# Sleep and Dreaming

How and why does the sleeping brain generate dreams?

Though the question is old, a paradigm shift is now occurring in the science of sleep and dreaming that is making room for new answers. From brainstem-based models of sleep cycle control, research is moving toward combined brainstem/forebrain models of sleep cognition itself. Furthermore, advances in philosophy, psychiatry, psychology, artificial intelligence, neural network modeling, psychophysiology, neurobiology, and clinical medicine make this a propitious time to review and bridge the gaps among these fields as they relate to sleep and dream research.

This book presents five papers by leading scientists at the center of the current firmament and more than seventy-five commentaries on those papers by nearly all the other leading authorities in the field. Topics include mechanisms of dreaming and REM sleep, memory consolidation in REM sleep, and an evolutionary hypothesis of the function of dreaming. The papers and commentaries, together with the authors' rejoinders, represent a huge leap forward in our understanding of the sleeping and dreaming brain, ultimately offering new and unique views of consciousness and cognition. They help provide new answers to both old and new questions, based on the latest findings in modern brain research. The book's multidisciplinary perspective will appeal to students and researchers in neuroscience, cognitive science, and psychology.

Edward F. Pace-Schott is Instructor in Psychiatry at Harvard Medical School.

Mark Solms is a Lecturer at the Royal London School of Medicine.

Mark Blagrove is Reader in Psychology, Department of Psychology, University of Wales Swansea.

Stevan Harnad is Professor of Cognitive Science in the Department of Electronics and Computer Science, University of Southampton.

# Contents

# Contents

# Preface

For centuries there have been theories about how and why sleep occurs. Since the discovery of REM sleep in 1953, science has also asked how and why the individual stages of sleep, such as REM sleep and slow wave sleep, occur and what relationship they have with dreaming. The relationship of dreaming to brain physiology and neurochemistry, and the possible functions, or lack of functions, of REM sleep and of dreaming have also been addressed. This book highlights the current debates, disagreements, and understandings among many of the world's leading researchers, from many different disciplines, on these questions, including both theoretical and experimental work. The book comprises a collection of target chapters, commentaries, and replies to commentaries that were first published as a special issue on sleep and dreaming of the journal *Behavioral and Brain Sciences* in December 2000.

These are currently areas of great ferment. Fifty years ago dreams seemed to occur almost exclusively in REM sleep, a few years later dreams were also shown to occur in non-REM sleep, and the debate continues today about whether these stages of sleep result in different types of dreams or whether dreaming can occur in all stages of sleep and how closely characteristics of dreaming, such as the illogicality of some dreams or the ease with which they are forgotten, are tied to the physiology and neurochemistry of the brain. Theories of the function of dreams have abounded, from the clearing out of memories to the linking and forming of memories to creative problem-solving. There have also been theories of the function of REM sleep, such as of brain maturation in the newborn and the consolidation of memories at all ages. This book addresses theories of the possible functions and causes of dreaming and REM sleep, with implications not only for our knowledge of two activities that take up much of the human life span, but also for the study of the relationship of conscious experience to the brain and of the possible functions of consciousness.

The target chapters and commentaries give examples of a wide range of scientific methodologies that aim to address these issues, including the phenomenology and neuroscience of conscious states, cognitive performance testing, and the relationship of dream content to waking life events. The book makes clear the relevance of the study of dreaming to neuroscience, psychology, psychoanalysis, cognitive science, neurology, philosophy, psychiatry, and other fields, and we wish to thank all the authors and commentators from the many sciences involved for their participation in this book.

For this book version of the *Behavioral and Brain Sciences* special issue, an introduction and a postscript have been added that provide updates on relevant papers published during 2000, 2001, and 2002. The necessity for these updates shows the current rapid expansion of investigations into sleep and dreaming, along with the growth of neuroscience and the study of consciousness. We hope that this volume will inspire further experiments and debate concerning the relationships between dreaming and the sleeping brain and the functions of sleep and dreaming.

Edward Pace-Schott, Mark Solms,
Mark Blagrove, & Stevan Harnad

# Introduction

**Mark Blagrove**

*Department of Psychology, University of Wales Swansea*

The target chapters in this book address three issues in the science of sleep and dreaming: the relationship of dreaming to brain physiology and neurochemistry and the possible functions, or lack of functions, of REM sleep and of dreaming. The target chapters provide detailed summaries of previous work and a background to these current issues. This introduction aims to summarize the main claims of each of the target chapters and to cite recent papers of relevance to those chapters that appeared around the same time as or after the production of the *BBS* special issue. A further update, by Edward Pace-Schott, specifically on the neuroscience of sleep and dreaming, is provided at the end of the book.

The first three target chapters of this book are concerned with the relationship between dreaming and brain physiology and neurochemistry, with particular reference to the relationship of REM sleep to dreaming. Hobson, Pace-Schott, and Stickgold detail their AIM model of the mind-brain during dreaming and other states of consciousness. This model describes three dimensions of brain neuromodulation, these being level of brain activity (A), internal or external source of stimulation for cognition (I), and mode of organization of cognition (M), which they relate to aminergic/cholinergic balance. This chapter emphasizes the importance of REM sleep to dreaming, reviews the comparison of dreaming to waking cognition [of relevance here is Kahn et al. (2000) on how character recognition occurs during dreams], and, in common with Nielsen's target chapter, reviews the history of investigations into the quantitative and qualitative differences and similarities between dreams in REM and NREM sleep. Hobson et al. then review recent studies on neuroimaging of the brain during dreaming and sleep, which indicate increased activation of limbic, diencephalon, and brain stem areas during REM sleep and the decreased activation of these areas during NREM sleep. The original activation-synthesis model of Hobson and McCarley is then updated with recent work on the relationship of cholinergic enhancement, and serotonergic and noradrenergic suppression, to REM sleep; there are also additional details from recent findings on the manner in which aminergic and cholinergic systems exert their influences on the REM-NREM cycle.

The AIM state space model that evolved from the activation-synthesis model holds that not only is the brain activated (A) with low external input (I) during REM sleep but that the modulatory factor (M) is related to the deficiencies of logic and memory found in dreams. Since the *BBS* special issue, Fosse et al. (2001a) argue that their finding of the reciprocal variation in thoughts and hallucinations between wake, sleep onset, NREMS, and REMS indicates the need for the M factor, in addition to A and I. This method of comparing the prevalence and reciprocal covariance of different cognitive variables within each dream is also used in Fosse (2000), with a reply to the analysis by Nielsen (2000), and operationalizes the classification of different types of dreams, such as thoughtlike and hallucinatory dreams. Dream experiences have also recently been quantified in terms of four factors, Emotionality, Rationality, Activity, and Impression, using the Dream Property Scale of Takeuchi et al. (2001b), who found that EEG power in some frequencies is correlated with Rationality (the dream being more ordinary and orderly), and at other frequencies with Impression (the dream being more clear and focused), indicating some brain-mind isomorphism using a questionnaire method that is answered on awakening and is not mediated by the experimenter, nor, apparently, confounded by dream length.

Solms's target chapter details that REM sleep and dreaming are dissociable states, in that each can occur in the absence of the other; one reason for this is that, although there are average differences between NREM and REM dreams, 5–10% of NREM dreams are indistinguishable from REM dreams, in terms, for example, of their vivid and dramatic content. Using neuroanatomical studies, loss of dreaming is shown in this chapter to be associated with lesions to the ventro-mesial quadrant of the frontal lobe and to the parieto-temporal-occipital junction. The former area is known to be important for goal-seeking and appetitive behaviors, the latter area for supporting mental imagery. Solms concludes that dreaming is controlled by forebrain, probably dopaminergic, mechanisms whereas REM sleep is controlled by cholinergic brain stem mechanisms. [Gottesmann (2000) also hypothesizes that the release of dopamine results in the "psychotic-like" fantasies and irrational mental activities of dreaming.] REM sleep is therefore held to be one of many possible activators of forebrain dreaming mechanisms; others include descending stage 1 at sleep onset and the rising morning phase of the diurnal rhythm. For an update on the dopaminergic and cholinergic influences on dreaming the reader should see Solms (2002) and Cicogna & Bosinelli (2001) for more on the interaction between top-down and bottom-up components of dream mentation and for a consideration of what aspects of consciousness remain or are diminished in dreaming. Reiser (2001) suggests that top-down (psychological, psychoanalytic) and bottom-up (neuroscience) approaches to dreaming and REM sleep are needed to account for the form and possible functions of dreams, with emotion being a bridge belonging to both domains. For example, to understand dreams one needs the top-down information derived from psychoanalysis of the dreamer's personal history and how it is incorporated into dreams together with the bottom-up PET studies of limbic areas that are activated during REM sleep.

Nielsen lists further examples where dreaming can occur in NREM sleep, such as when there is external sensory stimulation. He proposes that there are many levels of cognitive activity during sleep, from vivid intense dreams to thinking and fragmentary impressions to preconscious cognition, with the more complex or dramatic types of dreams more likely to occur in REM sleep. He shows that the liberalization (from around 1962) of what was accepted by dream researchers as a "dream" resulted in higher rates of dream recall being recorded from REM and NREM sleep and a diminution of the differences found between the dreams of the two states. The main purpose of the chapter is to review and reconcile evidence for the 1-generator view of dreaming, where similar processes of dream production are assumed to be occurring in REM and NREM sleep (but with the possibility that memory source activation and memory of dreams could be different in REM from NREM sleep), and the 2-generator view, where there are quantitative or qualitative differences between dreams from the two states. That the more vivid and even lucid types of dreams occur far more often in REM sleep is given as evidence for the 2-generator model.

Nielsen discusses, as do Hobson et al., the current debate about whether length of report should be controlled for in comparing dreams from REM and NREM sleep; such control has been found to eliminate almost all differences between dreams from the two states, but this statistical method is contentious because dream length may itself be a result of real differences in the characteristics of dreams from the two states. To explain why many dreams from NREM sleep are indistinguishable from REM sleep dreams, Nielsen then proposes that components of REM sleep can occur during NREM sleep but remain hidden ("covert") either because not all components are present at once, and so the sleep epoch is not scored as REM sleep, or because measures are not sensitive enough to detect the component. Nielsen proposes a probabilistic model whereby REM-like dreams can occur in NREM sleep during windows that are approximately 15 minutes prior to or following a REM period.

Of relevance to the discussion in these chapters of the comparison of dreams in REM and NREM sleep is a recent paper by Baylor and Cavallero (2001). In their reanalysis of three previously published studies they find that episodic memory sources occur significantly more often for NREM than REM dreams, and this stage effect remains even when report length is controlled for. Stickgold et al. (2001) found that dream report length increases with time in stage only for REM sleep, not for NREM sleep, and Takeuchi et al. (2001a) found that occurrence of dreams during sleep onset REM periods is related to amount of REM sleep, whereas occurrence of dreams during sleep onset NREM periods is related to number of arousals, but not to length of NREMS. However, note that Conduit et al. (2001) find that postsleep recall of events that occurred during NREM

sleep is worse than for REM sleep, and that therefore dreaming may occur equally in NREM sleep as in REM sleep, but just not be remembered. Similarly, Cicogna et al. (2000) ascribe the difference they find in REM and SWS dream recall rates to a mnemonic deficit, claiming that mental experiences may be produced almost continuously during sleep, this being a 1-generator theory. They did find a greater intensity of emotions during REM sleep, but level of self-participation and awareness did not differ between the two stages, and two significant differences between the stages, in number of characters and number of emotions, disappeared when dream length was partialled out.

The emphasis in these three chapters is on finding physiological and neurochemical correlates that explain the recall, lack of recall, and form of dreams across different brain states. For a review of psychological variables that affect individual differences in dream recall, see Blagrove & Akehurst (2000). As an indication that individual differences in length of dream reports may be a result of state-independent differences in reporting style, Stickgold et al. (2001) found intersubject differences in REM report lengths were correlated with similar differences in NREM, sleep onset, and waking report lengths.

In their target chapter Vertes and Eastman argue against REM sleep having a memory consolidation role, claiming that animal REM deprivation studies are divided equally in showing REM deprivation does or does not disrupt learning/memory. They account for REM deprivation studies that do show deficits in learning/memory as being due to the stress of the deprivation procedure, and a recent report by Spiegel et al. (1999) does find endocrine, metabolic, and stress effects of sleep restriction in humans. Vertes and Eastman claim that the majority of human REM deprivation studies find minimal or no effects on learning/memory, and they review work in humans on the suppression of REM sleep by antidepressants in which there are no, or negligible, effects on learning/memory. As a further context to their review, there has been debate over whether sleep restriction to less than 5 hrs per night, which can eliminate some REM sleep but generally leaves SWS intact, results in only minor or in large and cumulative cognitive deficits (Dinges et al. 1997; Pilcher & Huffcutt 1996). As Vertes and Eastman state in their reply to the commentaries, a recent series of experiments has compared memory after a period of early sleep (in which SWS predominates) with memory after a period of late sleep (in which REMS predominates). Readers may consult Wagner et al. (2001) for a recent example. Vertes and Eastman propose that REM sleep has a function of periodically activating the brain following each period of slow wave sleep. This view is supported by Horne's (2000) view that REM sleep tones up the sleeping cortex. Horne also accounts for the post-REM deprivation phenomenon of REM rebound, often taken as an indication of the importance of REM sleep to the organism, as instead being due to a default condition of loss of inhibition of REM sleep by non-REM sleep. Horne concludes that REM has little advantage over wakefulness in providing cerebral recovery or memory consolidation and that it can occur as a default condition when wakefulness is temporarily unnecessary to the organism, as would occur in situations of boredom or satiation.

Revonsuo's target chapter argues that during dreaming we rehearse threat perception and threat avoidance and that this mechanism has evolved because it increases the probability of reproductive success in threatening environments, such as those inhabited by ancestral humans. He argues that dreams may only infrequently act to solve intellectual problems (this being a current matter of debate, for example, Barrett 2001a, 2001b; Baylor 2001). He claims they do not have a function of solving emotional problems, as their function, like pain and fear, is to aid reproductive fitness rather than mental health or comfort, indeed Koethe and Pietrowsky (2001) find that nightmares induce anxiety and physical complaints.

One proposition that Revonsuo uses to arrive at his theory is that dreams are an organized and selective simulation of the perceptual world, that is, they are not random or disorganized or purely reflective of waking life. On his claim about dreams not being random, Stickgold et al. (1994) find sufficient coherence across dream reports for judges to distinguish intact from spliced reports. However, Roussy et al. (2000) demonstrate an inability of independent observers to detect a clear resemblance between participants' daily events and manifest dream content, although Revonsuo does note that in nonthreatening environments dreams may not show their threat simulation function. The evidence that dreams select for negative content is detailed in his chapter, such as the preponderance in dreams of misfortune over good fortune and the high frequency of aggressive strangers in dreams. Of relevance here is the finding of Bears et al. (2000), who, defining masochistic dreams as ones that involve negative self-representation, disappointment, loss, or lack or where the dreamer is lost, excluded, or ridiculed, report that depressed individuals tend to report masochistic dreams closer to morning than do nondepressed individuals, in whom masochistic dreams are equally distributed across the night. Revonsuo also proposes that waking threats activate the threat simulation system, and there is recent work relevant to this. The presence and intensity of the contextualizing (or central) images in dreams are higher among participants who report any abuse (physical or sexual, childhood or recent) compared to those who report no abuse (Hartmann et al. 2001). Furthermore, Zadra and Donderi (2000) found significant negative correlations between well-being and nightmare frequency and that the retrospective memory of frequency of nightmares and bad dreams greatly underestimates their frequency, in comparison to daily logs: a discussion of the directions of causality for the relationship between anxiety and disturbing dreams in adolescents is provided by Nielsen et al. (2000). Methodological problems with the assessment of emotions during dreams are described by Fosse et al. (2001b), who find that 74% of REM reports have at least one discrete emotion, and that the amount of positive or negative emotions in dreams does not change across the night.

Revonsuo then argues that stimulation of perceptual and motor skills during dreams, even when dreams are not explicitly remembered, leads to enhanced performance in waking life. On this point the theory is in accord with evidence that higher mental processes, such as those underlying social interaction, affect and evaluation, motivation and goal-setting, can occur without conscious choice or guidance (Bargh & Ferguson 2000). Revonsuo's view of the dream as an environment in which to act can be contrasted to the more metaphorical and symbolic view of dreaming (e.g., Lakoff 1993). Note, however, that Revonsuo states

that other functions may also be possible for dreams. Recent suggestions are that dreams aid attachment for people who are insecurely attached (McNamara et al. 2001) and that pleasant or moderately unpleasant dreams may be sources of personal insight during dream interpretation (Zack & Hill 1998).

The five target chapters now follow. It is hoped that the debates over the relationship between dreaming and brain physiology, and over the memory consolidation function of REM sleep and the proposed threat simulation function of dreaming, will enhance the wider fields of neuroscience and of the study of the nature of consciousness and the specific questions of how and why we dream.

## References

Bargh, J. A. & Ferguson, M. J. (2000) Beyond behaviorism: On the automatism of higher mental processes. *Psychological Bulletin* 126:925–45.

Barrett, D. (2001a) Comment on Baylor: A note about dreams of scientific problem solving. *Dreaming* 11:93–5.

Barrett, D. (2001b) *The committee of sleep: How artists, scientists and athletes use their dreams for creative problem solving.* Crown/Random House.

Baylor, G. W. (2001) What do we really know about Mendeleev's dream of the periodic table? A note on dreams of scientific problem solving. *Dreaming* 11:89–92.

Baylor, G. W. & Cavallero, C. (2001) Memory sources associated with REM and NREM dream reports throughout the night: A new look at the data. *Sleep* 24:165–70.

Bears, M., Cartwright, R. & Mercer, P. (2000) Masochistic dreams: A gender-related diathesis for depression revisited. *Dreaming* 10:211–19.

Blagrove, M. & Akehurst, L. (2000) Personality and dream recall frequency: Further negative findings. *Dreaming* 10:139–48.

Cicogna, P. & Bosinelli M. (2001) Consciousness during dreams. *Consciousness and Cognition* 10:26–41.

Cicogna, P., Natale, V., Occhionero, M. & Bosinelli, M. (2000) Slow wave and REM sleep mentation. *Sleep Research Online* 3:67–72. (http://www.sro.org/2000/Cicogna/67)

Conduit, R., Crewther, S. & Coleman, G. (2001) Poor recall of eye-movement signals from NREM compared to REM sleep: Implications for models of dreaming. *Sleep* 24 (Abstract supplement):A181.

Dinges, D., Pack, F., Williams, K., Gillen, K. A., Powell, J. W., Ott, G. E., Aptowicz, C. & Pack, A. I. (1997) Cumulative sleepiness, mood disturbance, and psychomotor vigilance performance decrements during a week of sleep restricted to 4–5 hours per night. *Sleep* 20:267–77.

Fosse, R. (2000) REM mentation in narcoleptics and normals: An empirical test of two neurocognitive theories. *Consciousness and Cognition* 9:488–509.

Fosse, R., Stickgold, R. & Hobson, J. A. (2001a). Brain-mind states: Reciprocal variation in thoughts and hallucinations. *Psychological Science* 12:30–6.

Fosse, R., Stickgold, R. & Hobson, J. A. (2001b). The mind in REM sleep: Reports of emotional experience. *Sleep* 24:947–55.

Gottesmann, C. (2000) Hypothesis for the neurophysiology of dreaming. *Sleep Research Online* 3:1–4. (http://www.sro.org/2000/gottesmann/1/)

Hartmann, E., Zborowski, M., Rosen, R. & Grace, N. (2001) Contextualizing images in dreams: More intense after abuse and trauma. *Dreaming* 11:115–26.

Horne, J. A. (2000) REM sleep – by default? *Neuroscience and Biobehavioral Reviews* 24:777–97.

Kahn, D., Stickgold, R., Pace-Schott, E. F. & Hobson, J. A. (2000) Dreaming and waking consciousness: A character recognition study. *Journal of Sleep Research* 9:317–25.

Koethe, M. & Pietrowsky, R. (2001) Behavioral effects of nightmares and their correlations to personality patterns. *Dreaming* 11:43–52.

Lakoff, G. (1993) How metaphor structures dreams: The theory of conceptual metaphor applied to dream analysis. *Dreaming* 3:77–98.

McNamara, P., Andresen, J., Clark, J., Zborowski, M. & Duffy, C. A. (2001) Impact of attachment styles on dream recall and dream content: A test of the attachment hypothesis of REM sleep. *Journal of Sleep Research* 10:117–27.

Nielsen, T. A. (2000) Dream mentation production and narcolepsy: A critique. *Consciousness and Cognition* 9:510–13.

Nielsen, T. A., Laberge, L., Paquet, J., Tremblay, R. E., Vitaro, F. & Montplaisir, J. (2000) Development of disturbing dreams during adolescence and their relation to anxiety symptoms. *Sleep* 23:727–36.

Pilcher, J. J. & Huffcutt, A. I. (1996) Effects of sleep deprivation on performance: A meta-analysis. *Sleep* 19:318–26.

Reiser, M. F. (2001) The dream in contemporary psychiatry. *American Journal of Psychiatry* 158:351–9.

Roussy, F., Brunette, M., Mercier, P., Gonthier, I., Grenier, J., Sirois-Berliss, M., Lortie-Lussier, M. & De Koninck, J. (2000) Daily events and dream content: Unsuccessful matching attempts. *Dreaming* 10:77–83.

Solms, M. (2002). Dreaming: Cholinergic and dopaminergic hypotheses. In: *Neurochemistry of consciousness: Transmitters in mind*, ed. E. Perry, H. Ashton, & A. Young. John Benjamins Publishing Company.

Spiegel, K., Leproult, R. & Cauter, E. V. (1999) Impact of sleep debt on metabolic and endocrine function. *The Lancet* 354:1435–9.

Stickgold, R., Malia, A., Fosse, R. & Hobson, J. A. (2001) Brain-mind states: I. Longitudinal field study of sleep/wake factors influencing mentation report length. *Sleep* 24:171–9.

Stickgold, R., Rittenhouse, C. D. & Hobson, J. A. (1994) Dream splicing: A new technique for assessing thematic coherence in subjective reports of mental activity. *Consciousness and Cognition* 3:114–28.

Takeuchi, T., Miyasita, A., Inugami, M. & Yamamoto, Y. (2001a) Intrinsic dreams are not produced without REM sleep mechanisms: Evidence through elicitation of sleep onset REM periods. *Journal of Sleep Research* 10:43–52.

Takeuchi, T., Ogilvie, R. D., Ferrelli, A. V., Murphy, T. I. & Belicki, K. (2001b) The Dream Property Scale: An exploratory English version. *Consciousness and Cognition* 10:341–55.

Wagner, U., Gais, S. & Born, J. (2001) Emotional memory formation is enhanced across sleep intervals with high amounts of rapid eye movement sleep. *Learning and Memory* 8:112–19.

Zack, J. S. & Hill, C. E. (1998) Predicting outcome of dream interpretation sessions by dream valence, dream arousal, attitudes towards dreams, and waking life stress. *Dreaming* 8:169–85.

Zadra, A. & Donderi, D. C. (2000) Nightmares and bad dreams: Their prevalence and relationship to well-being. *Journal of Abnormal Psychology* 109:273–81.

# 1

# Dreaming and the brain: Toward a cognitive neuroscience of conscious states

**J. Allan Hobson, Edward F. Pace-Schott, and Robert Stickgold**

*Laboratory of Neurophysiology, Department of Psychiatry, Harvard Medical School, Massachusetts Mental Health Center, Boston, MA 02115*
**{allan_hobson; edward_schott; robert_stickgold}@hms.harvard.edu**
**http://home.earthlink.net/~sleeplab**

**Abstract:** Sleep researchers in different disciplines disagree about how fully dreaming can be explained in terms of brain physiology. Debate has focused on whether REM sleep dreaming is qualitatively different from nonREM (NREM) sleep and waking. A review of psychophysiological studies shows clear quantitative differences between REM and NREM mentation and between REM and waking mentation. Recent neuroimaging and neurophysiological studies also differentiate REM, NREM, and waking in features with phenomenological implications. Both evidence and theory suggest that there are isomorphisms between the phenomenology and the physiology of dreams. We present a three-dimensional model with specific examples from normally and abnormally changing conscious states.

**Keywords:** consciousness, dreaming, neuroimaging, neuromodulation, NREM, phenomenology, qualia, REM, sleep

## 1. Introduction

Dreaming is a universal human experience that offers a unique view of consciousness and cognition. It has been studied from the vantage points of philosophy (e.g., Flanagan 1997), psychiatry (e.g., Freud 1900), psychology (e.g., Foulkes 1985), artificial intelligence (e.g., Crick 1994), neural network modeling (Antrobus 1991; 1993b; Fookson & Antrobus 1992), psychophysiology (e.g., Dement & Kleitman 1957b), neurobiology (e.g., Jouvet 1962) and even clinical medicine (e.g., Mahowald & Schenck 1999; Mahowald et al. 1998; Schenck et al. 1993). Because of its broad reach, dream research offers the possibility of bridging the gaps in these fields.

We strongly believe that advances in all these domains make this a propitious time to review and further develop these bridges. It is our goal in this target article to do so. We will study dreams (defined in the American Heritage Dictionary [1992] as "a series of images, ideas, emotions, and sensations occurring involuntarily in the mind during certain stages of sleep") and REM sleep, as well as the numerous forms of wake-state and sleep-state mentation. We will also review polysomnographically defined wake and sleep states. Our analyses will be based on comparisons and correlations among these various mental and physiological states.

### 1.1. An integrative strategy

Three major questions seem to us to be ripe for resolution through constructive debate:

1. Are the similarities and differences in the conscious experiences of waking, NREM, and REM sleep defined with sufficient clarity that they can be measured objectively? If so, do the measures establish clear-cut and major differences between the phenomenological experience of these three physiological states?

2. Are the similarities and differences between the brain

J. ALLAN HOBSON is Professor of Psychiatry at Harvard Medical School, Boston, MA, where he has been Director of the Laboratory of Neurophysiology since 1967. He is recipient of the 1990 Von Humbolt Prize for Science from the German government and the 1998 Distinguished Scientist Award of the Sleep Research Society. His major research interests are the neurophysiological basis of the mind and behavior, sleep, and dreaming and the history of neurology and psychiatry. He is author of *The dreaming brain* (1988), *Sleep* (1998), *Consciousness* (1998), *Dreaming as delirium* (1989) and, forthcoming in 2001, *The dream drug store* and *Out of its mind, Psychiatry in crisis* (with Jonathan Leonard).

EDWARD F. PACE-SCHOTT is Instructor of Psychiatry at Harvard Medical School, Boston, MA. His research interests include the cognitive neuroscience of dreaming, substance abuse, and sleep deprivation. He is also a psychotherapist.

ROBERT STICKGOLD is Assistant Professor of Psychiatry at Harvard Medical School, Boston, MA. His background is in biochemistry and cellular neurophysiology. He is a cognitive neuroscientist studying the roles of sleep and dreaming in off-line memory reprocessing, including off-line memory consolidation, transfer, integration, and erasure.

substrates of the states of waking, NREM, and REM sleep defined with sufficient clarity that they can be measured objectively? If so, do the measures establish clear-cut differences between these states at the level of brain regions, as well as at the cellular and molecular levels?

3. To the extent that affirmative answers can be given to the two preceding questions, can a tentative integration of the phenomenological and physiological data be made? Can models account for the current results and suggest experiments to clarify remaining issues?

Hoping to stimulate a useful debate, we will answer all three of the preceding questions affirmatively, documenting our responses with appropriate data drawn from our own work and from that of our colleagues. Referring to this ample literature, one can now identify numerous operationally defined psychological and physiological parameters with which to make such conscious state comparisons. In developing our answers, we will advance the thesis that the conscious states of waking, NREM, and REM sleep differ in three clear and important ways which are measurable at both the psychological and physiological levels. The three parameters will become the axes of a state space model that we introduce only briefly here but discuss in more detail in concluding this article.

### 1.2. A state space model of the brain-mind

In essence, our view is that the brain-mind is a unified system whose complex components dynamically interact so as to produce a continuously changing state. As such, any accurate characterization of the system must be multidimensional and dynamic and must be integrated across the neurobiological and psychological domains. Both neurobiological and psychological probes of the system must therefore be designed, applied and interpreted so as to recognize and clarify these features.

As a first step in that direction, we have created a three-dimensional state space model (AIM) that allows us to represent the system according to variables with referents in both the neurobiological and psychological domains as is shown in Figure 1. They are activation (A), information flow (I), and mode of information processing (M). Each of these terms has meaning both at the cognitive and neurobiological levels.

Roughly speaking, these dimensions are meant to capture respectively: (1) the information processing capacity of the system (activation); (2) the degree to which the information processed comes from the outside world and is or is not reflected in behavior (information flow); and (3) the way in which the information in the system is processed (mode).

The resulting state space model, while still necessarily overly simplistic, is nonetheless a powerful tool for studies of consciousness. It captures many aspects of the neurobiological, cognitive, and psychological dynamics of wake-sleep states, and is unique in several important respects that we will discuss in light of the controversial conceptual and empirical issues that have stymied the study of waking, sleeping, and dreaming.

### 1.3. Caveat lector

In setting the stage for a full explication of our integrative AIM model (sect. 4), we will review the evidence regarding the differentiation of brain-mind states at the levels of psychophysiology (sect. 2) and basic and clinical neuroscience

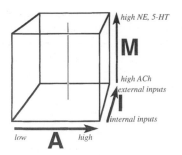

| Model Factor | Psychological | Neurobiological |
|---|---|---|
| **A-Activation:** Level of energy processing capacity | •Word count <br>•Cognitive complexity e.g., perceptual vividness, emotional intensity, narrative | •EEG activation <br>•Firing level and synchrony of reticular, thalamic and cortical neurons |
| **I-Information** Source internal or external. | •Real world space, time and person referents and their stability <br>•Real vs. imagined action | •Level of presynaptic and postsynaptic inhibition. <br>•Excitability of sensorimotor pattern generators. |
| **M-Mode:** Organization of data. | •Internal consistency? <br>•Physical possibility? <br>•Linear logic? | •Activity level of aminergic neurons |

Figure 1. The Activation-Input Source-Neuromodulation model (AIM). Illustration of three dimensional state space and the psychological neurobiological correlates of each dimension. See section 4 and also Hobson (1990; 1992a; 1997a).

(sect. 3). Although these reviews are extensive, they do not broach many of the fundamental questions of sleep research. For example, we do not consider the biological functions of REM sleep as we do elsewhere (Hobson 1988a) nor do we address the equally interesting question of how psychological and cognitive factors impinge upon sleep neurobiology, a subject which has been the focus of our most recent work (Stickgold et al. 1998a; 1999a; 2000a; Xie et al. 1996). As has often been shown, cognitive activity affects sleep as well as vice versa (e.g., Smith & Lapp 1991) reflecting, certainly, a reciprocal effect of psychological factors and their neural substrates. Additionally, we sidestep entirely the intriguing but difficult issue of whether dreaming itself, as a conscious experience, has a psychological function over and above the postulated benefits of sleep to homeostasis and heteroplasticity (Hobson 1988a). Finally, it is important to note that we deal here exclusively with what Chalmers (1995b) has termed the "easy problem" of consciousness, that is, the mechanisms of the cognitive components of consciousness, rather than the "hard problem" of how consciousness itself could arise from a neural system (see, e.g., Tononi & Edelman 1998; Woolf 1997).

## 2. The phenomenology and psychophysiology of waking, sleeping, and dreaming

In this section we discuss the evidence which has been gathered over the past 40 years in an effort to define the conscious states of waking, sleeping, and dreaming and to measure their formal features quantitatively. With respect to the first question raised by us in the introduction, we will defend the position that these three states *can* be defined, that their components can be analyzed and measured, and that they *are* significantly different from one another.

After presenting our justification for this claim, we will

address the claim made by many psychologists that differences between REM and NREM mentation – and even differences between REM and waking mentation – are much smaller than we believe. In the course of this discussion, we will identify several areas of disagreement and then suggest some new approaches to their resolution.

Definitions of dreaming have ranged from the broadest "any mental activity occurring in sleep" to the narrower one that we prefer:

> Mental activity occurring in sleep characterized by vivid sensorimotor imagery that is experienced as waking reality despite such distinctive cognitive features as impossibility or improbability of time, place, person and actions; emotions, especially fear, elation, and anger predominate over sadness, shame, and guilt and sometimes reach sufficient strength to cause awakening; memory for even very vivid dreams is evanescent and tends to fade quickly upon awakening unless special steps are taken to retain it.

We believe that this highly specified definition serves both folk psychology and cognitive neuroscience equally well. It captures what most people mean when they talk about dreams and it lends itself admirably to neurocognitive analysis as we now intend to show.

### 2.1. Early findings of distinct differences between REM and NREM mentation

Before proceeding, we provide definitions of "REM" and "NREM" sleep for those readers unfamiliar with these terms. These two clearly distinguishable types of sleep are defined, by convention, in terms of electrophysiological signs detected with a combination of electroencephalography (EEG), electroculography (EOG), and electromyography (EMG) whose measurement is collectively termed "polysomnography" (see Rechtschaffen & Kales 1968). First described by Aserinsky and Kleitmann in 1953, REM sleep (also known as "paradoxical," "active" or "desynchronized" sleep) is characterized by: (1) wake-like and "activated" (high frequency, low amplitude or "desynchronized") activity in the EEG; (2) singlets and clusters of rapid eye movements (REMs) in the EOG channel; and (3) very low levels of muscle tone (atonia) in the EMG channel. NonREM (NREM) sleep includes all sleep apart from REM and is, by convention, divided into four stages corresponding to increasing depth of sleep as indicated by the progressive dominance of the EEG by high-voltage, low-frequency (also termed "synchronized") wave activity. Such low frequency waves dominate the deepest stages of NREM (stages 3 and 4) which are also termed "slow-wave" or "delta" sleep. We refer the reader to Hobson (1989) for a comprehensive primer on sleep physiology.

Aserinsky and Kleitman's (1953) report of the correlation of REM sleep with dreaming began an intense period of research on the relation of brain to mind that lasted well into the 1970s. In the early days of the human sleep-dream laboratory era, much attention was paid to the specificity, or lack thereof, of the REM-dream correlation using the newly available sleep laboratory paradigm. Normal subjects, usually students, were awakened from either the NREM or REM phase of sleep in the sleep laboratory and asked to report their recollection of any mental experience preceding the awakening.

During this period, the similarities and differences in mentation between the brain states of waking, NREM, and REM sleep were lavishly documented (e.g., Foulkes 1962; Foulkes & Fleisher 1975; Goodenough et al. 1959; Herman et al. 1978; Monroe et al. 1965; Nielsen 1999; Pivik & Foulkes 1968; Rechtschaffen 1973; Rechtschaffen et al. 1963; Vogel 1991). We have summarized these REM-NREM differences in Table 1. Some of the important conclusions from this cross-sectional normative paradigm are:

1. Following REM sleep awakenings, variously defined dream reports are obtained much more frequently (Aserinsky & Kleitman 1953; 1955; Dement 1955; Dement & Kleitman 1957b; Kales et al. 1967; Wolpert & Trosman 1958) or at least substantially more frequently (Foulkes 1962; Goodenough et al. 1965a; Hobson et al. 1965; Molinari & Foulkes 1969; Rechtschaffen et al. 1963; Stoyva 1965) than after NREM awakenings. For reviews of this early work see Foulkes (1966; 1967), Herman et al. (1978), Nielsen (1999), Pivik (1991), Rechtschaffen (1973), and Snyder (1967). In an extensive review of 29 REM and 33 NREM recall rate studies, Nielsen (1999) found an average REM recall rate of 81.8 ($\pm$8.7)% compared to an average rate for NREM of 42.5 ($\pm$21.0)%.

2. The frequency of dream recall rapidly drops off as awakenings are delayed beyond the end of a REM period (Dement & Kleitman 1957b; Goodenough et al. 1965b; Wolpert & Trosman 1958), a finding which has recently been both supported (Stickgold et al. 1994a) and challenged (Rosenlicht et al. 1994). Subjects who are able to indicate that they are dreaming during sleep more often indicate dreaming during REM than during NREM (Antrobus et al. 1965).

3. There exists a positive relationship of both report word count and subjectively estimated dream duration with the length of preceding REM sleep (Dement & Kleitman 1957b) and this relationship has been recently replicated for word count (Stickgold et al. 1994a). Moreover, stimulus-incorporation studies suggest that there exists a positive relationship between the length of time dream events would occupy in real time and the duration of the preceding REM sleep epoch (Dement & Wolpert 1958).

4. Judges are able to distinguish unaltered REM mentation reports from NREM reports (Monroe et al. 1965), a finding that has been recently replicated (e.g., Herman et al. 1978; Reinsel et al. 1992). Furthermore, some dreamers can subjectively determine whether they themselves had been awakened from REM or from NREM (Antrobus & Antrobus 1967).

5. Reports from REM sleep awakenings are typically longer (Antrobus 1983; Casagrande et al. 1990; 1996b; Foulkes & Rechtschaffen 1964; Foulkes & Schmidt 1983; Stickgold et al. 1994a; Waterman et al. 1993), more perceptually vivid, more motorically animated, more emotionally charged, and less related to waking life than NREM reports (Antrobus et al. 1987; Cavallero et al. 1992; Foulkes 1962; Herman et al. 1978; Ogilvie et al. 1982; Rechtschaffen et al. 1963; see Nielsen, 1999 and Table 1 for summaries). In addition, there is linguistic evidence for greater consolidation of dream elements in REM (Salzarulo & Cipolli 1979).

6. In contrast to REM reports, NREM reports contain thought-like mentation and representations of current concerns more often than do REM sleep reports (Foulkes 1962; Rechtschaffen et al. 1963).

In a review of early data, Monroe et al. (1965) stated that "the high degree of success attained by the judges [in dis-

Table 1. Phenomenological differences between REM and NREM dream reports

| Study / Sleep Stage | # S's | # S's x # nights | # awakenings | % recall (any content) | % using more strict criteria | report length | bizarreness | visual vividness | emotionality | movement |
|---|---|---|---|---|---|---|---|---|---|---|
| **Antrobus (1983)** | 73 | 73 | | | | not compared | not compared | REM vs St. 2 n.s. when length controlled | not compared | not compared |
| REM | | | 73 | no report | | no data given REM >St. 2 p<.01 | | | | |
| St. 2 NREM | | | 73 | no report | | | | | | |
| **Aserinsky & Kleitman (1953)** | 10 | 14 | | | "dreaming" | | | | | |
| REM | | | 27 | 74 | 74 | | | | | |
| NREM | | | 19 | 22 | 11 | | | | | |
| **Casagrande et al. (1996)** | 20 | 40 | | | ≥1 sentence & ≥1 action" | using word count indices: | w. Antrobus et al., 1976 index: | w. Antrobus et al., 1976 index: | not compared | not compared |
| REM | | | | REM | | | | | | |
| early (in night) | | | 40 | early | 75 | early: REM> 2 & SO | early: REM> 2 & SO | early: REM> 2 & SO | | |
| late (in night) | | | 40 | late | 75 | late: REM & 2> SO | late: REM & 2> SO | late: REM & 2> SO | | |
| NREM (St. 2 abbreviated "2") | | | | NREM (2) | | | using a global rating: | using a global rating: | | |
| early | | | 40 | early | 50 | | REM always > 2 & SO | REM always > 2 & SO | | |
| late | | | 40 | late | 70 | | | | | |
| Sleep onset NREM St. 2 (SO) | | | | NREM (SO) | | | | | | |
| early | | | 40 | early | 50 | | | | | |
| late | | | 40 | late | 55 | | | | | |
| **Cavallero et al. (1992)** | 60 | 120 | | | not compared | temporal units | implausibility | not compared | % containing | not compared |
| REM | | | 60 | 89.2 | | 5.1 | 34 n.s. | | 62 p<.01 | |
| St. 3&4 NREM | | | 60 | 64.5 | | 1.88  p<.0001 | 50 | | 34 | |
| **Cicogna et al., 1998** | 36 | 72 | | | not compared | temporal units | implausibility* | not compared | number reported | body feelings |
| late spontaneous REM | | | 144 | 95 | | 7.3 | 84.2% | | .76 | 21.1% |
| late spontaneous St. 2 | | | 144 | 91 | | 6.0 | 79.6% | | .60 | 10.2% |
| **Dement (1955)** | 13 | ? | | not compared | "dreaming" | not compared | not compared | not compared | not compared | not compared |
| REM | | | 51 | | 88.2 | | | | | |
| NREM | | | 19 | | 0 | | | | | |
| **Dement & Kleitman (1957)** | 9 | 61 | | not compared | "dreaming" | not compared | not compared | not compared | not compared | not compared |
| REM | | | 191 | | 79.6 | | | | | |
| NREM | | | 160 | | 6.9% | | | | | |

| Study / Stage | n | (nights) | awakenings | recall % | dream content | length (word count / temporal units) | judged scene shift | judged visual | judged present/absent | judged present/absent (activity) |
|---|---|---|---|---|---|---|---|---|---|---|
| **Foulkes (1962)** | 8 | 56 | | | "vs. thinking" | subject-judged: REM>NREM p<.05 | judged scene shift: REM> NREM p<.02; subject-judged "distortion": R>N p<.05 | judged visual vs. not: REM> NREM p<.02; subject-judged "% visible": R>N p<.05 | judged present /absent: REM> NREM p<.02; subject-judged "present": R>N p<.05 | judged present /absent: REM> NREM p<.01; subject-judged "activity": R>N p<.05 |
| REM | | | 108 | 87 | 82 | | | | | |
| NREM | | | 136 | 74 | 54 | | | | | |
| St. 1 NREM | | | 32 | 69 | 56 | | | | | |
| St. 2 NREM | | | 32 | 74 | 51 | | | | | |
| St. 3&4 NREM | | | 32 | 70 | 51 | | | | | |
| **Foulkes & Rechtschaffen (1964)** | 24 | 48 | | | not compared | word count | subject-judged REM>NREM p=.01 | subject-judged REM>NREM p=.01 | subject-judged REM>NREM p=.01 | subject-judged REM>NREM p=.01 |
| REM | | | 143 | 88.8 | | 229.2 | | | | |
| NREM | | | 84 | 61.9 | | 158.6 | | | | |
| **Foulkes & Schmidt (1983)** | 23 | 69 | | | >1 temporal unit | temporal units | not compared | not compared | not compared | not compared |
| REM | | | 82 | 93 | 80 | 5.5 | | | | |
| NREM | | | 78 | 67 | 40 | 1.33 p<.01 | | | | |
| **Goodenough et al. (1959)** | 16 | 48 | | | not compared | not compared | not compared | not compared | not compared | not compared |
| REM | | | 91 | 69.2 | | | | | | |
| NREM | | | 99 | 34.3 | | | | | | |
| **Goodenough et al. (1965a)** | 10 | 98 | | | "dreaming" | word count | judged: REM > NREM p<.02 | judged visual imagery: REM >NREM p<.01 | not compared | judged activity: REM > NREM p<.01 |
| REM | | | 120 | 84 | 76 | 115 | | | | |
| NREM | | | 240 | 45 | 21 | 34 | | | | |
| **Hobson et al. (1965)** | 10 | 40–60 | | | "dreaming" | not compared | not compared | not compared | not compared | not compared |
| REM | | | 195 | 87.2 | 76.4 | | | | | |
| NREM | | | 102 | 37.2 | 13.7 | | | | | |
| **Kales et al. (1967)** | 3 | 40 | | dream+thinking | "dreaming" | not compared | not compared | not compared | not compared | not compared |
| REM | | | 134 | 83 | 81 | | | | | |
| NREM | | | 108 | 35 | 7 | | | | | |
| **Kamiya (1961)** | 25 | 250 | | | "dreaming" | not compared | not compared | not compared | not compared | not compared |
| REM | | | ? | ? | 85 | | | | | |
| NREM | | | 400 | 46 | 27 | | | | | |
| **Molinari & Foulkes (1969)** | 10 | 40 | | tonic / phasic; descend / ascend | not compared | not compared | not compared | subject judged tonic / phasic 60 100 / 79–80 | not compared | not compared |
| REM | | | 40 | 80 100; 72 75 | | | | | | |
| NREM | | | | | | | | | | |
| **Oglvie et al. (1982)** | 9 | 27 | | | not reported | not compared | judged: REM> St. 2 NREM p<.05 | judged: REM> St. 2 NREM "marginally" | judged: REM> St. 2 NREM "tendency" | not compared |
| REM | | | 54 | 79 | not reported | | | | | |
| NREM | | | 54 | | | | | | | |

(continued)

Table 1. (*Continued*)

| Study<br>Sleep Stage | # S's | # S's x<br># nights | # awak-<br>enings | % recall (any<br>content) | % using more<br>strict criteria | report<br>length | bizarreness | visual<br>vividness | emotionality | movement |
|---|---|---|---|---|---|---|---|---|---|---|
| Pivik & Foulkes (1968) | 20 | 40 | | | not compared | not compared | not compared | not compared | not compared | not compared |
| NREM total | | | 158 | 64.6 | | | | | | |
| NREM St. 2 | | | 74 | 71.6 | | | | | | |
| NREM St. 3 | | | 56 | 64.3 | | | | | | |
| NREM St. 4 | | | 28 | 46.4 | | | | | | |
| Rechtschaffen et al. (1963) | 17 | 30 | | | Ss say dreaming | not compared | subject judged | subject judged | subject judged | not compared |
| REM | | | | 86 | 87 | | 37% bizarre | 74% vivid | 74% emotional | |
| NREM | | | | 23 | 41 | | 6% bizarre | 24% vivid | 24% emotional | |
| Salzarulo & Cipolli (1979) | 8 | 80 | | | "contentful" | # sentences | not compared | not compared | not compared | not compared |
| REM | | | 240 | | 95 | 4.22 | | | | |
| NREM | | | 240 | | 85 | 3.48 | | | | |
| Stickgold et al. (1994) Nightcap | 11 | 110 | (spont.) | | > 100 words | | not compared | not compared | not compared | not compared |
| REM | | | 88 | 83 | 62 | 314 | | | | |
| NREM | | | 61 | 54 | 18 | 65 | | | | |
| Stoyva (1965) | 7 (deaf) | 28 | | | not compared | not compared | not compared | not compared | not compared | not compared |
| REM | | | 51 | 73 | | | | | | |
| NREM | | | 68 | 38 | | | | | | |
| Waterman et al. (1993) | 12 | 24 | 72 | not reported | not reported | REM>NREM | not compared | *w.Antrobus et<br>al., 1976 index<br>and length<br>partialed out:<br>REM>NREM* | not compared | not compared |
| Wolpert & Trosman (1958) | 10 | 51 | | | "dreaming" | not compared | not compared | not compared | not compared | not compared |
| REM | | | 54 | 90.8 | 85.2 | | | | | |
| NREM St. 2 | | | 26 | 3.8 | 0 | | | | | |

*Cicogna et al. 1998 actually found significantly more "space-time distortions" and a trend toward more "dimensional distortions" in Stage 2 versus REM reports, while the trend in global bizarreness (implausibility) went in the usual REM>Stage 2 direction. R = REM, N = NREM, spont. = spontaneous awakenings from identified sleep stage.

tinguishing REM from NREM reports] indicates that physiological sleep phase, REM or NREM, is highly diagnostic of the presence, amount, and quality of reported sleep mentation" (p. 456). In discussing the findings of this study, Rechtschaffen (1973) concluded that "these figures – discriminability ranging from about 70 to 90% – probably represent one of the best correlations ever discovered between psychological and physiological variables" (p. 163).

In REM sleep, the integrated conscious experience that is commonly referred to as dreaming is characterized by the following remarkably consistent set of features (see Hobson 1988b; 1994 for reviews):

1. Dreams contain formed hallucinatory perceptions, especially visual and motoric, but occasionally in any and all sensory modalities (Hobson 1988b; McCarley & Hoffman 1981; Snyder 1970; Zadra et al. 1998).

2. Dream imagery can change rapidly, and is often bizarre in nature (Hobson 1988b; 1997b; Hobson & Stickgold 1994a; Hobson et al. 1987; Mamelak & Hobson 1989a; McCarley & Hoffman 1981; Porte & Hobson 1986; Reinsel et al. 1992; Revonsuo & Salmivalli 1995; Williams et al. 1992). It has also been noted that dream reports contain a great many images and events which are relatively commonplace in everyday life (Dorus et al. 1971; Snyder 1970).

3. Dreams are delusional; we are consistently duped into believing that we are awake unless we cultivate lucidity (Barrett 1992; Hobson 1997b; Kahan 1994; LaBerge 1990; 1992; Purcell et al. 1986).

4. Self-reflection in dreams is generally found to be absent (Rechtschaffen 1978) or greatly reduced (Bradley et al. 1992) relative to waking and, when present, often involves weak, post hoc, and logically flawed explanations of improbable or impossible events and plots (Hobson 1988b; Hobson et al. 1987; Williams et al. 1992). It has been recently asserted, however, that self-reflection, self control and other forms of metacognition are more common in dreams than previously thought (Kahan 1994; Kahan & LaBerge 1994).

5. Dreams lack orientational stability; persons, times, and places are fused, plastic, incongruous and discontinuous (Hobson 1988b; 1997b; Hobson et al. 1987; McCarley & Hoffman 1981; Revonsuo & Salmivalli 1995; Rittenhouse et al. 1994; Stickgold et al. 1994b; 1997b; Williams et al. 1992).

6. Dreams create story lines to explain and integrate all the dream elements in a single confabulatory narrative (Blagrove 1992b; Cipolli & Poli 1992; Cipolli et al. 1998; Foulkes 1985; Hobson 1988b; Hunt 1991; Montangero 1991).

7. Dreams show increased and intensified emotions, especially fear-anxiety (Domhoff 1996; Merritt et al. 1994; Nielsen et al. 1991), which appear to integrate bizarre dream features (Merritt et al. 1994), and may even shape the narrative process (Seligman & Yellin 1987). Although the trend toward a predominance of negative emotion is prominent in most studies, other workers have found more balanced amounts of positive and negative emotion (for a good review, see Schredl & Doll 1998). Emotion also ranks as a prominent explanatory focus in functional theories of dreaming (e.g., Cartwright et al. 1998a; Greenberg et al. 1972; Kramer 1993; Perlis & Nielsen 1993).

8. Dreams show increased incorporation of instinctual programs (especially fight-flight), which also may act as powerful organizers of dream cognition (Hobson 1988b; Hobson & McCarley 1977; Jouvet 1973; 1999).

9. Volitional control is greatly attenuated in dreams

(Hartmann 1966b). The dreamer rarely considers the possibility of actually controlling the flow of dream events (Purcell et al. 1986) and, on those infrequent occasions when this does occur, the dreamer can only gain lucidity with its concomitant control of dream events for a few seconds (LaBerge 1990). Unlike the rarer form of dream control offered by lucidity, however, the more mundane self-control of thoughts, feelings and behavior may be fairly common in dreams (Kahan 1994).

All of these features can be found in REM dreams, and most REM dreams contain a majority of these features. Contrastingly, they are found relatively rarely in NREM reports (see Nielsen 1999). This is the empirical basis of our contention that all of these features will eventually be explainable in terms of the distinctive physiology of REM sleep.

We interpret the foregoing evidence as strongly supporting our conclusion that there are clear-cut and major differences among the states of waking, sleeping (NREM) and dreaming (REM) at the phenomenological level. We take the robust evidence for quantitative differences in amount of NREM and REM sleep mentation as convincing proof of the validity of an important role for not only activation (factor A) but for the two other factors, information source (I) and modulation (M) in our AIM model. In addition, we take the evidence that state transitions are gradual rather than discontinuous and the evidence that correlations between phenomenology and physiology are statistical rather than absolute as further support of this model.

### 2.2. Overview of the NREM-REM sleep mentation controversy

Although the discovery of REM sleep and its strong correlation with dreaming (Ascrinsky & Kleitman 1953) initially led to the strong hypothesis that dreaming occurred *only* during REM sleep (Dement & Kleitman 1957b), this hypothesis was clearly refuted by the discovery that reports of dreaming could be elicited from NREM sleep (Foulkes 1962) and that reports of dream-like mentation could also be obtained at sleep onset (Foulkes & Vogel 1965) and even from quiet waking (Foulkes & Fleischer 1975; Foulkes & Scott 1973). Given dreaming's lack of absolute state specificity, some investigators sought the psychophysiological correlates of specific dream features in the phasic events of REM and NREM sleep (Molinari & Foulkes 1969; see Kahn et al. 1997 and Pivik 1991 for reviews). Again, weak but consistently positive quantitative relationships were found (Kahn et al. 1997; Pivik 1991).

This lack of specificity led at least some investigators ultimately to conclude that investigations of REM sleep neurophysiology could provide no data helpful to understanding the genesis of dreaming (e.g., Bosinelli 1995; Foulkes 1990; 1991; 1993b; 1995; 1996a; 1997; Moffitt 1995). Such a view was encouraged by reports suggesting that in fact the differences between REM and NREM mentation were not nearly as great as had first been reported (e.g., Cavallero et al. 1992). In this section, we will present our reasons for rejecting these conclusions (see also Nielsen, target article).

How could the firm conclusions of the pioneer era (1955–1975) have apparently dissolved in the subsequent era of growing controversy (1975–1999)? In this section, we will analyze some of the scientific problems that led to the decline of the sleep-laboratory paradigm as this psychophysiological approach lost much of its initially enthusiastic support. In the subsequent section we will turn our attention to the concomitant development of cellular and molecular neurobiology and show how the findings of basic research provided an alternative approach.

### 2.2.1. REM sleep dreaming is not qualitatively unique.

While dream studies generally agree that REM reports are more frequent, longer, more bizarre, more visual, more animated and more emotional than NREM reports (Table 1), a pair of papers published in 1983 (Antrobus 1983; Foulkes & Schmidt 1983) led some researchers to the remarkable conclusion that the "characteristics [of dreaming] are pretty much the same throughout sleep" (Moffitt 1995) and that "dreaming in other sleep stages is not qualitatively different from REM dreaming" (Foulkes 1995). Because these papers are so central to the REM-NREM dreaming debate, we now offer a detailed review and critique of their findings and interpretations.

At the outset, it is important to point out that neither article actually concluded that REM and NREM dreams are indistinguishable, or even substantially the same, in either their quantitative or their qualitative features. In regard to qualitative features, Antrobus (1983) reported that when judges rated 154 REM and NREM reports for their relative "dreaminess" (using scales based on "visual imagery, bizarreness, hallucinatory quality and storylike quality"), they correctly identified 93% of the reports as either REM or NREM, indicating that REM dream reports were much more dreamlike than NREM reports. Similarly, Foulkes and Schmidt (1983, p. 276) concluded that "REM reports are likely to be significantly more dreamlike qualitatively (e.g., in character density, setting clarity) than typical NREM" reports, even when elicited after only five minutes of stage REM.

In regard to quantitative features, when Foulkes and Schmidt (1983) looked at 160 REM and NREM reports and characterized their lengths by the number of "temporal units" (narrative events), their data showed that temporal sequences (sequential events = temporal units − 1) were 14 times more common in REM reports than in NREM reports. In a similar way, Antrobus analyzed total recall frequency (TRF), which reflects the number of words in a report used to describe sleep mentation, and reported that word count significantly distinguished REM from NREM reports (F = 95.52). Using the same reports (J. Antrobus, personal communication), we have determined that the REM reports collected by Antrobus had a median length 6.4 times longer than their matched NREM reports, a number similar to the ratio of 7.0 obtained in a home study using reports from spontaneous awakenings (Stickgold et al. 1994a).

Since both Foulkes and Schmidt (1983) and Antrobus (1983) report such impressive differences between REM and NREM reports, one might wonder how and why these very authors have come to argue so strongly for a phenomenological sameness of these states. The critical question, raised by Foulkes and Schmidt and by Antrobus, pertains to the origin of the differences between REM and NREM reports, "whether there are . . . qualitative . . . differences as well as quantitative ones, and . . . whether such differences are merely attendant upon or are independent of the quantitative ones" (Foulkes & Schmidt 1983, p. 269). Or, as Antrobus wonders, whether "judges of Dreaming [dreaminess] implicitly rely on a dimension similar to the Total Recall Freq." (p. 562). It is this analysis that has led subsequent writers to claim that "when the quantitative characteristics of reports . . . from REM and nonREM . . . sleep are adjusted for length there are no differences in the characteristics of the reports" (Moffitt 1995, p. 19).

The normalization-for-length technique has been subsequently used to argue that bizarreness differences between REM and slow wave sleep (SWS) reports (Colace & Natale 1997), the number of dream-like features in a report (Fein et al. 1985; Rosenlicht & Feinberg 1997), memory sources of dreams (Cavallero et al. 1990) and even dream bizarreness itself (Bonato et al. 1991) are all directly and causally dependent on report length independent of sleep stage. Similar arguments have been advanced to explain correlations between dream bizarreness and creativity (Livingston & Levin 1991).

We will shortly reiterate our introductory arguments against this line of reasoning. Meanwhile, we emphasize some of these authors' own data that favor placing a strategic emphasis on the *differences* between REM and NREM mentation rather than using the similarities as a rationale for rejecting the cognitive neuroscience paradigm in favor of a purely cognitive description of mental states. (A similar critique of purely cognitive descriptions can be found in Nielsen 1999; and his target article.)

For example, Antrobus has recently shown that the REM/NREM distinction exerts a far greater effect on bizarreness than diurnal activation (Antrobus et al. 1995). He attributed the observed increase in bizarreness in REM reports to the increased activation seen in that state (Antrobus et al. 1995). It is also noteworthy that purely visual (versus verbal) imagery gave robust REM/NREM differences suggesting a differential sensory activation between the two states (Antrobus et al. 1995). And even when REM and NREM dreams were adjusted for length (a procedure we will shortly argue to be invalid), both Antrobus (1983) and Foulkes and Schmidt (1983) still found significant differences (e.g., in character density and setting clarity) between the two states. Notably, the persistence of a REM/NREM effect on bizarreness, visual imagery, and several other dream features in spite of normalization for report length has recently been confirmed (Casagrande et al. 1996b; Faucher et al. 1999; Nielsen 1999; and his target article; Raymond et al. 1999; Waterman et al. 1993). For example, when analysis of covariance (with report length as the covariate) is used to partial out the effect of report length on dream features, REM reports were still judged significantly more visual and bizarre than sleep onset or stage 2 reports (Casagrande et al. 1996b) and more visual than NREM reports (Waterman et al. 1993).

Even when dream features appear to be specifically linked to distinctive REM physiology, interpretations can still be cast toward either camp. Hong et al. (1997) reported an impressive correlation between visual imagery and REM density (r = 0.8), which we would argue as evidence for a dependence of dream imagery on a qualitative feature of REM sleep. But Antrobus et al. (1995) consider this to be another example of the simple dependence of dream content on levels of brain activation, arguing that rapid eye movements are not under strict brainstem cholinergic control, but come increasingly under the control of the frontal eye fields as general cortical activation increases.

Whatever one's assessment of the similarity versus difference argument, it is clear that none of the analyses in these two papers can distinguish between two competing hypotheses: (1) that dream features are dependent on report length; and its simpler converse (2) that report length is dependent on dream features. We now consider the arguments in favor of the second hypothesis, which we have adopted in our own work.

**2.2.2. The relationship between dream features and dream report length.** That report length depends on dream features was first implied by Hunt (1982) in his analysis of dreaming as fundamentally visuospatial versus verbal-propositional and was then explicitly proposed by Hunt et al. (1993). We agree with their logical assumption that reports with more dream features will require more words to describe them. For example, a report with such dream features as self-representation, visual hallucination, emotion, narrative plot, and bizarreness will almost certainly be longer than a report with none of these features. Similarly, it is highly unlikely that a report with a word count of only seven words, the median length of the Antrobus (1983) NREM reports (J. Antrobus, personal communication), could possibly have more than one of the above features.

Inexplicably, Antrobus (1983) and Foulkes and Schmidt (1983) both seem to regard word count and content as independent of each other. In doing so, each has emphasized a very different explanation. Although conceding that alternative explanations were "in no way excluded by these findings," Antrobus (1983) concluded that the NREM reports were shorter due to a defect in "the ability of the subject to recall and describe the [dream] events" (p. 567). In this view, the shorter reports failed to include dream features which were nonetheless present in the NREM dream itself. To us this seems, at best, a risky assumption. In contrast, Foulkes and Schmidt (1983) concluded that the shortened reports and the rarity of dream features reported resulted from differences in dream production. On this view, the differences reflected "the relative paucity and superficiality of mnemonic units active during NREM sleep" (p. 279) compared to REM sleep. The conclusion of Foulkes and Schmidt (1983) is strikingly similar to our position, which is that the relative brevity of NREM reports reflects a decrease in the types (superficiality) and number (paucity) of dream features present in the conscious experience reported in them. If Foulkes really agrees with us on this point, he cannot then also countenance controlling for word count in evaluating reports.

Analyzing the same data set used by Antrobus (1983) we have shown that REM/NREM differences can not be explained simply in terms of report length (Porte & Hobson 1986). Thus we agree with Antrobus when he pointed out that there is still a part of the REM/NREM variance that Dreaming (i.e., judges' idiosyncratic scales for "dreaminess") picks up better than a Total Recall Frequency factor.[1] Similarly, Foulkes and Schmidt (1983) reported that some residual REM/NREM differences in temporal unit composition (e.g., in character density) persist even after report length is controlled. Residual stage differences following normalization for report length in these as well as additional studies have recently been reviewed by Nielsen (1999).

In the face of such unambiguous statements, it is critical to try to understand why these results have been so frequently and so passionately misinterpreted. In part, the erroneous interpretations were encouraged by the original authors. For example, Antrobus (1983, p. 567) concluded that "although there are slight differences . . . it is quite clear that the global judgment of Dreaming adds little, if anything, to Total Recall [Frequency] with respect to the association with the sleep stages REM and NREM." Similarly, Foulkes and Schmidt (1983; p. 279) concluded that "*most* typically observed inter-stage differences in dream reports stem from different lengths rather than the different stages of the reports" (emphasis added). Because they have conflated causality with correlation, both Antrobus and Foulkes and Schmidt unjustifiably assume that most of the differences seen can be explained as correlates of report length. We disagree on the basis of the following studies.

Recent evidence provides strong support for Hunt's proposition that report length reflects the number and intensity of dreamlike features prior to awakening. Hunt et al. (1993) have argued "it is not the length of the dream that somehow makes bizarreness more likely, but . . . it is more parsimonious to conclude that episodes of bizarreness within the dream are one major determinant of overall dream length . . . making length a necessary consequence of bizarreness and not the other way around" (p. 180). In addition, Hunt et al. (1993) note that Hauri et al.'s (1967) factor analysis of dreams found that bizarreness and report length significantly load on the same factor (and therefore strongly co-vary), "which would make their enforced statistical separation highly questionable" (Hunt et al. 1993, p. 181). In other words, if quantity follows quality and is, in fact, caused by it, then longer reports are needed to describe dreamier dreams. On this view, word count is perhaps even a direct measure of dreaminess and might well be taken as such.

To support their position, Hunt et al. (1993) first demonstrated that awake subjects used more words to describe a visually bizarre picture than a mundane picture. They then showed that the bizarreness scores correlated positively with the number of words devoted to describing the bizarre episodes. Finally, they showed that normalizing dream features for report length actually eliminated the correlations of bizarreness with non-verbal imagination test scores. Hunt et al. therefore concluded that bizarreness directly determines a major component of report length and that controlling for total word count introduces an artifactual dilution of bizarreness scores.

In summary, a critical review of the papers of Antrobus (1983) and Foulkes and Schmidt (1983) reveals that these papers report significant quantitative differences in the features of REM and NREM dreams. Both papers also find features such as dreaminess or character density to differ significantly between REM and NREM dreams *even when report length is unjustifiably normalized.* Neither study reports data that argue against the contention that the strong correlation between report length and dream features occurs because reports with more dream features require more words to describe them (Hunt et al. 1993; Nielsen 1999). We urge the collection of additional data to further clarify the nature of these REM/NREM differences. Such data should include ample numbers of reports, collected longitudinally in naturalistic settings, which are obtained from home awakenings physiologically monitored with unintrusive devices such as the Nightcap (e.g., Rowley et al. 1998).

### 2.3. Methodological considerations in the study of dreaming

The study of mental states is replete with methodological shortcomings and conceptual confusions. We believe that some of these areas of confusion can be clarified in a manner that could increase consensus. In what follows, we address five methodological issues to point out the nature of the problems, offer clarifications, and suggest possible resolutions.

### 2.3.1. The reduction of psychological states to narrative reports.

The most profound problem in studying conscious states is the necessity of reliance on verbal reports. This method is problematic because these accounts are just *reports,* not the subject's experience of the states themselves. This reduction of conscious experience to prose has at least three important ramifications:

(1) A multimodal conscious experience including pseudo-sensory perceptual, emotional, and motoric dimensions is reduced to only one mode, that of narration. (To emphasize this point, we merely point out that if a picture is worth a thousand words, we certainly are not getting the whole picture with a seven-word report!)

(2) The narratives describing sleep state mentation are all generated during the waking state and are thus likely to mix, if not contaminate, the dreaming phenomenology with the phenomenology of waking (for a discussion of this point relative to dream meaning, see Hunt 1989, p. 9).

(3) Analysis of narrative dream reports is extremely limited in its power to recreate or model the true underlying mechanism of dream production at any fundamental, primordial level of explanation (be it cognitive-mnemonic, linguistic or neuropsychological) because narratives about experience display a high degree of what Pylyshyn (1989) terms "cognitive penetrability."

Pylyshyn's point can be applied to dreaming as follows. The behavior of the dream production system is highly malleable using the same cognitive processes invoked to explain its behavior such as the dreamer's goals and beliefs (see Pylyshyn 1989). For example, in the case of the dreamer's goals, the frequency of overall dream recall as well as lucidity can be greatly increased by auto-suggestion techniques that employ many of the same cognitive abilities (e.g., imagination and visualization) that most theorists believe contribute to dream production itself (see sect. 3.3). In the case of beliefs, the meaning of a dream experience *while it is occurring* is highly dependent on the dreamer's personal (and changeable) philosophy of what dreaming is (e.g., a message from a deity, a psychopathomimetic experience, "travel outside the body," etc.). According to Pylyshn (1989) such highly penetrable experiences, rather than illustrating primordial cognitive mechanisms, instead reflect "the nature of the representations and . . . cognitive processes operating over these representations" (p. 81), which, in the case of dream reports, is language itself. Given that Pylyshn (1989) asserts that cognitive penetrability can affect even highly objective and replicable psychological data (such as the visualized-image-size/image-scanning-time relationships described by Kosslyn & Koenig 1992), penetrability is all the more likely to influence the highly elaborated and individualistic phenomenon of dream reporting. The rendering of dream reports in conventional (wake state) grammar and syntax may, therefore, tend to obscure important differences between the actual experiences of waking and dreaming.

These considerations raise the concern that using the sentence or the word as a unit for quantifying mental activity may say more about language than about the multimodal nature of conscious experience. This is important because

so many researchers consider the quantification of report length as the single most salient feature of a dream. In this context, it is also worth noting that verbal retrospective reports are often considered inadequate to describe mental states that are closer to dreaming than to waking mentation. These states include religious conversion, near-death experience, functional psychosis, delirium, drug-induced conditions, and other altered states of consciousness.

This aspect of the REM physiology-dream mentation controversy may be particularly relevant to the current debate about self-representation and bizarreness in dreams of children aged 3 to 8 (see Foulkes 1990; 1993b; 1996a; 1996b; 1997; Resnick et al. 1994). Based upon an extensive longitudinal study (Foulkes 1982b) and a later cross-sectional study (Foulkes et al. 1990), Foulkes asserted that "dreaming is absent until ages 3 to 5 and does not assume the form of adult dreaming until ages 6 to 7" (Foulkes 1997, p. 4). Foulkes hypothesizes that, lacking or being deficient in their ability to consciously mentally represent their perceptuo-behavioral experience, young children (like animals) may not experience dreaming in spite of having an abundance of REM (Foulkes 1990; 1993c). He argues further that dreaming is "a high-level symbolic skill, a form of intelligent behavior with cognitive prerequisites and showing systematic development over time" (Foulkes 1993c, p. 120), and that dreaming has, as its prerequisite, conscious representational competence (Foulkes 1990; Foulkes et al. 1990). As evidence to support this, he cites studies in which he finds very low recall of dreaming and little bizarreness prior to age 5 (Foulkes 1982b; Foulkes et al. 1979), low rates of reporting at ages 5–8 (Foulkes 1982b; Foulkes et al. 1990), acquisition of kinetic versus static imagery only after age 6 (Foulkes et al. 1990), and acquisition of self-representation as an active dream participant as well as narrative continuity only after age 7 (Foulkes et al. 1990; 1991). Further, from his data showing correlation of report rate with measures of visuospatial versus verbal skills (Foulkes et al. 1990), Foulkes (1993b) suggests that "young children may fail to report dreams because they are not having them, rather than because they have forgotten them or are unable to verbalize their contents" (p. 201). For a recent review see Foulkes (1999).

Subsequent studies have shown that dream bizarreness does indeed increase over ages 3 to 8 (Colace et al. 1993; 1997; Colace & Tuci 1996; Resnick et al. 1994). However, other of Foulkes's findings have not been supported. For example, dream reporting rates in 4- to 5-year olds has been reported to be almost identical to that in 8- to 10-year olds (Resnick et al. 1994). In addition, active self representation in dreams of 4- to 5-year olds has been reported to occur in over 80% of their dream reports (Colace et al. 1995; Resnick et al. 1994). Finally, substantial occurrence rates for bizarre elements have been reported in the dreams of both 4- to 5-year olds (0.45 per 100 words) and 8- to 10-year olds (0.71 per 100 words) (Resnick et al. 1994).

Moreover, although rates of adult dream recall have been related to performance on tests of visuospatial skill (Butler & Watson 1985), rates of dream recall have also been correlated with individual differences in visual memory (Schredl et al. 1995). Therefore, any ontogenetic changes in visual memory would confound the effects of developmental changes in higher order visuospatial skills on dream reporting rates in children.

Overarching these conflicting data, however, is the theoretical point bearing on the current discussion: that is, that dream reports are given in waking and thus, of necessity, must be constrained by an organism's waking cognitive and linguistic abilities. At one extreme, it must be conceded that even if a cat had the most vivid of "dreams," it would not be able to report it. Similarly, if a toddler is variously unable (or unwilling) to conceive and verbalize a complex perceptual-emotional-motor REM experience, it does not mean it was not originally experienced in some form which, later in life, might be reported as a dream. In other words, we challenge here the assumption by Foulkes (e.g., 1990) and others (e.g., Bosinelli 1995) that "dreaming" is an experience that can occur only if it can be later reported by an organism possessing linguistic abilities. We recognize that verification of oneiric activity in organisms that are unable to report (or even, possibly, reflect upon) their experiences is currently impossible, although we do not rule out the possibility that new methods may someday provide hints as to the conscious experiences of nonverbal beings (e.g., see Marten & Psarakos 1995).

Nevertheless, as with many other psychological constructs such as emotional expression (e.g., Darwin 1873) or behavioral inhibition (e.g., Goldman-Rakic 1986), such inferences drawn between human developmental as well as mammalian phylogenetic levels has a long scientific tradition. It is, therefore, not inherently invalid to cautiously speculate from adult human oneiric experience to observed REM behavior in infants and animals, especially given the abundant behavioral correlates (e.g., ethologically meaningful oneiric behavior; for a full discussion see Jouvet 1999). Similarly, we specifically suggest that the human neonate, spending as it does more than 50% of its time in REM sleep (Hobson 1989), is having indescribable but nevertheless real oneiric experiences. An infant's waking experience remains essentially indescribable and speculative to us older persons but we do not doubt that infants enjoy some sort of waking conscious experience. For us, it is not at all difficult to imagine that an infant might be experiencing hallucinosis, emotions, and fictive kinesthetic sensations during REM sleep.

Given these caveats, we suggest that more effort be put into the development and use of other methodologies and scales such as the photo-response visual brightness and clarity scale (Antrobus et al. 1987; 1995; Rechtschaffen & Buchignani 1992), temporal unit analysis (Cavallero et al. 1990; Foulkes & Schmidt 1983), computerized content analyses (Gottschalk 1999), the analysis of dream drawings (Hobson 1988b), or the use of affirmative probes (e.g., Herman 1992; Merritt et al. 1994; Pace-Schott et al. 1997a; Stickgold et al. 1997a; see Herman 1992 and Hobson & Stickgold 1994a for further discussion). In other words, we need recourse to more diverse means to elicit detailed descriptions of salient aspects of conscious experience.

### 2.3.2. The sleep laboratory environment.
The sleep laboratory itself constitutes a second major methodological problem. Anyone who has ever slept in a sleep laboratory (as all of us have!) knows that it is an inhospitable and unnatural setting that makes sleep more difficult and less deep than is possible in more naturalistic settings. To appreciate this point, the reader need only imagine going to an unfamiliar place in an inner city neighborhood of dubious safety, encountering a technician who is a stranger and often of the opposite sex, having ten electrodes affixed to the scalp with cement that smells like airplane dope and then being bid "goodnight" and "pleasant dreams." Hence

the famous first night effect (objectively poor sleep owing to discomfort and anxiety) often extends to a second night, and may contribute to a constriction of dream experience (as in dreams of the sleep lab setting) over even longer times. The laboratory environment may even alter the content of dreams recalled from spontaneous awakenings in the laboratory at the end of a night's sleep as evidenced by the high frequency of laboratory references in morning spontaneous awakening REM and NREM laboratory dream reports (Cicogna et al. 1998).

Studies such as those of Dement et al. (1965), Domhoff and Kamiya (1964), Okuma et al. (1975) and Whitman et al. (1962) have shown substantial incorporation of the experimental situation into laboratory dream reports particularly on the first night in the laboratory but persisting, at a lower level, into subsequent laboratory nights (Dement et al. 1965; Domhoff & Kamiya 1964). Similarly, content differences have been noted between laboratory and home dreaming (Domhoff & Kamiya 1964; Domhoff & Schneider 1999; Hall & Van de Castle 1966), although it has been argued that these differences are very small (Domhoff & Schneider 1999). Although these early studies were confounded by spontaneous (home) versus instrumental (laboratory) awakening conditions (as has been noted by Foulkes 1979), later studies controlling for reporting conditions (Lloyd & Cartwright 1991; Weisz & Foulkes 1970) still found some content differences between the home and laboratory dreams of adults. Waterman et al. (1993) emphasize that home-laboratory differences can arise from both environmental factors and factors related to investigator expectancies and, therefore, both should be controlled. In our view, full adaptation to the sleep lab may take four days or longer (see Domhoff & Kamiya 1964) exceeding the length of most laboratory studies.

As in the case of NREM compared to REM dreaming, we are not arguing for a gross, qualitative distinction between home and laboratory dreams. Laboratory dreams are, undoubtedly, largely representative of many of the formal and content features of dreaming in naturalistic settings. Nevertheless, we suggest that quantitative constraints on the dreaming experience may be imposed by the laboratory setting so that the full potential expression of certain dream features is limited. Of additional concern is the finding by Antrobus et al. (1991) that REM-NREM differences in both word count and global judgment of dreamlike quality diminish over 14 nights in the sleep laboratory, an effect they attribute largely to motivational factors in dream reporting. Minimizing any such "laboratory-fatigue" confound constitutes further argument for longitudinal awakenings to be performed in the more comfortable environs of the home.

To overcome these problems, several options are possible. First, laboratory studies can simply be extended in time, perhaps recording each subject for a full week. This has obvious disadvantages including inconvenience, high cost, and the above noted motivational effects. A second option is to continue to run relatively short (1–4 night) paradigms, and accept the suppressive effects on sleep architecture and dream content. While perhaps no longer normatively valid, the data obtained would still be at least reliable. A third option, and the one that we have chosen, is to move recording into the home for extended longitudinal studies using the Nightcap (Ajilore et al. 1995; Mamelak & Hobson 1989b; Pace-Schott et al. 1994; Rowley et al. 1998; Stickgold et al. 1994a; 1998b).

**2.3.3. The question of "similarity" and "difference."** We have long thought that the argument over whether mentation in two states like REM and NREM sleep is more similar or different was specious. Thinking the dilemma to be false, we have ignored or minimized it in our previous writings. However, we now feel obliged to clarify for the reader how the debate over REM and NREM mentation has become inextricably entangled with the larger and more general question of the mind-brain problem. In doing so, we hope to elevate the debate from the parochial to the general level and to make our own position on mind-brain issues crystal clear.

In some ways, understanding the conflicting opinions that swirl around the sleep and dream mental content debate is relatively straightforward. One group of psychologists, exemplified by David Foulkes and the late Alan Moffitt, hypothesizes that the brain and the mind are so loosely linked that the study of the mind need not be constrained – or even informed – by the study of the brain (e.g., Bosinelli 1995; Foulkes 1991; 1993b; 1996a; 1997; Moffitt 1995). This group interprets the empirical data as indicating that mental content does not differ qualitatively across brain states. There is only one dream mentation production system that is more or less active during waking and sleep. In such theories, termed "One-Generator" models of sleep mentation by Nielsen (1999), it is only the fluctuating level of cognitive activation that determines differences between REM and NREM sleep in report length as well as in the broad range of dream features that co-vary with report length. By taking this position, these psychologists minimize the importance of physiology, which they assert to be irrelevant to the understanding of dreaming. How cognitive activation could be independent of brain activation is a question not addressed by these scientists.

Another group, consisting largely of psychophysiologists, holds that the mind and the brain form an integrated system, so tightly linked within and across states that detailed qualitative and quantitative distinctions at either level of analysis imply the existence of isomorphic distinctions at the other. This is the position that we take. For us, the cognition production system *is* the brain. And, of course, it is always the *same* brain. But we know that the brain's mode of information processing changes radically across states. So, therefore, must its mental products. Nielsen (1999) terms this point of view a "Two-Generator" model of sleep mentation. For us, the state-specific changes in brain function virtually guarantee concomitant changes in mental function, even if our psychological methodology may still be inadequate to identify these changes (just as for many years the physiological changes also eluded us!).

With respect, we suggest that the failure to demonstrate psychological differences concomitant with physiological ones must be laid at the door of inadequate psychological methodology. If psychology has so far failed to document the robust phenomenological differences between waking and dreaming that most people experience every day of their lives, then more vigorous and more creative psychological research is needed. Otherwise we are faced with the absurd and unacceptable conclusion that brain and mind have nothing to do with each other.

That even a single, "One-Generator" system (i.e., a "dream mentation production system") may show dramatically different features in different states is in no way a self-contradiction. To our way of thinking, states of the brain are analogous to other dynamic states of matter. Consider, for example, the way that liquid water changes state with changes in temperature: above 100° C it is steam; below 0° C it is ice. These states are analogous to the states of waking, NREM sleep, and REM sleep in the brain (as well as to less common mental states such as coma, hypnosis, and mania). No one would say that in the frozen state (ice) or in the vapor state (steam) that the material is not still water. Nor could any sentient person ignore the obvious differences in the properties and behavior of water across states. We believe that it is equally inappropriate to argue that since there is a single dream production system (i.e., the brain-mind), that the properties and behavior of its products, for example, dreams, must be identical or even similar across different states. Such an important error in scientific thinking would lead to minimizing or missing entirely the change in matter (in this case the brain) that underlies the change in its state-dependent properties (in this case, consciousness).

The question of whether REM and NREM mentation are the same or different has often devolved into a search for characteristics of mentation that are absolutely unique to REM sleep. We consider this quest to be a fool's errand and indeed no absolute qualitative distinction between the two states has yet been documented. Since the late 1950s, many sleep laboratory studies have shown substantial recall of mentation from NREM, thereby obviating an exclusive association of sleep mentation with REM (Cicogna et al. 1998; Foulkes 1962; 1966; Foulkes & Rechtschaffen 1964; Goodenough et al. 1959; 1965b; Kamiya 1961; Molinari & Foulkes 1969; Pivik & Foulkes 1968; Rechtschaffen et al. 1963; Salzarulo & Cipolli 1979; Stoyva 1965; Zimmerman 1970; see Foulkes 1967, Herman et al. 1978, and Nielsen 1999 for reviews). For example, among nine studies, the percentage of NREM awakenings yielding at least minimal recall varied from 23 to 74% (Foulkes 1967) and, as noted, Nielsen (1999) has found an average NREM recall rate of 42.5% over 33 published studies. Recall rates similar to those of NREM in general have even been obtained from stages III and IV of NREM (e.g., Bosinelli 1995; Cavallero et al. 1992; Goodenough et al. 1965b; Herman et al. 1978; Nielsen 1999; Pivik & Foulkes 1968; Salzarulo & Cipolli 1979; Tracy & Tracy 1974). In a review of eight studies of stages III and IV mentation, Nielsen (1999) found an average recall rate of 52.5 (+18.6)%, but also notes that a substantial percentage of subjects never recall stage III and IV mentation or require several nights of awakenings before reporting such mentation.

The findings of several studies have countered the hypothesis that NREM mentation is simply recall from previous REM (Foulkes 1962; 1967; Foulkes & Rechtschaffen 1964; Goodenough et al. 1965b; Rechtschaffen et al. 1963), although report length does drop precipitously following the end of REM periods (Stickgold et al. 1994a).

The fact that differences are *not absolute* does not mean however that *no* differences exist. Indeed, all the evidence shows that such differences *do* exist and we have already advanced good reasons to believe that these may have been seriously underestimated. For example, similarities in dream features such as bizarreness may be inflated when report length is controlled in REM and NREM reports (Hunt et al. 1993) and REM-NREM bizarreness differences may persist even when report length is partialled out (Casagrande et al. 1996b; Nielsen 1999; Waterman et al. 1993). In addition, recent work comparing sleep onset REM and NREM dreams using an experimental protocol which controlled for previous sleep and waking time has shown that sleep onset REM periods are specifically related to physiological signs of REM whereas NREM dreams were related to intrusions of waking into NREM (Takeuchi et al. 1999b). These authors conclude that the mechanisms underlying REM and NREM dreaming must, therefore, differ (Takeuchi et al. 1999b). We thus conclude that while *some* NREM dreams approach REM dreams in length, vividness, dreaminess, and bizarreness (Cicogna et al. 1998; Foulkes & Schmidt 1983; Herman et al. 1978; Nielsen 1999) and while "dream-like" versus "thought-like" mentation may predominate in some NREM reports (Foulkes 1962; Nielsen 1999; Rechtschaffen et al. 1963; Zimmerman 1970), NREM reports are far more likely than REM reports to be short, dull, and undreamlike (Nielsen 1999; Rechtschaffen et al. 1963).

Many of the above-noted problems inherent in assessing the similarity versus difference of two phenomena can be addressed with improved methodologies. For example, when two states (such as REM and NREM) are being compared in terms of specific parameters (such as bizarreness) to a third state (such as waking), the question of the similarity versus difference between the two states becomes much more tractable.

**2.3.4. The source and fate of dream memory.** A tendency to emphasize psychological similarity has also characterized recent studies on the memory sources of REM and NREM dreams. Using a modification of Tulving and Thomson's (1973) classification of memory sources and an experimental free association technique, Cavallero and his colleagues initially found a distinct difference in memory sources between early-night REM and NREM mentation (Bosinelli 1991; Cavallero & Cicogna 1993; Cicogna et al. 1986). Early-night NREM sources consisted primarily of discrete biographical episodes while REM sources were a mixture of episodic, abstract self-referential and semantic sources (Bosinelli 1991; Cavallero & Cicogna 1993; Cicogna et al. 1986). This observation fits with the commonly accepted distinction between NREM dreaming as a simpler and REM dreaming as a more complex state of consciousness.

However, when REM and NREM reports were collected later in the night and matched for "temporal unit composition" (a procedure akin to diluting bizarreness by controlling for word count), the same researchers emphasized the similarity of memory sources between REM and NREM (Bosinelli 1991; Cavallero & Cicogna 1993; Cavallero et al. 1988; 1990; 1992; Cicogna et al. 1991; Fagioli et al. 1989). Likewise, Cicogna et al. (1991) reported few REM/Stage 2 differences in number of temporal units, implausibility, self presence, settings or characters. Nonetheless, as in the case of dream content (Antrobus 1983; Foulkes & Schmidt 1983), some residual state-related memory source differences continued to be reported (Cavallero & Cicogna 1993; Cavallero et al. 1990; 1992; Cicogna et al. 1991) and these need to be explained.

The research on memory sources for mentation among the different behavioral states overlooks the far more robust difference in the overall functioning of memory processes that distinguishes sleep from waking. This is the notorious difficulty of recalling dreams or any other mental content following either instrumental laboratory or spontaneous awakening. Many dreamers are aware that recall actively eludes them as they awaken. And even when dream recall is confident and detailed, it is common for subjects to

assert that they are sure that there was much more antecedent dreaming that could not be recalled. One reason for the neglect of this robust phenomenon is that it is difficult to study something, in this case memory, that isn't there! But the very absence of recall is a datum which any dream theory must explain, especially in the face of the robust brain activation in REM sleep!

Freud's famous explanation was that dream forgetting was an active function of repression. We have instead attributed this prominent failure of recall to a state dependent amnesia caused by aminergic demodulation of the sleeping brain (Hobson 1988b). The waking level of aminergic modulation falls to 50% in NREM sleep and to nearly zero in REM (Hobson & Steriade 1986; Steriade & McCarley 1990a). It would appear that the intense activation of REM must overcome this demodulation and persist into subsequent waking in order for very vivid dreams to be remembered. In our view, the low level of production and recall of NREM mentation is due to the additive effects of inactivation and demodulation.

This hypothesis is consonant with subjective experience. For example, when one introspectively compares recall of a night's dreaming with that of a corresponding waking epoch, one of the most obvious differences lies in the far greater amount of detail that can be recalled in waking. Moreover, it is commonplace for long dreams to have complete scene shifts of which the dreamer takes no significant cognitive account. If such orientational translocations occurred in waking, memory would immediately note the discontinuity and seek an explanation for it. This intuitively convincing difference between memory for dreaming and memory of waking mentation is confirmed by several empirical studies (see below).

Although the frequent inability to recall dreamed experience in subsequent waking has been a robust finding in dream research (Goodenough 1991), there is also strong evidence of deficient memory for prior waking experience in subsequent sleep. For example, little continuity has been shown between pre-sleep stimuli and the content of REM dreaming when this phenomenon has been probed using the following paradigms:

1. Specific experimental pre-sleep stimuli in the form of films have little effect on dream content (Cartwright et al. 1969; DeKoninck & Koulack 1975; Foulkes et al. 1967; Foulkes & Rechtschaffen 1964; Goodenough et al. 1975; Karacan et al. 1966; Witkin 1969; Witkin & Lewis 1967).

2. Specific experimental pre-sleep stimuli such as static visual images or altered social milieu are rarely incorporated into dreams (Carpenter 1987; Orr et al. 1968; Shevrin & Fisher 1967).

3. Specific pre-sleep waking behavioral or thought experiences are not easily detectable in subsequent dreams (Bakeland 1971; Bakeland et al. 1968; Breger et al. 1971; Cartwright 1974b; Hauri 1970).

4. Presleep mentation is infrequently picked up by the dream process (Rados & Cartwright 1982; Roussy et al. 1996; 1997).

5. Naturalistic daytime events rarely enter dream content, casting grave doubt on the classical psychoanalytic concept of day residue as dream instigator (Epstein 1985; Harlow & Roll 1992).

6. Pre-sleep modification of biological drives or perceptual experience has very weak effects on dreaming (Baldridge et al. 1965; Bokert 1968; Dement & Wolpert 1958;

Roffwarg et al. 1978). (For reviews see Arkin & Antrobus 1978 and Cavallero & Cicogna 1993.)

It must, therefore, be concluded that because dreaming is so little shaped by pre-sleep experience, memory systems active during REM sleep have extremely poor access to recent waking memories. Even if dreaming is concerned far more with emotionally salient content than with current events, it is remarkable that the dream construction process fails to incorporate recent episodic memories, including emotionally salient ones, to any significant extent. Two experimental exceptions to this generality, however, should be noted. The first involves the practice of dream incubation whereby focused pre-sleep attention on a specific concern has been shown to increase its rate of occurrence in subsequent dreaming (Saredi et al. 1997). Dream incubation techniques, however, introduce substantial confounds in the form of artificially imposed practice effects as well as the focus on emotionally salient issues. The second involves the finding by Rosenblatt et al. (1992) that significantly more of cartoon segments viewed prior to sleep were recalled following REM versus Stage 2 NREM awakenings, a difference which disappears if a 30 second pre-reporting waking delay is interposed after awakening. Following the arousal-retrieval model of Goodenough (1991), Rosenblatt et al. attribute this REM-NREM difference to greater mnemonic capacity immediately following post-REM versus post-NREM awakenings resulting from greater immediately pre-awakening cortical arousal in REM versus NREM. Using the semantic priming task, we have recently reported a similarly positive mnemonic effect of pre-awakening REM versus NREM for associative memory processes (Stickgold et al. 1999b). Certain forms of memory, such as generating associations to weakly related word primes, may, in fact, be preferentially enhanced by both the activation and the neuromodulatory differences (see sect. 4) between REM and NREM (Stickgold et al. 1999b). In contrast, greater sleep inertia (Dinges 1990) following NREM awakenings (a phenomenon undoubtedly reflecting low pre-awakening brain activation) may less selectively impair a wide spectrum of mnemonic processes.

Even within sleep, memory appears impaired. If episodic experiences within sleep were to persist in the sleeper's memory, one would expect greater content and thematic continuity between contiguous REM periods than more distant REM periods. But despite the fact that content and thematic continuity of successive dreams is greater within the same night than across nights, continuity does not differ between contiguous and noncontiguous REM periods of the same night (Cipolli et al. 1987; Fagioli et al. 1989).

We have recently completed three preliminary studies that seek to quantify aspects of memory within sleep and to compare sleep memory to waking memory. In the first study, 27 subjects became aware of and could later recall three aspects of their memory functioning (semantic, recent, and remote episodic) more often during two waking experiences than during dreaming. Since both types of waking experience sampled were much shorter than the duration of a night's dreaming, results further support the concept of a mnemonic deficiency in dreaming compared to waking (Pace-Schott et al. 1997a).

A second study examined perceived duration of dreaming. The 22.5 minute median perceived duration of dreams by 54 subjects was associated with an unexpectedly large variation. Even ignoring the highest and lowest 10% still

left a 24-fold variation. Such wide variance in a basic memory function further suggests a profound alteration of memory processes in dreaming as compared to waking (Stickgold et al. 1997a).

In the third study, 11 subjects recorded the processes by which a total of 103 dreams were recalled. Fifty-two reports (50%) were recalled in "chunks" (i.e., entire dream segments were recalled as units). Another 38 reports (37%) were recalled all at once upon waking and 13 reports (13%) were recalled gradually. Nine of the 11 subjects reported at least one dream recalled in chunks, and there were often significant delays between the recall of different "chunks." These results point strongly to the presence of stored dream memories which cannot be readily accessed on awakening and further suggests both qualitative and quantitative alterations in basic memory processes during and after dreaming (Stickgold 1998; Stickgold et al. 1997a).

All of the above findings can be regarded as being caused by the failure of recent episodic memory (as defined by Tulving 1994) in sleep. And as we have noted, recent episodic memory is weak across wake-sleep and sleep-wake transitions as well as within sleep itself (Pace-Schott et al. 1997b). We believe that a deficiency of memory in dreaming may go a long way toward explaining such distinctive and robust dream phenomena as orientational instability, loss of self-reflective awareness, and failure of directed thought and attention.

**2.3.5. Type I versus Type II statistical analyses.** In analyzing studies of dream mentation, it is important to understand the nature of the statistical tests employed. In general, such tests calculate the probability that a specific null hypothesis – normally that there is no difference between two population samples – is or is not true. The most common statistical tests, that is, Student's t-test and ANOVA, measure Type I error, which determines the probability that the obtained results could be explained by the null hypothesis. When the probability is sufficiently low, normally less than 0.05, the null hypothesis is rejected and one concludes that the populations are different. Such analyses, however, provide no information on whether or not the null hypothesis is true. Thus, while a low $p$-value provides strong evidence that the null hypothesis is false, a high $p$-value does not necessarily indicate that it is true.

This is relevant to the conclusion of both of the papers we critiqued above. Antrobus (1983) concluded that "the global judgment of Dreaming adds little, if anything, to Total Recall Content with respect to the association with the sleep stages REM and NREM" (p. 567), although his statistics did confirm a significant contribution ($F(1,71) = 15.9, p < 0.01$). Nevertheless, this conclusion formed the basis of the wider interpretation that the differences between REM and NREM reports are merely a consequence of enhanced recall in REM.

In the second paper critiqued, Foulkes and Schmidt (1983) concluded that global discontinuity "is stage-invariant [and] *never* significantly discriminated reports from different stages of sleep, even in length-uncontrolled comparisons" (p. 277). Although this was true, it was also true that sleep onset reports contained 2.3 times more global discontinuity than NREM reports, a ratio that increased to more than 3 to 1 when normalized for report length (measured in "temporal units"), a fact that could lead to a conclusion quite different from the one drawn by the authors.

It thus appears premature to conclude, based on these early studies, that robust differences between REM and NREM sleep mentation do not exist. Until studies are carried out that measure Type II error and determine the likelihood that the null hypothesis is correct, it is only safe to say that these studies have failed to demonstrate either the presence or absence of differences between REM and NREM mentation. Under the circumstances, more recent studies reporting the presence of significant differences would appear more easily interpreted.

**2.3.6. The need for new approaches.** The conclusion that we draw from all these studies is that there are significant differences between the formal aspects of the states of consciousness associated with waking, NREM, and REM sleep. These differences, which are quantitative not qualitative, have not yet been adequately characterized for a variety of methodological reasons. Instead of continuing to argue over this issue, we urge our colleagues to join us in a more creative attempt to capture and measure the dimensions of conscious experience.

Basing the attempt to characterize dreaming solely on verbal reports of the poorly recalled subjective experience of subjects sleeping in unfamiliar, non-natural settings has led, not surprisingly, to a sterile and nonproductive controversy about whether the conscious correlates of waking, NREM sleep, and REM sleep are more similar or different, and to a very unfortunate split in what was once a unified field.

This mind-brain split is akin to the gulf that opened between psychiatry and neurology after Sigmund Freud abandoned the goals of his brain-based Project for a Scientific Psychology and declared brain science off limits to his psychology. To reunify two approaches that belong together, we call for a new neuropsychology of conscious states that integrates from the level of cellular-molecular events to the formal features of the mental states of which they form the substrate.

# 3. The cognitive neuroscience of waking, sleeping, and dreaming

We now turn our attention to the shifts in activation level, input-output gating processes, and the neuromodulatory balance of the brain that underlie the ultradian REM/NREM cycle in humans and in animals. We first enumerate the profound physiological differences that distinctively differentiate waking, NREM, and REM sleep and show that these differences are as robust as those shown above in the phenomenology of waking, sleeping, and dreaming. Then, we point out relationships between the physiological and phenomenological changes seen as the brain-mind shifts from one state to another, as a prelude to integrative modeling. Our overarching hypothesis is that for each phenomenological difference seen between conscious states it is possible to identify a specific physiological counterpart. The end result is a first approximation of a cognitive neuroscience of brain-mind states.

## 3.1. Recent findings in human neurobiology

**3.1.1. Neuroimaging studies.** The experimental study of human REM sleep dreaming has until recently been limited on the physiological side by the poor resolving power of the EEG. Even expensive and cumbersome evoked potential

and computer averaging approaches have not helped us to analyze and compare REM sleep physiology with that of waking in an effective way. This limitation has probably helped reinforce the erroneous idea that the brain activation of REM sleep and waking are identical or at least, very similar. However, recent technological advances in the field of human brain imaging have made it possible to document a highly selective regional activation pattern of the brain in REM sleep (Braun et al. 1997; 1998; Maquet et al. 1996; Nofzinger et al. 1997). At the same time, experiments of nature – in the form of strokes – have allowed a correlation of the locale of brain lesions with deficits or accentuations of dream experience in patients (Doricchi & Violani 1992; Solms 1997a).

Before discussing these intriguing new results, it is important to stress the methodological limitations of both the brain lesion and imaging techniques. We know from our long and relevant experience in basic sleep research that neither method can capture many significant mechanistic and functional details that emerge from cellular and molecular level neurophysiology (see Hobson et al. 1986 and Steriade & Hobson 1976 for a full discussion of these issues). For example, it is now clear that the lesion method, applied to the pontine brain stem, gave misleading results regarding both the general role of that region in state control and failed even to hint at the specific functions of its subcomponent nuclei. This is because the lesion method cannot discriminate between the effects of destruction and disconnection and cannot target specific neuronal groups in heterogeneous regions like the brain stem.

It is important to note that the preliminary regional functional neuroimaging studies that we review below suffer from such unavoidable limitations of new technologies as the following (see Rauch & Renshaw 1995 for a more complete discussion). First, one must consider whether or not more efficient functioning of an area might result in less versus more observed metabolism or whether glucose or oxygen uptake by inhibitory interneurons may produce local maxima in areas that are, in fact, less active due to inhibition. Second, there are statistical problems inherent in the small sample sizes used in some of these sleep studies (e.g., Braun et al. 1998; Nofzinger et al. 1997) as well as the repeated comparisons employed by the statistical parametric mapping technique (Friston et al. 1991), which is used by all these investigators. Third, global activation measures like electroencephalographic voltage averaging or cerebral blood flow cannot be expected to reveal mechanistic and functional details because they cannot identify small but influential neuronal populations like the locus coeruleus, the raphe nuclei and the pedunculopontine tegmental nucleus. Fourth, there is the potential of altered sleep physiology due to the sleep deprivation (Maquet et al. 1996) or REM deprivation (Braun et al. 1997; 1998) procedures used to maximize sleep stability and stimulate REM in these studies. And fifth, the functional activity of a brain area may vary with changes in its inputs as most dramatically illustrated by neuroplasticity involving recruitment of dedicated brain areas to subserve new modalities such as the visual cortex in Braille learning (e.g., Pascual-Leone 1999) or the reorganization of visual association cortex following V1 damage (e.g., Baeseler et al. 1999). Additionally, it is possible that normal functional disconnections, as occurs between V1 and visual association cortices in REM (Braun et al. 1998), result in the same neural structures performing differing, state-specific functional tasks.

In spite of these caveats, the widespread use of this technology and the broad agreement of the data with clinical neuropsychological findings argues strongly for the basic validity of neuroimaging as a tool in cognitive neuroscience (Cabeza & Nyberg 1997; 2000). Specifically in response to the fifth caveat above, strong suggestion that the functions of specific brain areas are similar between REM and wake is provided by the observable enactment of experienced dream movement in the REM sleep behavior disorder (Schenck et al. 1993). Moreover, wake-like function of regional brain areas is preserved in many abnormal states such as focal motor activity during seizures (Adams et al. 1997) or the recruitment of visual association cortex during visual hallucinations (Ffytche et al. 1998; Silbersweig et al. 1995). In future sleep research, many of these limitations may be overcome by the finer temporal and spatial resolution offered by functional MRI (fMRI) imaging (e.g., Ellis et al. 1999; Huang-Hellinger et al. 1995; Ives et al. 1997; Sutton et al. 1996; 1997; 1998; Lovblad et al 1999).

Our review of this new literature is undertaken with these shortcomings in mind. Three factors weighed heavily in our evaluation of these data: (1) their novelty and uniqueness in beginning to describe the role of forebrain subsystems; (2) the surprising concordance in the neuroimaging results that emerged from studies carried out simultaneously by three independent groups; and (3) the complementarity between the lesion and imaging studies that confer the value of a double dissociation on the validity of the inferences drawn.

### 3.1.2. PET studies indicating regional activation differences between REM sleep and waking.
Two very recent and entirely independent PET studies confirm the importance of the pontine brain stem in REM sleep brain activation (Braun et al. 1997; Maquet et al. 1996). This is an important advance because it validates, for the first time, the experimental animal data on the critical and specific role of the pontine brain stem in REM sleep generation. At the same time, these new studies also provide important new data for our understanding of dream synthesis by the forebrain. Instead of the global, regionally nonspecific picture of forebrain activation that has been suggested by EEG studies, all of these new imaging studies indicate a preferential activation of limbic and paralimbic regions of the forebrain in REM compared to waking (Braun et al. 1997; 1998; Maquet et al. 1996; Nofzinger et al. 1997). One implication of these discoveries is that dream emotion may be a primary shaper of dream plots rather than playing a secondary role in dream plot instigation.

### 3.1.2.1. The PET imaging findings of the Maquet group.
Maquet et al. (1996) used an $H_2^{15}O$ positron source to study REM sleep activation in their subjects who were then awakened for the solicitation of dream reports. In addition to the pontine tegmentum, significant activation was seen in both amygdalae and the anterior cingulate cortex (Table 2). Significantly, despite the general deactivation in much of the parietal cortex, Maquet et al. (1996) reported activation of the right inferior parietal lobe (Bredman area 40) – a brain region thought to be important for spatial imagery construction, an important aspect of dream cognition. The authors interpreted their data in terms of the selective processing, in REM, of emotionally influenced memories (see also Braun et al. 1997; Maquet & Franck 1997).

### 3.1.2.2. The PET imaging findings of the Braun group.
In another $H_2^{15}O$ PET study, Braun et al. (1997) largely replicated the Maquet group's findings of a consistent REM-

related brainstem, limbic, and paralimbic activation. In REM compared *individually* to delta NREM and to pre- and post-sleep waking (see Table 2), these authors showed relative activation of the pons, midbrain, anterior hypothalamus, hippocampus, caudate, and medial prefrontal, caudal orbital, anterior cingulate, parahippocampal, and inferior temporal cortices (Braun et al. 1997). Based on their observations, the Braun group then offered the following speculations which are relevant to the neurology of dreaming:

(1) Ascending reticular activation during REM as compared to waking may favor a more ventral cholinergic route leading from the brainstem to the basal forebrain over a more dorsal route via the thalamus.

(2) Activation of the cerebellar vermis in REM may reflect input to this structure from the brainstem vestibular nuclei. We note that these nuclei also constitute an important potential source of neuronal activation causing the unique vestibular features of fictive movement in dreams (Hobson et al. 1998c; Leslie & Ogilvie 1996; Sauvageau et al. 1998).

(3) Noting both a particularly strong REM sleep-related activation of the basal ganglia and the known connectivity of these subcortical structures, Braun et al. suggest that the basal ganglia may play an important role in an ascending thalamocortical activation network. They suggest that this network extends successively from the brainstem to the intralaminar thalamic nuclei, then to the basal ganglia, and back to the ventral anterior and ventromedial thalamic nuclei, and thence to the cortex.

This network contains multiple regulatory back projections including interconnections between the pedunculopontine tegmentum and the striatum further suggesting a possible role for the basal ganglia in the rostral transmission of PGO waves and the modulation of REM sleep phenomena. The extensive interconnections of the basal ganglia and the pedunculopontine area have recently been reviewed by Rye (1997) and Inglis and Winn (1995). The role of the basal ganglia in the initiation of motor activity may, in turn, be related to the ubiquity of motion in dreams (Hobson 1988b; Porte & Hobson 1996).

(4) The REM-associated increase in activation of unimodal associative visual (Brodmann areas 19 and 37) and auditory (Brodmann area 22) cortices contrasted with the maintained (NREM and REM) sleep-related deactivation of heteromodal association areas in the frontal and parietal cortex. Combined with findings of striate cortex deactivation in REM, this group (Braun et al. 1998) has subsequently theorized that, during REM, internal information is being processed between extrastriate and limbic cortices while they are functionally isolated from the external world both in terms of input (from the striate cortex) and output (via the frontal cortex).

(5) The prominent decrease in the executive portions of the frontal cortex (dorsolateral and orbital prefrontal cortices) contrasts with the REM-associated increase in activation of the limbic associated medial prefrontal area. This medial area region has the most abundant limbic connections in the prefrontal cortex, has been associated with arousal and attention, and disruption of this area has been shown to cause confabulatory syndromes formally similar to dreaming. (Note also the dream-wake confusional syndrome associated with anterior limbic cortical lesions reported by Solms 1997a.)

### 3.1.2.3. The PET imaging findings of the Nofzinger group.
Also confirming widespread limbic activation in REM

sleep, Nofzinger et al. (1997) described increased glucose utilization in the lateral hypothalamic area and the amygdaloid complex using an 18F-fluoro-deoxyglucose (FDG) PET technique (Table 2). The largest area of activation was, in their own words, " . . . an extensive confluent area along the midline that includes the lateral hypothalamic area, septal area, ventral striatum-substantia innominata, infralimbic cortex, prelimbic and orbitofrontal and the anterior cingulate cortex . . . Much of this is bilateral" (p. 198). The authors suggest that an important function of REM sleep is the integration of neocortical function with basal forebrain and hypothalamic motivational and reward mechanisms.

### 3.1.3. Selective deactivation of the dorsolateral prefrontal cortex in REM sleep.

Relevant to the cognitive deficits in self-reflective awareness, orientation, and memory during dreaming was the $H_2^{15}O$ PET finding of significant deactivation, in REM, of a vast area of dorsolateral prefrontal cortex (Braun et al. 1997; Maquet et al. 1996). A similar decrease in cerebral blood flow to frontal areas during REM has been noted by Madsen et al. (1991a) using single photon emission computed tomography (SPECT) and by Lovblad et al. (1999) using fMRI. Dorsolateral prefrontal deactivation during REM, however, was not replicated by an FDG PET study (Nofzinger et al. 1997) and this discrepancy, therefore, remains to be clarified by other FDG as well as $H_2^{15}O$ studies. (A potential cause of this discrepancy arising from differences between FDG and $H_2^{15}O$ methods is discussed further in sect. 3.3.5.2.)

Nevertheless, it seems likely that considerable portions of executive and association cortex active in waking may be far less active in REM, leading Braun et al. (1997) to speculate that "REM sleep may constitute a state of generalized brain activity with the specific exclusion of executive systems which normally participate in the highest order analysis and integration of neural information" (p. 1190).

Taken together, these results strongly suggest that the forebrain activation and synthesis processes underlying dreaming are very different from those of waking. Not only is REM sleep chemically biased but the preferential cholinergic neuromodulation is associated with selective activation of the subcortical and cortical limbic structures (which mediate emotion) and with relative inactivation of the lateral prefrontal cortex (which mediates directed thought). These findings greatly enrich and inform the integrated picture of REM sleep dreaming as emotion-driven cognition with deficient memory, orientation, volition, and analytic thinking.

The Maquet et al. (Maquet et al. 1996; Maquet & Franck 1997), Nofzinger et al. (1997), and Braun et al. (1997) groups all stress that their findings suggest assigning REM sleep a role in the processing of emotion (along with its cognitive and autonomic correlates) in memory systems via a limbic-cortical interplay. Additionally, PET researchers suggest the possible origin of dream emotionality in REM-associated limbic activation (Braun et al. 1997; Maquet & Franck 1997) and dream-associated executive deficiencies in REM-associated frontal deactivation (Braun et al. 1997; Maquet & Franck 1997). Although tantalizing correlations such as: (1) limbic activation and dream emotionality, (2) dream emotionality and affect-congruent dream narratives, and (3) frontal deactivation and dream bizarreness, are now becoming apparent in the sleep and dream literature, the precise causal sequence among these phenomena remains to be established by future research.

Two additional findings support this proposed cortico-limbic interaction. First, the anterior cingulate cortex has consistently shown increased activation in REM in other PET studies (e.g., Bootzin et al. 1998; Buchsbaum et al. 1989; Hong et al. 1995). Second, recent studies of human limbic structures with depth electrodes during REM sleep have shown distinctive rhythmic EEG patterns possibly related to the REM-associated hippocampal theta rhythms seen in animals (Mann et al. 1997; Staba et al. 1998). Human frontal midline theta has also been detected using scalp electrodes (Inanaga 1998).

### 3.1.4. Global and regional decreases in activation level in NREM sleep.

Neuroimaging studies also strongly support a distinction between REM and NREM sleep as states whose differing neuroanatomical activation patterns predict their observed phenomenological differences (Table 2). PET studies of NREM sleep generally show a decrease in global cerebral energy metabolism (i.e., $O_2$ or glucose utilization) relative to waking and REM (Buchsbaum et al. 1989; Heiss et al. 1985; Madsen & Vorstup 1991; Madsen et al. 1991b; 1999b; Maquet 1995; Maquet et al. 1990; 1992; 1997). The magnitude of this decline relative to waking has varied from 11% glucose utilization in stage 2 (Maquet et al. 1992) to 40% glucose utilization in stages 3 and 4 (Maquet et al. 1990). A similar pattern has usually been reported for global cerebral blood flow as measured by $H_2^{15}O$ PET, SPECT, near infrared spectroscopy or a modification of the Kety-Schmidt $O_2$ uptake technique (Braun et al. 1997; Hoshi et al. 1994; Madsen et al. 1991a; 1991b; Maquet et al. 1997; Meyer et al. 1987; Sakai et al. 1980), although some studies have failed to show this global hemodynamic change (Andersson et al. 1995; 1998; Hofle et al. 1997). In addition, cerebral energy metabolism decreases with progressively greater depth of NREM sleep (Maquet 1995) a result recently replicated with fMRI (Sutton et al. 1997). By contrast, in REM, global cerebral energy metabolism tends to be equal to (Asenbaum et al. 1995; Braun et al. 1997; Madsen et al. 1991b; Maquet et al. 1990) or greater than (Buchsbaum et al. 1989; Heiss et al. 1985) that of waking. Cerebral blood flow velocity measured in the middle cerebral artery similarly shows a slowing during NREM followed by values similar to waking during REM (Droste et al. 1993; Haiak et al. 1994; Klingelhofer et al. 1995; Kuboyama et al. 1997).

More striking than global patterns are the now well-replicated regional variations in cerebral energy metabolism over the wake-NREM-REM sleep cycle (Table 2). Earlier studies showing specific declines in thalamic glucose utilization in NREM relative to waking (Buchsbaum et al. 1989; Maquet et al. 1990; 1992) have been confirmed by recent oxygen utilization studies (Andersson et al. 1998; Braun et al. 1997; Hofle et al. 1997; Maquet et al. 1997). In addition to prominent thalamic deactivation, all three recent studies have found regional deactivation during NREM in the pontine brain stem, orbitofrontal cortex, and anterior cingulate cortex (Braun et al. 1997; Hofle et al. 1997; Maquet et al. 1997). NREM deactivation of lateral prefrontal cortex was also observed in some studies (Andersson et al. 1998; Braun et al. 1997). Thalamic activation was found to decline significantly concomitant with increased delta EEG activity and there was an additional decline associated with increased spindle-frequency activity when the decrements associated with delta were subtracted (Hofle et al. 1997). (For a very recent review see Maquet 2000.)

Table 2. *Review of relative activation of cortical and subcortical areas in REM and SWS noted in four recent PET studies (from Hobson 1998a; 2000)*

| SLEEP STAGE | REM | REM | REM | REM | NREM (3&4) | NREM (delta) | NREM (3&4) |
|---|---|---|---|---|---|---|---|
| STUDY | Maquet et al. 1996 | Nofzinger et al. 1996 | Braun et al. 1997 | Braun et al. 1997 | Maquet et al. 1997 | Hofle et al. 1997 | Braun et al. 1997 |
| TECHNIQUE | $H_2^{15}O$ | $^{18}FDG$ | $H_2^{15}O$ | $H_2^{15}O$ | $H_2^{15}O$ | $H_2^{15}O$ | $H_2^{15}O$ |
| RELATIVE TO | all other stages | waking | pre- (& post*)-sleep waking | NREM 3&4 | all other stages | change with increased delta | pre- or post-sleep waking |
| SUBCORTICAL AREAS | | | | | | | |
| brainstem | | | | | | | |
| pontine tegmentum | increase | | increase (R*) | increase | decrease | decrease:R | decrease |
| midbrain | increase | | increase* | increase | | | decrease |
| dorsal mesencephalon | | | | | decrease | | |
| diencephalon | | | | | | | |
| thalamus | increase: L | increase: R, Lat. | increase: A-POA | increase | decrease | decrease: M | decrease |
| hypothalamus | | | | increase: A-POA | decrease | | decrease: A-POA |
| basal forebrain | | | | | decrease | | |
| limbic system | | | | | | | |
| left amygdala | increase | increase | | | | | |
| right amygdala | increase | | | | | | |
| septal nuclei | | increase | | | | | |
| hippocampus | | | increase* | increase | | | |
| basal ganglia/striatum | | | | | | | |
| caudate | | increase: A, I, L | increase* | increase | decrease | | decrease |
| putamen | | increase | | increase | | | decrease: P |
| ventral striatum (n. accumbens, sub.innominata) | | | | increase | | | decrease |
| lenticular nuclei | | | | | decrease | | |
| cerebellum | | | incr. (vermis)* | increase (vermis) | | decrease | decrease: I |

| | REM | REM | REM | REM | NREM (3&4) | NREM (delta) | NREM (3&4) |
|---|---|---|---|---|---|---|---|
| **SLEEP STAGE** | REM | REM | REM | REM | NREM (3&4) | NREM (delta) | NREM (3&4) |
| **STUDY** | Maquet et al. 1996 | Nofzinger et al. 1996 | Braun et al. 1997 | Braun et al. 1997 | Maquet et al. 1997 | Hofle et al. 1997 | Braun et al. 1997 |
| **TECHNIQUE** | $H_2^{15}O$ | $^{18}FDG$ | $H_2^{15}O$ | $H_2^{15}O$ | $H_2^{15}O$ | $H_2^{15}O$ | $H_2^{15}O$ |
| **RELATIVE TO** | all other stages | waking | pre or post-sleep waking | NREM 3&4 | all other stages | change with increase delta | pre or post-sleep waking |
| **CORTICAL AREAS** | | | | | | | |
| **FRONTAL** | | | | | | | |
| dorsolateral prefrontal | decrease: L: 10,11,46,47 R: 8,9,10,11,46 | decrease: L, small areas increase: R | decrease: 46* | | | | decrease: 46 |
| opercular | | increase | decrease: 45* | | | | decrease: 45 |
| paraolfactory | | | | | | | |
| lateral orbital | | increase: 11,12 | decrease: 11* | | decrease: 11,25 | decrease: R 11 | decrease: 11 |
| medial orbital | | | | | | | decrease: R |
| caudal orbital | | increase | increase | increase | | | decrease |
| gyrus rectus | | | | | | | |
| **PARIETAL** | | | | | | | |
| Brodmann area 40 (supramarginal gyrus) | increase: R A 40 decrease: L 40 | | decrease: 40* | | | increase: L 40 | decrease: 40 |
| angular gyrus | | | decrease: 39* | | | | decrease: 39 |
| precuneus | decrease | | | | decrease: 7 | | |
| cuneus | | | | | decrease: 19 | | |
| pericentral | | | | | | increase: L 3/4 | |
| **TEMPORAL** | | | | | | | |
| mesiotemporal | | increase R | | | decrease: R 28 | incr: A R,L 21 | |
| middle | | | | | | increase: L 22 | |
| posterior superior | | | | increase: 22 | | | |
| inferior/fusiform | | | increase: 37,19 (post-sleep only) | increase 37,19 | | | |
| **OCCIPITAL** | | | | | | | |
| medial | | decrease: L, small areas | | | | incr: R 17/18 | |
| post-rolandic sensory | | | increase | | | incr: L 17 | |
| **LIMBIC ASSOCIATED** | | | | | | | |
| medial (prelimbic) prefrontal | | increase: R 32 | increase: 10 | increase: 10 | | | decrease: 10 |
| anterior cingulate | increase: 24 | increase: 24 | increase: 32* | increase: 32 | decrease: 24,32 | decrease: 24/32 | decrease: 32 |
| posterior cingulate | decrease: 31 | dec: R sm. areas | decrease* | | | | |
| infralimbic | | increase: 25 | | | | | |
| insula | | increase: L | decrease: P | increase: A I | | | decrease: A |
| parahippocampal | | increase | increase: 37* | increase: 37 | | | |
| entorhinal | increase | increase (in fusiform) | | | | | |
| temporal pole | | | | increase: 38 | | | decrease: 38 |

Abbr: L-left hemisphere; R-right hemisphere; A-anterior; P-posterior; C-caudal; M-medial; Lat.-lateral; I-inferior; S-superior; A-POA-anterior preoptic area; all numerals = Brodmann's area; sm.-small, dec.-decrease, inc.-increase.

Hofle et al. (1997) and Maquet et al. (1997) both interpret this pattern of decline as reflecting the progressive deactivation of the reticular activating system (RAS) that accompanies deepening NREM sleep. This deactivation leads to dysfacilitation of thalamocortical relay neurons, which allows the emergence of underlying thalamocortical oscillatory rhythms (Steriade & McCarley 1990a; Steriade et al. 1993a; 1993b; 1993c; 1993d; 1994; for recent reviews see Steriade 1997; 1999; 2000). GABAergic neurons of the thalamic reticular nucleus then further hyperpolarize and dysfacilitate thalamic relay neurons as NREM deepens (Steriade et al. 1994). In this hyperpolarized condition, thalamic neurons become constrained to burst firing patterns first in spindle (12–14 Hz) and later in delta (1–4 Hz) frequencies as NREM deepens from Stage 2 to delta sleep (Steriade et al. 1993a; 1993d). The cortex may further constrain these spindle and delta-wave-generating thalamocortical bursts within a newly described slow (<1 Hz) oscillation seen in cats (Steriade et al. 1993a; 1993b; 1993c; 1993d) and humans (Achermann & Borbely 1997). In conclusion, the metabolic decline seen during NREM is centered on the central core structures (brain stem, thalamus) which are known to play a role in generation of the slow oscillations of NREM sleep (Maquet 2000; Maquet et al. 1997).

The regional pattern of deactivation in NREM, therefore, sharply contrasts with the regional *activation* of these same regions (i.e., thalamus, pontine brain stem, anterior cingulate cortex) in REM (Braun et al. 1997; Maquet et al. 1996; Nofzinger et al. 1997). Details of these stage-related differences are shown in Table 2. Note that a recent cat study has shown a similar pattern of brain glucose metabolism in REM (Lydic et al. 1991a).

### 3.1.5. Interpreting the PET imaging results with respect to the psychophysiology of dreaming.
According to PET researchers, regional activation during REM may reflect a specific activation of subcortical and cortical arousal and limbic structures for the adaptive processing of emotional and motivational learning (Maquet et al. 1996; Nofzinger et al. 1997). Such processing may, in turn, account for the emotionality and psychological salience of REM dreaming (Braun et al. 1997). Some support for this comes from a PET (glucose) study showing correlation between content-analyzed dream anxiety and medial frontal activation (Gottschalk et al. 1991a).

In summary, the markedly differing physiology of wake, NREM, and REM cerebral activation should be reflected in the respective phenomenology of mentation reported from these three conscious states. More particularly, the specific phenomenology of REM mentation may reflect the neurobiologically specific brain activation pattern. Nofzinger et al. (1997) conclude that "the current findings of increased limbic and paralimbic activation during REM sleep . . . as well as global, regionally nonselective cortical deactivation and decreased metabolism during NREM sleep, are generally supportive of the traditional notion that more story-like affect-laden dreams are more attributable to the REM sleep, than NREM sleep behavioral state" (p. 199).

### 3.1.6. Brain lesions resulting in loss or alteration of dreaming.
#### 3.1.6.1. Solms's nosology for lesion-related disorders of dreaming.
A set of findings and conclusions which have proved remarkably complementary to the neuroimaging results have been reached following a neuropsychological survey of 332 clinical cases of cerebral lesions as well as a review of 73 extant publications on the dreaming-related sequelae of cerebral injury (Solms 1997a). Using these welcome and long overdue neuropsychological data, Solms proposes a new nosology for the brain-lesion related disorders of dreaming.

In one syndrome, "global anoneria," total cessation of dreaming in patients (whose normal waking vision is preserved) results from either posterior cortical or deep bilateral frontal lesions. The posterior global anoneria syndrome results from lesions of the inferior parietal lobes in either hemisphere, with lesions to Brodmann's areas 39 and 40 being the most restricted damage sufficient to produce the syndrome. The anterior variant of global anoneria results from deep medial frontal damage resulting in the disconnection of the mediobasal frontal cortex from the brain stem and diencephalic limbic regions. In this syndrome, bilateral damage to white matter in the vicinity of the frontal horns of the lateral ventricles was the most restricted site causing the syndrome.

The nosological distinction of a second syndrome, non-visual dreaming, from syndromes of global cessation of dreaming, was first systematically formulated by Doricchi and Violani (1992). In this syndrome, termed "visual anoneria" by Solms (1997a), bilateral medial occipito-temporal lesions produce full or partial loss of dream visual imagery (again with normal waking vision). Among his own patients, a decrease in the "vivacity" of dreaming was reported by two patients with damage to the seat of normal vision in the medial-occipital-temporal cortex (especially areas V3, V3a, and V4 but not V1, V5, or V6). Notably, a correlate of visual anoneria was visual irreminiscence, the inability to produce mental imagery in waking. In addition, partial variants of visual anoneria exist which involve selective loss of particular visual elements (e.g., "kinematic anoneria" or "facial anoneria").

In addition to these two disorders of attenuated dreaming, Solms reported another interrelated pair of symptom complexes that combined increased frequency and intensity of dreaming. He suggested that increased vivacity and frequency of dreaming was associated with anterior limbic lesions while recurring nightmares are associated with temporal seizures.

#### 3.1.6.2. Conclusions suggested by convergent PET and lesion findings.
We believe that these findings map particularly well onto the neuroimaging findings on REM. For example, extrastriate visual cortex is activated during REM (Braun et al. 1997; 1998) and lesions to this region produce the distinctive dream deficits of full or partial visual anoneria (Solms 1997a). In contrast, the striate visual cortex is deactivated during REM (Braun et al. 1998) while lesions to this region do not affect dreaming (Solms 1997a). Similarly, the seat of spatial cognition in the inferior parietal cortex (BA 40) is activated in the right (but not the left) hemisphere during REM (Maquet et al. 1996) while damage to this region, especially on the right, is sufficient to produce global anoneria (Solms 1997a). Moreover, much of the lateral prefrontal area is deactivated during REM (Braun et al. 1997; Maquet et al. 1996), while lesions to this region do not affect dreaming (Doricchi & Violani 1992; Solms 1997a).

Two exceptions to this general correspondence involve lesions of the brainstem (for which Solms reports no attenuation of dreaming) and lesions of the rostral limbic system (for which Solms reports an accentuation of dreaming). In the case of pontine lesions, we suggest that any lesion capable of destroying the pontine REM sleep generator mechanism would have to be so extensive as to eliminate consciousness altogether. We base this caveat upon the difficulty of suppressing REM by experimental lesions of the pons in animals. In the case of the rostral limbic system, we caution that lesions there could as well be irritative as destructive and that lesions in different areas of this functionally highly heterogeneous region (Devinsky et al. 1995) could produce dramatically different effects.

### 3.2. Reciprocal interaction: A neurobiological update

The discovery of the ubiquity of REM sleep in mammals provided the brain side of the brain-mind state question with an animal model (Dallaire et al. 1974; Dement 1958; Jouvet & Michel 1959; Jouvet 1962; 1999; Snyder 1966). While animal studies showed that potent and widespread activation of the brain did occur in REM sleep, it soon became clear that Moruzzi and Magoun's concept of a brain stem reticular activating system (Moruzzi & Magoun 1949) required extension and modification to account for the differences between the behavioral and subjective concomitants of waking and those of REM sleep (see Hobson & Brazier 1981).

**3.2.1. Implications for dream theory.** We take the theoretical position that it is the cellular and molecular level brain events to be discussed that bias the brain to produce the conscious state differences that contrast waking, NREM, and REM sleep. As we will point out in detail in section 4 when we develop the AIM model, the shift from aminergic dominance in waking to cholinergic dominance in REM lowers the probability that consciousness will be exteroreceptive, logical, and mnemonic while correspondingly raising the probability that consciousness will be interoceptive, illogical, and amnesic.

**3.2.2. Behavioral state-dependent variations in neuromodulation.** A conceptual breakthrough was made possible by the discovery of the chemically specific neuromodulatory subsystems of the brain stem (e.g., Dahlstrom & Fuxe 1964; for reviews see Foote et al. 1983; Gottesmann 1999; Hobson & Steriade 1986; Hobson et al. 1998; Jacobs & Azmita 1992; Lydic & Baghdoyan 1999; Mallick & Inoue 1999; Rye 1997; Steriade & McCarley 1990a) and of their differential activity in waking (noradrenergic and serotonergic systems on, cholinergic system damped) and REM sleep (noradrenergic and serotonergic systems off, cholinergic system undamped) (Aston-Jones & Bloom 1981; Cespuglio et al. 1981; Chu & Bloom 1973; 1974; Hobson et al. 1975; Jacobs 1986; Lydic et al. 1983; 1987; McCarley & Hobson 1975; McGinty & Harper 1976; Rasmussen et al. 1986; Reiner 1986; Steriade & McCarley 1990a; Trulson & Jacobs 1979).

**3.2.2.1. The original reciprocal interaction model: an aminergic-cholinergic interplay.** The model of reciprocal interaction (McCarley & Hobson 1975) provided a theoretical framework for experimental interventions at the cellular and molecular level that has vindicated the notion that waking and dreaming are at opposite ends of an aminergic-cholinergic neuromodulatory continuum, with NREM sleep holding an intermediate position (Fig. 2). The reciprocal interaction hypothesis (McCarley & Hobson 1975) provided a description of the aminergic-cholinergic interplay at the

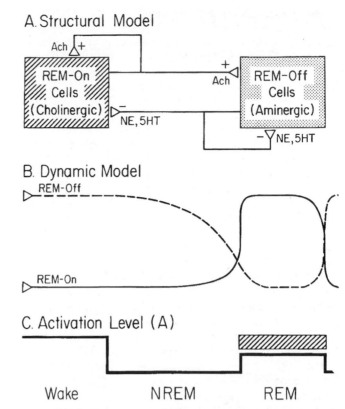

Figure 2. The original Reciprocal Interaction Model of physiological mechanisms determining alterations in activation level. A: Structural model of Reciprocal Interaction. REM-on cells of the pontine reticular formation are cholinoceptively excited and/or cholinergically excitatory (ACH+) at their synaptic endings. Pontine REM-off cells are noradrenergically (NE) or serotonergically (5HT) inhibitory (−) at their synapses. B: Dynamic Model. During waking, the pontine aminergic system is tonically activated and inhibits the pontine cholinergic system. During NREM sleep, aminergic inhibition gradually wanes and cholinergic excitation reciprocally waxes. At REM sleep onset, aminergic inhibition is shut off and cholinergic excitation reaches its high point. C: Activation level. As a consequence of the interplay of the neuronal systems shown in A and B, the net activation level of the brain (A) is at equally high levels in waking and REM sleep and at about half this peak level in NREM sleep. (Taken from Hobson 1992a.)

synaptic level and a mathematical analysis of the dynamics of the neurobiological control system (Figs. 2 and 3A). In this section we review subsequent work that has led to the alteration (Fig. 3B) and elaboration (Fig. 4) of the model.

Although there is abundant evidence for a pontine peribrachial cholinergic mechanism of REM generation centered in the pedunculopontine (PPT) and laterodorsal tegmental (LDT) nuclei (for recent reviews see Datta 1995; 1997b; 1999; Hobson 1992b; Hobson et al. 1993; Lydic & Baghdoyan 1999; Rye 1997), not all pontine PPT and LDT neurons are cholinergic (Kamodi et al. 1992; Kang & Kitai 1990; Leonard & Llinas 1990; 1994; Sakai & Koyama 1996; Steriade et al. 1988) and cortical acetylcholine release may be as high during wakefulness as during sleep (e.g., Jasper & Tessier 1971; Jimenez-Capdeville & Dykes 1996; Marrosu et al. 1995).

Recently, reciprocal interaction (McCarley & Hobson 1975) and reciprocal inhibition (Sakai 1988) models for control of the REM sleep cycle by brain stem cholinergic

## A. Original Model

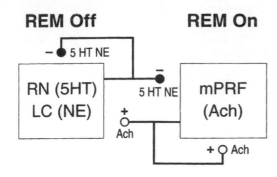

## B. Revised Model

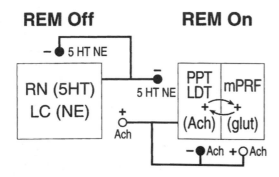

Figure 3. Synaptic modifications of the original reciprocal interaction model based upon recent findings. A: The original model proposed by McCarley and Hobson (1975) and detailed in Figure 2. B: Synaptic modifications of the original reciprocal interaction model based upon recent findings of self-inhibitory cholinergic autoreceptors in mesopontine cholinergic nuclei and excitatory interactions between mesopontine cholinergic and noncholinergic neurons (see Fig. 4 for more detail and references). Note that the exponential magnification of cholinergic output predicted by the original model (Fig. 2) can also occur in this model with mutually excitatory cholinergic-noncholinergic interactions taking the place of the previously postulated, mutually excitatory cholinergic-cholinergic interactions. In the revised model, inhibitory cholinergic autoreceptors would contribute to the inhibition of LDT and PPT cholinergic neurons, which is also caused by noradrenergic and serotonergic inputs to these nuclei. Therefore the basic shape of reciprocal interaction's dynamic model (illustrated in Fig. 2B) and its resultant alternation of behavioral state (illustrated in Fig. 2C) could also result from the revised model. *Abbreviations:* open circles, excitatory postsynaptic potentials; closed circles, inhibitory postsynaptic potentials; RN, dorsal raphe nucleus; LC, locus coeruleus; mPRF, medial pontine reticular formation; PPT, pedunculopontine tegmental nucleus; LDT, laterodorsal tegmental nucleus; 5HT, serotonin; NE, norepinephrine; Ach, acetylcholine; glut, glutamate.

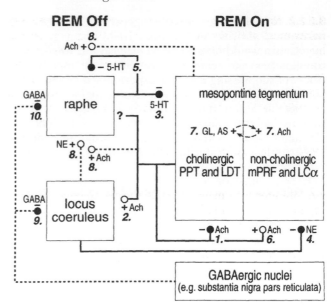

Figure 4. Additional synaptic details of the revised reciprocal interaction model shown in Figure 3B derived from data reported (solid lines) and hypothesized relationships suggested (dotted lines) in recent experimental studies (numbered on Figure and below). See text for discussion of these findings. Additional synaptic details can be superimposed on the revised reciprocal interaction model without altering the basic effects of aminergic and cholinergic influences on the REM sleep cycle. Excitatory cholinergic-non-cholinergic interactions utilizing Ach and the excitatory amino acid transmitters enhance firing of REM-on cells (6, 7) while inhibitory noradrenergic (4), serotonergic (3), and autoreceptor cholinergic (1) interactions suppress REM-on cells. Cholinergic effects upon aminergic neurons are both excitatory (2), as hypothesized in the original reciprocal interaction model and may also operate via presynaptic influences on noradrenergic-serotonergic as well as serotonergic-serotonergic circuits (8). GABAergic influences (9, 10) as well as other neurotransmitters such as adenosine and nitric oxide (see text) may contribute to the modulation of these interactions. *Abbreviations:* open circles, excitatory postsynaptic potentials; closed circles, inhibitory postsynaptic potentials; mPRF, medial pontine reticular formation; PPT, pedunculopontine tegmental nucleus; LDT, laterodorsal tegmental nucleus; LCα, peri-locus coeruleus α; 5HT, serotonin; NE, norepinephrine; Ach, acetylcholine; GL, glutamate; AS, aspartate; GABA, gamma-aminobutyric acid. *References:* (1) Baghdoyan et al. 1997; El Manseri et al. 1990; Kodama & Honda 1996; Leonard & Llinas 1990; 1994; Luebke et al. 1993; Roth et al. 1996; Sakai & Koyama 1996; Sakai et al. 1990. (2) Egan & North 1985; 1986b. (3) Horner et al. 1997; Leonard & Llinas 1994; Luebke et al. 1992; Thakkar et al. 1997. (4) Sakai & Koyama 1996. (5) Portas et al. 1996. (6) Sakai & Koyama 1996; Sakai & Onoe 1997; Vanni-Mercier et al. 1989; Yamamoto et al. 1990a; 1990b. (7) Greene & McCarley 1990; Leonard & Llinas 1994; Sakai & Koyama 1996. (8) Li et al. 1997. (9) Nitz & Siegel 1997; Datta 1997b; Datta et al. 1991. (10) Porkka-Heiskanen et al. 1997a (from Hobson et al. 1998b).

and aminergic neurons have been questioned (Leonard & Llinas 1994). Specifically, the self-stimulatory role of acetylcholine on pontine PGO-bursting neurons has not been confirmed in *in vitro* slice preparations (Leonard & Llinas 1994). For example, ACh has been shown to hyperpolarize cell membranes in slice preparations of the rodent parabrachial nucleus (Egan & North 1986a), LDT (Leonard &

Llinas 1994; Luebke et al. 1993), and PPT (Leonard & Llinas 1994). Similarly, LDT and PPT neurons with burst discharge properties most like those hypothesized to occur in PGO-burst neurons ("type I" neurons) may not be cholinergic (Leonard & Llinas 1990). Much evidence remains, however, that the reciprocal interaction model accurately describes essential elements of REM sleep cycle control even though some of its detailed synaptic assumptions need correction (Fig. 3B).

**3.2.2.2. New findings supporting the cholinergic enhancement of REM sleep.** Numerous findings confirm the hypothesis that cholinergic mechanisms are essential to the generation of REM sleep and its physiological signs (for recent reviews see Capece et al. 1999; Datta 1995; 1997;1999; Gottesmann 1999; Hobson 1992b; Hobson et al. 1986; 1993; Hobson & Steriade 1986; Lydic & Baghdoyan 1999; Jones 1991; 1998; Mallick & Inoue 1999; McCarley et al. 1995; 1997; Rye 1997; Sakai 1988; Semba 1999; Steriade & McCarley 1990a). A selection of the many recent examples follows:

1. Microinjection of cholinergic agonist or cholinesterase inhibitor into many areas of the paramedian pontine reticular formation induces REM sleep (Baghdoyan et al. 1987; 1989; Hobson et al. 1993; Vanni-Mercier et al. 1989; Velazquez-Moctezuma et al. 1989; 1991; Yamamoto et al. 1990a; 1990b). In addition to these short term REM induction sites, carbachol injection into a pontine site in the caudal peribrachial area has been shown to induce long-term (over 7 days) REM enhancement (Calvo et al. 1992; Datta et al. 1992; 1993).

2. Cholinergic (type II and III) PPT and LDT neurons have firing properties which make them well suited for the tonic maintenance of REM (Leonard & Llinas 1990).

3. PGO input to the LGB is cholinergic (Steriade et al. 1988) and can be antidromically traced to pontine PGO-burst neurons (Sakai & Jouvet 1980). Retrograde tracers injected into the thalamus label 50% or more of cholinergic PPT/LDT neurons (Oakman et al. 1999; Rye 1997). Moreover, stimulation of mesopontine neurons induces depolarization of cortically projecting thalamic neurons (Curro-Dossi et al. 1991).

4. PGO waves can be blocked by cholinergic antagonists (Hu et al. 1989) and neurotoxic lesions of pontomesencephalic cholinergic neurons reduce the rate of PGO spiking (Webster & Jones 1988).

5. PPT and LDT neurons show specifically c-fos and fos-like immunoreactivity following carbachol-induced REM sleep (Shiromani et al. 1995; 1996).

6. Low amplitude electrical stimulation of the LDT enhances subsequent REM sleep (Thakkar et al. 1996).

7. Electrical stimulation of the cholinergic LDT evokes excitatory post synaptic potentials (EPSPs) in pontine reticular formation neurons which can be blocked by scopolamine (Imon et al. 1996).

8. The excitatory amino acid, glutamate, when microinjected into the PPT dose-dependently increases REM sleep (Datta 1997a; Datta & Siwek 1997).

9. Microdialysis studies showed enhanced release of endogenous acetylcholine in the medial pontine reticular formation during natural (Kodama et al. 1990) and carbachol-induced (Lydic et al. 1991b) REM sleep.

10. Thalamic ACh concentration of mesopontine origin is higher in wake and REM than in NREM (Williams et al. 1994), a REM-specific increase of ACh in the lateral geniculate body has been observed (Kodama & Honda 1996), and both muscarinic and nicotinic receptors participate in the depolarization of thalamic nuclei by the cholinergic brainstem (Curro-Dossi et al. 1991).

11. Although *in vivo* cholinergic REM enhancement has been difficult to demonstrate in rats (Deurveiller et al. 1997), such enhancement has recently been reported (Datta et al. 1998; Marks & Birabil 1998) and a specific carbachol-sensitive site in the dorsal locus subcoeruleus of rats

has recently been described (Datta et al. 1998). Moreover, rats that are genetically supersensitive to ACh show enhanced REM sleep (Benca et al. 1996).

12. The new presynaptic anticholinergic agents have been shown to block REM (Capece et al. 1997: Salin-Pascual et al. 1995).

13. Muscarinic activation by carbachol has been shown to increase G-protein binding in brainstem nuclei associated with REM sleep (Capece et al. 1998).

14. Cholinergic PPT neurons have now been quantitatively mapped in the human pontine brainstem (Manaye et al. 1999).

It may not be an exaggeration to state that the evidence for cholinergic REM sleep generation is now so overwhelming and so widely accepted that this tenet of the reciprocal interaction model is an established principle. (For a recent review see Semba 1999.)

**3.2.2.3. New findings supporting the serotonergic and noradrenergic suppression of REM sleep.** But what about the essence of the theory: the idea that cholinergic REM sleep generation can only occur when the noradrenergic and serotonergic mediators of waking release their inhibitory constraint? The evidence for inhibitory serotonergic and noradrenergic influences on cholinergic neurons and REM sleep is now also quite strong. For example:

1. Serotonergic neurons have been shown to project to the LDT and PPT (Honda & Semba 1994; Steininger et al. 1997) and serotonin has been shown to hyperpolarize rat cholinergic LDT cells *in vitro* (Leonard & Llinas 1994; Luebke et al. 1992) and to reduce REM sleep percent *in vivo* (Horner et al. 1997).

2. Serotonin has been shown to counteract the REM-like carbachol-induced atonia of hypoglossal motoneurons (Kubin et al. 1994; 1996; Okabe & Kubin 1997).

3. Extracellular levels of serotonin are higher in waking than in NREM and higher in NREM than REM in the hypothalamus (Auerbach et al. 1989; Imeri et al. 1994), dorsal raphe (Portas et al. 1998) and frontal cortex (Portas et al. 1998) of rats, as well as the dorsal raphe (Portas & McCarley 1994) and medial pontine reticular formation (Iwakiri et al. 1993) of cats. And, the same pattern of extracellular serotonin concentration change over the sleep-wake cycle has recently been demonstrated in the human amygdala, hippocampus, orbitofrontal cortex, and cingulate cortex (Wilson et al. 1997).

4. Microinjection of the serotonin agonist 8-OH-DPAT into the peribrachial region impeded REM initiation in cats (Sanford et al. 1994b) and systemic injection of 8-OH-DPAT into serotonin-depleted rats also suppressed REM (Monti et al. 1994). However, localization of the serotonergic REM suppressive effect to the PPT/LDT has recently been challenged in favor of an amygdalar-pontine interaction (Morrison et al. 1999; Sanford et al. 1996; 1998b).

5. Microinjection with simultaneous unit recording has shown that 8-OH-DPAT suppresses the firing of REM-on but not REM-and-Wake-on cells of the cholinergic LDT and PPT (Thakkar et al. 1997; 1998).

6. *In vivo* microdialysis of serotonin agonists into the dorsal raphe nucleus (DRN) decreased DRN levels of serotonin (presumably via serotonin autoreceptors on DRN cells), which in turn increased REM sleep percent (Portas et al. 1996; Thakkar et al. 1998).

7. Electrical stimulation of the pons in the vicinity of the

(noradrenergic) locus coeruleus reduced REM sleep in rats (Singh & Mallick 1996) and locus coeruleus neurons have been shown to become quiescent during REM in the monkey (Rajkowski et al. 1997).

8. The alpha-2 noradrenergic agonist clonidine suppresses REM in human subjects (Gentili et al. 1996; Nicholson & Pascoe 1991) and the cat (Tononi et al. 1991) while the noradrenergic antagonist idazoxan increases REM when injected into the pontine reticular formation of cats (Bier & McCarley 1994).

9. There is near universal suppression of REM sleep in humans by acute dosage of serotonin and norepinephrine reuptake-inhibiting antidepressants (Gaillard et al. 1994; Nicholson et al. 1989; Vogel 1975; Vogel et al. 1990).

10. Mesopontine injection of a serotonin agonist depressed ACh release in the lateral geniculate body (Kodama & Honda 1996).

It can therefore also be stated that aminergic suppression of REM sleep is now an established principle (for recent reviews see Monti & Monti 1999 and Luppi et al. 1999a; 1999b).

**3.2.2.4. Modification of the original reciprocal interaction hypothesis to accommodate new findings.** Modifications of simple reciprocal inhibition or interaction models, which are consonant with recent findings, have been proposed for the brain stem control of REM sleep. For example, Leonard and Llinas (1994) suggest in regard to the McCarley and Hobson (1975) model that "indirect feedback" excitation via cholinergic inhibition of an inhibitory input or cholinergic excitation of an excitatory input or some combination of the two could replace direct feedback excitation in their model" (p. 327). A similar mutually excitatory or mutually inhibitory interaction between REM-on cholinergic and REM-on noncholinergic mesopontine neurons has also been proposed in the cat (Sakai & Koyama 1996). Such a mechanism is depicted in Figures 3B and 4.

From recent *in vitro* studies in the rat, the following modification of reciprocal interaction has been proposed proposed by Li et al. 1997 (see Fig. 4). During waking, presynaptic nicotinic facilitation of excitatory locus coeruleus noradrenergic inputs to the dorsal raphe enhances serotonergic firing. During REM, when the locus coeruleus is silent, the same presynaptic nicotinic input may facilitate serotonergic self-inhibition by raphe neurons themselves. *In vivo* microdialysis studies of GABA in the cat further suggests selective suppression of noradrenergic locus coeruleus neurons by GABAergic inhibition during REM (Nitz & Siegel 1997) as can be seen in Figure 4. Both of these modifications retain one or both of the major tenets of the reciprocal interaction model: cholinergic facilitation and aminergic inhibition of REM.

It is important to realize that many of the studies questioning reciprocal interaction or reciprocal inhibition (e.g., Egan & North 1986a; 1986b; Leonard & Llinas 1990; 1994; Luebke et al. 1993) have been carried out on *in vitro* rodent models, and the relationship of these findings to findings on the *in vivo* generation of REM sleep signs in the cat is only in its early stages (Datta 1995; Hobson et al. 1993; Sakai & Koyama 1996). Moreover, the hyperpolarization by ACh of cholinergic cells cited in these studies might be explained by recent findings suggesting the presence of ACh autoreceptors that contribute to homeostatic control of cholinergic activity (Baghdoyan et al. 1997; El Manseri et al. 1990; Ko-

dama & Honda 1996; Leonard & Llinas 1990; 1994; Roth et al. 1996; Sakai & Koyama 1996; Sakai et al. 1990). In contrast to the hyperpolarization of some mesopontine cholinergic neurons by cholinergic agonists, *in vitro* studies have shown the majority of medial pontine reticular formation (mPRF) to be depolarized by carbachol (e.g., Greene & McCarley 1990). This suggests that the exponential self-stimulatory activation which can be triggered by cholinergic stimulation in diverse meso- and medial pontine sites (Hobson et al. 1986; 1993; Hobson & Steriade 1986; McCarley et al. 1995; 1997; Steriade & McCarley 1990a) may involve noncholinergic excitatory intermediary neurons. Such cholinergic self-regulation combined with cholinergic-noncholinergic mutual excitation is illustrated in Figures 3B and 4.

We conclude that the two central ideas of the model are strongly supported by subsequent research: (1) noradrenergic and serotonergic influences enhance waking and impede REM via anticholinergic mechanisms; and (2) cholinergic mechanisms are essential to REM sleep and come into full play only when the serotonergic and noradrenergic systems are inhibited. Because many different synaptic mechanisms could mediate these effects, we now turn our attention to some intriguing possibilities.

**3.2.3. Other neurotransmitter systems.** Beyond the originally proposed cholinergic and aminergic neuronal populations, many additional neurotransmitter systems may participate in the control of REM sleep (see below). Since 1975, much progress has been made in the identification of other chemically specific neuromodulatory systems showing differential activation with particular behavioral states or with specific physiological signs within a behavioral state. We now discuss these new findings in terms of the way that they modify and extend the reciprocal interaction model.

In the brain stem and diencephalon, other neuromodulatory systems may interact with aminergic and cholinergic systems in the generation of REM sleep and its signs (for recent reviews see Jones 2000; Lydic & Baghdoyan 1999; Mallick & Singh 1999; Pace-Schott & Hobson, in press). In brief summary, these systems include:

1. GABAergic systems (Datta 1995; 1997b; Datta et al. 1991; Holmes & Jones 1994; Holmes et al. 1994; Jones 1991; 1993; Jones & Muhlethaler 1999; Luppi et al. 1999a; Nitz & Siegel 1997; Porkka-Heiskanen et al. 1997a; Sanford et al. 1998a; Steriade et al. 1990; Xi et al. 1997; for a recent review see Mallick et al. 1999);

2. Nitroxergic systems (Burlet et al. 1999; Datta et al. 1997; Leonard & Lydic 1997; Sippel et al. 1999; Williams et al. 1997; for recent reviews see Burlet et al. 1999 and Leonard & Lydic 1999);

3. Glutamatergic systems (Bartha et al. 1999; Datta 1997a; Datta & Siwek 1997; Holmes et al. 1994; Inglis & Semba 1996; Jones 1994; Lai & Siegel 1992; Onoe & Sakai 1995; Rye 1997; Sakai & Koyama 1996; Sanchez & Leonard 1996);

4. Glycinergic systems (Chase et al. 1989; Datta 1997b; Luppi et al. 1999a; Stevens et al. 1996; Yamuy et al. 1999);

5. Histaminergic systems (e.g., Lin et al. 1996; Saper et al. 1997; Shiromani et al. 1999);

6. Adenosinergic systems (Mackiewicz et al. 1997; Marks & Birabil 1998; McCarley et al. 1997; Porkka-Heiskanen et al. 1997a; 1997b; Portas et al. 1997; Rannie et al. 1994; 1997; Strecker et al. 1997a; 1997b);

7. A wide variety of neuropeptides such as: galanin

(Saper et al 1997; Sherin et al. 1998); orexin (Chemelli et al. 1999; Lin et al. 1999; Piper et al. 1999); vasoactive intestinal polypeptide (Bourgin et al. 1997; El Kafi et al. 1994; Murck et al. 1996; Obal et al. 1989; Prospero-Garcia et al. 1993; for a review see Steiger & Holsboer 1997) and nerve growth factor (Yamuy et al. 1995) (for a review of such substances see Inoue et al. 1999a); as well as numerous hormones including growth hormone releasing hormone (Zhang et al. 1999), prolactin (Morrison et al. 1999), and corticotropin releasing factor (Lai & Siegel 1999). (For a review of hormonal influences see Krueger et al. 1999; Obal & Krueger 1999.)

8. Dopaminergic systems (de Saint Hilaire et al. 1995; Gaillard et al. 1994; Gillin et al. 1973; 1978; 1994; Nicholson et al. 1989; Nishino & Mignot 1997; Olive et al. 1998; Post et al. 1974; 1978; Seidel et al.1997).

Numerous roles have been proposed for these neuromodulatory systems in the regulation of REM sleep and its physiological signs. Among the better known findings and hypotheses are the following:

1. In the initial stages of PGO wave generation, GABAergic and glycinergic cells may inhibit aminergic cells and thus release the cholinergic PGO-triggering or transmitting cells (Datta 1995; 1997b; 1999; Jones 1991; Nitz & Siegel 1997; for recent reviews see Mallick et al. 1999 and Luppi et al. 1999a; 1999b).

2. GABAergic afferents to the PPT and LDT originating in the substantia nigra pars reticulata (SNr) may exert direct inhibitory influences on PGO-related cells of these nuclei (Datta 1999; Datta et al. 1991; Kang & Kitai 1990; Leonard & Llinas 1990; Maloney & Jones 1997; Rye 1997) and the spike-bursting pattern in pontine PGO-burst cells may be the result of excitatory signals impinging on cells that are tonically inhibited by GABA (Datta et al. 1991; Sanford et al. 1998a; Steriade et al. 1990). Such excitatory signals may include corollary discharge from ocular premotor neurons commanding REMs (Steriade et al. 1990). In addition, GABAergic mechanisms may be involved in the medullary control of muscle atonia during REM (Holmes & Jones 1994).

3. Pontine glutamatergic cells may transmit REM sleep atonia-related signals to medullary sites (Lai & Siegel 1992; 1999; Rye 1997).

4. Medullary glycinergic cells may then affect the postsynaptic inhibition of somatic motoneurons during REM atonia (Chase et al. 1989). Glycinergic neurotransmission is also involved in the pre-motor functions of the pons (Gottesmann 1997; Stevens et al. 1996).

5. Adenosine may exert tonic inhibition over the glutamatergic excitatory inputs to the cholinergic cells of the LDT and PPT (McCarley et al. 1997; Rannie et al. 1994) and may contribute to the REM-related suppression of serotonergic raphe neurons (McCarley et al. 1997; Strecker et al. 1997a). Additionally, extracellular buildup of adenosine may constitute the sleep-promoting factor associated with prolonged wakefulness (McCarley et al. 1997; Portas et al. 1996).

6. Two very recent findings highlight the importance of neuropeptides in the regulation of sleep. The first is that inhibitory neurons in the ventrolateral preoptic area (VLPO) of the hypothalamus, a specifically sleep-active area (Sherin et al. 1996), utilize galanin as well as GABA to inhibit ascending arousal systems such as the the locus coeruleus (Saper et al. 1997). The second finding has come from studies on the genetic basis of narcolepsy using animal models. The neuropeptide orexin (or hypocretin), produced only by neurons in the lateral hypothalamus, may play a key role in

sleep regulation via its modulation of ascending cholinergic and monoaminergic arousal systems (Chemelli et al. 1999; Lin et al. 1999).

7. Because dopamine (DA) release does not vary dramatically in phase with the natural sleep cycle as do 5-HT, NE and acetylcholine (ACh) (Mamelak 1991; Miller et al. 1983; Trulson et al. 1981), dopaminergic agents have not been as extensively studied. It is often found, however, that REM sleep deprivation appears to enhance DA levels and DA receptor sensitivities (e.g., Brock et al. 1995; Nunes et al. 1994; Tufik et al. 1978). The effects of DA on sleep appear to be variable and are in need of further study. Studies on the administration of dopaminergic drugs have suggested that dopamine may play a role in dreaming, especially the induction and intensification of nightmares (Hartmann 1978; Hartmann et al. 1981; for recent reviews see Hobson & Pace-Schott 1999, and Thompson & Pierce 1999).

Two recent theories have proposed specific roles for DA in dreaming. First, Solms (1997a; 1999c) suggests that dreams are instigated by dopaminergically mediated appetitive drives from the ventral tegmental area (VTA) component of the mesolimbic reward system. Second, Gottesmann (1999) proposes that, during REM sleep, sustained dopaminergic modulation of the cortex in the absence of serotonergic and noradrenergic inhibitory influences but the renewed presence of cholinergic excitation contributes to the unique features of dream mentation such as its psychotomimetic quality. In keeping with the cholinergic hypothesis of REM and dreaming, mechanisms for dopaminergic enhancement of dreaming may involve mutual excitation by dopaminergic and cholinergic nuclei such as dopaminergic enhancement of cortical acetylcholine release (Moore et al. 1999; Smiley et al. 1999) and/or enhancement of mesolimbic dopamine release by cholinergic mesopontine neurons (Oakman et al. 1999).

Finally, as in much of neuroscience, research on behavioral state control is now beginning to extend its inquiry beyond the neurotransmitter and its receptors to the roles of intracellular second messengers (Capece et al. 1999) as well as intranuclear events (Bentivoglio & Grassi-Zucconi 1999; Prospero-Garcia et al. 1999; Schibler & Tafti 1999). Recent exciting results of a molecular genetic approach to sleep research includes the discovery of the role of orexin in sleep regulation (see above). In addition, molecular bases for consciousness are also now being proposed (e.g., Woolf 1996). Undoubtedly such inquiry, though beyond the scope of the present review, will increasingly enrich our understanding of sleep and dreaming.

### 3.2.4. REM sleep and other brain stem structures. In addition to this neurochemical diversity, a wide variety of brainstem structures other than the LDT, PPT, locus coeruleus, and raphe are crucially involved in the modulation of REM sleep and its distinctive physiological signs. These include diverse areas in the pontine reticular system such as noncholinergic areas within the pedunculopontine region (Rye 1997), the nucleus pontis oralis (Bourgin et al. 1995; Chase & Morales 1990), the locus coeruleus alpha and adjacent structures (Cespuglio et al. 1982; Sakai 1988; Shouse & Siegel 1992), peribrachial areas caudal to the LDT and PPT (Datta 1995; 1997b), as well as the midbrain central gray area (Maloney & Jones 1997; Sastre et al. 1996) and the medulla (Chase & Morales 1990; Gottesmann 1997). Figure 5 schematizes the generation of the various

physiological signs of REM at different levels of the CNS. Adding to the functional complexity of mesopontine cholinergic areas are their roles in other brain mechanisms such as motor control (Garcia-Rill et al. 1987; Inglis & Winn 1995; Rye 1997) as well as the cytoarchitectonic, cytochemical, and functional diversity within the PPT complex itself (Rye 1997). (For recent reviews on this functional neuroanatomy, see Datta 1995; 1997b; 1999; Gottesmann 1997; Hobson & Steriade 1986; Hobson et al. 1993; Jones 1991; Koyama et al. 1999; Pace-Schott & Hobson, in press; Rye 1997; Sakai 1988; Semba 1999; Siegel 1994; Steriade & McCarley 1990a; Vertes 1984.)

Therefore, even within the brainstem itself (i.e., pons, medulla, and midbrain) a diversity of structures and their neurochemical products modulate control of the REM sleep cycle by the aminergic and cholinergic nuclei. Exciting ongoing research in many laboratories now builds upon early findings summarized in the reciprocal interaction model and pursues the important goal of a more complete description of the complex brainstem mechanisms underlying REM sleep.

### 3.2.5. REM sleep and forebrain-brain stem interactions.
Other important contemporary research now extends the study of sleep-wake and REM sleep control mechanisms rostrally from the pontine brain stem to diencephalic structures in a manner consistent with connectivity studies (Morrison & Reiner 1985; Wainer & Mesulam 1990). In addition to the well described brainstem-thalamus-cortex axis, subcortical sleep control mechanisms intercommunicate with each other and with the cortex via an interconnected network of structures extending rostrally from the brainstem RAS to the hypothalamus, basal forebrain, and limbic system. Saper et al. (1997) classify three ascending arousal systems: the brainstem cortical projection system, the basal forebrain projection system, and the hypothalamic cortical projection system with the basal forebrain system projecting to topographically specific cortical areas and the other two systems projecting diffusely. Woolf (1996) has advanced an intriguing model of how these networks may interact in modulating memory and cognition. We now briefly summarize recent findings on this extended subcortical system that are pertinent to sleep-wake and REM sleep control. We will focus here on findings in the hypothalamus, basal forebrain nuclei, and amygdala.

### 3.2.5.1. The hypothalamus.
Histaminergic neurons originating in the posterior hypothalamus innervate virtually the entire brain (Panula et al. 1989) including brain stem structures such as the mesopontine tegmentum (Lin et al. 1996) and the vestibular nuclei (Tighilet & Lacour 1996). These brainstem regions, in turn, innervate both anterior and posterior hypothalamus (Abrahamson et al. 1997; Kumar et al. 1989; Steriade et al. 1980).

Anterior portions of the hypothalamus (preoptic area and adjacent basal forebrain) are known to be essential to sleep. Lesions here cause insomnia (Sallanon et al. 1989) while stimulation of this area promotes sleep (McGinty et al. 1994). In addition, stimulation of the locus coeruleus inhibits sleepactive neurons in this area (Osaka & Matsumura 1993).

Tonic firing of histaminergic neurons in the posterior hypothalamus play an important role in cortical arousal and the maintenance of wakefulness (Khateb et al. 1995; Lin et al. 1986; 1988; 1993; 1994; McCormick & Williamson 1991;

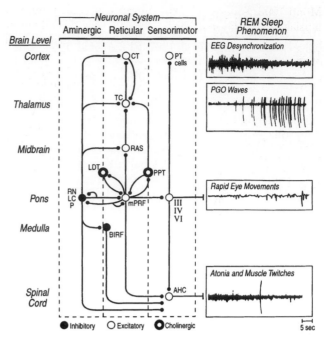

Figure 5. Schematic representation of the REM sleep generation process. A distributed network involves cells at many brain levels (left). The network is represented as comprising three neuronal systems (center) that mediate REM sleep electrographic phenomena (right). Postulated inhibitory connections are shown as solid circles; postulated excitatory connections as open circles; and cholinergic pontine nuclei are shown as open circles with darkened boundaries. It should be noted that the actual synaptic signs of many of the aminergic and reticular pathways remain to be demonstrated, and, in many cases, the neuronal architecture is known to be far more complex than indicated here (e.g., contributions of hypothalamic and basal forebrain systems). During REM, additive facilitatory effects on pontine REM-on cells are postulated to occur via disinhibition (resulting from the marked reduction in firing rate by aminergic neurons at REM sleep onset) and through excitation (resulting from mutually excitatory cholinergic-noncholinergic cell interactions within the pontine tegmentum).

The net result is strong tonic and phasic activation of reticular and sensorimotor neurons in REM sleep. REM sleep phenomena are postulated to be mediated as follows: EEG desynchronization results from a net tonic increase in reticular, basal forebrain, thalamocortical, and cortical neuronal firing rates. PGO waves are the result of tonic disinhibition and phasic excitation of burst cells in the lateral pontomesencephalic tegmentum. Rapid eye movements are the consequence of phasic firing by reticular and vestibular cells; the latter (not shown) directly excite oculomotor neurons. Muscular atonia is the consequence of tonic postsynaptic inhibition of spinal anterior horn cells by the pontomedullary reticular formation. Muscle twitches occur when excitation by reticular and pyramidal tract motorneurons phasically overcomes the tonic inhibition of the anterior horn cells. *Abbreviations:* RN, raphe nuclei; LC, locus coeruleus; P, peribrachial region; PPT, pedunculopontine tegmental nucleus; LDT, laterodorsal tegmental nucleus; mPRF, medial pontine reticular formation (e.g., gigantocellular tegmental field, parvocellular tegmental field); RAS, midbrain reticular activating system; BIRF, bulbospinal inhibitory reticular formation (e.g., gigantocellular tegmental field, parvocellular tegmental field, magnocellular tegmental field); TC, thalamocortical; CT, cortical; PT cell, pyramidal cell; III, oculomotor; IV, trochlear; V, trigmenial motor nuclei; AHC, anterior horn cell. (Modified from Hobson et al. 1986.)

Monti 1993; Saper et al. 1997; Shiromani et al. 1999; Szymusiak 1995) and neurons in this area may directly influence REM sleep (Reiner & McGeer 1987; Sallanon et al. 1989; Vanni-Mercier et al. 1984).

The tuberomammillary nucleus (TMN) plays a particularly important role in the posterior hypothalamic histaminergic arousal system (Saper et al. 1997; Sherin et al. 1996; Shiromani et al. 1999; Steininger et al. 1996; Vanni-Mercier et al. 1984). For example, Sherin et al. (1996) have proposed that a monosynaptic pathway in the hypothalamus may constitute a "switch" for the alternation of sleep and wakefulness. These workers have identified a group of GABAergic and galaninergic neurons in the ventrolateral preoptic anterior hypothalamus (VLPO) which are specifically activated during sleep and constitute the main source of innervation for the histaminergic neurons of the TMN. VLPO neurons may, therefore, specifically inhibit histaminergic neurons of the TMN in order to preserve sleep (Saper et al. 1997; Sherin et al. 1996; 1998).

A recent study has demonstrated extensive histaminergic innervation of the mesopontine tegmentum including the LDT (Lin et al. 1996). Suppression of slow wave activity and an increase in waking follows microinjection of histamine and histamine agonist into these areas (Lin et al. 1996). Recently, histaminergic projections from the TMN to the dorsal raphe as well as to areas of the basal forebrain involved in sleep-wake control have also been demonstrated in the cat (Lin et al. 1997). VLPO neurons have also been shown to innervate other components of ascending arousal systems such as the monoaminergic nuclei of the brainstem and there they may also exert a sleep-promoting inhibitory influence (Sherin et al. 1998). Moreover, also innervating most of the brainstem and diencephalic ascending arousal systems are the orexinergic cells of the lateral hypothalamus and these too may play a modulatory role in the sleep-wake cycle (Chemelli et al. 1999). Tying the hypothalamus to the pons in this dynamic manner may provide a critical link between the circadian clock and the NREM-REM sleep cycle oscillator (see also Liu et al. 1997; O'Hara et al. 1997). In this regard, it is notable that retinal input to the VLPO itself has recently been demonstrated (Lu et al. 1999).

### 3.2.5.2. The basal forebrain.

Basal forebrain (BF) nuclei have close anatomical connections with the locus coeruleus, raphe, and pontine nuclei (Butcher 1995; Jones & Cuello 1989; Szymusiak 1995) and, in turn, project to more rostral structures such as the cortex, thalamus, and limbic systems (Butcher 1995; McCormick 1990; Metherate et al. 1992; Steriade & Buzsazki 1990; Szymusiak 1995; Woolf 1996). In addition to its brain stem and cortical connectivity, the basal forebrain also has close anatomical connections with the anterior and posterior hypothalamus (Gritti et al. 1993; 1994; Szymusiak 1995), the amygdala, and the thalamus (Szymusiak 1995). (For a recent review of BF connectivity see Jones & Muhlethaler 1999.)

Neurochemically, acetylcholine plays a major role in BF control of behavioral state (Jones 1993; Jones & Muhlethaler 1999). For example, magnocellular cholinergic cells of the BF nuclei promote the activation of those cortical and limbic structures to which they project (Cape & Jones 1998; McCormick 1990; Metherate et al. 1992; Szymusiak 1995; Wainer & Mesulam 1990). For example, those of the Nucleus Basalis of Meynert activate topographically distinct areas of the cortex (Metherate et al. 1992; Szymusiak 1995;

Woolf 1996). Recent work in rats has also implicated BF magnocellular cholinergic neurons in the control of high voltage cortical slow waves such as are observed in NREM (Kleiner & Bringmann 1996; Nunez 1996). GABAergic BF cells may also interact with BF cholinergic cells in the regulation of oscillatory rhythms which accompany cortical activation (Jones & Muhlethaler 1999). Other BF cells, anatomically and neurochemically distinct from the cholinergic magnocellular neurons, function as sleep promoting elements (Szymusiak 1995), possibly by GABAergic inhibition of hypothalamic and brain stem arousal systems (Szymusiak 1995), the hippocampus (Mallick et al. 1997), or the cortex (Jones & Muhlethaler 1999).

There are extensive interactions between the brain stem structures (locus coeruleus, raphe nuclei, as well as the LDT and PPT) and the BF in sleep-wake control (Jones & Cuello 1989; Jones & Muhlethaler 1999; Semba 1999; Semba et al. 1988; Szymusiak 1995). Bidirectional interactions between the BF and sleep-related areas of the brainstem modulate behavioral state utilizing a variety of transmitter substances as illustrated by the following findings:

1. The cholinergic system of the mesopontine tegmentum communicates with the BF cholinergic system in a manner functionally relevant to sleep (Baghdoyan et al. 1993; Consolo et al. 1990). For example, simultaneous microinjection of carbachol into cholinoceptive regions of the BF suppresses the ability of carbachol to induce a REM-like state when injected into the pons (Baghdoyan et al. 1993).

2. Cholinergic BF structures, which activate the cortex, can be activated by brain stem glutamatergic cells (Rasmussen et al. 1994).

3. Glutamatergic systems of the BF can, in turn, affect behavioral state via projections to the mesopontine tegmentum (Manfridi & Mancia 1996).

4. Aminergic inputs to the BF nuclei from brainstem nuclei can influence behavioral state in a manner similar to their action in the pons. For example, the noradrenergic agonist isoproterenol increases wakefulness and suppresses REM when infused into the BF (Berridge & Foote 1996).

As in the brainstem, neuromodulatory systems interact within the BF itself. For example, BF cholinergic neurons may be under tonic inhibition by adenosine (Porkka-Heiskanen 1997b; Strecker et al. 1997b) while 5-HT can hyperpolarize cholinergic nucleus basalis neurons and decrease wake-associated gamma frequency oscillations in the cortical areas to which they project (Cape & Jones 1998). The BF nuclei, therefore, both directly participate in behavioral state-related functions and modify the activity of other areas involved in sleep such as the pontine REM generator.

### 3.2.5.3. The amygdala.

Of particular interest in view of the human neurobiology reviewed above (e.g., Maquet et al. 1996; Nofzinger et al. 1997), the amygdala has reciprocal connections with pontine regions involved in the control of REM sleep (Bernard et al. 1993; Calvo & Simon-Arceo 1999; Morrison et al. 1999; Sanford et al. 1995b; Saper & Loewy 1980; Semba & Fibiger 1992; Wainer & Mesulam 1990) and receives serotonergic innervation from the dorsal and medial raphe (Fallon & Ciofi 1992). For a recent thorough review of the amygdala in sleep regulation see Morrison et al. (1999).

Physiological signs of REM have been shown both to occur spontaneously and to be modifiable in the amygdala (see Calvo & Simon-Arceo 1999 for a review; see also Maquet 2000; Maquet & Phillips 1998; 1999 regarding recent human findings). For example, in the cat, PGO-like EEG activity has been detected in the basolateral amygdala (Calvo & Fernandez-Guardiola 1984). Moreover electrical stimulation of the cat amygdala significantly increased PGO number, spike density, and burst density (Calvo et al. 1987) as well as the amplitude and rate of acoustically elicited pontine PGO waves in the waking rat (Deboer et al. 1997; 1998), and burst firing of pontine cells in the rabbit (Morrison et al. 1999).

Aminergic and cholinergic stimulation of the amygdala has been shown to modify sleep in the directions predicted by reciprocal interaction for the action of these neurotransmitters in the pons. For example, cholinergic stimulation of amygdaloid sites in the cat enhanced REM sleep for several days, an effect akin to the long-term REM enhancement by cholinergic stimulation of the peribrachial pons (Calvo & Simon-Arceo 1995; 1999; Calvo et al. 1996). Furthermore, serotonergic stimulation of the amygdala in the cat caused short latency changes of state from either NREM or REM (Sanford et al. 1995b), while serotonergic antagonism during NREM increased PGO activity (Sanford et al. 1995a) and the relative amount of sleep (Sanford et al. 1995b). Similarly, noradrenergic stimulation of the amygdala suppressed sleep relative to wakefulness (Fuchino et al. 1996). Interestingly, the role of the amygdala in REM sleep control may differ between species (Deboer et al. 1997; Sanford et al. 1997a).

It has been suggested that serotonergic mechanisms in the amygdala constitute a mechanism whereby emotionally significant stimuli can influence the state of arousal (Sanford et al. 1995b). Such a role corresponds well with the proposed role of amygdala in the processing of emotional memory during REM (Maquet & Franck 1997).

### 3.2.5.4. Other subcortical structures.

Other diencephalic structures such as centralis lateralis nucleus of the thalamus possibly participate in the modulation of REM sleep (Mancia & Marini 1997; Marini et al. 1992). In addition, there are extensive striatal projections to the pedunculopontine region (Inglis & Winn 1995; Rye 1997) especially to glutamatergic cells of the midbrain extrapyramidal area (MEA) (Rye 1997). Interaction between the MEA and the basal ganglia may serve to modulate movement to accord with behavioral state (Rye 1997).

In addition to forebrain structures, brain stem structures rostral to the pons such as the ventrolateral periaqueductal gray (Sastre et al. 1996) may also be important in the modulation of REM sleep. Such rostral brainstem connections could facilitate ponto-limbic interactions in REM sleep generation and loss of this mechanism could account for loss of dreaming when such connections are severed by clinical lesions (Solms 1997a).

### 3.2.6. Neurophysiological evidence which supports the REM-NREM-waking distinction.

While the REM-NREM-waking distinction was first defined in standardized terms by the neurophysiological criteria of polysomnography (Rechtschaffen & Kales 1968), abundant additional physiological evidence has since accumulated which supports the biological differentiation of these three states. Although direct measurement of human CNS neuromodulators is still in its infancy, preliminary evidence points to a similar pattern of fluctuation across the sleep-wake cycle as is seen in animal models (Wilson et al. 1997). In addition, the following indirect evidence strongly supports the physiological distinction

between REM, NREM, and waking: (1) Autonomic activation is higher during NREM night terrors than during REM nightmares (Fisher et al. 1973). (2) While the locus coeruleus is active during waking and its noradrenergic output is associated with wake state anxiety responses (Bremner et al. 1996; Salzman et al. 1993), this region is quiescent in REM sleep (Hobson & Steriade 1986) despite the predominance of anxiety in the emotions of dreaming (Merritt et al. 1994). (3) Cholinergic activation of limbic structures probably underlies REM dream anxiety (Braun et al. 1997) whereas ACh is not prominently involved in waking anxiety (Salzman et al. 1993). (4) Nielsen (1999; and target article) notes additional physiological differences between REM and NREM sleep such as differing ERP patterns and external stimulus responses, which suggest differing cognitive processes taking place during these two sleep states.

### 3.2.7. Conclusions.

All of these findings indicate that the reciprocal interaction of cholinergic and aminergic systems may operate in areas other than the brain stem in ways that significantly amplify REM sleep generation or suppression. As has been hypothesized for learning and cognition (Woolf 1996), a subcortical medial ascending system of multiple nuclei, extensive reciprocal interconnections between nuclei, and system-wide sensitivity to neuromodulation controls behavioral state at a hierarchical level above that of specific subcomponent oscillators (e.g., the pontine REM generator). Furthermore, in view of the recent evidence of selective activation of the limbic lobe in human REM sleep (Braun et al. 1997; 1998; Maquet et al. 1996; Nofzinger et al. 1997), these new basic neurobiological findings have a particularly strong impact on the neurocognitive theory of dreaming.

We conclude that the essential tenets of the reciprocal interaction model have been strongly confirmed and that the interaction of the pontine structures with other brain structures can now begin to be studied in ways that will enrich our understanding of how the distinctive features of each conscious state are mediated and how their stereotyped sequencing is controlled.

### 3.3. Contemporary theories of conscious states

We now turn our attention to a review of theories on how conscious states are mediated. As the inadequacies of the Freudian model of dreaming have become more evident, many researchers have increasingly turned toward the establishment of a cognitive neuroscience of brain-mind states. Four major cognitive models of dreaming are discussed below. All four of these have been inspired by modern laboratory research but the degree to which they are deeply brain-based varies dramatically as we hope to make clear. In section 3.3.5, we address the ongoing debate on the relationship of REM eye movements to dream imagery. We do so because this controversy exemplifies both the basic differences between "top-down" (cortically driven) and "bottom-up" (subcortically driven) views on the origin of dreaming as well as the added complexity and realism offered by an approach to the biology of dreaming which takes into account the wide range of perspectives offered by contemporary neuroscience.

### 3.3.1. Activation models.

In 1970, Zimmerman advanced a theory in which dreaming (versus thinking or no mentation) occurred during sleep when "cortical arousal" exceeded a certain threshold, regardless of sleep stage. We will later describe various ways to measure cortical activation which we call factor "A" and take to be one of three critical factors in determining the probability of dreaming.

Antrobus and his colleagues have proposed an elaborated cortical activation-based model of mentation operating across all mental states (Antrobus 1986; 1990; 1991; Fookson & Antrobus 1992; Reinsel et al. 1992). According to Antrobus, the qualities of mentation in any state result from an interaction between the activation level of cortex and the current level of environmental stimulation as gated by current sensory thresholds. Interaction between cortical modules subserving various sensory, motor, and associative modalities create the dream narrative and integrate any cortical, subcortical or peripheral inputs via a "top-down" cortically controlled process (Antrobus & Bertini 1992). Antrobus and his colleagues describe the dynamics of this process in terms of parallel distributed process neural network models (Antrobus 1991; Fookson & Antrobus 1992). In our terms, the greater the value of "A," the greater the production and retrieval of associative trains of thought.

The Antrobus team theorizes that the high sensory thresholds of REM prevent interruption of ongoing mentation. In our terms, this process is measured as factor "I" which we see as shifted away from external sensory input, and correspondingly favoring internal, fictive sensory input. For Antrobus, the result is a more ongoing, story-like quality of REM mentation compared with wake mentation which, though similarly activated, is continually interrupted by external stimuli (Reinsel et al. 1986; 1992; Wollman & Antrobus 1986). In his model, dream bizarreness results when cortical networks, which are attempting to accurately reconstruct reality based on probabilities learned during waking, fail to fully integrate all of the various constructions being generated (Antrobus & Bertini 1992; Fookson & Antrobus 1992).

Antrobus implicitly rejects the role of aminergic-cholinergic neuromodulation (our model's factor M) in controlling the nature of dream mentation. Instead, he argues that since waking mentation can be dreamlike, this neuromodulatory shift is not necessary for dream mentation to occur and factor M of our three dimensional model is discarded. We invite Antrobus to explain the paradoxical memory defect and loss of self-reflective awareness and volition during dreaming on the basis of activation and sensory gating alone.

### 3.3.2. The cognitive psychological model of Foulkes.
Foulkes has advanced a cognitive, information processing model of dream production which questions the brain basis of conscious states and dream mentation (e.g., Foulkes 1982a; 1985; 1990; 1993b; 1997; Foulkes & Cavallero 1993). Instead, Foulkes describes dreams as resulting from the activation of mnemonic "systems" or "units." In his model, "activation" is conceived as the combination of both excitatory processes and the disinhibition of mnemonic systems previously inhibited by voluntary self-control (Foulkes 1985).

With the exception of general excitatory processes such as the cerebral activation of REM, Foulkes's model is explicitly a psychological, mentalistic construct which does not attempt to link psychological to physiological phenomena (Foulkes 1985; 1990). A similar position has been taken by Bosinelli (1995) and by Mancia (1995). Each of them asserts that mentalistic and physiological sleep phenomena cannot be explained from the same epistemological refer-

ents. As such, these models share with Freud's model a decision not to attempt to explain these mental functions in terms of brain actions.

Instead, Foulkes's earlier cognitive models emphasized similarity between the intermediate steps of a psycholinguistic model of language production and a "psychoneiric" model of dream production with the differences between the two processes occurring mainly at input and output stages of production (Foulkes 1982a). In more recent writings, Foulkes (1990) specifically equates the high level cognitive constructive processes which organize waking experience with those processes which organize dreaming. For example, he explains the consonance of dream emotion with dream plot as resulting from the primary narrative demands of the dream (Foulkes 1997; Foulkes et al. 1988b). Further, he specifically eschews any possible information-bearing role for subcortical stimuli in dream form or narrative. In his own words, "subcortical structures . . . simply turn on the light switch upstairs. They don't tell any of the creatures upstairs what to do or how to do it; they simply arouse them, enabling them to do whatever it is they characteristically do" (Foulkes 1997, p. 3).

Foulkes goes on to assert that if such higher level (and implicitly cortically based) cognitive processes cannot *consciously* construct an organized, episodically integrated, self-reflective account of waking (as in the case of an animal or a pre-operational child), they also cannot *unconsciously* construct a coherent dream narrative (Foulkes 1990). As previously noted, this model constrains the dream to adult human sleep mentation and does not account for conscious experiences during sleep which may be possible at a much lower level of integration. For example, given Foulkes's (1990) position, one might argue that severely cognitively regressed adults (e.g., with severe dementia or delirium) should lose much of *their* capacity to dream. However, this prediction is not supported by clinical findings (e.g., Cipolli et al. 1992; Doricchi & Violani 1992; Kramer et al. 1975). Instead, we see loss of dreaming associated with lesions to specific brain areas (for reviews see Doricchi & Violani 1992 and Solms 1997a), a finding which would be expected if specific circuits with a great degree of localization form the neural substrate of dreaming.

Although Foulkes's model cannot be specifically viewed in the context of our physiological AIM model, some hints of these concepts can be found in his work. For example, he does make a generalized claim that cortical activation by the brain stem (the "A" dimension of the AIM model) must be relatively high in dreaming (Foulkes 1997). In addition, he argues that the origin of dream scenarios comes from the quasi-random activation of a "mnemonic focus" (Foulkes 1985, p. 151), and specifically not from external stimuli. This corresponds to a value of low sensory input and high value of internal input on the "I" dimension. No position on the "M" dimension of our AIM model, however, can be inferred from his studies. We invite Foulkes to explain the several robust deficiencies of dream cognition, and especially the amnesia, in terms of his model.

### 3.3.3. The neuropsychological-psychoanalytic model of Solms.
Combining the clinical lesion studies described above in section 3.1.6 and the classical psychoanalytic theory of dreaming, Solms (1997a; 1999c) builds a neuropsychological model of normal dreaming, which is illustrated in Figure 6. Frontal dopaminergic mesolimbic reward cir-

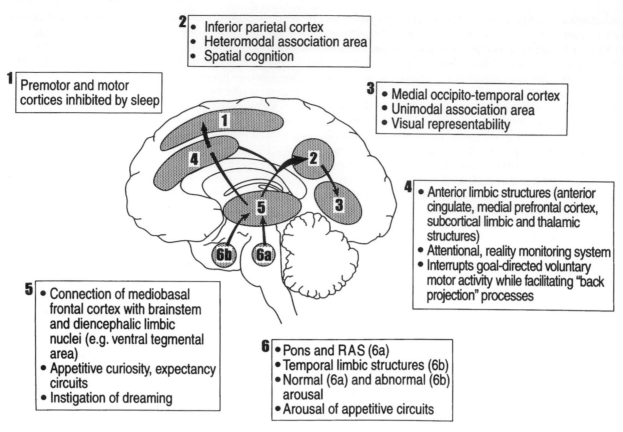

**1** Premotor and motor cortices inhibited by sleep

**2**
• Inferior parietal cortex
• Heteromodal association area
• Spatial cognition

**3**
• Medial occipito-temporal cortex
• Unimodal association area
• Visual representability

**4**
• Anterior limbic structures (anterior cingulate, medial prefrontal cortex, subcortical limbic and thalamic structures)
• Attentional, reality monitoring system
• Interrupts goal-directed voluntary motor activity while facilitating "back projection" processes

**5**
• Connection of mediobasal frontal cortex with brainstem and diencephalic limbic nuclei (e.g. ventral tegmental area)
• Appetitive curiosity, expectancy circuits
• Instigation of dreaming

**6**
• Pons and RAS (6a)
• Temporal limbic structures (6b)
• Normal (6a) and abnormal (6b) arousal
• Arousal of appetitive circuits

Figure 6. Forebrain processes in dreaming based upon a model proposed by Solms (1997a). Solms proposes that the dopaminergic mesolimbic reward circuits (region 5 in Fig. 6) produce an instigating impetus for dreaming when activated by arousing stimuli such as environmental input, ascending brainstem arousal in REM (region 6a in Fig. 6) or epileptiform discharge (region 6b in Fig. 6). He further hypothesizes that the posterior passage of this subcortical stimulus is gated by a reality monitoring process in anterior limbic areas (region 4 in Fig. 6) which both interrupt voluntary motor activity and facilitate back projection processes from the inferior parietal cortex (region 2 in Fig. 6) to medial temporal-occipital visual association areas (region 3 in Fig. 6). During this process, premotor and motor cortices (region 1 in Fig. 6) remain quiescent due to the combined effects of limbic blockage (region 4 in Fig. 6) of ascending impulses as well a sleep-related inhibition.

cuits produce an instigating impetus for dreaming when activated by arousing stimuli (e.g., ascending brainstem arousal in REM). The passage of this subcortical stimulus to posterior heteromodal association areas in the inferior parietal lobe is gated by a reality monitoring process mediated by anterior limbic areas. These anterior limbic areas also prevent this subcortical stimulus from activating the motor cortex as well as facilitating back projection of this stimulus to the posterior cortex. Back projection continues from the inferior parietal lobe (which contributes the capacity for spatial cognition) to visual association areas in medial occipito-temporal cortex (which contribute visual imagery) but not as far back as primary visual cortex. Solms speculatively assigns to the resultant network the sleep-protective function of Freud's classical dream work: appetitive subcortical impulses are "censored" by the anterior limbic system and then safely back-projected to posterior cortical representational mechanisms.

In support of the neuroanatomical details of this network Solms cites his findings on lesion-induced changes in dreaming. Loss of dream imagery (visual anoneria) is accompanied by an analogous waking deficit, visual irreminiscence, which involves the highly processed visual memory functions of unimodal association cortex and not the perceptual functions of the primary visual cortex. Since cortical area V1 lesions do not cause visual anoneria, Solms hy-

pothesizes that any back projection processes involved in dreaming do not extend all the way to primary visual cortex. On the basis of the findings that lesions in Brodmann areas 39 and 40 in either hemisphere appear to be the most restricted damage causing the posterior variant of global anoneria, he proposes that these heteromodal areas are the source of back projection to visual association areas. In support of this network's sleep-protective function, he notes that global anoneric patients report poorer sleep quality than non-cerebrally injured controls (Solms 1997a).

### 3.3.4. The activation-synthesis model

**3.3.4.1. The original-activation synthesis model.** Abundant studies in the 1960s and 1970s on the cellular neurophysiology of the sleep cycle as well as the functional reorganization of the visual system during sleep suggested a new conceptual approach to brain-mind states. First expressed as the activation-synthesis hypothesis of dreaming (Hobson & McCarley 1977), this model proposed the global mapping of brain states to mind states. This was the position taken by Freud in his famous *Project for a scientific psychology* (1895) but ostensibly abandoned in the *Interpretation of dreams* (1900). For a detailed discussion of this subject, see McCarley and Hobson (1977).

Enunciating the general principle of brain-mind isomorphism, the activation synthesis model placed emphasis on such aspects of the form of dreams which might be expected to have their roots traced to isomorphic forms of brain activity. In so doing, the new theory proposed some of the cellular and molecular mechanisms by which changes in activation, in stimulus origin and in neuromodulation could explain the state-dependent changes in perception, thinking and memory seen in shifts from waking to NREM and REM sleep (Flicker et al. 1981). The activation-synthesis hypothesis proposed that formal aspects of dream mentation reflected the outcome of attempts by sensorimotor and limbic regions of the forebrain to produce a coherent experience from the incomplete and chaotic inputs received from the brain stem. The specific formal features of dream mentation, it was proposed, could best be explained by examining the unique configuration of brain activity that occurs during REM sleep.

To illustrate how this global brain-to-mind mapping concept is articulated, we considered the probable consequences of a shift in visual system input source from the formed visual images on the retina in waking to the chaotic brain stem stimulation of REM sleep (Bizzi 1966a; 1966b; Callaway et al. 1987; Nelson et al. 1983; Pivik et al. 1977). This shift in input source occurs in the context of a concurrent cessation of activity in brain stem noradrenergic and serotonergic neurons (Hobson & Steriade 1986; Steriade & McCarley 1990a). The quiescence seen in these aminergic modulatory neurons results in the demodulation and disinhibition of the visual cortex (Evarts 1962), the lateral geniculate bodies (Bizzi 1966b) and brain stem oculomotor networks (Mouret et al. 1963).

As a result of the aminergic disinhibition, cholinoceptive peribrachial neurons become hyperexcitable and fire in bursts, causing phasic activation of the lateral geniculate bodies and visual cortex. This phasic activation is recordable in the REM sleep of cats as the PGO waves which, in turn, correlate with the direction of the rapid eye movements (Monaco et al. 1984; Nelson et al. 1983). We have speculated that this cholinergically mediated stimulation conveys information to the visual system about the direction of the eye movements which have become, in REM sleep, uncoupled from external sensory stimuli (Callaway et al. 1987).

The net result of these shifts is an activated brain stem and visual system which are (1) deafferentated, (2) aminergically demodulated, and (3) cholinergically auto-stimulated. But the brain stem signals still convey information about the direction of rapid eye movements to the deafferentated, demodulated forebrain. According to the activation-synthesis hypothesis, these changes in sensory input source and neuromodulation could contribute to such cognitive features of dreaming as (1) the hallucinatory visual imagery, (2) the frequent shifts and reorientations of attention, (3) the loss of voluntary control of both motor action and internal attention, (4) the emotional intensification especially of anxiety, elation, and anger, and (5) the memory loss within and after dreaming (Mamelak & Hobson 1989a).

**3.3.4.2. Evolution of the activation-synthesis model.** The original formulation of the activation-synthesis model of dream construction (Hobson & McCarley 1977) proposed that the phasic signals arising in the pontine brain stem during REM sleep and impinging upon the cortex and limbic forebrain led directly to the visual and motor hallucinations, emotion, and distinctively bizarre cognition that characterize dream mentation. In doing so, these chaotically gener-

ated signals arising from the brain stem acted as a physiological Rorschach test, initiating a process of image and narrative synthesis involving associative and language regions of the brain and resulting in the construction of the dream scenarios. Thus, it was the combination of this chaotic, bottom-up activation process and its resultant semi-coherent, top-down synthetic process which made up the overall process of dream construction.

Anticipating activation-synthesis by almost a decade, Molinari and Foulkes's (1969) application of Moruzzi's physiological tonic-phasic model to dream psychology first introduced the concept that the phasic events of sleep contribute hallucinatory raw material that was then secondarily elaborated during dream production. Using neurobiological data to support these concepts, the activation-synthesis model hypothesized that dreaming resulted from the interpretation by the cortex of information concerning eye movements and activated brain stem motor pattern generators. Seligman and Yellen (1987) added the consideration of emotional evaluation to the concepts of primary visual activation and secondary cognitive elaboration to generate a cognitive model of dream production, a suggestion strongly supported by recent PET studies showing preferential activation of limbic structures and adjacent cortices (Braun et al. 1997; Maquet et al. 1996; Nofzinger et al. 1997).

We have recently proposed that both cortical and limbic regions, when cholinergically activated by REM sleep events such as PGO waves, may synthesize their own information (Hobson 1988b; 1990; 1992a; 1997a; Hobson & Stickgold 1994a; 1994b; Mamelak & Hobson 1989a). For example, dream hallucinosis, while probably incorporating eye-movement information coded in PGO bursts, must also incorporate visual material from a variety of memory sources in an otherwise activated cortex. This aspect of the theory is very similar to Solms's suggestion of a "back projection" toward the visual cortex from the limbic forebrain (Solms 1997a) as the brain synthetically fits image to affect. Informing recent presentations of the activation-synthesis hypothesis are concepts from neural net modeling (Mamelak & Hobson 1989a; Sutton & Hobson 1994), self-organization theory (Kahn & Hobson 1993; Kahn et al. 1997), graph theory (Sutton et al. 1994a; 1994b), cognitive neuroscience (Hobson & Stickgold 1994a; 1994b) and, most recently and influentially, the new findings described above in section 3.3 on the functional neuroimaging of sleep and the clinical neuropsychology of dreaming (Hobson et al. 1998a; 1998b; 2000).

**3.3.4.3. Activation synthesis updated: An integrated model of REM sleep dreaming.** Integration of the original activation-synthesis model with new neuroimaging (Braun et al. 1997; 1998; Maquet et al. 1996; Nofzinger et al. 1997) and lesion (Solms 1997a) data allows the development of a more detailed activation-synthesis model of REM sleep dreaming (Hobson et al. 2000). Although the original activation synthesis model was necessarily weighted toward activation processes (e.g., PGO activation of thalamocortical circuits), these new findings allow us to begin to speculate on the neuroanatomical bases of the synthesis aspect of the model. In doing so, we present a neuropsychological model of dreaming differing substantially from that of Solms (presented above), which was based on lesion studies alone. This model is presented in Figure 7 and its components are described in more detail below.

In this model, dreaming consciousness results from pro-

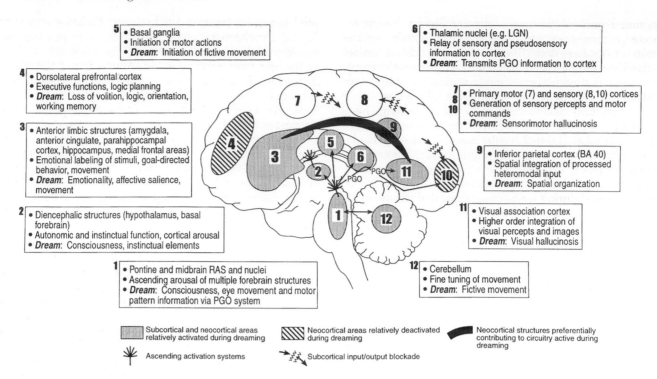

5 • Basal ganglia
• Initiation of motor actions
• *Dream*: Initiation of fictive movement

6 • Thalamic nuclei (e.g. LGN)
• Relay of sensory and pseudosensory information to cortex
• *Dream*: Transmits PGO information to cortex

4 • Dorsolateral prefrontal cortex
• Executive functions, logic planning
• *Dream*: Loss of volition, logic, orientation, working memory

7 • Primary motor (7) and sensory (8,10) cortices
8 • Generation of sensory percepts and motor commands
10 • *Dream*: Sensorimotor hallucinosis

3 • Anterior limbic structures (amygdala, anterior cingulate, parahippocampal cortex, hippocampus, medial frontal areas)
• Emotional labeling of stimuli, goal-directed behavior, movement
• *Dream*: Emotionality, affective salience, movement

9 • Inferior parietal cortex (BA 40)
• Spatial integration of processed heteromodal input
• *Dream*: Spatial organization

2 • Diencephalic structures (hypothalamus, basal forebrain)
• Autonomic and instinctual function, cortical arousal
• *Dream*: Consciousness, instinctual elements

11 • Visual association cortex
• Higher order integration of visual percepts and images
• *Dream*: Visual hallucinosis

1 • Pontine and midbrain RAS and nuclei
• Ascending arousal of multiple forebrain structures
• *Dream*: Consciousness, eye movement and motor pattern information via PGO system

12 • Cerebellum
• Fine tuning of movement
• *Dream*: Fictive movement

Subcortical and neocortical areas relatively activated during dreaming

Neocortical areas relatively deactivated during dreaming

Neocortical structures preferentially contributing to circuitry active during dreaming

Ascending activation systems

Subcortical input/output blockade

Figure 7. Forebrain processes in normal dreaming: an integration of neurophysiological, neuropsychological and neuroimaging data. Regions 1 and 2: ascending arousal systems; 3: subcortical and cortical limbic and paralimbic structures; 4: dorsolateral prefrontal executive association cortex; 5: motor initiation and control centers; 6: thalamocortical relay centers and thalamic subcortical circuitry; 7: primary motor cortex; 8: primary somatosensory cortex; 9: inferior parietal lobe; 10: primary visual cortex; 11: visual association cortex; 12: cerebellum. This figure serves as a visual model for section 3.3.4.3 ("Activation-synthesis updated: An integrated model of REM sleep dreaming") and each element of the figure is explained in detail in that section. Abbreviations: RAS, reticular activating system; PGO, ponto-geniculo-occipital waves; LGN, lateral geniculate nucleus; BA, Brodmann area. (From Hobson et al. 2000).

cesses of arousal impinging upon selectively facilitated, dysfacilitated or input/output-blockaded forebrain structures. The various elements of normal dreams are contributed by brain networks that include structures known to contribute to analogous processes in waking although, as the model suggests, dreaming is characterized by a deletion of certain circuits active in waking and, perhaps, the accentuation of others. The following text uses the enumerated brain areas in Figure 7 to present a model of the neuropsychological bases of dream phenomena.

*Ascending arousal systems (zones 1 and 2 in Fig. 7):* As in waking, activation of the forebrain occurs through ascending arousal systems located in the brainstem reticular activating system (Steriade 1996), the basal forebrain (Szymusiak 1995) and possibly the hypothalamus (Saper et al. 1997). Together these structures form an integrated ascending midline network (Woolf 1996) which includes ascending cholinergic systems. Braun et al. (1997) suggest that the ascending reticular activation of REM sleep may proceed relatively more via a ventral cholinergic route from the brainstem to the basal forebrain rather than via the dorsal route through the thalamus which is preferred in waking. This suggestion and the related idea of Solms (1997a), recall the early speculation of Jouvet (1962) that forebrain activation might proceed via the limbic midbrain circuit of Nauta.

The forebrain stimulation arising from such intrinsic arousal systems allows "consciousness" (as opposed to unconsciousness) to exist in dreaming. Such consciousness may be detected by the desynchronization of the traditionally measured cortical EEG frequencies (Hobson 1988b) as

well as by the appearance of gamma frequency oscillatory rhythms (Llinas & Ribary 1993; for reviews, see Hobson et al. 1998a; 2000; Kahn et al. 1997). Brainstem and diencephalic structures also contribute information in specific modalities via specific circuitries (such as the PGO network) resulting in distinctive dream features such as directionality of eye movement, distinctive motor pattern automata, and instinctive behavior and feelings such as rage, terror, or sexual arousal (Hobson & McCarley 1977).

*Thalamocortical relay centers and thalamic subcortical circuitry (zone 6 in Fig. 7):* The release of corticothalamic intrinsic oscillatory rhythms suppresses the experience of perception and mentation during NREM sleep (see above). During REM sleep, this process is reversed and the activated thalamic nuclei, which occupy key sites in sensorimotor relay as well as other brain circuits, contribute to the pseudosensory perceptual aspects of dream consciousness. For example, the lateral geniculate nucleus transmits PGO waves from the brainstem to the visual cortex. As an internal stimulus, PGO waves bear such information as the directionality of gaze shifts encoded in the form of corollary discharge from brainstem oculomotor nuclei (Hobson & McCarley 1977). Recent dipole tracing techniques in humans have shown PGO wave-like activity involving the pons, midbrain, thalamus, hippocampus, and visual cortex (Inoue et al. 1999b). Moreover, it has recently been shown that information encoded in the pattern of activation of geniculate neurons in the cat is sufficient to represent basic elements of natural scenes (Stanley et al. 1999).

As in waking, corollary discharge information from programmed instinctual motion commanded by brainstem motor pattern generators is transmitted rostrally via the thalamus (Hobson & McCarley 1977). In addition, nuclei within the thalamus participate in the subcortical circuitry of various motor pathways (Braun et al. 1997). Moreover, thalamic nuclei participate in the control of the sleep cycle itself (Mancia & Marini 1997) and recent findings have shown the ventrolateral thalamus may mediate the interaction of arousal and attention in humans (Portas et al. 1999).

*Subcortical and cortical limbic and paralimbic structures (zone 3 in Fig. 7):* As suggested by PET studies, medial forebrain structures, both cortical and subcortical, are selectively activated during REM sleep dreaming (Braun et al. 1997; 1998; Hobson et al. 1998b; 2000; Maquet et al. 1996; Nofzinger et al. 1997). Among these, limbic and paralimbic structures are consistently found to be active in REM and these contribute distinctive emotion-related dream features as follows.

As in waking (LeDoux 1996), amygdalar activation contributes emotional features, especially anxiety, to dreaming. Maquet emphasizes that those cortical areas activated in REM are rich in afferentation from the amygdala (anterior cingulate, right parietal operculum) while those areas with sparse amygdalar afferentation (prefrontal cortex, parietal cortex, and precuneus) were deactivated in REM (Maquet 1997; Maquet et al. 1996).

As in waking (Devinsky et al. 1995), anterior cingulate activation contributes additional emotional features to dreaming such as valence biases, the assessment of motivational salience, and the integration of dream emotion with fictive actions. Interestingly, in some PET studies, other elements of the rostral limbic and perilimbic circuits such as the ventral striatum and the orbitofrontal, insular, and medial prefrontal cortices have also been found to be activated during REM (Braun et al. 1997; Nofzinger et al. 1997). Such medial areas have the most abundant limbic connections in the prefrontal cortex (Barbas 1995; Braun et al. 1997) and their disruption is often associated with confabulatory or dream-wake confusional syndromes (Braun et al. 1997; Solms 1997a). Several recent findings also suggest the importance of medio-frontal, limbic-associated cortical areas to dreaming. First, during sleep, a scalp-recorded decrease in frontal alpha power and the persistence of waking frontal alpha asymmetry between hemispheres has been suggested to be linked to activation of underlying limbic structures during REM (Benca et al. 1999). Second, magnetic resonance spectroscopy has shown a sleep-related elevation of medial prefrontal glutamine (a glutamate precursor) to the unusually high levels seen in awake schizophrenics (Bartha et al. 1999). These authors go on to suggest that this elevation is linked to brain activity during dreaming.

Activated limbic circuits underlie the phenomenology of recalled dream emotion with its predominance of anxiety over other emotions (Domhoff 1996; Merritt et al. 1994; Nielsen et al. 1991). The finding that dream emotion is usually consistent with the dream narrative (Foulkes et al. 1988b) and that bizarre incongruities between emotion and narrative are rarer than incongruities among other dream elements (Merritt et al. 1994) can now be explained by viewing dream emotion as a primary shaper of plots rather than as a reaction to them (Seligman & Yellen 1987). Thus in a classic anxiety dream, the plot may shift from feeling lost, to not having proper credentials, adequate equipment or suitable clothing, to missing a train. These plots all sat-isfy the driving emotion – anxiety – while being only very loosely associated with one another in a category that we call "incomplete arrangements."

Two concerns arise when predicting that REM sleep dreaming is hyperemotional in comparison to other behavioral states. The first involves early findings of maximal galvanic skin response (GSR), an indicator of peripheral autonomic activity, in Stage 4 NREM rather than REM (Johnson & Lubin 1966) as well as the complementary findings of an "autonomic storm" accompanying Stage 4 night terrors (Fisher et al. 1973). It must be noted, however, that peripheral autonomic activity may be uncoupled from central autonomic activity in deep sleep. Thus we would not expect GSR to correlate with felt emotion in deep sleep. Moreover, if GSR did so correlate, it would constitute the sleep equivalent of the James-Lange hypothesis that emotion is the perception of peripheral autonomic changes, a hypothesis now felt to be inaccurate even in waking when the peripheral measures may themselves more faithfully reflect central autonomic activation. A second concern is the often reported lack of emotion-related physiological arousal accompanying dream events (e.g., violence) which would easily elicit such arousal in waking (Perlis & Nielsen 1993). Such emotional "numbing" in dreams could result both from a sleep-related dissociation of peripheral and central autonomic activity (as with peripheral arousal in Stage 4) combined with REM-related blockade of central readout to the periphery and peripheral sensory feedback to the CNS.

The amygdala is known to influence memory storage processes in the hippocampus (Cahill & McGaugh 1998). Such circuits could thus underlie the role of REM sleep and dreams in the processing of emotional memories that is often hypothesized by dream psychology theorists and by neuroimaging groups (Braun et al. 1997; Cartwright et al. 1998a; Hobson et al. 1998b; Kramer 1993; Maquet et al. 1996; Maquet & Franck 1997; Nofzinger et al. 1997; Perlis & Nielson 1993). For example, Nofzinger et al. (1997) suggest that an important function of REM sleep is the integration of neocortical function with basal forebrain hypothalamic motivational and reward mechanisms.

*Motor initiation and control centers (zone 5 in Fig. 7):* As in waking movement (Kolb & Whishaw 1996), the basal ganglia play a role in initiating fictive dream movement and their strong activation in REM relative to both waking and NREM (Braun et al. 1997) contribute to the ubiquity of hallucinated motion in dreams (Hobson 1988b; Porte & Hobson 1996). The cerebellum (zone 12 in Fig. 7) modulates these fictive movements and adds specific features such as vestibular sensations (Hobson et al. 1998c; Leslie & Ogilvie 1996; Sauvageau et al. 1998) via cerebellar connectivity with brainstem vestibular nuclei. It is interesting that pontine cholinergic neurons have recently been shown to project to the cerebellar vermis (Cirelli et al. 1998), a region of the cerebellum which has been found to be activated in REM (Braun et al. 1997). Moreover, the pons serves as a key intermediary structure in cortico-cerebellar and cerebello-cortical pathways (Schwartz & Thier 1999).

Braun et al. (1997) suggest a role for the basal ganglia in ascending thalamocortical activation (via their connectivity with the brainstem through the intralaminar thalamic nuclei) as well as a role for the basal ganglia in the rostral transmission of PGO waves (via their back-projections to the pedunculopontine tegmentum). Notably, the basal ganglia show extensive connectivity with regions of the pontine brainstem also known to regulate REM sleep phenomena (Inglis & Winn 1995; Rye 1997).

Motor input from cerebral levels rostral and caudal to the basal ganglia also contribute to the experience of movement in dreaming. Brainstem motor pattern generators (in zone 1 of Fig. 7) are stimulated along with the widespread pontine reticular activation of REM sleep and they could contribute to the frequent experience of programmed movement such as running in dreams (Hobson & McCarley 1977). The motor cortex (zone 10 in Fig. 7) also commands movement in dreaming as evidenced by the pathological expression of dreamed action in REM sleep behavior disorder (Schenck et al. 1993), although its output is normally blocked by the motor atonia of REM sleep (Chase & Morales 1990; Pompeiano 1967a). The premotor function of the anterior cingulate cortex (Devinsky et al. 1995) may also contribute to the experience of fictive movement in dreaming particularly in regard to emotionally motivated actions.

*Visual association cortex (zone 11 in Fig. 7):* Areas of the medial occipital and temporal cortices involved in higher order visual processing, as opposed to primary visual cortex, generate the visual imagery of dreams (Braun et al. 1998; Solms 1997a). Specific visual features of dreaming are generated by the same areas of the visual association cortex involved in their higher order processing during waking. For example, areas of the fusiform gyrus are both selectively activated in REM (Braun et al. 1997; 1998; Nofzinger et al. 1997) and are the portion of the ventral object recognition stream involved in face recognition (Kanwisher et al. 1997; McCarthy et al. 1997) which is a common, although often bizarrely uncertain and altered dream feature. Furthermore, in a very important recent finding, the same extrastriate ventral occipital areas are activated during waking hallucinations in patients with Charles Bonnet syndrome (Ffytche et al. 1998).

REM sleep combines the *activation* of visual association (e.g., Brodmann areas 37 and 19) and paralimbic cortices with the *deactivation* of primary visual and dorsolateral prefrontal cortices (Braun et al. 1997; 1998). The far lesser role of primary visual cortex (zone 10 in Fig. 7) in REM activation (Braun et al. 1997; 1998) and dream generation (Solms 1997) combines with the known sensory input and motor output blockade of REM sleep (Hobson 1988b; see zones 7, 8, and 10 in Fig. 7) to reinforce the concept that sensory information processing in dreaming may begin at levels downstream from primary sensory cortices (Braun et al. 1998).

*Inferior parietal lobe (zone 9 in Fig. 7):* The inferior parietal lobe, especially Brodmann's area 40, may generate the perception of a fictive dream space necessary for the global experience of dreaming (Solms 1997a). This is a brain region thought to be important for spatial imagery construction. Even with visual systems intact, destruction of this area in either hemisphere causes global cessation of dreaming (Solms 1997a). Other neuropsychological studies have suggested a vital role for this area in dreaming (Doricchi & Violani 1992). Turning to PET data, Maquet et al. (1996) note activation of the right parietal operculum despite general deactivation in much of the parietal cortex. Interesting to note, both lesion (Solms 1997a) and PET studies (Maquet et al. 1996) suggest a greater importance to dreaming of this area in the right versus the left hemisphere.

*Dorsolateral prefrontal executive association cortex (zone 4 in Fig. 7):* Neuronal modeling (Mamelak & Hobson 1989a) as well as neuroimaging (Braun et al. 1997; Maquet & Franck 1997) have suggested a possible origin of dream associated executive deficiencies in the REM-associated changes in frontal lobe functioning. The REM-associated activation of medial paralimbic frontal cortex contrasts with the prominent deactivation in the executive portions of the frontal cortex. The deactivation of the dorsolateral prefrontal cortices during sleep and their failure to then reactivate along with medial and parietal cortical structures in REM sleep underlies the prominent executive deficiencies of dream mentation.

The left dorsolateral prefrontal cortex has been shown to be selectively activated during human reasoning tasks (Goel et al. 1998). Its deactivation could account for the illogical ad hoc explanations offered for bizarre occurrences (Williams et al. 1992). Similarly, the dorsolateral prefrontal cortices have been consistently shown to activate during episodic and working memory tasks (Brewer et al. 1998; Cohen et al. 1997; Courtney et al. 1997; Fletcher et al. 1997; Tulving et al. 1996; Wagner et al. 1998); their deactivation in REM may contribute to the prominent mnemonic deficits in dreaming noted above in section 2.3.4. The other area found by PET to deactivate in REM compared to waking was the posterior cingulate cortex (Braun et al. 1997; Maquet et al. 1996; Nofzinger et al. 1997). This cortical area, especially its posterior-most retrosplenial portion, has been consistently implicated in episodic memory function with lesions to it resulting in episodic memory deficits (Maddock 1999).

Similarly, the dorsolateral prefrontal cortex is a structure specialized for the central executive function of working memory (Baddely 1998; Goldman-Rakic 1996); its deactivation in REM would thus result in the disorientation and bizarre uncertainties (Hobson et al. 1987) characteristic of dream mentation. Failures of working memory are prominent in dreaming. For example, scene shifts are experienced without reflection (Hobson et al. 1998b). In this sense, the dreamer could be seen as experiencing a frontal lobe dysfunction similar to "goal neglect" (see Baddely 1998; Duncan et al. 1996). Notable also is a recent PET study showing reduced working memory (WM) task-related activity in the right midfrontal gyrus in response to cholinergic enhancement with physostigmine (Furey et al. 1997). However, in this study, improved WM performance also resulted from cholinergic enhancement (Furey et al. 1997). Finally, Doricchi et al. (1993) present a convincing argument for an attenuation of frontal eye field inhibition of reflexive saccades during REM.

Interesting to note, hypoperfusion of the frontal cortex has been associated with pathological temporal limbic activation in epilepsy (Rabinowicz et al. 1997) and reciprocal inhibition between frontal and limbic areas has been hypothesized in theories on the etiology of schizophrenia (Weinberger 1995). REM sleep dreaming could thus be seen to involve a normal physiological state of the brain analogous to psychopathological conditions (Hobson 1994; 1997b; 1999b) in which limbic hyperactivation is combined with frontal hypoactivation.

*Hypothetical dynamic interactions of brain regions during normal dreaming:* In the view of modern cognitive neuroscience, component subsystems of global states of consciousness like dreaming are physically instantiated in networks or circuits each consisting of several to many discrete brain regions (e.g., Cummings 1993; Mesulam 1998; Nadel 1994).

Mesulam (1998) hypothesizes five global circuits each subserving a broad cognitive domain: spatial awareness language; explicit memory and emotion; face and object recognition; and working memory-executive function. In Mesulam's "selectively distributed processing" model of these networks, numerous brain regions participate in each cognitive function as opposed to there being functional brain "centers" for different aspects of cognition. The same individual brain region might participate in several functional networks which are differentiated by their component nodes (Mesulam 1998).

In a particular network, Mesulam suggests that certain multimodal nodes or "epicentres" serve to coordinate the functioning of (or to "bind") subsidiary nodes and are, therefore, key to determining this network's unique cognitive function. For example, epicenters in the transmodal posterior parietal cortex (e.g., Brodmann area 40) and the prefrontal cortex (e.g., Brodmann area 46) may coordinate nodes of a working memory-executive function network (Mesulam 1998). The same network can affect subcomponents of a more global cognitive function (e.g., explicit memory) by varying the relative levels of activation in the component nodes (Mesulam 1998).

We propose that during dreaming relative to waking, there is a relative dysfacilitation of the working memory-executive function network combined with relative facilitation of networks subserving emotional and memory consolidation processes. This echoes Braun et al.'s (1997) suggestion that "the 'limbic' loop connecting ventral striatum, anterior thalamus and paralimbic cortices, appears to be activated during REM sleep . . . However the prefrontal or 'association' loop, connecting the caudate, dorsomedial thalamus and prefrontal cortices . . . appears to be activated only in a partial or fragmentary way" (p. 1191). Given the sensory phenomenology of dreaming relative to waking (sect. 2), it might also be hypothesized that, during dreaming, the efficient functioning of spatial awareness and object recognition may be better preserved than the language networks resulting in predominance of visual versus auditory hallucinosis.

Flow of information between the regions localized by neuroimaging or lesion studies as crucial to dreaming is undoubtedly multidirectional with abundant re-entrant feedback and feedforward loops. At present, we propose three generalizations regarding this information flow: (1) Ascending arousal systems activate the forebrain regions involved in dream construction and do so in a manner chemically and anatomically different from that subserving waking arousal processes. (2) Cortical circuits activated in dreaming favor more medial circuits linking posterior association and anterior and posterior paralimbic areas (represented by central crescent in Fig. 7) versus circuits including the primary sensory cortex and/or frontal executive regions (see Braun et al. 1998). Such a predominance of medial circuitry in REM may underlie findings from lesion studies that features of dreaming are only weakly lateralized (Antrobus 1987; Doricchi & Violani 1992; Solms 1997a). (3) Subcortical circuits involving the limbic structures, basal ganglia, diencephalon, and the brainstem contribute strongly to regional brain activation in REM and, therefore, probably to the physiological substrate of dreaming.

Very promising new technologies, such as functional magnetic resonance imaging (e.g., Huang-Hellinger et al. 1995; Portas et al. 1999), transcranial magnetic stimulation (e.g., Cohrs et al. 1998), magnetic resonance spectroscopy (e.g., Bartha et al. 1999), receptor radio ligand PET (e.g., Sudo et al. 1998), near infrared spectroscopy (e.g., Tagaya

et al. 1999) and dipole tracing (e.g., Inoue et al. 1999b) are just now being applied to sleep science. Further research with such tools will undoubtedly further specify the key brain circuits and systems involved in the global experience and component elements of dreaming.

*Accommodation of NREM dreaming in an updated activation synthesis model:* As explained in detail in section 4, the AIM model of conscious state control predicts numerous gradations between states as well as possible dissociations of state characteristics during such transitions. This occurs because activation, input source, and modulation can, to some extent, vary independently.

Increased vividness of Stage 2 NREM dreaming near the end of the normal sleep period has been attributed to circadian increases in brain activation occurring at this time (Antrobus et al. 1995; Cicogna et al. 1998). Toward morning, activation (and perhaps also input source and modulation) may *differ the least* between Stage 2 periods and their adjacent REM periods compared to the other times of the night. Therefore, admixture of REM-like phenomena within Stage 2 NREM (including the brain activation accompanying REM) may be *maximal* late in the sleep bout and may sustain much longer and more vivid NREM dreaming. In other words, late night Stage 2 NREM dreaming may occur during a time when cortical and subcortical areas linked to dreaming (see Figs. 6 and 7) are becoming reactivated in anticipation of the next REM period. Alternatively, the activation of these areas may not as greatly diminish with the transition from late REM to late Stage 2 as it does earlier in the night during the descent from waking into slow wave sleep. (For a complete discussion of these possibilities see Nielsen's target article.)

Such transitional states might include the human equivalent of the well documented sleep stage termed SP (slow wave sleep with PGO waves) which heralds REM periods in the cat (Callaway et al. 1987; Datta 1995) and which has recently been hypothesized to occur in humans (Gottesmann 1999). In humans, recent experimental evidence has shown enhancement of visual imagery in Stage 2 NREM by acoustic stimuli below the threshold of awakening but of an intensity comparable to those triggering PGO waves in animals (Conduit et al. 1997; Drucker-Colin et al. 1983; Morrison et al. 1999). Therefore REM-like tonic (enhanced activation) as well as phasic (SP PGO waves) features may accompany late NREM and enhance dreaming at this time without in any way contradicting the assumption that REM sleep phenomena reflect the fullest expression of the physiological substrate of dreaming.

Nielsen (1999; and this volume) has recently proposed a very similar mechanism for the ubiquity of NREM dreaming which he terms "phantom" or "covert" REM sleep. According to this concept, elements of REM-like activation may commonly occur during NREM without, however, producing the full complement of signs necessary to score REM by Rechtschaffen and Kales's (1968) criteria. Nielsen suggests several examples of such partial expressions of REM physiology such as "missing" first REM periods with EEG desynchrony but lacking REMs or atonia, or NREM erections occurring with ultradian periodicity. Indeed, recent evidence has shown that the transition from NREM to REM sleep shows a typical order of appearance of the cardinal physiological signs of REM sleep as follows: atonia, saw-tooth waves, REMs (Sato et al. 1997).

Further candidate markers of "phantom REM sleep" include the numerous NREM events which investigators have correlated with mental phenomena ever since the lack of an exclusive sleep stage correlate to dreaming led them to seek physiological correlates of dreaming among the discrete phasic physiological events of sleep (Foulkes & Pope 1973; Molinari & Foulkes 1969; Ogilvie et al. 1980; Pivik 1991). For example, within NREM, phasic spinal reflex inhibition was associated with greater recall, auditory imagery, and hostility (Pivik 1991); PIPs (phasic integrated potentials) with enhanced recall (Rechtschaffen et al. 1972); and sleep onset theta bursts with discontinuity (Foulkes & Pope 1973). Such potential correlates continue to be identified and include the very rapid eye movements (VREMs) associated with K-complexes (Serafetinides 1991) as well as NREM imagery envoked by external stimuli (Conduit et al. 1997). As psychophysiological techniques in sleep research become increasingly sophisticated, it is likely that additional tonic and phasic correlates of sleep mentation will emerge in studies of both REM and NREM (e.g., Germain et al. 1999; Miro et al. 1999; Paiva & Guimaraes 1999; Rochlen et al. 1998; Takeuchi et al. 1999a; 1999b).

### 3.3.5. Comparison of activation-only to activation-synthesis models' explanations for the origin of dream imagery in relation to REM saccades and attentional processes.

Perhaps the greatest disagreement between "activation-only" models (sect. 3.3.1 above) and the activation-synthesis model (sect. 3.3.4 above) regards the origin of dream imagery in relation to REM sleep saccades and the dreamer's attentional processes. While the original activation-synthesis model argues that visual imagery and eye movements are largely initiated by chaotic brain stem activity transmitted to the cortex via ascending signals such as PGO waves (Hobson & McCarley 1977), Antrobus has argued for a primarily cortical origin for the visual imagery, REMs and even the PGO waves during dreaming (Antrobus 1990; Antrobus et al. 1995). A similar model for a cortical attentionally driven origin of REM saccades is presented as a revised scanning hypothesis (see below) by Herman (1992). We will address this controversy by integrating data from studies of neuroimaging, the neurophysiology of saccadic eye movement control and attentional processes. We will show that the relationship of dream imagery to REM saccades must involve the integrated activity of heterogenous brain mechanisms only some of which are initiated by exclusively top-down or bottom-up processes.

Before launching into this discussion it is important to situate its significance in a historical context. When REM sleep was first discovered and assumed to be a unique neurophysiological substrate of dreaming, it was logical to postulate a one-to-one correlation between the eye movements and the direction of hallucinated gaze in dreams. This "scanning hypothesis" (Roffwarg et al. 1962) was the strongest and most specific of the many theories of brain-mind isomorphism. In detailing the many difficulties that this theory has encountered, our goal is twofold: first, we want to emphasize that the field of dream research foundered because of its overinvestment in still unresolved arguments about scanning, and second, that promising alternative approaches to the psychophysiology of dreaming were overlooked because of this overinvestment. We will conclude our discussion by an appeal to keep the question of eye movement and dream imagery open until methods more adequate to its investigation are developed.

#### 3.3.5.1. Activation-only theories of a cortical origin for REMs and PGO waves.

Antrobus (1990) and Herman (1992) interpret the work of Herman et al. (1981; 1983; 1984) which shows partial confirmation of the scanning hypothesis (Roffwarg et al. 1962) as supportive of a largely cortical origin for the neural signals which initiate processes leading to dream imagery. Antrobus (1990) suggests that when cortical activation reaches a certain level due to the RAS-mediated arousal of REM sleep, the frontal eye fields are activated and begin to attempt to direct the eyes toward the virtual images being generated in a similarly activated posterior cortex.

In this model, REM saccades are the frontal eye fields' attempt to foveate on such fictive images and these cortical signals are transmitted to brainstem oculomotor nuclei via the same cortico-cerebellar pathways used in the fine-tuning of waking saccades (Antrobus 1990; Antrobus et al. 1995). PGO waves, in this model, are conceived as being similarly cortically evoked via cortico-cerebellar pathways connecting with the brachium conjunctivum, which, in turn, connects the cerebellum to pontine PGO elements (Antrobus 1990). In the Antrobus model, PGO waves may then provide secondary feedback to the frontal eye fields which remain the original instigator of both REMs and PGO waves (Antrobus 1990; Antrobus et al. 1995).

The failure of others (e.g., Jacobs et al. 1972; Moskowitz & Berger 1969) to replicate Roffwarg's original finding as well as the dissimilarities between waking and REM saccades are explained in various ways by current proponents of the scanning hypothesis. Herman (1992) emphasizes that early studies failed to take into account the dreamer's fictive head movements which, in dreaming, may coincide with cortically directed saccades and modify such saccades via the vestibuloocular reflex. Others suggest that visually guided, cortically initiated REM eye movements, in contrast to waking REMs, are saccadic movements toward stationary hallucinatory versus moving real targets (Hong et al. 1997). Although such explanations are plausible and are supported by some data (Herman 1992; Hong et al. 1997), much more work will be required to fully resolve the conflicting findings and daunting methodological challenges imposed by the various versions of the scanning hypothesis.

#### 3.3.5.2. Contributions from neuroimaging studies of REM sleep.

Recently, some investigators have suggested that neuroimaging technologies can shed new light on the scanning hypothesis. In particular, Antrobus et al. (1995) and Hong et al. (1997) cite a recent [18]fluorodeoxyglucose (FDG) PET study (Hong et al. 1995) as supporting their revised scanning hypothesis. Hong et al. (1995) showed that REM period eye movement number was positively correlated with glucose uptake in frontal cortical areas associated with saccadic eye movement control, the midline executive attentional system, and the visuospatial attentional system. Other authors have since interpreted these results as generally supporting visual scanning of the hallucinatory dream scene (e.g., Gottesmann 1997).

The major drawback of the Hong et al. (1995) study is that the measured variable was not REM activation relative to waking or NREM but rather the *within* REM and within waking correlations between eye movements and glucose uptake. Therefore, the only state dependent comparison here involves comparing the *degree of covariation* between REM counts and cerebral metabolism in regions of interest during waking as compared to during REM. In an earlier analysis of the same data set, this group had compared actual regional glucose metabolic rate between REM and waking reporting relatively fewer differences than did later PET studies (see below) although they did observe relatively greater activation of the anterior cingulate in REM (Buchsbaum et al. 1989).

Unlike the Hong study, later $^{15}$O PET studies found state-specific *negative* correlations between REM and cerebral blood flow in the dorsolateral prefrontal cortex with the positive correlations found instead in pontine tegmental, thalamic, and subcortical and cortical limbic structures (Braun et al. 1997; Maquet et al. 1996). Using the $^{18}$FDG PET method, Nofzinger et al. (1997) also found this thalamic, amygdala, and cingulate activation. Significantly for the scanning hypothesis, the $^{15}$O PET studies (Braun et al. 1997, 1998; Maquet et al. 1996) did not find relative activation during REM, as compared to waking or to NREM, in many of the saccade and attention-related cortical areas where Hong et al. (1995) found their positive correlations between eye movement number and glucose uptake (e.g., frontal eye fields, dorsolateral prefrontal cortex, left parietal operculum, precuneus).

It is important to note the significant methodological differences between the two PET imaging techniques (see Braun et al. 1997 and Nofzinger et al. 1997 for discussions). For example, $^{18}$FDG techniques integrate cortical activity over a much longer time than $^{15}$O PET (30 minutes versus 5 minutes) and thus $^{15}$O may better characterize shorter, more discrete PSG-defined sleep conditions (Braun et al. 1997). Therefore, although conclusions from both PET methods must acknowledge the limitations described above (sect. 3.1.1), activation of broader areas may be inherent to $^{18}$FDG compared to $^{15}$O PET. This difference is evidenced here by the greater area activated in $^{18}$FDG studies (Nofzinger et al. 1997) compared to $^{15}$O PET studies (Braun et al. 1997; Maquet et al. 1996) (see Table 2).

The utility of both methods for testing the scanning hypothesis is, therefore, limited because: (1) neither method can distinguish between tonic and phasic changes associated with REM sleep, and (2) neither can provide information on whether cortical activation precedes or follows REMs. Moreover, human PET studies could support either frontal eye fields and attentional systems being activated in response to brain stem activity or *vice versa*.

It seems quite likely to us that both possibilities will prove to be true. In other words, we suggest that some REM sleep eye movements are initiated in the brain stem, some in the frontal eye fields and, possibly, some in other nodes in the saccade-generation network (e.g., superior colliculus). Moreover, being elements of a network, these loci will robustly interact. Therefore, in the Hong et al. study, the similar patterns of correlation between metabolic activation and eye movement counts in both REM sleep and waking is not surprising given the approximately 30 minutes of $^{18}$FDG uptake during REM and waking saccade generation. Over this extended period, many nodes in saccade-generation networks may become activated in rough proportion to total eye movement counts.

### 3.3.5.3. Contributions from the neurophysiology of saccadic eye movement control.
A heterogeneity among the brain mechanisms controlling waking saccades in primates is a widely documented finding (Brooks 1999; Tehovnik et al. 1994) and certain of these circuits are independent of the

frontal eye fields (Tehovnik et al. 1994). Heterogeneity of REM saccadic eye movement control mechanisms was first suggested by an extensive series of lesion experiments in Jouvet's laboratory which showed that various forebrain structures add complexity to eye movements arising in the pons of cats (Jeannerod et al. 1965). Even the pontine cat, which lacked all the forebrain structures involved in eye movement control, still had some eye movements in REM (Jeannerod et al. 1965; Jouvet 1962). (For a thorough review and interpretation of these lesion studies see Doricchi et al. 1993.) Although citing those studies showing persistence of REMs and PGOs in decerebrate animals, Herman (1992) and Antrobus (1990) suggested that the decreased number, loss of bursting patterns, and stereotyped repetitiveness of REMs in such preparations indicates that the cortex controls the phasic components of REMs (presumably directing them toward internal hallucinatory stimuli). In their opinion, such purely pontine-generated REMs reflect only a tonic, repetitious baseline activation of the oculomotor nuclei while the cortex controls all potentially information-bearing REMs.

But additional findings must also be explained. For example, in the decerebrate cat, Pompeiano has been able to increase the frequency and clustering of REMs simply by increasing the cholinergic drive on the brain stem with physostigmine (Pompeiano 1980). Recent work in the cat has further demonstrated a diversity in neural mechanisms generating the saccades of REM and waking (Vanni-Mercier & Debilly 1998; Vanni-Mercier et al. 1994) with a specific region of the pons being implicated in the synchronization of REMs and PGO waves (Vanni-Mercier & Debilly 1998; Vanni-Mercier et al. 1996). This proves that the pons is not only necessary for all REM sleep eye movements but sufficient to generate many of them on its own. Under normal conditions, however, REM saccades, like those of waking, are very likely controlled by the final common pathway pontine generator whose output is modified by interactions with forebrain structures (Goldberg et al. 1991; Hepp et al. 1989; Ito 1987; Pierrot-Deseilligny et al. 1995), especially interactions between reflexively orienting attentional systems in the parietal cortex and superior colliculus as has been recently discovered and elucidated by Doricchi et al (1993).

### 3.3.5.4. The heterogeneity of attentional mechanisms.
The diversity of attentional mechanisms (see Posner 1994a and Kinchla 1992) further argues for a heterogeneity of attentional-oculomotor interaction among behavioral states. A widely distributed network of interconnected structures is known to participate in both attentional processes and the oculomotor control of saccades in waking (see, for example, Corbetta et al. 1993; Paus et al. 1993; Petit et al. 1996; Pierrot-Deseilligny et al. 1995; Sweeny et al. 1996; Wurtz & Munoz 1994). Such structures include those found by neuroimaging (e.g., Maquet et al. 1996) to be activated in REM such as the anterior cingulate cortex (Paus et al. 1993) as well as those shown to be deactivated in REM such as the prefrontal cortex (Boch & Goldberg 1989). An important dissociation between the frontally based attentional modulation of waking saccades and the lack of such frontal modulation in REM has been described by Doricchi et al. (1993; 1996) via the study of hemineglect patients.

### 3.3.5.5. Systems producing REM saccades with and without participation of cortical attentional structures.
Given the above-documented diversity and connectivity within

functional brain networks, it is likely that complex, reentrant interplay between cortical and subcortical structures will determine the relationships between REM saccades, dream imagery, and attentional processes (see Doricchi et al. 1993). In contrast, Antrobus's theory of an autogenous cortical origin of REM saccades predicts that phasic activity of the pontine generator, which must occur to produce any saccade (Goldberg et al. 1991), should always *follow* an initiating event in the cortex (the hallucinated, attended-to and then "saccaded-to" dream image). This can be termed a "top-down-only" mechanism. Contrary to this prediction, we now show that there are data indicating that pontine brain stem cells fire *prior* to REM saccades (a "bottom-up-only" mechanism) as well as *simultaneously* with REM saccades (a "mixed bottom-up and top-down" mechanism) in addition to *after* a saccade (as predicted by Antrobus's "top-down-only" mechanism).

*Evidence for bottom-up only mechanisms:* In the cat, pontine gigantocellular tegmental field (FTG) cells increase their firing rate 150 to 100 msec before eye movement (EM) onset in REM sleep (Pivik et al. 1977). Additional evidence for subcortical potentials anticipating REMs has recently been reviewed in Gottesmann (1997). Therefore, pontine PGO-triggering or transmitting cells may directly excite paramedian pontine reticular saccade burst cells within the pons and thereby initiate horizontal saccades whose directionality is conveyed to the occipital cortex by PGO waves to elicit visual imagery *following* the saccade (Hobson & McCarley 1977). The fact that the primary PGO wave is consistently ipsilateral to the directionality of a REM suggests that PGO waves can convey eye movement directional information to the posterior cortex (Datta & Hobson 1994; Monaco et al. 1984; Nelson et al. 1983). In this regard, it is also notable that, at the level of the pontine generation system, burst cells trigger saccades which are ipsiversive while at the level of the superior colliculus and above, control is contralateral (Goldberg et al. 1991). The impingement of ocular premotor excitatory corollary discharge on PGO bursting cells in the pons provides a mechanism whereby such directional information can be transferred from oculomotor neurons to rostral structures (Callaway et al. 1987; Nelson et al. 1983; Steriade et al. 1990).

*A collicular intermediary allows mixed bottom-up and top-down control of REMs:* The hypothesis that the superior colliculus can generate REM saccades independently of the frontal eye fields was first proposed and elaborated by Doricchi et al. (1993; 1996). Efferents from the PPT project to the superior colliculus (Beninato & Spencer 1986; Krauthamer et al. 1995; Rye 1997) and most cortical saccade-generating commands communicate with the brain stem saccade-generating system via the superior colliculus (Goldberg et al. 1991; Sparks & Hartwich-Young 1989). Moreover, the superior colliculus is able to initiate saccades even when frontal eye fields are damaged (Henik et al. 1994; Rafal et al. 1990; Tehovnik et al. 1994).

The potential importance of collicular mechanisms to the generation of REM sleep saccades is further suggested by the following three findings: (1) In REM sleep of the cat, superior colliculus damage decreases amplitude of saccades (Jeannerod et al. 1965). (2) In the albino rat, the superior colliculus is essential to the initiation of REM by the "lights-off" stimulus (Miller et al. 1997). (3) In humans, an extrageniculate or retinotectal orienting system centered in the superior colliculus has recently been extensively documented (Henik et al. 1994; Rafal & Robertson 1994; Rafal

et al. 1990; 1991; Sparks & Groh 1994; Wurtz & Munoz 1994). The failure of leftward hemineglect (i.e., right hemisphere parietal damage) patients to generate leftward REM-sleep saccades despite preserved (and rehabilitatively improvable) waking leftward saccades has led Doricchi et al. (1993; 1996) to propose the predominant involvement of reflexively orienting parieto-collicular circuits in the generation of REM saccades. Doricchi et al. (1993) go on to suggest that subcortically generated impulses (such as PGO waves) may constitute the endogenous stimuli to which the parieto-collicular system reflexively responds in REM.

If pontine PGO-triggering or transmitting cells directly excited collicular cells, then paramedian pontine reticular saccade burst cells could be excited and produce saccades without the involvement of cortical saccade-related centers. Under such conditions, PGO activation of the occipital cortex via the LGB and PGO-related initiation of saccades could occur *simultaneously*.

*Evidence for top-down only mechanisms:* At least some of the saccades of REM may be commanded by preceding activity of cortical structures (e.g., frontal eye fields), although even this possibility does not require that the dreamer is specifically orienting to hallucinated imagery from the posterior cortex. For example, although the Hong et al. (1995) PET data suggests that activation of certain cortical areas is temporally coincident with REM periods containing a high eye movement density, this correlation could either indicate causality or simply be secondary to intense PGO-associated activation of multiple cortical foci (see Amzica & Steriade 1996).

Additional evidence, however, suggests that cortical initiation of REM sleep saccades is in fact possible. For example: (1) REM density is reduced in patients with parietal damage (Greenberg 1966). (2) Hemi-inattention patients lose most REM-sleep saccades that are directed toward the visual field contralateral to their lesion (Doricchi et al. 1991; 1993; 1996) indicating the importance of parietal but not frontal cortices. (3) Directional eye movements can be voluntarily made during lucid REM dreaming (LaBerge et al. 1981). Again, however, none of these findings argue for an exclusively cortical initiation of REM saccades.

*The robust heterogeneity of mechanisms for REM sleep saccade generation suggests that REM sleep saccades might differ from waking saccades:* Behavioral state-related differences in saccade generation could arise either from an actual differential activation of brain regions or from differential contributions among the multiple cerebral saccade mechanisms (networks) in different behavioral states. And in fact such differences have frequently been described in both humans and in animal models (see Doricchi et al. 1993 and Gottesmann 1997 for recent reviews). For example, in humans, REM sleep saccades have been shown to be slower than those occurring during waking (Aserinsky et al. 1985; Fukuda et al. 1981; Jeannerod & Mouret 1963; Porte 1996). Moreover, saccades in the two states have been shown to possess a different velocity/amplitude relationship (Aserinsky et al. 1985; Fukuda et al. 1981). Studies of human eye movements in sleep predating the discovery of REM (reviewed by Gottesmann 1997) also revealed eye movements atypical in comparison to waking eye movements. In humans, another suggestion of neural control differences between REM and waking saccades in addition to their dissociation in hemi-inattention patients (Doricchi et al. 1991; 1993; 1996) are the amplitude-related constraints in a reported complementary relationship between experimentally controlled waking saccades and subsequent saccades in REM (DeGennaro et al. 1995). One final argument that REM-sleep saccades do not require the scanning of hallucinated dream imagery is the fact that such saccades are ubiquitous in the REM sleep of the congenitally blind who generally lack all visual dream imagery (Amadeo & Gomez 1966; Gross et al. 1965; see Weinstein et al. 1991 for a review).

In cats, REM saccades show a differing maximum velocity/amplitude (main sequence) relationship from that observed in waking (Vanni-Mercier et al. 1994). Moreover, in monkeys, REM saccades are disjunctive between the two eyes (Zhou & King 1997) and otherwise unlike those of waking (Fuchs & Ron 1968) while, unlike wake saccades, the REM saccades of cats are directionally asymmetrical (Vanni-Mercier et al. 1994). These results have led the authors of these three animal studies to argue against the scanning hypothesis. Studies such as these lead Vanni-Mercier et al. (1994) to conclude that REM and wake saccades do not share the same neural control circuits and that "eye movements of paradoxical sleep rather represent a stereotyped repeated pattern which is independent of dream content" (p. 1301). Authors of one cat study have, however, suggested that the REM saccades they observed are suggestive of scanning hallucinated imagery (Soh et al. 1992).

**3.3.5.6. Conclusion.** In conclusion, although some authors have interpreted the findings of Hong et al. (1995) as evidence for the scanning hypothesis (Antrobus et al. 1995; Hong et al. 1995; 1997), considerable improvement in temporal and deep structural resolution will be necessary before such evidence can be considered to be definitive. Such agnosticism is shared by the originator of the scanning hypothesis, Roffwarg (Roffwarg & Belenky 1996), who also emphasizes the need to visualize both cortical and subcortical structures simultaneously before assigning the initiation of REM sleep eye movements to either region. We therefore regard the question of exactly how the specific visual imagery of dreams is generated and attended to as being still entirely open at this time. One way to close this gap would be to compare cerebral blood flow patterns in subjects making directed visual images in waking with directed visual image-making in lucid REM sleep dreaming. In addition, it may soon be possible to temporarily deactivate specific cortical areas with transcranial magnetic stimulation during REM.

## 4. A new state space model: AIM

As the activation-synthesis model has evolved, it has metamorphosed into the three-dimensional framework of the AIM model. We now update the activation-synthesis concept as follows: (1) high levels of cortical activation (high values of "A") are a correlate of the mind's ability to access and manipulate significant amounts of stored information from the brain during dream synthesis; (2) the blockade of external sensory input and its functional replacement by internally generated REM sleep events such as PGO waves (internal sources of "I") provide the specific activation of sensory and affective centers that prime the cortex for dream construction; and (3) the shift of the brain from aminergic to cholinergic neuromodulation (low ratios of aminergic to cholinergic neuromodulation, "M") alters the mnemonic capacity of the brain-mind and reduces the reliability of cortical circuits, increasing the likelihood of

bizarre temporal sequences and associations which are uncritically accepted as waking reality when we are dreaming.

As the brain shifts from alert waking through drowsiness to NREM and REM sleep, a concerted set of physiological and chemical changes occur in the brain and periphery. Global changes are seen in all major physiological systems, including the nervous, respiratory, cardiac, renal, immunological, endocrine, and motor systems (Gottesmann 1997; Hobson 1989; Orem 1980; 2000). The changes in central neurophysiology include changes in gating of sensory input, inhibition of motor output and neuromodulation of widespread regions of the cortex (Gottesmann 1997; Hobson 1988b; Hobson & Steriade 1986; Steriade & McCarley 1990a). More specific neurophysiological changes involve both tonic and phasic activation of numerous brain regions, including, but not limited to, the medullary bulbar reticular formation, the pontine reticular formation, the hypothalamus, the lateral geniculate nucleus, the amygdala, the hippocampus, and the limbic and unimodal visual associative cortex, as well as regional *de*activation of the dorsal raphe, locus coeruleus, and multimodal association cortices (Amzica & Steriade 1996; Braun et al. 1997; Hobson & Steriade 1986; Maquet et al. 1996; Nofzinger et al. 1997; Steriade & McCarley 1990a). (See Table 2 and Fig. 7.) Not surprisingly, these changes are accompanied by dramatic shifts in the activity of the mind.

In the past, there has been a tendency to describe these shifting brain-mind states along a single axis, from wide awake to deeply asleep. The changes in mental state were perceived as dependent on variations in a single underlying parameter such as activity of the reticular activation system or overall brain activity as reflected in the EEG (e.g., Moruzzi & Magoun 1949). While conceptually useful at the time, it was clear from the outset that this activation concept was inadequate. And nowhere was this inadequacy more evident than in REM sleep, otherwise known as "paradoxical" sleep specifically because of the dissociation between level of behavioral arousal (low) and level of brain activation (high) (e.g., Jouvet & Michel 1959).

In response to this problem, researchers have recently suggested that the source of inputs for the brain-mind be considered a second dimension of brain-mind state (e.g., Antrobus 1991; Hobson 1990; 1992a). In their analysis of waking and dreaming, the neurophysiologists Llinas and Pare (1991) have ascribed all of the differences in subjective experience to the off-line status of the brain in REM. Likewise, the psychologist Antrobus has argued that sensory deprivation in the wake state produces dreamlike mentation because: (1) the brain is highly activated as it is in REM sleep (indicated by high frequency, low amplitude EEG patterns); and (2) the brain-mind has lost external sensory inputs and, again as in REM sleep, must turn to internal sources of input (Antrobus 1991; Reinsel et al. 1992). Although these two parameters tend to shift in concert, with brain activation and external input sources both decreasing as one moves from alert waking to deep sleep, such states as REM sleep (high brain activation and low external inputs) and sleep walking (low brain activation with some degree of preserved external inputs as evidenced by sleep walkers' ability to navigate) point out the potential independence of these two axes.

To this two-dimensional model we have added a critical third dimension which reflects the "mode" of information processing carried out by the brain-mind, a mode determined by the action of cortical neuromodulators (Hobson

1990; 1992a; 1997a). Within the brain, widespread cortical neuromodulation is effected by at least five specific neurotransmitters – acetylcholine, serotonin, norepinephrine, dopamine, and histamine (Cooper et al. 1996; Hobson & Steriade 1986; Saper et al. 1997; Steriade & McCarley 1990a) and probably others such as adenosine (McCarley et al. 1997) and orexin (Chimelli et al. 1999; Lin et al. 1999). With the exception of adenosine, each of the above neuromodulatory substances is produced by a highly localized group of subcortical neurons which project directly to widespread areas of the forebrain and are known to have powerful effects on mental state. Three of these – acetylcholine, serotonin, and norepinephrine – are known to play critical roles in the transitions from waking to NREM and then to REM sleep (Hobson & Steriade 1986; Steriade & McCarley 1990a).

Histamine and orexin also appear to be involved in sleep-wake transitions (Saper et al. 1997; Shiromani et al. 1999; Chimelli et al. 1999). Although dopamine does not appear to be a prime mover of normal conscious state regulation (Miller et al. 1983; Steinfels et al. 1983), it probably plays a major if perhaps secondary role in sleep regulation as evidenced by its interactions with other neuromodulatory systems (e.g., Kapur & Remington 1996; Mamelak 1991), its effects on normal sleep (Gillin et al. 1973; Olive et al. 1998; Post et al. 1974; Python et al. 1996; Trampus et al. 1993), and the effects of REM sleep deprivation on dopaminergic neurotransmission (Brock et al. 1995; Nunes et al. 1994; Tufik et al. 1978). It is thus not surprising that most of the psychopharmacological drugs used today which directly affect this neuromodulatory mode (Function M), often alter sleep and dreaming as well (e.g., Armitage et al. 1995; Lepkifker et al. 1995; Markowitz 1991; Pace-Schott et al. 1998; 1999; 2001; Sharf et al. 1978; Silvestri et al. 1998; in press; Vogel 1975; Vogel et al. 1990).

We have described this three-dimensional model of brain-mind state in our "AIM Model" (Hobson 1990; 1992a; 1997a; Hobson & Stickgold 1994b; Kahn et al. 1997). AIM makes three major claims:

1. AIM proposes that conscious states are in large part determined by three interdependent processes, namely the level of brain activation ("A"), the origin of inputs ("I") to the activated areas, and the relative levels of activation of aminergic (noradrenergic and serotonergic) and cholinergic neuromodulators ("M"). While these variables tend to vary in concert with one another, many paradoxical and dissociated mental states, both normal and abnormal, arise from the sometimes strikingly independent variation of these parameters as we will shortly illustrate.

2. The AIM Model proposes that the universe of possible brain-mind states can be construed as a three-dimensional state space, with axes A, I, and M (activation, input, and mode), and that the state of the brain-mind at any given instant of time can be described as a point in this space. Since the AIM model represents brain-mind state as a sequence of points, time is a fourth dimension of the model.

3. The AIM model proposes that while stable and reproducible mental states reflect the tendency of the brain-mind to occupy a small number of fixed locations in this state space, corresponding to such identified brain-mind states as alert wake or vivid REM sleep dreaming (see Kahn et al. 1997), all three parameters defining the state space are continuous variables, and any point in the state space can in theory be occupied. In the remainder of this section, we will discuss each of these three claims in detail.

## 4.1. The three dimensions of the state space

Experimental testing of the AIM Model requires that each of the three parametric axes of the brain-mind state space be directly measured and, ideally, manipulated. Toward this end, we have attempted to define the underlying parameters as well as to indicate how they can best be measured (see again Fig. 1). As we shall show below, reasonable measures of A and I can be readily obtained in both humans and animals. At the present time, M can only be measured directly in animals, but because its value can be manipulated experimentally in humans with pharmacological agents, its role in human conscious state determination can be indirectly assessed.

**4.1.1. Activation.** Conscious states show a clear-cut dependence on brain activation level. The production of conscious experience, as reflected in the length, intensity, and complexity of subjective reports of mental activity, as well as in levels of arousal and alertness, is generally greater in waking and in REM sleep than it is in deep NREM sleep and greater in alert waking than in quiet resting. The AIM model predicts that this physiological measure, "A," reflects the rate at which the brain-mind can process information regardless of its source (measured as "I") or its mode of processing ("M"). This activation parameter is based upon Moruzzi and Magoun's concept of a reticular activating system (Moruzzi & Magoun 1949; Steriade et al. 1980). Broad consensus already exists for the importance of this first dimension of the AIM Model.

In its simplest form, brain activation is defined as the mean firing frequency of brain stem neurons. It can be approximated in both humans and animals from the EEG spectrum, with increasing activation reflected by relatively high power in the high frequency range and relatively low power at low frequencies. In animals, the activity of the reticular activating system can be precisely quantified from the frequency of firing of neurons in the midbrain reticular formation (Huttenlocher 1961; Kasamatsu 1970; Steriade et al. 1980).

In humans, an alternative measure of overall brain activation might be the level of gamma frequency (30–70 Hz) oscillation in the brain (Llinas & Ribary 1993; Llinas et al. 1994). Although some recent work questions the association of gamma oscillation with REM sleep (Germain & Nielsen 1996), other work appears to confirm it (Uchida et al. 1997). Such gamma activity in humans has been shown to correlate with discrete cognitive events (Lutzenberger et al. 1995; Muller et al. 1996; Tallon-Baudry & Bertrand 1999; Tallon-Baudry et al. 1996; 1997; 1998) and to be measurable with depth electrodes in the human medial temporal lobe (Hirai et al. 1999).

**4.1.2. Input source.** Waking, NREM sleep and REM sleep represent states in which the sources of information processed by the brain differ dramatically. The second parameter of our AIM Model, input source (I), is a measure of the extent to which the brain-mind is processing external sensory data impinging upon receptors (as it is in waking) or from internal data sources (as in day dreaming or REM sleep). Because one component of sensory input is proprioceptive feedback reflecting the extent of motor activity, we also include the efficacy of such feedback in parameter I. Internally generated pseudosensory data can be produced by brain stem mechanisms (e.g., via PGO stimulation of visual cortex in REM sleep), it can be recalled from memory, or it can be intentionally created by directed mental imagery

In alert waking, the contents of our conscious experience (e.g., our thoughts and our feelings) tend to be driven by external stimuli and are predictive of subsequent motor behavior. During sleep, in contrast, conscious experience is normally driven by internally generated stimuli and has no apparent behavioral consequence. In the AIM Model, waking is characterized as both more exteroceptive and exteroeffective than either NREM or REM sleep, while REM sleep is markedly more interoceptive than NREM sleep but less exteroeffective than either waking or NREM sleep.

This second dimension of our AIM Model, though robust, has not been specified by many cognitive theorists who tend to regard internally generated signals as simply the phasic intensification of activation level. Such a view ignores what to us are very significant differences in such mental functions as vision, visual imagery, and visual hallucination. But while some seem to consider it an irrelevant factor, Llinas and Pare (1991) have suggested that this dimension by itself could be an adequate explanation of the phenomenological differences between such high activation states as waking and REM sleep (Llinas & Pare 1991). We agree with Llinas and Pare that both in waking and in sleeping, input source represents a major determinant of the nature of conscious experience. However, we do not regard the differences in input source to be an adequate explanation of the phenomenological distinction between waking and dreaming. How, for example, could it account for dream forgetting or the relatively low visual intensity and bizarreness of daydreams?

Physiologically, the input source axis of the AIM Model reflects both input-output gating and nonsensory activation of sensorimotor cortices. The activation of these cortical regions by external sensory stimuli can be directly measured in humans using evoked potential (ERP) techniques (e.g., Niiyama et al. 1997; Sallinen et al. 1996) or using stimulus threshold studies (see Arkin & Antrobus 1978 and Price & Kremen 1980 for reviews). In this regard, it is notable that Price and Kremen (1980) measured a rise in auditory stimulus threshold and Sallinen et al. (1996) observed a decreased ERP response in human phasic compared to tonic REM sleep. Similarly, the H-reflex can be used to measure motor blockade (Hodes & Dement 1964). In animals the same measures can be obtained and complemented by more refined assessments. For example, the amount of presynaptic inhibition of 1A afferent terminals (Bizzi & Brooks 1963; Pompeiano 1967b) specifically measures the sensory gate function while the amount of motoneuronal hyperpolarization (Chase & Morales 1990; Pompeiano 1967a) measures gating of motor activity. (For a recent review of such measurements see Gottesmann 1997.)

In humans and animals, eye movement density in REM sleep provides an estimate of the amount of internally generated pseudosensory data because eye movement density reflects brain stem PGO and motor pattern generator activity. In addition, the frequency of PGO waves (or the burst intensity of PGO waves) can be measured in animals to determine this parameter more directly. Currently, PGO waves cannot be easily or confidently recorded from humans although numerous suggestive EEG findings have been reported (McCarley et al. 1983; Miyauchi et al. 1987; 1990; Niiyama et al. 1988; Salzarulo et al. 1975 ) and new dipole tracing techniques show promise in identifying human PGO waves (Inoue et al. 1999b).

### 4.1.3. Modulation.

The third major and clear-cut physiological difference among waking, REM, and NREM is in the neuromodulation of the brain. In the AIM Model, we focus on the marked shift in modulatory balance seen from aminergic (noradrenergic and serotonergic) predominance in waking to cholinergic predominance in the REM sleep of animals. We call this modulatory factor M and define it as the ratio of aminergic to cholinergic chemical influence upon the brain.

It is our contention that this shift of neuromodulatory balance underlies the similar modal shifts in information processing (data processing, storage, and retrieval) seen as the brain shifts from one wake-sleep state to another. We propose that this modulatory factor M is involved in the regulation of such conscious state functions as directed attention, deliberate thought, self reflective awareness, orientation, emotion, memory, and insight. All of these functions are altered in the transition from waking to NREM sleep as a function of the diminished activation and sensory input level. But their even more marked dramatic alteration in dreaming, when the activation level is as high as in waking, must have another brain basis, which we think the changes in input-output gating alone are inadequate to explain. This element of our model has found little support among sleep psychologists who, we believe, either have failed to fully appreciate the extent of the alteration of cognitive features (such as the defective memory of REM sleep) or have simply rejected the concept of a neurophysiological description of psychological phenomenology (for one exception see Hartmann 1982).

Measurement of "M" is based on comparing the rates of firing or amounts of transmitter released by norepinephrine-containing locus coeruleus neurons and serotonin-containing raphe neurons to that of putatively cholinergic, PGO burst cells in the peribrachial region. State-dependent shifts in this parameter have been extensively documented in animal models (Datta 1995; 1997b; Foote et al. 1983; Hobson 1992b; Hobson & Steriade 1986; Hobson et al. 1986; Jacobs & Azmita 1992; Lin et al. 1994; Sanford et al. 1995b; Sherin et al. 1996; Steriade & Biesold 1990; Steriade & Hobson 1976; Steriade & McCarley 1990a; Szymusiak 1995). A more accurate measure of this parameter may be obtained by the simultaneous measure of release of the two classes of modulator using microdialysis techniques (e.g., Kodama & Honda 1996; Lydic et al. 1991b; Portas et al. 1998; Williams et al. 1994). Unfortunately, methodological constraints have so far largely prevented the measurement of this parameter in humans (although see Bartha et al. 1999; Sudo et al. 1998; Wilson et al. 1997). Evidence that such changes occur, and are significant, in humans is indirect but consistently confirmatory.

The role of this parameter in human conscious experience has been extensively studied in waking experiments using drugs known to alter neuromodulatory balance (see Perry & Perry 1995; Perry et al. 1999). In addition, cholinergic stimulation has been found to potentiate REM sleep (Berger et al. 1989; Gillin et al. 1991; Sitaram et al. 1976; 1978b) and dreaming (Sitaram et al. 1978a) while many aminergic agents are known to have REM suppressive and alerting effects (Gaillard et al. 1994; Nicholson et al. 1989) as well as effects on dreaming (Hobson & Pace-Schott 1999; Thompson & Pierce 1999). Reviews of psychopharmacological evidence suggests that the role of modulation in humans is homologous to that in experimental animals

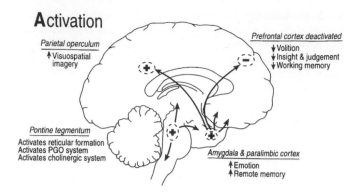

## Activation

**Parietal operculum**
↑ Visuospatial imagery

**Prefrontal cortex deactivated**
↓ Volition
↓ Insight & judgement
↓ Working memory

**Pontine tegmentum**
Activates reticular formation
Activates PGO system
Activates cholinergic system

**Amygdala & paralimbic cortex**
↑ Emotion
↑ Remote memory

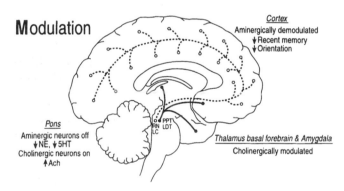

## Input Source

**PGO system turned on**
Fictive visual & motor data generated

Occipital cortex

Geniculate

Pons

**Sensory input blocked**
Real world data unavailable

**Motor output blocked**
Real action impossible

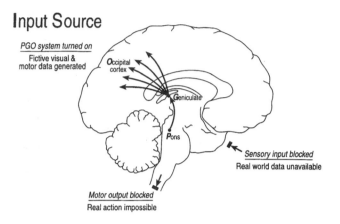

## Modulation

**Cortex**
Aminergically demodulated
↓ Recent memory
↓ Orientation

**Pons**
Aminergic neurons off
↓ NE, ↓ 5HT
Cholinergic neurons on
↑ Ach

RN PPT
LC LDT

**Thalamus basal forebrain & Amygdala**
Cholinergically modulated

Figure 8. Physiological signs and regional brain mechanisms of REM sleep dreaming separated into the activation (A), input source (I) and modulation (M) functional components of the AIM model. Dynamic changes in A, I, and M during REM sleep dreaming are noted adjacent to each figure. Note that these are highly schematized depictions which illustrate global processes and do not attempt to comprehensively detail all the brain structures and their interactions which may be involved in REM sleep dreaming (see text and Table 2 for additional anatomic details).

(e.g., Everitt & Robbins 1997; Hasselmo 1999; Perry & Perry 1995; Robbins & Everitt 1995).

An important aspect of the AIM model is its effort to mirror cognition's psychological features in its three physiological dimensions. Thus, "Activation" has a specific meaning at both the neurobiological and cognitive levels (see Anderson's ACT* model; Anderson 1983). Cognitivists also speak of information processing and thus share the concept of "input source" with neurobiologists, who express this dimension in terms of sensory thresholds, the excitability of motor pattern and efferent copy circuits, and the threshold for motor output. Finally, the mode concept is important to

cognitivists as a memory/amnesia dimension (as well as, possibly, an attention/inattention axis) while neurobiologists represent mode as the ratio of aminergic to cholinergic neuromodulator release. It is by these formal homologies between neurobiology and the cognitive sciences that the AIM model attempts to produce an integrated picture of the brain-mind.

An initial attempt to model the neuroanatomical structures participating in REM-state-dependent changes in activation, input source and neuromodulation is illustrated in Figure 8.

### 4.2. The AIM state space

The AIM model proposes that conscious states can be defined and distinguished from one another by the values of three parameters. These parameters can be considered as the axes of a three-dimensional state space. This state space can be represented visually as a cube where normal values for the parameters range along the three axes (Figs. 1 and 9). The model is not only useful in representing normal states but is also helpful as a heuristic tool to illustrate several critical issues in sleep research.

In quantitative renditions of the model (Hobson 1990; 1992a) the activation parameter (A) was derived from either the mean rate of firing of reticular formation neuronal populations that varies in animals from a low of 25/second in NREM sleep to 50/second in REM or from the inverse of the voltage amplitude of the EEG which varies from 25–50 μV in waking to 150–200 mV in Stage IV NREM sleep in humans. A four-fold range of values is assumed in visual representations of the model. The input source parameter can be derived from arousal threshold or H-reflex amplitude in humans or PGO wave frequency in animals. The range of these values is roughly the same order of magnitude as factor A. The modulatory parameter, M, is derived from the mean rate of neuronal population discharge of the aminergic populations (2–4 cycles/second in waking, 1–2 cycles/second in NREM, 0.01–0.1 cycles/second in REM) or from the concentration of norepinephrine, serotonin or acetylcholine in microdialysis studies which vary over a range of about 10-fold (Hobson & Steriade 1986; McCarley & Steriade 1990; Steriade & Hobson 1976).

All the parameters of the model are known to vary over the sleep cycle in a nonlinear manner. For example, factor M has a clearly exponential deceleration in the NREM-REM transition. Some aspects of this nonlinearity are embodied in earlier mathematical modeling of the reciprocal interaction model using the Volterra-Lotka equations (McCarley & Hobson 1975; McCarley & Massaquoi 1986) which yield ellipses as the graphical representation of the sleep cycle.

We acknowledge the tentative and necessarily speculative nature of our assumption of homology across mammalian sleep mechanisms, but point out that it is supported by abundant indirect evidence. And we recognize one important exception to this homology assumption: the relative complexity of the human forebrain gives rise to a greater complexity of EEG patterns in human NREM sleep compared to animals. We believe that this complexity is underestimated by currently available measures and that activation models of cognition likewise underestimate the differences between NREM states.

We do not pretend to have solved the problem of modeling conscious states, only to have proposed more realistic and heuristically valuable approaches to this problem. AIM constitutes only a simplified framework for modeling the physiology underlying changes of behavioral state and we in no way claim that it can fully account for the wide variety of human subjective experience, which includes thought, imagery, fantasy, and altered or pathological states as well as dreaming. Moreover, we recognize that the axes of the AIM state space are not independent. For example, at sleep onset a decline in general activation is likely to parallel a decline in aminergic modulation and a decline in the strength of external stimulus drive. Likewise at REM sleep onset the steep rise in cholinergic activity is likely to parallel the rise in internal stimulus drive and a rise in general activation level. But the axes of the model are uniquely capable of accounting for just the kinds of paradoxes that arise from an interactive system that changes its states paradoxically: that is, has high levels of activation in *both* waking and REM sleep; shifts from external to internal stimulus processing; and processes information differently in two equally activated states.

Current developments in basic and clinical neurobiology suggest the exciting possibility that the M dimension may become measurable in behaving (i.e., waking, thinking, performing, sleeping, dreaming) human beings. Already, microdialysis techniques with depth electrodes implanted to localize epileptic foci have shown fluctuations in serotonin across the wake-NREM-REM cycle paralleling those seen in animals (Wilson et al. 1997). Moreover, the newest PET techniques for radiolabeling receptor ligands as well as magnetic resonance spectroscopy (Rauch & Renshaw 1995) may yield further possibilities for the localization and quantitation of neuromodulatory dynamics in the human CNS.

One use of the AIM model is to depict the highly dynamic and variable nature of human consciousness, and thus to visually plot specific "states" of consciousness within the state space. As an example, normal consciousness, at the coarsest level, can be divided into the states of waking, REM, and NREM sleep. Each of these states can be characterized both by distinct physiologies and by distinct differences in mentation. To help the reader orient to the AIM state space, the positions of these three states in the AIM state space, as well as the trajectory from waking through NREM into REM sleep, are shown in Figure 9.

In this figure, the fully alert, wake state is depicted in the upper-right corner of the back plane of the cube. This corresponds to maximal levels of brain activation (right surface

of cube), maximal external input sources with minimal internal sources (back surface), and maximal aminergic and minimal cholinergic neuromodulation (top surface). Cognitively, this corresponds to alertness with attention focused on the outside world.

In the center of the cube lies deep NREM sleep, with low levels of brain activation, intermediate levels of both aminergic and cholinergic neuromodulation, and minimal levels of both external and internal input. In this state, the mind tends towards perseverative, non-progressive thinking with minimal hallucinatory activity, and this is reflected in the brevity and poverty of NREM sleep reports.

As cholinergic modulation increases and aminergic modulation decreases, the modulatory function falls to its low point. The brain-mind, however, regains waking levels of activation and moves from NREM into REM sleep. AIM (now referring to the brain's location in the AIM state space) moves to the bottom front edge of the cube, with input now internally driven (front surface) and neuromodulation predominantly cholinergic (bottom surface). We emphasize the paradox that instead of moving to the left surface of the cube – to a position diametrically opposed to waking (dotted line) – brain activation returns to waking level. This forces AIM to the right surface of the cube. As a result the mind is alert, but because it is demodulated and driven by powerful internal stimuli, it becomes both hallucinatory and unfocused. REM sleep's deviation from the main diagonal axis provides a visual representation of the distinctively unique phenomenology of REM sleep and shows why that state favors dreaming.

A second function of the AIM state space model is as a tool to clarify the concept of substates. While consciousness can be coarsely divided into waking, REM, and NREM sleep, these are only a few of many possible brain-mind states. For example, NREM sleep can be subdivided on physiological bases into substates: sleep onset, Stage II of NREM sleep, and deep Stages III and IV NREM sleep. Presumably, sleep mentation changes in concert with these physiological changes. Similarly, REM sleep can be subdivided physiologically into phasic and tonic REM or psychologically into lucid and nonlucid dreaming substates. Finally, the waking state can be subdivided into a vast multiplicity of substates, defined by attentive parameters (alert, attentive, vigilant vs. drowsy, inattentive, day dreaming), emotional parameters (calm, angry, sad, afraid), or even by information processing strategies (focused and goal directed vs. creative and freely associating). Other substates of waking can be produced by specific induction procedures, such as trance, hypnosis, sleep deprivation, and by the ingestion of psychoactive drugs.

For each of these substates, a subregion of AIM state space could, in theory, be defined which would characterize its physiological and psychological nature. However, as the distinctions between states become more subtle, these regions necessarily begin to overlap and blur. At the same time, the three dimensions of the AIM model quickly become inadequate. For example, the model is strained to account for differences between various emotional substates of waking. This could be partially resolved by adding a regional activation dimension to our model, such as the ratio of limbic to neocortical activation as suggested by neuroimaging studies (e.g., Maquet et al. 1996; Nofzinger et al. 1997).

Could the changes in regional activation of the brain be related to the shift in neuromodulatory balance that we have

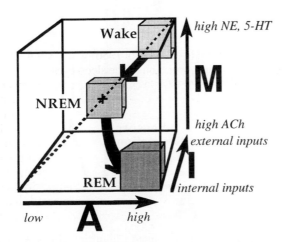

Figure 9. Normal transitioning within the AIM state space from wake to NREM and then to REM.

described? It seems likely to us that the changes in regional activation ($A_R$) are a combined function of changes in I and M so that, for example, it is the cholinergic pathway from pons to amygdala that is responsible for the selective activation of the limbic brain in REM sleep. Similarly, it could be that the deactivation of the frontal lobe is caused by the withdrawal of aminergic inputs to that region in REM sleep. These suggestions are not simply ways of saving the model's relative simplicity. Rather they demonstrate the capacity of the model to generate new, testable hypotheses about the cellular and molecular basis of regional brain activations.

**4.2.1. Dissociated states.** Given the multiplicity of parameters contributing to conscious states and the complex dynamics of their interaction, it is to the credit of evolutionary tinkering that the cardinal states of wake, NREM, and REM sleep appear so discrete and that their temporal sequence is normally so canonical. But this discreteness and canonical sequencing is only approximate. As the AIM state-space model attempts to make clear, any point within the state space can be occupied, and the parametric values which define the canonical states of waking, NREM, and REM sleep can be dissociated from one another. As a result, the appearance of dissociated states – states in which, for example, some parameters match their canonical NREM values while others match canonical REM or wake values – should be considered both natural and inevitable. Acknowledging this propensity of the conscious state system to dissociate enriches our view of both normal and abnormal neurological and psychiatric conditions.

These dissociations occur most commonly during the transition from one stable state to another as exemplified by state carry-over phenomena tapped by neurocognitive and psychological testing following the awakening of human subjects from NREM and REM sleep (Bonnet 1983; Doricchi et al. 1991; 1993; Fiss et al. 1966; Gordon et al. 1982; Lavie 1974b; Lavie & Giora 1975; Lavie & Sutter 1975; Rittenhouse et al. 1993; Rosenblatt et al. 1992; Stickgold et al. 1999b; Stones 1977), with perhaps the best known of these being the persistent lethargy termed as "sleep inertia" (Achermann et al. 1995; Dinges 1990). In such cases, the transitions of some parameters lag behind those of others and the dissociations are usually quite transient. But in other cases, they are more stable, as in sleep walking (Broughton 1968; Guilleminault 1987), where waking values of locomotor output are reached in NREM sleep. Interesting to note, recent PET data have shown persistence of selective deactivation, especially in the prefrontal and posterior inferior cortices, for more than 5 minutes post awakening from Stage 2 sleep (Balkin et al. 1999). Many of these dissociated states can be represented using the AIM state space model.

Thus, another function of the model is to organize and visually represent some of the conscious state dissociations seen in normal subjects, in patients with neurological and psychiatric symptoms, and in both groups when treated with drugs that affect brain neuromodulatory systems. The basic concept that we wish to convey is that while the three dimensions of AIM state space usually change synchronously as the brain-mind shifts between the three stable canonical states, genetic bias, life events, and pharmaceutical intervention can all conspire either to desynchronize the shifts occurring along the three axes or to create new stable states in which one or another dimension takes on an unexpected value.

The net result is a departure from the usual trajectory (shown in Fig. 9) or the creation of normal-hybrid states with mixtures of wake, NREM, and REM features as suggested in Figures 10–18. In these examples, dissociations along each of the three axes of the state space are examined. It should be emphasized that the discussion that follows is speculative and is intended to be heuristic rather than definitive. Although we have chosen examples that we believe to be realistic and have made assumptions that we hold to be reasonable, empirical tests of these hypotheses remain to be conducted.

**4.2.2. Activation.** To illustrate the vicissitudes of the activation function, we consider two normal phenomena, quiet waking and sleep onset, which are related to each other in ways that have a critical bearing on the issues discussed earlier in our target article. We will show how both quiet waking and the transition from wake to sleep may vary significantly depending upon the current level and the rate of change of the activation function. The transitional state of sleep onset has been extensively studied because of the unique mentation reports that can be obtained on arousal from this state. Yet the exact position of sleep onset in AIM state space is critically dependent on the precise temporal pattern of sleep onset.

*Quiet waking:* We first consider the period of quiet waking preceding sleep onset. Before lying down and closing his eyes, a subject is usually in an alert state (see again Fig. 9, "Wake"). Normally, on lying down and closing his eyes, he will shift into an alpha wave EEG pattern, reflecting a decrease in "A" and, because visual stimulation has been shut off, a decrease in "I" as well. At the same time, neuromodulatory shifts may begin to decrease aminergic output. Thus, he will begin to move along the main axis from Wake toward NREM, as indicated in Figure 9.

But when examined in detail, each individual will take a unique path through the state space from waking to NREM, depending on both the relative and absolute rates of decline of each of the three state space parameters. For example, if an individual is drowsy before retiring (Fig. 10, "Drowsy"),

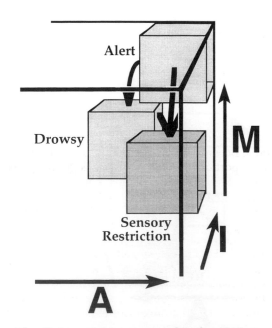

Figure 10. Quiet rest: Movement within the AIM state space prior to sleep onset depends on how sleepy the subject is as well as the extent of external sensory input.

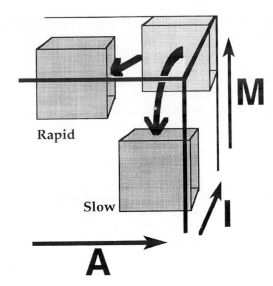

Figure 11. Sleep onset: With more rapid sleep onset, lowered activation precedes aminergic demodulation; with slow onset, the order is reversed.

values for "A" and perhaps also "M" will begin to drop well before the subject even goes to bed, while "I" remains high, placing one in the center of the back surface of the cube. In contrast, if an individual is quite alert when going to bed, "I" might drop before either "A" or "M" (not shown), followed by a small drop in "A" as alpha patterns appear in the EEG.

Under other conditions of quiet waking, such as when subjects were placed in a darkened, sound attenuated room by Antrobus in his "waking controls" for dream mentation (Reinsel et al. 1992), "I" would immediately shift because of the elimination of external sensory stimulation, and we expect that "M" would then slowly shift to relatively low values while "A" stayed high, placing one in the center of the right-hand surface of the cube (Fig. 10, "Sensory Restriction"). Under these conditions, the brain-mind state moves to a position midway between waking and REM sleep (cf. Fig. 9), rather than between waking and NREM. It is therefore not surprising to us that Reinsel et al. (1992) found that mentation became more dreamlike under these waking conditions.

We can use the AIM state space model to investigate the implications of Antrobus's paradigm. Since "I" falls virtually instantaneously upon being placed in the dark, AIM should initially occupy a position in the state space just in front of normal waking, with only "I" decreased. Then, over time, neuromodulatory shifts would move AIM lower in the state space, to the position shown in Figure 10 ("Sensory Restriction"). Because the AIM model hypothesized that "M" plays an important role in modulating cognitive processes, we would expect reports to become more and more dreamlike over the first 5 to 10 minutes in this condition. In contrast, Antrobus's activation-only model would seem to predict that reports should become *less* dreamlike with time, since activation would be expected to drop during quiet wake as EEG alpha increases. In fact, hallucinosis has been shown to increase over time as arousal diminishes during sensory deprivation protocols (Rossi et al. 1964). Indeed, it would be quite surprising to find mentation becoming more wakelike and less dreamlike with an increased period of waking sensory deprivation.

*Sleep onset:* As the subject moves from wake to sleep onset, further movement occurs within the state space (Fig.

11). The box labeled "Rapid" in Figure 11 represents a possible initial sleep onset state when the transition from waking to sleep is precipitous following sleep deprivation. In this case, the transition occurs before there is time for aminergic neuromodulatory levels to decrease. As a result, the "M" function remains on the top surface of the cube (modulation highly aminergic) while brain activation and external inputs diminish. In contrast, the box labeled "Slow" (Fig. 11) represents a gradual transition from waking to sleep as might be seen in situational insomnia. In this case, decreases in aminergic neuromodulation and external inputs might occur prior to the decrease in brain activation. In both cases, AIM would then move into the standard Stage NREM position (Fig. 9).

*Lucid dreaming:* Another dissociation along the "A" axis of the AIM cube may arise during lucid dreaming. Under normal circumstances, dreamers believe themselves to be awake – but occasionally individuals become aware that they are dreaming. In this state of "lucid dreaming" (Laberge 1990; 1992) waking insight combines with dream hallucinosis in an intriguing and informative dissociation. We assume that for lucidity to occur, the normally deactivated dorsolateral prefrontal cortex (DLPFC) must be reactivated but not so strongly as to suppress the pontolimbic systems signals to it. This dissociation is represented in the AIM model by splitting AIM so the portion representing the DLPFC can take a position dissociated from that of the rest of the brain (Fig. 12). When this partial reactivation of the DLPFC occurs, internally generated images are seen for what they are and are not misinterpreted as coming from the outside world.

The fact that lucidity can arise when the DLPFC is deactivated can also be explained using AIM. Lucid dreaming occurs spontaneously or can be cultivated by pre-sleep autosuggestion. Spontaneous lucidity indicates that the reduced amount of reflective self-awareness during dreaming is sometimes enhanced enough for the subject to recognize the dream state for what it is. Autosuggestion probably increases this probability by priming the brain circuitry – presumably in prefrontal areas – that subserves self-reflective awareness. In both cases, the phenomenon of lucidity clearly illustrates the always statistical and always dissociable quality of brain-mind states. AIM accommodates

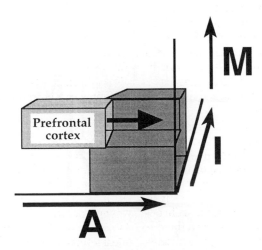

Figure 12. Lucid dreaming: Prefrontal cortical systems, which are normally inactive in REM sleep, shift toward higher, wake-like levels of activation, permitting conscious awareness of the dream state.

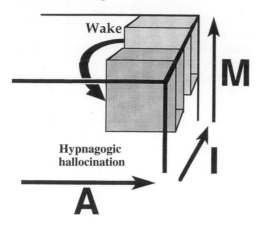

Figure 13. Hallucinosis: Internal stimuli shift the brain/mind forward along the "I" axis in AIM state space, with both internal and external inputs high. This condition may prevail during hypnagogic hallucinosis.

these features very well by proposing that lucid dreaming is a hybrid state lying across the wake-REM interface.

**4.2.3. Input source.** During waking, internal inputs are used mainly in the service of the ongoing sensorimotor integration of external signals. If, for any reason, internal signals became unusually strong, they could come to dominate the system with resulting hallucinosis. In this case, mentation would be driven by a combination of undifferentiated internally and externally driven imagery (see Mahowald et al. 1998).

*Hypnagogic and hypnopompic hallucination:* From the perspective of the AIM model, hypnagogic and hypnopompic hallucinations, associated with transitions into and out of sleep respectively, result from the REM-like enhancement of internal stimuli coupled with an activated, aminergically modulated waking brain (Figs. 13 and 14).

With internal and external inputs in an unstable balance (as occur during the hypnagogic period), AIM moves to a position half-way between the front and back surfaces of the cube (Fig. 13). But unlike NREM sleep, which is also at this midpoint of input source (with minimal internal and external inputs), both sources are being powerfully driven in hallucinosis. It is this unexpected combination of high internal and high external inputs that defines the functional

dissociation of these hallucinoid states. The frequency of this combination may be elevated by the abnormal physiology of narcolepsy, a condition in which the frequency of hypnagogic hallucinations is likewise elevated (Broughton et al. 1982; Mignot & Nishino 1999; also see Fasse 2000).

We can approximate a representation of the hypnopompic hallucinoid state by hypothesizing that while the brainstem signals continue to evoke internal representations in the cortex in the hypnopompic period, the blockade of external stimuli has broken down. As a result, the dissociated state results from a dissociation of the forebrain from the brainstem. This dissociation is represented in the AIM model by splitting the cube representing the brain-mind into forebrain (F) and brainstem (B) sections and showing their relative positions in AIM space (Fig. 14).

A more extreme example of this kind of dissociation is temporal lobe epilepsy in which abnormal phasic activation signals of limbic origin commandeer the cortex and force it to process external world data on limbic terms (e.g., Rabinowicz et al. 1997). Given the new findings on selective limbic activation in REM sleep (Braun et al. 1997; 1998; Maquet et al. 1996; Nofzinger et al. 1997), it seems reasonable to suppose that a similar, though normal, process may also drive the dreaming brain. By this we mean that the cortex of the dreaming brain is compelled to process internal signals arising from the pons and amygdala, as was originally suggested by the activation synthesis hypothesis. This epilepsy analogy is also cogent because the internal signals of REM sleep are spike and wave complexes arising in the pons and amygdala (Elazar & Hobson 1985). The limbic lobe may then direct the forebrain to construct dreams in a manner similar to that by which it creates the dreamy states of temporal lobe epilepsy (see Epstein 1995). Indeed, a recent study has shown more unpleasant and higher intensity emotions in the dreams of epileptics as compared to normals (Gruen et al. 1997).

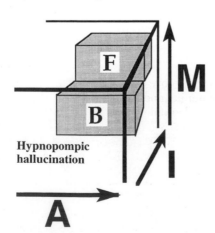

Figure 14. Hypnopompic hallucinosis: Forebrain (F) and brainstem (B) regions occupy different locations in the state space, with the brainstem initiating internal inputs while the forebrain continues to process external stimuli.

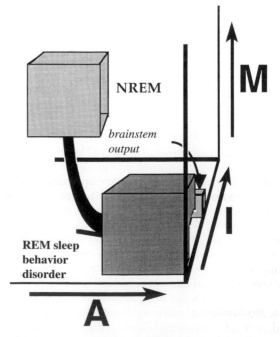

Figure 15. REM sleep behavior disorder: Brainstem inhibition of motor output is dissociated from other brain systems during REM sleep, moving toward waking values of the "I" parameter and leading to disinhibited motor output.

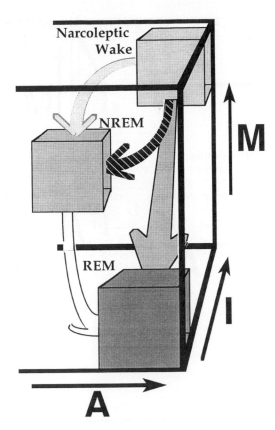

Figure 16A. Sleep onset in narcolepsy: the brain shifts down and forward in the AIM space prior to sleep onset, thereby inducing sleep onset hallucinations and direct entry into REM sleep at sleep onset.

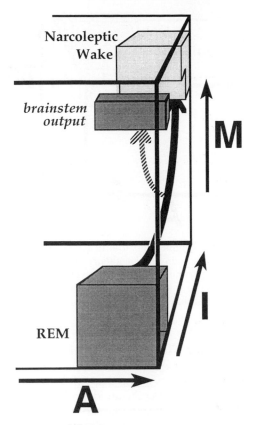

Figure 16B. Sleep paralysis in narcolepsy: Enhanced aminergic demodulation in narcolepsy increases inhibition of motor outputs, leading to dissociation of brainstem functions and continued motor inhibition after waking.

*REM sleep behavior disorder:* A particularly dramatic example of sensorimotor dissociation is seen in the REM sleep behavior disorder, in which the normal inhibition of motor output during REM fails (Mahowald & Schenck 1999; Schenck & Mahowald 1996; Schenck et al. 1993). Motor behaviors normally seen only in waking now arise completely involuntarily and automatically during REM, and patients physically act out their dreams (Mahowald et al. 1998). The historically oriented reader will recognize the similarity between this disorder and the dissociative phenomena that interested Charcot, Janet, and Freud.

During REM sleep, the motor cortex activation produces outputs similar to those seen in waking, but in response to exclusively internal inputs. Since the inhibition of spinal motorneurons usually occurs in concert with motor cortex activation, our single "I" parameter normally reflects the net inhibition of motor output. But in this case (as in the case of lucid dreaming) we represent this regional dissociation by a fragmenting of the AIM icon. In this case, the lower back quarter of the icon, representing brainstem output systems, has moved back in the state space toward a waking level of output (Fig. 15). It is this dissociation which produces the REM sleep behavior disorder.

**4.2.4. Modulation.** If aminergic modulatory power is weakened, as it is in narcolepsy (Mamelak 1991) and depression (Berger & Riemann 1993), and if cholinergic modulatory power is enhanced as it also appears to be in these two conditions (Berger & Riemann 1993; Mamelak 1991), then the value of M will decline. As a consequence, the ability of sub-

jects to maintain alertness may be compromised producing excessive daytime sleepiness. This would lead to a minor shift in the normal "alert" position in state space (Fig. 16A, "Narcoleptic Wake"). Moreover, REM sleep may be entered more rapidly or even directly from waking as in narcolepsy (Mitler et al. 1979). This shift in baseline values of M may also produce shortened REM latency (as in some forms of depression) or difficulty awakening fully from REM (as in narcolepsy).

These transitional abnormalities represent some of the clearest demonstrations of conscious state dissociation in sleep disorders medicine but they also instruct us about the normal phenomena, which they exaggerate. For example, narcoleptic subjects (Kayed 1995; Roth 1978) may hallucinate at sleep onset (Fig. 16A, striped arrow from Wake to NREM) as they move down and forward in the state space (more cholinergic modulation and hence more internal inputs) prior to sleep onset and its associated decrease in activation. This can be followed by normal entry into NREM sleep (striped arrow) or immediate entry into REM sleep without passing through NREM (gray arrow from wake to REM).

At the other end of the night, an inability to move, termed sleep paralysis (Mignot & Nishino 1999), which sometimes compounds the terror of hypnopompic hallucinations, represents a carry-over of the inhibition of spinal motorneurons into waking. This dissociation during narcoleptic awakening can be represented as a dissociation of brainstem motor activity along the "I" dimension secondary to a shift in "M" (Fig. 16B) as AIM moves toward the waking corner of the

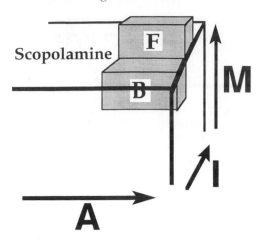

Figure 17. Scopolamine inhibition of REM sleep: Cholinergic inhibitors force the brain-mind to abnormally high ratios of aminergic to cholinergic neuromodulation, preventing entry into REM sleep and leading to simultaneous processing of external and internal inputs by forebrain (F) and brainstem (B) systems.

state space. This is the inverse of the dissociation seen in REM sleep behavior disorder (Fig. 15). The sleep abnormalities of narcolepsy, as well as those of depression, are relieved by drugs (e.g., tricyclic antidepressents and SSRIs) which enhance aminergic efficacy and suppress the cholinergic system (Gaillard et al. 1994; Nishino & Mignot 1997).

Other drugs that influence the M parameter produce "altered states of consciousness." Thus drugs which, like LSD, interfere with serotonergic neuromodulation (Aghajanian 1994), create dreamlike distortions of imagery and inhibit executive prefrontal cortical functions during waking, while anticholinergics (e.g., scopolamine) produce a delirious waking state with dream-like hallucinosis, disorientation, anxiety, and confabulation (Perry & Perry 1995). As seen in Fig. 17, scopolamine pushes AIM above the normal state space, pharmacologically reducing the levels of cholinergic neuromodulation below any normal physiological levels. At the same time, AIM splits as both external and internal inputs are activated.

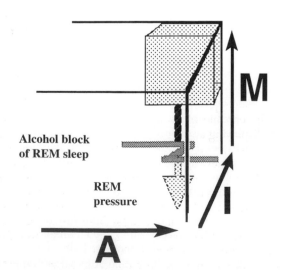

Figure 18A. Ethanol-induced suppression of REM sleep: Blockade of REM sleep leads to an increased biological pressure toward movement down in the state space, towards increased cholinergic modulation, but the blockade prevents movement.

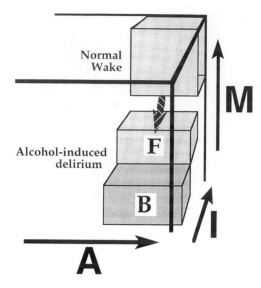

Figure 18B. Ethanol withdrawal: When the ethanol block is removed, the brain/mind shifts to abnormally high levels of cholinergic modulation, activating brainstem mechanisms for internal "sensory" inputs. This dissociates forebrain and brainstem systems and leads to alcohol-induced delirium.

**4.2.5. Dissociations.** In most of the cases described above, we have hypothesized that dissociation results from a fragmentation of normally unified neuromodulatory states. In short, the forebrain, midbrain, and brainstem fail to occupy a single position in the AIM state space. Instead, there is a split along the Activation or Input axis, with different brain regions occupying different positions in AIM space. Insight into how these dissociations might arise comes from the example of delirium associated with alcohol withdrawal.

Chronic alcohol usage blocks REM and upon withdrawal there is a REM rebound, marked by increased amounts and intensity of REM sleep (Pokorny 1978). It is during this period of REM rebound that delirium occurs. Presumably, the brain reacts dynamically to the alcohol-induced REM deprivation with an increased pressure towards REM sleep. We imagine this as pressure to move the brain lower in the AIM state space, towards lower aminergic and higher cholinergic neuromodulation. But while this pressure is exerted by the brain, the alcohol blocks the actual movement through the state space (Fig. 18).

When alcohol is withdrawn, the REM pressure forces AIM down in the state space causing increased REM sleep, but also causing hallucinations and delirium during waking (Fig. 18B). These symptoms of psychosis are caused by the release of brain systems which are normally inhibited except in REM sleep. In this case, it is an abnormal shift downward along the "M" axis of the state space which produces the splitting of AIM and causes its dissociation along the "I" axis. The net result is to move the brain-mind close to a position of REM sleep in waking.

### 4.3. Discrete conscious states and the continuous state space model

It is common, when discussing consciousness, to speak of "states" of consciousness. In doing so, it is often assumed that these are discrete brain-mind states with clearly definable boundaries; it is also assumed that at any given mo-

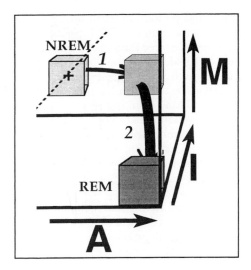

Figure 19. Time course of NREM to REM transition: Movement through AIM state space reflects the nonsynchronous shifts in EEG, neuromodulation, and muscle atonia. In this example, EEG desynchrony (1) occurs before this shift in neuromodulation and inputs (2). (Dashed lines as used in Fig. 9.)

ment the brain-mind is in one or another of these states. If this were true, then the transition between states would be absolute and instantaneous. As suggested by the examples presented above, the AIM state space model specifically rejects this conclusion. Rather, it proposes that although specific states of consciousness can be meaningfully described, shifts in consciousness reflect movements through a continuously varying state space, and not discontinuous jumps between discrete states. It also serves to demonstrate how a continuous state-space is compatible with the notion of discrete conscious states.

Specific states become defined because normal subjects tend to remain in a highly constrained region of the state space for long periods of time and then rapidly move to another similarly constrained region. Thus, after 16 hours of waking, the transition from waking to sleep can occur in less than one minute, and appears virtually instantaneous. Yet most researchers would agree that the transition is a continuous process rather than a sudden jump from one state to another; it is only the speed of the transition relative to the time spent in each "state" that makes it appear as a quantum shift.

Similarly, the transition from NREM to REM sleep, although rapid, shows a clear and finite time course (Fig. 19), with the typical REM signs of EEG desynchronization, muscle atonia, and rapid eye movements appearing in a variety of sequences over the course of 30 to 60 seconds – an observation familiar to all polysomnographers (see Butkov 1996; Rechtschaffen & Kales 1968; Sato et al. 1997). Recordings from single neurons in the cat brainstem further suggest that the shift in neuromodulation (the M axis in AIM state space) may be slower still as the shift from NREM to REM follows a continuous path from one state to the next (Hobson et al. 1975).

We emphasize that the AIM brain-mind state space is not a discontinuous collection of discrete states. Instead, any combination of values for A, I, and M is in theory possible, and although some ranges of these values are much more likely to be observed than others, movement from one sta-

ble state to another involves passing continuously along a path through the several state space domains.

A similar distinction is critical when the AIM state space is used to map both physiological states and states of consciousness. In its most specific description, AIM state space is mapped along three dimensions of physiology. When we map consciousness onto the three related dimensions of cognition, we achieve the same continuity and overlap of values that are seen in mapping physiology itself. And both domains thus achieve a realistic range of association-dissociation. Just as there is no absolute boundary between the waking, NREM, and REM domains in the physiological state space, there is no absolute boundary between the cognitive states determined by them. Thus, we do not claim that there can be *no* "NREM-like" mentation in REM sleep and *no* "REM-like" mentation in NREM or even in waking. Rather, we claim that there is a strong probabilistic relationship between positions in the physiological and cognitive state spaces; when a subject is in a given position in the physiological state space, he is most likely to occupy a nearly identical position in the cognitive state space. While we do believe that cognition and consciousness are totally determined by underlying physiological processes, we make no claim that we have more than begun to map the parameters (dimensions) of the state space which ultimately combine to define these psychological states.

### 4.4. Summary of the AIM model and the nature of conscious states

The AIM model describes a method of mapping conscious states onto an underlying physiological state space. In its strongest form, the AIM model relates not just to wake-sleep states of consciousness, but to all states of consciousness. It is limited by describing only three of what are undoubtedly numerous dimensions that must be specified to completely define this state space, but we have chosen those parameters that we feel are most critical for distinguishing among the basic wake-sleep states of consciousness.

By choosing activation, input source, and mode of neuromodulation as our three dimensions, we have selected *how much* information is being processed by the brain (A), *what* information is being processed (I), and *how* it is being processed (M). It is our belief that these three parameters are both necessary and sufficient to distinguish in a preliminary way among the basic wake-sleep states.

While the brain normally exists in specific regions of the AIM state space, only shifting from one area to another at relatively infrequent intervals, the brain is nonetheless theoretically capable of occupying any position in this state space, displaying any set of values of A, I, and M. As such, intermediate states and dissociated states are recognized as not only eminently possible but highly probable, and specific states of consciousness are seen more as convenient names for frequently occupied regions of the state space than as discrete, tightly bounded areas of the state space. In addition, transition from one stable brain/mind state to another involves moving along a continuous path through the state space, linking the two stable states.

Finally, although we believe that these three dimensions go a long way towards mapping what we know about the physiological processes underlying conscious states today, we believe that many more exist and as a result, our mapping from physiological state space to conscious states is an approximation that further experimentation can only serve to refine.

## 5. Conclusions

Our goal, as stated in the Introduction, was to begin to bridge interdisciplinary gaps in the study of sleep and dreaming; we have accordingly reviewed contemporary perspectives primarily from research psychology, neuropsychology, neurobiology, and clinical sleep medicine. Our AIM state-space model and a revised activation-synthesis theory of dreaming, summarized below, constitute our current and necessarily approximate synthesis of these data, which we hope will stimulate many future hypothesis-testing experiments. With regard to the areas not covered here, we refer the reader to published works (and we eagerly await future reviews) on dreaming and consciousness from scientists and scholars with specific expertise in clinical psychology, philosophy, literature, neural networks, artifical intelligence, as well as functional-evolutionary and molecular biology perspectives on sleep and dreaming.

We have shown that phenomenological differences between waking, NREM, and REM sleep are measurable. In our view, these differences are so great that they represent qualitative differences. A better understanding of the physiological processes underlying dream construction may be necessary before this issue can finally be laid to rest. But even when dream features appear to be specifically linked to distinctive REM physiology, interpretations can still be cast toward either camp. Hong et al. (1997) reported an impressive correlation between visual imagery and REM density ($r = 0.8$), which we would argue as evidence for a dependence of dream imagery on a qualitative feature of REM sleep. In contrast, Antrobus et al. (1995) consider this to be another example of the simple dependence of dream content on levels of brain activation, arguing that rapid eye movements are not under strict brainstem cholinergic control, but come increasingly under the control of the frontal eye fields as general cortical activation increases.

In the end, the issue may best be addressed in other forms. In the case of the major stages of sleep, it may be more useful to envisage psychophysiological continua, manifested at the levels of both the brain and the mind, whose various combinations define not only commonly experienced states of the brain-mind but uncommon ones as well. This is the strategy adopted by the AIM model with the dimensions activation (A), input source (I), and neuromodulation (M) representing three such continua.

Rather than fixed conditions, which must always show similar characteristics in order for brain-mind-body isomorphisms to be valid, behavioral states can be seen as relatively stable sets of values for these continua that have evolved as a result of adaptive benefit to the organism. Such multidimensional combinations can be influenced both at the level of the brain (as when we take a sleeping pill) and at that of the mind (as when we count sheep).

Along the dimension of Activation (A), neuroimaging studies strongly support an updated view of brain arousal in REM sleep as resulting from ascending influences from the brainstem and subcortex. The limbic subcortex and related cortex play a major part in the translation of this activation to associative, and perhaps even to sensorimotor areas of the cortex. Along the dimension of Input Source (I), newer research reinforces earlier findings on maximal sensorimotor blockade in REM. Along the dimension of modulation (M), recent research has confirmed the neuromodulation of conscious states by the interplay of cholinergic and aminergic influences arising from brainstem nuclei. This interplay is mediated and modulated by a diversity of cell populations and their neuromodulators in both the brain stem and the subcortical forebrain.

In a revised version of our activation-synthesis theory, the distinctive form of dream cognition may be explained at the level of the brain as follows:

1. The intense and vivid visual hallucinosis is due to autoactivation of the visual brain by pontine activation processes impinging, initially, at the level of unimodal visual association cortex and heteromodal parietal areas subserving spatial cognition.

2. The intense emotions, especially anxiety, elation, and anger are due to activation of the amygdala and more medial limbic structures. The emotional salience of dream imagery is possibly due to the activation of the paralimbic cortices by the amygdala and other subcortical limbic structures.

3. The delusional belief that we are awake, the lack of directed thought, the loss of self-reflective awareness, and the lack of insight about illogical and impossible dream experience are due to the combined and possibly related effects of aminergic demodulation and the selective inactivation of the dorsolateral prefrontal cortices.

4. The bizarre cognition of dreaming, characterized by incongruities and discontinuities of dream characters, loci, and actions, is due to an orientational instability caused by the chaotic nature of the pontine autoactivation process, its sporadic engagement of association cortices, the absence of frontal cortical monitoring, and episodic memory deficits that are, in part, due to failures of aminergic neuromodulation. We present a schematic explanation for the generation of these cognitive dream features which combines the above findings on state-dependent regional activation with the reciprocal interaction model for the neuromodulation of conscious states.

ACKNOWLEDGMENTS
This project was funded by the MacArthur Foundation Mind Body Network and NIH MH-48,832, MH13923, and MH01287. The authors wish to thank Eric A. Nofzinger, Allen R. Braun, James Quattrochi, David Kahn, Subimal Datta; Roar Fosse; also Jill Gustafson, Dorothea Abbott, Jennifer Holmes, Dawn Opstad, and April Malia.

NOTE
1. "Dreaming" was found to be better than "Total Recall Frequency" (TRF) (Table 3–8, $F[1,71] = 15.89$, $p < 0.01$), than TRF + "Waking Perception" (Table 3–10, $F[1,70] = 15.17$, $p < 0.01$), and than FRF + "Dreamer Participation" (Table 3–15, $F[1,70] = 13.70$, $p < 0.01$). In contrast, TRF + Dreamer Participation and TRF alone explained no significant amount of variance not already explained by Dreaming alone. In addition, judged Dreaming adds significantly to the REM/NREM variance when his "trichotomized' judges' scores versus their log-transformed scores are used in a step-wise analysis (Antrobus 1983).

# Dreaming and REM sleep are controlled by different brain mechanisms

**Mark Solms**

*Academic Department of Neurosurgery, St. Bartholomew's and Royal London
School of Medicine, Royal London Hospital, London E1 1BB, United Kingdom*
mlsolms@mds.qmw.ac.uk    www.mds.qmw.ac.uk

**Abstract:** The paradigmatic assumption that REM sleep is the physiological equivalent of dreaming is in need of fundamental revision. A mounting body of evidence suggests that dreaming and REM sleep are dissociable states, and that dreaming is controlled by forebrain mechanisms. Recent neuropsychological, radiological, and pharmacological findings suggest that the cholinergic brain stem mechanisms that control the REM state can only generate the psychological phenomena of dreaming through the mediation of a second, probably dopaminergic, forebrain mechanism. The latter mechanism (and thus dreaming itself) can also be activated by a variety of nonREM triggers. Dreaming can be manipulated by dopamine agonists and antagonists with no concomitant change in REM frequency, duration, and density. Dreaming can also be induced by focal forebrain stimulation and by complex partial (forebrain) seizures during nonREM sleep, when the involvement of brainstem REM mechanisms is precluded. Likewise, dreaming is obliterated by focal lesions along a specific (probably dopaminergic) forebrain pathway, and these lesions do not have any appreciable effects on REM frequency, duration, and density. These findings suggest that the forebrain mechanism in question is the final common path to dreaming and that the brainstem oscillator that controls the REM state is just one of the many arousal triggers that can activate this forebrain mechanism. The "REM-on" mechanism (like its various NREM equivalents) therefore stands outside the dream process itself, which is mediated by an independent, forebrain "dream-on" mechanism.

**Keywords:** acetylcholine; brainstem; dopamine; dreaming; forebrain; NREM; REM; sleep

## 1. Introduction

It is well established that humans spend approximately 25% of sleeping hours in a state of paradoxical cerebral activation, accompanied by bursts of rapid eye movement (REM) and other characteristic physiological changes (Aserinsky & Kleitman 1953; 1955). This state occurs in roughly 90–100 minute cycles, alternating with four well-defined stages of quiescent sleep known as non-REM (NREM) sleep (see Rechtschaffen & Kales 1968 for standardized definitions). In 70–95% of awakenings from the REM state, normal subjects report that they have been dreaming, whereas only 5–10% of NREM awakenings produce equivalent reports (Dement & Kleitman 1957a; 1957b; Hobson 1988b).[1] These facts underpin the prevalent belief that the REM state is "the physiological concomitant of the subjective experience of dreaming" (LaBruzza 1978, p. 1537) and that dreaming is merely "an epiphenomenon of REM sleep" (Hobson et al. 1998b, p. R12). The discovery of the brainstem mechanisms that control REM sleep (Jouvet 1962; McCarley & Hobson 1975) has led to the further inference that the same mechanisms control dreaming.[2]

This target article presents a body of evidence that substantially contradicts these prevailing assumptions. This evidence demonstrates that, although there is an important link between REM sleep and dreaming, they are in fact doubly dissociable states (Teuber 1955). That is, REM can occur without dreaming and dreaming can occur without REM. The evidence reviewed here suggests also that these two states are controlled by different brain mechanisms. REM is controlled by cholinergic brainstem mechanisms whereas dreaming seems to be controlled by dopaminergic forebrain mechanisms. This unexpected dissociation between REM sleep and dreaming – and the brain mechanisms that regulate them – requires a major paradigm shift in sleep and dream science.

## 2. REM sleep is controlled by pontine brain stem mechanisms

The conclusion that Jouvet (1962) drew from his pioneering ablation, stimulation, and recording studies – namely that REM sleep is controlled by pontine brain stem mechanisms – remains central to all major contemporary models of sleep cycle control (for reviews, see Hobson et al. 1986; 1998b). The *reciprocal interaction model* of McCar-

MARK SOLMS is a Lecturer in Psychology at University College London and Honorary Lecturer in Neurosurgery at the St. Bartholomew's and Royal London School of Medicine. He is author of *The neuropsychology of dreams: A clinico-anatomical study* (1997) Lawrence Erlbaum Associates.

ley and Hobson (1975) has dominated the field over the past two decades. According to this model, REM sleep – and therefore dreaming – is triggered by cholinoceptive and/or cholinergic "REM-on" cells, and terminated by aminergic (noradrenergic and serotonergic) inhibitory "REM-off" cells. The REM-on cells are localized principally in the mesopontine tegmentum and the REM-off cells in the nucleus locus coeruleus and dorsal raphe nucleus (Fig. 1). Although it is acknowledged that the complete network of nuclei contributing to and giving effect to this oscillatory mechanism is more widely distributed than initial findings indicated (Hobson et al. 1986), executive control of the REM/NREM cycle is still localized narrowly within the pontine brain stem (Hobson et al. 1998b).[3] The assertion therefore remains that "cholinergic brainstem mechanisms *cause* REM sleep and dreaming" (Hobson 1988b, p. 202).

## 3. REM sleep is not controlled by forebrain mechanisms

An important corollary of the hypothesis that REM sleep – and therefore dreaming – is controlled by pontine brainstem mechanisms is the hypothesis that it is *not* controlled by forebrain mechanisms. Jouvet (1962) classically demonstrated that the forebrain is both incapable of generating REM sleep and unnecessary for the generation of REM sleep: when cortex is separated from brain stem, it no longer displays the normal cycle of REM activation (which is preserved in the isolated brainstem). It is still widely accepted that the forebrain is a passive participant in the REM state. Even the once-popular notion that the eye movements of REM sleep are attributable to forebrain "scanning" of visual dream imagery has been questioned (Pivik et al. 1977). The dominant view seems to be that the eye movements, their associated ponto-geniculo-occipital (PGO) waves, and the resultant imagery – in short, all the visual events of REM sleep – are initiated by brain stem

neurons. The same applies to motor cortical events in REM sleep (Hobson 1988b; Hobson & McCarley 1977).

The brain stem localization of the mechanisms that regulate REM sleep physiology has become a springboard for far-reaching inferences about the mechanisms that regulate dream neuropsychology. An authoritative model of dream neuropsychology based on brain stem physiology is the *activation-synthesis model* (Hobson 1988b; Hobson & McCarley 1977). According to this model, which has dominated the field for the past two decades, dreams are actively generated by the brain stem and passively synthesized by the forebrain. The central tenet of this model is that the *causal* stimuli for dream imagery arise "from the pontine brain stem and not in cognitive areas of the cerebrum" (Hobson & McCarley 1977; p. 1347). The dream process is seen as having "no primary ideational, volitional, or emotional content" (p. 1347). Accordingly, the forebrain is assigned an entirely passive role: Its external input and output channels are blockaded by brain stem mechanisms, its perceptual and motor engrams are activated by brain stem mechanisms, and its memory systems merely generate "the best possible fit of [this] intrinsically inchoate data" (Hobson 1988b, p. 204). In this way it makes "the best of a bad job in producing even partially coherent dream imagery from the relatively noisy signals sent up from the brain stem" (Hobson & McCarley 1977, p. 1347).[4]

In the latest, admittedly speculative developments of this model (Hobson 1992; 1994; Hobson et al. 1998b), all the formal characteristics of dream psychology are accounted for by the above-described brainstem mechanisms. Dream hallucinosis, delusion, disorientation, accentuated affect, and amnesia are all attributed to the arrest of brain stem aminergic (noradrenergic and serotonergic) modulation of brainstem-induced cholinergic activation during REM sleep. It is even suggested that similar chemical mechanisms may underlie major psychotic symptoms that share formal features with dreaming (Hobson 1988b; 1992; 1994; Hobson & McCarley 1977). However, all of these propositions are questionable on several grounds.

## 4. Not all dreaming is correlated with REM sleep

Dreaming and REM sleep are incompletely correlated. Between 5 and 30% of REM awakenings do not elicit dream reports; and at least 5–10% of NREM awakenings do elicit dream reports that are indistinguishable from REM reports (Hobson 1988b). The precise frequency of NREM dreaming is controversial. However, the principle that *REM can occur in the absence of dreaming and dreaming in the absence of REM* is no longer disputed (Hobson 1988b; 1992; cf. Vogel 1978a).

The original source of controversy was Foulkes's (1962) observation that complex mentation can be elicited in more than 50% of NREM awakenings (Foulkes 1962). Subsequent studies have confirmed this observation – and suggested that an average of 43% of NREM awakenings elicit such reports (Nielsen 1999) – but the extent to which the reported mentation may legitimately be described as "dreaming" is still disputed (cf. Cavellero et al. 1992). This is due to the fact that there are qualitative differences between NREM and REM dreams: In short, the *average* NREM dream is more "thoughtlike" than the average REM dream. This appears to reaffirm the view that the physio-

mesopontine tegmentum

dorsal raphe nucleus

nucleus locus coeruleus

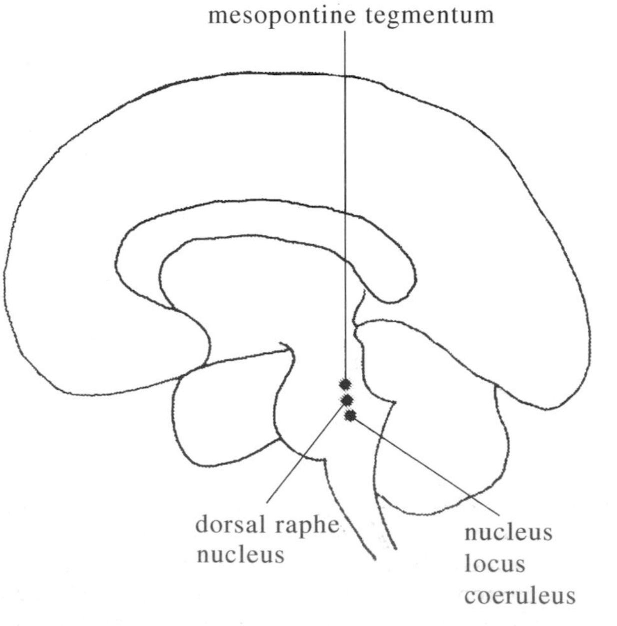

Figure 1. The major pontine brain stem nuclei implicated in REM/NREM sleep cycle control.

logical state differences between NREM and REM sleep are reflected in cognitive state differences between NREM and REM mentation. However, what is crucial for assessing the validity of the claim that dreaming is generated by the unique physiology of the REM state is not the question whether NREM "dreaming" occurs or not, but rather the extent to which NREM dreaming occurs that is *indistinguishable* from REM dreaming. This takes account of the problem of qualitative differences. It is generally accepted that NREM mentation that is indistiguishable from REM dreaming *does* indeed occur. Monroe et al.'s (1965) widely cited study suggests that approximately 10–30% of NREM dreams are indistinguishable from REM dreams (Rechtschaffen 1973). Even Hobson accepts that 5–10% of NREM dream reports are "indistinguishable by any criterion from those obtained from post-REM awakenings" (Hobson 1988b, p. 143). If we adjust this conservative figure to account for the fact that NREM sleep occupies approximately 75% of total sleep time, this implies that *roughly one quarter of all REM-like dreams occur outside of REM sleep.*

Moreover, REM-like NREM dreams are not randomly distributed through the sleep cycle; they cluster around specific NREM phases. As many as 50–70% of awakenings from sleep onset (descending NREM Stage I) yield reports that are not significantly different from REM dreams in all respects except for length (Foulkes et al. 1966; Foulkes & Vogel 1965; Vogel et al. 1972). Also, vivid REM-like reports are obtained with increasing frequency during the late NREM stages, in the rising morning phase of the diurnal rhythm (Kondo & Antrobus 1989).[5] This suggests that *these REM-like dreams are generated by specific NREM mechanisms.* In fact, within the reciprocal-interaction paradigm – where wakefulness and REM sleep are seen as terminal points on a continuum of aminergic demodulation – sleep onset and the rising morning phase have the opposite physiological characteristics to the REM state (Hobson 1992; 1994).

This is just one strand of the body of evidence that makes it difficult to retain the assumption that dreaming is generated by the unique physiological mechanism of the REM state.

In modifying the activation-synthesis model to accommodate these facts, the claim that all dreams are generated by the brain stem mechanisms that produce the REM state has recently been abandoned (Hobson 1992). This important shift in the dominant theory has passed almost unnoticed, however, because the closely related claim that all dreams are generated by *pontine brainstem* mechanisms has been retained (Hobson 1992; 1994). In the revised version of the activation-synthesis model (the Activation-Input-Mode [AIM] model), both REM *and* NREM dreams are attributed to reciprocal interactions between aminergic and cholinergic brainstem neurons (Hobson 1992; 1994). The formal characteristics of both REM *and* NREM mentation are therefore still described as "a function of the physiological condition of the reciprocally interacting brain stem neuronal populations that constitute the sleep-cycle control oscillator" (Hobson 1992, p. 228). Thus the doctrine of pontine brain stem control of dreaming has been retained, despite the fact that the assumption upon which it was explicitly based – the assumption of an isomorphism between REM sleep and dreaming (Hobson 1988b; 1992; Hobson & McCarley 1977) – has been disproved. The burden of evidence for the doctrine has thereby shifted from the phenomenological link between *REM sleep* and dreaming to the anatomical link between the *pontine brain stem* and dreaming.

## 5. Dreaming is preserved with pontine brain stem lesions

The assumption of an isomorphism between REM sleep and dreaming was important for the reason that the research program that isolated the brain mechanisms underlying REM sleep (ablation, stimulation, and recording studies) was conducted on infrahuman species in which concomitant effects on dreaming could not be monitored. The classical method for establishing brain-mind relationships in humans is the method of clinicoanatomical correlation in cases with naturally occurring lesions. If the assumption is correct that dreaming (like REM sleep) is controlled by brain stem mechanisms, it should be possible to demonstrate by this method that brainstem lesions in humans eliminate both REM sleep *and* dreaming.

Large lesions of the pontine brainstem eliminate all manifestations of REM sleep in domestic cats (Jones 1979), and this correlation has been confirmed in 26 human cases with naturally occurring lesions (Adey et al. 1968; Chase et al. 1968; Cummings & Greenberg 1977; Feldman 1971; Lavie et al. 1984; Markand & Dyken 1976; Osorio & Daroff 1980). However, elimination of REM (or near-elimination of REM) due to brainstem lesions was accompanied by cessation of dreaming in only one of these cases (Feldman 1971).[6] In the other 25 cases, the investigators either could not establish this correlation or they did not consider it (Adey et al. 1968; Chase et al. 1968; Cummings & Greenberg 1977; Lavie et al. 1984; Markand & Dyken 1976).[7]

Although cessation of dreaming has not been demonstrated in cases with elimination of REM due to brainstem lesions, the converse is also true: the preservation of dreaming in such cases has not been satisfactorily demonstrated (Solms [1997a] reported preserved dreaming in four patients with large pontine lesions, but polygraphic data was lacking). The paucity of evidence in this respect is at least partly due to the fact that pontine brain stem lesions large enough to obliterate REM usually render the patient unconscious (Hobson et al. 1998b).[8] Moreover, according to the revised version of the activation-synthesis model (the AIM model), dreaming is generated by both the REM and NREM components of the sleep-cycle control oscillator (Hobson 1992; 1994). This implies that dreaming can only be eliminated by very extensive brain stem lesions that obliterate *both* the REM and the NREM components of the oscillator. Such large lesions are almost certainly incompatible with the preservation of consciousness. It is therefore difficult to imagine how the assumption that dreaming is controlled by brainstem mechanisms can ever be refuted directly by lesion data. It can, however, be refuted indirectly via the corollary hypothesis that dreaming is not controlled by forebrain mechanisms. That is, the brain stem hypothesis would be falsified by clinicoanatomical methods if it could be demonstrated unequivocally that dreaming is eliminated by forebrain lesions that completely spare the brain stem.

## 6. Dreaming is eliminated by forebrain lesions which completely spare the brain stem

Subjective loss of dreaming due to a focal forebrain lesion was first reported more than 100 years ago. Wilbrand (1887; 1892) described a patient who dreamed "almost not at all anymore" (1887, p. 91) after suffering a bilateral occipital-temporal thrombosis. Müller (1892) documented a similar patient with bilateral occipital hemorrhages who "had no further dreams since her illness, whereas previously she not infrequently had vivid dreams and saw all sorts of things in them" (p. 868). Following these classical reports, 108 further cases with complete (or nearly complete) loss of dreaming in association with focal forebrain lesions have been published (Basso et al. 1980; Boyle & Nielsen 1954; Epstein 1979; Epstein & Simmons 1983; Ettlinger et al. 1957; Farah et al. 1988; Farrell 1969; Gloning & Sternbach 1953; Grunstein 1924; Habib & Sirigu 1987; Humphrey & Zangwill 1951; Lyman et al. 1938; Michel & Sieroff 1981; Moss 1972; Neal 1988; Nielsen 1955; Pena-Casanova et al. 1985; Piehler 1950; Ritchie 1959; Solms 1997a; Wapner et al. 1978). This clinicoanatomical correlation between subjective loss of dreaming and forebrain lesions has been confirmed repeatedly by the REM awakening method (Benson & Greenberg 1969; Brown 1972; Cathala et al. 1983; Efron 1968; Jus et al. 1973; Kerr et al.1978; Michel & Sieroff 1981; Murri et al. 1985) and by morning-recall questionnaires (Arena et al. 1984; Murri et al. 1984; 1985).[9]

In short, of the 111 published cases in the human neurological literature in which focal cerebral lesions caused cessation or near cessation of dreaming, *the lesion was localized to the forebrain – and the pontine brain stem was completely spared – in all but one case* (Feldman 1971). Critically, *the REM state was entirely preserved in all of the forebrain cases in which the sleep cycle was evaluated* (Benson & Greenberg 1969; Efron 1968; Jus et al. 1973; Kerr et al. 1978; Michel & Sieroff 1981). In view of the wide acceptance of the assumption that REM sleep is the physiological equivalent of dreaming, this lack of clinicoanatomical evidence correlating loss of REM sleep with loss of dreaming is striking.

The 110 published cases of loss of dreaming due to focal forebrain pathology fall into two anatomical groups (Fig. 2).[10] In 94 cases the lesion was situated in the posterior convexity of the hemispheres, in or near the region of the parieto-temporo-occipital (PTO) junction. The lesion was unilateral in 83 cases (48 left, 35 right) and bilateral in 11 cases. This localization has been confirmed repeatedly in substantial group studies (Arena et al. 1984; Cathala et al. 1983; Murri et al. 1984; 1985; Solms 1997a). In the other

16 cases, the lesion was situated in the white matter surrounding the frontal horns of the lateral ventricles. In these cases the damage was invariably bilateral. Of special interest is the fact that this lesion site coincides exactly with the region that was targeted in modified (orbitomesial) prefrontal leukotomy (Bradley et al. 1958). This association is confirmed by the fact that a 70–90% incidence of complete or nearly complete loss of dreaming was recorded in several large series of prefrontal leukotomy (Frank 1946; 1950; Jus et al. 1973; Partridge 1950; Piehler 1950; Schindler 1953). The many cases included in the latter series increases to almost 1,000 the number of reported cases of cessation of dreaming caused by focal forebrain lesions.

## 7. Dreaming is actively generated by forebrain mechanisms

It is not surprising that dreaming is lost with lesions in the PTO junction – a region that supports various cognitive processes that are vital for mental imagery (Kosslyn 1994). But why should it be lost with lesions in the ventromesial quadrant of the frontal lobes?

This region contains substantial numbers of fibers connecting frontal and limbic structures with dopaminergic cells in the ventral tegmentum (Fig. 3). These circuits arise from cell groups situated in the ventral tegmental area of Tsai, where the source cells for the mesolimbic and mesocortical dopamine systems are situated. They ascend through the forebrain bundles of the lateral hypothalamus via basal forebrain areas (synapsing on many structures along the way, including nucleus basalis, bed nucleus of the stria terminalis, and shell of the nucleus accumbens) and they terminate in the amygdala, anterior cingulate gyrus, and frontal cortex. Descending components of this system probably arise from the latter brain areas, and there is reason to believe that they are influenced strongly by cholinergic circuits (Panksepp 1985).

This system is thought to have been the primary target of modified prefrontal leukotomy (Panksepp 1985). Its circuits instigate goal-seeking behaviors and appetitive interactions with the world (Panksepp 1985; 1998a). It is accordingly described as the "SEEKING" or "wanting" command

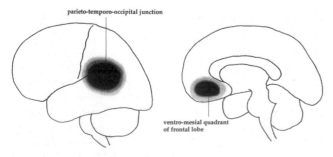

Figure 2. Lesion sites associated with loss of dreaming and preserved REM sleep.

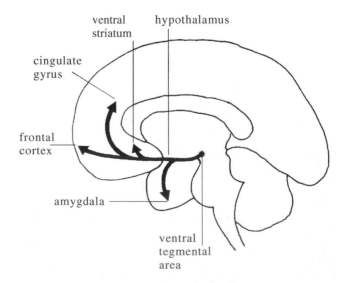

Figure 3. The mescortical/mesolimbic dopamine system.

system of the brain (Berridge, in press; Panksepp 1998a). It is considered to be the primary site of action of many stimulants (e.g., amphetamine and cocaine; see Role & Kelly 1991). The positive symptoms of schizophrenia – some of which can be artificially induced by l-dopa, amphetamines, and cocaine intoxication – are widely thought to result from overactivity of this system (Bird 1990; Kandel 1991; Panksepp 1998a). This system is also considered to be the primary site of action of antipsychotic medications (Role & Kelly 1991). A major psychological effect of antipsychotic therapy is loss of interactive interest in the world (Lehmann & Hanrahan 1954; Panksepp 1985). This underpins the popular view that antipsychotic medications – which block mesocortical-mesolimbic dopaminergic activity – yield "chemical leukotomies" (Breggin 1980; Panksepp 1985). Damage along this system produces disorders characterized by reduced interest, reduced initiative, reduced imagination, and reduced ability to plan ahead (Panksepp 1985). Lack of initiative or *adynamia* – where the patient does nothing unless instructed (Stuss & Benson 1983) – was a commonly observed side effect of orbitomesial prefrontal leukotomy (Brown 1985).

The following facts suggest that dreaming is generated by this dopamine circuit. First, dreaming ceases completely following transection of the forebrain component of this circuit (Frank 1946; 1950; Gloning & Sternbach 1953; Jus et al. 1973; Partridge 1950; Piehler 1950; Schindler 1953; Solms 1997a). These lesions have no effect on REM sleep. Transection or chemical inhibition of the same circuit reduces the positive symptoms of schizophrenia (Breggin 1980; Panksepp 1985), some formal features of which have long been equated with dreaming (Freud 1900; Hobson 1992; 1988b; Hobson & McCarley 1977). Second, adynamia (a common side effect of the surgical transection of this circuit) is a typical correlate of loss of dreaming following deep bifrontal lesions, and it statistically discriminates between dreaming and nondreaming patients with such lesions (Solms 1997a). Third, chemical activation of this circuit (e.g., through L-dopa) stimulates not only positive psychotic symptoms but also excessive, unusually vivid dreaming and nightmares (Nausieda et al. 1982; Scharf et al. 1978),[11] in the absence of any concomitant effect on the intensity, duration or frequency of REM sleep (Hartmann et al. 1980).[12] Fourth, drugs that block activity in this circuit (e.g., haloperidol) inhibit excessive, unusually frequent, and vivid dreaming (Sacks 1985; 1990; 1991) and other psychotic symptoms.

These facts suggest that the mesocortical-mesolimbic dopamine system plays a causal role in the generation of dreams. The relationship between this putative dopaminergic "dream-on" mechanism and the cholinergic REM-on mechanism of the reciprocal interaction model is discussed in the final section of this paper.

A further body of evidence strongly supports the view that dreaming can be *initiated* by forebrain mechanisms independently of the REM state. It is well established that nocturnal seizures – which typically occur during NREM sleep (Janz 1974; Kellaway & Frost 1983) – can present in the form of recurring nightmares[13] (Boller et al. 1975; Clarke 1915; De Sanctis 1896; Epstein 1964; 1967; 1979; Epstein & Ervin 1956; Epstein & Freeman 1981; Epstein & Hill 1966; Kardiner 1932; Naville & Brantmay 1935; Ostow 1954; Penfield 1938; Penfield & Erickson 1941; Penfield & Rasmussen 1955; Rodin et al. 1955; Snyder 1958;

Solms 1997a; Thomayer 1897). In 22 of the 24 published cases of this type, the recurring nightmares were caused by epileptiform activity in the temporal lobe, that is, by an unequivocally *forebrain* mechanism. (In the other two cases, the nightmares were associated with epileptiform activity in another part of the forebrain: the parietal lobe.) The causal link between the epileptic activity and the recurring nightmares in such cases was demonstrated by Penfield and his coworkers (Penfield 1938; Penfield & Erickson 1941; Penfield & Rasmussen 1955), who were able to reproduce the same anxious experiences artificially (in the form of waking "dreamy state" seizures) by stimulating the temporal lobe focus. This causal link between the forebrain seizures and the recurring nightmares was confirmed (in Penfield's and other cases) by the fact that both the underlying seizure disorder and the nightmares responded to anticonvulsant therapy and/or anterior temporal lobectomy (Boller et al. 1975; Epstein 1964; 1967; 1979; Epstein & Ervin 1956; Epstein & Freeman 1981; Epstein & Hill 1966; Solms 1997a). These observations demonstrate conclusively that *dreaming can be initiated by forebrain mechanisms (which are unrelated to REM sleep) and terminated by forebrain lesions (which spare the REM cycle).*

## 8. Dreams are generated by a specific network of forebrain mechanisms

In the activation-synthesis model, dream imagery was attributed to nonspecific forebrain synthesis of chaotic brainstem impulses. This conception of the neuropsychological mechanisms underlying the formal characteristics of dream imagery is incompatible with recent clinicoanatomical and functional imagery findings (Braun et al. 1997; 1998; Solms 1997a). Data derived from these two methods have produced a remarkably consistent picture of the dreaming brain (Hobson et al. 1998b). Both the clinicoanatomical studies (Solms 1997a) and the functional imagery studies (Braun et al. 1997; 1998; Franck et al 1987; Franzini 1992; Heiss et al. 1985; Hong et al. 1995; Maquet et al. 1990; 1996; Madsen 1993; Madsen & Vorstrup 1991; Madsen et al. 1991a; 1991b; Nofzinger et al. 1997) suggest that dreaming involves concerted activity in a *highly specific* group of forebrain structures. These structures include anterior and lateral hypothalamic areas, amygdaloid complex, septal-ventral striatal areas; and infralimbic, prelimbic, orbitofrontal, anterior cingulate, entorhinal, insular, and occipitotemporal cortical areas (Braun et al. 1997; Maquet et al. 1996; Nofzinger et al. 1997). Primary visual cortex and dorsolateral prefrontal cortex are deactivated during REM dreaming (Braun et al. 1998). The role of the parietal operculum is uncertain (Heiss et al. 1985; Hong et al. 1995; Maquet et al. 1996).

This differentiated pattern of regional activation and inactivation mirrors some striking neuropsychological dissociations that have been reported in the clinicoanatomical literature. For example, unimodal abnormalities of visual dream imagery occur only with lesions in visual association cortex (Solms 1997a), but lesions in primary visual cortex have no effect on dreams. That is, visual dream imagery is intact in cortically blind patients (with V1/V2 lesions) whereas patients with irreminiscence who are unable to generate facial and color imagery in waking life (due to V4 lesions) also cannot generate faces or colors in their dreams

(Adler 1944; 1950; Botez et al. 1985; Brain 1950; 1954; Charcot 1883; Grunstein 1924; Kerr et al. 1978; Macrae & Trolle 1956; Sacks 1985; 1990; 1991; Sacks & Wasserman 1987; Solms 1997a; Tzavaras 1967). Dream imagery is similarly unaffected by primary cortical lesions in the other modalities. Hemiplegic patients (with unilateral perirolandic lesions) experience normal somatosensory and somatomotor imagery in their dreams (Brown 1972; 1989; Grünstein 1924; Mach 1906; Solms 1997a). Similarly, aphasic patients with left perisylvian lesions experience normal audioverbal and motor speech imagery in their dreams (Cathala et al. 1983; Schanfald et al. 1985; Solms 1997a). These findings suggest that somatosensory, somatomotor, audioverbal, and motor speech imagery in dreams are generated outside of the respective unimodal cortices for these classes of perceptual and motor imagery (probably in heteromodal paralimbic or PTO cortex). This implies that perceptual and motor dream imagery does not isomorphically reflect the simple activation of perceptual and motor cortex during sleep, as was claimed by the authors of the activation-synthesis model (Hobson 1988b; Hobson & McCarley 1977). It also suggests that dream imagery is not generated by chaotic activation of the forebrain. Rather, it appears that *specific forebrain mechanisms are involved in the generation of dream imagery* and that *this imagery is actively constructed through complex cognitive processes.*

In addition, a detailed analysis of the known forebrain mechanisms implicated in dreaming accounts empirically (Solms 1997a) for the formal characteristics of dreams – such as hallucination, delusion, disorientation, negative affect, attenuated volition, and confabulatory paramnesia – which were previously attributed speculatively (Hobson 1992; 1994) to the arrest of brain stem aminergic modulation during REM sleep. Lesions in anterior thalamus, basal forebrain, anterior cingulate, and mesial frontal cortex cause excessively vivid and frequent dreaming, a breakdown of the distinction between dreaming and waking cognition, and other reality-monitoring deficits. This suggests that the hallucinated, delusional, disoriented, and paramnestic quality of dream cognition may be associated with inhibition of these structures during sleep. Discharging lesions in medial and anterior temporal cortex cause recurring nightmares during sleep and unpleasant hallucinatory experiences during waking life. This suggests that the typical emotional and complex episodic qualities of dreams are produced through activation of these structures during sleep. It also suggests that these structures participate causally in the generation of at least some dreams. Bilateral lesions in the ventromesial frontal white matter cause complete cessation of dreaming in association with adynamia and other disorders of volitional interest. This suggests that these motivational mechanisms are essential for the generation of dreams. Lesions in dorsolateral prefrontal cortex cause disorders of volitional control, self-monitoring, and other executive deficits, but they have no effect on dreaming. This suggests that dorsolateral prefrontal cortex is inessential for dreaming sleep, which might explain the attenuated volition and other executive deficiencies of dream cognition (and further account for the defective self-monitoring). Right-sided lesions in the PTO junction cause complete cessation of dreaming in association with disorders of spatial cognition. This suggests that normal spatial cognition is essential for dreaming. It also suggests that the concrete spatial quality of dreams is supported by right hemispheric PTO activation. Lesions in the same region of the left hemisphere convexity also cause cessation of dreaming in association with disorders of quasi-spatial (symbolic) operations. This suggests that quasi-spatial cognition is equally essential for dreaming, and that this aspect of dreaming is contributed by left PTO activation. Lesions in ventromesial occipito-temporal (visual association) cortex cause unimodal deficits of dream imagery, in association with identical deficits of waking imagery. This suggests that the visual imagery of dreams is produced by activation during sleep of the same structures that generate complex visual imagery in waking perception. It also suggests that these structures are activated in dreams by heteromodal structures that are downstream of these unimodal visual processes during waking perception. Lesions in other unimodal cortices have no effect on dream imagery, notwithstanding their marked effects on waking perceptual and motor functions. This accounts for the predominantly visual quality of dream hallucinosis. It also suggests that the "backward projection" process which presumably generates visual dream imagery (Kosslyn 1994; Zeki 1993) does not extend further back than visual association cortex (V3).[14]

These evidence-based clinicoanatomical inferences (which tally very closely with the available functional imagery data) place the neuropsychology of dreaming on an equivalent footing with that of other cognitive functions. This finally paves the way for a testable theory of the brain mechanisms underlying the complex psychology of dreaming (Solms 1997a).

A noteworthy disparity between the clinicoanatomical and functional imagery data is the involvement of the pontine brain stem in dreaming sleep in some of the functional imaging studies (Braun et al. 1997; Maquet et al. 1996) but not the clinicoanatomical studies (Solms 1997a). This disparity is readily attributable to the fact that dreaming sleep was equated with REM sleep in the relevant imaging studies, which precluded the possibility of comparing dreaming with nondreaming NREM epochs (cf. Heiss et al. 1985). Imaging studies of the dreaming brain at sleep onset, or during the rising morning phase of the diurnal rhythm (when the brainstem mechanisms that generate REM are uncoupled from the putative forebrain mechanisms that generate dreaming), would be enlightening on this point.[15]

## 9. The relationship between dreaming and REM sleep reconsidered

The high correlation between the REM state and dreaming has traditionally been interpreted as indicating that the brain stem mechanisms that generate REM simultaneously generate dreaming (i.e., that the REM state is intrinsic to and isomorphic with dreaming). However, the data reviewed above suggest that REM and dreaming are in fact doubly dissociable states, in both normal and pathological conditions, and that they are controlled by different brain mechanisms. The high correlation between REM and dreaming therefore requires an alternative explanation.

Perhaps the most reasonable possibility is suggested by the observation that the various brain states that correlate with vivid dream reports all involve *cerebral activation during sleep.* The most common of these is the "paradoxical" state of REM, in which the brain is simultaneously asleep and highly activated. Dream reports are also correlated with specific NREM states: descending Stage I (sleep onset) and the rising morning phase of the diurnal rhythm. These states are situated at polar ends of the sleep cycle, in the transitional phases between sleep and waking. The correlations between these states and dreaming have accordingly been interpreted as cerebral activation effects (Antrobus 1991; Hobson 1992). The same interpretation has been applied to the inverse correlation that exists between depth of NREM sleep (as measured by the sensory arousal threshold) and dreamlike mentation (Zimmerman 1970). Another state which triggers NREM dreaming is complex partial seizure activity, which could be described as a pathological form of cerebral activation during sleep. The fact that dreaming can be artificially generated by the administration of a variety of stimulant drugs, including both cholinergic[16] and dopaminergic agents, is open to a similar interpretation. Of crucial theoretical importance is the fact that dopaminergic agents increase the frequency, vivacity, and duration of dreaming without similarly affecting the frequency, intensity and duration of REM sleep (Hartmann et al. 1980). This observation, together with the equally important fact that damage to ventromesial frontal fibres obliterates dreaming but spares the REM cycle (Jus et al. 1973), suggests a specific dopaminergic dream-on mechanism that is dissociable from the cholinergic REM-on mechanism.

These observations show that dreaming is not an intrinsic function of REM sleep (or the brain stem mechanisms that control it). Rather, dreaming appears to be a consequence of various forms of cerebral activation during sleep. This implies a two-stage process, involving (1) cerebral activation during sleep and (2) dreaming. The first stage can take various forms, none of which is specific to dreaming itself, since reliable dissociations can be demonstrated between dreaming and all of these states (including REM). The second stage (dreaming itself) occurs only if and when the initial activation stage engages the dopaminergic circuits of the ventromesial forebrain. It is reasonable to hypothesize on this basis that these forebrain circuits are the final common path leading from various forms of cerebral activation during sleep (both REM and NREM) to dreaming per se. In this view, the high correlation between dreaming and the REM state merely reflects the fact that it is a regular and persistent source of cerebral activation during sleep. It is also possible that specific aspects of the REM state (e.g., noradrenergic and serotonergic demodulation) *facilitate* the primary dopaminergic effects. However, such facilitatory factors, which vary across the different sleep states associated with dreaming are not intrinsic to the dream process itself.

The biological function of dreaming remains unknown. This is at least partly attributable to the fact that the function of dreaming and the (equally unknown) function of REM sleep have been conflated for more than 40 years of research. Future studies of these functions should be uncoupled from one another. The statistical correlation between dreaming and REM sleep led early investigators to the understandable conclusion that they shared a single underlying mechanism. Subsequent research has demonstrated that this conclusion was erroneous: Dreaming and REM sleep are in fact doubly dissociable states, they have different physiological mechanisms, and in all likelihood they serve different functional purposes. The premise upon which the prevailing neuroscientific theories of dreaming

were based has therefore lapsed. Progress in this area will now be hampered if we do not acknowledge our initial error, and resist the temptation to compress our expanding knowledge of the dreaming forebrain into the initial REM-based theoretical framework.

NOTES

**1.** Reported dream recall rates vary, depending not only on the method of awakening and interview but also on the investigator's definition of "dreaming" (Foulkes 1966). The figures cited here are conservative (they are discussed in more detail in sect. 4). There is no generally accepted definition of dreaming. For our purposes, dreaming may be defined as *the subjective experience of a complex hallucinatory episode during sleep.* However, what is more important than an absolute definition of dreaming in the present context is the relative frequency with which dream reports obtained from REM and NREM sleep are considered indistinguishable by blind raters.

**2.** *Control* in this context implies activate, generate, sustain, and terminate.

**3.** The concept of "executive control" (Hobson & McCarley 1977, p. 1338; Hobson et al. 1998b, p. R7) implies that the distributed network of structures that contribute to and give effect to the various physiological manifestations of the REM state are *recruited* and *coordinated* by a cholinergic/aminergic oscillator that is "centered" in the mesopontine tegmentum (Hobson 1988b, p. 185). Accordingly, Hobson proposes that "the *on-off switch* is the reciprocal-interacting neuronal populations comprising the aminergic neurons and the reticular neurons of the brain stem" (p. 205).

**4.** "If we assume that the physiological substrate of consciousness is in the forebrain, these facts completely eliminate any possible contribution of ideas (of their neural substrate) to the primary driving force of the dream process" (Hobson & McCarley 1977, p. 1338).

**5.** These dreams are difficult to distinguish from REM dreams. The following are illustrative examples. The first is a sleep-onset dream (descending Stage I):

[It] had something to do with a garden plot, and I was planting seed in it. I could see some guy standing in this field, and it was kind of filled and cultivated, and he was talking about this to me. I can't quite remember what it was he did say, it seems to me as if it had to do with growing, whether these things were going to grow (Foulkes 1966, pp. 129–30).

The second example is a later NREM dream (25 minutes after the last REM episode):

I was with my mother in a public library. I wanted her to steal something for me. I've got to try and remember what it was, because it was something extraordinary, something like a buffalo head that was in this museum. I had told my mother previously that I wanted this head and she said, all right, you know, we'll see what we can do about it. And she met me in the library, part of which was a museum. And I remember telling my mother to please lower her voice and she insisted on talking even more loudly. And I said, if you don't, of course, you'll never be able to take the buffalo head. Everyone will turn around and look at you. Well, when we got to the place where the buffalo head was, it was surrounded by other strange things. There was a little sort of smock that little boys used to wear at the beginning of the century. And one of the women who worked at the library came up to me and said, dear, I haven't been able to sell this smock. And I remember saying to her, well, why don't you wear it then? For some reason or other I had to leave my mother alone, and she had to continue with the buffalo head project all by herself. Then I left the library and went outside, and there were groups of people just sitting on the grass listening to music (Foulkes 1996, pp. 110–11).

**6.** This was a case of closed head injury with traumatic occlusion of the basilar artery. Autopsy and relevant radiological data were lacking. The distinct possibility of forebrain damage in this case cannot be excluded.

**7.** In one report (Osorio & Daroff 1980) two patients recalled no dreams when awoken during atypical NREM epochs; this is not unexpected and does not constitute evidence of loss of dreaming.

**8.** However, this is not always the case. At least eight patients with cessation or near-cessation of REM have been reported who were sufficiently conscious to communicate meaningfully with an examiner (Feldman 1971; Lavie et al. 1984; Markand & Dyken 1976; Osorio & Daroff 1979).

**9.** The possibility that the reported loss of dreaming in these patients is attributable to amnesia for dreams rather than true loss of dreams has been excluded not only by REM awakening but also by neuropsychological examination of memory functions in dreaming versus nondreaming patients (Solms 1997a).

**10.** This analysis excludes the "several" cases of cessation of dreaming after cerebral commissurotomy reported by Bogen (1969), whose findings have never been replicated (Greenwood et al. 1977; Hoppe 1977).

**11.** Excessive, unusually frequent, and vivid dreaming (of the type stimulated by dopamine agonists) has also been described in association with lesions of the anterior cingulate gyrus, basal forebrain nuclei and closely related structures (Gallassi et al. 1992; Gloning & Sternbach 1953; Lugaresi et al. 1986; Morris 1992; Sacks 1995; Solms 1997a; Whitty & Lewin 1957). Similar phenomena have been linked with central visual deafferentation (Brown 1972; 1989; Grünstein 1924; Hécean & Albert 1978; Solms 1997a). In some of these cases, dreaming occurs continuously throughout sleep (Gallassi et al. 1992; Gloning & Sternbach 1953; Lugaresi et al. 1986; Morris et al. 1992; Sacks 1995; Solms 1997a; Whitty & Lewin 1957). These patients are unable to distinguish between dreams and real experiences, and reality monitoring in general is disturbed (Solms 1997a). Most striking are cases in which waking thoughts spontaneously transform into complex hallucinatory experiences, resulting in confabulatory delusional states (Solms 1997a; Whitty & Lewin 1957). This disorder has been interpreted (Solms 1997a) as indicating that basal forebrain nuclei and closely related structures – which are known to participate in discriminative cognitive processes – play a critical role in distinguishing between thoughts and perceptions (i.e., inhibiting hallucinosis). Accordingly, damage to these mechanisms results in excessive dreaming during sleep (when the visual system is deafferented) and the intrusion of dreamlike mentation into waking thought.

It is reasonable to assume that the normal alternations between thoughtlike and dreamlike mentation that occur throughout the sleep cycle are somehow related to these (largely cholinergic) forebrain mechanisms. However, they appear to exert this influence *in the opposite direction to that predicted by the activation-synthesis hypothesis.* The fact that damage to cholinergic forebrain structures (i.e., *reduction* in cortical acetylcholine) produces *excessive* dreaming and dreamlike mentation is consistent with the widely held view that cortical acetylcholine enhances discriminative cognitive mechanisms (Perry & Perry 1995). Likewise, it is well known that *anti*cholinergic agents (e.g., scopolamine or atropine), acting on the muscarinic receptors which predominate in the basal forebrain, produce dreamlike mentation and complex hallucinations in awake subjects (Perry & Perry 1995). These effects are enhanced by eye closure. Therefore, if the REM state is indeed partly mediated by basal forebrain cholinergic mechanisms, as has recently been suggested by proponents of the reciprocal-interaction hypothesis (Hobson et al. 1998b), then something else must be added to the cholinergic activation in order to account for the occurence and formal characteristics of dreamlike mentation during this state. What is proposed here is that this "something else" is provided by the putative dopaminergic mechanism discussed above, the stimulation of which correlates positively with the generation of complex hallucinations, delusions, and other dreamlike phenomena.

**12.** In view of the importance of these findings in the present context, Hartmann et al.'s (1980) study is briefly summarized here: 13 subjects slept in the laboratory on four occasions each. They were awakened at the end of the first and second REM periods and either l-dopa (500 mg) or placebo were administered, so that the action of the l-dopa would coincide with the third REM period. A study lasting 52 nights yielded 128 dreams, of which 90 were postmedication (42 l-dopa and 48 placebo). Each dream was scored by four blind raters on five dream content scales: dreamlikeness, nightmarelikeness, vividness, emotionality, and detail. The l-dopa condition dreams were significantly more dreamlike ($p < 0.01$), vivid ($p < 0.01$), detailed ($p < 0.01$), and emotional ($p < 0.05$; $t$-test for correlated samples) than the placebo condition dreams. The two treatment conditions did not differ significantly on any polygraphic measures, including REM frequency, duration, and density.

**13.** These are subjective experiences of complex hallucinatory episodes, not night terrors. Here is an example:

the patient [35 year old woman with idiopathic complex-partial seizures] reported a recurrent dream about her [dead] brother . . . which has reappeared several times. The dream is as follows: "I am walking down the street. I meet him. He is with a group of people whom I know now. I feel that I will be so happy to see him. I say to him, 'I'm glad you're alive,' but he'll deny that he is my brother and he'll say so, and I'll wake up crying and trying to convince him." (Epstein & Ervin 1956, p. 45)

Electroencephalography revealed a poorly defined right anterior temporal/right temporal spike focus, which appeared with the onset of drowsiness and light sleep.

**14.** This backward projection mechanism is apparently mediated in part by the cholinergic basal forebrain mechanism discussed previously.

**15.** The uncertain role of the parietal operculum in REM and NREM dreaming also awaits further investigation, but this question is unrelated to the main topic of the present paper.

**16.** Interesting to note, if cholinergic agents are administered prior to sleep onset they cause insomnia, if they are administered during NREM sleep they induce REM, and if they are administered during REM they provoke awakening (Sitaram et al. 1978b; Sitaram et al. 1976). This suggests a nonspecific activation-arousal effect.

# 3

# A review of mentation in REM and NREM sleep: "Covert" REM sleep as a possible reconciliation of two opposing models

**Tore A. Nielsen**

*Sleep Research Center, Hôpital du Sacré-Coeur de Montréal, Montréal, Québec, Canada and Psychiatry Department, Université de Montréal, Québec H4J 1C5, Canada*
**t-nielsen@crhsc.umontreal.ca**

**Abstract:** Numerous studies have replicated the finding of mentation in both rapid eye movement (REM) and nonrapid eye movement (NREM) sleep. However, two different theoretical models have been proposed to account for this finding: (1) a one-generator model, in which mentation is generated by a single set of processes regardless of physiological differences between REM and NREM sleep; and (2) a two-generator model, in which qualitatively different generators produce cognitive activity in the two states. First, research is reviewed demonstrating conclusively that mentation can occur in NREM sleep; global estimates show an average mentation recall rate of about 50% from NREM sleep – a value that has increased substantially over the years. Second, nine different types of research on REM and NREM cognitive activity are examined for evidence supporting or refuting the two models. The evidence largely, but not completely, favors the two-generator model. Finally, in a preliminary attempt to reconcile the two models, an alternative model is proposed that assumes the existence of *covert* REM sleep processes during NREM sleep. Such covert activity may be responsible for much of the dreamlike cognitive activity occurring in NREM sleep.

**Keywords:** cognition in sleep; dreaming; NREM sleep; REM sleep; sleep mentation

## 1. Introduction

### 1.1. The discovery of REM and NREM mentation

Initial reports of an association between REM sleep and vivid dreaming (Aserinsky & Kleitman 1953; Dement 1955; Dement & Kleitman 1957a; 1957b) inspired studies designed to clarify relationships between sleep physiology and dream imagery. A perspective emerged – referred to by many as the "REM sleep = dreaming" perspective (see Berger 1994; Foulkes 1993b; Lavie 1994; Nielsen & Montplaisir 1994; Rechtschaffen 1994 for overview) – from which dreaming was viewed as a characteristic *exclusive* to REM sleep. Mentation reported from NREM sleep was attributed to purportedly confounding factors, for example, recall of mentation from previous REM episodes or subjects' waking confabulations. Many subsequent studies cast doubt on the "REM sleep = dreaming" perspective (Foulkes 1962; 1966) primarily by demonstrating elevated levels of mentation recalled from NREM sleep stages. Although the REM sleep = dreaming belief did not disappear entirely, a debate over whether the *quality* of NREM and REM sleep mentation reports differ largely overshadowed it. Initially, qualitative differences in REM and NREM reports suggested that a different – possibly degraded – form of mentation occurs in NREM sleep. From these developments, two relatively distinct points of view concerning

REM/NREM mentation emerged and continue to influence the field. These points of view differ as to whether they consider NREM sleep mentation to stem from imagery processes that are fundamentally the same as or different from those that produce REM sleep mentation. I refer to these as the 1-gen (one-generator) and 2-gen (two-generator) models (reviewed in Nielsen 1999a); research supporting and/or refuting each model is reviewed in the following sections. The review concludes with the presentation of a third model, *the covert REM sleep processes*

TORE NIELSEN is Assistant Professor of Psychiatry at the Université de Montréal where he has been Director of the Dream & Nightmare Laboratory at the Hôpital Sacré-Couer de Montréal since 1991. He is a past Research Scholar of the Canadian Institutes of Health Research and presently a Senior Research Scholar of the *Fonds de la recherche en santé du Québec* (FRSQ or *Quebec Health Research Agency*). His main research interests are reflected in his many published papers on mechanisms of sleep mentation production, psychophysiological and EEG correlates of dreaming, hypnagogic imagery processes, disturbances of dreaming (including idiopathic and post-traumatic nightmares), and pain in dreams.

model, which combines aspects of both the 1-gen and 2-gen models in a way that may help to reconcile the two opposing points of view.

**1.1.1. The 1-gen and 2-gen models.** The 1-gen model stipulates that a single set of imagery processes produces sleep mentation regardless of the sleep stage in which it occurs. The model was suggested following demonstrations that reports of *cognitive activity* could be elicited from NREM sleep. Foulkes's (1962) application of more liberal criteria for identifying cognitive activity, as opposed to *dreaming activity,* allowed him and others to demonstrate a higher incidence of mentation during NREM sleep than was previously observed. Many others replicated these findings (see sect. 1.2.2.2).

Further support for 1-gen models came with the development of methods for effecting fair comparisons of mentation quality between reports of obviously different lengths. As REM sleep mentation reports were typically longer than their NREM equivalents, their qualitative attributes were thought to be confounded with quantitative attributes. Both Foulkes (Foulkes & Schmidt 1983) and Antrobus (1983) devised methods for removing quantitative differences and thus permitting – presumably – fair tests of residual qualitative differences. Both investigators found that when length of report was statistically controlled, qualitative differences diminished and often disappeared, a finding supporting the notion that all sleep mentation derives from *a common imagery source that is driven by different levels of brain activation.* Several models based upon the 1-gen assumption were subsequently elaborated (Antrobus 1983; Feinberg & March 1995; Foulkes 1985; Solms 1997a).

Foulkes's 1-gen model – the most influential – stipulates that mentation report from REM and NREM sleep arise from the same processes: (1) memory activation, (2) organization, and (3) conscious interpretation. Mentation differences stem primarily from differences in memory activation. When such activation is high and diffuse, during most REM but some NREM sleep, then organization is more intensely stimulated and conscious interpretation more probable and coherent. When memory activation is low and less diffuse, during most NREM but some REM sleep, then organization is less intensely stimulated and conscious interpretation less probable and coherent. It is thus the *diffuseness or availability of diverse memory elements* and not sleep stage physiology that determines the occurrence and form of sleep mentation.

Solms (1997a) adds some support to this model, primarily by refuting the physiological bases of Hobson's 2-gen model. He shows that lesions of the brainstem regions responsible for REM-related activation do *not* lead to loss of dreaming, whereas lesions in the forebrain ("anterior to the frontal horns of the lateral ventricles") or in the inferior parietal regions ("parieto-occipito-temporal junction"), lead to global cessation of dreaming. Mentation may occur in any state if these areas are active, even though it is most likely in REM sleep. Thus Solms, like Foulkes, views dreaming as largely independent of REM sleep-specific physiology. Unlike Foulkes, however, he does see dreaming to be associated with a neurophysiological substrate. The latter consists of a motivational-hallucinatory mechanism that is more akin to the Freudian psychoanalytical model than it is to a cognitive-psychological one (Solms 1995).

From the 2-gen perspective, REM and NREM sleep mentation reports stem from *qualitatively different* imagery generation systems. This difference was suggested by early findings that REM sleep reports are less thoughtlike, more elaborate, more affectively, visually and kinesthetically involving, and more related to waking life than are NREM sleep reports (Foulkes 1962; 1966; Monroe et al. 1965; Rechtschaffen et al. 1963a). The best-known 2-gen model was developed from the earlier activation-synthesis (A-S) hypothesis (Hobson & McCarley 1977) by Hobson's group (Hobson 1992a; Hobson & Stickgold 1994a; 1995; see also Seligman & Yellen 1987). McCarley (McCarley 1994; Steriade & McCarley 1990b) also updated the A-S hypothesis in different directions. A psycholinguistic 2-gen theory has also been proposed (Casagrande et al. 1996a).

Both the A-S hypothesis and its more recent variant (see Hobson et al., this issue) explain sleep mentation by combining (1) descriptions of the presumed physiological substrates of REM and NREM sleep (see Hobson 1988b; Kahn et al. 1997; McCarley & Hobson 1979 for reviews of the physiological findings) and (2) the assumption of formal mind-brain isomorphism. REM and NREM sleep physiological attributes determine the form of mental experiences and are isomorphic with them (Mamelak & Hobson 1989a). Dreaming mentation – characteristic of REM sleep – is distinguished from nondreaming mentation – characteristic of NREM sleep – according to the presence of six defining characteristics (Hobson & Stickgold 1994a): hallucinoid imagery, narrative structure, cognitive bizarreness, hyperemotionality, delusional acceptance, and deficient memory of previous mental content. Some of these features are embodied in newly proposed dream-content measures (e.g., emotional profile, visual continuity, thematic coherence; Baars & Banks 1994).

**1.1.2. Summary.** Both 1-gen and 2-gen models have had an important impact on sleep research over the last 40 years. That Foulkes's original findings were replicated and his model tested by so many researchers indicates that his cognitive-psychological framework and his 1-gen model have had a widespread influence. Solms's recent work further bolsters some of Foulkes's key assumptions while refuting others.

Until quite recently, the 2-gen model has been highly visible among the neurosciences and the popular press. The A-S hypothesis is today almost synonymous with dreaming. It has, nonetheless, been roundly criticized for various reasons (see below). How the model relates to dream content remains to be studied in greater depth, for example, discriminant validity of the index measures of the six proposed defining features of dreaming and non-dreaming mentation is still unknown.

As the use of cognitive methods has grown increasingly more popular in the brain and psychological sciences, both 1-gen and 2-gen models have continued to stimulate research within distinct subdisciplines. The result has been that the pros and cons of the two models have been scrutinized ever more closely, even though the two are only rarely compared directly one with the other.

### 1.2. Widespread evidence for cognitive activity in NREM sleep

**1.2.1. Distinguishing "dreaming" from "cognitive activity."** Distinctions between "dreaming" and "cognitive activity"

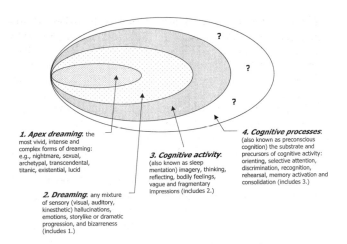

**1. Apex dreaming**: the most vivid, intense and complex forms of dreaming; e.g., nightmare, sexual, archetypal, transcendental, titanic, existential, lucid

**3. Cognitive activity**: (also known as sleep mentation) imagery, thinking, reflecting, bodily feelings, vague and fragmentary impressions (includes 2.)

**4. Cognitive processes**: (also known as preconscious cognition) the substrate and precursors of cognitive activity: orienting, selective attention, discrimination, recognition, rehearsal, memory activation and consolidation (includes 3.)

**2. Dreaming**: any mixture of sensory (visual, auditory, kinesthetic) hallucinations, emotions, storylike or dramatic progression, and bizarreness (includes 1.)

Figure 1. Four levels of specificity in defining sleep mentation. With an increasingly specific definition of sleep mentation, differences between REM and NREM mentation become more apparent. The two most specific levels (1 and 2) tend to occur much more exclusively in REM sleep. Cognitive activity (3) other than dreaming is predominant in NREM sleep. Beyond cognitive activity, there is likely an even more general level of cognitive processes (4) that consists of preconscious precursors to cognitive activity and that may be present in different degrees throughout REM and NREM sleep.

are key to appreciating differences between the 1-gen and 2-gen models. In general, dreaming – which is the object of study of most 2-gen theorists – is more specific than is cognitive activity (see Fig. 1). It is likely to be defined as imagery that consists of sensory hallucinations, emotions, storylike or dramatic progressions, and bizarreness, and that may exclude some types of cognition such as simple thinking, reflecting, bodily feeling, and fragmentary or difficult to describe impressions.

Nonetheless, there is currently no widely accepted or standardized definition of dreaming; definitions vary widely from study to study. There have been attempts to differentiate minimal forms of dreaming from more elaborate, vivid and intense forms, such as "everyday" and "archetypal" (Cann & Donderi 1986; Hunt 1989), "mundane," "transcendental," and "existential" dreaming (Busink & Kuiken 1996), "lucid" and "nonlucid" dreaming (Laberge et al. 1981), and ordinary versus "apex" (Herman et al. 1978) or "titanic" dreaming (Hunt 1989). In Figure 1, the term "apex" dreaming is adopted to refer to a subcategory of dreaming that is distinguished by exceptional vividness, intensity or complexity. Many of the forms mentioned above and other common types (e.g., nightmares, lucid dreams, sex dreams) fall into this category. The fact that such vivid dreaming occurs frequently during REM sleep but rarely during NREM sleep has led many to propose a qualitative difference between REM and NREM mentation, and thus to entertain a 2-gen perspective.

Cognitive activity is a more inclusive term than is dreaming. It is synonymous with the common term "sleep mentation" and refers to the *remembrance of any mental activity having occurred just prior to waking up* (Fig. 1). This may include static visual images, thinking, reflecting, bodily feeling, or vague and fragmentary impressions. However, the precise limits of this inclusiveness have not been clearly established. In a manner analogous to the model presented by Farthing for waking state conscious-

ness (Farthing 1992), cognitive activity during sleep could be viewed as a subset of an even more inclusive category (cognitive processes) that includes preconscious or "nonconscious" information processes (Fig. 1). Processes that are acknowledged building blocks of waking cognition, such as orienting, selective attention, sensory discrimination, recognition, rehearsal, memory activation, and consolidation, have also been shown to be active during sleep (see sect. 2.2) and are more or less accessible to consciousness. For example, most theorists presume that processes of memory retrieval are central to dream generation. In principle, such processes may be active whether or not they possess phenomenological correlates (e.g., sensory imagery) that can be recalled. However, many such processes can in principle become accessible to awareness if subjects are properly trained in self-observation and reporting (see Nielsen 1992; 1995 for examples). The fact that relaxation training (Schredl & Doll 1997) and probe-based interview techniques (Smith 1984) can enhance the amount and quality of recalled mentation illustrates this point. More research bearing on this question is needed.

Differences in definitions of "cognitive activity" and/or "dreaming" presumably account for much of the variability in levels of mentation recall from REM and NREM sleep that has been observed in previous studies. To illustrate, three different studies of NREM sleep mentation used three different definitions of content: a report of (1) "coherent, fairly detailed description of dream content" (Dement & Kleitman 1957b); (2) "a dream recalled in some detail" (Goodenough et al. 1959), and (3) "at least one item of specific content" (Foulkes & Rechtschaffen 1964). The different levels of stringency varied inversely with the number of awakenings with recalled NREM mentation, that is, 7, 35, and 62% respectively.

**1.2.2. Evidence for dreaming and cognitive activity in NREM sleep.** Numerous studies demonstrate cognitive activity during NREM sleep. How much of this activity qualifies as *dreaming* (or as *apex dreaming*) has been less clearly shown. Some of the strongest evidence for NREM mentation is the association of specific NREM contents with pre-awakening stimuli (Pivik 1991), for example, sleep talking (Arkin et al. 1970; Rechtschaffen et al. 1962) and experimental auditory and somatic stimuli (Foulkes & Rechtschaffen 1964; Lasaga & Lasaga 1973; Rechtschaffen et al. 1963b) that are concordant with NREM mentation. Similarly, presleep hypnotic suggestions often appear in mentation from all stages of sleep (Stoyva 1961).

An illustration of such incorporative "tagging" in NREM mentation is a report (Rechtschaffen et al. 1963a) of a subject who was stimulated during stage 2 sleep with a 500 Hz tone (7 sec) followed by a pause (27 sec), a second tone (7 sec), and then awakened 32 sec later:

> a little whistling tone was going on . . . and then it went off. And (the other person) said 'Oh, you had better get things over with quickly, because you may have to wake up soon' . . . I just said 'Oh!' to this, and I think I heard the whistling noise again. Then the same scene was there for some time, and I was just walking around trying to think of what was going on. (p. 412)

Some NREM parasomnias also demonstrate vivid mental experiences outside of REM sleep (Fisher et al. 1970; Kahn et al. 1991); sleep terrors arising from stage 3 and 4 sleep often result in reports of dramatic and frightening content. For some awakenings the content may be due to the arousal itself (Broughton 1968), for others there is some sign of a progression seeming to lead up to, and possibly to induce, the awakening. Fisher et al. also found stage 2 nightmares qualitatively similar to those from REM sleep.

**1.2.2.1. Sleep Onset (SO).** Perhaps the most vivid NREM mentation reports have been collected from SO stages. These include images from the Rechtschaffen and Kales stages 1 and 2 of sleep (Cicogna et al. 1991; Foulkes & Vogel 1965; Foulkes et al. 1966; Lehmann et al. 1995; Vogel 1991) as well as from the stages of a more detailed SO scoring grid (Hori et al. 1994; Nielsen et al. 1995). SO mentation is remarkable because it can equal or surpass in frequency and length mentation from REM sleep (Foulkes 1982b; Foulkes & Vogel 1965; Foulkes et al. 1966; Vogel 1978b; Vogel et al. 1966). Moreover, much SO mentation (from 31–76% depending upon EEG features) is clearly hallucinatory dreaming as opposed to isolated scenes, flashes or nonhallucinated images (Vogel 1978b).

**1.2.2.2. NREM sleep.** Many more studies of sleep mentation have concentrated on NREM stages of sleep other than those of SO. Although in many studies stages 2, 3, and 4 are indiscriminately combined, stage 2 sleep is by far the most frequently examined stage.

To summarize this literature, studies of REM and NREM mentation published since 1953 were consulted. Of these, 35 studies[1] were retained for the calculation of global estimates of mentation recall (Fig. 2). Excluded were studies of patients for whom an illness (e.g., depression, anorexia) may have affected mentation recall. To equally weight findings from all studies, only one estimate of recall from each study was included in the global average. If a study contained values for different subgroups (e.g., young vs. old, male vs. female), an average of the groups was taken. Estimates were also calculated separately for studies prior to Foulkes's (1962) work, which was the first to highlight the distinction between *dreaming* and *cognitive activity* (Table 1).

The overall difference in mean recall from REM (81.9 ± 9.0%) and NREM sleep (43.0 ± 20.8%) is close to 39%. However, this difference is much larger for the pre-1962 studies (i.e., 57.6%) than it is for the post-1962 studies (33.2%). Differences in median recall parallel those for the mean; total: 40%, pre-1962: 59%, post-1962: 37%. The present estimated NREM recall mean of 43.0% is very similar to that of 45.9% (± 15.8%) calculated from nine previous studies (Foulkes 1967). The present REM recall estimate of 81.9% also compares favorably with both (1) an estimate of 83.3% from over 200 subjects and 2,000 REM sleep awakenings (Dement 1965) and (2) an average of 81.7 ± 15.0% from 12 prior studies (Herman et al. 1978).

**1.2.2.3. Stages 3 and 4 sleep.** Some studies have found cognitive activity in stages 3 and 4 sleep (Armitage 1980; Armitage et al. 1992; Cavallero et al. 1992; Goodenough et al. 1965a; Herman et al. 1978; Pivik & Foulkes 1968). On average, recall from these stages is equal to that of stage 2 sleep; a tally of eight studies (Cavallero et al. 1992; Fein et al. 1985; Foulkes 1966; Lloyd & Cartwright 1995; Moffitt et al. 1982; Pivik 1971; Pivik & Foulkes 1968; Rotenberg 1993b) revealed an average recall rate of 52.5 ± 18.6%. The average stage REM recall rate in these studies was 82.2 ± 8.1%. The values for stages 3 and 4 are consistent with the finding that stage 2 and 4 mentation differences disappear for awakenings conducted at similar times of the night (Tracy & Tracy 1973). Three studies (Moffitt et al. 1982; Pivik 1971; Pivik & Foulkes 1968) found average recall rates to be higher in stage 3 (M = 56%) than in stage 4 sleep (M = 38%), a finding also true of children 9–11 years (42% vs. 26%) and 11–13 years (42 vs. 25%) (Foulkes 1982b). However, Pivik (1971) found nearly identical levels of recall of cognitive activity in stages 3 (41–56%) and 4 (38–58%).

Some subjects appear to have little or no recall of stage 3 and 4 sleep mentation. Ten of 60 subjects (17%) in one study (Cavallero et al. 1992) reported *no* mentation whatsoever after several nights of one awakening/night from stages 3 or 4 sleep; an additional 20 subjects (33%) required from one to five additional nights before recalling at least one instance of cognitive activity. These discrepancies have never been explained satisfactorily.

### 1.3. Summary

Numerous studies have replicated the finding of mentation outside of REM sleep as the latter is traditionally defined. All NREM sleep stages can produce some form of mentation. However, in accordance with the distinction between *dreaming* and *cognitive activity* discussed earlier, the more recent (post-1962) studies together indicate that about half of all NREM awakenings result in *no recall* of cognitive activity whatsoever. Further, about 50% of subjects appear to have noticeably degraded recall of mentation from NREM sleep, some (e.g., 17% of subjects in the Cavallero et al. 1992 study) have *no* recall after repeated awakenings. Further, because dreaming is a subset of cognitive activity, less than 50% of NREM awakenings produce dreaming. One liberal estimate is that only 25–50% of NREM reports bearing cognitive activity fulfill a minimal definition of dreaming (Foulkes 1962). Thus, at most 25%, but possibly as little as 12% of NREM awakenings in *susceptible* subjects will produce reports of dreaming. The more elaborate forms of ("apex") dreaming are even less prevalent. It has

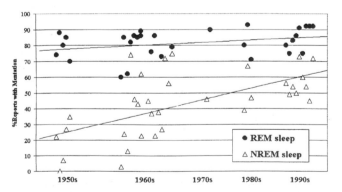

Figure 2. Summary of 35 studies of mentation recall from REM and NREM sleep over five decades. The percent of verbal reports that yielded some form of cognitive content after awakenings from NREM sleep increased from the 1950s to the 1990s, whereas the comparable percentage from REM sleep awakenings remained relatively constant. This difference is likely due to the widespread implementation in the 1960s of more liberal criteria for accepting reports as containing "cognitive activity" as opposed to simply "dreaming."

Table 1. *Summary of 35 studies of mentation recall from REM and NREM sleep (pre-1962 vs. post-1962)*

|  | N studies | Mean ± SD% | Median% | Range% |
|---|---|---|---|---|
| REM SLEEP RECALL |  |  |  |  |
| <1962 | 8 | 76.0 ± 11.5 | 77 | 60–92 |
| ≥1962 | 21 | 84.1 ± 6.7 | 86 | 71–93 |
| TOTAL | 29 | 81.9 ± 9.0 | 85 | 60–93 |
| NREM SLEEP RECALL |  |  |  |  |
| <1962 | 8 | 18.4 ± 15.4 | 18 | 0–43 |
| ≥1962 | 25 | 50.9 ± 15.5 | 49 | 23–75 |
| TOTAL | 33 | 43.0 ± 20.8 | 45 | 0–75 |
| REM/NREM SLEEP RECALL DIFFERENCES |  |  |  |  |
| <1962 | 8 | 57.6 | 59 | 60–49 |
| ≥1962 | 21 | 33.2 | 37 | 48–18 |
| TOTAL | 29 | 38.9 | 40 | 60–18 |

Recall of mentation from REM sleep has been consistently high in studies conducted from the 1950s to the present, whereas recall from NREM sleep has increased on average. This increase reflects liberalization (first operationalized by Foulkes in 1962) of the criteria for accepting a mentation report as a valid object of study: this marked the shift from studing the more delimited category of "dreaming" to studying the wider category of "cognitive activity."

been suggested (Herman et al. 1978) that vivid dreaming may occupy only 7% of recalled NREM mentation.

## 2. Experimental results bearing on the models

Resolving whether REM and NREM sleep mentation differ qualitatively is complicated by the thorny issue of whether the evaluation of sleep mentation conforms to commonly accepted psychometric principles of hypothetical construct validation, especially as these principles apply to psychophysiological studies. The validation of a hypothetical construct requires *several* criterion measures:

> It is ordinarily necessary to evaluate construct validity by integrating evidence from many different sources. The problem . . . becomes especially acute in the clinical field since for many of the constructs dealt with it is not a question of finding an imperfect criterion but of finding any criterion at all. (Cronbach & Meehl 1955, p. 285)

Further, the criterion measures under consideration should be as methodologically distinct from one another as possible to avoid "method artifact," that is, artifactual correlations among measures due to similarities in method (Strube 1990). Thus, solving the problem of qualitative differences in REM and NREM sleep mentation may require a construct validation approach sensitive to a wide range of methodologically diverse measures with probable or possible associations to sleep mentation. This is the principal justification for examining a variety of research methods in the following review.

How should a variable's "probable or possible associations" to sleep mentation be decided? Clearly, one's theoretical model is a determinant. Hobson's 2-gen model stipulates psychophysiological isomorphism; thus, the fact that REM and NREM sleep differ physiologically warrants investigation of physiological variables in relation to sleep mentation (Hobson & Stickgold 1995). Some proponents of the 1-gen model, on the other hand (Foulkes 1990), contend that mentation is psychologically driven. Physiological variables should be *excluded* from consideration. This assumption is supported by evidence that relationships between physiological variables and dream content have not been clearly demonstrated (see Pivik 1978; 1994; Rechtschaffen 1978, for reviews). However, as explained below, this assumption may not be completely justified on scientific grounds. To meaningfully compare the 1-gen and 2-gen points of view, a wide array of variables – including physiological variables – should be considered.

Foremost among the reasons for a lack of evidence for brain-mind relationships (Cacioppo & Tassinary 1990) may be the particular form of psychophysiological isomorphism proposed. One-to-one correspondences between a physiological ($\theta$) and a psychological ($\psi$) variable, such as those proposed by the 2-gen model, are not, in fact, common in the literature; more commonly, multiple $\theta$ responses accompany a $\psi$ variable or vice versa (Cacioppo & Tassinary 1990). To illustrate, EMG activity in the smiling muscle zygomaticus is associated with *both* positive dreamed affect and dreamed communication (Gerne & Strauch 1985). This problem can be resolved by evaluating a $\psi$ variable in relation to an appropriate *group* of $\theta$ measures ("spatial response profiles") or in relation to a combination of such spatial groups *over time* ("temporal response profiles"). Also grouping $\psi$ variables can give even greater specificity. Such procedures are rarely attempted for sleep mentation studies in part because of lack of computing tools, but also because of a dearth of theoretical frameworks for such work.

Another criterion for accepting a variable as a "probable or possible" correlate of sleep mentation concerns its existing status as a correlate of a waking state mental process. With much research demonstrating sleep mentation to be *continuous* with waking state experiences (see Schwartz et al. 1978, for review), it is reasonable to expect that physiological indicators of waking state experiences should also be valid during sleep. Such cross-state generalization of a measure's validity is, in fact, implicitly accepted whenever a measure (e.g., P300) that has been validated in one waking state (e.g., attentiveness) is applied during a different waking state (e.g., emotional arousal).

In summary, resolution of the debate about REM and NREM mentation is partly a problem of construct validation of the object of study. The debate was long ago widened to include *cognitive activity* as well as *dreaming* as dependent variables, and many pre-conscious *cognitive processes* may also belong in this category. It thus seems only fitting that a variety of process measures should be explored as potential markers of these objects of study. These measures should be methodologically diverse and have at least face validity as possible or probable correlates of the dependent measure. Thus, measures of cognitive content as well as accompanying physiological activity should be considered. In the review that follows, the measures considered are, for the most part, methodologically diverse and correlated with waking state cognitive processes. Even so, none involves the complex physiological profiles described earlier. Of the nine types of research examined, three (sects. 2.4, 2.6, 2.8) are closely tied to phenomenological features of sleep mentation. The others concern either physiological measures (sects. 2.3, 2.9), behavioral measures (sects. 2.1, 2.2, 2.5) or individual difference measures (sect. 2.7) that are presumed to index some critical aspect of cognitive activity during sleep mentation generation.

### 2.1. Memory sources inferred from associations to mentation

A 1-gen model might be expected to predict that REM and NREM reports of equivalent length derive from memory sources of equivalent type. This was supported in a study that used subjects' associations to dreams as a measure of their memory sources (Cavallero et al. 1990). Without controls for length, REM reports more frequently than NREM reports led to identifications of semantic knowledge sources, as opposed to autobiographical episodes or abstract self-references; with such controls – temporal unit weighting in this case – no memory source differences were found.

However, the 1-gen model is more often construed to be consistent with studies that *do* report qualitative differences in memory sources as a function of sleep stage. Comparisons of REM and NREM mentation reports do reveal differences in memory sources (Battaglia et al. 1987; Cavallero 1993; Cavallero et al. 1988; 1990; Cicogna et al. 1986; 1991; Foulkes et al. 1989). Compared with REM sleep mentation, memory sources of stage 2 mentation are more often episodic and less often semantic (see Cavallero 1993, for review) and more evidently connected to dream content (Foulkes et al. 1989). The memory sources of SO (1) are predominantly autobiographical and episodic (rather than an even mix of episodic memories, abstract self-references, and semantic knowledge as in REM sleep; Cavallero et al. 1988; 1990; Cicogna et al. 1986; 1991) and (2) more often have episodic sources referring to day residues than to earlier memories (as for REM sleep; Battaglia et al. 1987). Such results are taken to support the contention that "access to memory material is selective in SO, but probably undifferentiated in REM" (Cavallero & Cicogna 1993, p. 51).

**2.1.1. Problems with memory source experiments.** There are concerns with the notion that *diffuse mnemonic activation* is a precursor to sleep mentation (see sect. 2.9.1), because there are yet no valid correlates of such activation. Equally important is the question of whether memory activation should be considered to be distinct from the production of sleep mentation. If diffuse activation is dedicated exclusively to the production of sleep mentation and

is tightly and reciprocally coupled to this production, then might it not better be conceptualized as an integral, inseparable component of it? If so, qualitative differences in memory sources are in fact qualitative differences in mentation production processes.

Other explanations have been offered for some REM/NREM sleep mentation differences, for example, more frequent episodic memory sources for SO reports because of recency effects or a "carry-over" of episodic processes from immediately preceding wakefulness (Natale & Battaglia 1990). This reasoning is consistent with "carry-over" effects following awakenings from REM and NREM sleep as discussed under post-awakening testing (sect. 2.5); however, most of the latter research demonstrates differences for REM and NREM sleep, that is, supports a 2-gen model.

Qualitative differences in memory sources may be due to differential levels of engagement of the dream generation system, but few empirical findings speak directly to this issue. Some authors (Cavallero & Cicogna 1993) link changes in "levels of engagement" to levels of cortical activation, but cannot easily reconcile this explanation with the qualitative *differences* in physiological activation characterizing REM and NREM sleep. Others (Foulkes 1985) eschew links between psychological and physiological activation altogether.

### 2.2. Memory consolidation

Memory processes are central to both 1-gen and 2-gen models of mentation production. Of the several paradigms that have been used to investigate learning and memory consolidation during sleep, most have produced results consistent with the notion of different forms of cognitive processing during REM and NREM sleep (see Dujardin et al. 1990; McGrath & Cohen 1978; Smith 1995, for reviews). Although the evidence is not unanimous, most suggests that REM sleep is selectively implicated in learning new information.

Some studies have found discriminative responding during REM but not NREM sleep (Hars & Hennevin 1987; Ikeda & Morotomi 1997) or establishment of a classically conditioned response (e.g., hippocampal activity) selectively during REM sleep (Maho & Bloch 1992). Discriminatory cueing during REM sleep even enhances performance on a previously learned skill, whereas cueing during NREM sleep impairs it (Hars & Hennevin 1987). Smith and Weeden (1990) found that stimulation with 70 dB clicks that were previously paired with a learning task enhances later performance only when similar clicks are administered during REM, but not NREM, sleep. Further, stimulation of reticular formation only during REM sleep improves learning over 6 days (Hennevin et al. 1989); such stimulation enhances awake learning if applied after either training or cueing treatment (see Hennevin et al. 1995b, for review).

On the other hand, a few studies have demonstrated transfer of discriminative responding during NREM sleep (Beh & Barratt 1965; McDonald et al. 1975), for example, a second-order conditioned response can be entrenched during either REM or NREM sleep (Hennevin & Hars 1992).

Several types of perceptual, cognitive, and memory skills have been examined in relation to REM and NREM sleep using different types of procedures: selective REM/NREM deprivation, changes in REM/NREM sleep architecture after learning, retrospective assessment of sleep architecture differences in slow versus fast learners, and performance differences after REM and NREM awakenings.

Much of this research suggests *qualitative differences in the tasks that are dependent upon the integrity of REM and NREM sleep.* Some illustrative findings:

1. Disruption of REM, but not NREM, sleep diminishes performance on a basic visual discrimination task (Karni et al. 1994).

2. Deprivation of REM, but not NREM, sleep diminishes performance on procedural or implicit memory tasks, that is, Tower of Hanoi, Corsi block tapping, but not declarative or explicit memory tasks, that is, word recognition, paired associates (Smith 1995).

3. Training animals on a new, appetitive or aversive task is followed by an increase in REM, but not NREM, sleep (Hennevin et al. 1995b).

4. Successful intensive language learning is accompanied by increased %REM, but not %NREM (De Koninck et al. 1989).

5. Rearing in an enriched environment produces more dramatic increases in REM than in NREM sleep (Smith 1985).

6. Waking recall of stimuli presented during sleep is superior for stimuli presented just before awakenings from REM, but not NREM, sleep (Shimizu et al. 1977).

NREM sleep is associated with memory tasks only rarely; NREM sleep deprivation disrupts Rotor pursuit (Smith & MacNeill 1994) and the learning of lists of word pairs (Plihal & Born 1997). These findings nevertheless point to skills that are qualitatively different from those typically associated with REM sleep and are thus consistent with a 2-gen model.

### 2.2.1. Problems with memory consolidation experiments.
It remains unknown whether the memory processes essential to generating sleep mentation are the same as those shown to be associated with REM and NREM sleep. Almost invariably subjects in these types of experiments are never awakened to sample mentation in relation to learning. Some exceptions (Conduit & Coleman 1998; De Koninck et al. 1988; Fiss et al. 1977) unfortunately have not examined *both* REM and NREM sleep mentation to compare the two.

### 2.3. Event-related potentials

Different time-locked components of event-related potentials (ERPs) reflect different steps of perceptual and cognitive processing, steps that may be extrapolated to some extent to the various stages of sleep (see Kutas 1990; Salisbury 1994, for reviews). Short-latency auditory components – occurring within 10 to 15 msec of a stimulus – reflect sensory pathway integrity from receptors through to thalamus, and appear not to change in any sleep stage (Campbell & Bartoli 1986). Middle latency responses – 10 to 100 msec post-stimulation – reflect processes such as threshold detection associated with medial geniculate, polysensory thalamus, and primary cortex. Up to 40 msec, these components are largely unaffected by sleep/wake stage (Salisbury 1994). Beyond 40 msec, most studies show some reduction in amplitude and latency during sleep (Erwin & Buchwald 1986; Linden et al. 1985; Picton et al. 1974) although some show an increase in amplitude of potentials such as N1 and P2 (Nordby et al. 1996). These changes vary little from stage to stage, however. Long-latency components – typically later than 100 msec post-stimulation – are of particular interest because of their putative associations with cognitive processes such as selective attention (N1 or N100), sensory mismatch (N2-P3a), orienting (N2), surprise (P3b), novelty (P3a), and semantic processing (N400) (see Kutas 1990; Salisbury 1994, for reviews). Several studies (Addy et al. 1989; Nakano et al. 1995; Noguchi et al. 1995; Nordby et al. 1996; Roschke et al. 1996; Van Sweden et al. 1994) indicate that long-latency components from NREM sleep (vs. those from wakefulness), are both suppressed in amplitude and slowed in latency – independent of the sensory modality stimulated. Most studies find that these components in REM sleep resemble those of wakefulness to a greater extent than they do the more diminished potentials of NREM sleep.

Research pertinent to the critical question of whether P300, a presumed measure of complex cognitive processing, is differentially active during REM and NREM sleep has produced mixed results. Most studies find P300 in REM sleep and stage 1 NREM sleep but not in other NREM sleep stages (Bastuji et al. 1995; Côté & Campbell 1998; Niiyama et al. 1994; Roschke et al. 1996; Van Sweden et al. 1994) suggesting a distinctive mode of higher-order processing during the two sleep states with the most vivid imagery processes. Others have found either a diminished P300 in both REM and NREM sleep (Wesensten & Badia 1988) or no clear evidence of P300 in sleep (Nordby et al. 1996). These discrepant findings may be due, in part, to the large variability of this late component, a variability exacerbated in NREM sleep by the superimposition of endogenous K-complexes, as well as by the fact that oddball stimuli are often not sufficiently disparate (Salisbury 1994) or intense (Côté & Campbell 1998) to evoke the P300 response.

Both 1-gen and 2-gen models stipulate that the blocking of afferent information during sleep is a precondition for cognitive activity. Thus, early- and middle-latency results seem relatively irrelevant to differentiating the models. To the extent that higher-order cognitive functions are necessary for sleep mentation, long-latency ERP studies demonstrating degradation of these components in NREM, but not REM, sleep support the notion of *different* cognitive processes in the two states.

### 2.3.1. Problems with ERP studies.
It might be argued (from the 1-gen viewpoint) that long-latency ERP differences reflect only differences in degree – not quality – of mentation production processes in REM and NREM sleep. Diminished P300 amplitude in NREM sleep might simply index a reduction in memory diffuseness thought to occur (Foulkes & Schmidt 1983). This argument hinges in part on what transformations of the P300 waveform are ultimately found to be correlated with qualitative (and not simply quantitative) differences in REM and NREM mentation. One might expect that minor changes in amplitude or latency reflect only quantitative differences while more dramatic changes in ERP structure (e.g., absence of the waveform) reflect qualitative differences, but this remains an empirical question.

It might also be argued (from the 1-gen viewpoint) that the cognitive processing revealed by long-latency components does not reflect activity that is germane to mentation production. Such components may reflect processing occurring either so early or so late in production that they have no causal bearing on the outcome. Processes such as sensory mismatch recognition, or orienting/surprise to a stimulus could be simple affective *reactions* to unusual dreamed events, reactions with no real impact on imagery construction (Foulkes 1982c). Conversely, at least one well-articulated theory describes how orienting responses and related affective reactions *engender* sleep mentation (Kuiken & Sikora 1993). Moreover, many findings link P300 to emotional processes such as mood expectancy during reading (Chung et al. 1996) emotional prosody (Erwin et al. 1991) and emotional deficits (Bungener et al. 1996). On the other hand, the suggestion (Donchin et al. 1984) that P300 reflects processes of *creating, maintaining, and updating an internal model of the immediate environment* suggests that P300 underlies more basic representational processes.

### 2.4. Stimulation paradigms

The presentation of stimuli prior to sleep affects REM and NREM sleep mentation differentially, for example: (1) six hours of cognitive effort prior to sleep produces REM sleep mentation with less thinking and problem solving, and NREM sleep mentation with increased tension (Hauri 1970); (2) presentation of presleep rebus stimuli (e.g., image of a *pen* with a *knee* → penny association) has no effect on REM sleep mentation, but evokes conceptual references to the stimulus words (e.g., pencil, leg) in stage 2 mentation (Castaldo & Shevrin 1970); (3) auditory cues to picture learning leads to superior processing of higher order stimuli in stage 2 (Tilley 1979). These authors conclude that REM and NREM sleep are associated with different levels of cognitive organization – which squares with the notion that NREM sleep mentation is more conceptual or thoughtlike. However, auditory cues are also less impeded by sensory inhibition during stage 2 sleep than during phasic REM sleep (Price & Kremen 1980). On the other hand, superior processing of verbal materials during REM sleep was suggested in a study of associative learning (Evans 1972); such differences are not easily explained by elevated sensory inhibition during REM sleep.

**2.4.1. Problems with stimulation paradigms.** Many of these studies suggest sleep stage differences that are *opposite* in nature to those suggested by ERP studies, for example verbal stimulation preferentially influences stage 2 mentation, whereas REM sleep has more evident late ERP components of the type one might expect to index the registration of such verbal stimulation. Such ambiguities could be resolved by examining both sleep mentation and ERPs in the same study design.

### 2.5. Post-awakening testing

Post-awakening testing taps cognitive abilities immediately after awakening from REM or NREM sleep, and is based on the observation that cognitive and physiological components of a sleep state will "carry-over" and influence waking performance. Post-awakening testing has been used by at least six independent research groups in at least eight different studies (see Reinsel & Antrobus 1992, for review). Most studies concur that REM and NREM sleep awakenings produce different patterns of responding. The first demonstration of a "carry-over effect" (Fiss et al. 1966) was that thematic apperception test (TAT) stories generated following REM sleep awakenings were more "dreamlike" than those following NREM sleep. Subsequently, perceptual illusions, such as spiral after-effect and beta movement, were found to vary with preceding sleep stage (Lavie 1974a,

Lavie & Giora 1973; Lavie & Sutter 1975). Superior performance on right hemisphere (RH), primarily spatial tasks after REM sleep and on left hemisphere (LH), primarily verbal tasks after NREM sleep were also reported (Gordon et al. 1982; Lavie & Tzischinsky 1984; Lavie et al. 1984). Other studies (Bertini et al. 1982; 1984; Violani et al. 1983) demonstrated RH superiorities after REM sleep on a tactile matching task. Short-term memory is also better after REM versus NREM awakenings (Stones 1977).

One study (Reinsel & Antrobus 1992) did not replicate the reported stage differences, even though many of the same dependent measures were employed. The authors suggest that the discrepancies may be due to subtle methodological differences, for example, greater memory demands in the original studies (Reinsel & Antrobus 1992). Also, stage-related differences on trail-making and vigilance tasks were not found for REM and NREM awakenings (Koulack & Schultz 1974).

Most of these results support the interpretation that qualitatively different cognitive processes are active following and, by inference, just preceding awakenings from REM and NREM sleep. These include both lower-level (perceptual registration, stimulus matching) and higher-level (short-term memory, story generation) processes.

**2.5.1. Problems with post-awakening testing.** The replicability of post-awakening effects was questioned by at least one study (Reinsel & Antrobus 1992). There is also some concern about whether waking state measures are valid measures of preceding, sleep-related processes. Findings do support the "carry-over" construct, but the weight of evidence is not overwhelming. It is possible, for example, that post-awakening effects are due to different *changes of state* as opposed to "carry-over" of cognitive processes linked to a particular state.

### 2.6. Inter-relationships between mentation contents from different reports

The 1-gen model might predict that a single imagery generator would produce a great degree of thematic continuity between proximal REM and NREM reports within a night; the 2-gen model would predict different kinds of unrelated mentation. One study (Cipolli et al. 1988) supporting the 1-gen model found that low-level paradigmatic and lexical relationships (but not high-level syntagmatic and propositional relationships) between pairs of mentation reports were higher within the same night than they were between nights, regardless of whether the reports were REM-NREM pairs or REM-REM pairs. An earlier study (Rechtschaffen et al. 1963b) found that high-level themes were often repeated in REM and NREM reports from the same night.

**2.6.1. Problems with report inter-relationships.** If thematic similarity is an index of unified mentation production, then thematic difference may be construed as an index of two or more generators. In all likelihood, thematic differences would be more prevalent than similarities in any within-night REM/NREM mentation comparisons. Yet chance levels of thematic similarity in adjacent reports remain unknown. It may also be argued (from a 2-gen perspective) that similar themes nevertheless differ in some qualitative respects, for example, an interpersonal aggression may be more self participatory, affectively engaging and visual in a REM report than in a NREM report (cf Weinstein et al. 1991).

### 2.7. Subject differences in mentation content

Interactions between subject differences and stage-related cognitive activity may set limiting conditions on the generalizability of the two models, for example, they may suggest that one or the other model *is valid only for some types of subjects and under some circumstances.* Also, some prevalent subject variables linked to sleep mentation (e.g., age, insomnia, dream recall frequency) may determine subject self-selection for sleep studies and thus bias the estimated rates of mentation recall from REM and NREM sleep. Three variables illustrate this complexity.

**2.7.1. Light versus heavy sleepers.** Zimmerman (1970) first proposed that differences in *activation* may account for REM/NREM mentation differences. He classified subjects as either light or deep sleepers (based on auditory arousal thresholds) and awakened them twice each from REM and NREM sleep. Light sleepers reported *dreaming* after NREM awakenings more often (71%) than did deep sleepers (21%). REM and NREM mentation from these groups also differed qualitatively. For deep sleepers, NREM mentation was less perceptual, controlled, and distorted. For light sleepers, such differences did not obtain. If light-sleeping subjects are more cerebrally *aroused* than are deep-sleeping subjects during NREM sleep, then their NREM content may be much more REM-like. Thus, the 1-gen model may apply to light-sleeping subjects; the 2-gen model to deep-sleeping subjects.

**2.7.2. Habitual recall of dream content.** Mentation from REM and NREM sleep differs for subjects high and low in habitual dream recall. We (Nielsen et al. 1983; 2001) found that stage REM reports were higher on two measures of story organization (number of story constituents, degree of episodic progression) than were NREM reports, *but only for high frequency recallers.* The 1-gen and 2-gen models appear to describe low- and high-frequency recallers differentially.

**2.7.3. Psychopathology.** Measures of REM and NREM salience (i.e., recall and length) are correlated differentially with measures of psychopathology. For example, the MMPI L scale correlates with REM mentation recall whereas no scales correlate with NREM mentation recall (Foulkes & Rechtschaffen 1964). The two states are further differentiated by correlations between the MMPI Hy scale and REM word count and between several scales and NREM word count. NREM word count also correlates with Ego Strength and Hostility Control. A 2-gen model is favored by such results.

**2.7.4. Other studies of subject variables.** Many other subject variables are known to interact with sleep mentation although specific relationships remain to be clarified. Some include (1) the differential association of age with late night activation effects on REM and NREM mentation (Waterman et al. 1993), (2) large differences in recall of REM (but not NREM) related mentation for both insomniac (Rotenberg 1993b) and depressed (Riemann et al. 1990) patients versus normal controls, (3) the effects of introspective style on the salience of REM and NREM content (Weinstein et al. 1991) and elevated incorporation of laboratory characters

into REM (but not NREM) mentation for women, but not men (Nielsen et al. 1999). Other such correlates of dream recall have been reviewed (Schredl & Montasser 1997) and appear to be consistent primarily with the 2-gen model.

### 2.8. Residual differences in stage-related measures of mentation quality

Many authors feel that the fairest test of REM/NREM mentation differences is whether mentation reports differ on qualitative measures after report length has been controlled. However, many studies report qualitative REM-NREM stage differences even with such controls (Antrobus 1983; Antrobus et al. 1995; Cavallero et al. 1990; Cicogna et al. 1991; Foulkes & Schmidt 1983; Hunt et al. 1993; Porte & Hobson 1996; Nielsen et al. 1983). With length controls, REM and NREM mentation samples still differ on self-reflectiveness (Purcell et al. 1986), bizarreness (Casagrande et al. 1996b; Porte & Hobson 1986), visual and verbal imagery (Antrobus et al. 1995; Casagrande et al. 1996b; Waterman et al. 1993), psycholinguistic structure (Casagrande et al. 1996a), and narrative linkage (Nielsen et al. 1983). Strauch and Meier (1996) found fewer characters and lower self-involvement in NREM than in REM mentation, again, regardless of report length. Even Foulkes (Foulkes & Schmidt 1983) found more per-unit self-representation in REM than in SO mentation and more per-unit characterization in REM than in NREM mentation. Differences in characterization and self-representation are not trivial since they are two of the most ubiquitous constituents of dreaming.

Visual imagery is perhaps the most defining quality of dream mentation. Visual imagery word count and total word count both differentiate stage REM from stage 2 mentation reports – and a *significant predominance of visual words in REM over NREM reports remains even after total word count is controlled as a covariate* (Waterman et al. 1993). Antrobus et al. (1995) have replicated this finding, failing to replicate Antrobus's own earlier study (Antrobus 1983), as have Casagrande et al. (1996b).

A recent study (Porte & Hobson 1996) reports stage-related differences in fictive (imagined) movement, but also some support for the 1-gen model. Here, the subgroup of 10 subjects who produced the only motor reports in NREM sleep also had the longest mentation reports from both sleep stages. The authors suggest that some factor may have caused their NREM sleep to be influenced by REM sleep processes, for example, an increase in REM sleep "pressure" by REM deprivation, thus lengthening REM reports and raising the odds that a NREM awakening coincides with a pre-REM or post-REM sleep transitional window (Porte & Hobson 1996). I refer to this window as a type of *covert REM sleep* in a later section (see sect. 3).

The accumulation of findings of residual qualitative differences between REM and NREM sleep mentation after length control challenges the 1-gen argument that such controls cause qualitative differences to disappear (Foulkes & Cavallero 1993). Such differences are diminished by controlling length but they are not eliminated altogether.

### 2.9. Memory versus physiological "activation"

**2.9.1. Are memory activation and cortical activation isomorphic?** Foulkes's (1985) 1-gen model identifies memory activation as the instigating force of sleep mentation but excludes physiological activation as a determinant, even though known relationships between cerebral activation and sleep/wake stages might seem consistent with the model. For example, PET imaging studies of the brain have demonstrated that REM sleep is characterized by elevated and more widespread activation than is NREM sleep; higher levels of cerebral blood flow have been measured in most centrencephalic regions (cerebellum, brainstem, thalamus, basal ganglia, basal forebrain), limbic and paralimbic regions (hippocampus, temporal pole, anterior insula, anterior cingulate), and unimodal sensory areas (visual and auditory association; Braun et al. 1997). Note, however, that Foulkes's exclusion of neurophysiological correlates of brain activation in the development of 1-gen models is not supported by all 1-gen theorists.

Studies of whether cortical activation is indeed correlated with cognitive activation offer limited support for the notion of an association (see Antrobus 1991, for review). With EEG slowing and increased voltage there is an associated *decrease* in mentation recall (Pivik & Foulkes 1968, and there is more EEG slowing in NREM than in REM sleep (e.g., Dumermuth et al. 1983). In one study, both delta and beta amplitude predicted successful dream recall from REM sleep whether subjects were depressed or healthy (Rochlen et al. 1998). In our studies (Germain et al. 1999; Germain & Nielsen 1999) fast- and slow-frequency power was associated with recall of dreams from REM and NREM sleep respectively. If EEG-defined activation (delta) is statistically controlled, stage differences in mentation are still obtained (Waterman et al. 1993). At least one study (Wollman & Antrobus 1987) found *no* relationships between EEG power and word count of either REM sleep reports or waking imagery reports.

It is well known that both the recall (Goodenough 1978; Verdone 1965) and the salience (Cohen 1977a; Foulkes 1967) of sleep mentation increases in later REM episodes; these changes are likely due to activation associated with circadian factors (Antrobus et al. 1995). On the other hand, circadian factors appear to influence REM and NREM mentation equally (Waterman et al. 1993) – a finding that would seem to support the 1-gen model. However, when both stage and diurnal activation effects on variables such as visual clarity are assessed simultaneously, the effect size for time-of-night activation is only about 30% of the effect size for REM-NREM stage activation; this difference is interpreted to support the 2-gen, A-S model (Antrobus et al. 1995).

### 2.9.2. Partialling out activation: Problems with using report length.
Controls for report length are effected in different ways. Most studies estimate activation by total word count (TWC; Antrobus 1983), a tally, usually transformed by $\log_{10}(TWC+1)$ to remove positive skew, of all non-redundant, descriptive content words in the report. Length is then partialled out of correlations between variables or in some other way (Antrobus et al. 1995; Levin & Livingston 1991; Waterman et al. 1993; Wood et al. 1989). A procedure conceptually related to TWC is to weight dependent variables with a length estimate that is based upon report structure. Foulkes and Schmidt (1983) parsed reports for events that occurred contiguously, the so-called "temporal unit." Similarly, we (Nielsen et al. 1983; 2001) used the presence of story components (characters, actions, settings) to control for their organization – a REM/NREM difference was found in this study. We also used the proportional measures of the Hall and Van de Castle (1966) system to compare REM and NREM reports qualitatively – few REM/NREM differences were seen (Faucher et al. 1999).

Hunt's (1993) challenge to length-sensitive corrections is that variations in report length are an expected correlate of mentation that is qualitatively remarkable in some way, that is, that "more words are necessary to describe more bizarre experiences" (p. 181). To partial out report length from a given qualitative scale may be to partial out the variable from itself (p. 181) and may even "cripple our ability to study what is most distinctive about dreams by misleadingly diluting a key measure of the dreaming process" (p. 190). Even worse, using word frequencies to weight non-verbal variables (e.g., bizarreness) may arbitrarily transform findings and produce unpredictable and artificial effects (Hunt et al. 1993). Using report lengths and bizarreness ratings, Hunt demonstrated that a bizarre pictorial stimulus does indeed require more words to describe than does a mundane stimulus, and that the partialling out of TWC eliminates significant correlations between bizarreness and other measures. Weighting produced a significant loss of information related to the dependent variable.

### 2.10. Summary

Most of the research reviewed in the preceding nine categories tends to favor the 2-gen over the 1-gen model. The 2-gen model is supported particularly by evidence of REM/NREM differences in sleep mentation and by physiological measures, such as long-latency ERPs, that are valid correlates of waking cognitive processes. The principal claim of the 1-gen model, that qualitative differences are artifacts of quantitative differences, has been challenged by many studies demonstrating process differences and residual qualitative differences after length control, as well as studies questioning the assumptions underlying quantitative controls. Another argument, that residual qualitative differences are attributable to differences in memory inputs, has merit, but has not been supported by all attempts to quantify these inputs. There are also important questions about whether memory indeed functions in a diffuse manner as proposed, and whether memory source activation is not, in fact, an integral part of the dreaming process itself. Recent neuropsychological evidence favors the 1-gen model but has still not directly addressed the question of REM and NREM sleep mentation differences.

On the other hand, the evidence does not overwhelmingly support the 2-gen model either. Evidence for neurobiological isomorphism as currently defined is still slim, and leaves most of the conclusions of this model extremely speculative (Foulkes 1990; Labruzza 1978). The 2-gen model is also weak in describing the nature of REM and NREM mentation comparatively. As a model driven by physiological antecedents to cognition, it can also be criticized for not accounting for forebrain mechanisms that seem central to complex cognitive operations such as the narrative synthesis of dreaming (Antrobus 1990; Solms 1995; Vogel 1978a).

### 3. An alternative model: Covert REM sleep processes in NREM sleep

The literature presents an apparent paradox. On one hand, there is strong proof that cognitive activity – some of it dreaming – can occur in all sleep stages. On the other hand, there is evidence that REM and NREM sleep mentation and an array of their behavioral and physiological correlates differ qualitatively. The former evidence supports a 1-gen model, the latter a 2-gen model. How may this seemingly contradictory evidence be reconciled?

One possible reconciliation is that sleep mentation is, in fact, tightly coupled to REM sleep processes, but that some of these processes under certain circumstances may dissociate from REM sleep and stimulate mentation in NREM sleep in a *covert* fashion. This alternative conceptualization maintains a 1-gen assumption but couples it with an assumption of psychophysiological isomorphism. The same (REM sleep-related) processes are thought to be responsible for sleep mentation regardless of stage, even though in NREM sleep these processes may be activated in a piecemeal fashion and against an atypical neurophysiological background. Some REM sleep processes would thus combine in as yet unspecified ways with NREM sleep processes to produce unique profiles of NREM sleep physiology and intermittent occurrences of REM-like sleep mentation. The origin of these mechanisms in REM sleep events may explain observed similarities in REM and NREM mentation reports, while their dissociated nature may explain apparent qualitative differences. This model is in some respects similar to the 1-gen model in that it assumes commonality of processes for all mentation reports, but it differs in that it extends this commonality to physiological processes. The model is also similar in some respects to the 2-gen model in that it assumes psychophysiological isomorphism between sleep mentation and some features of sleep neurophysiology and in that it explains qualitative differences in REM and NREM mentation as a function of the dissociated quality of covert activation (e.g., piecemeal activation, atypical neurophysiological background).

This view leads to several straightforward and easily testable predictions about mentation in relation to sleep stage: *(1) mentation recalled from NREM sleep will be associated with factors linked to preceding and/or subsequent REM sleep.* For example, recall of mentation should be more likely, more abundant or more salient from NREM episodes that are in close proximity to a REM sleep episode, or from NREM episodes that are in proximity to particularly long or intense REM episodes. The former example is supported by several studies reviewed earlier and is described in more detail in the probabilistic model that follows. The latter example has not been systematically investigated. The covert REM sleep model also predicts that *(2) recall of mentation from NREM sleep will be more probable under conditions likely to stimulate covert REM sleep, for example, sensory stimulation during sleep, sleep deprivation and fragmentation, sleep onset, arousal during sleep, psychiatric and sleep disorders, medications.* Evidence supporting the preceding hypotheses is reviewed in more detail below. Finally, the model's isomorphism assumption leads to some predictions about the neurophysiological characteristics of REM and NREM sleep with and without mentation recall: *(3) the neurophysiological characteristics of NREM sleep* with *recall of mentation will differ from those of NREM sleep* without *recall, and (4) the neurophysiological characteristics of NREM sleep with the most vivid mentation will resemble the characteristics of REM sleep with typical mentation.* The former prediction we have supported to some extent with evidence that EEG spectral analysis differentiates between NREM sleep awakenings with and without recall of mentation (Germain & Nielsen 1999). The latter prediction we have supported to some extent with evidence of similarities in the EEG accompanying NREM imagery from sleep onset and that accompanying imagery from REM sleep (Nielsen et al. 1995). However, both predictions require testing with more refined multivariate methods.

Covert REM sleep is defined here to be *any episode of NREM sleep for which some REM sleep processes are present, but for which REM sleep cannot be scored with standard criteria.* This notion encompasses previous ideas that have been raised and expanded upon to varying degrees by different authors, but has never been elaborated into a systematic model. The following is therefore a synthesis and systematization of several existing ideas about covert REM sleep as well as a review of research findings that support these ideas. In brief, evidence is reviewed supporting the notion that covert REM sleep processes can occur in NREM sleep under many different circumstances. An easily testable model is then proposed that addresses two of these conditions: covert REM sleep occurring during NREM/REM transitions and that occurring during SO.

### 3.1. Covert REM sleep is suggested by "intermediate sleep"

Lairy et al. (1967) were among the first to identify atypical mixtures of REM and NREM sleep in human subjects. Their notion of "intermediate sleep" was of sleep that typically arises between REM and NREM sleep episodes but that consists of elements of both. Intermediate sleep was defined primarily by EEG configurations containing both REM and NREM sleep features, such as spindles or K-complexes separated by episodes of "EEG traces identical to that of REM sleep" (p. 277). Mentation elicited from intermediate sleep was noted to be less hallucinatory and more negative in feeling tone than that elicited from REM sleep. Intermediate sleep could also at times replace an entire REM sleep episode. In normal subjects, it was said to occupy 1–7% of sleep; in psychiatric cases, such as psychosis, from 10 to over 40% (Lairy et al. 1967). More recent clinical evidence (Mahowald & Schenck 1992) confirms that components of different sleep/wake states do indeed dissociate and combine in atypical patterns as a consequence of illness or other unusual circumstances. For instance, the violent dream-related outbursts of REM sleep behavior disorder seems to combine features of wakefulness (motor activity) with background REM sleep (Mahowald & Schenck 1994) whereas the cataplexy attacks of narcolepsy appear to combine aspects of REM sleep (muscle atonia) with background wakefulness.

### 3.2. Physiological processes anticipate REM sleep onset

Some studies suggest that covert REM sleep processes can occur during normal human sleep. First, the REM sleep-related shift in HR variability from predominantly parasympathetic to predominantly sympathetic can occur up to 15 minutes prior to the EEG-defined onset of REM sleep (Scholz et al. 1997). Second, the progressive suppression of REM-related sweating effector activity – an index of thermoregulation – anticipates REM sleep onset by 6–8 minutes (Dewasmes et al. 1997; Henane et al. 1977; Sagot et al.

1987). Fluctuations in this measure have been proposed to be due to occurrences of dreaming (Dewasmes et al. 1997; Ogawa et al. 1967). Third, the REM sleep-associated cortical process of N300 amplitude attenuation occurs several minutes prior to other REM sleep indices such as muscle atonia and eye movements (Niiyama et al. 1998). ·

### 3.3. Covert REM sleep during "missing" REM episodes

Covert REM sleep processes may be implicated in the atypical NREM sleep episodes for which the absence of one or more electrophysiological criteria prevents a score of REM sleep from being assigned. To polysomnographers, these episodes commonly, but not exclusively, appear as the troublesome "missed" REM sleep episodes early in the night. Their absence can lead to exceptionally long REM SO latencies being scored. During such episodes, most of the electrophysiological signs of REM sleep are present – for example, cessation of spindling, EEG desynchronization, changes occurring approximately 90 minutes after SO – but sometimes chin muscle tonus may remain high, or rapid eye movements may be slow or indistinct, or a brief waking arousal may occur. Such stages may be scored as stage 1 or 2 even though intuition strongly suggests that REM sleep is somehow present.

Other studies have reported the omission of REM periods at other times of the night. Nocturnal penile tumescence, a relatively robust correlate of REM sleep (e.g., Karacan et al. 1972), often occurs at the 90-minute junctures where REM sleep might be expected but is not scored because of missing criteria (Karacan et al. 1979). In Karacan's study, 12 of 19 erections occurring during NREM sleep were related to expected but incomplete REM sleep episodes; an additional four occurred during NREM sleep immediately after REM sleep awakenings. Their paper contains an illustrative hypnogram of three consecutive nocturnal erections overlying three corresponding covert REM episodes.

### 3.4. Proximity of NREM sleep awakenings to REM sleep

Recordings of spontaneous REM and NREM sleep awakenings in the home setting reveal that NREM mentation reports are longest if they occur within 15 min of a prior REM sleep episode, whereas REM mentation reports are longest if they occur 30–45 minutes into a REM episode (Stickgold et al. 1994a). In fact, in this study seven of the nine longest NREM reports occurred within 15 minutes of a REM episode. These findings replicate an earlier finding (Gordon et al. 1982) that NREM reports occurring within 5 minutes of previous REMs more often produce cognitive activity (81.8%) than do reports occurring more than 10 minutes post-REMs (3.8%). They also replicate the finding (Antrobus et al. 1991) that NREM reports occurring 5 minutes after a REM sleep episode contain more words per report than do those occurring 15 minutes post-REM. Stickgold et al. interpret these kinds of results as possibly supporting a covert REM sleep influence, that is, that "long NREM reports reflect transitional periods when some aspects of REM physiology continue to exert an influence" (p. 25). They also consider that reports from early in NREM sleep episodes might reflect recall of mentation from the preceding REM episode, a notion that has often been suggested as an explanation for dreaming during NREM sleep (Kales et al. 1966; McCarley 1994; Wolpert & Trosman

1958; and see Porte & Hobson 1996 for discussion). It should be noted that at least one study (Kamiya 1962) has found that NREM awakenings conducted prior to the first REM sleep episode of the night, when presumably no prior REM sleep influences could have occurred, nevertheless produced recall of cognitive activity (43%). Similarly, a study (Foulkes 1967) in which awakenings 30 minutes post-REM targeted the *middle* of NREM episodes – also found a sizable recall rate of 64.6%. These recall rates either equal or exceed the mean recall rate estimate for NREM sleep presented earlier. Both studies argue against the possibility of covert REM sleep processes. However, the reconsideration of SO as a possible source of covert REM sleep to some extent counters the first of these arguments (see sect. 3.5), whereas the substantial uncertainty associated with identifying the precise middle of NREM episodes responds somewhat to the latter (see sect. 4.1 below). These arguments are now considered in more detail.

### 3.5. Covert REM sleep during sleep onset (SO)?

Covert REM sleep processes may manifest during SO episodes. These brief wake-sleep transitions display many of the electrophysiological signs of REM sleep, for example, transient EMG suppressions and phasic muscle twitches, as well as extremely vivid sleep mentation. We have shown that the topographic distributions of fast-frequency EEG power for SO images and REM sleep are similar (Nielsen et al. 1995). REMs are less conspicuous at SO, but they are nevertheless observed (Vogel 1978b). However, the slow eye movements so characteristic of SO also occur frequently in REM sleep, suggesting that they may constitute an unrecognized marker of REM sleep (Porte 1997). It is thus possible that the vivid dreaming of SO derives from a brief, usually undetected passage through REM into descending stage 2 sleep. The sleep onset REM (SOREM) episodes observed frequently in both sleep disordered and normal individuals (Bishop et al. 1996) may be instances of covert REM sleep transitions that have been "unmasked" and thus *do* manifest all of the inclusion criteria for REM sleep. Such unmasking might be influenced by the build-up of REM pressure. For example, we found that SOREM episodes on the MSLT were twice as frequent in sleepy patients (with severe sleep apnea or idiopathic hypersomnia) than they were in non-sleepy patients (with mild sleep apnea or periodic leg movements without hypersomnia) (T.A. Nielsen, J. Montplaisir & A. Gosselin, unpublished results). The fact that reports of dreaming during MSLT naps are *not* good predictors of the presence of classical REM sleep (Benbadis et al. 1995) may reflect the difficulty of differentiating covert REM sleep from REM sleep as it is classically defined. Further evidence for covert REM sleep processes at SO is the variety of sleep starts commonly observed at SO among healthy subjects. Such starts consist of abrupt motor jerks and sudden flashes of visual, auditory, and some esthetic imagery; it has been suggested that they are intrusions of isolated REM sleep events into NREM sleep (Mahowald & Rosen 1990).

### 3.6. Covert REM sleep: A disorder of arousal?

Mentation is often reported after sleep terror awakenings, which occur in NREM sleep stages 3 or 4 (Fisher et al. 1973). Much of this mentation appears to be induced by the arousal itself, judging by the themes such as death anxiety associated with tachycardia and choking anxiety associated with respiratory difficulty. Other instances appear to be ongoing before the terror erupts although they too appear to be heavily influenced by stimuli from the laboratory (Fisher et al. 1973). In fact, it is possible to induce terrors by external stimulation, such as sounding a buzzer. Thus, it is possible that sleep terror mentation is also a type of brief covert REM sleep event induced by stimulation that arises either internally (autonomic arousal) or from the laboratory environment (electrodes, noise, etc.) during arousals from sleep (see also sect. 3.11 below).

Early studies that examined method of arousal as a determinant of mentation content reported that, relative to abrupt awakenings, prolonged awakenings increase the frequency of thoughtlike mentation reports from both REM and NREM sleep (Goodenough et al. 1965a; Shapiro et al. 1963; 1965). This may mean that the prolonged awakenings induced a type of covert REM sleep state regardless of whether the ongoing state was REM or NREM sleep; the thoughtlike mentation accompanying this sleep state parallels that of what is most commonly reported after NREM awakenings. Physiological evidence that prolonged awakenings produce covert REM sleep is scanty although "stage-1" sleep with rapid eye movements during arousals from NREM sleep have been observed in individual subjects (Goodenough et al. 1965a; Roffwarg et al. 1962). Further, Goodenough et al. report many occasions on which gradual awakenings from NREM sleep are accompanied by a REM sleep-like EEG profile but no rapid eye movements.

### 3.7. Covert REM sleep underlies the REM sleep "efficiency" concept

Polysomnographers applying the Rechtschaffen and Kales criteria have always accepted a certain degree of ambiguity in their scoring of REM sleep, especially in the notion of REM sleep "efficiency." Within the limits of a given REM sleep episode there can occur transitions into other stages – typically stage 2 or wakefulness – which reduce the efficiency of the REM episode. If this alternate activity does not exceed 15 minutes in length, then the stage is considered a temporary deviation of an otherwise continuous REM sleep episode. If it exceeds 15 minutes, it denotes the start of a new REM/NREM cycle, with a periodicity far short of 90 minutes, that is no longer factored into the efficiency score. Thus, the 15-minute criterion for REM sleep efficiency implies that the underlying physiological state of REM sleep is not *completely* suspended during intrusions by another stage for <15 minutes. Some factor continues to exert a "propensity" to express REM sleep, a factor that seemingly remains latent. In view of research reviewed here (see sect. 3.2), the choice of 15 minutes for calculation of REM sleep efficiency seems entirely appropriate.

### 3.8. Covert REM sleep "pressure" is augmented by REM sleep deprivation

Selective REM sleep deprivation is known to increase "pressure" to express REM sleep. This is measurable as an increased number of "attempts" to enter REM during NREM sleep (Endo et al. 1998), as well as an increased REM density, decreased REM sleep latency (Ellman et al. 1991) and REM sleep rebound on recovery nights. EEG changes on recovery have been observed, even up to three nights post-deprivation (Endo et al. 1998; Toussaint et al. 1997). The probability of covert REM sleep occurrences is thus likely to be increased during or after REM deprivation. This is in fact supported by three kinds of findings. First, REM deprivation produces an increase of ponto-geniculo occipital (PGO) activity during NREM sleep in animal subjects (Dusan-Peyrethon et al. 1967; Ferguson & Dement 1969). Second, REM deprivation destabilizes recovery sleep in some human subjects, producing mixtures of REM and NREM sleep events ("ambiguous" sleep; Cartwright et al. 1967). Third, REM deprivation increases the sensory vividness, reality quality, and dreamlikeness of NREM mentation reports (Weinstein et al. 1991). In fact, REM sleep-deprived subjects in Cartwright's study (Cartwright et al. 1967) were found to have high percentages of dream reports from pre-REM transitional sleep. For one sub-group of subjects in this study (the "substitutors"), the degree of REM rebound after deprivation was negatively correlated with dreamlike content from NREM sleep awakenings. These subjects appeared to "cope with the changed sleep cycle by substituting a pseudo-cycle in which a good deal of REM content comes into awareness during the preREM sleep" (p. 302). Porte and Hobson (1996) have also proposed that increased REM pressure may account for very dreamlike NREM sleep reports in laboratory studies.

### 3.9. Evidence of covert REM sleep from animal studies

Early animal studies (Gottesmann 1964; Weiss & Adey 1965) detected signs of covert REM sleep even before the observation of intermediate sleep in human subjects. Sleep characterized by combinations of high amplitude anterior spindles (a sign of NREM sleep) and low frequency, dorsal hippocampal theta (a sign of REM sleep) was observed in rats and cats. Jouvet (1967) described PGO activity during transitions from NREM to REM sleep and throughout the REM sleep period and thought that these reflected inputs relevant to the visual images of dreaming. Steriade et al. (1989) also described PGO-related discharges of lateral geniculate neurons during pre-REM sleep states in cats, finding their signal-to-noise ratios to far exceed those found during REM sleep. Steriade's findings suggest that "vivid imagery may appear well before classical signs of REM sleep, during a period of apparent EEG-synchronized sleep" (Steriade et al. 1989, p. 2228). McCarley (1994) further advanced this hypothesis in describing brainstem neuronal membrane changes associated with REM sleep that may begin well before either EEG or PGO signs of REM. The transition at the membranal level is "gradual, continuous, and of long duration" (p. 375); it may also continue after the offset of a REM episode (see also Kayama et al. 1992). McCarley, too, speculates that NREM dreaming takes place during such REM-active transitions. Recent work (reviewed by Gottesmann 1996) has described additional physiological characteristics of intermediate states, including a seeming deactivation of forebrain centers and an apparent link to the processes that generate REM sleep.

### 3.10. Drug-induced covert REM sleep

Many drugs have been found to influence covert REM sleep, primarily by increasing PGO activity during NREM sleep. Ketamine (Susic 1976), PCPA (Delorme et al. 1966), reserpine (Brooks & Gershon 1972; Delorme et al. 1965) and LSD (Stern et al. 1972) have all been found to augment the density of PGO spiking in NREM sleep in animal subjects. Other drugs have been found to affect intermediate sleep, such as the barbiturates and benzodiazepines, which prolong intermediate sleep at the expense of REM sleep (Gottesmann 1996), and nerve growth factor, which produces intermediate sleep ("dissociated" sleep) in addition to dramatically increasing REM sleep time (Yamuy et al. 1995).

### 3.11. Covert REM sleep induced by sensory stimulation

In addition to the many examples of spontaneously-occurring and drug-induced instances of covert REM sleep there are studies in which REM sleep-related processes have been experimentally activated during NREM sleep by simple sensory stimuli. In animal subjects, auditory stimuli reliably elicit PGO waves in all NREM sleep stages (Bowker & Morrison 1976; Hunt et al. 1998; Sanford et al. 1992b). Auditory stimuli also evoke phasic pauses in diaphragm activity during NREM sleep, another response typically associated with REM sleep (Hunt et al. 1998). There is a general tendency for PGO waves elicited in NREM sleep to have lower amplitudes than those from REM sleep (Ball et al. 1991b) although some studies fail to confirm this difference (Sanford et al. 1992a). In human subjects, combined auditory/visual stimulation during NREM sleep produces an *increase* in the amount of reported dream content (Conduit et al. 1997), a finding that prompted Conduit et al. to propose that the increase may be brought about by activation of REM sleep PGO activity during NREM sleep. Stimulation-induced covert REM sleep may even be exacerbated by REM deprivation because the latter reduces or eliminates inhibitory reactions to auditory stimulation during sleep (Mallick et al. 1991). Studies such as these indicate how easily covert REM sleep processes might be inadvertently triggered in (noisy) laboratory or home situations, and thereby produce elevated levels of sleep mentation reporting from NREM sleep. They may even help to explain instances of stimulus "tagging" in NREM sleep (see sect. 1.2.2) or instances of mentation recalled during sleep terror awakenings (see sect. 3.6).

### 3.12. Genetic factors

Studies of sleep in reptiles, birds, and rare mammals such as the echidna provide examples of apparent mixtures of REM and NREM sleep characteristics (Mukhametov 1987;

Siegel 1998; Siegel et al. 1996). Echidna sleep, for example, consists of high brainstem neuron discharge variability (similar to REM sleep) and high-voltage EEG (similar to NREM sleep) (Siegel et al. 1996). Similarities between such patterns and the sleep of neonates have been noted (Siegel 1998).

## 4. Summary

Evidence from human and animal studies suggests at least nine factors that might induce covert REM sleep to be activated during NREM sleep. These include (1) low-level transitional processes anticipating and following normal REM sleep, (2) sleep onset REM processes during NREM sleep, (3) arousal processes, (4) "omission" of expected REM sleep episodes, (5) sensory stimulation during NREM sleep, (6) REM sleep deprivation, (7) drug effects, (8) mental illness, and (9) genetic factors. Each of these factors and their many possible interactions can be assessed empirically with appropriate experimental designs. In the following section we examine a probabilistic model as it is applied to primarily the first two factors in the preceding list. However, similar probabilistic models could evidently be used to examine any of the factors.

### 4.1. Evaluation of a probabilistic model

Factors 1 and 2 in the preceding section provide the clearest basis upon which the probability of recalling sleep mentation from NREM awakenings can be modeled. If covert REM sleep is indeed linked to (1) NREM sleep immediately preceding and following REM sleep episodes, and (2) NREM sleep following sleep onset, then probabilities of recalling mentation may be calculated from normative architectural measures. To demonstrate this, I employ an average sleep episode calculated from a sample of 127 nights of sleep recorded from 111 healthy, medication-free subjects (55M; 56F; $M_{age}$ = 36.4 ± 14.5 years) in the Sleep Clinic of the Hôpital du Sacré-Coeur de Montréal. The ideal episode combines recordings from 25 first-night recordings and 102 second- or third-night recordings. Nights for which REM sleep onset latencies were greater than 150 minutes were excluded due to the possibility that these implicated "missing" REM sleep periods (see sect. 3.3). Subjects for whom any measure of REM or NREM time exceeded three standard deviations (SDs) of the mean were also excluded.

Table 2. *Descriptive statistics for six consecutive NREM and REM sleep episodes for 111 healthy non-medicated subjects (127 nights)*

| | NREM | | | | REM | | | | BOTH |
|---|---|---|---|---|---|---|---|---|---|
| | Duration | N | SD | % | Duration | N | SD | % | Duration |
| 1 | 84.4 | 127 | 24.8 | 85.7 | 14.1 | 127 | 7.8 | 14.3 | 98.5 |
| 2 | 85.4 | 127 | 22.0 | 78.5 | 23.4 | 127 | 11.4 | 21.5 | 108.8 |
| 3 | 84.0 | 126 | 20.7 | 76.6 | 25.7 | 124 | 13.4 | 23.4 | 109.7 |
| 4 | 68.4 | 116 | 21.8 | 71.1 | 27.8 | 106 | 14.2 | 28.9 | 96.2 |
| 5 | 56.5 | 67 | 19.5 | 68.8 | 25.6 | 49 | 14.8 | 31.2 | 82.1 |
| 6 | 52.3 | 21 | 21.4 | 66.3 | 26.6 | 7 | 13.7 | 33.7 | 78.9 |
| | 71.8 | 97.3 | 21.7 | 74.5 | 23.9 | 90.0 | 12.5 | 25.5 | 95.7 |

Table 3. *Probabilities of observing recall of sleep mentation assuming a 10-min (p-10) or a 15-min (p-15) covert REM sleep "window" around REM episodes (including sleep onset as a REM episode) for six consecutive NREM episodes. Window calculations are provided for mean NREM episode length and for ± 1 SD from this mean*

|  | MEAN | | | + 1 SD | | | − 1 SD | | |
|---|---|---|---|---|---|---|---|---|---|
|  | duration | p-10 | p-15 | duration | p-10 | p-15 | duration | p-10 | p-15 |
| 1 | 84.4 | 0.24 | 0.36 | 109.2 | 0.18 | 0.27 | 59.6 | 0.34 | 0.50 |
| 2 | 85.4 | 0.23 | 0.35 | 107.4 | 0.19 | 0.28 | 63.4 | 0.32 | 0.47 |
| 3 | 84.0 | 0.24 | 0.36 | 104.7 | 0.19 | 0.29 | 63.3 | 0.32 | 0.47 |
| 4 | 68.4 | 0.29 | 0.44 | 90.2 | 0.22 | 0.33 | 46.6 | 0.43 | 0.64 |
| 5 | 56.5 | 0.35 | 0.53 | 76.0 | 0.26 | 0.39 | 37.1 | 0.54 | 0.81 |
| 6 | 52.3 | 0.38 | 0.57 | 73.7 | 0.27 | 0.41 | 30.9 | 0.65 | 0.97 |
| All | 71.8 | 0.29 | 0.44 | 93.5 | 0.22 | 0.33 | 50.1 | 0.43 | 0.65 |

The duration of six consecutive REM and NREM sleep episodes were calculated and averaged over the 127 nights. No differences between men and women were noted so the two groups were combined. Descriptive statistics for these results appear in Table 2.

Probabilities of obtaining covert REM sleep (i.e., of recalling sleep mentation) in NREM sleep were calculated for a 10-min and a 15-min covert REM sleep window surrounding each REM sleep episode (Table 3). These two values were suggested by the literature reviewed above on the time course of covert REM sleep processes. They account for 20 and 30 min of each NREM episode respectively or a total of 120 and 180 min of total NREM sleep over the night. These numbers lead rather straightforwardly to probability estimates of finding covert REM in NREM sleep (Fig. 3). For the six NREM episodes, estimates ranging from 23–38% (mean: 29%) were found for the 10-min window and from 35–57% (mean: 43.5%) for the 15-min window. These percentages may be understood as probabilities of recalling sleep mentation with random awakenings from NREM sleep assuming either a 10- or a 15-min covert sleep window. Note that the 15-min window mean probability is strikingly similar to the average proportion of recall of mentation of 43.0% calculated from the 35 studies in Figure 2 (see also Table 1).

Calculations were repeated for the mean NREM episode length plus and minus 1 SD of this mean (Table 3). For longer NREM episodes (+1 SD), the 10- and 15-min window estimates dropped to 18–27% (mean: 22%) and 27–41% (mean: 33%) respectively. For shorter NREM episodes (1 SD), the two estimates climbed to 34–65% (mean: 43%) and 50–97% (mean: 65%) respectively. Thus, according to this model, with normal variations in NREM sleep episode length we might expect to observe large variations in the recall of sleep mentation – sometimes even exceeding the typical recall rate for REM sleep. This is, in fact, what we observed in the review of 35 studies. Across studies conducted after 1962, in particular, the recall of mentation from NREM sleep had a SD (15.5) that is over twice as large as that from REM sleep (6.7).

The prior calculations would suggest that the covert REM sleep window in human subjects is, on average, close to 15 min in duration. This may be an overly large estimate, given what is known about the time course of many processes preceding REM sleep. However, the value is based upon the assumption that mentation sampling takes place at random from any point in the entire NREM sleep episode. In practice (and in the 35 studies reviewed), researchers sample primarily stage 2 sleep, which tends to immediately precede and follow REM sleep. Calculated only for stage 2 NREM sleep, the probability of finding sleep mentation would be higher and the estimated REM sleep window would be correspondingly lower. In the present normative data set, 72.7% of NREM sleep was stage 2; weighting the 15-minute window by this proportion (.727) produces the more conservative estimate of 11 minutes.

Taken alone, the probabilistic model described here might

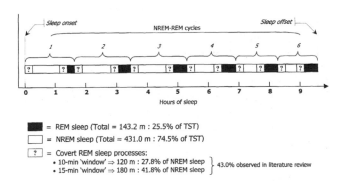

= REM sleep (Total = 143.2 m : 25.5% of TST)
= NREM sleep (Total = 431.0 m : 74.5% of TST)
= Covert REM sleep processes:
• 10-min 'window' ⇒ 120 m : 27.8% of NREM sleep } 43.0% observed in literature review
• 15-min 'window' ⇒ 180 m : 41.8% of NREM sleep

Figure 3. Probability model of covert REM sleep processes over six NREM-REM cycles: Normative results for 111 healthy nonmedicated subjects (127 nights). Illustration (to scale) of the normative sleep results listed in Table 2. The probability of obtaining covert REM sleep processes after a random awakening from NREM sleep may be calculated on a prototypical sleep episode with known architecture, here, a 9.5-hour night with six NREM-REM cycles. It is assumed in the model that covert processes (1) follow sleep onset and (2) precede and follow REM sleep episodes for a fixed duration or "window." The literature suggests a window of 10 to 15 min is possible. For a window of 10 min in length covert REM sleep accounts for 29.0% of NREM sleep. For a 15-min window, the value is 43.5% of NREM sleep. Random sampling of mentation during NREM sleep would thus fall upon covert REM sleep (where dreaming presumably occurs) 43.5% of the time for a 15-min window. Our literature review of mentation recall studies (see Fig. 2 and Table 1) revealed that overall 43.0% of NREM sleep awakenings are accompanied by mentation, a value similar to the postulated 15-min window. When weighted by the proportion of stage 2 sleep in the normative sample (.727), that is, by the stage most often sampled for mentation recall by researchers, the estimated window size can be adjusted to 11 min.

seem too simplistic to account for the numerous observations of mentation in NREM sleep. Evidence of mentation in stages 3 and 4 sleep is particularly difficult for this model to explain. Nevertheless, the large variability in NREM sleep episode length in the present normative sample illustrates the difficulty inherent in attempting to target the "middle" of NREM episodes to avoid possible covert REM sleep effects. One cannot be certain that covert processes anticipating the *next* REM sleep episode are not already active. Such attempts are clearly more likely to succeed from awakenings performed early in the night, but it is precisely at this time that less dreamlike mentation is observed.

In addition, this model does not bear on all factors thought to be associated with covert REM sleep processes, factors that might even trigger such processes unexpectedly in between the REM sleep windows. Studies reviewed earlier suggest that factors such as the intensity of prior REM episodes, extent of REM sleep deprivation, medication use and, especially, sensory stimulation during NREM sleep might evoke covert REM sleep processes. The laboratory itself influences many of these factors – as evidenced by the "first-night" (Browman & Cartwright 1980) and "second-night" (Toussaint et al. 1997) effects – and it may be an important determinant of the timing of covert REM sleep and, thus, of the chance of recalling mentation from NREM sleep. Research by Lehmann and Koukkou (1984) indicates that salient stimuli presented during all sleep stages may induce short-lasting brain states in the range of minutes, seconds or fractions of a second that are associated with discrete changes in cognitive process and EEG field potentials. They speculate that such "meaning-induced" changes in brain micro-state, whether evoked by internal or external stimuli, produce the typical characteristics of sleep mentation. Indeed, it is possible that closer attention to the phasic microstructure of EEG and other physiological variables may reveal measures by which covert REM sleep processes during NREM sleep can be quantified.

In conclusion, it is hoped that this exercise demonstrates how a new view of sleep stages as fluid and interactive, rather than as discrete and independent, may help reconcile a long-standing problem about one versus two imagery generators in sleep. As various phenomena of state overlap and intrusion among normal and sleep-disordered subjects are documented with increasing precision, their consequences for understanding sleep mentation will undoubtedly come into clearer focus. Obviously, not all recall of mentation from NREM sleep can be explained by the present probabilistic model. However, with further refinements, models of this type could account for a substantial portion of the variance in mentation recall. Several other factors, singly and in combination, remain to be more clearly defined, operationalized, and examined in systematic studies.

## ACKNOWLEDGMENTS
This work was supported by the Canadian Institutes of Health Research and the Fonds de la Recherche en Santé du Québec. Dominique Petit, Tyna Paquett, Sonia Frenette and Anne Germain were exceptionally helpful in correcting and commenting on the manuscript. I thank Jacques Montplaisir for permission to publish normative data from the Sacré-Coeur Sleep Center database.

## NOTES
**1.** Antrobus et al. 1995; Aserinsky & Kleitman 1953; Casagrande et al. 1996a; Castaldo & Holzman 1969; Conduit et al. 1997; Dement 1955; Dement & Kleitman 1957b; Fein et al. 1985; Foulkes 1962; Foulkes & Pope 1973; Foulkes & Schmidt 1983; Foulkes & Rechtschaffen 1964; Goodenough et al. 1959; 1965b; Hobson et al. 1965; Jouvet et al. 1960; Kales et al. 1967; Kamiya 1962; Kremen 1961; Lloyd & Cartwright 1995; Moffitt et al. 1982; Molinari & Foulkes 1969; Nielsen et al. 1998; Orlinsky 1962; Pivik & Foulkes 1968; Porte & Hobson 1996; Rechtschaffen et al. 1963a; Rotenberg 1993b; Slover et al. 1987; Snyder 1965; Stoyva 1965; Waterman 1992; Wolpert 1960; Wolpert & Trosman 1958; Zimmerman 1970.

# 4

# The case against memory consolidation in REM sleep

**Robert P. Vertes**

*Center for Complex Systems, Florida Atlantic University,*
*Boca Raton, FL 33431*
**vertes@walt.ccs.fau.edu**

**Kathleen E. Eastman**

*Department of Psychology, Northern Arizona University, Flagstaff, AZ 86011*
**k.eastman@nau.edu**

**Abstract:** We present evidence disputing the hypothesis that memories are processed or consolidated in REM sleep. A review of REM deprivation (REMD) studies in animals shows these reports to be about equally divided in showing that REMD does, or does not, disrupt learning/memory. The studies supporting a relationship between REM sleep and memory have been strongly criticized for the confounding effects of very stressful REM deprivation techniques. The three major classes of antidepressant drugs, monoamine oxidase inhibitors (MAOIs), tricyclic antidepressants (TCAs), and selective serotonin reuptake inhibitors (SSRIs), profoundly suppress REM sleep. The MAOIs virtually abolish REM sleep, and the TCAs and SSRIs have been shown to produce immediate (40–85%) and sustained (30–50%) reductions in REM sleep. Despite marked suppression of REM sleep, these classes of antidepressants on the whole do not disrupt learning/memory. There have been a few reports of patients who have survived bilateral lesions of the pons with few lingering complications. Although these lesions essentially abolished REM sleep, the patients reportedly led normal lives. Recent functional imaging studies in humans have revealed patterns of brain activity in REM sleep that are consistent with dream processes but not with memory consolidation. We propose that the primary function of REM sleep is to provide periodic endogenous stimulation to the brain which serves to maintain requisite levels of central nervous system (CNS) activity throughout sleep. REM is the mechanism used by the brain to promote recovery from sleep. We believe that the cumulative evidence indicates that REM sleep serves no role in the processing or consolidation of memory.

**Keywords:** antidepressant drugs, brain stem lesions; dreams; functional imaging; memory consolidation; REM deprivation; REM sleep; theta rhythm

## 1. Introduction

Although its origin is difficult to establish precisely, the view that memories are processed and consolidated in sleep, or specifically in REM sleep, dates back at least to the report of Jenkins and Dallenbach (1924) claiming that human recall improves following an intervening period of sleep. There was intense interest in the possible role of sleep in memory in the late 1960s to the 1980s as evidenced by the wealth of scientific papers on animals (and to lesser extent on humans) devoted to this issue. The position that memories are consolidated in REM has been championed by, among others, Pearlman (Pearlman 1971; 1978; 1979; Pearlman & Becker 1973), Fishbein (Fishbein 1970; 1971; Fishbein & Gutwein 1977; Gutwein & Fishbein 1980a; 1980b); Hennevin and colleagues (Bloch et al. 1979; Hars et al. 1985; Hennevin et al. 1995b; Leconte et al. 1974), and Smith (1985; 1995; 1996; Smith & Butler 1982; Smith & Kelly 1988; Smith & Lapp 1991; Smith & Rose 1996; 1997).

There was a marked decline in the number of studies devoted to this area beginning about the mid-1980s. As discussed below, the principal reason for this fall-off was that on balance the early work failed to convincingly demonstrate a relationship between sleep and memory. There were as many studies that failed to describe a link between sleep and memory as those that claimed such a relationship (Horne 1988; Horne & McGrath 1984; McGrath & Cohen 1978; Smith 1985).

There has been a renewed interest in the role of sleep and memory stemming in part from two complementary articles that appeared in *Science* in 1994: one by Wilson and McNaughton (1994) on rats and the other by Karni et al.

ROBERT P. VERTES is Professor of Neuroscience in the Center for Complex Systems and Brain Sciences at Florida Atlantic University. His main research interests include sleep, the anatomy and physiology of the brainstem, and subcortical systems controlling the theta rhythm of the hippocampus. He has written several articles and reviews on these topics. He is co-editor of *Brainstem Mechanisms of Behavior*. He currently holds a research career award from the National Institutes of Mental Health.

KATHLEEN E. EASTMAN is Assistant Professor of Psychology at Northern Arizona University. She has written publications in the areas of visual motion perception, pattern recognition, perceptual stability, and the neural basis of visual perception.

(1994) on humans. In a follow-up to a study by Pavlides and Winson (1989), Wilson and McNaughton (1994) reported that ensembles of hippocampal "place" cells tend to repeat patterns of activity of waking in subsequent episodes of slow wave sleep (SWS). Karni et al. (1994) showed that improvement on a visual task in humans depended on REM sleep. The two studies supported the view that memories are consolidated in sleep. It is interesting to note that, propelled by these reports, this area reached national public attention in the United States when Jonathan Winson and Matt Wilson appeared on the Charlie Rose television program explaining and promoting their shared belief that sleep is vital for memory consolidation.

This area has recently received a further boost from Allan Hobson and colleagues who have recently come out in favor of the hypothesis that memories are consolidated in REM sleep (Stickgold et al. 2000b). This recent position seems very much at odds with their earlier proposal, termed the activation-synthesis hypothesis (Hobson 1988b; Hobson & McCarley 1977), claiming that dreams (the cognitive component of REM sleep) represent the best cognitive fit (synthesis) to the undifferentiated and random action (activation) of the brain stem on the forebrain, and as such would have little value to the organism and presumably would not need to be remembered.

As indicated by our title, we do not subscribe to the view that memories are consolidated in REM sleep. This target article evolved from an earlier piece by Vertes (1995) which appeared as part of a series in *Sleep Research Society Bulletin* on the topic of sleep and memory. In the same series, Hennevin et al. (1995a), supporters of a role for sleep in memory consolidation, acknowledged why others may be skeptical of this position. They stated:

> The hypothesis of memory processing in sleep has always had to face criticism both from people working in the field of sleep, who predominantly consider that sleeping serves more basic biological functions, and from people in the field of learning and memory, who do not easily accept the idea that information processing can take place in a non-conscious state.

As researchers involved in both sleep (Vertes 1984; 1990) and memory work (Vertes 1986a; Vertes & Kocsis 1997), we remain skeptical on both counts, largely for the reasons put forth by Hennevin et al. (1995a); that is, sleep involves basic biological functions and memory requires consciousness.

## 2. Background

Memory consolidation refers to neural processing that occurs after information is initially registered, which contributes to its permanent storage in memory (Nadel & Moscovitch 1997). As mentioned, several reports appeared in the 1970s exploring the possible role of sleep in memory consolidation. These studies were of two basic types: (1) examinations of potential increases in REM sleep following heightened experiences in waking; and (2) examinations of the effects of REM sleep deprivation on previously learned tasks. A number of reviews (Dujardin et al. 1990; Fishbein & Gutwein 1977; Horne 1988; Horne & McGrath 1984; McGrath & Cohen 1978; Pearlman 1979; Smith 1985), including recent ones (Hennevin et al. 1995b; Rechtschaffen 1998; Smith 1995; 1996), have been devoted to the topic of sleep and memory. The following is not intended as a re-

view of this area, but rather is meant to serve as a general background and critical assessment of some important issues involving sleep and memory.

### 2.1. Effects of heightened experiences of waking on subsequent REM sleep

The rationale behind this set of studies is as follows: If REM serves to consolidate learning/memory, then exposure to enhanced learning situations or enriched environments in waking should result in increases in REM to process and consolidate these experiences. We will only briefly discuss this work for we do not believe that it represents a particularly powerful test of the REM consolidation hypothesis owing, among other things, to confounding effects of natural variations in REM sleep and the difficulty of establishing, at least for animals, that enriched experiences represent a significant departure from normal routines. Additionally, there is a certain degree of circularity in this position, in that enhanced learning experiences in waking presumably trigger increases in REM to consolidate them, yet they only become "learning experiences" after being processed and consolidated in REM sleep.

The findings of several reports in animals and humans using this paradigm have been mixed. In general, the majority of studies in animals have reported that heightened learning experiences or enriched conditions in waking produce increases in the amount of REM sleep (Horne 1988; Horne & McGrath 1984; McGrath & Cohen 1978; Smith 1985); on the whole, human studies have not shown this to be the case (Allen et al. 1972; Bowe-Anders et al. 1974; Horne 1976; Horne & Walmsley 1976; Zimmerman et al. 1978).

Horne and McGrath (1984) have raised objections to the animal work, pointing out, for instance, that in many of these reports: (1) increases in REM appeared to be an "artifact" of an overall increase in total sleep time (TST); that is, the proportion of REM to TST was not increased (Gutwein & Fishbein 1980a; 1980b; Kiyono et al. 1981; Krech et al. 1962; Mirmiran et al. 1982; Tagney 1973;); and (2) control animals were generally confined to impoverished environments, raising the possibility that differences between control and experimental animals involved decreases in REM (reflecting decreases in TST) for controls rather than increases for the experimental animals (Gutwein & Fishbein 1980a; 1980b; Krech et al. 1962; Tagney 1973).

McGrath and Cohen (1978) reviewed 15 studies in humans examining the effects of enhanced waking experiences on REM sleep (nondeprivation studies) and reported a lack of effect in 10 of the 15 reports. They concluded: "nondeprivation studies employing humans seemingly provide little support for a relationship between REM sleep and learning."

### 2.2. REM deprivation (REMD) studies in animals

REM deprivation (REMD) studies in animals and humans are of two types: prior REMD and post (or subsequent) REMD, reflecting whether the REM deprivation period precedes (prior) or follows (post) the learning situation.

**2.2.1. Post-REMD studies.** Specifically, the post-REMD procedure involves training animals to criterion on a task(s), depriving them of REM sleep for varying periods of time, and then retesting them on the task(s). If REM is critical for learning/memory, REMD should severely disrupt these functions; and if REM is not critical, REMD should have no effect on learning/memory.

The most widely used technique for depriving animals of REM sleep is the water tank (or pedestal) technique. In brief, animals are placed on top of a small pedestal (usually a small inverted flower pot) that is surrounded by water. As animals enter REM sleep, they lose postural tone (atonia), partially or fully slip from the pedestal into the water, and awaken. The procedure is thought to fairly selectively deprive animals of REM sleep. Controls are placed on larger diameter pedestals or allowed normal sleep in their home cages.

It is widely acknowledged that the pedestal technique introduces several spurious and uncontrolled variables that are generally recognized to confound results obtained with this method; these include isolation, wetness, heat loss, high levels of stress, muscle fatigue, and a significant loss of slow wave sleep as well as REM (Coenen & van Luijtelaar 1985; Ellman et al. 1978; Fishbein & Gutwein 1977; Grahnstedt & Ursin 1985; Horne & McGrath 1984; Kovalzon & Tsibulsky 1984; Youngblood et al. 1997). The pedestal technique is a severe method for REMD; alternatives are presently used such as the multiple platform and pendulum techniques (van Hulzen & Coenen 1980; 1982; van Luijtelaar & Coenen 1986) as well as the recently developed diskover-water method of Rechtschaffen and Bergmann (Rechtschaffen 1998; Rechtschaffen & Bergmann 1995).

It appears that the problems inherent in the pedestal technique have significantly clouded findings obtained with it. In fact, Bill Fishbein, an advocate of the REM consolidation hypothesis, recently acknowledged (Fishbein 1995) that he abandoned REMD work in mice because he was not able to respond adequately to criticisms leveled at the technique. He stated that he could not

> have anticipated all the flack that I received, in the years to come, about the "stress factor" produced by the mouse-on-the pedestal technique. I spent a great deal of time trying to prove that there was no stress factor. Despite my efforts to design experiments in a way that training and retention testing were not confounded by the pedestal procedure, it became clear that no matter what control experiment I did, I was never going to convince everyone. Eventually this controversy led me to completely abandon the REM deprivation procedure and look instead at the effects of learning on REM enhancement.

With the caveat, then, that many of the REMD studies supporting a role for REM in memory consolidation may lack validity based on the use of the pedestal technique, a review of the REMD work in animals shows studies to be about equally divided among those showing that REMD disrupted learning/memory (Fishbein 1971; Leconte et al. 1974; Pearlman & Becker 1973; 1974; Smith & Butler 1982; Smith & Kelly 1988) and those showing that this was not the case (Albert et al. 1970; Dodge & Beatty 1980; Joy & Prinz 1969; Miller et al. 1971; Shiromani et al. 1979; Sloan 1972; van Hulzen & Coenen 1979).

As discussed above, it is generally acknowledged that depriving animals of REM sleep with the pedestal technique or other means is debilitating. This has led to the view that the impairments seen following REMD are not true learning/memory deficits but merely performance deficits; that is, animals are simply unable to perform the required task(s), in large part owing to the physically debilitating effects of the deprivation. Attempts to separate learning from performance deficits primarily by looking at short term versus long term effects of REMD have largely shown that impairments are short term or, in effect, performance deficits (Fishbein 1970; 1971; van Hulzen & Coenen 1982).

For example, Fishbein (1971) trained groups of mice on a passive avoidance task, deprived them of REM sleep for 1, 3, 5, or 7 days using the pedestal technique, and then retested them on the task 30 min, 3 h, and 24 h following removal from the pedestal. The results showed that: (1) mice deprived of REM for 1 day showed no impairments at any of the three retest intervals (i.e., 30 min, 3 h, or 24 h) and (2) mice deprived for 3, 5, or 7 days showed marked deficits when retested at 30 min and 3 h but no impairments when retested at 24 h. In essence, mice deprived of REM for 3, 5, or 7 days were very impaired on short term but not on long term retest (i.e., 24 h), indicating that deficits were most likely performance and not learning/memory deficits.

**2.2.2. Prior REMD studies.** A number of reports (Bueno et al. 1994; Danguir & Nicolaidis 1976; Fishbein 1970; Hartmann & Stern 1972; Linden et al. 1975; Sagales & Domino 1973; Stern 1971; van Hulzen & Coenen 1982; Venkatkrishna-Bhatt et al. 1978) have shown that depriving animals of REM sleep prior to training (prior REMD) impairs acquisition/learning on a variety of tasks. These studies, however, do not seem to test the REM consolidation hypothesis since the deprivation period precedes training/acquisition and there is no potential carryover of information pre to post REMD as in the post-REMD design.

Aside from their intended purpose, we suggest that the prior REMD studies support the position that the deficits seen in post-REMD reports were performance and not memory deficits. With both paradigms (prior and post REMD) animals are impaired to similar degrees on the same types of tasks. In the post-REMD paradigm, however, the claim is made that deficits involve the inability of animals to use information learned prior to deprivation, as a direct result of the loss of REM; that is, animals perform poorly following REM deprivation because without REM they are unable to process, store, and utilize information acquired before deprivation to meet the demands of the task – a memory deficit. Although impairments are similar with the prior REMD paradigm, the claim could not be made that this involves a memory dysfunction. We suggest that in both cases the impairments are mainly performance deficits due in large part to the debilitating effects of deprivation procedures. The following report (van Hulzen & Coenen 1982) is consistent with this view.

van Hulzen and Coenen (1982) deprived two groups of rats of REM sleep for three days – one group with the pedestal (or water tank) technique and the other with the less stressful pendulum technique. Immediately following deprivation, both groups were trained on a two-way shuttle avoidance task (acquisition) and then retested six days later. Rats deprived with the pedestal technique showed severe impairments in acquisition but not on retest; those deprived by the pendulum method showed no deficits on acquisition or retest.

The results show that prior REMD by a stressful technique (pedestal), as opposed to a more moderate procedure (pendulum), affects immediate performance, while neither procedure impairs performance/learning when rats are fully recovered from REMD – that is, six days after deprivation. The findings suggest that stress (or other factors) associated with REMD and not necessarily the loss of a particular stage of sleep is largely responsible for the disruptive effects of REMD. This was indicated by the authors when they stated:

> shuttle box avoidance performance [was found] to be severely disrupted following 72 hrs of PS [paradoxical sleep] deprivation by means of the water tank technique. Similar effects could not be replicated in using the pendulum technique. Therefore, the possibility that these phenomena are not due to PS deprivation per se must seriously be considered. (van Hulzen & Coenen 1982)

**2.2.3. Summary and conclusions.** A review of REMD studies in animals shows that they are about equally divided in showing that REMD does or does not disrupt learning/memory. As developed above, it has been argued that reports claiming that REMD disrupts learning/memory are confounded by the use of very stressful deprivation procedures. It appears that stress (and associated factors) rather than the loss of sleep/REM sleep is responsible for the learning/memory deficits seen in these studies. While these reports are open to other interpretations, there appears to be no alternative explanation for studies that fail to show that REMD disrupts learning/memory.

Following a comprehensive review of the REMD literature, Horne (1988) concluded:

> The memory consolidation theories for REM sleep function are having increasing difficulty in handling REM sleep deprivation findings, as it is clear from both animal and human studies that even the longest periods of deprivation do not incapacitate memory, and at best only produce modest decrements.

And further, "In sum, and in relation to the memory consolidation hypothesis for REM sleep, I find the field of REM sleep deprivation and learning in animals unconvincing."

### 2.3. REM windows

Carlyle Smith, a foremost advocate of the REM consolidation hypothesis and a major contributor to this area, has put forth and provided supporting evidence for the existence of "REM windows"; that is, specific segments of REM sleep that are enhanced following learning and corresponding segments which when disrupted (REMD) impair learning/memory. According to the proposal, memories are selectively consolidated during the period of the REM windows (for review, see Smith 1985; 1995; 1996).

The REMD studies of Smith and coworkers focusing on REM windows appear subject to some of the same problems as other REMD studies, foremost of which is the inability to adequately control for the stress factor associated with the use of the pedestal technique for REMD. However, in defense of Smith and colleagues it should be noted that their work is less vulnerable to this criticism because their REMD periods are generally short, about 4–12 h.

On the other hand, there are difficulties with "REM windows" not encountered by other REMD studies. Of significant concern is the shifting nature of the REM window. As readily acknowledged by Smith (1985; 1996), the precise location of the window in REM varies widely, dependent on such factors as species and even strain of animals, the nature of the training tasks, and the number and distribution (concentrated or dispersed) of training trials per session

and/or per day. For instance, in separate reports, the times (post training) of the "REM window(s)" were: 9–12 and 17–20 h (Smith & Butler 1982), 48–72 h (Smith & Kelly 1988), 53–56 h (Smith & MacNeill 1993), 5–8 h (Smith & Rose 1996), and 1–4 h (Smith & Rose 1997). In fact, the last two studies (Smith & Rose 1996; 1997) involved virtually identical conditions (place learning with rats on the Morris water maze) yet the window shifted from 5–8 h in the earlier report to 1–4 h in the later one. Apparently, the only difference was a change from distributed (Smith & Rose 1996) to massed trials (Smith & Rose 1997).

It appears that REM windows (at least as defined for animals) are not present in humans. Smith and Lapp (1991) examined patterns of REM sleep (potential windows) in college students following an intense learning experience (post exams) compared to baseline periods (summer vacation), and reported that aside from an increase in the total number of (rapid) eye movements in test versus control conditions (most prominent in the fifth REM period), there were no changes in sleep/REM sleep under the two conditions. They stated: "No other REM-related measure (minutes of REM sleep, % REM sleep or latency from stage 2 onset to any of the five REM periods) was found to be significant. Further, there were no changes in any of the other sleep parameters measured" (Smith & Lapp 1991).

Finally, although there is some suggestion from recent work in humans that information is differentially processed in distinct phases of SWS and/or REM sleep (Plihal & Born 1997; Stickgold et al. 2000b), to our knowledge "REM windows" has not been independently demonstrated outside of the laboratory of Smith and colleagues (see Smith 1996). It seems that this potentially important phenomenon would be considerably strengthened if confirmed in other laboratories.

### 2.4. REMD studies in humans: Early reports

Compared to their numbers on animals, relatively few reports on humans have examined the effects of REMD on learning/memory. In contrast to the case with animals in which reports were about equally divided among those showing, or not, that REMD affects learning, the majority of studies in humans have described minimal or no effects of REM deprivation on learning/memory (Castaldo et al. 1974; Chernik 1972; Ekstrand et al. 1971; Lewin & Glaubman 1975; Muzio et al. 1972). If anything, complex tasks (Empson & Clarke 1970; Tilley & Empson 1978), as opposed to simple tasks (Castaldo et al. 1974; Chernik 1972), appear to be affected by REMD.

Following a review of early REMD studies in humans, Horne (1988) concluded:

> It is clear that, given before or after learning, REM sleep deprivation does not lead to any greater learning impairment on simple tasks, but difficult tasks are more affected. Whilst these latter findings can reach statistical significance, the effects are still relatively small, and not convincing enough to support any theory that REM sleep has a crucial role to play in the consolidation of memory.

### 2.5. REM sleep and memory consolidation in humans: Recent reports

Karni and Sagi (1993) initially showed that improved performance on a perceptual learning task required the passage of time; that is, subjects showed no improvement immediately following training but marked improvement 8–10 h following training. As discussed below, they have extended their original findings to sleep: performance was shown to improve not only with an intervening period of waking but also of sleep (Karni et al. 1994).

The task involved identifying the orientation of three diagonal lines (arranged either horizontally or vertically) embedded in a background of horizontal lines. The stimulus (target and background elements) was presented briefly (10 msec) in one quadrant of the visual field followed by a blank screen and then a patterned mask (100 msec). The interval between the onset of the stimulus and onset of the mask (stimulus-to-mask onset asynchrony, SOA) was varied, and the measure of performance was an 80% correct identification (threshold SOA) of the stimulus (horizontal or vertical lines) at a set interval. The index of improved performance was a decrease in threshold SOA (Karni & Sagi 1993; Karni et al. 1994).

In the sleep study, Karni et al. (1994) trained subjects on the task and then tested them after a normal night of sleep, sleep without SWS, or sleep without REM. They described significantly improved performance following a normal night of sleep as well as sleep that included REM but not SWS (SWS deprivation condition), but no gains in performance in the absence of REM sleep (REM deprivation condition). Karni et al. (1994) concluded that learning of this perceptual skill was a slow latent process requiring consolidation over time. The period of consolidation could be in waking or sleep, but if in sleep, it required REM sleep not SWS.

Using the identical visual display, Stickgold et al. (2000b) recently reported, like Karni et al. (1994), that subjects exhibited marked improvement on the task following sleep. Specifically, they reported: (1) no improvement on the task over the course of waking; (2) no improvement unless subjects obtained at least 6 h of sleep; (3) improved performance proportional to the total amount of sleep after 6 h of sleep; and (4) improved performance proportional to the amount of SWS in the first quartile of the night (SWS1) and to the amount of REM in the last quartile (REM4). They proposed that learning was a two-step process requiring both SWS (SWS1) and REM (REM4).

Although there are parallels between the two sets of findings (Karni et al. 1994; Karni & Sagi 1993; Stickgold et al. 2000b), there are several pronounced differences. A major difference involves the performance of subjects during waking. As discussed above, Karni and Sagi (1993) originally showed and subsequently confirmed (Karni et al. 1994) that performance significantly improved over time during waking. By contrast, Stickgold et al. (2000b) reported no improvement during post training waking behavior, even after 12 h, commenting: "12 hours of wake behavior was inadequate to produce reliable improvement while as little as 9 hours of sleep reliably produced improved performance."

Additional differences were as follows: (1) Stickgold et al. (2000b) demonstrated a direct relationship between improved performance and total amounts of SWS, particularly SWS1, whereas Karni et al. (1994) showed that depriving subjects of SWS did not alter performance; and (2) Stickgold et al. (2000b) reported that a minimum amount of sleep (6 h) was required for improved performance, and after 6 h gains were proportional to the total amount of sleep; neither was the case in the report by Karni et al. (1994). Un-

til these discrepancies are resolved, it is difficult to evaluate the reliability of the findings using this perceptual learning paradigm.

### 2.6. Theta rhythm and REM sleep

In a variation of the REM consolidation hypothesis, Jonathan Winson has proposed and provided supporting documentation for the position that certain types of memory, specifically memories that are critical for the survival of the species, are selectively processed and consolidated in REM sleep (Pavlides & Winson 1989; Winson 1985; 1990; 1993). The theta rhythm of the hippocampus figures prominently in this proposal (Greenstein et al. 1988; Pavlides et al. 1988; Winson 1972; 1978).

Winson (1972) reviewed the behavioral correlates of the theta rhythm of waking in several species and showed that theta was selectively present during certain behaviors characterized as species-specific behaviors that are critical for survival; for example, exploration in rats, defensive behaviors in rabbits, and predation in cats. In addition, theta is present throughout REM sleep (Vanderwolf 1969).

A number of recent reports (including those of Winson and colleagues) have shown that theta is directly involved in mnemonic functions of the hippocampus (for review, Vertes & Kocsis 1997). For example, it has been demonstrated that: (1) long term potentiation (LTP) is optimally elicited in the hippocampus with stimulation at theta frequency (i.e., 5–7 Hz or pulses separated by 170–200 msec) (Diamond et al. 1988; Greenstein et al. 1988; Larson & Lynch 1986; 1988; Larson et al. 1986; Leung et al. 1992; Rose & Dunwiddie 1986; Staubli & Lynch 1987); (2) stimulation delivered in the presence but not in the absence of theta potentiates population responses in the hippocampus (Bramham & Srebro 1989; Huerta & Lisman 1993; Pavlides et al. 1988); and (3) discrete medial septal (MS) lesions that abolish theta produce severe learning/memory deficits, as do MS lesions with unexplored effects on the hippocampal EEG (Berger-Sweeney et al. 1994; Dutar et al. 1995; Hagan et al. 1988; Hepler et al. 1985; Kesner et al. 1986; Leutgeb & Mizumori 1999; M'Harzi & Jarrard 1992; Mizumori et al. 1990; Poucet et al. 1991; Shen et al. 1996; Stackman & Walsh 1995; Walsh et al. 1996; Winson 1978).

In brief, then, Winson's position is that theta serves to encode survival-enhancing information during waking and to consolidate this information during REM sleep. In this scheme, theta is essential for the acquisition of skills for survival.

The primary focus of the research of the senior author is the theta rhythm of the hippocampus. In fact, the senior author was introduced to this area by Jonathan Winson and remains enormously grateful for the opportunity to learn from him. As is evident, however, we do not share Winson's view that theta is instrumental in consolidating memories in REM sleep.

We believe that the case is strong for the involvement of theta in mnemonic functions of waking but not of REM sleep (Vertes 1986a; Vertes & Kocsis 1997). This seeming discrepancy was recently addressed by Fishbein (1996) stating, "Robert Vertes has published a variety of studies that would lead one to assume he would be a leading champion of the theory of memory consolidation in REM sleep. Despite his important contributions he does not believe the collected evidence supports it."

Our position is that theta of REM is a by-product of the intense activation of the pontine region of the brainstem in REM sleep; theta merely reflects this activation and as such may not have any functional significance in REM or at least not the same functional significance as in waking. In a series of studies (Kocsis & Vertes 1994; 1997; Vertes 1979; 1981; 1988; 1992; Vertes & Martin 1988), we have shown that the theta rhythm is generated by a system of connections from the pontine reticular formation (PRF) to the septum-hippocampus. In brief, cells of nucleus pontis oralis of PRF fire tonically with theta and transfer this tonic barrage to the supramammillary nucleus of the hypothalamus where it is converted into a rhythmical pattern of discharge and then relayed to the GABAergic/cholinergic pacemaking cells of the medial septum to drive theta (Vertes & Kocsis 1997).

As previously described (Datta 1995; Jones 1991; Steriade & McCarley 1990a; Vertes 1984; 1990), pontine and lower mesencephalic regions of the brainstem contain discrete populations of cells that control individual events of REM sleep; when activated together these cell groups trigger each of the major indices of REM sleep (cortical EEG desynchronization, hippocampal theta, muscle atonia, PGO spikes, rapid eye movements, myoclonic twitches, and cardiorespiratory fluctuations), and hence the REM state. Part of this orchestration of activity of the pontine RF in REM involves excitation of nucleus pontis oralis and consequently theta. As argued above, theta of REM may simply reflect a highly activated brainstem in REM, and thus bear little functional relationship to its role in waking.

The presence of similar electrophysiological events in waking and sleep does not indicate that they serve the same (or even similar) physiological and/or behavioral function(s). For example, the cortical EEG desynchronization of waking and REM by no means signifies identical processes in the two states; that is, the EEG desynchronization of waking is associated with diverse sensory, motor, emotional, and cognitive processes that are notably absent in REM sleep.

As indicated, we favor the position that theta is critically involved in memory processing functions of waking (Vertes 1986a; Vertes & Kocsis 1997). Specifically, we propose that theta serves to gate and/or encode information reaching the hippocampus simultaneously with it from various external sources (e.g., the entorhinal cortex). In the awake state, the "information arriving with theta" is governed by the behavioral situation (context); that is, the sum of internal and external events relatively time locked to theta. If theta were involved in memory processing functions in REM, it should, in a similar manner, gate information to the hippocampus in that state. Unlike waking, however, in which the information reaching the hippocampus is dictated by behavioral circumstances, there appears to be no mechanism in REM for the selection and orderly transfer of information to the hippocampus from other sources. If the transfer of information in REM is not orderly, or is essentially chaotic, it would seem that there would be no functional value in consolidating or "remembering" this information. In effect, dream-like material might be presented to the hippocampus in REM, but there would be no purpose in storing or consolidating it during REM. This may be the reason that dreams (or other cognitive material of REM) are so poorly remembered.

In sum, the theta rhythm is present in waking and REM;

we believe that theta serves a mnemonic function in waking but not in REM sleep.

## 3. REM sleep and antidepressant drugs

It is well recognized that virtually all major antidepressant drugs suppress REM sleep (for review, Vogel et al. 1990) and it has, in fact, been proposed that the clinical efficacy of these drugs largely derives from their suppressant effects on REM sleep (Vogel 1975; 1983). The major classes of antidepressant drugs are the monoamine oxidase inhibitors (MAOIs), the tricyclic antidepressants (TCAs), and the recently developed and widely used selective serotonin reuptake inhibitors (SSRIs). A review of the actions of several members of these classes of antidepressants shows that they profoundly suppress REM sleep.

### 3.1. Monoamine oxidase inhibitors (MAOIs)

Of the antidepressants, the MAOIs have the strongest suppressive action on REM sleep. A number of early reports using normal and patient populations showed that MAOIs virtually completely (or completely) suppressed REM sleep for weeks to several months. In an initial study, Wyatt et al. (1969) reported that the MAOIs, isocarboxazid, pargyline hydrochloride, and mebanazine, reduced REM from about 20–25% of TST to 9.7, 8.6, and 0.4% of TST, respectively, and that in one subject REM was virtually eliminated for two weeks.

In a subsequent report in anxious-depressed patients, Wyatt et al. (1971b) described the remarkable findings that the MAOI, phenelzine (Nardil), given at therapeutic doses, completely abolished REM sleep in six patients for periods of 14 to 40 days. There was a gradual decline in amounts of REM sleep for the first two weeks on the drug and a total loss of REM after 3–4 weeks. In a complementary study with narcoleptic patients, Wyatt et al. (1971a) reported that phenelzine completely abolished REM in five of seven patients for the following lengths of time: 14, 19, 93, 102, and 226 days. They stated that: "The complete drug-induced suppression of REM sleep in these patients is longer and more profound than any previously described"; and further that "no adverse psychological effects were noted during the period of total rapid-eye-movement suppression."

Several other studies have similarly shown that MAOIs essentially abolish REM sleep. Akindele et al. (1970) reported that phenelzine completely eliminated REM sleep in four subjects (one normal and three depressed) for 2 to 8 weeks, and addressing possible behavioral consequences stated that, "Far from this leading to disastrous effects on mental functions, as some might have proposed, clinical improvement began." Kupfer and Bowers (1972) showed that phenelzine abolished REM in seven of nine patients, and drastically suppressed it in remaining patients from predrug values of 23.1 and 24.8% of TST to 1.4 and 0.5% of TST, respectively. Finally, Dunleavy and Oswald (1973) reported that phenelzine eliminated REM in 22 depressed patients.

If REM sleep were involved in memory consolidation, it would seem that the total loss of REM with MAOIs for periods of several months to a year (Dunleavy & Oswald 1973; Kupfer & Bowers 1972; Wyatt et al. 1969; 1971a; 1971b) would affect memory. As indicated above, the loss of REM did not appear to be associated with any noticeable decline in cognitive functions in these largely patient populations. These studies, however, made no systematic attempt to assess the effects of MAOIs on cognition.

Other reports, however, have examined the actions of MAOIs, primarily phenelzine, on cognition/memory and described an essential lack of impairment (Georgotas et al. 1983; 1989; Raskin et al. 1983; Rothman et al. 1962). For example, Raskin et al. (1983) observed no adverse effects of phenelzine on a battery of 13 psychomotor and cognitive tasks in a heterogeneous population of 29 depressed patients. Similarly, Georgotas et al. (1983; 1989) reported that elderly depressed patients given phenelzine for 2 to 7 weeks showed no alteration in several measures of cognitive function, and concluded that the lack of adverse effects with phenelzine suggests that it is preferable to TCAs (see below) in the treatment of depression in the geriatric population.

### 3.2. Tricyclic antidepressants (TCAs) and selective serotonin reuptake inhibitors (SSRIs)

As discussed below, virtually all of the commonly used TCAs and SSRIs significantly suppress REM sleep, but unlike the MAOIs, do not eliminate it. Also, the TCAs and SSRIs appear to exert immediate suppressive effects on REM (within the first few days of treatment); by contrast, the MAOIs produce maximal effects on REM about 2–3 weeks following the start of treatment.

An early report by Dunleavy et al. (1972) in normal subjects analyzed the effects on sleep of six TCAs and showed that four of them (imipramine, desipramine, chlorimipramine, and doxepin) markedly depressed REM, beginning with the first night of administration. Chlorimipramine had the strongest suppressive effect on REM sleep, producing a complete loss of REM for the first three nights and an approximate 50% reduction in REM for the remaining four weeks of the study.

Several subsequent examinations of the actions on sleep of these and other TCAs (amitriptyline, amoxapine, nortriptyline, imipramine, maprotiline, clomipramine) have demonstrated that, as a class, TCAs produce an immediate 40–70% reduction in REM and sustained 30–50% decreases in REM sleep (Brebbia et al. 1975; Hartmann & Cravens 1973; Kupfer et al. 1979; 1982; 1991; 1994; Mendlewicz et al. 1991; Nicholson & Pascoe 1986; Passouant et al. 1975; Roth et al. 1982; Shipley et al. 1984; Staner et al. 1995; Ware et al. 1989). Of the TCAs, clomipramine appears to be the strongest REM-suppressant (Passouant et al. 1975; Sharpley & Cowen 1995; Thase 1998).

The SSRIs, like the TCAs, produce an initial marked reduction in REM sleep that slightly abates with time. Examinations of the effects on sleep of several SSRIs (indalpine, fluvoxamine, fluoxetine [Prozac], paroxetine, and zimelidine) show that on average they produce an initial reduction in REM of 40–85% and long term decreases of 30–50% (Kupfer et al. 1991; Nicholson & Pascoe 1986; 1988; Nicholson et al. 1989; Oswald & Adam 1986; Saletu et al. 1991; Sharpley et al. 1996; Shipley et al. 1984; Staner et al. 1995; Vasar et al. 1994; Vogel et al. 1990).

In general, the SSRIs exert stronger suppressive effects on REM than do the TCAs. Staner et al. (1995) compared the actions on sleep of long term treatment with paroxetine (SSRI) and amitriptyline (TCA) in depressed patients, and

showed a 42% reduction in REM with paroxetine compared to a 30% reduction with amitriptyline. Similar findings have been described in other comparisons of these classes of antidepressants (Kupfer et al. 1991; Nicholson & Pascoe 1986).

Although the TCAs and SSRIs do not completely eliminate REM sleep, they significantly suppress it by as much as 75–85% in the short term (days) and 40–50% in the long term (weeks/months). As discussed for the MAOIs, if memories are consolidated in REM sleep, it would seem that the sustained reductions in REM with TCAs/SSRIs would alter memory.

There is a substantial literature describing the effects of TCAs and SSRIs on cognitive functions in normal and depressed subjects, including several reviews devoted to the topic (Amado-Boccara et al. 1995; Deptula & Pomara 1990; Knegtering et al. 1994; Thompson 1991; Thompson & Trimble 1982). Because these classes of antidepressants are in such widespread use, it is obviously important to know if they disrupt motor/cognitive functioning.

### 3.2.1. The effects of TCAs on cognition/memory.
Although there is conflicting evidence, mainly related to the diverse procedures used to evaluate the effects of antidepressants on cognition (Amado-Boccara et al. 1995; Deptula & Pomara 1990; Thompson & Trimble 1982), the general consensus is that some TCAs, primarily amitriptyline, impair memory, but most have minor or no effects on memory (for review, Amado-Boccara et al. 1995; Deptula & Pomara 1990; Thompson 1991; Thompson & Trimble 1982). Virtually all TCAs have some sedative and anticholinergic actions (Hardman et al. 1996), and if cognitive dysfunctions are present with TCAs they reportedly involve these properties (Curran et al. 1988; Deptula & Pomara 1990; Spring et al. 1992; Thompson 1991).

A number of studies have shown that amitriptyline disrupts memory – whether given acutely or long term, to the depressed or nondepressed, and across all age groups (Branconnier et al. 1982; Curran et al. 1988; Lamping et al. 1984; Linnoila et al. 1983; Spring et al. 1992; Warot et al. 1996). For instance, Spring et al. (1992) compared the effects of a four-week treatment with amitriptyline and clovoxamine (an SSRI) on psychomotor and memory tests in depressed outpatients, and reported that amitriptyline, despite alleviating depression, significantly impaired performance on the memory tasks. Clovoxamine, on the other hand, had no adverse effects of psychomotor/cognitive performance (see also below).

Spring et al. (1992) attributed the disruptive effects of amitriptyline on cognition to its anticholinergic actions, noting that, in general, anticholinergics (e.g., scopolamine) disrupt memory (Caine et al. 1981; Drachman & Leavitt 1974). They stated: "The decline in memory performance associated with amitriptyline apparently reflects the relatively high anticholinergic action of the drug, rather than a deficiency in its antidepressant action." And further, "Among the tricyclics, amitriptyline has the most pronounced anticholinergic effects, and would, therefore be expected to have the most adverse effect on memory."

Consistent with this interpretation, Curran et al. (1988) compared the effects on memory of four antidepressants (amitriptyline, trazodone, viloxazine, and protriptyline) that varied with respect to their sedative and anticholinergic properties, and showed that the sedating compounds (amitriptyline and trazodone) but not the nonsedating ones (viloxazine and protriptyline) impaired performance on a battery of memory tests, and that disruptive effects were considerably greater with amitriptyline (an anticholinergic) than with trazodone (no anticholinergic properties) (Gershon & Newton 1980).

In contrast to amitriptyline, most other TCAs have minimal or no adverse effects on memory/cognition. In a well-designed study, Peselow et al. (1991) examined the effects on learning/memory of a four-week treatment with the TCA imipramine (Tofranil) with 50 depressed outpatients, and reported that imipramine improved memory in these patients. Although the improvement in memory was attributed to the clinical efficacy of the compound (not to a memory-enhancing function for imipramine), Peselow et al. (1991) clearly demonstrated, as have several others (Amin et al. 1980; Friedman et al. 1966; Glass et al. 1981; Henry et al. 1973; Raskin et al. 1983; Rothman et al. 1962) that imipramine did not impair memory – even though imipramine is a powerful REM suppressant (Kupfer et al. 1994; Ware et al. 1989). For instance, Kupfer et al. (1994) showed that imipramine produced sustained 35–40% reductions in REM sleep for three years in depressed patients.

Finally, several other TCAs (doxepin, desipramine, nortriptyline, amoxapine, protriptyline, maprotiline, and chlorimipramine) that also suppress REM sleep reportedly produce little or no detrimental effects on memory (Allain et al. 1992; Curran et al. 1988; Georgotas et al. 1989; Liljequist et al. 1974; Linnoila et al. 1983; McNair et al. 1984; Pishkin et al. 1978).

### 3.2.2. The effects of SSRIs on cognition/memory.
As is well recognized, SSRIs are very widely used and currently the most prescribed treatment for depression. As a group the SSRIs do not appear to alter cognitive functions. For instance, there is no indication that any of the following SSRIs have any detrimental effects on psychomotor/cognitive functions in normal or patient populations: fluvoxamine, zimeldine, clovoxamine (an SSRI and partial noradrenergic reuptake inhibitor), sertraline, paroxetine, or fluoxetine (Curran & Lader 1986; Fairweather et al. 1993; 1996; Geretsegger et al. 1994; Hindmarch & Bhatti 1988; Hindmarch et al. 1990; Lamping et al. 1984; Linnoila et al. 1983; Saletu & Grunberger 1988; Saletu et al. 1980; Spring et al. 1992).

Kerr et al. (1992) recently examined the actions of paroxetine, alone or in combination with alcohol, on several psychomotor/cognitive tests in elderly nondepressed subjects with the goal of determining whether SSRIs, unlike compounds with anticholinergic and/or sedative effects, may alter cognitive functions. They speculated that SSRIs "are unlikely to have detrimental cognitive and psychomotor effects because of their unique pharmacological profile," and noted further that "patients often report that treatment with SSRIs leaves them feeling more able to think clearly." It was shown that paroxetine not only had no adverse effects on psychomotor and cognitive functions, but that it slightly ameliorated performance deficits produced by alcohol (Kerr et al. 1992).

Comparisons of the actions of amitriptyline and SSRIs on psychomotor/cognitive performance in healthy or depressed subjects (Curran & Lader 1986; Fairweather et al. 1993; Lamping et al. 1984; Linnoila et al. 1983; Spring et al. 1992) have demonstrated that amitriptyline but not SSRIs

produced significant impairments. Lamping et al. (1984) reported that even though amitriptyline and clovoxamine gave rise to comparable relief from depression, the two antidepressants differentially affected memory; that is, "an impairment of memory after chronic amitriptyline administration, as contrasted with an improvement in memory after chronic administration of clovoxamine." Spring et al. (1992) described virtually the identical findings using the same two compounds.

Finally, an early review of this area (Thompson 1991) concluded that: "Newer compounds devoid of antimuscarinic effects, particularly the serotonin reuptake inhibitors, if not sedative, have not been associated with memory impairment. Furthermore, a few more recent studies suggest that these drugs may exert a beneficial influence on memory processes in memory-impaired individuals"; while a recent review (Amado-Boccara et al. 1995) similarly concluded that: "antidepressants which inhibit serotonin reuptake seem to have no deleterious cognitive effects."

### 3.2.3. Summary of the effects of antidepressants on cognition/memory.
In summary, (1) MAOIs virtually abolish REM sleep but have no adverse effects on cognition/memory. (2) TCAs suppress REM by 30–70%. While amitriptyline, a strong anticholinergic and sedative compound, disrupts memory, most other TCAs produce minimal, or generally no, disruptive effects of cognitive/memory. (3) SSRIs suppress REM sleep by 40–85% but do not alter memory or other cognitive functions.

## 4. Brain stem lesions and REM sleep in humans

Although sizeable lesions at rostral, mesencephalic levels of the brainstem often result in persistent coma or death (Cairns 1952), those located more caudally within the pons are less severe and have been shown to give rise to a condition termed the "locked-in" syndrome. As originally described by Plum and Posner (1966), patients with this syndrome are fully conscious, alert, and responsive, but are quadriplegic and mute. Most of the patients retain the ability to make eye movements and very limited facial/head movements and some can communicate by small facial gestures. For instance, Feldman (1971) described a case of a woman with this syndrome who learned to communicate by Morse code using eye blinks and jaw movements.

A number of reports have examined sleep-wake profiles of these patients, and probably not surprisingly, have shown that most of them (or at least those with bilateral pontine lesions) completely lack REM sleep (Chase et al. 1968; Cummings & Greenberg 1977; Markand & Dyken 1976). For instance, Markand and Dyken (1976) reported that REM sleep was entirely absent in five of seven patients with the "locked-in" syndrome; SWS was present in essentially normal amounts. From case reports, the mental capacities of these patients, including memory for events and people, appear to be intact.

Although rare, there have been a few reports of patients with bilateral pontine lesions who are conscious, ambulatory, and verbally communicative (Lavie et al. 1984; Osorio & Daroff 1980; Valldeoriola et al. 1993). It appears that the lesions in these patients are less extensive than those with the locked-in syndrome. Nonetheless, like patients with the locked-in syndrome, they lack REM sleep (Osorio & Daroff 1980; Valldeoriola et al. 1993). Osorio and Daroff (1980) described two such patients. Both of them showed similar sleep deficiencies, the most prominent of which was a complete loss of REM sleep. It was further pointed out that aside from minor neurological deficits, the patients led normal lives. The authors stated: "Our two patients are the first awake and ambulatory humans in whom total absence of REM sleep has been demonstrated. These REM deprived patients behaved entirely appropriately and were by no means psychotic." The "psychotic" reference alludes to the early notion, subsequently dispelled (Vogel 1975), that long term REM deprivation produces psychosis.

Lavie et al. (1984) described the interesting case of a man who at the age of twenty suffered damage to the pontine region of the brainstem from shrapnel fragments from a gunshot wound. Following the injury, the man was comatose for 10 days, remained in critical condition for another two weeks and then recovered. An examination of his patterns of sleep at the age of 33 revealed that he essentially lacked REM sleep; that is, REM was absent on most nights and averaged 2.25% of TST on the other nights. Similar to the study by Osorio and Daroff (1980), Lavie et al. (1984) reported that despite the virtually total loss of REM sleep, the man led a normal life. For instance, following the injury the man completed college, then law school, and at the time of the study was a practicing attorney.

Although no systematic attempt was made to examine the cognitive capacities of these patients, the virtual total loss of REM sleep did not seem to result in any apparent cognitive deficits.

## 5. Functional imaging studies of brain activity in REM sleep

Recent functional imaging studies of human brain activity in REM sleep reveal patterns of activity that are consistent with dream processes but not with memory consolidation.

The mental/cognitive content of REM sleep is dreams. Although dreams are not restricted to REM, they are unquestionably a prominent feature of REM sleep. Dreams are the sole window to cognitive processes of REM sleep. Although the function(s) of dreams have been, and continue to be, strongly debated (see Revonsuo, this issue), a generally agreed-upon feature of dreams is that they are poorly remembered. Similar to its function, diverse explanations have been put forth to account for the amnesic quality of dreams.

Foulkes and coworkers (Foulkes 1982a; 1985; 1999; Foulkes & Fleisher 1975; Foulkes et al. 1989), leading proponents of the view that dreams are a meaningful extension of waking mental life, have suggested that the reason dreams are so easily forgotten is that the brain in REM sleep is in a reflective mode (akin to reminiscing about, or reflecting on, events during waking) rather than in an encoding mode. An important difference, however, between the reflections of dreams and waking is that during waking one can rapidly switch from the reflective to the encoding mode to integrate and possibly store information. This cannot readily be done in REM sleep and as a result the reflections/reminiscences of REM (dreams) are lost to memory (Foulkes 1985; Foulkes & Fleisher 1975).

At the opposite end of the spectrum to the position of Foulkes and others (Domhoff 1969; 1996; Domhoff & Schneider 1998; Hall & Van de Castle 1966; Van de Castle

1994) that dreams are logical and meaningful, Hobson and colleagues (Hobson 1988b; Hobson et al. 1998b) have argued that dreams can be defined by such characteristics as hallucinosis, bizarreness, delusion, and confabulation and have likened dreams to the "delirium of organic brain disease" (Hobson 1997b). Hobson et al. (1998b) have proposed a purely physiological explanation for the amnesia of REM, pointing to the likely correspondence between memory loss and underlying physiological changes in REM, stating: "The loss of memory in REM sleep makes dreaming consciousness much more difficult to recall than waking consciousness. This phenomenological deficit logically implies a physiological deficit: some functional process, present and responsible for memory in waking is absent, or at least greatly diminished, in REM sleep."

Independent of theories of dreams, recent functional imaging studies in humans during sleep have revealed patterns of activity in REM that appear to reflect dream processes, including its amnesic quality. Although differences exist among reports (Braun et al. 1997; 1998; Maquet et al. 1996; Nofzinger et al. 1997), a fairly consistent pattern of brain activity in REM sleep in humans has emerged from these studies. Some important findings are as follows: (1) the pontine reticular formation is highly active in REM sleep; (2) primary sensory areas (e.g., striate cortex for the visual system) are inactive in REM; by contrast, extrastriate (visual) regions (as well as other sensory association sites) are very active in REM; (3) limbic and paralimbic regions, including the lateral hypothalamus, the amygdala and anterior cingulate, and parahippocampal cortices, are intensely activated in REM; and (4) widespread regions of the frontal cortex including the lateral orbital and dorsolateral prefrontal cortices show marked reductions in activity in REM sleep (Braun et al. 1997; 1998; Maquet et al. 1996; Nofzinger et al. 1997).

This general pattern of activity in REM has been viewed as a "closed system" (Braun et al. 1998); essentially, an internal network disconnected from inputs and outputs. For instance, the suppression of activity in the primary visual cortex (input) is consistent with the well-characterized sensory blockade of REM, whereas the deactivation of the prefrontal cortex (output) parallels the failure of dreams to influence executive systems for behavior. With respect to the latter, Braun et al. (1997) stated: "REM sleep may constitute a state of generalized brain activity with the specific exclusion of executive systems which normally participate in the highest order analysis and integration of neural information."

In effect (and not unexpectedly), the brain in REM sleep mirrors the dreaming brain; that is, internally generated visual images are fed to (or recruited by) the limbic system. They are then incorporated into dreams but due to the suppression of activity of the prefrontal cortex dream scenarios are not often recorded and generally do not influence waking behavior. In this regard, in an article on the neural basis of consciousness, Jones (1998) commented that the recent demonstration in imaging studies (Braun et al. 1997; Maquet et al. 1996) that activity in the frontal cortex is depressed in REM suggests "an attenuation of processes important in episodic and working memory and perhaps explaining why unless awakened from a dream, a sleeping person has no memory of the dream."

Finally, if dream material is so readily forgotten in REM sleep (reflecting the state of the brain in REM), it seems unlikely that other mental phenomena that are not incorporated into dreams would be processed and permanently stored during REM sleep.

In summary, the pattern of brain activity in REM sleep is consistent with dreams but inconsistent with the orderly evaluation, organization, and storage of information which is the domain of attentive, waking consciousness.

## 6. A proposed function for REM sleep

It appears that the active state of the brain during REM has fueled claims that REM sleep is involved in complex, higher order functions, including memory (for review, Rechtschaffen 1998).

It is tempting to speculate, as several theories do, that magical processes occur during REM sleep; that is, that during the unconscious state of REM sleep some programmed or purposeful reordering of mental events occurs so that a nightly replay of daytime events during REM enhances the storage or consolidation of these events. In contrast to the view that the effects of REM extend beyond sleep to influence waking activities, we propose that REM can be entirely understood within the context of sleep without invoking mental phenomena or quasi-conscious processes (for review, Vertes 1986b). REM is a state of sleep; as such, it would seem that attempts to describe its function should look to sleep and not to waking.

As described in detail in our earlier theoretical paper (see Vertes 1986b), we propose that the primary function of REM sleep is to provide periodic endogenous stimulation to the brain which serves to maintain minimum requisite levels of CNS activity throughout sleep. REM is the mechanism used by the brain to ensure and promote recovery from sleep. We argued that the brain is strongly depressed in SWS, particularly in delta sleep, and incapable of tolerating long continuous periods of relative suppression. REM serves the critical function of periodically activating the brain during sleep without awakening the subject or disturbing the continuity of sleep. By analogy, the process of induction and recovery from general anesthesia is a delicate one requiring the special skills of highly trained medical professionals. The brain performs a very similar function daily and seemingly flawlessly. REM is an integral part of this process.

Our theory is consistent with sleep state organization; the main elements of which are that: (1) the percentage of REM sleep is very high in early infancy (about 50% of total sleep time) and declines sharply at 2–3 months of age; (2) sleep continuously cycles from light to deep sleep and back to lighter stages of sleep as the cycle repeats itself; and (3) REM sleep is quite evenly distributed throughout sleep (occurring about every 90 minutes) and the duration of REM periods become progressively longer throughout sleep.

Regarding this organization, we would suggest that the high percentage of REM sleep in neonates serves to offset equally high amounts of SWS in newborns (see also, Benington & Heller 1994); that sleep cyclically alternates between light and deep sleep to prevent the brain from dwelling too long in deep SWS; and that the progressively longer periods of REM throughout sleep serve to prime the brain for a return to consciousness as waking approaches. With respect to the latter, the disorientation experienced on sudden, unexpected awakenings from sleep (middle of the night), compared to natural awakening, may reflect an inadequate preparation of the brain for waking due to incomplete REM.

In line with the foregoing, reductions in REM, seen particularly with antidepressants, are generally accompanied by a reorganization of sleep; that is, marked increases in light SWS and corresponding decreases in deep SWS as well as frequent awakening (Cohen et al. 1982; Kupfer et al. 1989; 1991; Nicholson & Pascoe 1988; Saletu et al. 1983; 1991; Schenk et al. 1981; Shipley et al. 1984; Staner et al. 1995; Wyatt et al. 1971b). For the SSRIs, this has been referred to as the "alerting" effect on sleep of these antidepressants (Kupfer et al. 1989; 1991; Nicholson & Pascoe 1988; Saletu et al. 1983; 1991; Schenk et al. 1981; Shipley et al. 1984; Staner et al. 1995).

In accord with others (Benington & Heller 1995; Berger & Phillips 1995), we believe that the general purpose of sleep is restitution/recuperation for the CNS, and within this context, the primary function of REM sleep is to prepare the brain/CNS for recovery from sleep.

## 7. Conclusions

We believe that the evidence reviewed in this report disputes the claim that REM sleep serves a role in the consolidation of memory. Numerous studies have shown that depriving animals of REM sleep has no effect on learning/memory. Although other reports have shown that REM deprivation (REMD) disrupts memory, many of them have been questioned based on the use of the stressful pedestal technique for REMD leading to the view that reported deficits were performance and not learning/memory deficits. The majority of REM deprivation studies in humans have failed to show that REMD disrupts memory. Perhaps the strongest evidence against the memory consolidation hypothesis comes from the demonstration that antidepressant drugs or brain stem lesions profoundly suppress, or eliminate, REM sleep, yet neither appears to alter memory/cognitive functions. Finally, recent imaging studies in humans during sleep have described patterns of activity that are consistent with dreams, including their amnesic quality, but inconsistent with the orderly processing, evaluation, and storage of information that characterizes waking consciousness. In conclusion, we believe that the weight of evidence, as reviewed herein, fails to support a role for REM sleep in the processing or consolidation of memory.

ACKNOWLEDGMENTS
We thank Alison M. Crane for her critical reading and constructive comments on the manuscript. This work was supported by NIH Grant NS35883 and NIMH Research Career Scientist Award MH01476 to RPV.

# 5

# The reinterpretation of dreams:
# An evolutionary hypothesis
# of the function of dreaming

**Antti Revonsuo**

*Department of Philosophy, Centre for Cognitive Neuroscience, University of Turku, Turku FIN-20014, Finland*

**revonsuo@utu.fi    www.utu.fi/research/ccn/consciousness.html**

**Abstract:** Several theories claim that dreaming is a random by-product of REM sleep physiology and that it does not serve any natural function. Phenomenal dream content, however, is not as disorganized as such views imply. The form and content of dreams is not random but organized and selective: during dreaming, the brain constructs a complex model of the world in which certain types of elements, when compared to waking life, are underrepresented whereas others are over represented. Furthermore, dream content is consistently and powerfully modulated by certain types of waking experiences. On the basis of this evidence, I put forward the hypothesis that the biological function of dreaming is to simulate threatening events, and to rehearse threat perception and threat avoidance. To evaluate this hypothesis, we need to consider the original evolutionary context of dreaming and the possible traces it has left in the dream content of the present human population. In the ancestral environment human life was short and full of threats. Any behavioral advantage in dealing with highly dangerous events would have increased the probability of reproductive success. A dream-production mechanism that tends to select threatening waking events and simulate them over and over again in various combinations would have been valuable for the development and maintenance of threat-avoidance skills. Empirical evidence from normative dream content, children's dreams, recurrent dreams, nightmares, post traumatic dreams, and the dreams of hunter-gatherers indicates that our dream-production mechanisms are in fact specialized in the simulation of threatening events, and thus provides support to the threat simulation hypothesis of the function of dreaming.

**Keywords:** dream content; dream function; evolution of consciousness; evolutionary psychology; fear; implicit learning; nightmares; rehearsal; REM; sleep; threat perception

## Introduction

Dreaming is a universal feature of human experience, but there is no convincing explanation as to why we should experience dreams during sleep. Why do we have vivid, intense, and eventful experiences while we are completely unaware of the world that physically surrounds us? Couldn't we just as well pass the night completely nonconscious? The function of dreaming seems to be a persistent mystery, although numerous suggestions have been put forward about the possible functions it might serve. The leading neurocognitive theories, however, seem to have given up the hope of identifying any useful function for dreaming at all. They cannot provide us with an answer to the question "Why do we dream?" Instead, they seem to imply that we dream for no particular reason at all: Dreaming is biologically epiphenomenal. Dream consciousness is viewed as some sort of random noise generated by the sleeping brain as it fulfills various neurophysiological functions during REM (rapid eye movement) sleep.

Although the prospects for discovering useful functions for dreaming look rather bleak, the empirical evidence should be reevaluated once more from a truly multidisciplinary point of view, including dream content analysis, the neurophysiology of dream sleep, and evolutionary psychology. The exploration that I undertake in the present tar-

get article leads to the slightly surprising conclusion that dreaming does have a well-defined and clearly manifested biological function after all. In section 1, I clarify the nature of the basic question: What exactly is it that we want to understand when we inquire about the function of dreaming? The answer is that we need a clear idea of both what the phenomenon of dreaming is and of the sense in which we are using the word "function." In section 2, we review the currently dominant views on the function of dreaming in

ANTTI REVONSUO is a Fellow of the Academy of Finland at the University of Turku. He has published widely in cognitive neuroscience, neuropsychology, and consciousness studies. His research aims at understanding consciousness as a natural biological phenomenon and at fruitful interaction between philosophical and empirical research in the study of consciousness. He is co-editor of two books on consciousness, *Consciousness in philosophy and cognitive neuroscience* (Erlbaum, 1994) and *Beyond dissociations: Interaction between dissociated implicit and explicit processing* (Benjamins, 2000). Revonsuo is the European Editor of *Consciousness and Cognition*, and currently on the board of Directors of the Association for the Scientific Study of Consciousness.

the cognitive and neuroscientific literature as well as in the more clinically oriented dream psychology. The most common view in cognitive neuroscience is that dreaming has no function whatsoever. In clinical literature, the function of dreaming has been linked with problem solving and psychological adaptation, but the direct empirical evidence bearing on such functions remains scarce. In section 3 we point out that none of the previous theories have placed dreaming in the appropriate context for evaluating its possible biological functions: the human ancestral environment in which the dreaming brain was evolving for hundreds of thousands of years. If dreaming does have any biologically adaptive functions, they must have been effective in the evolutionary context, if anywhere.

In the rest of the article I argue that switching the context in such a way puts dreaming into an entirely new light, which suggests that the biologically adaptive function of dreaming is to simulate threatening events in order to rehearse threat perception and the appropriate threat-avoidance skills and behavioral programs. I emphasize that to claim threat simulation as the biological function of dreaming is not to claim that every single dream of every single individual should realize this function. It is only to claim that in certain adaptively important situations with certain ecologically valid cues, the system does become fully activated, and this is the principal reason why dreaming was selected for during our evolutionary history.

The threat simulation theory of dreaming is expressed here in the form of six propositions, each of which is empirically testable. The propositions can be summarized as follows:

1. Dream consciousness is an organized and selective simulation of the perceptual world.

2. Dream consciousness is specialized in the simulation of threatening events.

3. Nothing but exposure to real threatening events fully activates the threat-simulation system.

4. The threat simulations produced by the fully activated system are perceptually and behaviorally realistic rehearsals of threatening events.

5. The realistic rehearsal of these skills can lead to enhanced performance regardless of whether or not the training episodes are explicitly remembered.

6. The ancestral environment in which the human brain evolved included frequent dangerous events that constituted extreme threats to human reproductive success. They thus presented serious selection pressures to ancestral human populations and fully activated the threat-simulation mechanisms.

The empirical evidence relevant for the evaluation of each proposition is then reviewed (sect. 3). In the light of the currently available evidence, all of the propositions are judged as likely to be true, which consequently lends support to the threat-simulation theory of dreaming as a whole. In section 4, the dreams of hunter-gatherer populations and animals are considered in the light of the threat simulation theory. In section 5, new predictions are derived from the theory and the empirical testability of the theory is evaluated. Finally, the theory is elaborated upon and summarized in section 6.

After presenting the threat simulation theory, other theories that have taken an evolutionary perspective on dreaming are reviewed. Although some of them are related to the present view, none of them includes the idea that dreaming is a threat-simulation mechanism. In the final section, the theory is compared with neurocognitive theories of dreaming.

Taken together, this target article aims to show that the threat-simulation theory of dreaming integrates a considerable body of data from multiple sources in a theoretically meaningful way. The theory treats the conscious phenomenal experience of dreaming as a natural biological phenomenon best understood from the combined viewpoints of psychology, evolutionary biology, and cognitive neuroscience. This multidisciplinary treatment, I hope, manages to clarify the mystery of why we dream.

# 1. What is it that we want to understand when we inquire about the function of dreaming?

We should first make clear what it is we are asking when we inquire about the function of dreaming. We must explicate what we mean by dreaming and what we mean by function.

## 1.1. What is dreaming?

Dreaming refers to the subjective conscious experiences we have during sleep. We may define a dream as a subjective experience during sleep, consisting of complex and organized images that show temporal progression (Farthing 1992). Questions regarding the function of dreaming must be clearly distinguished from those regarding the function of REM sleep. Dreaming is a subjective conscious experience, while REM sleep is a physiologically defined stage of sleep. Furthermore, as is now clear, REM sleep is neither a necessary nor a sufficient physiological condition for dreaming, although it seems to be the typical and perhaps optimal physiological condition in which fully realized dreams are brought about (Pivik 1991). As Foulkes and Cavallero (1993, p. 9) emphasize, dreaming needs a level of explanation independent of the neurophysiological level at which REM sleep is defined, because "there almost certainly is REM sleep without dreaming and . . . there certainly is dreaming without REM sleep. No account of the distinctive physiology of REM sleep could provide either a necessary or a sufficient explanation of dreaming." Thus, the question we will be exploring is: Does it serve any useful function to have, during sleep, the sorts of conscious subjective experiences that dreaming consists of?

In order to make it clear that we distinguish the level of description at which dreaming proper resides from the levels of neurophysiological description, we may say that dreaming is realized at the experiential or *phenomenal* level of organization in the brain (Revonsuo 1997). We want to find out whether the realization of this level of organization during sleep serves any natural function. The specification of the functions that lower-level neurophysiological mechanisms serve during REM sleep does not constitute a specification of the functions that the realization of the phenomenal level serves, for the neurophysiological functions can be fully specified without ever mentioning the fact that subjective experience happens to be realized as well.[1]

## 1.2. What is it to be "functional"?

We must be clear about what we mean by "function" or "functional." The appropriate sense of "function" in this context is that of a biological, adaptive function. According to

Tooby and Cosmides (1995) the biological standard is the only standard of functionality that is relevant to analyzing why brain and cognition are organized in one fashion rather than another. A cognitive system is functional in the evolutionary sense if and only if it promotes the organism's inclusive fitness. That is, the biologically functional system must solve problems that will increase the probability that the organism possessing the system will produce offspring, or that the organism's kin will produce offspring. Evolutionary biology gives the concept of "function" a very specific content: The function of a system solely refers to how it systematically caused its own propagation in ancestral environments (Tooby & Cosmides 1995). If dreaming has an adaptive function, then dreaming must solve some adaptive problems whose solution tends to enhance survival and promote reproduction, thus causing the persistence of the brain's dream-production mechanisms and their spread in the population.

If dreaming does not have any adaptive function of its own, then it is likely to be coupled to properties that do. In that case, dreaming is a mere by-product, a nonadaptation that was not selected for (or against) during our evolutionary history but was dragged along because the features to which it was coupled were actively selected for. Flanagan (1995) makes an important distinction between "natural" and "invented" functions of dreaming. A similar distinction has been made by other dream theorists between what we do with dreams once we recall them, and what the dream can do itself (Blagrove 1996; Breger 1967). Natural functions are biological, adaptive functions in the sense defined above, whereas invented functions are derivative psychological or cultural functions. We can put our recalled dreams to a variety of personal or cultural uses,[2] but no matter how enlightening and meaningful such uses may be, they are invented by us, not by natural selection. It is doubtful that any truly natural function of dreaming could be based on the conscious recollection or verbal reporting of dream content, for the natural functions of dreaming, if any, must have been effective in such ancestral conditions and species in which self-reflective dream recollection or reporting were not likely to occur – thus, the natural functions of dreaming cannot have been dependent on them.

Now we are in the position to state our question more specifically. The question we are presently interested in is whether dreaming serves any natural functions: Does the realization of the subjective phenomenal level of organization (the experience of dreaming) solve any adaptive problems? That is, does phenomenal dreaming in any way enhance the prospects of the reproduction of the individual (and/or its close relatives); does dreaming increase the inclusive fitness of the individual?

## 2. Current theories of dream function

### 2.1. Theories in cognitive neuroscience

In cognitive neuroscience, recent theories and views on dreaming have led to the conclusion that dreaming as a conscious experience does not serve any useful biological function. Only the neurophysiological events associated with dreaming and REM sleep are assumed to be biologically functional, for they may serve important functions in the development of the brain and in periodically restoring the brain's neurochemical balance.

The Activation-Synthesis theory (Hobson 1988b; Hobson & McCarley 1977) emphasizes the randomness of dream imagery. During REM sleep, PGO waves originate in the pons and activate the forebrain. The forebrain attempts to make sense of this random activation and it synthesizes dream images to fit the patterns of internally generated stimulation. The forebrain selects images that isomorphically correspond to the patterns of eye movements and motor commands elicited during REM sleep. The images are loaded from memory, in which day residues are particularly salient. The theory delivers no answer to the question *why* the brain should generate any images at all during REM sleep; it is simply assumed to be an automatic process. The narrative content of dreams remains unexplained as well. More recently, Hobson (1994) has suggested that REM-dreaming might have a function in memory processing, and he specifically regards the rehearsal of motor programs as a possible function of dreaming during REM sleep. In Hobson's theory, however, dreaming *as an experience* with vivid phenomenal content is seen as a kind of random epiphenomenon that merely *reflects* some totally different events going on at other levels of organization where such events may serve useful neurobiological or mnemonic functions. The Activation-Synthesis theory suggests that the experiential dream imagery itself, the content of consciousness, is functionally as aimless as are the noises emitted by a computer when it processes information. The phenomenal level of organization is not regarded as biologically functional.

The theory presented by Crick and Mitchinson (1983; 1995) is related to Hobson's views, but contains some original ideas. In this theory, memory in the brain is compared to simple models of associative nets. When such a net gets overloaded, it easily starts to produce outputs that are combinations of actually stored associations. In order to make storage more efficient and avoid overloading, a process of reverse learning can be used. The net is disconnected from its normal inputs and outputs, and random input is given to it. The associations that this random input produces are consequently weakened, and the process is repeated many times with different kinds of random input. According to Crick and Mitchinson (1983; 1995) this is loosely analogous to what happens in the brain during REM sleep: the brain is disconnected from its usual inputs and outputs, and PGO waves provide it with more or less random input.[3] The theory explains why REM dreams are full of bizarre intrusions, consisting of mixtures of features previously stored in memory: these are the associations arising in an overloaded network and have to be unlearned. The reverse-learning theory does not even try to explain the narrative aspect of REM dreams, and it certainly does not assign any independent function to the phenomenology of dreaming; phenomenal dream images merely reflect the functioning of a memory-cleaning process.

David Foulkes (1985) has put forward a cognitive theory of dreaming. He proposes that dreaming originates in diffuse, more or less random activation of semantic and episodic memory during sleep: "Since it seems that the activation of mnemonic elements during dreaming and their selection for dream processing is random and arbitrary, it's not likely that the *particular content of our dreams* – in and of themselves – serve any adaptive functions" (Foulkes 1985, p. 200).

Foulkes, however, distinguishes dream content from dreaming as a process. Dreaming, unlike specific dream contents, has very predictable features. It involves an interrelated sequence of events occurring within a "world analog" (or a model of the world) composed of integrated multimodal sensory imagery; the dreamer participates in these events actively and personally; the contents and events depicted in the dream are related to the recent or distant past of the dreamer, not as a simple replay of a past experience but rather as a variation of the past as something that really could have happened to the dreamer. Foulkes suggests that, since the content of dreams seems to be random, what is important about the mnemonic activation is that it is in *some* way unique, not the precise way in which it is unique. In Foulkes's theory the phenomenal level of organization is not regarded as functional, apart from the general feature of producing novel and unique mnemonic configurations. Thus, Foulkes's theory is not essentially different from Hobson's as to the functionality of phenomenal dream content.

Solms (1997a) has recently defended the view originally proposed by Freud: the function of dreaming is to protect sleep. According to Solms, the dream process begins when external or endogenous stimuli activate "the curiosity-interest-expectancy circuits." Inhibitory mechanisms prevent the "appetitive interest," aroused by stimulation, from leading to motor activity; therefore the activity proceeds "regressively" in the direction of hallucinations. In anxiety dreams this mechanism of sleep protection fails. It is clear that this view does not attribute any functions to the specific content of dreams: Solms regards dreams simply as bizarre hallucinations that the weakened frontal reflective systems mistake for real perception.

Owen Flanagan (1995) explicitly denies that dreams as conscious experiences have any biological function. Dream experience, or *p*-dreaming (phenomenal dreaming) as Flanagan calls it, is "a likely candidate for being given epiphenomenalist status from an evolutionary point of view. *P*-dreaming is an interesting side effect of what the brain is doing, the function(s) it is performing during sleep. To put it in slightly different terms: *p*-dreams, despite being experiences, have no interesting biological function. I mean in the first instance that p-dreaming was probably not selected for, that *p*-dreaming is neither functional nor dysfunctional in and of itself" (Flanagan 1995, pp. 9–11). Flanagan argues that phenomenal experience during dreaming – dream consciousness – has no adaptive significance, because the functions of REM sleep and PGO waves, in early development of the visual system and in the restoration of neurochemicals for the next waking period, do not in any way require mentation of any sort. Furthermore, dream thoughts associated with such biological functions do not seem to be worth remembering. "The visual, auditory, propositional, and sensory-motor mentation that occurs is mostly noise" (p. 24). Antrobus (1993a) seems to agree with Flanagan's analysis. He says that since in REM sleep no sensory information is processed and no association-motor commands are executed, it makes no difference what the association cortex does. Dreaming has no maladaptive consequences, so it has survived.

In conclusion, theorists in cognitive neuroscience tend to regard the phenomenal content of dreaming as a biological epiphenomenon, although at least some of the (nonconscious) cognitive and/or neural activity during REM sleep are regarded as serving useful functions.

### 2.2. Theories in dream psychology

In psychological theories of dream function, the emphasis is on the individual person's psychological adaptation to his current waking life. The basic assumption behind this approach seems to be that dreaming is functional for the individual if the dream in some way helps the individual cope with his current waking concerns, solve current problems, and to promote psychological well-being. These views can be traced back to Jung (1933) who argued that dreaming helps to maintain the individual's psychic balance and Adler (1927) who believed that dreaming serves a personal problem-solving function.

These types of theories of the psychological function of dreaming can be divided into two categories. The first holds that dreaming has a problem-solving function in an intellectual or cognitive sense: The function of dreaming is to find solutions to (or to facilitate the solving of) intellectual problems. The second holds that the function of dreaming is related to emotional adjustment, not to intellectual problems. Any real-life event that can be considered an emotional concern for the dreamer can be seen as presenting a problem for psychological adjustment, and dreaming is assumed to contribute to the emotional or behavioral adjustment that is called for in order to solve the emotional problem (e.g., Breger 1967).

**2.2.1. Do dreams solve intellectual problems?** Some studies have directly addressed the question of whether we can solve intellectual problems in our dreams or with the help of them. Dement (1972) reports a series of experiments in which 500 undergraduate students were given a copy of a problem, and before going to bed the students were to spend exactly 15 min trying to solve the problem. In the morning, they wrote down any dreams they recalled from the previous night and, if the problem had not been solved, spent another 15 min trying to solve it. In 1,148 attempts, the problem was solved in a dream on only seven occasions. This means that less than 1% of the dreams were successful in solving the problem. Montangero (1993) reports a sleep laboratory experiment with six subjects. Four subjects were given a formal problem, while two were trying to solve an intellectual problem relating to their own professional careers. Although elements of the problems appeared in the dreams, none of the 29 reported dreams presented the solution to the problem. However, the subjects did find the solutions to the problems with relative ease during the first hour after awakening in the morning. Unfortunately, it remains unclear whether dreaming causally contributed to this problem-solving success at all. Cartwright (1974a) compared solutions to problems arrived at either after a period of REM sleep or an equivalent amount of waking. She concluded that "There is no evidence from this study that a period of sleep during which dreaming occurs is regularly followed by a better performance on intellectual tasks" (p. 454). In a study by Barrett (1993) the subjects were allowed to choose the problem that they tried to solve in their dreams. The results showed that problems of a personal nature were much more likely to find a solution through dreaming than problems of an academic or intellectual nature. The personal problems, however, lacked definitive criteria for what should count as a solution, raising the suspicion that at least some of the alleged solutions may have been attributed to the dream during retrospective reflection required during the reporting rather than having been arrived at within the dream itself.

Blagrove (1992a) presents a thorough review and critique of the problem solving paradigm of dream function. The assumption behind this paradigm is that the function of dreaming is to work actively and creatively toward solutions to actual current waking problems, thus going beyond what was known prior to the dream and causally contributing to the solution of a real-life problem. In order to evaluate the evidence for such claims, Blagrove distinguishes three types of problem-solving dreams: (1) Dreams that actually create a new and useful solution to a current problem in waking life; (2) Dreams that contain problem-solving activity that is internal to problems encountered in the dream world, but not relevant to waking problems; (3) Dreams that reflect solutions to waking problems, but for which there is no evidence that such solutions have not already occurred to the waking mind (i.e., the dream does not contribute to the solution, it merely reflects the solution once it has already been found during waking). Blagrove (1992a) argues that there is little evidence for problem-solving dreams of the first type; most of the dreams apparently solving problems either simply reflect solutions already known or solve problems only relevant in the context of the dream. Although a psychological change may be *correlated* with a dreamed solution to a problem, there is little reason to believe that there is a *causal* relationship between them. It is most likely that the actual solution first arises during waking, and the consequent dreaming merely reflects the solution, and thus becomes correlated with whatever the beneficial consequences of the solution were. The conclusion from Blagrove's (1992a) review is that whatever the function of dream experience is, it does not appear to be the finding of new and useful solutions to the problems we face in our waking reality.

### 2.2.2. Do dreams solve emotional problems?

Probably the most popular theory of dream function within psychology is the hypothesis that dreaming solves our emotional problems by helping us to adjust psychologically to, and maintain our mental health in, the real-life situations that trouble us emotionally and psychologically. There is an overwhelming amount of evidence showing that dream content indeed reflects the current emotional problems of the dreamer (Hartmann 1998; Kramer 1993). The question is: Does dreaming have an effect in reducing the negative affect and other negative psychological consequences induced by our real-life troubles and traumas?

Cartwright (1996) argues that the best way to test this hypothesis empirically is to study subjects who are undergoing a life event that creates genuine affect. She studied subjects undergoing marital separation. Seventy subjects were chosen from a group of 214 potential subjects. Forty of them were depressed as a consequence of the divorce. All subjects slept for three nights in the laboratory, and during the third night, REM dream reports were collected. The depressed subjects' dreams were emotionally more negative than those of the nondepressed subjects. Furthermore, the depressed subjects were more likely than the nondepressed to incorporate the about-to-be-former spouse as a character in the dreams. In a one-year follow-up, those depressed subjects who had dreamt about their spouse were better adjusted than those who had not. However, it remains unclear how this correlation should be interpreted; on the basis of this study no causal relationship between dream content and adjustment can be established.

Kramer (1991; 1993) argues that during REM sleep there is a surge of emotion, and that the function of dreaming is to contain or to attempt to contain this surge. If the dream is successful in fulfilling this function, it does not enter awareness or memory, but protects sleep. A successful pattern of dreaming first states and then works on and resolves the problem, which leads to a positive affective outcome and no dream recall. Kramer's (1993) studies show that a successful night's dreaming is associated with having more characters in the dreams and leads to increased happiness during the next waking period. If the problem is simply restated and not solved, as in repetitive nightmares, then the problem remains unsolved, emotions remain negatively toned, and the dream easily enters awareness. Nightmares and bad dreams are therefore seen as unsuccessful attempts at solving our emotional problems. This theory is called the selective mood regulatory theory of dreaming (Kramer 1993).

Hartmann (1995; 1996a; 1998) has recently argued that our dreams deal with our emotions and emotional concerns by making pictorial metaphors of them. Dreaming cross-connects or weaves in new material, which, according to Hartmann (1998), helps us adapt to future trauma, stress, and the problems of life. Thus, dreaming and psychotherapy fulfill somewhat similar functions. A stressful real-life experience can be processed in both cases in a similar way, essentially by "making connections in a safe place" – that is, by associating and integrating traumatic experiences with the rest of life in order to facilitate psychological healing. Dreaming "calms" the emotional "storm" going on in the mind. Hartmann calls the class of psychological adaptation views of dreaming consistent with his theory the "contemporary theory of the functions of dreaming."

Punamäki (1997; 1998) has recently tested the role of dreams and dream recall in protecting psychological well-being in traumatic conditions. She studied the dreams of a group of Palestinian children living in a violent area in Gaza and a control group living in a peaceful area in Galilee. She reports that traumatized children had better dream recall than nontraumatized ones, and the more the children were exposed to trauma, the more negatively emotional and the less bizarre were their dreams. Frequent dream recall was associated with depressive symptoms, whereas infrequent dream recall was associated with somatic and anxiety symptoms. Thus, the pattern of mental health effects associated with dream recall is not straightforward, for both good and bad dream recall were associated with some, although different, psychological symptoms. Furthermore, on the basis of this study it remains unclear whether dream recall was a cause or a consequence of these symptoms, as well as whether frequent or infrequent dream recall in any way serves a positive long-term mental health function in the recovery from trauma.

Thus, the literature on the possible mental health functions of dreaming is inconclusive as to whether dreams truly solve our emotional problems, protect our mental health, or help us to adjust psychologically and to recover from traumatic experiences. The empirical evidence for such psychologically adaptive functions appears to be relatively weak and correlational at best. Furthermore, it is not entirely clear what the predictions of such a theory really are and whether the empirical evidence confirms or disconfirms them. If the idea is that dreaming "protects" our mental health from negative emotional impact by turning the stressful emotional experience into something better and by integrating it with the rest of our lives, it is surprising how often this function deserts us when we need it most. Recurrent dreams during times of stress are accompanied by negative dream content, and are associated with a deficit in psychological well-being (Zadra et al. 1997–1998). When we live under constant emotional stress or have recently experienced trauma, our dream consciousness typically makes us suffer from intensive nightmares that constantly remind us of the trauma by reactivating powerful negative feelings and other elements from the trauma (see sect. 3.5). If the real function of dreaming is psychological healing, shouldn't we in fact expect exactly the opposite: pleasant, comforting, manifestly healing dreams – calming, not amplifying, the traumatic experience? Intuitively, reliving the emotional shocks over and over again in dreams would not seem to be exactly what traumatized people are psychologically in need of.

The usual explanation for this anomaly is that the assumed dream function has simply "failed"; nightmares are treated as "failures" of dream function (Kramer 1991). But if this is so, then dream function fails a little too regularly, and exactly when it would be needed most. In opposition to these psychological adjustment theories of dreaming, I shall argue that nightmarish dreams are not ones that failed to perform their function, but, by contrast, are prime examples of the kind of dreams that fully realize their biological function. The view that dreams solve our emotional problems and increase our happiness and psychological well-being seems to include the biologically misguided assumption that normal life is free of emotional pain and trauma. Biologically adaptive responses to danger, such as pain and fear, are not there in order to increase our happiness but to increase our reproductive success. Natural selection cares only about fitness, not our comfort (Nesse & Williams 1997). If dreams are biological adaptations, they may not care about our comfort either.

## 3. The biological function of dreaming

The discussion above shows that there is no convincing evidence that dreaming would causally contribute to the solving of either intellectual or emotional problems. We must look elsewhere to discover the biological function of dreaming.

### 3.1. Background assumptions

The construction of the appropriate context for discovering the biological function of dream consciousness requires clarification of the following two questions: (Q1) What is the level of organization to which we attribute a function when we attribute it to consciousness? (Q2) What was the biological context in which dream consciousness evolved? Here are brief answers to these questions:

(A1) Consciousness can be reconceptualized as the phenomenal level of organization in the brain (Revonsuo 1999a). A function attributed to consciousness concerns the causal powers and behavioral effects of events realized at the phenomenal level of organization. The phenomenal level forms the brain's real-time model of the surrounding world, of the organism's internal state, and of its external position in the environment. Dreaming as a subjective experience is realized at the phenomenal level.

(A2) The primary evolutionary context for considering

the possible adaptive function of dream consciousness is the prehistoric Pleistocene environment in which humans and their ancestors lived as hunter-gatherers for hundreds of thousands of years. If dream consciousness is biologically functional, it should have had adaptive value at least in that original environment, under the conditions in which human ancestral populations lived. Whatever the adaptive role of dream consciousness might have been in that long-gone original context, there is no guarantee that the average dreaming brain today, facing a completely different environment than the one in which it evolved, should fulfill any functions that we recognize as adaptive in the present environment.

I will simply take these answers as background assumptions that are reasonably well established; space does not permit a full defense of these views here (but for more on consciousness see Revonsuo 1995; 1997; and for an evolutionary perspective in cognitive neuroscience see Cosmides & Tooby 1995; Tooby & Cosmides 1995).

When put into the proper context in this manner, the question "Does dream consciousness have a function?" becomes: "Did the activation of an off-line model of the world in the ancestral human brain during sleep in some way enhance the probability of reproductive success of the individual living in the natural, original environment?"

My answer is in the affirmative: The off-line model of the world we call "dreaming" is specialized in the simulation of certain types of events that regularly and severely threatened the reproductive success of our ancestors, in order to enhance the probability that corresponding real events be negotiated efficiently and successfully.

### 3.2. Dream consciousness and threat simulation

We are now ready to formulate an evolutionary hypothesis on the function of dreaming. The hypothesis I am putting forward states that dream consciousness is essentially a mechanism for simulating threat perception and rehearsing threat-avoidance responses and behaviors. The threat simulation hypothesis of dreaming is presented below in the form of several independent empirically testable propositions. If each of these propositions is judged as probably true in the light of empirical evidence, then the threat-simulation hypothesis will receive considerable empirical support; but if most of them are not supported by empirical evidence, then the hypothesis will be falsified. I try to show that there are good reasons to believe that each of these propositions is actually true.

### 3.3. Proposition 1

Dream experience is not random or disorganized; instead, it constitutes an organized and selective simulation of the perceptual world.

The demonstration that something is a biological adaptation is always "a probability assessment concerning how likely a situation is to have arisen by chance" (Tooby & Cosmides 1992, p. 62). The content of dreams shows far too much organization to be produced by chance. Empirical dream research has shown that dream consciousness is organized along the same lines as our waking consciousness. All sensory modalities are involved in perceptual dream experience, and approximately with a frequency comparable to that of everyday waking experience (e.g., Foulkes 1985;

Strauch & Meier 1996; Zadra et al. 1998). The visual appearance of dreams is for the most part identical with that of the waking world (Rechtschaffen & Buchignani 1992). The dreaming brain constructs a complex, organized off-line model of the world in which there typically is an active dream self with a body-image much like the one we experience when awake, surrounded by a visuo-spatial world of objects, people, and animals, participating in a multitude of events and social interactions with other dream characters.

This highly predictable and organized form of dreaming presents a challenge to any view claiming that dream experience is merely an incidental by-product of neurobiological processes operating at a different level of organization. It is extremely implausible that a low-level neurochemical restoration process, for example, should produce as some sort of "noise" a complex and organized model of the world at a higher level of organization (cf. Foulkes 1985). If dreams truly were just noise, they should appear much more noisy and disorganized than they actually are. Random noise in the system is not likely to create organized perceptual wholes, nor is it likely to make a good story, or any story at all;[4] it would be expected to generate disorganized sensations and isolated percepts. True noise in the brain is produced in connection with an aura of migraine for example. It does not generate an organized perceptual world of objects and events; rather the contrary, it produces for instance white or colorful phosphenes, geometric forms, and scintillating and negative scotomata (Sacks 1992). The visual hallucinations connected with Charles Bonnet syndrome usually consist of static images of people, animals, buildings, and scenery (Schultz & Melzack 1991). Were our dreams closely to resemble these phenomena it would be easy to believe that dreams consist of nothing but random noise reflecting neurobiological processes at other levels of organization in the system.

It could, however, be argued that even random or disorganized processes might activate organized schemas and scripts and thus produce dreamlike phenomenology. For example, in Penfield's (1975) studies the direct electrical stimulation of temporal cortex produced vivid and realistic perceptual "flashbacks." Still, these experiences were in many ways dissimilar to dreams: they were short (a few seconds) and undramatic excerpts of the patients' previous experiences, like randomly chosen artificially activated memory traces: "The mechanism is capable of bringing back a strip of past experience in complete detail without any of the fanciful elaborations that occur in a man's dreaming" (Penfield 1975, p. 34). Thus, the activation of such traces would not produce dreams as we know them. Consequently, there is no evidence that any kind of essentially random activation could produce the phenomenology and narrative structure of fully developed dreams.

Dream phenomenology, therefore, is likely to be the consequence of an active and organized process rather than a passive by-product of disorganized activation. This process generates an organized world-model. Foulkes (1985) points out that dreams are coherently organized both momentarily and sequentially. The momentary phenomenal content of dream consciousness is comprehensible and conforms to the kinds of multimodal perceptual experiences that we have during waking perception. These momentary phenomenal contents cohere sequentially so as to constitute narrative stories or temporally extended episodes of experience of the same general form as our waking experience.

According to Foulkes, dreams are credible multimodal world analogs that are experienced as life: "The simulation of what life is like is so nearly perfect, the real question may be, why *shouldn't* we believe this is real?" (Foulkes 1985, p. 37).

Thus, all of the above shows beyond any reasonable doubt that dreaming is an organized simulation of the perceptual world; a virtual reality (Revonsuo 1995). Even granted this, it could still be the case that the phenomenal content of dreaming is simply a *random or indiscriminate sample* of the phenomenal content of waking consciousness (or the episodic memories thereof). However, this does not seem to be the case. There are certain experiences that are very frequent contents of consciousness during our waking lives but rarely or never enter our dreams. Hartmann (1998) describes two studies in which it was shown that even subjects who spend several hours daily reading, writing, or calculating virtually never dream about these activities. In the first study, two judges examined 129 written dream reports from several studies and found no instances of reading or writing and only one possible instance of calculating. In another study a questionnaire was mailed to 400 subjects who were frequent dreamers and interested in their dreams. They reported spending an average of six hours per day engaged in reading, writing, calculating, or typing, but answered that they dreamed "never" or "almost never" about any of these activities. They furthermore estimated on a seven-point scale how frequent different activities are in dreams compared with waking life. Their ratings showed a remarkable dissociation between waking and dreaming life: the average rating was at the "far more prominent in my waking life than my dream life" end of the scale as to the frequency of writing, reading, and typing.

This shows that dreaming is not only an organized but also a selective simulation of the world. Not every type of event or activity is simulated by the dream-production mechanisms, no matter how prominent they may be in our waking lives. Given that reading, writing, typing, and calculating are excluded from, or at least grossly underrepresented in, dream experience, what kind of phenomenal content is overrepresented in it? Which events is dream experience really specialized in simulating? This question leads us to Proposition 2.

### 3.4. Proposition 2

Dream experience is specialized in the simulation of threatening events.

**3.4.1. Dream content shows a significant bias toward representing threatening elements in dreams.** If dreams are specialized in simulating threatening events, then we ought to find that dream content is biased toward including various negative elements (reflecting threats) rather than positive elements. Several prominent features of dream content suggest that this bias indeed exists.

**3.4.1.1. Emotions in dreams.** In the normative study by Hall and Van de Castle (1966), 500 home dream reports from female and 500 from male college students, aged 18–25, were content analyzed. Of the more than 700 emotions expressed in the dream reports, about 80% were negative and only 20% positive. The figures remain similar when only the dreamers' own emotions are considered. About

91

half of the negative emotions experienced by the dreamers were classified as "Apprehension," the other half consisted of "sadness," "anger," and "confusion."

In the first normative laboratory study, Snyder (1970) collected 635 REM dream reports from students and found that more than two-thirds of the emotions mentioned in the reports were negative, fear being the most common and anger the next most common. Strauch and Meier (1996) report a sleep laboratory study in which they not only collected REM dream reports from 44 subjects but also asked them how they had felt during the dream. The emotions described in response to this question were analyzed. Specific emotions were mentioned in connection with every other dream. Negative emotions appeared twice as often as positive ones, with anger, fear, and stress being the most frequent types of negative emotions. In contrast to specific emotions, general mood states were found to be more often positively than negatively toned.

Foulkes et al. (1988a) and Revonsuo and Salmivalli (1995) have shown that emotions in dreams are in most cases appropriate to the dreamed situations in which they are experienced; therefore, the high proportion of negative emotions is a sign of frequent unpleasant dream events that should be expected to produce negative emotions if they were real. Emotions are evolved adaptations that increase the ability to respond appropriately in adaptively important situations. Negative emotions such as anxiety, fear, and panic, can be seen as adaptive responses that increase fitness in dangerous situations threatening a loss of reproductive resources (Marks & Nesse 1994). When emotions are experienced or expressed in dreams, they are much more likely to be negative than positive ones, and very likely to be appropriate to the dreamed situation. These findings are consistent with the hypothesis that dream content is biased toward simulating threatening events.

**3.4.1.2. Misfortunes in dreams.** "Misfortune" names a class of dream event in which a bad outcome happens to a character independent of anything the character has done (Hall & Van de Castle 1966). Misfortunes include, for example, mishaps, dangers, and threats. The opposite is called "Good Fortune." In the Hall and Van de Castle (1966) normative study, there were altogether 411 cases of Misfortune in 1,000 dream reports, and only 58 cases of Good Fortune. Thus, Misfortunes in dreams are seven times more frequent than Good Fortunes. Furthermore, about 70% of the misfortunes happen to the dream-self, and it is accidents, losses of possession, injuries or illnesses, obstacles, and threats from environment that comprise almost 90% of these misfortunes, whereas death and falling are rare types of misfortune (Domhoff 1996; Hall & Van de Castle 1966). Misfortunes, therefore, typically reflect situations in which the physical well-being or the resources and goals of the dream-self are threatened.

**3.4.1.3. Aggression in dreams.** Aggression is the most frequent type of social interaction found in dreams, the other classes in the Hall and Van de Castle (1966) scale being Friendliness and Sexual Interactions. About 45% of the dreams in the normative sample included at least one aggressive interaction. Dreamers are involved in about 80% of the aggressions in their dreams, and when they are involved they are more often the victim than the aggressor (Domhoff 1996; Hall & Van de Castle 1966).

**3.4.1.4. Summary.** Negative emotions, misfortunes, and aggression are prominent in dreams. These findings indicate that normative dream content frequently contains various unpleasant and threatening elements, which supports the view that dreams are specialized in simulating threatening events.

### 3.4.2. Dream content is consistent with the original evolutionary environment of the human species rather than the present one

**3.4.2.1. "Enemies" in our dreams.** Domhoff (1996) defines "Enemies" as those dream characters with which the proportion of aggressive encounters of all aggressive + friendly encounters is greater than 60%. This calculation on the Hall and Van de Castle (1966) normative sample reveals that animals and male strangers are the enemies in men's as well as women's dreams (Men vs. Animals 82%; Women vs. Animals 77%; Men vs. Male Strangers 72%; Women vs. Male Strangers 63%). Encounters with Female Strangers are not at all so aggressive, but predominantly friendly (Men vs. Female Strangers 40%; Women vs. Female Strangers 43%) (Domhoff 1996). According to Hall and Domhoff (1963), unknown males are responsible for the high proportions of victimization and physical aggressions with male characters.

Hall (1955) content analyzed 106 dreams of being attacked and found that the attacks predominantly represented situations in which the dreamer's life or physical well-being was at stake. The attacker was usually human or a group of humans (70%) but not infrequently an animal (21%). When the sex of the human attacker was identified it was virtually always male. The dreamer usually reacted to the attack by running, escaping, or hiding (unless she woke up). Hall and Domhoff (1963; 1964) analyzed aggressive and friendly interactions in more than 3,000 dream reports. They found that interaction was aggressive with 48% of the animal characters in men's dreams and with 29% of the animals in women's dreams.

Van de Castle (1983) compared college students' dreams (more than 1,000 dream reports altogether) in which humans were the dominating dream characters with those in which animals predominated. He found that dreams with animal figures typically take place in an outdoor setting, have a great deal of activity that is often of a violent nature, and that the dreamer typically experiences fear. If an animal figure initiates an interaction with the dream-self, the nature of the interaction is aggression 96% of the time and friendliness only 4% of the time. Van de Castle writes that "almost without exception, if the animal figure initiates any response to the dreamer, it is some form of threat or hostility" (p. 170).

Why are animals and male strangers our enemies in dreams? Ancestral humans lived in environments in which many animals (e.g., large carnivores, poisonous animals, parasite-carrying animals) presented an ever-present mortal threat for humans. Therefore, behavioral strategies to avoid contact with such animals and to escape or hide if attacked by them obviously were of high survival value. Some deep-rooted human fears and phobias of snakes, spiders, rats, and open spaces are indications that ancient threat avoidance programs still remain with us (Marks & Nesse 1994). Dreaming simulates and rehearses these ancestral threat avoidance programs in order to maintain their efficiency, because the costs of a single failure to respond appropriately when the danger is real may be fatally high, while the costs of repeated threat simulations during sleep are remarkably low.

Our present-day encounters with unfamiliar males in the waking life are not predominantly aggressive. In the ancestral human environment, however, intergroup aggression and the violent competition over access to valuable resources and territories is likely to have been a common occurrence. Since intergroup warfare and violence was and still is almost exclusively practiced by males (Wrangham & Peterson 1996; see also Campbell 1999), encountering male strangers is likely to have been a potentially threatening situation in the ancestral environment, comparable to the threats presented by dangerous animals. Indications that unfamiliar males often present a mortal threat to offspring come from other primates where infanticide by genetically unrelated males is common (Hrdy 1977). Furthermore, human infants universally develop stranger fear at about six months of age, and even in the modern world are much more likely to be killed or abused by genetically unrelated adults than by close kin (Daly & Wilson 1988). Thus, although an overwhelming majority of our current waking-life encounters with animals and male strangers are not particularly aggressive or threatening, dream content still reflects the ancestral conditions in which such encounters were potentially life-threatening. Dreams are biased toward simulating threats that were common in our ancestral environment.

### 3.4.2.2. Children's dreams.
If dreams are naturally biased toward simulating ancestral threats, then we should expect that the traces of these biases are strongest early in life, when the brain has not yet had the chance to adjust the biases in order to better fit the actual environment. This seems to be the case when it comes to the appearance of animals and aggressions in children's dreams. One of the most prominent differences between child and adult dreams is the much larger number of animal characters in children's dreams. Hall and Domhoff (1963; 1964) analyzed about 500 dream reports from children aged 2–12 years; Hall later increased the sample to 600 dreams and Domhoff (1996) reports the results from this larger sample. Animal characters make up about 25–30% of all characters in the dreams of children 2–6 years of age, and about 15% in 7–12 years of age, whereas the normative finding for adult dreams is about 5% (Domhoff 1996).

Van de Castle (1983) also reports studies of children's dreams. The 741 dream reports (one from each child) were written down by schoolteachers or directly reported by the pupils themselves. The general trend toward a decrease in the frequency of animal dreams as a function of age is clearly manifested. Two-year averages in the percentage of animal dreams for children 4–16 years old were 39.4% for 4–6 years olds, and 35.5, 33.6, 29.8, 21.9, and 13.7% for the next consecutive two-year age groups. In an earlier study on a smaller sample of dreams, Van de Castle (1970) reported closely similar figures (Fig. 1).

Surprisingly, in their dreams children often encountered animals that were seldom or never encountered in the waking world. Wild or frightening animals (e.g., snakes, bears, monsters, lions, spiders, gorillas, tigers, wolves, insects) comprised nearly 40% of all animal characters in children's dreams in this study, but less than 20% in college students'

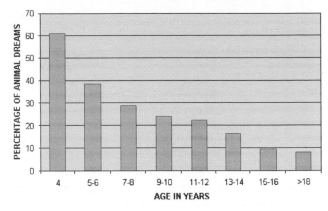

Figure 1. Percentage of animal dreams in relationship to age (modified from Van de Castle 1970).

dreams. Dogs, horses, and cats accounted for 28% of animals in children's dreams but 38% in college students' dreams (Van de Castle 1983). Thus, the proportion of domestic animals increases and that of wild animals decreases with age.

Due to the methods of collecting the dream reports, the studies mentioned above may have included a somewhat biased sample of dreams.[5] However, also in the laboratory study of Foulkes (1982b), animals were the major characters in the dream reports of children 3–5 and 5–7 years of age, appearing in 30–45% of the reports. Also the decrease in the number of animal characters with increasing age was confirmed. Strauch (1996) reports results from both home and laboratory REM dreams in Swiss children 9–11 years of age. Both types of dreams involved more animals than young adults' dreams, again confirming the decrease of dream animals with increasing age. Home dreams contained animals about twice as often as laboratory dreams, which was explained by dream report length: home dreams were longer and included more characters. Girls' dreams contained more animals than boys' dreams. In the REM dreams, 102 animals were found. In girls' dreams, tame animals and pets prevailed (63%) over wild native or exotic animals (37%), whereas in boys' dreams, wild animals were much more common than tame ones (61 vs. 39%). Taken together, on the average one out of two animals encountered in the children's dream world is an untamed wild animal. For boys around 10 years of age this is the most common type of dream animal.

Hall and Domhoff (1963) showed that children also have a higher rate of aggression in their dreams than adults. The greatest amount of aggression occurs in the dreams of children 2–12 years of age. According to Domhoff (1996), much of this larger amount of aggression is with animals and the child is usually the victim of an attack by the animal. In Strauch's (1996) data of combined REM and home dreams, about 30% of all the animals appearing in 10-year-old children's dreams were in the role of aggressors, compared to 10% in adults' dreams.

Levine (1991) studied the representation of conflicts in the dreams of 77 children who were about 10 years of age and came from three different cultures: Bedouin, Israeli, and Irish. Conflictual dreams accounted for about two-thirds of the reports and were reported about twice as often as nonconflictual dreams in all three cultures. The Bedouin children, who were living in a traditional seminomadic tribe, had dreams that were realistic and concerned with threats to physical survival, usually from the natural world.

Children's dreams thus show strong biases toward simulating a world that contains animals (especially wild animals), aggression, conflicts, animal aggressors, and victimization to a greater degree than does their own waking world or the dream world of adults. These biases decrease with age if the child's real environment is largely devoid of them. It seems unlikely that young children would have had more frequent real waking experiences of such things than teenagers or adults have had; therefore, it is difficult to explain these biases by referring to the waking lives of the children.[6] These biases seem to be another sign of the fact that the dream-production system is prepared to simulate threatening events consistent with and prevailing in the human ancestral environment. The biases decrease with age, as the perceptual world proves to be quite different from what was anticipated by the dream-production mechanisms.

### 3.4.2.3. Recurrent dreams and nightmares.

Robbins and Houshi (1983) asked 123 university students whether they had ever had recurrent dreams, and if so, they were asked to describe them. Sixty percent reported that they had had recurrent dreams, many beginning in childhood. A content analysis revealed that only one type of recurrent dream occurred with any frequency, an anxiety dream in which the dreamer was being threatened or pursued. The threatening agents were wild animals, monsters, burglars, or nature forces such as storms, fires, or floods. The dreamer was watching, hiding, or running away. The authors regarded these descriptions as reasonably close to nightmares (Robbins & Houshi 1983, p. 263). Recurrent dreamers reported more problems in their lives and more physical symptoms than those who did not report recurrent dreams, indicating that recurrent dreams may be related to increased levels of stress. Feldman and Hersen (1967) found that frequent recurring nightmares in adults were related to conscious waking concerns about death and to having experienced the death of a close relative or friend before the age of 10. Zadra et al. (1997–1998) reported that in both late teenagers and older adults recurrent dreams with negative content occur during times of stress.

Nightmares, or long, frightening dreams that wake the dreamer, are the paradigm cases of highly unpleasant dreams. It is estimated that almost everyone has had a nightmare, that children, especially from 3 to 6 years of age, very frequently experience some, and that adults quite commonly have occasional nightmares. In a study of 1,317 subjects, 5% reported having nightmares once per week and an additional 24% once per month (Feldman & Hersen 1967). The themes in the dreams of lifelong nightmare sufferers are remarkably similar to the themes of recurrent dreams, and the most frequent theme is, again, that of being chased or attacked (Domhoff 1996; Hartmann 1984). Such dreams usually begin in childhood and involve being chased by a monster or a wild animal. In adulthood, the chaser was more likely to be a large unfamiliar man, a group of frightening people, or a gang. These dreams can be frequent, seem very vivid and real, but still do not usually reflect any actual events that ever happened to the dreamer (Hartmann 1998).

Recurrent dreams and lifelong nightmares not directly connected with any real-life traumas appear to be very powerful simulations of rather primitive threats. Again, we should note that the origin of these simulations apparently is not in the real life of the dreamer. Where do these recurrent themes come from? In the light of the human ancestral environment, it makes great sense to simulate violent encounters with animals, strangers, and natural forces, and how to escape from such situations.[7] Therefore, these simulations are incorporated as default values in the threat simulation system, and they can be activated in almost anybody, at least occasionally. In lifelong nightmare sufferers the trigger seems to be the fact that, because of their highly sensitive personality ("thin boundaries"; Hartmann 1984), for them even everyday experiences may be highly stressful or traumatic (Domhoff 1996; Hartmann 1998) and, as we will see in later sections, such emotional triggers can have profound effects on subsequent dream content.

### 3.4.2.4. Absence of reading, writing, typing, and calculating.

One explanation for the fact that we do not dream about reading and writing is that they include little if any emotional charge for us. However, Hartmann (1998) found that walking, talking to friends, and sexual activity are represented in dreams about as often as in real life, although these activities differ considerably as to their emotionality. Therefore, the principal reason we do not dream about writing, reading, and doing arthmetic probably is that all these activities are cultural latecomers that have to be effortfully hammered into our evolved cognitive architecture. They were not present in ancestral environments nor are they neurally hardwired in the human brain in the way that other complex cognitive functions, frequently present in dreams, are (e.g., speech comprehension and production). Furthermore, they are highly dependent on abstract symbol systems rather than on the recognition or manipulation of concrete objects. Thus, they are in many ways activities fundamentally different from the ones that the human brain was selected for in its original environment.

### 3.4.2.5. Brain activation during REM sleep reflects the neural correlates of threat simulation.

If the essence of dreaming is threat simulation, then we should find that the brain areas active during REM sleep are ones involved in generating emotional and perceptual experience.

According to Hobson (1999a) PGO waves are believed to be the neural generators of the internal stimulation that results in dream phenomenology. They occur as bursts of waves during REM sleep, activating, in particular, the thalamocortical circuits involved in vision, but also radiating to the limbic lobe and amygdala. In the waking state PGO waves are triggered by strong, novel stimuli and are associated with surprise and fear: "PGO waves prepare us for fight or flight should these prove necessary. The startle reactions provoked in us by real or imaginary intruders are mediated by PGO-like signals" (Hobson 1999a, p. 169). Thus, the function of PGO waves during waking is clearly consistent with internal threat simulation during dreaming.

Research on emotionally charged memories and memory under stress has recently come up with the idea that there is a separable "hot" amygdala-centered emotional system distinct from the "cool" hippocampally centered episodic memory system (for a review, see Metcalfe & Jacobs 1998). The two systems work in cooperation, the "hot" system highlighting those species-specific or learned elements of memory traces that are highly emotional by nature. The "hot" system is believed to have a role in releasing species-specific behaviors such as fear or defensive responses to emotionally charged stimuli. As the stress levels of the organism increase, the "hot" memory system becomes increasingly activated. When a person is in a stressful and dangerous situation, the hippocampal "cool" system may not be optimal for responding to threat. Instead, the "hot" system may very efficiently process the threatening cues and immediately activate threat-avoidance mechanisms. The "hot" system is considered to be more automatic and primitive than the "cool" system, thus allowing the organism to realize rapid protective responses. In accordance with this view, a recent PET study suggests that the human amygdala modulates the strength of conscious episodic memories according to their emotional importance (Hamann et al. 1999).

Recent functional brain-imaging studies of sleep show that brain areas involved in the processing of emotionally charged memories are strongly activated during REM sleep and dreaming. The dream-production mechanisms thus seem to be in close interaction with the primitive "hot" memory system, preferably selecting memory traces with high emotional charge. A study of regional cerebral blood flow distribution showed that during human REM sleep, activation and functional interaction occurs between the amygdaloid complexes and various cortical areas, but the prefrontal cortices are deactivated (Maquet et al. 1996). The authors concluded that these interactions might lead to the reactivation of affective components of memories. A similar pattern was found in another study, concluding that pathways which transfer information between visual cortices and the limbic system are active during REM sleep (Braun et al. 1998).

In sum, neurophysiological studies and functional brain imaging reveal the dream-production mechanisms at work during REM sleep, searching for and processing emotionally charged memory traces in the evolutionarily ancient, "hot" memory system. The dream-production mechanisms, guided by the dominant emotional concerns of the dreamer, create the content of dreams in interaction with other long-term memory systems (Cavallero & Cicogna 1993) and perceptual cortical areas.

**3.4.2.6. Summary.** Many elements abundant in contemporary life (e.g., reading, writing) are absent from dreaming, whereas many such elements that are not common in waking life, but consistent with simulating primitive threats (e.g., aggressive interactions with animals and male strangers), are universally present in adults' dreams, children's dreams, recurrent dreams, and nightmares. Furthermore, brain activation during REM sleep is consistent with the activation of brain areas required to simulate emotionally charged, threatening events.

### 3.5. Proposition 3

Encountering real threats during waking has a powerful effect on subsequent dream content: real threats activate the threat simulation system in a qualitatively unique manner, dissimilar from the effects on dreaming of any other stimuli or experience.

**3.5.1. The effect of traumatic experience on dream content.** Real experiences of actual dangers or life-threatening events are very likely to be incorporated into dreams (Bar-

rett 1996; Hartmann 1984; 1996a). This is most clearly manifested in cases of post-traumatic nightmares. These nightmares are reported by people who have undergone, for example, wartime battles, natural catastrophies, terrible accidents, or assault, rape, or torture. The frequency of post traumatic nightmares depends, among other things, on the degree of threat perceived to be targeted at self and significant others. It appears that the greater the sense of threat created by the experience, the more likely it is that nightmares will follow (Nader 1996). For example, 100% of 23 children who were kidnapped and buried in a truck trailer; 83% of six children that underwent a life-threatening medical operation; 80% of 10 children witnessing their mothers being raped; 63% of children exposed to sniper fire on their schoolyard, and 40% of children whose suburb was exposed to radioactivity after a major nuclear power plant accident reported nightmares related to the respective incidents (for a review, see Nader 1996). Ninety-six percent of 316 Vietnam combat veterans described a combat nightmare in an interview (Wilmer 1996).

These are very impressive figures, especially in view of the fact that laboratory research has failed to find any strong determinants of dream content. Presleep stimuli, such as films depicting violence, are only marginally if at all incorporated into dreams. The conclusions of Vogel (1993), in a review of stimulus incorporation in dreams, are revealing:

> dream content is remarkably independent of external psychological and physical stimuli both before and during sleep and equally independent of currently measurable physiological processes during sleep. Therefore, the sources of dream content, that is, its themes and its specific elements, remain a mystery. (Vogel 1993, p. 298)

Laboratory research has failed to find the actual determinants of dream content, probably because it is practically and ethically impossible to expose experimental subjects to situations that evoke a deep enough sense of threat. The stimuli that are typically used in laboratory research on stimulus incorporation, such as films, never induce anything like a genuine sense of real threat to one's own life. Therefore, they do not function as ecologically valid cues for the dream-production mechanisms. We must turn to the cruel experiments inadvertently designed by wars, crime, and nature. The sense of severe personal threat probably is the most powerful factor we know of in the modulation of the content of dreams: the experience of a severe trauma can induce nightmares in almost anyone; the majority of people, especially children, involved in traumatic events do report nightmares; and traumatic nightmares can occur in several stages of sleep (Hartmann 1984).

Once the dream-production system encounters the memory of an event combined with a deep sense of threat, how does it handle that? There seems to be a more or less universal pattern involved in the ways in which post-traumatic dreams are constructed (Hartmann 1984;1998). In the first stage, immediately after the traumatic event, the frequency of trauma-related dreams and nightmares increases and the event is being replicated a few times in the dream world, in a form often closely similar but not exactly identical (Brenneis 1994) to the original experience. The first stage normally lasts a few days or weeks, but in severe post-traumatic disorder it may persist for years.[8] In a longitudinal study on children who were exposed to a sniper attack on the school playground, 42% continued to have bad dreams 14 months after the incident (Nader et al.

1990). In another study on a bushfire disaster, 18% of the children continued to have post-traumatic nightmares when studied 26 months after of the actual event (McFarlane 1987).

Gradually, the nightmares change into increasingly modified versions of the event. At this stage, the original experience is associated with and connected to other similar contents in memory. The resulting dreams may be small variations of the original threat, the original threat mixed with previously experienced ones, and with classical nightmare themes such as being chased or escaping powerful natural forces. Eventually, perhaps after a few weeks or months, the content of dreams returns to approximately normal. Even long after the original trauma, events that remind of it or also induce a deep sense of threat may trigger the recurrence of the trauma-related nightmares. In post-traumatic stress disorder, this normal development of the dream sequence does not occur; instead, replications and different variations of the original trauma continue to recur, even for years (Hartmann 1984; 1996a; Stoddard et al. 1996).

Ordinary as well as post-traumatic nightmares are especially frequent in children (Hartmann 1984). Nader (1996, pp. 16–17) mentions the following types of trauma-induced threat simulations in children's dreams. Kuwaiti and Croatian children exposed to war dreamed of being personally endangered by someone trying to kill them with a knife, a gun, or bare hands, and of being captured or tortured. Children from Los Angeles who had witnessed their mothers being raped dreamed of the rapist returning, of being threatened, of being severely physically harmed, of directly confronting the assailant, or of taking revenge. A girl who was chased and groped by an unfamiliar man had recurrent dreams in which people or animals chased her. Children who were in a cafeteria when a tornado knocked the wall down with serious consequences dreamed of the wall falling again, of houses being destroyed by a tornado, of branches falling, of being hit by glass, and of trying to find bandages for dead people. After a hurricane, both parents and children dreamed of being threatened by winds or tornadoes coming directly at them.

In a study of the dreams of Arab and Jewish children 11–13-years of age (Bilu 1989), all the dreams representing the "other side" were extracted and analyzed. In these 212 dreams, aggression appeared in about 90% of the interactions, while friendliness was virtually nonexistent (4%). Jewish children dreamt about Arab terrorist attacks and camouflaged detonating explosives in public places. In these dreams, the dreamer was usually the recipient of an unprovoked assault initiated by an adult adversary who was typically defeated in the end. Arab children living in a refugee camp dreamt about brutal physical aggression, which resulted in death on either side, in 25% of the dreams. The dreamers were typically harassed, expelled, arrested, beaten, injured, or killed. Bilu (1989, pp. 385–86) comments that the conflict between Arabs and Jews seems to have an even stronger presence in these children's dreams than it does in reality (i.e., it is *overrepresented* in dreams): "the intensity and pervasiveness of the conflict as reflected in the dreams cannot be taken for granted even by those well-acquainted with the situation."

Dreams after trauma reflect the dream-production system working at full capacity, producing a regular pattern that proceeds from near-identical replications to increasingly modified variations to gradual fading and possible re-

currence. Hartmann (1996a; 1998) suggests that dreams after trauma should be seen as the paradigm case of dream formation. He makes the important observation that:

> One hundred thousand years or so ago, when the human brain was gradually developing to its present form, our lives were considerably more traumatic; the after-effects of trauma may well have been an everyday reality. (Hartmann 1996a, p. 158)

According to Hartmann's (1996a; 1998) view, the content of dreams is greatly modulated by the current dominant emotional concern of the dreamer. Dreaming connects the trauma and the associated feelings and emotions to a wide variety of related images and memories in the dreamer's memory networks. Domhoff (1993) suggests that all dreams could be seen as dealing with traumatic experiences of differing degrees and regards recurrent dreams as "watered-down" versions of traumatic dreams, but otherwise basically within the same category of dreams. Domhoff (1996) treats nightmares, recurrent dreams, and dreams after trauma under the heading of "the repetition dimension" in dreams, and says that no theory of dreaming should be taken seriously if it cannot deal with this dimension. The present hypothesis explains this dimension as the paradigm case of threat simulation in dreams.

### 3.5.2. Real threats as cues that activate the threat simulation system.
The view that emerges can be summarized as follows: Experiences of real threats are the only ecologically valid cues for the threat simulation system. Encountering real threats powerfully activates the threat simulation system: first, they may intensify the neurophysiological events underlying threat simulations; second, they tend to render the threat simulations more realistic; and third, they may even influence the development of the dream-production system.

There is some evidence indicating that real threats may intensify REM sleep. In normal subjects the presence of stressful life events is associated with increased intensity of REM sleep (Williamson et al. 1995). One study (Ross et al. 1994) found that patients with post-traumatic stress disorder and frequent anxiety dreams showed elevated tonic REM sleep measures: they spent a higher percentage of total sleep time in REM sleep and their REM sleep periods were longer than those of control subjects. Furthermore, they had heightened phasic-event generation in REM sleep and manifested increased rapid eye movement activity. However, another study (Hurwitz et al. 1998) did not find any differences in polysomnographic sleep between Vietnam combat veterans and normal controls. Thus, more studies are needed to establish the relationship between stressful or life-threatening events and the intensity of REM sleep.

In a study on Palestinian children living in traumatic conditions it was found that the more the children were exposed to trauma, the more negatively emotional and the less bizarre their dreams were. The children exposed to trauma also had better dream recall than other children (Punamäki 1997). These findings indicate that the dream-production system creates especially vivid and realistic simulations of threatening events encountered in the real world.

Real threats might even trigger the ontogenetic development of dreaming. Foulkes (1999) argues that adult-like "true" dreaming appears relatively late in childhood, at about 7–9 years of age. By contrast, clinical case reports suggest that the earliest nightmares may be experienced as early as during the second year of life (Hartmann 1998). Some traumatized preschool children report fully developed nightmares, unlike those typical of the age (Nader 1996), suggesting that traumatic experiences may actually stimulate the development of the dream-production system, or, conversely, that a lack of real life-threatening events might hold it back, or at least preserve the dream-production system in a resting state although it would already be capable of generating threat simulations if only exposed to the ecologically valid cues (see also N. 5 on children's dreams).

It could be argued against Proposition 3 that both positive and negative real emotions are equally strong in activating the dream-production system. This alternative hypothesis is not supported by evidence. Hartmann (1998, p. 73) observes that "even when people experience a happy event, they are more likely to dream about problems associated with it than the pure happiness of the event itself." Thus, dreams tend to represent even happiness in the light of the possible threats that might endanger it.

According to the present hypothesis, the brain's dream-production system selects traumatic contents not because they represent unsolved emotional problems, but primarily because such experiences mark situations critical for physical survival and reproductive success. What from a psychological point of view is a "traumatic experience" is, from a biological point of view, an instance of threat perception and threat-avoidance behavior. Negative emotions, such as fear and terror, accompanying the perception of serious real-life threats, serve to label such events as critical to one's own survival and future reproductive success. The contents of the threat simulations are selected by the dream-production system from long-term memory, where recent memory traces associated with threatening emotional impact are the most salient ones to enter the dream-production mechanisms. The stronger the negative emotional charge, the more threatening the situation is likely to have been, and the more likely it is that it will be selected by the dream-production system as a recurrent theme for threat simulation. The dream-production system is highly sensitive to situations critical for the physical survival and future success of the individual: violent attacks, being chased by strangers or animals, finding intruders in one's private territory, losing valuable material resources, being socially rejected, encountering untamed natural forces or dangerous animals, being involved in accidents, and misfortunes. Such dream contents involve, from a biological point of view, threat perception, threat avoidance, antipredatory behavior, and coping strategies against threats.

### 3.6. Proposition 4

The threat simulations are perceptually and behaviorally realistic and therefore efficient rehearsals of threat perception and threat-avoidance responses.

**3.6.1. Perceptual realism and lack of insight.** So far we have shown that dreaming specializes in the repetitious simulation of threatening events. Next, we need to show that these events constitute realistic rehearsals of threat perception and avoidance, for otherwise they would not be useful simulations. First, dreams and especially nightmares consist of vivid images that seem perfectly real. Second, during dreaming we are in an uncritical, delusional, and isolated state of mind that very efficiently prevents us from re-

alizing it is all just a hallucinatory simulation (Rechtschaffen 1978). The relatively rare exception of lucid dreaming (Gackenbach & LaBerge 1988) notwithstanding, we take the dream world for real while it lasts, totally lacking insight into our true condition. Thus, these two factors, perceptual realism and delusional lack of insight, guarantee that the simulation is taken most seriously. If that were not the case, we might instantly recognize the dream world for what it is and not be motivated to defend ourselves against the simulated threats. Lucidity has in fact been recommended as a possible cure for recurring nightmares (e.g., LaBerge 1985; Zadra & Pihl 1997).

**3.6.2. Motor realism.** What about the dreamed action: What is its relationship to real motor behavior? It should be neurally realized in the same way as real actions are, otherwise it could not be regarded as an efficient rehearsal of what to do in a comparable real situation. Mental imagery of motor actions uses the same motor representations and central neural mechanisms that are used to generate actual actions (Decety 1996; Jeannerod 1995); moreover, dreamed action is experientially far more realistic than mere imagined action. Therefore we have good reasons to believe that dreamed action is equivalent to real action as far as the underlying brain mechanisms are concerned.

Classical neurophysiological studies in the 1960s (reviewed by Hobson 1988b) showed that the pyramidal-tract cells of the motor cortex increased their firing during REM sleep (compared to nonREM, NREM, sleep), having firing rates as high as those during waking with movement.[9] Thus, motor commands are generated during REM sleep at the cortical level but they are not realized in the periphery because of the operation of an inhibitory system that blocks the activity of motor neurons in the spinal cord, resulting in muscular atonia. According to Hobson (1999a), the *experience* of movement in dreams is created with the help of the efferent copying mechanism, which sends copies of all cortical motor commands to the sensory system. The brain thus receives internally generated information about issued motor commands and computes the expected consequences of those commands. The sensory system is not informed that these commands were not in actual fact carried out by the muscles, and therefore the illusion of movement comes about.

If the inhibitory mechanisms that produce atonia during REM sleep are malfunctioning, the result is a recently described sleep disorder called REM Sleep Behavior Disorder (RBD) (Schenck et al. 1986). These patients manifest violent behaviors during REM sleep, which are the acting out of the motor imagery being dreamt about. Thus, dreamed action corresponds to real action as far as the forebrain is concerned. The difference between dreamed and real motor action depends only on the inhibitory cell groups in the pons. Thus, within the forebrain, dreamed action has the same neural realization and the same causal powers as real action does. Dreamed action is experientially and neurophysiologically real. (For a similar view on all motor imagery, see Jeannerod 1994.)

Some illustrative cases of RBD have been reported in the medical literature. Dyken et al. (1995) describe the case of a 73-year-old man. During an episode of RBD, the patient leaped from his bed, fell, and struck the right side of his face on a corner of a chest, awakening him immediately. This resulted in subdural hemorrhage. He had dreamed of working on a loading dock and saw a man running. Someone yelled "Stop him!" and the patient had tried to do just that when he jumped out of his bed with the unfortunate consequences. In the sleep laboratory during REM sleep, the patient suddenly exhibited explosive running movements, followed by an arousal. The patient's actions again clearly corresponded to what he was dreaming about at the time of the observed behaviors. Boeve et al. (1998, pp. 363–70) describe a patient who, on one occasion, "held his wife's head in a headlock and, while moving his legs as if running, shouted: 'I'm gonna make that touchdown!' He then attempted to throw her head down toward the foot of the bed. When awakened, he recalled a dream in which he was running for a touchdown, and he spiked the football in the end zone." Comella et al. (1998) describe a group of patients with RBD. If these patients were awakened during an episode of abnormal sleep behavior, none of them realized that they had executed violent movements, although all recalled violent dreams at the time of awakening: being pursued by an enemy; trying to protect family members from unknown intruders; or fighting off unidentified assailants. Schenck (1993) describes a patient whose EEG, EMG, and EKG were polysomnographically recorded during an attack of violent behavior. The muscle tone was increased and the arms and legs showed bursts of intense twitching, accompanied by observable behavior. After a spontaneous awakening, the man reported a dream in which he was running and trying to escape skeletons that were awaiting him.

It is noteworthy that most cases of RBD involve intensive threat simulation dreams, and the behaviors manifested are (mostly adequate) responses to these threats. It may be that threat simulations are associated with increased cortical activation, leading to intensive motor imagery that breaks through the malfunctioning inhibitory mechanisms.

There are other sleep disorders that can be interpreted as an inappropriate activation of the threat-simulation system, leading to sleep-related behaviors. Night terrors,[10] sleepwalking (somnambulism), and nocturnal wandering appear to be, at least in some cases, threat simulations that take place during NREM sleep and lead to an altered state of consciousness that is a mixture of wakefulness and NREM sleep (Mahowald & Schenck 1992; Mahowald et al. 1998). In this state, one's subjective consciousness is focused on one internally generated, usually terrifying, image or belief. Appropriate threat-avoidance behavior is often realized automatically, violently, and efficiently in the absence of reflective thought – without an awareness of one's altered state, one's actions, or their actual consequences.[11] One patient described by Schenck and Mahowald (1995) once left the house in pajamas by running through a screen door, then entered his automobile and drove eight kilometers to his parents' home where he awakened them by pounding on their door. This episode of somnambulistic automobile driving was initiated by the subject's belief that someone was in the house and about to attack him. Another subject with nocturnal wandering once threw his wife on the floor, ran to his two children, took them into his arms and ran outside. Afterward he said he had believed that the house was on fire (Guilleminault 1995). When aged ten, one patient had risen from sleep, rushed into the sitting room where his parents were still sitting, and thrown the butter dish out of the window, believing it to be a bomb (Oswald & Evans 1985).

Thus, both RBD and NREM-related sleep disorders

show that threat simulation during sleep includes realistic and adequate motor activation in the brain in response to the perceived threats.

**3.6.3. Summary.** The evidence reviewed above shows that dreaming constitutes a realistic simulation that we tend to believe without questioning and that dreaming about an action is an identical process for cortical motor areas as actually carrying out the same action. In some pathological cases, the actions generated in the dream world are inadvertently performed in the real world. Thus, to dream about threat perception and threat-avoidance behaviors is to realistically rehearse these functions in a safe environment.

### 3.7. Proposition 5

Simulation of perceptual and motor skills leads to enhanced performance in corresponding real situations even if the rehearsal episodes were not explicitly remembered.

**3.7.1. Mental training.** It is a commonplace that training and repetition lead to enhanced performance. However, can actions only performed at the phenomenal level and not overtly executed lead to any kind of learning? Research on the effects of motor imagery and mental training to motor performance show that repeated motor imagery can lead to increased muscular strength (Yue & Cole 1992), improvement in the learning of new motor skills (Hall et al. 1992; Yaguez et al. 1998), and improved performance in sports (e.g., Lejune et al. 1994). These learning effects are thought to arise at the cortical programming levels of the motor system (e.g., through activation of Brodmann area 6 where the premotor and the supplementary motor areas reside), not from neural changes at the execution level (Yue & Cole 1992; Jeannerod 1994; 1995). Because even motor imagery and mental training can have these effects, there is every reason to believe that the intensive and thoroughly realistic motor imagery in our dreams can also lead to similar effects. Thus, repeated simulation of threat-avoidance behaviors should lead to enhanced threat-avoidance skills by increasing the efficiency of the programming and execution of motor activity required in the responses to perceived threats.

**3.7.2. Implicit learning and implicit memory.** There is one difference, however, between "mental training" and dreaming: We do not explicitly remember the learning and training episodes, nor do we have any idea of what the skills we are training in our dreams really are. Thus, doubts may be cast on whether it is possible to learn something in the absence of an intention to learn and memory of what one has learned. Extensive literature on implicit learning, however, confirms that many skills important for human performance are in fact learned without any conscious access to their nature (for reviews, see Berry 1994; Cleeremans et al. 1998; Lewicki et al. 1997). A person may have no idea that s/he uses certain types of acquired knowledge when performing a certain task. Even amnesic patients can learn motor skills despite their inability to remember having ever done the task before: Their performance becomes faster and more accurate, showing implicit skill learning in the absence of any conscious memory of the learning episode. Furthermore, amnesic patients can have implicit memory also for emotional experiences that they cannot remember explicitly (Glisky & Schacter 1988; 1989; Schacter 1996). Therefore, like any other skills, threat-avoidance skills also may be learned and rehearsed without explicit access to what has been learned.

Implicit learning is very sensitive to correlations and co-variations between different features of perceived objects. If two features are associated in our experience a few times, an initial coding rule can be acquired that biases perception to detect both features when only one of them is directly perceived (Lewicki et al. 1997). Dream experience might bias waking perception so that certain perceived features are automatically associated with certain other ones in order to be prepared for possible threats. Furthermore, we are predisposed to learn certain reactions to certain stimuli. Stimuli that reflect ancient threats easily come to be feared (Marks & Nesse 1994).

If the function of dreaming is realized through implicit learning and memory, then we should predict that REM sleep deprivation has a detrimental effect on tasks requiring implicit but not explicit memory. This is what in fact has been found: Smith (1995) reports that memory for explicit tasks is not affected by REM sleep loss, but memory for procedural or implicit tasks is impaired by REM deprivation.

**3.7.3. Summary.** I conclude that rehearsing threat-avoidance skills in the simulated environment of dreams is likely to lead to improved performance in real threat-avoidance situations in exactly the same way as mental training and implicit learning have been shown to lead to improved performance in a wide variety of tasks. It is not necessary to remember the simulated threats explicitly, for the purpose of the simulations is to rehearse skills, and such rehearsal results in faster and improved skills rather than a set of explicitly accessible memories. Furthermore, REM sleep physiology appears to selectively support implicit, procedural learning.

### 3.8. Proposition 6

The original environment in which humans and their ancestors have lived for more than 99% of human evolutionary history included frequent dangerous events that threatened human reproductive success and presented severe selection pressures on ancestral human populations. The ecologically valid threat cues in the human ancestral environment fully activated the threat-simulation system. Recurring, realistic threat simulations led to improved threat perception and avoidance skills and therefore increased the probability of successful reproduction of any given individual. Consequently, the threat-simulation system was selected for during our evolutionary history.

**3.8.1. Selection pressures and ancestral threats.** So far we have shown that dreams are specialized in threat simulations, effectively triggered by real-life threats and engaging the appropriate cognitive and neural mechanisms in ways that have been shown to lead to improved performance in other learning contexts. In order to complete the argument, we now need to show that the human ancestral environment was the kind of place that contained the relevant ecologically valid cues for constantly activating the threat-simulation system, and that there was likely to be a selectional advantage from improved threat-avoidance skills so that repeated threat simulations were likely to lead to increased reproductive success.

We need to show, first, that there was a high selectional pressure in the ancestral environment. How long did people live in those conditions? Which proportion of the population survived to reach the reproductive age? As far as we know, mean life expectancy was remarkably low compared with that of modern times, only 20–25 years. According to one estimation (Meindl 1992), of those who reached five years of age in ancestral hunter-gatherer populations, about 25% died before entering the reproductive period and about 70% died before completing it. Thus, mortality rates were high, and only a selected few ever got the chance to reproduce successfully.

Second, we need to show that the real threats in the ancestral environment were the kind of events the threat-simulation system is good at simulating. What were the most likely threats to survival in the ancestral environment? How severe were they? Some major causes of death in hunter-gatherer populations were probably exposure to predation by large carnivores, exposure to the elements, infectious disease, poor conditions and risky activities during hunting and gathering, and aggression or violent encounters, especially in defense of personal resources or group territories (Landers 1992; Meindl 1992).

These estimates render it quite obvious that the life of an average ancestral human was constantly at risk in the original environment.[12] The death or serious injury of close relatives and local group members was not an uncommon event. Confrontation with extremely dangerous or even life-threatening situations is likely to have been part of everyday life rather than a rare exception. In order to reproduce successfully under such conditions, an individual must have been quite skilled at perceiving and recognizing various threatening situations (e.g., predators, aggressive strangers, poisonous animals, natural forces, social rejection by own group members), at avoiding unnecessary dangers, and when a threatening situation could not be avoided, must have been able to cope with it by using efficient cognitive and behavioral strategies that promote survival.

**3.8.2. Activation of the threat simulation system in ancestral conditions.** The key question is: What was the dreaming brain dreaming about in those circumstances? In view of the extremely harsh conditions in which our ancestors lived, it is likely that every individual was continuously rather severely "traumatized," at least by modern standards. Therefore, their threat-simulation systems must have been repeatedly activated by the ecologically valid cues from threatening, real-life situations, resulting in a continuous flow of threat-simulation dreams. In effect, the dream-production system must have been in a more or less constant post-traumatic state. In fact, that probably was the *normal* state of the system then, although we who mostly live free of immediate threats to physical survival have come to regard it as a peculiar pathological state.

As Tooby and Cosmides note (1995, p. 1190), natural selection retained neural structures on the basis of their ability to create adaptively organized relationships between information and behavior; for example, the sight of a predator activates inference procedures that cause the organism to hide or flee. Threat simulation rehearses and improves performance in processing exactly such organized relationships, specifically between information interpretable as a threat to survival and efficient cognitive-behavioral procedures that need to be activated in response to such information. In the light of our present knowledge, it seems very likely that the dream-production system had more than enough threatening experiences to work with in the human ancestral environment. Therefore it was likely to simulate realistic threats thousands of times during an individual's lifetime, which was bound to result in improved threat-avoidance skills. Individuals with improved threat-avoidance skills were more likely to leave offspring. Since the neural basis of the dream-production mechanisms is innate, dreaming came to be selected for during our evolutionary history. Individuals without the threat-simulation system would have been in a disadvantageous position, and would have been selected against in the ancestral environment. Now that most humans live in environments far removed from the ancestral ones, and face threats completely unlike the ancestral ones, it may be that the threat-simulation system is not properly activated or not able to construct useful simulations of most of the present-day threats. But dreaming still is an important part of universal human experience, and its persistence and universality can now be explained by referring to the advantages in threat avoidance it provided our ancestors with.

## 4. The dreams of hunter-gatherers and animals

### 4.1. Threat simulation in the dreams of contemporary hunter-gatherers

If, as we have argued, the dreaming brain is a phylogenetically ancient threat-simulation system with default values reflecting ancestral rather than modern conditions, then we should expect to see this mechanism naturally activated in individuals who live in conditions closely resembling the ancestral ones. We should predict high levels of survival themes, threat simulation, and animal characters in the dreams of such individuals. Fortunately there are some studies of dream content in hunter-gatherer populations. Dreams from the Yir Yoront, an aboriginal society in Australia, were collected in the 1930s and later analyzed by Calvin Hall. Some of the results have now been published in Domhoff (1996). Compared to American males, the Yir Yoront males dream significantly more about animals, have a higher proportion of aggression with animals, and a very high percentage of physical aggression. They also often dream about sharing meat from the animals they have killed.

Gregor (1981) reported a content analysis of 385 dreams collected among the Mehinaku Indians in Central Brazil whose life had remained essentially traditional at the time of the study. The Mehinaku are an exceptionally fruitful group of informants for a study on dreams, for they have the habit of carefully recalling and often recounting their dreams in the morning. Gregor found that the dreams of the Mehinaku contain significantly more physical aggression, especially with animals, than dreams from the American normative sample. However, gender differences are similar in Mehinaku and American dreams: there is more aggression in men's than in women's dreams, and women are more likely than men to be the victims of aggression. The most frequent attackers are men and animals. There are many themes in the Mehinaku dreams that could be interpreted in the framework of evolutionary psychology: for example, women often dream about being the targets of sexually violent men, and men dream about having sex with women other than their own spouse, and consequently being attacked by jealous male rivals or angry female lovers.

In his paper, Gregor (1981) provides short summaries of the Mehinaku dream reports. Here are some examples of typical threats in the Mehinaku men's dreams:

A woman attempted to have sex with him, the jealous husband assaulted him;

Lost his belt and could not find it;
Desired and approached girl, struck by his jealous wife;
Attacked by a jaguar;
Stung by wasps while in woods;
Stung by ant;
Chased by snake, he turns and kills it;
Daughter almost drowned, rescued her;
Stung by bees;
Had sexual relations with girl friend, wife saw them and became angry;
Rescued drowning brother;
Attacked by a herd of wild pigs;
Shot at threatening jaguar but missed;
Killed a threatening snake.

The prediction from the threat-simulation hypothesis is that threatening events are overrepresented among dream events, and that nonthreatening, peaceful activities are underrepresented. In accordance with this prediction, peaceful and realistic nonthreatening and nonaggressive activities (e.g., "Went to river and saw birds"; "Worked in the forest"; Watched as the sun rose"; "Went to the garden"; "Went to the field to get corn") make up only about 20% of the 276 dreams reported by the Mehinaku men. In contrast, about 60% of the dreams have a threatening situation as a theme.[13] Even if their waking lives contained more threats than ours, it is unlikely that 60% of their waking time would consist of overtly threatening episodes; for that they would have to spend almost 10 hours per day in situations involving threats (i.e., 60% of the total estimated waking time of 16 hours). Therefore, the prediction holds in the Mehinaku dreams: Dream-production mechanisms selectively overrepresent threatening events and underrepresent peaceful activities.

Dream samples from contemporary hunter-gatherer groups are probably as close to ancestral dreams as it is possible for us to get, which is not to say that they would be identical. In any case, these studies show that threat simulation is very frequent in the dreams of such individuals, and that the dream-production system tends to generate fairly realistic threat simulations when the world it simulates is not very dissimilar from the ancestral human environment.

### 4.2. Dreaming among other mammals: Evidence for the rehearsal of survival skills

This reinterpretation of the function of dreaming is consistent with the inferences we can make concerning possible dream contents and the function of dreaming in other mammals. Although we cannot know with absolute certainty that other mammals have subjective experiences during sleep, we do know that they can manifest remarkably complex behaviors during REM sleep. In humans the comparable condition is the acting out of dream experience (REM sleep behavior disorder; see sect. 3.6.2). Therefore, we may assume that to the extent these animals have conscious phenomenal experience when awake, they are likely to have similar experiences, that is, dreams, when in REM sleep.[14]

REM sleep without atonia induces complex species-specific behaviors in the cat; for example, motions typical of orienting toward prey, searching for prey, and attacking (Morrison 1983a). In several species of mammals the hippocampal theta rhythm is associated with behaviors requiring responses to changing environmental information most crucial to survival: for example, predatory behavior in the cat and prey behavior in the rabbit (Winson 1990). The theta rhythm disappears in slow wave sleep but reappears in REM sleep. Winson (1990; 1993) suggests that information important for survival is accessed during REM sleep and integrated with past experience to provide a strategy for future behavior. Thus, there is empirical evidence that in other mammals the dreaming brain also rehearses species-specific survival skills, consistent with the present hypothesis that the human dream-production system is primarily a threat-simulation system.[15]

## 5. Testability and predictions

We can now summarize the central claims of the threat-simulation theory of dreaming, all of which are supported by the available evidence, and present some testable empirical predictions:

1. Dream consciousness is an organized and selective simulation of the perceptual world. *Predictions:* (1) The neural mechanisms directly underlying dream production and threat simulation function in a selective, orderly, and organized manner rather than randomly. (2) The triggering and construction of threat simulations are not random but, on the contrary, systematically modulated by the negative emotional charge attached to episodic memory traces in the amygdala-centered emotional memory systems.

2. Dream consciousness is specialized in the simulation of threatening events, especially the kind of events that our ancestors were likely to encounter frequently. *Predictions:* (1) If we define a new dream content category that specifically includes all the threatening events in dreams, we should find that such events are, in general, overrepresented in dreams. (2) The threatening events in our dreams should be found to include severe or mortally dangerous threats more often than our waking life typically does. (We are currently testing these two predictions in a content analysis study of threatening events in students' home-based dream reports.) (3) If activated by various kinds of real mortal threats, the threat-simulation system should be found to be capable of simulating ancestral threats (e.g., animal attacks, direct physical aggression, natural forces) more efficiently (i.e., with greater frequency or greater degree of realism) than modern fatal threats highly unlike ancestral ones (e.g., smoking, traffic accidents, explosives).

3. Only real threatening events can fully activate the dream-production system. Elements from such real events are regularly incorporated into the simulations. *Predictions:* (1) No class of nonthreatening, real-life events will be found that would activate the dream-production system in a manner comparable to real-life threats (i.e., propagating frequent dream simulations of the event after a single exposure to it and causing the simulation of such events to be overrepresented in dream life compared to waking life). (2) The activation of the threat-simulation system by real threats will be found to be a universal aspect of dreaming in humans, not dependent on any specific culture. (3) The intensity (i.e., frequency and persistence) of the threat sim-

ulations triggered by a real event will be directly related to the degree of personal threat that was experienced when the event took place in reality.

4. The threat simulations produced by the dream-production system are perceptually and behaviorally realistic rehearsals of real threatening events. *Predictions:* (1) When the dream-self is in mortal danger within the dream, the dream-self is more likely than not to display a reasonable and realistic defensive reaction. (We are currently testing this prediction in a content analysis study of threatening events in dreams.) (2) The direct neural correlates of subjective visual awareness in dreams will be found to be identical with the direct neural correlates of subjective visual awareness in waking experience. (3) If muscular atonia during REM sleep is completely removed in a controlled manner, then all the movements performed by the dream-self and realized in phenomenal dream imagery will be externally observed as fully realized by the physical body of the dreamer.

5. Perceptually and behaviorally realistic rehearsal of any skills, in this case threat-avoidance skills, leads to enhanced performance regardless of whether the training episodes are explicitly remembered. *Predictions:* (1) The kinds of threat perception and avoidance behavior that are employed in threat-simulation dreams can be shown to consist of such perceptual, cognitive, and motor skill components that become faster and more efficient through implicit (procedural) learning. (2) If exposed to threat-recognition or threat-avoidance tasks during waking, an amnesic person not able to remember the learning episodes explicitly will nevertheless become faster and more efficient in these tasks through repetitive rehearsals, showing implicit or procedural learning (i.e., implicit learning in amnesic patients during waking could be used as a model of implicit learning in normal subjects during dreaming).

6. The original environment in which humans and their ancestors have lived for more than 99% of human evolutionary history included frequent dangerous events that threatened human reproductive success and presented severe selection pressures on ancestral human populations. The ecologically valid threat cues in the human ancestral environment fully activated the threat-simulation system. Recurring, realistic threat simulations led to improved threat perception and avoidance skills and therefore increased the probability of successful reproduction. Consequently, the threat-simulation system was selected for during our evolutionary history. *Predictions:* (1) Children old enough to implement threat-recognition skills and threat-avoidance behavior during waking will be capable of threat simulation during dreaming if exposed to real ecologically valid threats. (2) Ontogenetically early exposure to experienced real (ancestral) threats will stimulate the threat-simulation system, leading to earlier, more frequent, and more intensive threat simulations, lasting throughout life. Conversely, if there is total isolation from exposure to real threats, the dream-production system will develop more slowly or stay in a resting state and threat simulations will remain less frequent and milder.

All of these predictions are empirically testable in principle, and most tests could be carried out in practice. What would primarily be needed to explore the threat-simulation hypothesis empirically is, first, content analysis methods with which to precisely quantify and describe threatening events in dreams and, second, systematically collected dream and nightmare report databases from various populations and age groups that have been recently exposed to threatening events varying in frequency and degree. Such studies would enable a precise description of the operation of the threat-simulation mechanisms in detail, and help us to conclude when and in whom and to what degree the mechanisms are typically activated.

In order to disconfirm the threat-simulation theory (or some part of it), it must be shown empirically that the above predictions are false. If it can be shown, for example, that dream generation is a truly random physiological process (as stated by several theories), or that even experiences completely different from threat-related ones regularly lead to intensive, recurrent simulations, or that there are cultures in which threatening experiences do not lead to threat simulations and nightmares, then the threat-simulation theory is in serious difficulty.[16]

As an evolutionary hypothesis, the threat-simulation theory of dreaming concerns historical events, and the historical events themselves of course cannot be subjected to observation or experimental manipulation. But it would not be correct to say, for example, that theories on what caused the mass extinction of the dinosaurs 65 million years ago, or why Asia has got the Himalayas, are not empirically testable because the original events cannot be observed or experimented on. Therefore, the threat-simulation hypothesis is open to empirical testing, confirmation, and disconfirmation to the same extent as any other hypotheses regarding the causal mechanisms at work in the past, leading the natural world to be as it is in the present.[17]

## 6. Threat simulation as a biological defense mechanism

This section summarizes and clarifies how the threat-simulation mechanism is hypothesized to operate in dream production, and in what sense this operation can be regarded as biologically functional.

### 6.1. Dream production and threat simulation

Dream production is an automatic, hard-wired, regularly activated feature of human brain function. The sources of dreams are selected from long-term memory by reactivating and recombining memory traces that are the most salient for the dream-production system. *Saliency* is a function of at least two factors: The degree of threat-related or negative emotional charge and the recency of the encoding or reactivation (or other priming) of the memory traces. Therefore, the most salient memory traces for dream production consist of the ones encoding the most threatening events most recently encountered (or whose memory traces have been otherwise most recently reactivated). The saliency of a set of memory traces gradually declines over time or may be overcome by that of other traces: memory traces compete for access to dream production through their saliency.

Simulations including elements of the selected threatening memories are then reconstructed by the dream-production mechanisms. In this process the dream-production mechanism tends to use dream settings and stereotyped scripts that are compatible with threats similar to ancestral ones (composing events that involve, e.g., attacks, fights,

pursuits, escapes, intrusions, losses of valuable resources, and events during which the dream self or close kin are endangered). Typical threat-simulation dreams, such as nightmares and recurrent dreams, are thus composed of a variable mixture of salient, episodic memory traces and suitable threat-simulation scripts. This mechanism is biologically functional (i.e., it solved adaptive problems for our ancestors) because in the original environment the dream-production system regularly generated simulations of such real events that directly or indirectly threatened the reproductive success of ancestral humans.

Simulating these events rehearses performance at two stages: *threat recognition* and *threat avoidance*. The simulation of threat recognition is supposed to proceed in the following way. Salient emotionally charged memory traces are first selected for dream production. The selected visual dream imagery is subsequently realized by the occipito-temporal ventral visual stream. When potentially threatening content is present in visual awareness, the amygdala is activated in order to evaluate the potential threat. Anatomically, the amygdala receives input from the late stages of visual object recognition in the temporal lobe but projects back to all stages of visual processing and has several connections to long-term memory networks (LeDoux 1998); the amygdala and the cortical regions with which it has strong connections are highly activated during REM sleep (Braun et al. 1998; Hobson 1999a; Maquet et al. 1996). Threat recognition simulation, therefore, primes this amygdalocortical network to perform the emotional evaluation of the content of visual awareness as rapidly as possible, in a wide variety of situations in which there is a visual object or event present that is potentially dangerous. The second stage, threat avoidance, consists of the rapid selection of a behavioral response program appropriate to the dangerous situation in question (e.g., fleeing, hiding, defending, attacking) and the immediate realization of this response. Threat-avoidance simulation primes the connections between specific perceptual-emotional content and specific behavioral responses, and rehearses the efficient release of these behavioral responses through the activation of cortical motor programs. The efficient, rapid functioning of these threat recognition and avoidance networks decreases the latency and increases the sensitivity and efficiency of responding to similar real threats during waking. Therefore, threat simulation during dreaming increases the probability of coping successfully with comparable real threats, leading to increased reproductive success.

### 6.2. Why do we dream about "current concerns"?

The capability for ancestral threat simulation is the essence of the biologically adaptive function of dreaming. However, the threat-simulation mechanisms automatically select *any* available memory traces with highest relative saliency and use them as a basis for simulation, regardless of the specific content selected; the mechanisms have no "insight" into what they are doing and whether it is biologically functional or not. When the individual's waking environment doesn't include any threat cues, the sources of dreams are quite varied and may be difficult to trace; almost any recently encoded or reactivated memory traces may become selected for dream production. Therefore, the present hypothesis is not in the least similar to Freud's (empirically untestable) thesis that *all* dreams are at bottom wish fulfillments. Not all dreams are threat simulations: the functionally crucial feature of dreaming is its *capability* for regular and efficient threat simulation in environments where the appropriate threat cues are constantly imminent.

The threat-simulation mechanisms operating in present-day humans who are living in safe environments rarely find salient memory traces corresponding to ancestral or mortal threats. Even the most salient traces typically represent only relatively mild threats. In the absence of truly dangerous threats, the threat-simulation system selects any recently encoded or reactivated emotionally charged memories that happen to have higher saliency relative to other traces. The selected traces in such cases are typically about the "current concerns" of the dreamer.

There is evidence that the saliency of current concern-related memory traces can be increased by presenting concern-related verbal stimuli during REM sleep (Hoelscher et al. 1981) or by giving concern-related waking suggestions (Nikles et al. 1998) to subjects. In these studies, concern-related topics led to dream incorporation significantly more often than nonconcern-related topics did; a finding well consistent with the current hypothesis of dream-production mechanisms.

Are there any ways to *separate* the predictions derived from the threat-simulation theory from current-concerns theories? The former, unlike the latter, predicts that threat simulations can sometimes be completely *dissociated* from the current concerns of the dreamer. Recurrent dreams and nightmares are often like this. Few people regularly worry about being chased by animals, monsters, aliens, or strangers, but they may nevertheless frequently dream about such events. The current-concerns theory cannot explain these kinds of dreams, whereas the threat-simulation hypothesis can explain both why we tend to dream about everyday current concerns (e.g., occupational or marital troubles) – they are mild emotionally charged threats that are more salient for dream production than emotionally completely neutral contents – and why we also dream about very severe and rather primitive threats ("ancient concerns") – they reflect the threat-simulation scripts embedded in the dream-production system as default settings, defining the types of threatening events that should be rehearsed most frequently. Consequently, the threat-simulation theory provides us with the most parsimonious explanation of dreaming because different kinds of dreams can be explained by referring to a single mechanism, the operation of the threat-simulation system. Different dream events can be ordered on a single continuum according to the different degrees of threat they contain, and their appearance in dreams can be explained by pointing to the relative saliency of the memory traces and threat scripts that the dreams are composed of.

Still, the simulation of the current concerns of modern humans probably has little if any biologically adaptive value. The threat recognition and avoidance programs, and especially the selection pressures and increases in reproductive success associated with current concerns, are hardly comparable to those associated with ancestral threats in the ancestral environment.

### 6.3. The mechanisms behind post-traumatic nightmares

Persistent post traumatic nightmares are produced by the threat-simulation mechanisms when a set of memory traces is associated with an overwhelming charge of threatening emotional content. This set of memory traces thus becomes overly salient for the dream-production mechanisms and, in the absence of serious competitors, tends to be selected over and over again. The saliency of the memory trace should normally slowly wear off, but in some cases the threat simulations themselves (as well as waking flashbacks) may reactivate the memory traces so often that they remain highly salient for extended periods of time.

Any procedure that decreases the emotional charge associated with the memory traces should render them less salient for dream production. There is evidence that recording one's nightmares and rehearsing them with a changed ending, or thinking about them in a relaxed state (desensitization) leads to significant decreases in nightmare frequency among chronic nightmare sufferers (Kellner et al. 1992; Krakow et al. 1995; 1996; Neidhart et al. 1992). These techniques probably decrease the negative emotional charge associated with the memory traces involved in nightmare generation, thus directly decreasing their saliency for dream production. In the preceding section we reviewed studies showing that the opposite effect, increasing the saliency of certain memory traces, can be achieved through current-concerns-related suggestions. Taken together, the indirect manipulation of dream content seems to be possible by directly increasing or decreasing the saliency of threat-related memory traces for dream production. Efficient methods for manipulating the saliency of the traces will obviously be clinically useful in the treatment of disturbing threat simulations (i.e., recurrent nightmares).

The threat-simulation hypothesis may seem to imply that, for example, war veterans suffering from PTSD and traumatic nightmares should be better adapted to the battlefield than those without any post traumatic nightmares. This prediction, however, does not flow from the threat-simulation theory. Frontline combat conditions undoubtedly create memory traces with the highest negative emotional charge, leading to post-traumatic nightmares, but the threats encountered in such conditions are hardly comparable to those in the human ancestral environment. There are few such skills among human threat-avoidance programs whose rehearsal would be of much help in an environment where one may at any moment get killed by shrapnel, the invisible sniper's bullet, nerve gas, hidden land mines, missiles shot from fighter planes, and so on. Only the ability to cope with threats that closely resemble ancestral ones should with any likelihood improve through repeated threat simulations. The threat-simulation system was useful in the ancestral environment, but it should not be expected to be useful in an environment where the original human threat-avoidance skills, no matter how well trained, are no guarantee of increased probability to survive and reproduce.

### 6.4. Threat simulation as a biological defense system

It is illuminating to compare the threat-simulation mechanism with other biological defense mechanisms. The immune system has evolved to protect us from microscopic pathogens, whereas the dream-production system (along with a number of other systems) has evolved to protect us from dangerous macroscopic enemies and events in the environment. When a pathogen has invaded the host, an appropriate immune response is elicited, and when the anti-gen has been removed from the system, the immune responses switch off, as they are no longer required, and the immune system is restored to a resting state. Certain parts of the immune system, however, "remember" the infectious agent and are now better prepared to fight it off next time. Analogously, when a threatening event is encountered in the real world, a threat-simulation response is elicited by the dream-production system, and when the response is completed after repeated threat simulations, the individual will be better prepared to cope with similar threats in the future. If real threats are completely removed from the individual's environment, the threat-simulation system gradually returns to a "resting state" where the content of dreams becomes more heterogenous and less troubled.

Even when the immune system is in the resting state, large numbers of leucocytes continue to be produced. For example, millions of granulocytes are released from the bone marrow every minute even in the absence of acute inflammation (Roitt et al. 1998). These cells only live for 2–3 days; thus, if the individual is saved from infections for some time, astronomical numbers of granulocytes live and die without ever realizing their biological function at all. One may ask: "But what is the biological function of all *those* granulocytes that never took part in any immune response? They must have some hidden function since they are so numerous and are produced so regularly." This question implies a misunderstanding of the biological functionality of the immune system. Similarly, to insist that all those dreams that do *not* simulate threats must have some hidden function of their own is to misunderstand the biological function of dreaming. Exactly as the evolved biological function of the immune system is to elicit appropriate immune responses when triggered by antigens, the evolved biological function of the dream-production system is to construct appropriate threat simulations when triggered by real threats. If no antigens are encountered and recognized within the organism, the immune system remains in a resting state but it nevertheless continues to produce leucocytes; if no traces of threat-related experiences are encountered during regular dream production, the system nevertheless always ends up producing some kind of dreams.

Furthermore, biological adaptations often have features that appear nonfunctional or even dysfunctional. Immune responses frequently occur in an exaggerated or inappropriate form (Roitt et al. 1998). Type I hypersensitivity – a typical allergic reaction – is an immune response caused by harmless antigens (e.g., pollen). In the worst cases it can lead to a generalized anaphylaxis and even death. Another example of extremely harmful immune responses is autoimmunity, where the immune system attacks the individual's own tissue. A highly efficient immune system may thus be prone to false alarms, but probably also more efficient when it is really needed. As long as the net result is that those of our ancestors equipped with an operational immune system were more likely to reproduce successfully than those without, the system is biologically functional, even if negative side effects sometimes occur.

Therefore, we should not be surprised to learn that efficient threat simulation sometimes may have harmful side effects. Subjects suffering from acute or chronic nightmares typically complain of sleep disturbance: for example, fear of going to sleep, awakening from sleep, restless sleep, insomnia, and daytime fatigue (Inman et al. 1990; Krakow et al. 1995b). However, in a group of Vietnam combat vet-

erans with PTSD and subjective complaints of disturbed sleep, no clinically significant sleep disorder could be found (Hurwitz et al. 1998). It is unclear to what extent ancestral humans might have suffered from sleep disturbances due to intensive threat simulations. If some of them sometimes did, that clearly would have been a negative side effect of threat simulation, but – like an allergic reaction – one whose costs would not typically have been too high compared with the benefits.

Natural selection can only take place if there is variation within the population with regard to the biological adaptation in question and if these differences can be genetically transmitted to offspring. Genetic factors have an important role in allergic hypersensitivity (Roitt et al. 1998), which can be regarded as an indication of the sensitivity of the immune system. The sensitivity of the threat-simulation system seems to show a similar pattern. Evidence from a recent study of 1,298 monozygotic and 2,419 dizygotic twin pairs (Hublin et al. 1999a) reveals that the tendency for children to have nightmares (an indication of the sensitivity of the threat-simulation system) has a substantial genetic basis, accounting for up to 45% of total phenotypic variance.

My conclusion is that the dream-production system can be seen as an ancestral defense mechanism comparable to other biological defense mechanisms whose function is to automatically elicit efficient protective responses when the appropriate cues are encountered.

## 7. Comparison with previous theories

### 7.1. Theories on dreaming and evolution

Theories of the evolutionary functions of dreaming are few, since the received view in contemporary cognitive neuroscience appears to be that dreaming has no such function. There are, however, a couple of exceptions. In a paper entitled "Toward an Evolutionary Theory of Dreaming," Frederic Snyder (1966) proposed that when it comes to mammals, sleep could be regarded as an adaptive mode of behavior for creatures that had to spend most of their time in hiding: sleep saves metabolic and energy resources and is conducive to longevity – early mammals used sleep to survive to the next period of activity and possible reproduction with costs as low as possible. Since the animals are highly vulnerable during sleep, a built-in physiological mechanism to bring about periodic awakenings would be called for, in order to scan the environment for possible dangers. According to Snyder (1966), virtually every REM period is followed by such a brief awakening, and this serves a "sentinel" or vigilance function. The REM period preceding the awakening serves a preparatory function, activating the brain in order to prepare it for possible fight or flight. The essence of dreaming as a biological phenomenon is endogenous perceptual activation that takes the form of a hallucinated reality such as the animal might be in danger of encountering at the time of awakening. A related hypothesis was earlier presented by Ullman (1959).

Snyder's theory is substantially different from the present one. He speculates that dream content, if it has any adaptive functions in addition to general activation of the central nervous system, in some way attempts to anticipate the particular situation that the individual might actually encounter immediately after having had the dream. This explanation of dream content is not particularly convincing, for the odds are obviously very much against having the dream-production system "guess" correctly what sort of danger might be approaching the sleeping organism. Furthermore, the idea that dream content should anticipate the immediately following waking experience is rather implausible in light of the fact of "dream isolation": that dream consciousness and the contents of dreaming are by and large isolated from, for example, stimulus input, reflective thought, autonomic activity, organismic state, and motor output (Rechtschaffen 1978). If the content of dreams were anticipatory of immediately following waking events, then one would expect external cues and stimuli to have a much greater effect on current dream content than is actually the case.

Michel Jouvet (1980) proposes that in mammals a periodic endogenous genetic programming of the central nervous system occurs during REM sleep. He argues that because the learning of epigenetic behaviors requires multiple repetitions of external stimulation in order to alter synaptic organization, we might expect that also endogenous behavior-regulating mechanisms need to be reprogrammed through repetitive endogenous stimulation in order to maintain, reestablish, or stabilize synaptic pathways. The programming requires temporary inhibition of perceptual inputs and motor outputs, but we are able to see the on-line results of the programming during REM sleep if postural atonia is removed. This can be done by lesioning the inhibitory mechanisms responsible for postural atonia during REM sleep, which reveals dramatic "oneiric" behaviors; for example, in the cat: "The cat will then raise its head and display 'orienting behaviour' towards some laterally or vertically situated absent stimulus. Afterwards, it may 'follow' some invisible object in its cage and even 'attack' it, or it may display rage behaviour, or fright. . . . Pursuit behaviour has been observed to last up to 3 min." (Jouvet 1980, pp. 339–40). Jouvet's theory is, however, presented purely as a theory of the function of REM sleep, and he does not comment on the content of dreams at all. More recently, Kavanau (1997) has suggested that, in order to maintain synaptic efficacy, repetitive spontaneous activation is needed in neural circuits that are in infrequent use. In REM sleep, patterns of activity including motor components would undergo this kind of "dynamic stabilization": memories involving motor circuitry are reinforced during REM sleep. However, Kavanau regards dreaming as biologically epiphenomenal.

As we mentioned in section 4.2., Winson (1990; 1993) suggests that in animals information important for survival is accessed during REM sleep and integrated with past experience to provide a strategy for future behavior. Although the theta rhythms relevant to his arguments have not as yet been recorded in humans, Winson nevertheless speculates that human dreaming during REM sleep may also reflect the integration of information that reflects the individual's strategy for survival. Thus, Winson's view comes quite close to the present one, and provides support to the hypothesis that the function of dreaming in simulating survival skills is not uniquely human. Still, I would not describe the result of dreaming as the forming of a "strategy" for survival. The essence of human dreaming is repeated threat simulation and the only strategy is to become as proficient as possible in coping with a variety of threatening situations without having to take unnecessary risks. Winson (1993) says that his theory actually encompasses the one emanating from

current dream research, "i.e., dreams reflect adaptation in the light of current experience" (p. 245). Thus, Winson sees his theory as closely related to the ones arising from clinically oriented dream psychology, pertaining to the *psychologically* adaptive function of dreaming, which we found not entirely convincing (see sect. 2.2).

All in all, there are previous theories on the evolutionary function of dreaming, but although many of them contain valuable insights and seeds of the present proposal, none of them has considered the human ancestral environment as the proper context of the dreaming brain.[18] Neither have they taken into account, within one unifying theory, the content of normal dreams, recurrent dreams, nightmares, children's dreams, post-traumatic dreams, the dreams of hunter-gatherer populations, and the dreams of nonhuman mammals.

### 7.2. Dreaming and daydreaming

We should still consider the possibility that it was daydreaming that was selected for in evolution as a safe method of virtual threat perception and avoidance, and that dreaming is only a nonadaptive consequence of this. Like night dreams, daydreams are often vivid and multimodal simulations of real experience and contain dreamlike features (Klinger 1990). Singer (1966; 1988) proposes that daydream and night-dream content are closely related: both typically have their sources in the current concerns of the dreamer. Daydreaming often reflects our attempts at exploring the future through trial actions or through positing a variety of alternatives.

However, there are also important differences between daydreams and dreams: daydreams very often contain interior verbal monologue and they are typically more pleasant than dreams. Findings from college students' daydreams suggest that, on the average, in daydreams we focus on anxiety-provoking or worrisome thoughts only about 3% of the time, and less than 1% of daydreams include violence (Klinger 1990, pp. 84–85). Furthermore, unlike in daydreams in dreams we invariably lose our self-reflectiveness: dream events happen to us without our control (Domhoff 1996). Thus, daydreaming appears to deal with the evaluation and setting of particular future goals, and charting the ways in which we might achieve such goals. Daydreaming is at least partly controlled voluntarily. By contrast, dreaming is a fully developed involuntary simulation of the perceptual world, tuned especially to simulate and rehearse the perception of, and immediate defensive reactions to, possible threatening events.

Both daydreams and night dreams consist of mental imagery, but the former tend to chart the *goals* we would like to achieve *in the future* (and we need to be reminded about), whereas night dreams tend to simulate the *dangers* we (or our ancestors) encountered *in the past* (and we would not particularly like to be reminded about). In light of these differences, it is unlikely that night dreaming should be only a nonadaptive consequence of what daydreaming was selected for. Their functions can rather be seen as complementary to each other.

### 7.3. Other theories on dreaming

The present hypothesis on the evolutionary function of dreaming is not seriously incompatible with many of the theories reviewed in the introduction (but see N. 16 for their predictions that conflict with the threat-simulation theory). Hobson (1994) suggests that the function of dreaming is memory consolidation and the linking of memory representations with motor programs. This is true, but it is an incomplete description of the real point of the system: Which memories are linked with which motor programs and why? Such questions can only be elucidated once we consider the evolutionary context of dreaming and the role of threatening experiences in the construction of dream content. In his latest book, Hobson (1999a) suggests in a remark made in passing that his current views are more or less consistent with the threat-simulation hypothesis. He writes (p. 170): "Waves of strong emotion – notably fear and anger – urge us to run away or do battle with imaginary predators. Flight or fight is the rule in dreaming consciousness, and it goes on and on, night after night, with all too rare respites in the glorious lull of fictive elation."

Foulkes (1985) regards the *form* of dream experience as the important factor. This is true as well: it is remarkable how closely the world-model created during dreaming corresponds to the one created during waking perception. The reason for such faithful replication (and perhaps also for the fact that we rarely recognize a dream for what it is) is the fact that if you want to simulate something in such a way that the simulation works as good training for the real thing, the simulation ought to be an exceptionally good copy of the real thing. This is true of dreaming: threatening elements in dreams do look and feel like the real thing. And better still, while inside the threat simulator (i.e., while we are dreaming) we take the simulation for the real thing and fight for our lives.

Foulkes (1985) regarded the novel combinations of memory representations as an important feature of dreams. It is indeed unlikely that having once encountered a threat, the same threat should be replicated in real life in *exactly* the way it was first experienced. Thus, in order to be prepared for all kinds of situations somehow reminiscent of the original event, it is reasonable to construct several possible variations on the theme rather than just one stereotyped original version. Blagrove (1992a; 1996) pointed out that dreams do not solve the problems of the waking world, although they might solve the problems internal to the dream world itself. This is a valid point: the dreaming brain is not adapted to solve problems such as finding a job, writing a thesis, or preventing pollution. Such problems did not exist in the ancestral environment; so they are not the kinds of problems that the dream-production system would recognize or know how to handle. By contrast, it does know how to handle problems that were abundant in the original environment but have become obsolete in most Western societies: escaping and fighting aggressors and predators, defending one's family and territory, and escaping natural forces. Furthermore, the activity of the dreaming brain is not appropriately described as "problem-solving." The specific *solutions* may not be as important as is the very *repetition* of the situations critical for survival. Hartmann (1998) and Domhoff (1996) are right in treating post traumatic dreams, recurrent dreams, and nightmares as the paradigm cases of dream functioning, although their interpretation of what this function is differs from the present view.

The present hypothesis is inconsistent with the theories presented by Freud (1900) and Flanagan (1995). Freud thought that wish fulfillment is the basic point of all dreams and he tended to dismiss anxiety dreams and post traumatic dreams as just exceptional. In the present view they are, on the contrary, the paradigm cases of the biological function of dreaming. Dreaming as threat simulation can be thought of as wish fulfillment only in the sense that dreams are expressions of the primeval "wish" to survive. Flanagan (1995) doesn't believe that dream consciousness has any survival value at all. However, his assertions are not based on any kind of review of the vast empirical literature on the phenomenal content of dreams, although that is exactly the empirical body of data relevant for the evaluation of his hypothesis. He furthermore neglects the proper evolutionary context of dreaming, the ancestral environment. Thus it is no wonder he ends up claiming that dream consciousness has no biological function. He has never even considered the relevant evidence seriously.

## Conclusion

Previous theories of dream function have not put dreaming into the evolutionary context in which it belongs: the ancestral human environment. If dreams have any biological function, any survival value at all, such functions must have been manifested in that context. The dreaming brain along with the rest of human cognitive architecture has its evolutionary history, and without an understanding of what that history was like, it may be extremely difficult to figure out what the brain is attempting to do when it dreams. The hypothesis advanced in the present paper states that we dream (i.e., the phenomenal level of organization in the brain is realized in its characteristic ways during REM sleep) because in the ancestral environment the constant nocturnal rehearsing of threat perception and threat-avoidance skills increased the probability of successful threat avoidance in real situations, and thus led to increased reproductive success.

ACKNOWLEDGMENTS
This research was supported by the Academy of Finland (projects 36106 and 45704). I am grateful to Mark Blagrove, Eric Klinger, Mark Mahowald, and the anonymous referees at BBS for their helpful and insightful comments on earlier drafts of this paper. I also thank Katja Valli for practical assistance during this project and Eira Revonsuo for checking my English.

NOTES
**1.** Owen Flanagan (1995) makes a corresponding distinction between the *p*-aspects (phenomenal) and *b*-aspects (brain) of dreaming. He says that these brain states are essential aspects or constituents of the conscious states. His view (to be discussed below) is that the phenomenal aspects of dreaming are biologically epiphenomenal.

**2.** An example of an invented function of dreaming is dream interpretation. Such a function may be quite meaningful and serve many useful purposes for those involved. For example, Hill et al. (1993) have shown that interpreting one's own dream led to deeper insight than interpreting another person's dream, indicating that dream reports include personally significant elements that may help in gaining self-understanding. Nevertheless, it is unlikely that this invented function of dreaming should be one that was selected for during human evolution, since the vast majority of dreams are totally forgotten and since our ancestors probably seldom recorded or communicated even the ones they might have recalled.

**3.** It has not been empirically tested whether or not the assumption that PGO spikes are "random" or that they induce "random" activation of the forebrain is in fact true.

**4.** This point has been made also by Rechtschaffen (1978, p. 106): "If there is any isomorphism between mental experience and brain activity, then one could hardly infer a disorganized brain from dream content because dream content is not especially disorganized. . . . dreams frequently take the form of definite stories. There is neither the kaleidoscope of unrelated images nor the cacophony of isolated thoughts and words that one might expect in truly disorganized consciousness."

**5.** There is an ongoing controversy about the nature of children's dreams and whether small children really have any dreams at all (e.g., Foulkes 1999; Resnick et al. 1994). These deep disagreements are due to the different results produced by different dream-collecting methods. Representative sampling of REM sleep in the laboratory suggests that dreaming is either not present at all or only very rudimentary in the preschool period, and only develops into full form from the ages of 5 to 9 (Foulkes 1999). This contradicts the earlier findings on children's home-reported dreams (e.g., Van de Castle 1970). Foulkes (1999) argues that small children's home-based dream reports are not reflections of subjective experiences during sleep at all, but, instead, are personal or social constructs of the waking reality: results from uncontrolled parental suggestion and active confabulation. His opponents have argued that the sleep laboratory situation somehow represses the natural flow of dream experience (e.g., Hunt 1989). My view is that neither database should be completely discredited. Foulkes's (1982b; 1999) data undoubtedly show that the REM periods of small children who are living in a safe modern environment are only rarely associated with conscious experiences that fulfill the criteria of a dream. However, these data do not show that such experiences are not possible, at least occasionally or in specific subgroups of children who are living in less safe environments or who have otherwise been exposed to various threatening events. It seems extremely implausible that the vast samples of children's home-based dream reports (e.g., Van de Castle 1970; 1983) would be nothing but products of suggestion and confabulation. It is hard to believe that parents would suggest to their children the topics that have been found to be prevalent in children's home-based dreams, such as the high proportions of aggression and victimization, since such dream content might easily be perceived by the parents as an indication of psychopathology or psychological disturbance in their children. Children's nightmares obviously are even less likely to be mere social constructions and confabulations: The American Sleep Disorders Association (1990) estimates that 10–50% of children at the age of 3–5 so frequently have nightmares as to alarm their parents. Furthermore, there are common features in the home-based and laboratory databases, such as the declining proportion of animal characters with increasing age, which suggests that both data flow from the same source. Foulkes (1999) advocates a highly contestable theory of consciousness on which his interpretations of the data are based: he takes consciousness to be "reflective consciousness" and argues that small children and animals lack it and therefore not only are unable to experience dreams but are in general like some kind of nonconscious zombies. Instead of accepting this view, the threat-simulation theory predicts that small children should be capable of having threat simulation dreams as soon as their perceptual and motor skills are at a level that enables threat recognition and avoidance in the waking state. However, this capability is only rarely realized if the child is not exposed to real threatening events that would activate the threat-simulation system properly. Children's home-based dream reports may thus largely reflect those relatively infrequent situations in which the threat-simulation system has become active and dreaming proper is experienced and consequently spontaneously remembered. This interpretation seems plausible in light of the fact that in home-based studies only one or a few dream reports at most, per child, were typically reported by a very large number of children (e.g., Van de Castle 1983), whereas in laboratory studies typically several REM-sleep awakenings were performed in a relatively small number of children but only a few reports that would qualify as dreams were obtained. The laboratory studies

primarily reveal that, in children, there is a lot of REM sleep without any dreaming. However, the home-based dream reports and clinical and parental observations reveal that, when dreams proper do occur in children, they remarkably often include threatening elements.

**6.** It could be claimed that children are exposed to fairy tales and cartoons that include animals, and therefore dream about them. However, when listening to fairy tales or watching cartoons, children never directly perceive the actual animals, but only rather poor representations of them. The child is never personally in danger; the threats are directed against some characters in the story. Furthermore, whereas the amount of animals and aggressions in dreams declines with age, exposure to increasingly violent stories, movies, games, and so on increases. Thus, if fiction were the main source of animal and aggressive content in children's dreams, we could expect, first, the simulations to be simulations of storytelling or watching- TV experiences ("I dreamt that father told me a frightening story about an angry bear"; "I dreamt that I was watching a very frightening TV-program about wild animals"), not of personal encounters with the threatening agents, and, second, the frequency of fiction-induced animal and aggression content in dreams to increase with increased exposure to all forms of fiction with age. Neither of these predictions is supported by the data. Furthermore, as Van de Castle (1970, p. 38) observes: "To say that [the high percentage of animal characters in children's dreams] would be attributable to the influence of the many animal characters that appear in children's books would be begging the question because one would then ask why are animals so frequently utilized in children's stories and what accounts for children's fascination with them?"

**7.** "Long-term, across-generation recurrence of conditions . . . is central to the evolution of adaptations. . . . Anything that is recurrently true . . . across large numbers of generations could potentially come to be exploited by an evolving adaptation to solve a problem or to improve performance" (Tooby & Cosmides 1992, p. 69).

**8.** Wilmer (1996, p. 88) mentions that 53% of 359 catastrophic dreams from Vietnam veterans suffering from post-traumatic stress disorder were "terrifying nightmares of the actual event as if it were recorded by cinema verité." These dreams portray "a single event in recurrent replays" and, according to Wilmer, "they are the only human dreams that define themselves in a completely predictable manner." Another 21% of the veterans' war nightmares contained plausible war sequences that conceivably *could* have happened but had not actually occurred. However, Brenneis (1994) argues that the relation between dreams and the original traumatic experience is not isomorphic: if trauma texts are paired with dream texts, at least some transformed elements can invariably be observed.

**9.** According to recent PET studies (e.g., Maquet et al. 1996), neither significant increase nor decrease of regional cerebral blood flow (rCBF) can be observed in the motor cortex (Brodmann area 4) or premotor cortex (Brodmann area 6) during REM sleep. However, such blood flow measurements reflect the actual neural activity only quite indirectly and with coarse spatial and temporal resolution. The increased neural activity during REM sleep in the selected population of pyramidal tract cells, verified by direct single-cell measurements in sleeping animals, evidently does not result in any measurable net change in rCBF in the motor areas (where there are also other types of cells that may behave differently during REM sleep). PET studies do show that there is a significant decrease of rCBF in a large area in the dorsolateral prefrontal cortex (Brodmann areas 10, 46, 9, and 8). These areas are believed to be involved in deliberate, "free-willed" actions and new choices that take place without the dictations of external cues but involve internal planning and voluntary decision-making (Passingham 1993). Thus such reflective planning and decision-making functions should not be supported by REM sleep. However, the performance of habitual, procedural actions in response to external cues is assumed to depend on premotor mechanisms alone (Passingham 1993), which are not suppressed during REM sleep. Threat-avoidance responses typically are externally cued (by the perceived external threat) and fairly "instinctive" actions whose efficiency the threat simulations aim to preserve or increase, and therefore the dorsolateral prefrontal cortex need not be involved in threat simulation.

**10.** A recent study by Hublin et al. (1999b) confirms that there is a strong correlation between the occurrence of nightmares and night terrors, supporting the present view that both phenomena may in fact reflect an increased level of activation in one underlying system, the threat-simulation mechanism.

**11.** This behavior is similar to panic, during which "Intense mental activity is focused on planning escape. When the overwhelming urge to flee is translated into action, all effort is concentrated on escape" (Nesse 1997, p. 77). Panic can be viewed as an adaptation that evolved to facilitate flight from life-threatening danger (Nesse 1997).

**12.** Meindl (1992) presents estimates of mortality based on three different hunter-gatherer cemetery sites in Africa and North America where hundreds of skeletons have been unearthed. The oldest of the communities dates back circa 10,000 years ago. Mean life expectation in each of the three populations is estimated to have been from 20 to 22 years. As Meindl (pp. 408–10) observes, "the relentless forces of mortality at every age assured that only a small proportion of a hunter-gatherer society was 'elderly' in our sense of the word"; instead, since "fertility must have been high to balance the annual death toll," "the paleodemographic data imply that the villages were rather like modern child day-care centers." Furthermore, the "demands of their economy may have compromised the health and safety of hunter-gatherers," "numerous healed long-bone fractures in the skeletons . . . as well as the higher mortality rates for males . . . suggest the perils of a foraging way of life" (Meindl 1992, pp. 408–10).

**13.** A content analysis of the 276 Mehinaku men's short dream summaries reported in Gregor (1981) was carried out. Two independent judges classified the dreams using the following mutually exclusive classes: (1) Threatening event (objective threat); (2) Subjective threat; (3) Peaceful activity; (4) None of the above or unclassifiable. The definitional criteria were refined and the use of the scale was practiced by first scoring the 109 Mehinaku women's dream summaries from Gregor (1981). The following definitional criteria were used:

1. Threatening event: Any event in the dream, which, if it were to occur in the waking life, would potentially decrease the probability of future reproductive success of the dream-self and close kin. Such events include the following: (a) Events that probably or potentially lead to immediate loss of life of the dream-self or close kin or local group members (i.e., any member of the local Mehinaku village of about 80 people); (b) Events that probably or potentially lead to physical injury of the dream-self or close kin or local group members; (c) Events that probably or potentially lead to loss or destruction of valuable physical or social resources of the dream-self or close kin. Physical resouces include all valuable possessions of and the territory controlled by the dream-self or close kin; social resources include membership and social status in the local group or society and access to desirable mates.

Examples of possible threatening events for the Mehinaku were outlined: Any local group member (including self and close kin) that is: (1) participating in an aggressive encounter with animal, human, or other malevolent characters (e.g., monsters, spirits) that can cause death, physical injury, or loss of territory or freedom; (2) encountering or perceiving dangerous animals in the vicinity (e.g., snake, wild pig, alligator, stingray, jaguar) even if the animal does not attack or show aggressive behavior; (3) being the victim of sickness or encountering animals or people or objects (e.g., parasite carrying animals, rotting food or corpses, feces) that carry or can otherwise cause disease; (4) victim of circumstances or natural elements (e.g., weather, coldness, heat, fire, rain) that can cause sickness or injury or prevent access to resources (e.g., making hunting, fishing, gathering difficult or im-

possible); (5) victim of accident or failure or misfortune that can cause death, physical injury, or loss of physical resources (getting lost, losing/breaking or not getting access to valuable possessions such as weapons, tools, prey, food, clothes). Dream-self or close kin is: (1) victim of social rejection or isolation that can cause loss of important social relationships and status in the group and/or loss of help and support from other group members; (2) taking part in risky activities (e.g., stealing, breaking rules/taboos) that can cause social punishment, isolation, shame, or loss of social status in the group.

2. Subjective threat: An event that does not fulfill the criteria of Threatening Event, but during which the dream-self nevertheless interprets the situation as threatening or experiences fear and anxiety.

3. Peaceful, everyday activity: An activity that is likely to be a part of the dreamer's everyday life, is realistic (nonbizarre), and involves no threatening or aggressive content.

4. None of the above: All such dreams that do not clearly fulfill any of the criteria of classes 1–3.

The results of the inter-rater agreement between the judges was 82.2% (i.e., 227 of the 276 dream summaries were scored identically). Disagreements were subsequently resolved through discussion.

The relative proportion of threatening events in Mehinaku men's dreams were as follows: Threatening events (objective threats): 56.2%; subjective threats: 7.6%; peaceful activities: 20.3%; none of the above: 15.9%.

To summarize: some kind of threatening elements are present in 63% of the dreams; events potentially threatening future reproductive success, were they real, ("objective threats"), make up the majority of these, accounting for 56% of all dream themes in the dreams of Mehinaku men.

**14.** Mammals had to live for at least 100 million years in the shadow of ferocious reptiles. Long periods of sleep allowed them to remain in hiding and save their strength for the brief active periods of finding food. Threat simulation or simulation of survival skills during REM-sleep may have been a valuable adaptation during this era, when mammals had to compete with the much larger and more numerous reptiles for resources. Dreaming may be just one more addition to the biological arms race whereby different species prosper in their different ecological niches. It may also be that other than mammalian brains simply cannot support the sort of multimodality simulations that dreaming consists of.

**15.** Thus, the hypothesis as applied to all mammals in general is: "Dreaming rehearses species-specific survival skills" – the exact nature of these skills, of course, varies from species to species depending on the niches that the species occupy. The hypothesis as applied to humans specifically is that dreaming rehearses threat perception and threat avoidance, particularly significant types of human ancestral survival skills.

**16.** We can contrast these predictions with those derived from other theories. All the theories claiming that dream production is based on fundamentally random processes (e.g., Crick & Mitchinson 1983; Foulkes 1985; Hobson & McCarley 1977) are of course inconsistent with predictions 1.1–1.3. All theories claiming that dreaming is specialized on some type of psychological content or effect (e.g., problem solving, emotional calming, mental health) other than threat simulation are in conflict with predictions 3.1. and 3.2. Such theories would need to show that the kinds of dreams they claim to be functional are generated by the dream-production mechanisms as reliably and effectively as threat simulations are and that having such dreams in the ancestral environment was likely to lead to increased reproductive success. Predictions 6.1 and 6.2 are inconsistent with Foulkes's theory of children's dreams (see N. 5). A central claim in Solms's (1997a) theory is that volitional motor activity is not possible during sleep and that the entire motor system is deactivated. These claims are inconsistent with Proposition 4 and the predictions derived from it. Furthermore Solms's theory is inconsistent with the data on high levels of activity in the corticospinal pyramidal tract neurons of the motor system during REM sleep (blocked downstream in the spinal cord to prevent dream enactment) and with the clinically well-documented phenomenon of REM sleep behavior disorder, a parasomnia disconfirming the claim that full motor activity and its execution would be impossible during REM sleep and dreaming (see also sect. 3.6.2 on the motor realism of dreams).

**17.** The starting point for the hypothesis was my view of dreaming as a natural virtual-reality simulation in the brain (first published in Revonsuo 1995). I subsequently asked: If dreaming is essentially a simulated perceptual world, what kind of simulations might be useful? I speculated that if flight simulators are used in order to train pilots to handle dangerous events that might arise during a real flight, perhaps the brain trains its own survival skills in a fight-or-flight simulator, specialized for extremely dangerous events that might be encountered in nature. The general prediction, that dream content should reflect this fact, was then made and, as the present paper documents, a surprising amount of evidence supporting the hypothesis has been found in the relevant literature. I was unaware of most of these empirical results when I began the project.

**18.** In his book *Affective Neuroscience* (which I happened to come across when the present paper was nearly finished) Jaak Panksepp makes an intriguing evolutionary suggestion on the origin of dreams: "Indeed, perhaps what is now the REM state was the original form of waking consciousness in early brain evolution, when emotionality was more important than reason in the competition for resources. This ancient form of waking consciousness may have come to be actively suppressed in order for higher brain evolution to proceed efficiently. This is essentially a new theory of dreaming" (Panksepp 1998a, p. 128). The threat simulation theory of dreaming is certainly consistent with Panksepp's proposal. [See also Panksepp "Toward a General Psychological Theory of Emotions" *BBS* 5(3) 1982.]

# Open Peer Commentary
and Authors' Responses

Table 1. *Commentators for special sleep and dreams issue*

| Commentators | Hobson et al. | Solms | Nielsen | Vertes & Eastman | Revonsuo |
|---|---|---|---|---|---|
| Antrobus, J. S. | JAH | MS | TAN | | AR |
| Ardito, R.B. | | | | | AR |
| Bednar, J.A. | | MS | | RPV | AR |
| Blagrove, M. | JAH | MS | TAN | RPV | AR |
| Borbély, A.A. & Wittmann, L. | JAH | MS | TAN | RPV | AR |
| Born, J. & Gais, S. | | | TAN | RPV | |
| Bosinelli, M. & Cicogna, P.C. | | | TAN | | |
| Cartwright, R. | JAH | MS | | RPV | AR |
| Cavallero, C. | | | TAN | | |
| Chapman, P. & Underwood, G. | JAH | | | | AR |
| Cheyne, J.A. | | | | | AR |
| Cipolli, C. | | | | RPV | |
| Clancey, W.J. | JAH | MS | TAN | RPV | AR |
| Coenen, A. | | MS | TAN | RPV | |
| Conduit, R., Crewther, S.G. & Coleman, G. | JAH | MS | TAN | RPV | |
| Domhoff, G.W. | JAH | MS | TAN | | AR |
| Doricchi, F. & Violani, C. | | MS | | | |
| Feinberg, I. | JAH | MS | TAN | RPV | |
| Fishbein, W. | | | | RPV | |
| Flanagan, O. | JAH | MS | TAN | | AR |
| Franzini, C. | | MS | TAN | | |
| Germain, A., Nielsen, T.A., Zadra, A. & Montplaisir, J. | | | | | AR |
| Gottesmann, C. | JAH | MS | TAN | RPV | AR |
| Greenberg, R. | JAH | MS | TAN | RPV | AR |
| Greene, R.W. | JAH | | | | |
| Gunderson, K. | JAH | | | | AR |
| Hartmann, E. | JAH | MS | TAN | | |
| Herman, J. | JAH | | | | |
| Hobson, J.A. | | MS | | | |
| Humphrey, N. | | | | | AR |
| Hunt, H.T. | JAH | MS | TAN | RPV | AR |
| Jones, B.E. | JAH | | | RPV | |
| Kahan, T.L. | [JAH] | MS | TAN | | |
| Khambalia, A., Shapiro, C.M. | | | | | |
| Kramer, M. | JAH | MS | TAN | | AR |
| Krieckhaus, E.E. | | | | | AR |
| LaBerge, S. | JAH | MS | | | |
| Lehmann, D. & Koukkou, M. | JAH | MS | TAN | | |

Table 1. *Commentators for special sleep and dreams issue*

| Commentators | Hobson et al. | Solms | Nielsen | Vertes & Eastman | Revonsuo |
|---|---|---|---|---|---|
| Levin, R. | | | | | AR |
| Lydic, R. & Baghdoyan, H.A. | JAH | | | | |
| Lynch, G., Colgin, L.A. & Palmer, L. | | | | RPV | |
| Mancia, M. | JAH | | | | |
| Marczynski, T.J. | JAH | MS | TAN | RPV | AR |
| Mazzoni, G. | | | | RPV | |
| Mealey, L. | | | | | AR |
| Montangero, J. | | | | | AR |
| Moorcroft, W.H. | JAH | MS | TAN | RPV | AR |
| Morgane, P.J. & Mokler, D.J. | JAH | MS | | RPV | |
| Morrison, A.R. & Sanford, L.D. | JAH | MS | TAN | RPV | AR |
| Nielsen, T.A. & Germain, A. | | | | | AR |
| Nofzinger, E.A. | JAH | MS | | | |
| Occhionero, M. & Esposito, M.J. | | MS | | | |
| Ogilvie, R.D., Takeuchi, T. & Murphy, T.I. | [JAH] | MS | TAN | RPV | |
| Pace-Schott, E.F. | | | TAN | | |
| Pagel, J.F. | JAH | MS | TAN | RPV | AR |
| Panksepp, J. | JAH | MS | TAN | RPV | AR |
| Perry, E.K. & Piggott, M.A. | JAH | MS | | | |
| Peterson, J.B. & DeYoung, C.G. | | MS | | | AR |
| Portas, C.M. | JAH | | | | |
| Porte, H.S. | | | TAN | | |
| Revonsuo, A. | JAH | | TAN | RPV | |
| Rotenberg, V.S. | | MS | TAN | RPV | AR |
| Salin-Pascual, R., Gerashchenko, D. & Shiromani, P.J. | | MS | | | |
| Salzarulo, P. | | | TAN | | |
| Schredl, M. | JAH | MS | TAN | RPV | AR |
| Shackelford, T.K. & Weekes-Shackelford, V.A. | | | | | AR |
| Shapiro | | MS | | | |
| Shevrin, H. & Eiser, A.S. | JAH | MS | TAN | | AR |
| Siegel, J.M. | | | | RPV | |
| Smith, C. & Rose, G.M. | | | | RPV | |
| Solms, M. | | | TAN | | |
| Steriade, M. | JAH | | TAN | | |
| Stickgold, R. | | | TAN | RPV | |
| Thompson, N.S. | | | | | AR |
| Vogel, G.W. | JAH | MS | TAN | | |
| Wichlinski, L.J. | | | | | AR |
| Zadra, A. & Donderi, D.C. | JAH | | | | AR |

# Open Peer Commentary

*Commentary submitted by the qualified professional readership of this journal will be considered for publication in a later issue as Continuing Commentary on these articles. Integrative overviews and syntheses are especially encouraged.*

## How does the dreaming brain explain the dreaming mind?

John S. Antrobus

*City College of the City University of New York, New York, NY 10025.*
john@psyche.socsci.ccny.cuny.edu

**Abstract:** Recent work on functional brain architecture during dreaming provides invaluable clues for an understanding of dreaming, but identifying active brain regions during dreaming, together with their waking cognitive and cognitive functions, informs a model that accounts for only the grossest characteristics of dreaming. Improved dreaming models require cross discipline apprehension of what it is we want dreaming models to "explain."

[HOBSON ET AL.; NEILSEN; REVONSUO; SOLMS]

The new brain imaging studies by Braun et al. (1997; 1998) and the new lesion-dreaming research by **SOLMS**, and the solutions to the exclusive REM sleep-dreaming model proposed by **NIELSEN** and **HOBSON ET AL.** afford a timely opportunity to evaluate our neurocognitive conceptions of dreaming sleep. Because it is impossible to use our most powerful experimental methods to study dreaming, models of cognitive processes ($\psi$) in sleep are more dependent on knowledge of neurobiological processes ($\Phi$) models of waking cognition. But neuroscientists' view of cognitive explanations of dreaming seem woefully simplistic, rather like a Manhattanite's picture of San Francisco from Manhattan in Steinberg's famous cartoon, where California is merely a small undifferentiated smudge on the other side of the Hudson River.

For some neuroscientists, a cognitive explanation is no more than a metaphor located in brain space and time. "Synthesis," "auto-activation," and "back projection" *imply* an explanation of the neural-cognitive process of dream construction that they do not deliver. It is the assumption by neuroscientists such as **HOBSON ET AL.** that these metaphors constitute explanatory models of cognitive processes that is at the heart of the controversy that **HOBSON ET AL.** attempt to resolve. In inviting us to respond to their revised AIM model, we thank **HOBSON ET AL.** for the opportunity to comment on this larger conflict between and "explanations" of what is, after all, the same process.

Despite their limitations, cognitive data provide us by far the most detailed information about dreaming. But we need brain scans, not to tell us that dreaming takes place in brain space and time, and not to confirm that dreams have visual imagery, but help us find out about those cognitive and affective characteristics and processes of dreaming that we do not *already* know. And that is a lot. Here, **HOBSON ET AL.** have provided an excellent account of what the recent imaging studies of Braun et al. (1997; 1998), Maquet et al. (1996), and Nofzinger et al. (1997). But as we move from cerebral cortex down to the brain stem, the specificity of the contribution of $\Phi$ to $\psi$ processes becomes increasingly diffuse, and correspondingly less informative to both $\psi$ and neurocognitive, $\psi(\Phi)$, models of dreaming. The revised pontine cholinergic-adrenergic model of REM-NREM sleep provided by **HOBSON ET AL.** provides an account for the location in *time* of most dreaming, and it shows how widespread cortical activation coupled with functional differentiation provides a *general* $\Phi$ basis for dreaming.

But the original Activation-synthesis and AIM models also claimed that PGO information and a cognitive "synthesis" process somehow constitute a $\Phi \rightarrow \psi$ explanation of dreaming. Although these proposals were, at the time, altogether plausible, they were found on careful study to be completely without empirical support. The detailed $\Phi$ account of the pontine mechanisms of the REM-NREM cycle has created, for neuroscientists and lay persons alike, the impression that the pontine-based $\Phi$ model of REM sleep constitutes also a model of dreaming – which it does not.

**HOBSON ET AL.** start with the thorny epistemological problem concerning what the mentation report can tell us about the actual dream. They review the familiar research literature that shows that the magnitude of every measure of sleep cognition is greater in REM than NREM, and acknowledge Pylyshyn's (1989) argument about the hazards of interpreting *anything* about private experience from verbal reports. These arguments notwithstanding, **HOBSON ET AL.** note that inasmuch as we all know from our private experience that we do dream, the failure to measure the private experience must be a limitation in psychological methodology. Then they proceed to take verbal report data pretty much at face value.

Several of our studies have successfully separated the dream report from evidence about the private mentation experience. Rosenblatt et al. (1992) showed that both log Total Recall and log Total Visual Imagery word counts of film clips shown before going to sleep were higher following REM than NREM awakenings. This evidence indicates that a portion of the REM-NREM differences in sleep mentation reporting difference may be attributed to superior recall and report processes operating *after* awakening from the respective states. Although I had proposed that the entire dream recall state difference might be attributed to enhanced REM recall in my 1983 REM-NREM word count paper (Antrobus 1983), after further research in our 1992 paper we clearly revised this suggestion, when we measured the magnitude of the recall-memory effect and showed that it was much smaller that the REM-NREM Total Recall effect. We reported that "a substantial part of the dream recall effect is the result of *pre*-awakening processes" (Rosenblatt et al. 1992, p. 223). **HOBSON ET AL.** fail to realize that even though there are real REM-NREM differences in mentation, some part of all REM-NREM differences is owing to a *REM-recall* advantage.

I also proposed in 1983 that the amount of dream content generated within REM sleep was more likely than in NREM "to be influenced by goals or motives established in the waking state" (p. 567), citing as evidence our finding that the dreamlike quality of REM reports dropped over 20 nights in the lab until it was indistinguishable from NREM reports (Antrobus et al. 1991). NREM dream reports were constant over time. This within-REM sleep motive to attend to dreaming appears to extend to the magnitude of visual imagery, holding total dream content constant. In the 1986 word count study (Antrobus 1986), although visual imagery words were strongly associated with the REM-NREM state difference, the visual effect was a subset of the stronger association REM-NREM association with total content count. But in the Antrobus et al. (1995) study where subjects not only gave a verbal mentation report but also matched their reported visual images to one of 16 photographs that varied by brightness and clarity, the proportion of visual imagery to recalled information increased substantially, so that the association of visual imagery with REM-NREM became *superior* to total recalled content. I assume that the emphasis in the report procedure on visual imagery motivated sleepers to attend more strongly to their visual imagery while asleep, and that this attention process operated more strongly during REM than NREM sleep – because as Braun et al. (1998) showed, both the limbic system and extrastriate visual cortex are activated in REM.

From this perspective, visual images and the dreamer's reaction to them may not be simply a set of fully-realized images, produced whether or not the dreamer is moved to notice them – as **HOBSON ET AL.** imply. Rather, the REM dream may be created by the process of attending to poorly structured information automatically generated (Antrobus 1991) in activated extra-pontine brain modules. In REM sleep, the attention process appears to be strongly determined by goal states established in the waking state. The memory of the dream is simply the functional residue of this attentional – construction process. This process of cognitive penetrability may not be, as Pylyshyn suggests, a post-awakening confound, or as Freud suggests, secondary revision of the original dream, but rather the primary process by which the dream is produced. This attentional process is a metaphor for the processes by which activated brain modules attempt to "interpret" or, (make sense of) the noisy activity in neighboring modules. These metaphors are made explicit in my neural network attractor model, DREAMIT2 model (Antrobus 1991). Upon awakening, the interpretive process accelerates as the dormant verbal and meaning modules of the left temporal and prefrontal cortices become active. Objects and persons become named and explained during the sleep-to-waking transition. Some aspects of the dream become more reasonable, while other relationships that were "acknowledged as seen" in the sleep-dream, are now judged bizarre. Conversely, other features that were implicitly understood but not visually imaged in the dream, for example, "I was in my car," are, upon awakening, reported as visual images. That is, reporters cannot recall any visual features.

My original $\Phi \rightarrow \psi$ activation model rested on assumptions about state differences in cortical activation and sensory thresholds that were basically those of Hobson et al. (1978), namely that pontine mechanisms produced widespread cortical activation in REM sleep, but the characteristics of the mentation itself were the best indication of localization of cortical activation. Twenty years later, sharing with **HOBSON ET AL.**, the monist assumption that the state differences in cognitive activation are largely confined to those regions identified by the brain scans, a model of dreaming stands to learn much from maps of regional brain activation in sleep. Of course, locating a cognitive feature, such as image brightness, in time and brain-space does not constitute a $\Phi$ explanation of $\psi$, but as I shall show below, it can certainly help.

But the heart of the controversy lies elsewhere. With the original Activation-synthesis model to the present updated AIM model, Hobson and his colleagues have consistently presented the detailed pontine generator model that accounts for the *when* of REM and NREM sleep as an explanation of *how* the cognitive characteristics of dreaming are produced. But their $\Phi \rightarrow \psi$ causal claims are highly speculative, and indeed, often contradicted by waking $\Phi \rightarrow \psi$ models upon which they are based. Aside from the original and important contribution – that the pons determines, or is at least one determinant of, the widespread activation of the cortex during REM sleep, these assumptions about *how* the pons determines the features of dreaming are completely without empirical support. The most problematic of these models, which I discuss below, concerns the assumption that the "chaotic nature of the pontine auto-activation process" constitutes a basis for the bizarre features of dreaming.

Given the powerful association of the *when* of the REM sleep with the *when* of dreaming, many of us have come to equate the *how* of REM sleep, $\Phi$, with the *how of* dreaming, $\psi$. **SOLMS** proposes that the temporal REM sleep = dreaming, $\Phi_{REM-NREM} \rightarrow \psi$, association is only indirectly related to the functional, or causal, $\Phi_{FRONTAL-DOPAMINERGIC} \rightarrow \psi$ relationship. That is: $\Phi_{REM-NREM} \rightarrow \Phi_{FRONTAL-DOPAMINERGIC} \rightarrow \psi_{DREAMING}$. If he is correct, and a substantial amount of data, including our 1995 diurnal rhythm paper, do support his position, the M = modulation vector of the AIM model loses its causal position in a $\Phi \rightarrow \psi$ model of dreaming.

This leaves us, once again, with a regional cortical-cognitive, $\Phi \rightarrow \psi$, model, where the pontine contribution of AIM is useful only for locating dreaming in time. As **HOBSON ET AL.** show, the accurate brain localization of activity in REM, NREM, and waking states (Braun et al. 1997; 1998), now allows us to attempt to map different features of dreams onto cortical and subcortical regions whose functions in the waking state have been identified – to the extent that these functions are invariant across states. But how far do, and can, these $\Phi \rightarrow \psi$ mappings take us toward an *explanation* of dreaming? It is noteworthy that almost all $\Phi \rightarrow \psi$ assumptions proposed from brain localization observations in sleep consist of mapping *already known* characteristics of dreams onto well-established functions of brain regions. For brain mapping to

tell us something about dreaming, however, it must identify new $\psi$ characteristics of dreaming or force us to change existing $\psi$ models of dreaming. The activation side of the Activation-Synthesis model did just that when it showed that the brain of the motorically-quiet REM sleeper was, in fact, quite active. The Braun et al. papers tell us much more. In particular, they show that REM sleep visual imagery cannot begin in the striate cortex, that many regions that participate cooperatively in waking are dissociated in REM sleep, and that the amygdala and limbic system that create the cognitive and affective characteristics of waking thought appear to contribute to the production of the REM dream.

The next question is, what can future $\Phi$ data contribute? Perhaps the most significant questions have to do with the relative influence of one cortical region over its neighbors, the magnitude of their interaction, and the ordinal character of these interactions. Both folk and psychoanalytic "interpretations" of dreams implicitly assume top-down, or what **SOLMS** calls "back projection" of information from meaning to visual-spatial functions, that is frontal to parietal and occipital locations. In 1991, I suggested that several other ordinal activation sequences are possible, particularly, visual-spatial $\rightarrow$ meaning $\rightarrow$ conation $\rightarrow$ motor (Antrobus 1991). That is, occipital-parietal regional may produce images that are interpreted by frontal structures, producing the surprise phenomena (" . . . and suddenly . . . ") of dreams. The subsequent imagined motor responses to these images then follow the same sequence they do in the waking state, except that, being imaged, they do not produce the same feedback as in waking, thereby eliciting additional cognitive and motor responses. Because the construction of the dream experience appears to continue even as the sleeper awakens and constructs a verbal description of his/her sleep experience, one cannot take at face value the ordered information in the report. Although it is well beyond the state of current brain imaging devices, such sequential brain scans could assist in the understanding of the ordinal causal effects in dream construction.

**SOLMS** takes this ordinal-spatial sequence for granted when he assumes limbic system effects are back projected to the visual cortex. But can, for example, a frontally-created goal – for example to seek out one's parent for protection – actually constrain the visual association and parietal cortices to construct an image of the parent? The assumption that dream motives can be "interpreted" from the sequence of the visual images is implicit and absolutely essential to the interpretation of dreams. Hasegawa et al. (1998) have shown that the retrieval of visual memories – in the temporal cortex – is under prefrontal cortex control. Braun et al. (1997) have shown that while the medial prefrontal cortex is more active in REM than NREM and wakefulness, the lateral orbital, dorsolateral and opercular prefrontal regions are *less* active. The evidence $\Phi$ is simply not clear enough at this time to determine whether top-down $\psi$, or back-projection, assumption is tenable. An equally plausible alternative to the top-down assumption is that image production is initially independent of limbic influence and that it precedes rather than follows the cognitive influence of the limbic structures. The latter might begin with the evaluation of the visual images ("Is that a friendly or unfamiliar face?") and be followed by imagined motor responses ("Shall I stay or run?"). More evidence on the prefrontal location of visual retrieval control, as well as the ordinal relation between these regions during dreaming will help to determine the strength of this key assumption about dream processes.

The next critical $\Phi \rightarrow \psi$ question is whether the pattern of dream features in a given state, such as REM sleep, is rigidly determined by the pattern of brain activation that is supported by subcortical structures, or whether the sleep state supports a general state-specific brain activation architecture that can be modified by the demands of the narrative dream sequence – as it is in waking perception – according to the demands of incoming information. For example, we assume that the dreamer's motor commands to run from an imagined strange man is accomplished in an activated motor cortex. Is the activation level of that motor cortex constant throughout the REM period, or is it *also* modified by the imagined demands of the dream? Support for the latter position comes from our finding of a decrement in REM dream reporting over 20 nights of lab awakenings, and the increment in visual imagery when photo scales were employed – as reported above.

HOBSON ET AL. show how the $\Phi \leftrightarrow \psi$ relations of a large number of studies can be mapped onto the three dimensions of their AIM model. But as the high dimensionality of these findings, particularly in brain imaging, expands, they acknowledge that to map multidimensional relationships onto three non-orthogonal dimensions tends to weakens the precision of the representation of process. Each of the three AIM "dimensions" is, in fact, an array of multi-dimensional input, $\Phi$, variables linked to another multi-dimensional set of output, $\psi$, variables that tell us little when they are concatenated into a single dimension. It is like mapping cities by their latitude. Although you can represent each city on a latitude line, you cannot locate it unless you also know its longitude. The real value of a model is to account in a systematic way for all the known evidence, and then use it to suggest new tests of the model until it can no longer represent the evidence and must be replaced by a better model. Although HOBSON ET AL. have indeed used AIM to attempt to account for a large amount of evidence and it has served them and us well, they have shown that it must now be superseded.

There are two problematic pontine-to-cortex issues that have persisted from the Activation-Synthesis to the AIM model. Brain imaging and lesion studies tell us what perceptual and cognitive processes are associated with particular cortical and subcortical regions. Nearly all of the evidence for these relationships comes from responses that are closely linked to antecedent *external* stimuli. Because our F $\Phi \rightarrow \psi$ knowledge is derived from the power of our stimulus response experimental procedures, we, as scientists, tend to think of the mind-brain as an organ whose every process, every thought and image, is initiated by an external stimulus. This poses a problem for a theory of dreaming, as well increment in visual imagery when photo scales were employed – as reported above as much of waking mentation, such as daydreaming, which is also independent of external stimuli (Antrobus 1991).

HOBSON ET AL. have long maintained that the pontine-generated PGO spikes are the extra-cortical information source whose information constitutes the origin of the dream, and they have proposed far-reaching implications from the fact that this information is generated subcortically. Although, as HOBSON ET AL. point out, I have long and most recently (Antrobus & Conroy 1999) argued that any active cortical region will create organized pattern out of chaotic neural activity so that *no* extra-cortical information is necessary to account for dream imagery (Antrobus 1991), they continue to claim a pontine origin for the dream.

Their PGO claim is also inconsistent with their claim to $\Phi \leftrightarrow \psi$ isomorphism. The assumption that PGO spikes carry eye movement *information* to the brain during REM sleep rests on the assumption that these spikes transmit this information in the waking state, so that having acquired this information in the waking state, the cortex makes the *same* interpretation in REM sleep. But no one claims to *know* the function of PGO spikes in *waking* visual perception, so there is no waking model to apply to sleep. More problematic for the model, the relationships between PGO and REMs is quite different in waking and REM sleep. While PGO spikes are materialized in REM sleep they are not materialized in the waking state. Further, while PGOs mark the *termination* of REMs in the waking state (they may tell the occipital cortex that the foveal image has now stabilized and striate cortex may proceed to analyze it), in REM sleep PGOs are *concurrent* with REMs. Monaco et al. (1984, p. 220) concluded that the dramatic PGO activity of REM sleep PGO seems to be due to "disinhibition resulting from the arrest of firing of diffusely projecting aminergic inhibitory neurons of the dorsal raphe and locus coeruleus." For a fuller statement of this argument, see Antrobus and Conroy (1999). In short, there is no obvious way in which the cortex could use such PGO misinformation. Left without this PGO input, the Activation-Synthesis model, and now the AIM model, leave the

cortex with nothing to *synthesize,* so that according to the AIM model, there can be no dream.

In their conclusions, HOBSON ET AL. continue to attribute bizarre cognition to chaotic pontine activation despite the fact that no experiments have supported this association, and furthermore, bizarreness mentation is frequently observed in states where PGO activity in minimal (Antrobus et a1. 1995; Reinsel et al. 1992). It is more likely that local chaotic neural activity represents neural attractors that cannot settle on a solution and so communicate nothing to neighboring neural locations, rather than, as HOBSON ET AL. propose, that chaotic neural activity in the pons produces bizarreness in the cortical dream process. Although there may be many ways to produce bizarreness in dream, the Braun et al. (1998) conception of the REM brain as one of dissociated regions of activation suggests a new possibility. Regions that collaborate in waking perception depend on each other for error correction so that when they are forced to function independently, as in acquired deep dyslexia, they produce strange errors, such as naming an apricot, a peach (Hinton & Shallice 1991). It is well known that the individual cerebral hemispheres function differently when separated and fully linked. Caution must be taken in assuming that each of the dissociated regions of the dreaming brain carries out the same functions that it does in waking – especially as they operate without the considerable assistance of the language cortex (Braun et al. 1998).

The dream-no dream dichotomy problem illustrates the rule that a neurocognitive theory cannot be better than the validity of its worst measure. The concept of dream comes to us from the vernacular. It is multi-dimensional: visual features of color, movement; it is thematic, bizarre, conative, and, at times, affective, and, more, verbal with a sense of self reflectiveness and control. Questions about across-sleep-state differences in dreaming carry assumptions about whether the pattern of these features is sustained across states or whether, for example, dreams in some state are more visual and, in other states, more verbal. Even if the underlying *pattern* of features is intact across states, do some features appear at low levels of cortical activation while the rarer features occur only at high levels of cortical activation (Antrobus et al. 1995)?

The SOLMS and NIELSEN analyses that are based only on the report of dreaming, not even scaled by magnitude, tell us far less about the questions they address than if they had used a multi-dimensional dependent variable scaled by magnitude. Forty years ago, Kamiya (1961) showed that the answer to whether dreaming occurs in both REM and NREM sleep or only in REM was a function of where, on the magnitude of dreamlike mentation scale, one draws the dividing line between dreaming and nondreaming. Because the point is absolutely arbitrary, it prejudices the answer to any question, including those of SOLMS and NIELSEN, about the relation between cognitive and neural processes.

This measurement problem compounds the fallacy of assuming that discretely-defined biological states imply discrete neurocognitive processes. Since the original Aserinsky and Kleitman (1953) discovery that proposed a discrete distinction between REM and NREM sleep, investigators have implicitly assumed that whatever $\Phi$ processes produced $\psi$ effects must also operate in an all or none fashion. It is gratifying, therefore, that both NIELSEN's alternate model and HOBSON ET AL., after years of prodding by Foulkes, have agreed that the sources of dreaming in REM and NREM sleep may be regarded as operating respectively in a high and low, rather than on-off mode. This position is consistent with SOLMS's model, except that the underlying $\Phi$ source of dreaming , $\psi$, is only indirectly tied to REM sleep.

Now that Braun et al. (1998) have provided us with evidence about the modular activation of different cortical regions in REM sleep, and SOLMS has shown that some forms of dreaming are accomplished in other sleep and waking states, and Antrobus et al. (1995) have shown that dreaming is also associated, during sleep, with the rising phase of the diurnal wake-sleep rhythm, we know that different brain structures may support different features to the cognitive dream. But we cannot determine the role of any given brain module unless it varies *independently* of the others. Conversely, if the activation of each brain region that participates in dreaming sleep covaries with the activation of the other regions, then even though brain activation is multimodal, it is nevertheless, one dimensional. And if it is one dimensional, we should expect the features of the cognitive dream to also be one dimensional, that is, they should covary – even though they consist of qualitatively different feature classes. For this reason, the lengthy quantitative-qualitative discussion by HOBSON ET AL. does not rise to the complexity of the questions they are trying to address. All qualitative differences are ultimately quantitative differences. The questions of interest are how the multidimensional quantitative $\psi$ patterns map onto the multidimensional quantitative $\Phi$ patterns – as described by brain state-specific maps.

SOLMS and NIELSEN avoid these issues by implicitly assuming that $\Phi$ state differences in $\psi$ all lie along a single dimension, namely dreaming or simply recall of any content (NIELSEN). That is, what is called dreaming in one state is assumed to have the same pattern, or profile of features, as dreaming in another. All features rise or fall together so that variation in the magnitude of dreaming describes the joint variation in all features. By restricting their measure of $\psi$ to one dimension their models are insensitive to possible qualitative differences that many vary across states of $\Phi$. For example, suppose that mentation is more verbal in NREM and more visual in REM sleep, but that when sleepers are asked, "Were you dreaming?" they answer in the affirmative in both cases. One would falsely conclude that the different $\Phi$ states produce the same quality of mentation, that is dreaming. This criticism is not evidence against their positions, but it renders less convincing SOLMS's conclusion that mentation produced in different $\Phi$ states is qualitatively the same.

At this point in time, however, we have little evidence to throw out the one dimensional $\Phi \leftrightarrow \psi$ model. HOBSON ET AL. attempt to address the question but since none of their analyses consider the *relationships among* the cognitive variables, their review of the literature simply does not speak to the issue. Rather, they report whether there are quantitative differences in each variable taken separately. Except for a few studies in our labs, tests for across-$\Phi$ state differences in $\psi$ *patterns* have not been carried out by any investigator, and oddly, our tests are not mentioned by HOBSON ET AL. That all variables increase in REM sleep does not speak to the question of whether the *rate of increase* is constant across all. In the absence of that evidence, one can say nothing about the dimensionality of $\Phi \leftrightarrow \psi$ dreaming relationships.

The only paper I know of that explicitly tests whether the pattern among cognitive relationships differs across two different $\Phi$ state *changes,* e.g., REM versus NREM, is our 1995 paper. It shows from the *pattern* of cognitive features that best discriminates between REM and NREM sleep reports is not different from the pattern that discriminates between two points in time along the rising phase of the diurnal activation cycle. That is, the pattern among the cognitive measures that describes the diurnal effect is unchanged in both REM and NREM sleep. It does not differentially magnify the REM or NREMsleep effect; rather it adds *equally* to both REM and NREM sleep mentation. It is important to note that this is not, as SOLMS suggests, a NREM effect, but rather an independent pattern of $\Phi \rightarrow \psi$ activation associated with the diurnal sleep-wake cycle.

In closing, I would like to say that the evolutionary hypothesis of the function of dreaming proposed by REVONSUO appears compatible with existing knowledge about mammalian evolution. My own disposition is to assume that the evolutionary value of $\Phi$ structures is determined by multiple interacting selection factors. We still have no hard evidence that dreaming has, or ever had, any behavioral function. If it did, it remains to be determined whether such value is incidental to some other more direct evolutionary function of REM sleep and others states (SOLMS). The study of dreaming has been driven by our curiosity about its dramatic strangeness rather than its function, and I think, many investigators feel that the attention they give to dreaming would be better

justified if it had a well documented functional value. REVONSUO reminds us not to lose sight of this question.

# Dreaming as an active construction of meaning

Rita B. Ardito

*Center of Cognitive Science, Department of Psychology, University of Turin, 10123 Turin, Italy.* ardito@psych.unito.it

**Abstract:** Although the work of Revonsuo is commendable for its attempt to use an evolutionary approach to formulate a hypothesis about the adaptive function of dreaming, the conclusions arrived at by this author cannot be fully shared. Particularly questionable is the idea that the specific function of dreaming is to simulate threatening events. I propose here a hypothesis in which the dream can have a different function. [REVONSUO]

REVONSUO deserves credit for exploring the possibility that dreams may have a specific adaptive function. His attempt to understand that function, by unifying in an evolutionary approach much of knowledge of both the phenomenological and neurophysiological aspects of our dreams, is laudable. That said, it seems some critical observations must inevitably be voiced.

Beyond the more or less marked differences among the various authors, the scientific debate on the nature of dreams has historically consisted of two opposing positions, one holding that the dream is a mere random byproduct of REM sleep physiology, the other that the dream is an organized subjective experience that performs a specific function. REVONSUO reintroduces this opposition in an original way aligning himself with the second position. I will suggest that these two positions need not be opposed.

The dream, notwithstanding the fact that it is based on neurophysiological processes that randomly activate particular cerebral structures, can nevertheless be understood as a structured subjective experience with a specific function. Just as our waking perceptions are amply guided by our expectations and by our interpretative model of the world, so in dreams non-structured stimuli activated by random neurophysiological processes take on meaning for dreamers who superimpose their own interpretative schemes and ways of conferring meaning onto the experiential flux of this stimulation. It is in this sense that the dream is the product of an active construction of meaning.

The dream can be understood as the guardian of sleep, but not in the sense intended by Freud (1900/1950). According to the founder of psychoanalysis, the dream eliminates the emotional stimulus that originates from an unconscious desire of the dreamer, by satisfying it in a hallucinatory manner. In the hypothesis that I am advancing, the processes of dream-construction preserve the sleep from the random cerebral stimulation that accompanies it. If it were not possible to give meaning to this stimulation, thanks to the formation of dreams, it would not be possible to sleep. Without dreams, the experiential state associated with sleep would be similar to a psychotic state that lacks precisely the possibility of assigning meaning to consciously perceived events. The dream can be seen as the solution to an adaptive problem: How to prevent the random neurophysiological stimulation accompanying sleep from impeding the organism in restoring itself. I suggest that this was the adaptive function for which dreaming was selected in our evolutionary history.

Dream images are not random, even while the neurophysiological processes that lead to them are. This position is distinct from that of Hobson (1988b) and Solms (1997a), because unlike them it explains why the brain generates the images that constitute the dream content, and why dreams have particular contents and narrative plots. Dreaming is the way our cognitive and signifying schemes give sense to stimulation that is in itself nonsense. The dream is therefore the reflection of our cognitive organization. Furthermore, this makes the dream a useful clinical and research tool with a specific psychological and cultural function that complements biological function.

From the above, it is apparent that I share with **REVONSUO** both the idea that the contents of a dream are not random, and the hypothesis that the dream has a particular adaptive function. I particularly embrace the idea that dream content is consistent with the original evolutionary environment. My essential divergence is with his idea that the dream has the specific function of simulating threatening events, and of rehearsing threat perception and threat avoidance. In reality, the process of dream construction could have the more general function described above; it would not be surprising that in the exercise of this function, there should sometimes be threat simulations as well, and that reading, writing, and calculating should be absent. As highlighted also by REVONSUO, this depends on the fact that our cognitive architecture, and hence the processes that are basic to the production of dreams as well, are the products of our evolutionary history.

Regarding the experimental studies and data cited by **REVONSUO** in support of his hypothesis, in particular proposition 2: First, the fact that two-thirds or more of the emotions expressed in dreams have negative connotations does not necessarily signify that these emotions are linked to events perceived in the dream as threats. In the waking state, negative emotions (for example, anger, boredom, disgust, and spite) manifest themselves independently of whether or not the one who experiences them is involved in a threatening situation. These emotions may present themselves in contexts very different from those hypothesized by REVONSUO. It is accordingly risky to cite these studies as indirect evidence that dream content is biased towards simulating threatening events, inasmuch as it leads to conclusions that are unjustified in the light of what we know from the psychology of emotions. Regarding "Misfortune" dreams, if it is true that some studies demonstrate the preponderance of these with respect to "Good Fortune" dreams, it is also true that Misfortune dreams often present situations in which events are uncontrollable or inevitable. In what way do dreams of this type help to avoid the real world threatening situations of the waking state?

Dreaming is not exclusively a specialized experience for the simulation of threatening events; it is the active construction of meaning in the state of sleep, for cerebral stimulation that has none. In this light, the simulation of a threat has its place as one aspect of our experiential lives, but not the only one. The alternative approach delineated here takes all dreams into account regardless of their content.

## ACKNOWLEDGMENT

This work has been supported by Ministero dell'Universita e della Ricerca Scientifica e Technologica of Italy (Confinanziamento 1999, Reasoning Processes and Mental Models).

# Internally-generated activity, non-episodic memory, and emotional salience in sleep

James A. Bednar

*Department of Computer Sciences, University of Texas at Austin, Austin, TX 78712.* **jbednar@cs.utexas.edu**
**www.cs.utexas.edu/users/jbednar/**

**Abstract:** (1) Substituting (as Solms does) forebrain for brainstem in the search for a dream "controller" is counterproductive, since a distributed system need have no single controller. (2) Evidence against episodic memory consolidation does not show that REM sleep has no role in other types of memory, contra Vertes & Eastman. (3) A generalization of Revonsuo's "threat simulation" model in reverse is more plausible and is empirically testable.

[HOBSON ET AL.; SOLMS; REVONSUO; VERTES & EASTMAN]

***One dream controller is as bad as another [Solms]*.** The **SOLMS** target article argues persuasively that not all dreaming can be

uniquely identified with activity in the REM generating areas of the brainstem (as was proposed in some earlier research). However, substituting the ventromesial forebrain as the single "controller" of dreams, as SOLMS proposes, seems like a step backward, even if lesion studies show that the area is important or even crucial for dreaming. As neuroimaging studies make clear, dreaming is a complex process occurring in a system of multiple interacting units distributed across the brain. In such a distributed system, lesion studies cannot provide any means for deciding on a single location as the controller, because in fact there need be no such clearly-defined module. (Thus in SOLMS's terms, no single brain area need be able to "activate, generate sustain, and terminate" all dreams.)

A more productive approach might be to focus on the essential aspects of dreaming, and only then to consider how the various brain areas might contribute to this process. The most obvious feature that distinguishes dreaming from waking is that dreaming relies on internally-generated inputs (Bednar & Miikkulainen 1998), while waking mentation can be traced, at least in part, to data from the senses. There can be many possible sources of this endogenous activity during sleep, all of which could be considered to "cause" dreaming in some sense. Among these, brainstem REM generators do seem to be "a regular and persistent source of cerebral activation during sleep," as SOLMS himself acknowledges. Thus there is no mistake at all in focusing on the REM state instead of searching for a single anatomical site to unlock the secret of dreaming.

A second minor point from the **SOLMS** article is worth mentioning in passing, because it involves a report most likely published after his article was written. SOLMS speculates that the cortical back-projections which appear to underlie mental imagery do not project as far back as primary visual cortex, which if true could explain why $V_i$ shows decreased activity during REM sleep. However, back projections to $V_i$ certainly seem to be present anatomically, and Kosslyn et al. (1999) have shown that mental imagery can in fact be measured in $V_i$, albeit only of a certain kind involving specific locations on the retina. Thus a theory of dreaming would have to explain why the back projections to $V_i$ do not typically take part in dreaming; it clearly is not because the connections do not exist.

***Non-episodic memory does not require consciousness [Vertes & Eastman]*.** Strong and valid reasons for discarding the idea that explicit episodic memories from the hippocampus are somehow consolidated during REM sleep are presented by **VERTES & EASTMAN**. However, in several instances they go much further than their cited data would support by concluding that "REM sleep serves no role in the processing or consolidation of memory." They appear to make this claim because of their unusually restrictive definition of memory, in which "sleep involves basic biological functions and memory requires consciousness."

Certainly, some types of memory are intricately linked with consciousness, in particular the episodic memory usually proposed for consolidation. However, memory is a very broad term that is applied to an enormous variety of wonderful phenomena, ranging from the strength of the connection between two Aplysia neurons (or even the state of charge of a certain capacitor in a computer chip), to vastly more complex processes. Rather than being a specific byproduct of consciousness, memory seems to be quite distributed, localized, and ubiquitous in the nervous system (Gilbert 1998). Indeed, it is arguably as much of a "basic biological function" as sleep is.

From this larger view, there is currently no reason to conclude that REM sleep serves no role in non-episodic memory processing, despite the lack of clinical impairments from REM deprivation. Given widespread plasticity combined with the strong activity found in many brain areas during REM sleep, the burden of proof is actually in the other direction: unless one can show a plausible mechanism by which the process of learning has somehow been disabled at each local synapse without abolishing activity, one must assume that the activity has the potential to modify those synapses. Plasticity of this sort would presumably underlie non-episodic memories, such as procedural/skill memory (Smith 1995) and limbic-system emotional associations (Maquet et al. 1996); such effects could be difficult to measure clinically. It is even possible that the episodic and working memory areas quiet during REM sleep might be suppressed precisely so that they would not undergo plasticity while the rest of the brain is processing other types of memory. Such processing could be very important to proposals such as REVONSUO's.

*How solid is the evidence that dreaming is an organized simulation of the world? [Revonsuo]* Even though I share with RE-VONSUO a suspicion that dreams are not as random as Hobson and McCarley (1977) proposed, I do not agree that his particular line of argument "shows beyond any reasonable doubt that dreaming is an organized simulation of the perceptual world." He correctly anticipates most of the argument's weaknesses, but a few important ones have been overlooked. For instance, he cites differences in narrative richness between dream reports and subjective reports from isolated electric stimulation in temporal cortex. However, these conditions are not comparable, since we know that REM sleep activation is a large-scale phenomenon with at least some spatial and temporal structure. An appropriate experimental control would thus require widespread and ongoing artificial brain stimulation, coupled with temporary deactivation of the same frontal-lobe areas suppressed in REM sleep. If such an experiment were practical, it might demonstrate that dream-like mentation could be generated from random activation; meanwhile we must at least consider it possible.

Similarly, when REVONSUO quotes Foulkes as saying "The simulation of what life is like is so nearly perfect, the real question may be, why shouldn't we believe this is real?" (Foulkes 1985), Dennett (1991) would probably point out that (1) there need be no "simulation" of life in dreams separate from the experiencing of the dream, and (2) any brain activity contributing to the experience need not be perfect or realistic at all, as long as the processing machinery treats it as realistic. Indeed, as REVONSUO acknowledges, there are many bizarre, non-realistic features of dreams that are obvious only in retrospect.

*What makes threat simulation so special? [Revonsuo]* REVON-SUO's general hypothesis for mammals, "Dreaming rehearses species-specific survival skills," seems much more defensible than his narrow version for humans: "dream consciousness is essentially a mechanism for simulating threat perception and rehearsing threat avoidance responses and behaviors." (Unfortunately, the version for humans is the one that is most clearly distinguishable from similar earlier theories, such as Winson 1990 and Jouvet 1978.) Threat simulation would seem primarily useful for species which are typically prey rather than predator, and humans clearly serve in both roles. Since REVONSUO acknowledges that "not all dreams are threat simulations," it seems arbitrary to assume that other commonly-cited dreams (such as flying) are mere side-effects.

Given that threats are not the only situations biologically important to humans, ancestral or otherwise, a much more intuitive hypothesis would be that dreams simulate biologically-significant situations in general. In humans that would presumably be approximated as emotionally-salient situations, in the absence of some other internal criterion for what is biologically significant. Threats would just turn out to be a particularly well-represented example of such situations, rather than the primary purpose of the system.

In making the case for threat simulation, REVONSUO dismisses most previous proposals for dream function because they do not systematically analyze dream content. Dream content analysis may be very helpful for formulating hypotheses, but by itself it cannot offer any definitive criterion for preferring one hypothesis over another because of the enormous and largely unknown biases involved in subjective dream content reports. Even during waking life, we focus disproportionately on emotionally salient events when reporting narratives, as a quick glance at the evening news or the movie listings will attest. Given the particular emotional salience of threatening events (again, witness the bizarre popularity of horror films), finding threatening events over-represented in reports does not necessarily indicate that they are over-represented in dreams, and finding them over-represented in dreams does not necessarily mean that they are specifically generated. At a minimum, to use dream content analysis one must compare dream reports with waking reports as opposed to actual waking life.

REVONSUO also seems to go too far in emphasizing the lack of adaptive function for threat avoidance in modern life. Although daily life for many people may be quite dull, certainly those who have served in war, who have lived in the inner city, who have played competitive sports, who have been assaulted, who have encountered a vicious dog, and so on are quite familiar with threatening situations. Despite being relatively safe, people still die every day owing to jealous lovers, natural disasters, and many other causes which have been presumably unchanged for millennia. And thus being alert and ready for quick, decisive action in threatening situations is surely not "obsolete," even if no longer as important as it once was.

*Revonsuo's dream model: Why not have emotion precede situation?* From data showing only a correlation between emotions in dreams and the situations in which they are experienced, REVONSUO assumes a specific direction of causality, that is that unpleasant dream situations cause the negative emotions through "threat recognition." However, common dream features such as emotional continuity in the face of narrative discontinuity (Seligman & Yellen 1987) would suggest precisely the opposite hypothesis: the brain may somehow activate a certain emotion, which prompts recall of events historically (or perhaps genetically) associated with that emotion.

Reversing the sequence in this way can simplify a key step in REVONSUO's neural model of dream generation. As originally formulated, his model requires some unspecified mechanism for initially selecting memories by their emotional salience. Such a mechanism is difficult to imagine because it would supposedly operate independently of the current emotional state of the brain, since limbic system areas like the amygdala are activated only later in his process. The model also requires another unspecified (and difficult to imagine) mechanism for deciding "when potentially threatening content is present in visual awareness," since no feedback for this judgment is available in the model.

Reversing the sequence leads to a simpler and more concrete approach similar to HOBSON ET AL.'s AIM model. This model would start with activation of an emotional state in the limbic system along with sparse random activation of the visual system. The initial activation would automatically activate (more or less at random) one or more emotionally salient episodic memories of waking experience. Such memories would presumably include specific patterns of activity in the sensory association areas and in motor cortex. In the simplest case, the process of activation could simply strengthen the already-present emotional association between the activated units through a simple connection-specific mechanism like Hebbian learning. This model would merely enhance emotionally salient memories at the expense of others which could be desirable.

Making this model slightly more extravagant to compare with the one proposed by REVONSUO, it could instead generate specific (rather than random) coarsely-determined visual input and/or specific motor cortex activations (Bednar, unpublished research proposal, June 1999). Such a system would amount to supervised training of an association between particular inputs (e.g., threatening situations, as in REVONSUO's model), particular outputs (e.g., fighting or fleeing), and a given emotional state (e.g., fear). The inputs, outputs, and emotional state would all need to be genetically specified somehow, which is what makes this hypothesis more extravagant than the simpler one above. However, the extravagance is no greater than REVONSUO's, and this model does not require the presentation of hypothetical scenarios on the input while hoping for the correct response from a brain that has no proposed feedback signal to guide the "threat recognition" process.

# Dreams have meaning but no function

Mark Blagrove

*Department of Psychology, University of Wales, Swansea SA2 8PP, United Kingdom.* **m.t.blagrove@swansea.ac.uk**
**http://psy.swan.ac.uk/**

**Abstract:** Solms shows the cortical basis for why dreams reflect waking concerns and goals, but with deficient volition. I argue the latter relates to Hobson et al.'s process I as well as M. A memory function for REM sleep is possible, but may be irrelevant to dream characteristics, which, contrary to Revonsuo, mirror the range of waking emotions, positive and negative. [HOBSON ET AL.; NIELSEN; SOLMS; REVONSUO; VERTES & EASTMAN]

**SOLMS** shows how dreams are dependent on cortical areas concerned with appetitive interactions with the world, which accords with previous work on dreams incorporating emotionally meaningful material (Hartmann 2000b), but it is interesting that dreams, which express our goals and concerns, seem also to have attenuated volition. **HOBSON ET AL.** also make the point that there is lowered volitional control in dreams, while referring to there being a current debate on its extent. More detail would be welcome on SOLMS's finding that lesions of the dorsolateral prefrontal cortex, involved with volitional control and self-monitoring in waking, do not have an effect on dreaming, because of the difficulty of measuring any such changes in volition and self-reflectiveness in dreams. Although there is evidence for deficient volition in dreams (e.g. Blagrove 1996; Kahan et al. 1997), Bargh and Chartrand (1999) show that there is a general lack of volition and self-reflection in waking life, and so goal seeking may be activated without conscious volition in both states. Furthermore, for waking life, Kirsch (1998) shows that behavior often follows the automatic activation of response sets, and that volition is often an erroneous attribution after an action. I suggest that in dreams the attenuated volition and self-reflection, and lack of surprise at bizarreness, may be a result of the lack of unexpected stimuli that are contrary to expectations, because, in an extension of their use in waking experience, response expectancies (Kirsch 1985) may be used to produce the successive contents of the dream.

For dreams the attenuated volition and self-reflection may thus be a result of the lack of feedback from an independent environment (which in waking life can cause surprises). That is **HOBSON ET AL.**'s process I, the sensory gating, although with the possibility, as **HOBSON ET AL.** state, that deficiencies in memory during dreaming are also involved. I would ask whether process M is any more than a measure of memory consolidation (their "memory/amnesia dimension"). For example, from Reinsel et al. (1986), mean total recall count for waking daydreams under minimal stimulation was 68 words, whereas REM recall was 34 words; such differences, although significant, raise the question of whether the low REM recall is a matter of low memory consolidation, memory retrieval in dreams, although usually outside conscious control, being quite resourceful and capable of some complexity.

**NIELSEN** makes the claim that NREM dreams may be dependent on the next REM sleep phase, as well as on the previous one, and of interest would be whether the window might have a different duration on either side of a REM sleep phase. I am concerned that some psychological REMS/NREMS differences may appear by chance. For example, in Foulkes and Rechtschaffen (1964) the MMPI L scale correlates with REM recall but not NREM, as **NIELSEN** reports, yet the 21 other MMPI scales had no relationship with REM or NREM recall. **NIELSEN** notes that after dream length is controlled for there are REM/NREM differences in visual imagery word count, number of characters, and self-involvement, and refers to these differences as "qualitative" although these differences are really quantitative, as **HOBSON ET AL.** state, and this would hold whether or not such differences could be eradicated by other methods of controlling for dream length. If the view that there are two systems of dream production arises because of the two physiological stages of sleep, should we be pos-

tulating a third system because mentation with dream-like characteristics can also be found briefly in waking daydreams (Foulkes & Fleisher 1975)? Parsimony suggests one system, with dream production turned on more frequently in REMS than NREMS.

**VERTES & EASTMAN** reason that as dream material is chaotic, and poorly remembered, then "the transfer of information in REM is not orderly," but these two processes may well be entirely independent. They argue that attention is needed for learning, but is absent in REM sleep, but even if attention is required in the first stages of learning, at input, and even that is questionable (cf. Lewicki et al. 1988), it would not be required at later stages of the storage of information, which would occur automatically and outside consciousness, as it does when one is awake. That Wilson and McNaughton (1994) found increased firing in hippocampal place cells during SWS after learning may indicate some effects of sleep on learning, as do the findings of Ambrosini et al. (1995) that SWS and REM sleep both have a memory consolidation function.

In response to their data on the reduction of REM sleep with antidepressants, it may be that if REM sleep is abolished then its functions can be fulfilled when awake; I am not clear whether the position **VERTES & EASTMAN** are attacking is that REM sleep has a function that cannot be fulfilled when awake, or that consolidation of memories can occur in REM sleep as well as when awake. There is also the problem of the alerting effect of REM deprivation (Nykamp et al. 1998) which could hide memory deficits, and in studying the effects of antidepressants, there are also problems if the baseline is the depressed state, during which there may also be memory deficits.

In their theory of REM sleep preparing the sleeping animal for waking, **VERTES & EASTMAN** do not account for the large amount of REM sleep in the fetus, for REM rebound, or for the possibility that REM sleep is evolutionarily earlier than SWS (Siegel 1997). The muscle atonia of REM sleep indicates against this arousal function, as would any confusion of waking up from a REM dream, and any difference in alertness between waking from REM sleep and from deep sleep may be too small to support this hypothesis. The comparison with slow recovery after anesthesia is not helpful, as there are biochemical reasons for this.

**REVONSUO**'s theory of the function of dreaming has similarities to Hartmann's (1996a) theory of the contextualizing of emotional concerns, with the addition of psychomotor practice to Hartmann's emphasis on forming connections in memory. Both authors use the extreme example of nightmares to argue about a function for dreams. **REVONSUO** asserts that in dreaming we "rehearse threat perception and threat avoidance," with the possibility, when threats are absent, of dreaming of emotionally charged memories, current concerns, or other mundane sources. The article is predicated on pain and fear being adaptive, and holds that dreams depict "ancient concerns" and have as "default values" the simulation of "violent encounters with animals, strangers, and natural forces, and how to escape from such situations." Which explains why REM behavior disorder frequently involves threatening actions, but on evolutionary terms would not positive reinforcement (e.g., dreaming of green fields, flowing water, success or friends) be as important as negative warnings? **REVONSUO** cites Hartmann's (1998, p. 73) finding that "when people experience a happy event, they are more likely to dream about problems associated with it than the pure happiness of the event itself," but Blagrove and Price (2000) found that happy skilled individuals tend to have happy dreams, and Kallmeyer and Chang (1998) found that particular positive (e.g. joviality, self-assurance) and negative (e.g. fear, sadness) waking emotions were associated with individuals who have positive and negative dreams, respectively.

**REVONSUO** claims that dream content is biased towards negative elements, yet although Strauch and Meier (1996, pp. 92–93) did find that negative emotions appear twice as often as positive ones, joy was the most common specific dream emotion (followed by anger, fear, interest, and stress), and of 500 REM dreams, as **REVONSUO** cites, general mood was more likely to be positive than negative. Strauch and Meier conclude that dreams are not predominantly influenced by fears, dismay or stress, but frequently display well-being and pleasant experiences (p. 94). Furthermore, Schredl and Doll (1998) found that although dreams rated independently had a preponderance of negative tone, when rated by the dreamer the ratio of positive to negative moods was balanced.

REVONSUO claims that reading, writing, and arithmetic do not appear in dreams because they are cultural latecomers, but they are usually unemotional activities, and I see no reason why writers would not dream of writer's block, or some other non-reproductive aspect of their professional life. It may thus be too narrow to claim that "the biological standard is the only standard of functionality." Also, SOLMS has dreams incorporating emotional and motivating stimuli, as with the traumatic events described by REVONSUO, but this does not show that the incorporation is functional, it may be a by-product of a system that incorporates positive and negative emotional stimuli and motivations into daydreams. However, HOBSON ET AL. remind us that dreams are highly penetrable cognitively, so the belief that dreams are concerned with threats may itself lead to such dreams occurring.

The test of this theory of an over-representation of threat simulation in dreams would surely be to find threat themes in people who are not hunter-gatherers or traumatized, because a theory of dreams as incorporating emotional events in general would similarly predict "high levels of survival themes, threat simulation and animal characters" in the dreams of hunter-gatherers. Rather than showing a mechanism of dream function this just shows the effects of being in those conditions. I note, however, that REVONSUO does state that today's changed environment may mean that dreams do not now have a function. The author is right however, that some type of selection is going on in the formation of dreams, but to study this selection, the frequency of threatening events in dreams should be compared to their frequency in autobiographical stories, or creative stories, rather than in real life itself.

REVONSUO reports changes in sleep due to PTSD but in the Williamson et al. (1995) paper cited dream variables were not measured, and in Ross et al. (1994) PTSD participants had more REM sleep and greater REM density, but of the 11 PTSD participants just one experienced an anxiety dream. Lavie et al. (1998) found that although PTSD patients had higher awaking thresholds than controls, and more aggressive and hostile dreams, the PTSD and control groups did not differ in dream recall frequency, and Dow et al. (1996) found no differences in dream recall or report length between Vietnam veterans with PTSD and major depression, veterans with depression alone, and veterans with neither PTSD nor depression, and for all groups dream anxiety was no more than mild. Anyway, although stress can increase nightmare frequency (Chivers & Blagrove 1999) and trauma can be represented in dreams and nightmares (Barrett 1996), the correlation of nightmare content with trauma, or even change in nightmare content with recovery from trauma, does not mean that dreaming has a causal role in that recovery (Blagrove 1992a).

REVONSUO states that on average one out of two animals in children's dreams are untamed wild animals, and "the proportion of domestic animals increases and that of wild animals decreases with age." And yet Foulkes (1985, p. 122) found that at ages 3–5 years dream animals "tended most often to come from two classes: domesticated farm animals or relatively familiar and unaggressive undomesticated animals" and Foulkes does not mention that children at ages 5–7 years dream of aggressive animals. Against the claim that children are likely to have infrequent actual experiences of animals, a source of there being so many animals in children's as opposed to adults' dreams may be present-day fairy tales and cartoons, rather than ancestral fears; why children are so interested in animals is then another matter: even if that interest has evolutionary origins dreaming about animals may not do so. Furthermore, Foulkes (1985) gives evidence to interpret strangers in children's dreams as a failed attempt to represent someone who is known, rather than an actual stranger, which is problematic for REVONSUO's claim that strangers in dreams may result from ancestral conditions in which encounters with strangers were po-

tentially life-threatening. REVONSUO asks why are male strangers our enemies in dreams, given that "present-day encounters with unfamiliar males in the waking life are not predominantly aggressive," yet surely such a view of male strangers is common in TV and newspapers, and hence is salient to us, even if exaggerated and unrepresentative of reality.

It is unclear why a dreamt simulation should help in the "perceiving and recognizing" of threatening situations. This surely requires real stimulation, and the analogy with flight simulators (n. 17) does not hold, because in using them the operator is highly conscious and attentive. REVONSUO argues that lack of consciousness is no problem to the function of dreams in the model, and if dreams have the role of providing practice for actions then this is true, but what of learning flexibility in actions'? Furthermore, the complex movements found to occur in cats during REM sleep without atonia have also been found to occur during wakefulness (Morrison & Bowker 1975), so Jouvet's widely cited result from sleeping cats without atonia may not be evidence for dreaming as motor practice.

REVONSUO is rightly not convinced by theories of "the psychologically adaptive function of dreaming," but dreams could be psychologically expressive of positive and negative emotions, as in work on the measurement of insight due to dream interpretation (e.g. Hill et al. 1993), and on the incorporation by divorcees of their former spouses (Cartwright 1991). It may be that day dreaming and imagery were selected for in evolution, with dreaming being an epiphenomenon. REVONSUO's argument against dreaming having resulted from the evolutionary selection of day dreaming is that dreaming has different features, such as in level of volition and type of moods, but these differences do not show that dreaming is not dependent, physiologically and evolutionarily, on day dreaming and on the ability to imagine and to have imagery.

## Sleep, not REM sleep, is the royal road to dreams

Alexander A. Borbély and Lutz Wittmann

*Institute of Pharmacology and Toxicology, University of Zürich, CH-8057 Zürich, Switzerland.* **borbely@pharma.unizh.ch**
**www.unizh.ch/phar/sleep**

**Abstract:** The advent of functional imaging has reinforced the attempts to define dreaming as a sleep state-dependent phenomenon. PET scans revealed major differences between nonREM sleep and REM sleep. However, because dreaming occurs throughout sleep, the common features of the two sleep states, rather than the differences, could help define the prerequisite for the occurrence of dreams.
[HOBSON ET AL.; NIELSEN; SOLMS; REVONSUO; VERTES & EASTMAN]

SOLMS provides an excellent summary of evidence that the REM sleep and dreaming are dissociable states. Although all authors in the present issue seem to agree that dreams occur throughout sleep, the temptation to associate them with REM sleep lingers on. Thus HOBSON ET AL. attempt to account for dreaming in nonREM sleep by invoking an "admixture of REM-like phenomena within stage 2," NIELSEN proposes the existence of "covert REM sleep processes" during nonREM sleep and sleep onset, and VERTES & EASTMAN state that "the mental/cognitive content of REM sleep and sleep is dreams." The conceptual dissociation of REM sleep and dreaming is still incomplete.

Dreaming occurs throughout sleep: it may be useful to focus on features that are common to both sleep states and different from waking. In PET scans they consist in the deactivation of hetromodal association areas in frontal and parietal cortex (Andersson et al. 1998; Braun et al. 1997). REVONSUO refers to SOLMS's view that dreams are "bizarre hallucinations that weakened frontal reflective systems mistake for real perception." Our recent study confirmed the deactivation of frontal areas in stage 2 and stage 4 of nonREM sleep as well as in REM sleep (Finelli et al. 2000). Another common feature appears to be the relative activation of unimodal cortical areas. In our study, parts of the occipital neocortex were more activated in stage 4 than in stage 2, and unimodal areas in the visual and parietalcortex were activated in REM sleep relative to waking. Hofle et al. (1997) showed for some of these areas a positive covariation with delta activity, an index of REM sleep intensity, and Braun et al. (1997) reported an activation of unimodal visual cortex in REM sleep.

The selective "deactivation" of frontal cortex described in PET studies seems to have an electrophysiological correlate. Thus in the initial nonREM sleep episodes, EEG slow wave activity shows predominance in the fronto-central derivation relative to caudal derivations (Werth et al. 1996; 1997). Moreover, brain mapping during and after prolonged waking revealed that frontal areas exhibit the largest increase of slow-wave activity in nonREM sleep and of theta activity in the waking EEG (Finelli et al., unpublished results; see also Cajochen et al. 1999a; 1999b).

Deactivation of heteromodal association areas, a feature common to nonREM sleep and REM sleep, could be a prerequisite for dreaming. However, its direct association with dream experience would have to be documented by comparing PET scans obtained for sleep periods with and without dreaming. If similar differences would emerge in nonREM sleep and REM sleep, then the pattern would deserve serious consideration as a physiological correlate of the dreaming process. Early studies have attempted to specify dream-related patterns of cerebral glucose metabolism (Gottschalk et al. 1991a; Heiss et al. 1985). However, the poor temporal resolution renders interpretation difficult. PET studies of regional cerebral blood flow using labeled water appear to be more propitious. The comparison of sleep periods with and without dreams would also be a useful approach in quantitative EEG analysis. Such a study would also be useful for testing the activation hypothesis of **SOLMS**.

A final comment pertains to **SOLMS**'s interesting proposition that the mesocortical-mesolimbic dopamine system plays a casual role in the generation of dreams. Neuroleptics such as haloperidol are powerful blockers of dopamine-D2 receptors and would be expected to eliminate dreaming. Awakenings from sleep on nights with and without neuroleptics would be a direct way to test the dopamine-dream hypothesis.

# REM sleep deprivation: The wrong paradigm leading to wrong conclusions

Jan Born and Steffen Gais

*Clinical Neuroendocrinology, Medical University of Lübeck, 23538 Lübeck, Germany.* born@kfg.mu.luebeck.de

**Abstract:** There are obvious flaws in REM sleep suppression paradigms that do not allow any conclusion to be drawn either pro or contra the REM sleep-memory hypothesis. However, less intrusive investigations of REM sleep suggest that this sleep stage or its adjunct neuroendocrine characteristics exert a facilitating influence on certain aspects of ongoing memory formation during sleep.

[NIELSEN; VERTES & EASTMAN]

REM sleep facilitates memory formation. Currently this is more a belief than a concept with convincing scientific support. Hence, **VERTES & EASTMAN**'s case against memory consolidation in REM sleep is a very timely contribution reflecting the true and persisting darkness in this area of sleep research. Unfortunately, **VERTES & EASTMAN** appear to be caught in similar misconceptions to those of researchers supporting a close link between REM sleep and memory consolidation. A great part of **VERTES & EASTMAN**'s review is devoted to studies evaluating recall of memories after a period of REM sleep suppression as compared to control situations, such as arousal from NonREM (NREM) sleep. Such stud-

ies did not provide evidence that REM sleep deprivation impairs recall of previously learned materials, under all circumstances, although changes, if they occurred after REM sleep deprivation, were always towards impairment rather than improvement of memory. It is very likely, however, that a stress response induced by REM sleep suppression is the principal factor responsible for recall deficits.

Recently, DeQuervain et al. (1998) demonstrated in rats that glucocorticoids, the release of which is a central marker of the stress response, have a distinctly impairing effect on the retrieval of long-term spatial memories. Hence the REM sleep suppression paradigm is not conclusive about what happens during REM sleep with ongoing consolidation. Moreover, a propensity for REM sleep must be assumed to persist during deprivation conditions, thereby contaminating the outcome of memory retrieval in an unpredictable manner. That is, by inducing nonspecific alterations of cognitive and emotional functions it may disturb or improve recall performance. Even more important, suppression of phenotypic REM sleep may miss those electrophysiological and neurochemical processes mediating memory consolidation.

Consonant with this view, **NIELSEN** in his target article on mentation in REM and NREM sleep proposes the concept of "covert REM sleep" as a kind of sleep that lacks some of the obvious signs of REM sleep, but shares underlying related processes. Traditional sleep scoring certainly does not focus on the phenomena determining memory during REM sleep which may persist (as reflections of propensity) even in the absence of the phenotypic signs of this sleep stage. This reasoning can be extended to all kinds of REM sleep suppression regardless of whether induced by behavioral techniques such as the pedestal method or by psychopharmacological intervention with antidepressant drugs. In addition, most of the latter work with antidepressant drugs, by focusing on changes during extended periods of treatment, is unable to distinguish drug effects on acquisition, consolidation, and retrieval. Also, the academic achievement of patients who have recovered from bilateral pontine lesions and do not reveal any common signs of REM sleep is clearly impressive. These data show that phenotypic REM sleep is not a prerequisite for memory consolidation, just as the occurrence of EEG desynchronization and theta activity is not restricted to REM sleep. Nevertheless, it cannot be concluded from these patients' performance whether processes are initiated during normal REM sleep, which facilitate certain aspects of a consolidation process.

In light of the apparent shortcomings of experimental procedures relying on suppressed REM sleeps it is amazing how little effort **VERTES & EASTMAN** devote to reviewing experiments that rely on less intrusive manipulations and, indeed, point to a supportive function of REM sleep on memory. **VERTES & EASTMAN** briefly mention the intriguing work of Stickgold et al. (2000b). The task used there (requiring a preattentive discrimination of visual textures) is remarkable as subjects, performance did not improve unless they obtained some hours of sleep after initial training. This suggests the presence of a slow continuous process of memory formation particularly sensitive to the influence of sleep (Karni et al. 1994; Karni & Sagi 1993). Stickgold and coworkers found that the improvement in texture discrimination was strongly correlated with the amount of slow wave sleep (SWS) in the first quarter of sleep time, and with the amount of REM sleep in the last quarter of sleep time. This pattern is of interest, because it rules out a one-to-one link between REM sleep and memory. Accordingly, Stickgold and coworkers proposed a two-step process of memory formation during sleep, with REM sleep becoming effective in a second step, strengthening associative connections at the neocortical level. However, correlations between the amount of REM sleep and recall performance do not necessarily reflect a relation between cause and effect, which limits respective conclusions.

Another interesting approach was developed by Ekstrands group in the 1970s (Barret & Ekstrand 1972; Ekstrand et al. 1977; Fowler et al. 1973; Yaroush et al. 1971), who compared retention rates across sleep periods of equal length but with different proportions of sleep stages. Ekstrand's group found greater improvement in recall of declarative memories after sleep during the SWS-rich early half of the night than after REM sleep-rich sleep during the late half of the night.

Recent studies in our laboratory have confirmed this (Plihal & Born 1997; 1999a). However, in extending the work of Ekstrand's group to several tasks of procedural memory (mirror tracing, word stem priming), we found greater improvement across late as compared to early sleep. This led us to suggest that some kinds of non-declarative memory not relying on the integrity of the hippocampus and adjacent temporal lobe structures particularly benefit from late sleep with predominant REM sleep.

In another study (Gais et al. 2000), we examined performance on the visual texture discrimination task mentioned above (Stickgold et al. 2000b; Karni & Sagi 1993). The comparison of retention intervals containing either a 3-hour period of early sleep or a 3-hour period of late sleep indicated early but not late sleep to be primarily necessary for the improvement in texture discrimination skills. It is interesting to note the improvement in task performance after an entire period of undisturbed nocturnal sleep containing early SWS as well as late REM sleep was on average more than 3-fold higher than after a period of early SWS-rich sleep alone. This outcome fits nicely with the two-step model of memory facilitation during sleep proposed by Stickgold et al. (2000b) and others (Giuditta et al. 1995) suggesting that REM sleep plays a role at a later stage of memory processing during sleep.

These experiments together provide evidence that REM sleep and associated processes can enhance memory formation. However, their role probably depends on the type of memory system and prior processing of the materials within this system. There are numerous processes (hormonal concentrations, temperature, etc.) changing in parallel with REM sleep that are candidates for explaining a sleep related memory enhancement as well as REM sleep per se. These processes may interact with cognitive functions in any sleep stage. However, they are neglected in **VERTES & EASTMAN**'s discussion of sleep associated memory formation and, notably, also in **NIELSEN**'s more general discussion of mentation during sleep.

Of utmost importance in this context is the release of corticosteroids from the pituitary-adrenal system, which in humans is at a minimum during early nocturnal sleep and reaches a maximum during late sleep. Glucocorticoids, that is, corticosterone in rodents and cortisol in humans, are potent modulators of ongoing EEG activity and memory function (e.g., DeKloet et al. 1999; Friess et al. 1994; Gronfier et al. 1997; Kirschbaum et al. 1996). In humans, infusion of cortisol during a period of early sleep completely blocked the improvement in declarative memory typically observed over this period (Plihal et al. 1999; Plihal &Born 1999b). Note that in the latter study the blocking effect of cortisol infusion on declarative memory consolidation during early sleep occurred without any concurrent reduction in signs of SWS. Thus, rather than the phenotype of SWS activity, the concurrent suppression of cortisol release turned out to be a crucial prerequisite facilitating declarative memory function during this period of sleep.

Comparable conditions may determine the putative memory process during late sleep when REM sleep prevails and glucocorticoid concentration is elevated. Studies in rodents indicated that memory of events that are emotionally highly arousing and aversive can be enhanced by glucocorticoid administration (Cahill & McGaugh 1996; DeKloet et al. 1999). Experimental improvement of memory ascribed to REM sleep might accordingly turn out to be a result of an accompanying elevation in glucocorticoid levels.

Another well-known example of neurohormonal processes modulating memory is sympathetic activity and the release of catecholamines. Through the activation of central-nervous adrenergic receptors, epinephrine can enhance storage of emotionally arousing events in humans (Cahill et al. 1994; van Stegeren et al. 1998). In humans, concentrations of epinephrine and norepinephrine in the blood are reduced during REM sleep as compared to SWS and wakefulness (Dodt et al. 1997). This could selectively

disfacilitate formation of emotional memory. Thus, neurohormonal processes only loosely linked to specific sleep stages may be more relevant for memory consolidation than a specific sleep stage. As an alternative view **VERTES & EASTMAN** propose that REM sleep serves to prime the brain for a return to consciousness as waking approaches. It is noteworthy that exactly the same function has been claimed for the release of pituitary-adrenal stress hormones increasing towards the end of sleep (Born et al. 1999).

# REM and NREM mentation: Nielsen's model once again supports the supremacy of REM

M. Bosinelli and P.C. Cicogna
*Department of Psychology, University of Bologna, Bologna, Italy.*
**cicogna@psibo.unibo.it**

**Abstract:** Nielsen's model presents a new isomorphic brain-mind viewpoint, according to which the sole dream generator is found in a REM-on (explicit or covert REM) mechanism. Such a model cannot explain the dreamlike activity during SWS (slow wave sleep), SO (sleep onset) and in the last period of sleep. Moreover the hypothesis contrasts with Solms's data, which show that dreaming is present also in case of destruction of the REM generator.
[**NIELSEN; SOLMS**]

In the fifties and sixties, a target article like the one by **NIELSEN** would not have been imaginable. The identification of REM sleep with dreaming had all the characteristics of an unshakable dogma. Today, **NIELSEN** documents the existence of rich production of NREM mental experiences; this has led to a divergence in researchers' theoretical views. On the one hand there are the supporters of a one-generator model, in which the REM and NREM dream production would be relatively autonomous from its physiological basis as related to sleep stage. On the other hand, there are the supporters of a two-generator model, according to which REM sleep would be responsible for an oneiric cognitive activity qualitatively different from the one generated by NREM sleep, regardless of stage.

**NIELSEN** attempts to reconcile the two models, assuming the existence of covert REM sleep processes in NREM sleep responsible for the concomitant NREM mentation. This attempt brings him back to the identification of dreaming and REM sleep, not dissimilar to the positions of Hobson and his group (1998b), in which an isomorphism between physiological background (REM or covert REM) and dream mentation is assumed. In this sense **NIELSEN**'s hypothesis does not seem to be a reconciliation between the two models, but a unique REM-one model of oneiric generation, either in its explicit or covert form. From a theoretical point of view, a unitary explanation is more parsimonious and therefore preferable to the two-generator model. However, **NIELSEN**'s arguments show some weak points that call for further investigation.

REM-sleep related processes would be responsible for NREM mentation, even though activated "in a piecemeal fashion and against an atypical neurophysiological background." (**NIELSEN** target article). Characteristic of this activation would be that it takes place near to the REM phase (10–15 min before or after), even though this is weakly supported by physiological data, making it very difficult to explain data on SWS mentation, as **NIELSEN** himself admits. SWS dream reports were collected in our lab (Cavallero et al. 1992; Occhionero et al. 1998) in cycle I, 40 min after SO (hence distant from a possible covert REM in SO and distant from REM I), as well as in cycle II 40 min after REM (unpublished data), yielding recall over 60% and differences in comparison with REM dreams with regard to length only. The same holds for spontaneous morning awakenings during NREM (about 70%) which are rich in oneiric activity and not near to any REM

(Cicogna et al. 1998). Moreover, if we were to accept NIELSEN's view, in sleep periods far from REM (when one could not impute covert REM activity) no mental activity would be present: which looks like a very hasty assertion. Only in deep coma, without any cortical activity, is there total absence of thought.

Another datum difficult to explain is oneiric activity in SO, which not only is very bright, but has high recall similar to REM. It seems difficult to maintain that this is covert REM, since physiological factors (transient EMG suppression, REMs, muscle twitches) that could provide evidence of a similarity to REM are almost non existent in SO-St2, the moment of SO in which experimental awakenings usually take place. Without any further data on the presence of REM-on physiological factors in SO-St2, it is difficult to interpret the oneiric richness of SO according to the model.

As to NIELSEN's very broad review of the literature, two remarks: (1) The scheme indicated in Figure 1 (NIELSEN) is not completely convincing with regard to the "apex-dream" typology that should be the most typical expression of a dream-like mentation (REM-like) and it in fact refers to rare situations or even to situations that violate the dream's hallucinatory quality (for example lucid dreams); (2) In our opinion the qualitative differences between REM and NREM mentation are overemphasized in the cases in which controls equating report length showed slight residual differences, to the point of making the authors infer that they were epiphenomena owing to quantitative aspects, explicable in terms of mnemonic spreading activation (Antrobus 1983; Cicogna et al. 1987; 1998; Foulkes & Schmidt 1983). These authors do not deny the influence of an underlying physiological background that may be responsible for the modulation of memory activity (for example EEG differences, differences in sensory thresholds), however they deny that this could be directly involved in the dream production and in the cognitive work typology.

There is also a general problem in research on sleep and dreaming, which is that of handling NREM sleep as a unitary and physiologically similar homogeneous entity, without considering the differences between stages, which are quite remarkable. In terms of radical isomorphism, one can think of as many dreaming generators as sleep stages. The limit of NIELSEN's model can be found in this "isomorphic" brain-mind view, in which the only oneiric generator is found in a REM-on mechanism (but it is unclear why). Among other things, this position contrasts with SOLMS's evidence (see target article) that dreaming continues in case of destruction of the REM generator and is absent in cases of forebrain lesions.

In our view as cognitive psychologists, there are higher cognitive processes which, after having been initiated by REM or NREM subcortical activation mechanisms, follow information processing rules that have no precise correspondence to neurophysiological areas or mechanisms. Even though one may want to find a correlation between cognitive and neurophysiological processes at all costs, the evidence adduced by NIELSEN himself as well as by Solms (1997a) shows that the association areas involved in the information processing are equally activated in REM and SWS, whereas the differences in the cerebral blood flow are found precisely in limbic and hippocampal areas affecting memory systems and emotions. The levels of physiologic activation do certainly modulate cognitive processes in terms of "amount of work," but they do not modify their operational modality.

# How and why the brain makes dreams: A report card on current research on dreaming

Rosalind Cartwright

*Department of Psychology, Sleep Disorder Service and Research Center, Rush-Presbyterian St. Luke's Medical Center, Chicago IL 60612.*
**rcartwri@rush.edu**

**Abstract:** The target articles in this volume address the three major questions about dreaming that have been most responsible for the delay in progress in this field over the past 25 years. These are: (1) Where in the brain is dreaming produced, given that dream reports can be elicited from sleep stages other than REM? (2) Do dream plots have any intrinsic meaning? (3) Does dreaming serve some specialized function? The answers offered here when added together support a new model of dreaming that is testable, and should revitalize this area of study.

[HOBSON ET AL.; NIELSEN; REVONSUO; SOLMS; VERTES & EASTMAN]

*Introduction.* The reader of these five BBS articles might come away with the impression that they have just witnessed another set of blind men describing an elephant. Although each does bring a different perspective to the problems associated with the phenomenon of dreaming, collectively, they make significant progress in clearing their way for further work. The first three papers (HOBSON ET AL., NIELSEN, and SOLMS) focus on activity within the sleeping brain to tackle the question of dream construction: How does the brain make dream experience happen? The last two papers (REVONSUO, VERTES & EASTMAN) address the function question: Why we do it? They both look into interactions between waking and REM sleep – one challenges the proposal that this sleep stage has any specific role in memory storage and the other champions a different function, that dreams make a contribution to our waking survival from threats.

The articles differ not only in where they look but how. The first two by HOBSON ET AL. and by SOLMS stay within the sleeping brain, tracing those pathways which are active and which blocked, in order to explain the variations in the cognition reported from sleep. HOBSON ET AL. focuses on the REM sleep system and its activation starting in the pons, SOLMS on the dream system he locates in a dopaminergic system within the forebrain. NEILSEN turns the problem the other way around, using the presence of a dream-like report, to predict the presence of REM sleep if only as fragments which previously have escaped traditional scoring. VERTES & EASTMAN look at REM sleep as both the independent and dependent variable in turn as they examine the evidence supporting one of its proposed functions: that REM sleep is involved in the storage of newly learned information. Is REM sleep enhanced by intensive pre-sleep learning and is performance post-sleep reduced following REM sleep deprivation? They find these data unconvincing. REVONSUO broadens the time frame for exploring a different function of dreaming by hypothesizing that the high proportion of negative affect characteristic of dream reports suggests these were developed from earliest times when waking life was more acutely dangerous. Perhaps they represent a genetically transmitted legacy of survival protocols retained and rehearsed during sleep for use in waking. This is the only paper in this collection that suggests some meaning for dream content.

*Background.* Following the discovery in the early 1950s of REM sleep, a flood of published reports confirmed that this stage of sleep was strongly associated with the presence of an ongoing dream. Using the reliable external indicators of REM sleep, awakenings were done to capture samples of the immediately preceding mentation. This allowed dreams to be studied systematically as never before possible. Thus the normative characteristics of REM-related dreams, their changes with age, and the effects on them of various manipulations and conditions were mapped out by the mid 1970s. Then came the drought when progress slowed to a trickle.

On reflection, three factors seem to be responsible for this turn of events. The first was the need to modify the initial brain/

behavior models to accommodate the finding that some of the awakenings made from NREM sleep stages also had dream-like characteristics. Before new schemes could be constructed to accommodate these data, the Activation-Synthesis hypothesis of dream construction was published. This effectively dismissed the importance of the dream as an object of study by accounting for its construction as a degraded effort by the sleepy cortex to interpret what were essentially random stimuli initiated from the pons. The third damper on the enthusiasm for the hunt to locate and explain dreaming came from the failure of REM deprivation studies to demonstrate any consistent effect of its loss on waking behavior.

Once dreaming lost its anchor to the REM state, and the questions of its meaning and function went unanswered, it is no wonder most serious investigators turned away. Now the problems raised by these challenges are addressed with new data and more elegant models, which are outlined in these five BBS target articles.

**1. The dream construction problem.** It was the reports of dream experience outside of' REM sleep, especially those collected from awakenings made shortly after sleep onset, before the criteria for the presence of REM sleep are met, that shook the assumption that REM sleep physiology was necessary for the production of dreams. If dreams can occur in descending Stage 1 when no rapid eye movements are visible, in Stage 2 following periods when REM sleep has been selectively suppressed, and if subjects can identify when they are experiencing dream imagery throughout all sleep stages by use of a signal and do so with more accuracy than experimenters were able to by using the REM markers (Brown & Cartwright 1978), then the activated brain state of REM may represent the best, but not the only, set of conditions under which dreams occupy awareness. The papers by HOBSON ET AL., NIELSEN, and SOLMS share the view that the relation of the sleeping mind to brain is more complicated than was originally described. No longer can we state with conviction: Every 24 hours we regularly cycle through three distinctively organized states of being: waking, NREM, and REM, each with its own physiological and psychological characteristics. Now we have to qualify this as more or less. Clearly there are differences in how firmly these states are separated from one another both within and between individuals. This permeability of the gates between states helps account for many anomalies in the sleep of some psychiatric and sleep disorder patients and the reports they give of their sleeping mental life.

All three of the papers that address dream production agree on three of the building blocks: (1) There must be a raised threshold for external sensory input. (2) This blockage from the periphery must occur in the presence of an activated brain which stimulates internal sources of stored sensory images, and (3) this source is biased toward the expression of basic motivational drives and negative affect. The three target article authors also call on the evidence from recent brain imaging studies showing the localized activity during REM sleep to differ from the activity level in those areas during NREM and in waking. The evidence of activation in the forebrain of the limbic and paralimbic system including the amygdala and hypothalamus supports that dream construction is emotion-driven. HOBSON ET AL. have worked out the conditions under which REM sleep is turned on starting in the pons. SOLMS has traced the dopaminergic mesencephalic tract and demonstrated the necessity of this being intact to sustain the experience of dreaming. NIELSEN offered one way to link the two, the pontine activation of REM sleep and that of the emotional- motivational dreaming system in the forebrain, with his concept of "covert REM" for dreams being experienced during periods of NREM sleep.

**2. The dream meaning problem.** The HOBSON ET AL. article is a heroic review of where we are in understanding dreams and how we got here. Having initially denied that these have intrinsic meaning these authors now propose a revision of the activation-synthesis model to account for cognition under many conditions

This new model leaves room for the study of the dream as more than an unplanned epiphenomenon of the "unthinking pons." SOLMS suggests that dreams may be experienced independently of REM sleep altogether as when the brain is activated by a seizure. After reviewing the difficulties in studying dreaming in the laboratory setting, HOBSON ET AL. build the case for home studies using the Nightcap system. This is a simplified two variable recording device based on the combination of rapid eye movements or no eye movements, in the presence of head movements or no head movements, without the EEG to distinguish sleep from wake and without the EMG of the submental muscle to distinguish quiet wakefulness from REM sleep. Thus, it would be difficult to test NEILSEN's concept of "covert REM" to account for NREM dreaming using this equipment. It would be daunting to distinguish sleep onset dreams in Stage 1 with REM intrusions from Stage 1 dreams with wake intrusions when there are eye movements present in the absence of head movement. Testing predictions based on the AIM model and NEILSEN's proposal will need further development of sensitive equipment to use in the home if experimenter and laboratory effects are to be avoided.

SOLMS's work gives more specificity to the areas of the brain and the connections responsible for the various aspects of the dream and the circuitry necessary to this activity. This suggests the possibility of developing a map of the circuits involved in contributing the various elements required to build a dream, perhaps equivalent to the five outlined to account for waking cognition by Mesulam (1998). This conversion of efforts to understand emotion- driven thought in waking and in sleep is to be applauded, and hopefully rapidly replicated. The next step would be to extend this brain mapping effort to the study of the sleep/wake transition phenomena such as the highly emotional states that are observed during night terrors and episodes of sleep walking with violence. These disorders highlight the difficulty posed by our reliance on the subject's report of the prior sleep mentation; these episodes are followed by nearly complete amnesia. Typically these subjects are unable to give an account of the perceptions responsible for their heightened drive-related emotional behavior. We need both a breakthrough in technology of more objective probes to illuminate what is happening centrally as well as more sensitive inquiry of the observers and patients to describe this experience. This will help to develop the maps of the brain areas that are functioning and not functioning during such episodes of dissociation.

For example, in two of three sleepwalking murder cases (Broughton et al. 1994; and Cartwright 2000) neither attacker recognized their victim. The face recognition pathway was not functioning while other visual pathways were operating that guided the perpetrators' spatial orientation. One man drove 15 kilometers to his mother-in-law's house, the other walked outside and assembled tools to begin work to repair a pool motor filter. Complex motor behaviors were intact. Neither responded to their victims' screams as did others who were more distant. Both were analgesic for a period following the attack; the first to pain inflicted in the struggle for the knife, the other to the cold water in the pool as he held his wife's head under water. Both had the genetic and personal history of a propensity to arouse abruptly from the delta sleep in the first cycle of sleep into a confused state that aborted REM. They clearly had a NREM to REM transition problem. When challenged during this state, both behaved as if under threat by initiating a fatal attack. Could this be explained as covert REM triggering a basic survival program? More likely it is the stress response of the neuroendocrine system that needs investigation.

NIELSEN's position that there are conditions under which the tight coupling of REM and dreaming is subject to dissociation is confirmed by the NREM dream reports of light sleepers who are in high arousal throughout all sleep and in others when there is a low threshold for arousal following sleep deprivation, and/or during acute stress. Both sleepwalking violence episodes reported above occurred followed periods of extended sleep loss and stress. Dissociation is also seen in REM sleep without atonia of those demonstrating the REM behavior disorder. This also represents a mixed state when the gates controlling movement during REM are lowered. It is interesting to note that this sleep disorder is sometimes the first symptom of a movement disorder, Parkinson's disease, in which dopamine production is low.

**3. The problem of REM function.** VERTES & EASTMAN deliver a devastating blow to the proposal that one of the functions of REM sleep may be to aid in the transfer of new learning from short term to long term storage. Certainly the evidence supporting this has been meager, and hard to replicate. Because the case is much stronger for an emotional-motivational function for REM/dreams, this is the next place to look.

This point is made when the REM suppression studies of the depressed are examined for their effect on mood rather than on memory consolidation. In the Vogel study (Vogel et al. 1975) REM suppression was carried out not by medication but by voice or hand awakenings at the first signs of REM sleep for six nights, followed by a night of uninterrupted sleep each seventh night for three or more weeks. On re-examination of the sleep in those who responded positively to this manipulation with a remission of depression, Vogel reported that it was not the REM deprivation that was responsible for the difference. The improvement in mood and increased drive behaviors was only seen in those who showed evidence of a build up of "REM pressure" on the intermittent nights without deprivation. Was this due to the appearance of covert REM? Vogel defines "REM pressure" as an increase in REM time and number of REM attempts that occur after the deprivation condition was lifted. This study suggests that depression, a state of low drive behaviors and mood, can be improved if there is rebound following a limitation of REM sleep. The fact that withdrawal from REM suppressing anti-depressants after long-term use can be followed by nightmares suggests an intensified rebound of REM/dreams, a heightening of experienced affect.

If waking mood and drive behaviors improve following the release from a period of REM deprivation, this suggests that there may also be a functional change in the nature of the dreams as well. Those who are severely depressed have little recall following REM awakenings or, at best, dreams with neutral affect. Cartwright et al. (1998b) reported a distinctive dream affect pattern within a night in those depressed volunteers who will later remit without treatment. Waking subjects to collect these dreams creates a night of reduced REM time and so constitutes a minor degree of deprivation. Remission could be successfully predicted when negative affect dreams dominated the first half of the night and more positive dreams were proportionally higher in the second half. This within-night dream affect pattern is also associated with an overnight improvement in depressed mood in normal subjects (Cartwright et al. 1998a).

REVONSUO's hypothesis that dreams involve rehearsal of fight and flight behaviors needed for survival from real life dangers reminds us that these are the behaviors that become dissociated from REM sleep in the sleep walking with violence cases and those with REM behavior disorder. Both these exhibit heightened aggression, the acting out of primitive drives including fighting, fleeing, and even inappropriate sexual behavior. These are expressed overtly in some adult sleep walkers in confusional states following arousal from the first cycle of NREM sleep before the muscle atonia of REM can confine this behavior to the safe expression of dreaming. In the RBD the aggressive behaviors occur when the loss of muscle atonia during REM allow these behaviors to be acted out in response to dreams they recall as having threatening content or which require their aggressive action. This argues that threat-avoidance programs of dreaming, if useful, may become malfunctioning in several ways: either too active or not active enough. In the depressed, this stress response may be attenuated until restimulated by some perturbing treatment. This new conception of dreaming calls for testing the relation of the survival dream scenarios and their adaptive function to the degree and length of prior waking stress and their effectiveness in terms of waking affect and coping behavior.

**Conclusion.** This group of papers set up a framework for research to fill out the picture of the mind asleep and its relation to pre-sleep and post-sleep waking behavior. After a long delay, we are moving toward a twenty-four hour picture of the brain/behavior relations as they vary around the clock, both in the normal mind and in the various disorders of the mind.

# REM sleep = dreaming: The never-ending story

Corrado Cavallero

*Dipartimento di Psicologia, Universita' degli studi di Trieste, 34134 Trieste, Italy.* **cavaller@univ.trieste.it**

**Abstract**: It has been widely demonstrated that dreaming occurs throughout human sleep. However, we once again are facing new variants of the equation "REM sleep = Dreaming." Nielsen proposes a model that assumes covert REM processes in NREM sleep. I argue against this possibility, because dream research has shown that REM sleep is not a necessary condition for dreaming to occur.
[NIELSON]

> Dream researchers face a paradoxical situation: Although a fairly large amount of evidence supports the idea that dreaming occurs during the whole night, irrespective of sleep stages, mental activity in nonrapid eye movement (NREM) sleep is still considered a second rate product in comparison with REM dreaming. And indeed, among scientists and the general public the old-fashioned – and wrong – equation "Dreaming = REM dreaming" is still widely accepted. (Cavallero et al. 1992, p. 562)

This was the start of the 1992 paper comparing REM and SWS (stage 3 and 4) reports and dream memory sources that I had thought might be substantial beyond the REM/NREM dichotomy. But I was wrong. Eight years later, notwithstanding new evidence strengthening the idea that equating REM sleep with dreaming is no longer viable, we are once again facing variants of the old REM = dream isomorphism.

Yet it has been amply demonstrated that dreaming occurs not only in REM but also during ordinary NREM sleep (including delta sleep) during sleep onset, and even during relaxed wakefulness (Foulkes 1985). It has also been shown (Antrobus 1983) that when length of dream report is partialled out, there are few if any qualitative differences between dreams collected in REM and NREM states. Moreover, a number of studies (Cavallero et al. 1990; 1992; Foulkes & Schmidt 1983) have found that when length of dream report is controlled, apparent qualitative differences between REM and NREM reports tend to disappear, suggesting that the same dream production mechanisms are involved across states. When time of night effect is controlled, narrative length is not proportional to time spent in REM prior to awakening; instead, prior sleep duration is a much more potent determinant of narrative length than time in REM (Rosenlicht et al. 1994). A number of studies on the mnemonic sources of dream content suggest that stage differences in dream recall appear more closely related to the level of mnemonic activation and to access to memory traces than to any special dream production mechanism unique to one stage of sleep (Cavallero 1987; Cavallero et al. 1990; Cipolli et al. 1988; Cicogna et al. 1986: 1991).

In general, these results suggest that the same cognitive systems produce mental activity irrespective of EEG sleep stage, as Foulkes proposed in 1985. Moreover, by comparing memory traces from day dreaming and sleep onset dreaming, Cicogna et al. (1986) found a similarity suggesting that "cognitive processes involved in the creation of original narrative sequences may be similar in sleep and waking." Further evidence comes from human neuropsychology, which has established that dreaming is coextensive with competence in mental imaging, a relatively late cognitive acquisition (Kerr 1993); and sleep-laboratory studies of children's dreaming, which indicate that dreaming is absent until ages 3 to 5, and does not assume the form of adult dreaming until ages 7 or 8 (Foulkes 1993c).

Given the above mentioned evidence, one might expect scientists will come to reject the idea, as appealing as it may be, that REM sleep is the brain correlate of the dream. On the contrary, we see a continuous quest for explanations of dreaming in physical events occurring just in REM sleep or just in REM sleep and its immediate temporal surroundings. NIELSEN's new model of covert REM processes in NREM sleep is a good example of these kinds of enterprise. He admits that NREM dreams exist and need to be taken into serious consideration (a good step forward in comparison with old-fashioned theorists who simply dismissed dreaming outside REM sleep as a kind of artifact). But then, instead of trying to develop a model that can account for dreaming as unitary phenomenon in terms of cognitive processes involved in its production, he goes back to the old idea that "real dreams" can be found only in REM sleep and hence one must find hidden REM features in NREM sleep to justify the existence of NREM dreaming. The idea in itself may be fascinating but it is the attempt to reduce dreaming to mechanisms found only in (or around) one state that is doomed to failure because, as I hope the evidence I have reported demonstrates, dreams can occur throughout human sleep and are not confined to a temporal window corresponding to REM sleep and its immediate surroundings. I must confess that I am rather skeptical about the possibility of discovering some covert REM processing underlying SWS dreaming.

# Mental states during dreaming and daydreaming: Some methodological loopholes

Peter Chapman and Geoffrey Underwood

*School of Psychology, University of Nottingham, Nottingham NG7 2RD, United Kingdom.* {peter.chapman; geoff.underwood}@nottingham.ac.uk
**www.psychology.nottingham.ac.uk**

**Abstract:** Relatively poor memory for dreams is important evidence for Hobson et al.'s model of conscious states. We describe the time-gap experience as evidence that everyday memory for waking states may not be as good as they assume. As well as being surprisingly sparse, everyday memories may themselves be systematically distorted in the same manner that Revonsuo attributes uniquely to dreams.

[HOBSON ET AL.; REVONSUO]

HOBSON ET AL. and REVONSUO use the difficulty people have remembering their dreams as key evidence in their model of conscious states. We would like to question their assumption that recall of mentation in waking states is so superior to that experienced during sleep. A critical difference in recall between waking and sleeping states may be the existence of an external narrative to which memories for internal events can be tied. Most recall of waking experiences is referenced by external events. When a subject performs any typical laboratory memory task, successful performance is predicated on the subject accepting and using the external temporal structure of the experiment. Recall of items which are internally generated, or of items presented from other learning episodes is regarded as an error (e.g., Roediger & McDermott 1995). If it is accepted that the perception of time during sleep is itself substantially distorted (Stickgold et al. 1997a) then this too may present a rather poor cue for recall. The very predictability of experiences in everyday life may provide both an illusion of memory for mundane events, and a structure with which to enhance memory for exceptions (cf. Reiser et al. 1985).

One of the most dramatic examples of memory failure for routine events is the time-gap experience (Chapman et al. 1999b; Reed 1972). The commonest example of a time-gap experience is when a driver suddenly realises that he or she has no recollection of some considerable part of the journey that is currently underway. The time-gap experience itself is characterised by a surprising failure to recall mentation. The very essence of the experience is this surprisingness. As Reed (1972) observes, a failure to recall significant mentation while spending half an hour sitting in the garden evokes little concern, while a similar failure to recall events during the drive from Bologna to Florence is perceived as a startling anomaly of attention. Two issues arise when considering such experiences – first, the question of whether the failure to recall specific episodic memories from routine, automatized tasks should in fact be surprising, and second, the question of the degree to which the "missing" mentations are internal or external in origin.

The first of these issues is the idea that the predictability of everyday experiences provides an illusion of episodic memory. Because it is clear that the only way to have progressed from Bologna to Florence in half an hour is to have driven along the road, we can confidently say that this has occurred without actually accessing new memory traces laid down by the experience. Moreover, the certainty that any interesting exceptions would have been stored allows us the knowledge that the journey passed in a routine manner in accordance with a general schema for such journeys. In fact, time-gaps may be absolutely routine aspects of most such journeys, the surprise that accompanies the occasional experience is simply brought about by some unanticipated event disrupting the normal flow of experience (Chapman et al. 1999a).

The second issue is the degree to which time-gaps may in fact be populated by internal events, task-related thoughts or daydreams. Our research, and the broader literature on daydreaming (e.g., Giambra 1995; Singer 1993) suggests that if subjects are interrupted during the performance of routine tasks they can often report task-unrelated images and thoughts (TUITs) which may be inaccessible after a delay. The reason that such TUITs or daydreams become inaccessible may simply be that because they are unrelated to the task, there is no retrieval cue available based on the normal structure of the remembered task. Without such an external narrative to impose on a sequence of events, the bizarreness of everyday cognition may itself be increased. Readers may well have had the experience of being engaged in a long and fascinating conversation when one participant suddenly exclaims "How did we ever get onto that topic?" A similar but stronger effect can be observed in one's own thoughts and daydreams – "How did I ever come to be thinking about that?" Without external events to tie previous thoughts to, the only way to answer this question may be a search for random associations between current thoughts and previous ones. We don't have direct access to what we were thinking ten minutes ago.

A common problem here for the investigation of both dreams and daydreams is the provision of any evidence (other than self-report) of their existence. Scientific evidence of any mental state can only come from systematic variation of response as the stimulus is varied. The dreams and daydreams that are hardest to recall may be the very mentations that are stimulus-free. Do they then exist? With time-gaps all we can report experimentally is the balance between internally and externally induced mentations summoned from memory. Although externally induced mentations may be subject to experimental manipulation, the opportunities for control over internally induced mentation are considerably reduced. As we argue below, even when we do know that the to-be-remembered event happened, recall can be notoriously distorted.

REVONSUO suggests that the over-representation of negative emotions, misfortunes, and aggression in dreams supports his hypothesis that dreams are specialised in simulating threatening events. An important issue here is that REVONSUO compares dream content with everyday life. Following our argument that memory for everyday life may not be as good as often is assumed, it is perhaps worth reflecting on the degree to which memory for everyday life is itself representative. A growing body of research suggests that autobiographical memory very substantially fails to represent everyday life. A recent study of ours looking at the poor recall of accidents and near-accidents (Chapman & Underwood 2000) not only demonstrates huge levels of forgetting for mundane events, but demonstrates selective retention of particular experiences in memory. Two key factors that determine the likelihood of events being represented in memory are precisely the degree of threat posed and the unpleasantness of the incident (in our study operationalised as the degree to which the participant felt they were to blame in the incident). Unpleasant, traumatic events are routinely over-represented in memory. Such findings are consistent with Wagenaar's (1986; 1994) extended analysis of his own autobiographical memory in which he reports heightened recall of highly unpleasant self-related events. Although we neither dispute nor support REVONSUO's analysis of the content of dreams, we suggest that memory, not real life, is the control condition to which the content of dreams must be compared.

REVONSUO cites Penfield's (1975) claim that random brain stimulation does not produce dreams, but instead produces memory traces. REVONSUO characterises these as short and undramatic excerpts of the patient's previous experiences. It is perhaps worth quoting Neisser's (1967) evaluation of the same data – "in short, the content of these experiences is not surprising in any way. It seems entirely comparable to the content of dreams, which are generally admitted to be synthetic constructions and not literal recalls. Penfield's work tells us nothing new about memory." (p. 169). Deciding how to characterise the reports from Penfield's patients is largely a subjective issue. We note one report of reliving the experience of childbirth (Penfield & Perot 1963). Surely it is not fair to characterise this as a short and undramatic excerpt from that patient's previous experience. More generally we suggest that the reports elicited from such stimulations may share many of the characteristics of dreams, but we stress that these characteristics may also be more representative of autobiographical memories than of real life.

## Play, dreams, and simulation

J. A. Cheyne

*Department of Psychology, University of Waterloo, Waterloo, N2L 3G1, Canada.* acheyne@watarts.uwaterloo.ca
www.watarts.uwaterloo.ca/~acheyne

**Abstract:** Threat themes are clearly over-represented in dreams. Threat is, however, not the only theme with potential evolutionary significance. Even for hypnagogic and hypnopompic hallucinations during sleep paralysis, for which threat themes are far commoner than for ordinary dreaming, consistent non-threat themes have been reported. Revonsuo's simulation hypothesis represents an encouraging initiative to develop an evolutionary functional approach to dream-related experiences but it could be broadened to include evolutionarily relevant themes beyond threat. It is also suggested that Revonsuo's evolutionary re-interpretation of dreams might profitably be compared to arguments for, and models of, evolutionary functions of play.

[REVONSUO]

The first part of REVONSUO's thesis, that dreams contain a disproportionate number of threat and predation themes, seems quite uncontroversial. As he points out, many studies have reported that a third or more of dreams contain negative emotions (see also, Merritt et al. 1994). Also reasonable is the claim that such figures seem substantially greater than would be likely in the waking lives of the subject populations of these studies, especially given the typical positivity bias (e.g., Cacioppo et al. 1999). Nonetheless, further studies such as those being carried out by REVONSUO and his colleagues are needed to further assess the degree of discrepancy. In particular, REVONSUO may wish to consider not only the relative incidence of threat and fear but also their intensity. Moreover, if simulated threat is what REVONSUO is truly interested in he might consider another common and often intensely frightening sleep-related REM phenomenon: sleep paralysis with hypnagogic and hypnopompic hallucinations (Cheyne et al. 1999a; 1999b). As many as 65% of people with such experiences give the maxi-

mum rating to their experienced fear. Although people who suffer from these sorts of nightmares may sometimes be experiencing stress in their waking lives, many volunteer that the level of fear experienced during the episodes exceeds anything they have ever experienced in their waking lives. Fear is often regarded as much too mild a word for the abject terror they experience. Also encouraging for REVONSUO's thesis is our finding of a substantial association between fear and the sense of a malevolent, unseen, threatening presence.

How do we now deal with the substantial remainder of non-threatening dream experiences? Do one-third, or half, or two-thirds of dreams have evolutionary significance and the remainder reflect random error? If the remainder of dreams were an undifferentiated morass, perhaps the narrowness of the threat simulation hypothesis would be less problematic. Dreams, however, are characterized by other themes of, for example, sex and/or flying. Floating and flying are also rather common hypnagogic and hypnopompic sleep-paralysis related experiences and these are more strongly associated with blissful feelings than with fear. We have argued that some of these phenomenal experiences are consistent with attempts to integrate conflicting vestibular and motor program activation during REM (Cheyne et al. 1999b). For example, as in ordinary dreams, activation of pontine vestibular nuclei, in the absence of feedback from compensatory head movements because of inhibition of motoneurons during sleep paralysis, may give rise to experiences of flying during sleep paralysis. Similarly, activation of motor programs, which are inhibited at the base of the spinal cord, continue to generate associated corollary discharge, which produces illusory and somewhat ethereal (because of the absence of feedback from the periphery) movements such as locomotion (Hobson & McCarley 1977; Hobson et al. 1998c).

REVONSUO's raising of the evolutionary thesis does suggest the interesting possibility that these temporary dissociations may also serve important integrative functions relating different aspects of the neural representation of bodily senses – and perhaps even the assembling of neural patterns underlying what Damasio (1999) refers to as the core self. Assembling and integrating neural maps of self representations seem at least as fundamental evolutionary functions as coping with external threats.

The evolutionary claims REVONSUO makes for dreaming are very similar to claims that have been made for play since the work of Karl Groos (1896). In one of the more rigorous versions of this sort of account, Fagan (1976) borrowed an interesting notion from engineering, arguing that the difference between practice play and "normal" functional activity was the difference between control and information functions. This analysis might equally be applied to dreaming in light of REVONSUO's suggestions. The information function operates in a manner similar to that suggested for the simulative mode in dreams.

Fagan draws upon aviation for illustrations in which the dynamic properties of aircraft and of their control may be optimized by putting aircraft through "unusual" and "exaggerated" maneuvers that would never be executed in the interest of efficient flight. In the biological example of the cat playing with a captured mouse, variations in amplitude of pouncing, for example, test the limits of the prey's reactions. Indeed, one might even understand that those limits might entail going so far as to permit the prey to escape. Such information may be important for efficient development of strategies that trade off speed, force, and accuracy.

It is intriguing that this way of thinking about dreams suggests that dreaming, as a practice mode, may have some advantages over play. One possible constraint on play (and practice modes more generally) is that it generally requires a "tension free field" or a "secure base." That is, because the informational requirements of practice test the limits of the organism's capacities (i.e., practice play is inherently dangerous), it is best to do this under relatively safe environmental conditions. Even here there are always inherent risks undertaken when one pushes any system to its limits – deliberately or not. Hence, a potentially strong point in favor of

REVONSUO's thesis is that dreaming allows even greater boldness in stretching at least the neural parameters of practice. The motor hallucinations and fictive movements of dreams seldom simply reproduce the mundane movements of everyday life (Hobson et al. 1998b). Rather they often have the unusual and exaggerated features of play. The inhibition of the peripheral motor system in REM would also allow motor programs greater latitude to experiment with (simulations of) extreme maneuvers. Parallel arguments may be made for the range of affect intensity. The attenuation of the somatic body-loop, especially motor reactivity, may allow for less constraint on the neurological components of terror and bliss. Thus arguments for the advantages of play as a practice mode may hold with even greater force for dreams.

# Iterative processing of information during sleep may improve consolidation

Carlo Cipolli

*Department of Psychology, University of Bologna, 40127 Bologna, Italy.*
**cipolli@psibo.unibo.it**

**Abstract:** The relationship between sleep and memory has been controversial since the 1950s. Studies on delayed dream recall and long-term retention of pre-sleep stimuli indicate that sleep may have a positive role in the consolidation of information. This positive indication counterbalances the negative one from the studies on the effects of REM deprivation. [VERTES & EASTMAN]

Periodically, a number of data are reinterpreted against, rather than in favor of, one or more of the three main hypotheses (interference, decay, consolidation) put forward for the relationship between sleep and memory. Such periodic reexamination is nevertheless useful to establish the value of the arguments brought by the various research strategies. The target article by VERTES & EASTMAN states that the evidence so far collected does not support any positive role of REM sleep in the consolidation of recently stored materials. This clear-cut conclusion is supported by two groups of complementary arguments, provided respectively by: (a) a thorough review of the data available on the effects of REM sleep deprivation induced using stressful laboratory techniques in animals and humans or by means of antidepressant drugs in humans; and (b) the interpretation of dream recall failure within the theoretical framework that information cannot be processed and consolidated during sleep as non-conscious state. However, the position of VERTES & EASTMAN cannot be considered conclusive, because the findings taken into account are not representative of the entire bulk of evidence available.

Concerning their first set of arguments, post-sleep retention of pre-sleep stimuli has been investigated by adopting two main strategies comparing respectively: (a) the retention rates after intervals of the same length, but characterized respectively by uninterrupted sleep and by selective sleep deprivation; and (b) the retention rates following sleep periods of similar length, but with different proportions of sleep stages (in particular, of REM and NREM sleep). By using the second strategy it has been shown that the capacity of enhancing retention is not exclusive to REM sleep; in particular, NREM sleep has a more positive effect than REM sleep on retention of simple stimuli such as paired words and sentences. These findings weaken the hypothesis of a superiority of REM sleep in determining long-term retention and also indicate that the sleep effect is influenced also by the characteristics of the materials to be retained.

As far as the second VERTES & EASTMAN's argument is concerned, several items of evidence support the possibility that (a) dream contents obtain a certain level of consolidation during sleep (Cipolli et al. 1992), and (b) stimuli externally delivered during sleep are retained in short-term memory in both REM and NREM sleep (Shimizu et al. 1977). The fate of oblivion of many dream experiences (which are quite ubiquitous during all sleep stages) is only apparent: after failure in spontaneous recall dream: subjects are capable of providing an accurate report if appropriately prompted by means of some content or sort of title they provided after night awakening. This means that dream contents are not decayed from long-term memory, but are not accessible because of interferences between the contents of dreams elaborated over the night.

The retention of stimuli delivered during (both REM and NREM) sleep in short-term memory makes them available for operations which may enhance the degree of consolidation. This possibility is crucial to understand whether consolidation also occurs during sleep for materials stored before sleep. Some data on cued recall (Smith & Weeden 1990) and dream organization (Cipolli 1995) indicate that pre-sleep stimuli can be repeatedly activated and processed in subsequent sleep stages and cycles (as it usually occurs in waking). Repeated auditory stimulation during REM sleep has proved to be capable of enhancing memory of a task previously learned in the presence of the same stimulation. This suggests that external stimulation initiates "recall" of the recently learned material and makes it available for further processing. Moreover, pre-sleep stimuli (such as sentences) are repeatedly incorporated into the contents of dreams elaborated during different stages and cycles of sleep. The similar incorporation rates in REM and NREM sleep and the iterative accessing to pre-sleep stimuli suggest that some processes of implicit memory are at work during all stages of sleep.

Finally, the retention rate of those contents of different dreams which share the same semantic features (the so-called interrelated contents) and, thus, derive from the same materials in memory, is higher than the retention rate of other contents. This suggests that iterative processing during sleep improves consolidation for materials internally accessed for insertion into dreams as well as for materials activated by external stimuli to which they have been associated before sleep. The evidence available, even if not conclusive, makes it plausible that the interactive access and processing during sleep has some consolidative effect for recently stored materials.

# Conceptual coordination bridges information processing and neurophysiology

William J. Clancey

*NASA/Ames Research Center, Computational Sciences Division, MIS 269-3, Moffett Field, CA 94035.* **bclancey@mail.arc.nasa.gov**
**www.ic.arc.nasa.gov/ic/clancey.html**

**Abstract:** Information processing theories of memory and skills can be reformulated in terms of how categories are physically and temporally related, a process called conceptual coordination. Dreaming can then be understood as a story-understanding process in which two mechanisms found in everyday comprehension are missing: conceiving sequences (chunking categories in time as a higher-order categorization) and coordinating across modalities (e.g., relating the sound of a word and the image of its meaning). On this basis, we can readily identify isomorphisms between dream phenomenology and neurophysiology, and explain the function of dreaming as facilitating future coordination of sequential, cross-modal categorization (i.e., REM sleep lowers activation thresholds, "unlearning").
[HOBSON ET AL.; NIELSEN; SOLMS; REVONSUO; VERTES & EASTMAN]

Now is a good time to bridge the different disciplines of the cognitive and neurosciences on the issue of dreams, with far-reaching implications for future theorizing across disciplines. But relating information processing theories to dream phenomenology and neurophysiology requires understanding the inherent, temporal basis of memory. In turn, a theory of consciousness can be developed that foregrounds how categories are constructed sequentially, cross-modally, and hierarchically in time (Clancey 1999) supported by **HOBSON ET AL.**'s analysis of REM neurophyisology.

From a connectionist perspective, the disinhibition of cross-modal activations suggests "reverse learning" (Crick & Mitchison 1983), by which the neural network is settling down to allow new associations to form or to lower the threshold required to coordinate experience sequentially. Thus procedural memory cannot be coordinated in REM sleep (ruling out any complex rehearsal of survival skills, contra **REVONSUO**). Instead, the function is to facilitate future learning in the awake state (contra **VERTES & EASTMAN**). Attempts to relate dreams to REM and NREM neurophysiology (**SOLMS** and **NIELSEN**) can be improved by characterizing how categories are related in sleep experiences.

What aspects of memory are missing? Dream phenomenology provides striking clues about the neurophysiology of REM sleep as well as aspects of memory and categorization that are essential for everyday consciousness. Perhaps of most importance are scene shifts and multi-modal discoordinations, which are taken for granted by the dreamer (**HOBSON ET AL.**). Freud (1900) characterized this phenomenology in terms of a rebus puzzle. For example in a dream I see a stick in the ground and say to myself "I have a lot at stake" and next am eating a steak. The meaning of the dream to me is revealed by my description, not the literal images or incidents (thus dream structure – the mix of images, sounds, and ideas – is organized by a verbal conceptualization of important concerns in my life, what I have "at stake").

However, to explain dream phenomenology in terms of neurophysiology, we need to characterize and relate both dream content and neurophysiology through an intermediate description of cognitive structure and temporal relationships (Clancey 1999). Finding an isomorphism between dream content and neurophysiology requires reformulating memory and learning in terms of categorization operating upon itself, eschewing notions of a random-access storehouse of beliefs and procedures (Clancey 1997).

Similarly, **HOBSON ET AL.**'s AIM model can be reformulated in terms of a categorization coordinating mechanism. The notion of "information" is characterized in neuropsychological terms as categorization (Edelman 1987); and "processing" is characterized as kinds of constructive operations by which multi-modal categories are physically and temporally related. Thus, I propose a three-layer analysis by which cognitive aspects of dreams and neurophysiology can be related:

1. Dream content (phenomenology).

2. Conceptual coordination analysis (structural and temporal relations of categorizing).

3. Neurophysiology analysis (neural activation between brain areas).

**HOBSON ET AL.** are right that a deficiency in memory goes a long way toward explaining orientational instability, loss of self-reflective awareness, and failure of directed thought and attention. However, the explanation is incomplete until we say more about what aspect of memory is relevant to these aspects of higher-order consciousness. This is the purpose of Level 2 in my analysis. What specific aspects of memory are missing?

Reformulating cognitive experience in terms of conceptual coordination, we find that higher-order consciousness (e.g., involving directed thought) requires three higher-order categorizing relations that are missing from REM sleep: (1) sequential correlation in multi-modal perceptual categorizing (e.g., relating sound and image), (2) holding a category active so it may be compared, counted, contrasted, etc., (3) categorizing a sequence of experience as a conceptual unit (chunking working memory). (See **HOBSON ET AL.** Figs. 4, 7, 8, 10.)

**HOBSON ET AL.**'s work shows nicely that the missing aspects of higher-order consciousness are due to aminergic demodulation. Or in conceptual coordination terms, the neurological mechanisms by which associations in different modalities are correlated or made consistent (sounds, images, and meanings correspond), by which categories are deliberately related, and by which episodes are held active (so that they may be objectified, named, and related) are not operating during REM sleep because of failures in aminergic neuromodulation.

In short, by viewing cognitive processes (information processing) in terms of how categories are formed and related (sequentially, hierarchically), the phenomenological structure of dreams can be explained. And by viewing these cognitive processes in terms of neurophysiology, their absence in REM sleep can be explained. This tripartite approach is essential because otherwise the phenomenology of dreams can only be loosely characterized in terms of "thought" or "episodes," and the role of the neurophysiolgical processes in everyday cognition will not be sufficiently articulated. As **HOBSON ET AL.** imply, psychology has heretofore failed to document the differences between waking and dreaming, just as it failed to document different kinds of consciousness among species, let alone between people and machines. This failure is rooted in a storage view of memory (with properties like copying and simultaneous multiple use of categories) and a verbally dominated view of thought (e.g., the assumption that visual thought and analogical reasoning only occurs by representing images as named objects and relationships, cf. Larkin & Simon 1987).

Relying on the information processing perspective of cognitive theory (such as a storage and retrieval view of memory), **HOBSON ET AL.**'s analysis is necessarily limited to talk about information instead of coordinated categorization in time. We can now reformulate AIM in conceptual coordination terms:

Activation (Information Processing Capacity) → perceptual categorization, scenes (coupled or synchronous categorizations), sequencing categorizations (episodes), holding a category active, holding a sequence of recent categorizations active (working memory), substitution within a sequence (e.g., saying chocaholic" by analogy to 'alcoholic"), categorizing a sequence (chunking, proceduralization), hierarchical activation of categories (bottom-up and/or top-down).

Input Source → perceptual categorization driven by: (external) sensory system, emotional correlation (e.g., dramatic theme such as "End of the World fear"), and/or conceptual (higher-order) categorization (e.g., verbal meanings influence imagery).

Modulation (Mode of IP) → how categories are conceptually coordinated, that is, how activation is modulated by other (higher-order) categorizations that are already active: correlating categories across modalities (especially sound, image, and meaning), counting, seeing-as, narrative, logical categorizing (e.g., implication, contradiction, identity), hierarchical goal-directed problem solving.

The changes in AIM during REM sleep involve an inability to hold a category or sequence of categorizations active (Activation), a mostly internally driven perceptual categorization (Input Source), and inabilities to conceptually coordinate across sensory systems and to categorize sequences (Modulation).

In summary, without the persistence enabled by sequential (and hence) temporal categorizing of the aminergic neurons, there is neither primary coordination sequencing required to follow and formulate procedural relations nor, consequently, secondary categorization (and awareness) of coordination that is occurring. Aminergic neurons are not categorizing sensorimotor activity over time (matrixing with cortical neurons is missing). With the shutdown of REM-off neurons, the reticular system is disinherited, contributing to the fantastic cross-modal activations of dreams, in which language, sounds, and images are freely associated. Attentional coordination is lost across systems, facilitated by the lack of feedback from sensorimotor interactions in the world. At the same time, the inability to hold non-sequential or non-synchronous categories active and relate them in time (which occurs in the conceptual coordination of higher-order consciousness) enables wild scene shifts.

How is activation of specific brain areas relevant? As we explicate how dreams are generated (what brain areas and paths are engaged), the conceptual coordination analysis can be mapped in more detail onto specific mechanisms involved in different aspects of categorization.

Insofar as dreaming occurs outside of REM sleep, as **SOLMS** argues, its story structure may be different from REM dreams. For instance, lucid dreams may combine disorientation and a capability to observe and comment on experience; whether these experiences are simultaneous or sequential is unclear. Building on SOLMS's (sect. 8) analysis, considering the kind of categorizing occurring in the person's experience may provide a clue about which areas are engaged and how they are relating to each other. For example, how are inabilities to hold a category active and to categorize sequences related to the deactivation of dorsolateral prefrontal cortex?

How does conceptual coordination differ during REM and NREM sleep? NIELSEN's effort to characterize the mentation in different forms of sleep may be improved by characterizing the organization of cognitive activity in terms of how categories are related. I suggest the order: perceptual categorization, scenes (simultaneous relation of multiple perceptual categories as in seeing a pen and a knee), sequencing (one scene/event follows another), correlation within a sequence (e.g., a sound is followed by a causally corresponding image), holding a categorization active (e.g., comparing ideas), and categorization of a sequence (consciousness of "what I'm doing now").

Thus, NIELSEN's Figure 1 might be improved by distinguishing the "cognitive processes" (item 4) that are higher-order categorizations missing in dreams (e.g., consolidation, rehearsal, plus forms of discrimination and selective attention) from the simpler relations found in dreams (e.g., perceptual memory activation, orienting/surprise). Aspects of conceptual coordination in sleep mentation can then be reordered (my Fig. 1) according to basic categorization (including NIELSEN's "preconscious precursors"), dreaming (scenes and narrative conceptualization), apex dreaming (protracted conceptualization of dramatic themes), and higher-order consciousness (sequentially coordinated ideas with causal and inferential relations, i.e., thinking). Because different kinds of conceptual coordination are occurring, it is too coarse to characterize NREM sleep as "more conceptual and thoughtlike." The question remains how thinking in NREM sleep and awake cognition differ.

What survival skills can be rehearsed without conceptual coordination? On a different level, REVONSUO has provided a broad-ranging, provocative account of the evolutionary function of dreaming. However, we must tighten up the notion of what is learned or reinforced and how what is learned relates to awake performance in the everyday environment. REVONSUO's analysis does not adequately distinguish between stimulus/response association and human inference. How could dreaming experiences, lacking basic aspects of goal-oriented attention, let alone reasoning by analogy and reinterpreting plans, constitute "training episodes" for skilled human performance in threat situations? The structure of dream experience, such as our inability to read text, reveals that conceptual coordination is impaired relative to awake cognitive activity, and hence we can rule out certain evolutionary benefits that require forms of logic, symbolic reference, and analogical reasoning. Although dream content reflects our everyday concerns, the primary function of dreaming, for humans at least, must be neurophysiological.

However, we must proceed carefully. Conceptualization of meaning occurs in dreams without the associated summarizing and encapsulating statements of meaning by which reasoning occurs when awake. The restricted consciousness of dreaming allows formation of new "dream thoughts," but without the elaborated structure of causally coherent narrative and planning that higher-order consciousness allows. The effect of such experience on the awake planning of humans is unclear. A both-and theory is required: Dream phenomenology is both "the consequence of an active and organized process" and "a passive byproduct of disorganized activation" (sect. 3.3). The coherence of dream drama is most definitely not like the sequence coherence of a narrative story or extended episode of experience. Although a dream may have an overarching theme or setting, the co-presence of dream elements (people, objects, and events) and the shifting story line is fundamentally unlike the coherence a person experiences (and indeed insists upon) when awake. Reading, writing, and calculating are absent because a dreaming person is unable to coordinate imagery and verbalization with a calculus. Such skills require procedural coordination (goal-directed, sequential behavior that is hierarchically organized with categorization "bindings" that may be substituted or generalized as behavior occurs; see Clancey 1999). Dream experiences are indeed multi-modal, but they are not sequentially coordinated and therefore cannot be simulations of real experience. Dream experience lacks higher-order consciousness ("insight into our true condition," sect. 3.6.l) – precisely what we rely upon to respond flexibly to threats in real life.

Human cognition is not just a stimulus-response system. Response to threats is not merely a matter of fight or flight. People anticipate (imagine what will happen next), plan (imagine what they might do next), make weapons (organize tools and get ready for some action). The complex behaviors involved in hunting and defending a habitat, especially in a social manner, are indeed skills. But they involve a kind of coordinated representation, reification, organization of materials, and behavior sequencing that are not possible during REM consciousness. What kind of simulations might be useful? Logical thinking!

Contrast the dream experience "stung by bees" with the skills of recognizing bee nests or areas where they might gather, methods of killing bees, getting honey from a bee hive, and interpreting how bee behaviors relate to climate and seasons. Aside from merely reinforcing a flight response, a dream about bees could at best reinforce a person's interest to learn more about bees or to attend to associated bee phenomena when awake.

One implication of REVONSUO's theory is that dreaming reinforces an unthinking way of responding to threat situations, merely based on reactive, perceptually, and emotionally driven behavior. If this was indeed an evolutionary advantage, it was originally conferred on other mammals, not Pleistocene man. Such learned associations, if any, are not like skilled human knowledge, because they are not procedurally integrated and flexibly controlled.

The presence of realistic imagery is not sufficient, there must be deliberate behavior, namely sustained attention that holds a goal in mind and orients interpretation and action in a coordinated way to accomplish the goal. In a daydream we can imagine a sequence of events and actions, with controlled behavior. But we lack this capability when dreaming.

Examples of "implicit learning" when awake merely show that

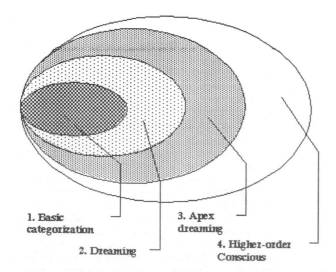

**1. Basic categorization**

**2. Dreaming**

**3. Apex dreaming**

**4. Higher-order Conscious**

Figure 1 (Clancey). Revision of NIELSEN's "levels of specificity" in terms of increasing conceptual coordination; each oval represents a form of consciousness in which simpler forms of categorization are temporally related in new ways.

correlations and sequential relations may be learned without reifying them into named objects and relations that are reasoned about (Clancey 1999). Nevertheless, the person is paying attention and performing with higher-order control. Indeed, dream experience lacks the correlation of incidents that is meaningful when awake, so how could a dream sequence produce a useful expectation of how events will unfold in awake life? REVONSUO cites evidence that REMD impairs memory for procedural or implicit tasks, which again supports the hypothesis that REM involves a relaxation/unlearning effect that facilitates later learning. So the benefit of REM would be to facilitate future learning in real-life situations, not to rehearse those situations during the dream itself.

At another level, if the average ancestral human were constantly confronted with threatening events (sect. 3.8.1), why would they need to be rehearsed? Everyday experience would surely provide enough practice to develop well-honed, adapted skills. Similarly, REVONSUO views survival skills too narrowly in terms of immediate physical dangers. Aren't "underrepresented peaceful activities," such as working in a forest, as important for survival as appropriate response to threats?

Does REM sleep facilitate future procedural learning? VERTES & EASTMAN's thesis is strongly supported by a conceptual coordination perspective that processing and consolidation of experience requires aspects of consciousness that are missing during REM sleep. But the conclusion that REM sleep could only serve to maintain CNS activity during sleep is not warranted.

VERTES & EASTMAN cite effects of post-REM deprivation on performance of an already practiced task (sect. 2.5) that appear consistent with the hypothesis that REM sleep "settles" the activation level of cross-modal coordination, thus lowering the threshold required for stimulation and hence improving performance. This in itself does not show that consolidation is operating. Instead, performance when awake could be enhanced by the clarity of mind that results from a lowered threshold required to coordinate behavior, and hence an ease in reconstructing practiced skills. For example, sitting down to play the piano in the morning you may find that the passages you labored over the night before are now effortlessly recollected. The practiced behaviors reactivate on an uncluttered path, as irrelevant relations, such as the conceptual context in which the practice sessions occurred, are not salient.

VERTES & EASTMAN cite other REMD studies supporting the hypothesis that REM prepares the brain for future multi-modal coordination learning (sect. 2.2.2). Further experiments might accordingly focus on learning involving multi-modal coordination such as sight-reading music, text comprehension involving visualization and calculation, or navigation involving multisensory cues and spatial orientation. (See also studies of REM sleep integrity and duration cited by NIELSEN, sect. 2.2, which provide related support.)

What does dreaming reveal about consciousness? Perhaps the most exciting result of this analysis is what it reveals about consciousness. First, we are conscious when we dream – a major shift from the idea that sleep is an "unconscious" state. Second, story comprehension – making sense of experience through narrative conceptualization – is more fundamental than logical thought (Donald 1991). Third, the essential coordination abilities of awake human consciousness are to hold a category active as a kind of anchor (e.g., to find a correlate and thus to have a basis of a higher-order relational categorization, such as "x is bigger than y") and to hold a sequence active and categorize it (e.g., to conceive of an episode, a procedure or method). These neuropsychological processes enable formulating goals and means for accomplishing them.

Now we may more fruitfully inquire about cognition in other animals. Do all primates have the conceptual coordination capabilities described here? Is counting possible without being able to hold a category active (e.g., scanning objects and incrementing the total)? Do other animals experience in their awake state the rapid scene shifts of human dreams? Does language confer a more stable way of holding a category active, so that cats may be goal directed, but be easily distracted and fooled because they do not name their intentions and reason about shifts in their attention? Can some personality dysfunctions (Rosenfield 1992) be reformulated in terms of inability to coordinate a protracted sequence of "what I'm doing now" (evidenced by Mr. T's rapid categorical shifts of "who I am" [Sacks 1987])?

Such questions are possible only because we no longer take for granted the conceptual coordination capabilities (binding, matching, storing, iterating) that procedural programming languages have given cognitive modelers for free. Reformulating memory, attention, and thought in terms of the neuropsychological mechanisms of consciousness is a dramatic breakthrough – perhaps the most important advance since the information processing revolution in psychology fifty years ago.

# The divorce of REM sleep and dreaming

Anton Coenen

*University of Nijmegen, Department of Psychology, Nijmegen, The Netherlands.* **coenen@@nici.kun.nl**

**Abstract:** The validity of dream recall is discussed. What is the relation between the actual dream and its later reflection? Nielsen proposes differential sleep mentation, which is probably determined by dream accessibility. Solms argues that REM sleep and dreaming are double dissociable states. Dreaming occurs outside REM sleep when cerebral activation is high enough. That various active sleep states correlate with vivid dream reports implies that REM sleep and dreaming are single dissociable states. Vertes & Eastman reject that REM sleep is involved in memory consolidation. Considerable evidence for this was obtained by REM deprivation studies with the dubious water tank technique.

[NIELSEN; SOLMS; VERTES & EASTMAN]

***Introduction.*** Few discoveries have provoked so much discussion as that of Aserinsky and Kleitman in 1953. An immediate association was established between REM sleep and dreaming: REM sleep was supposed to be the physiological sub-layer of the psychological phenomenon of the dream. A sensation was caused when it was discovered that when people were aroused from REM sleep, they always could recall a vivid dream (Dement & Kleitman 1957a). It was too good to be true! On the physiological side a high brain activity, without actual movements, and on the psychological side, the dream, a clear visual event accompanied by related emotions that simulate reality. After this firm association was established, a method appeared enabling the collection of a large number of dream reports. Researchers needed only to awaken REM sleeping subjects and inquire about their dreams. This dream recall research has been carried out countless times globally with similar results. It thus seemed that everyone undergoes several dream periods per night, and that most dreams concern normal everyday occurrences (Hall & Van de Castle 1966). It is quite striking, however, that during dreaming there is no form of critical awareness. We are not surprised about uncommon or impossible events, at the easy integration of external stimuli into the ongoing dream, such as, for example, the call of an alarm clock, and even less so about the combination of daily events that have no apparent relationship to each other.

The marriage of REM sleep with dreaming was so favored by researchers that facts not confirming this view were ignored as being insignificant. But gradually this intimate relationship was clouded. The papers of NIELSEN, SOLMS, and VERTES & EASTMAN are taken to discuss three main issues in REM sleep and dreaming research. First, the validity of the dream recall technique. Hypotheses and viewpoints on dreaming are mainly based on the results of this technique of which the validity is unknown. Second, theories on REM sleep are based on the outcomes of REM sleep deprivation, for which, in animals, the controversial water tank technique is often used. The supposed relationship of REM sleep and dreaming caused viewpoints on REM sleep and dreaming to become completely entangled. Third, and last but not least, the finding that dream mentation also occurred during slow wave sleep was ultimately decisive for the divorce of the two phenomena.

**The validity of the dream recall paradigm.** There is one point that deserves more attention and is much underexposed, although discussions about this point are not new (Cohen 1977b; Goodenough 1978). NIELSEN's paper only touches upon the validity of the dream recall paradigm. The critical point here is that dreaming occurs during sleep and is not directly observable by researchers. The actual dream cannot be studied, but only its reflection in the real world as told by the subject. No one can directly verify the accuracy of dream reporting. A dream is what someone describes upon awakening and researchers infer a one-to-one relationship between the dream and the way it is reported. But a dream report is once removed from an event or memory. It is therefore impossible to exclude such confounding factors as poor memory, overestimation, suppression or the effects of psycho-emotional factors on recall. Another problem can be that distractions at awakening impair dream recall. A further complicating matter is that there is only a short memory period after awakening in which the dream can be immediately recalled, while memory of a dream easily fades away in time. The depth of sleep from which one is aroused can also play a role in this process. Assume that awakening takes place from deep sleep; it takes time before recall can be made. Dreams seem to be recalled with ease only if the sleeper is awakened within seconds after the dream experience occurs. This short memory span for dreams is evidenced by the fact that so few people recall their dreams in the morning.

Thus, a main factor for recall is the accessibility of the actual dream, and among others is determined from the speed of awakening. This is smaller when one is aroused from light sleep compared to deep sleep with its high arousal threshold. In this way one can imagine a gradual course from NIELSEN's four levels of specificity in sleep mentation, running from apex dreaming, to dreaming, to cognitive activity, and finally to cognitive processes; a gradual dream scale ranging from most vivid intense dreams towards vague impressions. Sometimes awakening is facilitated by a frightening or bizarre dream, which then is vividly remembered. And why are dreams so illusory? Could it be that dream recall stories tell us the truth about what actually happened during the dream? So the actual question is what is the relation between the "primary" process (the actual dream) and the "secondary" process (the report or the memory of this experience). How large can the bias be? Is a dream report a reliable enough reflection of the actual dream so that we can base hypotheses on dreaming?

The "marker"-technique, introduced by Dement and Wolpert (1958), is a paradigm that may touch on this problem. These authors tried to mark a point in the dream by inserting an external stimulus into a REM period. The marker was a fine spray of cold water ejected from a syringe on the head of the sleeping subject. If this stimulus did not awaken the person, the subject was then allowed to sleep for a few minutes before being awakened by the experimenter and asked to report his dream recall. In most cases the spray of water was vividly incorporated into the dream story, for example, as a story about a leaking roof. In this way it is possible to get an idea of the relation of what actually happened and the reflection of it afterwards. However, despite the attractiveness of this marker technique, as far as I know, this experiment has not been frequently replicated in the literature. I only found one other experiment in which '"marking" or "tagging" took place. Rechtschaffen et al. (1963a) applied this technique in slow wave sleep, generally with the same results as Dement and Wolpert (1958). Perhaps this paradigm, when systematically applied may be a useful one to gain more insight in the relation of actual dream happenings and the manner in which it is reflected.

NIELSEN postulates an alternative model to explain the finding of dream mentation outside REM sleep. Covert REM sleep processes occur during slow wave sleep and these episodes are closely related to REM sleep in the sense that just before and just fol-

lowing a REM sleep episode such covert process occurs. In the view that I favour and that I explained before, this extra assumption is superfluous. I have to admit, however, that both views are not proven. Nevertheless, by systematically investigating the amount and nature of recall in relation to the nature of sleep, expressed in preceding EEG characteristics, the aforementioned views can perhaps be distinguished. It is now possible to link the complexity of the EEG of a given time period with methods derived from non linear dynamics which give a much better index for complexity than the classic visual analysis and fast Fourier transformation (Pradhan et al. 1995). A positive correlation is then expected between the degree of recall and the dimensional complexity of the foregoing EEG.

**The dubious role of the water tank technique of REM sleep deprivation.** VERTES & EASTMAN dispute the hypothesis that memory is consolidated during REM sleep and that REM sleep has an exclusive function in memory consolidation. In this view REM sleep deprivation should lead to a poor memory consolidation. A main factor in the considerations of VERTES & EASTMAN is the effects on cognition obtained when REM sleep deprivation was induced by the water tank technique. This technique is usually applied in animals, and particularly in rats. I agree with VERTES & EASTMAN that a thorough review of the literature regarding the cognitive function of REM sleep yields ambiguous results: some supportive, some ambivalent, and some negative. Together with van Hulzen and van Luijtelaar, I have carried out REM sleep deprivation studies for many years and personally experienced the debatable, controversial results of this type of research. Evidence accumulates that the effects of REM sleep deprivation produced by the pedestal technique are merely dependent on the technique of inducing REM sleep deprivation itself, instead of on the genuine effects of the lack of REM sleep. The most disturbing effects on cognitive functions are indeed obtained by using the water tank technique, fewer with the multiple platform technique, and the least with the pendulum and the selective awakening techniques. Furthermore, we determined that the effects on cognition and behavior ran more or less in parallel with the stress accompanying the applied technique (Coenen & van Luijtelaar 1985).

This strongly points in the direction of side effects of the stressful water tank technique as being responsible for the induced effects. Differential effects on behavioral indices were also found by Oniani (1984) using the platform technique together with the selective arousal method. He found behavioral changes only when animals remained on the platforms for the whole period, but not when the last part of the deprivation period was completed with hand-awakenings. Kovalzon and Tsibulsky (1984) replicated the enhanced locomotor activity and the increased intracranial self stimulation found after platform REM sleep deprivation, but could not replicate such changes when deprivation was induced by midbrain reticular formation stimulation, a variant of the selective arousal technique. Van Hulzen and Coenen (1979) demonstrated that consolidation of active avoidance is not reduced after selective deprivation of REM sleep, in contrast to the platform technique. Thus, storage of information acquired during avoidance learning is not dependent on the presence of REM sleep immediately following learning. It is concluded that learning deficiencies obtained after platform deprivation were not owing to REM sleep deprivation per se, but to adverse platform effects. Such a position is now strongly shared by Fishbein (1995).

Unfortunately, despite much research the platform factor responsible for the cognitive and behavioral changes is not yet identified, though the stress factor seems to play a pivotal role (Coenen & van Luijtelaar 1985). After publication of all these results, a drop in the number of animal REM sleep deprivation studies could be observed. Nowadays, the number of studies, however, is again firmly increasing. While the published studies seem to be a tip of the iceberg of unpublished studies, I can easily find more than 50 published studies over the last 5 years. It is likely that the simplicity and cheapness of the technique are the reason for this increase, as well as ignoring the older debates in the literature. A short survey of these studies shows a lot of diverging facts that are difficult to interpret. Moreover, studies are also directed toward a creation of a version of the classical water tank technique which induces even less stress than the multiple platform by placing more rats in the tank to overcome social isolation (Suchecki & Tufik 2000). Apart from such methodological studies, I would like to raise the question, in fact implicitly raised by **VERTES & EASTMAN**, of the acceptability of the water tank technique in sleep research. Thus, I challenge the international sleep society to thoroughly evaluate whether this technique is still acceptable according to international ethical guidelines, weighing the controversial effects that are difficult to interpret against exposing numerous animals to this technique. The scientific function of the flowerpot seems less adequate than the function for which it was originally designed!

Instead of a cognitive function for REM sleep, **VERTES & EASTMAN** propose a homeostatic function, reminiscent of the classic neural excitability hypothesis of REM sleep (Cohen & Dement 1965). A periodic endogenous stimulation maintains a requisite level of brain activation throughout sleep, and so promotes REM sleep, a faster recovery from sleep to wakefulness. Although I like this theory, the underlying evidence is still far from convincing. Based on this proposal it is now necessary that hypotheses on this proposal are formulated and adequately tested. Nevertheless in looking back on the results of the research of my group, in which all platform effects were disregarded and only effects obtained with the selective hand awakening and pendulum techniques were taken into account, a number of findings went in the same direction as the recent proposal of **VERTES & EASTMAN**. After deprivation, in some situations more behavioral activity was noticed (van Luijtelaar & Coenen 1985), a decrement in the amplitude of the evoked potential was found (van Hulzen & Coenen 1984), together with an increase in the number of beta-adrenoceptor sites in the cortex of the rat (Mogilnicka et al. 1986). All these effects, which were relatively small but significant, can be interpreted as belonging to a group of changes, all of which point to a small increase in the tonic arousal level as a result of deprivation (Coenen et al. 1986). It is inferred that REM sleep may be involved in regulating the arousal level in the waking state. However, I have to admit that all deprivation effects could also be ascribed to the drive of the brain to trigger REM sleep. To distinguish between these possibilities is a challenge for future research.

***REM sleep and dreaming: Double or single dissociable states?*** In his review **SOLMS** comes to convincing evidence for a relative independency of both phenomena. Before I comment on the paper of **SOLMS**, I will first make my own position in this matter clear. To this end I will first quote a passage of my paper (Coenen 1998):

> An important disappointment in dream research was that, now and then, but indeed consistently, non-REM sleeping subjects report dreams upon sudden awakening. This is a fundamental discrepancy that confounds the fixed relationship between REM sleep and dreaming. Although this finding has been abnegated as being insignificant, it cannot be refuted or overlooked. An opinion about dreaming could be that, if there is, for whatever reason, a sufficiently high brain activity during sleep, this may produce a dream. From this perspective, dreams are not the exclusive property of REM sleep; they are only the expression of a high brain activity during sleep. Accepting this explains the fact that an occasional dream recall during non-REM sleep can take place. One can be convinced that high brain activity that always accompanies REM sleep is at the core of dreaming, and causes the observer to mistake dreams rather than brain activity as the essential cause of REM sleep. The various dream-like phenomena that occur while one is falling into sleep, known as hypnagogic hallucinations, can also be declared as a mental expression of high brain activity. This type of dream event that occurs before one dozes off is unexplainable because the person's physiological state is not comparable to the REM sleep condition.

Based on supplementary evidence, **SOLMS**, to a large degree, agrees with the previous hypothesis. Dreaming may be the consequence of various forms of cerebral activation during sleep. He draws the conclusion that this implies a two-stage process. The first is cerebral activation during sleep and the second, the process of construction of a dream. In this respect Hobson and McCarley (1977) suggest that the cortex attempts to create a story from the bombardment from the brainstem and a dream story is the best fit the cortex could provide of this intense activity. They call this the activation-synthesis hypothesis of REM sleep. This view indeed implies a two-generator model. Firstly, cortical activation, which is for unexplained reasons of great importance; and secondly, a generator mechanism that creates a story based on this activation. Nevertheless, another view could also be that the dream is a mere by-product of this cortical activation. Perhaps, cerebral activation is the physiological basis underlying mental activity. Dreams could be merely the mental expression of intense activity in the brain that may be important for other reasons. In the same sense the noise of an automobile engine is merely a by-product of its running.

It is not completely clear what **SOLMS**'s viewpoint is on the previous models. He starts by accepting the statement that REM sleep and dreaming are double dissociable states: REM sleep can occur without dreaming and dreaming can occur without REM sleep. However, in the last part of his paper, in the reconsideration of the relationship between REM sleep and dreaming, **SOLMS** suggests an alternative explanation for the high correlation between REM sleep and dreaming. He mentions several examples of cerebral activation processes, such as induced by pathological processes and by stimulant drugs and also by REM sleep, and all are associated with dreaming. This thus implies that various brain states, which involve cerebral activation during sleep, are associated with dream reports. He thus shares my view on this, although his explanation of the one generator model (cerebral activation = dreaming) or the two generator model (where the brain itself creates a best fitting story for its own cerebral activity) is still unclear. Thus dreaming is not an intrinsic phenomenon of REM sleep, although dreaming always occurs during REM sleep. But I cannot see his often-mentioned point of the double dissociable states; the complete independency of REM sleep and dreaming. That dreaming can occur without REM sleep is now accepted, but the reverse is hard to accept. I also cannot find proof for this viewpoint. Given the unstable nature of memory for dreams, one can imagine that not every awakening from REM sleep results in a dream recall. In conclusion, rather than proposing, as **SOLMS** does, that REM sleep and dreaming are double dissociable states, it is perhaps better to regard them now as single dissociable states.

## Shedding old assumptions and consolidating what we know: Toward an attention-based model of dreaming

Russell Conduit,[a] Sheila Gillard Crewther,[a] and Grahame Coleman[b]

[a]*School of Psychological Science, La Trobe University, Melbourne 3083, Australia;* [b]*Department of Psychology, Monash University, Melbourne 3161, Australia.* {r.conduit; s.crewther }@latrobe.edu.au
grahame.coleman@sci.monash.edu.au
www.sci.monash.edu.au/psych/
www.psy.latrobe.edu.au

**Abstract:** Most current theoretical models of dreaming are built around an assumption that dream reports collected on awakening provide unbiased sampling of previous cognitive activity during sleep. However, such data are retrospective, requiring the recall of previous mental events from sleep on awakening. Thus, it is possible that dreaming occurs throughout sleep and differences in subsequent dream reports are owing to systematic differences in our ability to recall mentation on awakening. For this reason, it cannot be concluded with certainty that sleep cognition is more predominant or in any way different during REM compared to NREM sleep. It is our contention that REM sleep and ponto-geniculo-occipital (PGO) waves do not necessarily represent "pseudosensory" stimulation of the cortex in the generation of dreams, but might rather represent enhanced arousal of attention mechanisms during sleep, which results in the subsequent recall of attended mentation on awakening.

[HOBSON ET AL.; NIELSEN; REVONSUO; SOLMS; VERTES & EASTMAN]

**Background.** In 1953, Aserinsky and Klietman's discovery of a relationship between REM sleep and dream reporting reinforced growing biological reductionist concepts of brain-mind isomorphism. Such concepts also provided researchers with the impetus to study the biological mechanisms underlying REM sleep, with the hope that more general principles of hallucination could be established. This approach is now under threat, as evidence has mounted showing that REM sleep is not the exclusive domain of dreaming (e.g., **SOLMS**). In order to preserve underlying concepts of biological parallelism, researchers have hypothesized that processes underlying REM sleep could exist within NREM sleep (e.g., Pivik 1991; **NIELSEN**). However, a consistent relationship between underlying REM processes and dream reporting has not been found (see Pivik 1991). Regardless, animal-based PGO models have continued to dominate dream theory, despite the fact that a direct test of the relationship between PGO activity and dreaming has not been possible, as PGO activity cannot be directly measured in humans. Recent advances in PET neuroimaging techniques (e.g., Braun et al. 1997; 1998; Maquet et al. 1996), combined with brain lesion studies (Solms 1997a) have forced major modifications to the activation, inputs and modulation (AIM) model (**HOBSON ET AL.**). This model now suggests more modular PGO activation of association areas of the sensory cortex and limbic system in generating "pseudosensory" stimulation, rather than universal cortical activation or the activation of primary sensory areas suggested previously (e.g., Hobson & McCarley 1977; Stickgold et al. 1994a).

**HOBSON ET AL.** target article: Despite latest revisions to the AIM model, we believe some key aspects of its theoretical framework are still highly controversial for the following reasons:

1. Evidence against the notion that REM sleep is where dream mentation occurs and NREM sleep is predominantly a cognitive void: One central aspect of dream research often neglected is that psychological data regarding dreaming is collected retrospectively, requiring the recall of previous mental events on awakening. Therefore, in terms of strict scientific scrutiny, current evidence does not conclusively show that dreaming is more predominant or in any way different during REM compared to NREM sleep. It is possible that dreaming occurs throughout sleep and differences in subsequent mentation reports are due to differences in recall on awakening (e.g., Koukkou & Lehmann 1993). Most dream researchers assume that awake recall provides equal representation of previous REM and NREM sleep mentation. This may not be correct, as considerable data from sleep inertia research suggests cognitive performance parallels dream report frequency (best performance from REM to poorest performance from SWS; Dinges 1990). It is also interesting that poor dream report frequencies from NREM sleep are often treated as evidence for the absence of NREM dreams (e.g., **HOBSON ET AL.**, sect. 2), yet the absence of dream reports from REM sleep are often considered to be the poor recall of dreams (e.g., **HOBSON ET AL.** target article's sects. 2.2.1, 2.2.2). However, even if **HOBSON ET AL.**'s assumptions are accepted, there is considerable evidence against the proposal that REM sleep is a state of intense cognitive activity and NREM sleep is a relatively inactive brain state of low cognitive activity.

The results of recent PET studies are not always consistent with the AIM model of **HOBSON ET AL.** For example, EEG delta activity has been found to correlate positively with PET activation of the primary visual and secondary auditory cortex (Hofle et al. 1997). This led Hofle et al. (1997) to interpret this finding as reflecting "the occurrence of visual, auditory and perhaps verbal imagery during SWS" (p. 4806). Braun et al. (1997; 1998) have also demonstrated that the primary visual cortex consistently shows higher levels of activation during NREM sleep than during REM sleep. This led Braun et al. (1997) to observe that "SWS may not, as previously thought, represent a generalized decrease in neuronal activity" (p. 1173).

**HOBSON ET AL.**'s suggestion that the slow oscillatory rhythms of NREM sleep reflect decreases in brain activation are at odds with recent reports of Steriade and Amzica (1998), finding "frenzied" activity of cortical neurons during the depolarization phase of slow oscillations. "The frenzied activity of cortical neurons during the slow oscillation, occurring in natural sleep or deep anesthesia . . . during which consciousness is conventionally thought to be annihilated, prompts us to consider different roles played by the rhythmic bombardment of thalamic and cortical neurons upon their target" (Steriade & Amzica 1998, pp. 8–9).

**HOBSON ET AL.** also cite the observation of fast gamma frequency (30–70Hz) EEG and MEG oscillations during REM sleep (Llinas & Ribary 1993) as evidence for intense cognitive processing during REM sleep. However, gamma waves are also observable during NREM sleep (Llinas & Ribary 1993), SWS and deep anesthesia (Steriade & Amzica 1996). Such observations are inconsistent with the lack of PGO activity during these states. In acknowledging these inconsistencies, Kahn et al. (1997) state "the implications of finding the high frequency oscillations in NREM should be further investigated and the findings extended to human psychophysiology" (p. 23).

2. Evidence against the notion that PGO activity provides "pseudo-sensory" stimulation to the visual cortex: The original formulation of the Activation-Synthesis hypothesis was based on findings that pontine activation of eye-movements preceded activation of the cortex (Hobson & McCarley 1977). This hypothesis proposed that eye-movement and visual information was passed to the cortex from the pontine brainstem. This claim was also reinforced by the finding that patterns of PGO wave activity correlated with the direction of REMs during sleep (Nelson et al. 1983). However, this led to the claim that phasic PGO signals "led directly to the visual and motor hallucinations, emotion and distinctively bizarre cognition that characterize dream mentation." (**HOBSON ET AL.**, p. 41). Recently, evidence inconsistent with this "pseudosensory" nature of PGO waves has been derived from human neuroimaging and lesion studies. In fact, there is no consistent evidence supporting the notion that the primary sensory areas show enhanced metabolism during REM compared to NREM sleep (Braun et al. 1998). Nor is there any evidence that lesions to primary sensory areas eliminate dreaming (Solms 1997a). Possibly due to such findings, **HOBSON ET AL.** have revised their previous "pseudosensory" function of PGO activity. They have proposed that cortical and limbic regions may synthesize their own information when stimulated by PGO waves, claiming a similar induction of imagery to that of Solms's (1997a) concept of limbic back-projection to the visual association cortex. However, such hypotheses are still inconsistent with current imagery models of back projection to the striate cortex, which are based on PET data derived from subjects asked to view and imagine objects (Kosslyn & Thompson 2000). Imagery models based on awake subjects, are more scientifically sound, simply because we know with certainty that the PET data are derived from subjects engaged in visual imagery.

In support of their new version of PGO imagery generation, **HOBSON ET AL.** have cited findings that patterns of lateral geniculate nucleus (LGN) activity in waking cats are sufficient to represent basic elements of natural scenes (Stanley et al. 1999). However, earlier work showed that the occipital aspect of PGO waves was still present after LGN lesions (Hobson et al. 1969). It was then proposed that the thalamic aspect of PGO activity might not be entirely localized in the area of the LGN. In rats, PGO waves cannot be recorded in the LGN (Datta et al. 1998). Also, recent work by Marks et al. (1999) found that PGO innervation of the LGN in cats did not demonstrate the lamina specificity shown by retinal innervation of the LGN in visual processing. Marks et al. (1999) then conclude that the brainstem activation underlying PGO generation in the LGN controls neuronal activity in a different way to that of eye-specific, segregated retinal input to the LGN. In other words, the PGO influence on neuronal activity in the visual system is essentially different from that derived from visual experience.

However, despite **HOBSON ET AL.**'s new version of PGO imagery generation, the current model still reverts to the original activation synthesis concepts. For example: "Internally generated pseudosensory data can be produced by brainstem mechanisms (e.g., via PGO stimulation of visual cortex in REM sleep)" and "eye movement density in REM sleep provides an estimate of the amount of internally generated pseudosensory data because eye movement density reflects brain stem PGO and motor pattern generator activity" (HOBSON ET AL. pp. 55–56). We believe that to be accepted as a viable hypothesis, the "pseudosensory" role of PGO activity during REM sleep requires further clarification and investigation.

***PGO activity represents the arousal of attentional awareness during sleep.*** Based on the initial proposal that dreaming might represent a state of attentional awareness without volitional attentional control (Posner & Rothbart 1998), Conduit (1999) has put forth an attention-based model of dreaming. In the attention-based model, PGO activity is related to the arousal of attention mechanisms during sleep. This arousal produces heightened attentional awareness during sleep, allowing potential recall of attended sleep mentation on awakening. Several lines of evidence support this proposal of arousal of attention mechanisms during sleep.

***Orienting, attention, and PGO activity.*** Bowker and Morrison (1976) first raised the argument that the PGO wave was intimately linked to the startle response. They interpreted behaviors coincident with PGO activity as "alerting or orienting movements in response to some internal discharge, or as we suggest, 'startling' stimuli, that occur with each PGO spike appearance" (p. 188). However, years later, after extensive investigation of PGO and muscular variations in such things as timing, intensity and habituation, these researchers have come to a different conclusion: "neurons (we predict in the reticular formation) identify a signal that requires attention and that this requirement is passed via peribrachial neurons in the dorsal pons that respond to auditory stimuli and also generate PGO waves . . . thus, PGOs in the LGB could reflect a honing of neural mechanisms in the visual system to receive information" (Sanford et al. 1993, p. 443). Consistent with these findings, the pulvinar nucleus of the thalamus receives PGO outputs from the pons (Steriade et al. 1988). Furthermore; there are several lines of evidence showing the pulvinar has a central role in attention processes (Robinson & Peterson 1992). Thus, it is not unreasonable to suggest that phasic PGO activation of various regions of the thalamus could act to enhance the sensitivity and information gathering processes of a variety of sensory relay circuits (Sanford et al. 1994a), hence, heightening attention processes during sleep.

***Eye movements and attention.*** In an approach that we believe is more consistent with an attention-based model than AIM, **HOBSON ET AL.** suggest that the observation of bottom-up control of attentional eye movement (EM) mechanisms during sleep provides evidence in favor of pontine generation of dream imagery, and the observation of top-down control of EM attention mechanisms provides evidence for a scanning hypothesis. Using such an approach, these authors have literally used the activation of attention as an operational definition of dreaming. So, what do we really know? We know that EMs occur during REM. We know that either brainstem or cortical mechanisms can generate and modulate EMs. However, we also know that EMs are usually preceded by a shift in attention (Chelazzi & Corbetta 2000). Therefore, we are fairly certain that attention mechanisms are activated during sleep. We have no conclusive evidence that bottom-up EM control represents the activation of "pseudosensory" imagery.

***PET studies of REM sleep and attention.*** When one considers the possibility that REM may be a state of heightened attentional awareness during sleep, the amount of overlap in PET activation of brain areas during attention tasks in awake individuals and during REM sleep seems more than coincidental. PET studies of subjects during attention tasks have found activation of brain areas common to those activated during REM sleep. These include: the

brainstem (particularly the reticular formation), thalamus (particularly the pulvinar nucleus), anterior cingulate, hippocampus, parahippocampal gyri, anterior cingulate, and scattered association areas of the posterior occipital/parieto/temporal neocortex (particularly the parietal and extrastriate areas; Chelazzi & Corbetta 2000; Lockwood et al. 1997; Posner 1994b). Also, shifts in attention have little observable effect in the primary visual cortex (except maybe when the visual field is highly cluttered; Posner & Digirolamo 2000). This result fits well with the specific activation of extrastriate association areas rather than the striate cortex during REM sleep (Braun et al. 1998). The dorsolateral prefrontal cortex is heavily implicated in executive attentional control (Posner & Digirolamo 2000) and in conjunction with attentional awareness is proposed to be necessary for consciousness (Posner & Rothbart 1998). This is compatible with findings that the dorsolateral prefrontal cortex shows little activation during REM sleep (e.g. Braun et al. 1997; 1998).

***Electrophysiology during REM and attention.*** Electrophysiological studies support the proposal that REM is a sleep state of enhanced attentional awareness. A particular component of event related potentials (ERPs), the P300, is elicited in the waking state during the external orientation of attention in response to deviant stimuli or unexpected presentations. Sleep investigations have consistently produced the P300 during REM but not during other sleep stages (e.g., Cote & Campbell 1999). Occipital EEG alpha attenuation is also considered a physiological sign of the activation of visual attention. Recent research has found decreased occipital alpha spectral power during phasic REM periods compared to tonic REM (Cantero et al. 1999a).

Both electrophysiological and metabolic measures of neural activity during sleep can be interpreted as inconsistent with the AIM model. However, we believe such issues can only be resolved once the current temporal resolution of our investigative tools (PET and MRI) and the sleep scoring system we have adopted (Rechtschaffen & Kales 1968) are refined to adequately deal with sleep events lasting less than one second.

**SOLMS** target article: Some of the most challenging findings for the AIM model recently have been those derived from human lesion studies. Solms (1997a) essentially found that patients with brainstem lesions that eliminated REM sleep could still recall dreams, while patients with cortical lesions to areas such as the parieto-temporo-occipital (PTO) junction reported loss of dream recall with REM sleep intact. Hobson et al. (2000) dismissed the human brainstem lesion findings by stating that any lesion capable of eliminating the pontine REM sleep generator mechanism would eliminate consciousness altogether. After acknowledging this criticism, **SOLMS** has focused his latest review on the investigation of whether dreaming can be eliminated by forebrain lesions. From this, a large majority of cortical lesions resulting in the cessation of dreaming were located in or near the PTO junction (94/110). The small number of remaining lesions that eliminated dreaming were located near the ventro-mesial quadrant of the frontal lobe. **SOLMS** then argued that dreaming is driven by cortical back-projection, initiated from frontal DA circuits.

The brain lesion studies reviewed by **SOLMS** are also interpretable in terms of an attention-based model of dreaming. For example, frontal DA circuits have been implicated to play a key role in the regulation of attention processes (Granon et al. 2000), dysfunction of attention in schizophrenia (Swerdlow & Geyer 1998), and attention deficit disorder (Papa et al. 2000). Also, the underlying cognitive disorder of PTO lesions could be a deficit of visuo-spatial attention (see Posner 1994b). Solms's (1997a) findings that damage to extrastriate areas results in similar deficits in dreaming and waking perception is also consistent with an attention model. For example, when attending to the colour, form or motion of visual input, relative increases in neural activity occur within the same extrastriate areas that are believed to process such information (Posner 1994b). Solms (1997a) himself acknowledged his interpretations of the inhibitory function of the anterior cingulate and thalamus during dreaming were compatible with the proposed functional role of these structures during attention. Generally it is accepted that attention processes act to suppress unattended areas, resulting in a relative enhancement in activity of the cells coding for the attended stimulus (Posner 1994b). PET studies suggest that thalamic and anterior cingulate inhibitory projections enable the selective modulation of posterior parietal and extrastriate areas of the brain during attention (Posner 1994b). SOLMS highlights the finding that lesions to the dorsolateral prefrontal cortex have no effect on dreaming, but are implicated in significant deficiencies of executive control, and hence might explain the executive deficiencies of dream cognition. These findings are consistent with PET findings of dorsolateral prefrontal deactivation during REM (e.g., Braun et al. 1997), and support the proposal that dreaming is an example of heightened attentional awareness with deficient executive attentional control (Posner & Rothbart 1998).

SOLMS's evidence that dreams are cortically initiated is not necessarily conclusive. It could be argued that the functioning of a lesioned brain does not necessarily reflect the full neural circuitry utilized by an intact brain. For example, SOLMS argues that dreaming can only occur if the DA circuits of the ventromesial forebrain are aroused, and thus REM sleep is simply a state that reflects the effects of cerebral activation of this region during sleep. However, in a sleeping, intact brain, cortical arousal is essentially derived from the ascending reticular activating system and/or the PGO generator of the brainstem (Steriade 1996). If normal spontaneous arousal during sleep does not arise from the brainstem, where is its origin?

REVONSUO's target article has put forth a convincing argument that any biological theory regarding the function of dreaming should be accountable through concepts of evolutionary biology. However, the notion that dream consciousness is a unique state providing a mechanism for simulating threat perception and threat avoidance responses currently has inadequate empirical support.

The main evidence cited supporting this proposal has come from the interpretation that dream content shows a significant bias toward representing threatening events. However, most of this evidence comes from dream reports collected from home using dream diaries (e.g., Hall & Van de Castle 1966). Foulkes & Cavallero (1993a) have argued against the assumption that spontaneous dream reports collected from home provide a true representation of the nature of dreams. Human memory research suggests that events attracting attention by being more emotional or unusual are more easily recalled (Brown & Kulik 1977). Thus, vivid, emotional and/or bizarre dreams may be the majority that are reported simply because these are cognitive events we more reliably remember (Cohen & MacNeilage 1974; Van den Hout et al. 1989). If this is indeed the case, such recall biases might persist even in controlled laboratory awakenings. Foulkes and Cavallero (1993a) describe the results of research using systematic REM (and NREM) awakenings as "surprisingly mundane, built around relatively realistic situations" (p. 11). Considering such arguments, it is possible that the "over-representation" of threat in dreams may be due to the way we selectively attend and recall information, especially from spontaneous awakenings at home. This point is worth noting, particularly since REVONSUO is continuing to collect home-based dream reports in support of his theoretical claims (sect. 3. 3).

Even if we concede that "threatening" events are over-represented in dreams, rather than more easily remembered, REVONSUO's "threat simulation" proposal is not convincing. For example, if subjects were asked to recall events from their life history, emotionally significant or "threatening" events would probably be most prominent (Brown & Kulik 1977). In other words, all of the events we dream about must have an origin in memory. Therefore, it is the selective nature of attention and memory consolidation during waking that can provide an explanation for the proposed over-representation of "threatening" dream content, rather than a biased dream generation mechanism. This explanation can also be

put forth for why adults dream of "current concerns," like for example, divorce. People dreaming of current concerns are also thinking and attending to these problems in their waking life. Hence, their dreams reflect their current psychological state, and do not necessarily provide an overrepresentation of "threat."

Our attention-based model of dreaming is more compatible with Snyder's (1966) "Sentinel" hypothesis of dreaming than REVONSUO's model. Just as PGO mechanisms in an awake animal can heighten sensory awareness to deal with a possible approaching predator (Sanford et al. 1993), PGO activity during sleep might serve a similar function. Thus, it might be that phasic PGO waves act to periodically arouse the attention circuits of the brain enabling potentially threatening stimuli (such as novel or emotionally significant stimuli) to be perceived. If an external stimulus cannot be recognized as a "safe" stimulus (expected or of no emotional significance), the "novelty" or "emotional significance" of the stimulus should induce further attention and arousal to a point where a decision can be made as to whether there is a threat to survival.

VERTES & EASTMAN provide a strong case against the notion that memory consolidation occurs during REM sleep. However, a convincing alternative to the opposing consolidation model is not provided. We believe that if the activation model of VERTES & EASTMAN was developed to support the existence of heightened attentional awareness during sleep, a more consistent account of the existing data could be proposed in opposition to the consolidation model.

VERTES & EASTMAN offers an alternative to the memory consolidation function of REM sleep, arguing that REM provides "periodic endogenous stimulation to the brain" which maintains the "minimum requisite levels of CNS activity throughout sleep without awakening the subject or disturbing the continuity of sleep." If this is the case, what purpose does such stimulation serve? VERTES & EASTMAN begin to address this issue by stating that REM serves to "prime the brain for a return to consciousness as waking approaches" (sect. 6.5). Such interpretations are compatible with an attention-based model of REM sleep and dreaming, as the phasic arousal of attention mechanisms can be interpreted as a form of environmental monitoring in case of attack from predators, equivalent to Snyder's (1966) "sentinel" of sleep.

VERTES & EASTMAN provide a convincing critique of studies involving performance measures taken after REM deprivation, arguing that these studies are confounded by factors such as the stress and physically debilitating effects of deprivation procedures. However, VERTES & EASTMAN do not offer a strong explanation for observed increases in REM sleep after exposure to novel, enriched or enhanced "learning" environments. An attention-based model would predict that any altered environmental conditions will increase perceptions of possible danger or predation during sleep and thus result in REM enhancement and poorer sleep quality.

VERTES & EASTMAN refer to work relating hippocampal theta to long term potentiation and the observation that such activity is highly prominent during REM sleep (Winson 1993). They argue that the theta rhythm is generated as a by-product of the activation of brainstem mechanisms during REM and does not necessarily bear any functional relationship to its role in waking. However, such brainstem activation of the hippocampus is also present in waking during the engagement of attention (Buhusi & Schmajuk 1996). Curiously, VERTES & EASTMAN's proposal that theta "serves to gate and/or encode information reaching the hippocampus" (sect. 2.6) is analogous to other researchers interpretations that theta is involved in attentional processing (Buhusi & Schmajuk 1996). Under an attention-based model, hippocampal theta can be interpreted as a role in attention rather than memory consolidation, thus the unconvincing argument that theta indicates a different function depending on sleep state is not necessary.

An attention-based model of REM sleep and dreaming would maintain that cognition regarding salient memories previously consolidated during waking is attended to during REM/phasic sleep and thus may be recalled if the subject is awakened. On awakening, if there is any disruption, for example, a delay or distraction, recall will be impaired (Goodenough 1991), as in attentional blanking (Lawson et al. 1999). If consolidation of dream mentation occurs, it must occur following awakening; otherwise, dream mentation would be no more elusive to memory than any other waking event. Thus, from one source of insight into sleep cognition (dreaming), it seems that memory consolidation does not necessarily occur during sleep. Apart from the conditioning of reflexive physiological responses during sleep (e.g., Conduit & Coleman 1998), higher forms of learning requiring memory consolidation do not seem to be possible during sleep (Eich 1990).

**NIELSEN**'s proposal of "covert" REM sleep processes is an important and interesting one. It highlights the problems that modern sleep researchers have when attempting to investigate the possible existence of sleep phenomena which last for seconds (e.g., alpha blocking; Cantero et al. 1999a) or even milliseconds (e.g. PGO waves; **HOBSON ET AL.**), within a definition of sleep which has as its smallest unit, an epoch of 30 seconds (Rechtschaffen & Kales 1968). As new electrophysiology techniques approach submillisecond temporal resolution and PET/MRI scanning resolution is a matter of seconds, the current system of defining sleep must accommodate these advances if our understanding of sleep is to progress.

In many ways, **NIELSEN**'s model parallels previous tonic/phasic models of sleep mentation. Such models proposed the existence of phasic sleep processes (primarily PGO waves) underlying the recall of mentation from sleep (see Pivik 1991). Several lines of evidence presented as supporting **NIELSEN**'s model were originally cited as evidence for tonic/phasic models. These include: the proximity of NREM sleep awakenings to REM sleep (e.g., Stickgold et al. 1994a), REM deprivation effects on NREM recall (Foulkes et al. 1968), drug effects on NREM recall (e.g., Delorme et al. 1965) and sensory stimulation effects (e.g. Conduit et al. 1997).

However, since **NIELSEN**'s covert REM model is not necessarily dependent on the existence of PGO activity, this has allowed the incorporation of previous results relating sleep arousal to imagery reporting, where PGO activity is often absent. Findings such as sleep onset imagery (Vogel 1991), time of night effects (Rosenlicht et al. 1994) and sleep terrors (Broughton 1995) were previously considered inconsistent or incompatible with tonic/phasic PGO models (Pivik 1991).

**NIELSEN** states that the covert REM model is "similar to the one-generator model in that it assumes commonality of processes for all mentation reports, but it differs in that it extends this commonality to physiological processes" (sect. 3.2). The question that remains is what is this common underlying physiological process?

**Conclusions.** Previously, we have proposed that PGO activity might be indirectly related to dream reporting through the phasic activation of arousal, which then provides optimum conditions for the recall of ongoing mentation from sleep on awakening (Conduit et al. 1997). However, recent neuroimaging findings from REM sleep (Braun et al. 1997; 1998; Maquet et al. 1996), have shown PET activation of brain regions involved in attention. We now believe that heightened attentional awareness provides the conditions for subsequent recall of dreams on awakening and the unique characteristics of this recalled mentation. Thus, it might be the arousal of attention mechanisms that is the underlying physiological process of **NIELSEN**'s covert REM model, and might also better describe the "A" aspect of the AIM model of **HOBSON ET AL.** It might be damage to the brain mechanisms of attention that underlie the lack of dream reporting in patients suffering lesions to the parieto-temporo-occipital junction or the ventromesial quadrant of the frontal lobe, or the excessive dreaming of patients with damage to the anterior cingulate or thalamus (**SOLMS**). Arousal of attention mechanisms during sleep can be interpreted as supporting the case against memory consolidation during REM sleep, as memory is already consolidated and it is

heightened attentional awareness that is present during REM sleep. Finally, there is a sound evolutionary rationale for heightening attentional awareness during REM sleep, as it can be viewed as a mechanism of periodic environmental monitoring. However, if we are to even attempt to begin the testing of such new proposals, we must review our outdated sleep classification methods, so that at least our definition of sleep has the same temporal resolution of our current investigative tools.

## Needed: A new theory

G. William Domhoff

*Psychology Department, University of California, Santa Cruz, CA 95064.*
**domhoff@cats.ucsc.edu**

**Abstract:** Dream content is more coherent, consistent over time, and continuous with waking emotional concerns than most brainstem-driven theories of dreaming allow, but dreaming probably has no adaptive function. A new neurocognitive perspective focusing on the forebrain system of dream generation should begin with the findings on dream content in adults and the developmental nature of dreaming in children.
[**HOBSON ET AL.**; **NIELSEN**; **REVONSUO**; **SOLMS**; **VERTES & EASTMAN**]

***Introduction.*** A large body of findings with the Hall and Van de Castle (1966) coding system shows that dreams are more coherent, consistent over time for both individuals and groups, and continuous with past and present waking emotional concerns than **HOBSON ET AL.**'s emphasis on brainstem-driven bizarreness can accommodate (Domhoff 1996). In addition, Foulkes's (1982b; 1999) laboratory discovery of low levels of dreaming until ages 9–11 joins **SOLMS**'s (sect. 6) findings with brain-lesioned patients in demonstrating that REM sleep is insufficient for dreaming. A new neurocognitive theory of dreaming therefore should begin with the hypothesis that Foulkes's developmental findings may correlate with the maturation of the forebrain system of dream generation first uncovered through creative neuropsychological detective work by **SOLMS** (sect. 8). In addition, the findings with the Hall/Van de Castle system on the lifelong persistence of various kinds of negative dream content suggest there is a "repetition dimension" in people's dream life (Domhoff 1993; 1996) that may relate to the temporal-limbic and frontal-limbic origins of dreaming in **SOLMS**'s (sect. 7) model.

Contrary to **REVONSUO** (sect. 3.2), however, it is doubtful that dreams have any adaptive function. There are too many people, including children and brain-lesioned patients, who sleep adequately without them, and no evidence that either recalled or unremembered dreams have any functions (Antrobus 1993a; Foulkes 1985; 1993a). At best, people in some societies have invented uses for dreams, and in that sense dreams have an "emergent function that develops through culture" (Domhoff 1993, p. 315). Moreover, there is no evidence from systematic psychological studies that supports any psychotherapy-based dream theory claiming one or another function for dreams (Domhoff 1999a; Fisher & Greenberg 1977; 1996; Foulkes 1985).

***The brainstem/bizarreness commitment.*** Both **HOBSON ET AL.** (sect. 4) and **NIELSEN** (sect. 3) present interesting ideas that may explain away much of the "dreaming" in NREM sleep. However, they do concede there is enough dreamlike mental activity in NREM sleep to challenge the strict equation of dreaming and the REM stage of sleep, especially late in the sleep period. The empirical dream psychologists who abandoned the REM sleep/dreaming equation decades ago in the face of contradictory evidence summarized by Berger (1967; 1969), Foulkes (1966; 1967), and Hall (1967) did not ask for much more than what is now granted in these articles. **HOBSON ET AL.** are wrong to chastise psychologists (sect. 2.3.3) for focusing on the cognitive level when the constant changes in their own model show that their comprehensive mind-brain isomorphism is extremely premature (sect. 3).

NIELSEN (sects. 3.1–3.11) nicely demonstrates the arbitrariness of the "stages" of sleep agreed upon by Rechtschaffen and Kales (1968) in the face of great inconsistencies from laboratory to laboratory in analyzing sleep records. Considering the large number of situations that can lead to "missed" REM periods, "intermediate" sleep, stimulation-induced REM sleep, and transitions to Stage II during REM sleep, it would be interesting to know what percentage of a night's sleep is consistent with the scoring manual in a representative sample of uninterrupted nights of sleep from normal participants. A low percentage would strengthen NIELSEN's (sect. 3.14) call for a view of sleep stages as "fluid" and "interactive," which finds echoes in HOBSON ET AL.'s (sect. 4) emphasis on "dissociation" and "psychophysiological continua."

It is regrettable that HOBSON ET AL. took so long to broaden their theory in the face of contradictory evidence available long ago (Vogel 1978a), but it is possible that the "state" transition at sleep onset (sect. 4.2.2) and the greater activation late in a sleep period (sect. 3.3.4.3) explain much dreamlike NREM mentation. The disappointment is their continuing brainstem commitment, which is also preserved by NIELSEN (sect. 3) through his concept of "covert REM sleep." In the face of the new and old findings synthesized by SOLMS (sect. 6) to show that brainstem activation is not sufficient for dreaming, and in some unknown percentage of cases may not even be necessary, it would seem that research relating the forebrain system to many different aspects of dream content should now be the primary focus of mind-brain isomorphists.

HOBSON ET AL. (sect. 2.3.3) justify their desire to keep the brainstem at the forefront of their theory on the basis of a commitment to a mind-brain isomorphism. However, this insistence may also be due to their strong belief that dreams are bizarre and discontinuous, although one of their own studies reported "discontinuities" in only 34% of 200 dreams (Rittenhouse et al. 1994). Most others who have studied large samples of dream reports from groups and individuals see dreams as even more realistic (Dorus et al. 1971; Foulkes 1985; Snyder 1970; Strauch & Meier 1996). For example, Hall (1966) concluded that only 10% of 815 home and laboratory reports from 14 adult males had at least one "unusual element," using a scale that can be found in Domhoff (1996). In studies comparing REM reports to samples of waking thought collected from participants reclining in a darkened room, the waking samples were rated as more dreamlike (Reinsel et al. 1986; 1992).

To support their focus on brainstem activation and the bizarre nature of dream content, HOBSON ET AL. have to challenge several different sets of impressive findings. First, they reject (sect. 2.3.1) Foulkes's (1982b; 1999) conclusions on the low levels of REM dreaming in young children with the claim that these children are not able to communicate in words about their dreams. But Foulkes's data show that the rate of recall correlates with visuospatial skills, and that there are older children with good communication skills and poor visuospatial skills who do not recall very many dreams in the laboratory. It is more likely that young children do not dream often or well by adult standards, a conclusion favoring a cognitive theory of dreams.

HOBSON ET AL. (sect. 2.3.3) reject Foulkes's findings on the banality of the few dreams his young participants did report by saying the laboratory situation is not conducive to typical dreaming, but Foulkes (1979; 1996b; 1999) already has answered that claim very effectively. More generally, they overstate the differences between home and laboratory dreams. This is shown most recently in a reanalysis using effect sizes (Domhoff & Schneider 1999) with the original codings from the most comprehensive study of this issue, which was carried out by Hall (1966) with 11 young adult male participants who each spent three to four consecutive weeks sleeping in a laboratory bedroom in a house in a residential neighborhood.

HOBSON ET AL. (sect. 2.3.3) denigrate the findings on the everyday nature of most dream content by saying that psychological measurement has not been adequate, but they have not demonstrated that their evolving rating scales for the slippery

concept of bizarreness can be used reliably across laboratories. Furthermore, they ignore most of the findings with the Hall/Van de Castle system, which has shown high reliability when used by researchers in many different countries and produced results that have been replicated several times (Domhoff 1996; 1999b) However, HOBSON ET AL. (sect. 2.1) do note the Hall/Van de Castle findings on emotion in dreams, which anticipate their own findings of more negative than positive emotions, more reports of emotions in women's dreams, and no gender differences in the distribution of emotions (Merritt et al. 1994).

In their effort to emphasize differences between REM and NREM reports, HOBSON ET AL. (sect. 2.2.2) argue against any control for length of report. In so doing they do not seem to realize this problem is handled without loss of data by the indicators based on percentages and ratios that are now standard in the Hall/Van de Castle system (Domhoff 1999b; Schneider & Domhoff 1995).

NIELSEN (sect. 2.9.2) also discusses this issue, but does not come to any conclusion, perhaps because he did not make enough of a study in his laboratory using Hall/Van de Castle indicators with 20 REM and 18 Stage 2 NREM reports (Faucher et al. 1999). It showed the REM reports had higher rates of aggressive social interaction even with this small sample size, which is an impressive result because aggression is more sensitive to age, gender, culture, and home/laboratory comparisons than any other variable (Domhoff & Schneider 1999).

Strong support for the use of the Hall/Van de Castle content indicators in resolving disputes about the nature of REM and NREM reports is provided by a study Hall carried out three decades ago, but that was only recently reported by Domhoff and Schneider (1999). When NREM reports from early and late in the sleep period were compared with REM reports, several of the usual differences appeared. For example, the "cognitive activities percent" (the number of cognitive activities divided by the total number of all activities) was 20% in NREM reports, but only 11 % in REM reports. Conversely, the "verbal activity percent" was 37% in REM reports, but only 22% in NREM reports. However, the NREM reports from after the third REM period of the night were more similar to REM reports than early NREM reports on a summary measure for a wide range of Hall/Van de Castle categories. These results are consistent with the recent theorizing by HOBSON ET AL. (sect. 3.3.4.3).

HOBSON ET AL. (sect. 2.3.2) call for studies of dreams at home to obtain a more realistic sample of dream content, but they overlook the replicated longitudinal results with the Hall/Van de Castle system, which show that dream content can be constant for individual adults over years and decades, something that might not be expected if dreaming is as chaotic and bizarre as they claim (Domhoff 1996). One of these longitudinal studies showed that the dreams of "the Engine Man," used by Hobson (1988b) to show the bizzareness of dream structure, are highly consistent in content over just a three-month period. His dreams are also below the male norms on key social interactions, and continuous with his waking life in terms of the people and activities in his dreams (Domhoff 1996).

***Dream function dream negativism.*** REVONSUO (sect. 2.2) does a convincing job of critiquing rival functional theories, and his "threat simulation" hypothesis draws on an impressive array of ideas from many different kinds of studies (sect. 3.4). Unfortunately, several pieces of his complex argument are highly speculative, including his most crucial sleep/dream claim, the attribution of mental training and implicit learning (sect. 3.7) to REM sleep (VERTES & EASTMAN). It also seems unlikely that trauma could stimulate the development of dreaming (sect. 3.5.2), since Foulkes (1982b) found that children with tense home environments or personal problems did not report more dreams, or more negative content, than did other children. Nor is it possible to agree with the idea that the stereotypic movements of decorticated cats could be the acting out of dreams (sect. 4.2) because it is highly doubtful that animals dream (Foulkes 1983). Finally, it is hard to imagine that chase and attack dreams, which rarely contain successful defensive actions in any event, could make human beings any more primed for reacting to threat than they are due to the one-trial fear-conditioning system that is already found in reptiles (Le Doux 1996).

However, REVONSUO (sect. 3.5) is on to something when he links negative dreams and the "repetition dimension" (Domhoff 1993; 1996) to the vigilance/fear system centered in the amygdala. If this idea is placed within the context of ontogenetic development and SOLMS's (sect. 7) ideas on the forebrain mechanisms that activate dreaming, then REVONSUO has made a good case that the repetition dimension expresses a person's history of emotional concerns. Just as emotional memories can last a lifetime, so too can posttraumatic stress disorder dreams, recurrent dreams, recurrent themes in dreams, and heightened scores on Hall/Van de Castle indicators. Most generally, then, the available evidence suggests that dreams are both non-adaptive and psychologically revealing (Foulkes 1993a; 1999).

**Conclusion.** If the methodologically most sound descriptive empirical findings were to be used as the starting point for future dream theorizing, the picture would look like this: (1) dreaming is a cognitive achievement that develops throughout childhood (Foulkes 1999); (2) there is a forebrain network for dream generation that is most often triggered by brainstem activation (Hobson et al. 1998b; Solms 1997a); and (3) much of dream content is coherent, consistent over time, and continuous with past or present waking emotional concerns (Domhoff 1996). None of the papers reviewed in this commentary puts forth a theory that encompasses all three of these well-grounded conclusions. This suggests the need for a new neurocognitive theory of dreaming (Domhoff 2000).

# Mesolimbic dopamine and the neuropsychology of dreaming: Some caution and reconsiderations

Fabrizio Doricchi[a,b] and Cristiano Violani[c]

[a]*Centro Ricerche Neuropsicologia, Fondazione IRCCS S. Lucia, 1–00179 Rome, Italy;* [b]*Istituto di Psicologia "L. Meschieri," Università degli Studi di Urbino, Urbino I-61029 Italy;* [c]*Dipartimento di Psicologia, Università "La Sapienza," Rome I-00185, Italy.* {**Fdoricchi; violani** }**@uniromal.it**

**Abstract:** New findings point to a role for mesolimbic DA circuits in the generation of dreaming. We disagree with Solms about these structures having an exclusive role in generating dreams. We review data suggesting that dreaming can be interrupted at different levels of processing and that anterior-subcortical lesions associated with dream cessation are unlikely to produce selective hypodopaminergic dynamic impairments.
[HOBSON ET AL.; NIELSEN; SOLMS]

The cessation of dreaming after bilateral lesions of the deep white matter surrounding the tip of lateral ventricles is the relevant and original contribution of Solms (1997a) to the neuropsychology of dreaming. Starting from this evidence, which seems corroborated by brain imaging findings showing activation of several limbic structures in the medial basal forebrain, SOLMS now ascribes a fundamental and virtually exclusive role to the dopaminergic mesolimbic structures of the "reward-motivational system" in the generation of the dreaming state both within or outside REM sleep (provided a sufficient level of vigilance). But dreaming consists of a variety of concomitant neurocognitive operations; several hindbrain and forebrain mechanisms and several neurochemical systems maybe involved in a active construction and recall of dream experiences. We will critically review some of the arguments raised by SOLMS in favor of his hypothesis.

***Clinical and neurochemical evidence.*** The clinical correlates of global dream cessation documented by Solms (1997a) are not in themselves evidence that disruption of the reward-motivational system is playing a central role. Adynamia scarcely differentiated dreamers and non dreamers ($p < .1$), whereas measures of frontal function (preservation, $p < .001$) did. This points to the minimal specificity in the cognitive disorder induced by bilateral lesions of the deep white matter anterior to the tip of the lateral ventricular horns. In addition to their adynamia, these patients seem to suffer from a very severe deficit of attentional self monitoring. Solms (1997a) did not provide adequate measures of vigilance for these patients. (i.e., level of arousal was only clinically defined). Dream recall in frontal patients with lack of interest in initiating and sustaining actions should be reinvestigated with more adequate and specific tests to determine whether they suffer a general and diffuse deficit of intensive and/or selective attentional processing or whether their oneiric impairment arises from emotional-motivational deficit. In the latter case it should be further investigated whether the motivational impairment is a low level one (general hypoactivation) or whether it affects higher level processing (motivational learning and discrimination; Gaffan & Murray 1990). From a neuroanatomical point of view, bilateral lesions of the deep white matter anterior and inferior to the lateral ventricles are unlikely to interfere selectively with dopaminergic transmission because, for example, both noradrenegic fibers (Morrison et al. 1981) and cholinergic ones (Selden et al. 1998) traverse the same area to innervate very large sections of the cortical mantle. Hence damage to the deep frontobasal white matter probably has complex clinical and neurochemical effects that cannot be reduced to hypodopaminergic adynamia.

The idea that dopamine agonists and antagonists have opposite effects in increasing and decreasing hallucinatory activity is not completely convincing for at least two reasons: (1) It arbitrarily equates different hallucinatory phenomena endowed with different physiological and phenomenological qualities. Dopaminergic activation certainly plays a role in all these phenomena but the same phenomena cannot be exclusively defined by the level of DA activity. (2) It de-emphasizes the fact that cholinergic agonists can also induce dream-like activity (Sitaram et al. 1978a).

***Brain imaging evidence.*** In all the published activation studies, subjects underwent prior total sleep deprivation (36–48 hours). In their thorough discussion Braun et al. (1997) acknowledged the potential confounding effects produced by sleep deprivation. Here, we also recall that (1) Sleep deprivation first affects vegetative activities and the emotional section of the anterior cingulate (area 24 in the inferior genual area) is implicated in the regulation of vegetative responses (Devinsky et al. 1995) (2) Sleep deprivation (in particular REM sleep deprivation) enhances DA activity (Brock et al. 1995). Without denying the contribution of motivational-emotional activation to the shaping of dreams (although not all dreams are necessarily endowed with relevant emotional content; see target article by NIELSEN), one might suggest caution about the role attributed to dopaminergic activation of mesolimbic structures in the generation of dreams.

***Notes and conclusions.*** The assertion that published cases of loss of dream recall following stroke can be grouped as frontal ones (deep white matter) and parietal ones is incorrect and incomplete. In our (balanced) review of the literature (Doricchi & Violani 1992) we documented a consistent and clinically homogenous body of cases in which total dream cessation accompanied infero-mesial lesions producing visual-verbal disconnection. In the same review, which included 104 cases published in the neurological literature starting from 1883, we reported the nosology of preserved dream recall after frontal lesions with relevant involvement of the underlying white matter, the predominance of dream cessation after posterior lesions, and the loss of the visual component of dreaming. Solms (1997a) confirmed all these findings and provided further informative data on the locus of the lesion that suppresses the visual component of dreaming.

In the same paper, we formulated many specific and testable hypotheses on the relationship between left-hemisphere linguistic-semantic processing impairments and lack of dream recall as well as on the role of posterior parietal – temporal areas in the spatial shaping of dreams and the modulation of oculomotor activity in REM sleep (see also Doricchi et al. 1993; 1996). Unfortunately Solms (1997a) did not specifically test these hypotheses and his present target article rejects the role of posterior dorsal areas as supporting various processes involved in visual imagery. Citing our 1993 review (since all other relevant literature is reported) would have preserved the originality of SOLMS's contribution and made his review more balanced.

Current neuropsychological evidence clearly indicates that dreaming can be disturbed or interrupted at different levels of cognitive processing. The important developments in neuroimaging techniques and future clinical and experimental research will certainly provide a deeper picture of the various cognitive components leading to the construction of complex oneiric experience and the flow of information in the dreaming brain. At present, no model gives a satisfactory account of the pattern of neural activation and deactivation during dreaming. Some authors (see **HOBSON ET AL.**) simply and cautiously summarize available evidence from many other authors in a list of different neural structures and functions contributing to specific features of the dreaming experience. **SOLMS** (1997a) views mentalistic-psychoanalytic concepts as "censorship" or "hallucinatory backward projection;" in our view, this has little heuristic value and is incompatible with modern neurocognitive and biological approaches to the study of mental processes. Considerably more evidence and reanalysis is needed before we can assign to dopaminergic mesolimbic structures a selective and exclusive role in the generation of dreams.

# REM sleep: Desperately seeking isomorphism

Irwin Feinberg

*University of California, Davis and VA Northern California Health Care System, Davis, CA 95616.* **ifeinberg@ucdavis.edu**

**Abstract:** If reports given on experimental awakenings validly represent mental activity that was underway before the awakening, REM sleep is neither necessary nor sufficient for dreaming. Another intuitively attractive hypothesis for its function – that REM consolidates or otherwise modifies memory traces acquired while awake – is not supported by the preponderant evidence. There is growing acceptance of the possibility that REM functions to support sleep rather than waking brain processes.
[**HOBSON ET AL.; NIELSEN; SOLMS; VERTES & EASTMAN**]

***REM sleep and dreaming: Rosetta stone or red herring.*** As good a case as possible for the REM-dream isomorphism is made by **HOBSON ET AL.**, but one that ultimately fails to convince. It is not possible to review all the contrary evidence but I will cite some significant examples. **HOBSON ET AL.** note that Hong et al. (1997) found an "impressive" correlation of .8 between visual imagery and REM density and consider this "evidence for a dependence of dream imagery on a qualitative feature of REM sleep" (p. 138). However, this correlation was found in a single subjects (S); further experiments failed to demonstrate this relation in two additional Ss (Antrobus et al. 1995). Moreover, the early hypothesis that the rapid eye movements of REM sleep (REMs) indicate scanning of dream images has not been supported by subsequent, more careful studies (Moskowitz & Berger 1969). REMs are so dramatic a feature of REM physiology that it seemed obvious they must be functionally important. However, my colleagues and I proposed they are adventitious phenomena with no special relation to dream imagery (Feinberg et al. 1987). We suggested that REMs are overt but incidental manifestations of the intense, disinhibited neuronal firing sleep in many motor (and sensory) systems throughout the brain that occurs during REM. Whereas neuronal firing in motor centers that control the limbs must be blocked to prevent movements that would awaken the sleeper, the eyes can move without causing awakening. Nature did not estab-

lish an inhibitory pathway from the atonia centers of the brainstem to the oculomotor nuclei simply because none was needed. The physiologically important question then becomes why some brain structures are intensely active in REM sleep. A potential clue is that the structures showing this response tend to be "hard-wired" (Feinberg & March 1995).

**HOBSON ET AL.** recognize that Antrobus and his colleagues interpreted the eye movement density-visual imagery correlation reported by Hong et al. (1997) not as evidence of brain-behavior isomorphism but as "another example of the simple dependence of dream content on levels of brain activation." I agree with Antrobus's view and think it important to emphasize further the strong albeit circumstantial evidence that REMs density is proportional to within-sleep arousal level (or "activation") (Feinberg et al. 1987). This evidence includes: the reduction of REMs by total sleep deprivation the progressive increase in REMs density across successive REMPs; the further spectacular increase in REMs density when sleep is abnormally extended, becoming extremely "light"; and the strong suppression of REMs density by GABAergic hypnotics, drugs that specifically depress brain arousal.

To further support an isomorphism between REM sleep and dreaming, **HOBSON ET AL.** point to a "positive relationship" between length of preceding REM sleep and word count, citing Stickgold et al. (1994a). A positive relationship is not fully supported by the cited paper because word counts after 45–60 min of REM were about half as long as those in reports after 30–45 min of REM. Moreover, the results of a simple experiment in our laboratory, better controlled for time of night than that of Stickgold et al., challenge their findings (Rosenlicht et al. 1994). We awoke subjects (Ss) after 5 and 10 min from the second and fourth REMP of the night. Mentation was elicited with a standard protocol. The reports were tape-recorded, transcribed, and scored "blind" for word count by two raters. Word counts did not differ significantly in reports elicited after 5 versus 10 min of REM sleep (325 vs. 413; p = 0.114) but there was a highly significant difference between reports from the second fourth REMPs (264 vs. 474; p. < 001). We cannot fully explain the discrepancy between the results of the two studies. However, our experiment can easily be repeated and one hopes that it soon will be because its implications are substantial. If our findings are independently confirmed, they would demonstrate that the effects of REM sleep duration on word count are trivial compared to those of time of night. This point gains importance for **HOBSON ET AL.** because they now accept word count as a measure of "dreaming." So far as the underlying biology is concerned, we and many others have interpreted longer dream narratives later in the night as caused by higher within-sleep arousal ("activation") level. As discussed below, it is still not clear whether higher arousal level produces longer dreams, a wider span of recall for ongoing sleep mentation, or both.

Whether NREM and REM mentation differ qualitatively is the essence of the isomorphism issue. All of the experts in this *BBS* Special Issue on REM sleep and dreaming agree that dreamlike reports, qualitatively indistinguishable from those elicited from REM, can be obtained by awakenings from any stage of NREM sleep. Since the brain physiology of REM is massively different from that of NREM, this rules out a REM-dream isomorphism. At several points in their target article **HOBSON ET AL.** imply that failure to accept the REM-dream isomorphism is tantamount to rejecting the dependence of mental phenomena on the brain. This is hardly the case. One is rather rejecting the claim that a relationship exists between a particular psychological state (dreaming) and a particular physiological state (REM sleep). This rejection is not based on "mentalism" but on the strong contrary evidence. It is past time to accept the failure of this particular isomorphism and look elsewhere for the brain states that underlie dreaming. **SOLMS** does this in his interesting article.

***Solms on the neural substrate of dreaming.*** The isomorphism issue is tackled head on by **SOLMS**. Noting that disagreement remains on the precise frequency NREM dreaming, he emphasizes the general acceptance of "the principle that *REM can occur in* *the absence of dreaming and dreaming in the absence of REM*" (sect. 4). While even a rare instance of NREM dreams indistinguishable from those of REM would severely challenge isomorphism, NREM dreams are far from rare. SOLMS cites Hobson's (1988b) comment that "5–10% of NREM dream reports are indistinguishable by any criterion from those obtained from post-REM awakenings" (p. 143) and he points out that, since NREM sleep makes up 75% of total sleep time, "this implies that roughly one-quarter of all REM-like dreams occur outside of REM sleep" (SOLMS sect. 5, his emphasis).

SOLMS goes on to use clinicopathologic correlations to seek the neural substrate required for dreaming. His review finds that patients who have lost the ability to dream have suffered lesions in two forebrain areas: the "parieto-tempero-occipital junction" and the "ventro-mesial quadrant" of the frontal lobe. These observations are especially intriguing because several of these cases showed REM sleep when tested in the laboratory. However, without specific control for the memory impairment likely to accompany such brain lesions, one cannot know whether a patient has lost the ability to dream or the ability (and motivation) to recall and report dreams.

I strongly endorse SOLMS's conclusion that the REM sleep-dreaming relation is in need of so fundamental a revision as to constitute a paradigm shift. This shift is at least 10 years overdue. The REM-dream relation has not been a Rosetta stone but rather a red herring that has led us seriously astray. The failure of REM-dream isomorphism contains an ironic element. The irony lies in the fact that this failure, taken with our knowledge of the brain physiology of REM sleep, tells us a great deal about what the neuronal substrate for dream consciousness is *not*. SOLMS explicitly recognizes this. J. D. March and I arrived at a logically similar conclusion (Feinberg & March 1995). We reasoned that, since brain physiology is qualitatively different in NREM and REM, but the conscious experience of dreaming in the two states is not qualitatively different, "the striking NREM/REM differences in neuronal firing must not involve the neural systems that can affect the quality of conscious experience" (p. 106). Because it is almost certain that marked alterations of the firing patterns in these structures would affect waking consciousness, this conclusion implies that sleep involves disconnections within the brain, as well as a relative disconnection from the environment.

***Sleep and memory.*** Before commenting on VERTES & EASTMAN, whose paper deals mainly with this issue, I think it important to emphasize a fact that is too often overlooked: Virtually all modern sleep-dream research is based on the unproved assumption that narratives given by Ss when awakened from sleep represent mental activity that was going on prior to the awakening. Certainly this assumption is consistent with our subjective experiences of dreaming. Nevertheless, there are no data that rigorously exclude the possibility that dream reports are entirely constructed during the process of waking-up. A century ago Goblot (cited by Hall 1981) pointed to this possibility. Some of us who have been present (more recently) as Ss struggled to report their sleep mentation had the impression that a considerable process of reconstruction (construction?) was underway. On some of the infrequent occasions when Ss produced complex and elaborate dream narratives, I thought that the stories were being created *de novo*, while the S was in a fugue state intermediate between sleep and waking. A related point is that any quantitative or qualitative differences in the mentation elicited from the different stages of sleep might be caused by differences in the functional state of memory systems rather than in the mental activity produced during these stages (Feinberg & Evarts 1969).

VERTES & EASTMAN review the experimental literature on whether REM sleep promotes memory consolidation and show it to be unpersuasive. Many of these studies performed REM deprivation with the "flower pot" method that VERTES & EASTMAN rightly emphasize is contaminated by stress. Even in the presence of this stress, there are as many failures to show an impairment of learning and memory by REM deprivation as there are positive

studies. Carlyle Smith's evidence that memory "windows" exist during which REM sleep acts to consolidate memories has not been independently confirmed; moreover, the variability in the timing of these windows that Vertes and Eastman extensively document is disconcerting. Using another line of evidence, VERTES & EASTMAN cite data showing that monoamine oxidase inhibitors can virtually eliminate REM sleep without detriment to waking behavior. I agree with this point and, in fact, used it to support my arguments that REM serves a brain function intrinsic to sleep rather than (as does NREM) to waking (Feinberg 1974). Benington and Heller (1994) now also endorse a similar view.

In discussing the stress induced in rats by REM deprivation with the flower pot method, VERTES & EASTMAN note in passing that this criticism may not apply to deprivation with Rechtschaffen's yoked control-platform paradigm. This point is not essential to any of their main arguments. Nevertheless, because of the great theoretical importance currently placed on the physiological changes provided by the Rechtschaffen deprivation paradigm, it may be useful to draw the reader's attention to the fact that some investigators believe that these changes are due to stress rather than sleep loss. Thus, I noted (Feinberg 1999) that the pathophysiological changes produced in rats by prolonged total or selective sleep deprivation with the Rechtschaffen technique resemble the non-specific stress responses in Selye's General Adaptation Syndrome (Selye 1937). Rechtschaffen has strongly contested these arguments (Rechtschaffen & Bergmann 1999) and this issue remains controversial.

I would have emphasized more strongly than VERTES & EASTMAN the functional implications of the shut-down of memory consolidation systems during sleep. The degree of shut-down is roughly proportional to the level of high amplitude delta EEG, that is, it is maximal in stage 4 and least in stage REM. I have already emphasized that variations in arousal level might explain much of the variance in sleep mentation through its effects on memory function, a view previously proposed by several investigators (Antrobus 1991; Koulack & Goodenough 1976; Zimmerman 1970). Variations in memory function could also explain why Ss can produce non-random estimates of REM but not stage 4 durations (Carlson et al. 1978). If, as many of us assume, one function of sleep is to reverse certain effects of plastic neuronal activity during waking, it seems likely that memory systems would be involved. It makes intuitive sense that the systems being restored would be taken "off-line." The fact that memory consolidation systems are substantially disabled during sleep is therefore consistent with the possibility that one function of sleep is to permit recovery of these systems.

It is in their proposals for the function of REM sleep that I found VERTES & EASTMAN disappointing. Hypotheses similar to theirs have been advanced in the past and VERTES & EASTMAN offer no new evidence. As already mentioned, the hypothesis that REM serves a function intrinsic to sleep rather than to waking was advanced a quarter century ago (Feinberg 1974). A proposal similar to VERTES & EASTMAN suggestion that "the primary function of REM sleep is to provide periodic endogenous stimulation to maintain minimum requisite levels of CNS activity throughout sleep" was put forward by Ephron and Carrington (1966). Although VERTES & EASTMAN decry theories that propose "magical" processes for REM sleep, their own proposals seem vulnerable to similar criticism. Notions like "minimum requisite levels of CNS activity" or a brain "incapable of tolerating long continuous periods of relative suppression" could be viewed as vague and metaphorical.

***High levels of REM sleep in the neonate.*** Both HOBSON ET AL. and VERTES & EASTMAN cite the high neonatal levels of REM sleep to support their differing interpretations of REM's functional significance. But there are reasons to believe that the physiology of neonatal sleep differs fundamentally from that a few months later (Feinberg 1969). Brain wave patterns in the neonate are so rudimentary that one cannot distinguish the NREM from the REM EEG, making it necessary to distinguish sleep states behaviorally as "quiet" or "active" sleep (cf. Kahn et al. 1996)). Anatomical data can added to earlier arguments against accepting REM in the neonate as physiologically equivalent to that occurring later in infancy. Conel's atlases (Conel 1939) show cortical connectivity in the newborn human brain to be vastly limited compared to that present just a few months later, at which time REM levels are not greatly different from those in the adult. Frank et al. recently reported that "REM" sleep in the newborn rat differs pharmacologically from that in the more mature animal (Frank et al. 1997) and argued that this indicates a different physiological state. For these reasons, it seems hazardous to accept active sleep in the newborn as homologous with the REM of later infancy and adult life and to infer either functional or psychological significance from its high levels.

*Nielsen and the hypothesis of covert REM.* A masterful and objective review of the experimental literature on dreaming is provided by NIELSEN that should be useful for years to come. He accepts the "strong proof that cognitive activity – some of it dreaming – can occur in all sleep stages," and that the physiology of NREM and REM are qualitatively different. Nevertheless, he attempts to preserve an isomorphic REM-dream relation. To do so, he hypothesizes that "sleep mentation is tightly linked to REM sleep processes" and that these processes may dissociate from the REM state and "stimulate mentation in NREM sleep in a covert fashion."

Although I interpret differently much of the evidence NIELSEN cites in support of his hypothesis, I strongly agree that one encounters intermediate sleep states that have both NREM and REM features. Such states provoke considerable gnashing of teeth among the unfortunates whose task is to score sleep stages. However, it is a considerable leap from the sporadic occurrence of intermediate states to the notion that these states are invariably but covertly present whenever dreamlike narratives are elicited from NREM awakenings. This hypothesis seems particularly implausible in the case of stage 4 awakenings that give rise to vivid dreams since stage 4 physiology is the polar opposite of that in stage REM. But unless intermediate states with REM characteristics are *always* present when dreamlike narratives are elicited from sleep, covert REM cannot rescue the REM-dream isomorphism.

NIELSEN suggests that his hypothesis can be tested by simple and straightforward experiments. I do not agree that the tests he proposes would give unambiguous answers. NIELSEN proposes that dreamlike mentation reports will occur more frequently when elicited from NREM episodes in close proximity to REMPs, especially those that are lengthy. However, this result need not indicate the presence of covert REM. The one-stimulus model (see below) could parsimoniously interpret such findings as indicating that within-sleep arousal levels are higher at these points. Twenty years of research have shown that NREM sleep is not constant across a sleep cycle but shows the waxing and waning of delta intensity (cf. Fig. 1). Differences in the mentation elicited at the beginning and end of the cycle could reflect differences in the physiology of NREM at these points, independent of proximity to REM. A similar interpretation applies to NIELSEN's prediction that NREM mentation will be increased by sensory stimulation during sleep; such stimulation, already known to increase REM sleep (Drucker-Colin et al. 1983), could alter sleep mentation by raising arousal level. Experiments of the sort NIELSEN proposes would nevertheless be interesting. They could be strengthened if awakenings were performed in relation to points in the computer-quantified delta cycles rather than visually scored sleep stages. For example, it would be interesting if sleep mentation on the ascending limb of these curves differed from that elicited on the corresponding point of the descending limb (which would receive the same sleep stage scores and have similar proximity to REM).

In his discussion of "missing" REM episodes, NIELSEN seems unaware of some relevant literature. As NIELSEN notes, the "skipped" first NREMP leads to exceptionally long first NREMPs. This phenomenon is best understood on the level of basic sleep physiology. It has long been known that if one plots total EEG amplitude or spectral power or delta integrated amplitude across

sleep, one observes an irregular series of peaks and troughs (Church et al. 1975; Koga 1965; Lubin et al. 1973). The peaks correspond to visually scored stage 3–4, and the troughs are usually scored as stage REM, with stage 2 occupying the intermediate parts of the curve. However, in extremely deep sleep (e.g., in young normal children or young adults after total sleep deprivation – TSD), REM is frequently not scored in the first trough (Fig. 1). In these cases, application of curve smoothing methods

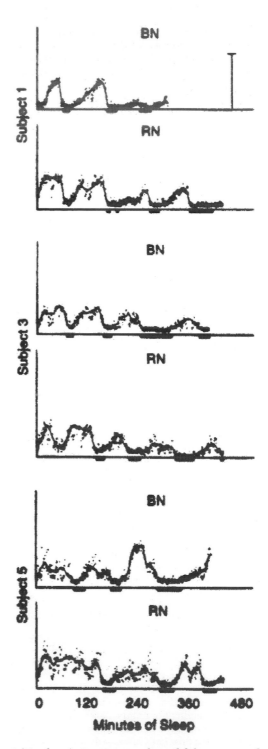

Figure 1 (Feinberg). Continuous plots of delta integrated amplitude on baseline (BN) and recovery (RN) nights following 24 h sleep deprivation. Because sleep is deeper, REM (black abscissa bars) is "skipped" in the first trough on the RNs, producing misleadingly long first NREMPs.

that objectively define successive peaks and troughs reveals that the duration of the first peak (NREMP 1) is not abnormally protracted in either the deep sleep of children (Feinberg et al. 1990) or in young adults after TSD (Feinberg & March 1988).

It will simplify the discussion if we now introduce the one-stimulus model of NREM and REM sleep that March and I proposed in 1988 and expanded in 1995. Briefly, this model holds that NREM and REM sleep occur at different points in the brain's response to a single inhibitory neuroendocrine pulse that occurs recurrently during sleep. This pulse is presumed to originate in the hypothalamus. This stimulus inhibits neuronal activity, reduces cerebral metabolic rate, induces EEG synchrony and depresses arousal level, that is, induces NREM sleep. Functional changes that occur in the inhibitory (NREM) state include relative sensory deafferentation and a shutting down of memory consolidation systems (see above). The intensity of the brain response to the stimulus parallels the waxing and waning of the EEG amplitude curves described above. When the strength of the inhibitory pulse falls below a critical threshold, escape from inhibition occurs. This neuronal escape *is* REM sleep which, as noted above, is characterized by intense, disinhibited firing in many neuronal systems. After a variable duration of REM, another pulse is released and the process repeats. The failures of REM to appear in the first trough of deeply sleeping Ss indicate that the critical arousal threshold for inhibitory escape has not quite been reached.

March and I have shown how the one-stimulus model, along with the homeostatic model of delta (Feinberg 1974) parsimoniously accounts for much of the known phenomenology of human sleep, including sleep architecture patterns and the effects of daytime naps on post-nap sleep (Feinberg et al. 1985; 1992). The model also explains the increased REM produced by (partially arousing) sensory stimulation during sleep, and Datta and Siwek's (1997) findings that low intensity stimulation of brainstem arousal centers converts NREM to REM and more intense stimulation converts REM to waking. Our model would also interpret the increased neuronal firing in cholinergic brainstem centers during REM as components of widespread disinhibition-release phenomena, rather than as specific stimuli for either the REM state or the cognitive events of dreaming.

## The case against memory consolidation in REM sleep: Balderdash!

William Fishbein

*Department of Psychology, The City College and Graduate School of The City University of New York, New York, NY 10031.* **wfatoffice@aol.com**

**Abstract:** Unfortunately, some researchers think a good scientific theory is one that has been repeatedly confirmed, and a bad theory is one that has not received consistent confirmation. However, confirmation of a theory depends on the extent to which a hypothesis exposes itself to disconfirmation. One confirmation of a highly specific, falsifiable experiment can have a far greater impact than the disconfirmation of twenty experiments that are virtually unfalsifiable. This commentary (1) counteracts misleading biases regarding the REM sleep/memory consolidation theory, and (2) demonstrates how chaotic cerebral activation during sleep is an essential component of long-term memory storage processes.
[**VERTES & EASTMAN**]

Most theories of the function of sleep – and REM sleep in particular – revolve around the idea that sleep serves an adaptational function for wakefulness. Four of the five authors of the target articles presented in this special journal issue – all except **VERTES & EASTMAN** – are of this view (as it is self-evident to most others). **VERTES & EASTMAN**'s idea is that the functional importance of REM sleep is solely neurobiological. They believe REM sleep is bound to the biological state of sleep itself, to the homeostatically upregulated depressed quietude of SWS (slow-wave sleep), in order to maintain obligatory levels of CNS activity throughout sleep.

The brain, they argue, is incapable of sustaining long periods of suppression produced by the delta activity of SWS and as such, requires endogenous stimulation to maintain neuronal homeostasis. However, as it has been for so many other theories ranging from that of Freud in 1900 to the present ones of the post-REM era, **VERTES & EASTMAN**'s theory is unfalsifiable. This is largely the reason the function of sleep (and dreaming) remains unknown. In any case, **VERTES & EASTMAN**'s theory is not under examination in this review; it is the REM-sleep/memory consolidation theory that is. This theory has received considerable attention because there are potential methods for evaluating the evidence relevant to the theory. That is, the theory includes the possibility data can be generated that will falsify it. Therefore some researchers have challenged its basic tenets.

***1. Two important caveats.*** (1) The principle of falsifiability (Popper 1959) has important implications for the way the theory of REM sleep and memory consolidation is evaluated. Many researchers think that a good scientific theory is one that has been repeatedly confirmed, and a bad theory is one that has not received consistent confirmation. They assume that the amount of confirming evidence is the critical factor. However, falsifiability implies that the number of times a theory has been confirmed (or not confirmed) is not the critical element; not all confirmations are equal. Confirmations are more or less important depending on the extent to which a hypothesis exposes itself to potential disconfirmation. One confirmation of a highly specific, potentially falsifiable experiment can have a far greater impact than the disconfirmation of twenty different experiments that are all virtually unfalsifiable. Therefore, it is necessary to look not only at the quantity of the confirming evidence, but also its quality. (2) Even with an earnest attempt at evaluation of the literature, the writer of a review organizes the material and emphasizes certain parts of it so as to persuade the reader to his view. For researchers who do not have a primary interest in the area of the review, and who do not peruse individual experiments reported, the reputation of the reviewer and his steadfast argument might convince the "outsider" the review represents the final word.

The purpose of the present commentary is to counteract the persuasive effects of honest, but nevertheless – what this writer believes – are misleading biases. Unfortunately, space limitations preclude a full critique, and therefore this commentary is limited to a few selected items of **VERTES & EASTMAN**'s target article.

***2. REM augmentation studies.*** One of the most consistent findings in the animal sleep literature (as overwhelmingly consistent as the evidence for cholinergic REM sleep generation) is the augmentation of REM sleep following heightened experiences. Yet, while human studies have not shown the same degree of consistency, the reader is led to think REM sleep augmentation is a bogus finding. Unfortunately, **VERTES & EASTMAN** provide no discussion of the differences between human and animal sleep cycle rhythmicity, nor is there any attempt to reconcile the differences in an effort to point out why REM augmentation may not be easily seen in human sleep. Given the imperfections of the disconfirming experiments in contrast to the consistency of the augmentation phenomenon, it is necessary to evaluate the extent and nature of the flaws in the disconfirming experiments. More emphatically, there is no evidence presented falsifying the confirming evidence that might lead to the theory being modified, or perhaps abandoned for an entirely new theory.

More dismaying is misinterpretation of research report findings. In the effort to build a case against the REM-sleep memory consolidation theory, **VERTES & EASTMAN** make reference to review articles and conclusions drawn by others, in particular the review paper of Horne & McGrath (1984). That paper refers to two publications from my laboratory (Gutwein & Fishbein 1980a; 1980b) in which we examined the effects of enriched and impoverished rearing on REM sleep. The original reports describe findings that are the opposite of what **VERTES & EASTMAN** state. **VERTES & EASTMAN** conclude that (1) REM sleep augmentation is an "artifact" of an overall increase in total sleep time (TST), and

(2) the differences between the experimental and control animals comes from a decrease in REM as a result of the controls being reared in impoverished conditions. A reader of the review is led to conclude that the research results were owing to a confounding variable, and therefore should be discarded. In fact, the reported findings in both research reports unambiguously show that REM sleep is significantly augmented as a result of enriched rearing compared to social control (not impoverished) animals and furthermore, the selective REM sleep augmentation in the enriched animals compared to the social control animals is not an artifact of SWS augmentation. Both papers, in considerable detail, report the statistical analyses with significance probability levels of $p < .001$ or greater.

**3. REM deprivation studies.** VERTES & EASTMAN then turn to the REM deprivation research that was so ubiquitously reported during the 1970s and 1980s and again the target article presents misrepresentations and omissions of significant experimental findings. This section begins with a quotation from my 1995 review paper (Fishbein 1995). I abandoned the REM deprivation work because I could not, it says: "adequately respond to criticisms leveled at the (REM deprivation) technique." This is gross distortion. To set the record straight, at the time the REM deprivation experiments were performed, I had reached the point whereby I had come to believe all the crucial experiments to demonstrate the relation between REM sleep and memory consolidation processes using the pedestal technique had been exhausted. To further pursue that line of research would produce experimental results that could only be seen as variations on a theme that, for the most part, had already been established. I believed it was time to move on and employ a new strategy.

**3.1. The pedestal technique.** The REM deprivation experiments from my laboratory were all designed with extraordinary care, primarily to handle the possible interpretation that the results would be seen as confounded by the "stress" of the pedestal technique. The view had emerged, at the time, that the pedestal technique produces results indistinguishable from experiments employing stress to produce impairments of learning or memory; interestingly, the REM deprivation/memory experiments reporting the effects of stress had all used rats (for a review, see Fishbein & Gutwein 1977). Our experiments, on the other hand, employed mice.

Certainly we knew that confirming evidence, based on reports from many laboratories examining the physiological effects of stress, and particularly those laboratories showing the pedestal technique not to be stressful, would not pacify skeptics wedded to the belief the pedestal technique is stressful. We also did not believe our descriptions of mice freely climbing about the underside of water filled cages, would pacify the skeptics either. Neither did we believe our open field activity experiments, showing no differences between REM deprived and normal animals, would placate the skeptics. However, we did think that if we could perform an experiment that would refute (falsify) the hypothesis that the pedestal technique is stressful by replicating the rat studies (i.e., restricting the activity of the mice while living on a pedestal in the midst of a pool of water), showing that restriction impairs (stresses) the mice, while nonrestriction (non stressed) does not, we would have direct confirmation of our hypothesis that the pedestal technique was not stressful (so long as the mice could climb about – and exercise – in the pedestal cages).

We performed that experiment; it is described in our 1977 paper (Fishbein & Gutwein 1977). The results indicated that the unrestricted mice (living on the pedestals for either 2 or 4 days) had identical activity scores (open field and passive avoidance step-through latencies) compared to controls (animals maintained in standard shoebox cages with wood shavings); whereas restricted animals (also on the pedestals for either 2 or 4 days) were considerably more active than controls and the unrestricted mice. With this experiment we clearly demonstrated why the rat studies had to be discounted as a result of the deprivation procedure, whereas the robustness of the mice demonstrated just the opposite. In

short, stress was not a factor we needed to concern ourselves with in our REM deprivation studies. The journal referees of our many manuscripts were convinced we had satisfactorily handled the stress criticism; they would not have allowed our reports to go to press if there was such a gross confound in our research. Yet despite all our efforts, there was always a small number of researchers (perhaps biased by their own theories of the function of REM sleep) who refused to be swayed by the data.

**3.2. Prior REMD studies.** In the fourth line of their brief discussion of prior REM deprivation studies, VERTES & EASTMAN dismiss this research out of hand. "These studies," they say, "do not seem to test the REM consolidation hypothesis since the deprivation period precedes training/acquisition and there is no potential carry over of information pre to post REMD." Period. That's it; and with that said, they select one study (van Hulzen & Coenen 1982) to drive home their point; prior REM deprivation produces only acquisition impairments. Of course, the stress factor is then resurrected and the whole matter is discharged without further ado.

However, this area of research is important and several experiments designed to examine the effects of prior REM deprivation on the conversion phase of memory consolidation have demonstrated that acquisition is unaffected by the prior deprivation. In fact, it is imperative to demonstrate in such experiments that subjects learn and remember normally for at least a brief period before amnesia sets in, otherwise – VERTES & EASTMAN would be correct – the experiments would be difficult to interpret because there would be no way to distinguish acquisition from retention impairments. The prior REM deprivation studies are centrally concerned with the role of REM sleep in the formation of newly acquired information.

In two experiments from my laboratory (cited in VERTES & EASTMAN's review, Fishbein 1970; Linden et al. 1975), we unequivocally demonstrate that mice deprived of REM sleep for 3 days prior to learning show perfectly normal retention up to 1 hour after learning compared to non-deprived controls. However, 3, 5, and 7 days after the animals had demonstrated normal retention, they are amnesiac. That is, a memory that had been established was now gone.

In the follow-up to this experiment we reasoned and predicted that if the prior REM deprivation impairs the permanent fixation of a long-term memory trace, the treatment induced its effect by altering the consolidation gradient of the memory fixation process. In this experiment the same experimental design was employed as in the previous one. Animals were deprived of REM sleep for three days and then trained. Electroconvulsive shock (ECS) was then administered at intervals varying from immediately after learning up to 6 hours afterwards. The animals were then tested for retention three days later, long after recovery from the deprivation and ECS. In this experiment we showed that 3 days after training the memory trace remains susceptible to disruption and furthermore a gradient of susceptibility was apparent as the animals recovered from the REM deprivation. The experimental results confirmed our hypothesis. Moreover, the experiment is without any stress confound. All animals were deprived for the same length of time; all animals received training at the same time after the REM deprivation. The only variable manipulated was the time between training and the administration of ECS. The experiment provided indisputable evidence that REM deprivation had sustained the memory trace in a labile form, thereby prolonging permanent consolidation of the memory trace. In short, REM sleep has an important role to play in the fixation phase of the memory consolidation process.

**3.3. Post learning REMD studies.** VERTES & EASTMAN similarly dispose of the REM deprivation studies in which the deprivation is inserted between learning and retention testing; in their view, these are all "performance" deficits. Considering the enormous number of publications that have examined the REM sleep-memory consolidation hypothesis from this perspective, it is hard to believe that so many experiments can be so easily dismissed. There are many confound-free experiments to choose from that underscore the important role of REM sleep in the mechanisms underlying the storage of long-term memory. One experiment, in particular, from my laboratory can serve as a model example. The experiment was one of our earliest (Fishbein et al. 1971) and is totally free of the "stress" factor. Mice were trained and immediately deprived of REM sleep by the pedestal method for 2 days and then separate groups were administered ECS varying from 5 minutes to 12 hours afterwards. Two days later, after recovery from deprivation and ECS the animals were tested for retention. As in the prior REM deprivation studies, ECS produced retrograde amnesia, with the amnesia gradient occurring two days after training. Important to note, the animals were not under any stress at the time of learning or retention testing, yet REM deprivation induced a brain change that led to a profound amnesia.

Moreover, contrary to **VERTES & EASTMAN**'s suggestion in all the experiments performed in my laboratory, the treatment effects were extremely large. Therefore, these experiments and many more that have followed, have provided exceptionally strong evidence that REM sleep has a major role to play in the processing and consolidation of long-term memories. The task of researchers is not to dismiss these powerful findings because other experiments have not systematically supported them, but to search for a pattern of flaws running through the research literature because the nature of the pattern may then provide reason to modify or possibly abandon the theory. However, that a confounding "stress" factor obscures the interpretation of an experimental result does not mean the theory need be abandoned.

Space constraints preclude further discussion despite other objections to be raised, yet it bears repeating that failure to reject Hebb (the null hypothesis) is not in itself evidence that the theory of REM sleep-memory consolidation should be abandoned. Nevertheless there is one abiding issue central to the REM sleep-memory consolidation hypothesis raised by **VERTES & EASTMAN** that must be addressed. The topic revolves around the role of theta activity involvement in memory processing.

### 4. Making dreams out of chaos.
Central to **VERTES & EASTMAN**'s objections to the REM-sleep-memory consolidation hypothesis is their view that theta waves – which, indisputably, involve memory processing during waking – merely reflect a highly activated brainstem during REM sleep, producing random (chaotic) activation of the cerebral cortex and limbic system (rather than reflecting memory processing activities).

Vertes and I have previously exchanged commentaries about these very points (Fishbein 1996; Vertes 1996). I suggested that the chaos may be the underlying basis of the dream itself. Information comes into the hippocampal system from the cerebral cortex, including the visual, auditory, sensory cortices and the motor cortex. It also receives information from the amygdala concerning odors, unsafe stimuli, and information about the person's emotional state: whether sexually excited, hungry, frightened, and so forth. Recent accumulated evidence suggests that the function of the hippocampus may be to tie together or relate all the things happening at the time the memory is stored. Many experiments suggest that the role of the hippocampus is to construct representational relationships between these various forms of experience, including the order in which events take place. The representations can be likened to a library card-catalogue filing system with essential information stored in neural networks distant from the hippocampus. The filing system performs the necessary work of filing and/or retrieving information. Disruption of the system will of course impair storage (as in Alzheimer's disease) or retrieval processes (as in schizophrenic hallucinations).

### 4.1. The relation of random excitability to memory.
**VERTES & EASTMAN** question whether random, unrelated events can have any functional value in the long-term memory encoding process occurring during REM sleep. They believe that since there is no mechanism in REM to select and orderly transfer the pontogeniculo-occipital (PGO) spike information to the hippocampus from the brainstem, or for that matter from various cortical regions, the information that does flow will be inherently random

and therefore "there would be no functional value in consolidating or 'remembering' this information."

However, chaos may be just the elixir needed to facilitate information storage. Information that is systematic, orderly, and time-locked to the behavioral experience, as **VERTES & EASTMAN** would have it, may not lead to strengthening of memory traces, but to its decline. D.O. Hebb (1949) was the first to point this out in his famous text, *Organization of behavior.* For example, there is no surprise about the fact that hearing a joke for the second time makes for poor entertainment – the better the details are remembered of the first telling, the less interest there is in the second. Similarly, in a prolonged training schedule, there are often periods when practice seems to have a negative effect. The more you try, the worse things get. Such mundane events – yet truisms – might suggest that connections involved in learning might actually be weakened by orderly repetition of the same sequence of events. This behavior is referred to as "habituation." And sometimes it is necessary to have a period of rest before functioning well can occur again ("spontaneous recovery"). In a similar vein, we know that distributed practice is superior to massed practice in learning a skill. Thus, the deteriorative effect of repetition (say, losing interest in repeating the same solution to a problem) leads to habituation.

Hebb's illustrations infer that a memory continually needs to be updated to maintain its organization and persistence. In the same way, perhaps a neuronal memory trace needs to be rearoused to be sustained, but the maintenance of the trace requires new combinations of excitation (Nick & Ribera 2000), which in turn might mean new cognitions or new ideas.

To quote Hebb, the "mere occurrence of a particular 'phase sequence' once, induces changes at the synapse (memory) that make it impossible for exactly the same sequence to occur again, unless the synaptic changes have disappeared with time" (1949, p. 228), or possibly the information needs to be introduced in a different form. In short, the chaos occurring during the REM sleep period that **VERTES & EASTMAN** believes serves no functional value in consolidating information may be the kind of excitation that neural networks require to consolidate and sustain information for prolonged periods.

### 5. The new findings of Braun and Solms.
Finally, **VERTES & EASTMAN**'s objections to the REM sleep-memory consolidation hypothesis must be abandoned in light of the recent brain imaging studies of Braun and colleagues (1997; 1998) and the new lesion-dreaming research of **SOLMS,** that presages renunciation of the brainstem activation-synthesis model of dreaming (and **HOBSON ET AL.**'s AIM update of the model), replacing it with a shared model in which forebrain cerebral activation during sleep and dreaming is either self-activated within the forebrain itself, or activated by either the random orchestration of the ascending brainstem cholinergic system (originating in the pons) primarily during REM sleep, or a non-REM activating system ascending through dopaminergic circuits originating in the midbrain ventral segmental area of Tsai, the origin of the mesolimbic and mesocortical dopamine systems. Therefore it is no longer necessary to look to the pontine brainstem as the sole source of endogenous cerebral activation. In short, (chaotic) activation of forebrain structures throughout the sleep cycle has the potential to reactivate neuronal circuits facilitating the consolidation of memory.

## Dreaming is not an adaptation

Owen Flanagan

*Department of Philosophy, Duke University, Durham, NC 27708.*
**ojf@duke.edu**

**Abstract:** The five papers in this issue all deal with the proper evolutionary function of sleep and dreams, these being different. To establish that some trait of character is an adaptation in the strict biological sense requires a story about the fitness enhancing function it served when it evolved and possibly a story of how the maintenance of this function is fitness enhancing now. My aim is to evaluate the proposals put forward in these papers. My conclusion is that although sleep is almost certainly an adaptation, dreaming is not.

[HOBSON ET AL.; NIELSEN; REVONSUO; SOLMS; VERTES & EASTMAN]

**Evolution and the dissolution of the hard problem.** Despite being orthodox naturalists and recommending whole-heartedly the view that mental processes are brain processes, the HOBSON ET AL. "dream team" (Flanagan 2000) buys into a bad philosophical idea. This is the idea that there is a "hard problem" of consciousness (Chalmers 1995a), one that they claim they do not treat or resolve with the AIM model, and one that they, like many others, seem to think is beyond current cognitive neuroscience, possibly beyond science, period. I want to convince the dream team that they do treat the hard problem.

What Chalmers (1995a) calls the "hard problem" of consciousness is the problem of explaining how subjectivity can arise from complexly organized material stuff. The hard problem truly exists in one sense, for we have at present nothing remotely approaching a complete theory of how the brain does everything it does, including how it produces consciousness. But Chalmers, as well other "new mysterians" (Flanagan 1991; 1992) like Colin McGinn (1989), have pressed the line that even if we are provided with a complete neurobiological theory of how the brain works, nothing will have been done to erase the intuition that there is an unbridgeable gap between the way the organized objective brain works and the first-person grasp I have of myself as a thinking-feeling creature. Knowing all the facts about how the brain works will fail to explain how the brain gives rise to subjective mental life.

My recommendation for HOBSON ET AL. is not to fall into the trap fostered by such intuitions as "It is amazing that my consciousness could emerge from brain processes" and "Thoughts don't feel as if they have neural texture." One falls for the trap when one allows the "gee-whiz" bug to get a grip and thinks that there is any harder, deeper, or further problem than explaining how the mind-brain works for each of the heterogeneous kinds of conscious mental state types. Explaining the mechanisms that give rise to the different types of waking consciousness, NREM, and REM-mentation, is all there is to solving the hard problem. There is no further hard problem that will remain once this labor is completed. It may still seem amazing that by explaining how the brain works we will have explained how the mind works. But so what? It may seem amazing, even incredible, that my solid maple dining table is actually a field in Hilbert space comprised mostly of empty space. But that is what it really is. How things seem, including the powerful intuition that there seems to be an unbridgeable gap between conscious experience and brain processes, has no evidentiary status whatsoever when it comes to how things are.

In fact, we have an evolutionary explanation for why the mind-brain relation seems so mysterious. Philosophers call such explanations "error theories," because they are designed to explain why otherwise intelligent people make grand reasoning errors, such as believing that there is evidence for God or that there are objective moral facts. In the case under consideration, Mother Nature designed us to be in touch first-personally with our brain states at a level of granularity that reveals nothing about their neural texture. But it is an inference to the best explanation that mental events just are brain events. The fact that they don't seem that way is irrelevant. Indeed, it was a wise evolutionary strategy to design us not to have first-personal touch with the deep structure of our mental states and processes, a case where more information would have been too much information. Awake consciousness in the five sensory modalities is an adaptation precisely because it allows us to detect reliably what is going on outside us and to use this information in fitness enhancing ways. There was nothing to gain and everything to lose had Mother Nature designed us to be in touch with our mental states at the level of granularity that neuroscience treats. So the alleged hard problem dissolves when we understand it in evolutionary terms.

What happens to dream consciousness when we think about it in evolutionary terms? HOBSON ET AL. once thought that dream consciousness functioned for the sake of memory fixation and consolidation. This hypothesis was motivated by the discovery that acetylcholine is implicated in fixing memories and in the discovery that acetylcholine levels are high during REM sleep. The more recent view that dreaming is probably an evolutionary epiphenomena is motivated by a clearer appreciation of a point that SOLMS (1997a) and I (Flanagan 2000) have pressed independently, that sleeping and dreaming are different phenomena, a claim also argued for (or implicit in) NIELSEN's, VERTES & EASTMAN's papers. High levels of acetylcholine may well support the idea that one function of REM-sleep is memory fixation and consolidation, but since we rarely dream about what we need to remember, the hypothesis that dreams themselves serve any memory enhancing function appears unwarranted. Furthermore, VERTES & EASTMAN present reason to worry about the memory consolidation hypothesis, favoring instead the view that REM sleep serves to maintain a level of CNS activity during sleep that assists the brain in recovering from sleep. I quite agree that the data on people with fine memories who do not REM is a problem. But one doesn't need to abandon the view (yet) that one normal function accomplished by REM is memory consolidation – perhaps the brain being plastic has other ways to accomplish the task if REM is interfered with, and it is entirely possible, indeed likely that REM sleep, like NREM sleep, serves multiple functions, perhaps including keeping CNS activity at a certain level.

In any case, it seems best, pending resolution of this debate, to think of awake consciousness as an adaptation and sleep as an adaptation (for a reason or reasons in need of further exploration), and to think of dreaming as a nonadaptive side-effect of what sleep produces in a brain designed to be conscious by the light of day. In saying this, however, the proposals of REVONSUO, SOLMS, and to some degree, NIELSEN, need to be addressed, because they explicitly or implicitly allow a proper evolutionary function for dreaming itself.

**A parsimonious account of threat simulation dreams.** It is true that in my argument for the thesis that "dreams are the spandrels of sleep" (Flanagan 1995; 1997; 2000), I did not consider any thing like REVONSUO's idea that dreams were selected for the purpose of simulating threatening situations and rehearsing appropriate responses. Three candidates for assigning an adaptive function to dreams that I did consider were Freud's (1900), HOBSON ET AL.'s (1988) and Crick and Michison's (1983; 1995). The ideas are, respectively, that dreaming was selected to express socially unacceptable wishes and thereby to preserve sleep, that dreaming functions to consolidate memories, and that dreaming functions to dispose of things not worth remembering.

In each case, my argument against the adaptationist proposal turned on the lack of support for the relevant hypothesis, given the actual content of what we dream about. Most dreams don't express wishes, most dreams don't involve entertaining things worth remembering, nor do they involve entertaining things worth forgetting. Freud aside, I do think that Hobson (1988a) and Crick and Michison (1983; 1995) are probably right that memory consolidation and brain-washing, SAVING and TRASHING, as it were, are one of the things that sleeping, especially REM-sleep, was selected to do (pace VERTES & EASTMAN). It is just that the phenomenology of dreams gives no support to the idea that dreaming contributes to this process. Despite their not having an evolutionary proper function, I claim that some dreams express things that our minds activate: emotions, worries, concerns, and memories that we have, and are in this way self-expressive, possibly and in some cases, even worth the effort of interpretation as sources of self-knowledge.

REVONSUO's proposal is that if we think carefully about the original evolutionary situation and at the same time examine the content of dreams, we will see that a plausible case exists for assigning dreams an adaptive evolutionary function. Dreaming was selected to simulate and rehearse threatening situations of the sort that we probably faced when we evolved tens of thousands of years ago. The argument turns on much interesting data: more dreams are unpleasant than pleasant, nightmares are common, fierce animals and threatening male strangers turn up much more frequently in dreams than they do in current environments, and much more frequently than the run-of-the-mill things we spend our time actually doing or worrying about doing.

These data then form the basis of REVONSUO's proposal that dreaming is not a spandrel after all but an adaptation. Here are some grounds for skepticism. First, we know thanks to work by Darwin (1873/1965) and Paul Ekman (1992) that humans did evolve with certain basic emotions. Ekman's list of basic and universal emotions now extends to seven emotions: fear, anger, sadness, disgust, contempt, surprise, and happiness. It seems utterly plausible to think that these emotions and the affect programs that govern them are adaptations, specifically adaptations that served awake humans who were up and about struggling to survive in harsh and threatening circumstances. It should be said, however, that it is entirely possible that the basic emotions and the affect programs they abide did not evolve among Homo sapiens but rather were handed over to us from earlier hominid ancestors.

This point is relevant because REVONSUO has us imagine a selection among Homo sapiens favoring those with heritable skills of simulating and rehearsing threatening events in dreams from those lacking the trait. Once the basic emotions are recognized as adaptations, a more parsimonious explanation for the threat simulation and rehearsal in dreams than the one REVONSUO offers is available. It is well known that the mechanisms activating sleep differentially activate the emotional centers of the brain. One widely accepted explanation is that this has to do with the proximity of the emotional centers to the brainstem from where the ins and outs of sleep are largely, but probably not exclusively, orchestrated. It would not be surprising, therefore, if the basic emotions and the associated affect programs were not thereby differentially activated. Since most of the basic emotions are negative, the associated dreams are also likely to be negative. Since the affect programs are attuned to be activated by stimuli or situations that operated in the original evolutionary context, it would not be at all surprising if minimal experiences with existing animals (canines, especially) and unrelated humans were sewn into the narratives that we know the brain tries hard to construct with the materials it is offered during sleep (probably by the forebrain as SOLMS points out).

It is even possible (although I remain agnostic on the matter) that the affect programs governing the basic emotions contain scenarios pre-loaded with content of threatening creatures and situations, so that we are primed to conjure up such scenarios once the relevant affect program is activated. This hypothesis, unlike REVONSUO's, is parsimonious because it requires no special selection pressures to have ever operated on dreaming, while nonetheless explaining why some of the data REVONSUO uses in making his argument might exist as he claims they do. However, one reason for preferring my proposal for explaining threatening dreams in addition to parsimony comes from a problem with thinking that dreams could be useful for practicing for threats that might occur in the light of day.

REVONSUO makes little mention of the fact that most researchers find REM dreams bizarre and disjointed, unlikely sites for realistic rehearsals of threatening scenarios to take place. Also we are not told whether the simulations he claims are the function of dreams occur mostly in NREM dreams, or in REM dreams. If in the former, there is no surprise, since no one, to the best of my knowledge, has denied that in NREM dreams we are often worried and anxious. The trouble is that in realistic NREM dreams we frequently do not think in particularly productive ways about what we are anxious or worried about, instead, we are caught in perseverative ruts. If, on the other hand, the threat simulation and rehearsal dreams are REM dreams then the worry about the bizarre and disjointed nature of REM dreams arises again with the attendant worry that close content analysis of these dreams will not un-

cover neat simulations and rehearsals conducive to preparation for real world threats.

**Freud ex machina.** I am extremely grateful to SOLMS for his wonderful, pathbreaking book (1997a), and especially for the empirical evidence he provides there and in his target article for distinguishing sleeping from dreaming. My argument (Flanagan 1995; 1997; 2000) was that sleeping is an adaptation, or better, a set of adaptations, whereas dreaming is a free rider on a system designed to be conscious while we are awake, and which is designed to sleep – during which time conscious mental states are serendipitously activated. This argument has been met surprisingly often by the objection that if sleep is an adaptation then so too is dreaming. The basic intuition behind this objection is that sleeping and dreaming are a unity, part of one and the same neurobiological process and thus not suited for separate analyses.

There are two common arguments for not trying to untangle sleeping from dreaming. One is that they universally co-occur. SOLMS, thankfully, has provided ample ammunition to quiet those who press this objection. There is a double dissociation. There are people who REM but do not dream and there are people who dream but do not REM. The second argument for not distinguishing sleep from dreams is that they are caused by one mechanism or two, one setting us into NREM sleep and its associated type of mentation, the other doing the same for REM sleep and its associated mentation. The exciting new evidence presented by SOLMS and NIELSEN for multiple mechanisms responsible for different stages of sleep and possibly still different, independent ones, for dreaming helps thwart the second argument in favor of identifying sleeping and dreaming. The brainstem may get us REM-ing but it is forebrain activation (probably of the dopaminergic system) that gets us dreaming.

Now the multiple generator models defended by SOLMS and NIELSEN do cause problems for one who like myself maintains that dreaming is likely not an adaptation, but an evolutionary epiphenomena. The reason is this: in deciding whether some mental process is an adaptation or a free rider on an adaptation (or set of adaptations) much turns on how it is caused. If dreaming is reliably initiated by the forebrain turning on dopaminergic circuits then the process is much more well specified and we can ask why evolution might have selected for a mechanism that reliably sets us to REM dreaming. Often, possibly usually, a reliable mechanism has an adaptationist explanation.

SOLMS, however, is careful to point out that "the biological function of dreaming [I would add, "if any"] remains unknown." But in his book he tentatively endorses an adaptationist proposal that is mentioned in three of the other target articles (HOBSON ET AL; REVONSUO; and NIELSEN). This is the Freudian view that dreams function as the protectors of sleep. One reason for thinking this is the fact that people who don't dream don't sleep well. It is also worth noting that people who dream but don't REM or suffer some form of REM suppression don't sleep well either. But all these people have other problems – they have suffered strokes, or take drugs that mess with their sleep and/or dreams. So people who sleep abnormally or who have abnormal dreams (excessively vivid dreams, REM-less dreams) don't sleep as well as normals. But I don't see why the abnormal cases SOLMS discusses leads to any view whatsoever on what the function of dreaming is. Especially, how it provides any evidence about what, if any, proper evolutionary function dreaming has.

Another idea that SOLMS toys with is that REM dreams actually serve the functions of wish-fulfillment, hallucinatory satisfaction, and involve censorship – essentially the orthodox Freudian view of the way in which dreams protect sleep and thereby preserve mental health (Solms 1997a, p. 174). SOLMS is careful to present this stronger view as a tentative but testable hypothesis. And I agree that it is. But I am happy to bet against it. The reason is this: We need an account of why the forebrain and dopaminergic systems are activated and in being activated produce REM dreams. We already know that there is differential activation of the aminergic and cholinergic systems during different parts of the sleep cycle and we are in possession of some decent hypotheses for why the brain is producing and/or stockpiling these neurotransmitters.

All these hypotheses – memory consolidation, trash disposal, stockpiling neurochemicals that are needed for attention and learning, even **VERTES & EASTMAN**'s proposal that REM sleep is designed to promote sleep recovery – require no story whatsoever about any biological function for the mentation itself. Meanwhile they can all avail themselves of exactly the same explanation why dreams occur and have the odd phenomenology they have, namely, that as sleep does what it is designed to do, it inevitably activates memories, emotions and so on, that are stored therein. So far, no hypothesis put forward requires that we think of dreaming as more than a side effect of the relevant functions of sleep.

Explanatory parsimony, and the expectation that dreaming will possess a unified (albeit complex) explanation lead me to expect that once we understand better why the brain needs dopamine, we will see that the activation of the dopaminergic system is just one other causal contributor to mentation that itself serves no fitness enhancing role. The issue is important, for the word on the street is that **SOLMS** has confirmed the orthodox Freudian view of dreams. To see this, consult Mortimer Ostrow's – President of the Psychoanalytic Research and Development Fund – contribution to the February 24 issue of *The New York Review of Books* (Ostrow 2000, p. 46). It would be good for **SOLMS** to explain where he stands on this important issue, and to explain his reason for not siding with me (assuming he doesn't) in betting that dreaming will, upon closer examination of the reasons for forebrain and dopaminergic activation, be further revealed to be caused by, but not itself be a contributor to, what the sleeping brain is designed to accomplish. Especially in light of the fact that he finds many mentally healthy souls who do not dream, I would have thought that the orthodox Freudian view would, on his own terms, not be thought to be much of a contender. Having read all five contributions to this special issue carefully, I am more rather than less convinced that dreams are the spandrels of sleep.

# Sleep, dreaming, and brain activation

Carlo Franzini

*Dipartimento di Fisiologia Umana, e Generale, Bologna 2, Italy.*
**franzini@biocfarm.unibo.it**

**Abstract:** Both Solms and Nielsen acknowledge the difficulty of accounting for the similarities between REM and NREM sleep mentation with a two-generator model, and each link dreams, either explicitly (Solms) or implicitly (Nielsen), to brain activation. At present, however, no data indicate that brain activation can be demonstrated whenever vivid dream reports are obtained.
**[NIELSEN; SOLMS]**

In reading and commenting on this series of articles on sleep and dreaming research, one is first and foremost impressed by the theoretical complexity of the field, a crossroads where different disciplines (epistemology, psychology, neurosciences) intersect, witnessing both the vitality of the research and the difficulty in attempting a unified theory. In fact all the proposed models involve tacit assumptions in each of these fields, which may outweigh the experimental evidence. The only antidote is the effort of spelling out the philosophic options underpinning the rules of the game agreed upon by author and reader; "a priori" implicit paradigms may hinder the theoretical debate. As the authors are neuroscientists, a preponderance of a radical reductionism identifying mind and brain is to be expected. However, the respectable old dualistic approach is still present. This makes for a mix of ideological (metaphysical) and scientific arguments in most theories of sleep and dream research.

**SOLMS**'s target article boldly fits into the thorny framework of the "mind-body problem" (sect. 5). The epistemological problems surface from the Introduction: "REM [sleep] is controlled by cholinergic brainstem mechanism, whereas dreaming seems to be controlled by dopaminergic forebrain mechanisms" where the author claims that different neural circuits (even neurochemically identified) underlie both a physiological state (REM sleep) and complex psychological activities (dreaming). That we inhabit the realm of extreme reductionism is confirmed by the comment on the activation-synthesis model: "the burden of evidence has shifted to the anatomical link between the pontine brainstem and dreaming" (sect. 4, para. 5). Psychology is at last reducible to anatomy. The "localizers" (neo-phrenologists) are alive and well.

**SOLMS** has the merit of tackling the relevant issues of sleep/dream research with a straightforward logical approach, making it easier for the commentator to explicate points of agreement and disagreement. Despite the suggestive title, the evidence that "Dreaming is preserved with pontine brainstem lesions" (sect. 5) is scanty. On the contrary, the data reported in section 6 are convincing. Common sense strongly supports the notion that mental activity during sleep is affected by forebrain lesions; the focal nature of these lesions is striking however. The disagreement here is confined to the author's open and honest reductionism. Any curious neuroscientist, used to browsing through the various chapters of a neuroscience textbook, must be impressed by the ubiquitous involvement of the dopaminergic system in accounting for diverse and behavioral disturbances. Hence the assertion that "dreaming is generated by this dopamine circuit" (sect. 7, para. 4) may well be correct in a reductionist paradigm, but then in the following statement (sect. 8, para. 1) "dreaming involves concerted activity in a *highly specific* group of forebrain structures" the term "specific" remains puzzling.

The final paragraph (sect. 9) of the **SOLMS** target article addresses the problem of the relationship between dreaming and brain activation. A wary adherent to some form of psychophysical parallelism or correlationism like myself should be content with the proposed correlation between dreaming and brain activation. However, experimental data do not support the conclusion that "dreaming appears to be a *consequence* of various forms of cerebral activation" (sect. 9, para. 3). Brain activation is an ill-defined term, generally implying cortical desynchronization and high levels of cerebral blood flow (CBF) and metabolism. However, Cavallero et al. convincingly demonstrated the existence of slow wave sleep (SWS) dreams indistinguishable from REM sleep dreams (Cavallero et al. 1992). Recent PET studies have shown the reduced metabolic cost of synchronizing modes of operation in the thalamocortical circuits (Maquet et al. 1997), and a negative correlation between delta activity and regional cerebral blood flows was found (Hofle et al. 1997).

On the other hand, continuous Doppler recordings of CBF changes during the night (Hajak et al. 1994) revealed a tonic, continuous drop of CBF, upon which phasic state-dependent changes are inscribed. As a result, late REM sleep episodes occur at lower CBF absolute values than SWS episodes occurring early in the night. Therefore indistinguishable mental activity during sleep can accordingly coexist with different degrees of cortical synchronization and different levels of energy consumption in the cerebral circulatory-metabolic machinery: No combination of the two indexes of brain activation (electroencephalographic or biochemical) can account for differences in mentation during sleep.

Can we draw some preliminary, operational conclusions from these data? The brain seems to be endowed with more degrees of freedom than we had thought possible. Redundancy is a general property of the central nervous system, which makes it extremely flexible in generating similar outputs through different internal operations (as a result, a correlational model is theoretically possible but extremely complex in practice). In the case of dreaming, neither cortical desynchronization nor metabolic level can be taken as obligate indexes for a specific type of mental activity during sleep. The search for the correlation goes on. Finally, I entirely agree with **SOLMS**'s conclusion that "the function of dreaming and the (equally unknown) function of REM sleep . . . should be un-coupled from one another" (sect. 9, para. 4).

In both the Introduction and in section 4 SOLMS states that "not all dreaming is correlated with REM sleep." This may be the starting point to confront his model with NIELSEN's. Nielsen recognizes the difficulty of establishing a rigid correlation between REM/NREM sleep on one hand and different types of mentation on the other hand (the two generator isomorphic model). His proposed solution is straightforward: Whenever we encounter mentation during sleep, REM sleep processes, manifest or covert, must be at work. It is therefore a one-generator model, which identifies in REM sleep processes the unique source of mental activity during sleep.

In my view, the model implicitly assumes that covert REM sleep processes are responsible for some form of brain activation, a feature that is hence shared by the two models. In fact, by disturbing the homeostatic condition of SWS, all physiological variables that connote the covert REM sleep processes may contribute to shifting the level of brain activation; NIELSEN specifically mentions cortical EEG desynchronisation in the "atypical NREM sleep episodes" (sect. 3.3) that may depend on covert REM sleep processes. Moreover, in the list of factors that might induce "convert REM sleep to be activated during NREM sleep" (sect. 3.2) quite a few (arousal processes, sensory stimulation, drug effects, sleep deprivation) are known to enhance the energy metabolism of the brain.

The two facets of brain activation (electroencephalographic and metabolic) are therefore prerequisites for the model. NIELSEN's hypothesis is based on well known physiological evidence, and has the merit of being experimentally testable. Polygraphic recordings show that the transition from NREM to REM sleep is not a clear-cut, abrupt event. Rather, different physiological variables change with different, contradictory time courses, and the macroscopic result may be the REM episode or an awakening or a return into NREM sleep. A single physiological variable can change alone, and may anticipate by many seconds, even minutes, the state change. It can be assumed that the complex process ultimately generating the full-blown REM sleep episode may have false starts and aborted outcomes; in this troubled transition (dynamic stage of train stem release in Parmeggiani's model, 1968); many physiological variables (increments in brain temperature and cerebral blood flow, heart rate, and blood pressure, Franzini 2000; motorneuron excitability changes, Nakamura et al. 1978) show a loose temporal link with the REM episode.

All this can be translated, in the terminology of NIELSEN's model, as "covert REM sleep processes" (sect. 3.2). As NIELSEN acknowledges, "evidence of mentation in stage 3 and 4 sleep (Cavallero et al. 1992) is particularly difficult for this model to explain" (sect. 3.14, para. 4). The difficult task of validating the model requires: (1) that some of the physiological markers of "covert REM sleep processes" be identified in the uniform and stable conditions of stage 3 and 4 sleep; and (b) that the apparently "deactivated brain" of SWS may show focal signs of metabolic brain activation linked to the same physiological markers.

# The prevalence of typical dream themes challenges the specificity of the threat simulation theory

Anne Germain,[a] Tore A. Nielsen,[b] Antonio Zadra,[a] and Jacques Montplaisir[b]

*Sleep Research Center, Hôpital du Sacré-Coeur de Montréal, [a]Department of Psychology, [b]Department of Psychiatry, Université de Montréal, Montréal, Québec H4J 1C5, Canada.* **a-germain@crhsc.umontreal.ca**

**Abstract:** The evolutionary theory of threat simulation during dreaming indicates that themes appropriate to ancestral survival concerns (threats) should be disproportionately represented in dreams. Our studies of typical dream themes in students and sleep-disordered patients indicate that threatening dreams involving chase and pursuit are indeed among the three most prevalent themes, thus supporting Revonsuo's theory. However, many of the most prevalent themes are of positive, not negative, events (e.g., sex, flying) and of current, not ancestral, threat scenarios (e.g., schoolwork). Moreover, many clearly ancestral themes (e.g., snakes, earthquakes) are not prevalent at all in dreams. Thus, these findings challenge the specificity of the threat simulation theory.
[REVONSUO]

REVONSUO's theory depends largely upon the observation that much dreaming is threatening in nature. But do the scenarios typically dreamed about today reflect the ancestral themes so central to the logic of this theory? Observations by our research group of the typical dream themes remembered by students and sleep-disordered patients are pertinent to evaluating the theory because they afford a global view of the scenarios most readily dreamed about over a lifetime within a given population. Our Typical Dreams Questionnaire (TDQ) includes 55 typical dream themes (cf. Griffith et al. 1958) that subjects check off if they have ever experienced them. We have administered the TDQ to over a thousand undergraduate students at different sites in Canada, the United States, and Japan and to close to a thousand sleep-disordered patients seen at the Sleep Disorders Center in Montreal. Consistencies between the lifetime prevalences of the most common TDQ items and those of Griffith et al. (1958) have been quite remarkable (Zadra & Nielsen 1997). Similarly, consistencies across our various samples have been very high (Nielsen et al. 1998; 1999c; Zadra & Nielsen 1999). In the case of three separate undergraduate student samples from McGill University (M age 20.3 ± 4.5 yrs; 113M; 228F), the most prevalent typical theme, endorsed by 78%, 86%, and 81 % of the student samples respectively (M = 82%) was the threat dream of being chased or pursued, but not physically injured (Zadra & Nielsen 1999). This theme was the second most prevalent typical dream of 233 Japanese undergraduates (M age 18.8 ± 2.3; 112F; 121M), 67% (Nielsen et al. 1999c), as well as the second and third most prevalent typical dream theme of two sleep-disordered patient samples much older in age (M age 44.9 ± 14.3 years; 249M; 235F), that is, 54% and 55% (M = 54%) (Nielsen et al. 1999b).

Such high lifetime prevalences of a threat theme are to some extent consistent with REVONSUO's theory of threat simulation during dreaming. However, the high prevalence of many other typical themes poses problems for the specificity of the theory. Two themes that fall consistently among the "top 4" in both our populations are of sexual experiences (undergraduates M = 76% and patients M = 55%) and falling (M = 72% and 47%). These are not obviously related to the ancestral threats described by REVONSUO. It might be argued that sexual dreams address issues of genetic transmission through sexual reproduction. However, threat is not the principal dynamic of such dreams. Falling dreams may echo long distant threats to the successful evolution of the upright stance in humans, but this is clearly not the type of evolutionary adaptation REVONSUO's model is attempting to explain.

Other highly prevalent themes pose similar problems. Dreams of flying or soaring through the air ranked 9th among both undergraduates (50%) and patients (38%) and typically reflect positive affect, not threat. Other prevalent themes deal less with ancestral sources of threat than with contemporary concerns, in the case of students, schools, teachers, studying (ranked 3rd; 73%), arriving too late, for example, missing a train (5th; 59%), trying again and again to do something (6th; 58%), and failing an examination 10th; 47%). It is not clear why dreaming should so often represent similar positive themes and/or contemporary concerns if its function is still geared toward dealing only with ancestral sources of threat.

Several of our least prevalent themes also do not support the theory because they are ancestral threat themes that occur in very few young or old subjects. Among the undergraduates, tornadoes or strong winds (ranked 45th; scored by 15% of sample) and earthquakes (48th; 12%) are rarely dreamed about. In fact, the likelihood of dreaming of these natural disasters is about the same as dreaming of being a member of the opposite sex (46th; 15%). Other natural disasters, such as fire (33rd; 23%) and threatening animals such as snakes (35th; 21%), wild, violent beasts (40th; 16%), or insects or spiders (23rd; 31%), also have low lifetime prevalences in our samples.

REVONSUO does offer some explanations for why such ancestral themes might be infrequent in dreams. First, some dream contents appear to change over time. For animals and aggressions at least, dreams appear to be more ancestral among children: "The brain has not yet had the chance to adjust the biases in order to better fit the actual environment" (sect. 3.4.2.2). The findings for children's dreams may well fit the threat simulation model, but it is not clear why the same pattern (i.e., high prevalence in the young, decreasing prevalence with age) should not hold true for other categories of threat, such as natural disasters. Nor is it clear why, in the case of children raised in environments relatively free from threat, the brain does not then adjust its simulations so as to be free of threat altogether. The notion of "change over time" in dream content (from ancestral themes to current themes) is problematic because such change would serve no obvious function. As described, it is only ancestral content that serves the (evolutionary) function stipulated by the theory. In sum, consistencies in the prevalences of typical dream themes in multiple study samples offer only limited support for the idea that dreaming is threat simulation. These findings would be more consistent with a less specific version of the theory that postulates simulations of positive, as well as negative, and of current, as well as ancestral, dream themes.

NOTE
Address correspondence to the first author.

## Each distinct type of mental state is supported by specific brain functions

Claude Gottesmann

*Laboratoire de Psychophysiologie, Faculté des Sciences, Université de Nice-Sophia Antipolis, 06108 Nice cedex 2, France.* **gottesma@unice.fr**
**www.unice.fr/psychophysiologie**

**Abstract:** Reflective waking mentation is supported by cortical activating and inhibitory processes. The thought-like mental content of slow wave sleep appears with lower levels of both kinds of influence. During REM sleep, the equation: activation + disinhibition + dopamine may explain the often psychotic-like mode of psychological functioning.
[HOBSON ET AL.; NIELSEN; REVONSUO; SOLMS; VERTES & EASTMAN]

**1. Brain support of mentation during sleep-waking cycle [Hobson et al.].** From a general point of view, it is difficult for a neurophysiologist to admit distinct modes of psychological functioning during waking and sleep unless they have different underlying brain states. Each 1/100 second of change, mentation probably involves thousands of variations in neuron activity in numerous complex circuits. This suggests that each mental state has to be sustained by some kind of specific brain state. We must be grateful to HOBSON ET AL. for carefully analyzing the psychological data in the literature to show that there are indeed general differences between slow-wave sleep and REM sleep mentation.

It is important to provide a model, as this generates hypotheses for future research. It is certainly bold to propose a unique functional schema to explain mentation during states as different as waking and the various stages of sleep. HOBSON ET AL.'s proposal is of high interest; they bring many convincing arguments forward to support different kinds of mental functioning. Nevertheless, two points are questionable. First, it is asserted that there seems to be an opposition between high noradrenergic and serotoninergic levels and a low acetylcholine levels and vice versa. This is in-

deed true for REM sleep at the cortical level. However, it is not the case for waking mentation where there is simultaneously a high release of noradrenaline (Aston-Jones & Bloom 1981; Hobson et al. 1975), serotonin (McGinty & Harper 1976; Rasmussen et al. 1984) and acetylcholine (Jasper & Tessier 1971; Marrosu et al. 1995). Second, it seems difficult to rule out an involvement of dopamine at least in REM sleep mentation. This transmitter is not taken into account in the "AIM" model.

Our view is that, as is generally accepted, consciousness is mainly generated in the cerebral cortex. Traditional EEG studies, gamma activity, neuron firing, blood flow, glucose uptake, and acetylcholine release all show that the cortex is in a different state during waking, slow wave sleep, and REM sleep, this last stage being defined by criteria very similar to those of attentive waking (for details, see Gottesmann 1999). All these data also demonstrate greater cortical activation in REM sleep than in slow wave sleep. However, inhibitory processes are also involved in cortex functioning. Dopamine, noradrenaline, serotonin (Reader et al. 1979), and histamine (Sastry & Phillis 1976) principally inhibit cortical neurons. The function of dopamine alone cannot explain differences in mentation in normal subjects, because studies in rats (Miller et al. 1983) and cats (Trulson & Preussler 1984) have shown that their neuronal firing rates do not change significantly during the sleep-waking cycle. Histamine neurons become silent as light sleep appears (Vanni-Mercier et al. 1984), and hence could potentially explain differences in cortical functioning during waking and sleep but not between slow wave sleep and REM sleep. In contrast, noradrenergic and serotoninergic neurons fire maximally during waking, decrease their activity during slow wave sleep, and become silent during REM sleep. Thus, they might control cortical functioning during the sleep-waking cycle. The importance of serotonin, at least, is established because decreased release induces the mental distortions associated with depression.

Our hypothesis accordingly is that during waking the cerebral structures involved in mentation are activated and thus able to generate mental activity, but that inhibitory processes in some way control or "normalize" this activation, thereby explaining reflective mentation. During slow wave sleep there is a decrease of both kinds of influence, explaining thought-like mental contents, because some controlled activation does persist. During REM sleep, the strong cortical activation occurs in a context of massive disinhibition, when all monoamines except dopamine are absent. This strong disinhibition alone could explain the original properties of mentation, which partly resemble psychotic symptoms, as described by Hobson et al. (1998b). We suggest that in this original activated and disinhibited state, the release of dopamine, strongly involved in psychosis, would reinforce this often schizophrenic-like mode of functioning. Indeed, an increased release of dopamine induces nightmares (Thompson & Pierce 1999) and psychotic disorders (Buffenstein et al. 1999). Moreover, it is well known that the reduction in the influence of dopamine by induced neuroleptics (Kinon & Lieberman 1996) alleviates schizophrenia.

**2. Experimental data to confirm the covert REM sleep hypothesis are still lacking [Nielsen].** The hypothesis of NIELSEN is highly important and confirms that dreaming only occurs in the physiological setting of REM sleep (Takeuchi et al. 1999b). This could explain why dreams have been described during slow wave sleep, in addition to thought-like activity. The problem is to determine the crucial physiological criterion to support REM sleep mentation. Dement wrote about thirty years ago (I do not remember where, nor does he) that REM sleep is like an orchestra playing a symphony: Several instruments (criteria of REM sleep) can be absent without suppressing playing (this sleep stage).

The arguments brought forward for sleep-onset dreaming are convincing. I have regularly such vivid life-like dreams currently and have wanted for months to record myself, being neither depressed nor narcoleptic. NIELSEN states that in addition to a possibly similar EEG, the same slow eye movements are seen at sleep onset as during REM sleep. To determine whether other criteria of REM sleep are found at sleep onset, particularly those linked to eye movements, it would be interesting to record phasic integrated potentials (PIPS) (Rechtschaffen et al. 1970), cortical waves (McCarley et al. 1983; Miyauchi et al. 1987), and the ear muscle activity (Pessah & Roffwarg 1972). During slow wave sleep, activation of the visual and, more lightly, of the secondary auditory cortex (Hofle et al. 1997) could be an index of REM sleep and explain dreams recorded during this stage (Foulkes 1962).

Prior to, and sometimes just after REM sleep, mice, rats, and cats show an intermediate stage with hippocampal criterion of REM sleep (for a review see Gottesmann 1996). Several other pieces of experimental evidence show REM sleep premises at this hinge-period, (McCarley & Hobson 1970 for cortical neuron firing; Morrison & Bowker 1975 for PGO waves; Morales & Chase 1978 for medulla neuron activity; Kanamori et al. 1980 for medulla oblongata neuron firing; Steriade & McCarley 1990a for mesopontine neuron activity; Sei et al. 1994 for blood pressure variations, etc.). In humans, as related by **NIELSEN**, Lairy et al. (1968) described an intermediate phase prior to and after REM sleep. It is characterized by interspersed criteria of slow wave sleep and REM sleep. However, the mental content does not correspond to the description of dreams.

First, it is difficult to establish a psychological contact with the subject behaviorally wakened from this stage. This is in contradiction to the good contact with the outside world observed on awakening from REM sleep, and with the sentry theory of Snyder (1968). Second, the verbal reports do not reveal visual contents but instead "a feeling of indefinable discomfort, anxious perplexity and harrowing worry" (p. 279). Although for Foulkes (personal communication, 1998) this result is debatable, Larson and Foulkes (1969) show that mental contents at this sleep time "are inconsistent with the hypothesis of an intensification of mental activity or cerebral vigilance at pre-REM EMG suppression. They seem, rather, to point to a reduction in reportable mentation and in efficient cognitive reactivity at this point of transition from NREM to REM sleep" (p. 552). These results are not in accordance with vivid visual dreaming activity. Despite its major interest, the hypothesis of **NIELSEN** needs to be confirmed by new psychophysiological studies performed particularly in the period preceding REM sleep.

**3. Is Revonsuo so far from Freud?** REVONSUO's is a rich hypothesis as it does not imply a systematic phenomenon: Its aim "is not to claim that every single dream of every single individual should realize this (threat simulation) function." This idea is original and supported by many dream contents reported in the literature. It is also true that the historical period of human life represents a ridiculously small part of the lifetime of the human species. Consequently, it is understandable that such ancestral fantasies should persist today above all because "dreaming has no maladaptive consequences, so it has survived." This is generally but not always the case (e.g., Huntington's chorea, which persists probably because reproduction occurs prior to the appearance of the disorder). REVONSUO comes close to Freud's (1900) position when he states "the content of dreams shows far too much organization to be produced by chance." Although he mentions "The Interpretation of Dreams," he overlooks subsequent work in which Freud mentions "primal fantasies" transmitted phylogenetically. More precisely, Freud (1918) described a famous dream "The man and the wolves," which fits with REVONSUO's thinking, for the patient was threatened by wolves. REVONSUO would deduce that "dreaming does have a well-defined and clearly manifested biological function (that is) to simulate threatening events," while Freud's psychological interpretation appeals to a hypothetical observation of parents' sexual "primal scene" and the ancestral danger of castration by the father, which is a threat to the subject's reproduction abilities; a concept also taken up in REVONSUO's theory. In fact, the two interpretations are in some ways complementary.

I am slightly more cautious about REVONSUO's use of physiological arguments. VERTES and I (see below) are doubtful about theta rhythm function during REM sleep. Perhaps in opposition

to currently accepted ideas (Hobson & McCarley 1977), I am also not totally convinced about the obligatory relation between PGO waves and dreaming. These spikes have to do with short-lasting activating influences (maximum 100 milliseconds, Miyauchi et al. 1987) that transiently reinforce cortical tonic activating processes of REM sleep. Although related eye movements appear in the pontine cat (Jouvet 1962), they are modulated by the cortex (Mouret 1964), which shows an activation of the saccadic eye movement system (Hong et al. 1995). Moreover, the cortical visual projection area is deactivated during REM sleep (Braun et al. 1998; Madsen et al. 1991a). Finally, rats seem to dream, as shown by behavioral characteristics (eye, vibrissae, ear, paws and tail movements) and pontine lesions (Mirmiran 1983). However, they display pontine spikes during REM sleep (Farber et al. 1980; Gottesmann 1967; 1969; Kaufman & Morrison 1981), but no REM-sleep-related spikes in lateral geniculate nucleus and cortex (Gottesmann 1967; 1969; Stern et al. 1974), despite a direct neuronal relationship between pons and visual cortex (Datta et al. 1998). It seems that dreams have a different time scale from that of short-lasting PGO waves, unless we accept that the successive spikes are responsible for the rapid changes of dream content, which currently seems doubtful.

**4. Sleep-waking stages are induced and mentation is supported by brain stem structures [Solms].** In his interesting contribution, **SOLMS** is right that dreaming is not initiated in the brain stem, contrary to the old, somewhat naïve theory of Hobson and McCarley (1977), rapidly discarded (Vogel 1978a). All his arguments in favor of the forebrain as generator of dreaming processes, and of mentation more generally, are convincing; today they seem self-evident. Nevertheless, his clinical arguments in support of the assertion that "dreaming is preserved with pontine brainstem lesions," thus that dreaming is able to occur without the brain stem inducing properties of REM sleep, are less convincing. Moreover, in the examples of forebrain-induced dreams it is not always easy to distinguish dreams from hallucinations; and in physiopathological conditions, hallucinations are often mistaken for dreams (Fischer-Perroudon et al. 1974). Through its midbrain and pontine structures the brain stem induces the different sleepwaking states and does not induce but supports correlative mentation.

It is somewhat difficult to agree entirely with **SOLMS**'s hypothesis concerning dopamine's almost exclusive influence on dreaming-generating processes. He is of course right when he emphasizes the role of dopamine. Indeed, as the species evolves this transmitter probably has an increasingly important influence on cortical functioning. Where dopaminergic projections end only at prefrontal level (Hökfelt et al. 1974) in the rat, in primates all cortical areas are concerned (Berger et al. 1991), and this is probably also the case in humans (Smiley & Goldman-Rakic 1993). However, in the normal subject, the absence of noradrenaline and serotonin cortical input seems to be the precondition for the participation of dopamine in the dream mentation of REM sleep. If it were not, we would always be in a dream state, for dopaminergic neurons fire similarly during all stages of the sleep-waking cycle. Moreover, several properties of dreaming seem to be explicable by the disinhibition process alone (Gottesmann 1999).

**5. Is it possible to speculate about theta rhythm function during REM sleep? [Vertes & Eastman].** The numerous arguments against the theory of memory consolidation during REM sleep by **VERTES & EASTMAN** are convincing, particularly those obtained in humans by pharmacology. Nevertheless, our own experience shows that it is possible to induce emitted potentials during REM sleep but not during slow wave sleep; this means that during REM sleep there is access to memory processes established during waking (Gauthier et al. 1986), although this result does not demonstrate memory consolidation.

An electrophysiological datum used by **VERTES & EASTMAN**, based on animal studies, appears debatable. They argue that although hippocampal theta rhythm occurs during waking and REM sleep, its properties are different in the two states: only during waking would it be involved in mnemonic function. As it has already been clearly shown by Grastyan et al. (1959), learning is associated with theta rhythm in animals. Green and Arduini (1954) were the first to show that this limbic rhythm is induced by activation of the midbrain reticular formation. This is probably owing in part to the stimulation of fibres passing nearby; Vertes (1981) has shown that the nucleus pontis oralis is the main origin of hippocampal synchronized activity. At first sight, it is hard to imagine that when the same (rather crude) basic structure induces similar theta activity during both waking and REM sleep, the target structure could function differently in the presence of the same activity during the two states. However, several pieces of experimental data support this dichotomy. First, the brain stem is not necessary for the induction of the theta rhythm. Acute intercollicular transected rats and cats (Gottesmann et al. 1980; 1984) show virtually continuous theta rhythm and it is difficult to assume that there are integrated functions in the hippocampus of a neocortically comatose animal. Second, serotoninergic innervation of the hippocampus, which is crucial for hippocampal functioning in memory processes (Matsukawa et al. 1997), becomes silent during REM sleep (Rasmussen et al. 1984); the influence of noradrenaline (Segal & Bloom 1974) is also suppressed (Aston-Jones & Bloom 1981). Therefore these two monoamines, which most often have inhibitory influences on higher brain structures, are important for mental processes. Third, by analogy, the neocortical EEG is similar during waking and REM sleep while consciousness is different. Consequently, as suggested by VERTES & EASTMAN, despite nearly identical theta rhythm, the hippocampus is in a different functional state during waking and REM sleep.

The function of REM sleep proposed by **VERTES & EASTMAN** seems to be a revival of Roffwarg et al.'s (1966) theory of the importance of endogenous brain activation in ontogenetic early brain maturation.

# Where is the forest? Where is the dream?

Ramon Greenberg

*Harvard University, Brookline, MA 02445.* **rgreenberg@hms.harvard.edu**

**Abstract:** In this commentary I discuss the importance of considering the isomorphism between the full richness of dreams and the great body of information about REM sleep that is amply documented in the five target articles. With this inclusive mode I point out the importance of looking at REM sleep as involving both pontine and cortical activity in an integrated network. We cannot have a full appreciation of sleep and dreaming (view of the forest) without taking both physiology and mental activity seriously. [**HOBSON ET AL., NIELSEN, REVONSUO, SOLMS, VERTES & EASTMAN**]

In discussing these five extremely detailed, scholarly, and thoughtful target articles, I will use two organizing principles: isomorphism and the forest and the trees. My commentary will consider some issues that involve all the articles and will also point out some problems that are specific to individual articles.

We are privileged, in this issue, to have extensive and quite complete reviews and summaries of much of the research that has emerged from the sleep research laboratories of the world during the last 50 years. Thus the reader is presented with many of the facts. However, the reader will also notice that different authors give different weight to certain facts in order to bolster their individual theories. I would liken these presentations to a wonderful description of sets of trees in a forest, but I am afraid, as I will try to demonstrate, the forest has been lost. One reason for this is that a major group of trees has been almost totally ignored. That has to do with a discussion of dreams. By this I mean dreams as seen in all their fullness and richness, rather than as collections of single units.

As **REVONSUO** quotes Farthing, "we may define a dream as a subjective experience during sleep, consisting of complex and organized images that show temporal progression" (Farthing 1992).

The loss here is as great as that which might occur if we studied Shakespeare by counting and tabulating words and sentences rather than considering the meaning of the units when assembled into story, plot, and poetry. An important reason for raising this issue is that we cannot have a really isomorphic or consistent picture (the forest) without a major part of the picture. These papers present an excellent view of the physiological and the cognitive aspects of sleep and dreaming but leave out what I think is the third leg of a three-legged stool, namely the dynamic and emotional meaning of dreams, which only two of the authors really touch on. Let me give an example of how helpful this can be.

In the 60s, as a psychoanalyst who was also involved in sleep laboratory research, I was troubled by the lack of fit between the total emphasis on the pons for the understanding of REM sleep and the richness and meaningfulness of the dreams that I worked with in the clinical situation. It seemed to me that the cortex must play a significant role in the process. Because of this I chose to study the REM sleep patterns in a group of patients with right parietal lesions and attention deficit visual disorders. The findings were clear and showed that the eye movements during REM sleep were missing in the direction of the affected visual fields (Greenberg 1966; **HOBSON ET AL.** cite this study but fail to note this basic finding. Dorrichi's study 25 years later is a replication of this study). This led to a review of Jouvet's oft cited study showing the persistence of REM sleep in decorticate patients. The REM sleep in these patients was not normal; the eye movements were isolated and lacked the bursts seen in normal sleep. This is certainly consistent with the idea that the pons might be firing away but without a cortex no visual imagery and dreams were being generated.

**SOLMS**'s contention that REM sleep can occur without dreaming and dreaming is separate from REM sleep seeks, on the other hand, to omit the role of the pons in the process. **SOLMS** can then maintain that it is the cortex alone that is responsible for dreams and that what we know about REM sleep has no relevance to understanding dreams. With Hobson's now partially abandoned but still frequently responded to idea that the pons alone is responsible for REM sleep with dreaming as an epiphenomenon, we have two starkly contrasting views which both fragment the process. In this commentary I wish to underline the fact that, as several of the authors suggest, we are dealing with a process which involves, in nonpathologic cases, the activity of a network of parts of the brain which must be working together to have a fully functional output. This includes the pons as a trigger plus various parts of the cortex which provide affective, imagistic, and motivational components to dreams.

An essential feature of this commentary depends on the idea that dreaming and the physiology of REM sleep are two aspects of the same process and the concept of isomorphism suggests that conclusions in each realm must be consistent with those from the other. Clearly the authors of these target articles disagree amongst themselves about the connection of dreams and REM sleep and the conclusions to be drawn from REM deprivation studies or from the exploration of dreams. I do think, however, that **HOBSON ET AL.** and **NIELSEN** present a very convincing case (the best I've yet seen) for the differences between REM and NREM mentation. I would like to add a few other bits of data. Although Foulkes is a major proponent of the difficulty in distinguishing REM from NREM mentation, a sample of 30 dreams he collected from a student from both REM and NREM awakenings revealed only one dream from NREM (Foulkes 1967). All the rest were from REM. Cartwright (Cartwright et al. 1967 in **NIELSEN**; and Cartwright 1972) has shown that the subgroup of normal subjects who do seem to generate dream-like material from NREM sleep awakenings are distinguished by various psychologic measures such as a high schizophrenia score on the MMPI. Also, it is mainly with gradual and not abrupt awakenings that mental activity is elicited from NREM awakenings (Goodenough et al. 1965a; Shapiro et al. 1963 and 1965 in **NIELSEN**), thus raising the question of whether the mentation is truly from NREM sleep. **NIELSEN** discusses this issue in more detail. Finally, as **REVONSUO** notes, the Penfield stimulation studies elicit visual flashbacks but these are "dissimilar from dreams" in various respects. Therefore, I will contend that REM sleep and dreaming co-occur and any theory must consider the isomorphism of theories derived from the too different manifestations of a single process.

Thus we have a process that includes well-studied physiologic activities that are associated with a very special type of mental activity. Can this association illuminate our understanding of these two seemingly different phenomena? That is, can what we have learned about REM sleep tell us something about what happens in dreams and can what happen in dreams orient us in our understanding of the physiologic findings? I have noted earlier how clinical experience with dreams led to a study of cortical activity during REM sleep. The role of the cortex is now, 30 some years later, much more clearly elaborated in the imaging studies noted by several of the authors and by the rich picture generated by **SOLMS**'s studies of patients with lesions.

Before considering how the physiologic findings have influenced our understanding of dreams, I must consider and respond to some of the observations the various target articles make about the function of REM sleep and especially the role of information processing. **VERTES & EASTMAN** raise a number of objections to the notion of the role of memory processing in REM sleep. They, in effect, suggest ignoring all the positive studies by proposing that stress is the major "villain." We should note here the very creative studies on cueing during REM sleep by Hars and Hennevin (1987; in **NIELSEN**) which **REVONSUO** fails to include. That nonarousing meaningful cues during REM but not NREM sleep can lead to improved learning indicates clearly that an effective consolidation process is occurring during REM sleep; one cannot invoke stress as issue.

**VERTES & EASTMAN** readily dismiss the fact that windows for sensitivity to REM deprivation occur, because in relation to different learning tasks or in different species the timing varies. This does not account for the fact that stress occurs equally in the animals deprived of REM during both the window and nonwindow periods, and yet there is the clearly demonstrable effect on memory consolidation only when REMD occurs during the critical period. Furthermore they do not deal with the fact that these time periods correspond to the periods of increase in REM seen after training trials during which there is an increased retention.

Finally, **VERTES & EASTMAN** seem unaware, in their claim that only Smith has shown windows (i.e., not replicated), that Pearlman (1973) had shown this phenomenon many years before. Another set of findings not considered is that different types of learning are or are not sensitive to REMD. We have formulated this as the difference between prepared and unprepared learning (Greenberg & Pearlman 1974). It is unprepared learning or complicated learning that seems sensitive to the REMD impairment of learning or that is followed by an increase in REM pressure after training. This holds for animals and for humans. (See Greenberg & Pearlman 1993 for full discussion.)

This idea is especially important in **VERTES & EASTMAN**'s description of the lack of effect on function of MAOI REM suppression in humans. A brief anecdote may speak most clearly to the kind of memory that is affected. Again it is not cognitive. A colleague was treating one of the subjects in Wyatt's study (personal communication). He did not know any of the sleep data or what drug the patient was taking. For months the patient seemed to be feeling better but also seemed to have no access to past emotionally meaningful experiences. Also no dreams. Then suddenly the patient began to bring in intense dreams and to be much more in touch with her past. The therapist then found out that an MAOI had been administered with complete suppression of REM and that the drug had been discontinued at the time things began to open up. We also have found a disconnection from past meaningful memories in subjects who have been REM deprived (Greenberg et al. 1983). Because of the lack of apparent cognitive deficits, **VERTES & EASTMAN** dismiss the use of REM suppressant drugs in animals to examine the effects of REM suppression, without stress, on learning. Yet the common denominator for the flower pot method and the drugs is REMD.

If we keep in mind, at this point, the idea of isomorphism, we might see that the observed affect of REMD on the integration of new experiences is consistent with some interesting "new" perspectives about dreams. I use the word "new" because in these target articles references to psychoanalytic ideas about dreams are exclusively references to Freud. Readers should be aware that much has been learned since "The Interpretation of Dreams" was published 100 years ago (see Greenberg & Pearlman 1999). Readers should become familiar with the work of Bonime (1962), French and Fromm (1964), and Reiser (1997) to see how these authors present a picture of dreams that is much more consistent (isomorphic) with much of the sleep lab research presented in these papers.

The picture of dreams that emerges is one in which one can see the current problems in adapting with which the dreamer is struggling and how solutions for problems are searched for. The manifest dream and affects become more important than Freud's theories have suggested. The dreams show the process that can only be inferred from the role seen for REM sleep in the processing of information. The dream can now be understood if one looks for problems rather than just for the categories of or numbers of individual items or events in the dream (see Greenberg et al. 1992).

This approach allows a somewhat different perspective from **REVONSUO**'s evolutionary theory. Yes, REM sleep and dreaming are important in man's and also other mammals' mastery of a threatening environment. By insisting on the separation of REM sleep and dreaming, **REVONSUO** can allow himself to look only at dreams. He goes even further in reducing his focus to dreams dealing with external dangers, that is, traumatic dreams. This allows him to miss the real import of the Hartmann (1984; 1998) work he cites. This has to do with the evolution of the traumatic dream from a replay of the trauma to one in which there is evidence of integration of the traumatic event into the dreamer's series of life experiences. As this evolution occurs, the clinical manifestations of trauma in both waking and dreaming life abate. The work of the dream is not just to deal with practicing for dealing with external dangers, but rather to enable the dreamer to deal with the complicated and unresolved feelings evoked by external events, be they wild animals, enemies, or other events that feel dangerous, such as abandonment, induced helplessness, humiliation, and so forth.

**REVONSUO** cites the appearance of threatening animals to support his idea that dreaming evolved to deal with external danger and continues to serve only that purpose. He fails to consider that the appearance of elements like wild animals may be the metaphoric or symbolic language with which the dreamer expresses more internal fears. He claims that dream content is independent of external psychological and physical stimuli. **HOBSON ET AL.** make the same claim. Here there is a failure to consider that dream content shows a great deal of what is emotionally significant. Breger et al. in "The Effect of Stress on Dreams" (1971) presented very clear evidence of the relationship of dream content to the stresses with which the subjects were dealing. In our study (Greenberg & Pearlman 1975) we also showed clear evidence of how the content of the dream was related to emotionally significant waking mental experience and Cartwright (1996; in **REVONSUO**) also illustrates this point. These studies demonstrate how the dreamer is struggling with life events.

**REVONSUO**'s theory is a special case for the role of dreaming. His idea that it is related to survival is discussed in a much more complete fashion by Rotenberg (1993a). Rotenberg's theory of search activity brings the role of REM sleep and dreaming into a very central position in our consideration of the question of survival. Of interest here is that Rotenberg's studies showed a role for search activity in the maintenance of the immune system. **REVONSUO** uses the analogy of the immune system to suggest that it is not always called into action. One must consider, however, the importance of it always being ready. One must also consider that external threats are not the only kind of problems that appear in dreams. Our study (Greenberg & Pearlman 1993) demonstrates that almost all dreams show evidence of problems and emotionally meaningful ones at that. What does vary is the appearance of solutions to the problems, varying from successful to unsuccessful. Just as the immune system is not always successful in fighting infection, so too the dream is not always successful, and in the traumatic dream, the system is clearly overwhelmed. What Rotenberg adds to our understanding is the idea of searching for solutions and this is what is of importance for survival. Thus I would suggest that while REVONSUO takes the dream very seriously, his focus is narrowed to a special kind of dream and in part this is the result of his failure to take seriously the idea of an isomorphism between the dream and what occurs in REM sleep in all species that have this process available.

Let us now turn to comments about the individual target articles. HOBSON ET AL. present a comprehensive picture in which they give reasons to believe that REM sleep and dreaming indeed go together. They reason that we are dealing, in this process, with a network that includes both the pons and cortical areas interacting. Where they fall short is the failure to include the richness of dreams and the implication that cortical lesions affect both the appearance and the quality of dreams. Thus, dreams provide an opportunity to learn more about the nature of the information that is processed during REM sleep. By taking the position that dreams do not show evidence of the inclusion of stimulating and emotionally important material from waking, they lose the meaningfulness of studies like Breger et al.'s (1971), Cartwright's (1996; in REVONSUO), or ours (Greenberg et al. 1992) to name a few. They also approach the forgetting of dreams from a purely physiologic perspective, rather than considering that the language of a dream is different from awake language and therefore harder to remember. Dreams do not just disappear. Note the frequent experience of the sudden remembering of a dream when a reminder during the day will bring a dream fully to mind. This is a frequent observation in the clinical situation but I suspect most readers have also had this experience. The dream experience is indeed a part of the memory system. How else could one explain the fact that the occurrence of REM sleep plays a role in learning? Something more permanent must be recorded. By continuing to view the forebrain as responding to brainstem input (I in AIM) rather than as a partner in the process, the Hobson group loses the opportunity for a truly isomorphic consideration where the richness of dreams is given a full hearing.

NIELSEN is one author who takes the phenomenology of dreaming into full account and his idea of covert REM mentation deserves consideration. I would raise one question for what is on the whole a plausible and well-documented presentation. Could it be that shifting from one to another kind of mentation could be analogous to the way the mind can wander during wakefulness? Thus, is the appearance of covert REM activity (dreams) during the NREM periods evidence of its constant activity or is there some shifting back and forth (mind wandering) which some subjects are more susceptible to (see Cartwright et al. 1967 and Cartwright 1972)?

SOLMS takes the role of the cortex very seriously and informs us of the special attributes of the cortical areas that are involved in dreaming and also in REM sleep. However, he would like to divorce the pons from any role in dreaming. I think he goes too far in arguing that the pons is not the cause of dreaming and ignores the possibility that it is the trigger for a process that involves networks in the cortex that are involved in dreaming. He then is able to totally ignore the connection between REM sleep and dreams and all the implications for our understanding of dreams which have emerged from studies of the function of REM sleep. For example, he refers to Panksepp's ideas about the "seeking" or "wanting" part of the brain but does not connect this with Rotenberg's ideas about the Search activity function of REM sleep. SOLMS also seems unaware that the "REM" sleep generated in the decorticate subject is not normal REM sleep (see above) and finally, as he

notes, the evidence is not in that subjects with pontine lesions abolishing REM sleep can actually still dream.

In conclusion, I have tried to provide a picture of the forest by adding a few trees, or groves, to the excellent descriptions provided by the authors of these articles. I would argue that only by integrating the information we have about dreams with what we know about REM sleep, considering that a network is activated through the whole brain, can we realize the parallels between what we have learned about dreaming and what we have learned about REM sleep. I do not know if this discussion will change any of the authors' minds but I hope that as they write their responses they will read some of the papers and books referred to in this discussion but not included in the target articles. I would also like to note that in a commentary in this journal entitled "The cortex finds its place in REM sleep" (Greenberg 1978), I was premature and only in the last few years has there been growing evidence of the role of the cortex. I hope that the place of the dream will not have to wait so long.

# State-dependent modulation of cognitive function

R. W. Greene
*Department of Psychiatry, HMS & VAMC, Brocton, MA.*

**Abstract:** The three introductory questions posed by Hobson et al. point toward further investigations of cellular, circuit, and systems mechanisms involved in cognitive function that include the effect of CNS-state related modulatory systems on these mechanisms.
[HOBSON ET AL.]

Two of three introductory questions posed by **HOBSON ET AL.** concern (1) the major differences between the phenomenological experience of these three physiological states (waking, REM, and NREM); and (2) measures that might establish clear cut differences between these states at the level of brain regions, as well as the cellular and molecular levels. These broad ranging questions might involve measures existing as the average levels of activity of the cholinergic and monoaminergic modulatory systems (Steriade & McCarley 1990a; Siegel 1990) referred to as "M" by the authors, but this has not been established with respect to cognitive function. The traditional view, based on studies reviewed by the authors, is of waking as a CNS state of high cholinergic and monoaminergic tone, of NREM sleep as a CNS state of low cholinergic and monoaminergic tone, and of REM sleep as a CNS state of high cholinergic and absent monoaminergic tone. Whether or not these simple measures are adequate may depend on questions like, "how do these simple changes affect the function of cortical and thalamic circuits and what might result in cognitive function?"

The answers are just beginning to surface, as for example, effects of modulatory tone on EEG activity (Buzsaki 1998; Herculano-Houzel et al. 1999; Steriade et al. 1993) and cholinergic modulation of visual systems interneurons (Xiang et al. 1998). However, the changes in circuit function that can derive from cholinergic and monoaminergic modulation depend on our understanding of the circuit function and its relationship to cognitive function (as for example: Eichenbaum 1999; Goldman-Rakic 1999; Hesselmo 1999; Lisman 1999; Wang 1999). Further, the magnitude of the complexity of this issue is emphasized by a recent study showing 25 different response combinations in CIA interneurons to noradrenaline, serotonin, muscarine, and mGluR agonists (Parra et al. 1998).

It would seem surprising yet it is conceivable that changes in circuit function in CA1 that results from a state specific in modulatory tone (restricted to just the cholinergic and monoaminergic systems) are sufficient to account for the state specific alterations in CA1 information processing. This provided that one also takes into account the effects of the change in cholinergic and monoaminergic modulatory tone on other afferent systems to the CA1 (a change in the pattern of septal gabaergic input for example; Dragoi et al. 1999). In other words the author's third question, "Can a tentative integration of the phenomenological and physiological data be made?" implies the challenge of both the investigation of cellular, circuit, and systems mechanisms involved in cognitive function as well as the effects of CNS-state related modulatory systems on these mechanisms.

# The dramaturgy of dreams in Pleistocene minds and our own

## Keith Gunderson

*Department of Philosophy, University of Minnesota, Minneapolis, MN 55455-0310.* **gunde002@maroon.tc.umn.edu**

**Abstract:** The notion of simulation in dreaming of threat recognition and avoidance faces difficulties deriving from (1) some typical characteristics of dream artifacts (some "surreal," some not) and (2) metaphysical issues involving the need for some representation in the theory of a perspective subject making use of the artifact.
[**HOBSON ET AL.; REVONSUO**]

Underlying the conceptual shape and contents of threat recognition and avoidance simulation in **REVONSUO**'s fascinating functional theory of dreams are, I believe, some metaphysical anomalies which make the view somewhat inchoate as it stands. I shall develop this diagnosis via heuristic analogies while drawing on observations contained within the theory of **HOBSON ET AL.**

**REVONSUO** proceeds on the bold metaphysical assumption that consciousness just *is* the organization of the brain at the phenomenal level. It includes the subjective experience of dreaming (described, quite interestingly as it turns out, as a mechanism) which in slumbering Pleistocene minds (PMs) contained simulations of threat perception and rehearsals of threat avoidance responses and behaviors. By "prepping" the PM for coping with real life counterparts of these simulations, the reproductive success of our ancestors was supposedly enhanced.

But how in general do we picture the contents of another's dreaming mind wherein **REVONSUO**'s simulations and rehearsals are said to have operated, when we reconstruct it using known or plausibly surmised historical facts coupled with seemingly parallel contemporary anthropological data involving dream reports? It will in significant ways depend on how we picture the contents of our own dreams when we recollect them.

Here are two very abbreviated and somewhat "cinematic" examples of how I think dreams often seem to us, and with slight permutations in content can be imagined as having occurred to PMs as well: (1) in our mind's eye, in memory, we imagine an image of our self (the dreamt self) spotting a dangerous animal bearing down on it, the dreamt self picking up a big stick and/or running away; (2) in our mind's eye, in memory, there is no image of oneself any more than there is in waking life when I run from a neighbor's nasty tempered mastiff and don't observe myself doing so.

How, then, might **REVONSUO**'s simulations utilize representations like those pictured, in either (1) or (2) so that we obtain a sense of how dream-production systems churning out thousands of realistic threats during a PM's lifetime is thereby "bound to result in improved threat-avoidance skills"? It is said to turn on "what the dream can do itself" (its "natural functions") where the notion of "the dream itself" (phenomenal or p-dreaming) explicitly excludes reference to any self-conscious recollections or "invented functions," that is, cultural and personal uses of dreams and dream reports (Flanagan 1995).

So how does the dream imaging prove to be functionally efficacious for **REVONSUO** (as it is not for Flanagan, nor "aimless" as in the Activation-Synthesis theory)? How can its utility as simulation rehearsal be depicted at that first person level? My concern

is that if it cannot, then the dramatic color and detail of simulation rehearsals as posited by **REVONSUO** washes off into the purely neurobiological domain, and signals not just a stylistic diminishment but a substantive one.

An analog suggests itself which might prove useful in that its outline parallels in interesting ways with **REVONSUO**, yet contains ingredients which his theory lacks, but may in some form need. This is Aristotle's influential theory that art involves imitation, is instinctual, instructive, and pleasurable. Aristotle, if we focus on the first three attributions and ignore the last, conjures up general, albeit inadvertent, likenesses to **REVONSUO**'s claims that ancestral dream consciousness was biologically functional with adaptive value. (Think of imitation and simulation, as going proxy for each other in numerous contexts, simulations and rehearsals as being instructive, and the species-specific non-conventional nature of the [Aristotelian] instinctual and **REVONSUO**'s biologically functional and adaptive.)

Consider, then, Aristotle and **REVONSUO** with respect to the following examples, any one of which to us looking in at them, as it were, might be called a *transparently instructive simulation* (TIS) of a real world thing or situation: (1) a detailed drawing of the anatomy of a cow for use in a course in veterinary medicine; (2) a video of a black belt karate instructor teaching a student a given kick; (3) a dress rehearsal of a play to iron out the kinks in it before actual production; (4) a cockpit of an airplane (*sans* the plane) wherein pilots can learn to master switches and buttons in preparation for flying a real plane.

In our "looking in" we assume a first person point of view of what it would be like for ourself to see such things or imagine ourself participating in such contexts, and we sense that if we had done so, we would have been edified. (We think of having looked at the cow picture and how that might have helped direct our dissection of a real cow; we think of having memorized the moves in the karate video and using them in self-defense, or, if we had been an actor in the play (the dreamer as other to itself in the dream) we might have noticed mistakes we had made and resolved to, correct.)

Any of these examples might prove friendly to Aristotle's claim that art as imitation (simulation) is instructive. Pursuing the parallel, we ask whether what Aristotle seems to have at hand – a range of TISes – has any kind of counterpart within the medium of dreams? If not, why? And does it matter?

Something *like* TISes seem to me demanded by **REVONSUO** but exactly what or how is so unclear as to make the alleged demand seem gratuitous. Consider some of the "remarkably consistent set of features" listed by **HOBSON ET AL.** as characterizing dreaming: "dream imagery can change rapidly and is often bizarre in nature;" self-reflection, often absent in dreams, "when present, often involves weak, post hoc, and logically flawed explanations of improbable or impossible events and plots"; dreams "lack orientational stability; times and places are fused, plastic, incongruous and discontinuous." Let me call these characteristics *serializing aspects of the dream as artefact of sleep* (or SER).

Dream dramas where it is as if a homuncular Salvador Dali reigned as dramaturge, strike me as unlikely material for TISes in somewhat the same way as a drawing of a cow's anatomy where its organs are distorted and melted into each other would seem less than useful in directing a dissection of a real one. Furthermore, there are other aspects of artefacts that contribute to their potential instructiveness: duration, intersubjective availability, potential for preservation, copying or replication, and so on that seem conspicuously absent from the phenomenal level of dreaming. Such "no shows," the density of SUR, and the exclusion, anyway, of all non-natural, "invented" or "cultural" features from consideration, make the budget of materials available for anything like counterparts to TISes in **REVONSUO**'s theory very skimpy indeed.

The problems I suggest might pose for **REVONSUO** primarily pertain to oddities attending the stuff of dreams and whether certain types of nocturnal "artifacts" could be constructed from it. (Compare: Could a rigid sculpture be made out of feathers and molasses?) These can be thought of as artifact problems. But there are other rather different difficulties which, though related to the foregoing, seem to me much more threatening to the dramaturgical coherency of REVONSUO's theory. These might be called perspective problems.

Part of what makes Aristotle's idea of instructive imitations (simulations) intelligible is our ability to find a locus of perspective or a subject who engages the simulations either from a standpoint outside of them, or through a presence within them: for example a person viewing or watching something (1 or 2) or being involved in some collective or singular action (3 or 4). But how does any perspective or participating subject inhabit REVONSUO's phenomenal "space" wherein its simulations reside? Which "whos" or "whats" constitute a simulation's dramatic personae and function vis-à-vis each other and the simulation viewed as artifact? It is difficult, indeed, impossible to conceive of either the dreamt self in a dream or the dreaming self (sans body-image) carrying out that role, for they, after all, belong to the "stuff of dreams" and cannot migrate literally and usefully into the real world in spite of anthropological reportage on peoples who may believe they can (Tedlock 1987a; Merrill 1987a). Nor can the real dreamer, in spite of impeccable ontic obduracy be the transporter of instructive dream "text." For this would involve exactly the sort of self-reflective sense of dreaming that REVONSUO regards as unlikely for a PM, and not germane in any case, since the dream's functional utility is attributed wholly to the simulation bearing dream experience itself. So who or what might audition better?

The only viable candidate remaining so far as I can tell is somehow the whole of the subjective experience of dreaming itself, described by REVONSUO as a "mechanism." If so, it is this mechanism – and I would suppose qua mechanism, a neurobiological one – that is also a subject with perspective, that is the locus of the dream, and furthermore whatever content the dream is defined by. If this interpretation is correct, REVONSUO's is not only a cognitive functional account of dreaming, but also a ticklish metaphysical position in which the subjective self which dreams is physically objectified as a neurobiological mechanism wherein the artifacts of dreaming coalesce with their artificers, and the content of dreams and the dreamer are one.

It is here (at last) where I believe the underlying metaphysical anomalies alluded to at the outset, to reside. To treat the brain mechanism as itself the needed subjective self with requisite perspective, is to create two more problems for REVONSUO (1) the complex and colorful idea of the dreamer-cum-rehearsal simulation is reduced to the very general idea of a person sleeping and, by dreaming about what was hair-raising that day, keeping in tune the neurobiological mechanisms needed to survive the morrow. (2) It assumes that the subjective self can be objectified in terms of a mechanism – a goal, to be sure, shared by virtually any physicalist account of the mind-body relationship. It also shares with these views the burden of coherency in the face of well known challenges from, for example, Nagel (1986), Jackson (1986), McGinn (1991), and many others.

# The waking-to-dreaming continuum and the effects of emotion

Ernest Hartmann

*Department of Psychiatry, Tufts University School of Medicine and Sleep Disorders Center, Newton-Wellesley Hospital, Boston, MA 02459.*
**ehdream@aol.com**

**Abstract:** The three-dimensional "AIM model" proposed by Hobson et al. is imaginative. However, many kinds of data suggest that the "dimensions" are not orthogonal, but closely correlated. An alternative view is presented in which mental functioning is considered as a continuum, or a group of closely linked continua, running from focused waking activity at one end, to dreaming at the other. The effect of emotional state is in-

creasingly evident towards the dreaming end of the continuum.
[HOBSON ET AL.; NIELSEN; SOLMS]

First of all, SOLMS's target article is about the control of dreaming – in other words, what portions of the brain are necessary for reports of dreaming (or visual dreaming) to occur. In these terms, SOLMS is certainly convincing in his demonstration that portions of the forebrain are involved – specifically the parieto-temporal-occipital junction, probably controlled or activated by a dopaminergic pathway in the ventral-mesial portions of the frontal lobes. SOLMS is right in pointing out that his data disprove the original Hobson and McCarley "Activation-Synthesis" view that dreaming is entirely dependent on REM-sleep activation of the forebrain by brainstem centers. HOBSON ET AL. do not exactly acknowledge this, but they do shift their emphasis from dreaming toward a broader attempt to explain the biology of the states of consciousness.

The target article by HOBSON ET AL. on the recent AIM model is impressively detailed, up-to-date in its references, and also creative and imaginative. It is a valiant effort to make sense of almost all current information on states of consciousness, organizing it according to three simple-sounding dimensions. The reach of the model, however, exceeds its grasp.

The AIM model makes most sense in terms of its ability to organize or "locate" three basic states – waking, NREM sleep, and REM sleep. I believe that the model is much less useful in its attempts to explain the psychological features of the states – including dreaming – even though HOBSON ET AL. use terms such as activation and information flow which are meaningful in cognitive as well as biological terms and thus could provide a bridge. For instance, one problem is that the waking state, which appears in the diagrams as a small, isolated box, actually supports a large array of cognitive activity, including overlap with the other states, as discussed below.

The chief problems I have with the three-dimensional AIM model are the following: First of all, an obvious question when looking at such a dimensional diagram is why *three* dimensions rather than more or less, and why these particular dimensions. The familiar spatial metaphor of a cube or box can delude us into thinking that three dimensions constitute an obvious number and that these three exhaust the possibilities. I do not question that activation (A), direction of information flow (I), and "modulation" (M) are of importance – though "modulation" is a very complex notion, defined at one point as "how information is handled," which is not convincingly a single dimension. I could think of some other relevant dimensions. For instance, do we simply want one dimension of "amount of activation?" What about a dimension of "focus," or "spread of activation," or perhaps a dimension involving velocity of activation, introducing time and measuring the speed with which different portions of the system are activated. I mention these dimensions because, like the others, they can easily be thought of in cognitive as well as purely biological terms. (In fact, although I was not thinking in terms of "dimensions," I made use of such variables years ago in what could be called an early "two-dimensional model." In formulating the principal characteristics of the three biological states we were beginning to recognize – waking, REM sleep, and NREM sleep – I summarized REM sleep overall as a state of "high-activity level," (similar to waking) but with "rough adjustment – poor feedback" (very different from waking) (Hartmann 1967, p. 149). (This formulation in fact may still be helpful in characterizing the physiological and psychological characteristics of REM sleep.)

Thus, three dimensions may not be the right number. However, my main concern is that the three dimensions of AIM may not truly be dimensions at all. It makes little sense to discuss in detail what happens in different regions of a three-dimensional cube unless one has reason to believe that the three axes are actually more or less orthogonal (independent). On the contrary, I believe there is considerable evidence suggesting that two of the important dimensions (I and M) and perhaps all three dimensions are intimately correlated.

Table 1 (Hartmann). *A continuum*

| | Focused<br>Waking thought | Looser,<br>Less-structured<br>thought | Reverie<br>Free Association<br>Daydreaming | Dreaming |
|---|---|---|---|---|
| What dealt with? | Percepts<br>Math symbols<br>signs, words | | fewer words, signs,<br>more visual-spatial<br>imagery | almost pure<br>imagery |
| How? | logical relationship –<br>If A then B | | less logic, more noting or<br>picturing of similarities,<br>more metaphor | almost pure<br>picture –<br>metaphor |
| Self-reflection: | highly self-reflective –<br>"I know I am sitting here<br>reading." | | less self-reflective,<br>more "caught up" in the<br>process, the imagery | in "typical"<br>(non-lucid) dreams<br>total *thereness*,<br>no self-reflection |
| Boundaries: | solid divisions,<br>categorizations,<br>thick boundaries | | less rigid categorization,<br>thinner boundaries | merging<br>condensation<br>loosening of<br>categories,<br>thin boundaries |
| Sequence of ideas or images: | A → B → C → D | A → B ⟨ C ↕ D | A → B ⟨ C ↓ D | A ✛ (B, C, D) |
| Processing: | | relatively serial; net functions chiefly as a<br>feed-forward net | | net functions more as an<br>auto-associative net |
| Subsystems: | | activity chiefly *within* structured<br>subsystems | | activity less *within*, more<br>across or *outside* of<br>structured subsystems |

Reprinted with permission from Hartmann (1998).

For instance, suppose we turn off the lights, close our eyes, and pull a blanket over our heads, but remain awake. Obviously we are moving from chiefly external input to internal input along dimension I, and as we do, we inevitably also move along the dimension M, and begin to process material, produce images, and so on, in a more "dreamlike" manner. A whole literature on "sensory isolation" supports this close association. Antrobus and others have shown that the longer a period of uninterrupted isolation (or internal processing) lasts, the more dreamlike cognitive activity becomes (see Antrobus 1990 for a review). All this occurs even if we are not tired. If we are tired and are "winding down" at the end of a day, our brains are presumably less activated. As "A" is reduced we concomitantly find ourselves paying less attention to the external world and more to the internal world (changes in "I"), and we also begin to think in a more imagistic and dreamlike way (changes in "M"). In fact, a huge body of work on relaxation states, meditation states, and hypnosis demonstrates such changes. This suggests that the "dimensions" are by no means independent, and that perhaps we need another way to conceptualize these changes in mental functioning.

My colleagues and I have in fact found it useful to think in terms of a continuum or a series of related continua, running from focused waking at one end through reverie and daydreaming, to dreaming at the other (Hartmann 1996a;1998; 2000a; Kunzendorf et al. 1997) (see Table 1).

This continuum, which we can call the "focused waking-to-dreaming continuum," obviously refers to states of the system capable of supporting conscious mental processing. It is clear that a certain level of activation is required to support consciousness, and thus all the points on the continuum in Figure 1 can be considered states of relatively high activation. Under these conditions we consider it useful to think of one continuum with many strands, rather than orthogonal dimensions.

This view emphasizes the continuity in mental functioning between the conscious states of waking and dreaming rather than the discontinuities. The work of Antrobus (1991), Foulkes (1990), Klinger (1990), Singer (1975), and others supports this continuum viewpoint. Material indistinguishable from dreams can be obtained from quiet waking, from sleep onset, and from NREM sleep. We have demonstrated that under certain conditions, daydreams and dreams are very similar and that depending on personality characteristics, the daydreams of some students are scored just as dreamlike and "bizarre" as the dreams of other students (Kunzendorf et al. 1997). Dreaming is mental activity (based on cortical activity) at the right end of the continuum that contains very little focused rapid-processing activity. For instance, we have shown that reading, writing, and arithmetic ("the three Rs") are extremely rare in dreams (Hartmann 1996b; 2000b). Activation or spread of activation is less focused and more diffuse, which I have related for many years (Hartmann 1973, p. 138, 155n) to low levels of norepinephrine in the cortex (among other factors).

Furthermore, the "qualitative" differences between dreaming

and waking mental activity are not as clear-cut as sometimes thought. For instance, if we consider the difficulty of recall, which HOBSON ET AL. frequently cite as a clear difference, I think it is useful to consider whether it is appreciably more difficult to recall last night's dreams than to recall the waking daydreaming or reverie we experienced this morning while washing or shaving.

In brief, our view is that at the dreaming end of the continuum there is more overlap of cortical activation patterns, or more bringing into conjunction elements often kept separate. Thus, we can say that presumably at the cortical level there is greater connectivity; connections are made more readily at the dreaming end of the continuum. As is well known, this can sometimes be useful artistically or creatively, but it is an everyday phenomenon as well. Five different women have told me a dream very close to the following: "I dreamt of Joe, my boyfriend, but in the dream he looked very much like my father" or "he turned into my father." I am not attempting any deep or Freudian analysis; my point is a simple one. In all five cases the women awakened and said something like: "Isn't that interesting. Of course Joe is like my father in three or four different ways; odd that I had never noticed that before." In other words the similarities were evident if one stopped to consider them but it took a dream – activity at the right hand end of the continuum – to bring the two networks together.

Thus, we consider connections to be made more broadly in dreaming – at the right end of the continuum – than in focused waking, but the process is not at all random; we have shown that it is guided by emotion. (Emotion is hardly mentioned in the models discussed in the target papers, although recent imaging studies – reviewed by HOBSON ET AL. and by SOLMS – have lent strong support to activation of the limbic system, especially the amygdala, during REM sleep.)

For my group, the paradigmatic dream is what we have come to call the "tidal wave dream." We have studied people – apparently normal adults – who have recently experienced an acute trauma: an escape from a fire in which others were killed, a rape, an attack, a sudden death of someone close. We believe this is an important starting point in the study of dreams, because here – unlike most of the time – we know clearly what is emotionally important in the dreamer's life. The dreamer, after an acute trauma, may or may not have a few dreams that repeat the actual trauma or aspects of it. Then, very frequently, one or more dreams such as the following occur:

> "I was walking along the beach when suddenly a huge tidal wave came and engulfed me. I was tossed around, I tried hard to get to the surface. I woke up terrified."

What is going on here? Obviously the person is not dreaming of his or her actual experience. Rather the dream is contextualizing (producing a picture-context for) the dominant emotion of terror/helplessness/vulnerability. We believe this is a paradigm, that the same process probably occurs in all dreams but is less easy to detect when we do not know of a clear dominant concern (for details, see Hartmann et al. 1998a). We have shown quantitatively that powerful images of the tidal wave type occur more frequently after trauma (Hartmann 1998, Hartmann et al. 1998b).

The view then is that the dream pictures or "contextualizes" the emotional state of the dreamer. Again, the continuum view suggests that emotion always influences our mental activity and imagery, but this is least evident at the left end of the continuum, when we are doing arithmetic or involved in focused thought. It becomes more evident as we move towards the right. There is evidence suggesting that emotion has a greater influence on our cognition and perception when we are relaxed or in a relaxed hypnotic state than in our ordinary waking state (for instance, see Kunzendorf & Maurer 1989; Klinger 1996).

It is of interest that Hobson's group continues its curiously negative view of the process of dreaming. They have previously spoken of dreaming as "delirium," (Hobson 1997b) and as "your brain on drugs" (Kahn & Hobson 1994). In the present target article

making an analogy with epilepsy where "activation signals of limbic origin commandeer the cortex and force it to process," they argue that "the cortex of the dreaming brain is compelled to process internal signal rising from the pons." All this is a very "focused-waking-centric" viewpoint. We consider dreaming in a more neutral manner as simply one end of a continuum of mental functioning. Dreaming, after all, is a widespread natural phenomenon, consuming a considerable amount of our time. It stands to natural-selection-based reason that dreaming probably has some function for the organism, and in fact several related functions have been proposed (Moffit et al. 1993; Hartmann 1998). Far from being "compelled," dreaming can be considered the least constrained type of mental activity. I find it natural to think of great portions of the cortex as an image generator. The system can be "constrained" to perform arithmetic or logic problems during focused waking, but it tends to "relax" into daydreaming or dreaming when there are no such constraints.

Viewing mental functioning along a continuum, as in Table 1, may also be useful in resolving the "one-generator versus two-generator" issues raised in the target article by NIELSEN. Our model certainly favors the idea of a single generator producing mental activity (thoughts and images) but a single generator whose products can vary along a continuum or a series of related continua as in Table 1. NIELSEN does not mention the varieties of waking mentation at all, but if one accepts the evidence that dreaming mental activity is not completely and qualitatively separate from daydreaming activity which in turn is not completely separate from other waking thought, one would hardly want to postulate a separate "generator" producing each of these related states. Nor is evidence presented by NIELSEN or others that would lead me to believe that NREM mentation is so qualitatively different from the entire continuum above that it would require a "generator" of its own.

Furthermore, from the point of view of brain anatomy, including the neuropsychological and brain imaging data reviewed, it makes a great deal of sense to consider a single image generating process, which, however, can be activated in a number of different ways, presumably using the cortical and subcortical pathways delineated in the target article by SOLMS.

The close relationship between dreaming and other forms of mental activity such as daydreaming or other waking imagery is also supported by ontogenetic studies by Foulkes, demonstrating that the ability to experience and report fully formed visuo-spatial dreams develops gradually, at around age 5–8, at about the same time as the development of full visuo-spatial abilities in the waking state (Foulkes et al. 1991).

The lesion studies by SOLMS may also be relevant to these questions, in the sense that it would be of great interest to know the precise status of daydream or reverie activities in the neurological patients who had lost the ability to dream, or to dream visually. SOLMS's study procedures include an extremely detailed dream interview covering 13 different areas (Solms 1997a, pp. 83–86). However, he does not specifically mention interview questions dealing with daydreams or reverie. If indeed SOLMS found a well-delineated group of patients who had stopped dreaming (or stopped dreaming visually) but continued to have clear visual daydreams and reverie exactly as before, I would take this as evidence against the continuum view I am discussing here. However, I do not believe this to be the case. No data are presented specifically on loss of daydreaming or reverie; however, SOLMS does mention in the present paper that in the large PTO (parietal-temporal-occipital) patient group other deficits were found. For instance, right-sided lesions in PTO were associated with not only cessation of dreaming but "disorders of spatial cognition." Left-hemisphere lesions were associated with disorders of "quasi-spatial (symbolic) operations." He also notes that lesions in visual association areas caused defects in visual dream imagery, "in association with identical deficits of waking imagery." SOLMS in fact suggests that "the visual imagery of dreams is produced by activation during sleep of the same structures that generate complex visual imagery in waking perception." Therefore I believe that SOLMS's results are consistent with a view of a single widespread cortical system or image generator (with several subsystems) generating what we usually call dreams, and that this system similarly generates a whole continuum of waking imagery.

# Reflexive and orienting properties of REM sleep dreaming and eye movements

John Herman

*Department of Psychiatry, University of Texas Southwestern Medical School, Dallas, TX 75235.* **joherma@childmed.dallas.tx.us**

**Abstract:** In this manuscript Hobson et al. propose a model exploring qualitative differences between the three states of consciousness, waking, NREM sleep, and REM sleep, in terms of state-related brain activity. The model consists of three factors, each of which varies along a continuum, creating a three-dimensional space: activation (A), information flow (I), and mode of information processing (M). Hobson has described these factors previously (1990; 1992a). Two of the dimensions, activation and modulation, deal directly with subcortical influences upon cortical structures – the reticular activation system, with regard to the activation dimension and the locus coeruleus and the pontine raphe neuclei, with regard to the modulation dimension. The focus of this review is a further exploration of the interaction between dreaming and the cortical and subcortical structures relevant to REM sleep eye movements.

[HOBSON ET AL. ]

The question of cortical versus subcortical control of saccadic eye movements during REM sleep is addressed by the target article and previous Hobson publications. **HOBSON ET AL.** review the controversy between brainstem-only models (bottom-up model) of control of REM sleep and combined brainstem-cortical models (top-down model). If REM sleep eye movements are exclusively regulated by brainstem mechanisms, then either they are totally independent of dream images, or both the eye movements and dream images of REM sleep are significantly governed by brainstem mechanisms. If they are under combined brainstem-cortical control, then dreaming could be of cortical origin and still linked to eye movements.

Lesion studies demonstrate that some form of REM-like eye movements occur during REM sleep in the decerebrate preparation: these and other studies demonstrate clearly that noncortical structures are necessary and sufficient for the generation of the eye movements of the REM sleep. The dilemma presented by such findings regards the commonly held assumption that visual dreaming during REM sleep is cortical phenomenon.

In 1975, McCarley and Hobson asserted that the occurrence of REM sleep and its timing are controlled by reciprocal interaction between cells in the pons (FTG cells) and cells in the nucleus locus coeruleus (LC cells). This assertion was instrumental in the neurophysiological and conceptual basis of the current modulatory (M) portion of **HOBSON ET AL.**'s three-dimensional AIM model. In 1977 McCarley and Hobson's activation-synthesis model (the A of the AIM model) literally turned the universally accepted model of dreaming on its head. They contend that brain stem neuronal mechanisms more significantly influence the timing, formal properties, imagery, and content of REM sleep dreaming and that the role of the cortex was secondary, synthesizing incoming volleys of corticofugal excitation (Hobson & McCarley 1977).

**HOBSON ET AL.** note that it is highly implausible that all REM sleep saccades are concordant with dream imagery, given the presence of eye movements in the congenitally blind. Also, animal studies in cats and monkeys, in which accurate measurements of the direction of both eyes and their positions are possible, indicate that the eyes are moving asymmetrically during REM sleep. **HOBSON ET AL.** conclude that even brainstem initiated REM sleep eye movements are most likely under the control of a final common

pathway which integrates brainstem generation and forebrain modification, a substantive modification of his earlier assertions of brainstem control (Nelson et al. 1983).

Such integration or cortical and non-cortical control of REM sleep eye movements is consistent with Doricchi et al. (1993; 1996), who observed that patients with left visual hemi-inattention, or visual neglect, showed dissociation between the direction of waking saccades and that of REM sleep. In contrast, during waking, exploratory saccades were present in both lateral directions, but confined to the right hemispace. Hence, the neglect patient is capable of executing saccades in both horizontal directions, but does not do so in REM sleep. Following two months of training, neglect patients increase leftward waking saccades, but none appear during REM sleep.

Therefore these authors (Doricchi et al. 1993) propose that two mechanisms control waking saccadic eye movements: one of a more voluntary, cognitive classification, typically employed in visual exploration, and the other more automatic and involuntary, related to reflexive-orienting. The total absence of leftward saccades during REM sleep indicates that *all saccadic eye movements during REM sleep are reflexive-orienting eye movements.*

Morrison and colleagues (Ball et al. 1991a; Bowker & Morrison 1976; Morrison et al. 1995; Sanford et al. 1992b; 1993), propose that REM sleep pontine geniculate occipital (PGO) waves constitute a response of central mechanisms to afferent information which, in the waking state, would be capable of eliciting an orienting response. Spontaneous PGO waves, similar to those accompanying the waking orienting response, are present throughout REM sleep, indicating that PGO waves are the REM sleep equivalent of an orienting response. They propose that *during REM sleep, higher cortical structures are in a state of virtually continuous orientation* (Ball et al. 1991b). This is a similar to Doricchi's conclusion that REM sleep eye movements are functionally equivalent to waking reflexive-orienting saccades.

Pompeiano and Morrison (1965; 1966; Morrison & Pompeiano 1966; 1970; Pompeiano & Valentinuzzi 1976) along with co-investigators, based upon a series of studies in the decerebrate preparation, have demonstrated a complete cessation of REM eye movements, following bilateral vestibular enucleation. They have identified the anatomical structures responsible for REM bursts as the vestibular nuclei, the oculomotor nuclei, and the oculo-orbital system.

Herman et al. (1983) have shown that the characteristic of eye movements during REM sleep in humans is more consistent with those observed with accompanying head movements than with those observed when the head is stationary. These observations, elicited by careful questioning, are not inconsistent with the concept that REM sleep eye movements are related to vestibular activity and reflective-orienting responses (Herman 1992).

It is proposed that a constant property of vivid REM dreaming is the sensation of orientation in the dreamt space, or the hallucinated impression that the dreamer is physically present in the dreamt scene, in three-dimensional space. The dreamer is directionally oriented and aware of a spatial relationship to the persons or objects present in the dream. The sensation of orientation is made possible by cortical-subcortical mechanisms, including the vestibular system. The associated eye movements are related to reflexive and orienting responses to the dreamt surround via feedback loop connecting cortex to vestibular nuclei. *It is postulated that the sense of orientation is the phenomenological equivalent in dreaming to the continuous vestibular activation during REM sleep.* This ubiquitous presence of self in the dream parallels Pompeiano and Morrison's finding that intact vestibular structures are required for the typical occurrence of REM sleep. In the same manner that the vestibular nuclei are necessary for the eye movements of REM sleep, so is the sense of self oriented in space an essential component of human dreaming.

# The ghost of Sigmund Freud haunts Mark Solms's dream theory

J. Allan Hobson

*Department of Psychiatry, Harvard Medical School, Boston, MA 02115.*
allan_hobson@hms.havard.edu

**Abstract:** Recent neuropsychological data indicating that an absence of dreaming follows lesions of frontal subcortical white matter have been interpreted by Solms as supportive of Freud's wish-fulfillment, disguise-censorship dream theory. The purpose of this commentary is to call attention to Solms's commitment to Freud and to challenge and contrast his specific arguments with the simpler and more complete tenets of the activation-synthesis hypothesis.
[HOBSON ET AL.; NIELSEN; SOLMS]

We have recently commented at length on SOLMS's neuropsychological data and dream theory (for scientific details and references see Hobson & Pace-Schott 1999). The following précis is drawn from a paper presented at the American Academy of Arts and Sciences in Boston on January 12, 2000, at which Mark Solms was the invited discussant. This commentary focuses sharply upon SOLMS's interpretation of his data, which clearly reveals his commitment to psychoanalytic dream theory. Only those data (and they are few) which can be retrofitted onto Freud's dream theory are deemed worthy of SOLMS's theoretical attention. The rest (and they are many) are either discredited or dismissed. While SOLMS does not mention Freud in his target article, it is clear from his book (Solms 1997a) and from his subsequent writings that his major theoretical orientation is psychoanalytic. The following commentary is thus designed to clarify our interpretive position and to contrast it with the Freud-Solms theory.

**Solms's attempt to resuscitate Freud's dream theory.** Until recently, I would have said that Freud's theory had no scientific status whatsoever and that activation-synthesis was its natural replacement (Hobson 1988b; Hobson & McCarley 1977). But recently SOLMS has interpreted his very interesting findings on the effects of brain lesions and brain disease upon dreaming as favoring Freud's dream-as-wish-fulfillment hypothesis (Solms 1997a). In essence, SOLMS has found that strokes or surgical interventions that damage two distinctive areas of the brain in the inferior parietal lobe and in the white matter tracts connecting subcortical structures to the medial prefrontal cortex are often followed by a loss of dreaming. The parietal region subserves spatial cognition and the white matter tracts connect important components of the limbic system that are thought to mediate emotion, motivation, and reward. This finding, interesting enough in itself, is complemented by a raft of brain imaging studies showing selective activation of these same structures and their directly associated cortical and subcortical regions during REM, the phase of sleep most capable of supporting dream consciousness. SOLMS also points out, quite correctly, that temporal lobe seizures produce dream-like states in waking and intensify dreaming in sleep.

Surely the new evidence indicates, as SOLMS claims, that Freud was on the right track in postulating wishes as dream instigators especially insofar as wishes are related to emotional arousal in humans. And so it might at first glance seem. But emotions, motivations, and rewards are not wishes in any unconscious, Freudian sense. Furthermore, positive dream emotions, like elation, joy, and erotic excitement do not qualify as unconscious Freudian wishes, and they certainly do not need disguise. More importantly, as we have shown in our formal analysis of dream emotion, they are not disguised in dreams. Why, after all, would we want to be unconscious of these emotions? Because they might wake us up? When I have dreams of flying or sex I hope they won't be censored and that I won't wake up!

What about negative emotions, like anxiety, fear, and anger which constitute a goodly portion of dream emotion? Are they wishes? I think not! I can see why I might want to disguise or censor them, but negative emotions are not disguised in my dreams either! The fact that I do wake up from these dreams is both a re-lief and proof that both the wish fulfillment and the guardian of sleep ideas that Freud concocted are just plain wrong!

What does activation-synthesis make of all this? Simply that in response to automatic and selective activation in sleep of the brain structures mediating the emotions, we consciously experience them undisguised and uncensored in our dreams. This is exactly what we said in our 1977 activation-synthesis paper.

> The new theory cannot yet account for the emotional aspects of the dream experience, but we assume that they are produced by activation of brain regions subserving affect in parallel with the better known sensorimotor pathways." (Hobson & McCarley 1977, p. 1336)

Recent PET imaging studies have confirmed this prediction and we have further elaborated this notion by suggesting that emotions may have a primary shaping force in determining dream plots.

If you want to interpret your anxiety dreams, or your elation dreams, or your anger dreams, go right ahead. You will find, as I do, that these emotions are associated with the same kind of plot content that they are in daily life (even if the dream plots are more humorously bizarre). For example, anxiety in my dreams is often associated with being lost (fear), with not having appropriate attire or credentials (anxiety), with being chased, or threatened (fear). Can this be read as the product of either wish fulfillment or disguise-censorship?

What I mean is that if you want to understand your dreams, the last person you would want to consult is Sigmund Freud or one of his psychoanalytic protégés! That is because unconscious wishes play little or no part in dream instigation, dream emotion is uncensored and undisguised, sleep is not protected by dreaming, and dream interpretation, via free association, still has no scientific status whatsoever. SOLMS's data-based and eloquent defense of Freud is nonetheless welcome. Why? Because SOLMS is so committed to the original text that he allows us to reframe the activation, synthesis theory in the light of the new neuropsychological findings and to show even more clearly the weakness of Freudian theory.

**Support for activation-synthesis by the new neuropsychological data.** If it is not unconscious wishes, then what does cause dreaming? Why are dreams bizarre? Why are dreams hyperemotional? And most interesting of all, why are dreams so quickly and thoroughly forgotten?

Here are activation-synthesis answers to these questions. Dreaming is caused by brain activation in sleep. Dreams are bizarre because the activation process differs in important ways from that of waking. Dreams are hyperemotional because the emotional brain is selectively activated in sleep. Dreams are forgotten because recent memory systems are disabled during sleep. Let us look at each of these issues in turn to see how well neurophysiology can explain them.

As soon as we fall asleep, the brain begins a dramatic transition from one activation state (waking) to another (REM sleep). The main features of this transition are the blockade of sensory inputs and motor outputs (which puts the activated brain in a closed-loop mode) and the chemical demodulation of the activated brain (which puts it in a distinctive processing mode). Associated with these changes, all of which are controlled by the brain stem, is the selective activation of internal information sources including pseudo-sensory data, which generate the dream perceptions and emotional data, which generate the dream affects.

Just after sleep onset, the brain is still activated enough to produce micro-dreams and it starts to do so as soon as external data can no longer have easy access to the still appreciably activated brain. Not only are the sensory and motor gates of the brain rapidly closing, but three of the specific chemical systems of the brain stem that support waking consciousness are also beginning to shut down. For these reasons the self-reflective awareness, directed thought, attention, and memory control of waking are rapidly slipping away. This concatenation of events is responsible for sleep onset dreaming.

Sleep onset dreaming is short-lived and evanescent because the activation status of the brain rapidly plummets and it ultimately falls to such low levels (in Stages III and IV of non-REM sleep) that mental activity of any kind is difficult to sustain. As sensory and motor gates continue to close and we become dead to the world, three of the brain stem chemical systems that help support waking consciousness decline even further. PET scans reveal that the lights have gone out in command central. Under these conditions, full-fledged dreaming is practically impossible. However, simple recall of at least some mental content from these stages is around 50 percent (Nielsen 1999).

But after about a 90 minute interval, the cerebral tables are turned and the brain is converted by REM sleep into an all but obligatory dream machine. This is because while two of the brain stem chemical systems regulating consciousness (those producing norepinephrine and serotonin) are completely shut off, the acetylcholine system is released from inhibition. The acetylcholine system not only reactivates the brain, but provides it with powerful, internally generated signals. These signals project to the sensory systems of the thalamus and cortex, triggering vivid dream perceptions, and to the limbic system of the forebrain triggering dream emotions. And, of course, those two forebrain systems communicate with each other. Under these conditions, dreaming is more vivid, more bizarre, and more sustained than in any other state.

Human PET scan studies of REM (see **HOBSON ET AL.** 1998a; 1998b; 2000; and target article, for reviews) confirm the main features of this picture and add an important, unexpected observation: the dorsolateral prefrontal cortex is selectively deactivated in REM sleep. The hallucinating, emotion-drenched brain is thus deprived not only of the chemicals it uses to tame such processes in waking but also loses the direction provided by its key top-down guidance system. The prefrontal cortex is generally considered to be the seat of executive cognitive functions like working memory, volition, self-reflective awareness, critical insight, and judgment, none of which work well in dreaming (see again **HOBSON ET AL.**)

No wonder that we are so hopelessly unable to get a grip on things in dreams, to know where we are or what exactly we are doing, or even to recognize that we are dreaming! On the contrary, we normally ignore all of the obvious internal evidence and conclude that we are awake! And no wonder we can't remember dream events either as they occur or after we wake up. Dream forgetting is not owing to repression as the Freudians would have us believe. It is simple, organically determined amnesia. No wonder, too, that dreams are bizarre. Improbable or downright impossible incongruities and discontinuities are synthesized by the hyperassociative brain only to pass unnoticed because the censor – far from being hyperactive – is sound asleep! It is in the accounting for dream bizarreness, that Freud's theory is not only incorrect, it is absolutely backwards!

Even **SOLMS** has admitted that disguise-censorship is the weakest part of Freud's theory. But I submit that this characterization of disguise-censorship is a euphemism, at best. Without disguise-censorship, Freud's dream theory is entirely negated because it cannot do what it was intended to do: that is, explain dream bizarreness. Even if one were to admit that dreaming is, in part, motivated by forces that include something like instinctual drives, it is clear on its face that these forces are undisguised and that dream bizarreness is entirely inadequate to the task Freud assigned to it.

Critiques of activation-synthesis, including **SOLMS**'s, have pointed to the tendency of late night NREM sleep to support dream mentation almost as well as REM. The obvious reason for this change is that as the night progresses, the general level of brain activation rises toward waking levels and NREM sleep becomes more and more REM-like. The EEG fluctuates between stages I and II (rather than III and IV). The input-output gates are still closed, and aminergic modulation is still weak enough to allow sleep to continue. Under the circumstances, dreaming is quite likely and, in essence, late night NREM dreaming is the other side

of the sleep onset dreaming coin. But REM sleep is still the optimal substrate for dreaming and that is why its neurophysiology is so informative. In addition, many REM-related physiological events probably occur in NREM sleep without producing the full complement of signs necessary to score REM sleep in traditional methods (Nielsen 1999).

***Can dreaming and REM sleep brain process be dissociated?***
Hoping to sidestep the devastating impact of modern sleep neurobiology upon Freud's dream theory, **SOLMS** and other critics of activation-synthesis have argued that since the forebrain is the acknowledged seat of dream consciousness and since dream consciousness can occur outside of REM sleep, then REM sleep physiology and its brain stem control systems are of no relevance to dream theory. In this view, the forebrain is free to act alone and can produce dreams independent of the activating, input-output gating and modulatory influences of the brain stem.

This argument is fatally flawed in three important respects:

1. The first fatal flaw is that there is now half a century of solid extensive evidence from laboratories around the world that the major determinant of the forebrain's neurophysiological state and of its propensity to dream is the set of brain stem core systems that run from the medulla up through the pons and midbrain to the basal forebrain, hypothalamus, and thalamus. Indeed **SOLMS**'s own speculative neurophysiological model of dream-instigating wishes invokes the dopamine system, which projects from the ventral tegmental area of the brain stem to the mesolimbic reward circuit in the ventral striatum, thalamus, and medial prefrontal cortex.

2. The second fatal flaw is that the brain is instantiating the neurophysiological conditions of REM sleep from the very moment of sleep onset and varying them in intensity continuously throughout the night until the moment of awakening. As their strength flows and ebbs, dreaming becomes more or less probable, more or less intense, and more or less emotional. In other words, REM physiology is relevant to dream generation even in so-called NREM sleep. I say "so-called" because we have shown that REMs occur in NREM sleep at an average level about one-third that of REM. So NREM sleep is a misnomer. It doesn't really exist! All of sleep is REM sleep (more or less). See **NIELSEN**'s target article.

3. The third fatal flaw is that REM sleep, by everyone's account, provides the optimal neurophysiological conditions for dreaming. The specific conditions of REM include cholinergic forebrain activation and stimulation, aminergic forebrain demodulation, and active input-output gating. These account for all the formal aspects of dreaming in a more economical manner than Freud's implausible theory.

As far as the notion that it is dopamine alone and not the mix of noradrenergic, serotonergic, and cholinergic neuromodulators that are critical to the propensity to dream, **SOLMS** is grasping at straws and the wrong straws at that. The medial forebrain lesions that lead to cessation of dreaming interrupt all of the neuromodulatory systems (and many other pathways) linking the brain stem and basal forebrain to the cortex. So why single out dopamine? Presumably because it is known to mediate motivation and reward. But how we get from motivation and reward to Freud's unconscious wishes is not at all clear. **SOLMS**'s choice of dopamine is all the more surprising because it is the one neuromodulatory substance that shows no tendency to change its output over the sleep-wake cycle being as active in NREM sleep as it is in REM as it is in waking (See Hobson & Pace-Schott 1999)! So, at best, dopamine could be as necessary to dreaming as it is to any other activated mental state but it could not possibly be either sufficient or specific enough to account for the distinctive aspects of dream consciousness.

## Dreaming as play

Nicholas Humphrey

*Centre for Philosophy of Natural and Social Sciences, London School of
Economics, London WC2A 2AE, United Kingdom.* **n.humphrey@lse.ac.uk**

**Abstract:** Dreaming can provide a marvelous opportunity for the "play-ful" exploration of dramatic events. But the chance to learn to deal with danger is only a small part of it. More important is the chance to discover what it is like to be the subject of strange but humanly significant mental states.
[REVONSUO]

At a time when theories of dreaming are tending to lose touch with psychological and biological reality, REVONSUO's target article comes as a welcome call for a return to common sense. Dreaming, REVONSUO reminds us, is about having dreams. Dreams tell stories in which the dreamer is an active protagonist. These stories can and often do leave lasting traces on the dreamer's mind. Hence, surely, the way to understand the evolutionary function of dreaming must be to consider the relevance of such stories to the kinds of survival problems that ancient humans had to face.

I have no doubt that this is the right way to go. And, as it happens, in the early 1980s I proposed a theory that is quite similar in spirit to the one described here (Humphrey 1980; 1983; 1986). I began by noting, as REVONSUO does implicitly, that there is an obvious analogy between dreaming and childhood play. Dreaming, like play, allows the subject to simulate his or her own participation in dramatic or dangerous events, without suffering the consequences these events would have in the real world. One of the chief functions of play is to provide an opportunity for the player to gain practice in exercising the relevant physical, intellectual and social skills. So, there is every reason to suppose this is a major function of dreaming too.

Now, REVONSUO picks up on just one aspect of this: he suggests that the main purpose of dreaming is the simulation of environmental threats, so that the dreamer is able to practice making his or her escape. I've no quarrel with this suggestion so far as it goes (and REVONSUO does make a good case for it). But, as a theory of dreams in general, it strikes me as being far too narrow – with regard to what it says both about the kinds of situation that are simulated and about the kinds of learning that take place.

To continue the analogy with play, even though childhood play does of course often centre around imaginary dangers, it is clearly not the case that learning to escape these dangers is play's main, let alone its only, function. Rather, play contributes in a major way to social and psychological development, especially through providing practice in role-playing and empathy.

> Play is a way of experimenting with possible feelings, possible identities without risking the real biological or social consequences. Cut! time for tea, time to go home – and nothing in the real world has changed, except perhaps that the child is not quite the person that he was before, he has extended just a little further his inner knowledge of what it can feel like to be human. (Humphrey 1986, p. 106)

But if this broad-band "sentimental education," as I have called it, is the functional rationale for play, surely we should expect something like it to be the rationale for dreams as well. In my own writings I've stressed in particular the key role that dreams can have in the education of a "natural psychologist" – through introducing him to introspectively observable mental states that are as yet unfamiliar in real life (and possibly beyond the scope of waking play).

> "Dreaming" represents the most audacious and ingenious of nature's tricks for educating her psychologists. In the freedom of sleep the dreamer can invent extraordinary stories about what is happening to his own person, and so, responding to these happenings as if to the real thing, he can discover new realms of inner experience. If I may speak from my own case, I have in my dreams placed myself in situations that have induced feelings of terror and grief, passion and pleasure, of a kind and intensity I have not known in real life. If I did now experience these feelings in real life, I should recognise them as familiar; more impor-

tant, if I were to come across someone else undergoing what I went through in the dream, I should have a conceptual basis for modeling his behavior. (Humphrey 1983, p. 85)

Neither is this mere arm-chair theorising. My interviews with people in psychologically-taxing situations have shown again and again that dreaming is indeed a recognised and valued resource for gaining insight into what it is like to be in another person's place. A young midwife, for example, revealed: "I think most mid-wives dream about giving birth when they start working in maternity units, and it was a fairly common experience among the students that I trained with. . . . I've never myself been pregnant. But my dreams have certainly made me more understanding, more relaxed and more confident in talking to mothers" (quoted in Humphrey 1986).

REVONSUO may object that this is all too rosy. It is all very well for me to point to the ways in which dreams can help with empathy-building and interpersonal understanding, in the relatively secure and sociable world that we now live in. But, for him, the true evolutionary context for dreaming was the harsh world of the Pleistocene, where human life was nasty, brutish, and short – and everyone lived in a constant state of post-traumatic stress.

I would answer that this Hobbesian vision of the EEA is simply much too bleak. Studies of contemporary hunter-gatherers such as the Kalahari Bushmen – those whom Sahlins (1977) has with good reason called "affluent savages" – have shown that, on the whole, their life is (and presumably has long been) remarkably easy, unstressful, and free of danger. In fact the main – if not the only – serious challenges these people face are precisely in the area of their human relationships (family politics, love affairs, status battles, jealousies).

Then why, to end with one of the stronger bits of evidence for REVONSUO's narrow view of what dreams are about, are there so many animal characters in children's dreams? And why, for that matter, so many animals in story books, in the play-room, in Walt Disney cartoons, and so on? What can these animals be doing, if it is not that they represent archaic threats? I believe the truth is that these play-animals are usually just what they seem to the child to be: simple, and indeed highly simplified, proxies for human beings – which, as it happens, are peculiarly well suited to the child's first tentative experiments in empathic projection and in applying a theory of mind. As Levi-Strauss (1962) once put it, animals are "good to think with." But this discussion is for another time.

## New multiplicities of dreaming and REMing

Harry T. Hunt

*Department of Psychology, Brock University, St. Catharines, Ontario L2S
3A1, Canada.* **hhunt@spartan.ac.brocku.ca**

**Abstract:** The five authors vary in the degree to which the recent neuroscience of the REM state leads them towards multiple dimensions and forms of dreaming consciousness (Hobson et al.; Nielsen; Solms) or toward all-explanatory single factor models (Vertes & Eastman, Revonsuo). The view of the REM state as a prolongation of the orientation response to novelty fits best with the former pluralisms but not the latter monisms.
[HOBSON ET AL.; NIELSEN; REVONSUO; SOLMS; VERTES & EASTMAN]

*1. Intimations of pluralism.* HOBSON ET AL.'s overview of REM dreaming as a state specific organization of consciousness represents an important and exciting expansion of Hobsons's previous attempts to synthesize the neuroscience and phenomenology of dreaming, with the potential for a still greater inclusiveness. Their addition of recent brain imaging and lesion studies showing hypothalamic, limbic, right parietal, and secondary visual activation, to his earlier model of pontine generation allows an expanded view of both a bottom-up and top-down REM state. The result is broadly consistent with Freud's suggestion of a "topo-graphical regression" in dreaming, as well as with views of the dreaming process that emphasize its basis in emotional activation (Lowy, Kuiken) and in cross-modal visual-spatial elaborations intrinsic to imaginative forms of dream bizarreness (Hillman, Hunt). It is also consistent with a specific phenomenology of a prototypical or background dream bizarreness related to delirium syndromes, with their shifting mix of cognitive confusion or clouding, emotionality, and intrusive, predominantly visual, hallucinosis (see also Hunt 1982; 1989), as well as separating that state specificity of REM dreaming from the more cognitive ruminations of NREM mentation.

HOBSON ET AL. make just the use of Hunt et al. (1993) for which I had originally hoped, showing that controlling dream bizarreness measures for report length, when novel events actually require more words for adequate description, will falsely dilute the defining feature of dream experience. Their organization of states of consciousness in terms of a three-dimensional space based on activation, source of input, and neuromodulation pathway is a significant contribution in its own right, and usefully differentiates prototypical REM and NREM mentation, as well as atypical variations in dreaming and sleep onset, although the number and level of specificity of the dimensions ultimately needed to classify discrete states of consciousness and forms of dreaming remains open (see Hunt 1982; 1989; Pekala 1991 for more phenomenologically-based attempts).

HOBSON ET AL.'s expanded approach might be further enriched by including the fuller implications of Morrison's (1983) demonstration that the positive activation of the REM state, with its cholinergic basis, motoric paralysis, and vestibular activation, is a prolonged form of the orientation or postural still response to novelty in wakefulness – Pavlov's "curiosity reflex." First of all, it suggests that pontine activation need not be considered as "random," but potentially already primed and/or self organized in terms of recent unassimilated novelty from wakefulness (Hunt 1989). The orientation response model is also consistent with less executive and volitional (higher forebrain) control, since a primary alerting and precognitive response is required in novel circumstances – which in our symbolic species will include not only intense emotionality (limbic) but also the recombinatory basis of creative symbolic operations (Geschwind's parietal zone of neocortical convergence). From this perspective the proper comparison is not REM state with generalized wakefulness but REM state with the cholinergic based orientation response to waking novelty, which might contradict the degree of REM state specificity posited by HOBSON ET AL.

Finally, the orientation response, as the fuller physiological context of the REM state, allows a more ready inclusion of non-normative but widely studied alternative forms of dreaming, not as "dissociations," but as augmentations and intensifications of REM dream cognition. In terms of psycho-physiology this seems clearest in lucid dreaming, which can involve an intensification of phasic REM features and parietal activation, consistent with its fully developed phenomenology of heightened bizarreness, enhanced kinesthetic and vestibular effects, and meditation-like self awareness and visual-spatial phenomena (see Hunt 1989; 1995). A prolonged orientation response to novelty is fully consistent not only with a fluctuating delirium as normative dream bizarreness but also with its potential for reorganization into the more specific forms of novelty "metabolism" found in lucid dreams, the imaginative dreams of such interest to Jung, nightmares, and the creative problem solving dreams documented by numerous scientists, writers, and artists. NIELSEN's term "apex dreaming" captures the way those forms of dreaming actually intensify REM dream features rather than dissociate them and reflect the limbic, vestibular, and parietalspatial activations that HOBSON ET AL. now include within their activation-synthesis model.

NIELSEN makes a plausible case that NREM mentation, especially in its more overtly dream-like aspect, involves brief phasic manifestations of the REM state. The question becomes whether this explanation works as well for more thought-like NREM mentation.

There may be a danger here of over-generalizing the specific form of cortical activation associated with the REM state, an acetylcholine mediated, pontine based orientation response (Morrison 1983), with other state specific forms of cortical activation. I have argued (Hunt 1989) that thought-like NREM mentation may reflect a dysfunctional cognitive activation in sleep, based on a defensive vigilant response to the artificialities and stress of the sleep laboratory setting (consistent with NIELSEN's recent finding of EEG power differences with versus without mentation) and as a subjectively disruptive response to depressive or obsessional issues. Certainly when we awaken in home settings having been "thinking all night" about ongoing waking concerns, it is associated with a sense of having slept poorly.

Such dysfunctional sleep rumination reflects a milder version of W. R. Bion's (1962) comment on some schizoid patients, that they suffer from a sense of not being able to dream, fully sleep, or fully awaken. Lairy's research on intermittent sleep, Lairy et al. 1967 cited by NIELSEN, may constitute a physiological reflection of this transitional state, but the presence of an EEG like that of the REM state, without other phasic REM indicators, need not indicate the actual REM state, since there is more than one pathway of physiological and biochemical activation of the cortex (HOBSON ET AL.). Correlations between thought-like NREM mentation, higher EEG activation, and waking conflict and psychopathology (see Hunt 1989) seem most consistent with a defensive hypervigilence, potentially distinct from the phasic bursts and after-effects of the pontine orientation response associated with more vivid, dream-like experience. NIELSEN, consistent with HOBSON ET AL. rejects a one generator model for all sleep mentation, which would be based on an overly generalized model of cortical activation, but then seems to put forward a one generator model for the physiological basis of NREM mentation. State-specific differences in phenomenology, coupled with the existence of different pathways of physiological activation, might be more consistent with a two generator model for the more thought-like and the more bizarre, dream-like patterns of NREM mentation.

SOLMS's evidence of the loss of dreaming with prefrontal damage, also linked to the loss of interest and affect after leukotomy, and with parietal lesions, also the region posited by Geschwind (1965) as the convergence zone for a cross modal translation capacity central to the symbolic capacity, is invaluable, and indirectly supportive of empirical phenomenologies of dreaming that center on affect and creative recombinatory imagery (bizarreness). However, while consistent with those cognitivists who have long sought to decouple a psychology of dreaming (seen as cortical) from the brain-stem physiology of the REM state, his own separation of the two may be premature.

On the one hand, there is the large literature linking dream bizarreness to pontine and related phasic features of the REM state, with both phenomenology and physiology correlated with levels of creative imagination in wakefulness (see Hunt 1989, for review). Similarly, while prefrontal and parietal mediators of REM dreaming can be artificially separated from brain stem features by lesions, it is important to note that they are normally conjoined in what amounts to a state specific patterning of consciousness in the sense posited by HOBSON ET AL.

Certainly, with SOLMS, dreaming, in the sense of a form of consciousness based on an attenuation of reflective thought and volition and a heightening of narratively organized imagery, can appear in conditions outside the REM state – as in sleep onset dreaming, NREM mentation, suggested dreams in hypnosis, waking daydreaming, and guided imagery. Although all these forms of "dream" will entail higher cortical processes, there would seem to be important differences in their form, consistent with very different pathways of subcortical and cortical activation. While, in one sense, SOLMS can say that the REM state is but one "arousal trigger" for a single dreaming process, the acetylcholine (and possibly dopamine) based pontine orientation response seems to produce a form of dream experience specific to the REM state (vivid and apex dreaming) and to similarly triggered and patterned altered states of consciousness in wakefulness (certain hallucinogenic drugs, visionary states). These seem, with **HOBSON ET AL.,** quite differently organized than most daydreaming or thoughtlike NREM mentation.

**SOLMS** is surely well supported in separating the REM state from a more general cognitive dreaming process, but a further equally supported step would be to distinguish within different forms of dream experience, so that prototypical REM dreaming would be a specific organization of cognitive-affective processes, normally entailing the entire circuit of cortical and subcortical activation that **SOLMS** has so carefully deconstructed through his neurological cases.

### 2. Assertions of monism

**Vertes & Eastman.** Why must the REM state, this exceedingly complex phenomenon of nature, have but one function? **VERTES & EASTMAN** seem to outline all possible skepticisms about memory consolidation research in order to advance their new version of the endogenous stimulation model also held by some of the pioneers of REM state research (Roffwarg) and more specifically developed by Ephron and Carrington (1966). However, Morrison's (1983) research, showing the REM state to be a specially sustained form of the orientation/still response to novelty, affords a view more consistent with multiple overlapping functions for the REM state (and REM dreaming). These include not only the endogenous stimulation model, but the equally plausible evidence for a separate role in fetal and neonatal maturation, as well as memory consolidation – especially for novel stimulation of high emotional significance.

More specifically, **VERTES & EASTMAN**'s demolition efforts with memory consolidation research seem narrowly conceived. The cortical activation of the pontine based orientation response operates from acetylcholine mediated pathways, distinct from the norepinephrine based activation of volitional executive control (Vanderwolf & Robinson 1981). Accordingly, the appropriate comparison for the mnemonic consequences of the REM state would be to the more recognitive encoding functions associated with emotional novelty and stress in wakefulness, which seems more consistent with the actual thrust of much of the mnemonic REM research reviewed, along with the lack of encoding for less vivid dreaming (coupled with life long recall for especially intense dreams; Knudson & Minier 1999), and a diffusely primed background for assimilating recent waking novelty that need not enter dreaming as such, yet is also consistent with empirical research on some unresolved "day residues" appearing in ordinary dreaming (see Hunt 1989).

Finally, to suggest that the high levels of REM sleep in neonates, still higher in the premature, is an endogenous compensation for slow wave sleep is not consistent with the near universality of the REM state late in gestation, the absence of true slow wave sleep until about three months, and the way that facial expressions in the neonatal REM state can be developmentally ahead of those shown in wakefulness – most consistent with a specific cortical maturation function distinct from the ontogenetically later one of endogenous compensation for NREM sleep (see Hunt 1989).

The REM state would seem to be one of the best illustrations for the economy and complex elegance of the natural order in which multiply nested functions can re-use a more basic organismic capacity.

**Revonsuo.** If the basic rule in scientific theory is still maximum parsimony given an unbiased consideration of the full range of evidence, along with at least potential testability, then **REVONSUO**'s model of threat simulation as the core of the REM state, based on an adaptation to the allegedly overwhelming threat situations of early Hominid evolution, ignores equally strong evidence for other forms and functions of dreaming and is as ad hoc, "extra," and inherently untestable as Freud's analogous hypothesis of the phylogenetic "primal horde."

Closer to home of what is directly researchable, threat simulation or nightmare dreaming does also follow as one major form of dreaming, especially prominent under major organismic stress, from Morrison's (1983) demonstration that the physiology of the REM state is identical to that of the orientation/still response to waking novelty. The orientation response as substrate is certainly consistent with nightmares as the predominant form of dreaming under system overload and with anxiety as the predominant dream affect, but it is equally consistent with other forms of dreaming such as lucidity, creative-imaginative dreaming, and the problem solving dreams attested by scientists, inventors, artists, musicians, and writers (see Hunt 1989). Not all novelty is stressful.

To consider only the most skeptical views on the existence of other dream forms, without a similarly close look at one's own all-explanatory model, bypasses the recent tendency of research in both the physiology and psychology of dreaming towards a new pluralism. Empirically then, **REVONSUO**'s use of anthropological evidence from contemporary hunter-gatherers is especially questionable, since in addition to "threat simulation," the literature here attests to the diverse forms of dreaming described in these complexly "dream centered" societies, including lucid, mythological-archetypal, social and personal problem solving, artistic creativity, and diagnosis and healing of illness (Hunt 1989; Tedlock 1987b).

On more theoretical grounds, if the most direct indication we have today of evolutionary threat simulation dreams are post-traumatic and stress based anxiety dreams, then it is difficult to see how our widely described paralyzed fears, slow motion running, and escape tactics based on absurd reasoning could be a rehearsal or simulation of anything adaptive. Pushing the origins of the model back into mammalian evolution seems even more questionable in terms of testability. However, on analogy to his use of hunter-gatherer evidence, my own domestic dogs have not only shown the expected growling and barking in their REM states, but also movements of drinking, sexuality, and tail wagging. Rapid paw movements (RPMs) can of course be as consistent with chase and play as with flight.

The view that nightmares and stress dreams show the essence of all dreaming is like saying that the underlying purpose of the vestibular system, with its compensatory eye movements restoring postural balance, is nausea and vomiting, because that is what happens when the system is overloaded in extreme dizziness. While vomiting is the extreme response, the predominant functioning of the vestibular system is to restore equilibrium as rapidly as possible. Helpless vomiting is no more "adaptive" to a vertigo sufferer climbing a tree than is the confused, panicked running we can do in our nightmares. Vestibular response, like REM sleep, is a complex multi-layered organismic process that can a final common pathway for potentially very different functions.

Empirically, it seems more and more that there is no single essence or core function to either dreaming or the REM state. They are complex and multiple, though with every possibility of our tracing the interrelations and interactions of their several dimensions once that complementarity is accepted.

# The interpretation of physiology

Barbara E. Jones

*Department of Neurology and Neurosurgery, McGill University, and Montreal Neurological Institute, Montreal, Quebec H3A 2B4 Canada.*
**mcbj@musica.mcgill.ca**

**Abstract**: Not at all self-evident, the so-called isomorphisms between the phenomenology and physiology of dreams have been interpreted by Hobson et al. in an arbitrary manner to state that dreams are stimulated by chaotic brainstem stimulation (an assumption also adopted by Vertes & Eastman). I argue that this stimulation is not chaotic at all; nor does it occur in the absence of control from the cerebral cortex, which contributes complexity to brainstem activity as well as meaningful information worth consolidating in the brain during sleep.
[HOBSON ET AL., VERTES & EASTMAN]

HOBSON ET AL.'s presentation of the isomorphism between REM sleep physiology and dreaming, has over interpreted the physiology. The physiology does not disprove any more than it proves, either Freud's theories (as put forward in *The interpretation of dreams,* 1900) or other more recent interpretations, to the effect that dreams originate in the cortex and contain highly meaningful information. I take issue with two fundamental points.

**1. Chaos reigns in the brainstem and determines the dream.** I do not know of any physiological evidence that the brainstem activation and stimulation of the forebrain is chaotic and would thus impose a chaotic influence on the cortex in dreams. On the contrary, circuits within the brainstem, as in the spinal cord, are highly ordered, so that specific motor patterns such as locomotion, chewing and vestibulo-ocular nystagmus, are generated there in a repetitive, rhythmic, and highly predictable manner. Complex organized behaviors are also generated in the brainstem, including sexual and rage behaviors that persist in decerebrate animals. Indeed, these are the very behaviors that are often unmasked in REM sleep without atonia and dreams. In addition to being highly organized behaviors, they are also instinctual and highly motivational, perhaps stimulating the wishes that emerge in dreams (see Jones 1991).

**2. The cortex has no control over the brainstem and the dream.** I do not know of any physiological evidence that the cortex has no control over the brainstem or over the central activities of dreams. On the contrary, corticofugal outputs reach the entire brainstem as well as the spinal cord, influencing the very neurons shown to be critical for the initiation and maintenance of REM sleep in the pontine reticular formation. Moreover, elimination of the corticofugal impulses by lesions of the cortex in cats were shown in Jouvet's laboratory to result in a complete impoverishment of both rapid eye movements and PGO spikes, reducing them to the very stereotypic, highly ordered and hence low-in-information-content pattern of purely brainstem driven activity (Jouvet 1975). The cortex thus appears both to control and introduce complexity to the brainstem activity and undoubtedly to dreams, which can accordingly also contain highly meaningful information.

Indeed, I might posit that the physiology of REM sleep provides considerable support for Freud's basic assumptions, according to which instinctual and highly motivational impulses arise from the brainstem and are in turn worked upon by the cortex where condensation, displacement, and symbol formation may control the continued activity of the brainstem and provide the complex and seemingly bizarre, though meaningful, content of dreams (Freud 1900).

VERTES & EASTMAN provide overwhelming evidence against memory consolidation in REM sleep. They also go on to provide convincing evidence for a critical role of theta in memory consolidation during waking. Theta occurs during active and attentive waking but also (and in the most robust, continuous way) during REM sleep. Yet these authors interpret this physiology differently during REM and waking, presenting theta as a mere epiphenomenon in REM sleep, because it would have no functional value occurring in association with the chaotic state stimulated by the brainstem in REM sleep. Either theta plays no role in memory consolidation or it does so in REM, as it does in waking.

I would agree with Winson (1990) that theta and REM are associated with species-specific highly motivational and orderly instinctive behaviors and underlying processes that are important for survival (see Jones 1998). In this framework, theta during REM, as during waking, is associated with maximizing, reinforcing, and potentiating the neural links underlying these behaviors as well as with reforming them in relation to the organism's changing world over a lifetime. Evidence to date has indicated that REM sleep is involved in consolidating procedural learning, a process that does not require or usually involve conscious awareness, (e.g., in walking, running or skiing). Such procedural learning can be highly practiced, with more and more efficiency, hence rapid responses and behaviors.

REM sleep in the fetus may provide procedural learning and preparation prior to birth for performing important behaviors, including locomotion and flight, by exercising the specific circuits. Such may be the case in the wildebeest, who begins to walk and then to run minutes after emerging from the womb in the dry season in order to escape predation and to begin a long migration of several hundred miles with its mother in search of a water supply. REM sleep during that migration may then facilitate the learning of speedy adjustments to changing terrain and rapid escape from real predators. Theta would be present during the waking experience and the dreamed replay to consolidate and potentiate the new sensory-motor associations and reactions with the old.

Such a process might also help in the learning and honing of nonessential skills by humans, such as more smoothly and rapidly negotiating the mogulled terrain on the second day down the ski slope following practice during dreaming through the intervening night. The importance of REM sleep for such highly practiced and unconscious processes and skills might be very difficult to document in the laboratory. Yet, as discussed by VERTES & EASTMAN, a few studies have recently shown effects of REM sleep on consolidation of procedural tasks. As in these experiments, the challenge has been to devise experiments and measures that are sensitive enough to reveal and confirm the full role of REM sleep in these processes. This role, though it may seem inessential, may provide a considerable advantage to the organism in strengthening associations and perfecting behaviors that have been important for the survival of each species in its evolutionary past.

# The "problem" of dreaming in NREM sleep continues to challenge reductionist (two generator) models of dream generation

Tracey L. Kahan

*Department of Psychology, Santa Clara University, Santa Clara, CA 95053.*
**tkahan@scu.edu**

**Abstract:** The "problem" of dreaming in NREM sleep continues to challenge models that propose a causal relationship between REM mechanisms and the psychological features of dreaming. I suggest that, ultimately, efforts to identify correspondences among *multiple* levels of analysis will be more productive for dream theory than attempts to reduce dreaming to *any* one level of analysis.

[HOBSON ET AL. ; NIELSEN]

NIELSON's "position," and a core issue in the debate between 1-gen and 2-gen theorists, turns on whether the differences in REM and NREM dreams are essentially *quantitative* (i.e., "relative"), or *qualitative* (i.e., "absolute"). The basic logic is that if REM and NREM dreams differ quantitatively, then the same qualities should describe both REM and NREM dreams, but to different degrees; this situation is consistent with 1-gen models (also see Foulkes & Cavallero 1993, p. 10). However, if REM and NREM dreams differ qualitatively, in that the defining qualities which characterize REM dreams (e.g., emotionality, vividness, bizarreness) do not characterize NREM dreams, then a 2-gen model is needed to explain these "absolute" differences (e.g., Hobson 1988b; Koukkou & Lehmann 1983).

It is notable that the debate between 2-gen and 1-gen theorists about the proposed differences between REM and NREM dreaming, as well as the extent to which the psychological experience of dreaming can be explained by neurophysiological mechanisms, is mirrored in the debate about whether sleep cognition is "discontinuous" or "continuous" with waking cognition (i.e., 1-gen model); (for recent discussions of this essentially parallel debate, see Purcell et al. 1993). For example, recent efforts to compare the incidence of high-order cognitive skills, such as self-reflection, decision making, and meta-attention, across REM dreaming and waking suggest that although these skills may be more characteristic of waking than dreaming, these metacognitive skills are, nevertheless, frequently associated with the recall of dream experiences (see, especially, Kahan & LaBerge 1996; Kahan et al. 1997). If one assumes a priori that dreaming involves, for example, a suspension of self-reflection (e.g., Hobson 1988b; Koukkou & Lehmann 1983), then one is not inclined to actively test this hypothesis by comparing the incidence of such skills across sleep and waking (also see Kahan 1994; Kahan & LaBerge 1994; Purcell et al. 1993 for discussion of this issue). In fact, many of the "discontinuity" theorists would claim that differences in cognition across waking and sleep are the result of the same physiological/psychological isomorphism that explains the differences in cognition across REM and NREM sleep (see **HOBSON ET AL.**).

In his balanced review of the evidence for 2-gen versus 1-gen models of dream generation, **NIELSEN** discusses the long and controversial research lineage which, ultimately, shows that REM is *not* required for vivid dreaming to occur, although vivid (and "apex") dreaming may occur *more often* in REM sleep. **NIELSEN** discusses a number of studies, some of them his own, which demonstrate that Stage 1 sleep at sleep onset, late-night Stage 2, and occasionally Stage 3/4 sleep, are associated with vivid dreaming that *cannot* be distinguished from vivid REM dreaming. Even **HOBSON ET AL.**, the consummate "2-gen" theorists, explicitly noted this fact when describing the early dream research: "Reports *qualitatively* indistinguishable from dreams were obtained from Stage 1 sleep at sleep onset, a phase of sleep without sustained eye movements . . . and some of the reports from non-REM sleep were indistinguishable *by any criterion* from those obtained from post-REM awakenings" (1988b, pp. 142–43, emphasis added). **NIELSON** also acknowledges the recent and persuasive neuropsychological evidence that shows an absence of dreaming even though REM sleep is intact (Solms 1997a). In other words, REM is neither necessary nor sufficient for vivid dreaming to occur (Cavallero & Cicogna 1993; FouIkes 1985; 1993c; Solms 1997a; this volume). This body of evidence is clearly contrary to the reductionist, 2-gen models (e.g., **HOBSON ET AL.** target article).

In light of his discussion of the dissociability of REM and dreaming, it is rather curious that **NIELSEN** ultimately concludes that the evidence supports a 2-gen model because of "residual *qualitative* differences" across REM and NREM dreams (sect. 2.8.4, emphasis added). Unquestionably, vivid dreaming occurs more often during REM than NREM sleep. Also, the reported incidence and qualities of dreaming sampled from NREM sleep is more unstable, in that the similarity between NREM and REM dreaming is influenced by the sleep stage and time of night when mentation is sampled (Kondo et al. 1989), by whether the participants are light/heavy sleepers (Zimmerman 1970) or high/low dream-recallers (Moffitt et al. 1972), and even by the theoretical predisposition of the investigators (Hermann et al. 1978). Critically, however, and as **NIELSEN** aptly points out, the reported frequency of NREM dreaming is intimately tied to how dreams are defined. As yet, there is no consensus among dream theorists on the formal features that define dreaming, and this seriously complicates the task of comparing across studies of purported sleep-stage differences in dreaming. It is not unusual for investigators to provide their own preferred list of the defining features and then proceed to compare REM and NREM mentation on these dimensions (e.g., Hobson 1988b; Koukkou & Lehmann 1983). Clearly, *which* qualities are considered the defining (or "formal") characteristics are often theoretically motivated and drive the comparison between REM and NREM dreaming, as well as, for that matter, comparisons between dreaming and waking cognition.

This question of what features are necessary to judge sleep mentation as a "dream" is central not only to comparisons of dreaming across different sleep stages, but also to theories claiming a formal isomorphism between sleep neurophysiology and the "formal" features of dreaming (e.g., **HOBSON ET AL.,** this volume; Koukkou & Lehmann 1983). If particular features of dreaming (e.g., bizarreness, emotionality, self-representation) are assumed to be *determined by* particular REM mechanisms, then agreement is

needed on just what features of dreaming are the consequence of REM neurophysiology. Further, if a formal isomorphism exists between REM and these dream features, then NREM mentation should not evidence these features; in other words, mentation during REM and NREM should differ in *kind* (i.e., qualitatively) rather than simply in *amount* (i.e., quantitatively)!

Unfortunately, **NIELSEN**'s efforts to "reconcile" the 1-gen and 2-gen models by adding to a 1-gen model the assumption of physiological isomorphism only complicates matters. The "one" system that **NIELSEN** proposes is the REM "system." His hypothesis is that variations in the incidence of dreaming occur in direct proportion to the involvement of REM mechanisms; whether "overt" in unequivocal REM sleep or "covert" in NREM sleep that seems to "mix" in some features of REM. NREM dreaming, then, is due to a (covert) carry-over of REM features into NREM sleep. However, I found this logic suspect; if a *subset* of REM mechanisms is observed in NREM sleep (a "mixed state"), and these REM mechanisms are responsible for NREM dreaming, then doesn't this first call into question the entire enterprise of sleep staging and, second, precisely *which* REM mechanisms are "responsible" for dreaming? **NIELSEN** does offer persuasive evidence that sleep stages are not always as discriminable as the literature sometimes implies; rather, sleep staging can be difficult and sometimes results in "mixed" states not clearly identifiable as REM or NREM. **NIELSON**'s "mixed states" model could account for some of the variability in the purported frequency of dream recall from NREM sleep, but this variability could also be accounted for by the sampling and individual differences factors mentioned earlier. More importantly, **NIELSEN**'s model does not offer a full accounting of the proposed *qualitative* differences that he considers pivotal to supporting 2-gen over 1-gen models. It is noteworthy that **HOBSON ET AL.** (this issue) offer a similar argument about lucid dreaming; that it occurs in a REM state that "mixes" in waking. However, Stephen LaBerge's work is contrary to this claim; he and his colleagues found that lucid dreaming tends to be associated with *intensified* REM, in which there is heightened brain activation and heightened muscle suppression (of the H-reflex)(Brylowski et al. 1989; LaBerge 1990).

Throughout this *BBS* special issue, an implicit question is: at what "level" dreaming is to be investigated and, eventually, explained: psychological (in terms of the cognitive skills of dreaming and the psychological or developmental conditions under which these skills manifest); neuropsychological (in terms of the consequences of brain damage for dreaming and other complex cognition); or neurobiological or evolutionary (in terms of the neurobiological evolutionary mechanisms that subserve the construction of complex cognitive experiences). Ultimately, we will need all of these – and other – levels of analysis to adequately account for what Harry Hunt called the "multiplicity" of dreaming (Hunt 1989).

At this point, it seems most constructive to consider all of the correlates of dreaming (and waking) cognition. Certainly, our neurophysiology and developmental stage constrain our dreaming and other cognitive experiences, but it seems to me it is the potential of the human cognitive system to generate *meaning* and to represent the self-in-world (e.g., Globus 1987) that makes "dreaming" possible. We thus need to understand the phenomenology of dreaming experience and, thus, must be willing to admit as data first-person, self-reports that describe not only the narrative of the dream events, but also the qualitative aspects and the personal or transpersonal meanings ascribed to that experience. In addition, we need to understand the psychological circumstances associated with dreaming, including the cognitive, developmental, motivational, and cultural conditions. And, we need to understand the neurobiological correlates of dreaming as well as what happens to dreaming (and its component cognitive skills such as memory, attention, imagery, language, organization, consciousness) when the brain is damaged. The puzzle of dreaming – how dreams are generated, why we dream, and how dreaming experience is related to waking experience – is complex and multidimensional, rather like a 3-D chess game; no one level of analysis is sufficient to solve the entire puzzle and all levels are inter-related. To attempt to reduce the multiplicity of dreaming to one "ultimate" level of analysis, whether neurobiological or phenomenological or psychological is to miss crucial puzzle pieces provided by the other levels of analysis and, hence, never to solve the puzzle and understand its intricacies.

# A new approach for explaining dreaming and REM sleep mechanisms

Amina Khambalia and Colin M. Shapiro

*Department of Psychiatry, University Health Network, Toronto Western Hospital, Toronto, M5T 2S8, Canada.* **amina_khambalia@hotmail.com**

**Abstract:** The following review summarizes and examines Mark Solms's article Dreaming and REM Sleep are controlled by different brain mechanisms, which argues why the understanding of REM sleep as the physiological equivalent of dreaming needs to be re-analyzed. An analysis of Solms's article demonstrates that he makes a convincing argument against the paradigmatic activation-synthesis model proposed by Hobson and McCarley and provides provocative evidence to support his claim that REM and dreaming are dissociable states. In addition, to situate Solms's findings in concurrent research, other studies are mentioned that are further elucidated by his argument.

[SOLMS]

SOLMS argues against the activation-synthesis model proposed by Hobson and McCarley. Refuting their theory that dream imagery is stimulated from the pontine brainstem, SOLMS claims that REM and dreaming are dissociable states controlled by different brain mechanisms. According to SOLMS, there is a separate forebrain mechanism responsible for turning on dreaming aside from that of the "REM-on" mechanism. Throughout the paper SOLMS cites many studies, including his own, as counter evidence to the Hobson and McCarley model. For instance, in section four, SOLMS refers to Vogel's "An alternative view of the neurobiology of dreaming." Vogel's paper is a direct response against Hobson's and McCarley's non-Freudian theory of dreaming, the activation-synthesis model. In his paper Vogel states that "all the evidence which modern neurophysiology can provide does not and cannot refute the Freudian dream hypothesis" (Vogel 1978a, p. 1531). In terms of Freudian scholarship, SOLMS himself has compiled several of Freud's articles, and has also published *The neuropsychology of dreams: A clinico-anatomical study*, which he frequently cites in his article (Solms 1997a).

In a well-organized format, SOLMS directs the reader through his developing argument. In his abstract, SOLMS's statement that the present assumption that REM sleep and dreaming are "physiologically synonymous" is false and in need of revision. Although the forebrain dopaminergic mechanism induces dreaming, the part of the brainstem controlling REM is only one of many activation pathways. According to SOLMS, the assumption that dreaming is an "epiphenomenon of REM sleep" was based on the finding that most dreaming occurs in REM sleep. This observation led to the assumption of dreaming and REM sleep being associable states; a conception that SOLMS intends to prove is incorrect.

Writing in a concise and logical manner, SOLMS develops his argument by reviewing past models and theories, and then suggesting new evidence, which challenges previous understandings and theoretical conclusions. For the past two decades Hobson and McCarley's reciprocal interaction model that states "cholinergic brainstem mechanisms cause REM sleep and dreaming" has been accepted by the neurobiology community at large. However, the hypothesis relies on the presupposition that REM sleep is not controlled by forebrain mechanisms, a point which proves essential to SOLMS's main argument, especially in section 6 where he discusses how forebrain lesions cause dreaming cessation and yet "spare the brainstem." Discussion in section 4 is dedicated towards presenting contradictory evidence that "dreaming is generated by the unique physiology of the REM state." SOLMS's suspicion of the activation-synthesis model is evident in section 3 when he states that Hobson's latest developments of the model are "admittedly speculative." Furthermore, SOLMS's confusion stems from the virtually unrecognized discovery that not all dreams are produced by the same brainstem mechanisms as REM sleep. SOLMS believes that the lack of acknowledgment of the latter is due to Hobson modifying his model. Although Hobson disclaims the previous physiological link between REM sleep and dreaming, he maintains an anatomical link between the brainstem and dreaming, a postulate, which, SOLMS feels, confirms the false conception of REM sleep and dreaming as associable states.

Further evidence suggesting that REM sleep and dreaming are controlled by different mechanisms is provided in section 4 which gives examples of when dreaming occurs in NREM sleep, and in section 5, which discusses research that demonstrates dreaming is unaffected by brainstem lesions. Section 5 however, as SOLMS points out, has unconvincing data because brainstem lesions large enough to "obliterate REM usually render the patient unconscious," a fact also mentioned by Hobson et al. (1998b). Consequently, it is unlikely that information refuting a correlation between dreaming and REM brainstem mechanism can be obtained using lesion data, a problem that SOLMS solves in section 6 by presenting evidence of forebrain lesions that eliminate dreaming without affecting the pontine brainstem. Citing a long list of references supporting his claim, SOLMS mentions that 108 cases have demonstrated a correlation between dreaming cessation and forebrain lesions confirmed by research using the REM awakening method and morning recall questions. In addition, the lesion site is precisely the same region as targeted in modified (orbitomesial) prefrontal leukotomy, a surgical process that results in complete or nearly complete loss of dreaming in 70–90% of recorded cases.

In section 7, SOLMS continues to argue that dreaming is actively generated by a forebrain mechanism. He states that the target zone of prefrontal leukotomy is the white matter of a ventral mesial quadrant of the frontal lobe. Transection of this area causes lack of initiative, or adynamia, reduced imagination or ability to plan ahead, and reduces positive symptoms of schizophrenia. In answer to the question of how any of this relates to dreaming, SOLMS cites three observations: features of schizophrenia have long been equated with dreaming, adynamia is a typical correlate of loss of dreaming, and the chemical activation of this circuit stimulates excessive, vivid, and unusually frequent dreaming. The chemical activation does not affect the intensity, duration or frequency of REM sleep, whereas drugs blocking the circuit eliminate excessive, vivid, and particularly unusual dreams. Further evidence is also mentioned relating nocturnal seizures to dreaming, events associated with the forebrain independently of REM sleep. SOLMS presents interesting findings that definitely support a fresh hypothesis that is still relatively speculative and in need of more evidence. In section 8, SOLMS returns to the activation-synthesis model and its explanation of dream imagery as a product of passive forebrain synthesis of brainstem impulses. SOLMS argues against the postulate by citing clinicoanatomical studies performed by himself and Braun et al. (1997). These findings suggest dreaming is caused by specific forebrain mechanisms, therefore the activation-synthesis model is incorrect because dreaming is not "isomorphically correlated with non-specific brainstem activation of perceptual and motor cortex during REM sleep." However, SOLMS does point out a disparity in the results between Braun et al.'s new functional imagery and his own clinicoanatomical data. In some functional imaging studies there is an involvement of the pontine brainstem during dreaming (Braun et al. 1997; 1998). SOLMS attributes the fact that "REM sleep was equated with dreaming in the imaging studies," owing to the focus on a comparison between dreaming and non-dreaming in NREM only. Hence, SOLMS feels the inclusion in some of the functional imagery studies of the pontine brainstem in dreaming accounts for the disparity between the imagery studies and the clinicoanatomical data.

Although **SOLMS** presents interesting findings that definitely support his relatively fresh hypothesis, his argument is still fairly speculative and in need of more subsidiary evidence. In his favor, **SOLMS** does suggest a further study to increase understanding: examining the dreaming brain during sleep onset, or during awakening. Most importantly, **SOLMS**'s article represents an alternative argument to the long accepted activation-synthesis model. Thus, **SOLMS** reminds his readers that we must constantly question and re-evaluate our knowledge and challenge our theories. Models by definition are simply representations and are liable to change and constant improvement. In effect, **SOLMS**'s article not only presents provocative information but it also encourages the reader to invite new ideas and consider changing what we presently accept as the truth. In considering **SOLMS**'s cogent argument, other findings related to dreaming, REM, and brain mechanisms are further elucidated. There are three studies concerned with sleep, which can be related to dreaming and REM as dissociable states. First, the review by Razmy and Shapiro (2000) discusses how dreams appear to occur outside of sleep in Parkinson's disease patients. The paper states that although less time was spent in sleep, "48% of PD patients presented altered dream phenomenon, while 26% displayed hallucinations" (Razmy and Shapiro 2000, p. 5). The distinction between dreaming and sleep in PD patients supports **SOLMS**'s claim that dreaming and REM can function independently. Second, a paper by McCarley and Sinton (2000) suggests there are different types of sleepiness in REM versus NREM. Moreover, efforts are being made to develop clearer neuropsychiatric elements of fatigue, "the possibility of multiple forms of fatigue being identified and defined seems likely" (Shapiro 1998). The plausible inference that different neurological circuits account for the distinctions between REM and NREM sleepiness, and different types of fatigue, makes it easier to conceptualize REM and dreaming as separate states.

A third clinical study was carried out on the airway resistance in adult asthmatics and controls in REM and non-REM sleep demonstrated that the "REM (dreaming) stage of sleep may be associated with bronchoconstriction" (Shapiro et al. 1986). In view of these findings, which suggest REM sleep "has a direct effect" on airway tone, and the observation that the content of the dream was significant in this regard, the existence of independent pathways and brain mechanisms performing various functions is conceivable.

Reference to these studies in relation to the differentiation between REM and dreaming as separate functioning states is both informative and enlightening. We anticipate that future investigations may further substantiate **SOLMS**'s argument and provide greater impetus to clarifying the neurology of dreaming.

# Dreaming has content and meaning not just form

Milton Kramer

*Clinical Professor of Psychiatry, School of Medicine, New York University, New York, NY 10024.* **milton.kramer@worldnet.att.net**

**Abstract:** The biological theories of dreaming provide no explanation for the transduction from neuronal discharge to dreaming or waking consciousness. They cannot account for the variability in dream content between individuals or within individuals. Mind-brain isomorphism is poorly supported, as is dreaming's link to REM sleep. Biological theories of dreaming do not provide a function for dreaming nor a meaning for dreams. Evolutionary views of dreaming do not relate dream content to the current concerns of the dreamer and using the nightmare as the paradigm dream minimizes the impact of poor sleep on adaptations.

[HOBSON ET AL.; NIELSEN; REVONSUO; SOLMS]

The current debates in the study of dreaming are reviewed from one perspective in this interesting series of articles. The focus is on the form and not the content of the dream with only one ex-

ception. The explanations for the form of the dream are biological rather than psychological. The explanations are essentially of a physiologically reductive nature and, in one case, in the form of an extreme brain-mind isomorphism. If a function for the dream experience is entertained, it is a biological or evolutionary one rather than a current concern or immediate psychological one. The possibility that the dream has meaning is either not addressed or rejected.

There is much to applaud and even more with which to agree in these highly informative essays, but the best use, I believe, that I can make of the space available to me is to focus, so to speak, on the other side of the debate and to raise question about various points raised in these reports.

**HOBSON ET AL.** present a magisterial integration of evidence attempting to explain the neuroscience basis for conscious states; based on demonstrating that waking, REM, and non-REM are physiologically different and that dreaming is REM based. As they appropriately point out, they do not address the "hard problem" of how consciousness could arise from a neuronal system. It is in this transduction that the explanation for consciousness lies. It is this crucial step that McGuinn (1999) believes is unlikely to be solved. A correlation, even if it could be shown between brain state and mental state, would not be an explanation. How does one go from neurons firing to the dream experience of being attacked, and what are the mechanisms to achieve this transformation? This crucial step is missing and seriously undermines the attempt to explain the mind in terms of the brain and even further raises questions about mind-brain isomorphism, even if it could be demonstrated.

There certainly are problems with trying to capture an experience such as dreaming in a verbal report, particularly as the experience occurs during one state and is reported in another (Winget & Kramer 1979). At this point in time, is there a meaningful substitute? That a picture maybe worth a thousand words does not necessarily negate the value of even a 7-word dream report. Certainly, picture drawing has its own limitations. The demand characteristics of probing for additional dream content adds another layer of problems to establishing what a dream experience was "really like." Arguing by analogy that if the physiology is the same in the adult and the infant, the experience must be the same is questionable. Because the authors can imagine the infant experiencing hallucinated emotions and fictive kinesthetic sensations does not demonstrate the relationship, it postulates the answer instead.

Does changing the setting by collecting dreams at home somehow move us closer to the " natural" dream experience? Granted we (Piccione et al. 1976) found little evidence of adaptation to the laboratory over 20 nights of dream collection from each REM period, the study, however, illustrates that extended laboratory studies are indeed feasible, have been done, and provide a wide ranging sample of the reported dream experience. The question remains whether evidence of the experimental situation appears in the dream collection experience at home? And if it does initially, does it adapt out over time? **HOBSON ET AL.** are of the opinion that state specific changes in brain function REM to non-REM virtually guarantee concomitant changes in mental functioning, and that the difficulty in demonstrating such changes is because of the inadequacy of our psychological measurements to identify the changes. The conclusion that the mind and brain have nothing to do with each other they find unacceptable. In searching for an explanation for aspects of the dream experience in a psychological framework, it does not follow that one has to deny a relationship between mind and brain, only that it may not be useful for a particular task.

The so-called inadequate psychological methodologies are more predictive night to night than are the more adequate, by implication, physiological ones (Kramer & Roth 1979). And the dream reports of a given night are related one to another so that one night's dreams are distinguishable from another (Kramer et al. 1976) which has not been shown for the physiological processes REM and non-REM). Conclusions are not a priori acceptable or unacceptable, but might better be based on the evidence available. It is not that mind and brain have nothing to do with each other but rather that the one to one correspondence may not be present and to assume it without evidence begs the question.

That the differences between REM and non-REM are not absolute raises serious question about the centrality of the state differences and the hypothetical mind-brain isomorphism. If the answer is that the significant aspects of the REM non-REM difference are not contained within the boundaries of their current definitions for example, PGO (ponto-geniculate occipital) waves in non-REM precede REM and continue after REM into non-REM and this then accounts for dream content in non-REM sleep. Then the shortcoming in demonstrating isomorphism is limited to the physiologic measurements. Foulkes may have early on attempted to show mind-brain isomorphism (Molinari & Foulkes 1969) but he was unable to replicate the work (Foulkes & Pope 1973). His efforts offer little or no support to the isomorphic position.

**HOBSON ET AL.** are surprised that direct representations of presleep experiences are not incorporated into the dream. The expectation is that such representation should occur. They see the same being true even for emotionally salient events not being incorporated. We found, as noted above (Piccione et al. 1976), that references to the laboratory continue to appear in the dream reports over 20 nights without much of an overall decrease. We found a significant relationship between the content of verbal samples before and after sleep and the, content of the intervening night's dreams (Kramer et al. 1981). We found that the interpersonal situation in which the dream was experienced and reported influenced the content of the dream report (Whitman et al. 1963a; 1963b; Fox et al. 1968). Pre-sleep themes are connected to the dreams that follow and the night's dreams are connected to the themes obtained after waking in the morning (Kramer et al. 1982). Traumatic experiences are reflected in nightmares but not as simple reproductions (Kramer et al. 1984). As memory is constructive rather than a reproductive enterprise, the expectation that waking events will be simply reproduced in dreams may be unrealistic.

It is unfortunate that a work that Freud never published and never wished to have published (a project for a scientific psychology; Freud 1895) is used as a source for his scientific thinking. Freud never declared brain science off limits as is reflected in *The interpretation of dreams* (Freud 1900/1955) in his discussing the stimuli and sources of dreams:

> Even when investigation shows that the primary exciting cause of a phenomena is psychical, deeper research will one day trace the path further and discover an organic basis for the mental event. But if at the moment we cannot see beyond the mental, that is no reason for denying its existence" (pp. 41–42).

The position that the biology sets limits on the psychology needs to be expanded so the psychology sets limits on the biology as the physiological reductionist is trying to explain a psychological experience. They need to account not just for the formal aspects of dreaming but the content as well. Whether the subject of the hallucination is an animal or my father needs to be explained.

The critique that **HOBSON ET AL.** provide of lesion and functional neuroimaging studies is most helpful. It serves to dampen the view that these studies provide the answer to understanding dreaming. The recognition that subjective states, emotion, may be the primary shaper of dream plot lines links their work to content focused dream studies (Kramer 1993). It may not be the emotion in the dream, but rather the emotional state prior to sleep which relates to the content of the night's dreams. Waking mood is related to the content of dreams, not to the latency to REM or the amount of REM sleep (Kramer et al. 1972). The non-REM aspects of sleep are related to the sleepy aspects of subjective state and the unhappy aspect of mood is related most particularly to the content of the dream.

Brain state modulation by various neurotransmitters contributes significantly to our understanding of REM-non-REM waking differences, but changes in norepinephrine, serotonin, and acetyl-

choline have not been linked to the experience of dreaming and their role in dream formation is not universally accepted (see target article by **SOLMS** who focuses on the role of dopamine in dreaming). The excessive emphasis on activation from the brain stem in earlier versions of the activation-synthesis hypothesis neglected the role of the higher centers for example, the cortex, in dream formation. The recognition that dreaming is not just a bottom-up process but has a top-down component was, I believe, a necessary corrective. The recall of dreaming is impaired by brain damage and aging in which the index of the damage is a gross measure of impairment (Kramer et al. 1975). Recall of night-reported dreams the next day follows the rules of classical memory theory; primacy, recency, and dramatic intensity are predictive of recallability (Trinder & Kramer 1975).

In their current version of the activation-synthesis hypothesis the brain fits the image to the affect. We have suggested that in the activated state the brain-mind responds to the pre-sleep affective state with an emotionally determined image (Kramer 1993). It is this view that is closest to this version of the activation-synthesis hypothesis. As dream emotion is not an inevitable accompaniment of dream reports (Strauch & Meier 1996), it is unlikely that dream emotion is the shaper of dream plots however the pre-sleep subjective state may be the shaper. The emotional numbing in dreaming fits well with the Freudian idea of the dream as an emotional dampener in its sleep protecting function. Patients who have had an insult to the brain and report no longer dreaming also report poorer sleep (Solms 1997a) and to my view of the dream as a selective affective regulator (Kramer 1993). The consequence of the dream experience is emotional numbing and not the cause of the experience or the dream plot.

In accounting for the formal features of the dream experience, **HOBSON ET AL.** neither address nor do they provide a way of understanding the gender, age, race, social status and marital status, differences in dream content (Kramer et al. 1971; Winget et al. 1972; Winget & Kramer 1979) or the dream content differences in various psychopathologic conditions (Kramer 2000). These differences are granted content but not form differences, and biologic approaches have not been invoked to account for these differences. The recognition of the need for a top-down aspect for the activation-synthesis hypothesis and recognizing the role of emotion in dreaming provides an overlap that may serve as a link between biological and psychological approaches to dreaming.

The burden remains on investigators, like **HOBSON ET AL.** to demonstrate the isomorphism they postulate and provide the transduction mechanisms that translate neural activity into dream consciousness. An advance in techniques and in conceptualizations of a high order will have to take place first. Whether the suggestion that REM sleep as currently defined needs to be redone in order to account for dreaming outside of REM as **NIELSEN** suggests remains to be tested. Without a marker for PGO waves in the intact human this may not yet be possible. The scanning hypothesis, as they note, remains an open question.

The AIM model calls attention to the three interacting aspects of the theory, activation, input, and modulation. These are the three aspects of the biology of dreaming that they feel are necessary to understand the dreaming process. The visual representation of AIM is interesting but I have trouble using it in a predictively testable manner, particularly as it does not deal with the content of the dream or the meaning of the dream. The model is used post-dictively and the multiplication of concepts that occurs when aspects of the model need to be split to fit the data weakens its possible explanatory power.

**SOLMS**'s elegant clinicoanatomical analysis of the relationship between REM sleep and dreaming, the hypothesis central to the activation synthesis theory of dreaming, points out that REM can occur without dreaming, and dreaming can occur without REM. He rejects the idea that dream imagery is isomorphically the consequence of activation of the perceptual and motor cortex. Dreaming is a response to activation, but not specifically from the brain stem; patients with forebrain lesions that spare the brain stem report the cessation of dreaming but continue to have REM sleep.

In addition to delineating the brain areas needed to generate dreaming, SOLMS is trying to explain the formal characteristics of dreaming from a biologic framework, but he does not deal with the specific content of dreams. Dreaming is the consequence of various forms of cerebral activation and occurs only if the activation engages the dopaminergic circuits of the ventromedial forebrain. He makes no statement as to the content of dreams or the function of dreaming either psychological or biological, although he does note that those who no longer report dreaming do not sleep as well as those who do. His contribution is a central critique of the REM = Dream equation and an elaboration of a dream formation mechanism that is not brain stem based but is cortical in nature, completely top down. Nevertheless, it is one in which the biology drives or determines the psychology and is therefore reductionistic in nature. As with all biological theorists SOLMS is unable to explain the transduction from neural activity to mental activity, nor can he account for the content of dreams, nor does he provide a basis to establish either a function for or the meaning of the dream experience. He does note that the loss of dreaming supports the Freudian position that dreaming protects sleep, however, he does not pursue this as a function of dreaming.

NIELSEN attempts to resolve the difference between those who see dreaming as independent of REM sleep and those who see it as the inevitable accompaniment of REM sleep. As the definition of the dream experience expanded to include so called cognitive mentation, the report of the experience is labeled sleep mentation rather than dreaming. There is no widely accepted standardized definition of dreaming because we are still doing the phenomenology of dreaming (Kramer 2000). There is no necessary reason why there should be. The standards applied to what will be counted as a dream, of course, alter the results. As we have no external criteria by which to judge what is and what is not a dream, all reports need to considered. Freud chose not to make a distinction reporting one word dreams such as the "Autodidasker" dream and a dream that took several hundred words in the telling "The Dream of Irma's Injection." Is there yet any necessary reason to chose one over the other? We found (Kramer et al. 1984), as did Fisher (Fisher et al. 1970a) that stage 2 nightmares were not different from REM nightmares in patients with post-traumatic stress disorder.

NIELSEN sees the evidence for neurobiological isomorphism as slim. He argues that the mental content in non-REM sleep is the result of the phasic aspects of sleep occurring in non-REM just prior to and immediately after REM sleep. These suggestions have not been tested and raise questions about the REM non-REM separation as it would attribute the mental phenomena to some sub-aspect of events occurring in REM sleep but not limited to REM sleep. The techniques are not available to test this hypothesis at the present time because we have no index in humans for the PGO waves. If this suggestion indeed is the case, NIELSEN would be confirming a biological explanation for the mental events in sleep and providing support for isomorphism. This position cannot account for the specific contents of the dream experience, nor is this its intent. It also does not deal with the transduction problem.

REVONSUO describes the biological function of dreaming as simulating threatening events and rehearsing threat perception and threat avoidance, which contributes to survival while awake and thereby increases the likelihood of reproductive success. The system operates in and out of awareness and is an alarm response so that all dreams at all times need not show evidence of threat perception and avoidance activity.

The nightmare is the paradigm for dreaming as it represents a threat perception and usually an avoidance. REVONSUO dismisses the poor sleep that accompanies this survival response. Nevertheless the 1991 National Sleep Foundation survey points out the negative consequences of insomnia (Roth & Ancoli-Israel 1999), an insomnia that would certainly accompany any traumatic night-

mare. The dreams of patients with post-traumatic stress disorder (PTSD) on the same night have only half their dreams related to the trauma (Kramer et al. 1984 ). The better adjusted patients with PTSD have decreased dream recall (Kamminer & Lavie 1991; Kramer et al. 1984) and an elevated arousal threshold (Schoen et al. 1984) which would make them more vulnerable to predators at night. If the dreaming system is threat perception sensitive, one might expect that strangers would be incorporated into dreams more easily than familiar persons, but the opposite is the case (Kinney et al. 1981). This view of an evolutionary function is rather limited and others have offered a broader, more encompassing way to think about evolutionary functions (Moffitt et al. 1993). An evolutionary theory of the sort proposed would not account for the demographically related content differences described above (Kramer et al. 1971; Winget et al. 1972).

The view REVONSUO offers of the emotionally focused dream theories is too narrow in its understanding. For example, in the mood regulatory theory of dreaming, it is the consequence of the improved mood in the morning that is the issue, not just having achieved a less unhappy state. This improved mood state has been shown to covary with a subsequent improvement in psychomotor performance (Johnson et al. 1990). It is the consequence of the threat perception that explains its alleged function.

# Papez dreams: Mechanism and phenomenology of dreaming

E. E. Krieckhaus

*Department of Psychology, New York University, New York, NY 10003.*
**krieck@worldnet.att.net**

**Abstract:** I agree with Revonsuo that dreaming, particularly about risky scenes, has a great selective advantage. Although the paleoamygdala system generally facilitates stress and alarm, the system which inhibits stress and alarm, initiates bold actions, and mediates learning in risky scenes is the arche, hippocampal system (Papez circuit). Because all thalamic nuclei are inhibited during sleep except arche, Papez probably also dreams in risky scenes.

[REVONSUO]

The mammillary bodies (MBs) which are the most ventral, caudal, and medial mammalian diencephalon, are unique among neurons whose axons constitute the chief input into one of the various thalamic nuclei, and thence to their corresponding cortices. The target of the MBs, anterior thalamic nucleus (ATN), is alone in not receiving inhibitory fibers from the thalamic reticular nucleus, which is usually active in sleep (Alonso & Llinas; Pare et al. 1991). Given that the backbone of the Papez circuit is the massive unidirectional mammillary projection to ATN, this lack of inhibition of ATN in sleep is the basis of my first thesis: Papez plays a critical role in dreaming.

I agree with REVONSUO who argues persuasively that:

1. Dreams are critical for survival;

2. There is considerable selective pressure to make the correct decision in threatening or risky real world contexts or scenes (Sparks 1998); and,

3. Dreams are imagined rehearsals for real scenes. This important idea is an extension of Tolman's vicarious trial and error (VTE) at the choice point (online) (Tolman 1938) to dreaming (offline), a complex sequence of events as scenes. But the same process is probably operative in both to let a physically possible sequence of events unfold and see what happens. If that doesn't work, then try another.

4. Given 1–3, dream VTE Dream vicarious trial and error (VTED) is thus presumably particularly critical in preparing for risky scenes.

I disagree with REVONSUO that:

1. The origin of risky VTED is to reduce the risk of predation in our earlier hunter-gatherer period; both the phenomenon of dreaming and its neural substrates are present throughout mammalia.

2. The mechanism of risk VTED is important primarily against predators; it is perhaps more import within a species, or within a tribe (e.g., when deciding whether to risk taking on the dominant male). Thus VTED learning in risky scenes plays a comprehensive and critical role within mammalia.

**Neurophysiology.** My second thesis is that two generally opposing forebrain functions are operative in VTED, just as in waking attention. During phylogeny, mammalian neocortex develops between the two allo pallia arche (hippocampus [HF]) and paleo (amygdala) (Nauta & Feirtag 1986). The allo system most discussed in mediation of action and affect in risky scenes is the paleo amygdala system which elicits specie-specific responses to threats (e.g. freezing and autonomic arousal) (LeDoux 1998). Amygdala also facilitates both stress (as excessive glucocorticoids) (Cullinan et al. 1993) and alarm (as excessive serotonin [5-HT]) (Maier et al. 1993).

However, Metcalfe and Mischel (1999) (discussed only too briefly by **REVONSUO**) introduce the opposite functions of a "cool" rational system (arche) which opposes the "hot" emotional system (paleo). Arche HF inhibits the activation of the stress cascade that amygdala excites (Cullinan et al. 1993; Krieckhaus 1999) and HF reduces alarm via mammillary inhibition of medial raphe (Kriekhaus 1999; submitted) that amygdala facilitates via excitation of dorsal raphe (Maier et al. 1993).

Given that arche inhibits stress and alarm, and arche is well known to mediate explicit learning (Aggleton & Brown 1999; Krieckhaus 1988; Squire et al. 1990) then it would be expected that arche lesions would uniquely interfere with learning to initiate bold action in threatening or risky scenes just the deficit seen following arche (MBs) lesions (Gabriel et al. 1995; Krieckhaus 1988; 1999). Finally, if **REVONSUO** is correct that dreaming is risky VTED, then arche should be critically involved in risky dreams just as it is in risky actions. Indeed, as we saw earlier: (1) Thalamic reticular nucleus can, in night dreams, inhibit all thalamic nuclei except arche ATN. (2) HF lesions severely disrupt choice point VTE and commensurate learning (Hu et al. 1997). Although VTE may be a different process from VTED, it is likely that HF mediates VTED as well. (3) Finally, given that dreams are for learning, then the neural system mediating learning should play a major role in dreaming; and since arche uniquely mediates explicit learning, arche Papez probably dreams explicitly.

**Discussion.** The diametrically opposite functions of fearful paleo and confident arche are fundamental to the mechanism of VTED or any other forebrain function. Whereas both paleo and arche receive high level invariant information (e.g., faces, grimaces) from posterior association cortex via entorhinal cortex (Krieckhaus et al. 1992; Ungerleider & Mishkin 1982), this information is consistently put to the opposite use (Cullinan et al. 1993; Graeff et al. 1996; Krieckhaus 1999; Metcalfe & Jacobs 1998). These two systems, arche and paleo, are the pallial two of four loops, which together with the two subpallial loops (somatic motor and visceral motor both worked out by Nauta 1966) constitute roughly half of the forebrain.

Because arche encourages rational explicit actions and can control negative affect, its functions are presumably what Freud characterized as secondary process thought (Freud 1911). The no less efficient and sophisticated functions of the paleo loop presumably correspond to what he called primary process thought. With this formulation of VTED, the survival value of Freud's "wish fulfillment" in dreaming is not to fulfill wishes but to instill in us hope and confidence by the responsible, reality oriented, arche dominating the scared and withdrawing paleo. Because these arche dominant dreams are rewarding they are more likely to lead to similar bold actions in later similar waking risky scenes, depending on the acumen of the organism's reality testing. Thus my third thesis: We mammals strive for a healthy equilibrium between arche and paleo, probably realized in a complete, continuous hegemony of arche over paleo.

Finally, as our understanding of neural mechanisms mediating the function of VTED increases, we are better able to distinguish the functions of experience in general, whether awake or dreaming, from the functions of the neural mechanisms that support these experiences. The general issue of the function of experience (given the sufficiency of its underlying neural substrate) reduces to the qualia problem of how neural (physical) activity can cause or be "mental" experience, an issue not pertinent here. More recently, the phenomenon of "blind sight" (relatively accurate adaptive actions but no commensurate experience) though still controversial, raises concrete questions about the functionality of experience, and makes more likely its emergence as epiphenomenal. Thus understanding of selective pressures for dreaming reduces to understanding the structure and function of the forebrain loops using verbal reports of experience simply as proxies for brain states.

**Conclusion.** Papez dreams, and, as argued by **REVONSUO,** dreaming is VTED learning, predominantly of risky scenes. The diametrically opposite functions of fearful paleo and confident arche determine the functions of dreaming and cognitive processes in general. The desired state of an adult mammal is for primary process paleo to be modulated by rational secondary process arche.

# Lucid dreaming: Evidence and methodology

Stephen LaBerge

*The Lucidity Institute, Stanford, CA 94309.* **slab@psych.stanford.edu**

**Abstract:** Lucid dreaming provides a test case for theories of dreaming. For example, whether or not "loss of self-reflective awareness" is characteristic of dreaming, it is not necessary to dreaming. The fact that lucid dreamers can remember to perform predetermined actions and signal to the laboratory allows them to mark the exact time of particular dream events, allowing experiments to establish precise correlations between physiology and subjective reports, and enabling the methodical testing of hypotheses.

[**HOBSON ET AL.; SOLMS**]

Just as dreaming provides a test case for theories of consciousness, lucid dreaming provides a test case for theories of dreaming. Although one is not usually explicitly aware that one is dreaming while in a dream, a remarkable exception sometimes occurs in which one possesses clear cognizance that one is dreaming. During such "lucid" dreams, one can reason clearly, remember the conditions of waking life, and act upon reflection or in accordance with plans decided upon before sleep. These cognitive functions, commonly associated only with waking consciousness, occur while one remains soundly asleep and vividly experiencing a dream world that is often nearly indistinguishable from the "real world" (LaBerge 1985).

Although lucid dreams have been reported since Aristotle, until recently many researchers doubted that the dreaming brain was capable of such a high degree of mental functioning and consciousness. Based on earlier studies showing that some of the eye movements of REM sleep corresponded to the reported direction of the dreamer's gaze (e.g., Roffwarg et al. 1962), we asked subjects to carry out distinctive patterns of voluntary eye movements when they realized they were dreaming. The prearranged eye movement signals appeared on the polygraph records during REM, proving that the subjects had indeed been lucid during uninterrupted REM sleep (LaBerge 1990; LaBerge et al. 1981). Figure 1 shows an example.

Our studies of the physiology of lucid dreaming fit within the psychophysiological paradigm of dream research that Hobson has helped establish. Therefore, I naturally agree with **HOBSON ET AL.** in believing it worthwhile to attempt to relate phenomenological and physiological data across a range of states including waking,

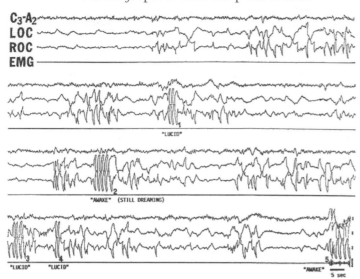

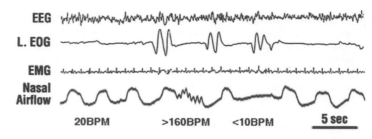

Figure 3 (LaBerge). Voluntary control of respiration during lucid dreaming. LaBerge and Dement (1982) recorded three lucid dreamers who were asked to either breathe rapidly or to hold their breath (in their lucid dreams), marking the interval of altered respiration with eye movement signals as shown in the figure. The subjects reported successfully carrying out the agreed-upon tasks a total of nine times, and in every case, a judge was able to correctly predict on the basis of the polygraph recordings which of the two patterns had been executed (binomial test, p < .002).

Figure 1 (LaBerge). A typical signal-verified lucid dream. Four channels of physiological data (central EEG [C3–A2], left and right eye-movements [LOC and ROC], and chin muscle tone [EMG]) from the last 8 min of a 30 REM period are shown. Upon awakening the subject reported having made five eye movement signals (labeled 1–5). The first signal (1,LRLR) marked the onset of luicidity. Skin potential artifacts can be observed in the EEG at this point. During the following 90 sec the subject "flew about" exploring his dream world until he believed he had awakened, at which point he made the signal for awakening (2,LRLRLRLR). This signal made in non-lucid REM shows that the precise correspondence between eye movements and gaze is not an artifact of lucidity. After another 90 sec, the subject realized he was still dreaming and signaled (3) with three pairs of eye movements. Realizing that this was too many, he correctly signaled with two pairs (4). Finally, upon awakening 100 sec later he signaled appropriately (5, LRLRLRLR). [Calibrations are 50 micro V and 5 sec].

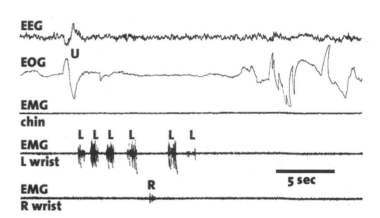

Figure 4 (LaBerge). Morse codes communication from the lucid dream. LaBerge et al. (1981) while testing a variety of lucidity signals found evidence of voluntary control of other muscle groups during REM. We observed that a sequence of left and right dreaming-fist clenches resulted in a corresponding sequence of left and right forearm twitches as measured by EMG. Here the subject sends a Morse code signal with left and right fist clenches corresponding to dots and dashes, respectively. Hence the message translates as "SL" (... .-..), the subject's initials. Note that the amplitude of the twitches bore an unreliable relationship to the subjective intensity of the dreamed action. Because all skeletal muscle groups except those that govern eye-movements and breathing are profoundly inhibited during REM sleep, it is to be expected that most muscular responses to dreamed movements would be feeble. Nonetheless, these responses faithfully reflect the motor patterns of the original dream.

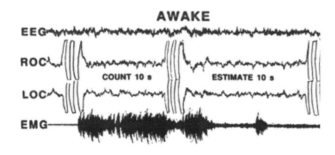

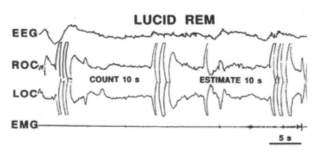

Figure 2 (LaBerge). Dream time estimations. We have straightforwardly approached the problem of dream time by asking subjects to estimate 10 sec intervals (by counting, "one thousand and one, one thousand and two, etc.") during their lucid dreams. Signals marking the beginning and end of the subjective intervals allowed comparison with objective time. In all cases, time estimates during the lucid dreams were very close to the actual time between signals (LaBerge 1980a; 1985).

NREM, and REM sleep. I also share HOBSON's view that REM sleep is unique in many ways; for example, stable lucid dreams appear to be nearly exclusively found in REM. As for the AIM model on which the HOBSON ET AL. article focuses, I regard it as an improvement on the earlier Activation-Synthesis model. The AIM model makes many plausible and interesting connections, but still doesn't do justice to the full range and complexity of the varieties of dreaming consciousness accompanying REM sleep.

One of the problems with AIM is that its three "dimensions" are actually each multidimensional. For example, from which brain area is "Activation" (A) measured? Obviously, A varies as a function of brain location. HOBSON ET AL. admit as much when they propose to locate lucid dreaming in a dissociated ATM space with PFC more activated than it usually is (see Fig. 12). If this is true, then non-lucid dreaming would have to be characterized by a low value of A. Incidentally, there is no evidence to support the idea that lucid dreaming is in any sense a dissociated state (LaBerge 1990). Still, the need for multiple A dimensions seems inescapable.

Similarly, the "Information flow" (I) dimension is more complex than at first appears. Experimental evidence suggests that it is possible for one sense to remain awake, while others fall asleep (LaBerge 1990). A further problem with the I "dimension" is the confounding of sensory input and motor output, as can be seen in several of HOBSON ET AL.'s examples (e.g., compare Figs.15, 16B, 19). Finally, "Mode of information processing" (M) attempts to reduce the vast neurochemical complexity of the brain to the global ratio of discharge rates of aminergic to cholinergic neurons. Is that really all there is to say about the neurochemical basis of consciousness? What about regional differences of function? What about the scores of other putative neurotransmitters and neuromodulators?

Perhaps due in part to the over-simplifications necessary to fit these multiple dimensions into an easy-to-visualize three, certain features of dreaming consciousness are misunderstood or exaggerated. For example, HOBSON ET AL. say "self-reflection in dreams is generally found to be absent (Rechtschaffen 1978) or greatly reduced (Bradley et al. 1992) relative to waking" However the two studies cited suffered from weak design and extremely small sample sizes. Neither in fact actually compared frequencies of dreaming reflection to equivalent measures of waking reflection. A study that did make direct comparisons between dreaming and waking (LaBerge et al. 1995) found nearly identical frequencies of reflection in dreaming (81%) as in waking (79%), clearly contradicting the characterization of dreams as non-reflective. Replications found similar results (Kahan & LaBerge 1996; Kahan et al. 1997). These studies were cited in Hobson's article but otherwise ignored.

Another unsubstantiated claim of HOBSON ET AL. is that "volitional control is greatly attenuated in dreams." Of course, during non-lucid dreams people rarely attempt to control the course of the dream by magic. The same is true, one hopes, for waking. But likewise, during dreams and waking, one has similar control over one's body and is able to choose, for example to walk in one direction or in another. Such trivial choice is probably as ubiquitous in dreams as waking and, as measured by the question "At any time did you choose between alternative actions after consideration of the options?" 49% of dream samples had voluntary choice, compared to 74% of waking samples (LaBerge et al. 1995). The lower amount of choice in dreams may be an artifact of poorer recall or a real difference, but choice is by no means "greatly attenuated."

While making the above claim, HOBSON ET AL. incorrectly attribute to me the false statement that "the dreamer can only gain lucidity with its concomitant control of dream events for a few seconds (LaBerge 1990)." In fact, lucid dreams as verified in the laboratory by eye-movement signalling last up to 50 minutes in length, with the average being about 2 minutes (LaBerge 1990). The relatively low average is partially due to the fact that subjects were carrying out short experiments and wanted to awaken with full recall. At the onset of lucid dreams there is an increased tendency to awaken, probably due to the fact that lucid dreamers are thinking at that point, which withdraws attention from the dream, causing awakening (LaBerge 1985).

The eye-movement signalling methodology mentioned above forms the basis for a powerful approach to dream research: Lucid dreamers can remember pre-sleep instructions to carry out experiments marking the exact time of particular dream events with eye movement signals, allowing precise correlations between the dreamer's subjective reports and recorded physiology, and enabling the methodical testing of hypotheses. We have used this strategy in a series of studies demonstrating a higher degree of isomorphism between dreamed actions and physiological responses than had been found previously using less effective methodologies For example, we found that time intervals estimated in lucid dreams are very close to actual clock time (see Fig. 2); that dreamed breathing corresponds to actual respiration (Fig. 3); that dreamed movements result in corresponding patterns of muscle twitching (Fig. 4); and that dreamed sexual activity is associated with physiological responses very similar to those that accompany actual sexual activity (see LaBerge 1985; 1990 for details).

These and related studies show clearly that in REM sleep, dreamed bodily movements generate motor output equivalent at the supraspinal level to the patterns of neuronal activity that would be generated if the corresponding movements were actually executed. Most voluntary muscles are, of course, paralyzed during REM, with the notable exceptions of the ocular and respiratory muscles. Hence, the perfect correspondence between dreamed and actual movements for these two systems (Figs. 1–3), and the attenuated intensity (but preserved spatio-temporal pattern) of movements observed in Figure 4.

These results support the isomorphism hypothesis (HOBSON ET AL.) but contradict SOLMS's notion of the "deflection" of motor output away from the usual pathways, and his speculation that it isn't only the musculo-skeletal system that is deactivated during dreams, but "the entire motor system, including its highest psychological components which control goal-directed thought and voluntary action" (Solms 1995, p. 58). I believe Occam's Razor favors the simpler hypothesis that the motor system is working in REM essentially as it is in waking, except for the spinal paralysis; just as the only essential difference between the constructive processes of consciousness in dreaming and waking is the degree of sensory input. See LaBerge (1998) for details.

Oddly, HOBSON ET AL. ignore the data on eye movements while appealing that we keep open the question of relationship between eye movement and dream imagery "until methods more adequate to its investigation are developed." There is no need to wait. Adequate methods have already been developed, as shown above (Figs. 1–4), and in our recent study showing smooth tracking eye movements during dreaming (LaBerge & Zimbardo 2000).

Memory is another area of inquiry upon which lucid dreaming can shed light. HOBSON ET AL. argue that memory during dreaming may be as deficient as it is upon awakening. They give the example of comparing one's memory of a night's dreaming to the memory of a corresponding interval of waking; unless it was a night of drinking being remembered, the dream will yield much less memory. But this is an example comparing episodic memory from waking and dreaming after awakening, and thus is not only unconvincing and vague, but irrelevant. Nobody disagrees that waking memory for dreams is sometimes extremely poor.

In the same vein, HOBSON ET AL. write that it is common for dreams to have scene shifts of which the dreamer takes little note. "If such orientational translocations occurred in waking, memory would immediately note the discontinuity and seek an explanation for it." Note the unquestioned assumption. In fact, recent studies suggest that people are less likely to detect environmental changes than commonly assumed (Mack & Rock 1998). For example, a significant number of normal adults watching a video failed to notice changes when the only actor in a scene transformed into another person across an instantaneous change in camera angle (Levin & Simons 1997).

Likewise, **HOBSON ET AL.** assert that "there is also strong evidence of deficient memory for prior waking experience in subsequent sleep." However, the evidence offered is always extremely indirect and unconvincing. A direct test requires lucid dreamers to attempt memory tasks while dreaming, as was done in a pilot study (Levitan & LaBerge 1993) showing that about 95% of the subjects could remember in their lucid dreams a key word learned before bed, the time they went to bed, and where they were sleeping. Subjects forgot to do the memory tasks in about 20% of their lucid dreams. That may or may not represent a relative deficit in memory for intentions.

A major methodological difficulty presented by dreaming is poor recall on awakening. The fact that recall for lucid dreams is more complete than for non-lucid dreams (LaBerge 1985) presents another argument in favor of using lucid dreamers as subjects. Not only can they carry out specific experiments in their dreams, but they are also more likely to be able to report them accurately. That our knowledge of the phenomenology of dreaming is severely limited by recall is not always sufficiently appreciated. For example, **HOBSON ET AL.** repeatedly substitute "dreaming" for "dream recall" (e.g., sect. 2.3.1). **SOLMS** (1997a) makes the same mistake, which in my view, is fatal to his argument. So when he writes "of the 111 published cases . . . in which focal cerebral lesions caused cessation or near cessation of dreaming" he is really saying "in which lesions caused cessation of dreaming or dream recall." To think otherwise would be to suppose that the dream is the report.

# All brain work – including recall – is state-dependent

Dietrich Lehmann[a] and Martha Koukkou[b]

[a]*The KEY Institute for Brain-Mind Research, University Hospital of Psychiatry, CH-8029 Zurich, Switzerland;* [b]*University Hospital of Clinical Psychiatry, CH-3000 Bern, Switzerland.* {dlehmann; mkoukkou}@key.unizh.ch www.unizh.ch/ch/keyinst/

**Abstract:** The continuous ongoing mentation is experienced as dreams in some functional states. Mentation occurs with high speed, is driven by individual memory, and uses state-dependent processing strategies, context material, storage options, and retrieval access. Retrieval deserves more attention. Multiple state-shifts owing to individual meaning as extracted also during sleep concatenate dream narratives and define access to segments for awake recall.

[**HOBSON ET AL.**; **NIELSON**; **SOLMS**]

Since the late 50s, dreams have been assumed to be the product of REM sleep. For dreams reported out of other sleep stages, **HOBSON ET AL.** argue now that there are "possible dissociations of state characteristics" that might permit "states in which some parameters match their canonical NREM values while others match canonical REM or wake values." **NIELSEN** makes a somewhat related proposal when speaking of "covert REM sleep processes during NREM sleep." **SOLMS** also postulates a specific dream state that in his view, however, can be independent of REM physiology.

We welcome this blurring of the earlier sharp distinction between two classes of sleep stages and between two classes of experiences. It opens the way for dream studies to take into account three major properties of the continuously ongoing mentation: its high speed, in the subsecond range (otherwise one couldn't even drive a car), its construction from individual memory, and its state-dependency (otherwise one would think in the same way when sober and inebriated, when awake or drowsy or asleep, as child or as adult) that is finely grained. Different modes of momentary mentation have been shown to be associated with distinctly different brain electric characteristics in the second time range (Lehmann et al. 1983; 1995) as well as in the mentation-relevant sub-second time range where the time trajectory of brain momentary states through state space consists of dwell times ("microstates") and shifts: that is, it is discontinuous (Lehmann et al. 1998). Mentation leads to the extraction of individual meaning that, if needed, initiates a state shift to optimize the conditions for the next step of brain work (Koukkou & Lehmann 1983; 1987); hence, the shifts of brain functional state during sleep need to be examined at a much higher time resolution (Cantero et al. 1999b) than in classical approaches to sleep psychophysiology.

The issue of retrieval is approached by **HOBSON ET AL.**, who say that "subjects assert that much antecedent dreaming could not be recalled," but he continues with "one reason for the neglect of this robust phenomenon is that memory isn't there!" Nobody can know for certain whether there was something – but there are compelling reasons to assume that there is always something in the mind during sleep, just as in wakefulness. The brain must continuously process information, from external and internal sources, not only during wakefulness, but also during sleep: otherwise sleeping people couldn't distinguish between relevant stimuli that require awakening or not (the sleeping mother who will awake at the whimper of her baby and sleep right on while the traffic roars by in front of the house; the subject who awakens to his name but not to names without biographical relevance). And, as in waking, if there are no external inputs that require state-shifts, the biographically generated memory has abundant material in waiting for review.

Well, maybe the current mind stuff is not stored for some reason? Again, nobody knows. But, absence of recall does not prove that nothing happened, it primarily shows that there is no recall. Experiences can be stored in a state too remote from the state during later attempts to recall (Koukkou & Lehmann 1983); but, when the original state is re-installed, the experiences become recallable again. There are numerous accounts portraying the effects of this mechanism, a very famous one described by Marcel Proust in "*La recherche du temps perdu*"; flashbacks in drug addicts are its modern version. A family classic is the no recall condition of the candidate when highly excited during an examination, and his "but I knew it all!" surprise when leaving the examination and relaxing. Events experienced in one state are optimally available for recall when the same state is reinstalled (Eich 1986).

We proposed the basic EEG measure of dominant "wave frequency" as first approximation of a metric for the state-dependency of type of dream mentation and in particular, of quality of dream recall (Koukkou & Lehmann 1983): the slower the dominant frequency and thus the further away the brain state at information experience is from the state at attempted information retrieval (i.e., wake), the poorer will be the recall – while arguing that this must zoom in on very brief time epochs (possibly "single waves"), not conventional EEG sleep stages. (But many more measurement dimensions of electric data should be added for comprehensive assessments of brain functional state, e.g., dimensionality of the embedding state trajectory, momentary coherency between intracerebral generator processes, or momentary spatial distribution Lehmann et al. 1998.)

A prime example of state-dependent non-availability of recall is sleepwalking with its goal-directed behavior that occurs in slow wave sleep (Jacobson et al. 1965). Not only is external input treated at high levels, also behavior is selected and implemented. But, the events that happen during sleep walking typically are not recallable after awakening. So-called childhood amnesia is another example: few events before the age of about 4 years are recallable by awake adults: because toddlers' EEGs are dominated by slower frequencies, and adults' EEGs are much faster. From pathological examples: behavior under scopolamine (that is associated with "slowing" of the EEG) cannot be recalled after the drug wore off (Bradley & Elkes 1957). Other classical, pathological conditions with EEG slow wave activity and unavailability of later recall of events are temporal seizures and head trauma.

In sum, we suggest that there is continual mentation during wakefulness and sleep, implemented always with the same basic, biography-driven brain machinery which leads to varying final products depending on the momentary global functional state of the brain: the momentary state is the fate of the information. The momentary state defines access to state-dependent processing strategies, context memory, storage procedures, and recall options; the numerous shifts of functional state during sleep accordingly concatenate very different mentation characteristics as segments of dreams (Koukkou & Lehmann 1983). Considering the state-dependency of all brain work including recall, and considering that the continual mentations and emotions are implemented in a split-second time range will help to clarify the ever-intriguing experiences during sleep.

## Nightmares: Friend or foe?

Ross Levin

*Ferkauf Graduate School of Psychology, Albert Einstein College of Medicine, Bronx, NY 10461.* **levin@mary.fordham.edu**

**Abstract:** Revonsuo's evolution-based theory places the nightmare as a prototype dream, which fully realizes its biological function. However, individuals who experience both repetitive (PTSD) nightmares and/or lifelong nontraumatic nightmares demonstrate impaired psychological functioning and attenuated information-processing. The importance of reconciling these discrepancies are addressed and ideas for providing stronger empirical tests of the model are presented.

[REVONSUO]

REVONSUO's evolution-based theory of dreaming is cogently presented in six empirically verifiable propositions and in many respects is quite plausible despite its tendency to overreach at certain points. revonsuo should be commended for drawing attention to the wealth of data indicating that dreaming is an organized and internally consistent simulation of the perceptual world. In addition, REVONSUO effectively supports his central assertion that dreaming is highly specialized in its selection of affect-charged memorial content, particularly of a threatening or dysphoric nature. However, his contention that nightmares are a prime example of dreams that fully realize their biological function is problematic and will be the focus of my comments.

If dreaming evolved to serve a threat simulation function, and if experiences of real threats which approximate the "primitive" human ancestral environment are the only ecologically valid cues for the activation of this system, why then are individuals who suffer from post-traumatic stress disorder (PTSD) so chronically overwhelmed by the very intrusive symptoms which should be helping them successfully resolve their problem? PTSD individuals demonstrate a marked hypersensitivity to overly activated fear-arousal networks (Foa & Kozak 1986) that interfere with their ability to process information effectively. Thus, it is difficult to understand how intrusive traumatic symptoms, whether they be repetitive nightmares or waking flashbacks, aid in successful coping, particularly as they may continue unabated for up to 50 years post-trauma with an absence of accompanying mastery or reduction in psychological distress (Lansky 1995). REVONSUO does address this tricky issue but only cursorily, concluding with the questionable assumption that modern-day trauma is too removed from the pre-technological daily traumas in which the dreaming system evolved to be adequately generalizable. While this may be true, modern living still provides numerous approximates of traumas that should activate threat simulator programs (e.g., incest, rape, physical beatings, natural disasters). REVONSUO's argument would be bolstered by data demonstrating that individuals who experience the latter types of psychological trauma recover quicker than those who are exposed to traumas involving insults for which the evolved threat-avoidance programs have not yet been incorporated.

REVONSUO also maintains that dreaming serves a similar bio-

logical defense mechanism function as do antigens in the pathoimmune system. We should then expect that exposure to early environmental threats should mobilize the dreaming immunological system and provide a psychological vaccine to deal more effectively with subsequent trauma. In fact, research has repeatedly demonstrated that exposure to early aversive environmental events is a primary pathogenic pathway to later psychopathology (Gershuny & Thayer 1999). Of course it could be argued that these experiences (parental loss, divorce, child abuse) are too disruptive to function effectively as a low dose for subsequent immunological functioning. It can also be argued that dreaming functions at a far more nomothetic and trans-species level and that its presumed protective qualities do not apply to individual cases in either of the above examples. In any case, these discrepancies should be fleshed out in greater detail.

REVONSUO does not address the issue of significant individual variation in the experience of lifelong non-trauma related nightmares. I believe this is crucial, as nightmares are a prototype dream within REVONSUO's model. Frequent lifelong nightmare sufferers demonstrate considerable psychological dysfunction and may be at increased risk for the development of schizophrenia-spectrum disorders (Hartmann 1984; Hartmann et al. 1987; Levin 1990a; 1994, 1998; Levin & Raulin 1991). It would be important to reconcile this data with his contention that such dreams were ancestrally selected for their adaptive function in preparing for fight-flight responses. It is interesting to note that in his influential paper on the development of schizophrenia, Meehl (1989) suggested that humans evolved from a schizotaxic genetic background. Could it be that nightmare sufferers retain a closer link to their ancestral vestige of the early fight-flight patterns than do other individuals? While highly conjectural, understanding better how these pieces fall together could develop our understanding of why nightmares occur.

Last, I believe that the author is too quick to dismiss alternative theories of dream function, particularly the problem-solving model. REVONSUO's claim that "there is no convincing" evidence that dreaming would casually contribute to the solving of either intellectual or emotional problems" runs counter to a large body of empirical data demonstrating otherwise (Cartwright 1986; Fiss 1993; Hartmann 1998; Koulack 1991; Moffitt et al. 1993; Levin 1990b; **VERTES & EASTMAN** 2000; Blagrove 1996 for opposing views). Furthermore, REVONSUO's claim that "the brain's dream production system selects traumatic contents not because they represent unsolved emotional problems but because such experiences mark situations for physical survival and reproductive success" is highly speculative and not directly supported by any empiric evidence. While it is difficult to predict the content of future dream production (Cartwright 1974b; Nikles et al. 1998), numerous studies by Cartwright (1986) and Kramer (1993) indicate that, "the dream is a selective affective regulator which functions as an 'emotional thermostat'" (Kramer, p. 182). In addition, REVONSUO fails to consider the literature on creative dreams (Dave 1978; Dreistadt 1971; Krippner 1981; Livingston & Levin 1991).

If dreams truly simulate phylogenetic threats, an important test of this model would be to determine what effect the presentation of stimuli that have known fear-relevant properties (those that meet the criteria for biological-preparedness for phobia acquisition such as spiders, snakes, angry human faces; McNally 1987) have on subsequent dream content. Furthermore, given REVONSUO's claim that the function of dreams may be optimally realized through implicit processing, such material could also be presented at subliminal activation levels and compared to supraliminal activation conditions. In order to provide a stronger test of his theory, levels of incorporation of the target stimuli in subsequent dream production could be compared to known current concerns of the dreamer (Nikles et al. 1998) to directly determine which stimuli have greater predictive utility. Further studies along these lines would be most helpful in providing new clues into the investigation of dream function.

# Koch's postulates confirm cholinergic modulation of REM sleep

Ralph Lydic and Helen A. Baghdoyan

*Department of Anesthesiology, The University of Michigan, Ann Arbor, MI 48109.* **rlydic@med.umich.edu**

**Abstract:** Robert Koch (1843–1910) discovered the causal agents for tuberculosis, cholera, and anthrax. The 1905 Nobel Prize acknowledged Koch's criteria for identifying the causal agent of an infectious disease. These criteria remain useful and the data reviewed below show that the cholinergic contributions to REM sleep control are confirmed by Koch's postulates.
**[HOBSON ET AL.]**

We congratulate **HOBSON ET AL.** for stimulating synthesis of cognitive science and sleep neurobiology. Their article demonstrates the unifying power of the localization-of-function concept so successfully advanced by nineteenth century German neurology. This commentary focuses on their proposal that the cholinergic hypothesis of REM sleep generation has been confirmed. We also draw from the nineteenth century, showing how data concerning the cholinergic modulation of REM sleep satisfy Koch's postulates (Brock 1999). Available data support Koch's postulates when one evaluates medial pontine reticular formation (mPRF) levels of acetylcholine (ACh) as a causal agent modulating the state of REM sleep.

> Postulate 1: The state must be reproduced when the agent is administered.

Microinjection of cholinergic agonists (Baghdoyan et al. 1984b) and ACh-esterase inhibitors (Baghdoyan et al. 1984a) into the mPRF causes REM sleep enhancement. REM sleep is inhibited by mPRF injection of the muscarinic cholinergic antagonist atropine (Baghdoyan et al. 1989; Lee et al. 1995) and by drugs that block the vesimacol receptor regulating the vesicular packaging of ACh (Capece et al. 1997). Normally, the mPRF is never exposed to cholinomimetics. How does cholinergically-induced REM sleep affect mPRF levels of the endogenous ligand ACh?

> Postulate 2: The agent is recovered during the experimentally-induced state.

Microdialysis data demonstrate significant enhancement of mPRF ACh release during the REM sleep-like state caused by contralateral mPRF administration of carbachol (Lydic et al. 1991b). Additional data satisfying this postulate include the finding that REM sleep is enhanced by electrical stimulation (Thakkar et al. 1996) of laterodorsal and pedunculopontine (LDT/PPT) neurons shown to regulate ACh release within the mPRF (Lydic &Baghdoyan 1993). A limitation of postulates one and two is that they are based on a REM sleep-like state produced by exogenous stimulation of the pons. The relationship between mPRF ACh levels and natural REM sleep is addressed by a third postulate.

> Postulate 3: The putatively causal agent should be present during every naturally occurring case.

Microdialysis data show that mPRF ACh release is significantly greater during spontaneous REM sleep than during waking or non-REM sleep (Leonard & Lydic 1995; 1997). Thus, levels of the putatively causal agent (ACh) are greatest in the mPRF during natural REM sleep.

> Postulate 4: Requires the isolation of the putatively causal agent from the host.

A ligand such as ACh is irrelevant without a functionally significant binding site. Therefore, in addition to ACh, cholinergic receptors also may be considered as agents to be isolated in the host (mPRF). Receptor mapping studies have identified M2 muscarinic cholinergic receptors (mAChRs) in the mPRF (Baghdoyan

1997; Baghdoyan et al. 1994). Functional data show that M2 muscarinic autoreceptors modulate ACh release in the mPRF (Baghdoyan et al. 1998) while mPRF M2 heteroreceptors contribute to REM sleep generation (Baghdoyan & Lydic 1999). All mAChRs are coupled to G proteins and in many brain regions M2/M4 mAChRs are linked to an inhibitory G protein ($G_i$). Pertussis toxin selectively ADP ribosylates $G_i$ proteins thereby preventing interaction with mAChRs. Cholinergic REM sleep enhancement is blocked by mPRF administration of pertussis toxin (Shuman et al. 1995). These data are consistent with G protein mediation of REM sleep, a conclusion supported by additional signal transduction studies showing cholinergic REM sleep modulation by mPRF adenylate cyclase, cAMP, and protein kinase A (Capece & Lydic 1997). Postulate four is supported by direct measurement of mPRF G protein activation by carbachol and inactivation by atropine (Capece et al. 1998).

We conclude that Koch's postulates have been satisfied for the cholinergic hypothesis of REM sleep generation. These advances concerning cholinergic neurotransmission in a defined LDT/PPT-to-mPRF network provide a solid basis for continued progress in sleep neurobiology.

ACKNOWLEDGMENTS
This work is supported by NIH Grants HL-40881, MH-45361, HL-57120, HL-65272, and the Department of Anesthesiology, The University of Michigan.

# "Spandrels of the night?"

Gary Lynch,[a] Laura Lee Colgin,[b] and Linda Palmer[c]

[a]*Department of Psychiatry and Human Behavior,* [b]*Institute for Mathematical Behavioral Sciences,* [c]*Department of Philosophy, University of California, Irvine, CA 92697.* **{glynch; lcolgin; lpalmer}@uci.edu**

**Abstract:** Vertes & Eastman argue against the popular idea that dreams promote memory consolidation and suggest instead that REM provides periodic endogenous stimulation during sleep. Although we suspect that much of the debate on the function of dreams reflects a too eager acceptance of the "adaptationist program," we nonetheless support the position of the authors and propose a specific advantage of periodic REM activity.
**[VERTES & EASTMAN]**

In the essay, "The spandrels of San Marco and the Panglossian paradigm: A critique of the adaptationist programme," Gould and Lewontin (1979) describe the remarkable aesthetic beauty of the spandrels of St. Mark's Cathedral in Venice. Their design is so "elaborate, harmonious, and purposeful" that one forgets that the spandrels are merely a necessary result of the original architectural purpose: mounting a dome on rounded arches. Similarly, the vivid intensity of dreams tempts investigators to believe that they evolved to serve a high-level cognitive purpose, such as memory consolidation (see article) or protection from repressed, "unconscious wishes" (Freud 1900). Although Freud's psychoanalytic theory of dreams has been abandoned for the most part, the memory consolidation hypothesis has survived, despite the lack of a convincing body of evidence to support it. Is it possible that the elaborate sensations of dreaming are analogous to the visual beauty of the spandrels of St. Mark's? Blinded by their beauty, we forget that dreams may not have directly evolved to serve a higher purpose but instead may be a necessary by-product of the basic structure of sleep.

REM sleep appears first phylogenetically in birds and is also present in almost all mammalian species. This has led researchers to conclude that REM sleep evolved to serve a high order cognitive function, namely memory consolidation. However, dolphins do not exhibit REM sleep (Mukhametov 1984), although dolphins are certainly capable of learning.

Bob Vertes's extensive contributions to the understanding of hippocampal theta rhythm production and possible function allow him certain insights into its role in REM sleep. We support his position that the theta rhythm of waking states does not have the same function as the theta rhythm of REM sleep. In fact we, along with others, suspect that each type of oscillation has a specific role in information processing. **VERTES** has previously proposed "that the theta rhythm may serve to gate or facilitate the transfer of information to the hippocampus, a process that may be involved in the long-term storage of that information" (Vertes & Kocsis 1997). Traub and colleagues have proposed that gamma oscillations in the cortex signify the details of a percept, while beta oscillations may reflect the occurrence of a stimulus with particular significance (Traub et al. 1999). The absence of this interplay of the different field potential oscillations during REM sleep may suggest that mnemonic functioning is absent, and theta is not functioning as it does during consciousness.

**VERTES & EASTMAN** are making wider claims than indicated by their title. Not only do they assert that there is no memory consolidation in REM sleep, they maintain that the function of REM sleep is to provide periodic stimulation to the brain to offset the depressed brain activity of slow wave sleep. This position is plausible, considering that REM sleep never occurs prior to episodes of slow-wave sleep, except in the case of narcolepsy.

A possible example of the importance of periodic stimulation involves the link between sleep and depression. A characteristic of depression is reduced latency of entry into REM sleep following sleep onset (Kupfer & Thase 1983). Sleep deprivation has an antidepressant effect (Wu & Bunney 1990, for a review), and the three major classes of antidepressants suppress REM sleep, as discussed by the authors in detail. In normal awake and slow wave sleep states, the raphe nucleus is releasing serotonin. At the onset of REM, raphe nucleus activity ceases, and serotonin release is suppressed. Serotonergic neurons have 5-HT1A autoreceptors, which regulate their function. Following REM sleep deprivation, these autoreceptors become less responsive to the effects of serotonin reuptake blockers, probably due to a desensitization of the autoreceptors, resulting in enhanced serotonergic transmission (Maudhuit et al. 1996). We would like to propose that REM activation serves to prevent the desensitization of autoreceptors that would occur if serotonin continued to be released. Similar needs for periodic activation via REM could involve endocrine functioning, as evidenced by body temperature increases (Wehr 1992) and hormonal changes (Obal & Krueger 1999) during REM sleep.

It may be premature to assert definitively that dreaming serves no higher cognitive function. Spandrels become an important artistic grammar in their own right. As Nietzsche wrote, "the cause of the origin of a thing and its eventual utility, its actual employment and place in a system of purposes, lie worlds apart" (Nietzsche 1992).

# Dream production is not chaotic

Mauro Mancia

*Istituto di Fisiologia Umana II, 20133 Milan, Italy.* **mauro.mancia@unimi.it**

**Abstract:** The AIM model proposed by Hobson et al. is interesting: We know the neurophysiological aspects of the activation process (A) and the external input (I), but very little about the internal input and neurocognitive process (M). Internal input could be an expression of unconscious experiences memorised by the subject containing his emotional and cognitive history. Therefore internal input could not be chaotic but might have an emotional and affective sense bound to the unconscious. The fact that dreams are present in the absence of REM sleep means that they may originate from other structures besides the pons. These structures may represent the archives of dreamer's affective history.

[HOBSON ET AL.; SOLMS]

My comment on the work of **HOBSON ET AL.** is essentially epistemological. My criticism concerns the concept of an isomorphism between the phenomenology of dreams and the physiology of the various phases of sleep. I agree that dreaming can be approached using a principle of ontological monism, (i.e., that every kind of mental activity, and hence also dreaming, is the result of processes taking place in the brain). Because we do not know the complex chain of events linking physiological and mental events, it is epistemologically incorrect to talk of mind/brain isomorphism as if mental events could be entirely identified with physiological events. I accordingly do not agree with Hobson (1988b) that dreaming is a physiological event. Dreaming is too complex a mental process to be directly explained using a model, albeit an interesting and sophisticated one, such as the AIM model proposed by **HOBSON ET AL.**

If we look at this model, we see that while we have neurophysiological evidence regarding the activation process (A), and we know the external input (I), we know little about what **HOBSON ET AL.** call internal input and even less about what they define as the cognitive neuromodulator process (M). Regarding the internal input, why not think of it as unconscious dynamics and experiences memorised by the subject, containing his emotional and cognitive history, rather than think in terms of chaotic input?

**HOBSON ET AL.**'s argument in favour of a double dream generator, organised in a profoundly different way on qualitative and quantitative levels during the phases of non-REM and REM sleep, comes up against equally convincing arguments proposed by other authors (Antrobus 1983; Bosinelli 1995; Foulkes 1997; 1999) in favour of a single dream generator relatively independent of the various biological phases of sleep itself. But even if the double organization of sleep were experimentally confirmed during the REM and non-REM phases, we would still need to explain the presence of dreams even in subjects with pontine lesions and no REM sleep (Solms 1995). **SOLMS**'s observation seems to have been greatly minimised by **HOBSON ET AL.** whereas they have stressed the importance of the pons in the process of neocortex and limbic system activation. In fact, it is reasonable to think that the visual cortex is activated in REM sleep to produce the visual hallucinations of dreams and that the amygdala and other limbic structures are activated to produce their emotions and anxieties. What is more difficult to accept, also owing to the absence of clear evidence, is the idea that this activation comes only from the pons and is chaotic, and that it activates the associative neocortices and the limbic system in a disorderly and unfocussed manner.

Since dreaming occurs even in the absence of pontine structures (Solms 1995), it might be thought that the source of the process of cortical and limbic dream activation is not only pontine but may also be found in other cortical and subcortical structures. Furthermore, the process may not be chaotic at all, but may instead retrieve from the memory archive emotional and cognitive experiences organised in the internal world of the dreamer, thus activating a neurocognitive system that takes into account possibly traumatic processes deposited in and removed from the unconscious. This retrieval may occur in a way which is not necessarily that linear, but distorted, so as to create manifest dream contents which, owing to condensation, symbolisation, oddities and absurdities, are different from the latent ones.

This neurocognitive hypothesis would bring the process of dreaming closer to that studied by psychoanalysis. In this line of thought, I believe the contribution of Eric Kandel (1998) is very important: rather than proposing a critical and destructive approach towards psychoanalysis, he proposes the constructive hypothesis that the speech and learning on which the psychoanalytical process is based may modify genic expression and therefore protein synthesis and consequently even the long-term functionality of certain synaptic structures.

In their review **HOBSON ET AL.** limit themselves to suggesting neurocognitive models with neurophysiological bases, and explain dreams using mental categories (hallucinations, thoughts, affection, emotions), without taking into consideration the contribution of psychoanalysis in the study of dreams and their significance. These authors apparently fail to recognise the simple historical fact that in the person of Freud psychoanalysis was dealing with dreams at least 50 years before neuroscientists and cognitivists were. I believe it is useful, as Kandel suggests, to bring neuroscience closer to psychoanalysis but for this to happen, it is not enough for psychoanalysts to confidently embrace neuroscience; it is also necessary for neuroscientists to know about psychoanalysis and the transformations that have taken place in the psychoanalytical method over the last 100 years, and to accept the extremely significant contributions that psychoanalysis has brought to the study and significance of dreams. I believe that every good scientist must know the limits of his method and must accept the possibility of integration from other disciplines, even those operating with different methods from his own, without falling victim to epistemological confusion.

# Novel concepts of sleep-wakefullness and neuronal information coding

Thaddeus J. Marczynski

*Department of Pharmacology, University of Illinois at the Medical Center M/C 868, Chicago, IL 60612.* **tadmar@uic.edu**
**www.uic.edu/depts/mcph/**

**Abstract:** A new working hypothesis of sleep-wake cycle mechanisms is proposed, based on ontogeny and functional/anatomic compression of two stochastic neuronal models of information coding that complement each other in a key/lock fashion: the axonal arbor patterns (AAP – "hardware") and the neuronal spike interval inequality patterns (SIIP – "software").
[HOBSON ET AL.; NIELSEN; REVONSUO; SOLMS; VERTES & EASTMAN]

Impressive analyses of clinical, behavioral, EEG and neuronal firing associated with sleep-wake cycles have been provided by **HOBSON ET AL., SOLMS, NIELSEN, VERTES & EASTMAN,** and **REVONSUO**. All five reviews ably described interactions between neuronal systems, but only in global terms of activation and inhibition. This conventional approach leaves out subtle important modes of neuronal interactions: the word-like intraburst timing of emitted action potentials, defined as spike interval inequality patterns (SIIP) that naturally, via axon arbor filters, seek their axonal arbor patterns (AAP). It is postulated that this communication system resulted from ontogenic compression of two stochastic processes: one linked to emission of SIIP and the other linked to shaping AAP, both aimed at efficient real-time information storage and retrieval.

In our studies of SIIP, the computer measured sequential spike intervals with 0.1 msec resolution and stored the data in sequential computer memory bins. Partially inspired by Norbert Wiener's (1921) criticism of mindless "infinitesimal" clockwork measures used in exploring brain cognitive functions and his favorable comments on the Weber's law that introduces measures derived from relative responses to sensory stimuli, we have explored the concept of SIIP as potential carriers of information emitted by brain neurons (Brudno & Marczynski 1977; Marczynski & Sherry 1971). A computer "window" for comparing sequential pairs of spike intervals moved one spike interval at a time; if the second interval in a pair was longer or shorter than the first interval, a (+) and (−) was entered respectively into another series of sequential computer memory bins. Excessively long intervals (>200 msec) were treated as "punctuation" gaps after which the inequality testing was resumed. Subsequently, the sequences of inequality signs were arranged into transition frequency matrices of various complexity. If the matri columns and the rows are labeled (+) and (−) respectively, the matrix cells tell how many times a (+) was followed by a (−) or by (+) and (−). In this manner, higher order matrices have been constructed that counted the occurrences of "words" composed of 3 through 6 inequality signs (trigrams through hexagrams). Based on the novel stochastic model (Brudno & Marczynski 1977), the probabilities were assigned to each SIIP permutation. The departures of SIIP occurrence from the model,

that is excessive emissions or deficits, were quantified using the chi square statistics.

Figure 1 summarizes the physiologic rationale of the SIIP concept. The arrow between SIIP-A and SIIP-B shows the direction of SIIP propagation in the main axon with seven collaterals. To keep the essential details simple, it was assumed that the geometric ratios at each axonal branching and the presence of the nodes of Ranvier (not shown) permit uninterrupted propagation of action potentials, although in reality most arbors work as electric filters that discriminate between SIIP (Deschenes & Landry 1980; Manor et al. 1991). In Figure 1, the inequalities between sequential pairs of spike intervals (moving one spike at a time) are expressed by signs (+) and (−). The mean spike rate and the mean spike interval are identical in SIIP-A and SIIP-B, yet their relative timing and therefore theoretical probabilities based on the stochastic model (Brudno & Marczynski 1977) are different. Thus, the timing of SIIP propagation into 7 axon collaterals must be different for SIIP-A and SIIP-B. Thus, these two SIIP must have different effects on functional dynamic "binding" in neuronal assemblies to which they project, despite that their SIIP Gaussian statistics are identical. There is a key/lock relationship and functional compression between each SIIP conceptualized as "software," and the corresponding axonal arbor patterns (AAP) conceptualized as "hardware." The term meta-organizing system (MOS) stands for the Hebb-like (1949) heteromodal association systems assumed to have "knowledge" of most sensory information and primary drives of the organism (MacKay 1965; Marczynski 1993), a system that operates mainly via dynamic interactions – "binding" among neurons and in real time (cf. von der Malsburg 1999).

In Figure 1 the ideas conveyed by SIIP-A and SIIP-B are presented as momentary "snapshots" disregarding the intermediate time frames. The SIIP-A is depicted at its most influential time frame. In contrast, SIIP-B is shown in the least influential time frame, and, due to its temporal structure, it could never achieve the effects of SIIP-A. If a condition represented by SIIP-B spike train would prevail for a longer time period and involve many neurons, a functional deafferentiation of cognitive systems would result, leading to a loss of consciousness and slow wave sleep (SWS).

***Origins of the SIIP information code.*** One can argue that the SIIP code, even though it ignores scalar data, should be acceptable as "hard" science. The SIIP code is most likely the product of unicellular organisms, because of its simplicity and reliability for selecting, in real time, adaptive cognitive/motor behavior. The SIIP code probably stems from the cell's ability to sense gradients, that is, inequalities of environmental stimuli, such as temperature and/or concentrations of attractant/repellent chemicals (Koshland 1974; Stock & Surette 1996). The brain ontogeny is one of the most complex processes malleable to environmental influences (cf. Aigner et al. 1995; Barinaga 1999; Benowitz & Routten-

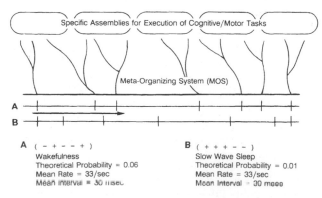

**Key / Lock Relations Between Spike Interval Inequality Patterns (SIIP) and Axonal Arbor Patterns (AAP)**

A ( − + − − + )
Wakefulness
Theoretical Probability = 0.06
Mean Rate = 33/sec
Mean Interval = 30 msec

B ( + + + − − )
Slow Wave Sleep
Theoretical Probability = 0.01
Mean Rate = 33/sec
Mean Interval = 30 msec

Figure 1 (Marczynski). Key/Lock relations between Spike Interval Inequality Patterns (SIIP) and Axonal Arbor Patterns (AAP).

berg 1997; Maurer et al. 1999; Numgung et al. 1997; Shatz 1990; 1992 Shea 1994; Smith & Skene 1997; Stirling & Dunlop 1995; Tessier-Levigne & Goodman 1996). Despite considerable computational capacities of cell proteins (Bray 1995), the ontological processes most likely disregard the physics-inspired infinitesimal clock-work formalism, the scalar data and the Gaussian statistics, and use instead biologically inspired sensing of gradients, that is, inequality judgments (cf. Korovkin 1975).

Even in adult humans, cognitive processes are largely based on inequality judgments, as revealed by experiments in which a subject is asked to compare two digit numbers. The reaction time, measured in milliseconds, referred to as symbolic distance effects, is significantly longer for numbers that are close together and require increased cognitive scrutiny, as compared to judging numbers that are far apart. This indicates that cognitive processes use abstract inequality concepts, even if they are in some way attached to sensory scalar values (Link 1990; Moyer & Landauer 1967; cf. Anderson 1995).

The simplicity of inequality judgments found the way to the commercially ubiquitous bar codes that label supermarket products. If scanned by a hand-held laser beam, the bar code retrieves, in real time, all pertinent information about the purchased product encoded in a few bar interval inequalities. The theoretically attractive-to-neurobiologists concept of brain look-up information tables had been tarnished by its use with convoluted Gaussian statistics and the radial basis function (Poggio 1990), which might increase the latency of real time brain responses, with disastrous consequences.

### The self-organizing properties of SIIP matrices.
The auto-associative memory is defined as a memory system in which every component or signal is encoded and is retrievable from other signal characteristics that have occurred simultaneously during storage (Kohonen 1984). Thus, in the model of auto-association SIIP matrices, a randomly selected subset of SIIP can be used to recover details of distributed memory comprising most of the remaining SIIP (Brudno & Marczynski 1977). This type of storage/retrieval of memory is failure-tolerant in the holographic system (Westlake 1970), which is deterministic and therefore useless for studying the mammalian brain in which the storage-retrieval of memories and other functions must be probabilistic to "protect" the living organism from mindless robot-like behavior. This conclusion is supported by the fact that the statistical distribution of SIIP in our model is probabilistic (cf. Brudno & Marczynski 1977). This is shown by the fact that in each SIIP matrix, the SIIP can be divided into two groups: (1) the essential, without which a matrix completion process in case "damage" would not be possible; and (2) the "redundant" SIIP whose statistical distribution can be deduced from distribution of the essential SIIP. As an example, in a matrix of 8 possible permutations of trigrams, that is, SIIP composed of three inequalities, 5 SIIP are essential and 3 are redundant. In a total of 64 possible hexagrams, 31 are essential and 33 SIIP are "redundant." When making cognitive/behavioral decisions, the mammalian brain "chooses" patterns from both pools (Brudno & Marczynski 1977; for in depth discussion of these topics, see Marczynski 1983).

Our biologically explored autoassociations in SIIP matrices seem to be powerful mechanisms for storing and retrieving memories, considering the availability of large numbers of neurons that may be recruited into cognitive functions of a healthy mammalian brain. These functions can be amplified by "training" the autoassociation network to handle heteroassociations, that is, input from the heteromodal sensory systems (cf. Churchland & Sejnowski 1992) which, in theory, have "knowledge" of virtually all cognitive sensory-behavioral transactions that occur in the brain (cf. MacKay 1966; cf. Marczynski 1993).

The autoassociative memory represented in statistical SIIP distribution differs from that of Anderson (1972) which is based on the mean neuronal firing rate. The SIIP autoassociative memory also differs from that of the celebrated Hopfield's network model (1982) and from the Boltzmann machine thermodynamic model

(cf. Hinton & Sejnowski 1986). These models seek the lowest energy level, a "motivation" which is hardly compatible with biological systems. Moreover, all operations are based on the mean neuronal firing rates. Our SIIP associative model also differs from that of Kohonen (1984) who uses learning rules based on physical deterministic laws of neuronal mean firing rates, a criterion that we rejected as misleading. Our biologically inspired SIIP model, by definition, uses inequalities of neuronal firing and therefore emphasizes nonlinear neuronal behavior as the carrier of information. On the other hand, the Hopfield model and the Boltzmann model regard the non-linear neuronal behavior as a "nuisance" to be ignored by "squashing" the non-linear data through the sigmoid function where the discrete temporal properties of spike trains are ignored and converted into a "static form of nonlinearity" (Hertz et al. 1991). Using the metaphor of a spoken language, this procedure is equivalent to trying to understand the meaning of a spoken word by averaging the pronunciation of its vowels and consonants!

### Spike interval inequality patterns (SIIP) correlate with subject's cognitive-motor functions.
For instance, in cat's transitions from an aroused state to a relaxed wakefulness, slow wave sleep (SWS) and REM sleep, are often not correlated with changes in the mean neuronal firing rate. However, the most interesting are the inversions in statistical distribution of patterns with reference to the stochastic model (Marczynski et al. 1984; 1992). This phenomenon is observed in about 6% of neurons monitored in the centrum medianum nucleus and in the nucleus reticularis of thalamus. The example from the latter region is shown in Figure 2 in the middle part of Figure 2, all 64 permutations of hexagram patterns are printed vertically and numbered from left to right. Each pattern should be "read" vertically from the bottom to the top sign. The chi square ordinates for each behavior measure pattern departures from the stochastic model. The black circle columns and the open circle columns represent respectively excessive emissions and deficits of patterns occurrences with reference to the stochastic model. The spike trains were monitored during cat's four behavioral states: vigilant, attentive, relaxed (REL), and slow wave sleep (SWS). The attentive state was caused by introduction to the experimental chamber of a transparent box containing a live mouse.

The overall impression from Figure 2 is that many patterns were emitted (filled circle columns) and others were suppressed (open circle columns) with reference to the stochastic model. The REL and SWS episodes show inversions in distribution of patterns, particularly obvious by comparing the attentive state with SWS, where the emitted and suppressed patterns changed to suppressions and emissions respectively. On the left of Figure 2, the four vertical scales of chi square values measure departures of pattern occurrences from the stochastic model. The legends on the right for each behavioral episode show: N = the number of spike intervals in a sample; MR = means neuronal firing rate; the chi square values without subscripts measure the sum total discordance of pattern distribution from the stochastic model. In the REL and SWS sample, the chi square values with subscripts i(28) and i(57) respectively represent values only for patterns that inverted their direction in deviating from the stochastic model. However, the most important message of Figure 2 is that the pattern inversion magnitudes are not random, but correlated, that is the larger emissions in an attentive behavioral state tend to be followed by proportionally large suppression of the same pattern in SWS, and vice versa. These relationships were quantified at the bottom of the figure by plotting the roots of chi square statistics. Plotting the "attentive" sample 18 emissions (Att)e versus SWS the same pattern suppressions (SWSs), a high degree of correlation was found ($p < 0.005$). An even more significant correlation was found by plotting (Att)e versus (REL + SWS)s which resulted in a correlation coefficient $r = 0.75$; $p < 0.005$. However, no significant correlations were found for (Att)s going to (SWS)e, nor (Att)s going to (REL+SWS)e.

The inversion phenomenon indicates that the occurrences of select SIIP are homeostatically controlled, most likely by the use-

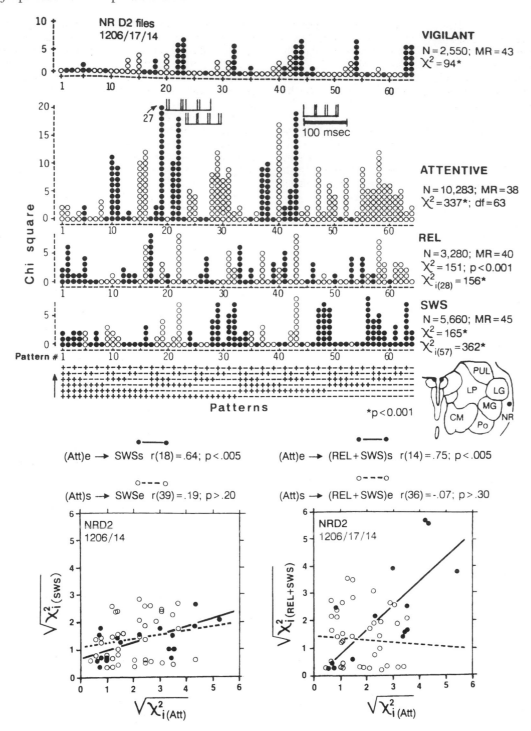

Figure 2 (Marczynski).    Single neuronal firing patterns during four behavioral states.

dependent desensitization of receptors to select transmitters/modulators which are specifically distributed on neuronal somata and dendrites, the latter, however, having most powerful influences on neuronal firing patterns (Mainen & Sejnowski 1966), as if modulating neuronal information coding by controlling SIIP "vocabulary." In many instances, the inversions in statistical distribution of patterns occurred without significant alterations in the mean neuronal firing rates (MR), indicating that SIIP seeking their AAP are more important than MR.

Another feature of the stochastic SIIP model is that it seems to be physiologically natural, because statistically significant and behaviorally correlated SIIP have been monitored in behaving felines from the CA1 region of the dorsal hippocampus, the pulv-

inar nucleus, the thalamic reticular nucleus, the centrum medianum, the visual cortex, and the feline nucleus abducens. The important finding was that during SWS there are always episodes during which single neurons generate virtually perfect SIIP stochastic distribution that has been conceived on the basis of theoretical assumptions of SIIP statistical distribution (Brudno & Marczynski 1977; Marczynski 1983; Marczynski et al. 1984).

As argued by Farley (1966), the main problems in constructing biologically inspired connectionist models of cognitive functions is to provide the system with the capacity to generalize and interpret newly encountered environments on the basis of previous experience. The Hopfield model (1982) and its extension, the Boltzmann machine of Hinton and Sejnowski (1986) have little or no a priori knowledge of what might be the consequences of particular "behavior." These models are "mindless" thermodynamic machines, which if turned on, begin to function from the "tabula rasa" baseline. A question arises of whether the stochastic SIIP distribution has memory of its own which could be utilized by neurons for encoding and transmitting information. The answer to this question is surprisingly positive and it is exemplified by the fact that the transition probability of a pentagram ($+-++-$) going to a hexagram ($+-++--$) equals 0.022024 which is greater than the probability of a pentagram ($--++-$) going to a hexagram ($--++--$) which equals 0.014987, despite the fact that the "history" of both pentagrams, going back four steps, is identical and differs only in the first event (Marczynski 1983; Marczynski et al. 1982). Less distant spike events have proportionally stronger influences on SIIP probabilities. Thus, the SIIP stochastic model of neuronal firing is sensitive to the history of events that can be formally defined as memory. Undoubtedly, this memory is generated by sequential inequality testing of spike intervals, because the process of comparing sequential intervals is advancing in a nonsaltatory manner, that is, one spike interval at a time. Whether or not brain neurons use this memory, remains to be investigated.

# Sleep can be related to memory, even if REM sleep is not

Giuliana Mazzoni

*Department of Psychology, Seton Hall University, South Orange, NJ 07079.*
**mazzoni@shu.edu**

**Abstract:** As reported by Vertes & Eastman, convincing evidence rules out any role for REM sleep in memory consolidation. However, they do not provide convincing evidence for their claim that sleep in general – as opposed to REM sleep per se – has no influence on memory consolidation. Recent correlational data suggest that the number of NREM/REM cycles is associated with performance on a verbal recall task.

[VERTES & EASTMAN]

The target article by **VERTES & EASTMAN** reviews an impressive amount of evidence that convincingly rules out any role of REM sleep on memory consolidation. It is correct, for example, to consider the results of deprivation studies in animals as biased by strong artifacts. In post-learning deprivation studies, animals deprived of REM sleep are overly stressed and their poor performance after REM deprivation might reflect a problem due to performance decrements, rather than to poor memory consolidation. Studies on REM deprivation in humans have the same flaw. Furthermore, compelling evidence about the role of REM sleep on memory consolidation cannot be derived from studies showing an increase in REM activity following a significant and enriched learning situation during waking. In this case the design does not provide a good test of the hypothesis that memory is consolidated during REM sleep, not only for the reasons put forward by **VERTES & EASTMAN**, but also because the method of testing the hypothesis is logically flawed. These are confirmatory tests; they do not consider potential alternative explanations of the data, some of which are accurately summarized by **VERTES & EASTMAN**.

Thus, overall, the article makes a good and important point, that is, there is no convincing evidence on the role of REM sleep in memory consolidation. But while the arguments concerning sleep are convincing, those concerning memory are much less so. It is arbitrary and wrong to assert that "memory requires consciousness." This is clearly not correct, since there is at least one type of memory (i.e., implicit memory) that does not require consciousness (for a review, Schacter et al. 1993). As for particular memory processes, studies on subliminal processing indicate that to a certain extent even encoding can occur successfully outside of consciousness (Draine & Greenwald 1998; Merikle et al. 1995). Retrieval typically requires consciousness, but the example of implicit memory shows that this is not always the case. Consolidation does not re-

quire consciousness. The idea that memory consolidates only through conscious rehearsal has long been abandoned. It is now accepted that consolidation can occur out of consciousness.

Let us distinguish then the concept of consciousness from the concept of a waking state. The claim of the authors should be rephrased as follows: Consolidation in memory cannot occur outside of the waking state, or, as they also claim, Sleep has no role in memory consolidation. But here they overstate their claim. Whereas the authors provide strong evidence against the role of REM sleep, they do not provide enough evidence on the lack of role of sleep in general on memory consolidation. The fact that REM sleep does not play a role in memory consolidation does not imply that sleep, as a whole, cannot play a role in memory consolidation.

Sleep cannot be studied as a juxtaposition of single stages, independent one of another. Sleep is a highly interconnected structure, or organization, where a modification in one stage can strongly influence the others. This has at least two consequences. First, it is difficult to conceive that by disrupting REM sleep the rest of the sleep activity in an individual remains unaltered, and this represents an additional criticism of the REM deprivation studies reviewed by **VERTES & EASTMAN**.

Second, and more important, the organization of the structure of sleep – rather than individual sleep stages – might affect the degree to which materials are consolidated in memory. Sleep cycles (NREM/REM cycles) can be an operational definition of sleep structure or sleep organization. An initial demonstration that the integrity of the structure of sleep as a whole plays a role in memory consolidation comes from a recent correlational study that still needs to be replicated in different populations (Mazzoni et al. 1999.) In this study, it was found that while REM sleep had no significant bearing on memory performance of a list of words learned just before sleep, two indices of sleep organization did. One was the total number of NREM-REM cycles, the other was the proportion of sleep spent in NREM/REM cycles over total sleep time (TST). A NREM/REM cycle was defined as a portion of sleep that contains Stage 1, 2, 3, and 4, followed by a period of REM sleep, without any sizable intervening spontaneous awakening. These data suggest that memory consolidation may not be a function of a single stage of sleep, but rather can be a function of the degree of sleep organization.

# The illusory function of dreams: Another example of cognitive bias

Linda Mealey

*Department of Psychology, College of St. Benedict, St. Joseph, MN 56374.*
**lmealey@csbsju.edu      www.employees.csbsju.edu/mealey**

**Abstract:** Patterns of dream content indicating a predominance of themes relating to threat are likely to reflect biases in dream recall and dream scoring techniques. Even if this pattern is not artifactual, it is yet reflective of threat-related biases in our conscious and nonconscious waking cognition, and is not special to dreams.

[REVONSUO]

**REVONSUO** presents an elegant argument for a functional model of dreaming consistent with the reasoning that form suggests function. However, when applying this type of reasoning to infer dream function, two problems exist. First, because of biases in dream recall and dream scoring techniques, we do not really know the true "form" of dreams, and without knowing their true form, we cannot usefully apply the argument of design. Second, even if we were to accept the premise that dream contents (and not just dream recall or dream scoring systems) emphasize threat-detection, we have no null model to test the hypothesis that threat-detection biases in dreaming cognition do not simply reflect threat-detection biases of our (conscious and nonconscious) waking cognition.

As REVONSUO and other evolution-minded theorists suggest, humans, like other animals, should have evolved a plethora of special threat-detection devices. The results of empirical research leave no doubt that this is, in fact, the case. Psychologists have documented a variety of threat-related cognitive biases at a variety of conscious and subconscious levels (e.g., Cosmides & Tooby 1992; Davey 1995; Hansen & Hansen 1988; 1994; Mathews & MacLeod 1985; McNally 1987; Mealey et al. 1996; Occhipinti & Siegal 1994; Ohman 1993; Shoemaker 1996; Spinks & Mealey, submitted). Survival-related threats are selectively perceived, attended to, remembered, and discussed.

This fact confounds our efforts to interpret dream content. Because dreams are elusive and dream recall is far from complete, dreams that are remembered are bound to be those that are particularly salient (Cohen 1974). Which dreams are most salient? Like waking stimuli, those that contain powerful emotions and elements of threat. Comparison of dream reports obtained under conditions of (1) immediate post-REM awakenings, (2) daily dream diaries, and (3) free recall after long delay, suggests that the least salient (least interesting and least emotional) dreams are forgotten first and progressively, until only the most salient dreams remain in memory. These particularly salient dreams are the ones most likely to be reported in the bulk of research studies, providing a highly selective and non-representative sample. If the function of dreams is to be deduced from the form and content of dreams, then a much less biased method of dream reporting must be used (see e.g., Foulkes 1999; Merrit et al. 1994).

Besides our selective recall of dreams, our existing cognitive biases for attending to threat also result in biases in categorizing dream content. Schredl and Doll (1998) showed that external raters impute relatively more negative emotion to dreams than do dreamers themselves, as a result of paying selective attention to the negative and ignoring the positive elements of dream reports. Furthermore, the most commonly used dream content scoring system (Hall & Van de Castle 1966) is itself biased for picking up on negative rather than positive emotions. The five emotion categories in this scoring system are anger, apprehension, happiness, sadness, and confusion, three of which are clearly negative and only one of which is clearly positive; the remaining "emotion" (confusion) is arguably more likely to be perceived by most people as negative than positive. It is hard to avoid the conclusion that dream content is full of threatening images and emotions when a majority of available scoring categories have negative valence.

Now, it may be that the Hall and Van de Castle system simply reflects the content of dreams rather than constructing it. Alternatively, perhaps the system reflects an extant bias in the English language. Spinks and Mealey (submitted) have shown that English (and, it seems, other languages) is biased to facilitate the labeling of threats: categorizing trait adjectives on dimensions of dominant-subordinate and prosocial-antisocial, they found far more adjectives describing dominant, antisocial people than people in any of the three other quadrants. Furthermore, it is quite possible that this bias of descriptive language in turn, reflects an actual bias of our emotions and our brain. There is strong consensus that the "primary" (cross-cultural and instinctive) human emotions are anger, fear, sadness, happiness, and disgust (e.g., Ekman 1971; Izard 1991; Panksepp 1982; 1999; Plutchick 1980). These are very close to the categories in the Hall & Van de Castle system and, of these, only one has positive valence while the other four are clearly negative. Indeed, anger, fear, and disgust seem to be phenomenological experiences designed specifically for threat-detection.

What this means in the context of a search for the function of dreams is that even if dreams are biased toward threat-related content, and this bias is significant in comparison to a baseline of everyday experience, it still may not reflect a special attribute of dreams. If that is the case, then the argument from design no longer holds.

Tooby (1999) asks "How do you test whether something is an adaptation?" His answer? "To establish something as an adaptation, all one needs to do is to collect evidence that justifies the re-jection of the hypothesis that the structure arose by chance (with respect to function.)" With dreams, we cannot do this. Even if there is a bias in dream content, as REVONSUO argues, we cannot reject the hypothesis that this bias is a by-product of other adaptive biases in our cognition, and is not specific to dreams.

Indeed, as Tooby continues "hypothesis testing is based on statistical inference, and the probability of obtaining the observations that support the hypothesis if the hypothesis were true, as compared to the probability of obtaining the same observations if the hypothesis were not true." With respect to testing a hypothesized function of an organ or process, "(t)his method involves comparing the problem-solving quality of a hypothesized adaptation with the problem-solving properties of other possible alternatives" (p. 3). I have not been convinced that the probability of observing biases in dream content is any different whether REVONSUO's model is true or whether REVONSUO's model is not true; we do not have an appropriate null model that includes the effects of known (non-dream related) cognitive biases. Furthermore, the problem-solving abilities of dreams (if they exist) are clearly inferior to the problem-solving abilities of other conceivable alternatives (Blagrove 1992a), suggesting that dreams are not a product of design after all (see also Flanagan 1995).

I suggest that dream recall is the end product of the serial treatment of REM-sleep neural processes through successive stages of cognitive processing, and as such, that it reflects all the biases of each of those sequential steps. This view is clearly consistent with HOBSON's activation-synthesis model as presented in this issue and elsewhere. I also suggest that our relentless desire to attribute function to dreams is simply one more manifestation of the same evolved cognitive processes that, as a byproduct of their otherwise effective heuristic functions (Gigerenzer & Todd 1999; Kahneman et al. 1982), attribute meaning to other meaningless patterns and create dreams in the first place.

## A more general evolutionary hypothesis about dream function

Jacques Montangero

*Department of Psychology, University of Geneva, CH-1211 Geneva 4, Switzerland.* **jacques.montangero@pse.unige.ch**

**Abstract:** Revonsuo's evolutionary theory of dream function is extremely interesting. However, although threat avoidance theory is well grounded in experimental data, it does not take other significant dream research data into account. The theory can be integrated into a more general hypothesis which takes these data into consideration.
[REVONSUO]

REVONSUO provides us with an original example of psychological dream analysis in terms of dream contents and behavioural adaption. His target article addresses a fundamental question related to dreaming – the function of our nocturnal representations – and proposes a very interesting theory, at once plausible and well grounded within theoretical arguments and experimental data from dream psychology and neurobiology. REVONSUO develops his arguments so skillfully that at first glance they seem convincing. Unfortunately, however, some of his assertions are questionable and his threat simulation theory remains conjectural and cannot explain the existence of the majority of our dreams. I accordingly suggest some complementary hypotheses.

REVONSUO's theory is based on two general ideas that I would be ready to accept. First, it is certain that if dreaming has an adaptive function, it should have enhanced biological fitness in the environment of our ancestors. Second, as Jung (1933) demonstrated with his idea of archetype, it is possible to establish a correspondence between the dream content of contemporary people and the needs and fear of humankind in the remote past. Unfortunately, however, the threat simulation theory of dreaming is based on other ideas that are not convincing. Let us consider the six propositions which summarise the theory.

Two of them are undoubtedly correct. First, the recall of our own dreams as well as the cognitive study of large samples of dreams show that they consist of an organised and selective simulation of the perceptual world (Proposition 1). Second, the threat simulations produced in dreams are indeed perceptually and behaviorally realistic rehearsals of threatening events (Proposition 4).

REVONSUO's other propositions are clearly overgeneralizations. It is true that dream consciousness is very adept at simulating threatening events. However, this does not mean that simulating threatening events is *the* specialised function of the dreaming process, as seems to be suggested in Proposition 2. If the frequency of topics dealt with in dreams is taken into account, it can also be hypothesised that dream consciousness is specialised for simulating human relationships, novel situations, highly desirable future events, and so forth. Another overgeneralization concerns the predominance of negative emotions in dreams. In fact, the majority of dreams are not accompanied by negative emotions. In the sample of 500 dreams studied by Strauch and Meier (1996), less than 30% of the dreams contained negative emotions. A similar overgeneralization can be observed in Proposition 6. It exaggerates the importance of threat representations activated by the experience of dangerous events, such as those that were frequent in primitive environment. Actually we know nothing about the frequency of nightmares experienced by our ancestors. As far as the dream content of Mehinaku Indians mentioned in the article is concerned, I suppose they were spontaneously remembered dreams. If this is the case, they constitute a biased sample of dreams. Everyone can remember a nightmare (at least for some time), while most people are unaware of an enormous quantity of more peaceful and mundane dreams that they could report if they were awakened during the night. For the same reason REVONSUO's numerous references to Hall and Van de Castle's studies (1996, etc.) are not pertinent.

The third proposition of the theory – according to which nothing but exposure to real threatening events can fully activate the threat simulation system – is false. Quite a number of people have frequent nightmares even though they have never been exposed to any particular danger. Depressed people, for example, are known to experience frequent nightmares which are due to inner psychological conditions rather than external causes. It must also be noted that many people drive too fast, cross streets outside pedestrian crossings, or practice dangerous sports, knowing that they may lose their lives during a moment of inattention. To my knowledge, these individuals do not have more frequent nightmares than cautious persons who avoid all danger. It is highly likely that those of our ancestors whose personality characteristics correspond to those of today's fast car drivers and dangerous sports lovers liked to go near wild animals, to swim through dangerous rivers, and attack enemies. They probably had no more nightmares than their remote descendants, however, and yet were adapted to these dangerous activities.

The most controversial proposition of threat avoidance theory, in my opinion, is No. 5, which states that the realistic rehearsal of threat avoidance skills in dreams can lead to enhanced performance. I quite agree with REVONSUO that motor actions represented in dreams might facilitate subsequent actual actions in the waking through implicit learning. However, no experimental data have shown that people's threat avoidance skills are improved after having nightmares. Second, mental images of motor activity can facilitate the subsequent performance of highly complex and non-instinctual movements like those involved in figure skating or golf. But threat avoidance "skills" represented in dreams, such as fleeing or hiding, are so elementary and instinctual that it is difficult to imagine how their representation could help to improve them. Nobody needs numerous rehearsals in order to know how to run away, to hide behind a rock, or to lie flat in the grass in presence of a danger.

In summary, the rehearsal function of dreaming threat avoid-

ance theory is interesting because it draws our attention to the relatively high frequency of archaic content and threatening situations in dreams. However it can be criticised on the following grounds:

1. Most dreams deal with non-threatening situations. REVONSUO's theory therefore cannot explain the functional significance of the majority of dreams.

2. Spontaneously remembered dreams constitute a biased sample of dream content.

3. Negative dream contents are not necessarily linked to actual dangers in real life.

4. The "skills" of threat avoidance in nightmares are so limited and instinctual that they hardly need any rehearsal.

I have suggested a more general hypothesis about the function of dreaming (Montangero 1999) which I would like to state here in slightly different terms. Dreaming is necessary in order to provide the mind with material to process during sleep. There are two reasons for this necessity to keep the mind active during the night. First, conscious reflection is so developed in the human species that if it were not busy with dream content, it might turn to external stimuli or to the current concerns of the sleeping person. This would tend to disrupt sleep. Dreaming thereby has the "guardian of sleep" function noted by Freud (1900/1955), but for different reasons. In this perspective, dreaming serves a biological function by permitting a full night's sleep, which in the long term favors the fittest physical condition in the daytime.

A second important benefit of dreaming is to maintain cognitive capacities such as encoding perceptions, making decisions, and planning actions. Specifically, if these capacities were not used for eight hours every twenty-four hours, they could be impaired in the long run and surely upon awakening. The threat avoidance function suggested by REVONSUO could therefore be included in this more general function of fundamental cognitive capacity preservations and mental vigilance.

Apart from this function of providing the mind with material to process, dreaming may have positive effects such as favoring the emotional balance by mastering or avoiding stress (Koulack 1991), or facilitating the discovery of novel solutions to problems upon awakening. However it must be admitted, as REVONSUO notes, that there are no conclusive experimental findings concerning these effects.

In conclusion, once it had endowed the human species with a high level of conscious reflection, nature had no choice. Cognitive processes involved in consciousness needed to produce evocations or simulations of reality when there was no need to encode perceptions or to plan actions (see, e.g., Foulkes & Fleisher 1975). The result was daydreaming, anticipation, and reminiscence in the daytime and dreaming at night.

## Sorting out additions to the understanding of cognition during sleep

William H. Moorcroft

*Laboratory of Sleep and Dreams, Luther College, Decorah, IA 52101.*
**moorcrwi@luther.edu**

**Abstract:** The target articles by Hobson et al., by Solms, and by Nielsen can be combined to further our understanding of the neurological basis of dreaming during REM and, notably, NREM sleep. Revonsuo adds to our understanding of the function of dreams from the perspective of behavioral biology but overstates its importance. Vertes & Eastman fail in their effort to discount memory enhancement as a function of REM sleep.
[HOBSON ET AL.; NIELSEN; REVONSUO; SOLMS; VERTES & EASTMAN]

HOBSON ET AL. do a thorough review of the disparate views of dreaming to show how new data from new technologies and approaches, including that of SOLMS using neuropsychological methods, is helping to resolve old questions, expand understanding, and show where more research is needed. NIELSEN offers a very reasonable explanation of the occasional presence of dreams during NREM sleep. In the past, too few scientists have been willing and able to work concurrently in the areas of the phenomenology and neural mechanisms of dreams. These target articles do a commendable job of attending to these areas.

HOBSON ET AL.'s overall goal of exploring how dreaming can be explained in terms of brain physiology is presented as three sub-goals. Although in such a long paper it is easy to find details that can be questioned or data that should have been included, tantamount is whether there are points of disagreement that might prove fatal to the theory or show a need for serious revision. Deficiencies that do not severely wound the author's intent are not as important, although some may be important in their own right. From this standpoint, I would like to comment on how well the authors succeed in achieving their stated sub-goals as relevant to their overall goal.

**Sub-goal 1:** That REM and NREM mentation differ from one another and, to lesser extent, from waking. The analysis of published data by HOBSON ET AL. rightly shows that it is at least premature at this time to consider REM dreaming and all NREM mentation identical. (NIELSEN concurs in this.) Likewise, the data are not sufficient to conclude that they are the same as waking mentation. While there may be some overlap in the characteristics of the mentation in the three stages, there are sufficient differences to at least allow the probability that they are fundamentally different. There is a real danger in proceeding as if REM and NREM mentation are the same, for which SOLMS seems to argue, if indeed they are not because of the difficulty of trying to discern the fundamental nature and purpose of a heterogeneous mix. Continuing to treat them as mostly unique would enable greater clarity in discovering more about their sources and purposes in the future. Should it turn out that they are indeed the same then it would be a relatively simple and unambiguous matter to combine the information gathered about them.

**Sub-goal 2:** That REM and NREM substantially differ physiologically (regionally, cellularly, and molecularly) from each other as well as from waking. This section of the paper also succeeds by marshaling evidence and drawing implications from it showing that REM, NREM, and waking are separable states.

Consistent with HOBSON ET AL.'s sub-goals 1 and 2 NIELSEN shows convincingly that dreaming sometimes occurs during NREM sleep concomitant with the occasional presence of certain physiological aspects of REM sleep. This (with acknowledgment of hints by others) he calls "covert REM sleep processes." This conclusion is given credence by HOBSON ET AL. and is compatible with SOLMS's activation of forebrain circuits as the final common path for dreaming. It is more comprehensive, comprehensible, and sensible than any other explanation yet put forward to explain the occasional occurrence of dreams during NREM sleep. This explanation implies that non-dreamlike mentation of NREM must emanate from brain mechanisms different from those that produce dreaming.

**Sub-goal 3:** That the phenomenological and physiological data about REM can be comfortably and usefully integrated (in this case as a three-dimensional model) leading to greater understanding of how dreams are generated. HOBSON ET AL. succeed here as well. In doing so they make a contribution toward the understanding of a "cognitive neuroscience of brain-mind state." It should be noted that along the way they show how activation of areas of the brain important in emotion play a significant role in the shaping of dream content.

Insofar as HOBSON ET AL. achieved their three sub-goals, it appears that they have achieved their overall goal. Specifically, this paper culminates with an updating of the activation-synthesis model, especially to the synthesis portion, by incorporating new findings. These revisions incorporate fresh data and the theoretical implications derived from them (including some important new insights from SOLMS based on his neuropsychological study of dreaming in brain damaged patients). These successful revi-

sions show that the activation-synthesis model is still the main contender among models seeking to show how dreams are generated. However, HOBSON ET AL. must make a better case of explaining SOLMS's data from patients who completely or nearly completely ceased dreaming following brain damage localized to two distinct forebrain regions. In over 99% of these cases the REM generators in pontine brainstem were spared and REM sleep was unaffected. Other patients show the apparent absence of any change in dreaming following pontine damage. SOLMS's data show REM sleep is not necessary for dreaming. However, as SOLMS points out, they do not eliminate the possibility that REM may be the sufficient and most favorable state for dreaming in the intact brain. Yet SOLMS still needs to explain how the forebrain areas become activated in order to instigate dreaming. It is possible that while these forebrain areas are preferentially activated by pontine influences during REM they may also be activated by non-pontine sources.

These models do what good models should do – summarize known relevant facts (including those most recently discovered), make some informed speculation, and point the way to future research. Additionally, the newer AIM model complements the activation-synthesis model by offering a testable explanation of the neurological basis of states of the brain/mind. Furthermore, the AIM model has broader implications for the understanding of human cognition beyond that of its focus on dreaming in the area of the phenomenology and neuronal basis of dreaming.

Nevertheless, I offer the following comments on two details in HOBSON ET AL., while not crucial to the main thrust of the paper, are important in their own right. In section 2. 3. 4, point 6 they confuse access to stored memories with utilization of them when they reason that because dreaming during REM sleep is so infrequently affected by manipulation of pre-sleep experience it must have extremely poor access to recent waking memories. However, because the dream content does not frequently show use of such stored memories does not mean that there is no access to them. In fact, studies of "dream incubation" (cf. Cartwright & Lamberg 1992; Delaney 1998), HOBSON ET AL.'s dismissal of them notwithstanding, suggest that there is indeed access to stored memories during the dream state. An alternative to the explanation of HOBSON ET AL. of why so little of pre-sleep experience finds its way into dreams may be that the agenda of dreams focusing on recent, waking emotional concerns ignores most experimenter-imposed pre-sleep experiences. As both SOLMS and REVONSUO point out, preferential activation of the limbic system during REM reactivates the neural networks containing these emotional memories that then play a major role in determining the content of the dream. It may be difficult to manipulate these emotional memories by artificial pre-sleep experience compared to naturally occurring, emotionally relevant experiences.

In section 4. 1. 1, paragraph 2, HOBSON ET AL. state that brain activation is defined as the "mean firing frequency of brain stem neurons." Cortical EEG intensity also serves to measure brain activation. By making their exclusive assertion they too easily eliminate any explanations of dreaming having a cortical origin. Allowing for cortical measures of arousal opens the possibility of finding non-brainstem origins for brain arousal, such as SOLMS proposes.

REVONSUO's paper about dreaming adds to our understanding of the functions of dreaming by showing its evolutionary roots. However it goes too far in claiming the "threat stimulation theory" is the sole explanation for dreaming. The argument presented in this paper hinges on its definition of function that emanates from the relatively new field of behavioral biology. The basis of this field is the belief that evolutionary success ("inclusive fitness") is the only primary function of any characteristic, including behavior, of any living organism. While this approach has yielded good insights into some of the ultimate reasons for the behavior of animals, it is not yet possible to fully accept the exclusiveness of its explanations.

Simply put, REVONSUO's thesis – that all dreaming stems from perceived threats to bodily welfare as were experienced by ancestral humans – is too narrowly wrought. In contrast, the hypothesis explains the same data if it is broadened to encompass the current emotional concerns of the individual as the focus of dream formulation. In ancestral humans such immediate, emotional concerns would indeed be bodily welfare and safety. So too, as this paper shows, for children and present day primitive people. However, in the contemporary Western world the emotional concerns of humans are more frequently psychological as shown by many studies (see REVONSUO's citations of Kramer 1993; Hartmann 1998). The attempts in this paper to discount such research are weak at best. For example, brushing off the research of Cartwright as simply correlation without causation does not hold because in some of her studies, Cartwright established causation when she actively trained some of her patients to change their dreams which resulted in significantly improved waking mood compared to the untrained subjects.

Portions of **REVONSUO**'s paper also reflect a misunderstanding of the application of some of the neurocognitive theories of dreaming such as those presented in **HOBSON ET AL.** That the source of dream generation is a random stimulation of brain structures involved in cognition does not mean that the resulting dreams are meaningless ("disorganized sensations and isolated precepts"). As stated in a later section of this paper, it is the brain areas that contain the individual's own thoughts, memories, emotions, sensations, motor movements, and patterns-of-cognitive-integration that are activated during dreaming. The initial activation may be random but the output has meaning for the individual. (The attempt to dismiss this possibility based solely on Penfield's memory research [Penfield 1975] does not work because Penfield's interpretation of his data as revealing true memories is no longer held to be valid.) Furthermore, as REVONSUO states, recent research has shown that the areas of the brain involved in emotions are preferentially activated during dreaming thus suggesting a primary role for emotions when dreaming. If we add the assumption, as stated in another section of the paper, that those brain networks most recently activated when awake are the ones that are most likely to be activated even by random inputs, then dreams are more likely to contain recent emotional concerns and "day residue" of relevance to the dreamer.

On a different note, **VERTES & EASTMAN** unconvincingly endeavor to directly dismiss the research that shows memory is enhanced during REM sleep by (1) pointing to some studies that fail to show this effect and (2) by showing how some of the earliest findings may have been owing to the stress of the procedures. Apparently the authors were not convinced by their efforts because they then devote most of its pages to attempting to show theoretically why memory cannot be enhanced by REM sleep. This later portion of the paper is akin to proving, using engineering principles, that hummingbirds cannot fly, in the face of reports that they sometimes do!

Since the most critical portion of **VERTES & EASTMAN**'s paper, then, is the first main section, I will focus most of my comments on it. This section fails to make its case that there is no valid research showing that an enhancement of memory can occur during REM. First, the listing of reviews is decidedly one sided, ignoring several reviews of the literature that conclude there is such an effect (for example: Cohen 1980; Dujardin et al. 1990; Smith 1993; Tilley et al. 1992). It also dismisses the positive findings by stating that there is roughly an equal number of negative findings. But negative findings easily result when looking in the wrong place – such as in the wrong REM window – not always because there is nothing to find.

Second, **VERTES & EASTMAN** focus their criticism on the methods used in some older studies rightly showing that they may be confounded by stress. Other, more recent research showing positive results using better methods (for example, Smith's REM window studies), is dismissed by stating that no one has endeavored to replicate these findings and that the explanation for part of the findings (in this example, the shifting nature of the REM window) is unknown. However, the history of science is replete with examples of how well established knowledge originated from a single source; examples that, for a time, were not fully explainable. (It should also be noted that stress cannot explain the results of REM window studies because there was no detriment of tested behaviors resulting from REM deprivation in general; memory deficits only occurred when the deprivation was during the REM window.)

Third, some recent research is ignored. For example, the study of people in intensive language learning situations (DeKoninck et al. 1990a) in which the more successful students had an increase in REM percent but no increase in total sleep time. These subjects also dreamt in the language more and had more verbal communication in their dreams. In other human research, positive results tended to be obtained for tasks with affective importance to the subject and tasks that required the learner to structure the material and use divergent thinking. Negative results occurred when the material was unimportant to the subject, or already structured, or required convergent thinking. An example of this kind of research is a study by Pirolli and Smith (1989). In this experiment, subjects learned a difficult logic task and a simpler paired word task. Subsequently, one group of subjects slept through the night, a second group was totally sleep deprived, a third REM deprived, and a fourth NREM deprived. One week later they were tested on both tasks. All groups performed equally well on the simple paired word task but only those subjects without REM (REM deprived and total sleep deprived) did worse on the difficult logic task.

There is another line of research not recognized by **VERTES & EASTMAN**. During the retention interval following new learning, some subjects are allowed to sleep (usually nap) when they would get much of one kind of sleep (REM or NREM) but little of the other kind. Other subjects remain awake. A problem for such research has been controlling successfully for time-of-day (e.g., circadian rhythm) confounds of when the sleep occurs. Nevertheless, some of this research has supported the notion that REM sleep is beneficial for memory consolidation, but a few studies have concluded that NREM is more beneficial. For example, Scrima (1984) administered a complex associative memory task to narcoleptics. They were then allowed a 20 minute nap or were to remain awake for 20 minutes. Since narcoleptics have a high amount of REM napping, many of the naps were mostly REM. Recall was best after REM and worse after remaining awake, with recall after NREM intermediate between the two.

In addition, **VERTES & EASTMAN** dismiss statistically significant but small gains in memory during REM because they are inconsequential. Yet a 10% enhancement in memory can be far from trivial especially if accumulated night after night. Consider, for example, what a 10% enhancement of exam scores would mean for a university student.

A problem for "proposed function for REM sleep" of **VERTES & EASTMAN** is the contradiction posed by the lengthening of successive REM periods. The shortest REM period (the first one of the night) follows the longest period of SWS. If this theory is correct then it would seem that the REM period at this time would instead be of considerable length if its function were to keep the brain aroused. On a related note, while it is possible that the REM periods get longer during the sleep period because the need for alertness becomes more likely with the increasing probability of waking as the end of the normal sleep period approaches, it should be noted that REM sleep shows a strong circadian propensity (peaking in the early morning hours) regardless of the timing of the sleep period (cf. Lavie & Segal 1989). Also REM sometimes occurs during naps following little NREM sleep. Finally, while it is possible that a function of REM is to maintain CNS arousal, this does not, necessarily, eliminate additional functions of REM such as memory consolidation. Indeed, most animal systems have multiple functions.

In the end, **VERTES & EASTMAN** show the need for continuing research on, rather than outright dismissal of, memory enhancement during REM sleep while **HOBSON ET AL., SOLMS, NIELSEN,** and **REVONSUO** make significant contributions to the understanding of the sources of dreaming.

# Dreams and sleep: Are new schemas revealing?

Peter J. Morgane and David J. Mokler

*Department of Pharmacology, University of New England, Biddeford, ME 04005.* dmokler@mailbox.une.edu    www.une.edu

**Abstract:** In this series of articles, several new hypotheses on sleep and dreaming are presented. In each case, we feel the data do not adequately support the hypothesis. In their lengthy discourse, Hobson et al. represent to us the familiar reciprocal interaction model dressed in new clothes, but expanded beyond reasonable testability. Vertes & Eastman have proposed that REM sleep is not involved in memory consolidation. However, we do not find their arguments persuasive in that limited differences in activity in REM and waking do not lend credence to the idea that memory consolidation occurs in one state and not the other. Solms makes an argument that dreams are generated from the dopaminergic forebrain based largely on pathological lesion studies in humans. We recognize that this argument has some intuitive appeal and agree with some of the tenets but we do not feel that the arguments are completely convincing due to the lack of anatomical controls, including symmetry and laterality. On the whole, there are interesting arguments put forward in these target articles but the evidence does not convince us that new vistas are opened. No Holy Grail of sleep here!

[HOBSON ET AL.; SOLMS; VERTES & EASTMAN]

***The new reciprocal interaction model-reverie or revelation? A Hobson's choice situation and déjà vu all over again: Hobson et al.*** In HOBSON ET AL.'s lengthy discourse attempting to move toward a cognitive neuroscience of conscious states, one comes away with a remorphed reciprocal interaction model dressed in new clothes but is, in reality, old wine in new bottles. The basic REM circuitry is draped in a multiple neurotransmitter type organization, including a dab of autoreceptor neurobiology. This "updated" reciprocal interaction scheme has now gone the way of most studies on regulatory systems, that is, the newer models of regulatory systems are widespread (distributed), involve multiple neurotransmitter pathways (not just, for example, the old "classic" aminergic/cholinergic "simpler" systems), and special newer families of receptors, including autoreceptors. Generally these complex, multilevel regulatory systems (brain substrates) are not reasonably testable and HOBSON ET AL. do not suggest experiments to clarify outstanding issues.

The "new" Hobson model presented has, in our opinion, only nebulous connection to reality so that the real versus virtual REM sleep/dreaming complex does not emerge. Why do HOBSON ET AL. not develop the dopamine aspects of dream sleep regulation as broached in his long discourse? This is especially relevant given SOLMS's views on a separate dopamine dream system entirely divorced from the REM brainstem mechanisms. And whatever happened to the hippocampal formation in this theorizing, given conclusive evidence of vigilance state-dependent gating of information flow through the hippocampal formation? How can this be totally ignored in the new "distributed model"? We should also remind Hobson that Hernandez-Peon et al. (1963) chemically mapped a cholinergic system that closely followed the limbic-forebrain pathways from the limbic midbrain. Since this issue is noted in the text (i.e., that Jouvet postulated such a pathway) we emphasize that Hernandez-Peon et al. actually mapped such a trajectory.

Let us give credit where is due. The original Hobson cellular neurophysiology of sleep cycle control was sound ground work for future studies. But that was only the scaffolding of the extended (distributed) system that has been coming into view over the past 20 or so years. It certainly has had its day, and invaluably so, but we cannot see much new in this re-review. Why codify the reciprocal interaction model so that the parts (pontine "generator," raphe, locus coeruleus) have become greater than the whole? The overall claim here is that the "essential tenets" of the reciprocal interaction model has been strongly confirmed which, to us, appears self-serving given that this component is such a limited part of the extended REM sleep/dreaming complex.

***To sleep, perchance to learn – aye, there's the rub! Vertes & Eastman.*** Two of the principal premises of the VERTES & EASTMAN article are, first, that the primary function of REM is the endogenous stimulation of the brain to maintain requisite levels of CNS activities throughout sleep. Secondly, that sleep involves basic biological functions whereas memory requires consciousness. Relative to the latter, we might well ask whether dreams are actually a form of consciousness. We don't see sound arguments to the contrary!

We do not find strong reasoning behind the view that the brain needs "requisite" levels of activity throughout sleep. VERTES & EASTMAN also postulate that theta serves memory function in waking but not in REM. Why so? We can imagine that without a great deal of extraneous "noise" seen in waking that such is tuned out in REM so that in this state processing could occur unencumbered. Might this help consolidation since theta is associated with selective diminished inhibition in the hippocampal formation in both waking and REM? Hence, why not assume theta involvement in memory functions in both waking and REM? The authors categorically state that the theta of REM is a "byproduct" of intense activity of the pontine region in REM sleep and thus may have no functional significance in REM. Why isn't waking theta a byproduct of similar activation? The authors also state, without proper documentation, that there is no mechanism in REM for selection and transfer of information to the hippocampal formation from other sources. Further, they state that since information in REM is chaotic, it can have no functional value. No clear distinctions are made as to theta quality so that it would serve memory function in waking but not in REM. Finally, the authors serve up, without adequate background reasoning and references, that REM is a mechanism to insure and promote recovery from sleep. This idea does not strike us as viable. Further, it is not intuitively appealing! The whole complex realm of REM components could not likely serve such a basic primary function as waking up the sleeping brain.

***Dopamine in the dream machine? Solms.*** Are we really ready for a major paradigm shift, that is, that REM is controlled by pontine brainstem mechanisms whereas dreaming seems to be controlled by dopamine forebrain mechanisms? Is this a form of blasphemy against long accepted views that REM sleep is the physiological equivalent of dreaming? How can dreaming be put forward as not an intrinsic function of REM? Only by separating it in space?

Evidence that dreaming is generated by dopamine circuits (particularly mesocortical and mesolimbic components), however, is somewhat soft. The author gives little or no proof of exactness of clinical lesions. Were these presumed to be bilateral (unlikely)? If unilateral, do they alter sleep only on the lesion side? Surely we cannot prove one-sided dreaming? SOLMS reviews work showing brainstem lesions leaving dreaming intact, whereas forebrain focal lesions result in cessation or near cessation of dreaming. The link between forebrain seizures and recurrent nightmares also does not constitute strong evidence that dopamine systems play causal roles in generation of dreams. It is suggestive and intriguing but certainly not causal!

SOLMS postulates that so-called "motivational mechanisms" (volition and adynamia) are essential for the generation of dreams. What reasoning is this based on? To us "motivation" is still the "phlogiston of psychology." SOLMS claims that the activation state "engages" the dopamine circuits of the ventromesial forebrain. What does this actually mean and how and where does such "engagement" occur? How do specific aspects of the REM state (NE and 5-HT demodulation) facilitate primary dopamine effects? Is the assumption that stoppage of NE and 5-HT activity activate the dopamine system(s) in the ventral tegmental area and, possibly, the substantia nigra? There is no direct neurophysiological evidence of this that these commentators are aware of.

The relationship of the putative dopamine "dream-on" mechanism and the cholinergic "REM-on" mechanism of the reciprocal interaction model is not developed to any extent, thus leaving us without any viable link. Dopamine may well be part of the dream machine but relations with physiological REM processes remain elusive.

# Critical brain characteristics to consider in developing dream and memory theories

Adrian R. Morrison[a] and Larry D. Sanford[b]

[a]*Laboratory for Study of the Brain in Sleep, Department of Animal Biology, School of Veterinary Medicine, University of Pennsylvania, Philadelphia, PA 19104-6045;* [b]*Division of Anatomy, Department of Pathology and Anatomy, Eastern Virginia Medical School, P.O. Box 1980, Norfolk, VA 23501.*
**armsleep@vet.upenn.edu      sanford@borg.evms.edu**

**Abstract:** Dreaming in sleep must depend on the activity of the brain as does cognition and memory in wakefulness. Yet our understanding of the physiological subtleties of state differences may still be too primitive to guide theories adequately in these areas. One can state nonetheless unequivocally that the brain in REM is poorly equipped to practice for eventualities of wakefulness through dreaming, or for consolidating into memory the complex experiences of that state.

[HOBSON ET AL., NIELSEN, SOLMS, VERTES & EASTMAN, REVONSUO]

**Dreams.** To discuss views on dreaming – its nature; whether dreams occur in both REM and NREM; and, if so, their degree of similarity – first requires discussion of the characteristics of the brain in sleep. We are of the mind, of course, that mental functioning requires brain functioning. We hasten to add, though, that the current view of how the brain functions may not match a future reality. The "computer" brain of today may be an analogy no closer to the brain's actual mode of operation than was the "telephone-line" brain of the past. Nevertheless, we believe that dreams will always be grounded in the physical workings of the brain.

The foregoing suggests an answer to the continuing debate on the nature of dreams or mental activity in non-REM and REM reviewed by **HOBSON ET AL., NIELSEN,** and **SOLMS. NIELSEN** suggests a way to resolve the conflict between the one-generator model of dreaming advanced by Foulkes and others and the two-generator model championed by **HOBSON** and colleagues: Covert REM processes can intrude into NREM to color mental activity like that in REM. Indeed, identifiable REM or "REM" events can appear in NREM as **NIELSEN** has reviewed. Foulkes (1997), however, has essentially rejected this idea, warning of a tautology if REM is defined to be present whenever dreaming occurs.

Returning to our introductory idea, may we suggest that both our traditional, "digital" staging of sleep/wake states and our understanding of just what neurophysiological processes equate with elements of cognitive processes are too primitive to resolve the debate one way or the other? Otherwise, it is impossible for us to believe that total absence of the pervasive aminergic activity seen in wakefulness and NREM, which is a hallmark of REM, as **HOBSON ET AL.** have reviewed, should not be reflected in a measurable difference in mental activity in the two sleep states. In our opinion, the AIM model developed by **HOBSON** and colleagues has considerable merit as a way to order one's thinking on the complexities of the state concept although we doubt that it will lead to an early truce.

A physiological difference between the two states that may ultimately bear on the problem, but ignored in the articles, is the profound alteration in hypothalamic regulation that is a feature of REM (Parmeggiani & Morrison 1990). One would think that such a dramatic change in regulation would, in some way, feed back into mental activity.

Readiness to move, although greatly suppressed, appears to be a feature of REM, while NREM is a quiescent state. **HOBSON ET AL.** noted that motor areas are highly active in REM although the background of atonia limits peripheral expressions to the intermittent brief muscle contractions that one sees in various striated muscles that result in limb and rapid eye movements. Pontine tegmental lesions in cats that eliminate the usual muscle atonia of REM reveal an impetus to move in REM (Henley & Morrison 1974; Jouvet & Delorme 1965). The movements are expressed as well organized behavior. However, no organized behavior emerges in NREM, another clear difference in the two states.

Yet, the elaborate behaviors observed in REM without atonia (REM-A) may not be regarded as a true expression of normal brain activity unencumbered by muscle paralysis, for the brain has been damaged. Furthermore, behaviors observed depend on the sites of the lesions in a very predictable way (Hendricks et al. 1982). Also, the animals in wakefulness have distinct abnormalities. In a study of activity during wakefulness of cats exhibiting REM-A in sleep (Morrison et al. 1981) found that all of them had a significant increase in exploratory locomotion.

Further, aggression during REM-A only appears in those with particular rostral lesions (Hendricks et al. 1982) and has the characteristics of predatory attack (Leyhausen 1979). In our study of 28 cats, eight expressed predatory attack behavior, not affective defense during REM-A; but the six that were also aggressive in wakefulness demonstrated their release of aggressive tendencies as affective defense (Morrison 1979). Unilateral lesions of the central nucleus of the amygdala also released aggressive behavior that differed in a similar way in wakefulness and REM-A (Zagrodska et al. 1998). Although these cats exhibited predatory attack during REM-A, they showed no increase in predation when tested during wakefulness but were very aggressive toward conspecifics. The different expressions of aggression most probably reflect the great reduction in sympathetic tone and hypothalamic control during REM in cats (Parmeggiani & Morrison 1990).

The characteristics of REM-A we have described, confounded as they are by brain damage, lead us to doubt that the behaviors observed serve as evidence that REM is a period when waking behaviors are being practiced. However, the behaviors certainly are consistent with the idea that the brain in REM is most like the brain in very alert wakefulness when an animal orients (Morrison 1979; Sanford et al. 1993). But practice for the realities of a stressful existence during REM, whether by cats, whatever their capacity for mentation, or by humans during dreams as **REVONSUO** proposes, would seem to be severely hampered by the absence or alteration of critical regulating systems of the brain during REM.

Both **HOBSON ET AL.** and **SOLMS** provide diagrams that suggest circuitry in the forebrain underlying dream elaboration. These are based on both lesion and imaging data. It is well to keep in mind that activity in the forebrain very likely plays a key role in maintaining, and most certainly in initiating, REM. Morrison and Reiner (1985) first emphasized that the important decerebrate experiments of Jouvet (1962) focused excessive attention on the caudal brain as the site of initiation of REM. Most certainly much has been learned as a consequence of this focus, but at the same time the forebrain was forgotten as a site also important for REM in *intact* individuals. Decerebrate cats are inordinately predisposed to enter a REM-like state following all sorts of strange stimuli: insertion of rectal thermometers, passing of stomach tubes, and pinches (Jouvet 1964). Morrison and Reiner (1985) reasoned that decerebration substituted for the processes in NREM that led to the suppression of hypothalamic control we have mentioned earlier. Now, much needed attention is being paid to the forebrain with regard to initiation and maintenance of REM (Morrison et al. 1999), which should feed in to further elaborations of dream theories.

**Memory.** The case for a role for REM in learning and memory consolidation appears, for the most part, to be built on somewhat tenuous correlational relationships between REM occurrence and indicators of performance. **VERTES & EASTMAN** present, in our minds, compelling arguments questioning the temporal relationship between the occurrence of REM and memory consolidation. While we make no claim to expertise in learning and memory, we have recently become interested in a learning paradigm, fear conditioning, and its effect on REM. In essence, this is a classical conditioning procedure training an animal to make an association between a neutral stimulus (cue) or situation (context) and an aversive stimulus (usually shock). Explicitly cued fear conditioning produces long-term potentiation-like changes in the lateral amygdala (Rogan et al. 1997), and contextual fear conditioning involves the hippocampus (Desmedt et al. 1998). Given arguments that learning is associated with increases in REM (Ambrosini et al. 1993; Smith 1985; 1995), one would have expected increased REM following fear conditioning. Far from resulting in enhanced REM, fear conditioning training selectively suppressed REM for 1 to 2 hours post-training (Sanford et al. in press). Adrien et al. (1991) utilizing a similar procedure reported a significant decrease in REM and no REM rebound during the subsequent 24 hours. From our perspective, then, it seems that a learning paradigm in which REM is selectively suppressed would be problematic for theories that REM is necessary for retaining the same learning. Interesting to note, Adrien et al. (1991) also reported an increase in NREM1 and we found an increase in NREM percent. These findings are consistent with suggestions that NREM may promote memory consolidation (e.g., Fowler et al. 1973; Wilson & McNaughton 1994).

The striking electrophysiological phenomena of REM are especially beguiling, leading researchers to search for special meaning or relevance for their occurrence. This has led to the inbuilt assumption for many theorists that neural activity specific to REM, as opposed to NREM or sleep in general, somehow aids in memory consolidation. That same activity would seem to us to pose potential problems for the processing of previous learning. For reasonably accurate memories to be formed, one would expect that reactivated traces (if such occur) would need to be free from internal and external interruptions. Alterations in hypothalamic function and the highly activated brain, as described in the previous section, would present possible sources of internal interference. In addition, brain processing may be almost as susceptible to external influences during REM as during wakefulness. Evoked potentials are similar during REM and wakefulness. This finding (among others) led to Llinás and Paré's (1991) suggestion that brain processing in REM and wakefulness is the same except for the elevated sensory threshold during REM. Actually, we demonstrated that cats in REM-A may behaviorally orient to simple external auditory stimuli of varying intensities in much the same way they do in wakefulness (Morrison et al. 1995). This suggests even more similarity between the way information is processed in wakefulness and REM. Indeed, these similarities do not rule out the possibility for rudimentary (S-R type) learning during REM itself, but in no way suggest that memory would be promoted. If so, such learning could pose problems for the idea that memory consolidation takes place during REM. According to interference theory, the formation of associations in the interval between learning and recall may be a factor in forgetting (Hulse et al. 1980).

One of the major problems we see with ascribing functional significance to neural activity in REM is the dramatically altered central orchestration of neural events. It seems to us that even theories that deal with specific processes must take into consideration the condition of the organism as a whole. In wakefulness, an extremely activated brain, irregular respiration, bouts of tachycardia, and twitching muscles coupled with potential extraneous interference from the environment would hardly be considered optimal for memory formation. We see no reason to think that some special quality of REM makes this same combination of factors conducive for consolidating information previously learned in another state.

## Post-traumatic nightmares as a dysfunctional state

Tore A. Nielsen and Anne Germain

*Sleep Research Center, Hôpital de Sacré-Coeur de Montréal, Montréal, Québec, Canada H4J 1C5; [a]Psychiatry Department, Université de Montréal, Québec, Canada.* **t-nielsen@crhsc.umontreal.ca**

**Abstract:** That PTSD nightmares are highly realistic threat simulations triggered by trauma is difficult to reconcile with the disturbed, sometimes debilitating sleep and waking functioning of PTSD sufferers. A theory that accounts for fundamental forms of imagery other than threat scenarios could explain the selection of many more adaptive human functions – some still pertinent to survival today. For example, interactive characters, a virtually ubiquitous form of dream imagery, could be simulations of attachment relationships that aid species survival in many different ways. [REVONSUO]

***PTSD as a dysfunctional dreaming state.*** The threat simulation theory would appear to suggest that nightmares, as exemplary threat simulations, are highly functional, for example, "nightmarish dreams are not ones that failed to perform their function, but, by contrast, prime examples of the kind of dreams that fully realize their biological function" (REVONSUO, sect. 2.2.8). Such a notion would be clearly at odds with the predominant psychiatric view that considers nightmares to be dysfunctional, as embodied in the Nightmare Disorder and Post-traumatic Stress Disorder categories of the DSM-IV (American Psychiatric Association 1994).

However, nightmare functionality in this model is limited primarily to a past, evolutionary function, not to a current regulatory function. REVONSUO likens nightmares to natural variations in a biological defense system such as the immune system. Like immune responses, which are sometimes overactive in susceptible, hypersensitive, individuals (e.g., allergy sufferers), acute or chronic nightmare sufferers may suffer merely from a "harmful side effect" of the threat simulation system – much like an allergic condition – but a side effect whose evolutionary costs (nightmare distress) nevertheless did not outweigh its benefits (survival). Further, such side effects are likely transmitted genetically, as natural selection of such variations would require. Thus, one cannot necessarily argue that the distress and impairment of Nightmare Disorder constitute evidence against the biological function of nightmares. Rather, they may simply be an inherited "cost" of the evolutionary necessity to avoid threat. This argument holds to the extent that Nightmare Disorder is inherited; there is at present only limited evidence supporting this possibility (Hublin et al. 1999a).

On the other hand, nightmares induced by trauma are much more directly pertinent to the predictions of the theory because they are less likely to be due to genetic dispositions than are idiopathic nightmares and because their severity is more likely to be due to trauma severity than to inherited factors (see Connor & Davidson 1997 for review). Rather, future PTSD susceptibility is increased by past exposures to trauma, particularly violent trauma; the more numerous the past exposures, the higher the likelihood that a future trauma will trigger PTSD (Breslau et al. 1999). Thus, if there is evidence that PTSD nightmares are associated with signs of dysfunctional adaptation to the environment, then the threat simulation theory is weakened.

REVONSUO acknowledges that PTSD nightmares do not necessarily facilitate adaptation to the trauma that incited them. The nightmares of war veterans with PTSD are not adaptive because their content does not deal with the real threats of the battlefield: "There are few such skills among human threat avoidance programs whose rehearsal would be of much help in an environment where one may at any moment get killed by shrapnel, the invisible sniper's bullet, nerve gas, hidden land mines . . . and so on" (REVONSUO, sect. 6.3, para. 3). It appears that only current threats that correspond to ancestral threats may benefit from the "rehearsals" of threat simulation. Nonetheless, one may question this reasoning in the case of war trauma (where a strategy of "combat avoidance at any cost" could well help to save a soldier's life), as well as for rape and assault trauma (where avoidance of the perpetrator and/or the crime scene could well prevent worse injuries), for motor vehicle trauma (where avoidance of driving could enhance survival), or for any number of other, somewhat predictable, trauma. It is not clear why these types of trauma would *not* benefit from the threat simulations proposed by the theory whereas other similar, or even less predictable ancestral types of trauma, such as natural disasters, would.

Furthermore, PTSD may well be a dysfunctional, if not completely debilitating condition, which can hinder rather than facilitate adaptation. REVONSUO does not review a rather large body of evidence describing the dysfunctional aspects of PTSD. He thus leaves the impression that PTSD would not be likely to be an impediment to the goal of survival. It is our impression, however, that the accumulating mass of evidence characterizing PTSD as dysfunctional supports the notion that it may work counter to the evolutionary pressures described by REVONSUO. First, and perhaps most obviously, the nightmares of PTSD can often disrupt sleep and engender dysfunctional reactions in the daytime. In severe cases, such reactions can be worse than those induced by Nightmare Disorder. Moreover, many studies have found abnormalities in REM sleep latency, REM sleep amount, and REM density (see Benca 1996, for review), evidence favoring the hypothesis that PTSD is a function of disturbed REM sleep (Ross et al. 1989). Studies of PTSD sufferers have also found anomalies of breathing (Krakow et al. 2000), arousal regulation (Mellman 1997), sleep efficiency (Mellman et al. 1997), body and limb movements (Mellman et al. 1995), and NREM sleep awakenings (Kramer & Kinney 1988), among others. These, and numerous studies assessing perturbations in waking state variables as diverse as memory (Moradi et al. 1999; Wolfe & Schlesinger 1997), visual imagery (Bryant & Harvey 1996), startle (Orr et al. 1997), P300 (Metzger et al. 1997), and corticotrophin-releasing hormone (Baker et al. 1999) all indicate severe abnormalities in PTSD sufferers. Such global perturbations of key cognitive and physiological systems would seem to decrease an individual's chances of survival significantly. Whereas the threat simulation theory would predict that PTSD nightmares are evolutionary remnants that are, at worst, non-functional in nature, the evidence together suggests that they reflect a more generally disturbed, dysfunctional state that is induced by traumatic, much more than genetic, factors.

**The polyvalence of successful evolution.** His limited characterization of dreaming as threat simulation leads REVONSUO to consider only one specific adaptive function pertinent to human evolution. For example, the evolutionary advantage afforded by dreaming dealt with *"behavioral* strategies to avoid contact with such animals and to escape or hide if attacked by them" (sect. 3.4.2.1, para. 4, emphasis added). Presumably, detouring, running fast, hiding, and the like were the behaviors that gave humans a reproductive edge in this case. However, in prehistoric times there were also naturalistic events that led to the selection of highly advanced, cognitive, social, and emotional skills that were not necessarily organized around threat. Why were such skills also not simulated during dreaming so that waking-state adaptation could be facilitated on several fronts at once'?

Such a notion seems more consistent with the wide variety of very common themes and structures seen in dream reports (see commentary by Germain et al. this issue). In fact, it could be argued that any dream content with a high overall prevalence is a candidate for supporting a biological function analogous to that of threat simulation. For instance, the observation that interactive character imagery is virtually universal to dreaming could lead forthright to a theory of dreaming as simulation of attachment relationships. Attachment relationships (Bowlby 1969) are also fundamental to survival and may have been as essential to threat mitigation as were the behavioral strategies of running from predators and disasters. Strong interpersonal bonds could have ensured strong tribal structures which, in turn, could have enabled organized defenses against predators and cooperative problem-solving skills more generally. Perhaps more important, such a socio-emotional function for dreaming would still have clear adaptive significance for dreams occurring today. For example, family and group cohesion remain essential ingredients in many aspects of health and survival (e.g., Albert et al. 1998; King 1997).

Similar arguments might be made for different ubiquitous classes of dream imagery such as self-imagery and place-imagery. For example, self-imagery may facilitate functions related to ego and self-state development (Fiss 1986) or the learning of new motor competencies; place-imagery may facilitate functions related to spatial learning and orientation (Winson 1993). All such functions may have evolved much in the way that REVONSUO describes for threat perception and avoidance, with the important difference that these are more polyvalent cognitive and socio-emotional functions that are pertinent to the continuing evolution of our species today.

# Insights from functional neuroimaging studies of behavioral state regulation in healthy and depressed subjects

Eric A. Nofzinger
*Western Psychiatric Institute and Clinic, Pittsburgh, PA 15213.*
**nofzingerea@msx.upmc.edu**

**Abstract:** New data are presented showing excellent replicability and test-retest reliability of REM sleep findings from functional brain imaging studies in healthy subjects on which newer brain-based models of human dreaming have been constructed. Preliminary region-of-interest findings related to bottom-up versus dissociable brain systems mediating REM sleep and dreaming are also presented.
[HOBSON ET AL.; SOLMS]

The field of dream research is indebted to the efforts of each of these groups of investigators in their tireless efforts to formulate synthetic models of brain function that underlie the experience of dreaming. SOLMS has provided an intriguing challenge to the basic conceptualization of dreaming as a bottom-up phenomena and the work of HOBSON ET AL. reviews an astonishing array of preclinical, experiential, and cognitive neuroscience data in their most recent formulation of a brain-state model of consciousness. I can only add a few observations from our functional brain imaging studies across the behavioral states of waking, NREM, and REM sleep in healthy and depressed subjects that may have relevance to these areas of inquiry (Nofzinger et al. 1997; 1998; 1999; 2000).

A concern in human brain imaging studies of sleep is whether the findings are replicable both across and within subjects. This is important, since isolated disparate findings should not direct models of brain function as conceptualized by each of these groups of authors. This is an appropriate concern, since most studies have relied on statistical methods involving thousands of statistical comparisons across all brain pixels in relatively small sample sizes. Our group has now replicated in an independent group of four subjects our original findings of brain structures that have increased relative glucose metabolism in REM sleep when compared with waking. Additionally, in the new sample, we performed a test-retest reliability study in which the waking to REM sleep

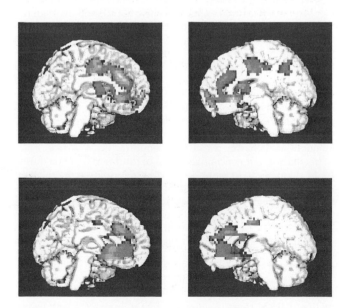

Figure 1 (Nofzinger). Bilateral mid-sagittal sections showing REM sleep minus wake activations. Two figures on top demonstrate regions activated in four healthy controls at each of two time-points separated by 12 weeks. Two figures on bottom demonstrate regions activated in six independent healthy subjects from a prior study.

functional brain imaging study was repeated on two occasions separated by 12 weeks. The figures on the previous page show that the pattern of activation present in the original subjects is also present in an independent group of subjects at each of two time-points. The areas of activation are remarkably similar to the areas of interest diagrammed by the **HOBSON ET AL.** in their Figure 7: "Forebrain processes in normal dreaming – integrated model." They include activation of their anterior paralimbic group, Zone 3, beginning in the ventral striatum and continuing into sub-, pre-, and supragenual anterior cingulate cortex with some medial prefrontal cortex; bilateral activation of the basal ganglia, Zone 5; bilateral activation of inferior parietal cortex (not shown), Zone 9; and perhaps difficult to see, activation of basal forebrain, Zone 2. These additional findings support the original findings on which the HOBSON ET AL. integrated model are built.

Even in this replication study, however, we did not see clear evidence for a change from waking to REM sleep in the pontine reticular formation, an important structure in a bottom-up approach to REM sleep. Perhaps the absence of a change simply reflects that this region is similarly active between waking and REM sleep, both states of cortical activation. Perhaps this is a limitation of the method in which there are spatial resolution constraints of scanners and in which there is significant smoothing of images that is performed across subjects to control for inter-subject regional variations in brain morphology. These factors would preclude our ability to see small pontine nuclei in the PET images. In an attempt to address some of these issues, we then drew regions-of-interest (ROIs) in the pontine reticular formation across several axial planes, then compared the activity in these regions across behavioral states. As expected, a clear drop in functional activity in the pontine reticular formation was noted from waking to NREM sleep, then a return of the waking level of activity was noted following entry into REM sleep, although not to levels exceeding that of wakefulness. This pattern of change paralleled changes in global metabolism across the entire forebrain, supporting at a metabolic level distinctions in global forebrain function and pontine reticular formation function across these unique behavioral states.

Still, we remain puzzled by some preliminary observations of our REM sleep imaging data when we explored the relationships between the pontine ROIs and other brain structures thought to play a role in an integrated model of forebrain processes in dreaming. We were reassured to find a positive correlation between relative metabolism in the pontine reticular formation and that of the thalamus (G). We also found a positive correlation between relative metabolism in the pontine reticular formation (P) and that in (0) primary visual cortex (pons × left occipital cortex correlation = .928). Do these relationships represent a metabolic correlate of PGO wave generation? Also of interest is that functional activity in the amygdala (A) paralleled activity in the pons, thalamus, and occipital cortex. This would be supportive of more recent efforts to more directly link amygdala function, and presumably its role in emotional behavior, with REM sleep.

In contrast, our conceptual model of forebrain function during REM sleep began to break down when we explored the relationships between this PGO-A system and that of the anterior paralimbic REM activation axis that is becoming a signature of forebrain function during REM sleep in human functional brain imaging studies. Relative activity in the anterior paralimbic system was negatively correlated with that in the PGO-A REM system (pons × left pregenual anterior cingulate correlation = −.77; pons × right pregenual anterior cingulate cortex correlation = −.945). How can this be? Shouldn't functional activity in the anterior paralimbic system parallel that in the pontine reticular formation if REM sleep is generated by the brainstem with consequent forebrain manifestations? Similar preliminary studies in depressed subjects help clarify this to some degree. When we additionally looked at the relationship between basal forebrain and hypothalamus function in relation to the pontine reticular formation and anterior paralimbic system in depressed patients, we found that functional deficits in basal forebrain and hypothalamus

paralleled those deficits in the anterior paralimbic system. In contrast, in healthy subjects, there was more of a direct relationship between functional activity in the basal forebrain and hypothalamus and the reticular formation consistent with the notion that these structures may be rostral extensions of an ascending activation system. The findings in depressed patients suggests that there may be unique functional roles served by a more generalized posterior ascending activation system from the pontine reticular formation, through thalamus and on to cortex and a more specific anterior paralimbic activation system from the basal forebrain and perhaps hypothalamus in the service of mediating adaptive or motivational behavior. Given the more selective activation of anterior paralimbic structures over occipital cortex during REM sleep in human imaging studies, it may be that the more selective anterior activating system is preferentially activated during normal, healthy REM sleep.

Does the pontine reticular formation play a role in triggering function in this anterior ascending system? If the primary forebrain function during REM sleep is the maintenance of anterior paralimbic forebrain activity, it may be that the inverse relationship between pontine reticular formation activity and the anterior paralimbic system represents the efficiency of the system. The easier it is for the ascending system to engage anterior paralimbic activity, the less work it has to do. In support of this, with increasing severity of depression, depressed patients show increasing difficulty in activating the anterior paralimbic system from waking to REM sleep. Concurrently, they also demonstrate increasing relative activity in the pontine reticular formation during REM sleep with increasing depression severity. This raises the possibility that the increased REM sleep production in depressed patients reflects a compensatory brainstem drive of more posteriorly located ascending activation in response to behavioral deficiencies in anteriorly located ascending activation mediated through the basal forebrain and into anterior paralimbic structures. These preliminary observations, however, await confirmation in larger sample sizes as well as replication as we have now done for the findings in healthy subjects.

In closing, we agree that the recent sleep imaging work in humans is important and we are glad that the findings have helped shape models regarding forebrain processes in dreaming. We also recognize the inherent limitations of these methods, primarily in terms of spatial and temporal resolution in more clearly identifying the temporal sequencing of regional brain activity in proximity to the time that the functional activity is occurring at the electrophysiological level. We feel that future developments in this area will only come via collaborative interchanges between the preclinical research labs and the labs performing the human studies as each can amplify, extend, and provide meaningful interpretation of the others' data. Future refinements in imaging technology providing increased spatial and temporal resolution will undoubtedly make these early human sleep imaging studies obsolete and leave us with richer datasets on which to further refine these models of human brain function in relation to dreaming.

# Toward a new neuropsychological isomorphism

M. Occhionero and M. J. Esposito

*Department of Psychology, University of Bologna, Bologna, Italy.*
**{occhione; esposito}@psibo.unibo.it**

**Abstract:** The deactivation of the dorsolateral prefrontal cortex is likely to be essential for generating some characteristics of the dream. The heterogeneous nature of NREM sleep makes it difficult to assume that there are different NREM dream triggers. Different cortical and subcortical neurophysiological conditions modulate mentation both in waking and in sleeping without any specific direct triggering factor.
[SOLMS]

SOLMS's is a target article of unquestionable elegance. The conclusions he draws from the analysis carried out on manifold sources allow us to free oneiric activity from the neurophysiological processes of REM sleep. Besides "rehabilitating" the cortex in the production of sleep mentation, SOLMS's model gives dreams back the dignity of thought. Dreaming becomes the active product of cognitive operations taking place through the intervention of areas that are normally delegated to control complex functions. The dream is no longer an epiphenomenon derived from the assignment of meaning by the forebrain to casual stimuli coming from the mechanisms that regulate the REM/NREM cycle alternation. What is puzzling is that SOLMS is still looking for specific anatomic-functional mutual relationships, even in dealing with mental processes as complex as oneiric thought.

A second point on which we do not agree is SOLMS's conclusion about the deactivation of the dorsolateral prefrontal cortex. This cerebral area carries out the very highest cognitive processes, regulating self-monitoring and strategic control over functions. Neuroimaging studies have shown that this area is deactivated during sleep and neuropsychological observations have shown that a lesion here does not affect oneiric activity. SOLMS concludes from this that "the dorsolateral prefrontal cortex is inessential for dreaming sleep." Some peculiar characteristics of dreaming (e.g., inefficiency in reality testing, the lack of strategic control over the course of thought, the frequent presence of temporal-spatial distortion) might be the consequence of weak and ineffective control by the executive functions of the dorsolateral prefrontal cortex. The hypoactivation of this area may subserve the specific cognitive organization of oneiric activity: this is hardly "inessential."

Dream production may accordingly be the result of complex patterns of cerebral reorganization depending on CNS's specific functional equilibrium (activation-hypoactivation) and to specific dream content. A point that requires further clarification from SOLMS is "the necessary presence of triggers" for dream production. If the pontine cholinergic mechanism (REM-on) triggers a sudden modification of the electro-encephalographic pattern, the NREM trigger is not well specified. NREM sleep in fact consists of heterogeneous stages with no possible precise boundary between them; changes are progressive rather than abrupt.

SOLMS should clarify what he means by NREM trigger. At present, one frequent question answerable only approximately, concerns the continuity/non continuity of oneiric activity during sleep. If separate, stage-specific NREM triggers were documented, one would expect from SOLMS's model, that the oneiric activity in NREM depends on them. This would support the hypothesis that sleep mentation is not continuous. In contrast, the absence of specific state-dependent triggers would free dream like activity from specific neurophysiological layers, favoring the hypothesis that it may occur in any sleep stage, even in a continuous way.

SOLMS does a broad review of neuroanatomical, neurochemical, and neuropsychological data; however, he disregards psychological data somewhat. He uses only a small portion of REM-like dreams in NREM to be able to support his hypotheses. To identify a neurophysiological layer that acts as NREM trigger, SOLMS emphasizes the high percentage of dream recall in SO-stage 1 and in the morning awakenings. These involve some physiological activation that is close to wakefulness in both conditions. He further adds that there should be a negative correlation between depth of sleep (as measured by the acoustic threshold for awakening) and the presence of dream-like activity. However, the author he cites, Zimmermann (1970), only did the awakenings in stage 2. SOLMS does not take into account documented dream-like activity (>60%) in SWS, in which the threshold for awakening is notoriously high. (Cavallero et al.1992; Occhionero et al. 1998).

From a psychophysiological point of view, interpreting activity SO-Stage 1 as dream-like is not very convincing. Sleep onset is a condition in which a gradual transition from waking to sleep takes place. Its electrographic boundaries stretch from relaxed waking to Stage 2. The mental activity present in this period has special characteristics. SO-Stage 1 is not strictly oneric; rather it is a grad-

ual disorganization of voluntary thought. Stage 2 (I cycle) does unquestionably exhibit more dream-like features than the preceding stage. This suggests that there is a gradual modification of cognitive organization in a dream-like direction, while the signals indicating an ongoing state of sleep are present in the recording (Bosinelli 1991). Furthermore in a study in which reports obtained in sleep onset (Stage 2) and in morning spontaneous awakening were compared, the morning stage 2 reports were more dream-like than those obtained in SO Stage 2 (Cicogna et al. 1998).

In our opinion, triggers are not necessary for dreaming; cortical and subcortical neurophysiological mechanisms differently modulate mentation throughout wakefulness and sleep, yet they do not seem to directly generate mentation. Therefore, we propose that mentation during wakefulness and sleep is functionally autonomous. Its relationship with physiological background is interactive rather than one of cause and effect.

# Expanding Nielsen's covert REM model, questioning Solms's approach to dreaming and REM sleep, and reinterpreting the Vertes & Eastman view of REM sleep and memory

Robert D. Ogilvie, Tomoka Takeuchi, and Timothy I. Murphy

*Psychology Department, Brock University, St. Catharines, Ontario L2S 3A1, Canada.* **rogilvie@brocku.ca; {ttomoka; tmurphy}@spartan.ac.brocku.ca www.psyc.brocku.ca/~rogilvie/sleep.html**

**Abstract:** Nielsen's covert REM process model explains much of the mentation found in REM and NREM sleep, but stops short of postulating an interaction of waking cognitive processes with the dream mechanisms of REM sleep. It ranks with the Hobson et al. paper as a major theoretical advance. The Solms article does not surmount the ever-present problem of defining dreams in a manner conducive to advancing dream theory. Vertes & Eastman review the REM sleep and learning literature, but make questionable assumptions in doing so.
[HOBSON ET AL.; NIELSEN; SOLMS; VERTES & EASTMAN]

NIELSEN *mentation in REM and NREM sleep: A review and possible reconciliation of two models* HOBSON ET AL. *Dreaming and the brain.* The target articles by **HOBSON ET AL.** and by **NIELSEN** provide very well-constructed arguments. Both point out current problems in the methodological and theoretical aspects of mind-brain relationships in a logical and fair manner based on enormous evidence including many new findings using neuroimaging techniques.

The AIM model of **HOBSON ET AL.** is noteworthy because it tries to capture the entire concept of our consciousness, using three dimensions that are supported by neurophysiological findings. It is unique to postulate "the state of the brain-mind at any given instant of time can be described as a point in this space." Under this proposition, we could predict the dynamics among different manifestations in our consciousness such as normal, altered, and abnormal states. We believe that their AIM model incorporating the "evolved" Activation-Synthesis model will inspire dream researchers to drive forward, further challenging concepts of consciousness in the same manner that their original A-S model has influenced current dream research.

The covert REM model by **NIELSEN** gives dream researchers positive perspectives on the controversy regarding whether mentations from REM and NREM sleep originate from the same process. His review represents the most creative theoretical formulation of sleep mentation since Foulkes's (1962) discovery that dreams occur outside REM sleep. **NIELSEN** cites compelling evidence for REM sleep intrusions or precursors at both physiological and psychological levels. Covert REM sleep processes may account for a considerable amount of sleep mentation variance and will generate testable hypotheses including several outlined in his review. It is particularly useful in terms of his recognizing the fluidity of the REM/

NREM boundaries. He shows that setting aside the convenience of the conventional R&K system appears quite productive when studying dreams and other states associated with sleep behavior. Thus NIELSEN's model of covert REM sleep processes has succeeded in merging previously incompatible and controversial findings represented by two models "the 1-gen model and the 2-gen models" into a more flexible and accountable model of sleep mentation.

Nevertheless, the commentators wondered if more phenomena could be explained by introducing *waking cognitive behavior* including micro-arousals into the covert REM sleep process as a second primary determinant of sleep mentation. Thus, this different type of 2-gen model – one in which one generator subserves waking cognitive activity with another serving REM dream processes – might allow one to predict mentation throughout waking and sleeping. Such an attempt might provide a means of broadening NIELSEN's very useful model to include some mentation that could not be fully explained by covert REM mechanisms: It would allow his model to explain NREM and REM instances of non-dream-like activity as well.

Waking cognitive behaviors might influence sleep mentation more than we think when one recalls how often micro-arousals appear during sleep (Mathur & Douglas 1995). Subjects are often unaware of these arousals even though their EEG clearly shows arousal during sleep (Ogilvie et al. 1989). During such arousals, some "wake-type" cognition or mentation may occur. When people wake up, they might recall or amend these mentations as if they had been experienced during sleep. For example, our data obtained from sleep-onset NREM and REM periods in normal subjects showed an interaction of brief arousals on REM and NREM dream recall rate (Takeuchi et al. 1999a; 1999b). This enables us to postulate different dream production processes for REM and NREM sleep, that is, the arousal process promotes NREM dreaming but blocks REM dreaming. Adapting waking cognition to the covert REM model would enable one to explain more fully the variety of mentations that seem to appear outside covert REM windows, such as mentations at the initial sleep onset in the first NREM-REM cycle and slow wave sleep as follows.

1. *Sleep onset mentation:* NIELSEN tries to explain sleep-onset mentation by his covert REM model. However, it is difficult for us to imagine covert REM mechanisms underlying sleep onset mentation in the first NREM-REM cycle in healthy nocturnal sleep considering the circadian nature of sleep-onset REM periods (Sasaki et al. 2000). Sleep-onset REM periods appear specifically when the sleep cycle is disrupted (Carskadon & Dement 1980; Fukuda et al. 1987) but do *not* appear during initial sleep onset in the first NREM-REM cycle in healthy nocturnal sleep. Hence, it seems unlikely that REM mechanisms function during the initial sleep onset in normal individuals. Thus, it seems more parsimonious to postulate that the initial sleep onset mentation may be explained by waking cognition rather than covert REM processes.

2. *Slow wave sleep mentation:* NIELSEN finds it difficult to explain mechanisms underlying slow wave sleep mentation by his model. There are a number of NREM phenomena (including parasomnia and mentation related to pre-awakening stimuli) which are difficult for the covert REM model to explain.

We feel that expanding NIELSEN's probabilistic model to include waking cognition might explain virtually all sleep mentation. Further, his covert REM model and HOBSON ET AL.'s AIM model would complement each other in terms of their direction. These models – both bottom-up and top-down approaches – will provide theoretical direction for future dream studies.

### Solms. Dreaming and REM sleep are controlled by different brain mechanisms.

A useful basis is provided by SOLMS for continuing the debate about REM sleep and dreams. However, much of this debate may be semantic in nature. Dreaming is a somewhat ambiguous concept and his distinctions may add to the ambiguity. Differing definitions of "dreaming" create different results. The increase in NREM dream reports over time is owing to a changing definition of dreaming (NIELSEN). SOLMS argues that dreams are generated by the forebrain. However, some of the evidence for this proposition comes from people with seizures, brain damage, or drug use. These *dreams* are described by the author as "nightmares," "unusually vivid," and so on; in other words, atypical. If we limit ourselves to a more typical, intrinsic definition of "dreaming" then REM sleep and brain stem activity become crucial to dreaming.

SOLMS claims that the increasing frequency of dream recall during the late NREM stages in the rising morning phase of the diurnal rhythm suggests that these REM-like dreams are generated by specific NREM mechanisms. However, these longer NREM mentations are typically obtained within 15 minutes of prior REM sleep (Stickgold et al. 1994a) and could stem from covert REM windows (NIELSEN). Considering the circadian influence on REM pressure, it is logical that more mentation would be obtained from wider covert-REM windows in parallel with higher REM pressure in the morning. SOLMS claims that "cessation of dreaming has not been demonstrated in cases with elimination of REM sleep due to brain stem lesions." However, the elimination of observable REM sleep is not proof that no REM mechanisms are active. He also claims that the brain stem is not solely responsible for producing dreams by using the indirect argument that frontal lobe damage does eliminate dreaming. This can be debated on at least two levels.

First, forebrain damage may affect dream recall processes. The frontal lobes are involved in working memory (Smith & Jonides 1999). Working memory is required to keep thoughts active for an extended period of time such as during the arousal process. Thus, dream cessation after forebrain damage does not mean that the forebrain "produces" dreams. These people may have dreams, but do not recall them. SOLMS claims that 70–90% of people with prefrontal leukotomy experience a "complete or nearly complete loss of dreaming." Hence, 10–30% still dream. If he accepts that "25% of NREM dreams are indistinguishable from REM dreams" as proof that REM and dreaming are not the same, then having 10–30% of patients with frontal leukotomies still dreaming is equally compelling evidence that the forebrain is not the generator of dreaming. The hypothesis that dreams are generated by the forebrain after cerebral activation is worthy of further study. However, at this time we remain unconvinced that this explanation can negate the data showing such a strong relationship between REM sleep and dreaming. This is especially true if atypical dreams associated with various abnormal brain states (seizures, medications, etc.) are not considered.

### Vertes & Eastman. The case against memory consolidation in REM sleep.

Last, we would like to comment on the VERTES & EASTMAN article. In particular, we would like to raise several questions about the assumptions and logic used in their critique of the REM sleep and memory literature:

1. *Competing theories and assumptions:* We agree with VERTES & EASTMAN that an important function of REM sleep is to provide CNS activation periodically through the night. Their formulation is an interesting blend of the Roffwarg et al. (1966) Ontogenetic Hypothesis and Snyder's (1966) Sentinel Hypothesis. But the authors seem to imply that if REM serves to activate the brain periodically during sleep, it cannot also be involved in other activities – particularly memory consolidation, which they feel must take place during wakefulness. "Sleep involves basic biological functions and *memory requires consciousness*" (sect. 1, para. 5). This is a *huge* unsubstantiated assumption, following which they begin their review of the literature. To assume that waking and sleeping processes do not interact is inconsistent with the evidence, particularly in light of the papers of HOBSON ET AL. and NIELSEN.

Another difficulty is that by simply counting "for" and "against" studies, they weigh studies which fail to reject the null hypothesis equally with those which *do* reject it, overlooking the powerful difference in the logical strength of these two positions, that is, one can never prove the existence of "no differences" – in this case that memory consolidation does not take place in REM.

2. *Animal REMD studies:* VERTES & EASTMAN of course, are right in saying that the flower pot method of REM deprivation in-

duces stress, but it is also a very effective means of almost totally eliminating REM (Smith & Gisquet-Verrier 1996). They are right in recommending the use of multiple platforms (pots) or Recht-schaffen's rotating disc-over-water/yoked control apparatus for REM deprivation studies. Both provide better controls for stress and are preferable to the flowerpot technique, but are not used routinely in REMD studies because they are much more costly and tedious to use.

But here an important judgment call must be made: Does one categorically reject all studies using the flower pot method, as **VERTES & EASTMAN** do, or should one interpret such studies with caution, being cognizant of the stress effect, but looking for con-vergent validation across labs and approaches? The latter ap-praisal leads to very different conclusions from those reached by **VERTES & EASTMAN**. As they note, the REM window studies are less susceptible to the stress criticism because the animals are de-prived for shorter periods.

**VERTES & EASTMAN** are concerned that "windows" appear to move as a function of different tasks and species. However this may be a rather elegant demonstration of the specificity of mem-ory processing. To our knowledge, this is the first behavioral evi-dence that parallels the superb neuroimaging work, which shows how dramatically individually patterned is the neural activity ac-companying a number of behavioral tasks *and* a variety of differ-ent types of memory. That being so, is it so strange that the con-solidation of these demonstrably different memory processes should take place at different rates?

Recent studies using the post-REMD design have almost unan-imously found that REMD following learning produces deficits in the days or weeks to follow. **VERTES & EASTMAN** have incom-pletely reviewed this literature. The series of papers by Smith and colleagues is apparently dismissed because of their use of the pedestal technique. We would prefer to consider these experi-ments, noting the stress element.

3. *Human studies:* In the reviewed recent human studies of REM sleep and memory, there is no mention of Smith's work, though he and his coworkers are one of the most active groups working in this area. This exclusion seems unwarranted. After all, Smith does not place his *human* participants on flower pots.

In conclusion, we agree that there is insufficient evidence to ac-cept that REM is solely for memory consolidation or that memory consolidation occurs only during REM sleep. However, there does appear to be sufficient evidence to confirm the existence of a link between REM sleep and memory consolidation.

# Nielsen's concept of covert REM sleep is a path toward a more realistic view of sleep psychophysiology

Edward F. Pace-Schott

*Department of Psychiatry, Harvard Medical School, Boston, MA 02115.*
**edward_schott@hms.harvard.edu**

**Abstract:** Nielsen's concept of "covert REM sleep" accounts for more of the complexity in sleep psychophysiology than its conceptual predecessors such as the tonic-phasic model. With new neuroimaging findings, such concepts lead to more precise sleep psychophysiology including both tra-ditional polysomnographic signs and neuronal activity in greater proxim-ity to the actual point sources and distributed networks which generate dreaming.

[HOBSON ET AL.; NIELSEN]

**NIELSEN**'s current reviews comparing REM and NREM sleep mentation (see also Nielsen 1999) provide a much-needed com-prehensive review and incisive analysis of the extant psychophys-iological data bearing on this controversy. His attention to the de-tailed physiology of sleep, and his concept of "covert REM sleep" offers an empirical path around the conceptual impasse imposed by over-reification of the cardinal sleep states as unitary, non-dissociable physiological "black boxes" whose subjective manifes-tations must be compared dichotomously. **NIELSEN**'s considera-tion of conceptually and experimentally diverse comparisons between REM and NREM associated cognitive processes (e.g., state-dependent differences in the effects of sleep manipulation on memory consolidation, ERP responses to stimulation, and post awakening performance) is essential in that it moves this debate away from the conceptually flawed (see the **HOBSON ET AL.** and **NIELSEN** discussions of Hunt et al. 1993) and inherently limited method of normalization for report length which has dominated the REM/NREM controversy to date. Equally important is his ap-preciation for and analyses of the difficulties inherent in achieving construct validity for sleep mentation variables and obtaining un-ambiguous psychological physiological correlations in any state.

In one sense, **NIELSEN**'s idea of covert REM sleep allows a re-consideration of the tonic-phasic model (Molinari & Foulkes 1969; Ogilvie et al. 1980; Pivik 1991) originally proposed to explain the lack of an exclusive association between dreaming and REM sleep. The tonic-phasic model itself encountered difficulties due to the very weak (although consistently positive) associations be-tween the phasic events of REM and NREM sleep and the details of associated mentation reports (see Pivik 1991 for a comprehen-sive review). However, as we have argued elsewhere (Kahn et al. 1997), the polysomnographically measurable phasic events of REM and NREM sleep are most realistically viewed as integrated, attenuated "surface readouts" of complex and varied brain events and, therefore, their lack of a tight temporal relationship to the re-ported features of mentation is fully predictable. **NIELSEN** has now taken a fresh and more in-depth look at the psychological and physiological manifestations of these CNS events.

This new consideration comes at a time when the underlying brain events producing the psychophysiologically measured pha-sic and tonic events of sleep are also beginning to be revealed by functional neuroimaging studies of both sleep (e.g., Andersson et al. 1998; Braun et al. 1997; 1998; Bootzin et al. 1998; Hofle et al. 1997; Hong et al. 1995; Kajimura et al. 1999; Lovblad et al. 1999; Maquet et al. 1996; 1997; Nofzinger et al. 1997) and dream-like intrusions on waking (e.g., Ffytche et al. 1998; Rabinowicz et al. 1997; Silbersweig et al. 1995). The neuroimaging data leads to an updated concept of the neural circuitry underlying dreaming in REM sleep which emphasizes activity in limbic circuits under conditions of decreased activity in executive cortical regions (e.g., Braun et al. 1998; Hobson et al. 1998a; 2000, and this volume; Ma-quet & Franck 1997; Nofzinger et al. 1997).

These new findings provide a first look at how **NIELSEN**'s hy-pothesized generators might be physically instantiated. For exam-ple, the comparative activation patterns of REM and NREM sleep suggest that one significant difference between the REM sleep and NREM sleep generators lies in the much greater activity of limbic structures during REM (for details, see **HOBSON ET AL.** 2000 and this volume). PET findings and their interpretation by Braun et al. 1998 suggest that an additional difference between the two gener-ators may involve the neural sources of visual hallucinosis. NREM imagery may be initiated or generated further upstream in visual processing networks such as in the striate cortex (see also Hofle et al. 1997) whereas REM imagery may arise more from activity of vi-sual association cortex (Braun et al. 1998) and thus perhaps be more similar to waking hallucinosis (see Ffytche et al. 1998).

When viewed in the light of abundant new neuroimaging data on a multitude of waking cognitive functions and subjective expe-riences (see Cabeza & Nyberg 1997; 2000; Gazzaniga 2000), the question of one versus two generators of sleep mentation seems inadequate to encompass the diversity of distributed and point sources of neural activity which must occur in multiple CNS net-works in order to generate the complex phenomena of sleep men-tation. For example, in addition to the distributed sensory and associative cortical modules hypothesized or implied by "one-generator" theories (e.g., Antrobus 1990; Foulkes 1993b), dream phenomenology indicates that subcortical-cortical networks subserving emotion, instinct, and motivation (see Cummings 1993; Kalivas & Barnes 1993) must contribute at least as powerfully to the neural basis of sleep mentation. In addition, the wide difference in affective tone between individual dreams must reflect differential activation of limbic circuits subserving different emotional states. For example, nightmares compared to euphoric dreams may respectively reflect greater relative activation of amygdalar fear circuits versus mesolimbic reward circuits.

Toward the goal of dissecting the neural mechanisms that comprise sleep mentation generators, several of the ways in which NIELSEN compares cognitive processes in REM and NREM point toward specific neural networks and suggest hypotheses that might allow us to examine sleep mentation at a level of specificity comparable to current waking state paradigms. For example:

1. Comparison of memory sources between REM and NREM suggests the differential cortico-hippocampal information transfer in different sleep states that has been demonstrated in animal models (Buzsaki 1996). Such state-dependent differences in information processing have recently been hypothesized to underlie some of the state-dependent differences in memory sources available for the elaboration of sleep mentation (Stickgold 1998; Stickgold et al. 1999b)

2. Normal and abnormal variations in the extent or stability of lateral prefrontal cortex deactivation during sleep may contribute to subject differences in dream recall from REM and NREM sleep. The lateral prefrontal cortices have been shown to be essential to both encoding and retrieval of episodic memory (Fletcher et al. 1997). Therefore, for example, the greater NREM mentation recall in light sleepers (see NIELSEN's discussion of Zimmerman 1970) might result from lesser NREM associated deactivation of prefrontal structures subserving episodic memory encoding, retrieval or both. Similarly, Braun (1999) has suggested that sleep-related prefrontal deactivation may degrade working memory resulting in encoded dream memory traces which then become relatively inaccessible to retrieval processes owing to a paucity of simultaneously encoded contextual cues.

Relative regional prefrontal activation also becomes a useful dependent variable for neuropsychological hypothesis testing of mechanisms for dream lucidity (i.e., relative engagement of executive functions) or for the variations in the degree of interrelationship between mentation content from different reports of the same night, which NIELSEN discusses (reflecting, perhaps again, hypothetical individual and temporal variations of encoding and retrieval). The great advantage of such specific hypotheses over the vaguer theories of "sleep mentation generation systems" proposed in the past is that they are eminently testable with current and developing technologies in cognitive neuroscience such as functional MRI or transcranial magnetic stimulation. Moreover, hypothesis formation in dream research can now be rapidly informed by new wake-state findings as they are reported.

3. The interrelationships between psychopathology and dreaming can be tested at both the psychological and physiological level using validated psychological instruments (such as the MMPI scales NIELSEN discusses) in combination with the ability to visualize activity in brain regions which respond to sleep-based treatments such as the anterior cingulate (Smith et al. 1999; Wu et al. 1999). Pioneering work in this area investigating dream anxiety among normals has already been reported by Gottschalk et al. 1991a; 1991b.

NIELSEN's model of covert REM sleep and his enumeration of nine factors by which it might be evoked suggests new and powerful ways of identifying and differentiating neural mechanisms underlying the known features of sleep mentation. For example:

1. The notion of a stage of "intermediate sleep" in the transition of REM to and from NREM has been well described in animals by Gottesmann and colleagues (see Gottesmann 1996, for a review) but has been under-investigated by students of the psychophysiology of sleep mentation (e.g., the only specific investigations cited by NIELSEN or Gottesmann appear to be early work by Lairy and colleagues; see also Larson & Foulkes 1969). In seeking specific

and relatively isolated physiological correlates of sleep mentation, this polysomographically identifiable period should be a prime target for combined electrophysiology and neuroimaging studies especially given the apparent predictability in the non-simultaneous appearance of the cardinal PSG signs of REM (Sato et al. 1997).

2. NIELSEN notes that sleep onset similarly represents a prime candidate for a natural transitional state in which the dissociated psychophysiological correlates of REM may appear non-synchronously. As with REM onset, there may exist a degree of predictability as to the order in which different types of dream-like mentation may arise (Rowley et al. 1998).

3. Pathological and/or experimentally manipulated transitional or dissociated states suggested by NIELSEN (e.g., cataplexy; see Asenbaum et al. 1995, REM-deprivation induced sleep onset REM, pharmacologically altered sleep, auditorily evoked NREM mentation) provide an abundance of potential paradigms for combining electrophysiology and neuroimaging.

Studies such as the above might not only better define the neural substrate of physiological signs which are known to correlate with REM and/or dreaming (e.g., gamma-frequency oscillations: Gross & Gotman 1999; Llinas & Ribary 1993), but may identify previously unmeasured depth events with an even higher degree of correlation to psychological events such as the elusive (in humans) PGO wave. The current development of event-related fMRI techniques (e.g., Jessen et al. 1999) is particularly promising in this regard. Moreover, NIELSEN's empirically tested model of probability for encountering covert REM sleep processes in NREM based upon temporal proximity to REM periods illustrates the straightforward methodology which could be employed in such studies (e.g., temporal correlation of regional activations with the incidence of mental events).

In summary, NIELSEN's paper represents an important theoretical advance in the psychophysiological study of dreaming. In the future, such ideas may be viewed as important steps in a transition from viewing dreaming as a manifestation of a defined behavioral state to directly linking its components to the ebb and flow of neuronal processes which themselves define the behavioral states. In this sense, what we now refer to as "REM sleep" may come to be defined as that configuration of possible states in varied neuronal networks during sleep which results in the concurrent distinct experience of emotion, hallucination, and movement in synchrony with maxima of diverse physiological processes. In that sense, there is no covert REM sleep, only various degrees of dissociation or synchrony in the functioning of these modules. In turn, an overriding ultradian oscillatory mechanism may normally cause these interacting networks to reach maximal activity in phase with one another, thus producing the synchronous rise and fall of physiological processes and subjective experiences which we now characterize as the REM-NREM cycle.

# Dreaming is *not* a non-conscious electrophysiologic state

J. F. Pagel

*Dream Section, American Academy of Sleep Medicine, University of Colorado Medical School, Pueblo, CO 81005.* pueo34@juno.com

**Abstract:** There has been no generally accepted cognitive definition of dreaming. An electrophysiologic correlate (REM sleep) has become its defining characteristic. Dreaming and REM sleep are complex states for which the Dreaming = REMs model is over-simplified and limited. The target articles in this *BBS* special issue present strong evidence for a dissociation between dreaming and REM sleep.
[HOBSON ET AL.; NIELSEN, REVONSUO; SOLMS; VERTES & EASTMAN]

*The vision of dreams is this against that, the likeness of a face confronting a face.*

Sira, Dead Sea Scrolls, 180 BC

Dreaming is a cognitive state. If no one experienced the recall of sleep mentation on awakening, dreaming would not exist as a phenomenon for discussion. Instead we would be discussing the perceptual isolation of sleep, and the non-conscious significance of REM sleep, NREM sleep, PGO spikes, hippocampal theta and sawtooth waves. In the last thirty years these electrophysiologic (particularly REMS = Dreaming) have become the defining characteristics of dreaming. That period may be ending. The target articles in this *BBS* special issue present strong evidence for a dissociation between dreaming and REM sleep. The scientific study of dreaming is in the midst of a millennial year's change.

**What is a dream?** Aristotle, with characteristic conceptual clarity, defined dreaming as the mental activity of the sleeper insofar as he is asleep (Aristotle 235/1952). For Aristotle, dreaming was an idealized state with a reality independent of waking experience. This definition has survived more than two thousand years, to be found almost verbatim in current dictionary definitions: a dream is a series of thoughts, images, or emotions occurring (passing through the mind) during sleep (Webster's Dictionary 1993; Random House Dictionary 1983). In this definition, dreaming is a state occurring during sleep, not accessible to study during the wakefulness. It suggests that dreaming may be best defined by its nonconscious correlates. The cognitive dream, in other words, the recall waking of a dream, is not part of the Aristotelian definition.

Scientific methodology was developed as a formal approach in order to analyze external reality and hence to differentiate it from dreams (Descartes 1637). The modern age of the scientific study of sleep can be dated back to the staging of sleep based on EEG, EMG, and EOG criteria from the mid 1950s and 1960s (Aserinsky & Kleitman 1953; Rechtschaffen & Kales 1968). Although sleep has well defined electrophysiologic correlates, sleep onset is still generally defined behaviorally rather than electrophysiologically. Sleep is defined as a reversible behavioral state of perceptual disengagement from and unresponsiveness to the environment (Kryger et al. 1994).

Archeological research has shown that human interest in dreaming dates back over at least 4,000 years. Codices recovered from the Sumerian city of Ladak (2500 B.C.) describe King Gudea's orientation of his new temple based on insight obtained from a dream experience. Despite such a prolonged interest, limited progress has been made towards understanding dreaming. Freud wrote on page 1 of *The interpretation of dreams*, "In spite of many thousands of years of effort, the scientific understanding of dreams has made very little advance . . . little or nothing that touches upon the essential nature of dreams." Freud's contribution was to identify a role for dreaming in the diagnosis and therapy of psychiatric illness (Freud 1900). Psychoanalysts, psychologists, and psychiatrists continue to use dreaming in diagnosis and therapy. However, modern researchers and therapists often allude to dreams without definition. The tomb of psychiatry (the Diagnostic and Statistical Manual of Mental Health, DSM) avoids definition of a dream, mentioning only the nightmare – a frightening dream (DSM-IV 1994). The meaning, content, allusions, and associations of dream (undefined) are regularly the focus of discussion and publication. A topic may be REM sleep, personal moti-

vation, neuroanatomy, film, hallucinations, literature, neuropharmacological effects of medication, real estate, the unconscious mind, or anthropology; and dreaming (undefined) is often the title. Check any library catalog system for its listing on dreams.

The issue is compounded by the personal nature of dreaming. Between two and four years of age most children develop a conceptual definition of a dream (Olson et al. 1988). This private dream reality remains outside study and definition, being made available only as the individual dreamer desires. In the words of Shakespeare, "The eye of man hath not heard, the ear of man hath not seen, man's hand is not able to taste, his tongue to conceive, nor his heart to report, what my dream was" (Shakespeare 1595/1986). When we gather to study dreams, we each bring to the table our personal definitions. Dreaming is not the only aspect of mental activity that has such loose attributes. Imagination, creativity, intelligence, and even consciousness are poorly defined cognitive concepts that suffer the same diffuse social attributions. Dreaming has an advantage in that it has a long history of studies that have described the dream personally, physiologically, and psychologically. There is, however, a generalized tendency in the field to avoid defining dreaming (Pagel 1999a).

Different fields have widely varying definitions for dreaming. Psychoanalytic definitions of dreaming date from Freud – unconscious wish fulfillment (Freud 1900). One current psychiatric definition of dreaming can be considered: bizarre or hallucinatory mental activity occurring during a continuum that extends through stages of sleep and waking. The newer field of sleep medicine often defines dreaming as mental activity (images, feelings or emotions) occurring during sleep (Mahowald et al. 1998). Psychology has generally avoided a definition of dreaming, concentrating on defining methodology and study populations. A wide spectrum of fields from anthropology to literature characterize dreaming by its associations and attributes. In popular culture, the most generally accepted definition is loosely Freudian: dream marriages, dream homes – the projected image of conscious wish fulfillment.

The topic of one study in which dreaming is mentation occurring during sleep may be very different from hallucinatory or delusional thought content. Results are likely to be comparable only when comparable definitions of the topic of study are used. This has produced conceptual confusion that has contributed to a lack of structural rigor in the field of study. Single structural definition has not been possible owing to the diverse spectrum of fields and an historic multiplicity of definitions applied to the study of dreaming. A classification schema for definitions of dreaming has been described for incorporation into scientific studies of dreaming so the results of studies across epistemologically diverse fields can be compared (Table 1) (Pagel 1999b).

As Descartes points out, "Now the principal and most frequent error that can be found in judgments consists in the fact that I judge that the ideas, which are in me, are similar to, or in conformity with certain things outside me." (Descartes 1641/1980). The present target articles reflect the disarray in the fields of dream study that have occurred because of the fields' inability or disinclination to cognitively define the topic of study – dreaming. It is the electro-

Table 1 (Pagel). *AASM dream definition paradigm*

| (a) | Sleep | Sleep Onset | Dreamlike States | Routine Waking | Alert Wake |
|---|---|---|---|---|---|
| | * | * | * | * | * |

| (b) | No Recall | Recall | Content | Associative Content | Written Report | Behavioral Effect |
|---|---|---|---|---|---|---|
| | * | * | * | * | * | * |

| (c) | Awareness of Dreaming | Day Reflective | Imagery | Narrative Story | Illogical Thought | Bizarre Hallucinatory |
|---|---|---|---|---|---|---|
| | * | * | * | * | * | * |

physiologic correlates of dreaming, especially REM sleep, that have been used to define its presence or absence. This Aristotelian conceptual approach (Dreaming = REMS) has achieved almost mythological status despite extensive criticism. A cognitive definition of dreaming should incorporate the awake recall of dreaming. Sleep, a state with clearer electrophysiologic correlates than dreaming, is still defined by its behavioral and cognitive correlates. The reproducibility of results is based on clear limits of one's object of measure. In part, that is why the view that Dreaming = REMS has been so attractive. It would make this loosely defined cognitive state much easier to study and understand.

SOLMS's paper is the shortest but perhaps most important in this series owing to the clarity with which research and clinical findings have been applied to the question of whether dreaming is "an epiphenomenon of REM sleep." Since not all dreaming is correlated with REM sleep, SOLMS asks whether what has been defined as the REM sleep type of dreaming occurs outside of REM sleep. He presents evidence that the REM sleep can occur without dreaming and dreaming without REM sleep. A consistent body of literature suggests that dreaming correlates better with cerebral activation during sleep than with the occurrence of REM sleep. SOLMS's argument is that REM sleep and dreaming are doubly dissociable states with different physiological mechanisms in all likelihood serving different functional purposes. He suggests that the REMS = Dreaming premise upon which most prevailing neuroscientific theories of dreaming are based, is invalid.

VERTES & EASTMAN consider the lack of evidence for memory processing and consolidation in REM sleep. This too reflects some of the difficulties rising from prevailing REMS = Dreaming theories. As the authors state, "The mental/cognitive content of REM sleep is dreams . . . Dreams are the sole window to cognitive processes of REM sleep." Their reviews of imaging brain studies of brain activity during REM sleep are accordingly extrapolated to dreaming: "in summary, the pattern of brain activity in REM sleep is consistent with dreams, but inconsistent with the orderly evaluation, organization and storage of information which is the domain of attentive, waking consciousness." It may very well be, as they suggest, that REM sleep can be understood within the context of sleep without invoking mental phenomena or quasiconscious processes – the cognitive process of dreaming.

REVONSUO presents an eloquent and well researched argument for the role of dreaming as a cognitive process in the evolution of threat avoidance. He defines dreaming as conscious experiences during sleep: "We may define a dream as a subjective experience during sleep, consisting of complex and organized images that show temporal progression." He argues that consciousness can be reconceptualized as the phenomenal level of organization of the brain with dreaming a subjective experience realized at the phenomenal level. Based on prevailing REMS = Dreaming theories he concludes, "the phenomenal level of organization of the brain is realized in its characteristic ways during REM sleep." He incorporates research from both REM sleep studies and cognitive dream research, considering that both fully reflect dreaming. It is suggested that dreaming and REM sleep may have complementary yet different evolutionary roles.

NIELSEN acts as an apologist for the Dreaming = REMS Paradigm, redefining REM sleep as a state in which REMS type dreaming occurs: hallucinoid imagery, narrative structure, cognitive bizarreness, hyperemotionality, delusional acceptance, and deficient memory of previous mental content. He distinguishes "dreaming" (REMS type mental processes as described above) from other cognitive activity occurring during sleep which is not considered to be "dreaming." REM sleep is characterized by a general motor atony, bursts of sciatic eye movements, cardiac and respiratory irregularity, PGO spikes, hippocampal theta rhythms, genital tumescence, and low frequency "awake-like" EEG activity which may include sawtooth waves (Kryger et al. 1994; Rechtschaffen & Kales 1968). NIELSEN redefines REM sleep as any of these REM sleep associated events occurring in association with "dreaming" as defined above. It is suggested that REM sleep may

occur even in states which are not clearly sleep (the disorganized EEG drug effects induced by katamine and LSD) because "dreaming" is associated with these states. This paper concludes that if both dreaming and REM sleep are radically redefined the REMS = Dreaming model can be preserved.

HOBSON ET AL.'s important paper comes from the group that first advanced the REM = Dreaming hypothesis. Although they persist in correlating dream features with distinctive REM sleep physiology, the authors suggest that the correlation should be loosened, with cognitive states better viewed as inter-related psychophysiological continua manifested at the levels of both the brain and the mind.

Dreaming is initially defined as a series of images, ideas, emotions, and sensations occurring involuntarily in the mind during certain stages of sleep and REM sleep as well as during the numerous forms of wake-state and sleep-state mentation. Later in the paper, dreaming is redefined by the narrower constraints HOBSON ET AL. prefer: "mental activity occurring in sleep characterized by vivid sensorimotor imagery that is experienced as waking reality despite such distinctive cognitive features as impossibility or improbability of time, place, persons, and actions; emotions, especially fear, elation and anger predominate over sadness, shame and guilt and sometimes reach sufficient strength to cause awakening; memory for even very vivid dreams is evanescent and tends to fade quickly upon awakening unless special steps are taken to retain it." The focal argument, however, is that dreaming and other states of consciousness can be best defined by a model that includes: (1) the information processing capacity of the system (activation); (2) the degree to which the information processed comes from the outside world and is or is not reflected in behavior (information flow); and (3) the way in which the information in the system is processed (mode). This model corresponds to the various axes of the AASM dream definition protocol (Table 1) in which the activation axes can be equated with the awake/sleep axis, the information axis with the recall axis, and the mode axis with the content axes (Table 1).

This model is applied to the description of a series of cognitive states viewed as brain-body-mind isomorphisms varying in multidimensional combinations. With this model, HOBSON ET AL. Dreaming = REMS controversy, utilizing the dream axis definition criteria to model "dreaming" and other cognitive states of both waking and sleep. It is the authors' contention that their model is both necessary and sufficient to distinguish in a preliminary way among the basic wake-sleep states: the resulting state space model, while still necessarily overly simplistic, is nonetheless a powerful tool for studies of consciousness."

***Comparing definitions of dreaming.*** In these papers, the REMS = Dreaming model – until now so pervasive in the fields of sleep and dream study – is useful for comparing studies that are based on irreconcilable cognitive definitions of dreaming. These papers use a spectrum of definitions of dreaming. NEILSEN uses a narrow content based definition for dream in his presentation: mentation characterized by hallucinoid imagery, narrative structure, cognitive bizarreness, hyperemotionality, delusional acceptance, and deficient memory of previous mental content. His definition excludes other cognitive activity occurring during sleep which is not considered to be "dreaming." The Awake/Sleep or Activation axis of his definition is broad and undefined with dreaming considered to occur in sleep, wake, and drug induced states (Table 2). The dreaming that REVONSUO considers is very different, including a broad spectrum of content (subjective experience realized at the phenomenal level), specific recall (evolutionary effects), and specific Awake/sleep axis (occurring only during sleep). VERTES & EASTMAN (the mental/cognitive content of REM sleep is dreams) are also considering an unlimited range of content occurring out of an even narrower segment of the Awake/Sleep axis – REM sleep. VERTES & EASTMAN and REVONSUO are studying dreaming that NIELSEN defines as sleep cognitive activity (non dreaming). SOLMS and HOBSON ET AL. define dreaming as an epiphenomenom of REM sleep in order to argue for and against the REMS =

Table 2 (Pagel). *Dream definition paradigm comparing definitions used in these papers*

| (a) | Sleep | Sleep Onset | Dreamlike States | Routine Waking | Alert Wake |
|---|---|---|---|---|---|
| | * | * | * | * | * |

| (b) | No Recall | Recall | Content | Associative Content | Written Report | Behavioral Effect |
|---|---|---|---|---|---|---|
| | * | * | * | * | * | * |

| (c) | Awareness of Dreaming | Day Reflective | Imagery | Narrative Story | Illogical Thought | Bizarre Hallucinatory |
|---|---|---|---|---|---|---|
| | * | * | * | * | * | * |

**NEILSEN:** mentation characterized by hallucinoid imagery, narrative structure, cognitive bizarreness, hyperemotionality, delusional acceptance, and deficient memory of previous mental content.

| (a) | Sleep | Sleep Onset | Dreamlike States | Routine Waking | Alert Wake |
|---|---|---|---|---|---|
| | * | * | * | * | * |

| (b) | No Recall | Recall | Content | Associative Content | Written Report | Behavioral Effect |
|---|---|---|---|---|---|---|
| | * | * | * | * | * | * |

| (c) | Awareness of Dreaming | Day Reflective | Imagery | Narrative Story | Illogical Thought | Bizarre Hallucinatory |
|---|---|---|---|---|---|---|
| | * | * | * | * | * | * |

**REVONSUO:** subjective sleep experience realized at the phenomenal level affecting evolutionary process.

| (a) | Sleep | Sleep Onset | Dreamlike States | Routine Waking | Alert Wake |
|---|---|---|---|---|---|
| | * | * | * | * | * |

| (b) | No Recall | Recall | Content | Associative Content | Written Report | Behavioral Effect |
|---|---|---|---|---|---|---|
| | * | * | * | * | * | * |

| (c) | Awareness of Dreaming | Day Reflective | Imagery | Narrative Story | Illogical Thought | Bizarre Hallucinatory |
|---|---|---|---|---|---|---|
| | * | * | * | * | * | * |

**VERTES & EASTMAN:** the mental/cognitive content of REM sleep is dreams
**SOLMS:** dreaming is an epiphenomenon of REM sleep.

| (a) | Sleep | Sleep Onset | Dreamlike States | Routine Waking | Alert Wake |
|---|---|---|---|---|---|
| | * | * | * | * | * |

| (b) | No Recall | Recall | Content | Associative Content | Written Report | Behavioral Effect |
|---|---|---|---|---|---|---|
| | * | * | * | * | * | * |

| (c) | Awareness of Dreaming | Day Reflective | Imagery | Narrative Story | Illogical Thought | Bizarre Hallucinatory |
|---|---|---|---|---|---|---|
| | * | * | * | * | * | * |

**HOBSON ET AL.:** mental activity occurring in sleep characterized by vivid sensorimotor imagery that is experienced as waking reality despite such distinctive cognitive features as impossibility or improbability of time, place, person, and actions; emotions, especially fear, elation, and anger predominate over sadness, shame, and guilt and sometimes reach sufficient strength to cause awakening; memory for even very vivid dreams is evanescent and tends to fade quickly upon awakening unless special steps are taken to retain it.

Dreaming paradigm. In creating their cognitive state models, **HOBSON ET AL.** use a definition of dreaming that specifically defines all three axes (Table 2).

Both **SOLMS** and **HOBSON ET AL.** address the diversity of definitions for the dream state, applying specific definitions to compare data while maintaining an awareness that adaptive models are required to address the different phenomenon that researchers call dreaming. If interdisciplinary studies are to be compared, such a specific yet adaptive multi-axis approach to dreaming state definition, as **HOBSON ET AL.** suggest, provides us with a powerful tool that can be applied to cognitive study of other conscious states.

**Dreaming is not a non-conscious electrophysiologic state.** These are exciting times in the field. The papers in this special issue of *BBS* describe a major paradigm shift in our understanding of the association between sleep and dreaming. **SOLMS** argues cogently that REM sleep and dreaming are doubly dissociable states with different physiological mechanisms in all likelihood subserving different function. Support comes from **VERTES & EASTMAN** who argue that REM sleep can be understood within the context of sleep without invoking mental phenomena or quasi-conscious processes – the cognitive process of dreaming.

As often seems to be the case, the greatest support for this paradigm shift away from REMS = Dreaming comes from its defenders. **NIELSEN** shows how the paradigm could be saved by radically defining both REM sleep and dreaming. In order to preserve the model, cognitive activity occurring during sleep must be excluded from the definition of dreaming, and REM sleep must be re-defined to occur throughout the Awake/Sleep axis. In order to preserve REMS Dreaming, we are required to restructure the entire sleep and dream study. Perhaps the greatest support for this paradigm shift comes from **HOBSON ET AL.**, the developers of the REMS = dreaming model, who have chosen to side-step defense of their paradigm and move to a multi-dimensional cognitive state model for dreaming in which REM sleep is but one point on an Activation = Awake/Sleep continuum.

Suggestions that the REMS = Dreaming premise upon which most prevailing neuroscientific theories of dreaming are based is no longer valid, **SOLMS** proposes new neuroanatomical and clinical approaches to the study of REM sleep and dreaming. **VERTES & EASTMAN** have set the framework for the analysis of dreaming using memory paradigms independent of the states' association with REM sleep. **REVONSUO** demonstrates that an evolutionary model is applicable to both REM sleep and dreaming as independent yet complementary phenomenon. **HOBSON ET AL.** apply a multi-axis paradigm for dreaming that can be used to describe other cognitive states, and perhaps even model non-biologic systems.

Descartes asserted that scientific method could be applied to reality and not to dreams (Descartes 1637/1980). But science can use multi-axis definitions to study the specific cognitive and electophysiologic correlates of dreaming that are reality. Dreaming and REM sleep are complex states for which the Dreaming = REMS model has become excessively simple and limited. Progress in a scientific field occurs when structural models change. As this collection of papers suggests, the paradigm has shifted.

## "The dream of reason creates monsters" . . . especially when we neglect the role of emotions in REM-states

Jaak Panksepp

*Department of Psychology, Bowling Green State University, Bowling Green, OH 43403.* **jpankse@bgnet.bgsu.edu**
**http://caspar.bgsu.edu/~neuro/Faculty/Faculty_jpanksepp.html**

**Abstract:** As highlighted by Solms, and to a lesser extent by Hobson et al. and Nielsen, dreaming and REM sleep can be dissociated. Meanwhile Vertes & Eastman and Revonsuo provide distinct views on the functions of REM sleep and dreaming. A resolution of such divergent views may clarify the fundamental nature of these processes. As dream commentators have long noted, with Revonsuo taking the lead among the present authors, emotionality is a central and consistent aspect of REM dreams. A deeper consideration of emotions in REM dreams may serve as the conceptual salve to help heal the emerging rifts in this field of inquiry.
[HOBSON ET AL.; NIELSEN; REVONSUO; SOLMS; VERTES & EASTMAN]

I use Goya's epigram for one of his final engravings in my title to highlight how our widespread failure to deal with basic emotional processes of the brain may have led to many conundrums in our search for the adaptive functions of dreaming and REM sleep. The same can be said about our aspirations to understand many other mind-brain relations, from consciousness to intrinsic organismic values.

**SOLMS** has done a great service in highlighting, more dramati-

cally than ever before, the increasingly evident fact that REM mechanisms and dreaming are not strictly isomorphic. Clearly, REM sleep reflects very ancient aspects of neural organization while the richness of human dreams is a comparatively recent brain development. As **SOLMS** discusses, distinct types of arousal during sleep may activate dreams in the absence of REM, and fairly restricted damage to neocortex can lead to the disappearance of dreams while REM persists. As a result of his challenge, we must no longer equate REM sleep and dreaming as casually as in the past. How these distinct but interactive brain-mind processes can be understood without doing injustice to either, is the central theme of the present discussion.

**HOBSON ET AL.** provide us with a masterful update of neural and psychological aspects of REM sleep and dreaming, and a new synthetic model of mind to boot. Even though they mention emotions repeatedly and at times foreshadow **REVONSUO**'s intriguing proposal emphasizing anxiety over other emotions as the prime "shaper" of dream plots, **HOBSON ET AL.** fail to cultivate emotional concepts in any comprehensive way. Since their AIM model of mental space "strained to account for differences between various emotional substates," it was good to have **REVONSUO** develop an emotional theme elegantly as an exemplar of where we need to go. The analyses of **NIELSEN** and **VERTES & EASTMAN** also remained regrettably devoid of emotional concerns. Despite their rich tapestries of evidence and argumentation, a judicious consideration of brain emotional systems remains a most promising path to understand the role of REM in dreams as well as brain-mind evolution. Thus, I would encourage all participants to consider brain emotional issues further in their final comments, and also provide additional predictions (especially disconfirmatory ones) for their various viewpoints.

**NIELSEN** attempts mightily to bind up the wound that **SOLMS** has opened up. The one- and two-generator models of dreaming can be reconciled in various ways, but how, without circularity, might the "covert REM" hypothesis deal with the fact that elimination of overt REM sleep does not necessarily eliminate dreaming? If we accept the traditional idea that REM is a uniquely powerful, though not an obligatory trigger for dreaming (and most studies of the intact brain-mind do affirm that conclusion), I do not see how we can escape some variant of a two generator point of view. Perhaps the truth is still rather close to what most have been led to believe – REM dreams are fundamentally emotional while NREM dreams are more strictly cognitive. In other words, the two major types of dreaming may emerge from measurably distinct albeit interactive neuromental spaces.

Despite the dissociations between REM mechanisms and dreaming in broken brains, many of us – including perhaps **HOBSON ET AL., NIELSEN,** and **REVONSUO** – will continue to believe that the relationships between the two (especially the most vivid emotional dreams) in normal physiology are as dynamically interwoven as those of conductor and orchestra (or horse and carriage for the more blatantly behavioristic). However, as implied by both **SOLMS** and **HOBSON ET AL.**, we should remain open to the possibility that several functionally distinct conductors can guide the dream orchestra – each with different tempos, neuropsychological consequences, and state spaces in the brain-mind. In this view, the distinction between the emotional dreams of REM and the non-emotional ones of NREM, as well as their varied functions, may still prove to be physiologically and psychologically meaningful. However, **SOLMS**'s point would still hold – the cognitive contents of dreams are not choreographed in any detail by REM processes; rather, REM is simply the most emotionally minded conductor of one dream symphony.

The classic Freudian division between latent and manifest contents may also help us parse these issues in reasonable ways. For instance, while REM sleep arises substantially from lower brainstem processes, providing latent energies for many dream possibilities, the manifest contents require the participation of higher, heteromodal cortical processes. **SOLMS**'s neuropsychological work jibes remarkably well with modern brain imaging data on REM dreams as detailed comprehensively by **HOBSON ET AL.** – the dorsolateral prefrontal working-memory systems remain asleep in REM while basomedial appetitive-emotional systems and most other limbic areas become remarkably aroused. In contrast, those emotional brain areas remain fairly quiescent during NREM sleep. Whether REM characteristic brain patterns accompany NREM dreams remains unknown, but I hope **SOLMS** will clarify whether he would predict massive neurophysiological congruences between the two? If not, would he accept, in principle, the likelihood of at least two distinct types of dream life?

Although **SOLMS** and **VERTES & EASTMAN** seem to imply that we may have overvalued REM-dreams, I would not discard the possibility that REM dreams are of substantial functional importance in normal brain-mind homeostasis. The REM process appears quite privileged in breathing primary-process affective life into the cognitive flow of dreams within the intact brain. REMs may sequentially arouse various basic emotional tendencies during paradoxical sleep, providing latent neurodynamic support for reprocessing affective experiences encountered during waking within the manifest dreams percolating in higher brain areas. Indeed, it may be that only the less emotional dreams survive damage to REM circuits. Considering this, the emotional analysis of dreams of individuals whose REM mechanisms are impaired may be informative, and perhaps **SOLMS** can enlighten us on such issues.

Despite the striking dissociations highlighted by **SOLMS,** I suspect that the most emotion laden dreams remain so only because of the sustaining influences of REM arousal. Indeed, since the co-occurrence of REM and emotional dreams may be the only way we shall ever decode the biological, neuroadaptive nature of our most vivid dreams, we should continue to pursue the strategy that a detailed study of REM related processes provides our best opportunity to understand the nature and functions of our most memorable (perhaps archetypal) dreams. Despite **VERTES & EASTMAN'**s well-reasoned skepticism about any global information-processing hypothesis, REM dreams may still have specific functions in the integration of affective information, perhaps along the lines outlined by **REVONSUO.**

**REVONSUO** shared his ideas concerning dreams at a Tucson III Plenary panel on the "Neural Correlates of Consciousness" (May 2, 1998). I inquired from the floor whether he or other panel members had an explanation for the perplexing neurological fact that REM arousal mechanisms, and hence presumably the rudiments of dreaming, are more ancient in brain evolution than the waking circuits of the Extended Reticulo Thalamic Activating System (ERTAS). All seemed perplexed by this prickly question. It has also received inadequate attention from the sleep-research community. My puzzlement was prompted by the recognition that the epicenter for REM sleep is slightly more caudal in brainstem tissues than the epicenter for waking, at least as affirmed by the classic pre-trigeminal preparation.

A resolution of this apparent paradox may yield a coherent understanding of REM mechanisms and the related affective dream contents. My personal solution was offered in passing several years ago (Panksepp 1993) and, as **REVONSUO** notes, also (Panksepp 1998a, pp. 134–35): The REM process may be the functional residue of an ancient form of waking – a simple-minded form of emotional arousal that was "reined in" through the evolution of REM-atonia as higher "pos-trigeminal" ERTAS systems prevailed over primordial pre-propositional forms of waking. This would help explain why REM still arouses basic emotional processes and infuses affect into cognitively manifested dream deliberations. Might it be that people with damage to cortical areas that abolish manifest dreams still experience affective states but no longer have memorial residues of those experiences as is the case in amnesic temporal lobe patients (Damasio 1999)?

From an evolutionary perspective, it should be obvious that basic waking mechanisms have a long history, and certain simple-minded solutions may have gradually been superseded by more sophisticated ones. Perhaps older forms could not be eliminated but, because of adaptive constraints, had to be integrated, often in

perplexing ways, with subsequent layers of neural control. According to this view, REM arousal may reflect an ancient form of waking arousal that was devoted largely to activating genetically ingrained emotional subroutines, which guided behavioral actions in ancestral species long before the behavioral flexibility provided by higher cerebral evolution. Those ancient, value-coding processes may still provide background operations that help higher brain mechanisms sift and integrate fundamental survival concerns from the Niagara of cognitive information flowing in from newly evolved forebrain regions.

Although we cannot be privy to the cognitive contents of animal dreams, we do have some behavioral access to their emotional contents. Following destruction of their atonia mechanisms, cats exhibit at least four "archetypal" REM-dreams – predatory intent, fearful withdrawal, angry assertiveness, and licking/grooming (Sastre & Jouvet 1979). These categories reflect primal emotional concerns of all mammals, elaborated by subcortical systems which may have had a more important role in the emergence of consciousness than is commonly recognized (Damasio 1999; Panksepp 1998b; 2000). In unpublished work, I have repeatedly found these four primal themes to prevail in human dream reports.

If REM merely reflected some type of homogeneous, nonspecific cerebral arousal as **VERTES & EASTMAN** suggest, these oneiric behaviors would remain puzzling. However, these emotional expressions provide excellent clues for unraveling the nature of the neural "conductors" that may guide REM-dreams in all species. Perhaps a search for the psychological functions of REM sleep should be premised on an understanding of these "archetypal" psycho-behavioral themes. Would **NIELSEN** and **VERTES & EASTMAN** consider including such affective themes in their analysis of REM functions?

Any comparable analysis of NREM dreams will remain more problematic until someone learns to extract the cognitive contents of sleeping animals, a project that is becoming conceivable with neurophysiological measures of brain representational activities: It has been surprising to many that spatial data processed during waking are better represented in hippocampal circuitries during slow-wave sleep than in REM sleep (Shen et al. 1998; Wilson & McNaughton 1994). However, perhaps this simply affirms that NREM dreams are more laden with past cognitive experiences. I trust future investigators will eventually find predominating neuronal footprints of past emotional experiences in REM sleep

In accepting the connectedness of REM dreams and the ancient emotional force of fear, **REVONSUO** clarifies why threat perception and harm avoidance lie at the heart of so many of our dreams. However, one empirical difficulty for **REVONSUO**'s fear-constrained analysis may be the tendency of predatory mammals to exhibit more REM sleep than their prey. Surely the latter should exercise their defensive capacities more than the former. One could, of course, suggest that predators have a continual fear of starvation, but that would be taking the analysis in circular directions that have traditionally yielded more confusion than clarity in our attempt to scientifically understand the nature of psychological processes.

In any event, fear has been a most compelling way to bring basic emotional issues to the attention of the scientific community. It will be especially interesting to know whether humans have fear dreams more in anticipation of harm than its aftermath. Perhaps a study of well-motivated students or sick people before difficult mental and bodily examinations, as well as dream analyses of winning and losing teams around closely contested championships, could shed light on such issues.

Of course, the other basic emotional systems also deserve study in such an "experience expectant" dream framework. Since Freud focused on wish fulfillment as one common theme of dreams, let me briefly follow the lead of **SOLMS** on the possibility that dopamine arousal helps to mediate such urges in all animals (Ikemoto & Panksepp 1999). There are remarkable similarities between the emotional energies of REM and dopaminergic lateral hypothalamic urges reflected in self-stimulation behaviors (Panksepp

1998a, pp. 142, 163); it is also established that dopamine neurons do not shut down during REM as do norepinephrine and serotonin ones (Steinfels et al. 1983; Trulson & Preussler 1984). Unfortunately, as **HOBSON ET AL.** emphasize, we don't really know whether dopamine is vigorously released during activated sleep. Indeed, with all the many other neurochemical participants in REM sleep, dopamine may only be one conductor of the dream symphony.

In any event, if dopamine is a major player in certain dreams, we might anticipate dopamine neurons to begin exhibiting "bursting" activity during REM sleep. Such changes, have not yet been empirically evaluated to my knowledge, although the DA supersensitivity following REM deprivation, as noted by **HOBSON ET AL.,** support that view. From simple ethnological observations we do know that rats exhibit lots of sniffing – an excellent indicator of dopamine arousal and appetitive engagement – during REM. In **REVONSUO'S** perspective, we might also note that avoidance of danger may operate through dopaminergic "wish-fulfillment" type processes since "approach to safety" rather than mere "escape from danger" may be the self-centered dynamic around which many avoidance behaviors are constructed in the brain (Ikemoto & Panksepp 1999). With such ideas, the views of **SOLMS** and REVONSUO could be integrated nicely.

In any event, REVONSUO has provided an impressive evolutionary scenario that can help us make sense of the chaotic evidence for the information-integration views of REM sleep that **VERTES & EASTMAN** criticized. If one restricted such information-integration inspired analyses of REM functions to difficult emotional realms, rather than including all cognitive problems animals need to solve, the existing evidence for REM-promoted memory consolidation may be less chaotic than VERTES & EASTMAN portrayed.

For example, the largest and most consistent effects of post-training REM deprivation ever described in animal studies are those that have employed devilishly complex tasks such as the two-way shuttle avoidance (Smith 1985). No animal has been prepared by evolution to continually run between fluctuating danger and safe zones, and it takes a great deal of training for animals to master such emotionally horrendous tasks (Greenberg & Pearlman 1974). The severe deficiency of REM-deprived animals (using the stressful island method) in the acquisition of shuttle avoidance supports **REVONSUO**'s hypothesis that complex harm avoidance strategies may be strengthened by REM sleep. As Smith (1985) has highlighted, such effects can be obtained with short periods of REM deprivation and many studies have employed rather good controls to help rule out generalized stress effects – one of the most, compelling being that of Leconte et al. (1974). The failure of van Hulzen and Coenen (1982) to obtain such effects with their kinder form of REM deprivation (rocking), may have been owing to their use of a pre-learning deprivation paradigm on learning of rather mild shock avoidance, which may have aroused less fear than is present in many other studies. Clearly, more work is needed on the issue before we have definitive conclusions. Still, **VERTES & EASTMAN**'s assertion that REM-deprivation technique typically produces a great deal of stress having distinct behavioral effects is certainly correct (e.g., Kovalzon & Tsibulsky 1984).

With regard to the information-processing dilemma that **VERTES & EASTMAN** highlight, from the present vantage, REM deprivation should selectively impair emotionally loaded tasks that truly challenge animals' coping resources toward the breaking point. REM dreams may operate effectively on the statistical contingencies present in such difficult situations, providing opportunities for new emotion-relevant "insights," or at least coping adjustments, to emerge from REM inspired information juggling in higher regions of the brain. However, rather than REM merely consolidating information in a cognitive realm (which may be much more of a NREM function as highlighted by Buzsaki [1998] as well as the aforementioned work from the McNaughton group), REM may help create novel psycho-behavioral connections within the subconscious emotional habit-structures of animals. As **REVONSUO** and others have emphasized, dreams may allow organisms to deal better with emotionally charged situations in novel ways. De-

spite VERTES & EASTMAN's compelling challenge, this alternative view has yet to attract adequate empirical investigations.

To digress, the fact that antidepressants which eliminate REM have few cognitive effects may not be pertinent to the present argument: First, these agents allow animals to cope better with emotionally stressful situations independently of their REM suppressing effects. Second, since there is good reason to believe that REM sleep normally sustains the synaptic efficacy of transmitters such as serotonin and norepinephrine, the utilization of antidepressive agents may ameliorate some of the neurochemical deficits normally produced by REM deprivation (Panksepp 1998a, p. 140). Regrettably, the emotional tendencies of patients with brainstem damage disrupting REM sleep remain to be adequately evaluated, but the present view would predict problems in long-term emotion/cognition integrations. Also, **VERTES & EASTMAN** fail to note that the absence of apparent waking-up deficits in such individuals might be a negation of their suggestion that REM helps sleeping brains to recover from the psychological torpor of deep somnolent states. Likewise, I anticipate the antidepressant and many other psychological effects of REM-deprivation might be difficult to explain with the minimalist view espoused by VERTES & EASTMAN.

If most REM-dreams reflect forward directed, experience-expectant emotional processes, then they may not be the epiphenomenal or psychologically irrelevant flotsam (as **REVONSUO** puts it "random noise generated by the sleeping brain as it fulfills various neurophysiological functions") that many investigators are coming to believe. If REM-dreams are truly laden with self-referential configurations and permutations of emotional problems to be solved, while NREM dreams are laden with less affective contents, then we still have a relatively straightforward conceptual solution to several dilemmas highlighted by this excellent series of papers: A careful consideration of our fundamental emotional nature. which like REM itself emerged in brain evolution long before sophisticated cognitive abilities, may be essential to make sense of the most activated phases of mammalian sleep and the dreams they energize. Slow wave sleep, and its duller dreams, may be more important for dealing with the less passionate cognitive deliberations and adjustments of the brain and body.

# Neurotransmitter mechanisms of dreaming: Implication of modulatory systems based on dream intensity

E. K. Perry and M. A. Piggott

*Medical Research Council Neurochemical Pathology Unit, Newcastle General Hospital, Newcastle upon Tyne NE4 6BE, United Kingdom.*
**e.k.perry@ncl.ac.uk**

**Abstract:** Based on increasing dream intensity and alterations in neurophysiological activity from waking, through NREM to REM sleep, dreaming appears to correlate with sustained midbrain dopaminergic and basal forebrain cholinergic, in conjunction with decreasing brainstem 5-HT and noradrenergic neuronal activities. This, model, with features in common with the modulatory transmitter models of Hobson et al. and Solms, is consistent with some clinical observations on drug induced alterations in dreaming and transmitter correlates of delusions.
[HOBSON ET AL.; SOLMS]

The growing realization that dreaming occurs, to a greater or lesser extent throughout sleep, and essentially independent of REM, requires – as **SOLMS** suggests – a paradigm shift regarding neurobiological mechanisms. Linking dreaming to the function of individual transmitter systems is no doubt simplistic, given the multiplicity of transmitters that – as in waking consciousness or cognition – are likely to be involved. Nevertheless a hypothetical framework of selective neuromodulation provides the focus for investigations ranging from disease to drug effects. The original evidence that brainstem cholinergic neuronal activation underpins REM sleep has for example been linked to REM sleep abnormalities in Lewy body disease in which pedunculopontine neuropathology occurs and REM behavioural disorders may be presymptomatic (Boeve et al. 1999). Objective measures of REM sleep, not yet established for dreaming, have no doubt provided much of the attraction of the original hypothesis that REM sleep and dreaming are equivalent in terms of brain mechanisms.

Because dreaming occurs throughout the sleep cycle, hypotheses regarding neurobiological mechanisms need to be based on patterns of transmitter activity that, rather than distinguish REM from NREM sleep, distinguish sleep from wakefulness. Moreover, if as evidence suggests, dreaming intensity is greater in REM than NREM sleep, transmitter correlates of dreaming should vary accordingly.

Based on activation or deactivation during REM and NREM (reviewed, Gottesman 1999, **HOBSON ET AL., SOLMS**) the following neuromodulatory systems have been implicated in dreaming: brainstem and basal forebrain cholinergic, brainstem 5-HT and noradrenaline, and midbrain dopaminergic. As dreaming normally ranges from none to more intense in parallel with changes for wakefulness to NREM to REM sleep, candidate transmitter correlates would be those demonstrating a similar gradation in activation or deactivation. Activity of dopaminergic neurons in the ventral tegmental area, implicated by **SOLMS**, is no doubt essential but not alone sufficient to account for dreaming, because these neurons are active throughout the entire sleep-wake cycle (Miller et al. 1983). Similarly basal forebrain cholinergic neurons are, to judge by cortical release of acetylcholine, active to a greater or lesser extent throughout (Marrosu et al. 1995). Brainstem cholinergic neuronal activation, originally highlighted by Hobson, cannot, as **SOLMS** concludes be essential since this does not generally occur during NREM. The only modulatory systems so far identified with the required variation in firing patterns are apparently the 5-UT and NE brainstem neurons which both demonstrate a gradation between wakefulness, NREM, and REM from active through less active to inactive (Hobson et al. 1975; McGinty & Harper 1976). Histaminergic neurons in the hypothalamus are incidentally equally silent during both NREM and REM sleep (Vanni-Mercier et al. 1984). As Gottesman (1999) has concluded, disinhibition in target projection areas including cerebral cortex, associated with declining 5-UT and noradrenergic activities, may underpin the generation of dreaming and explain the delusional, altered affective and other characteristics associated with information processing that is repressed during waking. It is thus suggested that dreaming is essentially associated with decreased 5-HT and noradrenergic activity in conjunction with the maintenance of VTA dopaminergic and basal forebrain cholinergic activities – a model that includes aspects of both models of transmitter modulation proposed by **HOBSON ET AL.** and **SOLMS**.

Continued activation during NREM and REM of midbrain dopaminergic and basal forebrain cholinergic neurons may provide the basis of, respectively, motivation or drive (albeit non-volitional), and selective attention or conscious awareness (variably reduced) during dreaming. Variations in cholinergic function between REM and non-REM, which include brainstem neuronal activation in REM, and increased forebrain activation during REM compared to NREM, may contribute to the increased frequency and/or intensity of dreaming reported during REM. Since pontine cholinergic neurons activate VTA dopaminergic neurons (Gronier & Rasmussen 1998), enhanced dopaminergic function may also contribute to dream intensification during REM. However variations in reported intensity could reflect reduced levels of conscious awareness during NREM compared to REM. Reduced release of acetylcholine, which has been specifically implicated in conscious awareness (Perry et al. 1999), occuring during NREM compared to REM and alert wakefulness could contribute to decreased registration of dreams during NREM.

The proposed association between dreaming, on the one hand, and a distinct activation/deactivation pattern in specific modula-

tory transmitter pathways may not have the merit of simplicity, but is consistent with some aspects of dream physiology, pharmacology, and pathology. It is for example not easy to identify a single system which alone can account for the broad range of brain areas activated during dreaming, to judge from the lesion and in vivo neuroimaging data reviewed by **HOBSON ET AL.** and **SOLMS**. Innervation by dopaminergic axons, suggested by **SOLMS**, may correlate less well than innervation by dopamine and acetylcholine together and/or by 5-HT and noradrenaline. Reports regarding the dopaminergic innervation of primate cortex are variable, depending on the transmitter indices and, based on tyrosine hydroxylase, indicate that the motor cortex is particularly densely innervated (Berger et al. 1991; Parnavelas 1990), and yet this area is not reported to be activated during dreaming. Activated cortical areas, including the amygdala, limbic cortex, and also hypothalamus receive dense cholinergic in addition to dopaminergic innervations. Areas where activity is decreased during dreaming, such as primary visual cortex contains the highest density of 5-HT innervation and 5HT2 receptors (Parnavelas 1990; Pazos et al. 1987).

Other aspects of dreaming which can be examined in the context of the multitransmitter hypothesis proposed in this commentary in response to the models of **HOBSON ET AL.** and **SOLMS**, are alterations in dreaming induced by drugs or related to disease. Alterations in dreaming are potentially relevant to clinical practise. Thus for example in Parkinson's disease, dreaming abnormalities (vivid dreams, nightmares, night terrors) have been reported to precede the occurrence of hallucinations, delusions and delirium induced by levo-dopa (Factor et al. 1995; Pal et al. 1999), and REM abnormalities have been linked to drug induced hallucinations in Parkinson's disease (Comella et al.1993). In relation to increased dreaming intensity associated with anti-Parkinsonian medication, it is interesting that pramipexole (a dopamine agonist with preference for the D3 and no affinity for the Dl receptor) is more likely to induce this as a side effect than l-dopa (Pal et al. 1999), and the distribution of the D3 receptor may provide information on the role of dopamine in dreaming. In Alzheimer's disease, treatment with cholinesterase inhibitors can induce dreaming abnormalities including nightmares (Ross & Shua-Haim 1998) that may restrict the use of this type of drug; similar increased dreaming has been reported in accidental cases of organophosphate toxicity (Warburton 1979).

There have been few reports so far on abnormalities in dreaming, independent of drug treatment, associated with diseases of the brain which involve degeneration of specific neuronal nuclei. Alzheimer and Lewy body types of diseases affect, to varying degrees all of the modulatory systems discussed above and alterations in dreaming are likely to provide new insights in disease mechanisms and prognosis. There is one report of increased dream intensity in patients with Parkinson's disease which was not correlated to medication (Van Hilten et al. 1993) and more recently of one patient with cessation of dream recall independent of medication (Sandyk 1997). It might be predicted that basal forebrain cholinergic and/or VTA dopaminergic pathology would be associated with decreased dreaming, locus coeruleus, and/or dorsal raphe pathology with increased dreaming, and various degrees of combined pathology with intermediate effects. More attention has been focused on the parallel between dreaming and delusions associated with various cerebral disorders especially schizophrenia. Dopamine receptor antagonists reduce delusional symptomatology in a variety of disorders. Delusions or psychosis in Alzheimer's disease have been related to reductions in 5-HT and 5-HIAA in the subiculum (Zubenko et al. 1991). In dementia with Lewy bodies delusions have been linked to elevated muscarinic Ml receptor binding in temporal cortex (Ballard et al., submitted). Delusions can be induced by organophosphate cholinesterase inhibitors in normal individuals (Warburton 1979), although in Alzheimer's disease delusions are reduced by such drugs as metrifonate. Pathology in the locus coeruleus and raphe nuclei may in this instance contribute to a differential drug response in the disease condition. If delusions during waking are closely related to a dream-like state, then these observations implicating dopamine, 5-HT, and acetylcholine are generally consistent with the transmitter model of dreaming suggested in this commentary.

In conclusion, original contributions to understanding neurobiological mechanisms of dreaming, such as those of **HOBSON ET AL.** and **SOLMS,** will no doubt continue to generate not only new models but also new directions for research in neuropsychiatric disease.

## Metaphoric threat is more real than real threat

Jordan B. Peterson and Colin G. DeYoung

Department of Psychology, University of Toronto, Toronto, Ontario, Canada
M5S 3G3. peterson@psych.utoronto.ca
www.psych.utoronto.ca/~peterson/welcome.htm

**Abstract:** Dreams represent threat, but appear to do so metaphorically more often than realistically. The metaphoric representation of threat allows it to be conceptualized in a manner that is constant across situations (as what is common to all threats begins to be understood and portrayed). This also means that response to threat can come to be represented in some way that works across situations. Conscious access to dream imagery, and subsequent social communication of that imagery, can facilitate this generalized adaptive process, by allowing the communicative dreamer access to the problem solving resources of the community.

[**REVONSUO; SOLMS**]

**REVONSUO** believes that dreaming enhanced adaptive fitness in the ancestral environment and that dreams provide "perceptually and behaviorally realistic rehearsals of threatening events." His notion of "realistic," however, appears shaped by the implicit presumption that environmental selection mechanisms can best be considered as the array of things and situations that "leap out at us" and "cry out to be named" (Brown 1965, p. 478), and not as something more generally conceptualized. In REVONSUO's scheme, these are the particular and namable elements of the Pleistocene savannah – the specific dangers that lurked there (predators, enemies, disasters). It is not clear, however, that the ancestral and current human environments are fundamentally different, or that threat can best be mastered as a consequence of its "basic-level" representation.

On the surface, this appears contentious. How can the environment of the prototypical African human progenitor be considered reasonably equivalent to that of the modern individual? The problems we face while sitting at our computers seem very much unlike those of the more "natural" world. What constitutes threat, however, or even "environment," depends on level of abstraction, and there are levels that allow for representations of danger that are isomorphic across all conceivable frames of reference (Peterson 1999).

Let us first determine just what "threat" means, in the broadest possible sense. Humans are goal-directed (Adler, in Ansbacher & Ansbacher 1956; Gray 1982; Oatley 1999). Emotions, including anxiety, signal the interruption of specific goal-directed schemes of conceptualization and patterns of action. Anxiety signals threat, to be sure (Gray 1982) – but more generally indicates the emergence of the unknown or the anomalous (which is initially nothing but undifferentiated evidence for the insufficiency of current plans) (Peterson 1999). This means that the concrete dangers of the natural world may be most usefully considered specific exemplars of a more general category. This more general category – the anomalous – lurks everywhere; it is a universal constituent element of experience. This is because we dramatically simplify the world (Miller 1956), while engaging in our goal-directed processes, and because these simplifications may constantly be revealed as insufficient, in the real environment. It is such revelation that constitutes the most basic and universal threat (Binswanger 1963).

This implies that the dream may represent threat most usefully at the highest level of abstraction – that level allowing for most

cross-situational generalization. Once this is understood, the relationship between the dream, consciousness, and the adaptive activity of cultural construction can be explicitly comprehended. Consider an actual dream, as exemplar – the production of a highly verbal five year old boy, about to leave his family and join the novel world of kindergarten. He was happy during the day, although deeply immersed in a pretend world: He spent much of his time dressed as a knight, with a plastic helmet and sword. He was not sleeping well, however, and frequently screamed for his mother late in the evening. One morning he described a nightmare. Armless, greasy, dwarf-like beaked creatures had been jumping on and biting him. Each creature had a cross shaved on the top of its hairy head. In the background loomed a fire-breathing dragon. The dragon exhaled smoke and fire, which promptly transformed itself into more biting beaked dwarves. Everyone who heard his dream report was fascinated and shocked.

It was clear that this boy had never really encountered biting dwarves or dragons. What possible purpose could such representation therefore serve? Well, after the boy had recited his tale, he was asked a question: "What could you do about this dragon?" This seems something simple, but it is not. It is instead the sort of utterance that allows a lawyer to "lead" a witness. It is a question full of "triggers" (Bruner 1986) or implicit information. The question says as much as it asks. It says, "something can be done about dragons," for example, and "small boys like you can do that something." This leading question therefore puts forth in exceedingly compressed form the plot and character elements necessary to successfully complete the narrative of the dream – that is, to solve the problem it poses.

The boy said, excitedly: "I would take my dad, and go after the dragon. I would jump on its head and poke its eyes out with my sword. I would go down its throat to the fire. I would cut out the box the fire came from, and make a shield from it." This is a complete and spontaneous recreation of a traditional hero myth – and hero myths detail successful encounters with the unknown (Peterson 1999). It is not necessary, however, to posit the derivation of this tale from the "collective unconscious" (Jung 1959). This boy had seen many movies, heard many stories, and had observed patterns of successful (and unsuccessful) real-world behavior. So the pattern for the "hero" was something thoroughly embedded in his social world. But he had never conceptualized himself as heroic. One leading question, however, provided sufficient motivation for that. His dream represented him as threatened by "archetypal" dangers – not so much by particular threats (in the form of the dwarves), but by threat itself (in the form of the dragon). When he reconceptualized himself, therefore – as a consequence of social prompting – he came to understand that he was more than someone who could face particular threats: He was someone who could overcome the class of threatening things itself. This is a far more useful conceptualization because of its cross-situational generalizability (and one that did in fact eliminate his nightmares).

The fact that it was social prompting that led to such reconceptualization also sheds light on an additional mystery. Why communicate dream content? There is a simple answer to this question: Two heads – or two thousand – are better than one. Traumatized individuals experience intrusive thoughts about the threatening occurrence (Tait & Silver 1989) and need to talk about their experience (Ersland et al. 1989; Rime 1995). Those denied opportunities to engage in social-mediation of such experiences tend to suffer more, in the aftermath (reviewed in Petrie et al. 1998). Why? What good does talking do? Well, ability to report on internal states in a communicable manner also means capacity to draw on the problem-solving resources of the community to deal with threat. This capacity to communicate dreams could have been selected for after the rise of language. So – threatening dreams become memorable and compel communication (become nightmares) precisely when they represent a threat so profound that it exceeds the current adaptive capacities of the dreamer. Such dreams are then reported, in a dramatic and intrinsically fascinating fashion. Then the community helps solve the problems they pose.

The global significance of this process should not be underestimated. Dreams are part of the lengthy, historically-elaborated process by which threats, as a class, come to be metaphorically represented – as something reptilian, for example, unpredictable, chaotic, devouring (Eliade 1978) – and then, as a class, come to be mastered (Peterson 1999). The construction of protective culture itself can reasonably be regarded as a consequence of this motivated process – not so much to escape from the specific dangers of the Pleistocene environment, but to alleviate the total consequences of human vulnerability, across all conceivable contexts.

**SOLMS**'s observations on the potential dopaminergic mediation of dreams are interesting in this regard and help tie the threat representation capacity of the dream to its evident facility for bizarre conceptual portrayal. We know, for example, that dopaminergic activation is associated with exploration (Gray 1982) and with increased categorical flexibility (Ashby et al. 1999; Lubow 1989). These two phenomena are logically related: Categorical flexibility should increase during exploration, so that current schemes of apprehension may be modified as a consequence of learning. Adaptation to threat means either reconceptualization of self and the acquisition of new and relevant skills, or reconceptualization and recategorization of the feared object (Foa & Kozak 1986; Williams et al. 1989; 1997). Finally, we have the fact that general mood states in dreams tend to be positive (**REVONSUO**) – something in keeping with the first two phenomena, as dopaminergic activation is associated with positive emotion (Gray 1982; Ashby et al., 1999). This all implies that dreams may be positive, exploratory, creative play, when they are not dealing specifically with an anomaly intense enough to be traumatizing. So it seems reasonable to posit that dreams may be considered more broadly part of the process of adjustment to novelty, and that their facility for dealing with threat might be considered as something subsidiary to that broader function.

# One machinery, multiple cognitive states: The value of the AIM model

C. M. Portas

*Wellcome Department of Cognitive Neurology, Institute of Neurology, University College London, London, WC1N 3BG, United Kingdom.*
**cportas@fil.ion.ucl.ac.uk    www.fil.ion.ac.uk/groups/frith/chiara.html**

**Abstract:** The AIM model represents an original and comprehensive example of how changes in conscious states can be reconciled with specific neurophysiological factors. However, further elucidation of the biological parameters necessary to define a specific space-state relationship should be considered.
[**HOBSON ET AL.; SOLMS**]

**HOBSON ET AL.** and **SOLMS** review an impressive amount of phenomenological and physiological data in relation to waking, NREM, and REM sleep. The major objective of these authors is to find an integrative model of conscious experience in which distinct cognitive states may be quantified and tracked down to specific neurobiological events. The importance of the activation information processing modulation (AIM) model formulated by this group is owing not to this integrative approach only but especially to the understanding of consciousness as a multidimensional and dynamic process. Hence, normal and abnormal cognitive states find their own definition on the basis of the neurobiological parameters taken in account (only three in the AIM model but arguably many more). Thus the AIM-type unitary approach may be applied when studying cognitive states very different in appearance (e.g., hallucination, anesthesia, coma, etc.). In this regard the value of the AIM model goes far beyond the confines of sleep physiology.

Despite my overall appreciation of the AIM model, there are in my view several points of oversimplification that should be further discussed.

***Activation.*** In the activation domain, it appears that the sleep-

wake related brain activity shown by functional imaging studies has been overstressed and only one of the many possible interpretations of the data is offered to the reader. This commentary does not provide the space to address detailed criticisms of the functional imaging studies, however, some elements of misinterpretation should be considered. I will limit my comments to a couple of examples.

1. It has been suggested that the higher activation of the brainstem during REM sleep is dependent on intense neuronal activity of the REM sleep generator system. This is a likely possibility and one that I happen to favor, as do the authors, our shared opinion being that the brainstem generates REM sleep's phasic and tonic phenomena. However, I am interested in why brainstem activity should be higher in REM sleep than in wakefulness. It is known that the neuronal populations responsible for arousal are located in the brainstem. For instance, the locus coeruleus noradrenergic neurons, the raphe dorsalis serotonergic neurons, the reticular activating system neurons, and so on, all lie in the proximity of the REM sleep generator system (McCormick & Bal 1997). Also it is known that in the lateral dorsal tegmentum (LDT) and ponto pedunculum tegmentum (PPT), the number of neurons selectively active during REM sleep are outweighed by the number of neurons active during both REM sleep and waking or waking alone (Kayama et al. 1992). From this observation I would expect to see similar patterns of brainstem activation in both wakefulness and REM sleep. This point remains unresolved.

2. Another example is given by the higher activation of the anterior cingulate cortex during REM sleep compared to NREM sleep or waking. This effect has been discussed by the authors in the following way: "As in waking, anterior cingulate activation contributes additional emotional features to dreaming such as valence biases, the assessment of motivational salience, and the integration of dream emotion with fictive actions" (from Subcortical and Cortical limbic and perilimbic structures in the Activation Synthesis, sect. 3.4.4, **HOBSON ET AL.**). This interpretation is one possibility amongst several others. However, I question how we can reconcile this view with the large amount of data showing that the anterior cingulate cortex is a crucial part of the executive attentional and executive system and activates in tasks requiring performance monitoring and error detection (e.g., Awh & Gehring 1999; Carter et al. 1998). Such complex cognitive features do not easily fit with the authors' proposition that the dreaming brain lacks of self-awareness, judgment capability, volitional control, and so on. Finally, the higher activation of the anterior cingulate cortex contrasts with the lesion studies reviewed by **SOLMS** which show a correlation between anterior cingulate lesions and increased frequency and vivacity of dreaming (Solms 1997a). In conclusion, I suggest that neither the functional imaging nor the lesion studies results should be overemphasized at this stage. These results are not conclusive. In addition, several discrepancies (more than are usually considered) are present among different functional imaging studies (see Table 2 of **HOBSON ET AL.** article).

***Information processing.*** The information processing domain of the AIM model implies a blockade of information flow during NREM and REM sleep in particular. Despite the fact that threshold for awakening is higher in NREM and REM sleep, there is evidence that sensory inputs are processed at the thalamo-cortical level during sleep (Mariotti & Formenti 1990; Pare & Llinas 1995) and a recent study has shown differential processing of relevant and irrelevant auditory stimuli during sleep (Portas et al. 1999). In addition, paradoxical phenomena like sleepwalking imply a certain degree of sensory processing (being sensory processing necessary for ambulation). Thus, the possibility of residual sensory processing and therefore cognitive functionality during sleep should be acknowledged and the concept of sensory blockade should be drastically reviewed in the AIM model.

***Modulation.*** Another point worth discussing is the necessity of experiments that may address more directly the modulator mechanisms of the AIM model. The long, tedious (and necessary!) list of single cell recording, microdialysis, receptor binding, and so on, experiments used by the authors to support the validity of the Reciprocal Interaction model, loses its strength when applied to explain the highly dynamic aspects of the AIM model. In fact, it is my belief that the dynamic functional models call for more dynamic experimental evidence. In other words it would be necessary to test on-line neuromodulatory interactions which are able to manipulate the space-state relationship of the AIM model (and consequently the conscious state). Ungerstedt and collaborators have made attempts in this direction. The Swedish groups, using microdialysis, systematically monitor the level of several neurotransmitters and metabolites in distinct areas of the human brain after traumatic or ischemic injuries (e.g., in humans: Hillered & Persson 1992; Nillson 1999; Persson & Hillered 1992; in animal models: Nillson et al. 1990). The aim of their study is to correlate patterns of decreased/increased level of neurochemicals with cognitive outcomes. More experiments of this type are required to test the neuromodulatory requirements of the conscious states.

# Neural constraints on cognition in sleep

Helene Sophrin Porte

*Department of Psychology, Cornell University, Ithaca NY 14853–7601.*
**hsp2@cornell.edu**

**Abstract:** Certain features of Stage NREM sleep – for example, rhythmic voltage oscillation in thalamic neurons – are physiologically inhospitable to "REM sleep processes." In Stage 2, the sleep spindle and its refractory period must limit the incursion of "covert REM," and thus the extent of REM-like cognition. If these hyperpolarization-dependent events also inform Stage NREM cognition, does a "1-gen" model suffice to account for REM-NREM differences?

[**NIELSEN**]

To a debate that persists (despite the field's outgrowing it), **NIELSEN**'s target article proposes a singularly constructive and clearly stated resolution. This commentary examines the concept of covert REM sleep, which **NIELSEN** defines as "any episode of NREM sleep for which some REM sleep processes are present but for which REM sleep cannot be scored with standard criteria." Of the questions this definition raises, I shall address three: In which NREM episodes does "covert REM" *not* reside? Which "profiles of NREM Sleep physiology" *do* admit "REM sleep processes" and may thus generate "intermittent REM-like mentation"? Does that intermittence account robustly, as **NIELSEN** claims, for differences between Stage REM reports and Stage NREM reports?

1. In which NREM episodes does "covert REM" not reside? Omitting the K complex, Stage 2 sleep (on which I shall focus) is epitomized by the *sleep spindle*. In a neural network that topographically interconnects (GABAergic) reticular thalamic neurons and (glutamatergic) thalamocortical neurons, spindling indexes a repeating cycle of (1) slow membrane depolarization, mediated by the low voltage-dependent calcium current $I_T$, and (2) at action potential threshold, rhythmic burst discharge (for a review, see McCormick & Bal 1997). This cycle waxes and wanes at variable intervals; between spindles, thalamocortical neurons are depolarized. Do "REM sleep processes" take advantage of this intermittent depolarization to insinuate themselves – if only here and there – into NREM sleep?

Not necessarily. In my view, the spindle wave refractory period – the portion of the interspindle interval mediated by the hyperpolarization-dependent action current $I_H$ – must prevent a ubiquitous invasion of covert REM into the interspindle interval. To be sure, the refractory period does inactivate $I_T$, replacing the rhythmic burst with the single spike as the unit of thalamic network neural discharge (McCormick & Bal 1997). But does the refractory period (1) excite the reticular activating system, both tonically and phasically; (2) phasically inhibit reticular thalamic neurons; and (3) depress the release of serotonin, norepinephrine, and histamine? I suggest that insofar as any NREM episode

merely concatenates sleep spindles, that episode will exclude these REM sleep processes.

2. Which "profiles of NREM sleep physiology" do admit Stage REM processes? The interspindle interval in Stage 2 NREM does not always exclude covert REM. For one thing, standard sleep stage scoring rules, acknowledging both the necessity of spindling and its intermittence, allow an interspindle interval of up to 3 minutes in Stage 2. For another, the duration and constitution of the interspindle interval will change across the sleep period.

As **NIELSEN** rightly suggests, sleep's circadian architecture must influence the variability (and viability) of covert REM sleep. Indeed, in accord with the principle of "delta homeostasis," neural network state can be instantaneously quantified as the sum of (1) globally diminishing "delta power" (Feinberg & March 1995) and (2) periodically alternating hyperpolarization and depolarization in the thalamic network (the NREM-REM cycle; cf. Kahn et al. 1997). Therefore in the "missed REM period" – or in the first REM period, where Stage REM is visible but often mixes with spindling – network hyperpolarization may dominate emergent Stage REM at the surface electrode, but cannot obliterate it altogether in the thalamic neural network.

Thus question No. 2 may be answered as follows: "Covert REM" may reside in the interspindle interval in Stage 2 NREM, but only inside the 3 minute limit for scoring Stage 2, and only outside the spindle wave refractory period. On one hand, "REM sleep processes" may become increasingly salient in the interspindle interval as (cortically instigated) network hyperpolarization diminishes across the sleep cycle. On the other hand, the effect of diminishing delta power on the duration of the $I_h$-mediated spindle refractory period remains to be elucidated. Could waning delta power in fact aggrandize the refractory period?

3. Does the "intermittence" of REM-like mentation account robustly, as **NIELSEN** claims, for differences between Stage REM reports and Stage NREM reports? Clearly, the structure of the Stage 2 NREM episode can accommodate "intermittent REM-like mentation." Nonetheless, a "1-gen" model does not account for all of the differences between Stage REM reports and Stage 2 NREM reports. In my view the sleep spindle alone can support a robust "2 gen" model. It is well known that spindling opposes thalamic neurons' ability to respond articulately to stimuli in their receptive fields. In this and other ways, spindling is like its bigger, $GABA_B$ receptor-mediated cousin, the "spike and wave" network discharge that characterizes absence epilepsy. The symptoms of absence seizure run a gamut from "mental confusion" to "complete blackout" joined to a blank, staring facial expression (Penfield & Erikson 1941). Might "spindling mentation" display similar symptoms? Support exists for this comparison. For example, a set of Stage 2 mentation reports previously rejected as uninformative (Antrobus 1991) in fact contains many absence-like reports: "I blanked out," or – consonant with subjects' typical inability to recall absence – "I can't remember." In the same sample, significantly fewer Stage REM reports (in fact, virtually none) qualify as absence-like (Porte 2000).

If the sleep spindle does not produce dreamlike mentation, and if the spindle wave refractory period does not accept "REM sleep processes" but does (theoretically) produce mentation, a "second generator" of cognition in sleep – possibly, even a third – cannot be rejected out of hand.

In conclusion, I would like to ask whether *any* NREM episode displays the phasic, structured, endogenous network excitation that we measure at the surface electrode as (for example) eye movement, or by virtue of the system's adaptive response to strong phasic somatic motor excitation – as muscle atonia. Certainly NREM Stage 1, as **NIELSEN** would agree, is a candidate. But doesn't Stage 1 (minus the embryonic spindling that heralds Stage 2) occupy a physiologic and cognitive continuum with Stage REM? Take eye movement, for example. Might higher levels of serotonin, norepinephrine, and histamine – and thus greater reduction of "leak" potassium currents (McCormick & Bal 1997) – contribute to slower eye movement velocities in Stage 1 than in stage REM?

Likewise, do Stage 1 levels of those neuromodulators permit some somatic motor excitation, but not enough to warrant adaptive paralysis? In this light, I suggest that "REM-like" processes are not covert at all in Stage 1 NREM. Rather – owing to an old accident of stage classification, they are called by the wrong name.

By raising issues such as these, NIELSEN's target article does much to move us beyond an old debate, clearing the ground for questions of greater current importance: What is the relevance of cognition in sleep to the structure of cognition in waking? What are the adaptive uses of spontaneous neural activity in sleep?

# The contents of consciousness during sleep: Some theoretical problems

Antti Revonsuo

*Department of Philosophy, Center for Cognitive Neuroscience, University of Turku, FIN-20014 Turku, Finland.* **revonsuo@utu.fi**
**http://www.utu.fi/reasearch/ccn/consciousness.html**

**Abstract:** The approach of Hobson et al. is limited to the description of global states of consciousness, although more detailed analyses of the specific contents of consciousness would also be required. Furthermore, their account of the mind-brain relationship remains obscure. Nielsen's discussion suffers from conceptual and definitional unclarity. Mentation during sleep could be clarified by reconceptualizing it as an issue about the contents of consciousness. Vertes & Eastman do not consider the types of memory (emotional) and learning (implicit) that are relevant during REM sleep, and therefore dismiss on inadequate grounds the possibility of memory functions associated with REM sleep.

[HOBSON ET AL.; NIELSEN; VERTES & EASTMAN]

Dreaming provides us with a unique window to the way the phenomenal level (consciousness) is generated and organized in the brain. The study of dreaming and its neural correlates is an important source of empirical data for consciousness research: The dreaming brain can be regarded as a model system that isolates consciousness from the normal sensory and motor interactions and reveals consciousness in a very basic form (Revonsuo 1995). Progress in understanding dreaming and consciousness requires conceptual clarity and tight interaction between theory and empirical research. In this commentary I point out some conceptual and theoretical issues that have been inadequately dealt with in the target articles by HOBSON ET AL., NIELSEN, and VERTES & EASTMAN. The resolution of these issues would clarify considerably the positions advocated by these theorists.

HOBSON ET AL. present a detailed model of the states involved in waking and sleep. Such cross-disciplinary work is a step in the right direction, for it integrates the study of dreaming with mainstream cognitive neuroscience. However, some difficulties with the approach of HOBSON ET AL. remain. They wish to make their own position crystal clear on the more general mind-brain problem (sect. 3), but their view remains rather obscure philosophically. They say that distinctions at either level (mind or brain) imply the existence of isomorphic distinctions at the other: changes in brain function virtually guarantee concomitant changes in mental function and, conversely, for each phenomenological difference it is possible to identify a specific physiological counterpart. While the latter claim (that there can be no mental difference without a corresponding physical difference) is commonly accepted by philosophers as the mind-brain supervenience relation (e.g., Kim 1998), the former claim is more problematic. Even if we accept supervenience between the mental and the physiological, it does not follow that all distinctions at the physiological level are necessarily accompanied by phenomenological ones: There can be physiological distinctions, realized at completely nonconscious levels, that have no direct counterparts at the level of conscious experience. HOBSON ET AL. seem to deny this. It remains unclear how exactly they construe the mind-brain relation: do they adopt the standard supervenience relation or not?

At the empirical level, HOBSON ET AL.'s model concentrates on global states instead of specific contents of consciousness. There are compelling methodological reasons for this: it is much easier to measure and model the physiological and phenomenological changes at a coarse-grained level, where only global states need to be distinguished from each other, than at a fine-grained level, where specific contents of consciousness and their neural correlates should be defined. The psychological and neurophysiological methods currently available may be inadequate for the latter task. However, studies at a fine-grained phenomenological level would seem to be the most revealing ones when we try to understand the mechanisms of consciousness. Consider the detailed analysis of dream bizarreness (to which Hobson's group has made important contributions, e.g., Rittenhouse et al. 1994).

The bizarreness of dreams is of particular interest for the study of consciousness, because the concept of "bizarreness" in dream research is closely related to the concept of "binding" in cognitive neuroscience and consciousness research (Revonsuo 1999b). Indeed, what dream researchers have conceptualized as different forms of bizarreness can in many cases be reconceptualized as referring to different forms of binding, or, to be precise, to different types of deviations or aberrations in the binding of dream images coherently together (Kahn et al. 1997; Revonsuo 1995; Revonsuo & Salmivalli 1995). For example, one specific form of bizarreness, the incongruity of dream images, can be characterized in the following way: Incongruous dream elements are ones that either have features that do not belong to corresponding elements in waking reality or that appear in contexts in which the corresponding elements would not appear in waking reality. Thus, seeing a blue banana, finding a banana growing in an apple-tree, or seeing the President of the United States in one's home, would all be examples of incongruous elements in dreams. These bizarre elements can be characterized in more detail with the help of the concept of binding. A blue banana is a good example of erroneous feature binding: the representation of "banana" in our semantic memory should primarily associate with bananas the color yellow and to a lesser degree the color green, but not the color blue. And a banana growing in an apple-tree, or the President having a cup of coffee in my kitchen, are cases of erroneous contextual binding: even though the individual elements of such dream images are internally coherent, the images do not fit together in the light of our semantic knowledge of the world. Another variety of dream bizarreness is the discontinuity of dream elements: for example, a banana may suddenly appear, disappear, or be transformed into an apple in a manner not possible in the waking world.

Discontinuity seems to be a case of defective binding across time: Successive dream images do not always retain or update individual phenomenal representations in a consistent manner, which results in sudden and inexplicable appearance, disappearance or transformation of objects, persons, and places in dreams. Explaining the different types of bizarreness in dream images thus turns out to be the task of explaining how the mechanisms of binding and the unity of consciousness operate during sleep.

Hence in addition to a general model about the global states of waking and dreaming, we need detailed phenomenological descriptions of the specific contents of dreams and neurocognitive theories to explain how the different kinds of dream images are generated in the brain. This implies further methodological improvement both in dream content analysis and brain imaging. Even more important, there should be fruitful conceptual integration between dream research and cognitive neuroscience. Phenomena described in traditional dream research by a set of more or less folk-psychological concepts (e.g., "bizarreness," "anxiety dream") should be reconceptualized in such a manner that their detailed description can be given in a cognitive neuroscience framework (e.g., "deviations of visual binding caused by the spreading of activation in a neural network," "threat simulation response"). This kind of cross-disciplinary conceptual integration will be necessary to reach HOBSON ET AL.'s ultimate goal, the reunification of the psychological and neuroscientific approaches in the study of consciousness.

NIELSEN attempts to answer the questions: What generates "mentation" during sleep? What are the causal mechanisms involved? He divides "sleep mentation" into two categories, "dreaming" and "cognitive activity." It seems to me that these basic concepts could be defined much more clearly and precisely, by emphasizing the fact that we are actually talking about varieties of subjective conscious experience during sleep. "Sleep mentation" includes all kinds of subjective (conscious) experiences during sleep. The crucial division is between contents of consciousness that fulfill the criteria of dreaming and those that do not. Reconceptualizing the issues with concepts that directly refer to the content of consciousness would give us the following, much clearer distinctions: (1) Complex, temporally progressing contents of consciousness during sleep (dreaming); (2) Other contents of consciousness during sleep (non dreaming; NIELSEN's "cognitive activity"); (3) Nonconscious information processing during sleep (NIELSEN's "cognitive processes").

Because NIELSEN does not make such clarifications, various conceptual and definitional difficulties abound in the target article. According to him, dreaming is likely to be defined as imagery that consists of sensory hallucinations, emotions, story-like or dramatic progressions, and bizarreness. NIELSEN's definition of "dreaming" thus comes dangerously close to the definitions typically given by the proponents of 2-gen models and criticised by the proponents of 1-gen models.

Although some dreams are emotional, dramatic and bizarre, most are not, and it would be a mistake to uncritically include these features in the definition of a dream. In order for us to ever resolve the empirical issues about the neural correlates of dreaming, it would be crucially important to first find a theoretically more neutral definition for "dreaming," for example, "dream" = subjective conscious experience during sleep that involves complex organized mental images which show temporal progression or change (Farthing 1992). Another conceptual confusion can be found in Figure 1 where NIELSEN says that "cognitive activity" is included in the class of "cognitive processes," which seems odd since the latter category consists of nonconscious or preconscious processes and the former of the subjective contents of consciousness. How could the contents of consciousness be a part of nonconscious/preconscious processes? It would seem more logical to treat the categories "dreaming," "cognitive activity," and "cognitive processes" as forming different levels of description. Phenomena at the lower nonconscious levels are probably necessary for the processes at higher levels of description, but the higher-level phenomena (e.g., dreaming) cannot be coherently depicted as "a part" of the nonconscious levels.

After the reconceptualization suggested above, the central question of the target paper could be formulated much more clearly: "Is there a qualitative difference between contents of consciousness during REM sleep and NREM sleep, and must this possible difference be explained by referring to two distinct causal mechanisms or just one?" It is difficult to see how any evidence that does not directly reflect the contents of consciousness could resolve this issue one way or the other. In spite of this fact, only three of the nine lines of evidence that NIELSEN considers are like that. However, it does not suffice to show that REM and NREM differ with respect to physiology or nonconscious information processing. Such differences are not at the correct level of description and reveal little if anything about possible qualitative differences in the contents of consciousness.

NIELSEN introduces the concept of "covert REM sleep": an episode of NREM sleep for which some REM sleep processes are present, but for which REM sleep cannot be scored with standard criteria. These REM sleep processes are responsible for bringing about the "imaginal experiences," that is, contents of consciousness (dreams and non-dreams) during NREM as well as during REM sleep. It remains unclear how exactly NIELSEN's model disagrees with the paradigmatic 1-gen model by Foulkes who says that "we have one dream production system, rather than two (or many). It explains non-REM dreaming as involving some measure

of degradation in the operations of the same dreaming system that is generally working at full steam during REM sleep" (Foulkes 1985, p. 61). This position seems to be entirely consistent with NIELSEN's model; the only difference is what Foulkes calls "degradations in the operations of the dream production system," NIELSEN calls "covert REM processes." This raises the suspicion that the novelty of postulating "covert REM processes" may be more verbal than substantial.

VERTES & EASTMAN claim that brain activity in REM sleep is "inconsistent with the orderly evaluation, organization and storage of information." Such a conclusion is, however, not quite implied by the evidence. VERTES & EASTMAN fail to consider the possibility that memory systems other than working memory and explicit episodic memory might be at work during REM sleep. The neurophysiological evidence appears to be consistent with the hypothesis that memory systems specialized in the processing of emotionally charged memories are very active during REM sleep. Limbic areas (especially the amygdala) are very active in REM sleep. Recent evidence by Hamann et al. (1999) shows that the human amygdala modulates the strength of conscious episodic memories according to their emotional importance. Other researchers (e.g., Metcalfe & Jacobs 1998) have recently proposed that there is an amygdala-centered "hot" emotional memory system, separate from the "cool" hippocampal system. Thus, areas very central to the evaluation and processing of episodic memory traces are highly active in REM sleep, which shows that, contrary to VERTES & EASTMAN's claims, REM sleep is consistent with the orderly evaluation, organization, and storage of emotionally charged information. Although this processing does not typically lead to the consolidation of explicit episodic memories, this does not exclude the possibility that other types of memory traces might be formed or strengthened. There are in fact good reasons to believe that REM sleep and dreaming implicitly strengthen memory patterns related to threat perception and avoidance (see my target article in this issue). VERTES & EASTMAN dismiss the possibility of any memory processing or consolidation in REM, although they fail to consider all the different kinds of memory systems that exist in the brain.

Furthermore, if the principal function of REM sleep is to decrease the harmful side effects of prolonged low brain activation associated with slow wave sleep, as VERTES & EASTMAN claim, then shouldn't it follow from this that total sleep deprivation (at least in small doses) – that is, avoiding the harmful low-level activation altogether so that REM is not needed to fight its side-effects – is in fact less harmful than pure REM-deprivation? Should we not be better off totally without sleep than wakened after a couple of hours of SWS but with little if any REM? Such an implausible prediction would seem to flow from their hypothesis.

ACKNOWLEDGMENT
The author was supported by the Academy of Finand (project 45704).

## Search activity: A key to resolving contradictions in sleep/dream investigation

V. S. Rotenberg
*Sleep Laboratory of the Abarbanel Mental Health Center, Bat-Yam, Israel.*
vadit@post.tau.ac.il    www.zionet.co.il/hp/evadim/book4.html

**Abstract:** The target articles on sleep and dreaming are discussed in terms of the concept of search activity integrating different types of behavior, body resistance, REM sleep/dream functions, and the brain catecholamine system. REM sleep may be functionally sufficient or insufficient, depending on the dream scenario, the latter being more important than the physiological manifestation of REM sleep. REM sleep contributes to memory consolidation in the indirect way.
[NIELSEN, REVONSUO; SOLMS; VERTES & EASTMAN]

*General introduction.* My approach to all problems discussed in this *BBS* special issue is based on the concept of search designed to change the situation (or the subjects attitude to it) in the absence of certainty about outcome but with constant monitoring of outcomes at all stages of the activity. Search activity, manifesting itself in flight, fighting, creativity, orienting behavior, and so on, increases body resistance, especially in a stressful situation (Rotenberg & Arshavsky 1979; Rotenberg 1984). Search activity is absent in stereotyped behavior.

Renouncing search (giving up) manifests itself in freezing, panicky behavior, helplessness, depression, neurotic anxiety; this is maladaptive, regressive, and decreases body resistance. It is important to emphasize that the positive effects of search activity on survival depend on the process of searching per se, not its pragmatic results. Search activity may diminish not only after continuous failures but also after very meaningful achievements, causing the diseases attending achievement. In functionally sufficient REM sleep dreams, represent a special form of search activity which compensates for the renouncing search in prior wakefulness and restores search activity in the subsequent wakefulness (for details see Rotenberg 1993a). With this general introduction, let us now turn to comments on particular papers.

**1. The differentiation of functionally sufficient versus insufficient REM sleep [Solms].** A very clear and convincing differentiation is presented by **SOLMS** between brainstem mechanisms responsible for REM sleep as a physiological condition for dream mentation, and forebrain mechanisms responsible for dream mentation, as a psychological phenomenon. Although previous investigators understood that forebrain mechanisms are obligatory for the subjective dream experience, two contradictory and restrictive approaches to this topic have dominated the scientific literature. According to the first approach, REM sleep performs purely physiological functions not requiring any subjective experiences, while the dream is a side effect of general brain activation in REM sleep. As a result, the dream represents some occasional combinations of images free from any special psychological function. The second approach, although more common, is less definitively expressed in the literature. This approach acknowledges the importance of psychological functions in dreams, yet implicitly connects dream experience to REM sleep as its natural, obligatory, and inseparable component. According to this point of view, which has caused many paradoxes and misunderstandings, forebrain mechanisms responsible for dream production are considered to be a part of the REM sleep system. If REM sleep is considered a physiological mechanism isomorphic to dream mentation, it is difficult to reconcile with the fact that in mental and psychosomatic diseases, physiological REM sleep mechanisms are often strained but idle, being "empty" of dream content. Thus, sleep structure is not in itself sufficient to indicate whether REM sleep is efficient. It is also necessary to go on and investigate dream reports.

A few years ago, I suggested distinguishing functionally sufficient and functionally insufficient REM sleep (Rotenberg 1988). The latter is characterized by impoverished dream content or by the absence of dream reports after awakenings during REM sleep. I suggested that dream experience is responsible for the adaptive capacity of REM sleep. Based on **SOLMS**'s concept, functionally insufficient REM sleep is characterized by dissociation between the preserved brainstem mechanisms controlling REM sleep and the disturbed forebrain mechanisms responsible for dream experience. **SOLMS** presents strong arguments for the role of forebrain mechanisms dream generation. I would add only that a correlation between forebrain focal epileptic seizures and recurring anxious dreams does not confirm that dreams are produced by forebrain mechanisms. It shows only that dream content can be influenced by forebrain activity.

Although animals do not report dreams, they probably do have dreams based on their behavior in REM sleep without muscle atonia. If so, they may also demonstrate the dissociation between REM sleep and dreams, and presumably in such cases REM sleep will likewise be functionally insufficient. This may be pertinent to the contradictions in the data on REM sleep alterations under stress or sleep deprivation. I would conjecture that the predomi-

nance of hippocampal theta rhythm in REM sleep indicates an integration of REM sleep and dreams in animals.

I highly appreciate **SOLMS**'s hypothesis that dreaming is generated by the dopamine circuit which instigates goal-seeking behavior. This hypothesis accords concepts of search activity (Rotenberg 1984; 1993a), which starts in the presence of a certain critical level of brain monoamines utilized in the course of search activity. However, in parallel with utilization, brain monoamines are continuously restored in the process of search activity, providing the circuit with positive feedback (Rotenberg, 1994a). In the state of renouncing search, this positive feedback is blocked. This agrees with **SOLMS**'s suggestion that dreaming is generated by the same dopamine circuit that stimulates goal-seeking behavior. However, I expect that other monoamines such as norepinephrine may also participate in this mechanism.

It is important to stress that, although in pathology both systems – brainstem system for REM sleep, and forebrain system for dream generation – may be separated from one another; in healthy subjects both systems function in a highly integrated way. REM sleep provides the best physiological condition for dream generation: on the one hand, it causes a massive general activation of the forebrain; on the other hand, contact with the outside world is substantially reduced. At the same time, the relationship between the two systems is reciprocal: Dreams may have a secondary influence on REM sleep physiology. For instance, there is evidence for a directional correspondence between dream imagery and rapid eye movements in REM sleep (Herman et al. 1984). It has also been shown that in healthy subjects eye movement density correlates with the active participation of the dreamer in dream events (Rotenberg 1988). In his attempts to show the independence of two systems, **SOLMS** underestimates some of these facts. Nonetheless functional interrelationship between the systems does not diminish the centrality or importance of his conclusion that these are two separate systems which can be disintegrated in some special conditions.

**2. Behavioral attitude in dreams: The main variable. [Revonsuo].** The topic of **REVONSUO**'s target article is especially interesting and intriguing if we consider the evolutionary context and biological function of dreaming. In most investigations, biological functions are ascribed to REM sleep, while dream mentation is considered to belong to the domain of human psychology. However, the behavior of animals in REM sleep without muscle atonia (the model of Jouvet and Morrison) allows speculation that animals may also have dream experience (see Jouvet & Delorme 1965; Morrison 1982). Thus, an integrative theory of dream function must be relevant to humans as well as animals, and to psychology as well as biology. There are many inner contradictions in REVONSUO's paper, however, which are not sufficiently discussed, even though the author seems aware of some of them.

**2.1. Revonsuo's main idea** is that the biological function of dreaming is to simulate threatening events in order to rehearse threat perception and the appropriate threat avoidance skills. He stresses that in order to perform this function, dreams must contain recurring realistic threat simulations (perceptual realism) and must activate realistic avoidance skills relevant to these threats. Rehearsing threat avoidance skills in the simulated environment in dreams is likely to lead to improved performance in real threat avoidance situation in wakefulness, in the same way that mental training for specific movement skills improves these skills in real behavior.

**REVONSUO** supports his hypothesis with data from numerous investigations of dream content containing threatening situations critical for physical survival, for example, confrontations with wild animals, snakes, aggressive strangers, and so on. One can agree with the author's conjecture that such dream contents mirror the threatening environment of our ancestors. However, in modern society the real threat in most cases is psychological rather than physical. Precise behavioral strategies that are relevant for the ancestral environment are not relevant for modern life. They are not applicable to modern threatening events such as failures, injuries, and conflicts in different social relationships. Such failures threaten not physical survival, but survival of the self. The special skills trained by fighting or escaping aggressive animals in dreams are not useful in coping with modern social threats and frustrations. In this context, the ancestral dream content has no sense for the modern life and can be considered a psychobiological atavism.

The concept of dream scenario preparing the subject for waking behavior does make sense, however, in the context of the search activity concept discriminating two opposite types of behavior in a stressful situation. Surprisingly, REVONSUO has missed this concept in his very substantial review of psychological and biological functions of dreams, although it is very relevant and avoids the fundamental contradiction noted earlier. If the task of the dream is not to prepare the subject for a special threatening situation, but to restore search activity as a general adaptive mechanism, then dream events have no special meaning. They may reflect recent modern problems or problems more typical of the ancestral environment. *The only important feature of the dream content is the dreamer's behavior in the dream scenario: search activity versus renunciation of search.* I suppose that even in ancient times, dreams performed the same function because the restoration of search activity after giving up is more important for adaptive waking behavior and survival than even the training of special skills.

**2.2. In considering the dream** as a simulation of the threatening system, REVONSUO does not acknowledge that such a simulation may evoke different forms of behavior – not only successful coping strategies but also giving up (helplessness). As a result, he comes against another very serious inner contradiction. In his view, nightmares and recurring threatening dreams are typical and efficient threat stimulation dreams. Thus, they must have high adaptive value and increase adaptive skills in wakefulness. However, for clinicians, nightmares (a typical symptom of post-traumatic disorder, PTSD) are a pathological sign. Sleep disturbance associated with nightmares is difficult to consider as a side effect of the positive outcome of anxiety dreams because there are no positive effects on waking behavior. In nightmares, the subject is usually a helpless victim of threatening events, which means that the dream is functionally insufficient and unable to restore search activity.

Space limitation does not permit me to discuss other contradictions, such as those related to the nature of images of wild animals in the dreams of children. The author ignores a well known fact that for children events in fairy-tales are as real as objective reality, and this may be the source of such images, especially since they are absent in dreams of Bedouin children who have no experience with such tales.

**3. Covert REM state: Possible physiological and psychological manifestations [Nielsen].** A covert REM sleep process in NREM sleep would explain dream-like reports after awakenings in NREM. Many studies on humans as well as animals confirm this. This idea is also free of internal contradictions compromising the 1-gen and 2-gen models. Aside from all the contradictions cited by NIELSEN, I wish to stress that the qualitative similarity between REM and NREM mentation implied by the 1-gen model does not correspond to different effects of REM and NREM deprivation on mental state and behavior in humans. On the other hand, as has been shown, the 2-gen model has problems in explaining the similarity between REM and NREM reports.

Among data confirming REM state intrusion in NREM there is the forgotten investigation of Toth (1971). Using special subtle electrodes connected directly to eyelids, Toth detected, in different stages of NREM, short periods of electrical activity very similar to that accompanying small eye movements. When awakened in these NREM periods, including SWS, healthy subjects reported dream-like states. This has not been replicated so far as I know. However, if it is valid, this method can be used in psychophysiological investigations of NREM mentation, allowing quantitative measurement of covert REM states.

In our work on galvanic skin reaction (GSR) distribution in night sleep 25 years ago (Rotenberg et al. 1975), we detected a phenomenon presumably related to covert REM states in NREM.

It is well known that in healthy subjects, GSR is usually less prominent in the first cycle than in the second (first cycle just after the first REM sleep episode). It is usually very visible, as if REM sleep were changing some physiological conditions allowing GSR to express itself. At the same time, in the first cycle, GSR is greater in stage 2 following SWS than in stage 2 before SWS. In the second cycle there is no such difference. Moreover, in the first cycle in SWS alone, GSR often does not increase for some period and then increases suddenly. Sometimes it looks like an explosion of GSR. It looks as if A is a critical point in the first cycle (not necessary in SWS) after which GSR begins to increase. In view of the regular increase of GSR in the second cycle after the first REM sleep period, one can conjecture that such a critical point in the first cycle corresponds to the covert REM state.

The idea of a covert REM state in NREM sleep may be helpful in explaining paradoxes of total sleep and REM sleep deprivation in animals. There is a discussion in the literature about whether NREM sleep and pre-REM state can partly repay REM sleep debt (see *Sleep*, 1999, n. 8). If NREM sleep includes covert REM-like states, such a compensation may be possible. It is also possible that the shift of REM sleep towards sleep onset (decrease of REM sleep latency) in depressed patients reflects the unmasking of covert REM sleep episodes. Although covert REM sleep in NREM sleep can explain a substantial portion of the data on NREM mentation, it does not explain it all. Not all mental experiences in NREM sleep are caused by covert REM sleep episodes in NREM sleep, and NIELSEN realizes this. Thus, his model does not reduce the importance of distinguishing mentation during "pure" NREM and during REM sleep incorporation in NREM. As NIELSEN stresses in his review, self-reflectiveness and self-involvement are especially typical of REM reports. Self-representation maybe a singular feature of dream mentation. In humans, the dream serves to undo repression (Grieser et al. 1972) and is an important part of psychological defense mechanisms protecting the self from disintegration. Self-representation in dreams may be related to this dream function and perhaps can be used as a sign of true dream experience. On the other hand, active self-participation in a dream scenario is crucial for the restoration of search activity as a primary biological function of dreaming (Rotenberg 1993a).

I propose that there is also another feature of true dream mentation separating it from any other form of sleep mentation. This is related to the right-hemispheric nature of dream experience. In contrast to the left-hemispheric monosemantic way of thinking, which is dominant in wakefulness and in NREM, typical dreams are polysemantic (see Rotenberg 1994b). This is evident in subjects' reflections on their own dream reports after awakenings. While remembering all details of dream images and events, subjects feel that their report is missing something very important, the thing that made the dream experience so meaningful and affective. Subjects feel that something very substantial and crucial has been lost during the presentation of the dream content, and that the dream story they are reporting differs significantly from their actual experience. There are numerous relationships between dream events and images forming the polysemantic context of the dream; these are lost during a verbal monosemantic presentation.

**4. REM sleep and memory consolidation [Vertes & Eastman].** I agree with VERTES & EASTMAN's main conclusion that REM sleep does not in itself play a crucial role in memory consolidation. However, my approach differs enough to make it worthwhile to discuss this point in detail. The studies have been performed on animals, and the authors present a vast amount of experimental data. They emphasize that the pedestal REM deprivation technique includes stress, which may cause the negative outcome of REM deprivation on memory and behavior. This is a reasonable assumption. However, one must go on to discuss the particular nature of this stress. Stress on the pedestal includes the relative restriction of activity, isolation, deprivation of search activity, regular frustration of the biological need for REM sleep, and punishments for any attempt to achieve REM sleep. Such regular frustration and punishment usually causes helplessness. Stress evoked by the combination of these factors increases the requirement for REM sleep in order to compensate for the lack of search activity (Rotenberg 1993a). However, REM sleep is prevented. As a result, helplessness is not abolished, and this state itself may cause failures in testing memory function (Rotenberg 1992). Thus, I agree that REM sleep alone does not have a special function in memory consolidation. However, REM sleep may have indirect positive influence on memory as well as on many other functions in compensating for a state of helplessness (renouncing search) which is harmful for any mental and biological functions and even for survival (see General Introduction). Performance deficits are only a part of this general impairment of all functions caused by REM deprivation on pedestal.

In order to induce this state of search renuciation, a stressful situation combined with REM sleep deprivation has to last long enough. This can explain the data of Fishbein (1971), that the marked deficits in performance appeared only after REM deprivation for longer than one day. The state of helplessness caused by REM deprivation prior to training can also explain later impaired learning. Thus, I agree with **VERTES & EASTMAN** that performance deficits caused by REM deprivation can explain the data memory consolidation in REM sleep. However, the performance deficit in itself and the disturbance of mental functions is an outcome of renouncing search (helplessness). This is not caused by stress alone, but by the combination of stress and ensuing helplessness with the parallel suppression of REM sleep. If REM sleep is not prevented and if it is functionally sufficient, it compensates for the state of helplessness. If stress is less prominent, such as in the rotating water tank, the development of this state will take more time.

The same line of reasoning can explain the results of the work on humans. First, not every learning situation is stressful enough to produce a state of helplessness. In the human studies, experimental paradigms which can evoke this state are rarely used. It is obvious that complex tasks are more available to produce this state than simple tasks, and it is exactly these complex tasks as opposed to simple tasks that are affected by REM deprivation

**VERTES & EASTMAN** present an interesting discussion of the function of hippocampal theta rhythm in wakefulness and sleep. I agree that theta rhythm is involved in the mnemonic functions of waking and not those of REM sleep. However, I cannot agree that theta rhythm in REM sleep is a byproduct of the intense activation of the pontine region and has no function. Moreover, I assume that in wakefulness and in REM sleep this rhythm displays the same state-search activity which can manifest itself in different forms of overt behavior and in dreams. This is based on a balance between theta rhythm in wakefulness and REM sleep: The more pronounced it is in wakefulness, the less it is in subsequent REM sleep, and even REM sleep itself is diminished (Oniani & Lortkipanidze 1985). In wakefulness, mnemonic functions (selection and encoding of information) require search activity and correlate with the theta rhythm. In REM sleep, search activity does not relate to mnemonic functions but compensates the lack of search activity in the preceding wakefulness.

I agree that the fact that REM sleep reduction with antidepressants is not accompanied by memory disturbances is good evidence against memory consolidation in REM sleep. I have presented some additional supporting arguments (Rotenberg 1992): (1) Activating drugs such as amphetamine have a beneficial effect on memory, although they suppress REM sleep; (2) REM sleep has a tendency to increase with neuroleptic treatment and with reserpine, although this does not have a beneficial effect on memory.

My final comment is related to **VERTES & EASTMAN**'s hypothesis that REM sleep is a mechanism used by the brain to ensure and promote recovery from sleep. I disagree, and not only because I ascribe a function to REM sleep in restoring search activity. This hypothesis is also inconsistent with certain facts: (1) In humans, REM sleep is concentrated in the last third of the night when NREM sleep is superficial; (2) On the first night of rebound effect after sleep deprivation, sleep in psychologically stable subjects is deeper than usual and contains increased SWS but not increased REM sleep. According to **VERTES & EASTMAN** it would be reasonable to expect a compensatory increase of REM sleep. In addition, there are special mechanisms protecting the brain from unlimited increases in SWS – spontaneous shifts to the more superficial sleep stages, and shifts after body movements. Both types of stage shifts, and especially shifts after body movements, cause decreases in SWS.

***General conclusion.*** By taking into consideration (1) the fundamental difference between two types of behavior in wakefulness and dreaming, and (2) the role of the functionally sufficient REM sleep in the restoration of search activity, one can explain many contradictions in the data related to REM sleep functions in dream experience, resistance to stress, and memory consolidation.

ACKNOWLEDGMENT
The Sleep Laboratory is affiliated with Tel Aviv University.

# Some myths are slow to die

Rafael Salin-Pascual, Dmitry Gerashchenko,
and Priyattam J. Shiromani

*West Roxbury VA Medical Center and Harvard Medical School,
West Roxbury, MA 02132.* **pshiromani@hms.harvard.edu**

**Abstract:** Solms and the other authors in this series of *BBS* target articles accept the findings that the executive control of the REM/NREM cycle is still localized within a narrow region of the pontine brainstem. However, recent findings challenge this notion. We will review the recent data and suggest instead that the hypothalamus is the primary regulator of states of consciousness. If the hypothalamus indeed controls all the fun stuff, such as sex, eating, drinking, sleeping, and so on, then one can more easily accept Solms's argument that dreams are also generated from the forebrain.
[SOLMS]

Nauta (1946) found that the hypothalamus regulates sleep and wakefulness. However, his findings were set aside because Moruzzi and Magoun's work (1949) began to emphasize that an ascending reticular activating system emanating from the brainstem activated the cortex. The idea of an ascending brainstem arousal system took firm hold, once it was established in Dahlstrom & Fuxe (1964) that noradrenergic neurons were localized to specific regions of the brainstem. During that period, Jouvet also demonstrated changes in the electroencephalogram following transections and lesions of the brainstem, which prompted him to propose the monoaminergic theory of sleep-wake regulation (Jouvet 1969). Galvanized by the neuroanatomical and physiological data, investigators began to monitor the firing patterns of neurons within the brainstem to determine whether neuronal firing could be associated with specific sleep-wake states. For half a century, the transection, lesion, and electrophysiology studies have been the driving force that has shaped theories regarding sleep-wake regulation.

Although much of the sleep research community focused on the brainstem, a group of investigators at UCLA, headed by Carmine Clemente and Barry Sterman, who continued to investigate the hypothalamus. Studies from that group supported Nauta's findings that the preoptic area was important for sleep (reviewed in Szymusiak 1995). For instance, lesion of the preoptic area produces long-lasting insomnia while electrical or pharmacological stimulation and warming induces sleep. Electrophysiology studies have identified sleep active neurons in this region (Szymusiak et al. 1998). Recent studies (Sherin et al. 1996) have identified a specific neuronal group within the preoptic area that projects monosynaptically to wake-active neurons in the tuberomammillary nucleus (TMN) in the posterior hypothalamus. These neurons are located in the ventral lateral preoptic (VLPO) area and are sleep-active based on electrophysiology and c-Fos studies. Neurons in the VLPO are GABAergic (and also contain the inhibitory peptide galanin) (Sherin et al. 1998), and their activation would release inhibitory agents onto wake-active posterior hypothalamic wake-active cells and sleep would ensue (Shiromani 1998). Lesions of the VLPO produces long-lasting insomnia (Lu et al. 2000).

Recently, another group of neurons were identified in the posterior hypothalamus as being important for the regulation of wakefulness (De Lecca et al. 1998; Sakurai et al. 1998). These neurons contain the neuropeptide hypocretin, also named orexin. These neurons project to virtually the entire brain and spinal cord, providing especially heavy innervation to regions implicated in the regulation of wakefulness such as the TMN and the locus coeruleus (Nambu et al. 1999; Peyron et al. 1998). This peptide exerts an excitatory influence on target neurons (Hagan et al. 1999; Horvath et al. 1999) and some believe that this peptide is important for wakefulness and given the innervation of neurons implicated in wakefulness (Peyron et al. 1998). Consistent with this possibility it has been found that canine narcolepsy is associated with a mutation in the hypocretin-2 receptor (Lin et al. 1999), and another study found that orexin knockout mice sleep more at night, have more REM sleep, and also experience cataplexy (Chemelli et al. 1999).

The sleep disorder narcolepsy has provided insight into the neurobiology of sleep-wakefulness. That the disorder is associated with a neuronal population located in the hypothalamus is one more reason to stop looking to the pons and the brainstem for all the answers. The hypothalamus also contains neurons regulating other vital bodily functions, such as feeding, drinking, sexual behavior, and temperature. Moreover, the suprachiasmatic nucleus, which represents the master clock, also resides in the hypothalamus. Thus, it is not at all surprising that neurons regulating another fundamental behavior, sleep, are also present here.

# Time course of dreaming and sleep organization

Piero Salzarulo

*Department of Psychology, University of Florence, 50125 Florence, Italy.*
**salzarulo@psico.unifi.it**

**Abstract:** The complexity and mysteriousness of mental processes during sleep rule out thinking only in term of generators. How could we know exactly what mental sleep experience (MSE) is produced and when? To refer to REM versus NREM as separate time windows for MSE seems insufficient. We propose that in each cycle NREM and REM interact to allow mentation to reach a certain degree of complexity and consolidation in memory. Each successive cycle within a sleep episode should contribute to these processes with a different weight according to the time of night and distance from sleep onset. This view would avoid assuming too great a separation between REM and NREM functions and attributing psychological functions only to a single state.

[**NIELSEN**]

The experimental approach to dreams (or to mental sleep experience (MSE; Cipolli et al. 1981; Salzarulo & Cipolli 1979; Salzarulo et al. 1973) underwent enormous development in the sixties, mainly because of the heuristic value of the discovery of REM sleep by Aserinski and Kleitman (1953). For the next two decades (see Schulz & Salzarulo 1993 ), however, much less work was devoted to dreams. The theoretical work of **NIELSEN** in this issue is accordingly welcome and much appreciated. What follows are a few remarks and a proposal concerning problems raised by the complex **NIELSEN** paper.

One of the main arguments developed by **NIELSEN** concerns whether sleep mentation is generated by a single or a double source. Sleep mentation is assessed from reports obtained after REM and NREM sleep. This raises some questions. The term "generator," referring here to mental activity, reminds us of physiological terminology and concepts (e.g., the use of the term generator for PGO activity, Siegel 1989). This use of physiological paradigms for the study of mentation has led to equating dreaming

with REM sleep, which has been criticized (Lairy & Salzarulo 1975). Now, NIELSEN is speculating about 1 versus 2 generators, that is one versus two forms of sleep mentation. Why look for two generators? First, there are two "containers," that is two sleep states, REM and NREM. Second, in a well known experimental work, Foulkes (1962) showed that there could be a substantial recall of sleep mental activity not only in REM sleep (Dement & Kleitman 1957b) but also in NREM sleep. Once it was established that mental activity could also occur in NREM, it was question of comparing NREM and REM mental activity. Psychological measures were used by some researches, while others (including the NIELSEN paper) used mainly physiological measures (see for discussion Salzarulo et al. 1973).

The complexity and the mysteriousness of mental processes during sleep rule out thinking solely in term of generators. How can we know precisely what MSE is produced and when? Are we sure that the indicators of physiological events, with their specific time-windows, are useful for "localizing" mental processes in time? What kinds of events should be included? To refer constantly to REM versus NREM is in our opinion insufficient. There is another possibility partially connected with the last part of the NIELSEN paper.

Retrospective evaluation of mental functions from post-awakening recall has revealed prominent differences in memory processes between NREM and REM (Cipolli 1995). Analyzing memory to understand sleep mentation is one of the methods used by some researchers in recent years. These studies not only explained differences between REM and NREM reports (see Antrobus 1983; Salzarulo & Cipolli 1979) but also found various degrees of consolidation of MSE using psycholinguistic indicators (Salzarulo & Cipolli 1974; 1979). Cipolli et al. (1998) further showed that memory consolidation improves across the night sleep cycles thanks to an iterative process. Indeed, a role for sleep cycles in memory processes, rather than sleep states per se, has been demonstrated recently (Ficca et al., 2000; Mazzoni et al. 1999; Salzarulo et al. 1997).

We proposed that in each cycle NREM and REM interact to allow mentation to attain a certain degree of complexity and consolidation in memory (Mazzoni et al. 1999; Salzarulo 1995). Each successive cycle within a sleep episode should contribute to these processes with a different "role" (weight) according to the time of night and distance from sleep onset. This avoids assuming too great a separation between REM and NREM functions and attributing psychological functions to a single state; it also emphasizes the temporal dimension and sleep organization.

To emphasize the usefulness of taking into account the temporal dimension in the comparison between physiological and psychological processes, we show in Figure 1 the time course of EEG activity (Dijk et al. 1991) and memory processes involved in dream recall (Cipolli et al. 1998). The increasing number of units consolidated in memory parallels the decreasing amount of slow wave activity. Hence, the physiological S process declines in parallel with increasing of memory consolidation.

In conclusion, I prefer not to speak about (or to seek) dream generators; instead I see physiological sleep activities as conditions (frames) within which sleep mentation can be elaborated and consolidated in memory. Consolidation can be achieved by iterative access to contents, possibly related to a single production system. Beyond this, how the production system functions, and why it starts to function during sleep, still remains a mystery.

ACKNOWLEDGMENTS
I thank G. Ficca for redrawing data from original sources shown in Figure 1.

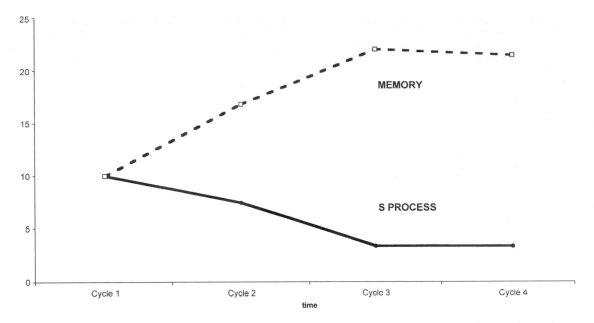

Figure 1 (Salzarulo).  Trends of night-time memory and the S (sleep) process. Abscissa: rank of NREM-REM cycles Ordinate: Relative changes of memory process (scores) (Cipolli et al. 1998) and S process (SWA: total power density in 0.5–4 HZ frequency bin) (Dijk et al. 1991) across cycles. First cycle value is conventionally assigned 10.

## Dream research: Integration of physiological and psychological models

Michael Schredl

*Sleep Laboratory, Central Institute of Mental Health, 68072 Mannheim, Germany.* **schredl@as200.zi-mannheim.de**
**www.zi-mannheim.de/schalab/abteilung.html**

**Abstract:** All five target articles are of high quality and very stimulating for the field. Several factors such as dream report length and NREM/REM differences, may be affected by the waking process (transition from sleep to wakefulness) and the recall process. It is helpful to distinguish between a model for REM sleep regulation and a physiological model for dreaming. A third model accounting for cognitive activity (thought-like dreaming) can also be of value. The postulated adaptive function of dreaming in avoidance learning does not seem very plausible because the two major basic assumptions (specificity of dream content and benefit of negative dreams) are not clearly supported by modern dream research: The critique of studies investigating memory consolidation in REM sleep is justified. Future studies integrating the knowledge of memory processes and sleep research will shed more light on the role of sleep, especially REM sleep in memory consolidation.

[HOBSON ET AL.; NIELSEN; REVONSUO; SOLMS; VERTES & EASTMAN]

The five target articles discuss important aspects of dreaming and REM sleep, that is, physiological mechanisms of dreaming, the function of dreaming, and the role of REM sleep in memory consolidation. In view of the vast range of topics addressed in the articles, these following comments are restricted to a few selected points. First, some basic issues of contemporary dream research will be discussed.

***Basics of dream research.*** First of all, it seems important to look at the definition of dreaming. Although it is not always explicitly mentioned, it is obvious that only the dream report, that is, the part of the mental activity during sleep that can be reported upon awakening, is available to the researcher. This process involves the transition of at least two thresholds: (1) recall of previous experiences and (2) sleep/wake transition. Both factors may be important considering the issue of dream length and the REM/NREM differences in dream content. Rosenlicht et al. (1994) found a relationship (r = .23) between dream report length and the report length of a re-narration of a previously shown video

film. This finding indicates that reporting style may well affect dream length.

Because dreams are not perceived as films, it will be interesting to use standardized situations which are recounted afterwards and to test whether the length of these reports correlates substantially with dream report length. From experience with recording dreams and listening to dream reports, it seems obvious that a variety of factors in addition to report style can affect dream report length. Dreams (at least REM dreams) are experiences similar to those of waking life, that is, a dream scene can be described in a few words if one focuses on the major dream action. On the other hand, the report of the same scene will be much longer if all the small visual details are recounted (if they were remembered).

The effect of motivation and in-depth inquiries about specific dream details on dream length was seldom investigated in a systematic way, but other studies, for example, investigating the experience of tactile sensations (Strauch & Meier 1996) or emotions (Schredl & Doll 1998) have shown that the dream report comprises only parts of dreamed experiences. A study of dreaming among the elderly (Schredl et al. 1996) revealed a marked relationship between dream report length and verbal short-term memory, whereas visual memory and overall cognitive performance were not correlated. This finding points to the fact that specific abilities are necessary in the recall process after awakening to record or report the dream experience in a temporal sequence. These investigations make clear that the assumption of HOBSON ET AL. that dream report length is largely determined by the quantity of dream experiences should be regarded with caution since at least the above mentioned factors must also be taken into account.

Second, the sleep/wake transition may also affect the generation of the dream report. A comprehensive model was proposed by Koukkou and Lehmann (1983). The basic assumptions of their model are the following:

Cognitive activity can be assigned to different functional states which are associated with differences in processing modes, memory stores, and EEG pattern. They assume that information of lower levels of activation (NREM sleep, REM sleep) can only be sparsely recalled by functional states of high activation (wakefulness) and that the closer the two functional states are, the better is the transference of information. On the other hand, information from higher memory stores is available for lower states. Although the findings of EEG studies are inconsistent (cf. Schredl & Montasser 1996; 1997), the model offers a plausible interpretation of the differences in recall rates of NREM and REM awakenings, since NREM sleep is associated with a lower functional state than REM sleep. This may be comparable to the activation component of the AIM model (**HOBSON ET AL.**).

The functional state-shift model stresses the fact that both functional states, that is, prior and after awakening, are important in the process of recalling the dream experience. Spitzer et al. (1993), for example, have proven so-called carry-over effects, that is, after REM awakenings semantic priming in a word association task was significantly more pronounced than after NREM awakenings and in the waking state. The authors interpreted their finding as inhibition of the semantic associative network.

Similarly, **HOBSON ET AL.** pointed out that the cholinergic-aminergic modulation of the brain did not change abruptly. In view of these findings regarding the carryover effect, the effect of the awaking process and the differences between the functional states prior to and after awakening on dream report length and other dream characteristics should be investigated in detail.

In the future it may be possible to utilize brain imaging techniques. It may be that the NREM/REM differences in dream content (**NIELSEN**) are partly owing to the different states of the brain after the awakening. Another aspect of dream recall is the large inter-individual variability of dream recall frequency and the fact that dream recall frequency can easily be enhanced by training (cf. Schredl 1999). It seems implausible that a purely physiological model (cholinergic-aminergic modulation of the brain) can explain why some persons recall a dream almost every morning whereas others have not remembered a dream for years. Similar patterns were detected for low dream recallers in the sleep laboratory (e. g., Meier Faber 1988). Recall after REM awakenings differed considerably between subjects in this study. In this context, it would be very interesting to conduct studies testing whether dream recall frequency after NREM awakenings could be enhanced by training.

Another important issue is the definition of dreaming, which was attempted by **NIELSEN** (Fig. 1). From this it seems clear that only a part of the brain activity can be recalled. Dreaming is defined in contrast to cognitive activity as a mixture of sensory hallucinations, story-like or dramatic progression, and bizarreness (similar to the definition of **HOBSON ET AL.**). This definition, however, is based on dream reports with all the above-mentioned limitations.

The formal characteristics of dream reports also vary considerably from person to person, for example, the "thin versus thick boundaries" personality dimension is related to dream bizarreness (Schredl et al. 1999). Since a strict differentiation between the types of mental activity during sleep (apex dreaming, dreaming, cognitive activity, cognitive processes) based on dream reports is difficult, it will be necessary to carry out studies that elaborate on a clear definition and link these types to physiological processes. The investigation of inter- and intra-individual differences will be very promising because, all types of mental sleep experiences (except cognitive processes) can be obtained after REM awakenings as well as after NREM awakenings.

The authors themselves have pointed it out, but it nevertheless seems important to stress that the AIM model provides an explanation for the formal characteristics of the dream, for example, bizarreness. The model does not, however, provide anything about dream content. For this reason, researchers focus on the so-called continuity hypothesis (e.g., Domhoff 1996; Hartmann 1998), which states that emotional concerns and preoccupations of the waking life are reflected in dreams. However, the findings regarding the dream-lag effect, the temporal references of dream elements, and the effects of experimental manipulation on dream content are not consistent in a way to formulate a more specific model explaining dream content.

The AIM model describes the states of consciousness using three dimensions. **HOBSON ET AL.** pointed out that this is an oversimplification, and this is evident when the authors apply the model to other states of consciousness, e.g., lucid dreaming, and REM behavior disorder, since they have to introduce other dimensions. It seems not clear whether the model is useful beyond the descriptive aspect. On the other hand, the model contains a modulation component, as an explanatory aspect, because the preponderance of a specific neuro-transmitter affects the mode of information processing.

In this context, studies as carried out by Hartmann et al. (1980) with L-dopa investigating the effect of drugs such as muscarinic cholinergic agonists and acetylcholinesterase inhibitors on formal dream characteristics, such as bizarreness, will allow a specific test of the M component of the AIM model. Preliminary data of a pilot study (Schredl et al. 2000) carried out in our laboratory indicate that donepezil, an acetylcholinesterase inhibitor, intensifies dreaming but has only a small effect on REM sleep percent although REM latency is shortened and REM density is heightened under medication. It will be interesting to test whether such drugs affect dream content in the predicted way. In addition, the investigation of interindividual differences in dream content in relation to parameters of neurotransmission will also be of value.

The findings of **SOLMS** suggest that it may be helpful to develop two models, one to explain the physiology of REM sleep regulation and the second for the physiology of dreaming which explains the formal characteristics of dreams. In order to do this, it will be necessary to have a clear and exact definition of dreaming to correlate mental activity with brain physiology beyond sleep stages. The next step could be a search for a link between the two models (cf. **NIELSEN**). This differentiation may add to the explanation of other states of consciousness (sensory deprivation, daydreaming, narcosis, near-death experiences).

Another problem which was discussed at the Third International Congress of the World Federation of Sleep Societies (held 1999) is the definition of sleep. What does it mean for a single cell or groups of cells whether the brain as a whole is sleeping? This argumentation is also evident in the article of **NIELSEN** who has shown that distinct REM features which are present during NREM sleep are related to dream recall. Therefore, a differentiation between REM sleep physiology and physiology of dreaming seems promising. It may be that a third model is necessary to explain the occurrence of cognitive activity (thought-like dreaming) if one does not agree with Hartmann's hypothesis (1998) in which a continuum ranging from bizarre dreams to focused waking thought is conceptualized.

***Functions of dreaming.*** Quite a few hypotheses about the function of dreams were formulated over the years. Some of them are reviewed by **REVONSUO**. First, it is important to draw attention to the fact that the function of dreaming cannot be studied empirically in a direct way, because dream content can not be elicited without the involvement of the waking mind, for example, dreams have to be recalled in order to test the hypothesis whether they are helpful in problem solving. It may also be possible that thinking about the dream causes subsequent changes in waking life and not the dream per se.

A few arguments which should lead to a critical evaluation of **REVONSUO**'s theory will be briefly enumerated. First, it seems implausible to assume that dreams are specialized in replaying threatening experiences. The widely acknowledged continuity hypothesis states that emotional concerns and preoccupations are reflected in dreams (e.g., Domhoff 1996; Hartmann 1998). A component which **REVONSUO** has not considered is the fact that not only real experiences of the waking-life can be found in dreams but also thoughts, emotions, things seen on TV, movies, and so forth. If one examines this input, for example, TV news, it may be possible that one finds a preponderance of negative themes which affect dream content. In order to compare waking life and dreaming, it will be necessary to elicit the inner world of the person as completely as possible.

The second point is related to the assumed preponderance of

negative dream emotions. Schredl and Doll (1998) have pointed out that several methodological issues (recalling dreams which occurred a long time ago, only focusing on explicitly mentioned emotions within the dream report, the using of raters instead of self-ratings) limit the generalizability of many studies in this field. These arguments are also valid for the study of Mehinaku dreams which was cited by **REVONSUO**, that is, very vivid and negatively toned dreams which could be remembered easily were reported to the researcher. Dreams recorded immediately upon awakening show a balanced ratio of negative and positive emotions (Schredl & Doll 1998).

Another example to support the continuity hypothesis and not the specificity for threatening contents is the occurrence of the male stranger as major dream aggressor; almost all murderers and soldiers are male, that is, the pattern of male aggressiveness is reflected in dreams (e.g., Schredl & Pallmer 1998). There is evidence from studies in the field of learning that some learning occurs abruptly, for example, classical conditioning in avoidance tasks (avoiding an aversive stimulus), acquiring an aversion that is very resistant to extinction to specific food (e.g., Margraf 1996). This kind of learning makes a repetition of negative experiences unnecessary. In addition, the adaptive function of PTSD or anxiety disorders seems not to be very reasonable. **REVONSUO** pointed out that these negative effects may only be present in war-related PTSD, but investigations of rape victims and persons who experienced natural disasters did not support the hypothesis that war-PTSD is specific (cf. Barrett 1996). The marked relationship between psychopathology and negative dream emotions (e.g., Schredl & Engelhardt 2000) did not support the idea that threatening dreams serve an adaptive function; on the contrary, depressive patients or patients with anxiety disorders suffer from their dreams. Similarly, the correlation between low life satisfaction and negative dream emotions in the elderly (Schredl et al. 1996) can be interpreted in a way that negative dreams are associated with poor psychological adjustment. Hartmann (1991) has investigated persons with thin boundaries who often suffer from frequent nightmares. He described these persons as not sensitive to possible danger in foreign cities; an observation which does not fit within REVONSUO's framework.

The question of utmost importance is whether threat simulation during dreaming increases the probability of coping successfully with comparable real threats. As mentioned above, this hypothesis could not be tested empirically in a direct way since the remembering of nightmares or post-traumatic re-enactments may stimulate waking thoughts which affect subsequent behavior and not the nightmares themselves. It will be very interesting to explore whether or not animals experience nightmares (strongly negative dreams which end by an awakening). It may be possible to traumatize animals and measure the physiological anxiety responses during REM sleep. Animals with nightmares should learn an avoidance task more easily. To my knowledge, such studies have yet to be carried out. If – as shown for example by Hublin et al. (1999) – a genetic factor plays an important role in the etiology of nightmare, one might argue that the group of frequent nightmare sufferers should be increased by natural selection. Since this group is small (about 5 percent, e.g., Bixler et al. 1979), this seems not to be very plausible.

To summarize, the two basic questions (selectivity for threatening experiences, benefit of replaying these experiences in dreams) are in my view not supported by the presented evidence. On the other hand, one can follow the assumption of Kuiken and Sikora (1993) that dreaming serves multiple purposes.

**Function of REM sleep.** In contrast to **REVONSUO** who outlined a possible function of dreaming, **VERTES & EASTMAN** discussed the function of REM sleep in memory consolidation. A major problem of previous research, at least in humans, is in my view the lack of integration between memory research and sleep research (Schredl et al. 1998). Modern research has shown that different types of memory processing (explicit vs. implicit, declarative vs. procedural memory) are associated with different brain areas

(e. g., Markowitsch 1996). Recent studies in humans (Karni et al. 1994; Plihal & Born 1997) have shown that procedural memory may be consolidated during REM sleep but declarative memory performance is enhanced after undisturbed slow wave sleep. It may be expected that findings will be more consistent if the knowledge from the two disciplines are combined.

On the other hand, it is important to consider the fact that memory consolidation also takes place during the waking state. Mandai et al. (1989), for example, have formulated a two-step model with memory storage and consolidation during waking and additional memory consolidation in REM sleep. The findings cited by **VERTES & EASTMAN** have shown that REM deprivation (i.e., disturbed sleep, stress, etc.) did not affect simple memory tasks and, thus, the hypothesis of memory consolidation during the waking state is supported. Similarly, the findings regarding drugs which suppress REM sleep, can be interpreted in this way. However, these studies did not retest material or skills which have been trained the previous day (or days) for the first time. A pilot study (Schredl et al. 2000) has shown that the amount of REM sleep in nights with donepezil (an acetylcholinesterase inhibitor) is strongly related to the increase of performance in a memory task from the evening training session to the morning test session. Therefore, more sophisticated studies including training of new knowledge or skills and retest after at least one sleep period have to be carried out to evaluate the effect of REM sleep on memory consolidation.

Buzsaki (1998) has suggested a model which indicates that both slow wave sleep as well as REM sleep play an important role in memory consolidation during sleep. A recent study showed that the diurnal cortisol profile (minimum in the first part of the night) affect the consolidation of declarative memory but not the occurrence of specific sleep stages (Plihal et al. 1999). From a methodological viewpoint, it must be said that modern literature reviews use the technique of meta-analysis. With this technique, effects of the deprivation technique (pedestals, multiple platforms, pendulum technique) can be tested statistically (in this case, a sufficient number of studies allowing the computation of effect sizes were published). In addition, it will be necessary to carry out EEG studies with animals to estimate the bias introduced by stressful REM sleep deprivation techniques.

To summarize, the database regarding the role of REM sleep and sleep in memory consolidation is indeed not very solid because of the limitations pointed out by **VERTES & EASTMAN** and the lack of integration between memory research and sleep research. But recent studies are promising and support that sleep plays a role (in addition to processes during waking) in memory consolidation.

**Future directions.** In the future, imaging techniques (e.g., MRI) will offer the option to investigate the relationship between brain processes in different sleep stages and dream content. In order to do this properly it will be necessary to improve the present imaging techniques (e.g., time resolution) and to develop precise instruments for measuring dream content. It will be helpful to use two or three models explaining (1) REM sleep regulation, (2) dreaming, and (3) cognitive activity (thought-like dreaming). In addition, it will be important to investigate the awaking process in a more detailed way in order to evaluate the relationship between dream report and original dream experience. Last, a theory modeling the relationship between waking life and dream content should be formulated which goes beyond the simple statement of continuity between these two states of consciousness.

# Threat simulation, dreams, and domain-specificity

Todd K. Shackelford and Viviana A. Weekes-Shackelford

*Division of Science-Psychology, Florida Atlantic University, Boca Raton, FL 33314.* {tshackel; vwee9812}@fau.edu

**Abstract:** According to Revonsuo, dreams are the output of a evolved "threat simulation mechanism." The author marshals a diverse and comprehensive array of empirical and theoretical support for this hypothesis. We propose that the hypothesized threat simulation mechanism might be more domain-specific in design than the author implies. To illustrate, we discuss the possible sex-differentiated design of the hypothesized threat simulation mechanism.

[REVONSUO]

REVONSUO proposes that dreaming is the output of an evolved "threat simulation mechanism." According to the author's argument, the hypothesized mechanism was selected for because ancestral humans whose psychology included this mechanism experienced dreams in which threats were simulated and thereby more efficiently and effectively dealt with in waking life. Those early humans who had dreams in which threats to survival and reproductive success were simulated were better able to solve similar classes of threats in waking life and, therefore, out-reproduced conspecifics whose psychology did not include the threat simulation mechanism. We appreciate this argument, and believe that REVONSUO has done an excellent job of synthesizing an impressive array of empirical and theoretical support for the argument. We propose, however, that the hypothesized threat simulation mechanism is more domain-specific than the author presents.

**The hypothesized threat simulation mechanism may be too domain-general.** According to REVONSUO, dreams serve as a means for the dreamer to rehearse events that would have threatened survival or reproductive success in ancestral environments. The author proposes two primary threatening events or episodes – one in which the dreamer is being chased or attacked by an unfamiliar adult male, and one in which the dreamer is being chased or attacked by a wild and dangerous animal. REVONSUO provides sound theoretical arguments why unfamiliar adult males and wild animals were key threats to ancestral humans and, therefore, why they are prominent in the self-reported dreams of modern humans. We believe this is an excellent starting point for an evolutionary psychological analysis of dreams. We propose, however, that the threat simulation mechanism that generates dreams may be far more domain-specific. Instead of generating general threat dreams that include the two key events proposed by REVONSUO, we suggest that perhaps the dreams of modern humans might reveal greater domain specificity. Might the threat simulation mechanism generate dreams that are more specific to the adaptive problems faced recurrently by humans over human evolutionary history? For example, might the threat simulation mechanism generate different classes of threatening dreams when it is operating in a child living with one stepparent and one genetic parent than when it is operating in the psychology of a child who lives with two genetic parents (see Daly & Wilson 1996)? As another example, might the threat simulation mechanism generate different classes of threat scenarios when it is operating in male psychology than in female psychology? We discuss the latter example in the remainder of this commentary.

**Is the threat simulation mechanism sex-differentiated?** An overwhelming collection of theoretical and empirical work suggests that males and females faced different adaptive problems recurrently over human evolutionary history (see Buss 1994, for a review). For this class of adaptive problems, modern evolutionary psychologists expect the evolution of sex-differentiated psychological mechanisms. One such adaptive problem recurrently faced by ancestral humans is a long-term partner's infidelity. Because fertilization occurs internally to females, females always can be certain that they are the genetic parent of any offspring they produce. Males, in contrast, never can be certain that they are the genetic parent of the offspring produced by their partner. Males, but not females, risk cuckoldry – unwittingly investing in offspring to whom they are genetically unrelated. Although both sexes are upset by a partner's infidelity, males are more upset by a partner's sexual infidelity than by a partner's emotional infidelity – infidelity in which resources such as social support and material wealth are channeled to another person. Females, in contrast, are more upset by a partner's emotional infidelity, which places them at risk of losing to another woman the investment their partner would otherwise channel to them and their children (see Buss 2000, for a review of this work).

If the threat simulation mechanism generates dreams that simulate ancestral threats to survival and reproductive success, we propose that sex-differentiated ancestral threats will have selected for a threat simulation mechanism that generates sex-differentiated dreams. The mechanism might be sensitive to and triggered by different classes of infidelity cues when situated in male psychology than when situated in female psychology. Relative to a partner's emotional infidelity, sexual infidelity presented a graver adaptive problem for ancestral males than for ancestral females. We therefore hypothesize that the threat simulation mechanism will generate in males relative to females more dreams about a partner's sexual infidelity. In addition, we hypothesize that dreams about a partner's sexual infidelity will be more upsetting for males than for females. A partner's emotional infidelity presented a graver adaptive problem for ancestral females than for ancestral males. We hypothesize that the threat simulation mechanism will generate in females relative to males more dreams about a partner's emotional infidelity. In addition, we hypothesize that dreams about a partner's emotional infidelity will be more upsetting for females than for males. According to this argument, the threat simulation mechanism is the same in males and in females, but the design features of the mechanism – the class of information that triggers the mechanism, and the output generated by the mechanism, may be sex-differentiated.

In summary, REVONSUO has provided us with a wonderful example of the heuristic value of an evolutionary psychological perspective. The target article significantly advances our understanding of dreams by proposing that dreams are generated by an evolved threat simulation mechanism. Although we find the core of REVONSUO's argument compelling and convincing, we suggest that the hypothesized mechanism may be more domain-specific than the author implies. We propose that the dreams of males will more frequently include a partner's sexual infidelity, whereas the dreams of females will more frequently include a partner's emotional infidelity. In addition, we propose that dreams of a partner's sexual infidelity will be more distressing for males than for females, whereas dreams of a partner's emotional infidelity will be more distressing for females than for males. We hope that future work might investigate the domain-specificity of the hypothesized threat simulation mechanism. We suggest as a starting point the investigation of the possible sex-differentiated design features of this mechanism that might be revealed with an analysis of dreams about a partner's infidelity.

# Continued vitality of the Freudian theory of dreaming

Howard Shevrin[a] and Alan S. Eiser[b]

[a]*Department of Psychiatry, University of Michigan, Ann Arbor, MI 48105;* [b]*Sleep Program in Psychiatry, University of Michigan Hospitals, UH-8D8 702, Ann Arbor, MI 48109-0117.* {shevrin; aeiser }@umich.edu

**Abstract:** A minority position is presented in which evidence will be cited from the Hobson, Solms, Revonsuo, and Nielsen target articles and from other sources, supporting major tenets of Freud's theory of dreaming. Support is described for Freud's view of dreams as meaningful, linked to basic motivations, differing qualitatively in mentation, and wish-fulfilling.

[HOBSON ET AL.; NIELSEN; REVONSUO; SOLMS]

In this commentary we will be taking a decidedly minority position. We will point out that there is considerable evidence, some of it provided by the articles themselves, that supports Freud's theory of dreaming. With the exception of **SOLMS**'s paper, the **HOBSON ET AL.**, **REVONSUO**, and **NIELSEN** papers either discard Freud's theory or ignore it. Our commentary will be divided in six sections, each devoted to a proposition from Freud's theory and evidence in its support.

**Proposition 1.** Despite their bizarre and illogical appearance, dreams are organized on the basis of certain principles and are the outcome of specifiable mental processes. Dreams are not random or epiphenomenal.

The articles by both **HOBSON ET AL.** and **SOLMS** spell out the consistency and complementarity of findings obtained from brain imaging studies of REM sleep and lesion studies of dreaming. From the PET studies **HOBSON ET AL.** conclude that REM sleep may involve "a specific activation of subcortical and cortical arousal and limbic structures for the adaptive processing of emotional and motivational learning." Similarly, **SOLMS**'s neuropsychological findings indicate that dreams depend upon the concerted activity of a very particular set of mental functions located in various areas of the forebrain. These findings are quite consistent with a view of dreams as specifically organized to carry out particular functions.

Nevertheless, **HOBSON ET AL.** continue to maintain that important features of dreaming are determined by random, chaotic noise from the brainstem impinging on the forebrain. Further, the data **HOBSON ET AL.** cite concerning the rich bidirectional interactions, reciprocal connections, and feedback loops between brainstem and forebrain structures do not support a view of the brain as functioning in a nonintegrated fashion, with brainstem activity impinging disruptively and randomly on the forebrain. **HOBSON** characterizes dreams in terms of cognitive deficiencies, and draws an analogy between dreams and epileptic seizures; the possibility is not considered that dreams are organized in a different, rather than deficient, fashion that has its own particular logic.

**REVONSUO** also views dreams as organized phenomena providing a selective simulation of external threats in a perceptually realistic way so as to evoke rehearsal of skills in the efficient recognition and response to these threats. However, this differs considerably from our view of the specific function of dreams, as discussed in Propositions 3 and 5.

**Proposition 2.** The occurrence of dreaming is determined by two causes (1) the state of sleep itself, and (2) basic motivations such as sex and aggression whose persistence creates conflict in both the waking and sleep states.

1. It has been clear since the discovery of REM sleep and its association with dreaming that the first part of this proposition requires further specification. Although there is a range of views on the differences between REM and NREM mentation, there is little doubt that REM sleep is particularly facilitative of dreaming. Indeed, Freud as early as 1895 hypothesized that a motor paralysis was a necessary condition for dreaming to occur; such a paralysis is found only in REM sleep. We find persuasive both **HOBSON ET AL.** and **NIELSEN**'s marshalling of evidence that there are important qualitative differences between REM and NREM mentation, and here would differ with **SOLMS,** who views REM as only one of many arousal triggers of dreaming. The very high correlation between REM sleep and dreaming suggests to us a more integral, even if not absolute, connection, as does the very different pattern of brain activation in REM versus NREM sleep reported in the imaging studies and consistent with **SOLMS**'s lesion findings about the neuroanatomy of reported dreams. We note, with **HOBSON ET AL.**, that **SOLMS**'s data have not established that dreaming is preserved with brainstem lesions that eliminate REM sleep. We would characterize the differences between REM and NREM mentation in terms of a greater predominance of, in Freudian terms, primary-process versus secondary-process thinking (see Proposition 4) during REM sleep.

2. Freud's view of instinctual drives and their development implies an ongoing source of conflicted sexual and aggressive wishes that may serve as the motive force for dreaming. We find evidence compatible with this understanding in several of the findings discussed in these papers. **SOLMS** has reported that one of the two major forebrain areas associated with loss of dreaming in lesion studies is the parietal-temporo-occipital junction, which is associated with "appetitive interactions with the world," "the 'SEEKING' or 'wanting' command system of the brain." Similarly, the imaging findings consistently indicate the important participation of areas of the brain involved with emotion, motivation, and reward, "quite in accordance with older, more general views that REM sleep, and specifically dream content, is associated with internally generated, or instinctual behaviors that subserve adaptive mechanisms" (Nofzinger et al. 1997, p. 199). The concept of wishes that are conflicted also finds support. A hallmark of conflict in Freudian theory is that wishes associated with danger arouse anxiety, and the amygdala, which is especially associated with anxiety, is found to be activated in REM sleep.

There is another set of findings that is strikingly consistent with the association of dreams and persistent basic motivations. We refer to studies on the effects of REM sleep deprivation on motivation in the waking state. Dement and colleagues (1967; Dement 1969) carried out a series of prolonged REM deprivation procedures in cats and reported that REM-deprived cats became hypersexual and hyperphagic. The sexual behavior reported involved unusual, persistent efforts on the part of a number of male cats to mount other male cats. These findings were not presented in fully quantified form, and must be regarded as exploratory and suggestive. However, additional credence is lent to them by a number of more formal experimental reports in the literature with similar findings. Some examples: Ferguson and Dement (1969) reported that in REM-deprived rats primed with amphetamine, stereotyped aggressive and sexual behaviors were seen in the absence of the usual releasing stimuli, the aggressive behaviors in virtually all experimental rats and the sexual behaviors in a subset. Peder et al. (1986) found that REM deprivation alone resulted in increased aggressive behavior in rats, and increased genital exploration among female rats. Morden et al. (1967) reported that REM deprivation led to increased shock-induced fighting behavior in rats, which persisted beyond the recovery period. Conversely, electrical stimulation of hypothalamic defence (rage) reactions in cats led to reductions in subsequent REM sleep and, in REM deprived cats, in subsequent REM rebound (Putkonen & Putkonen 1971), a finding suggestive of reciprocal elements in the relation between drive and REM sleep. This sampling of results provides intriguing evidence from animal research of a relationship between REM sleep/dreaming and the expression of sexual and aggressive drives/wishes (see also Vogel 1979).

**Proposition 3.** The dream itself is an attempt to deal with these conflicts with the means available during sleep, just as these conflicts are dealt with in the waking state with the means available in that state

The major disagreement with this proposition derives from **REVONSUO**'s position that dreams function to deal with external threats as experienced by our ancestors. However, a closer examination of **REVONSUO**'s data leads to a different conclusion more in keeping with the Freudian conflict hypothesis.

First, let us consider the physical threats our ancestors had to guard against which figure so prominently in **REVONSUO**'s theory. These threats were encountered because of the need to venture out in order to hunt and gather food, or seek a mate. Threats would have been considerably diminished by simply staying put in a safe cave. Thus conflict at least between the need for food and the need for a mate and running the risk of injury or death in their pursuit must have existed among our ancestors. If during sleep a desire for food or sex was aroused this would bring with it the threat posed by these desires.

Observations of primate groups reveal that the most frequently experienced threats come from other conspecifics rather than from predators. The constant jockeying for alpha status among males, the need to select an appropriate mate among females, are not life threatening but rank high as sources of frequent conflict. It is the internal pressure to be dominant among males and to mate by choice among females that results in conflict and threat. Threat cannot be defined solely externally, but is defined by the internal state of the individual and determined by specific motivations. This is entirely consistent with the Freudian theory of dreams.

**Proposition 4.** Freud's theory is a two generator, or two mentation type theory which Freud called the primary and secondary process.

We will cite experimental evidence that REM mentation is organized along primary process lines and NREM along secondary process lines which fits with a two mentation model. The study by Shevrin and Fisher (1967) cited by **HOBSON ET AL.** is to our knowledge the only sleep-dream study in which the effects of a waking subliminal stimulus on REM and NREM mentation has been investigated. The relevance of the study to the current controversy over one versus two mentation theories lies in the fact that it introduces operationally through subliminal stimulation the place of unconscious influences on sleep mentation, an issue of great importance to a Freudian theory of dreaming, and we believe an overlooked factor in sleep-dream research. The subliminal stimulus was designed in accord with Freud's hypothesis that the unconscious mentation underlying dreams was rebus-like in nature, by which he meant that the elements of a dream were juxtaposed and combined not in terms of their customary logical and conventional relationships but in terms of what he referred to as "superficial" associations, borrowing a term from Wundt. In contemporary terms, these would be seen as distant associates. However, in no current linguistic or cognitive theory would the prediction be made that the distant associates would be combined to form a new entity with its own associates. It was to this process that Freud gave the name condensation, one of the primary process mechanisms, in which two quite different unrelated elements are combined to form a new entity.

The stimulus was a picture of a pen and a knee forming the rebus for the word penny. The effects of the stimulus could then be measured along two dimensions: secondary process (logical, conventional) associates of pen (e.g., ink, paper) and knee (e.g., leg, bent) and primary process associates of penny (e.g., cent, money). Scoring was based on association norms collected prior to the experiment. Following awakenings from REM and NREM sleep the subjects' accounts of the immediately prior sleep events were obtained as well as two minutes of free associations. These free associations revealed that the rebus effect (penny associates) was significantly greater following REM awakenings, and the secondary process associates (pen and knee associates) were significantly greater following NREM awakenings. These results support Freud's view that it is necessary not only to know the sleep content but also the unconscious process giving rise to the sleep content which may differ depending on sleep state along primary process (REM) and secondary process (NREM) lines.

In view of the fact that in the Shevrin and Fisher study the results were found in a waking response following sleep awakenings, it is important to note that **NIELSEN** (p. 15) cites eight studies in which cognitive and physiological components of the sleep state carry over and influence waking performance, and one study which failed to find such differences. He concludes that most of the results "support the interpretation that qualitatively different cognitive processes are active following and, by inference, just preceding awakenings from REM and NREM sleep."

In a second study conducted by Castaldo and Shevrin (1970), it is important to note that the pictorial rebus stimulus was not presented subliminally prior to sleep, but was presented as words through earphones during REM or NREM sleep. Measures were based on the waking accounts of sleep mentation rather than on free associations. Despite these considerable differences from the Shevrin and Fisher study, Castaldo and Shevrin found that pen and knee associations were significantly more frequently found in

accounts of NREM mentation and significantly less frequently found in accounts of REM mentation. The rebus effect was not replicated and may depend on such factors as pictorial and subliminal presentation.

In their reference to the Shevrin and Fisher study, **HOBSON ET AL.** do not cite the association findings which we believe are of special relevance to understanding the role of unconscious processes in sleep mentation. The Shevrin and Fisher results support **HOBSON ET AL.**'s position on the qualitatively different nature of sleep mentation associated with REM sleep, and fail to support **SOLMS**'s conclusion that there is no qualitatively different mentation related to REM activation.

**Proposition 5.** The dream will sometimes succeed in providing an hallucinated gratification of sexual and aggressive motivations and thus will be wish-fulfilling; at other times the dream will fail, resulting in anxiety dreams and awakenings. When optimally functioning the dream is thus a protector of sleep

**REVONSUO** directly challenges the wish-fulfilling aspect of dreaming proposed by Freud, replacing it with an evolutionary explanation based on rehearsal of threat. It is thus of interest that when **REVONSUO** provides some detailed description of dreams from the Mehinaku Indians that the role of threat is cast in a different light. Of the 14 dreams described, 7 can be construed as having manifest wish-fulfilling implications (e.g., desired and approached girl, struck by jealous wife). Our kinship with the Mehinaku Indians resides in the sexual and aggressive desires we share. Dreams from this standpoint are not outmoded rehearsals of prehistoric threats, but serve important current psychological purposes rooted in our evolutionary past.

**Proposition 6.** Exigent motives are highly activating or, in Freud's terms, cathected with considerable psychic energy and thus can provide the impetus for dreaming.

The evidence previously cited that REM deprivation can cause animals to become hypersexual and hyperaggressive suggests that these motives are highly activating. We have also noted that **SOLMS**'s lesion findings indicate that appetitive circuits are essential for the dreaming process, and that imaging studies point to the crucial involvement of motivational and emotional centers in the limbic and paralimbic system during REM sleep. It is intriguing to consider in addition the possible role of the deactivation of prefrontal cortical areas involved in executive functions, such as volitional control and self-monitoring. We would see this as resulting in a shift in the balance between wishes on the one hand and controlling/inhibitory functions on the other in favor of the former, resulting in wishes being relatively stronger and more exigent during REM sleep. This view also seems consistent with the aspect of **HOBSON ET AL.**'s AIM model that stresses a different balance in neuromodulation (his M factor) during REM sleep, with the preference for cholinergic over aminergic modulation favoring structures which mediate emotion over those which mediate directed thought.

Despite the insurance provided by motor paralysis that entertaining powerful motives would not result in acting on them, Freud reasoned that even becoming directly aware of them would disturb sleep and awaken the dreamer. For this reason there was a need for disguise which employed the primary process mechanisms described earlier and supported by the Shevrin and Fisher study. But the Shevrin and Fisher study did not establish the disguising function of the primary process. Further work is needed on this hypothesis, although nothing thus far reported would necessarily be inconsistent or contradict this hypothesis.

## Phylogenetic data bearing on the REM sleep learning connection

J. M. Siegel

*Department of Psychiatry, School of Medicine and University of California at Los Angeles, Neurobiology Research 151A3, Sepulveda VAMC, North Hills, CA 91343.* **jsiegel@ucla.edu**     **www.bol.ucla.edu/~jsiegel/**

**Abstract:** The phylogenetic data are inconsistent with the hypothesis that REM sleep duration is correlated with learning or learning ability. Humans do not have uniquely high amounts of REM sleep. The platypus, marsupials, and other mammals not generally thought to have extraordinary learning abilities have the largest amounts of REM sleep. The whales and dolphins (cetaceans) have the lowest amounts of REM sleep and may go without REM sleep for extended periods of time, despite their prodigious learning abilities.

[VERTES & EASTMAN]

The idea that REM sleep with its elaborate associated dream mentation has a role in memory consolidation is a very attractive one. It is a pity that the evidence supporting this beautiful idea is so weak. VERTES & EASTMAN marshal impressive evidence inconsistent with a major role of REM sleep in learning. To this I add the mammalian phylogenetic data.

Much of the evidence that has been advanced as supporting a positive relationship between REM sleep and learning derives from reported increases in REM sleep after learning and blockade of learning after reduction in REM sleep with deprivation procedures. As VERTES & EASTMAN point out, most if not all of this evidence does not withstand careful scrutiny. Nevertheless if we pursue the logic of this approach, one would predict that animals with greater learning capacity would have greater amounts of REM sleep.

Indeed, learning and memory theorists often imply that humans have large amounts of REM sleep. In fact, humans follow the general trends within the animal kingdom. Being large animals they share the inverse relationship between size and total sleep amount (Zepelin 1994). Humans sleep less than most smaller mammals. If REM sleep is calculated as a percentage of total sleep time, humans appear to have a lot of REM sleep, though not a uniquely large amount. However, the animals with the largest amounts of REM sleep are not the primates. The animal with the most REM sleep is the duckbilled platypus, which has, depending on how the calculation is done, approximately 7–8 hours of REM sleep a day (Siegel et al. 1997; 1999). REM sleep in the platypus has some unusual features. Perhaps of most significance is the lack of the low voltage EEG that characterizes REM sleep in other adult mammals. If we put aside the platypus data, the next contenders for the REM sleep championship are the black-footed ferret and the armadillo (Marks & Shaffery 1996; Prudom & Klemm 1973; Van Twyver & Allison 1974). What intellectual attribute do these three animals have in common? Is it intelligence or stupidity? Without disparaging the beauty and role of these animals in the ecosystem, they are largely instinct driven. Clearly they can learn as can all mammals, but they do not appear to be unique in their mental skills. In general, the marsupials and monotremes have more REM sleep than the placentals (Zepelin 1994).

How about the other end of the spectrum? The mammals with the least REM sleep are the cetaceans (whales and dolphins). Early reports in captive animals did not detect any clear episodes of REM sleep (Mukhametov et al. 1977; Mukhametov 1987; Shurley et al. 1969; Oleksenko et al. 1992; Flanigan 1974). Clearly if dolphins have any REM sleep at all, they can go without it for days or weeks. A more recent study in a captive gray whale demonstrated occasional twitches during sleep (Lyamin et al. 2000). The most generous estimates of the REM sleep total in these animals would be less than 15 minutes a day. How does the learning ability of dolphins and whales, animals with the largest brains ever to exist on earth, compare with that of the platypus, ferret, and armadillo? It would be difficult to defend the notion that the latter are smarter than the cetaceans. Across mammals, REM sleep time is negatively correlated with brain weight (Zepelin 1994).

Work by Jouvet-Mounier (Jouvet-Mounier et al. 1970) and a survey of the literature by Zepelin (1994) led to the conclusion that REM sleep time was correlated with immaturity at birth. Our recent findings in the platypus at the high end of the REM sleep scale and cetaceans at the low strongly support this conclusion. The immaturity of the platypus, hatching from an egg and remaining attached to its mother for an extended period after birth is consistent with its high level of REM sleep. The maturity at birth of the cetaceans, which can swim free of the mother and defend themselves immediately after birth is consistent with their low level of REM sleep. Neither the platypus nor the cetacean data is consistent with a relation to intellectual function or memory.

One way out of this dilemma for the learning-REM supporters is to argue that amount of REM sleep is not an informative variable; that REM sleep in the platypus may be less intense or efficient than that in the cetacean. However, this post hoc reasoning is not persuasive. There is no evidence that the very short REM sleep periods in the cetaceans are more intense or that those in the platypus are less intense. In fact the best evidence in terms of phasic event intensity argues just the reverse. The platypus has more than 6,000 phasic events during sleep/24 hours while the Gray whale has fewer than 10. As to the contention that time in REM sleep is not the important variable; this is the very basis of the claim of a relation between REM sleep and learning. The learning theorists cannot convincingly argue this point both ways.

In conclusion, the phylogenetic data provide additional evidence for the case against a key role for REM sleep in memory consolidation or intellectual function.

## Evaluating the relationship between REM and memory consolidation: A need for scholarship and hypothesis testing

Carlyle Smith[a] and Gregory M. Rose[b]

[a]*Department of Psychology, Tarent University, Peterborough, Ontario, Canada K9J 7B8;* [b]*Memory Pharmaceuticals Corporation, New York, NY 10032.* **csmith@trentu.ca**     **rose@memorypharma.com**

**Abstract:** The function of REM, or any other stage of sleep, can currently only be conjectured. A rational evaluation of the role of REM in memory processing requires systematic testing of hypotheses that are optimally derived from a complete synthesis of existing knowledge. Our view is that the large number of studies supporting a relationship between REM-related brain activity and memory is not easily explained away.

[VERTES & EASTMAN]

The article by VERTES & EASTMAN illustrates nothing so well as the difficulty of unraveling the mechanisms underlying memory consolidation and the current dismaying trend of reducing complex topics to sound bites. In the case of REM sleep, it cannot be said that, without REM, absolutely no memory consolidation of any kind will occur. However, the difference between this statement and the authors' thesis, that REM plays no role in memory processing, is vast. VERTES & EASTMAN's conclusion is possibly the result of their lack of direct experimental work examining the relationship of sleep to learning and memory, or, for that matter, with learning and memory at all. (VERTES is rightly recognized as a foremost expert in the neurophysiology of paradoxical sleep and theta rhythm generation in rodents.) We acknowledge that there is no simple guide through the complex and seemingly contradictory literature that deals with REM and memory consolidation. In the limited space allowed, we will attempt to rescue VERTES & EASTMAN from some of the conceptual traps into which they have fallen.

***A general comment.*** Science is an empirical discipline; understanding of a subject is always the consequence of a procession of discoveries. VERTES & EASTMAN seem to ignore this fact in assigning equal value to all REM/memory studies, whether they were performed 30 years ago or 5 years ago, and the authors exhibit a distinct preference for early (1970s) work (Fishbein & Gutwein 1977; McGrath & Cohen 1978) to support their hypothesis. It is germane that more recent studies, utilizing more quantitative methodology, better experimental design (including appropriate control groups), and an appreciation of the existence of multiple memory systems, have clearly and consistently shown that REM deprivation impairs the consolidation of some types of memories, but not others (Plihal & Born 1997; Smith 1985; 1995; 1996). Interesting to note, this work has also revealed that other stages of sleep play a role in consolidating some types of memory for which REM is not involved (Smith 1995; Smith & MacNeill 1994).

***Why animal studies using REM deprivation do not always affect learning.*** First, to reiterate a statement made above, it is clear that REM-based memory consolidation processes are involved with only some, and not all, types of learning. More generally, it must be considered that some tasks are so simple (i.e., that the subject is so genetically and/or experientially prepared to solve the problem) that no consolidation mechanism will be activated to a level that can be detected by experimental intervention (Pearlman 1979; Smith 1985). Second, the existence of REM Windows (Smith 1985; 1995; 1996) can explain why mis-timed periods of REM deprivation would not impair consolidation for a particular task. We are nonplused by **VERTES & EASTMAN**'s lack of appreciation that REM windows could be different in different learning situations (e.g., for different tasks or as the consequence of changing task demands). The idea of having a rigid REM window occurring at the same time after training, no matter what the task or number of training trials, is neither intuitively consistent, based upon how most biological systems are known to function, nor supported by systematic experimental work.

***Is stress the mechanism through which REM deprivation affects consolidation?*** One of the most persistent complaints about the REM-memory literature is that stress associated with techniques related to REM deprivation has not been adequately controlled. It is interesting that the stress argument is repeatedly invoked, despite the general absence of data indicating that stress has a detrimental effect on learning or memory. As has been argued in detail elsewhere, several types of studies appear to counter the argument that stress plays a role in REM-deprivation induced memory impairments: (1) pharmacological blockade of REM is not stressful, yet blocks consolidation (Smith 1995; Smith et al. 1991); (2) REM window experiments expose animals to procedures that produce equivalent stress, but produce memory deficits when brief (4 hour) REM deprivation occurs only at a specific time following training (Smith 1985; 1995; 1996); (3) blockade of all sleep except during REM window periods, a procedure known to be stressful, does not impair learning (Smith & Butler 1982); and (4) under certain conditions, REM deprivation has been shown to enhance memory (Kitahama et al. 1976; Smith & Gisquet-Verrier 1996). Taken together, this work makes it clear that stress has no simple role in REM deprivation studies, and parsimony suggests that it plays none at all.

***Studies of REM and learning in humans by Vertes & Eastman.*** The treatment of this literature is painfully incomplete. Studies in humans to date clearly support the idea that REM is at best a modulator of memory and that it plays a role in the consolidation of certain types of memories, but not others. One of the most salient findings of these studies is that REM is not involved in declarative/explicit type memory tasks, at least of the simple type, exemplified by word list remembering, that are often used (Smith 1995). By contrast, tasks that can be classified as procedural/explicit are sensitive to REM sleep loss, which results in impairments ranging from 20–50%. This knowledge undercuts the author's argument that humans with REM-eliminating lesions or REM-depriving pharmacological treatments are normal, because this cannot be established if the subjects are not tested in tasks in which REM is known to be involved. The REM sleep variables in humans are augmented with learning in the tasks examined so far and can be either the amount of REM sleep (min.) (DeKoninck et al. 1990b; 1989; DeKoninck & Prevost 1991) or the number

and density of REMs (Smith 1993; 1999; Smith & Lapp 1991). REM windows have been reported both within a single night (Stickgold 1998) and over several nights (Smith 1993). Although there are fewer human than animal studies, recent results support the REM-memory hypothesis and suggest involvement of a phasic REM component (Karni 1994; Smith 1995; 1999; Stickgold 1998).

In conclusion, we wish to state that while the precise role of REM sleep in memory consolidation is far from being completely defined, **VERTES & EASTMAN**'s conclusion that there is no connection is not supported by a balanced evaluation of the existing data. Advancement of scientific knowledge depends upon scholarship and hypothesis testing, both of which are incomplete in **VERTES & EASTMAN**'s target article.

# The mechanism of the REM state is more than a sum of its parts

Mark Solms

*Academic Department of Neurosurgery, St. Bartholomew's Royal London School of Medicine, Royal London Hospital, London E1 1BB, United Kingdom.* **mlsolms@mds.qmw.ac.uk**

**Abstract:** Nielsen has not demonstrated that NREM dreams are regularly accompanied by fragments of the REM state. However, even if this hypothetical correlation could be demonstrated, its physiological basis would be indeterminate. The REM state is a configuration of physiological variables, the basis of which is a control mechanism that recruits and coordinates multiple sub-mechanisms into a stereotyped pattern. The diverse sub-mechanisms underlying each individual component of the REM state do not have an intrinsic relationship with the REM state itself. [NIELSEN]

**NIELSEN** observes that the frequent occurrence of "almost dreams" during NREM sleep does not invalidate the original observation that most *actual* dreams occur during REM sleep. His attempt to explain by recourse to the notion of "almost REM" (**NIELSEN**'s "covert REM"), the subsequent discovery that some actual dreams (indistinguishable by blind raters from REM dreams) do also occur during NREM sleep, is therefore inconsistent. "Almost REM" is just as problematical a concept as "almost dreams." This would be true even if it were possible to demonstrate that REM-like physiological events routinely accompany NREM dreams (which **NIELSEN** has not in fact demonstrated, and which is contradicted by some of the available evidence; see Larson & Foulkes 1969; Pivik 1991; Rechtschaffen 1973; Rechtschaffen et al. 1972).

REM sleep is defined as a state in which diverse physiological factors covary in a distinctive pattern (Rechtschaffen & Kales 1968). Research into the neural basis of the REM state has accordingly focused on attempts to isolate "executive control" centres which "recruit and coordinate" the multiple component factors that constitute this distinctive pattern (Hobson 1988b; **HOBSON ET AL.** 1986; 1998b; ). This is because the very existence of a stereotyped pattern of physiological variables suggests the existence of an underlying control mechanism which generates (recruits and coordinates) that pattern. The mechanism in question has no special relationship with each of the individual variables that participate in the pattern. Each of the individual physiological variables has its own mechanism (or, more usually, mechanisms); but these diverse mechanisms are not the issue here. The mechanism by which each of the various factors might be activated (or suppressed) in isolation is therefore of comparatively little consequence for our understanding of the REM state. This is because the state *is* the pattern. What is at issue with respect to the REM state is something else, namely the executive mechanism that recruits and coordinates the various factors into a distinctive configuration. It is *this* mechanism that is said to generate dreams; and it is *this* claim (i.e., the claim that the executive control mechanism of the REM state is also the mechanism whereby dreaming is generated) that is under dispute.

A useful analogy might be drawn with the concept of clinical syndromes. When each of the component elements of a syndrome appears in isolation, the diagnostic implications are indeterminate. It is only when all the elements appear simultaneously in a known, pathognomonic configuration, that they have definite diagnostic implications (i.e., they imply that a known pathophysiological mechanism is operative).

> The essence of a medical syndrome is that a collection of signs or symptoms, *when all present*, indicate the presence of a specific disease. The correlation between elements in a syndrome may be high, low or in-between . . . Indeed, a syndrome is most useful as a diagnostic tool precisely when the elements usually are not found together. When they are found together, this strongly points to some special pathological process. (Strub & Geschwind 1983, pp. 317–18; emphasis added).

A clinician who interprets isolated symptoms or signs as "covert" expressions of a known pathognomonic mechanism would commit serious diagnostic errors (cf. the fallacy of "partial syndromes"; Kinsbourne 1971). By the same reasoning, when individual elements of the REM state appear in isolation, the physiological implications are indeterminate; they have no necessary relationship with the REM state itself, or with its known physiological mechanism. It would be an error to infer that the REM state is "covertly" present, and that its control mechanism has somehow been "partially" activated, when individual features of the state appear in isolation or in bits of the stereotyped configuration.

**NIELSEN** commits precisely this error when he construes isolated instances or couplings of saccadic eye movement (rapid or slow), or muscle atonia, or EEG desynchronization, and the like, as somehow implying a partial or covert expression of the REM state. The physiological meaning (the underlying mechanism) of these isolated events is indeterminate, and has to be established in its own right in each instance. The events in question might well have nothing to do with activation (partial or otherwise) of the known executive control mechanism that recruits and coordinates the known pattern of events that constitute the REM state.

For example, a burst of EEG desynchronization during a NREM period might reflect a type of forebrain activation derived from a source quite different from and unrelated to the pontine cholinergic/aminergic oscillator that is thought to generate such desynchronization in the REM state. EEG desynchronization comes in many varieties, reflecting a wide range of different states, generated by diverse physiological mechanisms. EEG desynchronization, by itself, therefore, can mean almost anything. One can only "diagnose" the causal presence of the known pontine mechanism of the REM state if the burst of EEG desynchronization in question occurs within the context of the known REM "syndrome." This is true whether the desynchronization co-occurs with dreaming or not. Accordingly, if it can be demonstrated that NREM dreaming is regularly accompanied by phasic EEG desynchronization, then it is not at all justifiable to infer that the dreaming was causally triggered by partial or covert activation of the control mechanism of the REM state. It may well be that it was generated by an entirely different mechanism. This same principle applies to all the hypothetical "covert REM" events that **NIELSEN** refers to.

The pertinent question, therefore, still remains: *What generates those NREM dreams that are "indistinguishable by any criterion" from REM dreams* (Hobson 1988b)? For the reasons outlined in my target article, I do not believe that they are generated by reciprocal interactions between pontine cholinergic and aminergic mechanisms. If **NIELSEN** is suggesting that they are, then his thesis lacks conceptual coherence (and empirical support!). If, on the other hand, he is suggesting merely that the REM state shares scattered variables in common with other physiological states that are also productive of dreams, then he is making a very different claim – and a far weaker one: one which begs the main question that is at stake here. I agree that the REM state and certain NREM states which are productive of REM-like dreams are likely to share some physiological properties in common. The questions

then become: (1) What are those shared properties? and (2) *What control mechanism recruits and coordinates them, and thereby generates the dreams?* To my mind, the search for these common properties and this underlying control mechanism should start from the observable fact of the dreams, not from theoretical preconceptions derived from our understanding of *REM* physiology.

One plausible empirical approach to the problem, then, is the classical clinico-anatomical approach; that is, to ascertain what anatomical structures are essential for dreams to occur. (This is the approach that I have taken.) To date, only two such structures have been identified: the parieto-temporo-occipital cortical junction and the ventromesial frontal white matter. (Nobody has ever demonstrated that pontine brainstem structures are essential for dreaming [dreaming in particular, as opposed to consciousness in general].) The task now is to verify whether the control mechanism we are seeking is indeed localizable to one of the identified structures. For the reasons set out in my target article, I believe that the best candidate for this role (in the present state of our knowledge) is a dopaminergic pathway that courses through the ventromesial frontal white matter. Now we need to establish whether and how activation of (and influences upon, and effects of activity in) this pathway (and other pathways in the ventromesial frontal quadrant) correlate with the actual experience of dreaming.

The old master of clinical neurology, Charcot, is reputed to have once said: "*La théorie c'est bon, mais ça n'empêche pas d'exister*" [Theory is good, but it does not prevent facts from existing]. (Freud 1893, p. 13). The psychophysiological theory of REM/dream isomorphism is, I suspect, preventing **NIELSEN** from acknowledging the existence of some unexpected clinico-anatomical facts which are difficult to reconcile with that theory.

## Neuronal basis of dreaming and mentation during slow-wave (non-REM) sleep

M. Steriade
*School of Medicine, Laval University, Quebec G1K 7P4, Canada.*
**mercea.steriade@phs.ulaval.ca**

**Abstract:** Although the cerebral cortex is deprived of messages from the external world in REM sleep and because these messages are inhibited in the thalamus, cortical neurons display high rates of spontaneous firing and preserve their synaptic excitability to internally generated signals during this sleep stage. The rich activity of neocortical neurons during NREM sleep consists of prolonged spike-trains that impose rhythmic excitation onto connected cells in the network, eventually leading to a progressive increase in their synaptic responsiveness, as in plasticity processes. Thus, NREM sleep may be implicated in the consolidation of memory traces acquired during wakefulness.
[**HOBSON ET AL.; NIELSEN; VERTES & EASTMAN**]

Our experimental evidence on reciprocally related activities of neocortical and thalamic neurons supports some concepts in the target articles by **HOBSON ET AL.** and by **NIELSEN**. One of the ideas in the paper by **HOBSON** and colleagues is that NREM dreaming contains thought-like mentation whose content is much less dissimilar than is commonly believed when compared to that occurring in waking; in other terms, NREM reports are related to waking life. According to **NIELSEN**, the recall rate of dreaming mentation in NREM sleep is quite high. And, both **HOBSON ET AL.** and **NIELSEN** mention the vividness of dreaming near the end of the normal sleep period that may correspond in humans to the period of cat NREM sleep with incipient ponto-geniculo-occipital (PGO) waves, which heralds REM periods.

In this commentary, I would like to present some data that help to understand the neuronal basis of the association between NREM sleep and mentation. This may be surprising in view of previous postulates regarding this sleep state as accompanied by a global inhibition of cortex and subcortical structures (Pavlov 1923), which would underlie the "abject annihilation of consciousness" (Eccles 1961). The two major claims arising from our recent experiments, using intracellular recordings from cortical and thalamic neurons under anesthesia as well as during the natural waking-sleep cycle of behaving cats, are as follows. (1) Despite the fact that the cerebral cortex is deprived in NREM sleep of signals from the external world because of their blockade within the thalamus, the gateway to neocortex, cortical neurons continue to entertain during this sleep stage a vivid dialogue, which is reflected in the high rates of spontaneous firing and preserved synaptic excitability to internally generated signals. Such aspects were unexpected during a behavioral state which is conventionally qualified as "passive," "resting" or "inactive." (2) The rich activity of neocortical neurons during NREM sleep consists of prolonged spike-trains that impose rhythmic excitation onto connected cells in the network, eventually leading to a progressive increase in their synaptic responsiveness, as in plasticity processes. Thus, we postulate that NREM sleep may be implicated in the consolidation of memory traces acquired during wakefulness. Our data are only indirectly related to the basic assumption in the target paper by **VERTES & EASTMAN**, denying the role of REM sleep in memory consolidation, as we champion the role of NREM sleep in this process.

The limit of our hypothesis is that, at the present time, intracellular recordings are performed, by necessity, in animals with a rather limited behavioral repertoire. And, terms such as "memory" are used to describe the occurrence, after rhythmic and prolonged testing volleys, of *spontaneous* neuronal activity displaying exactly the same patterns as those exhibited by stimulus-locked responses evoked during the prior period (see Fig. 6B in Steriade 1999). On the other hand, the advantage of our approach is that, for the first time, dual intracellular recordings from cortical or cortical and thalamic neurons are obtained *in vivo* in animals under an anesthetic that best simulates natural NREM sleep (Steriade et al., 1996) and, furthermore, intracellular recordings are performed during the natural waking-sleep cycle in behaving animals (Fig. 1). There is no need to elaborate on the advantages of intra-

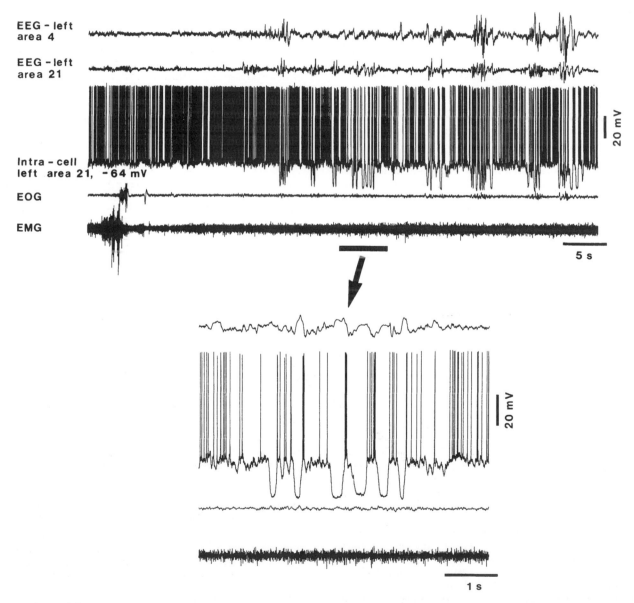

Figure 1 (Steriade). *Intracellular recording of neocortical neuron during natural waking and sleep states in behaving cat.* Upper panel: five traces depict (from top to bottom): EEG from left cortical areas 4 and 21, intracellular activity of neuron from area 21, electro-oculogram (EOG), and electromyogram (EMG). The panel illustrates the transition from wakefulness to NREM sleep (see onset of low-frequency, high-amplitude EEG waves). Part marked by horizontal bar is expanded below (arrow); only EEG from area 21 was illustrated. Note cyclic hyperpolarizations (downward deflections) of neuron during NREM sleep, but brisk firing during the depolarizing periods of the slow sleep oscillation. Unpublished data by M. Steriade, I. Timofeev, and F. Grenier.

cellular recordings from multiple sites, the only method that provides information on subthreshold membrane potential fluctuations during shifts in the state of vigilance and that can reveal the dissociation between exaltation of activity in cortical neurons and simultaneous postsynaptic inhibition in related thalamocortical cells, an inhibition mediated by the activation of GABAergic thalamic reticular neurons driven by corticothalamic projections (Steriade & Contreras 1995).

Let me start with the rich activity of neocortical neurons during NREM sleep, a state when these neurons are deprived of thalamic inputs that originate in pathways from the brainstem and the external world. That the thalamus is the first brain relay station where incoming signals are deeply inhibited from the very onset of drowsiness and even more inhibited during NREM sleep, without any change in the magnitude of the afferent volley that monitors the input reaching the thalamus, was documented by simultaneous recording of pre- and postsynaptic field potentials (Steriade 1991) and intracellular activities (Timofeev et al. 1996). Then, how is it that, without thalamocortical inputs, neocortical neurons still remain spontaneously active and responsive during NREM sleep? The fact is that corticocortical synapses exceed by far those made by thalamocortical axons. The major cortical rhythmic activity that characteristically defines NREM sleep, is a slow oscillation with a frequency between 0.5 and 1 Hz in both cats (Contreras & Steriade 1995) and humans (Achermann & Borbély 1997; Amzica & Steriade 1997). This oscillation is generated intracortically, as it survives an extensive thalamectomy (Steriade et al. 1993) and it disappears in the thalamus following decortication (Timofeev & Steriade 1996).

The distinctive feature of the cortical slow oscillation during NREM sleep is the alternation between prolonged periods of depolarization leading to spike-trains and long lasting periods of hyperpolarizations associated with neuronal silence. This stands in contrast with the tonic depolarization and firing of cortical neurons during both waking and REM sleep (Fig. 1). Although neuronal discharges are absent during the hyperpolarizing periods of the slow sleep oscillation, the overall firing rates of cortical neurons are quite close in NREM sleep and in wakefulness ($\sim$12 Hz and $\sim$14 Hz, respectively, in electrophysiologically identified regular-spiking neurons). This is due to the fact that, during the depolarizing phase of the slow oscillation, cortical neurons discharge vigorously, in many instances above the level observed in waking or in REM sleep (Steriade et al. 1999). Not only do cortical neurons fire spontaneously at high rates during NREM sleep, but their synaptic excitability to intracortical signals is enhanced during this stage. In contrast to the synaptic inhibition of thalamic neurons during NREM sleep and particularly during sleep spindles (which explains the thalamic blockade of incoming signals), earlier extracellular recordings in behaving monkeys showed an increased responsiveness of neocortical neurons to callosal volleys in NREM sleep compared to waking (Steriade et al. 1974) and recent intracellular recordings demonstrate that the cortically-evoked excitatory responses of cortical neurons are not diminished during the prolonged hyperpolarization of the slow sleep oscillation (Timofeev et al. 1996).

The result of intracellular recordings in naturally sleeping animals, showing high rates of spontaneous neuronal firing in neocortex during NREM sleep, raised the obvious question: what is the functional role of rhythmic spike-trains fired by neocortical neurons during the slow sleep cortical oscillation (0.5–1 Hz)? With the corollary: What may be the influence of rhythmic spike-bursts fired by thalamocortical neurons during sleep spindles (7–15 Hz) upon cortical neurons? Both these types of sleep rhythms (spindles and slow oscillation) have an impact on neocortical neurons and may change their responsiveness and even induce structural changes in their dendritic arbor that could have important consequences for the consolidation of traces produced by events occurring in other states of vigilance.

We started to work out the above hypotheses by simulating a major sleep oscillation, thalamically-generated spindles, using trains of thalamic stimuli applied at $\sim$10 Hz, while simultaneously recording cortical and thalamic neurons intracellularly (Steriade et al. 1998). The responses to pulse-trains at $\sim$10 Hz grow progressively in size, from the second stimulus in the train and, therefore, are termed *augmenting*. Both the thalamus and the cerebral cortex have the neuronal machinery that is necessary to generate augmenting responses, as shown by the fact that such responses can be recorded in the neocortex of athalamic animals and in the thalamus of decorticated animals. However, the full consequences of the augmenting phenomenon, which include self-sustained oscillations and plastic changes in network activities, require intact-brain preparations. This need for an intact brain, including generalized modulatory systems, is also shown by the state-dependency of augmenting responses which have maximal amplitudes in NREM sleep and lowest amplitudes during full alertness and REM sleep.

Augmenting responses are associated with plasticity processes, that is, decreases in inhibitory responses and persistent and progressive increases in excitatory synaptic responses. Such changes can lead to self-sustained oscillations due to resonant activities in closed loops, as in memory processes (Steriade 1999). The repeated circulation of impulses in reverberating circuits between the cortex and thalamus, may lead to synaptic modifications in target structures, which favor alterations required for memory processes. This hypothesis was also proposed (Buzsáki 1989) and tested experimentally (Wilson & McNaughton 1994) in the hippocampus. The hippocampal "place cells" were found to display higher firing rates and had a tendency to discharge synchronously during sleep, as if neuronal states are played back as part of the memory consolidation process.

Finally, NIELSEN's model, implicating covert REM processes before the full-blown REM sleep, follows experimental data published more than a decade ago in which we attempted to link the dreaming mentation to the appearance of PGO waves well before muscular atonia and EEG activation, during EEG synchronization of the final NREM period. In that paper (Steriade et al. 1989), we proposed that "vivid imagery may appear well before the classical signs of REM sleep, during a period of apparent EEG-synchronized sleep." This idea was based on the fact that visual thalamic neurons fire spike-bursts related to PGO waves (the robust bursts are owing to the fact that thalamic neurons are still hyperpolarized during that final period of NREM sleep); and also because the spontaneous firing rates of thalamic neurons is low, the signal-to-noise ratio during the PGO-related spike-bursts is very high, thus possibly underlying the vivid mental experiences outside REM sleep.

# Inclusive versus exclusive approaches to sleep and dream research

Robert Stickgold

*Department of Psychiatry, Harvard Medical School, Boston, MA 02115.*
**rstickgold@hms.harvard.edu**

**Abstract:** By assuming that REM sleep either plays a critical role in all memory consolidation or no role in any, Vertes & Eastman have chosen to reject, rather than explain, robust experimental findings of a role for sleep in memory and learning. In contrast, Nielsen has attempted to integrate conflicting findings in the dispute over REM versus NREM mentation. Researchers must trust the data more and the theories less, and build integrative rather than exclusionary models if they hope to resolve these knotty problems effectively.
[NIELSEN; VERTES & EASTMAN]

***Introduction.*** It is striking that 100 years after Freud (1900), there is absolutely no agreement as to the nature of, function of, or brain mechanism underlying dreaming. There is even disagreement as to what constitutes a legitimate approach to the question. In this environment, it is perhaps not surprising that various researchers have staked out strikingly different positions that are often presented as incompatible with one another. An alternative interpretation, one that I will argue here, is that we are discovering, but in many cases ignoring, the rich complexity of sleep and dreaming. In what follows, I will argue that the paper of **VERTES & EASTMAN** has fallen into this trap, choosing to look at only some of the available data to draw a conclusion that rejects, rather than explains, other robust experimental findings. In contrast, I will support **NIELSEN**'s attempt to integrate conflicting findings in the dispute over REM versus NREM mentation. In the end, I will conclude that we should probably trust the data more and the theories less, and build integrative rather than exclusionary models. Only by moving in this direction will we be able to resolve these knotty problems effectively.

***Vertes & Eastman: The case against memory consolidation in REM sleep.*** The argument against REM sleep having a role in memory consolidation by **VERTES & EASTMAN** is a disappointingly anachronistic and one-sided review of this rich literature. At the end of their introduction, they give the overarching reasons that lead them to conclude that REM sleep can have no possible role in learning and memory: "Sleep involves basic biological functions and memory requires consciousness." What could they possibly mean? Are learning and memory consolidation not basic biological functions? Are these "psychological" phenomena and hence not of any basic importance? Second, assuming for the moment (as they clearly do) that we are unconscious when we dream in REM sleep, are old memories not activated, associated, and integrated to form dream imagery and narrative? And if we remember our dreams on awakening, haven't new memories been laid down during sleep? It is as if they want to say that memory is too complex for sleep and sleep is too important for memory. These attitudes could have been forgiven if they were followed by a good, critical review of the literature. Unfortunately, they were not.

***Memory systems.*** One problem with **VERTES & EASTMAN**'s discussion is that they have a view of memory systems that is at least seven years out of date (Schacter & Tulving 1994). While vaguely acknowledging that there is data supporting the concept that REM sleep is more important for learning complex tasks than simple tasks and making passing reference to suggestions that "information is differentially processed in distinct phases of SWS and/or REM sleep," they never mention procedural or declarative memory systems and only refer to working and episodic memory in a quote from Barbara Jones (1998) about the possible role of sleep-related decreases in frontal lobe activity on subsequent dream recall. Yet for the last decade much of the work on sleep and memory has focused specifically on the differential effects of various sleep stages on procedural versus declarative versus motoric memory systems (Plihal & Born 1997; Smith & MacNeill 1994; Smith and Rose 1997; Smith et al. 1998; Stickgold 1998; Stickgold et al. 2000b). **VERTES & EASTMAN** seem to take the position that either REM sleep is absolutely critical for all forms of learning and memory consolidation or else it has no role whatsoever in any form of these processes. Obviously, this is an unjustifiable dichotomy.

***Biased reviewing.*** A more worrisome problem is how they choose to cite the literature. Whether it is noting the appearance of Jonathan Winson and Matt Wilson on the Charlie Rose TV program "promoting their shared belief" in a role for sleep in learning, or suggesting that our own findings of sleep-dependent memory consolidation is "very much at odds" with our activation-synthesis model of dreaming, or quoting the frustration of Elizabeth Hennevin and Bill Fishbein with the resistance of the sleep community to their empirical findings (Fishbein 1995; Hennevin et al. 1995a), **VERTES & EASTMAN** often seem to be spending more time criticizing the researchers than their data. Beyond this, their discussion of the actual data is frustratingly skewed. In their discussion of REM deprivation effects on memory, they pay considerable attention to the legitimate concern over the interpretation of cognitive tests given 30 min or 3 hr after several days of exposure to stressful REM deprivation techniques, but they ignore

findings of equally strong effects that are seen a week after a relatively brief 4-hr period of REM deprivation (Smith & MacNeill 1994) or after a single night of sleep deprivation in humans (Aubrey et al. 1999; Smith & MacNeill 1994). Again, they note that both Karni et al. (1994) and our laboratory (Stickgold et al. 2000b) have reported strong relationships between REM sleep and learning on a visual discrimination effect (we reported an *r*-value of 0.89 between post training sleep parameters and subsequent improved performance, a correlation significant at $p < 0.0001$), but then they focus on what we and Karni agree are relatively minor differences between our findings and conclude that "until these discrepancies are resolved, it is difficult to evaluate the reliability of the findings." No such critical lens is held up to the work of those who find no correlation between sleep and learning!

***Alternative explanations.*** Having rejected all findings of effects of REM deprivation on learning under the rubric of "very stressful deprivation procedures," they go on to conclude that there must be no effect at all of REM on memory because "there appears to be no alternative explanation for studies which have failed to show that [REM deprivation] disrupts learning/memory." Yet as all researchers should know, the failure of a test to reach statistical significance only indicates a failure to demonstrate the presence of a difference. It does not demonstrate the absence of such a difference. Beyond the issue of statistics, there is another very simple "alternative explanation." Since all proponents of REM-dependent memory consolidation agree that REM is not involved in consolidation of declarative memories such as those formed in paired associates training, the failure to observe REM-dependent consolidation may simply reflect the testing of a memory system that is not REM-dependent. The failure of the reviewers to address this issue, despite reviewing articles that directly discuss it (Plihal & Born 1997; Smith & MacNeill 1994; Smith & Rose 1997; Smith et al. 1998; Stickgold 1998; Stickgold et al. 2000b), is a frustrating disappointment.

***Theta rhythms.*** A role for theta rhythms is acknowledged by **VERTES & EASTMAN** in the processing of memories during wake, but they reject any similar role in REM: "If the transfer of information in REM is not orderly, or is essentially chaotic, it would seem that there would be no functional value in consolidating or 'remembering' this information" (sect. 2.6). Fair enough. But there is very often order in nature that we fail to see. Let me suggest an example of how REM sleep might be processing stored information in a less obvious manner. One critical task of the mammalian brain is to search for useful new connections between previously stored memories, most usefully memories stored in neocortical association networks, rather than the hippocampal, declarative system. While not critical on a minute by minute or even day by day basis, these new connections allow organisms to identify classes of related precepts, superclasses of causal relationships, and insights into novel relationships among sets of stored information. These are precisely the types of new associations that people are seeking when they "sleep on a problem." The problems one "sleeps on" are not trying to remember phone numbers. No one believes that sleep will help recall lost episodic or declarative memories. Rather, one sleeps on problems that involve assessing alternative explanations of past events or possible outcomes of future events.

The physiology of REM sleep would seem to support such information processing. Information outflow from the hippocampus to neocortex is shut off (Buzsáki 1989; 1996), cortical association nets are loosened (Stickgold 1998; Stickgold et al. 1999b), and cortical memory formation is enhanced (Hasselmo 1999; Hasselmo & Bower 1993) during REM sleep. Together, these allow the brain to (1) ignore the predictable interpretations driven by the replay of episodic memories (noticeably absent in REM dreaming), (2) seek out and test novel associations within the cortex, and (3) strengthen them as appropriate. The consolidation of procedural memories, also residing in neocortical rather than hippocampal memory systems, could follow a similar strategy. Interestingly, the flow of information back into the hippocampus during REM sleep might then serve to signal the appropriateness of "forgetting" episodic memories (Crick & Mitchison 1983) after cortical integration. Evidence for such erasure via LTD has been proposed by Poe et al. (1997).

If this is the role of sleep in memory consolidation, or if this even touches on sleep's role, it is not surprising that we find such variable results in published studies. These are not easy forms of consolidation to quantify. Indeed, these processes probably involve memory integration even more than they do simple memory consolidation. So it is not surprising that simple cognitive and psychomotor memory tests fail to show any obvious impairment of performance after administration of drugs that disrupt REM sleep. These tests classically measure working memory and declarative memory systems that we would not expect to be affected by REM deprivation. We know of no cases in which anyone, for example, tested the effects of these drugs on complex perceptual procedural learning.

*Magical processes.* Near the end of their article, **VERTES & EASTMAN** note that many have been tempted to believe that "magical processes occur during REM sleep." Indeed they do, and not just in REM sleep. We do not understand how dreaming comes about, how memories and concepts are so intriguingly woven to form the narrative of the dream. We do not understand how, in waking, the brain comes to understand language, or color, or beauty. But this does not mean that they do not occur. It merely means we do not know how they are produced. When we find a correlation between sleep and learning that explains 80% of intersubject variance and is significant at the 0.0001 level, something is happening, magical or not.

In the end, I am honestly baffled by **VERTES & EASTMAN**'s decision to reject all of the work that points toward a role for sleep in memory consolidation and learning. It is as if, finding the cup half full (or half empty), they insist that it must be either completely full or completely empty. Surely, the best reading of the literature would say that sleep plays an important, if not necessarily critical, role in some forms of memory consolidation and learning. Drawing such a conclusion is not a compromise. It is scientific fact.

*Nielsen: Mentation in REM and NEM sleep: A review and possible reconciliation of two models.* The question of how to characterize the role of sleep states in the control of sleep mentation remains a thorny one. In this context, **NIELSEN**'s offering is a welcome breath of fresh air. He presents an impressively complete review of the extant literature and provides a valuable theoretical framework for the critical analysis of this literature. **NIELSEN** has attacked the dichotomous approach to REM and NREM sleep even more effectively than we did in our article (**HOBSON ET AL.**). By offering the concept of "covert REM sleep" to describe the spread of REM sleep physiology into NREM states, he emphasizes one essential aspect of our state space model: the continuity, overlap, and dissociability of state features.

There are two main points I would make about his model. First, while the early part of his paper points strongly toward qualitative differences between REM and NREM mentation, **NIELSEN** seems in the end to suggest that all NREM mentation is due to the covert intrusion of REM state processes. If this is to explain the qualitative differences between REM and NREM, one would have to assume that different REM state processes add different features to the mentation process. While this may, indeed, be true, I still would argue that there is at least a basal tendency toward sleep mentation even when all of these REM state processes are inactive. This could explain the mentation reports of SWS and would further simplify the explanation of qualitative differences; as more and more REM state processes are activated, the mentation shifts more and more from a "pure" NREM phenomenology toward a "pure" REM phenomenology.

My second point is that I believe **NIELSEN** could have taken his model even further. I would propose three basic tenets of such an expanded model:

1. Sleep mentation and dreaming are products of the brain, and are determined at any given time by the levels of activation and interaction of disparate brain systems. These include anatomically defined systems such as the brainstem, amygdala, and frontal cortex, cognitively defined systems such as attention, emotion, and memory, and neuromodulatory systems such as the cholinergic, noradrenergic, and serotonergic systems.

2. All of these systems show fluctuations in their levels of activation and functional connectivity across the sleep cycle.

3. The use of polysomnography to define sleep stages represents a crude division of these rich and complex physiological fluctuations.

Given these assumptions, perhaps Rechtschaffen and Kales's sleep stages (Rechtschaffen & Kales 1968) should be seen as only a first attempt at defining the rich heterogeneity of sleep. For many purposes it is a completely adequate description of the system. For other purposes it will clearly prove only marginally adequate or even totally inadequate. Hobson (1992a) has proposed a three dimensional model of sleep state space based on levels of sensory input, brain activation, and neuromodulation. In our paper in this issue (**HOBSON ET AL.**), we begin to move toward expanding this model so that different brain regions can be in different portions of this state space. Eventually the number of dimensions will necessarily increase as will the number of brain regions that need to be viewed separately. **NIELSEN** is hinting at this complexity, but I suspect that before too long we will have to accept that REM and NREM are useful concepts when looking grossly at sleep phenomena, but inadequate as we refine our investigations of these phenomena.

In conclusion, I suggest that we need to go even beyond **NIELSEN**'s idea of covert REM sleep processes in NREM sleep and, of course, covert NREM sleep processes in REM sleep, and instead accept a rich, complex, and confusing panoply of brain processes that show robust, but nonetheless only statistical, probabilities of co-occurring and of being sustained for periods of minutes or tens of minutes.

---

# Evolutionary psychology can ill afford adaptionist and mentalist credulity

Nicholas S. Thompson

*Department of Psychology and Biology, Clark University, Worcester, MA 01610.* **nthompson@clarku.edu**

**Abstract:** The idea that dreams function as fright-simulations rests on the adaptionist notion that anything that has form has function, and psychological argument relies on the mentalist assumption that dream reports are accurate reports of experienced events. Neither assumption seems adequately supported by the evidence presented.

[**REVONSUO**]

**REVONSUO**'s core idea – that dreams must be a biological threat simulation system because dream reports are highly structured narratives uniquely sensitive to threatening circumstances – provoked in me both biological and psychological misgivings. On the biological side, the argument fails because it requires the false premise that all the things that have form have function. The form of many recognizable structures (e.g., the human nose, the armpit, the pseudopenis of the female hyena, etc.) is determined by selection on other features of the organism and has no role in their determination (Glickman et al. 1993; Gould & Lewontin 1979). Even the association of "randomness" with "formless" is inappropriate. If you and I were to bump into each other in the street, we might very well speak of a random meeting. We could truthfully characterize the meeting in this way, but not because the behavior that led to it was formless or without direction, nor even because the meeting was not entirely predictable. We would say that the meeting was random because the factors that determined the meeting were not governed by it. To show that the meeting was non-random, we would have to show that our behavior was in some sense designed to produce a meeting.

Similarly, to show that dreams function to improve performance in threatening situations, we must show not only that dreams have this consequence but also that they are designed for it. Even the evidence offered by the author that dreams produce improvements in performance is shaky enough: it seems to be based solely on an analogy with motor imaging experiments. The evidence that dreams are designed to produce these consequences seems even shakier. (Thompson 1981): (1) the comparative method, (2) the engineering method, and (3) the examination of reproductive consequences. Of the three classes of evidence, the only evidence consistent with design is related to the engineering criterion, in this case evidence that dreaming is sensitive to the occurrence of threatening situations in the dreamer's life and that dream content is related to threatening occurrences. No comparative evidence is provided – no evidence that across species, animals dream more that are subject to hazards: nor is there any evidence that non-dreamers live particularly short and unfecund lives in hazardous environments.

On the psychological side, **REVONSUO** seems unaware of the sort of misgivings that might be entertained by behaviorists concerning his research program. The author speaks of dreaming as if it posed no special philosophical or methodological difficulties. According to his account, the dream is a shoe the dreamer puts on for himself because it helps him practice for dire situations. Waking up a dreamer is sort of like interrupting a person who is watching TV drama. Once we get his attention, we can ask him about the nature of the "program" he has been watching, and he can report luridly concerning his experience. Only the dream is problematized. Remembering the dream and reporting on that memory are taken as unproblematic. This stance is unlike any that we would take with reports of emotion-laden events obtained in non-dreaming contexts. Consider, for instance, the stance a therapist would take toward a client who reported an argument he had with his spouse during the previous eight hours. Unlike the dream researchers, the therapist would not presume that the client has exact recall of the argument or that the client would give a faithful report of such intimate and troublesome events.

Such methodological credulity with respect to dream recall and dream report is particularly troubling because dreams are notoriously ephemeral. Dreaming, or being aware of our dreams, or remembering our dreams, or telling coherent accounts of our dreams are not skills that the people of my acquaintance possess equally. Some people seem to expand a lot of effort in rehearsing and relating their dream reports: others to be hesitant to report their dreams, bad at remembering them, or, perhaps, dream rarely if at all. Even the best dreamers around me confess that dreams are so evanescent that they will be lost or altered in memory if they are not written down or related immediately upon awakening.

In fact, why do we assume that there is anything that is dream apart from the subject's reporting of it? Even if I grant that I, too, awake in the morning with the sense that I have had experiences while sleeping, what is the reason for believing that these experiences correspond to any facts of the matter whatsoever? The supporting evidence used to be that the REM sleep syndrome was a necessary concomitant of dream reports, but I gather that even that evidence is no longer credible (see **REVONSUO**, sect. 1.1). Is the ontology of dreaming so firm that we cannot even imagine a skeptical account of dream reports that does not require the existence of dreams? Whatever we conceive sleep to be, forced awakening certainly provokes a massive reorganization of neural activity. Is it such a wonder that such a profound intrusion into ongoing neural activity should not knock off shards of memory and send them hurtling into awareness? Is it such a wonder, storytelling creatures that we are, that these shards should be assembled into fragile narratives? Is it so strange that we should perceive these stories as occurring prior in time, during the previous sleep? Finally, given that these narratives are commonly assembled in the context of an interruption of ongoing activity, is it such a wonder that they should have a threatening tone? Surely, the dream report literature should be subjected to the same sort of skeptical assault that tested the hypnosis literature in the 60s and 70s (Sarbin 1981) Can experimenters reliably distinguish fake from "real" dreamers? Does dream fakery get better if the subject is drowsy when asked to generate a fake dream? What if the drowsy subject is forbidden to tell a dream but allowed only to relate an imaginative story (cf. Fiss et al. 1966)? Or to give an account of something that "happened yesterday"? Is a recently awakened subject more likely to relate a negative story than a randomly probed subject? Wouldn't you?

As evolutionary psychology gains attention and respect, the question of what kind of a psychology it is to be – mentalist or behaviorist, adaptionist or selectionist – becomes more important. Is evolutionary psychology to explain feelings, thoughts, beliefs, desires, and other events "within the head" by reference to vaguely conceived benefits? Alternatively, is it to explain patterns in the activities of humans in terms of a history of differential reproduction of individuals enacting those patterns? Insensitivity to these crucial issues will slow the development of the field.

ACKNOWLEDGMENT
Thanks are due to my colleagues Joe deRivera, Jim Laird, Patrick Derr, and Michael Addis for their kind assistance.

# Critique of current dream theories

Gerald W. Vogel

*Sleep Research Laboratory, Emory University, Atlanta, GA 30306.*
**gvogel@emory.edu**

**Abstract:** Modern lab research has found that contrary to the suggestions of Hobson et al., Nielsen, and Solms, dreams are organized, mundane stories. Hence, their theories to explain the distortions and bizarreness of dreams are misdirected. Hobson et al. propose that REM sleep processes are responsible for dreams. But dreaming occurs in absence of REM sleep and REM sleep is often accompanied by no dreaming. Hence, REM sleep is not necessary or sufficient for dreaming.
[**HOBSON ET AL.; NIELSEN; SOLMS**]

**HOBSON ET AL., SOLMS,** and **NIELSEN** propose dream theories to explain distortion, disorganization, and bizarreness in dreams. They assume that these are the salient characteristics of typical dreams, a nineteenth century view also held by Freud. We now know that this view is wrong. It is based mostly upon reports of dreams that are spontaneously recalled upon waking in the morning. These dreams are likely the most dramatic, bizarre dreams and are not representative of dream life in general. The collection of large dream samples from throughout the night from large samples of ordinary people, both children and adults, has shown that most dreams are mundane, organized, everydayish stories (Dorus et al. 1971; Snyder et al. 1968). Though novel, they are not bizarre. Though often vivid, they do not often express inappropriate or extraordinarily intense emotion. Though unreflective, they are not disorganized. **HOBSON ET AL., SOLMS,** and **NIELSEN** ignore this massive evidence about the nature of dreams. Thus their dream theories are inappropriately directed and miss the main point – and mystery – about dreams, namely, that even during sleep, the mind/brain produces organized, coherent, understandable mentation.

**HOBSON ET AL.**'s dream theory, the activation-synthesis hypothesis (ASH) (Hobson & McCarley 1977), is also inconsistent with other twentieth century findings. The hypothesis asserts that during REM sleep, random neuronal discharge of the hindbrain REM sleep generator activates the forebrain to produce the conscious dream. According to this hypothesis, distortion, disorganization, and bizarreness are dream characteristics because the forebrain can only perform "the best of a bad job" in synthesizing its random brain stem stimulation. The hypothesis predicts that dream distortion and bizarreness will increase during more intense "random" stimulation by REM generator (Hobson & McCarley 1977). Tested in several studies, this prediction has been refuted although with some minor exceptions (for review, see Pivik 1991; Rechtschaffen 1973). The hypothesis also predicts that dreaming will occur only in REM sleep. Many studies found that dreams were reported from more than 50% of awakenings from nonREM sleep during periods far separated from waking and REM sleep and during which the REM generator neurons are silent (for review, see Pivik 1986). And many studies found that dreams were reported from the 70% of awakenings from sleep onset when the REM generator is silent (for review, see Vogel 1991). Most important, two independent studies found that, equated for length, REM and nonREM reports were indistinguishable (Antrobus 1983; Foulkes & Schmidt 1983; ). These findings indicate that REM sleep is not necessary for dreaming and in particular for dreaming with the formal characteristics (bizarreness, distortion, disorganization) that Hobson and others claim are the distinctive formal characteristics of REM dreams. HOBSON's response to this critique is that REM-like nonREM mentation is produced by REM sleep processes during nonREM sleep.

HOBSON ET AL. propose that several findings support this requirement of the ASH. Here is a point by point refutation of their proposals.

1. HOBSON ET AL. propose that circadian rhythm based increases in late night stage two activation sustain larger and more vivid nonREM dreaming. They then suggest that brain activation during nonREM sleep is a "REM-like phenomenon" (p. 56). In rebuttal, this is about brain activation, which is not a distinctively REM sleep process. Brain activation also occurs when awake. Thus, dreaming during late night Stage 2 activation is not associated with a distinctive sleep process as required by ASH. Furthermore even HOBSON ET AL.'s proposed association between nonREM sleep activation and dreaming is contradicted by evidence. For example, (a) dreams are reported from nonREM stages 3 and 4, which, based on metabolic studies, are periods of low brain arousal (Pivik 1986; Pivik & Foulkes 1968; Rechtschaffen 1973); and (b) Pivik and Foulkes (1968) found that stage 3 dream recall was similar to stage 2 dream recall.

2. HOBSON ET AL. propose that nonREM dreaming occurs during pre REM slow wave sleep in association with preREM sleep PGO waves, which are prominent phenomena during REM sleep. In rebuttal, this proposal is hypothetical. Though preREM sleep PGO waves have been observed in the cat, there is no direct evidence of their occurrence in humans. Second, even if they do occur, contrary to HOBSON ET AL.'s proposal, Pivik's review (1991) indicates that nonREM dreaming is not increased during preREM period; and Larson & Foulkes (1969) found fewer, less dream-like reports during the REM EMG drop than during control periods. Although HOBSON has theorized prodigiously about the mental correlates that accompany the initial activation of REM mechanisms, he has conducted absolutely no psychophysiological research to verify his accounts. Others, however, have studied the phenomenological correlates of immediately pre-REM and early REM physiology. Both Larson and Foulkes (1969) and Foulkes et al. (1980) found that early signs of REM sleep activation (e.g., the sudden loss of muscle tone) were accompanied by reliable decreases in the dreamlike quality of sleep mentation, a remarkable and direct refutation of the heart of HOBSON ET AL.'s theory.

3. HOBSON ET AL. propose that during nonREM sleep external stimuli produce PGO spikes which then cause visual imagery, the hallmark of dreams. This is speculation. HOBSON ET AL.'s work indicates that in humans acoustic stimuli during stage 2 enhance visual imagery and they cite other work indicating that in cats sound stimuli stimulates PGO. But they present no evidence of the requirement of the ASH, namely, that in humans external stimuli during nonREM sleep reliably stimulate REM-like activity in association with reported imagery. In fact, in cats, external stimuli produce isolated PGO spikes, not the long trains of PGO spikes that ASH would require of dream substrates.

4. HOBSON ET AL. cite a Pivik result that nonREM phasic spinal reflex inhibition was associated with greater recall, auditory imagery, and hostility. In contrast, Rechtschaffen's (1973) analysis of this study concluded that there was no significant relationship

of phasic variables studied (including dreamlike fantasy, sexuality, morality, thought, emotion, etc.). only hostility and auditory imagery were significantly related to phasic events. It appears that by selective reporting, HOBSON ET AL. were trying to give a positive spin to what was essentially a negative result.

5. HOBSON ET AL. cite a Rechtschaffen result that nonREM PIPs (phasic integrated potentials), were associated "with enhanced recall" of mentation. PIPs are extraocular spikes recorded from surface electrodes in humans. These spikes have morphological and distributional features like the extraocular spikes associated with PGO potentials in cats. Thus, they have been interpreted as surface indicators of PGOs in humans. But HOBSON ET AL. are wrong in their assertions about the relationship between PIPs and recalled nonREM mentation. In Rechtschaffen et al.'s work (1972), 51% of PIP awakenings and 53% of control awakenings showed recalled content, clearly no substantial differences in percentage of nonREM awakenings with recall. As Rechtschaffen (1973) pointed out, "PIPs do not appear responsible for dreaming per se" and "there is more dream activity in nonREM sleep than spike activity." Finally, PIP bursts in nonREM sleep typically last only a few seconds. They are never long enough to account for a long nonREM dream.

The bottom line is that long, detailed dreams – visually indistinguishable from the most elaborate of REM dreams – can be elicited on awakenings from nonREM sleep with little or no evidence of distinctive or near distinctive REM-like phenomena, including (1) low voltage, mixed frequency EEG; (2) rapid eye movements; (3) tonic or phasic EMG suppression; (4) saw-tooth waves; (5) PIPs (possibly representing PGO); and (6) penile erections. Therefore, dreaming is not dependent on REM sleep physiology, and the characteristics of dreaming in terms of REM-like intrusions into nonREM sleep have speculatively relied on relatively subtle indications of such intrusions with little evidence that they are related to nonREM dream reports. This evidence indicates that REM sleep phenomena are not necessary for dreaming.

The ASH also predicts that REM sleep is sufficient for the production of bizarre, distorted, disorganized dreams. This is because the hypothesis leads us to expect that excitation of the brainstem REM sleep generators will produce bizarre, distorted dreams. The evidence contradicts this prediction. In a large collection of studies, 17% of awakenings from REM sleep produced no dream reports. Also, dream reports may be sparse or absent on awakenings from REM periods of young children and when present, they are usually mundane and undistorted (Foulkes 1982b). Finally, as mentioned above, short REM reports are like typical nonREM reports – thought-like, mundane, undistorted (Foulkes & Schmidt 1983).

In short, contrary to the ASH, REM sleep is not necessary or sufficient for dreaming. Thus, unique REM sleep processes, such as discharge of the brainstem REM sleep generator, cannot be causes of dream production. Thus, the ASH is wrong.

HOBSON ET AL.'s neurophysiological theory of dream generation rests on what they call a mind/brain isomorphism. The theory is based in the proposition that dream distortion is caused by "random" discharge of neurons in the REM sleep generator. But we have no evidence that "disordered" meaning is caused by a disordered (temporally random) discharge pattern of individual neurons. In more general terms, we do not know the neural correlates of particular sequences of mentation, ordered or disordered. We do not also know that the orderliness of mental sequences is determined by the pattern of hindbrain stimulation of the forebrain. It is possible that mentation is relatively independent of hindbrain stimulation and is primarily determined by the response characteristics of the forebrain. In any case, our ignorance about these issues is very great. In view of such ignorance, the claim that disordered (incoherent) meaning is caused by disordered (random) discharge of individual pontine cells appears to be based on verbal similarity rather than empirical findings (Vogel 1978a). Indeed, the empirical findings listed above do not support this claim.

Aside from the particular dream theories, a major premise of

**HOBSON ET AL., SOLMS,** and **NIELSEN** is that brain physiology can tell us about the psychology of dream experiences.

The idea is that the nocturnal activation of previously identified "centers" for emotionality, symbolism, integration, etc., will implicate, or fail to implicate, these physiological processes in dreaming. But the evidence that these brain "centers" actually are substrates of psychological phenomena rests on the correlation of their activity with those phenomena in the first place. Thus, if dreams prove generally to be well integrated sequences of mental activity, while brain scans tell us that alleged integrative brain centers are quiescent, then we have found, not that dreams are disorganized, but rather that contemporary neurophysiology has not yet adequately identified the neural substrates of ideational integrity. This use of brain sciences to "prove" or "disprove" psychological findings or theories is, in principle, erroneous, and since that is the main premise of this account of dreaming, the account is fatally flawed. (Foulkes, personal communication)

Finally, let us apply these methodological comments about mind/brain to Freud's psychological dream hypothesis. This asserts that dreams are instigated by – and express – unconscious, unacceptable wishes. A test of this hypothesis will measure the empirical correlation between dream reports and unconscious, unacceptable wishes disguised in these reports. Neurophysiological findings cannot provide these data. Only reliable psychological data can. The fact that 100 years have passed without such a test suggests that Freud's hypothesis is empirically untestable and hence outside the realm of science.

# The pharmacology of threatening dreams

Lawrence J. Wichlinski

*Department of Psychology, Carleton College, Northfield, MN 55057.*
**lwichlin@carleton.edu**

**Abstract:** The pharmacological literature on negative dream experiences is reviewed with respect to Revonsuo's threat rehearsal theory of dreaming. Moderate support for the theory is found, although much more work is needed. Significant questions that remain include the precise role of acetylcholine in the generation of negative dream experiences and dissociations between the pharmacology of waking fear and anxiety and threatening dreams.

[REVONSUO]

**REVONSUO** has generated a provocative and persuasively argued theory for the biological function of dreaming. The author has limited this case (not unreasonably so) to the phenomenal level. What I propose to do in this commentary is to extend the hypothesis to biological and pharmacological levels of analysis and to evaluate whether the author's case is consistent with this literature.

Several predictions reasonably derive from the threat rehearsal hypothesis: (1) biological mechanisms should be found that subserve dreams of a threatening nature – if not in fact, then at least in principle; (2) agents which suppress those biological mechanisms should inhibit such dreams – either yielding dreams of a less threatening nature and/or eliminating dreaming altogether; (3) the withdrawal of such agents following repeated administration should lead to a reinstatement, perhaps even an exacerbation, of threatening dreams; and (4) there should be significant overlap between drugs that inhibit fear and anxiety in the waking life and those that inhibit fear and anxiety during dreams.

I will now briefly review the pharmacological evidence that is both consistent and inconsistent with **REVONSUO**'s hypothesis. Importantly, I will operate on his assumption that REM sleep is the typical, perhaps even optimal (though not exclusive) phase of sleep for dreams involving threat rehearsal. Because of its association with night terrors (Fisher et al. 1973) deep slow-wave sleep (SWS) will also play a prominent role in my analysis.

First of all, a number of agents with anti-anxiety properties suppress stage 4 slow-wave sleep, rapid-eye movement (REM) sleep,

or both. Tricyclic antidepressants (TCAs) (Vogel et al. 1990) MAO inhibitors (Vogel et al. 1990), and selective serotonin reuptake inhibitors (SSRIs) (Nicholson & Pascoe 1988) reduce REM sleep. Barbiturates and benzodiazepines suppress deep SWS and either suppress or delay REM onset (Declerck & Wauquier 1990; Hartmann 1976; Kay et al. 1976). Compared to the other classes of agents listed above, the benzodiazepines are less likely to reduce REM time (Hartmann 1968; Itil 1976).

More directly relevant to this paper, both phenelzine (an MAO inhibitor) and imipramine (a tricyclic antidepressant) have been found to be useful in reducing nightmares in patients with post-traumatic stress disorder (PTSD) (Kosten et al. 1991; Ross et al. 1994). However, the antidepressant nefazodone has no effect on REM sleep, and yet it is useful in reducing nightmares in PTSD sufferers (Gillin et al. 1999).

Benzodiazepines (e.g., diazepam) and TCAs (e.g., imipramine) suppress night terrors (Cooper 1987; Fisher et al. 1973). However, Fisher et al. (1973) found that benzodiazepines could suppress stage 4 sleep in some patients at doses that did not reduce night terrors. Also, Cooper (1987) reports a failure of diazepam in relieving night terrors in two clinical cases that, nevertheless, responded to imipramine.

A large percentage of patients with narcolepsy report vivid, frightening dreams (Lee et al. 1993). Narcolepsy is treated with TCAs and amphetamines, both of which suppress REM (Rechtschaffen & Maron 1964; Vogel et al. 1990). Since this syndrome is characterized by an atypically early onset of REM, the sleep mechanisms responsible for REM onset may also influence bad dreams. Alterations in monoamine function have been suggested in human narcolepsy. A canine model of narcolepsy implicates supersensitivity of muscarinic cholinergic receptors, specifically pontine M2 receptors (Tononi & Pompeiano 1995). TCAs are known for their anticholinergic properties (Baldessarini 1991), whereas the amphetamines have mixed effects at best on cholinergic neurotransmission (Cheney & Costa 1978). The above findings are significant since muscarinic cholinergic mechanisms are implicated in REM sleep generation and presumably dream induction as well (Baghdoyan & Lydic 1999; Hobson 1988b).

On the reverse side are agents that enhance REM and elicit bad dreams. For example, the antihypertensive reserpine enhances REM and induces nightmares in humans (Hartmann 1970). Along the lines of the previous paragraph, reserpine has been found to elevate acetylcholine, at least regionally in some species (see Palfai et al. 1986, for references).

The anxiogenic inverse agonist beta-CCE, which reduces chloride conductance at GABA/benzodiazepine receptors, substantially enhanced REM and SWS duration when administered to cats (Kaijima et al. 1984). In humans, a closely related inverse agonist, FG 7142, produced intense anxiety in two volunteers (Dorow et al. 1983). The authors did not investigate the dreams of subjects the night after taking this agent, so we know nothing about the subjective dream experiences produced by this class of agents.

Theory predicts that withdrawal from agents that suppress REM should lead to REM rebound, accompanied by an increase in the number and/or intensity of bad dreams. Consistent with this theory, Adams and Oswald (1989) found a five-fold increase in bad dreams accompanying REM rebound after withdrawal from the benzodiazepine triazolam (25 nights at 0.5 mg/night). Both Kales and Jacobson (1967) and Oswald and Priest (1965) showed that barbiturate withdrawal is accompanied by negative dream experiences. However, REM rebound following cessation of TCAs and low-potency phenothiazines (which also have anticholinergic properties) is not consistently accompanied by negative dream experiences (Kales & Vgontzas 1995).

A number of neurochemical systems have been implicated in anxiety and fear in animals, and in anxiety disorders in humans. These include (but are not limited to) norepinephrine (NE), serotonin, GABA, and cholecystokinin (CCK) (Charney & Bremner 1999). Any hypothesis attempting to link daytime anxiety with sleep-related anxiety must reconcile significant contradictions, such as the suppression of locus coeruleus (NE) firing during REM sleep (Aston-Jones & Bloom 1981) with the apparent hyperactivity of NE systems in waking anxiety and fear (Charney & Bremner 1999). Some degree of consistency, on the other hand, appears to exist for GABAergic mechanisms in this context. Blockade of GABA-mediated chloride conductance is linked to anxiety and fear (Charney & Bremner 1999) and there is abundant evidence, albeit indirect, for a GABA-negative mechanism in REM sleep onset, perhaps one involving an endogenous inverse agonist (Wichlinski 1996). This system may occur either upstream or downstream (or both) from the REM generator mediated by acetylcholine, for which there is compelling evidence (Hobson 1992b). This cholinergically mediated REM generator may itself be inhibited by some classes of antidepressants, including the tricyclic antidepressants and the MAO inhibitors.

It is clear that REM, stage 4 SWS, negative dream experiences and daytime anxiety are dissociable, viewed either phenomenologically or pharmacologically. Suppression of REM and/or stage 4 SWS is neither necessary nor sufficient for the elimination of negative dream experiences or daytime anxiety and fear. Moreover, those agents that are successful at eliminating daytime anxiety do not always suppress negative dream experiences. Nevertheless, pharmacological inhibition of stage 4 SWS and REM sleep tends to go hand in hand with inhibition of negative dream experiences and suppression of daytime anxiety.

The mechanisms that underlie threatening dreams most likely overlap those responsible for REM and stage 4 SWS. Moreover, it is plausible that the premature onset of normal REM mechanisms, including REM related threatening dreams, may account for night terrors seen in stage 4 SWS (Arkin 1978). Although the pharmacological data are limited and do not directly address REVONSUO's hypothesis, the preponderance of findings thus far are consistent with it. Nevertheless, significant questions remain:

1. What is the precise role of acetylcholine in generating dreams, especially threatening dreams? Why is it that antimuscarinic agents have hallucinatory properties when theory predicts the exact opposite? Is it due to differential muscarinic subtypes or differing responses to associated GABAergic neurons (Perry & Perry 1995)?

2. Why aren't centrally acting anticholinergic agents useful in treating both daytime anxiety disorders and parasomnias with negative content (e.g., dream anxiety attacks, night terrors)? Or have they simply not been adequately tested for these capacities?

3. How can activation of noradrenergic systems account for both daytime anxiety and nighttime threatening dreams if the locus coeruleus shuts down during REM sleep?

It is likely that the neural mechanisms underlying threat perception – like most behaviors studied – involve a number of different neurotransmitter systems. Perhaps it is asking too much of the brain to expect it to use the same mechanisms to encode the experience of fear and anxiety in the waking state as in the sleeping state. Finally, it is useful to be reminded that anxiety and fear are not unitary states – either across or within species; therefore, attempts to generate models of brain-behavior relationships which fail to discriminate the various types of fear and types of anxiety in different species under specific conditions are bound to invite confusion (Kagan 1998).

REVONSUO has offered a provocative hypothesis for the biological function of dreaming. Much, though certainly not all, of the pharmacological data on sleep and negative dream experiences are consistent with it. Significant questions remain and further work is needed on the pharmacology of threatening dreams before a comprehensive evaluation can be undertaken on this aspect of the theory. Nevertheless, this hypothesis represents a crucial step toward an enhanced understanding of the basic mechanisms involved in dreaming and in clinical conditions associated with negative dream experiences.

# Threat perceptions and avoidance in recurrent dreams

A. Zadra[a] and D. C. Donderi[b]

[a]Department of Psychology, University of Montreal, C.P. 6128 succ. centre-ville, Montreal, Quebec, Canada H3C 3J7; [b]Department of Psychology, McGill University, Montreal, Quebec, Canada I I3A 1B1.
zadraa@psy.umontreal.ca    donderi@hebb.psych.mcgill.ca

**Abstract:** Revonsuo argues that the biological function of dreaming is to simulate threatening events and to rehearse threat avoidance behaviors. He views recurrent dreams as an example of this function. We present data and clinical observations suggesting that (1) many types of recurrent dreams do not include threat perceptions; (2) the nature of the threat perceptions that do occur in recurrent dreams are not always realistic; and (3) successful avoidance responses are absent from most recurrent dreams and possibly nightmares.
[HOBSON ET AL.; REVONSUO]

REVONSUO agrees with Domhoff's (1996) position that to be taken seriously, a theory of dreaming must account for the repetition dimension in dreams and asserts that his evolutionary hypothesis of the function of dreaming explains this dimension as the paradigm case of threat simulation in dreams.

Dreaming is viewed as both an organized and selective simulation of the world. A particularly well-organized form of dream content is the recurrent dream which is distinguished by its complete repetition as a remembered experience (Brown & Donderi 1986). Many kinds of dream theories (e.g., Gestaltist, object-relations, Jungian) converge in their view that recurrent dreams are associated with a lack of progress in recognizing and resolving conflicts in the dreamer's life and that the cessation of a recurring dream indicates that the conflict has been successfully dealt with. Research results support the generic clinical dream theory that recurrent dreams are associated with the presence of unresolved stressors. Studies have shown that in both late teenagers and older adults, recurrent dreams are accompanied by negative dream content in everyday dreams and that they are associated with a relative deficit in psychological well-being (Brown & Donderi 1986; Robbins & Houshi 1983; Zadra et al. 1998). Moreover, the cessation of a previously recurrent dream in adulthood is associated with a positive rebound effect on well-being (Brown & Donderi 1986).

REVONSUO cites a study (Robbins & Houshi 1983) showing that the most frequently reported theme in a student sample of recurrent dreams is one in which the dreamer is being chased (43% of all dreams). However, most of these recurrent dreams had first appeared in childhood. Zadra (1996) conducted content analyses of 110 recurrent dreams from adulthood (i.e., reported as having first occurred after the age of 18) and 53 recurrent dreams from childhood (i.e., ceased to recur before the age of 12). Although chase and pursuit dreams were the most frequently reported theme in both samples, they represented less than 15% of the adult recurrent dreams and 42% of the childhood ones. A broader category encompassing all themes in which the dreamer is in some kind of danger has been found to characterize approximately 40% of recurrent dreams (Cartwright & Romanek 1978; Robbins & Houshi 1983). What is the thematic content of the remaining 50 to 60% of recurrent dreams? The other themes include having difficulties with house maintenance, losing one's teeth, discovering or exploring new rooms in a house, driving a car that is out of control, being unable to find a private toilet, and flying. In addition, 10% of all recurrent dreams contain only pleasant emotions while another 5% are described as containing no affect (Zadra 1996). Our content analyses have also shown that about 25% of negatively-toned recurrent dreams contain emotions other than fear and apprehension (e.g., sadness, anger, confusion, guilt). Finally, the content of over 30% of all recurrent dreams is idiosyncratic and in great part unrelated to any threats. Based on the range of thematic content and affective expression represented in recurrent dreams, it may be misleading to conclude that most recurrent dreams are dissociated from the dreamer's current concerns. It would appear that a great many recurrent dreams are not realistic rehearsals of a threatening event but rather pictorial metaphors of current concerns.

Nightmares are highly unpleasant dreams which, by definition, awaken the sleeper. Although we have not conducted a systematic content analysis of the several hundred nightmares we collected from non-traumatized adults, it is our clear impression that the overwhelming majority of nightmares contain threat perceptions but not evolutionarily adaptive threat-avoidance programs. In fact, many contain no appropriate behavioral response beyond a relatively straightforward fear reaction: dreamers awaken either while trying (unsuccessfully) to escape or at the moment they are caught or attacked. Furthermore, nightmares have been described where subjects experience the total destruction of their body. These dreams appear to simulate failure rather than any form of adaptive response. As for the nature of the threats themselves, many nightmares do not contain real-life threats based on the human ancestral environment but involve unrealistic and unusual circumstances (e.g., Zadra & Pihl 1997). Nightmares may very well be one type of dream which reveals the prime importance of threat simulation mechanisms. However, **REVONSUO**'s theory does not adequately account for the bizarre and unrealistic content of many nightmares and the fact that most do not contain reasonable and realistic adaptive behaviors. A detailed content analysis of the threatening events in nightmares as well as of the ensuing responses is needed to clarify this issue. Such an analysis is now being conducted by **REVONSUO** as well as our own research group.

In sum, the data indicate that many recurrent dreams, and possibly nightmares, do not include situations critical for physical survival and reproductive success. Even in those cases where they do, these dreams rarely contain constructive threat avoidance behavior or coping strategies against threats. The simulation of threat recognition during REM sleep may very well fulfill the goal of priming an amygdalocortical network to perform rapid and appropriate emotional evaluation of the potential danger. However, the second stage (i.e., rapid selection of an appropriate behavioral response and its instantiation) appears to be lacking.

That dreams can be realistic is not in doubt. Following her medal-winning performance at the 1988 Winter Olympics, Canadian figure skater Elizabeth Manley explained that she had been having difficulty with her routine, and was worried about it. The night before, she dreamed that she did the entire routine flawlessly, and when it came time to perform the routine before the judges, it was flawless enough for second place. Tholey (1991) provides evidence for the use of lucid dreaming in sports training. Real rehearsal under lucid or semi-lucid control clearly plays a role in dreams. **HOBSON ET AL.** discuss neuromodulation of the limbic-prefrontal axis (sects. 3.2.4, 3.2.7) as an important determinant of dream emotion and direction. They suggest that relative inhibition of the prefrontal cortex and activation of the limbic and paralimbic cortex leads to non-directed, emotionally valent

imagery in dreams; lucid dreaming, by contrast, involves relatively less inhibition of the prefrontal cortex, relatively greater control over dream imagery, and self-awareness of dreaming. These are conditions where volition can guide imagery and produce conscious rehearsal. This is the raison-d'être of dreams, according to **REVONSUO**'s theory; we see it as one, and not the most frequent, process that occurs during dreaming sleep.

**REVONSUO** supposes that preliterate humans dreamed realistic threat dreams that rehearsed their control of a dangerous physical environment. With rare, athletic exceptions, the threat environment we now face is more symbolic than physical. Most threats involve language. Through metaphor and simile, language allows us to both manipulate our own mental images, and to try and control the mental images of others. Freed from goal-directing prefrontal control by the inhibitory processes associated with REM sleep, it may be that language mechanisms act on emotionally valenced memories to create the unpredictable metaphors and similes of dreams: a visual form of metaphoric and symbolic emotional free association. Many recurrent dreams and nightmares are unrealistic because they are imaginally metaphoric and free-associative; just like our own language when freed from goal-directed constraint. If we regard Elizabeth Manley's performance anxiety about figure skating in the winter Olympics as the modern equivalent of an evolutionary threat, then her confirmatory example fits neatly into **REVONSUO**'s theory. It was literally a perceptually and behaviorally realistic rehearsal of a threatening event. **REVONSUO**'s theory explains Elizabeth Manley-type performance dreams, but does not explain the far more numerous anxiety dreams in which the dream images are unrealistic. Perhaps **REVONSUO**'s answer will be that, in this type of dream, the system does not become fully activated. This answer dismisses most dreams as evolutionarily useless epiphenomena that just waste time until a dreamer, under the immediate threat of losing a medal, losing a job, or losing out to a rival, realistically rehearses skating a program, confronting the boss, or carrying out a murder.

Three questions not solved by **REVONSUO**'s theory are: (1) why are realistic threat perceptions absent from many if not most recurrent dreams? (2) Why are efficient or successful avoidance responses absent from most recurrent dreams and possibly nightmares? and (3) Why is the balance of mental activity during dreaming more like metaphoric and symbolic free association, and the balance of mental activity during waking more like goal directed thought? The answer to (3) may be **HOBSON ET AL.**'s distinction between limbic and forebrain-directed dreaming; differences which lie along the neuromodulation (aminergic-cholinergic) dimension of their three-aspect model of physiological changes in dreams. According to **HOBSON ET AL.** the prefrontal cortex, vitally involved in the goal-directed planning of behavior and therefore in the streaming and directing of mental content towards a goal, is relatively inhibited during normal dream states. Limbic system activation of remembered mental content, unconstrained by goal needs, fills dream content with emotionally loaded but behaviorally non-sequential imagery.

# Authors' Responses

## Dream science 2000: A response to commentaries on *Dreaming and the brain*

J. Allan Hobson, Edward F. Pace-Schott, and Robert Stickgold

*Department of Psychiatry, Harvard Medical School, Boston, MA 02115.*
{allan_hobson; edward_schott; robert_stickgold}@hms.harvard.edu
www.earthlink.net/~sleeplab

**Abstract:** Definitions of dreaming are not required to map formal features of mental activity onto brain measures. While dreaming occurs during all stages of sleep, intense dreaming is largely confined to REM. Forebrain structures and many neurotransmitters can contribute to sleep and dreaming without negating brainstem and aminergic-cholinergic control mechanisms. Reductionism is essential to science and AIM has considerable heuristic value. Recent findings support sleep's role in learning and memory. Emerging technologies may address long-standing issues in sleep and dream research.

## HR1. Introduction

"Why are academic debates so acrimonious?" After a long pause the joke teller then answers his own question. "Because there is so little at stake!"

In the case of *Dreaming and the brain,* it is obvious that the answer must be different. Debates in this domain can be acrimonious. But it is because there is so much at stake! No interested party could or should easily cede the adequacy of any general theory of brain-mind states and especially those that propose to reduce aspects of dream phenomenology to neurobiology. The stakes are indeed high in this game. So it is no surprise that the interests and passions of the players are intense.

When, in 1986, we wrote our *BBS* target article on REM sleep neurobiology (Hobson et al. 1986) we were astonished by the depth and range of the responses. That experience taught us that we had much more to learn from our critics and our friends than we had ever imagined. This time, that principle has been even more solidly upheld, in part because in addition to our own target article, *Dreaming and the brain,* there are four other papers on closely related topics. The field of sleep and dream research is obviously undergoing an upheaval; a millennial repositioning that we see as a renaissance; and we are happy to be a part of it. It is humbling to realize just how much creative, thoughtful, and passionate dialogue has been generated by our work. Humbling, because so many talented people have taken the time to comment. And humbling because we know our response to the commentaries can hardly do them justice. But precisely because there is so much at stake, we will do our best.

We begin our response by thanking especially those commentators who have taken such strongly critical positions against our arguments. They give us the chance to explain and clarify even if we cannot convince. We also appreciate the support of many commentators who offer alternative hypotheses. They give us the chance to amplify and extend the theories we have put forth and to broaden their experimental testing ground. We hope it is not too immodest to suggest that this special issue of *BBS* shows clearly that Dream Science in the year 2000 has a strongly established beachhead. We trust that the discussion, debate, and even the controversy that our article has inspired is both healthy and constructive of the future growth of our field.

We have organized our responsive comments in what seems to us a logical order. First (and foremost) we address definitional issues: What is dreaming and how can it be characterized and measured? Second, we take up the debate regarding the location of dreaming in the sleep-wake cycle to reiterate and clarify our recognition and understanding of the strong correlation of dream features with REM. Third, we discuss the commentaries concerning brain underpinnings of this correlation as a prelude to developing our fourth theme, the concept of isomorphism and its instantiation in the activation-synthesis and AIM models. Fifth (and most future oriented), we address the question of function with special reference to learning and memory enhancement.

## HR2. Formal approach to phenomenology

### HR2.1. Defining dreaming

The definition of dreaming remains unclear. As pointed out in the target article by NIELSEN as well as the commentaries by **Antrobus** and **Pagel,** there is no clearly agreed upon definition of what a dream is, and as was made clear in a panel discussion at the 1999 meetings of the Association of Professional Sleep Societies, we are not even close to agreement. Our own preference is to reserve the term "dream" for mentation during sleep which has most, if not all, of the following formal features: hallucination, delusion, narrative structure, hyperemotionality, and bizarreness. This corresponds to NIELSEN's innermost circle of dream definition, "apex dreaming." We prefer to use the term "sleep mentation" for mental activity during sleep that does not meet this criterion. In doing so, we are not attempting to bind others to our definitions. Instead, we only wish to make clear what we are referring to by these terms. As will be clear in what follows, we do not restrict our discussion to just this "apex" dreaming, or even to sleep mentation. We can instead measure more detailed formal features of mental activity – like hallucinatory imagery or thinking – and map them onto our brain measures across wake-sleep states.

As we have emphasized in our writings on consciousness, we feel strongly that there is no way for cognitive neuroscience to sidestep first-person accounts of subjective experience (Hobson 1999b). If the psychophysiology and neuropsychology of mental life are to advance, we must develop the means of characterizing and quantifying the subjective experience of conscious states. The recent spate of brain imaging articles makes it clear that no amount of technical sophistication can compensate for neglect of exactly what psychological features the neurobiological data are asked to explain. For example, now that we can subtract blood flow measures in NREM and REM from those in waking, it is essential to characterize and measure the experiential aspects of waking and the two sleep states so that an isomorphic subtraction can be performed at the psychological level as well. But just as it is no longer acceptable to equate REM sleep with dreaming, it is also no longer possible to regard waking as a single state. Here again, we suggest examining discrete components of waking, like self-reflective awareness.

### HR2.2. Self-reflective awareness and dream lucidity

An example that appeals to us is **LaBerge** and **Kahan**'s favorite subject, lucid dreaming. We suggest testing the hypothesis that self-reflective awareness will fluctuate with rises and falls in blood flow to the dorsolateral prefrontal cortex (DLPFC). In a PET or fMRI study, we would thus predict the following general result: both measures (i.e., self-reflective awareness and DLPFC blood flow) are high in waking, low in NREM sleep, and lowest in REM sleep. We would further predict that, within waking or REM sleep, increases in self-reflective awareness will correlate with increases in blood flow to the DLPFC and vice versa. To do this experiment, valid and reliable quantitative measures of self-reflective awareness akin to our bizarreness measure, such as the SR scale (Darling et al. 1993; Purcell et al. 1986; 1993) or the DRS scale (Kahan 1994) would need to be applied. In our view, scales based on affirmative probes (focused inquiries about the presence of subtle psychological features) are most likely to have the requisite sensitivity to produce success. Indeed, Kahan (1994) has shown that the self-rated DRS produces a higher incidence of metacognitive activity in dreaming than does the judge-rated SR scale. Using lucid dreaming to increase self-reflective awareness during REM, and fantasies (Williams et al. 1992) to decrease it during wake, would permit testing of this hypothesis.

We like this example not only because it identifies a feasible experiment and provides guidelines about how to do it but also because it illustrates two critical aspects of scientific phenomenology: (1) the focus on the form rather than the content of mental activity, and (2) the precise and explicit definition of those formal aspects of mental state which have a reasonable chance both of being measurable and of correlating well with brain activity as currently measured.

### HR2.3. Dream bizarreness

Because these criteria (formal focus, measurable variables) are so rarely fulfilled, it is easy to understand why so many investigators cannot confidently distinguish between the mental content of waking, NREM, and REM sleep. To our mind, **Clancey**'s commentary is clearly focused on the formal level: Compared to waking, dreaming suffers from difficulties in conceiving sequences (which we measure as plot discontinuity; Stickgold et al. 1994b) and in coordinating across modalities (which we measure as dream incongruity; **HOBSON ET AL.** 1987; Rittenhouse et al. 1994; Sutton et al. 1994a; 1994b). Most of the studies cited by **Vogel, Domhoff,** and **Kramer** don't even try to achieve the goal of defining and differentiating formal features. For example, Vogel's assertion that dreaming is "banal" is based on Snyder's early work (Snyder 1970). Because one of us (J.A.H.) was Snyder's collaborator at the time that study was done, it can be confidently asserted that no effort was made to define, identify, or quantify bizarreness at the microscopic level, an effort that is necessary to achieve sensitivity. While it may well be true that many dreams are concerned with mundane, everyday themes, they are interconnected in an incongruous and discontinuous manner. This feature of dreams has been recognized for hundreds of years, as the following poem, written by John Dryden in 1700, makes clear:

> Dreams are but interludes which fancy makes
> When monarch reason sleeps, this mimic wakes.
> Compounds a medley of disjointed things
> A mob of cobblers and a court of Kings.

This mob of cobblers and court of kings that are compounded in the dream are banal enough (especially at the time when Dryden wrote his poem). However, we just don't expect to find them together since they were normally segregated by social station. This is classic dream bizarreness.

Notice, too, that Dryden – being a scientific rationalist – speculated presciently about the cause of dream bizarreness, namely a decline in waking rational capacity (monarch reason sleeps), and the rise in mimicry (the fanciful imitation of reality). Now, 300 years later, we can go a step further and equate "monarch reason" with top-down frontal lobe control of the waking brain and the "mimic" as bottom-up subcortical subversion of the dreaming brain.

**Vogel**'s use of the dream data of the 1960s is a bit like Cajal's predecessors staining the brain with the wrong reagent, looking at it under the low power of the microscope, and concluding that all of the cells are connected to one another in a syncytium! The brain's cellular details and the mind's psychological details only reveal themselves when special techniques, like Golgi's stain for cell processes and the orientational instability index of bizarreness (Sutton et al. 1994b), are applied. These two approaches are isomorphic in that they are microscopic and seek to identify the structural and functional units of brain (neurons and circuits) and mind (items of content and their association). It is worth noting in regard to the alleged similarities between wake and sleep mentation that in a recently reported analysis of 1595 reports, only 2 of 872 wake reports were incorrectly judged as sleep mentation reports, with an additional 9 judged as sleep onset reports, giving an overall accuracy of 99% for judging wake reports (Stickgold et al. 1998b). Similarly, only 7 or 236 REM reports were misjudged wake mentation reports, for an accuracy of 97%. Clearly, it is not so difficult to distinguish the two types of reports.

### HR2.4. Dream emotion

Regarding **Panksepp**'s suggestion, the same formal and quantitative criteria must be applied to dream emotion. It is not enough to say that dream emotion is an important dimension of dreaming or that it has been neglected. We need to define and measure it. Having always felt that emotion was a crucial formal feature of subjective experience, we have worked hard – though admittedly not yet hard enough – to identify and measure emotion in reports of subjective experience (Merritt et al. 1994; Sutton et al. 1994a). We have found that affirmative probes for dream emotion are ten times more sensitive than unstructured inquiries. Continuing the microscopic power analogy, this is like switching from low to high power. Now we need to go further in pursuit of the microscopic character of dream emotion and its relationship to dream plot. We thank Panksepp for his suggestion of a "seeking" impulse linking cognition and emotion. This idea fits nicely with the finding that REM sleep dreams are always animated (McCarley & Hoffman 1981; Porte & Hobson 1996). But it doesn't explain why the dreamers are so often lost, that robust formal dream feature of which the orientational instability factor is explanatory.

### HR2.5. Dream content and dream meaning

Our interest in dream emotion gives us common cause with commentators like **Greenberg, Hartmann, Kramer, Mann** cia, **Shevrin & Eiser,** and **SOLMS** who chide us for ignoring what they take to be the personal meaningfulness of dreams. We have always regarded dreaming as emotionally salient even before we knew that the amygdala, the adjacent parahippocampal cortex, and the anterior cingulate were selectively activated in REM sleep. But because we do not yet see a way of actively measuring salience (not to mention the even more slippery notion of meaning), we regard it as still outside the reach of dream science. We thus find ourselves entirely comfortable with **Vogel**'s emphatically negative conclusion about the scientific status of psychoanalysis. Vogel asserts that only a reliable psychological test could confirm or refute Freud's theory. "The fact that 100 years have passed without such a test suggests that Freud's dream hypothesis is empirically untestable and hence outside the realm of science."

We nonetheless believe that neurobiology can – and does – inform and weaken many tenets of psychoanalytic dream theory by providing simpler and quite different explanations of many formal dream features such as:

1. Instigation of dreaming (forebrain activation in sleep vs. release of unconscious wishes).

2. Vivid visual imagery (activation of visual association cortex vs. regression to the sensory side).

3. Memory deficits (organic amnesia vs. repression).

4. Diminished cognitive capacity (inactivation of dorsolateral prefrontal cortex and aminergic demodulation vs. disguise and censorship of unacceptable unconscious wishes).

5. Bizarreness (disorientation and organic amnesia vs. disguise-censorship).

6. Intense emotion, especially fear/anxiety (activation of the amygdala vs. symptomatic defensiveness).

As we map these dream features to their possible neurobiological roots, the very existence of explanations at the most complex psychological level becomes uncertain, and the likelihood that these psychoanalytic explanations have any true validity becomes greatly diminished. Readers who are interested in a more detailed discussion of these issues are referred to our recent NeuroPsychoanalysis target article and response to the commentaries of Braun, Solms, Reiser, Nersessian, and Gilmore (Hobson & Pace-Schott 1999).

We appeal to our psychodynamically inclined colleagues to develop measures of emotional salience that can put their psychological theories to the experimental test using neurobiological tools in conjunction with quantitative phenomenology. **Cartwright**'s focus on emotional adaptation, **Greenberg**'s emphasis on the transparent meaningfulness of manifest dream content, and **Hartmann**'s construct of thick and thin boundaries are all steps in the right direction but because they only partially fulfill the two criteria noted above (i.e., focus on form vs. content; definable, measurable indices) they cannot yet be fully integrated into a program of brain-mind integration.

### HR2.6. Methodological problems

The difficulty with memory that we consider to be a fundamental formal feature of dreaming does indeed pose a formidable obstacle to the development of a science of dream phenomenology that depends upon dream reporting. Having attempted our own study comparing dreams and fantasies in the same subjects (Williams et al. 1992), we certainly agree with **Chapman & Underwood** when they

say that even the recollection of waking fantasies is not all that good. We believe that one reason for this difficulty is that fantasy is a background mental state which has to compete with foreground input-output processing. Lacking the context of external space-time may contribute to the problem of recall in both day dreaming and night dreaming just as it contributes to other dream features (**Blagrove** and **Bednar**). But night dreaming is unlike day dreaming because it has nothing with which to compete. As Rechtschaffen (1978) has shown, dreaming is impressively single-minded so that competition and attentional allocation cannot be the issue. To us, it is more likely that the limiting factors are the memory deficit within sleep and the disruption of the ongoing dream experience by the state change that is required to generate reports. As an example of a within-sleep deficit, Braun (1999) suggests that the poor memory for dreams results from a lack of contextual cues encoded simultaneously with the dream experience, a deficiency which is compounded by sleep-related deficits in working memory.

### HR2.7. Sleep inertia and motivational factors

It is **Feinberg** who puts his finger on an important variation on this theme – the difficulty of awakening subjects, especially from non-REM sleep, early in the night. We know, from personal experience, that subjects awakened from stage IV may generate long rambling "dream" reports while their EEG shows persistent high voltage slow wave activity (unpublished observation). Some of our subjects, having spontaneously awakened from REM and given long, elaborate dream reports, reentered REM directly on going back to sleep (unpublished observation). Were they really awake? Was their memory of antecedent mentation veridical? Or did they too confabulate from a dissociated state combining, in this case, REM and wake state features. One safeguard against the inclusion of this very probably confabulated material is to insist that only those laboratory awakenings which have EEG evidence of full arousal be included in data sets aiming to compare mental content between states like NREM and REM.

Another approach, too rarely used, is subject training. The longer a subject is studied – either at home or in the sleep lab – the more that subject can easily awaken and report reliably and confidently. Subject and scorer bias is easily controlled by double-blind designs. When this approach was used in our 14-day home-based Nightcap controlled study, the 16 subjects generated 1,800 reports that blindly scoring judges correctly identified as descriptions of subjective experience in waking, sleep onset or REM sleep (Stickgold et al. 1998b). The only state that posed difficulty was sleep onset and that, not surprisingly, tended to be confused with wake (Stickgold et al. 1998b). This third-person accuracy raises questions regarding **Antrobus**'s assertion that REM dreams increasingly come to resemble NREM mentation due to reduced reporter motivation in longer sleep laboratory protocols and again suggests that the naturalistic, home-based collection of reports yields a more accurate longitudinal sampling of sleep state-related differences in mentation (**HOBSON ET AL**). Given all of the above problems, it is little wonder that there are discrepancies in the data. Our way of avoiding discouraging and erroneously negative conclusions is to use a very large, repeated measure database with each subject as his or her own control and focus on such internal consistency as emerges. As our most recent studies indicate, this approach pays off handsomely (Stickgold et al. 1998b, and submitted; Fosse et al., in press).

### HR2.8. Novel approaches

To those commentators who suggest novel ways of conceptualizing dream phenomenology, we offer thanks and encouragement. **Conduit et al.**'s proposal to investigate attentional aspects of dreaming seem to us particularly attractive. We think that a scale that included an affirmative probe for surprise and for subtle shifts in attention would almost certainly be rewarding. This is because shifts in attention are predictable from **Morrison & Sanfords**'s PGO-orienting response theory (Sanford et al. 1993) and from the high scores of our own orientational instability measure (Sutton et al. 1994b). As a corollary, we would also suggest finding a way to ask subjects if they can attend to dream images or if, as we might hypothesize, their attention is instead being seized by those intrusive images.

We also like **Humphrey**'s whimsical commentary suggesting that dreaming is playful or even amusing, an idea already entertained by us (Hobson 1988b). We quote here Victor Turner's (1950) definition of play and ask readers if it sounds anything like their experience in dreaming:

> (Play's) metamessages are composed of a potpourri of apparently incongruous elements: products of both hemispheres are juxtaposed and intermingled. Passages of seemingly wholly rational thought jostle in a Joycean or surreal manner . . . Play is neither ritual action nor meditation, nor is it merely vegetative, nor is it "just having fun"; it also has a good deal of ergotropic and agonistic aggressivity.

Notice the key words which denote dreamlike bizarreness (B) and emotionality (E): potpourri (B); incongruous (B); juxtaposed (B); intermingled (B); jostle (B); Joycean (B); surrealist (B); ergotropic (E); agonistic (E); and aggressivity (E).

## HR3. The NREM-REM controversy

Many commentators have again raised the question of dreaming in REM versus NREM sleep. Why does this question continue to rise, phoenix-like, from its ashes? There are several reasons, many made apparent by our commentators. We will try to structure some of the claims and clarify our position on them.

### HR3.1. Does dreaming occur during all stages of sleep?

Perhaps we should simply answer this question, Yes. But it is an ill-formed question, rife with ambiguity. Aside from the question of the definition of dreaming (e.g., do dreams, as we define them, occur in stage IV?), the very structure of the sentence is ambiguous. For example, it is very likely that English is spoken in every country of the world. For this claim to be true, only one person in each country would have to speak English. Still, many groups would take umbrage at this statement, because it has a tinge of suggesting that English is an important language or even the main language of their country. Likewise, what does it mean that we dream in all sleep stages? If this statement simply means that dreaming, even by our restricted definition, is capable of occurring in all stages of sleep, and, with at least a mea-

surable frequency does occur in all stages, then we agree, and know of no researcher who disagrees. Thus we explicitly reject **Vogel**'s claim that our theories predict that "dreaming will occur only in REM sleep." At the same time, we absolutely reject, as we believe do most other authors and all research, any suggestion that there are not major and important differences in normative sleep mentation across stages.

### HR3.2. On average, are mentation reports from REM and NREM sleep substantially different from each other?

Again, we answer, Yes. All studies show that REM sleep mentation, averaged across the night, is reported more frequently, at greater length, and with a higher prevalence of the formal dream features used in our definition of dreaming than is NREM mentation. Indeed, whether judged by length, bizarreness, narrative style, or just generic "dreaminess," judges can readily distinguish a collection of REM reports from a collection of NREM reports (see **HOBSON ET AL.**).

**Antrobus** notes that few studies have looked for differences in frequency of cognitive features across *both* states and features, correctly pointing out that "questions about across-state differences in dreaming carry assumptions about whether the pattern of these features is sustained or whether, for example, dreams in some states are more visual and in others more verbal."

We have now begun to report the results of our 1,800 report database that examines over 25 dream features across waking, sleep onset, REM, and NREM sleep in 16 subjects studied for 14 consecutive days and nights using our Nightcap sleep monitor and experiencing sampling techniques (Fosse et al., in press; Stickgold et al. 1998b; submitted). In our second report on this database, Fosse et al. (2001), we have shown that the frequency of reported thoughts in mentation reports decreases from waking to sleep onset and then further in NREM and finally REM sleep, while the frequency of hallucinations follows a reciprocal path (Fig. R1).

These data clearly indicate that a cardinal dream feature, hallucinatory imagery, and a cardinal wake feature, thinking, when considered together all but qualitatively distinguish the extreme ends of the state spectrum and co-vary negatively across that spectrum. We ask rhetorically: What clearer proof could anyone seek for the validity of the brain-mind state paradigm?

### HR3.3. Are some NREM reports every bit as dreamy as REM reports?

Again, we answer with a qualified, Yes. Some NREM reports are clearly as dreamy as the average REM report. But as the most "dreamy" REM report? On this we are agnostic. We have collected REM mentation reports 3,000 words in length. We have never seen a NREM report close to this long. In our more recent study, the longest of 229 REM reports had a TRC ("total recall count," Antrobus 1983) of 1,723 words, while the longest of 165 NREM reports was only 623 words (Stickgold et al., submitted). Perhaps NREM reports will never reach these exceptional lengths. While saying this, we would like to point out the literature suggesting that as the night progresses and REM reports become

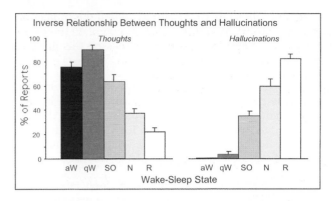

Figure HR1. Prevalence of thoughts and hallucinations in reports collected across wake-sleep states. As subjects moved from quiet wake through sleep onset and NREM into REM, the frequency of thoughts in mentation reports dropped 4-fold while the frequency of hallucinations rose more than 10-fold. Thus, separate mechanisms must underlie their appearance across the wake-sleep cycle. aW - active wake; qW - quiet wake; SO - sleep onset; N - NREM; R - REM. Error bars = s.e.m. across subjects (From Fosse et al., 2001).

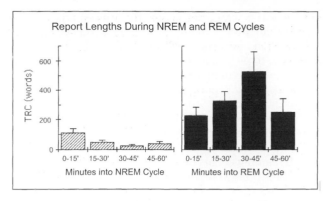

Figure HR2. Temporal distribution of report lengths in NREM and REM cycles. While NREM reports peaked in length at the start of NREM periods, REM reports were of maximal length when collected 30–45 min. into the REM period. Thus, the differences observed between REM and NREM dreams will be minimal when reports are collected early into the cycle, and maximal when collected 30–45 min. into each period. Times indicate minutes after the start of a NREM (left) or REM (right) period. Error bars = s.e.m. (From Stickgold et al. 1994a).

longer and more dreamy, so do NREM reports (Antrobus et al. 1995). In a recent study (Fosse et al., in preparation), we have shown that the prevalence of thoughts and hallucinations in NREM reports from late in the night is nearly identical to that found for REM reports from early in the night. **Antrobus** (see, Antrobus et al. 1995) also raises this issue, and it is intimately connected to **Borbély**'s (1982) two factor model in which both a circadian process (C) and a sleep process (S) contribute to brain-mind state control. This is an important point, because the outcome at the cellular and molecular level of the interaction between these two processes is largely unpredictable. The reason for this unpredictability is that most of the relevant data has been collected from cats, who are notoriously non-circadian in their sleep patterns. Thus, while we frequently act as if it were not the case, the physiological and neuromodulatory studies done in the cat provide no hint of how the underlying physiology and neuromodulation of REM and NREM sleep might vary across the 8 hr of consolidated sleep seen in humans.

A second point here involves the tendency to assume that both physiology and psychology are constant across a given REM or NREM period. This clearly is not the case. Indeed, the change in firing rates at the REM-NREM border in pontine neurons comes closer to describing a sinusoidal curve than a step function (HOBSON ET AL. 1975), and we have reported that sleep mentation report lengths were shortest during the first 15 min in REM, but longest during the first 15 min in NREM (Fig. R2; Stickgold et al. 1994a) These results are, in fact, in agreement with those of **Feinberg** who reports in his commentary a 27% increase in REM report lengths (325 to 413) between 5 and 10 min into REM. For comparison, in an analysis of 88 REM reports, we found a 45% increase between the first and second 15 min of REM (Stickgold et al. 1994a) while in our more recent study, with 229 REM reports, we found a 53% increase (Stickgold et al. 2001).

Together, these findings make clear the importance of defining the times at which reports are collected for comparison. If the physiological transitions between REM and NREM are gradual, and if the physiological characteristics

of the states vary across the night, it should not be surprising to find similar changes in mentation reports.

### HR3.4. Is intense dreaming confined largely to REM sleep?

We believe that for our definition of dreaming and for the first 4 to 6 hrs of the night, the answer, again, is Yes. **Antrobus**'s classic study of REM and NREM reports from 77 subjects (Antrobus 1983) yielded REM reports with a median length (TRC) of 44 words and NREM reports with a median length of only 7 words. All told, 46% of the NREM reports had TRCs of 5 words or less, and over 70% had TRCs of a dozen words or less. It seems highly unlikely that any of these would meet our definition of a dream.

Figure R3 shows the distribution of report lengths (as TRC) from these reports (Antrobus, personal communication). Median lengths for the two sets of reports differed by

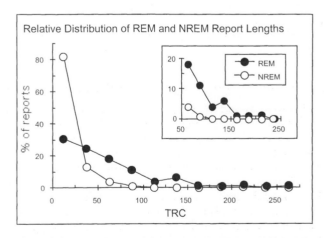

Figure HR3. Relative numbers of REM and NREM with different TRCs. Insert: All told, 43% of REM reports but only 4% of NREM reports were over 50 words in length. Thus most NREM reports appear too short to describe as classically intense REM dreams (J. S. Antrobus, personal communication).

a factor of 6.4, with two-thirds of the NREM reports being shorter than 83% of the REM reports. Similar differences were seen in our own study (Stickgold et al. 1994a). Thus, if we were to pick an arbitrary cutoff of 50 words as a crude estimate of the minimum length in Antrobus's (personal communication) reports that would reflect a dream meeting our definition, 43% of the REM reports and only 4% of the NREM reports would qualify as dreams.

In summary, we state our substantial agreement with the following conclusions:

1. Dreaming occurs during all stages of sleep, and
2. Some NREM reports are every bit as dreamy as REM reports, but
3. On average, mentation reports from REM and NREM sleep are substantially different from one another, and
4. Intense dreaming is largely confined to REM sleep.

Having said all this, we must add a caveat. The second, third, and fourth points above assume that there is no recall bias across sleep states. In theory, all of the reported differences in sleep mentation – between REM and NREM, between early and late in the night, between early and late in a REM or NREM period, between normal and depressed subjects, and even among normal controls, might simply reflect recall and reporting bias. Indeed, several of the commentaries in this issue, including those by **Antrobus, Conduit et al., Feinberg, Lehmann & Koukkou,** and **Schredl,** raise this question. While we suspect that such biases do play some role in these differences, we find no basis for assuming that they explain the bulk of the differences. Certainly, there is no data even remotely suggesting that all people have the same intensity and quality of dreaming at all times of the night. And such findings as the reciprocal relationship seen between the frequency of hallucinations and thoughts across wake sleep states are difficult to explain in terms of recall phenomena.

### HR3.5. Is dreaming the subjective correlate of REM sleep physiology?

First, we should look closely at the question itself. Unfortunately, it is another poorly framed question. First of all, "REM sleep physiology" is not a unitary state (**NIELSEN**). The first REM period of the night may have no rapid eye movements at all, as may the first minute or two of later REM periods. And how much REM physiology do you need for dreaming? This is the question so clearly posed by **NEILSEN** in his article. Again, we would like to make our position clear.

**HR3.5.1. Dreaming can occur outside of Rechtschaffen and Kales-defined REM sleep.** By any definition of dreaming, we take this to be true, even if relatively rarer and less dreamlike the further from REM sleep one goes.

**HR3.5.2. Substantial components of REM sleep physiology can occur outside of Rechtschaffen and Kales-defined REM sleep.** We take this to be similarly clear-cut. Aside from the transitional period, when polysomnographic signs of REM can be observed in stage II sleep (**Gottesmann** 1996), evidence from cat studies shows relatively slow transitions in the firing patterns of brainstem neuronal populations associated with the much more punctate state transitions (**HOBSON ET AL.** 1975). In the human, no comparable data are available. In addition, there are clearly

identifiable "aborted" REM periods, when the subject begins to go into REM and then fails to complete the transition (Butkov 1996). **NIELSEN** has discussed this in much greater detail than we can go into here.

**HR3.5.3. REM sleep dreaming becomes more intense later in the night, when REM sleep shows more intense physiological manifestations as well.** The dream data are quite clear, as are the increases in REM period duration, REM density, and REM magnitude (Antrobus et al. 1995; Fosse et al., submitted; Stickgold et al., 2001).

**HR3.5.4. NREM sleep becomes less deep later in the night, as NREM mentation becomes more dreamlike.** We take the disappearance of SWS later in the night, as well as the shortening of the NREM periods as indications of this lightening of NREM sleep. To what extent the physiology of these late NREM periods become more REM-like is unknown. But as alluded to earlier, mentation reports from late NREM periods are longer, more hallucinatory, and less thought-like than are reports from earlier NREM periods (Antrobus et al. 1995; Fosse et al., submitted).

**HR3.5.5. The unique physiology of REM sleep biases the brain toward the production of dreamlike mentation.** We believe that everyone should be prepared to accept this statement, with an understanding that the bias factor can be larger or smaller. There must be something about the brain physiology underlying REM sleep that makes mentation reports from this sleep stage so much more frequent, so much longer (seven times longer in both our 1994 study and that of Antrobus), and so much more bizarre, emotional, and scenario-like.

Many authors and commentators on this issue have pointed to the importance of new brain imaging studies in mapping changes in dreaming back onto the underlying physiology of the brain. Indeed, **Borbély & Wittmann** raise the important question of how changes in regional brain activation during REM and NREM are similarly dif-

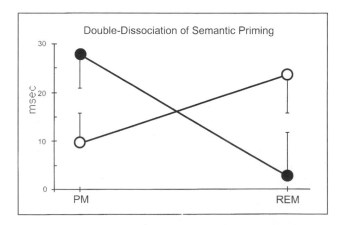

Figure HR4.    Double dissociation of semantic priming. The relative efficacy of strongly related (filled circles) and weakly related (open circles) semantic primes in a semantic priming task are reversed from normal, following awakenings from REM sleep. This suggests that the rules for activating associated memories are altered during REM sleep to favor normally weak associations. PM - priming measured during the afternoon; REM - priming measured immediately following awakenings from REM sleep. Error bars are s.e.m. (from Stickgold et al. 1999b).

ferent from waking. Answering this question will help provide us with an understanding of what produces both the similarities and differences in sleep mentation across sleep stages.

Another exciting new approach, originally championed by Antrobus (1991), is to attempt to take a more cognitive approach to this question. We have recently suggested, for example, that associative mental processes are less tightly constrained in REM sleep than in NREM (Stickgold et al. 1999b).

By waking subjects from these two states and immediately testing them on a semantic priming task, we showed that the relative efficacy of weakly and strongly related primes in enhancing responses to target words was reversed following awakenings from REM sleep (p = 0.01, Fig. HR4). These results suggest that the neural networks underlying associative processes are altered during REM sleep compared to both waking and NREM sleep, so as to enhance the activation of weakly, as opposed to strongly related semantic links. Such a shift could be the result of parallel shifts in aminergic and cholinergic modulatory systems that are known to affect these systems at both the physiological and cognitive levels (Beversdorf et al. 1999; Clark et al. 1989). The consequence of this shift, in turn, would be an increase in the frequency of occurrence of unusual and unexpected associations in dreams, precisely the increase in bizarreness that has been frequently reported.

## HR4. Neurobiology and neuropsychology

As we had hoped, our necessarily preliminary speculations on the neural basis of dream experience have elicited thoughtful commentary and critique from investigators whose specific expertise and perspectives differ from our own as well as some previews of exciting new findings. Given the burgeoning technology and efforts in both neurocognitive and sleep science, we can only imagine what a review on this topic might look like in another five years! As possible augury of things to come, one might note that, in a 1997 review on the neuroanatomy of cognition, Cabeza and Nyberg reviewed 73 PET articles published prior to December, 1995, while in their 2000 edition of this review (extended to December 1998), they assessed a full 275 PET and fMRI studies (Cabeza & Nyberg 2000)! Therefore, as **Morrison & Sanford** suggest, it is likely that different aspects of our current models will be both vindicated and contradicted by future findings.

In this spirit, we address the specifically neurobiological commentaries by focusing on the following issues:

1. How and to what extent do currently measurable tonic physiological conditions (e.g., dorsolateral prefrontal deactivation) and phasic events (e.g., REMs) of sleep reflect specific phenomenological aspects of dreaming?

2. What is the comparative contribution of the forebrain (diencephalon, limbic subcortex, basal ganglia, and cortex) versus the brainstem in the initial, causal events of dream instigation?

3. What is the relative importance of neuromodulators other than acetylcholine, serotonin and norepinephrine (e.g., dopamine) to dreaming?

4. What are the limitations of current measures in detecting correlates of dream phenomenology and what new opportunities do emerging technologies offer?

### HR4.1. Currently measurable sleep physiology and dream phenomenology

**HR4.1.1. Hypofrontality and limbic activation.** We agree that before the neurobiology of sleep stage differences can be understood, the broad differences between wake and sleep psychophysiology need to be addressed (**Borbély & Wittmann, Perry & Piggott**). Clearly sleep-related deactivation of heteromodal association areas, and especially the dorsolateral prefrontal cortex, is key to the phenomenology of dreaming. This hypothesis is supported even by those who strongly disagree with our assignment of certain other dream features to subcortical substrates (e.g., **Occhionero & Esposito**). Borbély & Wittmann's observation that the onset of slow wave activity in sleep following sleep deprivation is greatest in frontal EEG derivations fits well with recent neuroimaging findings of altered prefrontal function following sleep deprivation (Drummond et al. 1999; 2000) and relatively delayed prefrontal reactivation following awakening (Balkin et al. 1999).

What, however, differentiates a general sleep-associated frontal deactivation from conditions specific to REM is the selective re-activation of limbic regions in REM sleep. As noted by **Nofzinger,** the relative activation of anterior limbic cortex and subcortex during REM has now been widely replicated in human neuroimaging studies.

**Portas** presents a particularly valuable commentary on the descriptive model we propose in our target article, noting that to limit the role of the anterior cingulate in dreaming to its salience detection, valence labeling, and premotor functions leaves out its demonstrated waking roles in executive attentional systems, performance monitoring, and error detection. Portas points out that our model hypothesizes that such executive functions are lacking in dreaming and also that enhanced dream vivacity and dream-wake confusion may result from lesions to this area (SOLMS 1997a). These are certainly valid critiques and we offer the following possibilities as a first attempt to resolve this apparent contradiction.

First, the anterior cingulate is only part of the frontal executive and attentional networks active in waking, and the key function of working memory (subserved by the dorsolateral prefrontal cortex) remains relatively "off-line" in REM (Braun et al. 1997; 1998; Maquet et al. 1996). Although the dreamer's attention is undoubtedly engaged by sometimes quite surprising dream events (**Conduit et al.**) and a certain amount of self reflective awareness and self-control is normally present (Kahan & LaBerge 1994), the nature of this executive control remains quite deficient compared to waking, perhaps in part due to deficient memory, especially working memory, and consequent deficits in orientation (HOBSON 1999b).

Second, as we note in our target article, the anterior cingulate is functionally a highly heterogeneous structure with subdivisions into at least, affective, cognitive, and premotor areas as well as connectivity with many functionally and anatomically differing frontal circuits (Bush et al. 2000; Devinsky et al. 1995). Therefore, current neuroimaging techniques may lack the resolution to distinguish these subdivisions. This is important, because lesions to different regions of the anterior cingulate can produce very different behavioral consequences (Devinsky et al. 1995).

Third, an issue closely related to **Portas**'s points regarding the executive role of the anterior cingulate in error detection is the role of this area (along with medial aspects of other prefrontal areas such as the orbitofrontal cortex) in decision making, social cognition, and social judgment (Adolphs 1999; Bush et al. 2000; Damasio 1996). This concept has been most clearly enunciated in Damasio's somatic marker hypothesis (Damasio 1996). This hypothesis ascribes to the ventromedial prefrontal cortex the role of "marking" complex stimuli (such as social situations) with records of past autonomic and emotional responses evoked by such situations, thereby making possible high level social judgments such as behavioral inhibition or assessment of complex social cues (Damasio 1996; Tranel et al. 2000). Portions of these ventromedial prefrontal cortices are among those areas re-activated in REM following their deactivation in NREM sometimes to levels exceeding those of waking (Braun et al. 1997; 1998; Maquet et al. 1996; Nofzinger et al. 1997) and, in this regard, it is notable that dreams are highly "social" experiences (Domhoff 1996; Hall & van de Castle 1966; Kahn et al. 2001). Therefore such activation of ventromedial prefrontal cortices may contribute to the ubiquity of salient social scenarios.

Finally, Cohen et al. (2000) have recently proposed a complex feedback loop between the anterior cingulate and prefrontal cortex which controls performance through the adjustment of cognitive control based on task demand. In their model, this loop is driven by activity in the locus coeruleus and aids in the redirecting of attention in our dreams, the sustained deactivation of dorsolateral prefrontal areas with consequent deficiencies of working memory and orientation may account for some of the bizarreness in these dreamed scenarios relative to specific task demands. It is striking that with the locus coeruleus shut down during REM sleep and the dorsolateral prefrontal cortex deactivated as well, the brain would appear to be running Cohen's circuit in an open-loop mode. Such open-loop activation might contribute phenomenologically to the failure of attentional systems and the inappropriateness (i.e., incongruity) of dream imagery, and may also contribute functionally to the testing of associative networks for emotional salience without feedback and adjustment. As such, it may contribute to some of the more complex putative memory functions of sleep which we discuss below.

**HR4.1.2. NREM activation patterns and mentation.** Another phenomenon noted by **Borbély & Wittmann** is the relatively greater activation of striate cortex in NREM compared to REM (Braun et al. 1998; Hofle et al. 1997) which Hofle et al. speculate may underlie NREM imagery. **Steriade** provides one possible mechanism by which this imagery might occur. He first gives us a view into the dynamic electrical oscillatory milieu occurring among neurons within the cortex during NREM which takes the form of a slow (0.5 to 1 Hz) alternation of a quiescent hyperpolarization phase followed by intense spike trains during a depolarized phase. Steriade proposes that in NREM just prior to REM, renewed thalamic input to the cortex (which is absent during most of NREM) in the form of PGO waves (which are known to precede the full complement of REM features) impinges on cortical neurons which are still in the slow oscillation mode and are thus primed to emit spike trains which, in turn, are experienced as vivid NREM mental imagery. Notably, this could account both for the observed activation of the striate cortex and for the transient, static but intense imagery ascribed to NREM mentation (for a recent elaboration, see Steriade 2000).

**Porte** provides a thoughtful discussion of NREM physiology and its possible constraints on the occurrence of "covert REM" and its concomitant mental manifestations hypothesized by **NIELSEN**. She suggests that, during Stage 2 NREM, the sleep spindle and its refractory period may intermittently prevent REM-like processes and associated mentation but that these processes can occur during prolonged interspindle intervals. It should be noted, however, that recent lesion and stimulation studies of human thalamic nuclei suggest that the linkage between electrographically measured sleep spindles and thalamocortical activity in humans may be more complex than in the cat (Arnulf et al. 2000; Santamaria et al. 2000).

**HR4.1.3. Parallel neurochemical and cerebral blood flow data.** It is noted by **Perry & Piggott** that the striate cortex contains relatively greater serotonergic innervation than limbic areas which are heavily innervated by cholinergic and dopaminergic projections. This observation chemically differentiates the same regions differentiated by relative regional cerebral blood flow in PET studies comparing REM and SWS (Braun et al. 1997; 1998). Similarly, Braun (1999) cites a primate study showing that relative cortical distribution of cholinergic synthetic and degradative enzymes – high in limbic and paralimbic regions and low in prefrontal and striate regions – roughly parallels the areas relatively activated in PET studies of REM. Such regional differences in innervation patterns may combine with state-dependent variations in neuromodulatory cell activity to produce the observed state-dependent patterns of regional brain activation and thus the changes in dream quality and frequency across states.

**HR4.1.4. REM saccades and dream imagery.** **Herman** gives careful consideration to the still mysterious relationship between REM saccades and dream imagery. His current analysis differs from our own in only minor ways. For example, while we do acknowledge that a brainstem oculomotor final common pathway integrates cortical input for many saccades, this is not necessarily the case for all saccades. Direct connectivity between subcortical nuclei (vestibular, collicular) and the oculomotor complex exists and these structures could interact independently of the cortex. The loss of leftward saccades in hemineglect patients with right parietal lesions suggests cortical mediation of their REM saccades but does not necessarily imply that the initial instigation of these saccades originates in the intact cortical hemisphere. For example, as **Herman** notes, Doricchi et al. (1993) suggest that, unlike some frontally commanded waking saccades, the reflexive saccades of REM sleep, although parieto-collicularly commanded at an early stage, are probably in response to endogenous signals of subcortial origin such as PGO waves. Similarly, while we agree that a vestibulocortical circuit is likely to contribute to the dreamer's sense of existing in three-dimensional space and that the inferior parietal cortex is a likely cortical node for spatial integration (Solms 1997a), it is open to question whether all dreams accompanied by REMs (e.g., those of the congenitally blind) are accompanied by such a sense. An experimental paradigm which might be able to address many of these issues would be the study of lucid dreamers with event-related fMRI techniques and instrumental awakenings as we have already suggested in section 1 of this response.

### HR4.2. Contribution of forebrain versus brainstem in dream initiation

**HR4.2.1. Sleep control systems include both brainstem and forebrain.** In our recent overview of the neural mechanisms subserving sleep (Pace-Schott & Hobson, in press) we have fully acknowledged and reviewed findings on the role of diencephalic and other forebrain structures in regulating the complex interactions of intrinsic neuronal oscillators whose frequencies vary over a range of at least 86,400 (from the once per day circadian to the 30- to 80-persecond gamma EEG). In our article we concluded:

> Interactions of diverse neuromodulatory systems operate in widespread subcortical areas to amplify or suppress REM sleep generation as well as to facilitate onset and offset of the control of behavioral state by the pontine REM/ NREM oscillator. An ascending medial brainstem and diencephalic system of multiple nuclei with extensive reciprocal interconnections and system-wide sensitivity to neuromodulation controls the regular alternation and integration of the sleep-wake and REM/ NREM cycles. (Pace-Schott & Hobson, in press)

Even before the new PET studies appeared, we had stressed that the initiation of certain REM signs as well as dreaming itself must integrate multiple forebrain inputs (e.g., Kahn et al. 1997). However, we continue to be often misperceived as holding that both derive for the most part from the pontine REM/NREM oscillator (e.g., **SOLMS**). For example, while Solms argues that EEG desynchronization need not arise from the pons, we too have stressed that ascending tracts from the basal forebrain constitute the cholinergic component of cortical activation. (**HOBSON ET AL.**; Pace-Schott & Hobson 2000). Nonetheless, the pontomesencephalic cholinergic system provides a major excitatory input to the cholinergic basal forebrain (although even this circuit carries the added complexity of a glutamatergic intermediary within the basal forebrain [Semba 1999]). Similarly, given the distribution of ascending cholinergic tracts, even that component of cortical activation derived from thalamic nuclei which are activated by the pontomesencephalic cholinergic system is not itself cholinergic (Pace-Schott & Hobson 2000).

**HR4.2.2. New models of neuroanatomy subserving the sleep-wake and REM/NREM cycles.** We appreciate the commentaries of **Morrison & Sanford** and **Salin-Pascual et al.** emphasizing new research showing important roles of forebrain mechanisms in the mediation of sleep-wake and REM/NREM cycles. We ourselves have stressed the involvement in sleep regulation by forebrain regions such as the hypothalamus, amygdala, and basal forebrain as well as the roles of adenosine, histamine, galanin, and orexin in the initiation of NREM sleep (**HOBSON ET AL.**; Pace-Schott & Hobson 2000). In these papers, we extensively cite recent reviews on the hypothalamus (Shiromani et al. 1999), amygdala (Calvo & Simon-Arceo 1999; Morrison et al. 1999) and basal forebrain (Jones & Muhlethaler 1999; Szymusiak 1995) and attempt to integrate this plethora of new data in a set of connectivity schematics which recognize the complexity and plurality of the brain system represented.

We find very plausible the model of Shiromani et al. (1999) for the control of the wake-NREM transition and its link to the ultradian REM-NREM cycle. In summary, Shiromani et al. (1999) suggest that:

1. During prolonged wakefulness, accumulating adenosine inhibits specific GABAergic anterior hypothalamic and basal forebrain neurons which have been inhibiting the sleep-active, hypothalamic ventrolateral preoptic (VLPO) neurons during waking.

2. Thus, the disinhibited sleep-active GABAergic neurons of the VLPO and adjacent structures then inhibit the wake-active histaminergic neurons of the tuberomammilary nucleus (TMN) in the posterior hypothalamus as well as pontine aminergic (DR and LC) and cholinergic (LDT/PPT) ascending arousal systems thereby initiating NREM sleep.

3. Forebrain activation by these ascending arousal systems is thus disfacilitated.

4. Once NREM sleep is established, the executive networks of the pons initiate and maintain the ultradian REM/ NREM cycle described by reciprocal interaction.

Regarding the role of amygdalar input in the initiation of REM (Calvo & Simon Arceo 1999; Morrison et al. 1999), we have noted that early work on the pontine cat provides evidence that REM/NREM cycling can be independent of the forebrain (Pace-Schott & Hobson, in press). **Morrison & Sanford** address this concern noting that decerebration may have made the pontine cat abnormally predisposed to enter REM. We remain agnostic on this point noting, however, that other investigators have identified potential REM triggers in the pontine caudolateral peribrachial area (Datta 1995; 1997b).

**HR4.2.3. Multiple ascending cholinergic systems and pontine activation.** With regard to ascending cholinergic activation in REM and waking, an important distinction is made by both **Nofzinger** and Braun et al. (1997) between a posterior pontine reticular system which activates thalamocortical circuits and an anterior system, originating in the basal forebrain and hypothalamus, which activates limbic and paralimbic structures. Both suggest that forebrain activation by the anterior system may predominate in REM compared to waking. It is important to note, however, that pontine nuclei have anatomical connections with both the basal forebrain (Szymusiak 1995) and hypothalamus (Kumar et al. 1989) and the PPT/LDT provides a major source of excitatory input to the basal forebrain (Semba 1999). The functional divergence of these two cholinergic systems may, therefore, occur rostral to the pons. Given these two cholinergic systems, it seems quite plausible, as **Perry & Piggott** suggest, that NREM dreaming is sustained by cortical cholinergic activation from the basal forebrain plus dopaminergic activation from the ventral tegmental area (VTA) with the addition of brainstem cholinergic drive accounting for the greater intensity of REM dreaming.

**Nofzinger** further notes the lack of relative pontine activation in REM compared to waking in PET studies. However, as noted by **Portas** and also by Semba (1999), much of the ascending brainstem reticular arousal systems are shared between REM and waking. Therefore, as noted by Portas, one would expect a similar degree of pontine activation in REM and wake. In contrast, the pontine tegmentum does show relatively increased activation in REM when compared to slow wave sleep (Braun et al. 1997) or to the average of NREM and waking (Maquet et al. 1996).

A major problem still confounds our efforts to integrate the cellular and molecular level analyses of pontine activity in animal models with human neurobiology. That is the uncertain relationship of PET documented blood flow changes with neuronal activity in general and the impossibility of resolving individual nuclei (like the LC, DRN, and

PPT) not to mention different cellular subcomponents of these nuclei. One gap-closing strategy is higher resolution imaging. Another is imaging in the animal model itself. Both are desirable.

**HR4.2.4. The functional complexity of the frontal medio-basal forebrain.** As noted by **Doricchi & Violani,** the detailed connectivity of mediofrontal subcortical and cortical regions as well as the projection patterns of aminergic and cholinergic nuclei in primates argue against Solms's (1997a; 1999c; **SOLMS**) assignment of dream instigation solely to mesocortical dopaminergic projections from the VTA. There are at least four observations supporting this assertion:

First, as **Doricchi & Violani** point out, SOLMS's clinical description of anterior mediobasal lesion patients does not effectively rule out attentional or motivational deficits contributing to their apparent global cessation of dreaming.

Second, **Doricchi & Violani** emphasize the multiple reciprocal subcortical-cortical connections that traverse the mediofrontal white matter area in which SOLMS claims leucotomy or stroke selectively disrupt dopaminergic corticopetal pathways (Heimer & Alheid 1999; Kalivas & Barnes 1993; Rempel-Clower & Barbas 1998; Salloway et al. 1997). In addition to mesocortical dopaminergic projections from the VTA (Roth & Ellsworth 1999), these tracts contain cholinergic projections from magnocellular neurons in the nucleus basalis (Mesulam et al. 1992), histaminergic projections from the posterior hypothalamus (Rempel-Clower & Barbas 1998) and the corticopetal prefrontal projections of noradrenergic and serotonergic projections exiting the medial forebrain bundle (Gaspar et al. 1989; Jacobs & Azmita 1992).

Third, if lesions in the white matter of the ventral mesial quadrant of the frontal lobe which cause global cessation of dreaming (Solms 1997a; 1999c; **SOLMS**) were to extend just slightly posteriorly, they would damage actual basal forebrain and hypothalamic nuclei as well as aminergic tracts within the medial forebrain bundle itself.

Fourth, the thalamocortical tracts connecting the mediodorsal nucleus of the thalamus to the prefrontal cortex, which were targeted by leucotomy, integrate many "upstream" limbic inputs. In addition to mesolimbic, reward-related input from the ventral striatum, the mediodorsal nucleus receives input from areas associated with fear, rage, and other aversive responses (e.g., LeDoux 1996). Such inputs might account for the fear and anxiety (Merritt et al. 1994) and instinctual behaviors (Jouvet 1999) which are at least as prevalent as wishing/seeking experiences in dreams.

For example, **Krieckhaus** describes a thalamo-subcortical system possibly involved in the construction of an evolutionarily advantageous adaptation involving the opposing, homeostatic interaction of a stress-reactive amygdala-based circuit with a subcortical system for adaptive inhibition involving the hippocampus, mammilary bodies and anterior thalamic nucleus. The amygdala-based circuit promotes reactive defense or flight whereas the hippocampus-based system inhibits such overt behavior. The relative balance of these two systems provides a substrate for evolution to fine tune adaptive behaviors such as predator avoidance.

Notably, **Krieckhaus**'s system involves the neural substrate of instinctual defensive behaviors and their experiential correlates such as fear-anxiety rather than the appetitive circuits that are emphasized by **SOLMS.** As we have pointed out elsewhere (Hobson & Pace-Schott 1999),

SOLMS's exclusive focus on the dopamine system can be understood as an effort to find a neurobiological substrate for Freud's highly dubious and discredited theory of dream instigation by unconscious wishes.

These facts in no way diminish our interest in the dreaming-related roles of this midline limbic region which, after all, consists of just those areas shown to be selectively activated during REM in PET studies (**Nofzinger**). We simply suggest, as do **Doricchi & Violani,** that the limbic contribution to dreaming is far more complex than just dopamine-driven reward-seeking impulses.

### HR4.3. The relative importance of different neuromodulators

**HR4.3.1. Neuromodulation *must* affect cognition.** Although some scientists still minimize the role of state-dependent neuromodulatory changes in the differential phenomenology of dream and wake mentation (e.g., **Antrobus**), others (e.g., **Morrison & Sanford**) agree with us that it is illogical to assume that such massive physiological changes as noradrenergic and serotonergic demodulation in REM (and, to a lesser extent in NREM) would not cause equally massive mental effects. One need only consider the profound psychological effects of minuscule amounts of the mixed serotonin agonist/antagonist LSD (Aghajanian 1994) or the mental effects of hypercholinergic toxicity caused by exposure to organophosphates (see **Perry & Piggott**) to appreciate the dependency of mental status on physiochemical conditions (see Hobson 1999b).

Among researchers who do accept the importance of state dependent neuromodulation to dream phenomenology, there continues to be a debate as to which neuromodulators are of greatest importance (e.g., Solms 1999c). Notwithstanding the wide variation in emphasis placed on the association between REM sleep and dream quality or frequency, their strong quantitative association (see above) suggests that the physiology of REM (and of NREM) should, at the very least, be considered and well understood. In this regard, **Lydic & Baghdoyan**'s analysis of the cholinergic contribution to REM generation using Koch's postulates of causality adds to the overwhelming experimental evidence supporting a key role for acetylcholine in the executive networks controlling the REM/NREM cycle. It is notable that Lydic & Baghdoyan's conclusion that an early stage of REM generation involves the cholinergic stimulation of the pontine reticular formation by mesopontine nuclei is exactly the conclusion drawn by Semba (1999) in her comprehensive and critical review of this literature.

**HR4.3.2. Dopamine and dreaming.** Contrary to some commentaries (e.g., **Gottesmann**), we by no means rule out the involvement of dopamine in REM sleep mentation. In recent reviews (Hobson & Pace-Schott 1999; **HOBSON ET AL.;** Pace-Schott & Hobson 2000) we have explored possible roles of dopamine in normal REM sleep and dreaming. Our finding is that "dopamine is as often a dream destroyer as it is a dream creator and that dreaming is enhanced as well as impeded by dopamine deficiency" (Hobson & Pace-Schott 1999, p. 215). A few examples illustrate this point as well as the likely involvement of cholinergic and other aminergic systems in the generation of normal dreaming.

**HR4.3.2.1. Dopaminergic stimulation does not always enhance dreaming.** There are many drugs which enhance dopaminergic neurotransmission via reuptake inhibition which, in both experimental studies and clinical lore, are widely known to delay or disrupt sleep but are not reported to produce vivid dreams. This absence of reported dream effects is especially striking given that the psychostimulants methylphenidate, pemoline, and dexedrine have been widely used in the treatment of ADHD and narcolepsy, while the abused psychostimulants such as cocaine and methamphetamine have been widely studied by investigators concerned with the sleep effects of these agents (e.g., Brower et al. 1992; Cottler et al. 1993; Gillin et al. 1973; Kowatch et al. 1992; Lukas et al. 1996; Post et al. 1974; Roehrs et al. 1998; Watson et al. 1992). A 1966–1999 MED LINE text word search for dream effects of dopaminergic drugs yielded citations for only L-DOPA, some DA agonists, some atypical amphetamines (Thompson & Pierce 1999), and bupropion (Becker & Dufresne 1982; Posner et al. 1985). Similarly, perusal of the 1999 PDR revealed low incidences of dream abnormalities and nightmares associated with dopamine agonists but virtually no such effects for the DA reuptake inhibitors. Moreover, at least one dopamine agonist, piribedil, has been shown to decrease dream recall (Passouant et al. 1978).

### HR4.3.2.2. Dopaminergic dream enhancement is non-specific.
The well-known dream enhancing effects of the dopamine agonists such as L-DOPA (e.g., Moskovitz et al. 1978) which are stressed by SOLMS (1997; 1999) are also noted in patients receiving the noradrenergic beta-receptor blockers (Thompson & Pierce 1999) and the cholinesterase inhibitors (Ross & Shua-Haim 1998), both effects fully predicted by REM enhancement in accord with the reciprocal interaction model. Two other observations are also consonant with reciprocal interaction and AIM. First, as also noted by **Doricchi & Violani**, experimental administration of cholinergic agonist can induce REM sleep with dreaming (Sitaram et al. 1978a). Second, recent findings from our laboratory have described serotonergic suppression of dream recall frequency combined with a pattern of enhancement of dream intensity consistent with the notion of cholinergic rebound (Pace-Schott et al. 1999; 2000; 2001).

### HR4.3.2.3. Dopamine deficiency does not diminish dreaming.
As noted by **Borbély & Wittmann**, SOLMS's dopamine hypothesis predicts a paucity of dreaming in dopamine deficient conditions such as Parkinson's disease or neuroleptic therapy. However, the dream narratives of L-DOPA-treated Parkinson's disease patients did not vary as a function of drug dosage (Cipolli et al. 1992) and, as **Perry & Piggott** note, enhanced dreaming has been observed in Parkinson's disease independently of dopaminergic medication. Moreover, Gaillard and Moneme (1977) showed near 100% dream recall in REM awakenings of both sulpiride (a D1 antagonist) treated and placebo-matched subjects with only a small, nonsignificant reduction in recall and report length in the sulpiride group. Similarly sulpiride increased dream recall in schizophrenics (Scarone et al. 1976) and, as also noted by **Wichlinski**, a catecholamine depleting agent, reserpine, has been associated with excessive, bizarre dreaming (Hartmann 1966a) as well as being associated with elevation of acetylcholine levels (Wichlinski).

### HR4.3.3. Resolving the dopamine-acetylcholine conflict.
There are several possible resolutions of this apparent conflict between two models (i.e., cholinergic pontine versus midbrain dopaminergic instigation of dreaming). The first is to recognize, as pointed out by **Greene**, that "the changes in circuit function that can derive from cholinergic and monoaminergic modulation depend on our understanding of the circuit function and its relationship to cognitive function." That is, neuromodulatory substances modify functioning neural circuits and the interactions of different neuroactive substances may vary from antagonistic, to neutral, to interchangeable, to synergistic depending on the circuitry in question. Such neuromodulatory interactions may explain the variability of sleep and dream alterations among patients with the same neurodegenerative disease diagnosis (Christos 1993; Comella et al. 1993; Perry et al. 1999), since the specific affected nuclei can vary as is the case with serotonergic and noradrenergic deficiencies in Parkinson's disease (Nausieda et al. 1982; Valldeoriola et al. 1997). An additional source of variability in sleep and dream effects by similar pharmacological agents may be the differential brain distribution of specific receptor subtypes such as the D3 receptor (**Perry & Piggott**).

The second possible resolution is to note that the normal sleep and dream effects of DA may be mediated by its interactions with other neuromodulatory systems. For example, DA interacts with brainstem systems involved in the executive control of the REM/NREM cycle and with basal forebrain cholinergic cells which activate cortical and limbic structures (Hobson & Pace-Schott 1999; **Perry & Piggott**). In addition, dopamine has been shown to enhance cortical acetylcholine release (Moore et al. 1999; Smiley et al. 1999) while cholinergic mesopontine neurons have been shown to enhance mesolimbic dopamine release (Oakman et al. 1999). Braun (1999) has suggested that the (cholinergic) lateral habenula may play a role in coordinating serotonergic and dopaminergic interactions during sleep. Mutual facilitation between cholinergic and dopaminergic systems may serve to maintain or intensify REM sleep and/or dreaming (Hobson & Pace-Schott 1999; Perry & Piggott) especially given the continued activity of dopamine neurons during REM (Miller et al. 1983; Trulson et al. 1981).

### HR4.4. Opportunities for dream studies in emerging neuroscience technologies

In our target article, we have speculated on the promise to dream psychophysiology of new techniques in neuroscience such as transcranial magnetic stimulation (TMS), event-related functional magnetic resonance imaging, magnetic resonance spectroscopy, receptor radio ligand PET, near infrared spectroscopy, and dipole tracing (**HOBSON ET AL.**). Following up on our above discussion of REM saccades and dream imagery, added power might be conferred on the study of lucid dreamers with event related fMRI techniques and instrumental awakenings by transiently deactivating the frontal eye fields using TMS. Indeed, TMS has recently been used to demonstrate involvement of striate cortex in visual imagery (Kosslyn et al. 1999).

In our target article, we cite one microdialysis study on CNS neuromodulation across the human sleep-wake cycle (Wilson et al. 1997). It is exciting to learn from **Portas** that Ungerstedt and his colleagues in Sweden are now actively combining cognitive neuroscience with human microdialysis techniques.

Depth electrode recording and stimulation in epilepsy

or Parkinson's disease patients offer yet another possible source of discoveries on the brain basis of dreaming. For example, subdural recordings from epilepsy patients have been used to study human theta oscillations during virtual maze navigation (Kahana et al. 1999) while the sleep effect of direct thalamic stimulation used in the treatment of tremor have recently been described (Arnulf et al. 2000).

## HR5. AIM/isomorphism

### HR5.1. Brain-mind isomorphism

**HR5.1.1. Philosophical issues.** We need to remind our philosophically minded commentators (**Franzini, Mancia, Flanagan, Antrobus, Revonsuo**) of our commitment to **Antrobus**'s concept of the brain-mind as a unified system. For us, the scientific study of such a system entails the mapping of features in the domain of subjective experience onto features in the domain of neurobiology and vice versa. We use the term isomorphism to denote similarity of form in the two domains (McCarley & Hobson 1977). A simple and easily understandable example would be the universally accepted isomorphism of cognitive activation and of brain network activation. This is clearly visible as factor A in our AIM model.

Because it is at present difficult to know exactly which cognitive measures to map onto which neurobiological measures, we have adopted the conscious state paradigm – with all of its imperfections and problems – as a start point for the mapping effort. We greatly appreciate **Flanagan**'s encouragement regarding the so-called "hard problem" of understanding how the two domains are linked. Although we can see how one might disagree (and, indeed, we disagree among ourselves), for us the hard problem is not so much imagining how the brain could give rise to consciousness as it is imagining how it could not. We think that a system that is capable of representing the complexity of the outside world as a set of dynamic neural codes and is also possessed of self-activation (factor A), input-output gating and internal stimulus generation (factor I), and a vast and differential chemical modulator system (factor M) presupposes consciousness and the differentiation of that consciousness in the multiplicity of states of which such a system is agreeable. The astronomical numbers arrived at by multiplying neuronal numbers by neuronal connections emphasize that the human brain has an almost infinite capacity for information encoding, processing, and storage and that emergent properties, as consciousness may be, are inevitable in so complex a system. No one doubts that the brain encodes the visual world or that it encodes personal history. Why, then, could not the brain use these representations as a self who sees and sees that he sees?

### HR5.2. The bogeyman of reductionism

To our psychologically minded prosecutors, we plead guilty to the charge of reductionism. Indeed we take it as a compliment. This is because, following Churchland (1986), we take reductionism to be the very essence of science. No reductionism, no science. Reductionism is nothing more or less than the explanation of events at a macro level (like eye color or memory loss) by events at a micro level (like gene expression and shifting levels of neuromodulators). Reductionism is not eliminative materialism

because the explanation does not make eye color or memory go away.

Reductionism can be successful within domains (as for example when dream bizarreness reduces to plot discontinuity and incongruity and these formal features further reduce to deficient executive functions including working memory), as well as across domains as when the working memory deficits of REM sleep dreams are reduced to aminergic demodulation and diminished blood flow to the dorsolateral prefrontal cortex. Within the neurobiological domain, we can even go a reductionistic step further and propose that regional brain activation and regional blood flow will reduce to changes in neuromodulatory balance because it is very likely that both activation and blood flow are controlled by neuromodulators!

We remind our psychologist critics that no one was more reductionistic than Sigmund Freud. But just because Freud could not accomplish the cross-domain reduction of mind to brain that he envisaged in 1900 does not mean that no one in the twenty-first century can.

The stepwise chain of reductionism we have just traced simply explains each level of data with the data from a lower level. This paradigm also allows us to unify events at regional level of the brain – in this case, the cortex – with events at a lower level, in this case the brain stem.

By reducing dream bizarreness to changes in neuromodulatory ratio we are neither denying nor eliminating the dream as a subjective experience, nor are we denying the forebrain's role as dream instantiator when we show its state to be dependent on (or in **REVONSUO**'s terms, supervenient upon) the state of the brain stem. And we are not saying that REM is superior and NREM second class. Not at all. We are only saying that dreaming favors REM sleep. REM sleep just happens to be the brain state when those physiological and neurochemical conditions which underlie the occurrence of dream bizarreness and other dream phenomenology reveal themselves most strongly, clearly, and frequently. Each of these conditions, and sometimes all of them, can express themselves in other states.

### HR5.3. Sleep as the substrate of dreaming

Thus we agree with **Borbély & Wittmann** that sleep itself entails the neurobiological substrate of dreaming. This substrate is now a more specifiable set of changes in the forebrain; as sleep proceeds through the NREM phase, these conditions intensify to become maximal in REM. Because all of these conditions are brain stem and diencephalically mediated and because sleep onset and NREM sleep are progressive steps on the way to REM, we can also agree with **NIELSEN** when he attributes extra-REM dreaming to REM-like physiology. Repeating our ultra-reductionist claims for dreaming, the one neurobiological variable that best distinguishes wake from sleep and NREM from REM is the progressive decline in aminergic neuromodulation. That is factor M in our AIM model.

We regret that there is no psychophysiological dissociation for physiologically phobic dream psychologists to take comfort in! On the contrary, what we see is an increasing degree of association between REM brain state variables and dream mind state variables. Ultimately, we suppose the dreaming brain will be seen to be so inseparable from the dreaming mind as to force the conclusion that they are one and the same thing, as a unified system seen from the vantage point of the third person (the brain) and from the vantage point of the first person (the dreamer). In this sense we can envisage the mind body problem, the hard problem, and the subject-object problem to be nonexistent. Rather than claiming the problem to be "solved" we will see that there is no such problem.

**HR5.3.1. The AIM model.** Some of our physiologically inclined colleagues appreciate the AIM model (**Gottesmann, Nielsen & Germain, Ogilvie et al., Perry & Piggott**). Others do not (**Hartmann, Morgane & Mokler,** and **Solms**). We ourselves like it, but we are dissatisfied by it as well. We like it because it is an advance from two-dimensional models like traditional sleep graphs and our own reciprocal interaction structural and mathematical models because all of AIM's three dimensions are realistic and measurable, at least in animals. We are dissatisfied for some of the same reasons as our critical commentators.

Of course we may need more dimensions and when we do, we will add them. But please note that with time represented as the interval between the succession of dots in AIM state space the model already has four, not just three dimensions, so it is already at least twice as realistic as its predecessors. To critics like **Morrison & Sanford,** who believe that any model is premature because we do not know enough about the brain, we say that so far the reciprocal interaction has served us and our pharmacological colleagues very well. We now hope that AIM will help to close the gap between basic and clinical sleep research by showing that all conditions of the brain-mind are points in a multidimensional state space.

As with dimensions, AIM may also need more neuromodulators (like histamine or dopamine) or neuropeptides (like orexin). And we will add them if the release of those chemicals changes in a state-dependent manner that is different from the chemicals we have already modeled. AIM is not yet the Model of Everything (only of the conscious state). We say that ironically and self-effacingly, while freely admitting to **Hartmann**'s charge that AIM's reach exceeds its grasp. While the model does a pretty good job with increasing hallucinatory propensity and its reciprocal deficits in memory and with thought, we cannot yet see how to deal with important conscious state variables like emotion which will almost certainly need a brain region or subregion dimension unless, as we hypothesize, regional activation is a combined function of factors I and M.

The AIM model, originally proposed in 1990 (Hobson 1990) and revised first in 1992 (Hobson 1992a) and now in our target article, is an evolving model. Its original goal is, we are convinced, still a valuable one, namely to point out that wake-sleep mentation is a continuously varying phenomenon dependent on several underlying and similarly varying neurophysiological variables. The fact that the brain tends to remain in a small region of the AIM state space for prolonged periods of time permits the identification of canonical brain-mind states, such as waking, REM, and NREM sleep. But we have assumed all along that the brain-mind is much more complex than any three dimensions might suggest. Already in our target article we have begun to talk about the differences in AIM space position of various brain regions. Indeed, all the brain imaging studies make clear that we cannot talk about a single value for level of brain activation. Similarly, neuromodulation is not constant across these regions. It would be highly inaccurate to

suggest that a single ratio of aminergic to cholinergic neuromodulation describes the micromodulation of each and every brain region at a given time

But as our understanding of the physiology and phenomenology of dreaming continues in its exceptional growth, we believe that the AIM model retains heuristic value. Physicists and chemists continue to talk of matter being made up of electrons, neutrons, and protons at the same time that they realize that this grossly underestimates the complexity of the system. Indeed, we feel as if we are in a period similar to the recent one in physics when the number of fundamental particles seemed to be increasing without limit, or the current period in neurobiological research in which the number of known neurotransmitters similarly seems to be increasing without limit. Thus in response to the friendly criticism of commentators such as **Morrison & Sanford,** who suggest that our AIM model is premature, we respond that it may almost be too late to develop such a model.

But we still believe that the AIM model is important heuristically because it reinforces the view that: (1) mentation and physiology are strongly linked to one another; (2) that each varies in a continuous manner with various "local attractors" where the brain and mind spend most of their time; and (3) that the three neurophysiological parameters of brain activation, modulation, and input source explain more of the variance in mental content than any comparable parameters.

One aspect of mentation that we suspect might be underrepresented in our model is that of emotion, and we assure our impatient critics like **Panksepp, Kramer,** and **Greenberg** that we are fully committed to the premise of emotional salience. This premise flows from the fact that dream emotion is not bizarre but always appropriate to the plot even when the plot itself is full of discontinuities and incongruities (Foulkes et al. 1988b; Merritt et al. 1994)! In this regard, we are struck by the PET evidence of activation of limbic and immediately paralimbic structures suggesting, à la Penfield (see Tees 1999), that emotion and its directly associated cognition are evoked by limbic autostimulation. Penfield's observation and the dream-like mentation associated with some temporal lobe seizures show that REM-like activations of the forebrain can indeed evoke dreamlike mentation. Fortunately, most of us do not need either seizures or Penfield to stimulate our temporal lobes. We have a built-in mechanism for that purpose. It is REM sleep.

In the interest of theoretical economy, we assume that the temporal lobe activation of REM is itself a function of the drop in M and the corresponding increase in interoceptive I, of which PGO activity is the clearest example. In fact, PGO waves are notably epileptic (Elazar & Hobson 1985). Furthermore, they are recordable in the amygdala during REM (Calvo & Fernandez-Guardiola 1984). Finally, the pontine cholinergic system which triggers PGO activity projects heavily to the amygdala (Morrison et al. 1999).

Here again, we see the power of reductionism in explaining emotional salience. Through a chain of processes having their origin in the brain stem (but, of course, recruiting the midbrain as well as the basal forebrain and other diencephalic structures), cholinergic phasic activation is projected selectively to the limbic lobe where it triggers emotion and its cortically based cognitive correlates. These cognitive correlates may be intrinsically bizarre (owing to aminergic demodulation) but still salient because of the connectivity to associated image producing structures in the temporal cortex. Obviously, REM is not the only way to set these processes in motion. It is just our favorite way to do so!

## HR6. The functions of sleep and the learning-memory controversy

In our target article, we spoke only briefly of the possible function of sleep. But in light of the target articles of **VERTES & EASTMAN** and **REVONSUO** addressing possible cognitive functions of sleep, we would like to address this issue here. We begin with the hypothesis that sleep clearly plays a role in neuronal plasticity. The alternative theory, that the brain is left unchanged after a night's sleep, is unacceptable on its face. Indeed, we suppose that an immense number of plastic changes occurs each night. It seems self-evident to us that since so much of the night contains some form of mentation, and since this mentation is inevitably built up from images and concepts stored in one's memory, that either the entire evolved mechanism which creates and supports this activation and synthesis of memories and associations, along with its accompanying mentation, either does so for no purpose at all or else that some aspect of these memories and associations becomes altered by the process. Our scientific intuition is that whereas our subjective awareness of this processing may be gratuitous, the processing itself is not.

### HR6.1. Synaptic plasticity

The real question then is what types of plastic changes occur during sleep and toward what ends. Sleep related changes could be (1) controlled solely at the level of individual cells and synapses, (2) involve calculations across local neuronal networks, (3) extend across the brain regions, or (4) require conscious dreaming as part of the algorithmic calculation of the appropriate plastic changes in neuronal connectivity. Similarly, the goals of these sleep-dependent changes might be to (1) erase memories, undoing plastic changes that occurred during the day and restoring the brain to an earlier state, (2) consolidate memories, stabilizing changes produced during the day, (3) fine tune discrete

networks, strengthening some synapses and weakening others, or (4) introduce totally novel changes, creating neuronal configurations that had never existed before.

It is difficult to identify precisely where the target articles lie among these options. **VERTES & EASTMAN** are clearly addressing the concept of memory consolidation, probably conceptualized as occurring primarily at the levels of single neurons and local neural networks. In contrast, **REVONSUO** appears to be proposing that dreams serve to create novel connectivity within the brain. If we add the proposal of Crick and Mitchison (1983), we have examples of each of the options proposed above.

Is it conceivable that functional neuronal plasticity is occurring during sleep at all of these different levels of mechanistic and functional complexity? Without arguing that it does act at all of these levels, we do argue that it is not unreasonable to think so. Table HR1 compares possible mechanisms of synaptic plasticity during sleep with known mechanisms for modulating enzymatic activity within single cells.

On the left side of this table, one can see the range of conceptual and biochemical levels of enzymatic control. Evolution clearly found advantageous the ability to exert control at numerous points in the process. Thus, even in bacteria, enzyme activity is modulated at several levels. When product accumulates, it binds to the enzyme and inhibits further production of the product. This is an unavoidable consequence of how enzymes work. But at a more sophisticated level, enzymes evolve separate binding sites for other related molecules that can also control the enzyme. Thus, if an enzyme, $\alpha$, catalyzes the coversion of "A" into "B" as the first step in the pathway, $A \rightarrow B \rightarrow C \rightarrow D \rightarrow E \rightarrow F$, it is not uncommon for the enzyme $\alpha$ to contain an inhibitory binding site for the final product, "F," in order to slow down the biosynthetic chain at the start when the final product builds up to adequate levels. But even more long range control systems have evolved to control the actual level of the enzyme in the cell. Thus, in our example above, "F" might be transported into the nucleus and actively inhibit the transcription of the gene coding for the enzyme $\alpha$, thus stopping further synthesis of the enzyme when it is not needed.

As another example, disparate levels of regulation can be described for the regulation of blood glucose level, which

Table HR1. *Hierarchical regulatory systems*

| Regulation of Enzyme Activity | Sleep and Neuronal Plasticity |
| --- | --- |
| Enzyme kinetics | Synaptic kinetics |
|   Substrate activation at enzymatic site |   Newly activated synapses strengthened |
|   Product inhibition at enzymatic site | |
| Allostery | Strengthening weak associations |
|   "Relevant" molecules activate or inhibit at |   "Relevant" synapses identified and strengthened or |
|   regulatory subunit of enzyme |   weakened |
| Enzyme relocation | Memory relocation |
|   For example, internalization of membrane bound enzymes |   For example, hippocampal memories moved to neocortex |
| Nuclear control | Inter-regional control |
|   Transcription and translation controlled by |   Emotional conflicts and unresolved issues resolved |
|   "relevant" molecules |   through new or strengthened connectivity |

Since the regulation of enzyme activity (left) is of great evolutionary value, a number of different levels of control appeared during evolution. If sleep-dependent neuronal plasticity (right) is similarly valuable, it is reasonable to assume that a comparable range of regulatory mechanisms would have evolved.

involves mechanisms ranging from single glucose transport molecules to a global hunger drive. It thus seems reasonable to us that evolution, having identified the value of sleep-dependent modifications of synaptic plasticity, would end up producing multiple levels at which these modifications occur.

### HR6.2. Developmental plasticity

But how many of these putative levels of synaptic plasticity during sleep can be supported by data? At the most basic level, Shaffery et al. (1996; 1999) have proposed that REM sleep is critical for the experience-dependent wiring of binocular cells in the young kitten's striate cortex. In this case, REM sleep serves to enhance plastic changes that are initiated during waking visual activity, as part of an evolutionarily designed developmental process. The changes occur only during a brief developmental period, and the sleep-dependent process serves a very specific purpose, namely, to reduce the susceptibility of the lateral geniculate nuclei to undesirable plastic changes during normal CNS growth and maturation. In this case, sleep-dependent neuronal plasticity would most likely be classified as involved in neither learning nor memory processes. Since selective blockade of PGO waves without disruption of REM sleep produced similar effects, they proposed that PGO waves during sleep act to minimize "undesirable variability" during an intense period of experience-dependent plastic changes. **Jones** has also noted this possible function of REM sleep.

### HR6.3. Experience dependent plasticity

At the next higher level are the studies of Karni et al. (1994) and **Stickgold** et al. (2000a; 2000b), showing a REM sleep and SWS requirement for consolidation of a visual discrimination learning task. While this clearly qualifies as perceptual skill learning, changes may be occurring simply at the level of orientation specific neurons in striate cortex (Karni et al. 1995) and may not involve consolidation of memories as normally construed. In this case, evidence for a sleep dependency is seen in (1) the failure of any improved performance to develop over 12 hours of wake, coupled with the appearance of robust improved performance (Fig. HR5A) after a night of sleep (Stickgold et al. 2000b), (2) the absence of improvement after a night's sleep with instrumental REM, but not SWS, disruption (Karni et. al. 1994), (3) the strong correlation (Fig. HR5B) of overnight improvement with amounts of early SWS and late REM sleep (r = 0.89, p < 0.0001; Stickgold et al. 2000a; 2000b), and (4) the absence of improvement after a night of full sleep deprivation, despite two subsequent nights of recovery sleep (Stickgold et al. 2000c). In this last study, we have demonstrated that when subjects are sleep deprived the night after the training and then retested at 72 hours, after two full nights of recovery sleep, no significant improvement is seen (p > 0.3), with the average improvement being only one-fifth that of controls (controls > deprived, p = 0.014). Thus, we can now go beyond the correlational findings of our first study to state that the improved performance requires sleep within 24 hours of training.

### HR6.4. Procedural memory

Contrary to **VERTES & EASTMAN**'s claims, considerable evidence does support the concept of sleep-dependent mem-

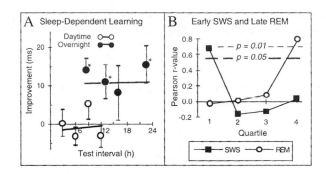

Figure HR5.    Visual discrimination learning. A: Time course of improvement. No improvement was seen until after a night of post-training sleep. Subjects were tested either the day of training with no intervening sleep (open circles) or on the day after training, following a night's sleep (filled circles). Each point in the graph represents a separate group of subjects. Means and error bars (s.e.m.) are shown for each test interval. Solid lines are linear regression fits to data points. B: Correlation of learning with SWS and REM across the night. Overnight improvement showed significant correlations with the amount of SWS in the first 2 hours of the night and the amount of REM sleep in the last 2 hours. For each quartile of the night, the Pearson correlation coefficient between SWS% and overnight improvement (filled squares) and between REM% and overnight improvement (open circles) was calculated. Heavy dashed line: correlations are significant (p < 0.05) for Pearson r-values greater than 0.57; light dashed line: correlations are more significant (p < 0.01) for Pearson r-values greater than 0.71. *Individual groups showing significant improvement. (From Stickgold et al. 2000b).

ory consolidation in various appetitive and aversive training paradigms. Specific response to **VERTES & EASTMAN**'s article can be found in the commentaries of **Bednar, Born & Gais, Cipolli, Fishbein, Jones, Mazzoni, Morgane & Mokler, Ogilvie et al., Smith & Rose,** and **Stickgold**. We conclude that in fact the only reasonable reading of the totality of the animal sleep and memory consolidation literature leads inevitably to the conclusion that sleep, and most clearly REM sleep, plays an important role in procedural memory consolidation, including spatial memory tasks. Much of this has been reviewed elsewhere (Smith 1985; 1995).

### HR6.5. Declarative memory

Moving to yet more complex systems, evidence has now begun to accrue arguing for a role of sleep, although not necessarily REM sleep, in the consolidation of episodic memories. The role of NREM sleep in the consolidation of declarative memories has been discussed in the commentaries of **Born & Gais, Smith & Rose, Jones,** and **Stickgold**. Further support for this role comes from the hippocampal recordings of Wilson and McNaughton (1994) and Buzsáki's model of a hippocampo-neocortical dialogue (Buzsáki 1996). A general discussion of the issues of REM versus NREM consolidation of procedural versus declarative memories can be found in the commentary of Stickgold (1998).

### HR6.6. Memory transfer

As we move to yet higher levels, issues of information transfer between brain regions arise. Wilson and McNaughton

(1994) have shown that ensembles of rat hippocampal "place cells" (Otto et al. 1991) which are uniquely activated during exploration of a new environment are selectively reactivated during subsequent SWS. Again, it is difficult to conceive of an explanation of this replay of activity patterns that does not involve memory consolidation, transfer, or integration. Buzsaki (1996) has shown that, during this SWS, hippocampal activation is played back through the entorhinal cortex to the neocortex. Such replay, according to the model of McClelland et al. (1995), serves to transfer hippocampal episodic memories to neocortical memory systems, where they are integrated into general semantic memory networks.

More interesting, this flow of information between the hippocampus and neocortex is state specific. Information flows out of the hippocampus and into the neocortex during SWS, but out of the neocortex and into the hippocampus during REM sleep. Buzsaki (1996) has referred to this back-and-forth communication as a "hippocampo-neocortical dialogue."

The upshot of all these findings is a model of sleep and memory consolidation (Stickgold 1998) in which (1) information flows out of, but not into, the hippocampus during NREM sleep, when dreams are indeed more thought-like and linked to episodic memories, and (2) information flow out of the hippocampus is shut off during REM sleep, when associations are weakened, limbic brain regions are reactivated, and dreams become bizarre, emotional, and isolated from waking memories. Within such a model, a prime function of NREM sleep is the consolidation of hippocampal, episodic memories and their transfer to cortical semantic memory systems, while REM sleep serves to integrate these and older memories within semantic networks. Of course, in the larger context of this special issue on sleep, we must reconfirm that all of these processes vary across the night and across individual REM and NREM periods, and that none of the features listed is absolutely constrained to any specific polysomnographically defined sleep stage. In this context, we have recently shown in humans that memories of recent waking activities can be reactivated at sleep onset in hypnagogic imagery, even in densely amnesiac patients with extensive bilateral hippocampal and medial temporal lobe damage who are otherwise unable to recall these waking activities (Stickgold et al. 2000a; 2000d).

### HR6.7. Problem solving

At still higher levels, we encounter theories such as that presented by **REVONSUO,** where sleep and possibly dreaming, are seen as part of a very high level of process of memory consolidation and integration that can lead to problem solving, conflict resolution, and creativity. Here there seems to be little agreement. Part of the problem has to do with difficulty in developing adequate paradigms for studying these issues, although both **Cartwright** et al. (1998a) and **Smith** (1993) have approached this question, albeit from very different angles. As mentioned earlier, we have begun to look at this question from still another vantage point by looking at changes in semantic priming following awakenings from REM and stage 2 NREM sleep (Stickgold et al. 1999b), and showing that REM awakenings lead to an unprecedented enhancement of the priming ability of weakly related words (Fig. HR4). These results suggest that the functional connectivity among associations in the neocortex are altered

during REM sleep, as if to facilitate the identification and possible strengthening of weak, "unexpected associations."

In summary we have shown that there are many potential roles for sleep in the functional modulation of synaptic plasticity, and that different phases of sleep may contribute differently to each of these functions. Certainly, the vast differences in electrophysiology, regional brain activation, hippocampo-neocortical communication, neuromodulators and even hormones such as the corticosteroids (see **Born & Gais**), point to different functional roles for different portions of the sleep cycle. From our analysis, there is now substantial evidence for the role of sleep in more basic forms of synaptic plasticity, up to the level of both animal and human learning paradigms, and tantalizing hints at all of the higher levels described. Finally, it is our very strong belief that the continued study of this question, borrowing heavily on the paradigms of cognitive neuroscience, will produce breathtaking breakthroughs in our understanding of this critical function of sleep over the next decade.

## HR7. Conclusion

In the year 2000 Dream science has just completed its first century of life. The proto-scientific efforts of Sigmund Freud at the turn of the twentieth century were theoretically brilliant but empirically inadequate (Freud 1900). The field has developed particularly strongly in the last fifty years because the brain sciences have supplied some of the fundamental neurobiological material that Freud needed to accomplish his failed purpose: to create a dream psychology that was perspicacious and free from doubt. Although they are far from free of doubt, brain based theories of dreaming are certainly perspicacious in that they can account for global mental processes in cellular and molecular terms.

Now, at the turn of the twenty-first century, brain science has supplied another key piece of information: knowledge about the relative strengths of activation (and deactivation) of brain regions that regulate many of the critical aspects of cognition that are altered in dreaming. In this response, we have tried to explain, extend, and further justify our assessment of the current state of knowledge in five important areas of dream science as follows:

1. *Phenomenology:* More widespread and more detailed attention to defining and quantifying first person data is needed. We showcase the formal approach and urge its wider adoption.

2. *Psychophysiology:* In the debate regarding the ability of the NREM and REM states to produce dreaming, the emphasis should shift and be placed on the goodness of fit between the psychological and physiological levels of analysis. To our mind the emphasis on dissociation is both unjustified and counterproductive.

3. *Neurobiology and Neuropsychology:* The success of cellular and molecular level analyses of waking, NREM, and REM sleep provide an embarrassment of explanatory riches which now comprise many subcortical brain regions including the spinal cord, medulla, pons, midbrain, hypothalamus, and thalamus. The advent of imaging should stimulate an upsurge of attention to other subcortical regions (e.g., the amygdala) and many specialized regions of the cerebral cortex including the dorsolateral prefrontal cortex, the inferior parietal lobe, anterior cingulate and other limbic-related cortices and visual associative areas.

4. *Models and philosophy:* Dream science is now poised to advance from a descriptive to a hypothesis testing phase. Models are essential to this welcome coming-of-age. To be effective, such models must be simple enough to be understood and tested, complex enough to be realistic, and sophisticated enough to anticipate cross-domain reductionism. AIM is such a model.

5. *Learning and Memory:* It is a plain fact that the brain is processing information during sleep and empirical support for progressive, elaborative, and adaptive aspects of this processing is robust. We anticipate explosive growth of informative evidence through the combination of cognitive, cellular, and molecular neuroscience with neuroimaging techniques.

## Forebrain mechanisms of dreaming are activated from a variety of sources

Mark Solms

*Academic Department of Neurosurgery, St. Bartholomew's and Royal London School of Medicine, Royal London Hospital, London E1 1BB, United Kingdom.* mlsolms@mds.qmw.ac.uk    www.mds.qmw.ac.uk

**Abstract:** The central question facing sleep and dream science today seems to be: What is the physiological basis of the subset of NREM dreams that are qualitatively indistinguishable from REM dreams ("apex dreams")? Two competing answers have emerged: (1) all apex dreams are generated by REM sleep control mechanisms, albeit sometimes covertly; and (2) all such dreams are generated by forebrain mechanisms, independently of classical pontine sleep-cycle control mechanisms. The principal objection to the first answer is that it lacks evidential support. The principal objection to the second answer (which is articulated in my target article) is that it takes inadequate account of interactions that surely exist between the putative forebrain mechanisms and the well established brainstem mechanisms of conscious state control. My main response to this objection (elaborated below) is that it conflates nonspecific brainstem modulation – which supports consciousness in general – with a specific pontine mechanism that is supposed to generate apex dreaming in particular. The latter mechanism is in fact neither necessary nor sufficient for apex dreaming. The putative forebrain mechanisms, by contrast, *are* necessary for apex dreaming (although they are nor sufficient, in the limited sense that *all* conscious states of the forebrain are modulated by the brainstem).

## SR0.  Introduction

I very much appreciate the confirmation, amplification, and elaboration of my views that the commentators have provided. For obvious reasons, however, I will focus in this response on the outstanding areas of *disagreement* between the commentators and me. This provides an opportunity not only to remove misunderstandings but also to reconsider some aspects of the proposals outlined in my target article.

Considering the first three target articles together (**HOBSON ET AL., SOLMS, NIELSEN**) through the lens of the commentaries, it is apparent that we have all attempted to answer the same basic question: namely, what is the physiological basis of NREM dreams which are "indistinguishable by any criterion" (Hobson 1988b, p. 143) from REM dreams?

### Qualitative differences between REM and NREM dreams

I should point out immediately, in view of continuing misunderstanding by some commentators (**Antrobus; Clan-**

cey; **Moorcroft; Panksepp; Shevrin & Eiser**),[1] that this question does not concern normative differences between *average* NREM and REM dreams; it concerns only the *subset* of NREM dreams that are indistinguishable from REM dreams. Few people would disagree that the average NREM dream is more "thoughtlike" than the average REM dream. The existence of thoughtlike dreams presents no difficulties for the REM = dreaming doctrine. What presents problems for that doctrine is the fact which Hobson (1988b) acknowledges in the quotation cited above, namely that *some* NREM dreams *are* REM-like. (In the remainder of this response, I use **NIELSEN**'s term "apex dreams" in this broad sense to denote all REM-like dreams.)

## SR1.  What generates NREM apex dreams?

**HOBSON ET AL.** and **NIELSEN**, after some equivocation, provide essentially the same answer to this question: They claim that NREM apex dreams (like REM ones) are caused by REM sleep control mechanisms. In other words, they claim that *all* apex dreams are generated by the pontine oscillator that generates the REM state. I am grateful to **Hobson** for stating the assumption underlying this new viewpoint so baldly: "REM physiology is relevant to dream generation even in what is called NREM sleep . . . [because] *all of sleep is REM sleep* (more or less)" (item 3 from the last section of his commentary; my emphasis). This claim, to my mind, is the only way in which the REM = dreaming doctrine can still be defended in the light of all the evidence showing that NREM apex dreams occur quite frequently, in both health and disease.[2] There could be no other valid basis for continuing to assert (as Hobson does) that "the REM state is the optimal substrate for dreaming and therefore uniquely informative about its physiology," and that "REM sleep and dreaming co-occur . . . [and are] two different manifestations of a single process" (**Greenberg**, para. 3). Fortunately, it is not difficult to test **HOBSON ET AL.** and **NIELSEN**'s new claim. **Feinberg** stipulates the simple criterion by which we may do so: "Unless intermediate states with REM characteristics are *always* present when [apex] dreamlike narratives are elicited from sleep, covert REM cannot rescue the REM-dream isomorphism" (sect. 5, para. 2 of his commentary). The evidence currently available (some of it summarized by **Bosinelli & Cicogna; Ogilvie et al.; Perry & Piggott; Porte; Vogel**) indicates that this criterion is unlikely to be met.

I would go further than **Feinberg**, and suggest (with **Conduit et al.; Gottesmann;** and **Panksepp**) that there should be *some common physiological factor* which is always present in the diverse "intermediate" states hypothetically associated with NREM dreaming. Only this common factor (presumably some component *or correlate* of the classical REM state) could legitimately be described as the "generator" of apex dreaming. If such a factor can be identified, we will need to ascertain whether or not it is produced by the classical pontine sleep-cycle control oscillator in sporadic instances of NREM apex dreaming. The "covert REM" hypothesis leaves wide open the possibility that apex dreaming is generated by some other mechanism – including "top-down" mechanisms (of the kind postulated by **Nofzinger; Salin-Pascual et al.; Vogel,** and others). This reintroduces the original question in a different form: What generates "covert REM"? (see **Solms** commentary on **NEIL-**

**SEN**). In short, if one really believes that "all of sleep is REM sleep" (**Hobson**), then the critical question as to what generates apex dreaming remains unanswered.

In my target article, I provided a different answer to the original question. I argued that the fact that apex dreams frequently occur outside of REM sleep (as classically defined by Rechtschaffen & Kales 1968) suggests that the pontine sleep-cycle control oscillator probably does *not* control dream generation. If one nevertheless assumes (with **Gottesmann; Perry & Piggott,** and others) that a mental state as distinctive as apex dreaming is likely to have a distinctive neurophysical correlate (and I note that some commentators do appear to dispute this assumption; e.g., **Ardito**) then the question becomes: What other than the REM/NREM oscillator generates apex dreams?

## SR2. Apex dreams in relation to other varieties of dreams

Before answering this question, I wish to emphasize again that it concerns only apex dreaming. It certainly is possible (indeed likely) that the core mechanisms of dream generation vary somewhat, in tandem with the many varieties of dream experience that actually occur in nature. Thus, although I can see why **NIELSEN** classified me as a "one generator" theorist, I can only accept this classification with respect to *apex* dreams (both REM and NREM); I do not claim that *all* varieties of dream experience are generated by a single, invariable mechanism (cf. **Antrobus; Hunt; Panksepp**).

## SR3. Role of brainstem arousal

One obvious way to begin answering the question "What generates apex dreams?" is to isolate the neuroanatomical structures and/or neurophysiological functions that are essential for apex dreaming to occur. There are various ways of doing this, the now-conventional ones being lesion studies, functional imaging studies, and pharmacological studies. The available evidence arising from such studies, summarized in my target article, focuses attention on the mesocortical/mesolimbic dopamine fibres coursing through the ventromesial quadrant of the frontal lobes. On the basis of this evidence, I suggested that the distinctive neural generator (i.e., initiator) of apex dreaming is (1) mesocortical/mesolimbic dopaminergic innervation, (2) in the context of nonspecific cerebral activation, and (3) during the state of sleep.

Those commentators who assert that I "maintain that it is the cortex alone that is responsible for dreams" (**Greenberg**, para. 2) and that I "divorce the pons from any role in dreaming" (Greenberg, second para. from end), overlook the second element in this equation. The same applies to **HOBSON**'s suggestion that I think "the forebrain is free to act alone and can produce dreams independent[ly] of the activating, input-output gating and modulatory influences of the brainstem" (final section of his commentary).

**Feinberg** points out that **HOBSON ET AL.** conflate rejections of their conception of dream physiology with rejection of the very idea of dream physiology. Likewise here, a rejection of **Hobson**'s view of the brainstem's role in dreaming is conflated with a rejection of *any* role for the brainstem in dreaming. I do not deny that nonspecific

cerebral activation (which is usually brainstem mediated) is a necessary condition for apex dreaming to occur.[3] What I dispute is the more specific claim that the brainstem's contribution to apex dreaming is provided by the unique properties of *REM* activation (i.e., only by aminergically demodulated, pontine cholinergic activation of the forebrain). **Vogel** presents precisely the same argument. The essential empirical basis for this opinion is that apex dreaming not infrequently occurs independently of the REM state (classically defined; i.e., the state that Hobson & McCarley's reciprocal interaction model was meant to explain).

Any theory of dream generation which excludes a role for the consciousness sustaining structures of the brainstem is certainly incomplete. The conscious forebrain is *always* modulated by the brainstem. However, we must recall that apex dreaming is not merely a state of consciousness, it is a *particular variety* of consciousness; and it is the particularities of the state that require explanation. I agree with **Gottesmann** that nonspecific brainstem arousal mechanisms *support* rather than *cause* dreaming (cf. **Hunt**). I therefore do not agree (with **Coenen,** and others) that dreaming is *solely* a function of high brain activation during sleep. This latter viewpoint cannot explain why dreaming so often does *not* occur when the brain is highly activated (e.g., during REM sleep in cases with bilateral ventromesial frontal lesions; Jus et al. 1973 – this being one "proof" of double dissociation, of the kind that Coenen asks for). Likewise, it cannot explain why L-DOPA intensifies the "apex" properties of REM dreams (vivacity, complexity, affectivity) without also intensifying the concomitant REM activation (Hartmann et al. 1980).

It is for these reasons, among others, that I suggest dreaming requires nonspecific cerebral activation during sleep *plus* mesocortical/mesolimbic dopamine innervation (*contra* **Franzini** and **Hobson**). For the same reason, I cannot accept **Panksepp**'s suggestion that "emotion laden dreams remain so only because of the sustaining influences of REM arousal." It appears that apex dreaming is obliterated only by damage to Panksepp's dopaminergic SEEKING system (notwithstanding the preservation of REM arousal in these cases) and not by damage to any of the other basic emotion command systems described by him (Panksepp 1998a) *nor by the basal forebrain cholinergic systems that* **HOBSON ET AL.** *now implicate in dream generation* (see below).

## SR4. Loss of REM with preservation of dreaming

I acknowledge that *preservation* of dreaming with loss of REM sleep has not been unequivocally demonstrated (cf. **Franzini; Greenberg**) but in view of the fact that cessation of REM *is* sometimes accompanied by the preservation of reticulo-thalamic extended activating mechanisms (and therefore consciousness; see note 8 of my target article), I expect that this clinicoanatomical dissociation can be confirmed in principle. Unfortunately, such cases are few and far between. They are also gravely ill, with the result that ethical considerations add to the difficulty in studying them comprehensively. Lack of evidence in this regard cuts both ways, of course. One might equally well ask why no one has ever been able to demonstrate an unequivocal case of cessation of REM sleep associated with *loss* of dreaming.

We also need to acknowledge that literally hundreds of non-dreaming cases with focal forebrain lesions have been documented.

## SR5. Late morning effect in relation to REM/ NREM distinction

I did not mean to imply in my target article that **Antrobus**'s "late morning effect" is a specific NREM effect in the sense that it is generated by the known physiological mechanisms of NREM sleep control. When I described it as a non-REM effect, I meant that it was *not* mediated by REM control mechanisms. Antrobus's clarification (echoed by **Cavallero** and amplified by **Feinberg**) to the effect that the late morning effect is orthogonal to both REM and NREM mechanisms (i.e., that apex dreaming increases across both REM and NREM stages toward the end of the night) actually underscores the point I was trying to make: namely, that apex dreaming cannot be reduced to classical pontine sleep cycle control mechanisms. (I note that **Hobson** and **Ogilvie et al.** still construe the late morning effect as if it *were* reducible to REM mechanisms).

## SR6. Dopamine in relation to other neurotransmitters

**Gottesmann** and **Perry & Piggott** accept that mesocortical/mesolimbic dopamine plays an essential role in apex dreaming, but they suggest that this role should be contextualized in a broader formula which also incorporates noradrenergic and serotonergic demodulation and cholinergic activation (**Panksepp** and **Rotenberg** make similar suggestions). As already stated in my target article, I accept that NA and 5HT demodulation probably facilitate the "dream-on" effects of mesocortical/mesolimbic DA activation during sleep. However, in view of the fact that apex dreaming apparently also occurs during sleep stages traditionally characterized by relatively high levels of NA and 5HT activation (e.g., sleep onset and the rising morning phase of the diurnal rhythm), I do not consider these particular facilitatory effects to be *obligatory* for the appearance of apex dreaming.

As to the role of ACh: I cited evidence in my target article to show that dreamlike mentation is facilitated by *anticholinergic* drugs and by the *destruction* of the (cholinergic) basal forebrain nuclei that **HOBSON ET AL.** now implicate in dream generation. (On these grounds I must reject **Doricchi & Violani** and **Hobson**'s suggestion that cessation of dreaming following ventromesial frontal white matter lesions might be attributable to the interruption of cholinergic fibres.) It is doubtful that ACh activation (at the forebrain level, at least) contributes to the generation of apex dreaming in any simple fashion. It is possible that ACh release during sleep facilitates dreaming indirectly via the mediation of DA mechanisms. It is even possible that basal forebrain nuclei are actually inhibited during sleep by ascending pontine cholinergic projections (see Braun 1999).[4] As **Panksepp** points out, REM arousal mechanisms are more primitive in brain evolution than the waking circuits of the extended reticulo-thalamic activating system. It is almost certainly the case that the nonspecific activating effects of the REM state on the forebrain are mediated indirectly, perhaps largely by noncholinergic mechanisms (see

Braun 1999 for a summary of evidence supporting this view).[5] Either way, a cholinergic mechanism that is obligatory for the appearance of apex dreams has never yet been demonstrated.

A number of commentators draw attention to the fact that ventral tegmental DA activity does not co-vary in a simple fashion with classical sleep cycle stages in rats and cats, thereby calling into question the role I have assigned to mesocortical/mesolimbic DA fibres in dream generation. This reasoning assumes that dream generation is a simple function of the classical sleep stages, which I do not accept.[6] Moreover, it assumes that sleep architecture bears the same relationship to apex dreaming in rats and cats that it does in humans. This assumption might well be incorrect – especially in view of known differences in DA innervation of the forebrain across the species in question (Berger et al. 1991), and also in view of the central role that the PTO junction appears to play in the maintenance of dream consciousness. (It is questionable whether Brodmann's areas 39 and 40 have direct homologues in other primates, let alone in rats and cats; Creuzfeldt 1995.)

**Cavallero**'s reminder that apex dreaming appears to be a relatively late cognitive acquisition in human development is also relevant here. I cannot emphasize enough that what I am attempting to isolate when I refer to the role of DA in dreaming is the neural correlate of the *mental state* of apex dreaming itself. What is required, therefore, are studies which correlate mesocortical/mesolimbic dopamine activity with fluctuations in *mental state*, which (sadly) can only be done in humans. How, for example, does one begin to interpret the implication for *dreaming* in the observation that ventral tegmental DA cells in rats fire with a more variable interspike interval during REM than during slow wave sleep (Miller et al. 1983)?

In this connection, I strongly agree with **Lehmann & Koukkou**'s remarks to the effect that shifts in brain state need to be studied at a higher temporal resolution than in classical approaches to sleep physiology. However, I also think that "state" encompasses more relevant physiological dimensions than can be measured by EEG; we need to distinguish momentary shifts in the dynamic states of *different neuromodulatory systems* (cf. **Franzini**; **Hartmann**; **Nofzinger**).

I accept that if mesocortical/mesolimbic DA were to play an exclusive role in dream generation, then VTA dopaminergic activity (in humans) should co-vary in some direct way with dream frequency and/or intensity. However, along the lines of the formulae advanced by **Gottesmann** and **Perry & Piggott**, it seems more likely that what is decisive for the appearance of apex dreaming is not the absolute level (or pattern) of DA activation but rather the *ratio* of mesocortical/mesolimbic DA activity to some other modulatory variable(s). Critical tests (which **Panksepp** calls for) of the role of dopamine in such dynamic equations can begin from the experimental hypothesis that anything which increases or decreases mesocortical/mesolimbic DA activity (perhaps at D3 receptors in particular) should result in increased and decreased frequency and/or intensity of dreaming, respectively. This hypothesis could be operationalized in various ways, including lesion studies,[7] advanced functional imaging, and pharmacological studies (cf. **Borbély & Wittmann**).

**Bednar, Doricchi & Violani,** and **Hobson** ask why I have assigned an exclusive role to mesolimbic dopamine circuits in dream generation. As should be evident from what

I have just said, and also from section 8 of my target article which summarizes the network of forebrain structures that generate dreams, I do not assign an exclusive role to the mesolimbic dopamine circuit in apex dream generation (although I do venture to assign it an obligatory role). The emphasis on the "control" of dreaming in the structure of my target article derived more from the structure of the model I was arguing against than from my own way of thinking about dreams. Doricchi and Violani (1992) correctly point out that dreaming can also be interrupted at other levels of processing (e.g., at the level of visuospatial representation).

## SR7. Backward projection

**Antrobus** complains that I take for granted the folk psychological notion that information is "projected backwards" in dreaming, from the level of meaning generation to that of visuospatial representation (i.e., from limbic-frontal to parieto-occipital stuctures). He argues that an equally plausible alternative is that (parieto-occipital) image production is initially independent of (limbic-frontal) conative influences, and that the latter might consist in the subsequent evaluation of the visual images ("Is that a friendly or unfamiliar face?") followed by the imagined motor responses ("Shall I run?"). This alternative seems unlikely in the light of the clinicoanatomical evidence. Lesions in the occipitotemporal structures which I place at the terminal (representational) end of the dream process are associated with dreams that are normal in every respect, apart from their lack of visual imagery (or specific components of visual imagery). If the process started with the visual images, there should be nothing for the frontal-limbic structures to evaluate (and therefore no dreams) in such cases. Conversely, lesions in the deep frontal-mesencephalic structures which I situate at the initiating end of the process are associated with cessation of dreaming, not with visual dreams *sans* conative associations.

It is also noteworthy in this connection that disorders of dream imagery reverse the hierarchy that characterizes disorder of waking visual perception. In waking perception, early visual (striate) cortical lesions obliterate visual perceptual consciousness; late unimodal visual (extrastriate) cortical lesions affect selective aspects of complex visual perception (colors, faces, movement, etc.); heteromodal posterior cortical lesions spare primary visual perception altogether but disturb visuospatial cognition. By contrast, early visual cortical lesions have no effect on dream imagery, late unimodal visual cortical lesions disturb selective aspects of complex visual dream imagery (color, faces, movement, etc.), and heteromodal posterior cortical lesions obliterate dream imagery completely. Nevertheless, I freely concede that this whole issue needs further study.

**Rotenberg** makes a point similar to that of **Antrobus** when he says that the correlation between some complex partial seizures and recurring nightmares does not imply that dreams are *produced* by forebrain mechanisms but only that they are *influenced* by forebrain activity. This argument overlooks the fact that the same nightmares can actually be produced experimentally – in the form of "dreamy state" seizures – by stimulating the temporal lobe focus (Penfield 1938; Penfield & Erickson 1941). Complex partial seizures are of course circumscribed forebrain events by definition.[8] (Incidentally, the powerful affective tone of these epileptic nightmares further contradicts **Panksepp**'s intimation that only REM dreams are emotional.)

**Bednar** and **Conduit et al.**'s remarks about the role of V1 in normal visual imagery does not bear any necessary relationship to its role in apex dreaming, as the mental phenomena in question are quite different. Normal mental imagery is not hallucinatory and is highly constrained by dorsolateral prefrontal structures (among other things) which are deactivated during most apex dreaming. On the other hand, **Hartmann**'s intuition to the effect that waking imagery and fantasy are affectively bland and, in other respects, deficient in anoneiric patients does seem to be correct (see Frank 1950; Hoppe 1977).

## SR8. Importance of converging evidence

**Doricchi & Violani** correctly point out that I (Solms 1997a) did not directly test their hypothesis to the effect that visual-verbal disconnection underpins reported cessation of dreaming in some (left occipito-temporal) cases. This was due to the fact (mentioned in the preface to my book) that I had already completed my research some 12 months before the appearance of their review (Solms 1991). They also believe that I now "reject" the role of the PTO junction. One reason for de-emphasizing the role of the PTO junction in my target article was the fact (mentioned briefly in note 15 of the article) that the role I originally assigned to the PTO junction in dreaming was called into question by the subsequent findings of Braun et al.'s (1997) PET study which, unlike Hong et al. (1995) and Maquet et al.'s (1996) earlier studies, suggested that this region was deactivated during REM sleep. In other words, by the time I wrote my target article, the involvement of this region in REM dream generation had become controversial. (I note from **Nofzinger**'s commentary that he has since confirmed the earlier observation to the effect that this region is highly activated in REM sleep.) All of the PET evidence published after I completed my study, by contrast, confirmed my conclusion (based on clinicoanatomical evidence) to the effect that the ventromesial quadrant of the frontal lobes (and associated basal forebrain structures) are strongly implicated in apex dream generation. (Incidentally, the convergence of the clinicoanatomical and functional imaging evidence in these respects is one reason to doubt Doricchi & Violani's suggestion that the anterior cingulate activation noted in the imaging studies might be a sleep-deprivation artefact.) I hope it is clear, therefore, that I have focused on the mesocortical/mesolimbic DA system for the reason that *converging lines of evidence* (derived from various methods), summarized in my target article, point to the conclusion that this is the component of the ventromesial frontal white matter that best explains loss of dreaming with lesions in this site. However, this should not detract from the fact that occipitotemporal (and perhaps inferior parietal) tissues are implicated in the representation of dream cognition and perception.

A number of commentators have criticized other individual aspects of my evidential base. I cannot deal with each of these specific points in detail here. I can only say that while I accept that some aspects of the evidence I cite are stronger than others, criticisms of my general conclusions have to be measured against the evidential base *as a whole*.[9] This applies especially to the very important question as to whether "cessation of dreaming" actually means "loss of recall for dreams" (**Feinberg; LaBerge; Ogilvie et al.**). I have considered this possibility from every conceivable methodological angle (Solms 1997a) and concluded that reported cessation of dreaming due to focal brain lesion cannot be attributed to a disorder of recent memory (see note 9 of my target article). The commentators in question seem to be unaware of this fact.

## SR9. Role of dorsolateral prefrontal cortex

The convergence of clinicoanatomical and functional imaging findings also contradicts **LaBerge**'s speculation that the cortical motor system functions in REM essentially as it does in waking. (This, incidentally, has important implications for **REVONSUO**'s target argument.) Findings from both methods indicate that this region most likely plays no active role in (apex) dreaming. The apparent lack of involvement of dorsolateral frontal cortex in apex dreaming also make lucid dreamers poor subjects for clarifying the role of executive mechanisms in normal dream cognition (cf. LaBerge) for the reason that these atypical subjects are likely to be exceptional with respect to precisely the mechanisms at issue. More useful investigations of lucid dreamers in the present context than what LaBerge suggests might be to determine whether the capacity for lucid dreaming survives DLPFC lesions. (This underlines the point made by **Occhionero & Esposito** to the effect that the hypoactivation of the DLPFC makes a specific – albeit negative – contribution to dream cognition; I agree that my description of it as "inessential" to apex dreaming was an unfortunate choice of word.)

I take it as given that every method has its strengths and limitations, and that the most reliable conclusions are based on evidence derived from a range of different methods. The same applies, in my view, to the distinction between investigations of the subjective experience of dreaming and the neural correlates of dream experience (a distinction which a surprising number of commentators refer to: see **Ardito; Hartmann; Kramer; Mancia; Pagel; Vogel**). I take it for granted that some questions about dreams are best addressed by psychological methods and others by neurological ones. Nevertheless, the two approaches ultimately study one and the same "thing" and therefore must – in principal – be reconcilable with one another (see Solms 1997b, *pace* **Franzini**'s attribution to me of "extreme reductionism").

## SR10. The specter of psychoanalysis

The previous point allows me to respond pertinently to **Hobson**'s puzzling comments regarding my "commitment to Freud."[10] I do not think it appropriate to rule out *a priori* any method or hypothesis on the basis of simple prejudice. (cf. also **Doricchi & Violani**'s anti-psychoanalalytic opinion.) My interest in Freud's dream theory in this regard is two-fold. First, early research findings which pointed to a REM-based (autochonous pontine brainstem) mechanism of dream generation were widely interpreted as disproving Freud's theory (including by HOBSON himself: Hobson 1988b; Hobson & McCarley 1977). If more recent research now casts doubt on the validity of the earlier findings, then it is necessary – and only fair – to re-evaluate those criticisms in the light of the new evidence. This entails a commitment to scientific truth and fair play, not to Freud.

The new clinicoanatomical and functional imaging findings do indeed seem to suggest that Freud's hypothesis to the effect that dreaming is instigated by the arousal of instinctual mechanisms during sleep is more tenable than we previously thought. Whether Freud's view is actually "confirmed" (**Flanagan**) by the new data is a different matter. I do not think this data is any more confirmatory in a definitive sense of Freud's theory than it is of some of the other interesting hypotheses advanced by commentators in this volume, linking dreams with various attentional, orienting, curiosity, and search mechanisms (e.g., **Hunt; Conduit et al.; Jones; Morrison & Sanford; Panksepp; Rotenberg**). Nevertheless, by seeking to establish the neuropsychological correlates of Freud's theory, we do at last open it to rigorous experimental testability (cf. **Vogel**). This is the second aspect of my interest in Freud's dream theory. For example, now that we have identified a group of patients who cannot dream as the result of focal brain lesions, it should finally be possible to test his hypothesis that dreams function to protect sleep. It is not possible to answer Flanagan's question in this regard (or indeed his broader question as to whether or not dreams serve any function at all) until such experiments have been performed.

**Flanagan**'s view that abnormalities associated with loss of dreaming in brain damaged cases cannot cast light on the function of normal dreaming for the reason that these cases have abnormal brains overlooks the conventional rationale for comparing control and experimental groups who are alike in all respects other than the dependent variable. If, for example, two groups of patients with PTO junction lesions are compared, and the dreamers sleep more soundly than the non-dreamers, then this says something about the function of dreaming in particular rather than the effects of PTO junction abnormality in general. (**Kramer** is therefore wrong to say that I do not provide a basis for establishing the function of the dream experience.)

## SR11. Concluding remarks

In closing, I want to restate my essential theoretical conclusions as clearly as I can:

1. *Dreams are generated by forebrain mechanisms.* The functional anatomy of these mechanisms has been laid bare (in broad outline) by strongly convergent clinicoanatomical and functional imaging findings. The neurochemistry of these mechanisms has not yet been adequately characterized, but converging lines of evidence *suggest* that mesocortical/mesolimbic dopamine is an essential element in an emerging dynamic equation. The psychology of these mechanisms has not yet been adequately studied.

2. *The forebrain "dream generator" can be activated from a variety of sources.* REM arousal is undeniably the most common source. However, the intrinsic mechanisms of the REM state are external to dream generation itself because *other sources of activation produce the same forebrain effects.* These sources of activation apparently include "top down" influences, some of which may be intrinsic to the dream generator. However, it is of course likely that all such influences recruit brainstem activating mechanisms (and these may include some components or correlates of the REM generator).

I would like to thank my colleagues most sincerely for the careful attention they have given to evaluating my findings and hypotheses concerning the neuropsychology of dreaming. It appears that we truly are in the midst of a paradigm shift in sleep and dream science, and I consider myself fortunate to be part of it.

NOTES

1. I am referring to comments such as this: "There is a real danger in proceeding as if REM and NREM mentation are the same, which Solms seems to argue" (**Moorcroft**, para. 4).

2. **Ogilvie et al.** appear to think that this happens only in pathological cases.

3. This issue is obviously relevant to **Conduit et al.**'s question: If spontaneous arousal during sleep does not arise from the brainstem, where is its origin? Cf. **Moorcroft**'s implicit answer: "it is possible that while these forebrain areas are preferentially activated by pontine influences during REM they may also be activated by non-pontine sources" (para. 7).

4. Likewise, when **Portas** draws attention to the apparent discrepancy between my observation that anterior cingulate *lesions* are associated with increased frequency and vivacity of dreaming and the functional imaging data which show that this region is highly *activated* during "dreaming sleep" (REM sleep), she neglects the possibility that the observed activation is inhibitory.

5. Braun (1999) also summarized numerous "viable links" (of the kind requested by **Morgane & Mokler**) between the cholinergic REM-on mechanism and the putatively dopaminergic dream-on mechanism.

6. Cf. **Feinberg**'s pregnant remark: "We reasoned that, since brain physiology is qualitatively different in NREM and REM, but the conscious experience of [apex] dreaming in the two states is not qualitatively different, 'the striking NREM/REM differences in neuronal firing must *not* involve the neural systems that can affect the quality of conscious experience'" (emphasis added).

7. Here is a critical test of the *obligatory* involvement of DA in apex dreaming: cases with suitably located, complete lesions of the ventromesial frontal dopamine pathways and *preserved* apex dreaming would disconfirm my hypothesis. Incidentally, **Morgane & Mokler** seem to be unaware of the "unlikely" fact that all reported cases of cessation of dreaming with pure ventromesial frontal lesions did indeed sustain *bilateral* damage (Solms 1997a).

8. **Occhionero & Esposito** ask for specific examples of NREM triggers of dreaming. Complex partial seizures (which are not "stage specific" but usually occur during NREM sleep) provide an excellent example. Normal equivalents may be inferred. Incidentally, I do not see a basis for the distinction that **Gottesmann** makes in this connection between "dreams" and "hallucinations." Are apex dreams not hallucinations?

9. For example, **Doricchi & Violani** point to the weak statistical correlation between cessation of dreaming and adynamia in a small group of deep ventromesial bifrontal cases reported in my (1997a) study, but make no mention of the ubiquity of this symptom in the vast psychosurgical literature. (Cf. **Morgane & Mokler**'s questions concerning the putative link between dreaming and motivational mechanisms.)

10. I have responded elsewhere to his detailed criticisms of Freudian dream theory in relation to recent neuroscientific findings (cf. Hobson 1999c; Solms 1999c; 2000) and therefore will not address them again.

# Covert REM sleep effects on REM mentation: Further methodological considerations and supporting evidence

Tore A. Nielsen

*Sleep Research Center, Hôpital du Sacré-Coeur de Montréal, Montréal, Québec,Canada and Psychiatry Department, Université de Montréal, Montréal, Québec H4J 1C5 Canada,* **t-nielsen@crhsc.umontreal.ca**

**Abstract:** Whereas many researchers see a heuristic potential in the covert REM sleep model for explaining NREM sleep mentation and associated phenomena, many others are unconvinced of its value. At present, there is much circumstantial support for the model, but validation is lacking on many points. Supportive findings from several additional studies are summarized with results from two new studies showing (1) NREM mentation is correlated with duration of prior REM sleep, and (2) REM sleep signs (eye movements, phasic EMG) occur frequently in NREM sleep. The covert REM sleep model represents one class of explanatory models that combines the two assumptions of mind-body isomorphism and a 1-gen mentation generator; its future development will depend largely upon a more detailed understanding of sleep state interactions and their contribution to mind-body isomorphisms.

## NR0. Introduction

Reactions to my target article varied from the extremely skeptical to the highly supportive with as many commentators favoring it as doubting its conclusions. Eight principal themes addressed by various authors are listed in Table NR1; these are dealt with in turn in the sections that follow.

## NR1. The definition of dreaming is inadequate

Some authors (**Antrobus; Clancey; Kahan; Pagel; Revonsuo**) expressed dissatisfaction with the definition of sleep mentation adopted in my target article. This dissatisfaction is justified to the extent that the classification scheme proposed in Figure 1 illustrates only in very broad strokes distinctions existing in the REM- NREM mentation literature that are central to my review, rather than providing a detailed classification system *per se*. However, as the covert REM sleep model has evolved, I have found it increasingly imperative to develop criteria to discriminate among very brief and minimal forms of mentation. To contribute to this goal, I have revised my previous Figure 1 to incorporate several concerns raised in the commentaries (see Fig. NR1).

I agree that a more theoretically neutral definition of dreaming is desirable (**Revonsuo; Kahan**), that is, that a definition of dreaming should be based as much as possible upon the *contents* of subjective experience.[1] At the very least, such a definition would allow investigators of different theoretical orientations to study the same phenomenal objects in a convergent fashion. A chronic lack of agreement on the definition of dreaming has contributed much to the current confusion in the 1-gen versus 2-gen debate (cf. **Pagel**). Revonsuo is therefore justified in questioning my inclusion of "cognitive processes" in the classification of sleep mentation. Cognitive processes are, indeed, a theory-laden descriptor whose superordinate position in relation to other categories in Figure NR1 is based upon the hypothetical notion (e.g., Dixon 1981; Freud 1900) that most activity supporting subjective awareness occurs outside of

Table NR1. *Commentaries on Nielsen target article: Main themes*

| Theme | Commentaries |
|---|---|
| 1. The definition of dreaming is inadequate | **Antrobus, Clancey, Kahan, Pagel, Revonsuo** |
| 2. Authors add new information that supports the model | **Borbély & Wittmann, Born & Gais, Cartwright, Feinberg, Gottesmann, Greenberg, Lehmann & Koukkou, Pace-Schott, Rotenberg, Salzarulo, Steriade** |
| 3. Waking state processes need further consideration | **Greenberg, Hartmann, Ogilvie et al., Schredl** |
| 4. Dreaming occurs during stages 3 and 4 sleep | **Blagrove, Bosinelli & Cicogna, Cavallero, Feinberg, Moorcroft, Ogilvie & Koukkou, Stickgold** |
| 5. The model links dreaming exclusively to brainstem activation in REM sleep | **Bosinelli & Cicogna, Domhoff, Porte, Solms, Salin-Pascual et al.** |
| 6. Evidence for isomorphism is lacking | **Hunt, Kramer, Morrison & Sanford, Panksepp, Solms, Vogel** |
| 7. Elimination of REM sleep does not eliminate dreaming | **Panksepp, Solms, Shevrin** |
| 8. The model needs validation | **Blagrove, Coenen, Conduit et al., Franzini, Gottesmann** |

that awareness. Although I signaled the tentativeness of this category with question marks in my original Figure 1, its predominance in the diagram cannot be justified on observation alone. I therefore clarify in Figure NR1 that these processes (unobservable cognitive activity) are not necessarily associated with the other categories of mentation. I also describe a second type of cognitive activity that is normally unobservable but accessible through introspective effort. Justification for the category is given below.

**Revonsuo** proposes an alternative definition of dreaming. "Complex, temporally progressing content" is suggested to be a relatively theory-free feature that distinguishes dream-

ing from other types of cognitive activity during sleep. **Clancey** also proposes an alternative classificatory system that includes the sequencing or progression of perceptual categories. Temporal progression corresponds to the well-known criterion of "dramatic" quality that Freud (1900) borrowed from Spitta (1882) to define dreams, that is, dreams construct a *situation* out of hallucinatory images (Freud 1900, p. 114). While temporal progression may indeed be a common feature of much dreaming, and especially the dreaming common to most REM sleep, it is not likely a defining feature of all dreaming. For example, the criterion of temporal progression would exclude many of the uni-

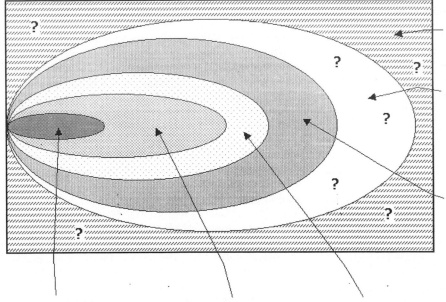

**6. Unobservable cognitive activity:** also known as pre-conscious or unconscious activity (distinct from, but likely implicated in, 1-4)

**5. Cognitive activity requiring introspective effort:** activity not available to awareness without reflective effort, e.g., orienting, selective attention, discrimination, recognition, rehearsal (distinct from, but likely implicated in, 1-4)

**4. Cognitive activity:** absence of sensory hallucinations: thinking, reflecting, bodily feelings, vague and fragmentary impressions (distinct from, but likely implicated in, 1-3)

**1. Apex dreaming:** vivid and narratively complex progression of sensory (visual, auditory, kinesthetic) hallucinations, e.g., sexual, archetypal, transcendental, titanic, existential, and lucid dreams; nightmares

**2. Typical dreaming:** any progression of sensory hallucinations (includes 1)

**3. Minimal dreaming:** any isolated sensory hallucination (includes 1-2)

Figure NR1.   Levels of specificity in defining sleep mentation – revised version of Figure 1 from target article. See text for details.

modal, static hallucinatory images typically reported in our studies of sleep onset mentation (Germain & Nielsen 1997; Nielsen et al. 1995), and this on a seemingly arbitrary basis. Arbitrary because the studies, including my self-observational studies of brief hypnagogic images (Nielsen 1992; 1995) (http://www.crhsc.umontreal.ca/dreams/TNmodeling .htm), suggest that such static images are often endowed with a hallucinatory quality that renders them quite dream-like. The hallucinatory quality is unmistakable, even for "fleeting" images and "sleepiness" sensations that occur prior to the more fully formed hypnagogic images themselves. Hallucinatory quality is associated with the seeming *sensory* nature of the imagery and appears to involve a degree of *apparent orientation* to ("self-participation" in) the imagery (e.g., Bosinelli et al. 1974; **Herman**). Apparent orientation here refers to illusory sensations of a spatial distribution of objects, including, and sometimes consisting only of, the self, the apparent vertical, apparent depth, and/or apparent motion. Hallucinatory quality was to Freud as important a defining attribute as was dramatic quality, the purported "transformation of ideas into hallucinations" (Freud 1900, p. 114). In Figure NR1, hallucinatory quality defines a minimal dream, whereas temporal progression distinguishes minimal dreaming from more complex and typical forms of dreaming.

This revision in Figure NR1 also responds somewhat to **Shevrin & Eiser**'s comment that Freudian theory is ignored by the covert REM approach. It may also respond to **Antrobus**'s point that a unidimensional measure of mentation recall/non-recall is inferior to a multidimensional approach in making fair comparisons of REM and NREM mentation. The criterion of "hallucinatory quality" might be applied equally well to mentation in all sensory dimensions, and possibly also to emotion, pain, and other organic sensations. If so, fair *uni*dimensional comparisons of "minimal dreaming" could still be made across sleep states using this criterion.

More generally, I believe that the continued disagreement over defining dreaming is based upon at least two methodological shortcomings. First, there is not only disagreement over how best to accomplish an accurate phenomenology of subjective experience (e.g., Dennett 1991), but all too often available phenomenological methods (e.g., Busink & Kuiken 1996; Husserl 1965) are disregarded in research. The result is that definitions are proposed without much reference to methods of deriving them (cf. **Pagel**), and no standardization is possible. Second, subjects in sleep mentation experiments, on whose responses definitions of subjective experience are often based, are typically naïve to the exigencies of introspective reflection. This issue goes beyond the concerns voiced in commentaries by **Antrobus** and **Schredl** that mentation reports have uncertain validity. Rather, the point is that introspectively untrained subjects simply cannot accurately report upon all microstructural constituents of hallucinatory quality that might be crucial in identifying a subjective experience as a dream. Conversely, there is today very little support for introspective approaches that involve training subjects and/or investigators to access these microstructural levels of subjective experience precisely and reliably. To reflect this concern, Figure NR1 distinguishes a type of cognitive activity that is available to awareness only with some degree of introspective effort.

In sum, although I agree that definitions of dreaming

should be theory-free, I doubt that such approaches can be developed without a more concerted emphasis on introspective and self-observational methods of study that involve the training of both subjects and experimenters. Therefore, in lieu of importing definitions from consciousness research or elsewhere, the most reasonable course of action in the short-term may simply be to refine terminology that has evolved over the years *within* the discipline of dream research and whose connotations and nuances are thus understood more or less consensually by a large number of researchers active in the area. However, a long-term strategy for addressing this basic issue is clearly needed.

## NR2. Authors add new information that supports the model

At least 12 commentaries (**Borbély & Wittmann; Born & Gais; Cartwright; Feinberg; Gottesmann; Greenberg; Lehmann & Koukkou; Pace-Schott; Porte; Rotenberg; Salzarulo; Steriade**) described research and/or theory consistent with or supportive of the covert REM sleep model. An important paper by Toth (1971), which was suggested by Rotenberg (1982) and another by Schwartz (1968), which was mentioned by Gottesmann, were not referred to in my target article but contain evidence fairly directly supporting the covert REM sleep model. I will briefly summarize both. Toth (1971) devised miniature electrodes which, when glued to the eyelids overlying the cornea, more than doubled the sensitivity of standard EEG recordings. This innovation allowed him to quantify very small amplitude eye movements occurring in NREM sleep (cited in Rotenberg 1982). Although this study urgently needs replication, the report suggests both a straightforward method for measuring covert REM sleep processes in NREM sleep and, if confirmed, that such processes may be more present in NREM sleep than has been appreciated.

Schwartz (1968) observed "indeterminate sleep" in both hypersomnolent patients and control subjects shortly after sleep onset during afternoon naps. Distinguishing among *very slow* eye movements, *medium fast* eye movements, and *rapid* eye movements he found that medium fast eye movements could be observed in all patients and controls at each sleep onset and that they were more common than very slow eye movements. Medium fast movements were recorded consistently in stage 1B and especially in stage 2, and then decreased in quantity and amplitude as slow waves predominated. They were rare in stage 3, but nevertheless often accompanied K-complexes. He noted that the voltage of these eye movements varied with electrode distance and individual differences in anatomy, thus standard EOG recordings may be insufficient to identify them under routine recording conditions. He also identified phasic EMG activity occurring immediately after the onset of EEG-defined sleep stage 1Band. These consisted of small movements or twitches of the face, hands, feet, head, shoulders, and even the abdomen, and were indistinguishable from the phasic movements of REM sleep. Schwartz noted that medium fast eye movements occur also in REM sleep, especially just before the onset of rapid eye movement bursts. Finally, he found dream recall after spontaneous awakenings from stages 1B and 2 sleep that had been accompanied by medium fast eye movements. He also cites a study by Kuhlo and Lehmann (1964) in which eye movements similar to his medium fast eye movements were studied in conjunction with hypnagogic imagery. We also report these types of events in preliminary study 2 reported in section NR8.2 (see Figs. NR3–8). Although Schwartz's study also requires replication with a larger sample of healthy control subjects, his findings concerning REM sleep-like eye movements, phasic EMG activity, and dreaming at sleep onset are strongly supportive of the covert REM sleep model. Together, our results, the findings of both Toth and Schwartz, and the neurophysiological observations concerning sleep onset eye movements contributed in the **Porte** commentary, all bolster two points I make in the target article: (1) rapid eye movements may not be particular only to REM sleep and (2) slow eye movements may also be a correlate of REM sleep. If so, sleep onset may be considered to be a kind of short-lived or fragmentary episode of (covert) REM sleep, and sleep onset imagery a type of brief (covert) REM dream.

Other commentators discuss findings from sleep deprivation research that are consistent with the covert model. **Born & Gais** and **Cartwright** both emphasize that REM sleep propensity is heightened after REM sleep deprivation. This covert propensity may be a critical factor in studies of deprivation effects on memory because of continued effects of covert REM sleep processes on memory consolidation, despite the apparent absence of the REM sleep state itself (Born & Gais). The improvement in mood and increased drive behaviors produced by sleep deprivation in depressed subjects may also be due to covert REM sleep (Cartwright). We have observed that healthy subjects undergoing sleep deprivation sometimes manifest REM sleep signs in their NREM sleep polysomnograms during recovery sleep (Nielsen & Carrier 2000, unpublished). To illustrate, Figure NR2 shows the sleep onset tracing and hypnogram of a 31-year-old healthy female following 40 h of sleep deprivation. The tracing contains distinct rapid, medium fast, and slow eye movements in conjunction with a background of stage 1 sleep.

**Cartwright** also suggests that the covert REM sleep model is supported by studies demonstrating a coupling of REM sleep and dreaming under dissociated circumstances such as the NREM dream reports of light sleepers who are in high arousal throughout sleep, and in other sleepers for whom there is a low arousal threshold following sleep deprivation or acute stress. Violent sleepwalking episodes also occur following periods of extended sleep loss and stress. Finally, sleep state dissociation is seen in subjects with REM sleep behavior disorder in which there are REM sleep signs but lapses of muscle atonia. There is a wide range of phenomena that involve dreamlike mentation in NREM sleep (see review in Nielsen & Zadra 2000) whose closer study could shed light on whether dissociated REM sleep processes are implicated in the mentation. Dissociation of REM sleep processes is discussed in greater depth in section NR5.

Several commentators suggested ways that EEG or other brain imaging methods might be harnessed to quantify covert REM processes. A figure in the **Feinberg** commentary illustrates very nicely how delta EEG power could serve as such an index. Delta power normally drops sharply at the onset of REM sleep episodes and then rises again with the start of the following NREM episode and repeats this variation across the night. The Feinberg figure illustrates three types of commonly observed events that are consistent with the covert model (see also Dijk et al. 1995; Landolt et al. 1996):

1. *Sleep onset REM processes:* Not only is delta power low during REM episodes, but it is similarly low at sleep onset, when dissociated REM sleep processes are hypothesized to occur.

2. *"Skipped" first REM episodes:* Delta power estimates during the first 90 min of subjects' 1 and 3 recovery nights (RN) drop sharply even though the expected REM sleep episodes are not scored. **Feinberg** indicates that these episodes are often not scored during RN while they are scored during baseline nights (BN). Such findings support the existence of covert REM processes during "skipped" REM episodes as discussed in the target article and further suggest that they may be more likely during recovery from sleep deprivation. Delta power analyses reveal that such tendencies toward skipped REM episodes are more striking in children and young adolescents than young or middle-aged adults (Gaudreau et al., in press) and confirm that the exceptionally long REM onset latencies (up to 3–4 h) seen in young children are often likely due to such skipped REM episodes (Benoit 1981; Bes et al. 1991; Dement & Fisher 1964; Palm et al. 1989; Roffwarg et al. 1966; 1979). Palm et al. (1989), for example, found in a sample of 8–12-year-olds that on 67% of nights the first sleep cycle lacked REM sleep as traditionally scored; in 88% of these, "an abortive EEG sleep pattern was found with traits specific to REM as well as to non-REM" (p. 306). The main anomaly observed in their study was a *lack* of rapid eye movements during the anomalous REM episode. Other research (e.g., Carskadon et al. 1987) has suggested that long REM latencies (i.e., skipped REM sleep episodes) may interact with both the "first-night" effect" (with REM latencies higher on the first night) and gender (with REM latencies decreasing over laboratory nights 1 to 3 for girls and nights 1 and 2 for boys). Skipped first REM periods also appear in adults who are under conditions of sleep loss (Berger & Oswald 1962).

3. *Pre- and post-REM covert effects:* The gradient with which delta power decreases and increases before and after REM sleep varies from subject to subject, within nights, and over experimental conditions. Subject 1's BN plot shows that power increased moderately after the first REM episode but remained very low after the second and third. Such profiles correspond to a predominance of stage 2 sleep in the subject record. Are covert REM sleep processes more likely to manifest during these lulls in delta power? Possibly. Waterman (1992) found delta power, but not other frequency bands, to be negatively correlated with dream recall (word count) and to account for a significant portion of the REM-NREM and time of night differences in dream recall. Furthermore, these findings held for young, but not older, subjects. **Salzarulo** also emphasized an inverse relationship between delta power (slow-wave activity or SWA) and cognitive processing in sleep – in this case, the number of statements that comprise each dreamed "story event." Salzarulo goes further, however, to suggest that SWA reduction across the night reflects diminution of the more general process S, and that this reduction serves as a physiological condition for cognitive experience irrespective of sleep stage. Such a 1-gen notion is, in fact, consistent with studies demonstrating increases in dream intensity later in the night (e.g., Antrobus et al. 1995), but the effect appears to be much smaller than the REM-NREM sleep mention in dream intensity (**Antrobus** et al. 1995).

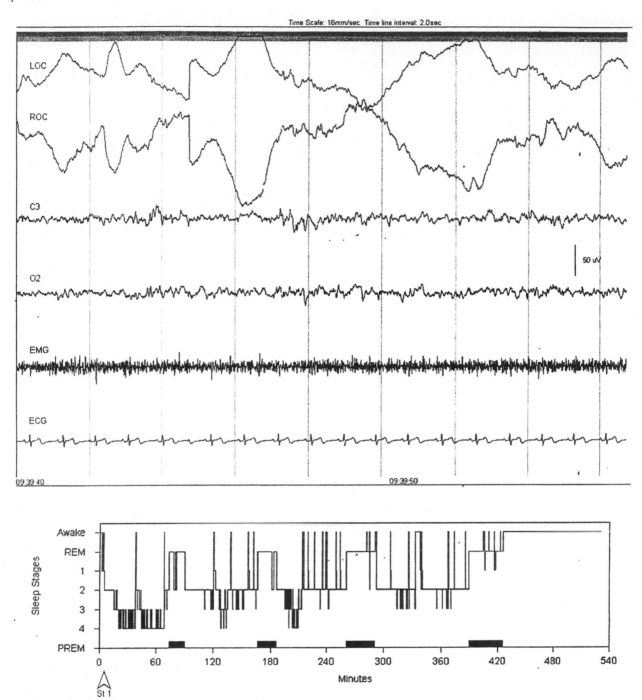

**Figure NR2.** Hypnogram and polysomnographic (PSG) tracing from a healthy 31-year-old female subject on her first recovery (daytime) sleep after enduring a 40-hour constant routine. Rapid, medium fast, and slow eye movements are clearly visible against a background of stage 1 EEG and EMG.

The commentators considered various other brain imaging measures in relation to hypotheses about covert REM sleep and dream production. The suggestion that episodes of covert REM sleep are equivalent to lapses of attentional control during the waking state (**Greenberg**) is conceptually similar to the hypothesis that a basic dream production mechanism depends upon activation of attentional mechanisms (**Morrison & Sanford; Conduit et al.**), e.g., the PGO wave, and that such mechanisms may be activated sporadically in NREM sleep. Such processes may be indexed by more detailed measures of spontaneous EEG during REM and NREM sleep or by various evoked potential techniques. The dissociation of REM sleep processes into other sleep states also corresponds well with **Lehmann & Koukkou**'s (1984) notion of *momentary brain states,* that is, very brief (in the order of seconds or less) changes in brain state within a sleep stage. Their work suggests that evidence of such momentary state changes may be "hidden" in rapidly changing EEG parameters, but that their decodification may be forthcoming with more sophisticated methods of quantifying the EEG. Alternatively, covert REM sleep processes may parallel rises and falls in mechanisms of brain *synchrony* (**Pace-**

Schott), presumably a measure derivable from EEG coherence. We have found that some features of dream content are associated with generalized cortical coherence in REM sleep (Nielsen & Chénier 1999) but we have yet to examine NREM mentation for the same correspondences.

**Steriade** points to work he published over a decade ago that supports the covert REM sleep model in suggesting that increases in the signal-to-noise ratio of PGO-related spike bursts in visual thalamus is high during pre-REM sleep transitional periods, a change that might underlie the generation of vivid mental experiences. Other brain indicators of covert REM sleep processes may be tied to deactivation of heteromodal association areas, as indicated by recent brain imaging studies (**Borbély & Wittmann**). Such studies implicate structures and mechanisms in covert REM events that may be beyond the capacity of present-day EEG methods to quantify.

**Porte** points to the need for further investigation of EEG spindle characteristics in relation to REM sleep signs and describes how the neurophysiological structure of NREM stage 2 sleep could, in fact, be compatible with the intermittent appearance of such signs. Specifically, covert REM processes may be more likely to occur between distantly spaced sleep spindles because of an inhibitory influence during the interspindle wave refractory period. This notion is consistent with our own observations in study 2 (see sect. NR8) of medium fast and rapid eye movements occurring between spindles in stage 2 sleep. However, in our study some eye movements were also observed to occur in close proximity to, if not simultaneous with, sleep spindles (see Fig. NR7), suggesting that any inhibitory influence of the spindle generator on intermittent REM sleep events may be variable and transitory. It must also be noted that non-cortical REM sleep processes such as muscle twitches, penile tumescence, heart rate variability, and other autonomic fluctuations that may manifest in NREM sleep are not likely to be affected by the spindle wave refractory period.

Of course, the development of new forms of sleep monitoring need not be restricted to the EEG. To illustrate, REM and NREM sleep are distinguished by autonomic changes, most notably an increase in sympathetic activation during REM sleep (Berlad et al. 1993). The description of such changes has until recently been severely restricted by a lack of appropriate recording methods. It is therefore noteworthy that a recently developed plethysmographic method for quantifying peripheral vasoconstriction during sleep has found that vasoconstriction is highly characteristic of REM sleep, and that its increase can be detected at least 30 minutes before the beginning of REM sleep as it is traditionally scored (Lavie et al. 2000). This finding is entirely consistent with the covert REM sleep model and suggests that the "window" around the REM sleep state during which covert processes might influence NREM sleep mentation could be larger than the 10–20 min window discussed in the **NIELSEN** target article.

In sum, by directing attention to both micro- and macrostructural dissociations of REM sleep processes into NREM sleep, the covert REM sleep model highlights potentially fruitful directions in which biosignal imaging and interpretation methods may be developed. These methods may lead to more precise definitions of sleep stages and their relationships.

## NR3. Consideration of waking processes in the model

Some commentators (**Greenberg; Hartmann; Ogilvie et al.; Schredl**) expressed dissatisfaction that the covert REM sleep model does not deal with potential incorporations of waking state processes into sleep. They viewed this as either a weakness in the model or as a potential avenue for its further elaboration. On the one hand, some authors pointed to the immediate post-awakening state as a factor that could potentially influence REM/NREM mentation differences. For instance, Greenberg emphasized that gradual awakenings from NREM sleep can lead to reporting of more dream content (Goodenough et al. 1965a). Goodenough believed that this accounted for some but not all instances of NREM mentation. However, it remains an open question whether such "gradual awakenings" involve the intermingling of waking state processes with NREM sleep mentation or the brief activation of REM sleep processes during transition to full awakening. There may occur a substantial degree of secondary elaboration during awakening as Freud (1900) suggested, or content may be produced as part of the arousal process as in the case of some sleep terrors (Fisher et al. 1973). In the target article I deal at greater length with the possibility that brief or fragmented episodes of REM sleep occur unnoticed in the course of waking up. It is important to emphasize that even a minor elaboration or generation of content at this time would be sufficient for a report of genuine dreaming to be "identified." As studies of both hypnagogic imagery and "disorders of arousal" demonstrate, *even fleeting experiences of hallucinatory content are sufficient to generate* bona fide, *albeit diminutive, reports of dream mentation.* Subject differences even further complicate the picture, because some factors unique to subjects may enhance REM/NREM differences (**Schredl**). Since more elaborate mentation reports may be given by subjects who have a more verbose verbal style, who have superior verbal short-term memory, or whose recall is "enhanced" by training, the degree of elaboration of even brief mentation samples may also be increased.[2] Subjects who are introspectively inclined and verbally confident may well find it a simple task to elaborate a single fleeting image into a coherent, multi-propositional, narrative episode.

A study by **Herman** et al. (1978) illustrates the subtlety of the problem. This work demonstrates clearly that mentation reports from NREM (but not REM) sleep are rendered more "dreamlike" (as measured by Foulkes's *dreamlike fantasy* scale) when experimenters or subjects themselves are systematically biased to believe that this is the expected result. Herman et al. even suggested that "a possible major source of variance in NREM recall studies is the predisposition of the investigator" (p. 91). Factors such as experimenter influence are methodological obstacles to conducting fair and unbiased comparisons between REM and NREM mentation. The covert REM sleep model helps to bring many of these methodological issues into focus and it suggests novel means for controlling them. It is, in one sense, a methodologically driven model whose stance in the face of acknowledged shortcomings in the definitions of REM and NREM sleep is to advocate that these definitions be more precise and their presumed cognitive correlates be more thoroughly studied.

Other authors consider waking state processes as a means

of extending the covert REM sleep model. For example, **Ogilvie et al.** take issue with the notion of covert REM mechanisms underlying sleep onset mentation in the first NREM-REM cycle, this based upon the presumably circadian nature of sleep onset REM periods (Sasaki et al. 2000). It is argued that waking state processes are more likely to be incorporated into sleep onset mentation that are REM sleep processes. This suggestion is feasible and consistent with some work on sleep onset mentation (e.g., Cicogna 1994) and some results from study 2 reported in section NR8.2. However, the covert REM sleep explanation cannot be ruled out in light of several studies previously described. For example, the study by Schwartz (1968) and our own preliminary findings (sect. NR8) are consistent with the assertion that REM sleep events occur at sleep onset. I agree that REM sleep processes are influenced by circadian factors, but such factors do not necessarily *preclude* the occurrence of extremely brief, if not fragmented, REM sleep processes at sleep onset and elsewhere. In fact, if a REM sleep potential does exist early in sleep, a very weak circadian pressure might be *expected* to fragment, dissociate, or diminish it rather than simply to impede its expression in an all-or-none fashion.

**Hartmann** suggests that dreaming mentation should be seen as part of a continuum with daydreaming and other varieties of waking mentation, and that the components of this continuum are not different enough to warrant considering them products of different mentation generators. It is true that some comparative studies of waking and sleep mentation find evidence of structural similarity (Kahan et al. 1997; Kahan & Laberge 1996) but there are in my view too few comparative studies of such features and their physiological correlates to elaborate a definitive model. *The evidence in support of a REM-NREM sleep mentation continuum is controversial enough!* Nevertheless, Hartmann does take some constructive steps toward specifying a global structure for one possible wake-sleep mentation continuum and of proposing factors that might describe how dreaming and waking vary on this continuum.

## NR4. Demonstrations of dreaming during stages 3 and 4 sleep and their implication for the existence of mentation unique to NREM sleep

Several authors suggest that the covert REM sleep model cannot explain reports of dreamlike mentation in stages 3 and 4 sleep (or slow-wave sleep; SWS). Supporters of this notion point to, among other evidence, a study by **Cavallero** et al. (1992) that involves direct sampling of SWS mentation. There is much evidence reviewed in the target article and in the present reply that provides a basis for at least questioning the definitiveness of this and other such studies of SWS cognition (Cicogna et al. 2000). In general, I question how many of the mentation reports collected from SWS occurred under conditions which, according to the covert model, were demonstrably free from the potential influence of covert REM sleep? These include variables such as time from preceding REM sleep periods, time prior to next REM sleep periods (which, with today's instruments, may be impossible to calculate with any certainty), partial sleep deprivation (producing increased REM sleep pressure), sources of sensory stimulation during sleep (which are potentially numerous in a laboratory), the effects of drugs or alcohol

and/or withdrawal from these, and so forth. This might seem like an exorbitant list of criteria to exclude but the approach is not unlike how a clinician proceeds in excluding possible alternative diagnoses of a sleep problem. In fact, a partial remedy to the caveats posed by the covert REM sleep model may be to routinely evaluate (and publish) pertinent details of subjects' sleep states along with the usual reporting of sleep mentation characteristics. For example, analyses of NREM sleep hypnograms or sleep tracings from the pre-awakening interval could exclude the presence of sleep fragmentation, eye movements, motor activation, and other possible REM sleep signs. Further, quantified measures of sleep state transitions, sleep efficiency, and so forth could provide valuable information about how "dissociable" a subject's sleep is. Subjects could also be screened for frequency of nightmares and other parasomnias, especially because such subjects may be particularly inclined to participate in studies of sleep mentation. Our findings from study 2 (see sect. NR8) suggest that covert REM processes might be more prevalent or more active among nightmare sufferers. One post-traumatic nightmare patient from our sample who demonstrated a very high number of REM sleep signs in NREM sleep also had an extremely variable hypnogram on both recording nights and reported dreaming vividly throughout the night (see Figs. NR6 and NR7).

In addition to these concerns, the **Cavallero** et al. study and others like it should be interpreted with caution for at least two methodological reasons. First, several subjects (17%) in the Cavallero et al. study recalled *no mentation from SWS whatsoever* and were excluded from the study sample. Other subjects required more than one night in the laboratory to achieve a recall of mentation from SWS. Had such observations been made for awakenings from REM sleep, they would likely have caused a significant stir and provoked further investigation to determine their clinical implications. However, for NREM sleep such a finding raises no eyebrows, is readily dismissed, yet remains completely inexplicable to a model that proposes regular SWS dreaming. Second, it is not stated whether the experimenters in this study were naive to the nature of the hypotheses. Subjects could have been pressured inadvertently by experimenters to produce mental content, as **Herman** et al. (1978) so clearly demonstrate. As noted in the previous section, the amount of mental activity during SWS that is stimulated either by covert REM sleep or wakefulness processes could be quite small while still seeming to produce a somewhat elaborate mentation report from SWS. Cavallero et al.'s work on SWS dreaming has made an important contribution to research in the area but it is not without its methodological limitations.

Some commentaries (**Bosinelli & Cicogna; Cavallero**) reiterated the argument that studies of REM/NREM mentation that have controlled for the length of the mentation report (with, for example, total word count as a covariate) have found that apparent REM/NREM stage differences are diminished or disappear altogether. The finding of residual differences that are discussed in the target article are thus seen to be artifactual, for example, the result of differences in the spreading of mnemonic activation in the two sleep states. Such research findings are interpreted as supporting the view that dreaming occurs in both REM and NREM sleep but not because of any link to possible covert REM sleep processes. Although more studies would seem to be called for, two points should be reiterated: (1) The widespread use of report-length correction methods over the last decade may well be in doubt (see discussions in **NIELSEN** and **HOBSON ET AL.** target articles). Thus, the seeming diminution of stage differences with length-control may be a dramatically over-stated phenomenon. (2) My review of the literature on REM/NREM mentation comparisons in the target article resulted in no less than a dozen studies that report residual differences, *despite* the implementation of such report-length controls. In fact, in this literature I have found little evidence that stage differences are ever entirely eliminated with length controls.

**Blagrove** adds to this debate the observation that purportedly qualitative residual differences are nevertheless quantitative in nature (e.g., number of characters, visual imagery word count); there are thus no qualitative differences *per se* between REM and NREM reports, and a 1-gen hypothesis is supported. This observation points out an important problem: measurements are quantitative (usually), whereas features themselves are qualitative (usually). So a seemingly quantitative difference between groups could belie what is, in fact, an important qualitative difference. For example, it would be foolish to suggest that a group of subjects each bearing three eyes was only *quantitatively* different from a group of normal two-eyed subjects. Yet an eye-count measure would lead to just such a conclusion. Such comparisons must be informed by the normative context of the measurements. One solution to this type of methodological problem is discussed in the **HOBSON ET AL.** target article (disallow length controls). Another is discussed by **Antrobus** (compare mentation reports on a multidimensional measure). Alternatively, if the use of report-length controls is justifiable, then a fair approach would seem to be to evaluate all quantitative measures in the same units as the weighting factor, for example, word count of all bizarreness text weighted by total word count (cf. **Hunt** et al. 1993). Such an approach could also lend itself to multidimensional comparisons because all measures would be based upon the same metric. This approach is similar to one employed by Antrobus et al. (1995).

## NR5. The model links dreaming exclusively to brainstem activation in REM sleep

Several commentators (**Bosinelli & Cicogna; Domhoff; Porte; Solms; Salin-Pascual et al.**) suggest that the covert model implies a particular view of REM sleep as governed exclusively by brain stem sources of activation. This "bottom-up" interpretation of the model derives from the early reciprocal interaction model of REM sleep (McCarley & Hobson 1979) that places control of REM sleep in pontine "REM-on" neurons. The **Solms** commentary provides a clear definition of this view of REM sleep state and thus allows useful comparisons with the covert REM sleep model. Solms defines REM sleep to be synonymous with an executive mechanism that recruits various physiological events (e.g., EEG desynchronization, muscle atonia, rapid eye movements) and coordinates them into "a distinctive configuration." He identifies the brainstem as this executive mechanism and he disputes whether it can, in fact, be responsible for the generation of dreaming. The **SOLMS** target article further addresses this claim. This view, the separation of REM sleep into a specific control mechanism and its coupled components, can be compared with the covert REM model by posing the following three key questions about the definitional concepts.

### NR5.1. Are all aspects of REM sleep control located in the brainstem?

There is still disagreement as to the extent of involvement and, ultimately, of the importance to REM sleep generation of pontine brainstem regions. **Salin-Pascual et al.** review several studies that challenge the notion and that implicate a major role for the hypothalamus. **Morrison & Sanford** and **Feinberg** also qualify this notion with reference to forebrain structures, such as the hypothalamus, which may influence brainstem activity. **Jones** calls into doubt brainstem control by referring to Jouvet's critical experiments that eliminated REM sleep by eliminating corticofugal influences on brainstem. **Nofzinger** describes new brain imaging findings that support forebrain involvement and that cast doubt on the specificity of brainstem involvement. **Lydic & Baghdoyan**, on the other hand, support the notion of brainstem control quite vigorously. This small sampling of diverse opinions reveals the wide disagreement about whether pontine brainstem should be accorded the status of a *unique* control mechanism for REM sleep. It also underlines the importance of distinguishing among types of executive control; for example, between mechanisms that *trigger* REM sleep onset and those that *maintain* REM state integrity over time. Pontine brainstem may well be a primary determinant of REM sleep onset (although this notion is still contested) while forebrain may affect REM sleep intensity, consolidation, or duration. Consistent with this possibility, there is evidence (Montplaisir et al. 1995) that among patients with Alzheimer's disease, which affects basal forebrain but not pontine brainstem, REM sleep timing is normal, but REM sleep episodes are shorter than normal in duration. To reiterate the preceding, there is disagreement as to whether brainstem is the only, or even the most important, controller of REM sleep; this is largely because there are so many features of REM sleep that must be controlled.

### NR5.2. Do isomorphic correlates of dreaming exist only at the level of REM sleep executive control?

Notwithstanding the previous problem, it may be premature to conclude that REM sleep control and dreaming control are isomorphic. This is because little if any research has studied the isomorphism question at these corresponding levels of complexity. In fact, most studies seeking to find isomorphic relationships in sleep have concentrated exclusively on what **Solms** refers to as the individual "components" of the REM sleep state. As I argue in the next section, there is in fact evidence that isomorphic relationships exist between isolated physiological variables and specific attributes of dream content. On the other hand, there are no studies that have yet managed to directly assess whether the pontine "REM-on" neurons and their presumed executive control structure are associated with dreaming.

In contrast to **Solms**'s view, I think it is feasible that some essential processes of dream organization occurring at a *microstructural* level may be found to be associated with components of the REM sleep state. By microstructural organization I mean processes governing the ordered and coherent presentation to awareness of a sequential flow of inter-connected multisensory images. To achieve this, it seems likely that the dream production system depends upon a great degree of autonomy in the *local* organization of image elements such that the integrity of every part of the (arguably complex) imagery sequence does not hinge upon the fidelity of a single, central control mechanism. Image elements may have mechanisms of attraction and repulsion that allow them to dissociate and regroup into larger units much as basic physical elements combine to create more complex molecules and substances. Elsewhere (NIELSEN 1995) (www.crhsc.umontreal.ca/dreams/TNmodeling.htm), I describe a mechanism referred to as *transformative priming* that could fulfill such a local control function over information contained in a wide variety of modalities. Transformative priming involves one image or image element activating a conceptually related image or element (priming) and then combining with it into a completely novel form (transformation). The process, which unfolds over a time span of milliseconds, could account for the local coherence of minimal dreaming and of more complex forms of dreaming as well.

### NR5.3. Can REM sleep events dissociate from the REM sleep configuration?

According to **Solms**'s commentary, even individual physiological events that may be correlated with dreaming should not be identified with the REM sleep state if they occur outside of that state because they are not part of the presumed brainstem control mechanism; their source is "indeterminate." On the other hand, the notion of the covert REM sleep model is that REM sleep events that occur outside of REM sleep are somehow *dissociated* from the state and can continue to exert an influence; their source is somehow still "linked" to REM sleep. In fact, to the extent that the frontal and parietal structures identified by Solms are typically implicated in dreaming and are also typically associated with REM sleep, I would view his findings as completely consistent with, if not splendidly supportive of, my own model. The action of these structures Solms considers to be *independent* of REM sleep; the covert model would describe them as a *dissociation* of REM sleep processes into another sleep state. The solution to this discrepancy may lie in whether state dissociation can be proven to be a valid construct.

A substantial body of literature in fact supports the concept of sleep state dissociation (Mahowald & Schenck 1991) and thus also supports the related notion of dissociated or covert REM processes. State dissociation purportedly explains a variety of bizarre clinical phenomena involving mentation, such as the symptoms of narcolepsy, REM sleep behavior disorder, disorders of arousal (e.g., sleep terrors, sleepwalking, sleep drunkenness), automatic behavior, and "out-of-body" experiences. In most of the cases discussed by Mahowald and Schenck, however, the state *into* which intrusions occur is of more importance in defining the phenomenon than is the state *from* which the isolated intrusions originate. For example, in the case of REM sleep behavior disorder, there is very little doubt that the REM sleep state is involved, whereas the precise origin of the isolated, intruding event (absence of muscle atonia) is of less importance to the definition of the syndrome. It may be a waking-state intrusion or some unspecified type of motor activation. In the case of covert REM sleep, identification of the state *from* which intruding events arise is of primary importance. Thus, the REM sleep processes that may in-

trude upon other states vary in complexity from, on the one extreme, the *absence* of a single defining component (as in the absence of eye movements during "skipped" first REM periods) to, on the other extreme, the *presence* of a single component in a NREM sleep state (as in the presence of eye movements during stage 2 sleep). It is validation of the latter type of event, involving the intrusion of single components, that is most at issue in **Solms**'s commentary; instances of the former type are more obviously variations of a known state. The problem of validating many such isolated physiological events as bona fide REM sleep dissociations will require more detailed scrutiny of the events' characteristics. To illustrate, Lavie (1990) describes episodes of penile tumescence without REM sleep in a patient with shrapnel fragments lodged in his left cerebellar hemisphere and prepontine cistern. Over five recording nights, this patient had a total lack of REM sleep on three nights, and only a single REM episode on each of the two others (REM% = 0.6 and 5.9%, respectively). The episodes of tumescence might thus seem to be "indeterminate," that is, completely unrelated to REM sleep. Nevertheless, closer scrutiny reveals that episodes of penile tumescence were recorded (1) that followed the expected temporal REM sleep rhythmicity of about 90 min (e.g., erections were spaced 82, 150, and 101 min apart on three recording nights), (2) that occupied portions of total sleep time that were similar to typical REM sleep times (35.5, 22.9, and 26.2% on the three nights), and (3) that were coincident with REM sleep on the two occasions that REM sleep was, in fact, detected. Lavie even concluded that "in spite of the drastic reduction of REM sleep, there was an indication of a 'REM-like' cyclicity in penile erections" (p. 278). To Lavie, the finding "supports the notion that nocturnal penile erections can be dissociated from REM sleep" (p. 278), a notion proposed earlier by Karacan and colleagues (Karacan 1982; Karacan et al. 1976).

To extend this notion even further, the *dissociability* of physiological processes during REM sleep may be speculated to be a basic feature of the state. **Antrobus** points out that the imaging results of Braun et al. (1998) reveal a high degree of dissociation among normally associated brain structures in REM sleep. The same is true of a wide variety of autonomic systems (Parmegianni 1994). Much cognitive literature (e.g., Hecker & Mapperson 1997; Livingstone & Hubel 1987) demonstrates how components of perception and memory can be experimentally dissociated, revealing that such information is processed in parallel along anatomically separate channels in the CNS. Dissociation of information may just be a necessary condition of dreaming which, as Foulkes (1985) proposes, must draw upon a diffuse pool of "dissociated elements of memory and knowledge" (p. 27). If REM sleep is at least partly about the dissociation of normally coupled systems in the service of reorganizing them for dream formation, then perhaps we should not be surprised to see such dissociations also occurring outside of the state.

Arguments about organization and isomorphism aside, differences between **Solms**'s model and my own may only constitute a difference in *interpretation* of findings. If a given process is reliably associated with a given sleep state, say with a concordance of 85–100%, and if that relationship is highly specific to that sleep state, then it would seem appropriate to consider the attribute as a biological marker of the sleep state. But if the relationship is not specific to the sleep state, then its role as a marker is cast in doubt. It is the *degree* of specificity of the process to the state that will determine whether it is trusted to be a valid marker of the state. The covert model is an attempt to more precisely identify that degree of specificity for REM sleep.

To summarize, until isolated REM sleep signs occurring in NREM sleep can be confidently *excluded* as (1) being "linked" to REM sleep initiation or maintenance or (2) bearing some isomorphic relationship to sleep mentation variables, I am comfortable in viewing them as "dissociated" rather than "indeterminate" events. The interpretation of these signs depends heavily upon how the REM sleep state is conceptualized as well as upon what specific and/or general features of REM sleep prove to be isomorphic with sleep mentation.

## NR6. Lack of evidence for isomorphism

At least six commentators (**Hunt; Kramer; Morrison & Sanford; Panksepp; Solms; Vogel**) referred to the lack of evidence for isomorphic relationships between physiological variables and sleep mentation, evidence that is critical in evaluating the covert REM sleep model. Although authoritative reviews of psychophysiological isomorphism such as those by Pivik (1991) are often taken as evidence that strongly refutes isomorphism, such reviews in fact offer ample evidence supporting some types of isomorphic relationships, and even some evidence supporting the covert REM sleep model. First, whereas there is inconsistency in many findings that bear on different classes of physiological variables in relation to mentation, some classes (e.g., autonomic) appear particularly strongly associated with sleep mentation variables. Variability in respiration rate has been observed to correlate with both quantitative (Shapiro et al. 1964) and qualitative (Hobson et al. 1965; Kamiya & Fong 1962; Van de Castle & Hauri 1970) aspects of sleep mentation. Hobson et al. (1965) even observed such relationships in both REM and NREM sleep. Other autonomic indicators, such as sudden penile erections, have also been found to be associated with increased recall (Karacan 1966) and erotic content (Fisher 1966). In NREM sleep, including stages 2, 3, and 4, both the recall and hallucinatory quality of mentation has been found to be higher on awakenings that follow brief phasic inhibitions of the H-reflex (Pivik 1971). Sleep onset has also yielded associations between EEG theta bursts on the one hand and visual imagery and discontinuity on the other (Pope 1973). The physiological measures in NREM sleep (respiration variability, H-reflex inhibition, theta bursts), by virtue of their similarity to REM sleep phenomena, are good candidates for indicators of covert REM sleep processes. Note that this holds true for both stage 2 sleep and SWS. As I specified in the target article, one reason that isomorphic relationships between physiological and sleep mentation variables have not been more robust may be because methods for analyzing combinations of such variables in coherent groupings have not been available. Studies that are able to simultaneously consider variations in respiration, penile tumescence, EMG inhibition, and other autonomic indicators may well prove to demonstrate more reliable isomorphic relationships with sleep mentation at different levels of complexity.

## NR7. Elimination of REM sleep does not eliminate dreaming

Two commentators (**Bosinelli & Cicogna; Panksepp**) and a target article (**SOLMS**) suggest that the covert REM sleep model is inconsistent with the demonstration (Solms 1999b) that elimination of REM sleep does not necessarily eliminate dreaming. This contention depends crucially on whether REM sleep can, in fact, be eliminated as claimed. **HOBSON ET AL.** suggest in their target article that it cannot. They suggest, on the basis of proven difficulties in experimentally suppressing REM sleep with pontine lesions in animals, that any lesion capable of destroying the pontine REM sleep generator in humans would have to be so widespread so as to eliminate consciousness altogether. Solms (1999b) himself conceded this point at a recent symposium on the neurophysiology of sleep.

Repeated polysomnography over many nights would be crucial to determining the presence or absence of REM sleep or, more precisely perhaps, the degree of presence of REM sleep. This was amply demonstrated by the case of purportedly suppressed REM sleep described in section NR5 (Lavie 1990). The subject of this case study had severely reduced REM sleep, but it was found to be totally absent on only three out of five recording nights. Experimental awakenings from sleep in subjects like this, who suffer from brainstem lesions and reduced REM sleep, could serve as a critical test of the covert REM sleep model. Subjects' sleep records could be examined for evidence of residual REM sleep events, even in the absence of stage REM sleep as traditionally scored. As Lavie's paper demonstrated, REM sleep signs can be detected in the absence of the full-blown REM sleep state.

## NR8. The model needs validation

I agree wholeheartedly with commentators (**Blagrove; Conduit et al.; Franzini; Gottesmann**) calling for validation of the covert REM sleep model. I think that the **NIELSEN** target article, many of the excellent points raised in the commentaries, and this reply to the commentaries together suggest straightforward ways in which such validation could proceed:

1. Replication of early unreplicated findings demonstrating state overlap in NREM sleep (Schwartz 1968) and at sleep onset (Toth 1971).

2. Extension of previous studies that have examined percent and type of NREM mentation recall as a function of preceding REM sleep characteristics. Time since previous REM sleep has been evaluated in several studies, but time in previous REM sleep, intensity of previous REM sleep, propensity for previous REM sleep, and so on, have not (although see results of Study 1 in sect. NR8.1).

3. Assessment of clinical phenomena in which vivid NREM dreaming occurs (e.g., stage 2 nightmares) for evidence of covert REM processes.

4. Replication of recent findings concerning the effects of during-sleep stimulation on dreaming, for example, Conduit et al.'s (1997) finding that stimulation in NREM sleep increases recall of mentation.

5. Examination of EEG parameters for evidence of brief state shifts (Lehmann & Koukou 1984) and REM sleep-like

intrusions, for example, brief EEG desynchronizations in NREM sleep.

6. Use of topographic mapping to determine simultaneous activation of NREM and REM signs in NREM sleep (e.g., central vs. frontal leads).

7. Examination of continuous delta power profiles for evidence of reduced delta and/or rapid delta fluctuations during the covert REM sleep of "missing" first REM periods (cf. **Feinberg**).

8. Exploration of covert REM sleep signs during REM sleep deprivation (cf. **Cartwright**).

9. Effects of measurements taken at home versus in the laboratory on NREM mentation; does the laboratory environment induce covert REM sleep processes?

10. Architectural assessment of covert REM signs (e.g., penile tumescence, eye movements, EMG bursts) in relation to mentation recall: do they conform to a 90-min ultradian rhythm? Is their duration from 20–25% of TST? Are they in close proximity to an overt REM sleep episode? Are they concordant with other REM signs (eye movements, phasic muscle activity, heart rate or respiratory variability, etc.)?

11. Assessment of REM-NREM content differences in subjects highly trained in introspection.

12. Effects of experimenter bias, subject verbosity, speed of awakening, and so on, on frequency and complexity of NREM mentation reports.

I undertook preliminary validation of the model in two studies that address the first three of these considerations. One study was designed to assess correlations between the amount of mentation recalled following awakenings from stage 2 sleep and the simple duration of immediately preceding REM and NREM sleep stages. The second study was an exploratory assessment of a sample of sleep records from both normal and sleep-disordered subjects for evidence of signs of covert REM sleep in NREM sleep. I briefly describe these studies below.

### NR8.1. Study 1: Is stage 2 mentation associated with prior duration of REM and NREM sleep?

To test whether the amount of mentation recalled from stage 2 sleep would be associated with longer durations of prior REM and/or NREM sleep, we drew upon a sample of 26 healthy control subjects (20W, 6M; Mean age = 25.7 ± 6.5 years, range: 18–42) who in a previous study (Faucher et al. 1999) had been awakened from REM and stage 2 sleep to report mentation. We identified all stage 2 awakenings for which there had also occurred a preceding, uninterrupted REM sleep episode (N = 74). A trained polysomnographer scored the sleep records for two variables: (1) time in prior REM sleep, and (2) time in prior stage 2 sleep (stage 2 onset to point of awakening), according to the standard criteria (Rechtschaffen & Kales 1968). Another judge counted the number of relevant, nonredundant words in each mentation report from which total word count (TWC) and log (TWC + 1) were calculated. Correlations were calculated for the entire sample of 74 (N = 26 subjects) and for a reduced sample of 34 reports (N = 18 subjects) that excluded all TWC scores that were equal to zero.

TWC and log (TWC + 1) scores gave similar patterns of results (Table NR2). Correlations only partly supported the hypothesis that proximity to a prior REM episode ("prior stage 2 duration") would be associated with lengthier stage

Table NR2. *Correlations between total word count (TWC) and duration of prior REM and NREM sleep episodes*

| | TWC r ($p$) | $Log_{10}$ (TWC+1) r ($p$) |
|---|---|---|
| Reports with WC≥0 (N=74) | | |
| Prior REM duration | +0.380 (*.001*) | +0.335 (*.004*) |
| Prior stage 2 duration | −0.138 (*.243*) | −0.033 (.789) |
| Reports with WC>0 (N=34) | | |
| Prior REM duration | +0.373 (*.030*) | +0.255 (*.145*) |
| Prior stage 2 duration | −0.315 (*.069*) | −0.420 (*.014*) |

2 mentation reports. Duration of prior stage 2 sleep correlated negatively with TWC $r = −.315$, $p = .069$) and log (TWC + 1) ($r = −.420$, $p = .014$) when zero-recall reports were excluded, but not when they were included (both $p$ = NS). Further, duration of the prior REM sleep episode was positively correlated with report length whether zero-recall reports were included ($r = .380$; $p = .0008$) or not ($r = .373$, $p = .030$). This did not seem to be due to a circadian phase effect (i.e., longer REM episodes occurring later at night) because correlations between the clock time of REM episode onset and TWC were negligible ($r = .097$ and .118) for the two samples (both $p$ = NS).

These analyses thus partly support predictions of the covert REM sleep model and replicate the findings of several previous studies demonstrating greater recall with closer proximity to REM sleep (see **NIELSEN** sect. 3.4 "*Proximity of NREM sleep awakenings to REM sleep*"). They are also the first to suggest that parameters of a prior REM sleep episode *other* than its proximity might influence NREM mentation. Whether the REM duration measure in the present study reflects heightened REM pressure (due to awakenings for mentation recall from other REM episodes) or some other factor has yet to be determined. However, the findings together are consistent with the possibility that the presence and degree of elaboration of stage 2 sleep mentation is affected by interactions between prior REM and stage 2 sleep processes. Specifically, the present results suggest that the *duration* of a prior REM episode may determine whether or not content will appear in a subsequent stage 2 episode, but that the stage 2 episode's *proximity* to this REM episode may determine the degree of elaboration of that content, given that it is present.

### NR8.2. Study 2: Do signs of covert REM sleep appear in NREM sleep?

To examine whether REM sleep signs appear at sleep onset and in NREM sleep more generally, a polysomnographer with six years of full-time experience using the Rechtschaffen and Kales (1968) criteria evaluated a series of 35 records from 20 subjects (11W, 9M; mean age = 32 ± 11.6) for evidence of rapid eye movements and other signs in NREM sleep. Eight of these subjects (5W, 3M; mean age = 29 ± 12.5) were healthy controls, seven (3W, 4M; mean age = 27.6 ± 5.4) were patients consulting for idiopathic nightmares (INM), and five (3W, 2M; mean age = 44.6 ± 8.4) were patients consulting for post traumatic nightmares (PTNM). The polysomnographer used Schwartz's (1968) criteria for scoring slow, medium fast, and rapid eye move-

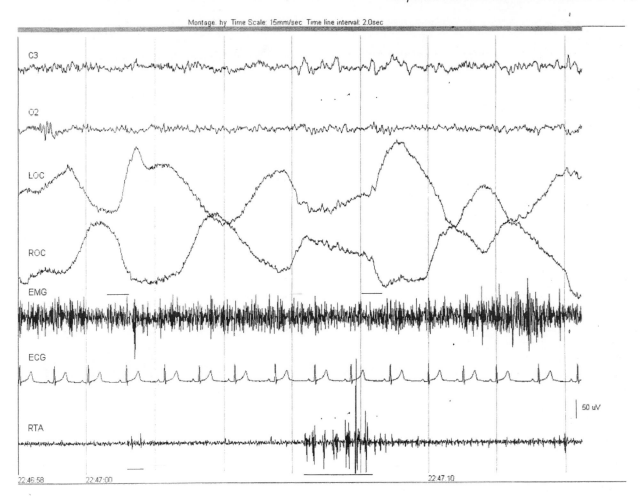

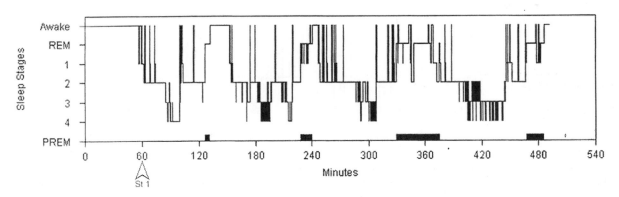

Figure NR3. Hypnogram and polysomnographic (PSG) tracing from a 24-year-old male patient with long-term idiopathic nightmares (INM). Medium fast and rapid eye movements are visible in this sleep onset stage 1 epoch, with phasic tibialis activation occurring between two bursts. C3: C3/linked ears; O2: O2/linked ears; LOC: left ocular; ROC: right ocular; EMG: chin muscle activity; ECG: bipolar cardiac; RTA: right tibialis anterior. Vertical grey lines indicate 2 second intervals.

ments as a guide only, since the latter criteria were not found to be precise enough to apply systematically. For example, the duration criteria for the three types are *slow:* 1.0 to 4.0 sec; *medium fast:* 0.25 to 2.0 sec; and *rapid:* 0.2 to 1.5 sec.

Of the 20 subjects, 12 (60%) showed at least one clear example of covert REM signs either at sleep onset (No. events = 13) or during later stage 2 or 3 sleep (No. events = 16). Examples were noted in 4 of 8 (50.0%) control subjects, 4 of 7 (57.1%) INM patients, and 4 of 5 (80.0%) PTMN patients. They occurred in 6 of 11 (54.5%) women and 6 of 9

(66.7%) men. Events were found more often in stage 2 sleep (17/30 or 56.7%) than in stage 1 sleep (12/30 or 40.0%), stage 3 sleep (1/30 or 3.3%) or stage 4 sleep (0/30 or 0.0%). More events occurred shortly after (23/30 or 76.7%) rather than before (2/30 or 6.7%) an episode of wakefulness, and before (4/30 or 13.3%) rather than after (1/30 or 3.3%) an episode of REM sleep. Some examples of these REM sleep events with their corresponding hypnograms appear in Figures NR3 to NR8 (see also Fig. NR2).

Figures NR3 and NR4 are taken from a 24-year-old male

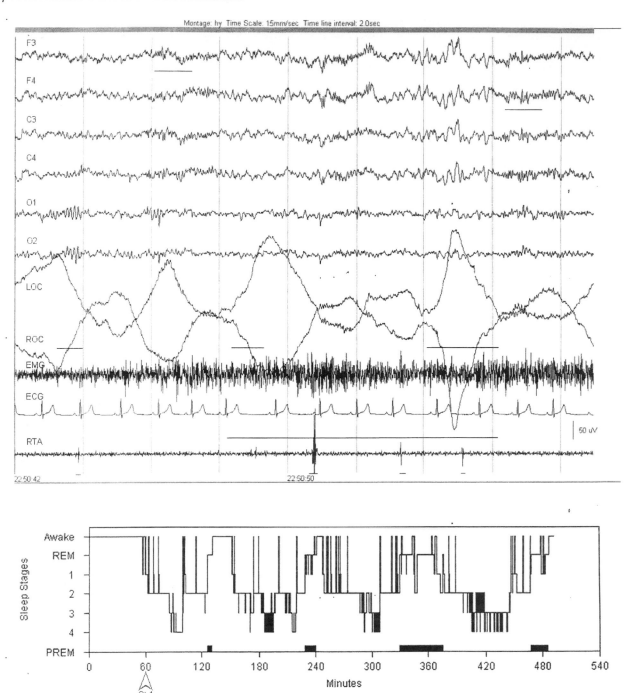

Figure NR4. Hypnogram and PSG tracing from same patient as in Figure NR3. The tracing occurred within 4 min of the previous one. A mixture of slow, medium fast, and rapid eye movements can be seen. Phasic tibialis EMG is also evident as is REM sleep-like cardiac variability on the ECG. Spindles are clear in the EEG. Legend as in Figure NR3 with addition of F3, F4, C4, and O2 all referenced to linked ears.

INM patient. These tracings occurred within 4 min of each other only minutes after initial sleep onset. They illustrate a mixture of slow, medium fast, and rapid eye movements occurring within the same eye movement bursts. A given eye movement may be medium fast or rapid in one direction yet slow in the other. Further, these eye movement bursts are accompanied by REM sleep-like phasic tibialis muscle bursts (both Figures) and abrupt cardiac variability, as well as by spindling in the EEG (Fig. NR4).

Figure NR5 is taken from a 25-year-old female patient with INM. It displays a section of stage 1 sleep shortly after

a long episode of wakefulness in the sleep onset period. Rapid and medium fast eye movements again occur in the same eye movement burst. Spindles are also present.

Figure NR6 is a section of late night stage 2 sleep from a 43-year-old male PTNM patient. This patient had the highest number of identified REM sleep signs (3 at sleep onset; 9 in late night NREM) out of the entire sample and had a highly fragmented hypnogram in general. He also reported dreaming vividly throughout the night, every night. A phasic EMG burst of chin muscle activity and a single rapid eye movement occur amidst several stage 2 sleep spindles in the

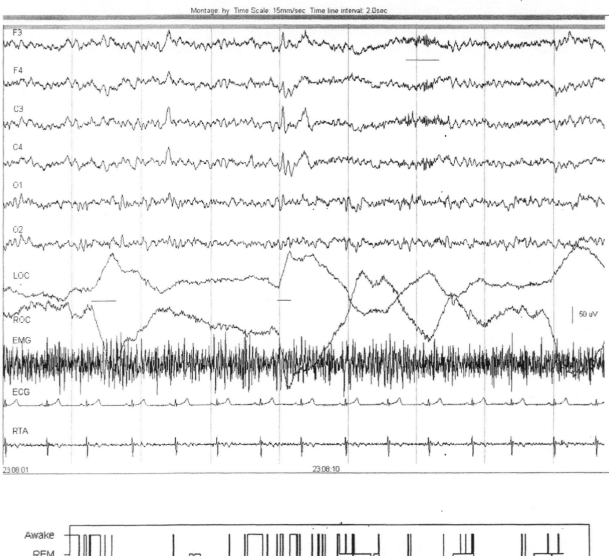

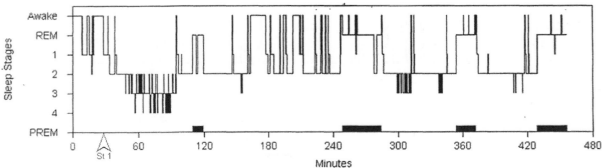

Figure NR5. Hypnogram and PSG tracing from 25-year-old female with INM. A section of stage 1 sleep with spindling at sleep onset contains both medium fast and rapid eye movements in the same eye movement burst. Legend as in Figure NR4.

tracing. This patient displayed a second such event 9 min later, just prior to an apparently aborted REM sleep episode.

Figure NR7 is a section of stage 2 sleep from the same patient as in Figure NR6 but on the following night and transpiring less than 10 min after a lengthy REM sleep episode. The tracing shows medium fast and rapid eye movements, one of which occurs in exact synchrony with a sleep spindle. This type of synchrony suggests that inhibitory influences associated with sleep spindles (see **Porte** commentary) may be less generalized than is thought.

Figure NR8 is taken from a 30-year-old female INM patient. It illustrates a burst of medium fast-to-rapid eye

movements coincident with a 5-sec burst of chin muscle activity against a background of relatively quiescent EMG in stage 3 sleep. This event occurred several minutes prior to a brief awakening.

This study was not undertaken to prove that eye movements and other REM sleep signs observed in NREM sleep are frequent enough to account for all the observed sleep mentation reported in this stage, although the correspondence between the fact that 50% of normal subjects had such signs and that recall of NREM sleep mentation is about 50% on average (see target article) should be noted. Rather, it was intended simply to raise doubts in a concrete

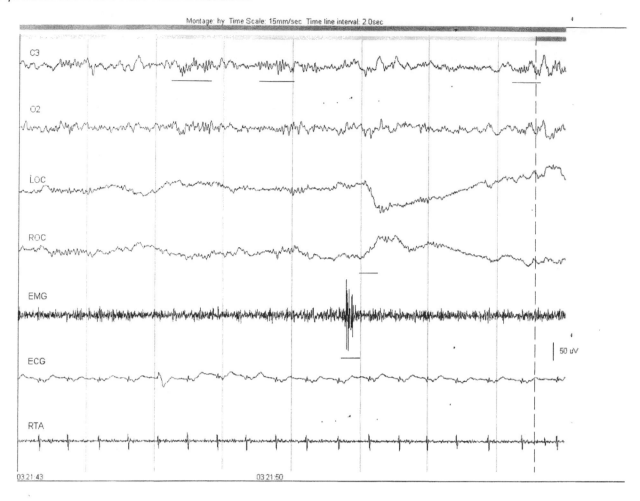

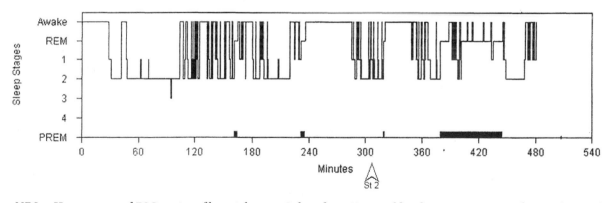

**Figure NR6.** Hypnogram and PSG tracing of late night stage 2 sleep from 43-year-old male post-traumatic nightmare (PTNM) patient. This patient had the most REM sleep signs of the entire sample and a fragmented sleep hypnogram on both nights (see Fig. NR7). He also reported dreaming vividly throughout the night, every night. A phasic EMG burst of chin muscle activity and a single rapid eye movement occur with stage 2 sleep spindles. A second similar event occurred 9 min later, just prior to an apparently aborted REM sleep episode. Legend as in Figure NR3.

fashion as to whether REM and NREM sleep states are as completely distinct as commonly thought. The findings together do suggest that: (1) REM sleep events are common enough in NREM sleep that they warrant more careful study with more sensitive recording equipment (e.g., higher sensitivity eye movement detectors); (2) sleep onset, in particular, often resembles REM sleep, if only for brief intervals, with some of the standard scoring criteria absent; (3) covert REM signs occur in normal subjects but more frequently in sleep-disordered patients; and (4) covert

REM signs are closely linked to prior wakefulness, and to *subsequent* (more so than to *preceding*) REM sleep. The importance of the last point is that subsequent REM sleep episodes are technically very difficult to predict and thus are very likely to affect NREM mentation reports.

If, as this study suggests, readily measurable peripheral signs of REM sleep occur with some regularity in NREM sleep, then there should be even more reason to suspect that *less* easily measurable peripheral and central signs of REM sleep may also be active outside of their normal

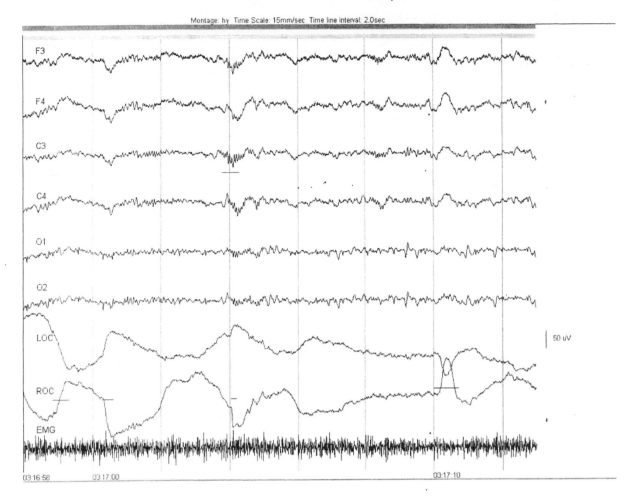

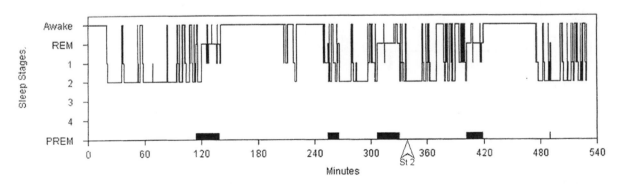

Figure NR7. Hypnogram and PSG tracing of stage 2 sleep from same patient as in Figure NR6 but on the following night. The epoch occurs less than 10 min after a lengthy REM sleep episode. Medium fast and rapid eye movements are visible; one of these occurs in exact synchrony with a sleep spindle. Legend as in Figure NR4 minus RTA.

boundaries. There is a multiplicity of physiological systems participating in the chaos of REM sleep but only a fraction of these are ever monitored. In fact, many such processes may manifest sporadically during NREM sleep even when *none* of the standard criteria for REM sleep are visible. In particular, important changes in a variety of autonomic effector systems in REM sleep (Parmeggiani 1994) are often technically difficult to measure, yet these seem particularly pertinent to assessing emotional features of sleep mentation that might become dissociated from REM sleep (cf. **Panksepp**).

## NR9. Conclusion

The covert REM sleep model can be seen to be an instance of one of four alternative viewpoints on the sleep mentation question, each of which makes a different combination of assumptions concerning (1) mind-body isomorphism and (2) the presence of one versus two mentation generators (see Table NR2). Isomorphism with a 1-gen assumption describes the covert REM sleep processes model. Isomorphism with a 2-gen assumption describes the activation-synthesis and AIM models, while non-isomorphism with

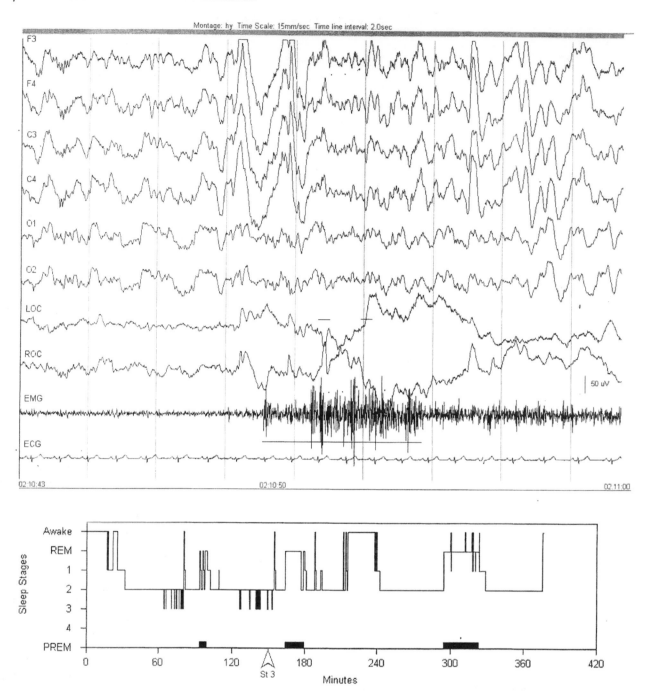

Figure NR8. Hypnogram and PSG tracing of stage 3 sleep from a 30-year-old female INM patient. A burst of medium fast-to-rapid eye movements coincides with a 5-sec burst of chin muscle activity against a background of quiescent EMG. A brief awakening occurred several minutes later. Legend as in Figure NR4 minus RTA.

1-gen and 2-gen assumptions describe Foulkes's model and models such as that proposed by Casagrande, respectively. There is in all likelihood room for models that take intermediate positions on these two basic assumptions. For example, commentators such as **Cavallero, Bosinelli & Cicogna,** and **Feinberg** acknowledge a limited role for cortical activation in initiating sleep mentation, but they do not appear to subscribe to isomorphism beyond this general level. Because so little is known about mind-body isomorphism, it would be premature to exclude consideration of such models.

If *both* strict isomorphism and a 1-generator mechanism are true assumptions, then so also is the covert REM sleep model true *in some form.* By this I mean that some uniform set of physiological isomorphs exists that is reliably correlated with sleep mentation – regardless of sleep state. In fairness to the most adamant critics of the covert model, such physiological variables *need not* be the dissociated REM sleep processes that I propose. They may prove to be much subtler patterns of neural coding that have little to do with the overt measures that we routinely record from surface electrodes. Some examples are discussed in Helekar (1999). They may even be active during much of the waking state. Then again, it may prove to be convenient to adopt a REM sleep-related nomenclature if only because these variables will likely be more typical of REM than of NREM sleep, that is, they will be more prevalent, more fre-

Table NR3. *Models of sleep mentation necessitated by different assumptions about isomorphism and number of mentation generators*

| | 1-generator true | 2-generator true |
|---|---|---|
| Isomorphism false | A. One factor mnemonic activation model (Foulkes and others) or equivalent | B. Two-factor psycholinguistic model (Casagrande and others) or equivalent |
| Isomorphism true | C. Covert REM sleep processes (Nielsen and others) or equivalent | D. Activation-synthesis and AIM models (Hobson, McCarley, and others) or equivalent |

quent, and more intensely activated in REM sleep than they will in NREM sleep – or in the waking state for that matter. This fact, the regular association of vivid imagery with REM sleep, still remains as the legacy of last century's neurobiologically driven dream research, regardless of the convincing demonstrations of sleep mentation in NREM sleep. However, a definitive explanation of dreaming awaits a much more detailed understanding of what constitutes REM and NREM sleep, and of precisely how mind and body are inter-related as these states surge, recede, dissociate, and blend together across the sleep/wake cycle.

NOTES

1. I prefer the term "subjective experience" (cf. Helekar 1999) to "conscious experience" and especially to "subjective conscious experience" in the case of sleep mentation because the manner in which dreaming is "conscious" vis-à-vis waking consciousness has not been clearly articulated (although cf. Kahan & Laberge 1996).

2. This kind of explanation is very difficult to evaluate because verbatim mentation reports are only rarely published.

## REM sleep is not committed to memory

Robert P. Vertes[a] and Kathleen E. Eastman[b]

[a]Center for Complex Systems, Florida Atlantic University, Boca Raton, FL 33431; [b]Department of Psychology, Northern Arizona University, Flagstaff, AZ 86011. **vertes@walt.ccs.fau.edu    k.eastman@nau.edu**

**Abstract:** We believe that this has been a constructive debate on the topic of memory consolidation and REM sleep. It was a lively and spirited exchange – the essence of science. A number of issues were discussed including: the pedestal technique, stress, and early REMD work in animals; REM windows; the processing of declarative versus procedural memory in REM in humans; a mnemonic function for theta rhythm in waking but not in REM sleep; the lack of cognitive deficits in patients on antidepressant drugs that suppress or eliminate REM sleep; the disposition of conscious (dreams) and nonconscious material of REM sleep; and finally our theory of REM sleep. Although our position was strongly challenged, we still hold that REM sleep serves no role in the processing and consolidation of memory.

### VR0.  Seeds of our target article

Several years ago I (**VERTES**) carried out a series of studies in behaving rats examining the relationship between the activity of cells of the pontine reticular formation (PRF) and the theta rhythm of the hippocampus. I showed that the discharge of a subset of PRF neurons was highly correlated with theta rhythm of waking and REM and subsequently

that these PRF cells are directly involved in the generation of the theta rhythm.

Prior to recording, I deprived rats of REM sleep in order to increase the amount of time spent in REM sleep (i.e., REM rebound) during subsequent recording sessions. Rats were deprived of REM for 24–36 hours using the pedestal technique. Although my sole purpose for using REMD was to boost REM during recording periods, I was surprised to observe that even 24 h of REMD produced severe detrimental effects on the rats. The rats were cold and often still wet from having fallen in the water, physically fatigued from balancing on the small diameter surface of the inverted flower pot, tired from a considerable lack of sleep (mostly REM, but both SWS and REM), and generally debilitated (much like we would be without sleep for 1–2 days). Although rats are reportedly hyperactive following REMD, I found that they were essentially immobile for at least 6 h post REMD. This experience led me to question the validity of experiments examining the effects of REMD on learning and memory; that is, if rats were so severely incapacitated following this procedure how could they adequately perform on behavioral tasks following REMD?

In 1995, Peter Shiromani asked me to participate in a forum on sleep and memory for Sleep Research Society (SRS) Bulletin. I agreed and indicated that I would be taking the "con" position: no relationship between REM sleep and memory. Of eight participants in the forum, I was the only one taking this position. Possibly based on my article in SRS Bulletin, Mike Chase invited me to participate in a debate with Carlyle Smith on this same topic at an international workshop on sleep and cognitive function sponsored by the World Health Organization in Cancun, Mexico, in 1999. The debate was fruitful and further fueled my interest in the issue of memory consolidation and REM sleep. The target article by my colleague and me developed from this background.

### VR1.  Early REMD studies in animals, the pedestal technique, and stress

As we discussed in our target article, there was an intense interest in the role of REM sleep in memory consolidation in the 1960–1970s, interest waned in the 1980s, and has recently resurfaced. This is now a lively topic in the sleep field. As we previously indicated, our coverage of the early REMD work in animals was not meant to serve as a detailed analysis of this area, but rather to convey a general sense of the net contribution of this work to an understanding of the possible involvement of REM sleep in memory consolidation. We reached two main conclusions: (1) the early studies in animals were quite contradictory with as many reports opposing as favoring a role for REM in memory consolidation; and (2) the "stress" associated with the use of the pedestal technique for REMD was a confounding factor in many studies using this technique.

It appears that, on balance, most commentators agreed with these conclusions; that is, the early work was contradictory and much of it was methodologically flawed (**Born & Gais; Cartwright; Coenen; Feinberg; Mazzoni; Ogilvie et al.; Panksepp; Rotenberg; Siegel**). This was well put by Born and Gais: "There are obvious flaws of REM sleep suppression paradigms which do not allow for any conclusion, either pro or contra the REM sleep-memory hypothesis."

It was generally agreed that the stress associated with the pedestal technique confounded findings obtained with it. Some however, felt that we overplayed the "stress card" (**Fishbein; Greenberg; Moorcroft; Smith & Rose**) or as Greenberg stated, we laid everything at the feet of the "villain" stress.

Although several commentaries addressed this issue, perhaps the most insightful was that of **Coenen**, who has extensively examined the learning abilities of rats using the pedestal technique and has compared its use to other less disruptive forms of REMD. He essentially concludes that the detrimental effects of REMD on learning/memory primarily involve the stress of REMD procedures rather than the loss of REM, per se, pointing to a direct relationship between degree of stress and extent of learning/memory impairments. Coenen made reference to their work (van Hulzen & Coenen 1979) showing that a relatively stressful procedure (platform technique) for REMD, but not a mild one (selective hand awakening), disrupted active avoidance learning in rats. Finally, Coenen questions the use of the pedestal technique in sleep research on ethical grounds; that is, given the controversial findings obtained with this procedure, is it appropriate to continue to expose rats to it?

**Fishbein** argues that he ruled out stress as a possible confounding factor in pilot REMD studies in mice by the use of a modified version of the pedestal technique that produces little or no stress (Fishbein & Gutwein 1977). With this method, mice are not restricted to the pedestal as with conventional techniques, but are able to latch onto and climb about the underside of a wire mesh lid positioned above the enclosure containing the pedestal. Fishbein concluded that his demonstration that "stress-free" mice still showed learning deficits negates stress as a major contributor to the deficits. While this interpretation is possible, it seems that stress was only assessed by a single measure (open field activity), and to our knowledge, this procedure has not been used with mice outside of Fishbein's laboratory, thereby providing no independent verification of the claim that the mice are stress-free.

**Smith & Rose** counter the "stress argument" with the following: (1) memory deficits are present following the non-stressful pharmacological suppression of REM sleep; (2) learning impairments are seen only following four hours of REMD (i.e., REM windows); (3) animals deprived of total sleep (excluding windows) which is presumably more stressful than depriving them of four hours of REM sleep (window) show no learning/memory deficits; and (4) in some instances, REMD has been shown to improve mem-

ory. With respect to point 1, it seems unlikely that the pharmacological blockade of REM is stress free, and whatever the outcome of these studies, they do not address the question of whether stress was a major contributor to the deficits seen in reports using the pedestal technique. Regarding points 2 and 3, in our target article, we acknowledged that Smith's use of relatively short REMD periods rendered his studies less vulnerable to the stress factor, but not invulnerable. It is quite possible that even short periods of REMD are stressful to rats; that is, sufficiently stressful to disrupt learning. Regarding point 4, the findings that animals show no impairments or even improved performance following REMD would seem to indicate that in some circumstances animals can override the effects of stress.

## VR2. REM windows

In our target article, we discussed three issues related to REM windows: (1) several conditions (e.g., species and even strain of animals, type of training tasks, number and distribution of training trials per session and/or per day) can affect the post-learning position of the window (i.e., shifting windows); (2) the phenomenon has only been demonstrated by Smith and colleagues; and (3) REM windows have not been described in humans.

Regarding the first point, several commentators (**Greenberg; Moorcroft; Ogilvie et al.; Smith & Rose**) questioned our "lack of appreciation" of the shifting nature of the REM window. In essence, many expressed surprise that we would expect anything less than fluctuating windows reflecting very different initial learning conditions. According to Smith & Rose, the notion of a rigid window occurring at the same time after training would be intuitively inconsistent. Borrowing from Smith & Rose, we *intuitively* have great difficulty with the notion of REM windows. For argument's sake, if REM were involved in memory consolidation, why would information be selectively consolidated in one shifting four-hour block of REM-dominated time and no other. In separate studies, Smith and co-workers reported post-training REM windows of 53–56 h (Smith & MacNeill 1993) and 48–72 h (Smith & Kelly 1988). What is the utility to animals of consolidating information 2–3 days after it has been learned? Might not the information be needed in the interim? Finally, on a practical level, the shifting nature of the windows allows for the possibility that studies showing no effect of REMD on memory could be dismissed on the grounds that they missed the window.

Regarding the second point, **Greenberg** indicated that Pearlman described REM windows several years before Smith did, but **Smith & Rose** did not dispute our statement that the effect originated with Smith. Finally, with respect to the last point, other than a brief reference by Smith & Rose, no one disputed our claim that REM windows are not present in humans.

## VR3. Human studies, declarative and procedural memory

### VR3.1. REM sleep is involved in procedural but not in declarative memory

A number of commentators felt that we did not do justice to (or ignored) recent human studies on memory and REM sleep (**Born & Gais; Schredl; Smith & Rose; Stickgold**). In our defense, a section was devoted to this, and much of the important work in this area has only recently appeared. For instance, Born & Gais admonished us for giving little attention to this work, but then cited the example that we "briefly mentioned the intriguing work of Stickgold et al. (2000b)." This report was not published at the time we submitted our target article. If our previous coverage was inadequate, we hope to rectify the situation with the present treatment.

**Born & Gais** point to a very useful approach to the study of memory and sleep in humans, which is to compare rates of learning following either of two equally long periods of sleep; one with high amounts of SWS (SWS-rich) and the other with high amounts of REM sleep (REM-rich). Using this approach, Ekstrand and co-workers (Barret & Ekstrand 1972; Fowler et al. 1973; Yaroush et al. 1971) and subsequently Plihal and Born (1997; 1999a) showed that there was significantly greater improvement in the recall of declarative memories following SWS-rich (first half of night) than REM-rich (second half of night) (see also Born & Gais). In addition, Born and colleagues (Plihal & Born 1997) recently reported significantly improved recall in various procedural (nondeclarative) memory tasks (mirror tracing, word stem priming) following late (REM) but not early (SWS) sleep. These findings indicate that SWS is involved in the processing/consolidation of declarative memories; REM in procedural memories.

In our target article, we described a visual discrimination task developed by Karni and colleagues that involves the identification of the orientation (horizontal or vertical) of three diagonal lines embedded in background of horizontal lines. The task has been described as a procedural learning task (Karni et al. 1994; Stickgold 1998). To date, three laboratories (Karni, Born, and Stickgold) have used this task, and have described relatively consistent findings with it. Karni et al. (1994) reported that subjects showed improved performance on the task following a period of sleep that contained REM sleep (SWS deprived) but not one lacking REM sleep (REM deprived), and concluded that REM was critical for the learning of this procedural task. In like manner, Stickgold et al. (2000b) demonstrated significant increases in amounts of SWS in the first quarter of sleep and REM in the last quarter of sleep in subjects showing improved performance on the task, while Gais et al. (2000) (see **Born & Gais**) reported a three-fold improvement in performance on the task in subjects with both SWS and REM sleep compared to those with only SWS (SWS-rich sleep). All three studies, then, show that REM sleep is directly involved in the acquisition of this procedural learning task, and together with those reviewed above, indicate that REM serves a critical role in procedural learning.

In line with the foregoing, **Smith & Rose** indicate that REM serves to consolidate some types of memories but not others; that is, according to them, "REM is not involved in declarative/explicit type memory tasks"; but serves a clear role in procedural tasks producing "impairment ranging from 20–50% in these tasks." **Stickgold** fully concurs stating: "Since all proponents of REM-dependent memory consolidation agree that REM is not involved in declarative memories such as those formed in paired associates training, the failure to observe REM-dependent consolidation may simply reflect the testing of a memory system that is not REM-dependent."

In summary, leading proponents of the memory consolidation hypothesis seem to have recently come to a very similar conclusion which is that REM sleep is involved in procedural memory but not in declarative memory.

### VR3.2. Contesting the position that REM sleep is involved in procedural memory

The position that REM may be uniquely involved in procedural memory is late developing and appears to largely rest on recent work using the perceptual discrimination task of Karni and associates. In our target article, we pointed out differences between the results of Karni and Stickgold using this task. As indicated above, Born and colleagues are also now using this task in studies on sleep and memory in humans. There are, however, inconsistencies among the findings of the three groups. For instance, Born (Gais et al. 2000) described a role for SWS (early sleep) but not REM sleep (late sleep) in the task; Karni et al. (1994) reported the opposite (a role for REM but not SWS) and **Stickgold** contends that both are involved (SWS and REM sleep).

**Stickgold** suggests that we unfairly highlighted minor differences between his work and that of Karni using this perceptual task. We disagree, particularly with respect to one very important difference between the findings of the two laboratories; that is, the improvement (Karni & Sagi 1993; Karni et al. 1994) or lack of improvement on the task (Stickgold et al. 2000b) in the waking state.

In an initial study devoted entirely to waking, Karni and Sagi (1993) described the important findings that improved performance on the perceptual task simply required the passage of time; no gains were seen immediately after learning but only after a period of 8–10 hrs following learning. In a follow-up report, Karni et al. (1994) replicated their earlier effect (improvement over time in waking) and further demonstrated a comparable improvement with time in REM sleep. Taken together these results indicate that the acquisition of this procedural skill requires consolidation over time but this time need not be in REM sleep. In marked contrast to these findings, Stickgold et al. (2000b) clearly indicated that there was no improvement on the task during waking, commenting: "12 hours of wake behavior was inadequate to produce reliable improvement while as little as 9 hours of sleep reliably produced improved performance." We believe that this is a very important difference in the findings of the two groups (Karni and Stickgold), especially considering that they used an identical perceptual task.

In summary, the results of the three groups using the Karni perceptual task significantly conflict; that is, gains in performance on the task have been variously attributed to waking (Karni & Sagi 1993; Karni et al. 1994), SWS (Gais et al. 2000), REM (Karni et al. 1994), and both SWS and REM (Gais et al. 2000; Stickgold et al. 2000b). We believe that these differences undermine the findings of each of the groups and need to be resolved, especially since this task seems to be evolving as the standard for examining the role of sleep in procedural memory. At the very least, the above results indicate that REM serves no unique role in the acquisition of this procedural learning task.

**Jones** supports an involvement of REM in procedural memory, arguing that the rehearsal of motor sequences in REM could enhance efficiency of motor performance in waking. For instance, Jones suggests that the high amounts of REM in the fetus may be used by animals to prepare them for meeting contingencies after birth such as locomotion and flight, and cites the example of the wildebeest who is fully mobile upon birth and immediately begins long treks with its mother in search of water. **Siegel**, however, describes the opposite; that is, an inverse relationship between degree of maturity at birth and amounts of REM sleep: the less mature the species, the more REM sleep; the more mature the species, the less REM sleep. Siegel refers to his recent work with the platypus and cetaceans (whales and dolphins) stating: "The immaturity of the platypus, hatching from an egg and remaining attached to its mother for an extended period after birth is consistent with its high level of REM sleep. The maturity at birth of the cetaceans, which can swim free of the mother and defend themselves immediately after birth is consistent with their low level of REM sleep."

In summary, there appears to be general agreement that REM sleep is not involved in declarative memory. The case for a REM involvement in procedural memory seems to largely rest on the recent work of three groups (**Born,** Karni, and **Stickgold**) using the visual discrimination task of Karni. As indicated, there are marked differences in the findings of these groups, which until resolved, make it difficult to evaluate the reliability of these results.

## VR4. A memory processing function for the theta rhythm in waking but not in REM sleep

We surprisingly received more comments than expected on the theta rhythm. Although comments were directed to several issues related to theta, most involved our contention that theta does not serve the same function in waking and REM; that is, a mnemonic function in waking but not in REM sleep. To paraphrase **Jones**: if theta plays a role in memory consolidation, it should do so in REM as well as in waking. Commentators were divided on this issue; some supported our position (**Gottesmann; Lynch et al.; Rotenberg**), others did not (**Conduit et al.; Fishbein; Jones; Morgane & Mokler; Stickgold**). Gottesmann expressed some skepticism that a target structure could function differently in the presence of the same activity (i.e., theta) in two different states, but then described circumstances in which this happens. He cited work of his laboratory with animals with transections at the intercollicular level of the brainstem. This is a unique preparation in that animals are comatose yet show a continuous theta rhythm. Gottesmann argued that it is very unlikely that theta serves a memory processing function in comatose animals.

In further support of our position, **Gottesmann** and **Lynch et al.** argue that the theta rhythm cannot be viewed in isolation from events that occur with it and which undoubtedly affect its functional role in either waking or REM. For instance, Gottesmann points to well documented findings that monoaminergic neurons fire at their highest rate in waking and lowest in REM, indicating that the monoaminergic drive to the hippocampus is very different in waking and REM. This would seem to have obvious functional consequences for the hippocampus as well as for the role of theta in the hippocampus in waking and REM sleep. In like manner, Lynch et al. note that during waking there is an exquisite interplay among various field oscillations (theta, beta, gamma) which signal diverse aspects of

the environment, and, also important, that: "the absence of this interplay of the different field potential oscillations during REM sleep may suggest that mnemonic functioning is absent, and theta is not functioning as it does during consciousness." Finally, Gottesmann points out, as did we, that even though the neocortical EEG is the same in waking and REM there are significant differences in the state of the animal in the two states, notably, differences in consciousness.

**Jones** has difficulty with our position that theta does not serve a mnemonic function in REM sleep. Interestingly, Jones also supports a role for REM in the consolidation of procedural memories (see earlier discussion). Procedural memory, however, is not thought to involve the hippocampus (see **Born & Gais; Stickgold**) and by extension would not likely involve the hippocampal theta rhythm. The developing view that REM is integral to the processing/consolidation of procedural memories would seem to rule out both theta and the hippocampus in these functions in REM sleep. In summary, waking and REM sleep are obviously very different states. It would seem likely that electrophysiological events common to waking and REM would serve different rather than the same functions in these two states.

## VR5. Lack of effects on memory of brainstem lesions or antidepressant drugs that profoundly suppress or eliminate REM sleep

As indicated in our target article, perhaps the strongest evidence against the memory consolidation hypothesis involves the demonstration that brainstem lesions or antidepressant drugs significantly suppress or eliminate REM sleep, but do not, on the whole, adversely affect cognition/memory. These findings strongly challenge the view that REM serves a critical role in memory consolidation.

Although we described this work in considerable detail, particularly that dealing with antidepressants, few commentators addressed it. **Born & Gais** remarked, however, that the demonstration that the loss of REM with brainstem lesions does not noticeably impair cognition shows, at minimum, that "phenotypic REM is not a prerequisite for memory consolidation."

While acknowledging the general lack of adverse actions of antidepressants on learning/memory, a number of commentators argued that these compounds may selectively affect certain kinds of memory and not others; that is, they may alter types of memories that were not previously tested for in patients taking antidepressants (**Bednar; Greenberg; Panksepp; Revonsuo; Smith & Rose; Stickgold**). For example, Greenberg described anecdotal evidence from a colleague who treated one of Wyatt's patients on long-term MAOIs. According to the account, while being treated with MAOIs, the patient seemed to have no access to past meaningful emotional experiences and had no dreams (and no REM sleep), but upon removal of the MAOIs the patient became more connected with her emotional past and experienced intense dreams. From this, Greenberg concluded that REM suppression with MAOIs affects emotional but not "cognitive" memories. We question whether "emotional memories" exist separate from their obvious cognitive content.

**Smith & Rose** and **Stickgold** similarly argue that the apparent lack of learning/memory deficits in patients with brainstem lesions or on antidepressants may have involved the failure to examine certain types of memory; that is, mainly procedural memory. For instance, Smith & Rose contended that by not distinguishing between different types of memories, we undercut our argument that "humans with REM-eliminating lesions or REM-depriving pharmacological treatments are normal, since this cannot be established if the subjects are not tested in tasks in which REM is known to be involved." In like manner, Stickgold states

> It is not surprising that simple cognitive and psychomotor memory tests fail to show any obvious impairment of performance after administration of drugs that disrupt REM sleep. These tests classically measure working memory and declarative memory systems that we would not expect to be affected by REM deprivation. We know of no cases in which anyone, for example, tested the effects of these drugs on complex perceptual procedural learning.

Although **Smith & Rose** and **Stickgold** seem to acknowledge that antidepressant drugs do not alter declarative memory, we are not willing to concede that they affect procedural memory. As we indicated, the widespread use of antidepressants has prompted a close examination of their possible side-effects, not only cognitive, but motor. With few exceptions, most of the commonly used antidepressants seem to have little or no adverse actions on motor functions – indirectly indicating a lack of an effect on procedural memory. We nonetheless agree with Stickgold that the possible effects of these drugs on specific procedural tasks have not been examined. This should be done.

## VR6. Dreams and recent imaging studies of the brain in REM sleep

In our target article, we indicated that the sole window to the cognitive content of REM sleep is dreams. We would be more sympathetic to the position that REM serves a memory consolidating function if dreams more or less faithfully reproduced waking experiences. They obviously do not. Freud (1900) grappled with this issue in *The interpretation of dreams,* considering, but then dismissing, the possibility that dreams merely replicate waking experiences and thus serve to store them. Freud speculates that:

> It might perhaps occur to us that the phenomenon of dreaming could be reduced entirely to that of memory: dreams, it might be supposed, are a manifestation of a reproductive activity which is at work even in the night and which is an end in itself.

And continuing:

> But views of this sort are inherently improbable owing to the manner in which dreams deal with the material to be remembered. Strümpell rightly points out that dreams do not reproduce experiences. They take one step forward, but the next step in the chain is omitted, or appears in altered form, or is replaced by something entirely extraneous. Dreams yield no more than *fragments* of reproductions; and this is so general a rule that theoretical conclusions may be based on it.

Finally, Freud remarked that only in very rare instances do "dreams repeat an experience with as much completeness as is attainable by our waking memory."

**Flanagan** reached a similar conclusion that dreams are not a mechanism for the storage of information from waking, stating: "since we rarely dream about what we need to remember, the hypothesis that dreams themselves serve any memory enhancing function appears unwarranted."

It would seem that most proponents of the memory con-

solidation hypothesis would agree that waking experiences are not faithfully reproduced in dreams or committed to memory through dreams. In a twist of logic, however, that we find difficult to understand, it appears that proponents of the consolidation hypothesis seem willing to acknowledge that conscious material of REM (dreams) is not stored in REM sleep, while at the same time holding that material that never reaches dream consciousness (whatever its nature) is somehow magically processed and consolidated in REM sleep.

In our target article, we reviewed recent human imaging studies of the brain in REM sleep. As indicated, the findings show a pattern of activity in REM that is consistent with dreams; that is, a suppression of major sensory inputs and motor outputs, reflecting, as termed by Braun et al. (1997) a "closed system"; a highly activated limbic system reflecting the rich emotional architecture of dreams, and strongly dampened activity within the frontal cortex corresponding to a lack of a higher order processing and integration of information in REM sleep.

As also discussed, activity within memory processing systems of the brain appears to be attenuated in REM as evidenced by the amnesia of that state, or as **Jones** (1998) observed: "an attenuation of processes important in episodic and working memory and perhaps explaining why, unless awakened from a dream, a sleeping person has no memory of the dream." Reviewing the same (imaging) data, Hobson et al. (1998b) similarly concluded that: "some functional process, present and responsible for memory in waking is absent, or at least greatly diminished, in REM sleep." We would argue that the "absent or greatly diminished" mnemonic capacity of the brain in REM affects both conscious (dreams) and nonconscious material of REM sleep.

In line with the foregoing, **Morrison & Sanford** observed that dreaming in REM and cognitive processing in wakefulness represent very different functional states of the brain, and that the "dreaming brain" is ill-equipped to deal with the requirements of wakefulness, including memory. According to them, "one can state unequivocally that the brain in REM is poorly equipped to practice for eventualities of wakefulness through dreaming, or for consolidating into memory the complex experiences of that state."

In summary, the foregoing indicates that the brain is in a non-encoding mode in REM sleep which accounts for the amnesic quality of dreams, and in our view, amnesia for all other cognitive material, conscious or not. We find difficulty with the position that acknowledges, on the one hand, that material reaching awareness in REM (dreams) is lost to memory, while at the same time claiming that material that does not reach dream consciousness is faithfully stored in memory in REM sleep.

## VR7. Our theory for the function of REM sleep

Our theory for the function of REM sleep received considerable attention; both positive (**Coenen; Lynch et al.; Ogilvie et al.**) and negative (**Blagrove; Clancey; Conduit et al.; Feinberg; Fishbein; Gottesmann; Hunt; Moorcroft; Morgane & Mokler; Panksepp; Revonsuo; Rotenberg**). We expected that the theory would be challenged, for, among other reasons, few are willing to concede that any current theory can fully account for the intricacies of REM sleep.

Before considering specific comments on the theory, we believe it is important to address two general issues related to our theory: (1) our intent, as **Hunt** suggests, was not to discredit the memory consolidation hypothesis in order to advance our theory of REM sleep; and (2) the theory did not originate with the target article but as we indicated was an abbreviated summary/restatement of a theory previously published by Vertes (1986b).

To elaborate, our target article was not meant to serve a forum for our theory, but having completed a critical analysis of the memory consolidation hypothesis, we felt it important to present our hypothesis for the function of REM sleep. **Morgan & Mokler** stated that we did not provide necessary background material to our theory, while others indicated that we did not give due credit to earlier theories that were forerunners to ours (e.g., Cohen & Dement 1965; Ephron & Carrington 1966; Roffwarg et al. 1966; Snyder 1965). This was done in the original, complete version of the theory by Vertes (1986b).

A number of comments were directed to our statement/position that the activation of REM serves to offset the inactivation of SWS, thereby, as we proposed, preventing the brain from dwelling too long in SWS and preparing the brain for a return to consciousness. Specifically, commentators questioned our position that REM reverses the effects of SWS, pointing to well documented mismatches between SWS and REM. For instance, (1) REM is present in significantly greater amounts than is SWS in the fetus and in newborns (**Blagrove; Feinberg; Hunt**), and (2) relative amounts of SWS and REM do not precisely co-vary throughout the sleep cycle: amounts of SWS are high and REM relatively low in early sleep; the reverse in late sleep (**Moorcroft; Rotenberg**).

We address the very high amounts of REM sleep in the fetus/newborn in a theoretical treatment on the sudden infant death syndrome (SIDS) (Vertes & Perry 1993). In brief, we discussed evidence showing that the respiratory system is undeveloped in newborns and hence abnormally sensitive to the effects of hypoxia in SWS. We proposed that REM exerts pronounced stimulatory actions not only on the CNS but on the respiratory system in newborns which in part serves to prevent hypoxia-induced respiratory failure during sleep in early infancy. We cited the early work of Baker and McGinty (1977) showing that kittens exposed to hypoxic conditions for several days showed extreme irregularity and slowing of respiration in SWS that led to death unless reversed by the activation of REM sleep. Baker and McGinty (1977) drew parallels between their findings with kittens and SIDS, speculating that active sleep (AS) (REM sleep) may serve to protect human infants from SIDS. They stated: "The predominance and tenacity of the AS state in the newborn period may account for the paradoxical immunity to SIDS in the first month of life. The peak incidence for SIDS coincides with the rapid decrease in AS time between 2 and 3 months of age." In essence, then, the need for more REM in neonates may be related to potentially greater detrimental effects of SWS at this age.

With respect to the relative differences in amounts of SWS/REM throughout sleep, we proposed that REM serves two complementary functions in sleep: it offsets SWS and promotes recovery from sleep. The shorter REM periods in early sleep would seem to be of sufficient length to periodically activate the CNS in sleep, while the increasingly longer REM periods throughout sleep would pro-

gressively prepare the brain for a return to consciousness as waking approaches.

In summary, to restate our theory, we propose that the brain/CNS is strongly depressed in SWS, particularly in delta sleep, and the function of REM is to provide periodic endogenous stimulation to the brain which serves to maintain minimum requisite levels of CNS activity throughout sleep. REM is the mechanism used by the brain to ensure and promote recovery from sleep. We further believe that theories of REM should contain two important elements: (1) the function of REM should remain constant throughout the life span and (2) as a state of sleep, the function of REM should be described entirely within the context of sleep. Our theory meets these criteria.

## VR8. Other evidence supporting our position

As pointed out by **Morrison & Sanford**, the case for memory consolidation in REM sleep has largely been built on the rather "tenuous correlational relationship between REM occurrence and indicators of performance" and that we presented compelling arguments questioning this temporal relationship. They discussed recent findings of their laboratory that run counter to the memory consolidation hypothesis. They reported a suppression of REM sleep in rats for 1–2 h following training on a conditioned avoidance task (fear conditioning), and concluded that these findings present problems for the position that REM is instrumental in consolidating information following learning.

In an interesting comparison with waking, **Morrison & Sanford** posed the following question: Would the constellation of REM events transferred to waking be conducive for memory consolidation in that state? That is, would a state of wakefulness characterized by a hyperactive brain, irregular respiration, tachycardia, muscle twitches, and the uncontrolled intrusion of extraneous mental images (dreams) be optimal for the consolidation of memory?

**Siegel** describes phylogenetic data inconsistent with the memory consolidation hypothesis. Siegel argues that if REM were involved in memory consolidation, one might expect that learning ability would be directly correlated with amounts of REM across species, but this is not the case. Siegel reviews evidence showing that: (1) humans do not have uniquely high amounts of REM sleep; (2) animals with the largest amount of REM sleep (7–8 h/day) such as the duckbilled platypus, black-footed ferret, and armadillo are not noted for their intelligence; and (3) species with the lowest amount of REM sleep (less than 15 minutes/day) such as whales and dolphins exhibit "prodigious learning abilities."

Finally, supporting our position, **Rotenberg** reports that some drugs (e.g., amphetamine) have a beneficial effect on memory but suppress REM, while others such as reserpine or neuroleptic agents have no noticeable effects on memory but increase REM sleep.

## VR9. Is slow-wave sleep (SWS) involved in memory consolidation?

Several commentators raised the possibility that SWS alone (**Steriade; Morrison & Sanford**) or in combination with REM (**Blagrove; Born & Gais; Cipolli; Mazzoni; Moor-**

croft; **Schredl; Smith & Rose; Stickgold**) may be involved in memory consolidation. As discussed, **Born & Gais** described a role for SWS in declarative memory and REM in procedural memory, while **Stickgold** implicated both SWS and REM in procedural memory.

Commentators drew attention to the recent work of McNaughton and colleagues as well as Buzsáki implicating SWS in memory consolidation. Specifically, McNaughton and co-workers (Wilson & McNaughton 1994; Skaggs & McNaughton 1996) have shown that ensembles of hippocampal place cells tend to repeat patterns of activity of waking in subsequent episodes of SWS, while Buzsáki (1989; 1998) has proposed that hippocampal sharp waves and associated high frequency (200 Hz) bursts (ripple), that are prominent in SWS, serve to transfer information from the hippocampus to the neocortex in SWS.

**Steriade** pointed out that SWS is not generally viewed as a state of mentation but rather one involving a global inhibition of subcortical and cortical structures. However, recent findings of his laboratory suggest that SWS may not be the "restful" state that it is commonly thought to be. He reports that, despite an absence of external input, neocortical neurons fire at high spontaneous rates and respond to internally generated signals in SWS, indicating that they may be involved in higher order, possible mnemonic, processes in SWS.

As indicated by our title, the focus of our case is memory consolidation in REM sleep. At this time, we take no position on the possible role of SWS in memory consolidation. We do, however, believe that a role for SWS in this process is far from proven, and doing so will involve re-tracing most (or all) of the steps taken by advocates of the memory consolidation in REM hypothesis.

## VR10. Conclusion

We believe that this has been a very fruitful debate on the topic of memory consolidation in REM sleep. This is obviously not the final word on this topic. We conclude with an observation of **Born & Gais** that we believe reflects the current state of affairs in this very important area. They stated: "REM sleep facilitates memory consolidation. Currently this is more a belief than a concept with convincing scientific support. Hence, **VERTES & EASTMAN**'s case against memory consolidation in REM sleep is a very timely contribution reflecting the true and persisting darkness in this area of sleep research."

## Did ancestral humans dream for their lives?

Antti Revonsuo

*Department of Philosophy, Center for Cognitive Neuroscience, University of Turku, Turku FIN-20014, Finland.*
**revonsuo@utu.fi     www.utu.fi/research/ccn/consciousness.html**

**Abstract:** The most challenging objections to the Threat Simulation Theory (TST) of the function of dreaming include such issues as whether the competing Random Activation Theory can explain dreaming, whether TST can accommodate the apparently dysfunctional nature of post-traumatic nightmares, whether dreams are too bizarre and disorganized to constitute proper simulations, and whether dream recall is too biased to reveal the true nature of dreams. I show how these and many other objections can be accommodated by TST, and how several lines of new supporting evidence are provided by the commentators. Accordingly TST offers a promising new approach to the function of dreaming, covering a wide range of evidence and theoretically integrating psychological and biological levels of explanation.

## RR0. Overview

I am grateful for the stimulating and thoughtful commentaries on the Threat-Simulation Theory of dreaming (TST). As expected, the reactions to this reinterpretation of dreams vary greatly. I took note of the following general lines of criticism emerging in a number of commentaries: First, some commentators seem to believe that all the data presented in the target article can in fact be explained without assuming any adaptive biological functions for dreaming, or at least without accepting TST. Second, several commentators suggest that TST implies a restricted view of dreaming and, while TST may contain a partial truth, it ignores many other forms and functions of dreaming. A number of commentators also presented new data or reinterpretations of old data, that lend support to TST and lead to further empirically testable hypotheses and predictions. This, more than anything else, shows that a fruitful and productive research program on dreaming could be established on the theoretical foundations defined by TST. Even if the hypothesis would eventually have to be modified or discarded, it seems that TST will be able to contribute to the progress we are making in the scientific understanding of the dreaming mind-brain.

I will proceed in this response in the following order:
RR1. Random Activation Theory (RAT) does not explain the form and content of dreams
RR2. Alternative explanations of the data
RR3. Negative effects of PTSD and nightmares
RR4. Evidence from typical dreams
RR5. Evidence from recurrent dreams and nightmares
RR6. Other forms and functions of dreaming
RR7. Support for TST from multiple independent sources
RR8. How to test TST properly
RR9. TST and the philosophy of consciousness
RR10. Conclusion: Why do we dream?

In section 1, I discuss the Random Activation Theory and explain why it cannot account for the data in a convincing manner. In section 2, I reply to a variety of counter arguments against TST, and I explain why the alternative explanations of the data are not as convincing as those offered by TST. In section 3, I consider the challenge posed by the negative effects of post-traumatic stress disorder, and I explain why it is unlikely that ancestral humans would have suffered greatly from them. In sections 4 and 5, I discuss the studies about typical and recurrent dreams that were described in the commentaries, and I show that the results lend strong support to TST. In section 6, I reply to those commentators who argued that TST includes too narrow a view on dreaming and its functions. I emphasize that TST does not deny that other forms of dreaming exist; it is just doubtful that any of these forms are biologically functional. In section 7, I summarize the new lines of evidence supporting TST contributed by several commentators. In section 8, I try to describe briefly what kind of empirical tests and findings are most critical for testing TST, and what kind of tests would be ambiguous or irrelevant. In section 9, I briefly comment on the philosophy of consciousness on which TST is erected. In the final section, I summarize the outcome of this highly valuable open peer commentary. I conclude that the Threat Simulation Theory of dreaming seems to survive all the major challenges raised by the commentators and will be, I hope, regarded as one of the least implausible attempts to explain why we dream.

## RR1. Random Activation Theory (RAT) does not explain the form and content of dreams

The view that dreams are nonfunctional biological epiphenomena brought about by random neurophysiological stimulation seems to be the most popular alternative to TST. It appears in various formulations in the commentaries, by, for example **Ardito, Antrobus, Moorcroft, Mealey,** and **Flanagan.** The advocates of this view argue that the content of dreams is just a haphazard by-product of brain activation during sleep. No special biological design needs to be invoked in order to explain why dreams are generated and why they have the form and content that they do.

The explanation offered by these theorists is something like this (e.g., **Flanagan**): Emotional centers in the brain contain specific affect programs corresponding to our basic emotions (most of which are negative). In the brain's memory networks, the most recently activated traces are more salient than other ones. The brainstem and other mechanisms actively generating sleep just happen to be so close to the memory and emotional networks that the activation related to sleep somehow by accident leaks over to them. This results in random activation of the emotional and memory networks. Since most of our basic emotions are negative, random input is more likely to activate negative than positive emotional programs. And since in our memory networks recent memories are primed, they are more likely to be activated than remote memories. This random activation is then sewn into a narrative as the forebrain tries to interpret and make sense of the emotions and images that happened to emerge in dream consciousness. The organization and apparent meaningfulness of the dream is actively constructed by the forebrain giving sense to stimulation that is nonsense by its nature. Dreams are haphazard contents of consciousness that the brain was never designed to produce, but their accidental generation does not really make any difference for the brain.

### RR1.1. Dreams are organized simulations, not random activation patterns

The Random Activation Theory (RAT) of dreams does not explain several prominent features of dreaming. It does not explain the organized form of dreams. Dreams are not noisy patterns of activation, but organized as simulations of "self-in-world": When we dream, we find ourselves as an experientially embodied self in the center of a visuo-spatial world of objects, persons, and events. This extremely predictable and organized form of dreams requires highly coordinated interaction between several cognitive modules in the brain that are in charge of the perception of objects, faces, places, color, motion, speech, emotion, and the overall space experienced around us. The activation of multiple cognitive modules is evident also in PET images of REM sleep (as pointed out by **Antrobus**). The careful coordination of complex modularly organized activity is a sign that dream-

ing is biologically designed to be a sophisticated simulation of waking experience rather than an accidental haphazard collection of emotions and images activated entirely by chance. The central disagreement can be formulated as follows: TST conceptualizes dreaming as an "organized simulation" whereas RAT depicts it as (a result of) "random activation patterns" and "noise." RAT theorists simply take the form of the phenomenal dream world for granted as if such remarkable organization would inevitably or self-evidently follow from sleep-related random processes in the brain and therefore would require no further explanation. The burden of proof is on the advocates of RAT to show how the random, disorganized, noisy activation that their theory assumes can regularly produce an organized multimodal phenomenal self-in-world where events show temporal progression.

**Bednar** suggests that a proper test of RAT would require widespread and ongoing artificial brain stimulation coupled with temporary deactivation of the frontal lobes, paralleling the activation patterns in REM sleep. Such an experiment may be difficult to carry out in practice. However, epileptic discharges in – and electrical stimulation of – temporal lobe structures induce patterns of excitation and inhibition in widely distributed neuronal networks (Gloor 1990). In these cases, the patient typically reports seeing a scene, a face, or hearing a voice or a piece of music being played. There may be actual recall or an illusory feeling of familiarity. Fear is the most common affective response, but any other emotion may be experienced as well. These experiences are clearly different from dreams and threat simulations in several ways:

> Thus the experiences do not move forward in time, with the single exception of the hallucination of hearing music. Scenes are static; they do not evolve, there is no story to be told. The patient remains passive and does not feel that he actively participates in the hallucinated scene. Perceptual detail may be fragmentary or lacking. . . . The auditory hallucination of hearing a voice is almost always without semantic content, even though the voice may sound familiar and may be identifiable. (Gloor 1990, p. 1675)

More interesting, Gloor (1990) reports an observation of direct electrical stimulation of the right amygdala in one patient, who re-experienced a threatening event from his childhood as if he had been there again: he was at a picnic in Brewer Park in Ottawa, a kid was coming to push him into water, and he experienced being pushed into the water. Fragments of this experience were first experienced with 2–3 seconds of stimulation, but the memory emerged during the third and longest (4.4. seconds) stimulation. However, the patient

> described it in a fragmentary manner, stressing its emotionally laden features. He did not tell a detailed story; the recollection of the event was reduced to a few of its essential elements: its locale and time and the fact that he had been physically abused by submersion into water by a stronger boy. There was no other detail and yet the experience had a subjective immediacy similar to that of the original event. (Gloor 1990, p. 1677)

It remains unclear whether there was any visual hallucination involved. When questioned, the patient denied having seen himself being chased by the big fellow. The experience seems to have consisted of the affective impact of the remembered experience and a feeling of being there and being chased, with little perceptual detail. Thus, this experience was a re-invoked emotional memory image (as would be expected when the amygdala is stimulated) rather than a complex progressive threat simulation in which one actively participates. Although the random activation of temporal areas can sometimes produce fragmentary images and remembrances, the elicited experiences do not constitute organized and variable (threat) simulations as we know them from our dreams and nightmares. Furthermore, we should remember that these experiences occur only in seizures arising from the temporal lobe, and by no means in all of them, and that in some patients they can be elicited by stimulation in the temporal lobe but never by stimulations applied to other regions of the cerebral cortex (Gloor 1990). Therefore, truly random activation all over the cortex would most of the time elicit no experiences at all, and at best a mixture of static mental images or fleeting affective states.

In the light of these facts RAT should predict that, as a result of random activation of temporal lobe areas during sleep, we will experience various emotional states and various visual and auditory sensations or images. There is no reason why the random activation patterns should show any specific organization. This kaleidoscope of randomly activated experiential states, actually predicted by RAT, is a far cry from the organized simulation of self-in-world that constitutes dreaming. The explanation offered by RAT is that although the activated images are random, the brain somehow can "make sense" of them and construct the organization and narrative of the dream. The organization is not assumed to be in the input itself, but somehow "constructed" later on when trying to interpret the chaotic input.

This explanation is not entirely convincing either. If it were true that the brain can impose remarkable organization on random, noisy activation patterns, then we should expect that other forms of random brain activation (e.g., epileptic or migraine auras, Charles Bonnet Syndrome hallucinations) also result in the brain "making sense" of the chaos by imposing a structure and a narrative to these experiences. However, people in such cases report truly random phenomenology (varying images and sensations that are not organized to form a perceptual world or a temporally progressing narrative). Why does the brain not make sense out of these types of random activation although it succeeds so remarkably well in the case of dreams? The TST answer is that in the case of dreams the brain activation was organized to begin with and therefore resulted in well-structured phenomenology: a simulation of self-in-world. Dreaming consists of a transparently meaningful and organized perceptual world, not of random noise that the brain struggles to make sense of somehow.

### RR1.2. Dreams show unique preference for threatening events

Furthermore, RAT does not explain the special status that threatening events have in the generation and the content of dreams. Why should random activation elicit simulations of animals, aggressive encounters, and pursuits or escapes, especially in children's dreams? Why should the memory of a brief but life-threatening event be activated by random activation so regularly and for such a long time after the incident really happened? Why exposure to events that aroused strong emotions other than threat does not lead to similar repetition of the same emotional event in dreams? It seems that there are no recurrent "post-triumphant" dreams after

great successes have been experienced in sports or business or war, or following great luck in gambling or lottery. There do not appear to be recurrent dreams specialized in simulating waking events where intensive but other than threat-related basic emotions were experienced (e.g., happiness, surprise). Since the activation is random and should not have any preference for threatening events, **Flanagan**'s version of RAT predicts that following positive emotional experiences, the positive "affect-programs" should be activated during dreaming as frequently as, following negative experiences, the threat-related ones are. The evidence reviewed in the target article indicates that dreams show high preference only for threatening content and are strongly modulated only by real threatening events, not by other emotional events. RAT cannot account for these systematic biases and causal relationships between real threats and dreaming.

### RR1.3. Random activation of "affect programs" or systematic activation of threat simulations

It is apparent that **Flangan** is aware of the deficiency in his version of RAT; to counter this he is prepared to accept that the "affect programs" that are randomly activated might in fact contain "scenarios pre-loaded with content of threatening creatures and situations, so that we are primed to conjure up such scenarios once the relevant affect program is activated." Note especially that **Flanagan** thus attributes to "affect programs" what in TST are called "threat scripts." Note also that a concept resembling "threat-simulation" has thus been smuggled into the RAT by talking about "scenarios" pre-loaded with content of threatening creatures and situations. But if the affect programs contain complex "threat-scenarios" that get activated whenever the program is activated, then should we not hallucinate plausible threat scenarios whenever the program is activated in the waking state?

As **Flanagan** explicitly mentions, the affect programs are adaptations that served awake humans who were up and about struggling to survive. It would not seem to serve reliable threat detection to have pre-loaded threat scenarios that might interfere with veridical perception; the postulated threat scenarios would not seem to serve any adaptive purpose for the awake individual. Even if such threat scenarios existed in the affect programs, RAT would not be able to explain why they are activated not randomly but in a systematic and organized manner during sleep, depending on, for example, personal history or position in sleep cycle. Children's nightmares and recurrent dreams include ancestral threats; traumatized people have recurrent nightmares of the traumatic event; repetitive, realistic post-traumatic dreams occur early in the sleep cycle whereas other nightmares occur late (van der Kolk et al. 1984) as if these two types of threat simulations had separate slots within the schedule of the threat simulation mechanism.

These regularities would seem to reveal an organized system rather than the aimless effects of the random activation of affect programs during sleep. **Flanagan**'s suggestion about the affect programs containing threat scenarios seems a bit like an ad hoc assumption that must be added to RAT in order to explain the data. TST offers a much less contrived explanation: there is a separate dream production mechanism that contains threat scripts designed by natural selection to be regularly released in specific conditions. The dream production system is capable of constructing complex, organized simulations of threatening events whenever it is activated by ecologically valid stimulation. This "threat-simulation response" is the core of TST. Any theory of dreaming that does not explain it is like a theory of the immune system that gives no account of the immune response.

### RR1.4. Flanagan's dreams: Random gurgling of the brain or threat simulation?

Incidentally, in his recent book, Flanagan (2000) reports some of his own dreams that seem to lend support to TST: When he was five years old, he had a dream of a pack of wolves chasing him; he was terrified and could not run away fast enough. This is a classical theme in children's threat simulation dreams, and if this kind of a dream does not constitute a realistic simulation of an ancestral threat, I do not know what would. Flanagan (2000) believes that the origin of this dream was in the stories of the Three Little Pigs that were terrorized by the huffing and puffing wolf, and Little Red Riding Hood, who was deceived by the clever wolf that dressed up as the girl's grandmother. I do not find this explanation particularly plausible: the wolves in the stories are manifestly bizarre and unrealistic whereas in the dream they are extremely realistic; the wolves in the stories behave like intelligent humans, not in the least like the pack of hunting predators encountered in the dream. If fairy-tales were the real origin of children's threat simulation dreams, we should find that children's dreams contain utterly bizarre animals that huff and puff and dress up as grandmothers to cheat the poor children.

By contrast, we find that, even if a child's only experience about wolves is in the form of these bizarre representations in fairy tales, the threat simulation mechanism knows better how real wolves behave and what should be done if they were ever encountered. In one of his recent dreams, Flanagan (2000) was involved in a military manoeuver where he was expected to risk his life in battle, but he tried to resist and he insisted that the government's orders should not be followed. This threat simulation obviously flows from his later real experiences that involve threats very unlike ancestral ones and therefore resulted in a complex but not terribly realistic or ancestral threat simulation. **Flanagan** however does not recognize his own dreams for what they are – threat simulations – because he has made a prior commitment to the view that dreams are noisy, disorganized, bizarre and random gurglings of the sleeping brain and do not have any biological adaptive functions. In the light of TST, Flanagan's dreams are not all that random, but practically paradigm examples of the threat simulation mechanisms at work.

### RR1.5. Dreams are more organized than they are bizarre

RAT theorists (e.g., **Flanagan**) are eager to emphasize that REM dreams are highly bizarre and disjointed and therefore unlikely sites for realistic threat simulations to take place (this point was also brought up by **Gunderson**). The bizarreness of dreams is regarded as good evidence for the disorganization of dreams and the randomness of brain activation during dreaming. Unfortunately, the high level of bizarreness and disorganization in dreams is simply assumed, although what would be needed, before any judgment is passed, are detailed quantitative studies and evaluations about the frequency and quality of bizarreness in dreams. Thus, we need to ask: How bizarre exactly are dreams? Is there more organization than disorganization in dreams, or vice versa? Are threat simulation dreams as bizarre as, or more realistic than, other dreams? Are all forms of bizarreness necessarily dysfunctional?

Revonsuo and Salmivalli (1995) conducted a detailed quantification of nearly 8,000 normal and bizarre features in dream reports in order to find out how great a proportion of all dream elements are somehow deviating from the expectations of the real world (note that this is a remarkably low criterion for bizarreness because it includes all kinds of deviations, not only physically impossible ones). The elements most commonly described in dream reports were objects, actions, persons, and places, covering 67% of all elements described. Only 15–20% of these descriptions were classified as bizarre, indicating that the vast majority (75–80%) of the dominating perceptual elements in our dreams are entirely normal. In another study on the bizarreness of human characters in dreams (Revonsuo & Tarkko, currently in preparation), we have found that about half of the human characters we encounter in our dreams are entirely in accordance with the expectations of our waking perceptual world: they look the same, feel appropriately familiar or unfamiliar to us; if familiar, we remember who they are; they appear and disappear in a physically possible manner; they are encountered in contexts and places where they are not unlikely to appear, and they speak and behave in a manner not out of the ordinary in the waking world.

On the basis of these detailed quantifications of dream bizarreness, we may say that dreams are remarkably organized. No random process could ever create such a complex simulation of the waking perceptual world. Bizarreness certainly is a regular feature of dreams, but it constitutes a relatively mild deviation against the solid background of sophisticated organization; a little amount of noise within a highly organized signal. RAT can only explain the bizarreness – the small degree of noise – but it cannot explain the high degree of organization in which the bizarreness is embedded. Dreams are much more organized than they are disorganized. A proper theory of dreaming should explain both the organization and the disorganization of dreams.

As **Flanagan** (2000) himself notes, all that natural selection really cares about are mechanisms that work pretty well, but not necessarily optimally:

> It would be a mistake to assume optimal functioning for any human capacity even at the times and in the places for which such capacities were designed to do their job. And it is an even greater mistake to complain that a capacity is not functioning well when it doesn't need to (Flanagan 2000, p. 38).

Thus, to complain about the fact that our present-day ordinary dreams contain bizarre features is a mistake, for in the absence of ecologically valid cues, the dream production system does not need to function particularly well. However, when the threat simulation mechanism is activated by the proper cues, bizarreness decreases and realism increases: as for example, recurrent dreams of ancestral threats and post-traumatic dreams show, threat simulation dreams tend to be more realistic than ordinary dreams, exactly as predicted by TST, revealing that threat simulation is a special function of the dream production system.

Furthermore, as **Cheyne** suggests, some forms of bizarreness may not be noise at all but functional exaggerations that test the limits of the system or that make the simulation more demanding than a real situation. He makes the important observation that dreaming as a practice mode may have some advantages over play. One purpose of the practice mode is to test the physical limits of the system's capacities, which would be risky during ordinary play. However, dreaming allows exaggerated emotions, reactions, and motor programs to be simulated without any of the risks that would be involved with them during wakefulness. Therefore, it is possible that some forms of bizarreness are not simply nonfunctional noise but functional exaggerations that are used during enhanced threat simulations in order to practice extreme manoeuvers by pushing the system to its limits.

A fairly common observation in chase dreams is that you suddenly have heavy and slow legs that don't seem to take you anywhere just when you should run for your life at full speed, which makes you struggle: you try harder and harder. This is reminiscent of some forms of sports training where the athlete runs carrying a heavy backpack, sometimes on soft and sinking ground or uphill to make the task even harder. The hypothesis that threat simulation uses exaggerated manoeuvers to make the rehearsal more efficient could be tested, for example, by finding out how frequent the "heavy legs syndrome" is in dreams and whether it selectively appears only in connection with threatening events. If it appears frequently and selectively in escape situations, it could be considered as enhanced threat simulation in the sense that **Cheyne** suggests.

To summarize, RAT disregards the organized form and content of dreams and emphasizes bizarreness in order to make dreams seem as "just noise, like the gurgling of the stomach" (Flanagan 2000, p. 24). But an earnest exploration of the empirical evidence reveals that, in actual fact, dreams are highly organized simulations of the perceptual world, specialized in the simulation of threatening events. Therefore, dreams are beyond the explanatory powers of RAT.

## RR2. Alternative explanations of the data

The commentators raised several questions about the ability of TST to account for the data and they suggested alternative explanations that they deemed more plausible. In the following I will explain how these points can be explained by TST and why the alternative explanations are not all that plausible.

### RR2.1. Cognitive biases and the unreliability of dream recall

There are a few skeptical commentators who try to cast doubts on the validity of the data that TST tries to explain. **Thompson, Mealey, Bednar,** and **Conduit et al.** argue that we do not really know what dreams are like because dream recall is inherently unreliable and heavily biased. **Thompson** takes this line even further and asks "Why do we assume that there is anything that is the dream apart from the subject's reporting it?" He says that I seem to be unaware that these sorts of criticisms could be entertained by behaviorists against TST. On the contrary, I am well aware of behaviorist and verificationist attacks on dreaming. In fact, such theories used to be very influential in the

philosophy of mind: Malcolm (1956; 1959) denied that there is any phenomenon of consciousness behind the reported dream, and more recently Dennett (1976) argued that dreams may not be experiences at all, but only false memories activated at the time of awakening. I did not bother to mention these theories in the target article, because in an earlier paper (Revonsuo 1995) I have shown that the results of modern dream research have made Malcolm's and Dennett's views utterly implausible. (I have to admit that they were not terribly plausible to begin with.) Furthermore, the philosophical starting point of the research program I advocate is that consciousness is a real natural biological phenomenon (Revonsuo 2000), and I have explained elsewhere why the elimination of consciousness or Dennett's neobehaviorism are not compatible with such a research program (e.g., Revonsuo 1994a; 1994b). Hence I am not particularly worried about behavioristic attacks against TST, for behaviorists have in any case placed themselves outside any reasonable research program on subjective phenomenal consciousness.

**Mealey** and **Conduit et al.** stress the unreliability of dream recall and claim that because of biases in dream recall and dream scoring techniques, we do not know the true form of dreams. This general criticism of course is not a problem only for TST but for all dream research in general. However, it is one of which dream researchers are well aware. There are a number of studies comparing the content of dreams recalled in different situations (home/laboratory/Nightcap) and from different stages of sleep and different conditions of awakening. All of the methods have their own problems. Home-reported dreams show more aggression than laboratory reports, whereas laboratory reports often show elements related to the experimental situation (Weisz & Foulkes 1970). The differences between dreams reported at home and in the sleep laboratory are, however, not all that great, and in any case they have been well documented. Therefore, I think it is an exaggeration to say that we do not know what the true form of dreams is like. A wide ranging survey of the dream research literature, such as the one in the target article, including many different kinds of studies, surely gives us a reasonably good picture of what dreams are like. If that is not the case, then the content of dreams must be beyond the reach of systematic empirical investigations altogether. **Mealey**'s plea for ignorance about the "true" content of dreams is thus not particularly persuasive, nor does it appear to me as a particularly fruitful approach to the study of dreams.

**Mealey**'s more specific criticism is that, first, we are biased to remember only highly salient dreams and that any dreams containing survival-related threats will be highly salient because threats are perceived, attended to, and remembered better than other types of material. Second, cognitive biases in attending to threat result in biases in dream content analyses (also **Schredl** claims that dream emotions just seem negative because of biases in the studies; **Conduit et al.** claim that selective memory and attention provide an explanation for the overrepresentation of threatening content in dreams; and **Montangero** argues that spontaneously remembered dreams constitute a biased sample and that therefore the Hall & Van de Castle (1996) results are not pertinent). Mealey claims that we do not have an appropriate null model that includes the effects of these cognitive biases. In contrast to Mealey, I think that it is not all that difficult to find an appropriate null model or control condition against which we can compare the frequency of threatening events in dreams. As **Chapman & Underwood** point out in their insightful commentary, an excellent baseline is provided by everyday memory for waking states, because everyday memories also are systematically distorted in the same manner as dream recall is. According to Chapman & Underwood, there are huge levels of forgetting for mundane events and, conversely, selective retention of events including high degrees of threat and unpleasantness. Therefore, in order to find out whether there is a bias for threatening events in dreams as compared to waking life, we simply have to find out the quantity and quality of threatening events in dream recall as compared to everyday memories.

It would thus not be impossible (or even very difficult) to demonstrate empirically that threatening events are truly overrepresented in dreams in a way completely independent of biases in dream recall or other cognitive processes. We can define a category of severe threats where the dreamer's life or physical well-being is in danger. Such events are very rare in the everyday lives of most Westerners. Let us assume that during a specified period of, for example, a couple of weeks, a group of subjects does not report having been involved in any such events in their waking lives. Again, if their dream reports from that period (or another period of comparable length) do include such severely threatening events, then severe threats undeniably are overrepresented in dreams. Even if the subjects have forgotten some of their more mundane dreams (and waking events), it would not change the fact that during the period of time covered by the study, the subjects have been involved in highly dangerous events in the dream world more frequently than in the waking world. If in my dreams I have to run for my life to escape an aggressive attacker once per week or once per month, but in the real world, maybe at most once in a decade, then the threatening event "being chased and running for my life" surely is overrepresented in my dreams. The same applies to any other life-threatening events. They are so rare in real life that if they occur with almost any frequency at all in dreams, they are very likely to exceed the frequency of occurrence in real life. The possible biases in memory cannot confound this result, for the biases help us to remember all highly threatening events, regardless of whether they happened during dreaming or waking. We can, thus, safely assume that we have access to a representative sample of life-threatening events in both the dream and the waking world, and any comparison we make between them should reveal conclusively which world contains a greater amount of serious dangers. In this comparison, it is irrelevant how many mundane dream and waking events have been forgotten: a direct comparison between the frequency of threats during dreaming and waking is sufficient to reveal the predicted overrepresentation.

A study we have recently conducted sheds some light on this question. We found that highly threatening events occur with considerable frequency in the home-based dream diaries of Finnish students (Revonsuo & Valli 2000). When the data set consisting of nearly 600 dream reports was content-analyzed for life-threatening or physically highly dangerous events, about 150 such events were found, and 79% of the subjects reported at least one such event in their dreams. Although we do not yet have quantitative data that would reveal how many life-threatening or highly dangerous events similar subjects typically

encounter during a comparable period of their everyday lives, it is virtually certain that average Finnish students are very seldom if ever assaulted, shot at, pursued by strangers or animals, or participate in combat. Neither do most of them face life-threatening events once per month in their everyday lives, although that is what happens in their dream world.

Furthermore, although all subjects had at least a few mildly threatening events in their dreams, 34% of the 600 dream reports in this study did not include any threatening events, which speaks against **Mealey**'s and **Conduit et al.**'s view that nonthreatening dreams would not be adequately remembered or reported. In sum, there is no reason to believe that the bias to simulate threatening events could not be properly quantified and compared to waking life, and, although additional studies are needed about the frequency and severity of dream threats, all the facts that we currently have show that threatening events truly are overrepresented in dreams compared to waking life.

**Schredl** attempts to explain away Gregor's (1981) results by implying that the study on Mehinaku dreams suffers from methodological weakness such as recalling dreams that occurred quite a long time ago so that very vivid and negatively toned dreams which could be memorized more easily were reported to the researcher. **Schredl** probably did not check how the reports were collected in the original study of Gregor (1981), because Gregor's method of dream collecting does not suffer from the methodological weakness assumed by **Schredl**: the Mehinaku dreams were actually recalled and recorded every morning immediately after awakening:

> "According to the Mehinaku, dreams occur when the soul leaves its home in the iris of the eye to wander about through a nocturnal world peopled by spirits, monsters, and the souls of other sleeping villagers. In the morning the villagers are careful to recall the adventures of their souls, since they are a clue to the future . . . Each morning, the Mehinaku remember their dreams and often recount them to their families and housemates. This penchant for recall and immediate verbalization is ideally suited for research, as it insures that the night's dreams are not lost in the cloud of amnesia that follows waking for most Westerners. During my research, the villagers made superb informants as I circulated through the houses each morning to harvest the previous night's crop of dreams." (Gregor 1981, p. 354)

Gregor admits that the dream narratives were collected under less favorable conditions than Calvin Hall's data for example, but he believes that the reports nevertheless accurately reflect the dream themes, settings, characters, and emotional tone. Therefore, the high proportions of threatening events in the Mehinaku dreams cannot be dismissed on methodological grounds. **Schredl**'s assumption that dreams recorded immediately upon awakening should show a balanced ratio of negative and positive emotions probably applies to Western people living in nonthreatening environments, but there is no reason to believe that it should apply to hunter-gatherer populations that live in an environment much closer to the ancestral one.

### RR2.2. The resort to dream symbolism

An alternative explanation as to why (especially children's) dreams contain an abundance of animals and strangers is offered by **Blagrove**, **Humphrey**, and **Greenberg**; perhaps these elements are actually symbolic representations of something else. Animals in dreams might well be simplified proxies for human beings (Humphrey), wild animals are in fact a symbolic expression of internal fears (Greenberg), and strangers are symbols for failures to recognize familiar people (Blagrove). In my view this kind of resort to dream symbolism is unwise, for once we refuse to take the dream images for what they are, it is impossible to determine in any objective manner which dream images should be regarded as "symbols" and what exactly they might symbolize. One can always get rid of dream elements that do not fit in with one's favorite theory by saying that they actually stand for something else (most likely for something that the favorite theory happens to be able to explain very neatly). If TST could be justifiably criticized by saying that some of the threatening dream elements that seem to be in accordance with the theory in fact lend no support to the theory because they are symbols for something else entirely, then TST could be defended as justifiable by claiming that some of the nonthreatening elements in dreams are in fact symbols for threatening elements, which neatly supports TST. This absurd scenario makes it clear that any serious scientific theory of dreaming must explain the contents of dreams as they are actually experienced and reported, and not enter the slippery slope of dream symbolism which only too easily leads to theoretically baseless distortion of the actual data.

### RR2.3. Dreams are not proper simulations

Dreams are not appropriate for the simulation of threat perception and avoidance, **Clancey** claims, because they lack goal-directed sustained attention that orients interpretation and action, and dream experiences are not sequentially coordinated. Again, these characterizations of dreams are, unfortunately, presented without providing any supporting data or references. However, if we look at typical threat-simulation dreams, we see that these characterizations are simply not true.

Consider the following threat simulation dream by a Finnish university student:

> "I was in a bus. I had my schoolbag, another large bag, and a plastic bag. The bus stopped at a bus stop where I was supposed to get off. I had so much to carry that I decided to take the plastic bag and the large bag out first. When I was about to take my schoolbag, a big crowd of schoolkids were leaving the bus so that I could not fetch my bag. I saw the bus leaving the bus stop where I was standing. I remembered that in the school bag there was my wallet with $100 in it, and all my anatomy textbooks.
>
> Next, I was at home and I called the lost and found and some other place, but my bag was not there. I asked my mother what other places I should call, and we decided to call the bus company."

This dream, contrary to **Clancey**'s assertions, does show sequential coordination, goal-directed attention, and action; very reasonable measures are taken by the dreamer to fight the threatening event (the loss of valuable personal property). This dream also does not support Clancey's claim that threat simulation dreams only reinforce an unthinking way of responding to threat situations. Therefore, it is unwise to assert what threat simulations in dreams are like without first carefully looking at the actual data. Such assertions may reveal what the widely held beliefs about the nature of dreams are, but they do not necessarily tell us what threat simulations during dreaming actually are like. In order to find that out, we need systematic content analyses of threatening events in dreams. Only careful empirical investigations, not our personal hunches, can reveal what is and what is not possible (or typical) in threat simulation dreams.

### RR2.4. Threat simulation would have been unnecessary

Another kind of critique against TST is made by **Clancey, Schredl,** and **Montangero.** They suggest that there was no need for training threat perception and avoidance in the ancestral environment. **Schredl** (and also **Domhoff**) says that classical conditioning in avoidance tasks occurs immediately and therefore no training is needed. However, it is clear that threat perception and avoidance are much more complex skills than what is involved in classical conditioning. We not only need to learn the relationship between, say, a certain perceptual cue and expected negative consequences, but we need to recognize and classify threatening objects accurately and rapidly, and immediately activate the kind of response that is most adequate to the situation in question. We need to rehearse the speed, accuracy, and intensity of employing these complex skills, because such features are the currency for survival in real situations.

**Clancey** insists that rehearsal during dreaming is futile since average ancestral humans were in any case constantly confronted with threatening events, and consequently their everyday experience surely provided enough practice to develop well-honed, adapted skills. It is not difficult to see why this proposition cannot be taken very seriously. It ignores the very reason why all behavioral simulations are valuable: a simulation, unlike the real event, does not involve the high risks and costs that real situations always entail. If a skill can be made faster, more reliable, and more efficient without taking any of the risks that the real event necessarily involves, the individual is much better off every time it needs the skill in reality. An ancestral human who only trained his threat avoidance skills in real situations probably did not live long enough to leave any offspring, whereas those who had gone through repeated threat simulations during sleep every night were much better prepared to begin with, and were not as easy to catch as their unfortunate and untrained fellow.

To demonstrate the usefulness of training, we may think about it in the context of sports: would you cheerfully enter the boxing ring without having had any training because, instead, you have booked for yourself so many prize fights in the near future that surely you will get all the training you'll ever need? The problem of course is that your boxing career will be far too short to allow you to get any better. Analogously, the career of those ancestors who only trained their threat avoidance skills in real situations was so short that they never had the opportunity to get significantly better. That is why the tendency to simulate dangerous events without any costs during dreaming (and play) are built into present-day humans.

### RR2.5. The continuity hypothesis

**Schredl** argues that the data presented in the target article could be accounted for by the "continuity hypothesis" and biases in dream reporting. He thinks that if we examine the total input that an individual receives (e.g., through the

TV), we might find the basis for the negative bias in dream content. Both **Blagrove** and **Rotenberg** suggest that the origin of threat-related themes and biases is not in the ancestral environment, but in fairytales, cartoons, TV, movies, and newspapers. In the target article, I showed why this is an unlikely explanation for the content of children's dreams (see n. 6 in the target article. See also my comments on **Flanagan**'s explanation of his childhood dream above). To repeat: Children have dreams where they personally encounter realistic threats (wild animals, aggressive male strangers, pursuits) corresponding to ancestral ones. These biases become weaker as the child ages. Unlike threatening dream animals, the animals in fairy-tales and cartoons are often human-like and otherwise highly unrealistic; unlike the concrete threats in dreams that typically victimize the dreamer, all forms of fiction and media present rare and exotic threats that happen to distant strangers or story characters; unlike the frequency of ancestral-like threats in children's dreams, the exposure to films and news and other materials with a high degree of aggressive and violent content increases with age. All of these facts speak against the explanation that threat simulations would originate in the exposure to fiction and media. If that were the case, small children would mostly dream about friendly TeleTubbies or Moomins going about their peaceful business, or at worst some bizarre cartoon animals chasing each other, but there would not be any fierce wild animals, monsters, or male aggressors on the point of attacking the dreamer. Furthermore, with age the biases in dreams should change in step with the growing exposure to more and more violent types of fiction and mass media programs, which it does not do.

Fortunately, this is an issue that could easily be investigated by comparing dream reports from children with different degrees of exposure to fairytales or the mass media. For example, a study on typical anxiety dreams conducted in the 1940s (Harris 1948) shows that the very same theme, being chased or attacked or threatened by a dangerous environmental object, was by far the most typical theme in the anxiety dreams of both children and adults at a time when people were not exposed to mass media to the same extent as they are today. (Similar results from modern studies are reviewed by **Germain et al.**) **Rotenberg** claims that images of wild animals are absent from Bedouin children because they have no experience of fairytales. If that were truly the case, it would constitute remarkable evidence against TST. Unfortunately, Rotenberg provides no data or references to back up his claim. In fact, the only published study on Bedouin children's dreams that I am aware of (Levine 1991), shows that they dreamed about nonhuman opponents (including animals) in their dreams significantly more often than children from the two other cultures that were studied. In Levine's (1991) study, even a sample dream report from a Bedouin child includes an aggressive encounter with a wolf. Rotenberg's assertion thus simply flies in the face of the available evidence.

Another way to test empirically whether fiction and media are the main sources of threat simulations is to do a content analysis of threatening events in dreams and try to evaluate what the source of the depicted threats might be. We have actually done exactly this kind of a study (Revonsuo & Valli 2000). We classified 672 threatening events in dreams according to the likely source of information in real life to learn about the kind of event that the dream represents. The vast majority (63%) of these threatening events were based on the personal life of the dreamer; that is, they were the kind of events that the dreamer or a person in a similar position (an average Finnish student) could in principle encounter in his or her own life. The mass media accounted for 33% of the threats, whereas fiction and fantasy covered only 4%. Thus, the more unrealistic and distant from our personal life a threat is, the less probably it will be included into threat simulations during dreaming. This result makes it even more unlikely that children's dreams would be primarily constructed on the basis of fairytales, cartoons, or TV.

**Schredl** further argues for the Continuity Hypothesis that since almost all murderers are male, it is no wonder that this pattern of male aggression is reflected in dream content. First, we should note that the preponderance of male aggression is itself rooted in evolution (see, e.g., Campbell 1999). Second, the real issue with dream content was not that most aggression is with male characters, but that most interactions with male strangers (and animals) in dreams are aggressive, although that is not the case in waking life (see Domhoff's [1996] concept of "dream enemies"). Schredl confuses here the statement "most aggressors are male" (true about the waking world) with the statement "most males are aggressors" (true about unknown males in the dream world but false in the waking world). The point that argues against the continuity hypothesis is that the latter statement is not true in our waking world: we may encounter hundreds of unknown males as we do business or participate in various social situations, without a single aggressive encounter with any of them, whereas in dreams an interaction with an unknown male is much more likely to be aggressive than friendly (Domhoff 1996). The same goes for animals: in the waking world we mostly encounter harmless wild animals or friendly pets, but in dreams the animals we encounter are much more likely to be aggressive than friendly towards us (Domhoff 1996). Yet, not many of us adults (let alone small children) have regularly been targets of aggression by the animals or the male strangers we have met in the real world, which speaks against the continuity hypothesis.

### RR2.6. Do animals practice survival skills during REM dreaming?

Doubts are expressed by **Domhoff, Morrison & Sanford,** and **Blagrove** that the behavior of cats during REM sleep with atonia is evidence for dreaming and survival skill practice during sleep. Domhoff simply asserts that it is highly unlikely that animals dream and refers to Foulkes's (1999) view. I have explained in Note 5 in the target article why I do not agree with Foulkes's dismissal of animal consciousness, so I will not repeat it here. Morrison & Sanford and Blagrove base their doubts on some of the experimental findings, such as that the site of the lesion modulates the type of oneiric behavior observed, and that some of the cats also exhibited altered waking behaviors. Morrison & Sanford admit that the oneiric movements are expressions of well-organized behavior and that the brain in REM is most like the brain in very alert wakefulness, although they do not believe that waking behaviors are being practiced in this state because they see REM as a state not optimal for the formation or consolidation of memory traces. Oneiric behavior has been studied mainly in two laboratories, Morrison's and Jouvet's. The latter originally described the phenomenon in 1965. In a recent book, Jouvet (1999) offers a summary and an interpretation of these studies, providing a slightly different view from that of Morrison & Sanford.

According to Jouvet (1999), after the brainstem lesions there is no obvious motor or behavioral disturbance in the cats during waking. The oneiric behaviors begin at the same instant when the first PGO waves appear: the cat opens its eyes, raises its head, and looks around. The behavioral patterns that follow are stereotypical but unpredictable: hunting, stalking, running as if chasing imaginary prey, predatory aggression, aggressive attack (fighting with an imaginary enemy), fear and defensive posture, rage. By contrast, sexual behaviors, shivering, panting, vomiting, or sneezing have never been observed during oneiric behavior. Jouvet (1999) points out that it is very difficult to find out what the relationship is between different behavioral patterns and the PGO activity, because we do not know whether the random-looking PGO activity contains some kind of a "code" that could be deciphered. Jouvet's findings on the PGO waves show that there is simultaneous motor programming and excitation of sensory areas. He concludes:

> Thus, the hypothesis that a cat dreams of actions characteristic of its own species (lying in wait, attack, rage, flight, fright, pursuit) during its paradoxical sleep is quite plausible. (Jouvet 1999, p. 92)

Directly relevant to TST is the observation that when we humans are in a corresponding state to the one that the cats were in these experiments (REM Sleep Behavior Disorder, see target article), we do experience vivid threat simulation dreams and our bodies do act out the dreamed actions. This undeniable correspondence between dream content and externally observable oneiric behavior renders Jouvet's hypothesis even more plausible. What would be needed are more studies on the oneiric behaviors of other mammals besides cat (if technically possible). TST is directly concerned only with the function of human dreaming, but the evidence we currently have about oneiric behaviors in cats and humans supports the idea that animals also dream, and during dreaming they practice species-specific behaviors that are important for their reproductive success in their natural environment.

Thus, there seems to be considerable evolutionary continuity in the function of dreaming in different species. **Panksepp** offers some support to this view by pointing out that he has repeatedly found similar archetypal REM dream themes in humans to those expressed by the oneiric behaviors in cats. However, he also poses the question as to why predatory mammals should exhibit more REM sleep than their prey. Part of the correct answer probably has to do with prey-predator relationships: large prey who lack hiding places must remain vigilant at all hours. Thus, horse and giraffe have a total daily sleep time of only 2–3 hours, but 0.5 hours from that is REM sleep, which makes the proportion of REM sleep similar to other mammals with longer sleep times. Rabbit spends 1 hour, rat 2.5 hours and hamster, cat, and dog 3 hours in REM sleep per day (Zepelin 1993). There should thus be plenty of time for survival skill simulation during REM sleep in different mammalian species, but the exact amount of REM sleep (and total sleep time) probably depends on multiple factors other than the pressure to practice defensive reactions or threat recognition.

### RR2.7. People without dreams

It is suggested by **Domhoff** that there are people who sleep adequately without dreaming and therefore it is doubtful that dreams have any adaptive function. However, the issue is not whether people without the ability to have dreams can sleep adequately or survive in the present-day world but whether or not such people would have been worse off as to reproductive success in the ancestral environment. We could just as well argue that since people who are physically unable to run (due to overweight, knee problems, etc.) survive in our society, running has no adaptive function. But if such people should suddenly be transported 100,000 years back in time, they would quickly find out that running does serve important functions, and that people unable to run will not be as successful in the original environment as those who are fast runners and have physical stamina. According to TST, the same applies to dreaming: in the original environment, people who were unable to have dreams were thereby unable to rehearse important threat perception and threat avoidance programs during sleep, and therefore were less likely to leave offspring.

**Domhoff**'s argument ignores the ancestral environment and thus, is not an adequate test of TST. Furthermore, he argues that it is unlikely that trauma could stimulate the development of dreaming, since children with tense home environments or personal problems did not report more dreams or more negative dreams in one study. "Tense home environment" or "personal problems," however, hardly constitute ecologically valid cues for the threat simulation system and therefore this study is not a proper test of the hypothesis that real threats may trigger dream development. Clinical observations by Nader (1996) of children who have experienced true trauma do support it, but new systematic studies on this question certainly will be demanded.

### RR2.8. Are threat simulations useful and adaptive?

**Domhoff** asserts that threat simulation dreams "rarely" contain successful defensive actions and therefore cannot be adaptive. In the same spirit, **Hunt** says that it is difficult to see how our "widely described" paralyzed fears, slow motion running, and escape tactics based on absurd reasoning could be a rehearsal of anything adaptive. Unfortunately neither of these commentators back up their assertions with any systematic or quantified data; therefore their claims appear premature. In order to be able to pass any judgments on the matter, we would first have to have some criteria as to what should count as an adaptive response in a certain situation. Hunt seems to think that "paralyzed fears" are inherently nonadaptive, but actually the proper behavioral response to fear includes an initial freezing response which is adaptive in situations where a danger (e.g., a predator) has been noticed at a distance (Panksepp 1998a). Furthermore, we need to find out how often threat simulation dreams actually do contain successful defensive actions and how often not.

In a recent study, we tried to estimate this by conducting an analysis of the reactions of the dream self to threatening events in dreams (Revonsuo & Valli 2000). First, we found out that if the dream self reacts in some way to the threatening event, the reaction includes in almost all cases (94%) an appropriate action, whereas irrelevant reactions and physically impossible reactions were rare (6%). Second, we found that there was a statistically significant relationship between the severity of the threat and the probability of the dream self taking relevant action. A relevant reaction to the threat was significantly more likely than no reaction at all (67% vs. 33%) if the event was life-threatening for the dream self, whereas if the event was not life-threatening, relevant reaction and no reaction were about equally likely (47% vs. 53%). The results reported in Revonsuo & Valli (2000) apply to students' home-reported dream threats, but more studies would be needed to find out what happens in the threat simulation dreams of special populations (e.g., children's dreams, post-traumatic dreams). In any case, we should not make any hasty judgments on issues that have not yet been properly investigated. In the light of the data that are presently available, threat simulation dreams may well contain useful and adaptive defensive reactions.

### RR2.9. The connection between actual dangers and threat simulation

It is suggested by **Montangero** that the fact that some people have frequent nightmares without having encountered any real threats constitutes evidence against Proposition 3 (that real threats activate the threat simulation system in a qualitatively unique manner, dissimilar from the effects on dreaming of any other stimuli or experience). He further points out that many people deliberately expose themselves to dangers (e.g., they speed although they know it's dangerous) but do not suffer from nightmares. The point of Proposition 3 was not to state that the threat simulation system could never be activated without encountering a real danger; in fact TST postulates "threat scripts" that sometimes can be activated in almost anyone, especially during times of stress. The point was to deny that there is any other type of experience apart from real threats that would influence dream content as dramatically as threats do.

Real threats, if severe enough, are immediately incorporated into dreams and frequently repeated during dreaming even for years after a single exposure to such an event. No other type of experience will be incorporated in this manner into dreams. The reason why speeding by car or crossing the street does not cause threat simulations is that such events, although dangerous in the light of traffic accident statistics, do not constitute ecologically valid cues for threat recognition or simulation. Such events do not naturally invoke in us a sense of severe threat or intense fear; therefore they are not salient for threat simulation. Instead, our threat recognition systems are naturally tuned to recognize ancestral threats. People are much more fearful, for example, of snakes, or being alone in dark woods at night, than traffic, although in the modern world it is much more probable to die in a traffic accident than be killed by a snakebite or by nocturnal predators in the woods. However, involvement in a real traffic accident where one's life is in danger will most likely trigger threat simulations about similar situations.

## RR3. Negative effects of PTSD and nightmares

In their thoughtful commentary, **Nielsen & Germain** correctly observe that the function of nightmares is a past evolutionary function rather than a current regulatory one. They argue, however, that although idiopathic nightmares

could plausibly be viewed as evolutionary adaptations, nightmares induced by trauma (PTSD nightmares) are not, for the latter is a dysfunctional state with debilitating pathology (e.g., insomnia), and therefore could not have been adaptive during our evolutionary past (also **Levin** and **Kramer** present this argument). This is an important issue and deserves careful consideration. I think that the resolution will be found by examining different kinds of post-traumatic dreams and by trying to find out what would have been the case in the ancestral environment.

The question is: Can threat simulation be biologically adaptive if it results, after trauma, in a state involving insomnia, daytime sleepiness, and impaired daytime performance? To answer this question we need a cost-benefit analysis about the relative advantages and disadvantages of threat simulation, as well as an estimation as to how probable it is that ancestral humans frequently suffered from severe PTSD.

First, I will argue that ancestral humans were not particularly prone to suffer from the adverse effects of PTSD, for they were adapted to higher levels of stress throughout their lives, and the threats in the ancestral environment were, for them, mostly familiar and predictable, and did not involve such strong and prolonged fear and terror inducing stimulation as do the worst threats (e.g., wartime combat, imprisonment in concentration camps) that cause the most severe and persistent PTSD in our world (e.g., Rosen et al. 1991; Wilmer 1996). Therefore, our ancestors arguably had frequent nocturnal threat simulations about threats from the natural world, which nevertheless were not likely to severely disturb their daytime activities. This seems to be the case also for present hunter-gatherer populations such as the Mehinaku Indians (Gregor 1981): they have very frequent threat simulations in their dreams, but are not in general inflicted with PTSD.

Second, even if ancestral humans sometimes did suffer from PTSD-like symptoms, it is unlikely (as I argued in the target article) that this cost would overweigh the benefits. If an ancestral human broke his leg, the pain made him suffer and decreased his ability to move about efficiently for some time, but it also gave time for the leg to heal; if he was suffering from an infection, a high fever may have made him temporarily unable to forage but eventually he may have got rid of the infection. Biological defense mechanisms often make us temporarily disabled and vulnerable, but in the long run and on the average it is worth while. Therefore, immediately after a severe trauma, one may sleep less well for a few days or weeks, but after the most intensive threat simulations are accomplished, the individual is, on the average, better off now, equipped with more efficient threat perception and avoidance programs in case similar threats should return. I will now present these arguments in more detail.

### RR3.1. Were ancestral humans prone to suffer from PTSD?

We should remember that not all post-traumatic nightmares indicate PTSD: there is a crucial difference between acute post-traumatic nightmares and the repetitive nightmares of chronic PTSD. The former type of nightmares occur after a single traumatic event and at first they contain repetitions of the original trauma, but the nightmares change with time and gradually become more infrequent and less disturbing (Hartmann 1998). PTSD nightmares, by contrast, are chronic, repetitive, and their content does not show substantial change through time. They are typically seen in war veterans who were exposed to extremely threatening conditions for a prolonged period of time. Most of the studies on PTSD-related sleep disturbances have been made on war veterans. There is no consistent pattern of sleep laboratory findings that has emerged in all or most studies (Hurwitz et al. 1998; Ross et al. 1989, 1994; van der Kolk et al. 1984). Therefore, it is difficult at this stage to say whether, and exactly how, the objective sleep patterns are altered in PTSD, and whether the effects of other factors (substance abuse, drugs, alcoholism, aging) have been adequately controlled. Let us assume, however, that at least some PTSD patients do show altered and abnormal sleep patterns and experience severe insomnia, and that this is due to the PTSD itself rather than other confounding factors. What are the implications for TST? Does this mean that ancestral humans suffered from severe chronic PTSD and therefore were less adapted to their environment?

There are good reasons to believe that ancestral humans did not suffer from the adverse effects of PTSD nearly to the same extent as some present-day humans do. Ancestral humans lived in a world which was full of dangers from the natural world. The constant presence of predators, natural forces, rival bands, and risky activities was part of their everyday life, and the death or physical injury of local group members must have occurred rather often. Therefore the average Pleistocene human must have been adapted to significantly higher levels of trauma and stress from early childhood on than we are. By contrast, modern humans in the Western world are adapted to low levels of stress and a low frequency of life-threatening events in their everyday lives. Furthermore, some threats in the modern world, such as combat activities or concentration camp conditions, are probably even more terrifying than ancestral threats were, because these highly non-natural threats involve prolonged periods of simultaneous exposure to several extreme stressors, for example, sleep deprivation, physical exhaustion and injury, undernutrition, imprisonment, torture, victimization, observing mass murder and other atrocities, sounds of deafening explosions, and so on. Typical ancestral threats, no matter how dangerous they were, did not reach such extremes of prolonged and severe fear stimulation. Thus, in the modern world the baseline level of stress is considerably lower than in the ancestral environment, whereas the conditions known to induce the worst PTSD symptoms in our times involve a higher degree of more prolonged stress and fear than any natural condition ever did.

There is evidence indicating that only individuals with preexisting low levels of stress tolerance continue to have PTSD nightmares decades after exposure to combat (Brill & Beebe 1955, quoted in van der Kolk et al. 1984). Thus, the difference between the stress levels we are adapted to in our everyday lives and the peak levels induced by the highly traumatizing conditions prevalent in our world is much greater than what was the case for ancestral humans. I suggest that post-traumatic nightmares can more easily develop into a severe and chronic dysfunction in our environment where individuals who are used to very low levels of stress are suddenly exposed to extreme threats, and, consequently, the dream production system will get stuck, even for years, with a few inordinately salient memory traces (see

sect. 6.3 in the target article). Paradoxically, the modern world contains both much more protected and much more horrifying environments than any ancestral one, and an individual who moves from one extreme to the other is in special danger of later suffering from chronic PTSD nightmares. It is unlikely that our ancestors would have become severely dysfunctional with PTSD after encountering threats that always had been a natural part of their world and their relatively short lives. It is more likely that they had acute post-traumatic nightmares that constituted a temporary, functional threat-simulation response but did not develop into chronic PTSD with severe insomnia.

**Levin** points out that modern living still provides numerous approximates of ancestral threats (e.g., rape, physical beatings, natural disasters). He suggests that TST would be supported by evidence showing that people recover quicker from traumas resembling ancestral threats than ones completely unlike them. This prediction is worth testing, but it should be noted that even these cases are far from perfect models of what was going on in the ancestral world. Unlike most modern humans, for our ancestors the ecologically valid threat cues were not rare exceptions that intruded into an otherwise peaceful existence. Instead, they were a part of the ever-present everyday reality.

### RR3.2. Are some features of PTSD adaptive in a dangerous environment?

PTSD includes some symptoms which, although dysfunctional in our environment, may have been useful in a more dangerous one. Individuals with PTSD are more sensitive to threatening stimuli and contexts. They show increased autonomic responses to sudden stimuli, which is a defensive response that increases the organism's preparation for motor activity. Hypervigilance, another typical PTSD symptom, is a manifestation of heightened readiness to respond to threatening stimuli or situations (for a review, see Pitman et al. 2000). These characteristics would have been adaptive and useful in conditions with continuous exposure to natural threats. In exaggerated form they are also part of PTSD symptoms, but they become dysfunctional especially after returning from an environment with high levels of threat to a completely safe environment (which did not exist for ancestral humans). A relatively high arousal level and sensitive responses to threats probably were useful rather than harmful in the ancestral environment.

### RR3.3. Does threat simulation necessarily cause disturbing nightmares?

The clinical definition of "nightmare" includes the idea that the terrifying dream awakens the sleeper from REM sleep, and thus disrupts sleep. This definition may be clinically valuable, but theoretically we would need to find out how frequently threat simulation dreams occur after traumatic events, regardless of whether the dreams awaken or otherwise disturb the sleeper. That is, we would need to have an estimation of the frequency of threatening content both in dreams that do not radically disrupt sleep and in disturbing nightmares. Then we could compare the frequency and content of nondisturbing nightmares to ones that actually wake up the dreamer and cause insomnia. The prediction derived from TST is that the nightmares that radically disrupt sleep are only the tip of the iceberg; the exception rather than the rule. The frequency of nightmarish dreams with threat simulations but without considerable debilitating consequences should be higher than the frequency of such nightmares that severely disrupt sleep and cause grave insomnia. This would ensure both that the threat simulations will accomplish, without premature interruptions, rehearsal of the relevant threat avoidance skills and that the individual will not be unduly deprived of sleep. This prediction should hold primarily for threats that are reminiscent of ancestral ones as to their nature and severity, and for individuals who have been used to living in environments where such threats are constantly present. We should remember that the Mehinaku Indians seem to have nightmarish content appearing very regularly in their dreams (see target article, sect. 4.1), but Gregor (1981) who stayed in the Mehinaku village collecting the dreams does not mention that these people would have suffered from any kind of sleep disturbances or other symptoms of PTSD. There is no reason to believe that our ancestors would have been constantly debilitated by PTSD, although they were likely to suffer from occasional nightmares as side effects of the efficient threat simulation in which their dream production system was continually engaged.

### RR3.4. The correlation between nightmares and psychological problems

It is pointed out by **Schredl** that there is a correlation between psychopathology, low life satisfaction, and negative dream emotions. **Levin** notes that frequent lifelong nightmare sufferers demonstrate considerable psychological dysfunction. Thus, people who are more troubled and who have mental health problems tend to have more unpleasant and negative dreams. Schredl argues that negative dreams are associated with poor adjustment and therefore the idea that threatening dreams serve an adaptive function is not supported. However, this argument is not entirely convincing, for it ignores two crucial issues: First, TST does not imply that threat simulation dreams should promote psychological or social adjustment to present-day environments, or increase an individual's happiness or satisfaction. TST only claims that threat simulation dreams increased the inclusive fitness of ancestral humans in the ancestral environment: it says nothing at all about any other type of adaptation or adjustment than biological, nor about dream functionality in any other type of environment than the ancestral. Hence, the facts presented by Schredl are not proper tests of the actual claims included in TST. Second, a correlation between threat simulation dreams and psychological problems or psychopathology is just that: a correlation. It is not the same as causation: there is no evidence that frequent threat simulations would actually cause psychopathology, although they may be indicative of the presence of psychological problems.

Then why does threat simulation co-occur with psychological troubles? The simplest explanation is that people who suffer from high levels of anxiety, phobias, fearfulness, depression, stress, and so on, feel extremely threatened, whether by imagined or real threats. Therefore, it is no miracle that the threat simulation system is intensely active in such circumstances, trying to simulate the perceived threats and prepare the individual to overcome them (although perhaps to little avail in some of these cases). **Schredl's** ar-

gument that the correlation between psychological problems and negative dreams is evidence against the adaptive function of nightmarish dreams could be compared to an argument about the immune system: there is a correlation between immune responses and multiple injuries and infections present in an individual. In fact, the less physically healthy the individual is, the more likely it is to find that massive immune responses are taking place in his body. Thus, since there is a high correlation between ongoing immune responses and low degrees of physical well-being, the immune response cannot be adaptive. I guess anyone can see what the problem with both arguments is: if the immune response and the threat simulation response are biological defense mechanisms, they are biologically programmed to co-occur with physical and psychological problems, respectively, because they fight the adverse effects of such conditions. But the regular correlation is hardly any evidence against their biological functionality; on the contrary, it is evidence for it.

### RR3.5. The costs and benefits of threat simulation

**Levin** suggests that nightmare sufferers "retain a closer link to their ancestral vestige of the early fight-flight patterns than do other individuals." I agree, but would rather say that the threat-simulation system is more sensitive or more highly activated in individuals with chronic nightmares. **Schredl** argues that the group of frequent nightmare sufferers should have increased by natural selection, but since in actual fact this group is small, threat simulation has not been selected for. This argument disregards the fact that genetically transmitted phenotypic features usually show variation within the population so that selection can work on them. Most features involve both benefits and costs, relative to the environment where they are expressed. Thus, individuals with a very sensitive threat simulation mechanism perhaps enjoy the benefit of having extremely primed threat perception and threat avoidance programs but suffer the cost of somewhat disrupted sleep or severe PTSD more easily than individuals with less sensitive threat simulation systems. By the same token, individuals with an extremely sensitive immune system may more easily fall prey to severe allergies or autoimmune diseases. Natural selection does not exclusively favor sensitive immune systems or threat simulation mechanisms over less sensitive ones, for at some point or in some environments the costs of increased sensitivity will exceed the benefits. Therefore, the fact that chronic nightmare sufferers represent only a small proportion of the total population is in accordance with the normal distribution in the population of the sensitivity of the threat simulation mechanism.

## RR4. Evidence from typical dreams

Interesting data on the most prevalent themes in dreams are provided by **Germain et al.** In order to be able to interpret the results properly, it is important to understand what these studies reveal and what still remains unknown. The data were collected by having the subjects fill in a questionnaire where they check off from a list of 55 typical dream themes the ones that they have experienced. Thus, this method basically measures what kind of dream themes the subjects can remember having experienced at least once in their lives. To check off a theme, it is sufficient to have experienced it only once or a few times, but in such a way that it has had a lasting effect on long-term memory. Therefore, the results do not reveal anything about the absolute or relative frequency of the themes that are checked off. We cannot tell which dream themes perhaps occurred only very rarely but were highly memorable and which ones perhaps recur regularly almost every night. Thus, we should keep in mind that this measure cannot distinguish the memorability from the frequency of dream themes. Once this reservation is granted, we will see that none of the results pose serious problems for TST, on the contrary, the better part of them seem to directly support it.

The results reviewed by **Germain et al.** reveal that as much as 67–86% of undergraduate students remember dreaming about being chased or pursued, and that this is the leading remembered theme in their dreams. The result is of course very well in accordance with the predictions of TST. It could be interpreted as strong evidence for the existence of universal threat scripts in the human dream production system, and for the occasional activation of these scripts in almost everyone, regardless of whether the person has ever actually encountered any ancestral threats comparable to the content of the threat scripts.

The results show that contemporary concerns are also among the most prevalent dream themes. **Germain et al.** say that it is not clear why dreaming should so often represent contemporary concerns if its function is geared toward dealing only with ancestral sources of threat. However, as I explained in section 6.2 of the target article, TST does explain why and how current concerns get incorporated into our dreams: they are mild threats and have higher saliency for dream production than completely neutral or positively charged themes. All memory traces with negative emotional charge and/or recent activation are salient for the dream production mechanisms and therefore, in modern populations, also the traces representing current concerns frequently end up being themes of threat simulation. The high prevalence of current concerns cannot be regarded as evidence against TST.

Still, some dream themes with high prevalence are not related to threat simulation: especially sexual experiences and flying. Falling is more ambiguous and could well be part of a threat simulation dream, since falling from cliffs or other high places obviously constituted a serious risk to physical health in the ancestral environment. How much serious evidence against TST do these results constitute? Not very serious, for we should now remember that prevalence is not the same as frequency. While threatening events in dreams have both high prevalence and high frequency, sexual experiences and flying have only high prevalence but very low frequency. In a recent study, we calculated the frequency of all kinds of threatening events in 592 home-based dream reports from 52 Finnish university students (Revonsuo & Valli 2000). We found that there were altogether 672 threatening events described in these reports, on the average 1.2 threatening events per dream report, although some reports contained several threats and one third of the reports contained none. All the subjects had at least some threatening events in their dream reports.

These results confirm that threatening events not only have high prevalence but also high frequency in dreams.

By contrast, this does not seem to be the case for sexual dreams. In the Hall & Van de Castle normative sample, only 4% of the dreams reported by female subjects and 12% by male subjects contained any sexual activity. Furthermore, as Domhoff (1996, p. 60) notes, "sexual interactions in dreams are far more likely to be fraught with negative emotions than popular stereotypes suggest." Therefore, the presence of sexual activity in a dream by no means excludes the possibility that it is a threat simulation dream depicting for example, threat of being raped, other types of sexual harassment, or current concerns about one's sexual potency or desirability. The frequency of flying in dreams is even lower, with about 0.5–1% of dream reports containing it (Domhoff 1996). The high prevalence in this case obviously reflects the high memorability rather than high frequency of the theme.

**Germain et al.** mention that only about 20–30% of undergraduates remember ever dreaming about snakes, wild violent beasts, or insects and spiders, but they correctly note that this can be explained by the general trend, with age, towards lower frequency of animals in dreams if the actual environment contains few. And of course, we should not expect undergraduates to recall what kind of animals they were dreaming about when they were small children. However, Germain et al. then raise the question of why natural disasters have low prevalence in dreams, and (although we need more data on this) it appears that even children do not very frequently dream about natural disasters. Our own studies on the frequency of threatening events in dreams confirm this (Revonsuo & Valli 2000). Of the 672 threatening events that were identified in the students dream reports, only 2.5% represented "catastrophies" (uncontrollable natural, technological or social forces that pose a threat). Why might this be the case? I think we need to consider the original ancestral environment once again.

There is wide agreement that the most important geographical area in hominid and human evolution is Africa, especially sub-Saharan, eastern and southern Africa. This area is not located at the margins of the earth's crustal plates where structural stresses may be released as frequent earthquakes and intense volcanicity. Therefore, earthquakes and volcanic eruptions have probably been too infrequent to pose a constant selective pressure on ancestral human populations. Furthermore, it is unclear whether there would be any effective responses available against the threats posed by large-scale disasters, such as earthquakes, that could have been rehearsed during dreaming. Tornadoes, strong winds, forest fires, and tidal waves have probably been more frequent than earthquakes, but still much more infrequent than encounters with the ever-present predators, other dangerous animals, or aggressive humans. There has been a considerably higher selective pressure to train the skills needed in perceiving and avoiding the latter types of dangers than natural disasters. Furthermore, there are several readily available behavioral patterns that can be used and rehearsed for efficient avoidance of predators, dangerous animals, and aggressive attacks. Therefore the threat scripts and other biases in our threat simulation system primarily reflect that kind of ancestral threat. We can even formulate a prediction from this: if paleoecological studies identify an ancestral threat that occurred with high frequency in the ancestral environment and posed a significant selective pressure on ancestral humans and was the kind of threat where skilled avoidance behavior did make a difference for the inclusive fitness of the ancestral individual, then we should find that the same theme is frequently simulated by the threat simulation mechanism during dreaming. Conversely, if such ancestral threats are discovered but we never dream about them, it would constitute disconfirmatory evidence against TST.

**Germain et al.** wonder why the threat simulation system should adjust its simulations to better fit the actual environment; they say that such change would not serve any obvious function. On the contrary, I would say that it does serve a rather obvious function. It would be advantageous to have a somewhat flexible threat simulation system that is sensitive to possible changes in the environment and to the threats it contains. For example, in ancestral populations climatic changes or migration to new habitats may have introduced new types of threats that involved great selective pressures for the population but had not been encountered previously and did not belong to the innate repertoire of the threat simulation system. A flexible threat simulation system will make an attempt to simulate these new types of threats as well, whereas an inflexible one would continue simulating only the threats contained in the original repertoire. It appears that the flexible system would be able to solve a greater number of adaptive problems than the inflexible one. Germain et al. also wonder why children raised in environments completely free from threat do not cease to have threat simulation dreams altogether. The answer is that the threat simulation system is a biological defense mechanism that is always ready to react to possible threats. It evolved in an environment where certain types of threats recurred over and over again for hundreds of thousands of years. This very long evolutionary history left a lasting mark in our dream production system that the present, "non-natural" environment cannot wipe out, at least it has not been able to do it in the very short period of historical time (less than 1% of the whole human evolutionary history) during which ancestral threats have ceased to present major selective pressures on humanity. The threat simulation system, just like the immune system, cannot be turned off for good even if no appropriate enemies are presently in sight. Our threat simulation system will make an attempt to simulate whatever current concerns are encountered, and the immune system attacks whatever invaders are discovered within us (sometimes only harmless pollen). From our point of view it might be more adaptive if these defense systems could simply be turned off by will, but they cannot, and they will stubbornly try to do the job that natural selection designed them to do, even if it appears that in the modern world they have sometimes been sent on a fool's errand.

In sum, I think **Germain et al.** have done a valuable service by reviewing the data from the prevalence of typical dream themes. However, since the study is based on what people remember about the dreams that they have had, not on the quantitative content analysis of systematically collected dream reports, the results must be taken with proper caution and combined with studies of the frequencies of such themes in reported dreams. When this is done, the data offers remarkably strong support for TST. The worries that Germain et al. have about the low prevalence of natural disaster dreams, the high prevalence of other than ancestral themes in dreams, and the change of dream content with time did not prove to be difficult to explain by TST.

## RR5. Evidence from recurrent dreams and nightmares

The content of recurrent dreams is informatively reviewed by **Zadra & Donderi** (see also Zadra 1996). There are several features in this data that lend support to TST. First, being chased is the most common theme in the recurrent dreams of both children and adults. Negative affect was far more common than positive affect (about 80% vs. 10%), fear or apprehension was the most frequently reported emotion (occurring in 79% of childhood recurrent dreams), and misfortune far more frequent than good fortune (42% vs. 4%). All of this is in accordance with the view that the most typical recurrent dreams are powerful threat simulations. Furthermore, there were interesting differences between childhood and adult recurrent dreams: in childhood dreams the dreamer was more often in danger than in adult dreams (65% vs. 42%). The most frequent childhood recurrent dreams contained severe threats that would have been highly relevant to a child's survival in the ancestral environment (being chased; death of family members; being alone and stuck or trapped; facing natural forces). The recurrent dreams of adults contained a lower degree of ancestral threats but a higher degree of current concerns for example about house maintenance and personal competence. This again demonstrates the malleability of the threat simulation system to incorporate other than ancestral threats that frequently occur in the present environment. Ancient concerns, the default values in the threat simulation system, are slowly (but never completely) replaced by milder current concerns.

In the light of these results, there is no doubt that most (though not all) recurrent dreams can be regarded as threat simulations. Recurrent dream themes that may have other roots are: flying, losing one's teeth, and being unable to find a private toilet. It should be noted that the latter two themes only occur in adults', not in children's recurrent dreams, and although both themes depict rather peculiar threats, there are plausible physiological explanations for both. Stimulation from a full bladder during REM sleep might sometimes get incorporated into a dream, and thus modulate dream content, creating this type of recurrent dream where one feels the need to find a private toilet. Bruxism, or toothgrinding, occurs in 70% of all people during sleep, it is more severe during times of stress, and it occurs in all stages of sleep (Stepanski 1993). I am unaware whether there are any relevant studies in this area, but I would hypothesize that dreams about losing one's teeth are triggered by episodes of bruxism during sleep when the sensations elicited in the mouth are incorporated into the dream. Thus, the origin of these peculiar recurrent themes is most likely physiological, and the existence of these themes does not change the fact that most recurrent dreams are threat simulations of ancestral or current concerns.

**Zadra & Donderi** report that in their database of several hundred nightmares, they find threat perceptions but not adaptive threat avoidance programs, because the dreamers awaken either while trying to escape or at the moment they are caught or attacked. However, they have not made any quantitative analysis of the data yet. Furthermore, they say that many nightmares contain bizarre and unrealistic elements. As Zadra & Donderi note, to evaluate these issues we would need to conduct quantitative content analyses of nightmares. I would add that such analyses should be made on dreams from different populations, especially children's nightmares and nightmares from populations that live in environments that contain ecologically valid cues of ancestral threats. It is only such studies that could tell us whether the threat simulation system might have worked appropriately in the ancestral environment. Furthermore, we need to analyze not only nightmares that woke the dreamer up but also completed threat simulations that were not disrupted by awakening. It is clear that we will not find appropriate responses to threats in nightmares if the dream was disrupted before there was any time to respond. Thus, we need to make content analyses of threatening events in such dreams and nightmares that were not disrupted but were experienced and reported in full, and we also need to find out how often there are threat simulations that do not disrupt sleep compared to those that do.

We have recently reported a study on threatening events in the dreams of university students (Revonsuo & Valli 2000) that sheds some light on the issues raised by **Zadra & Donderi**. Of the 672 threatening events that we identified in the dream reports, 17% were disrupted and 14% discontinuous, which means that it was not possible to properly evaluate the response to, or the consequences of, the threatening event. In some dreams, the dream self was not present in the scene of threat or otherwise did not react to the threat. However, in 54% of the threatening events the dream self was reported to react somehow to the threatening event. Ninety-four percent of these reactions were appropriate and relevant to the threatening situation, whereas irrelevant reactions (4%) and physically impossible reactions (2%) were rare. Furthermore, a relevant reaction by the dream self (as opposed to no reaction) was significantly more likely to be associated with life threatening situations than less dangerous ones, showing that when the threat simulation system is properly activated by a highly dangerous threat, then the dream self probably reacts in some way that would be relevant in a corresponding real situation. Also this issue should be studied in detail in children's dreams and in the dreams of people exposed to ancestral threats; TST would predict an even higher proportion of relevant reactions in their dreams.

Another issue raised by **Zadra & Donderi** is that the consequences that the dream self suffers because of threat simulation may be severe, such as total destruction of the body. Our study (Revonsuo & Valli 2000) does not support such a conclusion. Most threat simulations are not reported to have any consequences. Severe consequences were extremely rare: there were no cases among the 672 threatening events where the dream self would have suffered fatal injuries or death, and only in less than 2% was there serious injury or death of significant others. Threatening events do not often simulate severe losses; in fact, the threat simulation system does not seem to show that much interest in simulating what happens after the immediate threat is over. Neither do our data support Zadra & Donderi's view that threat simulations would usually be bizarre or unrealistic. When we evaluated the sources of the threatening events, only 4% could be traced back to unrealistic sources (e.g., fantasy, science fiction, horror stories, mythology), whereas 63% depicted events that belong in the dreamer's everyday world.

**Zadra & Donderi** do not seem to be satisfied with the answer that the threat simulation system becomes fully active only under some special circumstances, because they

think that it dismisses most dreams as evolutionarily useless epiphenomena that just waste time. However, we should remember that in the ancestral environment (for which the dream production system was designed) humans could rarely enjoy the luxury of periods completely free from threats. Modern human populations now have artificially created environments free from ancestral selective pressures where the simulation of ancestral threats is no longer really required. The dream production mechanisms (or TST) should not be blamed for that. We ourselves have made dreams less important by removing ancestral threats from our environments. If one day humankind will be able to remove all infectious micro-organisms from its environment, the immune system might become largely useless as well, because it is not essential in a completely sterile environment. But this will not make the immune system a biological epiphenomenon since it is an evolved adaptation in origin. Furthermore, there are several biological adaptations that are useful only during quite limited periods of time. For example, female breasts are not particularly useful unless the woman has recently given birth to a child. If we take a "representative sample" of females of all ages, we will probably find that, at the time of the study, breasts serve no function for most of them. Even so, we would not be justified to conclude that most breasts are therefore mere biological epiphenomena. Some of them may already have ceased to serve their function, some may one day become fully active and start to serve their function, if certain conditions are met. The same applies to human sperm cells: countless billions of them never have the slightest chance to fulfill their biological function, fertilization, yet the wasted sperm cells are not biologically epiphenomenal.

Analogously, a representative (and indiscriminate) sample of dreams may not reveal what the dream production system is tuned to do under certain special circumstances: many biological adaptations only become active under specific circumstances. Dreams that do not simulate threats are not "evolutionarily useless epiphenomena" to any greater extent than female breasts that do not give milk, or the billions of male sperm cells that never fertilize an egg cell. The real question about biological functionality is whether these biological systems would react otherwise if certain conditions were met (if the need to prime threat perception, to feed a baby, or the opportunity to enter an egg cell, should arise). To point out, for example, as **Montangero** does, that since most dreams deal with nonthreatening situations, the function of the majority of dreams cannot be threat simulation is to misunderstand the nature of biological functionality. Furthermore, the claim that "most dreams deal with nonthreatening events" may turn out to be incorrect: our recent study shows that most dreams (66% of dream reports in our sample) do include threatening events (Revonsuo & Valli 2000).

## RR6. Other forms and functions of dreaming

Several commentators regard TST as implying too restricted a view of dreaming and dream function. **Hunt** complains that TST ignores or denies many other forms of dreaming, such as lucidity, creative-imaginative dreaming, and problem-solving dreams. He also protests that there are dream-centered societies that have different kinds of dreams and cultivate dreaming for other purposes than the

Mehinaku Indians. In order to reply to these accusations, let me explain what kind of claims TST actually implies and what it does not. TST is a theory about the biological functions of dreaming and it identifies the kinds of dreams and dreaming that would have had adaptive value for ancestral humans in the original environment. TST does not claim that other forms of dreaming than threat simulation do not exist. That would be outright silly, for obviously other forms of dreaming do exist: there certainly are lucid dreams, dreams certainly have inspired creative thinking, perhaps in a few rare cases dreaming has contributed to scientific advances (in fact, some of my own threat simulation dreams contributed to the formulation of TST, so I can testify that dreams have helped in the discovery of this particular theory!). In some cultures these types of dreams may be regarded as more interesting or significant or valuable than threat simulation dreams.

Still, these other forms of dreaming have no biological function. The culturally invented functions of dreaming are independent of the biological function. The Mehinaku Indians do not regard their own dreams as threat simulations, but, instead, as clues to the future, symbols of events to come, that can be decoded with the help of complex metaphoric equivalencies of color, shape, and action (Gregor 1981). There is no doubt that other cultures have invented other uses for their dreams; that is by no means questionable, or denied by TST, as **Hunt** seems to believe. However, I am not aware of any other present-day dream-centered societies – that could be regarded as reasonably good models of ancestral hunter-gatherer societies – the theoretically most crucial source of data for testing TST. A dream-centered society that leads a life completely different from the hunter-gatherer way of living may not have any particular relevance to testing TST (which of course is not to say that it would not be interesting to study such societies and their dreams for other purposes).

Many commentators were prepared to accept the view that dreams are simulations, but they argued that it would be useful to simulate also other types of activities than threat perception and avoidance. **Humphrey**, **Cheyne**, and **Peterson & DeYoung** all regard dreaming as closely similar to play.[1] Peterson & DeYoung suggest that dreams may be positive, exploratory, creative play and thus part of the process of adjustment to novelty. Humphrey argues that dreams can help in the development of social skills, empathy, and interpersonal understanding; a child can apply the developing theory of mind to the characters encountered in dreams and can introduce introspectively observable mental states that are as yet unfamiliar in real life. **Nielsen & Germain** argue that dreaming might simulate multiple, cognitive, and socio-emotional functions that are pertinent to our species today: the simulation of interactive character imagery would constitute adaptive simulation of attachment relationships; simulation of self-imagery may facilitate functions related to ego development or motor competence; place imagery may facilitate spatial learning and orientation. **Blagrove** argues that rehearsing positive things is as important as threat simulation, and **Clancey** suggests that peaceful activities (such as could be found in the Mehinaku dreams) are as important for survival as threats so perhaps it would be useful to simulate them as well. **Peterson & DeYoung** propose that since nightmares are easily recalled and communicated to others, perhaps the function of dreaming is to let the whole community

know about these current problems and threats so that they could be solved with the help of the whole social group.

**Rotenberg** (and also **Greenberg**) advocates the idea that REM sleep and dreams are related to the restoration of search activity. The concept is somewhat vague, characterized as including for example fight, flight, and creativity but excluding "stereotyped behavior," "freezing" or "panicky behavior." However, it is difficult to figure out exactly how "search activity" is related to threat perception and avoidance, since adaptive fight or flight responses to threats may include initial freezing (to avoid being detected by a potential enemy), panicky escapes from mortal dangers, or stereotyped defensive actions. The concept of "search activity" should be clarified substantially before it is possible to evaluate its relation to TST. **Shevrin & Eiser** support the Freudian view that dreams deal not with external threats but with internal conflicts, and sometimes succeed in their function of wish fulfillment; they regard bad dreams as failures of dream function. **Shevrin & Eiser** (and also **Ardito** and **Montangero**) support the idea that the function of dreams is to protect sleep.

First and foremost I want to emphasize that TST certainly does not show or claim that other functions of dreaming than threat simulation do not exist or are not possible. Having said this, I hasten to add that, although I find some of the above proposals fascinating, unfortunately the case for them does not seem to be very strong. If these suggestions are offered as serious candidates for being biological functions of dreaming, then they will have to be subjected to a similar evolutionary cost-and-benefit analysis as threat simulation. These suggestions suffer from one serious weakness: it is difficult to imagine what the selective pressures would have been in the ancestral environment that would have increased the inclusive fitness of individuals with the kinds of alternative dream simulations suggested by the commentators. The crucial difference between threat simulation and all other cognitive or social simulations is that real threats involve fatally high risks, and therefore encounters with them have costs too high to allow much rehearsal in the real world. All the other activities and skills that were mentioned as possible functions of dreaming involve little cost if explored and practiced in the real world. Therefore, it is difficult to see what the advantage, in terms of increased fitness, would have been to simulate them during dreaming if the real thing was available all the time, and with little cost, during waking (or, conversely, what the disadvantage would have been of the failure to simulate these things).

I would welcome any carefully formulated hypotheses on other biological functions of dreaming, as long as they are backed up by convincing empirical evidence. Those who claim that dreaming has other biological functions than threat simulation have the burden of proof to show that their hypothesis is supported by a wide range of evidence from normative dream data, typical dreams, recurrent dreams, post-traumatic dreams, children's dreams, hunter-gatherer dreams, and so on. The task for TST is not to prove that other biological functions of dreaming do not exist. TST only shows that at least threat simulation is a biological function of dreaming. So much the better for dream theory if other plausible hypotheses can be formulated, but I am not tempted to give it a try, nor do I believe it to be an easy task. So far as I can see (which may not be very far), it is not viable to make as strong a case for any other biological function of dreaming as for threat simulation. The sheer variety of the different proposals, the disagreements between them, and the fact that none of them covers as wide a range of data as TST does, make it difficult to see which proposals, if any, should really be taken seriously.

## RR7. Support for TST from multiple independent sources

So far, I have concentrated on responding to the critical points raised against TST in the commentaries. However, I was happy to note that many commentators also provided evidence supporting TST and formulated new hypotheses that could be tested in future studies. Whether or not TST will be accepted by the research community in the long run, at least it is capable of generating new testable predictions at multiple levels of description, and therefore seems to be a hypothesis well worth taking seriously.

**Humphrey**'s and **Panksepp**'s views on the function of dreaming appear to be in broad agreement with TST. Both acknowledge the importance of an evolutionary approach, and Panksepp emphasizes the role of emotions in the explanation of REM sleep and dreaming. His view of the REM state as an ancient form of waking consciousness, an emotionally guided state of mind, is not in conflict with TST. **Bednar** suggests that a modification of the model by reversing the sequence of perceptual content leading to emotional response would simplify the model without giving up the basic idea of the theory. I am open to the possibility that Bednar's proposal might work just as well as the original one. The difficulty is in devising empirical experiments that could contrast these two hypotheses with each other. In any case, we do need more precise explications of the learning mechanisms involved in threat simulation, and Bednar's suggestion is a viable option.

**Shakelford & Weekes-Shackelford** accept the evolutionary psychological perspective of TST and they propose further testable predictions concerning the content of dreams. Their hypothesis is that male and female psychology might generate different kinds of threat scenarios for the dream production mechanisms to simulate. That is a definite possibility that will have to be addressed in future studies. In our recent study (Revonsuo & Valli 2000), we did not find significant gender differences in the relative distribution of different kinds of threats, apart from the higher proportion of direct physical aggression in the threatening events reported by male subjects (this gender difference has been reported in numerous studies before). However, we did not look into dreams about sexual infidelity separately. My impression is that such a theme was not frequent in our sample of threatening dream events. A more common threat theme in female dreams appeared to be being involved in sexual activity with, or having an intimate commitment to, an undesirable partner. This might well be a sex-specific threat theme with a clear evolutionary rationale, but it is too early to say for sure until a proper quantitative analysis of such threats is conducted in a sufficiently large sample of dreams.

As discussed above, the results from studies on typical dreams (**Germain et al.**) and recurrent dreams (**Zadra & Donderi**) provide substantial positive evidence for TST, especially if they are combined with detailed studies, such as our own, on the frequency of threatening events in dreams (Revonsuo & Valli 2000). Furthermore, Zadra &

Donderi provide a beautiful example of a fully realistic dream a figure skater had during the Winter Olympics. In this case, the dream apparently constituted a useful rehearsal of complex motor performance in the face of a threatening failure. **Cheyne** supports the idea expressed in the target article that several sleep disorders can be better understood if they are interpreted as inappropriate activation of the threat simulation mechanisms. Garfield interprets depression as involving decreased activity of dream production and threat simulation, whereas in REM sleep behavior disorder and in some cases of violent behavior during sleepwalking we are dealing with an overactive or inappropriately timed activation of the threat simulation mechanisms. Cheyne points out that sleep paralysis associated with hypnagogic hallucinations very frequently involves intense fear and an experience of being assaulted by an intruder or at least the sense of a malevolent unseen threatening presence. In these cases the threat simulation mechanisms appear to be strongly activated as well. Taken together, TST seems to unify this new data from a variety of sources in a theoretically meaningful way.

**Wichlinski** explores the implications of TST at the neurochemical and pharmacological level of description, and provides several significant observations that seem to support TST. Most pharmacological agents that reduce REM sleep also reduce nightmares and, conversely, many agents that enhance REM induce nightmares. The REM rebound that occurs after pharmacological REM suppression can dramatically increase the frequency of bad dreams. Thus, it would appear that pharmacological REM suppression and enhancement correlate with the level of activity of the threat simulation mechanism.

**Wichlinski** is rightly rather cautious in the interpretation of this data, since many questions about the complex neurochemical mechanisms that mediate these effects remain open. In any case, it is encouraging that TST is consistent with data from a completely independent source dealing with the level of pharmacological and neurochemical mechanisms. **Krieckhaus** seems to accept the core assumptions of TST with some minor modifications. He argues that the neuroanatomical basis of the threat simulation system is to be found in the Papez circuit, and the competition between hippocampus and amygdala. I agree that the interaction between amygdala and hippocampus seems to have an important role in dream generation, but it is probably too early to anatomically localize the threat simulation mechanisms with any certainty.

## RR8. How to test TST properly?

TST can explain (or at least is consistent with) a vast amount of data from a variety of sources. However, it is clear that the theory must be rigorously tested in future studies in order to find out its strengths and potential weaknesses. New falsifiable empirical hypotheses and predictions, derived from the theory, must be formulated. (This is indeed possible, contrary to **Hunt**'s assertion that the theory is "inherently untestable"). Here we must be very clear about the claims and the predictions of the theory in order to find out what would constitute disconfirmatory or confirmatory evidence for the theory.

TST postulates that there is a dream production mechanism that brings about organized simulations of the perceptual world during sleep. The biological function of this system is to produce organized threat simulations. These are constructed from two sources: threat scripts depicting ancestral dangers, and emotionally charged, recently activated memory traces of real threatening events encountered in the environment. The threat simulation system has default values consistent with the ancestral environment and ancestral threats. The default values dominate threat simulations in children's dreams, but if there are no ancestral threats in the real environment, then whatever other types of threats there are will slowly take the lion's share of the threat simulations. The threat simulation system can be activated in different degrees. The lowest level of activation should occur in an environment that is completely safe and where the individual is free from stress and fear. The highest level of activation (a strong threat simulation response) should occur in an environment where the individual's life and physical well-being is seriously threatened. The optimal functioning of this system occurs in individuals whose life resembles the life of ancestral humans in that there are threats from the natural world corresponding to ancestral ones, and that they have been constantly present from early childhood on so that the individuals are well adapted to a constant level of stress. Since the life expectancy of ancestral humans was barely 25 years, we should expect the threat simulation mechanism to work optimally in children and adolescents for whose survival the system made a difference in the ancestral conditions. There was no age-group there that would resemble the elderly people of our world; therefore, the threat simulation system may not be activated in the elderly people to the same extent as in people under 20 years of age.

How do we test whether the threat simulation mechanism really exists and whether it behaves the way the theory postulates? What are the most crucial tests and what kind of tests are not so revealing? How to separate the predictions of TST from the predictions of competing theories? TST postulates that threatening content constitutes special, unique input for the dream production system, whereas the random activation theory (RAT) and the continuity hypothesis deny this. By contrast, they postulate that all emotions are represented in dreams to the same extent that they are represented in the waking world (or in autobiographical memory thereof). No type of emotion has a special status in dream production. The central tenet of TST is the existence of a specific threat simulation response to perceived external threats. TST predicts that if an individual experiences real events, each one of which elicits different (positive and negative) strong emotions, the events that represent serious threats for the individual or significant others (e.g., physical assault, serious accident) will be incorporated into dreams as threat simulations significantly more often, and such dreams will persist significantly longer than any dreams incorporating the events involved in nonthreatening emotional events. If threatening and nonthreatening events that both involve strong emotions were found to be incorporated into dreams to the same extent, then we would have evidence that threatening real events do not have any special status in the construction of dream simulations, that is, there is no special threat simulation response as postulated by TST. If one really desires to falsify TST once and for all, the best way to do it would be to show that there is no such thing as a specific threat simulation response.

Blagrove and Price (2000) report that happy skilled individuals tend to have happy dreams. Does this constitute falsifying evidence against TST? First, we should note that the data in this study were collected with a questionnaire, which gives a limited view of the variety of dreams that people have. Second, the correlation between waking happiness and dream happiness, although significant at the $p < 0.05$ level, was relatively low, only .19. The correlation between traumatic life-events and threat simulation dreams (nightmares) would appear to be much stronger. In any case, it is quite likely that happy, skilled people also live in an environment that is relatively free of severe threats, therefore it would not be surprising to find that their threat simulation system is not particularly active. Thus, this study does not properly test TST, for TST would not predict a high frequency of threat simulation dreams in such a population. A proper test of TST would be along the lines described in the previous paragraph: the responses of the dream production system to similar intensities of different kinds of emotions should be compared. If the system does not show a special response to threatening content, then the prediction derived from TST would be disconfirmed and those derived from competing theories would be confirmed.

**Blagrove** argues that a proper test of the theory would be to find threat themes in people who are not hunter-gatherers or traumatized. But what can we expect to find in such cases where TST predicts that the threat simulation system is not highly activated? TST postulates that there are ancestral threat scripts dormant in everyone's dream production system, even in people who never were traumatized or practiced the foraging way of life. These scripts may become active every now and then in almost anyone. The most convincing evidence of the existence of these scripts are the uncontestable and highly replicable findings (see the commentaries by **Germain et al.** and **Zadra & Donderi**) that as much as 80–90% of ordinary young adults (most of whom probably had not been traumatized and certainly none were hunter-gatherers) have had dreams of being chased or pursued and that over 40% have had recurrent dreams of this type, especially during childhood. This is the most frequently reported dream theme in studies of both typical dreams and recurrent dreams. The explanation provided by TST is that this reflects the activation of the ancestral threat scripts built into the threat simulation mechanisms. Competing theories, RAT and the continuity hypothesis, are unable to account for these findings, for there is nothing random in these highly organized threat simulations, and very little that forms a continuum with the everyday life of school children or typical college students in the modern world.

Thus, any study that properly tests TST must carefully take into account the actual predictions that can be derived from this theory before some intuitively plausible counterexamples which nevertheless are not in conflict with the theory. For example, as I have tried to explain in the target article and in this response, the following facts do not constitute disconfirmatory evidence against TST:

1. There are many dreams that do not simulate threats.

2. There are some threat simulations that have negative effects and disturb sleep.

3. Many dreams contain bizarre features.

4. Having nightmares or recurrent dreams correlates with psychological problems.

I will not here recapitulate why these facts do not disconfirm TST; the arguments can be found elsewhere in this treatment. However, the following facts, if confirmed, would pose problems for the theory:

(a) There is no special or unique threat simulation response that would be any different from the effects of non-threatening, positive events: all highly emotional real events elicit a similar dream simulation response where there are frequent dreams about the emotionally charged event.

(b) There were ancestral threats which posed serious and continuous selection pressures on humans and for which they had potentially efficient behavioral avoidance strategies that were rehearsable, but we never dream about such threats. Here TST would thus predict the existence of a corresponding threat script.

(c) The dreams of children and adolescents who have been exposed to ancestral-like threats since early childhood do not simulate the threats or efficient responses to them.

(d) There is a population of hunter-gatherers that is a good model of ancestral human way of life, but whose dreams only infrequently simulate ancestral threats.

(e) There is a culture whose members never have threat simulations in their dreams or never experience nightmares.

(f) Threat simulations make no difference to perceptual and motor skills and do not support or improve later performance in similar threatening situations.

The most important empirical task now is, as noted by several commentators, to conduct detailed quantitative content analysis studies that describe in detail the frequency and nature of threatening events in the dreams and nightmares of different populations, especially children, traumatized individuals, and hunter-gatherer populations. Our own recent study was carried out in this spirit (Revonsuo & Valli 2000): we developed a detailed rating scale to identify and describe threatening events in dreams. The first results on students' dreams are consistent with the predictions of TST. We will continue similar studies with dreams collected, for example, from traumatized children.

The most difficult aspect of TST to test empirically is the claim that threat simulations are causally responsible for better performance in comparable situations. We can test this claim indirectly by investigating the role of REM sleep in implicit learning. As mentioned in the target article, there is evidence that REM sleep deprivation impairs implicit learning. **Panksepp** furthermore points out that REM-deprived animals are deficient in the shuttle avoidance task, which supports TST. **Schredl** suggests that it may be possible to traumatize animals and see whether nightmares can be induced in them and whether such animals learn an avoidance task more easily. We will certainly need many imaginative ideas like this before the issue about threat simulations and improved threat perception and avoidance skills can be settled. We should be able to somehow measure the performance (e.g., reaction times and error rates) in threat perception and threat avoidance tasks that are more or less naturalistic (requiring, e.g., visual search for cues of potential threats, sustained attention, and rapid response selection). Consequently, performing such a task should induce similar threat simulation dreams. This may be very difficult to achieve and, even if achieved, such an experiment may be ethically problematic as it might induce persistent recurrent nightmares in the participants.

Perhaps some sort of virtual-reality game that presents threatening events and requires fast and correct responses would be technically possible. If it is realistic and immersive enough, perhaps it would elicit true emotions in the subject and result in true threat simulation dreams. If the level of performance improvement in such a game correlates with the frequency of threat simulations about the game in an individual's dreams, then TST would gain support. It remains to be seen whether any such threat simulation game will ever be a practically feasible option as an experimental method to test TST.

## RR9. TST and the philosophy of consciousness

**Gunderson** raises some philosophical issues that TST must face, since TST is not a philosophically innocent theory of dreaming but includes metaphysical assumptions and commitments concerning the nature of consciousness in general. I am grateful to Gunderson for bringing these questions to the forefront, for I believe that the philosophical implications of a theory on dreaming are at least as important as the empirical ones. Although this is not the place to develop a full-scale philosophical theory (some initial ingredients can be found in Revonsuo 1995; 2000), I will briefly clarify my position. Gunderson argues that TST suffers from what he calls "perspectival problems." He says that instructive simulations need a participating subject or perspective, but that it cannot be identified in the dream, at least not any such subject that would also exist in the real world. My view is that, as empirical evidence shows (e.g., Foulkes & Kerr 1984), the vast majority of dreams is experienced from the first-person point of view, as a personal embodied presence in the dream world. The experience we have of being present and embodied in the waking world is practically identical. Therefore, the phenomenal body image that provides us with the experienced presence in a world is, I believe, the same, but during dreaming it is modulated by a short causal chain that originates in the activation of the body representations in the brain, whereas during waking it is modulated by a longer causal chain that originates in the sensory organs and sensory nerve endings all over our actual flesh-and-bones body. Thus, it is the same phenomenal subject, me, both in the dream world and in the waking world. It is an uninteresting grammatical question who, in that case, is the person or self or subject who dreams. The facts are that sometimes I find myself present in the waking world, sometimes in the dream world. Perhaps it would be best to say that it is the brain that is either awake or dreaming, but in both cases I myself am simply present in a world, whether a dream world or a real one. As Rodolfo Llinas once put it in a talk: "Life is but a dream guided by the senses."

**Gunderson** is on to something when he says that I treat the whole of subjective experience or the phenomenal level of organization as a mechanism; a neurobiological mechanism that is also a subject with a perspective. I have expressed this metaphysical position elsewhere in more detail (e.g., Revonsuo 2000); the core idea is that consciousness is a real, natural biological phenomenon and should be reconceptualized as the phenomenal level of organization in the brain. I am painfully aware that this position needs to be worked out in much more detail than I have been able to do so far. However, I do not think that this position will lead to the difficulties that Gunderson points out at the end of his commentary. He seems to think first that if TST regards consciousness as a neurobiological mechanism, then we have ceased to discuss in terms of phenomenal consciousness altogether, and second, that we have erroneously objectified the subjective self in terms of a mechanism.

Let me put it this way: By treating consciousness as a biological phenomenon, my aim is not to objectify it as a standard neurobiological mechanism, but, on the contrary, to subjectify some neurobiological phenomena to such an extent that there is room for phenomenal consciousness amongst the complex and layered micro-macro levels of organization in the brain. Since the different levels of organization in the brain are closely related and dependent on each other, dreaming of course has complex nonconscious mechanisms hidden beneath the surface of the subjective phenomenal dream world. However, when threat simulation is rehearsed, the threatening event must be realized at the phenomenal level of organization, because threatening agents, the emotional intensity of threat they pose, the potential escape routes and the defensive movements can only be represented in a perspectival system: there is a self-in-world in whose presence also the threatening agent is. The actual perceptual and motor programs that will be modulated and primed through repeated threat simulations reside at lower levels of organization, but that does not render the phenomenal level irrelevant.

The philosophical point is precisely that we need to get rid of the categorical opposition between consciousness and neurobiological mechanisms and, instead, learn to treat subjective phenomenal consciousness as an integral level of organization in the brain, densely interconnected with and constituted by the lower nonconscious levels of organization. The brain requires perspectival, embodied representations of self-in-world when trying to guide the flesh-and-bones body in the world towards prosperous trajectories and away from disastrous ones. Dreams, threat simulations, and the rest of the contents of consciousness are just the phenomenal surface of what in actual fact are complex biological phenomena at multiple levels of organization in the brain, trying to perform this formidable task. The important point is that we should not treat consciousness as just another ordinary biological phenomenon, we should treat it instead as a very special one with several puzzling and amazing features, but a biological phenomenon all the same.

## RR10. Conclusion: Why do we dream?

The commentators have put forward important questions and objections to TST, which have been of enormous help in the more precise formulation of the theory. Although some of the questions have been challenging, it seems to me that TST has been able to solve them without serious difficulties. The commentators have also provided the theory with significant support and revealed several new sources of supportive evidence, which has made the theory even more appealing. The various competing views on the function of dreams, advocated by some of the commentators, do not seem to be able to match TST in explanatory power, the range of evidence covered, or the degree of theoretical integration of several different fields achieved. Therefore, I come to the conclusion that TST is a fruitful and scientifically respectable new theory of dreams that should be developed, rigorously tested, and its theoretical and clinical implications explored. TST not only provides a definite answer to the questions "Why do we have dreams?" and "What is the function of dreaming?", it also clarifies the meaning of these questions in such a way that answers based on empirical facts can be given. If TST by and large proves to be on the right track, as seems to be case, then we are for the first time in a position to truly understand why we dream.

ACKNOWLEDGMENT
A.R. is supported by the Academy of Finland (project 45704).

NOTE
1. The adaptive functions of human and animal play behavior remain unclear, although a number of hypotheses have been put forward: For example rehearsal of aggressive behavior for real fights or real predator avoidance, general physical exercise, testing dominance positions, reaffirming social relationships, reducing the fear to launch attacks, modification of the neuromuscular system during a sensitivity period in young mammals (for recent papers, see Bekoff & Byers [1998]). There is some evidence for all of these hypotheses, but no single explanation seems to apply across all species or types of play behavior. Playing and dreaming may have some complementary functions in threat avoidance skill rehersal, because dreaming is perceptually and emotionally more realistic than play, whereas the actual execution of motor programs and muscular movements may be the essence of play. There is one crucial difference between play and dreaming in age distribution: in all species, play occurs predominantly during a brief period in the young, after which it is "turned off" (Byers 1998), whereas dreaming and REM sleep occur throughout the whole lifetime of an individual.

# References

**Letters "a" and "r" appearing before authors' initials refer to target article and response, respectively.**

Abrahamson, E. E., Card, J. P. & Moore, R. Y. (1997) Afferent connections of the posterior hypothalamic area. *Society for Neuroscience Abstracts* 23:2133. [aJAH]

Achermann, P. & Borbély, A. A. (1997) Low frequency (<1 Hz) oscillations in the human sleep electroencephalogram. *Neuroscience* 81:213–22. [aJAH, MSt]

Achermann, P., Werth, E., Dijk, D. J. & Borbély, A. A. (1995) Time course of sleep inertia after nighttime and daytime sleep episodes. *Archives Italiennes de Biologie* 134:109–19. [aJAH]

Adams, I. & Oswald, I. (1989) Can a rapidly-eliminated hypnotic cause daytime anxiety? *Pharmacopsychiatry* 22:115–19. [LJW]

Adams, R. D., Victor, M. & Ropper, A. H. (1997) *Principles of neurology, sixth edition.* McGraw-Hill. [aJAH]

Addy, R. O., Dinner, D. S., Luders, H., Lesser, R. P., Morris, H. H. & Wyllie, E. (1989) The effects of sleep on median nerve short latency somatosensory evoked potentials. *Electroencephalography and Clinical Neurophysiology* 74:105–11. [aTAN]

Adey, W., Bors, E. & Porter, R. (1968) EEG sleep patterns after high cervical lesions in man. *Archives of Neurology* 24:377–83. [aMS]

Adler, A. (1927) *The practice and theory of individual psychology.* Harcourt. [aAR]

(1944) Disintegration and restoration of optic recognition in visual agnosia: Analysis of a case. *Archives of Neurology and Psychiatry* 51:243–59. [aMS]

(1950) Course and outcome of visual agnosia. *Journal of Nervous and Mental Diseases* 111:41–51. [aMS]

Adolphs, R. (1999) Social cognition and the human brain. *Trends in Cognitive Sciences* 3:460–79. [rJAH]

Adrien, J., Dugovic, C. & Martin, P. (1991) Sleep-wakefulness patterns in the helpless rat. *Physiology and Behavior* 49:257–62. [ARM]

Aggleton, J. P. & Brown, M. W. (1999) Episodic memory, amnesia and the hippocampal-anterior thalamic axis. *Behavioral and Brain Sciences* 22:425–89. [EEK]

Aghajanian, G. K. (1994) Serotonin and the action of LSD in the brain. *Psychiatric Annals* 24:137–41. [arJAH]

Aigner, L., Arber, S., Kapfhammer, J. P., Laux, T., Schneider, C., Botteri, F., Brenner, H.-R. & Caroni, P. (1995) Overexpression of the neural growth-associated protein GAP-43 induces nerve sprouting in the adult nervous system of transgenic mice. *Cell* 83:269–78. [TJM]

Ajilore, O. A., Stickgold, R., Rittenhouse, C. & Hobson, J. A. (1995) Nightcap: Laboratory and home-based evaluation of a portable sleep monitor. *Psychophysiology* 32:92–98. [aJAH]

Akindele, M. O., Evans, J. I. & Oswald, I. (1970) Mono-amine oxidase inhibitors, sleep and mood. *Electroencephalography and Clinical Neurophysiology* 29:47–56. [aRPV]

Albert, I., Cicala, G. A. & Siegel, J. (1970) The behavioral effects of REM sleep deprivation in rats. *Psychophysiology* 6:550–60. [aRPV]

Albert, M., Becker, T., McCrone, P. & Thornicroft, G. (1998) Social networks and mental health service utilisation – a literature review. *International Journal of Social Psychiatry* 44:248–66. [TAN]

Allain, H., Lieury, A., Brunet-Bourgin, F., Mirabaud, C., Trebon, P., Le Coz, F. & Gandon, J. M. (1992) Antidepressants and cognition: Comparative effects of moclobemide, viloxazine and maprotiline. *Psychopharmacology* 106:S56–61. [aRPV]

Allen, S. R., Oswald, I., Lewis, S. A. & Tagney, J. (1972) The effects of distorted visual input on sleep. *Psychophysiology* 9:498–504. [aRPV]

Alonso, A. & Llinas, R. R. (1992) Electrophysiology of the mammillary complex in vitro. II. Medial mammillary neurons. *Journal of Neurophysiology* 68(4):1321–31. [EEK]

Amadeo, M. & Gomez, E. (1966) Eye movements, attention and dreaming in subjects with lifelong blindness. *Canadian Psychiatric Association Journal* 11:501–507. [aJAH]

Amado-Boccara, I., Gougoulis, N., Poirier Littre, M. F., Galinowski, A. & Loo, H. (1995) Effects of antidepressants on cognitive functions: A review. *Neuroscience and Biobehavioral Reviews* 19:479–93. [aRPV]

Ambrosini, M. V., Mariucci, G., Bruschelli, G., Colarieti, L. & Giuditta, A. (1995) Sequential hypothesis of sleep function. V. Lengthening of post-trial SS episodes in reminiscent rats. *Physiology and Behavior* 58:1043–49. [MBl]

Ambrosini, M. V., Mariucci, G., Colarieti, L., Bruschelli, G., Carobi, C. & Giuditta, A. (1993) The structure of sleep is related to the learning ability of rats. *European Journal of Neuroscience* 5:269–75. [ARM]

*American Heritage Dictionary, third edition* (1992) eds. A. H. Soukhanov et al. Houghton Mifflin. [aJAH]

American Psychiatric Association (1994) *Diagnostic and statistical manual of mental disorders, fourth edition (DSM-IV).* American Psychiatric Association. [TAN]

American Sleep Disorders Association (1990) *The international classification of sleep disorders. Diagnostic and coding manual.* American Sleep Disorders Association. [aAR]

Amin, M. M., Khan, P. & Lehmann, H. E. (1980) The differential effects of viloxazine and imipramine on performance tests: Their relationship to behavioral toxicity. *Psychopharmacological Bulletin* 16:57–58. [aRPV]

Amzica, F. & Steriade, M. (1996) Progressive cortical synchronization of ponto-geniculo-occipital potentials during rapid eye movement in sleep. *Neuroscience* 2:309–14. [aJAH]

(1997) The K-complex: Its slow (<1 Hz) rhythmicity and relation with delta waves. *Neurology* 82:952–59. [MSt]

Anderson, J. A. (1972) A simple neural network generating an interactive memory. *Mathematical Biosciences* 14:197–220. [TJM]

(1995) *An introduction to neural networks.* A Bradford Book/MIT Press. [TJM]

Anderson, J. R. (1983) *The architecture of cognition.* Harvard University Press. [aJAH]

Andersson, J., Onoe, H., Hetta, J., Broman, J. E., Valind, S., Lilja, A., Sundin, A., Lindstrom, K., Watanabe, Y. & Langstrom, B. (1995) Regional changes in cerebral blood flow during sleep as measured by positron emission tomography. *Journal of Cerebral Blood Flow and Metabolism* 15:S871. [aJAH]

Andersson, J., Onoe, H., Hetta, J., Lindstrom, K., Valind, S., Lilja, A., Sundin, A., Fasth, K. J., Westerberg, C., Broman, J. E., Watanabe, Y. & Langstrom, B. (1998) Brain networks affected by synchronized sleep visualized by positron emission tomography. *Journal of Cerebral Blood Flow and Metabolism* 18:701–15. [AAB, aJAH, EFP-S]

Ansbacher, H. L. & Ansbacher, R. R. (1956) *The individual psychology of Alfred Adler: Selections from his writings.* Harper Torchbooks. [JBP]

Antrobus, J. S. (1983) REM and NREM sleep reports: Comparison of word frequencies by cognitive classes. *Psychophysiology* 20:562–68. [JSA, MBo, CCa, arJAH, MM, aTAN, EFP-S, PS, GWV]

(1986) Dreaming: Cortical activation and perceptual thresholds. *Journal of Mind and Behavior* 7:193–212. [JSA, aJAH]

(1987) Cortical hemisphere asymmetry and sleep mentation. *Physiological Review* 94:359–68. [aJAH]

(1990) The neurocognition of sleep mentation: Rapid eye movements, visual imagery and dreaming. In: *Sleep and cognition,* ed. R. R. Bootzin, J. F. Kihlstrom & D. L. Schacter. American Psychological Association. [EH, aJAH, aTAN]

(1991) Dreaming: Cognitive processes during cortical activation and high afferent thresholds. *Psychological Review* 98:96–121. [JSA, IF, EH, arJAH, aTAN, HSP, aMS]

(1993a) Dreaming: Could we do without it? In: *The functions of dreaming,* ed. A. Moffitt, M. Kramer & R. Hoffman. SUNY Press. [GWD, aAR]

(1993b) The dreaming mind/brain: Understanding its processes with connectionist models. In: *Dreaming as cognition,* ed. C. Cavallero & D. Foulkes. Harvester Wheatsheaf. [aJAH]

Antrobus, J. S. & Antrobus, J. S. (1967) Discrimination of two sleep stages by human subjects. *Psychophysiology* 4:48–55. [aJAH]

Antrobus, J. S., Antrobus, J. S. & Fisher, C. (1965) Discrimination of dreaming and nondreaming sleep. *Archives of General Psychiatry* 12:395–401. [aJAH]

Antrobus, J. S. & Bertini, M. (1992) Introduction. In: *The neuropsychology of sleep and dreaming,* ed. J. S. Antrobus & M. Bertini. Erlbaum. [aJAH]

Antrobus, J. S. & Conroy, D. (1999) Dissociated neurocognitive processes in dreaming sleep. *Sleep and Hypnosis* 2:105–11. [JSA]

Antrobus, J. S., Fein, G., Jordan, L., Ellman, S. J. & Arkin, A. M. (1974) Measurement and design in research on sleep reports. In: *The mind in sleep,* ed. A. M. Arkin, J. S. Antrobus & S. J. Ellman. Erlbaum. [JSA]

(1991) Measurement and design in research on sleep reports. In: *The mind in sleep: Psychology and psychophysiology,* ed. S. J. Ellman & J. S. Antrobus. Wiley. [JSA, aJAH, aTAN]

Antrobus, J. S., Hartwig, P., Rosa, D., Reinsel, R. & Fein, G. (1987) Brightness and clarity of REM and NREM imagery: Photo response scale. *Sleep Research* 16:240. [aJAH]

Antrobus, J. S., Kondo, T., Reinsel, R. & Fein, G. (1995) Dreaming in the late morning: Summation of REM and diurnal cortical activation. *Consciousness and Cognition* 4:275–99. [JSA, IF, arJAH, arTAN]

Arena, R., Murri, L., Piccini, R. & Muratorio, A. (1984) Dream recall and memory in brain lesioned patients. *Research Communications in Psychology, Psychiatry and Behavior* 9:31–42. [aMS]

Aristotle (235/1952) On dreams and their interpretation. In: *Great books of the Western world,* vol. 8, ed. R. Hutchings. Encyclopaedia Brittanica. [JFP]

Arkin, A. M. (1978) Night-terrors as anomalous REM sleep component manifestation in slow-wave sleep. *Waking and Sleeping* 2:143–47. [LJW]

Arkin, A. M. & Antrobus, J. S. (1978) The effects of external stimuli applied prior to and during sleep on sleep experience. In: *The mind in sleep: Psychology and psychophysiology,* ed. A. M. Arkin, J. S. Antrobus & S. J. Ellman. Erlbaum. [aJAH]

Arkin, A. M., Toth, M. F., Baker, J. & Hastey, J. M. (1970) The frequency of sleep talking in the laboratory among chronic sleep talkers and good dream recallers. *Journal of Nervous and Mental Disease* 151:369–74. [aTAN]

Armitage, R. (1980) Changes in dream content as a function of time of night, stage of awakening and frequency of recall. Masters thesis, Carleton University, Ottawa, Canada. [aTAN]

Armitage, R., Hoffmann, R. & Moffitt, A. (1992) Interhemispheric EEG activity in sleep and wakefulness: Individual differences in the basic rest-activity cycle (BRAC). In: *The neuropsychology of sleep and dreaming,* ed. J. S. Antrobus & M. Bertini. Erlbaum. [aTAN]

Armitage, R., Rochlen, A., Fitch, T., Trivedi, M. & Rush, J. (1995) Dream recall and major depression: A preliminary report. *Dreaming* 5:189–98. [aJAH]

Arnulf, I., Bejjani, B.-P., Garma, L., Bonnet-A.-M., Damier, P., Pidoux, B., Dormont, D., Cornu, P., Derenne, J.-P. & Agid, Y. (2000) Effect of low and high frequency thalamic stimulation on sleep in patients with Parkinson's disease and essential tremor. *Journal of Sleep Research* 9:55–62. [rJAH]

Asenbaum, S., Zeithofer, J., Saletu, B., Frey, R., Brucke, T., Podreka, I. & Deecke, L. (1995) Technetium-99m-HMPAO SPECT imaging of cerebral blood flow during REM sleep in narcoleptics. *Journal of Nuclear Medicine* 36:1150–55. [aJAH, EFP-S]

Aserinsky, E. & Kleitman, N. (1953) Regularly occurring periods of eye motility and concomitant phenomena during sleep. *Science* 118:273–74. [JSA, AC, RCo, aJAH, aTAN, JFP, PS, aMS]

(1955) Two types of ocular motility during sleep. *Journal of Applied Physiology* 8:1–10. [aJAH, aMS]

Aserinsky, E., Lynch, J. A., Mack, M. E., Tzankoff, S. P. & Hurn, E. (1985) Comparison of motion in wakefulness and REM sleep. *Psychophysiology* 22:1–10. [aJAH]

Ashby, F. G., Isen, A. M. & Turken, A. U. (1999) A neuropsychological theory of positive affect and its influence on cognition. *Psychological Review* 106:529–50. [JBP]

Aston-Jones, G. & Bloom, F. E. (1981) Activity of norepinephrine-containing locus coeruleus neurons in behaving rats anticipates fluctuations in the sleep-waking cycle. *Journal of Neuroscience* 1:876–86. [CG, aJAH, LJW]

Aubrey, J. B., Armstrong, B., Arkin, A., Smith, C. T. & Rose, G. (1999) Total sleep deprivation affects memory for a previously learned route. *Sleep* 22:S246. [RS]

Auerbach, S. B., Minznberg, M. J. & Wilkinson, L. O. (1989) Extracellular serotonin and 5 Hydroxyindolacetic acid in hypothalamus of the unanaesthetized rat measured by in vivo dialysis coupled to high performance liquid chromatography with electrochemical detection: Dialysate serotonin reflects neuronal release. *Brain Research* 499:281–90. [aJAH]

Awh, E. & Gehring, W. J. (1999) The anterior cingulate cortex lends a hand in response selection. *Nature Neuroscience* 2(10):853–54. [CMP]

Baars, B. J. & Banks, W. F. (1994) Special issue: Dream consciousness: A neurocognitive approach. *Consciousness and Cognition* 3(1):1–128. [aTAN]

Baddely, A. (1998) Recent developments in working memory. *Current Opinion in Neurobiology* 8:234–38. [aJAH]

Baeseler, H. A., Morland, A. B. & Wandell, B. A. (1999) Topographic organization of human visual areas in the absence of input from primary cortex. *Journal of Neuroscience* 19:2619–27. [aJAH]

Baghdoyan, H. A. (1977) Location and quantification of muscarinic receptor subtypes in rat pons: Implications for REM sleep generation. *American Journal of Physiology* 273:R896-R904. [RLy]

Baghdoyan, H. A., Fleegal, M. A. & Lydic, R. (1997) Acetylcholine (ACh) release in the medial pontine reticular formation is regulated by M2 muscarinic autoreceptors. *Society for Neuroscience Abstracts* 23:2131. [aJAH]

(1998) M2 muscarinic autoreceptors regulate acetylcholine release in the pontine reticular formation. *Journal of Pharmacology and Experimental Therapy* 286:1446–52. [RLy]

Baghdoyan, H. A. & Lydic, R. (1999) M2 muscarinic receptor subtype in the feline medial pontine reticular formation modulates the amount of rapid eye movement sleep. *Sleep* 22:835–47. [RLy, LJW]

Baghdoyan, H. A., Lydic, R., Callaway, C. W. & Hobson, J. A. (1989) The carbachol-induced enhancement of desynchronized sleep signs is dose dependent and antagonized by centrally administered atropine. *Neuropsychopharmacology* 2:67–79. [aJAH, RLy]

Baghdoyan, H. A., Mallios, J. V., Duckrow, R. B. & Mash, D. C. (1994) Localization of muscarinic receptor subtypes in brain stem areas regulating sleep. *NeuroReport* 5:1631–34. [RLy]

Baghdoyan, H. A., Monaco, A., Rodrigo-Angulo, M., Assens, F., McCarley, R. W. & Hobson, J. A. (1984a) Microinjection of neostigmine into the pontine reticular formation of cats enhances desynchronized sleep signs. *Journal of Pharmacology and Experimental Therapy* 231:173–80. [RLy]

Baghdoyan, H. A., Rodrigo-Angulo, M. L., McCarley, R. W. & Hobson, J. A.

(1984b) Site-specific enhancement and suppression of desynchronized sleep signs following cholinergic stimulation of three brain stem regions. *Brain Research* 306:39–52. [RLy]

(1987) A neuroanatomical gradient in the pontine tegmentum for the cholinoceptive induction of desynchronized sleep signs. *Brain Research* 414:245–61. [aJAH]

Baghdoyan, H. A., Spotts, J. L. & Snyder, S. G. (1993) Simultaneous pontine and basal forebrain microinjections of carbachol suppress REM sleep. *Journal of Neuroscience* 13:229–42. [aJAH]

Bakeland, F. (1971) Effects of pre-sleep procedures and cognitive style on dream content. *Perceptual and Motor Skills* 32:63–69. [aJAH]

Bakeland, F., Resch, R. & Katz, D. D. (1968) Pre-sleep mentation and dream reports. *Archives of General Psychiatry* 19:300–11. [aJAH]

Baker, D. G., West, S. A., Nicholson, W. E., Ekhator, N. N., Kasckow, J. W., Hill, K. K., Bruce, A. B., Orth, D. N. & Geracioti, T. D. (1999) Serial CSF corticotropin-releasing hormone levels and adrenocortical activity in combat veterans with posttraumatic stress disorder. *American Journal of Psychiatry* 156:585–88. [TAN]

Baker, T. L. & McGinty, D. J. (1977) Reversal of cardiopulmonary failure during active sleep in hypoxic kittens: Implications for the sudden infant death syndrome. *Science* 199:419–21. [rRPV]

Baldessarini, R. J. (1991) Drugs and the treatment of psychiatric disorders. In: *The pharmacological basis of therapeutics,* ed. A. Goodman Gilman, T. W. Rall, A. S. Nies & P. Taylor. Pergamon Press. [LJW]

Baldridge, B. J., Whitman, R. M., Kramer, M. A., Ornstein, P. H. & Lansky, L. (1965) The effect of external physical stimuli on dream content. *Psychophysiology* 4:372–73. [aJAH]

Balkin, T. J., Braun, A. R., Wesensten, N. J., Varga, M., Baldwin, P., Carson, R. E., Belenky, G. & Herskovitch, P. (1999) Bidirectional changes in regional cerebral blood flow across the first 20 minutes of wakefulness. *Sleep Research Online* 2 (Supplement 1):6. [arJAH, EFP-S]

Ball, W. A., Hunt, W. H., Sanford, L. D., Ross, R. J. & Morrison, A. R. (1991a) Effects of stimulus intensity on elicited ponto-geniculo-occipital waves. *Electroencephalography and Clinical Neurophysiology* 78:35–39. [JH]

Ball, W. A., Sanford, L. D., Morrison, A. R., Ross, R. J., Hunt, W. H. & Mann, G. L. (1991b) The effects of changing state on elicited ponto-geniculo-occipital (PGO) waves. *Electroencephalography and Clinical Neurophysiology* 79:420–29. [JH, aTAN]

Barbas, H. (1995) Anatomic basis of cognitive-emotional interactions in the primate prefrontal cortex. *Neuroscience and Biobehavioral Reviews* 19:499–510. [aJAH]

Bargh, J. A. & Chartrand, T. L. (1999) The unbearable automaticity of being. *American Psychologist* 54:462–79. [MBl]

Barinaga, M. (1999) New clues to how neurons strengthen their connections. *Science* 284:1755–57. [TJM]

Barrett, D. (1992) Just how lucid are lucid dreams? *Dreaming* 2:221–28. [aJAH]

(1993) The "committee of sleep": A study of dream incubation for problem solving. *Dreaming* 3(2):115–22. [aAR]

Barrett, D., ed. (1996) *Trauma and dreams.* Harvard University Press. [MBl, aAR, MSc]

Barrett, T. R. & Ekstrand, B. R. (1972) Effect of sleep on memory: III. Controlling for time-of-day effects. *Journal of Experimental Psychology* 96:321–27. [JB, rRPV]

Bartha, R., Williamson, P. C., Drost, D. J., Malla, A. K. & Neufeld, R. W. J. (1999) Medial prefrontal glutamine and dreaming. *British Journal of Psychiatry* 175:288–89. [aJAH]

Basso, A., Bisiach, E. & Luzzatti, C. (1980) Loss of mental imagery: A case study. *Neuropsychologia* 18:435–42. [aMS]

Bastuji, H., Garcia-Larrea, L., Franc, C. & Mauguière, F. (1995) Brain processing of stimulus deviance during slow-wave and paradoxical sleep: A study of human auditory evoked responses using the oddball paradigm. *Journal of Clinical Neurophysiology* 12:155–67. [aTAN]

Battaglia, D., Cavallero, C. & Cicogna, P. (1987) Temporal reference of the mnemonic sources of dreams. *Perceptual and Motor Skills* 64:979–83. [aTAN]

Becker, R. E. & Dufresne, R. L. (1982) Perceptual changes with Bupropion, a novel antidepressant. *American Journal of Psychiatry* 139:1200–201. [rJAH]

Bednar, J. A. & Miikkulainen, R. (1998) Pattern-generator-driven development in self-organizing models. In: *Computational neuroscience: Trends in research,* ed. J. M. Bower. Plenum Press. http://www.cs.utexas.edu/users/nn/pages/publications/abstracts.html#bednar.cns97.ps.Z [JAB]

Beh, H. C. & Barratt, P. E. H. (1965) Discrimination and conditioning during sleep as indicated by the electroencephalogram. *Science* 147:1470–71. [aTAN]

Bekoff, M. & Byers, J. A. (1998) *Animal play.* Cambridge University Press. [rAR]

Benbadis, S. R., Wolgamuth, B. R., Perry, M. C. & Dinner, D. S. (1995) Dreams and rapid eye movement sleep in the multiple sleep latency test. *Sleep* 18:105–108. [aTAN]

Benca, R. M. (1996) Sleep in psychiatric disorders. *Neurologic Clinics* 14:739–64. [TAN]

Benca, R. M., Obermeyer, W. H., Larson, C. L., Yun, B., Dolski, I., Kleist, K. D., Weber, S. M. & Davidson, R. J. (1999) EEG alpha power and alpha power asymmetry in sleep and wakefulness. *Psychophysiology* 36:430–36. [aJAH]

Benca, R. M., Overstreet, D. E., Gilliland, M. A., Russell, D., Bergman, B. M. & Obermeyer, W. H. (1996) Increased basal REM sleep but no difference in dark induction or light suppression of REM sleep in Flinders rats with cholinergic supersensitivity. *Neuropsychopharmacology* 15:45–51. [aJAH]

Beninato, M. & Spencer, R. (1986) A cholinergic projection to the rat superior colliculus demonstrated by retrogade transport of horseradish peroxidase and choline acetyltransferase immunohistochemistry. *Journal of Comparative Neurology* 253:525–38. [aJAH]

Benington, J. H. & Heller, H. C. (1994) Does the function of REM sleep concern non-REM sleep or waking? *Progress in Neurobiology* 44:433–49. [IF, aRPV]

(1995) Restoration of brain energy metabolism as the function of sleep. *Progress in Neurobiology* 45:347–60. [aRPV]

Benoit, O. (1981) Le rhythme veille-sommeil chez l'enfant. I. Physiologie. [Waking-sleep cycle in children. I. Physiology]. *Archives Francaises de Pediatrie* 38:619–26. [rTAN]

Benowitz, L. I. & Routtenberg, A. (1997) GAP-43: An intrinsic determinant of neuronal development and plasticity. *Trends in Neurosciences* 20:84–91. [TJM]

Benson, D. F. & Greenberg, J. (1969) Visual form agnosia: A specific defect in visual discrimination. *Archives of Neurology* 20:82–89. [uMS]

Bentivoglio, M. & Grassi-Zucconi, G. (1999) Immediate early gene expression in sleep and wakefulness. In: *Handbook of behavioral state control: Molecular and cellular mechanisms,* ed. R. Lydic & H. A. Baghdoyan. CRC Press. [aJAH]

Berger, B., Gaspar, P. & Verney, C. (1991) Dopaminergic innervation of the cerebral cortex: Unexpected differences between rodents and primates. *Trends in Neuroscience* 14:21–27. [CG, EKP, rMS]

Berger, M. (1994) Dreaming and REM sleep – commentary. *World Federation of Sleep Research Societies Newsletter* 3:13. [aTAN]

Berger, M. & Riemann, D. (1993) REM sleep in depression – an overview. *Journal of Sleep Research* 2:211–23. [aJAH]

Berger, M., Riemann, D., Hochli, D. & Spiegel, R. (1989) The cholinergic REM-sleep-induction test with RS86: State or trait marker of depression? *Archives of General Psychiatry* 46:421–28. [aJAH]

Berger, R. J. (1967) When is a dream is a dream is a dream? *Experimental Neurology* (Supplement) 4:15–27. [GWD]

(1969) The sleep and dream cycle. In: *Sleep: Physiology and pathology,* ed. A. Kales. Lippincott. [GWD]

Berger, R. J. & Oswald, I. (1962) Effects of sleep deprivation on behavior, subsequent sleep, and dreaming. *Journal of Mental Science* 108:457–65. [rTAN]

Berger, R. J. & Phillips, N. H. (1995) Energy conservation and sleep. *Behavioural Brain Research* 69:65–73. [aRPV]

Berger-Sweeney, J., Heckers, J., Mesulam, M. M., Wiley, R. G., Lappi, D. A. & Sharma, M. (1994) Differential effects on spatial navigation of immunotoxin-induced cholinergic lesions of the medial septal area and nucleus basalis magnocellularis. *Journal of Neuroscience* 14:4507–19. [aRPV]

Berlad, I., Shlitner, A., Ben-haim, S. & Lavie, P. (1993) Power spectrum analysis and heart rate variability in stage 4 and REM sleep: Evidence for state-specific changes in autonomic dominance. *Journal of Sleep Research* 2:88–90. [rTAN]

Bernard, J. F., Alden, M. & Besson, J. M. (1993) The organization of the efferent projections from the pontine parabrachial area to the amygdaloid complex: A phaseolus vulgaris leucoagglutinin (PHA-L) study in rat. *Journal of Comparative Neurology* 329:201–28. [aJAH]

Berridge, C. W. & Foote, S. L. (1996) Enhancement of behavioral and electroencephalographic indices of waking following stimulation of nonadrenergic betareceptors within the medial septal region of the basal forebrain. *Journal of Neuroscience* 16:6999–7009. [aJAH]

Berridge, K. (in press) Pleasure, pain, desire, and dread: Biopsychological pieces and relations. In: *Understanding quality of life: Scientific understanding of enjoyment and suffering,* ed. D. Kahnemann, E. Diener & N. Schwartz. Russell Sage Foundation. [aMS]

Berry, D. C. (1994) Implicit learning: Twenty-five years on. In: *Attention and performance XV,* ed. C. Umiltá & M. Moscovitch. MIT Press. [aAR]

Bertini, M., Violani, C., Zoccolotti, P., Antonelli, A. & Di Stefano, L. (1982) Performance on unilateral tactile test during waking and upon awakening from REM and NREM. In: *Sleep,* ed. W. P. Koella. Karger. [aTAN]

(1984) Right cerebral activation in REM sleep: Evidence from unilateral tactile recognition test. *Psychophysiology* 21:418–23. [aTAN]

Beversdorf, D. Q., Hughes, J. D., Steinberg, B. A., Lewis, L. D. & Heilman, K. M. (1999) Noradrenergic modulation of cognitive flexibility in problem solving. *Journal of Cognitive Neuroscience* 11:2763–67. [rJAH]

Bier, M. J. & McCarley, R. W. (1994) REM-enhancing effects of the adrenergic antagonist idazoxan infused into the medial pontine reticular formation of the freely moving cat. *Brain Research* 634:333–38. [aJAH]

Bilu, Y. (1989) The other as a nightmare: The Israeli-Arab encounter as reflected in children's dreams in Israel and the West Bank. *Political Psychology* 10(3):365–89. [aAR]

Binswanger, L. (1963) *Being in the world.* Basic Books. [JBP]

Bion, W. R. (1962) *Learning from experience.* Heinemann. [HTH]

Bird, E. (1990) Schizophrenia. In: *An introduction to neurotransmission in health and disease,* ed. P. Riederer, N. Kopp & J. Pearson. Oxford University Press. [aMS]

Bishop, C., Rosenthal, L., Helmus, T., Roehrs, T. & Roth, T. (1996) The frequency of multiple sleep onset REM periods among subjects with no excessive daytime sleepiness. *Sleep* 19:727–30. [aTAN]

Bixler, E. O., Kales, A., Soldatos, C. R., Kales, J. D. & Healey, S. (1979) Prevalence of sleep disorders in the Los Angeles metropolitan area. *American Journal of Psychiatry* 136:1257–62. [MSc]

Bizzi, E. (1966a) Changes in orthodromic and antidromic response of optic tract during the eye movements of sleep. *Journal of Neurophysiology* 29:861–70. [aJAH]

(1966b) Discharge patterns of single geniculate neurons during the rapid eye movements of sleep. *Journal of Neurophysiology* 29:1087–95. [aJAH]

Bizzi, E. & Brooks, D. C. (1963) Functional connections between pontine reticular formation and lateral geniculate nucleus during deep sleep. *Archives Italiennes de Biologie* 101:666–80. [aJAH]

Blagrove, M. (1992a) Dreams as the reflection of our waking concerns and abilities: A critique of the problem-solving paradigm in dream research. *Dreaming* 2:205–20. [MBl, LM, aAR]

(1992b) Scripts and the structuralist analysis of dreams. *Dreaming* 2:23–37. [aJAH]

(1996) Problems with the cognitive psychological modeling of dreaming. *The Journal of Mind and Behavior* 17:99–134. [MBl, RL, aAR]

Blagrove, M. & Price, C. (2000) Happiness in dreams: Associations with waking happiness, skills, and challenges. *Sleep* 23 (Suppl. 2):A100-A101. [MBl, rAR]

Bloch, V., Hennevin, E. & Leconte, P. (1979) Relationship between paradoxical sleep and memory processes. In: *Brain mechanisms in memory and learning: From the single neuron to man,* ed. M. A. Brazier. Raven Press. [aRPV]

Boch, R. A. & Goldberg, M. E. (1989) Participation of prefrontal neurons in the preparation of visually guided eye movements in the rhesus monkey. *Journal of Neurophysiology* 62:1064–84. [aJAH]

Boeve, B. F., Silber, M. H., Ferman, T. J., Kokmen, E., Smith, G. E., Ivnik, R. J., Parisi, J. E., Olson, E. J. & Petersen, R. C. (1998) REM sleep behavior disorder and degenerative dementia: An association likely to reflect Lewy body disease. *Neurology* 51:363–70. [EKP, aAR]

Bogen, J. (1969) The other side of the brain, II: An appositional mind. *Bulletin of the Los Angeles Neurological Society* 34:135–62. [aMS]

Bokert, E. (1968) The effects of thirst and a related verbal stimulus on dream reports. *Dissertation Abstracts* 28:4753B. [aJAH]

Boller, F., Wright, D., Cavelieri, R. & Mitsumoto, H. (1975) Paroxysmal "nightmares": Sequel of a stroke responsive to diphenylhydantoin. *Neurology* 25:1026–28. [aMS]

Bonato, R. A., Moffitt, A. R., Hoffmann, R. F., Cuddy, M. A. & Wimmer, L. (1991) Bizarreness in dreams and nightmares. *Dreaming* 1:53–61. [aJAH]

Bonime, W. & Bonime, F. (1962) *The clinical use of dreams.* Basic Books. (Reprinted: Da Capo Press, 1982). [RG]

Bonnet, M. H. (1983) Memory for events occurring during arousal from sleep. *Psychophysiology* 20:81–87. [aJAH]

Bootzin, R. R., Hubbard, T. L., Reiman, E. M., Bandy, D., Yun, L. S. & Munzlinger, T. (1998) Brain regions preferentially affected during different stages of sleep and wakefulness. *Sleep* 21 (Supplement):272. [aJAH, EFP-S]

Borbély, A. (1982) A two-process model of sleep regulation. *Human Neurobiology* 1:195–204. [rJAH]

Born, J., Hansen, K., Marshall, L., Mlle, M. & Fehn, H. L. (1999) Timing the end of nocturnal sleep. *Nature* 397:29–30. [JB]

Bosinelli, M. (1991) Recent research trends in sleep onset mentation. In: *The mind in sleep, 2nd edition,* ed. S. J. Ellman & J. S. Antrobus. Wiley-Interscience. [aJAH, MO]

(1995) Mind and consciousness during sleep. *Behavioral Brain Research* 69:195–201. [aJAH, MM]

Bosinelli, M., Cicogna, P. & Molinari, S. (1974) The tonic-phasic model and the feeling of self-participation in different stages of sleep. *Giornale Italiano di Psicologia* 1:35–65. [rTAN]

Botez, M., Olivier, M., Vézina, J.-L., Botez, T. & Kaufman, B. (1985) Defective revisualization: Dissociation between cognitive and imagistic thought. Case report and short review of the literature. *Cortex* 21:375–89. [aMS]

Bourgin, P., Escourrou, P., Gaultier, C. & Adrien, J. (1995) Induction of rapid eye movement sleep by carbachol infusion into the pontine reticular formation of the rat. *NeuroReport* 6:532–36. [aJAH]

Bourgin, P., Lebrand, C., Escourrou, P., Gaultier, C., Franc, B., Hamon, M. &

Adrien, J. (1997) Vasoactive intestinal polypeptide microinjections into the oral pontine tegmentum enhance rapid eye movement sleep in the rat. *Neuroscience* 77:351–60. [aJAH]

Bowe-Anders, C., Herman, J. & Roffwarg, H. (1974) Effects of goggle-altered color perception on sleep. *Perceptual and Motor Skills* 38:191–98. [aRPV]

Bowker, R. M. & Morrison, A. R. (1976) The startle reflex and PGO spikes. *Brain Research* 102:185–90. [RCo, JH, aTAN]

Bowlby, J. (1969) *Attachment and loss. Vol. I. Attachment.* Hogarth Press. [TAN]

Boyle, J. & Nielsen, J. (1954) Visual agnosia and loss of recall. *Bulletin of the Los Angeles Neurological Society* 19:39–42. [aMS]

Bradley, K., Dax, E. & Walshi, K. (1958) Modified leucotomy: Report of 100 cases. *Medical Journal of Australia* 1:133–38. [aMS]

Bradley, L., Hollifield, M. & Foulkes, D. (1992) Reflection during REM dreaming. *Dreaming* 2:161–66. [aJAH]

Bradley, P. B. & Elkes, J. (1957) The effect of some drugs on the electrical activity of the brain. *Brain* 80:77–117. [DL]

Brain, R. (1950) The cerebral basis of consciousness. *Brain* 73:465–79. [aMS]
    (1954) Loss of visualization. *Proceedings of the Royal Society of Medicine* 47:288–90. [aMS]

Bramham, C. R. & Srebo, B. (1989) Synaptic plasticity in the hippocampus is modulated by behavioral state. *Brain Research* 493:74–86. [aRPV]

Branconnier, R. J., DeVitt, D. R., Cole, J. O. & Spera, K. F. (1982) Amitriptyline selectively disrupts verbal recall from secondary memory of the normal aged. *Neurobiology of Aging* 3:55–59. [aRPV]

Braun, A. R. (1999) Commentary on Hobson's "The new neuropsychology of sleep: Implications for psychoanalysis." *Neuropsychoanalysis* 1:196–201. [rJAH, EFP-S, rMS]

Braun, A. R., Balkin, T. J., Wesenten, N. J., Carson, R. E., Varga, M., Baldwin, P., Selbie, S., Belenky, G. & Herscovitch, P. (1997) Regional cerebral blood flow throughout the sleep-wake cycle – an (H$_2$O)-O-15 PET study. *Brain* 120:1173–97. [JSA, AAB, RCo, FD, WF, arJAH, AK, aTAN, EFP-S, arMS, arRPV]

Braun, A. R., Balkin, T. J., Wesensten, N. J., Gwadry, F., Carson, R. E., Varga, M., Baldwin, P., Belenky, G. & Herscovitch, P. (1998) Dissociated pattern of activity in visual cortices and their projections during human rapid eye movement sleep. *Science* 279:91–95. [JSA, RCo, WF, CG, arJAH, AK, rTAN, EFP-S, aAR, aMS, aRPV]

Bray, D. (1995) Protein molecules as computational elements in living cells. *Nature* 376:307–12. [TJM]

Brebbia, D. R., Branchey, M. H., Pyne, E., Watson, J., Brebbia, A. F. & Simpson, G. (1975) The effects of amoxapine on electroencephalographic stages of sleep in normal human subjects. *Psychopharmacologia* 45:1–7. [aRPV]

Breger, L. (1967) Function of dreams. *Journal of Abnormal Psychology Monograph* 72(5):1–28. [aAR]

Breger, L., Hunter, I. & Lane, R. W. (1971) The effects of stress on dreams. *Psychological Issues* 7 (3), Monograph 27. [RG, aJAH]

Breggin, P. (1980) Brain-disabling therapies. In: *The psychosurgery debate*, ed. E. Valenstein. Freeman. [aMS]

Bremner, J. D., Krystal, J. H., Southwick, S. M. & Charney, D. S. (1996) Noradrenergic mechanisms in stress and anxiety: II. Clinical studies. *Synapse* 23:39–51. [aJAH]

Brenneis Brooks, C. (1994) Can early childhood trauma be reconstructed from dreams? On the relations of dreams to trauma. *Psychoanalytic Psychology* 11(4):429–47. [aAR]

Breslau, N., Chilcoat, H. D., Kessler, R. C. & Davis, G. C. (1999) Previous exposure to trauma and PTSD effects of subsequent trauma: Results from the Detroit area survey of trauma. *American Journal of Psychiatry* 156:902–907. [TAN]

Brewer, J. B., Zhao, Z., Desmond, J. E., Glover, G. H. & Gabrieli, J. D. E. (1998) Making memories: Brain activity that predicts how well visual experience will be remembered. *Science* 281:1185–87. [aJAH]

Brill, N. & Beebe, G. (1955) *A follow up study of war neuroses.* VA Medical Monograph. US Government Printing Office. [rAR]

Brock, J. W., Hamdi, A., Ross, K., Payne, S. & Prasad, C. (1995) REM sleep deprivation alters dopamine D2 receptor binding in the rat frontal cortex. *Pharmacology, Biochemistry and Behavior* 52:43–48. [FD, aJAH]

Brock, T. D. (1999) *Robert Koch: A life in medicine and bacteriology.* University of Wisconsin Press. [RLy]

Brooks, B. A. (1999) Saccade. In: *Encyclopedia of neuroscience, vol. II*, ed. G. Adelman & B. H. Smith. Elsevier. [aJAH]

Brooks, D. C. & Gershon, M. D. (1972) An analysis of the effect of reserpine upon ponto-geniculo-occipital wave activity in the cat. *Neuropharmacology* 11:499–510. [aTAN]

Broughton, R. J. (1968) Sleep disorders: Disorders of arousal? Enuresis, somnambulism, and nightmares occur in confusional states of arousal, not in "dreaming sleep." *Science* 159:1070–78. [aJAH, aTAN]
    (1982) Neurology and dreaming. *The Psychiatric Journal of the University of Ottawa* 7:101–10. [aJAH]
    (1995) Parasomnias. In: *Sleep disorders medicine. Basic science, technical*

considerations and clinical aspects, ed. S. Chokroverty. Butterworth-Heinemann. [RCo]

Broughton, R. J., Billings, R., Cartwright, R., Doucette, D., Edmeads, J., Edwardh, M., Ervin, F., Orchard, B., Hill, R. & Turrell, G. (1994) Homicidal somnambulism: A case report. *Sleep* 17:253–54. [RC]

Brower, K. J., Maddahian, E., Blow, F. C. & Beresford, T. P. (1992) A comparison of self-reported symptoms and DSM-III-R criteria for cocaine withdrawal. *American Journal of Drug and Alcohol Abuse* 14:347–56. [rJAH]

Browman, C. P. & Cartwright, R. D. (1980) The first-night effect on sleep and dreams. *Biological Psychiatry* 15:809–12. [aTAN]

Brown, J. (1972) *Aphasia, apraxia, agnosia: Clinical and theoretical aspects.* Thomas. [aMS]
    (1985) Frontal lobe syndrome. In: *Handbook of clinical neurology*, ed. P. Vinken, G. Bruyn & H. Klawans. Elsevier. [aMS]
    (1989) Essay on perception. In: *Neuropsychology of visual perception*, ed. J. Brown. Erlbaum. [aMS]

Brown, J. & Cartwright, R. (1978) Locating NREM dreaming through instrumental response. *Psychophysiology* 15:35–39. [RC]

Brown, R. (1965) *Social psychology.* Free Press. [JBP]

Brown, R. & Kulik, J. (1977) Flashbulb memories. *Cognition* 5:73–99. [RCo]

Brown, R. J. & Donderi, D. C. (1986) Dream content and self-reported well-being among recurrent dreamers, past recurrent dreamers, and nonrecurrent dreamers. *Journal of Personality and Social Psychology* 50:612–23. [AZ]

Brudno, S. & Marczynski, T. J. (1977) Temporal patterns, their distribution and redundancy in trains of spontaneous neuronal spike intervals of the feline hippocampus studied with a non-parametric technique. *Brain Research* 125:65–89. [TJM]

Bruner, J. (1986) *Actual minds, possible worlds.* Harvard University Press. [JBP]

Bryant, R. A. & Harvey, A. G. (1996) Visual imagery in posttraumatic stress disorder. *Journal of Traumatic Stress* 9:613–19. [TAN]

Brylowski, A., Levitan, L. & LaBerge, S. (1989) H-reflex suppression and autonomic activation during lucid REM sleep: A case study. *Sleep* 12:374–78. [TLK]

Buchsbaum, M. S., Gillin, J. C., Wu, J., Hazlett, E., Sicotte, N., Dupont, R. M. & Bunney, W. E. (1989) Regional cerebral glucose metabolic rate in human sleep assessed by positron emission tomography. *Life Science* 45:1349–56. [aJAH]

Bueno, O. F. A., Lobo, L. L., Oliviera, M. G. M., Gugliano, E. B., Pomarico, A. C. & Tufik, S. (1994) Dissociated paradoxical sleep deprivation effects on inhibitory avoidance and conditioned fear. *Physiology and Behavior* 56:775–79. [aRPV]

Buffenstein, A., Heaster, J. & Ko, P. (1999) Chronic psychotic illness from amphetamine. *American Journal of Psychiatry* 156:662. [CG]

Buhusi, C. V. & Schmajuk, N. A. (1996) Attention, configuration, and hippocampal function. *Hippocampus* 6:621–42. [RCo]

Bungener, C., Lehouezec, J., Pierson, A. & Jouvent, R. (1996) Cognitive and emotional deficits in early stages of HIV infection – an event-related potentials study [review]. *Progress in NeuroPsychopharmacology and Biological Psychiatry* 20:1303–14. [aTAN]

Burlet, S., Leger, L. & Cespuglio, R. (1999) Nitric oxide and sleep in the rat: A puzzling relationship. *Neuroscience* 92:627–39. [aJAH]

Bush, G., Luu, P. & Posner, M. I. (2000) Cognitive and emotional influences in anterior cingulate cortex. *Trends in Cognitive Sciences* 4:215–22. [rJAH]

Busink, R. & Kuiken, D. (1996) Identifying types of impactful dreams – a replication. *Dreaming* 6:97–119. [arTAN]

Buss, D. M. (1994) *The evolution of desire.* Basic Books. [TKS]
    (2000) *The dangerous passion.* Free Press. [TKS]

Butcher, L. L. (1995) Cholinergic neurons and networks. In: *The rat nervous system*, ed. G. Paxinos. Academic Press. [arJAH]

Butkov, N. (1996) *Atlas of clinical polysomnography.* Synapse Media. [arJAH]

Butler, S. F. & Watson, R. (1985) Individual differences in memory for dreams: The role of cognitive skills. *Perceptual and Motor Skills* 61:823–28. [aJAH]

Buzsáki, G. (1996) The hippocampo-neocortical dialogue. *Cerebral Cortex* 6:81–92. [rJAH, EFP-S]
    (1998) Memory consolidation during sleep: A neurophysiological perspective. *Journal of Sleep Research* 7 (Supplement 1):17–23. [RWG, JP, MSc, rRPV]
    (1989) Two-stage model of memory trace formation: A role for "noisy" brain states. *Neuroscience* 31:551–70. [MSt, RS, rRPV]
    (1996) The hippocampo-neocortical dialogue. *Cerebral Cortex* 6:81–92. [RS]

Byers, J. A. (1998) Biological effects of locomotor play: Getting into shape, or something more specific? In: *Animal play*, ed. M. Bekoff & J. A. Byers. Cambridge University Press. [rAR]

Cabeza, R. & Nyberg, L. (1997) Imaging cognition: An empirical review of PET studies with normal subjects. *Journal of Cognitive Neuroscience* 9:1–26. [aJAH, EFP-S]
    (2000) Imaging cognition II: An empirical review of 275 PET studies and fMRI studies. *Journal of Cognitive Neuroscience* 12:1–47. [arJAH, EPS]

Cacioppo, J. T., Gardner, W. L. & Berntson, G. G. (1999) The affect system has parallel and integrative processing components: Form follows function. *Journal of Personality and Social Psychology* 76:839–55. [JAC]

Cacioppo, J. T. & Tassinary, L. G. (1990) Inferring psychological significance from physiological signals. *American Psychologist* 45:16–28. [aTAN]

Cahill, L. & McGaugh, J. L. (1996) Modulation of memory storage. *Current Opinion in Neurobiology* 6:237–42. [JB]

(1998) Mechanisms of emotional arousal and lasting declarative memory. *Trends in Neuroscience* 21:294–99. [aJAH]

Cahill, L., Prins, B., Weber, M. & McGaugh, J. L. (1994) Beta-adrenergic activation and memory for emotional events. *Nature* 371:702–704. [JB]

Caine, E. D., Weingartner, H., Ludlow, C. L., Cudahy, E. A. & Wehry, S. (1981) Quantitative analysis of scopolamine-induced amnesia. *Psychopharmacology* 74:74–80. [aRPV]

Cairns, H. R. (1952) Disturbances of consciousness with lesions of the brain stem and diencephalon. *Brain* 75:109–46. [aRPV]

Cajochen, C., Foy, R. & Dijk, D. J. (1999a) Frontal predominance of a relative increase in sleep delta and theta EEG activity after sleep loss in humans. *Sleep Research Online* 2:65–69. http://www.sro.org/pdf/2536.pdf [AAB]

Cajochen, C., Khalsa, S. B., Wyatt, J. K., Czeisler, C. A. & Dijk, D. J. (1999b) EEG and ocular correlates of circadian melatonin phase and human performance decrements during sleep loss. *American Journal of Physiology* 277:R640–49. http://ajpregu.physiology.org/cgi/reprint/277/3/R460.pdf [AAB]

Callaway, C. W., Lydic, R., Baghdoyan, H. A. & Hobson, J. A. (1987) Ponto-geniculo-occipital waves: Spontaneous visual system activation occurring in REM sleep. *Cellular and Molecular Neurobiology* 7:105–49. [aJAH]

Calvo, J. M., Badillo, S., Morales-Ramirez, M. & Palacios-Salas, P. (1987) The role of the temporal lobe amygdala in ponto-geniculo-occipital activity and sleep organization in cats. *Brain Research* 403:22–30. [aJAH]

Calvo, J. M., Datta, S., Quattrochi, J. J. & Hobson, J. A. (1992) Cholinergic microstimulation of the peribrachial nucleus in the cat. Delayed and prolonged increases in REM sleep. *Archives Italiennes de Biologie* 130:285–301. [aJAH]

Calvo, J. M. & Fernandez-Guardiola, A. (1984) Phasic activity of the basolateral amygdala, cingulate gyrus, and hippocampus during REM sleep in the cat. *Sleep* 7:202–10. [arJAH]

Calvo, J. M. & Simon-Arceo, K. (1995) Long-lasting enhancement of REM sleep induced with carbachol micro-injection into the central amygdaloid nucleus of the cat. *Sleep Research* 24A:17. [aJAH]

(1999) Cholinergic enhancement of REM sleep from sites in the pons and amygdala. In: *Handbook of behavioral state control: Molecular and cellular mechanisms*, ed. R. Lydic & H. A. Baghdoyan. CRC Press. [arJAH]

Calvo, J. M., Simon-Arceo, K. & Fernandez-Mas, R. (1996) Prolonged enhancement of REM sleep produced by carbachol microinjection into the amygdala. *NeuroReport* 7:577–80. [aJAH]

Campbell, A. (1999) Staying alive: Evolution, culture, and women's intrasexual aggression. *Behavioral and Brain Sciences* 22:203–52. [arAR]

Campbell, K. B. & Bartoli, E. A. (1986) Human auditory evoked potentials during natural sleep: The early components. *Electroencephalography and Clinical Neurophysiology* 65:142–49. [aTAN]

Cann, D. R. & Donderi, D. (1986) Jungian personality typology and the recall of everyday and archetypal dreams. *Journal of Personality and Social Psychology* 50:1021–30. [aTAN]

Cantero, J. L., Atienza, M., Salas, R. M. & Gomez, C. (1999a) Alpha power modulation during periods with rapid oculomotor activity in human REM sleep. *NeuroReport* 10:1817–20. [RCo]

(1999b) Brain spatial microstates of human spontaneous alpha activity in relaxed wakefulness, drowsiness period, and REM sleep. *Brain Topography* 11:257–63. [DL]

Cape, E. G. & Jones, B. E. (1998) Differential modulation of high frequency gamma electroencephalogram activity and sleep-wake state by noradrenaline and serotonin microinjections into the region of cholinergic basalis neurons. *The Journal of Neuroscience* 18:2653–66. [aJAH]

Capece, M. L., Baghdoyan, H. A & Lydic, R. (1998) Carbachol-stimulated [35S] guanosine 5-(gamma-thio)-triphosphate binding in rapid eye movement sleep related brainstem nuclei of rat. *The Journal of Neuroscience* 18:3779–85. [aJAH, RLy]

(1999) New directions for the study of cholinergic REM sleep generation: Specifying pre- and post-synaptic mechanisms. In: *Rapid eye movement sleep*, ed. B. N. Mallick & S. Inoue. Marcel Dekker. [aJAH]

Capece, M. L., Efange, S. M. N. & Lydic, R. (1997) Vesicular acetylcholine transport inhibitor supresses REM sleep. *NeuroReport* 8:481–84. [aJAH, RLy]

Capece, M. L. & Lydic, R. (1997) Cyclic AMP and protein kinase A modulate cholinergic rapid eye movement (REM) sleep generation. *American Journal of Physiology* 273:R1430–40. [RLy]

Carlson, V. R, Feinberg, I. & Goodenough, D. R. (1978) Perception of the duration of sleep intervals as a function of EEG sleep stage. *Physiological Psychology* 6:397–500. [IF]

Carpenter, K. A. (1987) The effects of positive and negative pre-sleep stimuli on dream experiences. *The Journal of Psychology* 122:33–37. [aJAH]

Carskadon, M. A. & Dement, W. C. (1980) Distribution of REM sleep on a 90 minute sleep-wake schedule. *Sleep* 2:309–17. [RDO]

Carskadon, M. A., Keenan, S. & Dement, W. C. (1987) Nighttime sleep and daytime sleep tendency in preadolescents. In: *Sleep and its disorders in children*, ed. C. Guilleminault. Raven Press. [rTAN]

Carter, C. S., Braver, T. S., Barch, D. M., Botvinik, M. M., Noll, D. & Cohen, J. D. (1998) Anterior cingulate cortex, error detection, and the online monitoring of performance. *Science* 280(5364):747–49. [CMP]

Cartwright, R. D. (1972) Sleep fantasy in normal and schizophrenic persons. *Journal of Abnormal Psychology* 80:275–79. [RG]

(1974a) Problem solving: Waking and dreaming. *Journal of Abnormal Psychology* 83(4):451–55. [aAR]

(1974b) The influence of a conscious wish on dreams: A methodological study of dream meaning and function. *Journal of Abnormal Psychology* 83:387–93. [aJAH, RLe]

(1986) Affect and dream work from an information processing point of view. *Journal of Mind and Behavior* 7:411–28. [RLe]

(1991) Dreams that work: The relation of dream incorporation to adaptation to stressful events. *Dreaming* 1:3–9. [MBl]

(1996) Dreams and adaptation to divorce. In: *Trauma and dreams*, ed. D. Barrett. Harvard University Press. [RG, aAR]

(2000) Sleep-related violence: Does the polysomnogram help establish the diagnosis? *Sleep Medicine* 1:331–35. [RC]

Cartwright, R. D., Bernick, N., Borowitz, G. & Kling, A. (1969) Effect of an erotic movie on the sleep and dreams of young men. *Archives of General Psychiatry* 20:263–71. [aJAH]

Cartwright, R. D. & Lamberg, L. (1992) *Crisis dreaming: Using your dreams to solve your problems.* Harper Collins. [WHM]

Cartwright, R. D., Luten, A., Young, M., Mercer, P. & Bears, M. (1998a) Role of REM sleep and dream affect in overnight mood regulation: A study of normal volunteers. *Psychiatry Research* 81:1–8. [RC, arJAH]

Cartwright, R. D., Monroe, L. J. & Palmer, C. (1967) Individual differences in response to REM deprivation. *Archives of General Psychiatry* 16:297–303. [RG, aTAN]

Cartwright, R. D. & Romanek, I. (1978) Repetitive dreams of normal subjects. *Sleep Research* 7:174. [AZ]

Cartwright, R. D., Young, M., Mercer, P. & Bears, M. (1998b) Role of REM sleep and dream variables in the prediction of remission from depression. *Psychiatry Research* 80:249–55. [RC]

Casagrande, M., Violani, C. & Bertini, M. (1996a) A psycholinguistic method for analyzing two modalities of thought in dream reports. *Dreaming* 6:43–55. [aTAN]

Casagrande, M., Violani, C., Lucidi, F., Buttinelli, E. & Bertini, M. (1996b) Variations in sleep mentation as a function of time of night. *International Journal of Neuroscience* 85:19–30. [aJAH, aTAN]

Casagrande, M., Violani, C., Vereni, F., Lucidi, F. & Bertini, M. (1990) Differences between SO, Stage 2 and REM reports assessed by a psycholinguistic scale. *Sleep Research* 19:133. [aJAH]

Castaldo, V. & Holzman, P. (1969) The effects of hearing one's own voice on dreaming content: A replication. *Journal of Nervous and Mental Disease* 148:74–82. [aTAN]

Castaldo, V., Krynicki, V. & Goldstein, J. (1974) Sleep stages and verbal memory. *Perceptual and Motor Skills* 39:1023–30. [aRPV]

Castaldo, V. & Shevrin, H. (1970) Different effects of auditory stimulus as a function of rapid eye movement and non-rapid eye movement sleep. *Journal of Nervous and Mental Disease* 150:195–200. [aTAN, HS]

Cathala, H., Laffont, F., Siksou, M., Esnault, S., Gilbert, A., Minz, M., Moret-Chalmin, C., Buzaré, M. & Waisbord, P. (1983) Sommeil et rêve chez des patients atteints de lésions pariétales et frontales. *Revue Neurologique* 139:497–508. [aMS]

Cavallero, C. (1987) Dream sources, associative mechanisms, and temporal dimension. *Sleep* 10:78–83. [CCa]

(1993) The quest for dream sources. *Journal of Sleep Research* 2:13–16. [aTAN]

Cavallero, C. & Cicogna, P. (1993) Memory and dreaming. In: *Dreaming as cognition*, ed. C. Cavallero & D. Foulkes. Harvester Wheatsheaf. [aJAH, TLK, aTAN, aAR]

Cavallero, C., Cicogna, P. & Bosinelli, M. (1988) Mnemonic activation in dream production. In: *Sleep '86*, ed. W. P. Koella, F. Obal, H. Schultz & P. Visser. Gustav Fischer Verlag. [aJAH, aTAN]

Cavallero, C., Cicogna, P., Natale, V., Occhionero, M. & Zito, A. (1992) Slow wave sleep dreaming: Dream research. *Sleep* 15(2):562–66. [MBo, CCa, CF, aJAH, arTAN, MO, aMS]

Cavallero, C., Foulkes, D., Hollifield, M. & Terry, R. (1990) Memory sources of REM and NREM dreams. *Sleep* 13:449–55. [CCa, aJAH, aTAN]

Cespuglio, R., Faradji, H., Gomez, M. E. & Jouvet, M. (1981) Single unit recording in the nuclei raphe dorsalis and magnus during the sleep-waking cycle of semi-chronic prepared cats. *Neuroscience Letters* 24:133–38. [aJAH]

Cespuglio, R., Gomez, M. E., Faradji, H. & Jouvet, M. (1982) Alterations of the sleep wake cycle induced by cooling of the locus coeruleus area. *Electroencephalography and Clinical Neurophysiology* 54:570–78. [aJAH]

Chalmers, D. J. (1995a) *The conscious mind.* Oxford University Press. [OF]
(1995b) The puzzle of conscious experience. *Scientific American* 273:80–86. [aJAH]

Chapman, P., Ismail, R., Avellano, T. & Underwood, G. (1999a) Time-gaps while driving. In: *Behavioural research in road safety IX*, ed. G. Grayson. Transport Research Laboratory. [PC]
(1999b) Waking up at the wheel: Accidents, attention and the time-gap experience. In: *Vision in vehicles VII*, ed. A. G. Gale, I. D. Brown, C. M. Haslegrave & S. P. Taylor. Elsevier Science B. V. [PC]

Chapman, P. & Underwood, G. (2000) Forgetting near accidents: The roles of severity, culpability and experience in the poor recall of dangerous driving situations. *Applied Cognitive Psychology* 14:31–44. [PC]

Charcot, J. M. (1883) Un cas de suppression brusque et isolée de la vision mentale des signes et des objets, (formes et couleurs). *Progrès Médical* 11:568–71. [aMS]

Charney, D. S. & Bremner, J. D. (1999) The neurobiology of anxiety disorders. In: *Neurobiology of mental illness*, ed. D. S. Charney, E. J. Nestler & B. S. Bunney. Oxford University Press. [LJW]

Chase, M. H. & Morales, F. R. (1990) The atonia and myoclonia of active (REM) sleep. *Annual Review of Psychology* 41:557–884. [aJAH]

Chase, M. H., Soja, P. J. & Morales, F. R. (1989) Evidence that glycine mediates the post synaptic potentials that inhibit lumbar motorneurons during the atonia of active sleep. *Journal of Neuroscience* 9:743–51. [aJAH]

Chase, T., Moretti, L. & Prensky, A. (1968) Clinical and electroencephalographic manifestations of vascular lesions of the pons. *Neurology* 18:357–68. [aMS, aRPV]

Chelazzi, L. & Corbetta, M. (2000) Cortical mechanisms of visuospatial attention in the primate brain. In: *The new cognitive neurosciences*, ed. M. S. Gazzaniga. MIT Press. [RCo]

Chemelli, R. M., Willie, J. T., Sinton, C. M., Elmquist, J. L., Scammell, T., Lee, C., Richardson, J. A., Williams, S. C., Xiong, Y., Kisamuki, Y., Fitch, T. E., Nakazato, M., Hammer, R. E., Saper, C. B. & Yanagisawa, M. (1999) Narcolepsy in orexin knockout mice: Molecular genetics of sleep regulation. *Cell* 98:437–51. [aJAH, RS-P]

Cheney, D. L. & Costa, E. (1978) Biochemical pharmacology of cholinergic neurons. In: *Psychopharmacology: A generation of progress*, ed. M. A. Lipton, A. DiMascio & K. F. Killam. Raven Press. [LJW]

Chernik, D. A. (1972) Effect of REM sleep deprivation on learning and recall by humans. *Perceptual and Motor Skills* 34:82–94. [aRPV]

Cheyne, J. A., Newby-Clark, I. R. & Rueffer, S. D. (1999a) Sleep paralysis and associated hypnagogic and hypnopompic experiences. *Journal of Sleep Research* 8:313–18. [JAC]

Cheyne, J. A., Rueffer, S. D. & Newby-Clark, I. R. (1999b) Hypnagogic and hypnopompic hallucinations during sleep paralysis: Neurological and cultural construction of the nightmare. *Consciousness and Cognition* 8:319–37. [JAC]

Chivers, L. & Blagrove, M. (1999) Nightmare frequency, personality and acute psychopathology. *Personality and Individual Differences* 27:843–51. [MBl]

Christos, G. (1993) Is Alzheimer's disease related to a deficit or malfunction of rapid eye movement (REM) sleep? *Medical Hypotheses* 41(5):435–39. [rJAH]

Chu, N. S. & Bloom, F. E. (1973) Norepinephrine containing neurons: Changes in spontaneous discharge patterns during sleeping and waking. *Science* 179:908–10. [aJAH]
(1974) Activity patterns of catecholamine containing pontine neurons in the dorsolateral tegmentum of unrestrained cats. *Journal of Neurobiology* 5:527–44. [aJAH]

Chung, G., Tucker, D. M., West, P., Potts, G. F., Liotti, M., Luu, P. & Hartry, A. L. (1996) Emotional expectancy – brain electrical activity associated with an emotional bias in interpreting life events. *Psychophysiology* 33:218–33. [aTAN]

Church, M. W., March, J. D., Hibi, S., Benson, K., Cavness, C. & Feinberg, I. (1975) Changes in frequency and amplitude of delta activity during sleep. *Electroencephalography and Clinical Neurophysiology* 39:1–7. [IF]

Churchland, P. S. (1986) *Neurophilosophy: Toward a unified science of the mind-brain.* MIT Press. [rJAH]

Churchland, P. S. & Sejnowski, T. J. (1992) *The computational brain.* A Bradford Book/MIT Press. [TIM]

Cicogna, P. (1994) Dreaming during sleep onset and awakening. *Perceptual and Motor Skills* 78:1041–42. [rTAN]

Cicogna, P., Cavallero, C. & Bosinelli, M. (1986) Differential access to memory traces in the production of mental experience. *International Journal of Psychophysiology* 4:209–16. [CCa, aJAH, aTAN]

(1991) Cognitive aspects of mental activity during sleep. *American Journal of Psychology* 104:413–25. [CCa, aJAH, aTAN]

Cicogna, P., Cavallero, C., Bosinelli, M., Battaglia, D. & Natale, V. (1987) A comparison between single- and multi-unit dream reports. *Sleep Research* 16:228. [MBo]

Cicogna, P., Natale, V., Occhionero, M. & Bosinelli, M. (1998) A comparison of mental activity during sleep onset and morning awakening. *Sleep* 21(5):462–70. [MBo, aJAH, MO]
(2000) Slow wave and REM sleep mentation. *Sleep Research Online* 3:67–72. [rTAN]

Cipolli, C. (1995) Sleep, dreams and memory: An overview. *Journal of Sleep Research* 4:2–9. [CCi, PS]

Cipolli, C., Baroncini, P., Fagioli, I., Fumai, A. & Salzaruo, P. (1987) The thematic continuity of mental sleep experience in the same night. *Sleep* 10:473–79. [aJAH]

Cipolli, C., Bolzani, R., Massetani, R., Murri, L. & Muratorio, A. (1992) Dream structure in Parkinson's disease. *Journal of Nervous and Mental Disease* 180:516–23. [arJAH]

Cipolli, C., Bolzani, R. & Tuozzi, G. (1998) Story-like organization of dream experience in different periods of REM sleep. *Journal of Sleep Research* 7:13–19. [aJAH, PS]

Cipolli, C., Fagioli, I., Baroncini, P., Fumai, A., Marchi'o, B. & Sancini, M. (1988) The thematic continuity of mental experiences in REM and NREM sleep. *International Journal of Psychophysiology* 6:307–13. [CCa, aTAN]

Cipolli, C., Fagioli, I., Baroncini, P., Fumai, A., Marchi, B., Sancini, M., Tuozzi, G. & Salzarulo, P. (1992) Recall of mental sleep experience with or without prior verbalization. *American Journal of Psychology* 105:385–407. [CCi]

Cipolli, C. & Poli, D. (1992) Story structure in verbal reports of mental sleep experience after awakening in REM sleep. *Sleep* 15:133–42. [aJAH]

Cipolli, C., Salzarulo, P. & Calabrese, A. (1981) Memory processes involved in morning recall of REM-sleep mental experience: A psycholinguistic study. *Perceptual and Motor Skills* 52:391–406. [PS]

Cirelli, C., Fung, S. J., Liu, R.-H., Pompeiano, O. & Barnes, C. D. (1998) Cholinergic neurons of the dorsal pontine tegmentum projecting to the cerebellar vermal cortex of the kitten. *Archives Italiennes de Biologie* 136:257–71. [aJAH]

Clancey, W. J. (1997) *Situated cognition: On human knowledge and computer representation.* Cambridge University Press. [WJC]
(1999) *Conceptual coordination: How the mind orders experience in time.* Erlbaum. [WJC]

Clark, C. R., Geffen, G. M. & Geffen, L. B. (1989) Catecholamines and the covert orientation of attention in humans. *Neuropsychologia* 27(2):131–39. [rJAH]

Clarke, L. P. (1915) The nature and pathogenesis of epilepsy. *New York Medical Journal* 101:522, 567–73, 623–28. [aMS]

Cleeremans, A., Destrebecqz, A. & Boyer, M. (1998) Implicit learning: News from the front. *Trends in Cognitive Sciences* 2:406–16. [aAR]

Coenen, A. M. L. (1998) Is dreaming a characteristic of REM sleep? In: *Sleep-wake research in The Netherlands, vol. 9*, ed. D. G. M. Beersma, A. L. van Bemmel, H. Folgering, W. F. Hofman & G. S. F. Ruigt. Dutch Society for Sleep-Wake Research. [AC]

Coenen, A. M. L., van Betteray, J. N. F., van Luijtelaar, E. L. J. M. & van Hulzen, Z. J. M. (1986) Should the neural excitability hypothesis of paradoxical sleep be revised? *Abstracts of the Eighth European Congress of Sleep Research, Szeged,* 67. [AC]

Coenen, A. M. L. & van Luijtelaar, E. L. J. M. (1985) Stress induced by three procedures of deprivation of paradoxical sleep. *Physiology and Behavior* 35:501–504. [AC, aRPV]

Cohen, D. B. (1974) Toward a theory of dream recall. *Psychological Bulletin* 81:138–54. [LM]
(1977a) Changes in REM dream content during the night: Implications for a hypothesis about changes in cerebral dominance across REM periods. *Perceptual and Motor Skills* 44:1267–77. [aTAN]
(1977b) Sources of bias in our characterization of dreams. *Perceptual and Motor Skills* 45:98. [AC]
(1980) The cognitive activity of sleep. *Progress in Brain Research* 53:307–24. [WHM]

Cohen, D. B. & MacNeilage, P. F. (1974) A test of the salience hypothesis of dream recall. *Journal of Consulting and Clinical Psychology* 42:699–703. [RCo]

Cohen, H. B. & Dement, W. C. (1965) Sleep: Changes in threshold to electroconvulsive shock in rats after deprivation of "paradoxical" phase. *Science* 150:1318–19. [AC, rRPV]

Cohen, J. D., Botvinick, M. & Carter, C. S. (2000) Anterior cingulate and prefrontal cortex: Who's in control? *Nature Neuroscience* 3:421–23. [rJAH]

Cohen, J. D., Perstein, W. M., Brauer, T. S., Nystrom, L. E., Noll, D. C., Jomides, J. & Smith, E. E. (1997) Temporal dynamics of brain activation during a working memory task. *Nature* 386:604–608. [aJAH]

Cohen, R. M., Picker, D., Garnett, D., Lippel, J., Gillin, J. & Murphy, D. L.

(1982) REM sleep suppression induced by selective monoamine oxidase inhibitors. *Psychopharmacology* 78:137–40. [aRPV]

Cohrs, S., Tergau, F., Reich, S., Kastner, S., Paulus, W., Ziemann, U., Ruther, E. & Goran, H. (1998) High-frequency repetitive transcranial magnetic stimulation delays rapid eye movement sleep. *NeuroReport* 9:3439–43. [aJAH]

Colace, C., Doricci, F., Di Lorento, E. & Violani, C. (1993) Developmental qualitative and quantitative aspects of bizarreness in dream reports of children. *Sleep Research* 22:57. [aJAH]

Colace, C. & Natale, V. (1997) Bizarreness in REM and SWS dreams. *Sleep Research* 26:240. [aJAH]

Colace, C. & Tuci, B. (1996) Early children's dreams are not bizarre. *Sleep Research* 25:147. [aJAH]

Colace, C., Tuci, B. & Ferendeles, R. (1997) Bizarreness in early children's dreams collected in the home setting: Preliminary data. *Sleep Research* 26:241. [aJAH]

Colace, C., Violani, C. & Tuci, B. (1995) Self representation in dreams reported from young children at school. *Sleep Research* 24:69. [aJAH]

Coleman, R. M., Roffwarg, H. P., Kennedy, S. J., Guilleminault, C., Cinque, J., Cohn, M. A., Karacan, I., Kupfer, D. J., Lemmi, H., Miles, L., Orr, W. C., Phillips, E. R., Roth, T., Sassin, J. F., Schmidt, H. S., Weitzman, E. D. & Dement, W. C. (1982) Sleep-wake disorders based on a polysomnographic diagnosis. A national cooperative study. *Journal of the American Medical Association* 247:997–1003. [rTAN]

Comella, C. L., Nardine, T. M., Diedrich, N. J. & Stebbins, G. T. (1998) Sleep related violence, injury, and REM sleep behavior disorder in Parkinson's disease. *Neurology* 51:526–29. [aAR]

Comella, C. L., Tanner, C. M. & Ristanovic, R. K. (1993) Polysomnographic sleep measures in Parkinson's disease patients with treatment-induced hallucinations. *Annual Reviews in Neurology* 34:710–14. [rJAH, EKP]

Conduit, R. (1999) A phenomenological investigation of the relationship between PGO waves and dream reporting in humans: Toward an attention-based model of dreaming. Unpublished Ph. D. dissertation, La Trobe University, Melbourne, Australia. [RCo]

Conduit, R., Bruck, D. & Coleman, G. (1997) Induction of visual imagery during NREM sleep. *Sleep* 20:948–56. [RCo, aJAH, arTAN]

Conduit, R. & Coleman, G. (1998) Conditioned salivation and associated dreams from REM sleep. *Dreaming* 8:243–62. [RCo, aTAN]

Conel, J. L. (1939/1963) *The post-natal development of the human cerebral cortex.* Harvard University Press. [IF]

Connor, K. M. & Davidson, J. R. (1997) Familial risk factors in posttraumatic stress disorder. *Annals of the New York Academy of Sciences* 821:35–51. [TAN]

Consolo, S., Bertorelli, R., Forloni, G. L. & Butcher, L. L. (1990) Cholinergic neurons of the pontomesencephalic tegmentum release acetylcholine in the basal nuclei complex of freely moving rats. *Neuroscience* 37:717–23. [aJAH]

Contreras, D. & Steriade, M. (1995) Cellular basis of EEG slow rhythms: A study of dynamic corticothalamic relationships. *Journal of Neuroscience* 15:604–22. [MSt]

Cooper, A. J. (1987) Treatment of coexistent night-terrors and somnambulism in adults with imipramine and diazepam. *Journal of Clinical Psychiatry* 48:209–10. [LJW]

Cooper, J. R., Bloom, F. E. & Roth, R. H. (1996) *The biochemical basis of neuropharmacology, seventh edition.* Oxford University Press. [aJAH]

Corbetta, M., Miezin, F. M., Shulman, G. L. & Peterson, S. E. (1993) A PET study of visuospatial attention. *Journal of Neuroscience* 13:1202–26. [aJAH]

Cosmides, L. & Tooby, J. (1992) Cognitive adaptations for social exchange. In: *The adapted mind: Evolutionary psychology and the generation of culture,* ed. J. H. Barkow, L. Cosmides & J. Tooby. Oxford University Press. [LM]

(1995) From function to structure: The role of evolutionary biology and computational theories in cognitive neuroscience. In: *The cognitive neurosciences,* ed. M. S. Gazzaniga. MIT Press. [aAR]

Côté, K. A. & Campbell, K. B. (1998) Event-related potential evidence for consciousness during REM sleep. *Sleep* 21(Supplement):273. [aTAN]

(1999) P300 to high intensity stimuli during REM sleep. *Clinical Neurophysiology* 110:1345–50. [RCo]

Cottler, L. B., Shillington, A. M., Compton, W. M., Mager, D. & Spitznagel, E. (1993) Subjective reports of withdrawal among cocaine users: Recommendations for DSM-IV. *Drug and Alcohol Dependence* 33:97–104. [rJAH]

Courtney, S. M., Ungerleider, L. G., Keil, K. & Haxby, I. V. (1997) Transient and sustained activity in a distributed neural system for human working memory. *Nature* 386:608–10. [aJAH]

Creuzfeldt, O. (1995) *Cortex cerebri: Performance, structure and functional organization of the cerebral cortex.* Oxford University Press. [rMS]

Crick, F. (1994) *The astonishing hypothesis.* Scribners. [aJAH]

Crick, F. & Mitchison, G. (1983) The function of dream sleep. *Nature* 304:111–14. [WJC, rJAH, aAR, RS]

(1995) REM sleep and neural nets. *Behavioural Brain Research* 69:147–55. [OF, aAR]

Cronbach, L. J. & Mechl, P. E. (1955) Construct validity in psychological tests. *Psychological Bulletin* 52:281–302. [aTAN]

Cullinan, W. E., Herman, J. P. & Watson, S. J. (1993) Ventral subicular interaction with the hypothalamic paraventricular nucleus: Evidence for a relay in the bed nucleus of the stria terminalis. *Journal of Comparative Neurology* 332:1–20. [EEK]

Cummings, J. L. (1993) Frontal-subcortical circuits and human behavior. *Archives of Neurology* 50:873–80. [aJAH, EFP-S]

Cummings, J. L. & Greenberg, R. (1977) Sleep patterns in the "locked in" syndrome. *Electroencephalography and Clinical Neurophysiology* 43:270–71. [aMS, aRPV]

Curran, H. V. & Lader, M. (1986) The psychopharmacological effects of repeated doses of fluvoxamine, mianserin and placebo in healthy human subjects. *European Journal of Clinical Pharmacology* 29:601–607. [aRPV]

Curran, H. V., Sakulsriprong, M. & Lader, M. (1988) Antidepressants and human memory: An investigation of four drugs with different sedative and anticholinergic profiles. *Psychopharmacology* 95:520–27. [aRPV]

Curro-Dossi, R., Pare, D. & Steriade, M. (1991) Short-lasting nicotine and long-lasting muscarinic depolarizing response of thalamocortical neurons to stimulation of mesopontine cholinergic nuclei. *Journal of Neurophysiology* 65:393–406. [aJAH]

Dahlstrom, A. & Fuxe, K. (1964) Evidence for the existence of monoamine-containing neurons in the central nervous system. I. Demonstration in the cell bodies of brain stem neurons. *Acta Physiologica Scandinavica* 62:1–55. [aJAH, RS-P]

Dallaire, A., Toutain, P. L. & Ruckebusch, Y. (1974) The periodicity of REM-sleep: Experimental and theoretical considerations. *Physiology and Behavior* 13:395–400. [aJAH]

Daly, M. & Wilson, M. (1988) Evolutionary social psychology and family homicide. *Science* 242:519–24. [aAR]

(1996) Violence against stepchildren. *Current Directions in Psychological Science* 5:77–81. [TKS]

Damasio, A. R. (1996) The somatic marker hypothesis and the possible functions of the prefrontal cortex. *Philosophical Transactions of the Royal Society of London* 351:1413–20. [rJAH]

(1999) *The feeling of what happens: Body and emotion in the making of consciousness.* Harcourt Brace. [JAC, JP]

Danguir, J. & Nicolaidis, S. (1976) Impairments of learned aversion acquisition following paradoxical sleep deprivation in the rat. *Physiology and Behavior* 17:489–92. [aRPV]

Darling, M., Hoffman, R., Moffitt, A. & Purcell, S. (1993) The pattern of self-reflectiveness in dream reports. *Dreaming* 3:9–19. [rJAH]

Darwin, C. (1873/1965) *The expression of the emotions in man and animals.* Appleton. [aJAH] University of Chicago Press, 1965. [OF]

Date, Y., Ueta, Y., Yamashita, H., Yamaguchi, H., Matsukura, S., Kangawa, K., Sakurai, T., Yanagisawa, M. & Nakazato, M. (1999) Orexins, orexigenic hypothalamic peptides, interact with autonomic, neuroendocrine and neuroregulatory systems. *Proceedings of the National Academy of Sciences USA* 96:748–53. [RS-P]

Datta, S. (1995) Neuronal activity in the peribrachial area: Relationship to behavioral state control. *Neuroscience and Biobehavioral Reviews* 19:67–84. [arJAH, aRPV]

(1997a) Brainstem cholinergic cells in wakefulness and sleep. *Sleep Research* 26:10. [aJAH]

(1997b) Cellular basis of pontine ponto-geniculo-occipital wave generation and modulation. *Cellular and Molecular Neurobiology* 17:341–65. [arJAH]

(1999) PGO wave generation: Mechanism and functional significance. In: *Rapid eye movement sleep,* ed. B. N. Mallick & S. Inoue. Marcel Dekker. [aJAH]

Datta, S., Calvo, J., Quattrochi, J. & Hobson, J. A. (1992) Cholinergic microstimulation of the peribrachial nucleus in the cat: I. Immediate and prolonged increases in ponto-geniculo-occipital waves. *Archives Italiennes de Biologie* 130:264–84. [aJAH]

Datta, S., Curro-Dossi, R., Pare, D., Oakson, G. & Steriade, M. (1991) Substantia nigra reticulata neurons during sleep-wake states: Relation with ponto-geniculo-occipital waves. *Brain Research* 566:344–47. [aJAH]

Datta, S. & Hobson, J. A. (1994) Neuronal activity in the caudo-lateral peribrachial pons: Relationship to PGO waves and rapid eye movement. *Journal of Neurophysiology* 71:1–15. [aJAH]

Datta, S., Patterson, E. H. & Siwek, D. F. (1997) Endogenous and exogenous nitric oxide in the pedunculopontine tegmentum induces sleep. *Synapse* 27:69–78. [aJAH]

Datta, S., Quattrochi, J. & Hobson, J. A. (1993) Effect of specific muscarinic M2 receptor antagonist on carbachol induced long-term REM sleep. *Sleep* 16:1. [aJAH]

Datta, S. & Siwek, D. F. (1997) Excitation of the brain stem pedunculopontine tegmentum cholinergic cells induces wakefulness and REM sleep. *Journal of Neurophysiology* 77:2975–88. [IF, aJAH]

Datta, S., Siwek, D. F., Patterson, E. H. & Cipolloni, P. B. (1998) Localization of pontine PGO wave generation sites and their anatomical projections in the rat. *Synapse* 30:409–23. [RCo, CG, aJAH]

Dave, R. (1978) Effects of hypnotically induced dreams on creative problem solving. *Journal of Abnormal Psychology* 88:293–302. [RLe]

Davey, G. C. L. (1995) Preparedness and phobias: Specific evolved associations or a generalized expectancy bias? *Behavioral and Brain Sciences* 18:289–325. [LM]

Deboer, T., Sanford, L. D., Ross, R. J. & Morrison, A. R. (1997) Electrical stimulation in the amygdala increases the amplitude of elicited PGO waves. *Society for Neuroscience Abstracts* 23:1847. [aJAH]

(1998) Effects of electrical stimulation in the amygdala on ponto-geniculo-occipital waves in rats. *Brain Research* 798:305–10. [aJAH]

Decety, J. (1996) Do imagined and executed actions share the same neural substrate? *Cognitive Brain Research* 3:87–93. [aAR]

Declerck, A. C. & Wauquier, A. (1990) Influence of antiepileptic drugs on sleep patterns. In: *Epilepsy, sleep, and deprivation, 2nd edition,* ed. R. Degen & E. A. Rodin. Elsevier. [LJW]

DeGennaro, L., Casagrande, M., Violani, C., DiGiovanni, M., Herman, J. & Bertini, M. (1995) The complementary relationship between waking and REM sleep in the oculomotor system: An increase of rightward saccades during waking causes a decrease of rightward eye movements during REM sleep. *Electroencephalography and Clinical Neurophysiology* 95:252–56. [aJAH]

DeKloet, E. R., Oitzl, M. S. & Joels, M. (1999) Stress and cognition: Are corticosteroids good or bad guys? *Trends in Neurosciences* 22:422–26. [JB]

De Koninck, J., Christ, G., Hebert, G. & Rinfret, N. (1990a) Language learning, dreams and REM sleep: Converging findings. *Sleep Research* 19:134. [WHM]

(1990b) Language learning efficiency, dreams and REM sleep. *Psychiatric Journal of the University of Ottawa* 15:91–92. [CS]

De Koninck, J., Christ, G., Rinfret, N. & Proulx, G. (1988) Dreams during language learning: When and how is the new language integrated? *Psychiatric Journal of the University of Ottawa* 13:72–74. [aTAN]

De Koninck, J. M. & Koulack, D. (1975) Dream content and adaptation to a stressful situation. *Journal of Abnormal Psychology* 84:250–60. [aJAH]

De Koninck, J. M., Lorrain, D., Christ, G., Proulx, G. & Coulombe, D. (1989) Intensive language learning and increases in rapid eye movement sleep: Evidence of a performance factor. *International Journal of Psychophysiology* 8:43–47. [aTAN, CS]

De Koninck, J. M. & Prevost, F. (1991) Le sommeil paradoxal et le traitement de l'information: Une exploration par l'inversion du champ visuel. *Canadian Journal of Psychology* 45:125–39. [CS]

Delaney, G. (1998) *All about dreams.* Harper Collins. [WHM]

De Lecea, L., Kilduff, T. S., Peyron, C., Gao, X., Foye, P. E., Danielson, P. E., Fukuhara, C., Battenberg, E. L., Gautvik, V. T., Bartlett, F. S., Frankel, W. N., Van den Pol, A. N., Bloom, F. E., Gautvik, K. M. & Sutcliffe, J. G. (1998) The hypocretins: Hypothalamus-specific peptides with neuroexcitatory activity. *Proceedings of the National Academy of Sciences USA* 95:322–27. [RS-P]

Delorme, F., Froment, J. L. & Jouvet, M. (1966) Suppression du sommeil par la p-chloro-métamphétamine et la p-cholorphenylalanine. *Comptes Rendus des Séances de la Société de Biologie et de des Filiates* 160:2347–51. [aTAN]

Delorme, F., Jeannerod, M. & Jouvet, M. (1965) Effets remarquables de la reserpine sur l'activité EEG phasique ponto-génico-occipitale. *Comptes Rendus des Séances de la Société de Biologie et de des Filiates* 159:900–903. [RCo, aTAN]

Dement, W. C. (1955) Dream recall and eye movements during sleep in schizophrenics and normals. *Journal of Nervous and Mental Disease* 122(45):263–69. [aJAH, aTAN]

(1958) The occurrence of low voltage, fast, electroencephalogram patterns during behavioral sleep in the cat. *Electroencephalography and Clinical Neurophysiology* 10:291–96. [aJAH]

(1965) An essay on dreams: The role of physiology in understanding their nature. In: *New directions in psychology II,* ed. F. Barron, W. C. Dement, W. Edwards, H. Londman, L. D. Phillips, J. Olds & M. Olds. Holt, Rinehart, & Winston. [aTAN]

(1969) The biological role of REM sleep (circa 1968). In: *Sleep physiology and pathology,* ed. A. Kales. Lippincott. [HS]

(1972) *Some must watch while some must sleep.* Norton. [aAR]

Dement, W. C., Henry, P., Cohen, H. & Ferguson, J. (1967) Studies on the effect of REM deprivation in humans and in animals. In: *Sleep and altered states of consciousness,* ed. S. S. Kety, E. V. Evarts & H. L. Williams. Williams & Wilkins. [HS]

Dement, W. C., Kahn, E. & Roffwarg, H. P. (1965) The influence of the laboratory situation on the dreams of the experimental subject. *Journal of Nervous and Mental Disease* 140:119–31. [aJAH]

Dement, W. C. & Kleitman, N. (1957a) Cyclic variations in EEG during sleep and their relation to eye movements, body motility, and dreaming. *Electroencephalography and Clinical Neurophysiology* 9:673–90. [AC, aTAN, aMS]

(1957b) The relation of eye movements during sleep to dream activity: An objective method for the study of dreaming. *Journal of Experimental Psychology* 53(3):339–46. [aJAH, aTAN, PS, aMS]

Dement, W. C. & Wolpert, E. (1958) The relation of eye movements, body motility, and external stimuli to dream content. *Journal of Experimental Psychology* 55:543–53. [AC, aJAH]

Dennett, D. C. (1976) Are dreams experiences? *Philosophical Review* 73:151–71. [rAR]

(1991) *Consciousness explained.* Little, Brown. [JAB, rTAN]

Deptula, D. & Pomara, N. (1990) Effects of antidepressants on human performance: A review. *Journal of Clinical Psychopharmacology* 10:105–11. [aRPV]

DeQuervain, D. J., Roozendaal, B. & McGaugh, J. L. (1998) Stress and glucocorticoids impair retrieval of long-term spatial memory. *Nature* 394:787–90. [JB]

de Saint Hilaire, Z., Python, A., Blanc, G., Charnay, Y. & Gaillard, J. M. (1995) Effects of WIN 35,428, a potent antagonist of dopamine transporter on sleep and locomotion. *NeuroReport* 6:2182–86. [aJAH]

De Sanctis, S. (1896) *Il sogni e il sonno.* Alighieri. [aMS]

Descartes, R. (1637/1980) *Discourse on method,* trans. D. A. Cress. Hackett. [JFP]

(1641/1980) *Meditations on first philosophy,* trans. D. A. Cress. Hackett. [JFP]

Deschenes, M. & Landry, P. (1980) Axonal branch diameter and spacing of nodes in the terminal arborization of identified thalamic and cortical neurons. *Brain Research* 191:538–44. [TJM]

Desmedt, A., Garcia, R. & Jaffard, R. (1998) Differential modulation of changes in hippocampal-septal synaptic excitability by the amygdala as a function of either elemental or contextual fear conditioning in mice. *Journal of Neuroscience* 18:480–87. [ARM]

Deurveiller, S., Hans, B. & Henneoin, E. (1997) Pontine microinjection of carbachol does not reliably enhance paradoxical sleep in rats. *Sleep* 20:593–607. [aJAH]

Devinsky, O., Morrell, M. J. & Vogt, B. A. (1995) Contributions of anterior cingulate cortex to behaviour. *Brain* 118:279–306. [FD, arJAH]

Dewasmes, G., Bothorel, B., Candas, V. & Libert, J. P. (1997) A short-term poikilothermic period occurs just after paradoxical sleep onset in humans: Characterization changes in sweating effector activity. *Journal of Sleep Research* 6:252–58. [aTAN]

*Diagnostic and statistical manual of mental disorders* (1994) *Fourth edition,* ed. A. Francis. Chairperson task force on DSM IV. American Psychiatric Association. [JFP]

Diamond, D. M., Dunwiddle, T. V. & Rose, G. M. (1988) Characteristics of hippocampal primed burst potentiation in vitro and in the awake rat. *Journal of Neuroscience* 8:4079–88. [aRPV]

Dijk, D. J. (1995) EEG slow-waves and sleep spindles: Windows on the sleeping brain. *Behavioural Brain Research* 69:109–16. [rTAN]

Dijk, D. J., Brunner, D. P. & Borbely, A. A. (1991) EEG power density during recovery sleep in the morning. *Electroencephalography and Clinical Neurophysiology* 78:203–14. [PS]

Dinges, D. F. (1990) Are you awake? Cognitive performance and reverie during the hypnopompic state. In: *Sleep and cognition,* ed. R. Bootzin, J. Kihlstrom & D. Schachter. American Psychological Association. [RCo, aJAH]

Dixon, N. F. (1981) *Preconscious processing.* Wiley. [rTAN]

Dodge, A. & Beatty, W. (1980) Sleep deprivation does not affect spatial memory in rats. *Bulletin of the Psychonomic Society* 16:408–409. [aRPV]

Dodt, C., Breckling, U., Derad, I., Fehm, H. L. & Born, J. (1997) Plasma epinephrine and norepinephrine concentrations of healthy humans associated with night time sleep and morning arousal. *Hypertension* 30:71–76. [JB]

Domhoff, G. W. (1969) Home dreams versus laboratory dreams: Home dreams are better. In: *Dream psychology and the new biology of dreaming,* ed. M. Kramer. Charles C. Thomas. [aRPV]

(1993) The repetition of dreams and dream elements: A possible clue to a function of dreams. In: *The functions of dreaming,* ed. A. Moffitt, M. Kramer & R. Hoffman. SUNY Press. [GWD, aAR]

(1996) *Finding meaning in dreams: A quantitative approach.* Plenum Press. [GWD, arJAH, arAR, MSc, aRPV, AZ]

(1999a) Drawing theoretical implications from descriptive empirical findings on dream content. *Dreaming* 9(2/3):201–10. [GWD]

(1999b) New directions in the study of dream content using the Hall and Van de Castle coding system. *Dreaming* 9(2/3):115–37. [GWD]

(2001) A new neurocognitive theory of dreams. *Dreaming* 11:13–33.

Domhoff, G. W. & Kamiya, J. (1964) Problems in dream content study with objective indicators: I. A comparison of home and laboratory dream reports. *Archives of General Psychiatry* 11:519–24. [aJAH]

Domhoff, G. W. & Schneider, A. (1998) New rationales and methods for quantitative dream research outside the laboratory. *Sleep* 12:398–404. [aRPV]

(1999) Much ado about very little: The small effect sizes when home and laboratory collected dreams are compared. *Dreaming* 9:139–51. [GWD, aJAH]

Donald, M. (1991) *Origins of the modern mind: Three stages in the evolution of culture and cognition.* Harvard University Press. [WJC]

Donchin, E., Heffley, E., Hillyard, S. A., Loveless, N., Maltzman, I., Ohman, A., Rosler, F., Ruchkin, D. & Siddle, D. (1984) Cognition and event-related potentials. II. The orienting reflex and p300. *Annals of the New York Academy of Sciences* 425:39–57. [aTAN]

Doricchi, F., Guariglia, C., Paolucci, S. & Pizzamiglio, L. (1991) Disappearance of leftward rapid eye movements during sleep in left visual hemi-inattention. *NeuroReport* 2:285–88. [aJAH]

(1993) Disturbances of the rapid eye movements (REMs) of the REM sleep in patients with unilateral attentional neglect: Clue for the understanding of the functional meaning of REMs. *Electroencephalography and Clinical Neuropathology* 87(3): 105–16. [FD, JH, aJAH]

(1996) REMs asymmetry in chronic unilateral neglect does not change with behavioral improvement induced by rehabilitational treatment. *Electroencephalography and Clinical Neurophysiology* 98:51–58. [aJAH]

Doricchi, F. & Violani, C. (1992) Dream recall in brain-damaged patients: A contribution to the neuropsychology of dreaming through a review of the literature. In: *The neuropsychology of sleep and dreaming,* ed. J. S. Antrobus & M. Bertini. Erlbaum. [FD, aJAH, rMS]

(1996) REMs asymmetry in chronic unilateral neglect does not change with behavioral improvement induced by rehabilitation treatment. *Electroencephalography and Clinical Neurophysiology* 98(1):51–58. [FD, JH]

Dorow, R., Horowski, R., Paschelke, G., Amin, M. & Braestrup, C. (1983) Severe anxiety induced by FG 7142, a B-carboline ligand for benzodiazepine receptors. *Lancet* 2:98–99. [LJW]

Dorus, E., Dorus, W. & Rechtschaffen, A. (1971) The incidence of novelty in dreams. *Archives of General Psychiatry* 25:364–68. [GWD, aJAH, GWV]

Dow, B. M., Kelsoe, J. R., Jr. & Gillin, J. C. (1996) Sleep and dreams in Vietnam PTSD and depression. *Biological Psychiatry* 39:42–50. [MBl]

Drachman, D. A. & Leavitt, J. (1974) Human memory and the cholinergic system: A relationship to aging? *Archives of Neurology* 30:113–21. [aRPV]

Dragoi, G., Carpi, D., Recce, M., Csicsvari, J. & Buzsaki, G. (1999) Interactions between hippocampus and medial septum during sharp waves and theta oscillation in the behaving rat. *Journal of Neuroscience* 19(4):6191–99. [RWG]

Draine, S. C. & Greenwald, A. G. (1998) Replicable unconscious priming. *Journal of Experimental Psychology: General* 127:286–303. [GM]

Dreistadt, R. (1971) An analysis of how dreams are used in creative behavior. *Psychology* 8:24–50. [RLe]

Droste, D. W., Berger, W., Schuler, E. & Krauss, J. K. (1993) Middle cerebral artery blood flow velocity in healthy persons during wakefulness and sleep: A transcranial Doppler study. *Sleep* 16:603–609. [aJAH]

Drucker-Colin, R., Bernal-Pedraza, J., Fernandez-Cancino, F. & Morrison, A. R. (1983) Increasing PGO spike density by auditory stimulation increases duration and decreases the latency of rapid eye movement REM sleep. *Brain Research* 278:308–12. [IF, aJAH]

Drummond, S. P. A., Brown, C. G., Gillin, J. C., Stricker, J. L., Wong, E. C. & Buxton, R. B. (2000) Altered brain response to verbal learning following sleep deprivation. *Nature* 403:655–57. [rJAH]

Drummond, S. P. A., Brown, C. G., Stricker, J. L., Buxton, R. B., Wong, E. C. & Gillin, J. C. (1999) Sleep deprivation-induced reduction in cortical functional response to serial subtraction. *NeuroReport* 10:3745–48 (also published in *Nature,* February 10, 2000). [rJAH]

Dujardin, K., Guerrien, A. & Leconte, P. (1990) Sleep, brain activation and cognition. *Physiology and Behavior* 47:1271–78. [WHM, aTAN, aRPV]

Dumermuth, G., Langer, B., Lehmann, D., Meier, C. A. & Dinkelmann, R. (1983) Spectral analysis of all-night sleep EEG in healthy adults. *European Neurology* 22:322–39. [aTAN]

Duncan, J., Emslie, H., Williams, P., Johnson, R. & Freer, C. (1996) Intelligence and the frontal lobe: The organization of goal-directed behavior. *Cognitive Psychology* 30:257–303. [aJAH]

Dunleavy, D. L. F., Brezinova, V., Oswald, I., Maclean, A. W. & Tinker, M. (1972) Changes during weeks in effects of tricyclic drugs on the human sleeping brain. *British Journal of Psychiatry* 120:663–72. [aRPV]

Dunleavy, D. L. F. & Oswald, I. (1973) Phenelzine, mood response, and sleep. *Archives of General Psychiatry* 28:353–56. [aRPV]

Dusan-Peyrethon, D., Peyrethon, J. & Jouvet, M. (1967) Etude quantitative des phenomènes phasiques du sommeil paradoxal pendant et après sa deprivation instrumentale [Quantitative study of phasic phenomena of paradoxal sleep during and after its instrumental deprivation]. *Comptes Rendus des Séances de la Société de Biologie et de Ses Filiates* 161:2530–33. [aTAN]

Dutar, P., Bassant, M. H., Senut, M. C. & Lamour, Y. (1995) The septohippocampal pathway: Structure and function of a central cholinergic system. *Physiological Reviews* 75:393–427. [aRPV]

Dyken, M. E., Lin-Dyken, D. C., Seaba, P. & Yamada, T. (1995) Violent sleep-related behavior leading to subdural hemorrhage. *Archives of Neurology* 52:318–21. [aAR]

Eccles, J. C. (1961) Chairman's opening remarks. In: *The nature of sleep (CIBA Foundation Symposium),* ed. G. E. W. Wolstenholme & M. O'Connor. Churchill. [MSt]

Edelman, G. M. (1987) *Neural Darwinism: The theory of neuronal group selection.* Basic Books. [WJC]

Efron, R. (1968) *What is perception? Boston studies in the philosophy of science.* Humanities Press. [aMS]

Egan, T. M. & North, R. A. (1985) Acetylcholine acts on M2-muscarinic receptors to excite rat locus coeruleus neurons. *British Journal of Pharmacology* 85:733–35. [aJAH]

(1986a) Acetylcholine hyperpolarizes central neurons by acting on an M2 muscarinic receptor. *Nature* 319:405–407. [aJAH]

(1986b) Actions of acetylcholine and nicotine on rat locus coeruleus neurons in vitro. *Neuroscience* 319:565–71. [aJAH]

Eich, E. (1990) Learning during sleep. In: *Sleep and cognition,* ed. J. Bootzin, J. Kihlstrom & D. Schacter. American Psychological Association. [RCo]

Eich, J. E. (1986) Epilepsy and state specific memory. *Acta Neurologica Scandinavica* 74 (Supplement 109):15–21. [DL]

Eichenbaum, H. (1999) The hippocampus and mechanisms of declarative memory. *Behavioural Brain Research* 103(2):123–33. [RWG]

Ekman, P. (1971) Universals and cultural differences in facial expressions of emotion. In: *Nebraska Symposium on motivation, vol. 19,* ed. J. Cole. University of Nebraska Press. [LM]

(1992) Are there basic emotions? *Psychological Review* 99:550–53. [OF]

Ekstrand, B. R., Barrett, T. R., West, J. N. & Maier, W. G. (1977) The effect of sleep on human long-term memory. In: *Neurobiology of sleep and memory,* ed. R. R. Drucker-Colin & J. L. McGaugh. Academic Press. [JB]

Ekstrand, B. R., Sullivan, M. J., Parker, D. F. & West, J. N. (1971) Spontaneous recovery and sleep. *Journal of Experimental Psychology* 88:142–44. [aRPV]

Elazar, Z. & Hobson, J. A. (1985) Neuronal excitability control in health and disease: A neurophysiological comparison of REM sleep and epilepsy. *Progress in Neurobiology* 25:141–88. [arJAH]

Eliade, M. (1978) *A history of religious ideas. Vol. I: From the stone age to the Eleusinian mysteries.* Chicago University Press. [JBP]

El Kafi, B., Cespuglio, R., Leger, L., Marinesco, S. & Jouvet, M. (1994) Is the nucleus raphe dorsalis a target for the peptides possessing hypnogenic properties? *Brain Research* 637:211–21. [aJAH]

Ellis, C. M., Monk, C., Simmons, A., Lemmens, G., Williams, S. C. R., Brammer, M., Bullmore, E. & Parkes, J. D. (1999) Functional magnetic resonance imaging neuroactivation studies in normal subjects and subjects with the narcoleptic syndrome. Actions of modafinil. *Journal of Sleep Research* 8:85–93. [aJAH]

Ellman, S. J., Spielman, A. J., Luck, D., Steiner, S. S. & Halperin, R. (1978/1991) REM deprivation: A review. In: *The mind in sleep,* ed. S. J. Ellman & J. S. Antrobus. Erlbaum edition, 1978; Wiley 1991. [aTAN, aRPV]

El Manseri, M., Sakai, K. & Jouvet, M. (1990) Responses of presumed cholinergic mesopontine tegmental neurons to carbachol microinjection in freely moving cats. *Experimental Brain Research* 83:115–23. [aJAH]

Empson, J. A. C. & Clarke, P. R. F. (1970) Rapid eye movements and remembering. *Nature* 227:287–88. [aRPV]

Endo, T., Roth, C., Landolt, H.-P., Werth, E., Aeschbach, D., Achermann, P. & Borbely, A. A. (1998) Selective REM sleep deprivation in humans: Effects on sleep and sleep EEG. *American Journal of Physiology* 274:R1186–94. [aTAN]

Ephron, H. S. & Carrington, P. (1966) Rapid eye movement sleep and cortical homeostasis. *Psychological Review* 73:500–26. [IF, HTH, rRPV]

Epstein, A. W. (1964) Recurrent dreams: Their relationship to temporal lobe seizures. *Archives of General Psychiatry* 10:49–54. [aMS]

(1967) Body image alterations during seizures and dreams of epileptics. *Archives of Neurology* 16:613–19. [aMS]

(1979) Effect of certain cerebral hemispheric diseases on dreaming. *Biological Psychiatry* 14:77–93. [aMS]

(1985) The waking event-dream interval. *American Journal of Psychiatry* 142:123–24. [aJAH]

(1995) *Dreaming and other involuntary mentation.* International Universities Press. [aMS]

Epstein, A. W. & Ervin, F. (1956) Psychodynamic significance of seizure content in psycho-motor epilepsy. *Psychosomatic Medicine* 18:43–55. [aMS]

Epstein, A. W. & Freeman, N. (1981) The uncinate focus and dreaming. *Epilepsia* 22:603–605. [aMS]

Epstein, A. W. & Hill, W. (1966) Ictal phenomena during REM sleep of a temporal lobe epileptic. *Archives of Neurology* 15:367–75. [aMS]

Epstein, E. & Simmons, N. (1983) Aphasia with reported loss of dreaming. *American Journal of Psychiatry* 140:109. [aMS]

Ersland, S., Weisaeth, L. & Sund, A. (1989) The stress upon rescuers involved in an oil rig disaster: "Alexander Kielland" – 1980. *Acta Psychiatrica Scandinavica* 80:38–49. [JBP]

Erwin, R. & Buchwald, J. S. (1986) Midlatency auditory evoked responses: Differential effects of sleep in the human. *Electroencephalography and Clinical Neurophysiology* 65:383–92. [aTAN]

Erwin, R., Van Lancker, D., Guthrie, D., Schwafel, J., Tanguay, P. & Buchwald, J. S. (1991) P3 responses to prosodic stimuli in adult autistic subjects. *Electroencephalography and Clinical Neurophysiology* 80:561–71. [aTAN]

Ettlinger, G., Warrington, E. & Zangwill, O. (1957) A further study of visual-spatial agnosia. *Brain* 80:335–61. [aMS]

Evans, F. J. (1972) Hypnosis and sleep: Techniques for exploring cognitive activity during sleep. In: *Hypnosis: Research developments and perspectives,* ed. E. Fromm & R. E. Shor. Aldine/Atherton. [aTAN]

Everit, B. J. & Robbins, T. W. (1997) Central cholinergic systems and cognition. *Annual Review of Psychology* 48:649–84. [aJAH]

Evarts, E. V. (1962) Activity of neurons in visual cortex of the cat during sleep with low voltage fast EEG activity. *Journal of Neurophysiology* 25:812–16. [aJAH]

Factor, S. A., Molho, E. S., Podskalny, G. D. & Brown, D. (1995) Parkinson's disease: Drug-induced psychiatric states. *Advances in Neurology* 65:115–38. [EKP]

Fagan, R. (1976) Modeling how and why play works. In: *Play: Its role in development and evolution,* ed. J. S. Bruner, A. Jolly & K. Sylva. Penguin. [JAC]

Fagioli, I., Cipolli, C. & Tuozzi, G. (1989) Accessing previous mental sleep experience in REM and NREM sleep. *Biological Psychology* 29:27–38. [aJAH]

Fairweather, D. B., Ashford, J. & Hindmarch, I. (1996) Effects of fluvoxamine and dothiepin on psychomotor abilities in healthy volunteers. *Pharmacology Biochemistry and Behavior* 53:265–69. [aRPV]

Fairweather, D. B., Kerr, J. S., Harisson, D. A., Moon, C. A. & Hindmarch, I. (1993) A double-blind comparison of the effects of fluoxetine and amitriptyline on cognitive function in elderly depressed patients. *Human Psychopharmacology* 8:41–47. [aRPV]

Fallon, J. H. & Ciofi, P. (1992) Distribution of monoamines within the amygdala. In: *The amygdala: Neurobiological aspects of emotion, memory and mental dysfunction,* ed. J. P. Aggleton. Wiley-Liss. [aJAH]

Farah, M., Levine, D. & Calviano, D. (1988) A case study of mental imagery deficit. *Brain and Cognition* 8:147–64. [aMS]

Farber, J., Marks, G. A. & Roffwarg, H. P. (1980) REM sleep PGO-type waves are present in the dorsal pons of the albino rat. *Science* 209:615–17. [CG]

Farley, B. G. (1960) Self-organizing models for learned perception. In: *Self-organizing systems, 2nd edition,* ed. M. Yovits & S. Cameron. Pergamon Press. [TJM]

Farrell, B. (1969) *Pat & Roald.* Hutchinson. [aMS]

Farthing, W. G. (1992) *The psychology of consciousness.* Prentice-Hall. [RG, aTAN, aAR, AR]

Faucher, B., Nielsen, T. A., Bessette, P., Raymond, I. & Germain, A. (1999) Qualitative differences in REM and NREM mentation reports using a standardized coding system. Paper presented at the Thirteenth Annual Meeting of the Association of Professional Sleep Societies, Orlando, Florida, June 19–24, 1999. [aTAN] Published in *Sleep* 22 (Supplement 1):S177. [GWD, aJAH, rTAN]

Fein, G., Feinberg, I., Insel, T. R., Antrobus, J. S., Price, L. J., Floyd, R. C. & Nelson, M. A. (1985) Sleep mentation in the elderly. *Psychophysiology* 22:218–25. [aJAH, aTAN]

Feinberg, I. (1969) Effects of age on human sleep patterns. In: *Sleep: Physiology and pathology. A symposium,* ed. A. Kales & J. B. Lippincott. [IF]

(1974) Changes in sleep cycle patterns with age. *Journal of Psychiatric Research* 10:283–306. [IF]

(1999) Delta homeostasis, stress, and sleep deprivation in the rat: A comment on Rechtschaffen et al. *Sleep* 22:1021–24. [IF]

Feinberg, I. & Evarts, E. V. (1969) Changing concepts of the function of sleep: Discovery of intense brain activity during sleep calls for revision of hypotheses as to its function. *Biological Psychiatry* 1:331–48. [IF]

Feinberg, I. & Floyd, T. C. (1979) Systematic trends across the night in human sleep cycles. *Psychophysiology* 16:283–91. [rTAN]

Feinberg, I., Floyd, T. C. & March, J. D. (1987) Effects of sleep loss on delta (0.3–3Hz) EEG and eye movement density: New observations and hypotheses. *Electroencephalography and Clinical Neurophysiology* 67:217–21. [IF]

Feinberg, I., Maloney, T. & March, J. D. (1992) Precise conservation of NREM period 1 (NREMP1) delta across naps and nocturnal sleep: Implications for REM latency and NREM/REM alternation. *Sleep* 15(5):400–403. [IF]

Feinberg, I. & March, J. D. (1988) Cyclic delta peaks during sleep: Result of a pulsatile endocrine process? [letter]. *Archives of General Psychiatry* 45:1141–42. [IF]

(1995) Observations on delta homeostasis, the one-stimulus model of NREM-REM alternation, and the neurobiologic implications of experimental dream studies. [Review]. *Behavioural Brain Research* 69:97–108. [IF, aTAN, HSP]

Feinberg, I., March, J. D., Flach, K., Maloney, T., Chern, W.-J. & Travis, F. (1990) Maturational changes in amplitude, incidence and cyclic pattern of the 0 to 3 Hz (Delta) electoencephalogram of human sleep. *Brain Dysfunction* 3:183–92. [IF]

Feinberg, I., March, J. D., Floyd, T. C., Jimison, R., Bossom-Demitrack, L. & Katz, P. H. (1985) Homeostatic changes during post-nap sleep maintain baseline levels of delta EEG. *Electroencephalography and Clinical Neurophysiology* 61:134–37. [IF]

Feldman, M. H. (1971) Physiological observations in a chronic case of "locked-in" syndrome. *Neurology* 21:459–78. [aMS, aRPV]

Feldman, M. J. & Hersen, M. (1967) Attitudes toward death in nightmare subjects. *Journal of Abnormal Psychology* 72:421–25. [aAR]

Ferguson, J. & Dement, W. (1969) The behavioral effects of amphetamine on REM deprived rats. *Journal of Psychiatric Research* 7:111–18. [aTAN, HS]

Ffytche, D. H., Howard, R. J., Brammer, M. J., David, A., Woodruff, P. & Williams, S. (1998) The anatomy of conscious vision: An fMRI study of visual hallucinations. *Nature Neuroscience* 1:738–42. [aJAH, EFP-S]

Ficca, G., Lombardo, P., Rossi, L. & Salzarulo, P. (2000) Morning recall of verbal material depends on prior sleep organization. *Behavioral Brain Research* 112:159–63. [PS]

Finelli, L. A., Landolt, H.-P., Buck, A., Roth, C., Berthold, T., Borbély, A. A. & Achermann, P. (2000) Functional neuroanatomy of human sleep states after zolpidem and placebo: A H$_2$O -PET study. *Journal of Sleep Research* 9:161–73. [AAB]

Fischer-Perroudon, C., Mouret, J. & Jouvet, M. (1974) Sur un cas d'agrypnie (4 mois sans sommeil) au cours d'une maladie de Morvan. Effet favorable du 5-hydroxytryptophane. *Electroencephalography and Clinical Neurophysiology* 36:1–18. [CG]

Fishbein, W. (1970) Interference with conversion of memory from short-term to long-term storage by partial sleep deprivation. *Communications in Behavioral Biology* 5:171–75. [WF, aRPV]

(1971) Disruptive effects of rapid eye movement sleep deprivation on long-term memory. *Physiology and Behavior* 6:279–82. [VSR, aRPV]

(1995) Sleep and memory: A look back, a look forward. *Sleep Research Society Bulletin* 1:53–58. [AC, WF, aRPV, RS]

(1996) Memory consolidation in REM sleep: Making dreams out of chaos. *Sleep Research Society Bulletin* 2:55–56. [WF, aRPV]

Fishbein, W. & Gutwein, B. M. (1977) Paradoxical sleep and memory storage processes. *Behavioral Biology* 19:425–64. [WF, CS, arRPV]

Fishbein, W., McGaugh, J. L. & Swarz, J. R. (1971) Retrograde amnesia: Electroconvulsive shock effects after termination of rapid eye movement sleep deprivation. *Science* 172:80–82. [WF]

Fisher, C. (1966) Dreaming and sexuality. In: *Psychoanalysis: A general psychology; Essays in honor of H. Hartmann,* ed. R. M. Loewenstein, L. M. Newman, Schur, M. & A. J. Solnit. International Universities Press. [rTAN]

Fisher, C., Byrne, J. V., Edwards, A. & Kahn, E. (1970a) A psychophysiological study of nightmares. *Journal of the American Psychoanalytic Association* 18:747–82. [MK]

(1970b) REM and NREM nightmares. *International Psychiatry Clinics* 7:183–87. [aTAN]

Fisher, C., Kahn, E., Edwards, A. & Davis, D. M. (1973) A psychophysiological study of nightmares and night terrors: I. Physiological aspects of the Stage 4 night terror. *Journal of Nervous and Mental Disease* 157:75–98. [aJAH, arTAN] Also in: *Archives of General Psychiatry* 28:252–59. [LJW]

Fisher, S. & Greenberg, R. (1977) *The scientific credibility of Freud's theories and therapy.* Basic Books. [GWD]

(1996) *Freud scientifically appraised.* Wiley. [GWD]

Fiss, H. (1986) An empirical foundation for a self psychology of dreaming. *Journal of Mind and Behavior* 7:161–91. [TAN]

(1993) The "royal road" to the unconscious revisited: A signal detection model of dream function. In: *The function of dreaming,* ed. A. Moffitt, M. Kramer & R. Hoffmann. State University of New York Press. [RLe]

Fiss, H., Klein, G. S. & Bokert, E. (1966) Waking fantasies following interruption of two types of sleep. *Archives of General Psychiatry* 14:543–51. [aJAH, aTAN, NST]

Fiss, H., Kremer, E. & Litchman, J. (1977) The mnemonic function of dreaming. *Sleep Research* 6:122. [aTAN]

Flanagan, O. (1991) *The science of the mind, 2nd edition.* MIT Press. [OF]

(1992) *Consciousness reconsidered.* MIT Press. [OF]

(1995) Deconstructing dreams: The spandrels of sleep. *The Journal of Philosophy* 92(1):5–27. [OF, KG, LM, aAR]

(1997) Prospects for a unified theory of consciousness, or what dreams are made of. In: *Scientific approaches to consciousness*, ed. J. D. Cohen & J. W. Schooler. Erlbaum. [OF, aJAH]

(2000) *Dreaming souls: Sleep, dreams, and the evolution of the conscious mind.* Oxford University Press. [OF, rAR]

Flanigan, W. F. (1974) Nocturnal behavior of captive small cetaceans. II. The beluga whale, *Delphinapterus leucas. Sleep Research* 3:85. [JMS]

Fletcher, P. C., Frith, C. D. & Rugg, M. D. (1997) The functional neuroanatomy of episodic memory. *Trends in Neuroscience* 20:213–18. [aJAH, EFP-S]

Flicker, C., McCarley, R. W. & Hobson, J. A. (1981) Aminergic neurons: State control and plasticity in three model systems. *Cellular and Molecular Neurobiology* 1:123–66. [aJAH]

Foa, E. & Kozak, M. (1986) Emotional processing of fear: Exposure to corrective information. *Psychological Bulletin* 99:20–35. [RLe, JBP]

Fookson, J. & Antrobus, J. (1992) A connectionist model of bizarre thought. In: *The neuropsychology of sleep and dreaming*, ed. J. S. Antrobus & M. Bertini. Erlbaum. [aJAH]

Foote, S. L., Bloom, F. E. & Aston-Jones, G. (1983) Nucleus locus ceruleus: New evidence of anatomical and physiological specificity. *Physiological Review* 63:844–914. [aJAH]

Fosse, R. (2000) REM mentation in narcoleptics: An empirical test of two neurocognitive theories. Consciousness and Cognition

Fosse, R., Stickgold, R. & Hobson, J. A. (2001) Brain-mind states: Reciprocal variation in thoughts and hallucinations. *Psychological Science.* 12:30–36. [rJAH]

(in preparation) Overnight changes in thoughts and hallucinations. [rJAH]

Foulkes, D. (1962) Dream reports from different stages of sleep. *Journal of Abnormal and Social Psychology* 65:14–25. [CG, aJAH, aTAN, RDO, PS, aMS]

(1966) *The psychology of sleep.* Charles Scribner's. [GWD, aJAH, aTAN, aMS]

(1967) Nonrapid eye movement mentation. *Experimental Neurology* 19 (Suppl. 4):28–38. [GWD, RG, aJAH, aTAN]

(1979) Home and laboratory dreams: Four empirical studies and a conceptual reevaluation. *Sleep* 2:233–51. [GWD, aJAH]

(1982a) A cognitive-psychological model of REM dream production. *Sleep* 5:169–87. [aJAH, aRPV]

(1982b) *Children's dreams: Longitudinal studies.* Wiley. [GWD, aJAH, aTAN, aAR, GWV]

(1982c) REM-dream perspectives on the development of affect and cognition. *Psychiatric Journal of the University of Ottawa* 7:48–55. [aTAN]

(1983) Cognitive processes during sleep: Evolutionary aspects. In: *Sleep mechanisms and functions in humans and animals: An evolutionary perspective*, ed. A. Mayes. Van Nostrand Reinhold. [GWD]

(1985) *Dreaming: A cognitive-psychological analysis.* Erlbaum. [JAB, MBl, CCa, GWD, aJAH, TLK, arTAN, aAR, AR, aRPV]

(1990) Dreaming and consciousness. *European Journal of Cognitive Psychology* 70(2):39–55. [EH, aJAH, aTAN]

(1991) Why study dreaming? One researcher's perspective. *Dreaming* 1:245–48. [aJAH]

(1993a) Data constraints on theorizing about dream function. In: *The functions of dreaming*, ed. A. Moffitt, M. Kramer & R. Hoffman. SUNY Press. [GWD]

(1993b) Dreaming and REM sleep. *Journal of Sleep Research* 2:199–202. [aJAH, aTAN, EFP-S]

(1993c) Children's dreaming. In: *Dreaming as cognition*, ed. C. Cavallero & D. Foulkes. Harvester Wheatsheaf. [CCa, aJAH, TLK]

(1995) Dreaming: Functions and meanings. *Impuls* 3:8–16. [aJAH]

(1996a) Dream research 1953–1993. *Sleep* 19:609–24. [aJAH]

(1996b) Misrepresentation of sleep-laboratory dream research with children. *Perceptual and Motor Skills* 83:205–206. [GWD, aJAH]

(1997) A contemporary neurobiology of dreaming? *Sleep Research Society Bulletin* 3(1):2–4. http://www.srssleep.org/publications.html [aJAH, MM, ARM]

(1999) *Children's dreaming and the development of consciousness.* Harvard University Press. [GWD, aJAH, MM, LM, arAR, aRPV]

Foulkes, D., Bradley, L., Cavallero, C. & Hollifield, M. (1989) Processing of memories and knowledge in REM and NREM dreams. *Perceptual and Motor Skills* 68:365–66. [aTAN, aRPV]

Foulkes, D. & Cavallero, C. (1993a) Introduction. In: *Dreaming as cognition*, ed. C. Cavallero & D. Foulkes. Harvester Wheatsheaf. [RCo, aJAH, aTAN, aAR]

Foulkes, D. & Cavallero, C., eds. (1993b) *Dreaming as cognition.* Harvester Wheatsheaf. [TLK]

Foulkes, D. & Fleisher, S. (1975) Mental activity in relaxed wakefulness. *Journal of Abnormal Psychology* 84:66–75. [MBl, aJAH, JM, aRPV]

Foulkes, D., Hollifield, M., Bradley, L., Terry, R. & Sullivan, B. (1991) Waking self-understanding, REM-dream self representation, and cognitive ability at ages 5–8. *Dreaming* 1:41–52. [EH, aJAH]

Foulkes, D., Hollifield, M., Sullivan, B., Bradley, L. & Terry, R. (1990) REM dreaming and cognitive skills at ages 5–8. *International Journal of Behavioral Development* 13:447–65. [aJAH]

Foulkes, D. & Kerr, N. H. (1984) Point of view in nocturnal dreaming. *Perceptual and Motor Skills* 78:690. [rAR]

Foulkes, D., Pivik, T., Ahrens, J. & Swanson, E. M. (1968) Effects of dream deprivation on dream content: An attempted cross-night replication. *Journal of Abnormal Psychology* 73:403–15. [RCo]

Foulkes, D., Pivik, T., Steadman, H. F., Spear, P. S. & Symonds, J. D. (1967) Dreams of the male child: An EEG study. *Journal of Abnormal Psychology* 72:457–67. [aJAH]

Foulkes, D. & Pope, R. (1973) Primary visual experience and secondary cognitive elaboration in Stage R. *Perceptual and Motor Skills* 37:107–18. [aJAH, MK, aTAN]

Foulkes, D. & Rechtschaffen, A. (1964) Presleep determinants of dream content: Effects of two films. *Perceptual and Motor Skills* 19:983–1005. [MBl, aJAH, aTAN]

Foulkes, D. & Schmidt, M. (1983) Temporal sequence and unit composition in dream reports from different stages of sleep. *Sleep* 6:265–80. [MBo, CCa, aJAH, aTAN, GWV]

Foulkes, D. & Scott, E. (1973) An above-zero baseline for the incidence of momentary hallucinatory mentation. *Sleep Research* 2:108. [aJAH]

Foulkes, D., Scott, E. A. & Pope, R. (1980) The tonic-phase strategy in sleep mentation research: Correlates of EEG theta bursts at sleep onset and during nonREM sleep. *Richerche di Psicologia* 16:121–32. [GWV]

Foulkes, D. & Shepherd, J. (1972) Stimulus incorporation in children's dreams. *Sleep Research* 1:119. [aJAH]

Foulkes, D., Spear, P. S. & Symonds, J. D. (1966) Individual differences in mental activity at sleep onset. *Journal of Abnormal Psychology* 71:280–86. [aTAN, aMS]

Foulkes, D., Sullivan, B., Kerr, N. H. & Brown, L. (1988a) Appropriateness of dream feelings to dreamed situations. *Cognition and Emotion* 2(1):29–39. [aAR]

(1988b) Dream affect: Appropriateness to dream situations. In: *Sleep '86*, ed. W. P. Koella, F. Obal, H. Scholz & P. Vizzer. Gustav Fisher Verlag. [arJAH]

Foulkes, D. & Vogel, G. (1965) Mental activity at sleep onset. *Journal of Abnormal Psychology* 70(4):231–43. [aJAH, aTAN, aMS]

Fowler, M. J., Sullivan, M. J. & Ekstrand, B. R. (1973) Sleep and memory. *Science* 179:302–304. [JB, ARM, rRPV]

Fox, R., Kramer, M., Baldridge, B., Whiman, R. & Ornstein, P. (1968) The experimenter variable in dream research. *Disease of the Nervous System* 29:698–701. [MK]

Franck, G., Salmon, E., Poirrier, R., Sadzot, B., Franco, G. & Maquet, P. (1987) Evaluation of human cerebral glucose uptake during wakefulness, slow wave sleep and paradoxical sleep. *Sleep* 16:46. [aMS]

Frank, J. (1946) Clinical survey and results of 200 cases of prefrontal leucotomy. *Journal of Mental Science* 92:497–508. [aMS]

(1950) Some aspects of lobotomy (prefrontal leucotomy) under psychoanalytic scrutiny. *Psychiatry* 13:35–42. [arMS]

Frank, M. G. (1999) Phylogeny and evolution of rapid eye movement (REM) sleep. In: *Rapid eye movement sleep*, ed. B. N. Mallick & S. Inoue. Marcel Dekker. [aJAH]

Frank, M. G., Page, J. & Heller, H. C. (1997) The effects of REM sleep-inhibiting drugs in neonatal rats: Evidence for a distinction between neonatal active sleep and REM sleep. *Brain Research* 778:64–72. [IF]

Franzini, C. (1992) Brain metabolism and blood flow during sleep. *Journal of Sleep Research* 1:3–16. [aMS]

(2000) Cardiovascular physiology: The peripheral circulation. In: *The principles and practice of sleep medicine, third edition*, ed. M. H. Kryger, T. Roth & W. C. Dement. W. B. Saunders. (in press). [CF]

French, T. & Fromm, E. (1964) *Dream interpretation: A new approach.* Basic Books. [RG]

Freud, S. (1893) Charcot obituary. In: *The complete psychological works of Sigmund Freud, vol. 3*, ed. J. Strachey. Hogarth Press. [MS]

(1895) Project for a scientific psychology. In: *The origins of psychoanalysis. Letters to Wilhelm Fliess, drafts and notes: 1887–1902*, ed. M. Bonaparte, A. Freud & E. Kris. Basic Books. [aJAH, MK] Also in: *The standard edition of the complete psychological works of Sigmund Freud, vol. 1*, ed. J. Strachey. Hogarth Press. [HS]

(1911/1950/1953/1955/1965) *The interpretation of dreams*, trans. and ed. J. Strachey. 1955, Basic Books. [WF, OF, arJAH, BEJ, MK, JM, rTAN, rRPV] Also published as: *Standard Edition of the Complete Psychological Works of Sigmund Freud, vols. 4 & 5.* 1953, Hogarth Press. [CG, EEK, GL, JFP, aMS] 1950, Random House edition. [RBA, aAR, RS] 1965, Avon Books. [WJC]

(1918) The dream and the primary scene. In: S. Freud, *From the history of an infantile neurosis.* Hogarth Press. [CG]

Friedman, A. S., Granick, S., Cohen, H. W. & Cowitz, B. (1966) Imipramine

(Tofranil) vs. placebo in hospitalized psychotic depressives (a comparison of patients' self-ratings, psychiatrists' ratings and psychological test scores). *Journal of Psychiatric Research* 4:13–36. [aRPV]

Friess, E., van Bardeleben, U., Wiedemann, K., Lauer, C. J. & Holsboer, F. (1994) Effects of pulsatile cortisol infusion on sleep-EEG and nocturnal growth hormone release in healthy men. *Journal of Sleep Research* 3:73–79. [JB]

Friston, K. J., Frith, C. D., Liddle, P. F. & Frackowiak, R. S. J. (1991) Comparing functional (PET) images: The assessment of significant change. *Journal of Cerebral Blood Flow and Metabolism* 11:690–99. [aJAH]

Fuchino, S. H., Sanford, L. D., Ross, R. J. & Morrison, A. R. (1996) Effects of microinjections of the alpha-1 agonist, methoxamine, into the central nucleus of the amygdala on sleep-wake states. *Sleep Research* 24:6. [aJAH]

Fuchs, A. & Ron, S. (1968) An analysis of rapid eye movements of sleep in the monkey. *Electroencephalography and Clinical Neurophysiology* 25:244–51. [aJAH]

Fukuda, K., Miyasita, A. & Inugami, M. (1987) Sleep onset REM periods observed after sleep interruption in normal short and normal long sleeping subjects. *Electroencephalography and Clinical Neurophysiology* 67:508–13. [RDO]

Fukuda, T., Wakawura, M. & Ishikawa, S. (1981) Comparative study of eye movements in the alert state and rapid eye movement sleep. *NeuroOpthalmology* 1:253–60. [aJAH]

Furey, M. L., Pietrini, P., Haxby, J. V., Alexander, G. E., Lee, H. C., VanMeter, J., Grady, C. L., Shetty, U., Rapoport, S. I., Schapiro, M. B. & Freo, U. (1997) Cholinergic stimulation alters performance and task-specific regional cerebral blood flow during working memory. *Proceedings of the National Academy of Sciences, USA* 94:6512–16. [aJAH]

Gabriel, M., Cuppernell, C., Shenker, J. I., Kubota, Y., Henzi, V. & Swanson, D. (1995) Mammillothalamic tract transection blocks anterior thalamic training-induced neuronal plasticity and impairs discriminative avoidance behavior in rabbits. *Journal of Neuroscience* 15(2):1437–45. [EEK]

Gackenback, J. & LaBerge, S., eds. (1988) *Conscious mind, sleeping brain.* Plenum Press. [aAR]

Gaffan, D. & Murray, E. A. (1990) Amygdalar interaction with the mediodorsal nucleus of the thalamus and the ventromedial prefrontal cortex in stimulus-reward associative learning in the monkey. *Journal of Neuroscience* 10:3479–93. [FD]

Gaillard, J. M. & Moneme, A. (1977) Modification of dream content after preferential blockade of mesolimbic and mesocortical dopaminergic systems. *Journal of Psychiatric Research* 13:247–56. [rJAH]

Gaillard, J. M., Nicholson, A. N. & Pascoe, P. A. (1994) Neurotransmitter systems. In: *Principles and practice of sleep medicine, 2nd edition,* ed. M. Kryger, T. Roth & W. Dement. W. B. Saunders. [aJAH]

Gais, S., Plihal, W., Wagner, U. & Born, J. (2000) Early sleep triggers memory for early visual discrimination skills. *Nature Neuroscience* 3:1335–9. [JB, rRPV]

Gallassi, R., Morreale, A., Montagna, P., Gambetti, P. & Lugaresi, E. (1992) "Fatal familial insomnia": Neuropsychological study of a disease with thalamic degeneration. *Cortex* 28:175–87. [aMS]

Garcia-Rill, E., Houser, C., Skinner, R., Smith, W. & Woodward, J. (1987) Locomotion-inducing sites in the vicinity of the pedunculopontine nucleus. *Brain Research Bulletin* 18:731–38. [aJAH]

Gaspar, P., Berger, B., Febvret, A., Vigny, A. & Henry, J. P. (1989) Catecholamine innervation of the human cerebral cortex as revealed by comparative immunohistochemistry of tyrosine hydroxylase and dopamine-beta-hydroxylase. *Journal of Comparative Neurology* 279:249–71. [rJAH]

Gaudreau, H., Carrier, J. & Montplaisir, J. (in press) Age-related modifications of NREM sleep EEG: From childhood to middle age. *Journal of Sleep Research.* [rTAN]

Gauthier, P., Hamon, J. F. & Gottesmann, C. (1986) Etude de la variation contingente negative (VCN) et des composantes tardives des potentiels évoqués au cours du sommeil chez l'Homme. *Revue d'EEG et de Neurophysiologie Clinique* 15:323–30. [CG]

Gazzaniga, M., ed. (2000) *The new cognitive neurosciences, second edition.* MIT Press. [EFP-S]

Gentili, A., Godschalk, M. F., Gheorghiu, D., Nelson, K., Julius, D. A. & Mulligan, T. (1996) Effect of clonidine and yohimbine on sleep in healthy men – a double blind, randomized, controlled trial. *European Journal of Clinical Pharmacology* 50:463–65. [aJAH]

Georgotas, A., McCue, R. E., Reisberg, B., Ferris, S. H., Nagachandran, N., Chang, I. & Mir, P. (1989) The effects of mood changes and antidepressants on the cognitive capacity of elderly depressed patients. *International Psychogeriatrics* 1:135–43. [aRPV]

Georgotas, A., Reisberg, B. & Ferris, S. (1983) First results on the effects of MAO inhibition on cognitive functioning in elderly depressed patients. *Archives of Gerontology and Geriatrics* 2:249–54. [aRPV]

Geretsegger, C., Bohmer, F. & Ludwig, M. (1994) Paroxetine in the elderly depressed patient: Randomized comparison with fluoxetine of efficacy, cognitive and behavioural effects. *International Clinical Psychopharmacology* 9:25–29. [aRPV]

Germain, A. & Nielsen, T. A. (1996) Spectral analysis of global 40Hz EEG rhythm during sleep onset imagery and wakefulness. *Sleep Research* 25:135. [aJAH]

(1997) Distribution of spontaneous hypnagogic images across Hori's EEG sleep onset stages. *Sleep Research* 26:243. (Abstract). [rTAN]

(1999) EEG power correlates of mentation recall in REM sleep and Stage 2 sleep. Paper presented at the Sixteenth International Conference of the Association for the Study of Dreams, Santa Cruz, CA, July 6–10, 1999. [aTAN]

Germain, A., Nielsen, T. A., Khodaverdi, M., Bessette, P., Faucher, B. & Raymond, I. (1999) Fast frequency EEG correlates of dream recall from REM sleep. Paper presented at the Thirteenth Annual Meeting of the Association of Professional Sleep Societies, Orlando, Florida, June 19–24, 1999. *Sleep Research* 22 (Suppl. 1):131. [aJAH, aTAN]

Gerne, M. & Strauch, I. (1985) Psychophysiological indicators of affect patterns and conversational signals during sleep. In: *Sleep '84,* ed. W. P. Koella, E. Ruther & H. Schulz. Gustav Fischer Verlag. [aTAN]

Gershon, S. & Newton, R. (1980) Lack of anticholinergic side effects with a new antidepressant – trazodone. *Journal of Clinical Psychiatry* 41:100–104. [aRPV]

Gershuny, B. & Thayer, J. (1999) Relations among psychological trauma, dissociative phenomena, and trauma-related distress: A review and integration. *Clinical Psychology Reviews* 19:631–57. [RLe]

Geschwind, N. (1965) Disconnection syndromes in animals and man. *Brain* 88:237–94, 585–644. [HTH]

Giambra, L. M. (1995) A laboratory method for investigating influences on switching attention to task-unrelated imagery and thought. *Consciousness and Cognition* 4:1–21. [PC]

Gigerenzer, G. & Todd, P. (1999) *Simple heuristics that make us smart.* Oxford University Press. [LM]

Gilbert, C. D. (1998) Adult cortical dynamics. *Physiological Reviews* 78(2):467–85. http://physrev.physiology.org/cgi/content/abstract/78/2/467 [JAB]

Gillin, J. C., Post, R., Wyatt, R. J., Goodwin, F. K., Snyder, F. & Bunney, W. E., Jr. (1973) REM inhibitory effect of L-DOPA infusion during human sleep. *Electroencephalography and Clinical Neurophysiology* 35:181–86. [arJAH]

Gillin, J. C., Pulvirenti, L., Withers, N., Golshan, S. & Koob, G. (1994) The effects of lisuride on mood and sleep during acute withdrawal in stimulant abusers: A preliminary report. *Biological Psychiatry* 35:843–49. [aJAH]

Gillin, J. C., Smith-Vaniz, A., Zisook, S., Stein, M., Rapaport, M. & Kelsoe, J. (1999) Effects of nefazodone on sleep, nightmares and mood in PTSD. *Sleep* 22 (Supplement):S277–78. [LJW]

Gillin, J. C., Sutton, L. & Ruiz, C. (1991) The cholinergic REM induction test with arecholine in depression. *Archives of General Psychiatry* 8:264–70. [aJAH]

Gillin, J. C., vanKammen, D. & Bunney, W. E., Jr. (1978) Pimozide attenuates effects of d-amphetamine in EEG sleep patterns in psychiatric patients. *Life Science* 22:1805–10. [aJAH]

Giuditta, A., Ambrosini, M. V., Montagnese, P., Mandile, P., Cotugno, M., Grassi, Z. G. & Vescia, S. (1995) The sequential hypothesis of the function of sleep. *Behavioural Brain Research* 69:157–66. [JB]

Glass, R. M., Uhlenhuth, E. H., Hartel, F. W., Matuzas, W. & Fischman, M. W. (1981) Cognitive dysfunction and imipramine in outpatient depressives. *Archives of General Psychiatry* 38:1048–51. [aRPV]

Glickman, S. E., Frank, L. G., Holekamp, K. E., Smale, L. & Licht, P. (1993) Costs and benefits of androgenization in the female spotted hyena: The natural selection of physiological mechanisms. *Perspectives in Ethology* 10:87–118. Also in: *Behavior and evolution,* ed. P. P. G. Bateson, P. H. Klopfer & N. S. Thompson. Plenum Press. [NST]

Glisky, E. L. & Schacter, D. L. (1989) Extending the limits of complex learning in organic amnesia: Computer training in a vocational domain. *Neuropsychologia* 27:107–20. [aAR]

Globus, G. (1987) *Dream life, wake life: The human condition through dreams.* SUNY Press. [TLK]

Gloning, K. & Sternbach, I. (1953) Über das Träumen bei zerebralen Herdläsionen. *Wiener Zeitschrift für Nervenheilkunde* 6:302–29. [aMS]

Gloor, P. (1990) Experiential phenomena of temporal lobe epilepsy. *Brain* 113:1673–94. [rAR]

Goel, V., Gold, B., Kapur, S. & Houle, S. (1998) Neuroanatomical correlates of human reasoning. *Journal of Cognitive Neuroscience* 10:293–302. [aJAH]

Goldberg, M. E., Eggers, H. M. & Gouras, P. (1991) The oculomotor system. In: *Principles of neural science,* ed. E. R. Kandel, H. H. Schwartz & T. M. Jessell. Appleton and Lange. [aJAH]

Goldman-Rakic, P. S. (1986) Circuitry of the prefrontal cortex and the regulation of behavior by representational knowledge. In: *Handbook of physiology: The nervous system. Higher functions of the brain,* ed. F. E. Bloom. American Psychological Society. [aJAH]

(1996) Regional and cellular fractionation of working memory. *Proceedings of the National Academy of Sciences, USA* 93:13473–80. [aJAH]

(1999) The "psychic" neuron of the cerebral cortex. *Annals of the New York Academy of Sciences* 868:13–26. [HTH]

Goodenough, D. R. (1978) Dream recall: History and current status of the field. In: *The mind in sleep: Psychology and psychophysiology,* ed. A. M. Arkin, J. S. Antrobus & S. J. Ellman. Erlbaum. [AC, aTAN]

(1991) Dream recall: History and current status of the field. In: *The mind in sleep, 2nd edition,* ed. S. J. Ellman & J. S. Antrobus. Wiley. [RCo, aJAH]

Goodenough, D. R., Lewis, H. B., Shapiro, A., Jaret, L. & Sleser, I. (1965a) Dream reporting following abrupt and gradual awakenings from different types of sleep. *Journal of Personality and Social Psychology* 2:170–79. [RG, aJAH, arTAN]

Goodenough, D. R., Lewis, II. B., Shapiro, A. & Sleser, I. (1965b) Some correlates of dream reporting following laboratory awakenings. *Journal of Nervous and Mental Disease* 140:365–73. [aJAH, aTAN]

Goodenough, D. R., Shapiro, A., Holden, M. & Steinschreiber, L. (1959) A comparison of "dreamers" and "nondreamers": Eye movements, electroencephalograms, and the recall of dreams. *Journal of Abnormal and Social Psychology* 59:295–302. [aJAH, aTAN]

Goodenough, D. R., Witkin, H. A., Koulack, D. & Cohen, H. (1975) The effects of stress films on dream affect and on respiration and eye-movement during rapid eye movement sleep. *Psychophysiology* 15:313–20. [aJAH]

Gordon, H. W., Frooman, B. & Lavie, P. (1982) Shift in cognitive asymmetries between wakings from REM and NREM sleep. *Neuropsychologia* 20:99–103. [aJAH, aTAN]

Gottesman, C. (1964) Données sur l'activité corticale au cours du sommeil profond chez le rat. *C. Royal Société de Biologie* 158:1829–34. [aTAN]

(1969) Etude sur les activités électrophysiologiques phasiques chez le Rat. *Physiology and Behavior* 4:495–504. [CG]

(1996) The transition from slow-wave sleep to paradoxical sleep: Evolving facts and concepts of the neurophysiological processes underlying the intermediate stage of sleep. *Neuroscience and Biobehavioral Reviews* 20:367–87. [CG, rJAH, arTAN, EFP-S]

(1997) Introduction to the neurophysiological study of sleep: Central regulation of skeletal and ocular activities. *Archives Italiennes de Biologie* 135:279–314. [aJAH]

(1999) Neurophysiological support of consciousness during waking and sleep. *Progress in Neurobiology* 59:469–508. [CG, aJAH, EKP]

Gottesmann, C., ed. (1967) *Recherche sur la psychophysiologie du sommeil chez le Rat.* Presses du Palais Royal. [CG]

Gottesmann, C., Gandolfo, G. & Zernicki, B. (1984) Intermediate stage of sleep in the cat. *Journal de Physiologie* 79:365–74. [CG]

Gottesmann, C., User, P. & Gioanni, H. (1980) Sleep: A physiological cerveau isolé stage? *Waking and Sleeping* 4:111–17. [CG]

Gottschalk, L. A. (1999) The application of a computerized measurement of the content analysis of natural language to the assessment of the effects of psychoactive drugs. *Methods and Findings in Experimental and Clinical Pharmacology* 21:133–38. [aJAH]

Gottschalk, L. A., Buchsbaum, M. S., Gillin, W. C., Reynolds, C. A. & Herrera, D. B. (1991a) Anxiety levels in dreams: Relation to localized cerebral glucose metabolic rate. *Brain Research* 538(4):107–10. [AAB, aJAH, EFP-S]

Gottschalk, L. A., Buchsbaum, M. S., Gillin, W. C., Wu, J., Reynolds, C. A. & Herrera, D. B. (1991b) Positron-emission tomographic studies of the relationship of cerebral glucose metabolism and the magnitude of anxiety and hostility experienced during dreaming and waking. *Journal of Neuropsychiatry and Clinical Neurosciences* 3:131–42. [EFP-S]

Gould, S. L. & Lewontin, R. C. (1979) The spandrels of San Marco and the Panglossian paradigm: A critique of the adaptationist programme. *Proceedings of the Royal Society of London B: Biological Sciences* 205(1161):581–98. [GL, NST]

Graeff, F. G., Guimeraes, F. S., De Andrade, T. G. & Deakin, J. F. (1996) Role of 5-HT in stress, anxiety, and depression. *Pharmacology, Biochemistry and Behavior* 54:129–41. [EEK]

Grahnstedt, S. & Ursin, R. (1985) Platform sleep deprivation affects deep slow wave sleep in addition to REM sleep. *Behavioral Brain Research* 18:233–39. [aRPV]

Granon, S., Passetti, F., Thomas, K. L., Dalley, J. W., Everitt, B. J. & Robbins, T. W. (2000) Enhanced and impaired attentional performance after infusion of D1 dopaminergic receptor agents into rat prefrontal cortex. *Journal of Neuroscience* 20:1208–15. [RCo]

Grastyan, E., Lissak, K., Madarsz, I. & Donhoffer, H. D. (1959) Hippocampal electrical activity during the development of conditioned reflexes. *Electroencephalography and Clinical Neurophysiology* 11:409–30. [CG]

Gray, J. A. (1982) *The neuropsychology of anxiety: An enquiry into the functions of the septo-hippocampal system.* Oxford University Press. [JBP]

Green, J. D. & Arduini, A. (1954) Hippocampal electrical activity in arousal. *Journal of Neurophysiology* 17:532–57. [CG]

Greenberg, R. (1966) Cerebral cortex lesions: The dream process and sleep spindles. *Cortex* 2:357–66. [RG, aJAH]

(1978) The cortex finds its place in REM sleep. *Behavioral and Brain Sciences* 3:490–91. [RG]

Greenberg, R., Katz, H., Schwartz, W. & Pearlman, C. (1992) A research based reconsideration of the psychoanalytic theory of dreaming. *Journal of the American Psychoanalytic Association* 40:531–50. [RG]

Greenberg, R. & Pearlman, C. (1974) Cutting the REM nerve: An approach to the adaptive function of REM sleep. *Perspectives in Biology and Medicine* 17:513–52. [RG, JP]

(1975) A psychoanalytic dream continuum: The source and function of dreams. *International Review of Psychoanalysis* 2:441–48. [RC]

(1993) An integrated approach to dream theory: Contributions from sleep research and clinical practice. In: *The functions of dreaming,* ed. A. Moffitt, M. Kramer & R. Hoffman. State University of New York Press. [RG]

(1999) The interpretation of dreams: A classic revisited. *Psychoanalytic Dialogues* 9:749–65. [RG]

Greenberg, R., Pearlman, C., Schwartz, W. & Youkilis, H. (1983) Memory, emotion, and REM sleep. *Journal of Abnormal Psychology* 92:378–81. [RG]

Greenberg, R., Pillard, R. & Pearlman, C. (1972) The effect of dream (stage REM) deprivation on adaptation to stress. *Psychosomatic Medicine* 34:257–62. [aJAH]

Greene, R. W. & McCarley, R. W. (1990) Cholinergic neurotransmission in the brainstem: Implications for behavioral state control. In: *Brain cholinergic mechanisms,* ed. M. Steriade & D. Biesold. Oxford Science. [aJAH]

Greenstein, Y. J., Pavlides, C. & Winson, J. (1988) Long-term potentiation in the dentate gyrus is preferentially induced at theta rhythm periodicity. *Brain Research* 438:331–34. [aRPV]

Greenwood, P., Wilson, D. & Gazzaniga, M. (1977) Dream report following commissurotomy. *Cortex* 13:311–16. [aMS]

Gregor, T. (1981) A content analysis of Mehinaku dreams. *Ethos* 9:353–90. [arAR]

Grieser, G., Greenberg, R. & Harrison, R. (1972) The adaptive function of sleep: The differential effects of sleep and dreaming on recall. *Journal of Abnormal Psychology* 80:280–86. [VSR]

Griffith, R. M., Myiago, O. & Tago, A. (1958) The universality of typical dreams: Japanese vs. Americans. *American Anthropologist* 60:1173–79. [AG]

Gritti, I., Manville, L. & Jones, B. E. (1993) Codistribution of GABA with acetycholine synthesizing neurons in the basal forebrain of the rat. *Journal of Comparative Neurology* 329:438–57. [aJAH]

(1994) Projections of GABAergic and cholinergic based forebrain and GABAergic preoptic-anterior hypothalamic neurons to the posterior lateral hypothalamus of the rat. *Journal of Comparative Neurology* 339:251–68. [aJAH]

Gronfier, C., Luthringer, R., Follenius, M., Schaltenbrand, N., Macher, J. P., Muzet, A. & Brandenberger, G. (1997) Temporal relationships between pulsatile cortisol secretion and electroencephalographic activity during sleep in man. *Electroencephalography and Clinical Neurophysiology* 103:405–408. [JB]

Gronier, B. & Rasmussen, K. (1998) Activation of midbrain presumed dopaminergic neurons by muscarinic cholinergic receptors: An in vivo electrophysiological study in the rat. *British Journal of Pharmacology* 124:455–64. [EKP]

Groos, K. (1896) *The play of man.* Appleton-Century-Crofts. [JAC]

Gross, D. W. & Gotman, J. (1999) Correlation of high-frequency oscillations with the sleep-wake cycle and cognitive activity in humans. *Neuroscience* 94:1005–18. [EFP-S]

Gross, J., Byrne, J. & Fisher, C. (1965) Eye movements during emergent stage 1 EEG in subjects with lifelong blindness. *Journal of Nervous and Mental Disorders* 141:365–70. [aJAH]

Gruen, I., Martinez, A., Cruz-Olloa, C., Aranday, F. & Calvo, J. M. (1997) Caracteristicas de los fenomenos emocionales en las ensonaciones de pacientes con epilepsia del lobulo temporal. *Salud Mental* 20:8–15. [aJAH]

Grünstein, A. (1924) Die Erforschung der Träume als eine Methode der topischen Diagnostik bei Grosshirnerkrankungen. *Zeitschrift für die gesamte Neurologie und Psychiatrie* 93:416–20. [aMS]

Guilleminault, C. (1987) Disorders of arousal in children: Somnambulism and night terrors. In: *Sleep and its disorders in children,* ed. C. Guilleminault. Raven Press. [aJAH]

Guilleminault, C., Moscovitch, A. & Leger, D. (1995) Forensic sleep medicine: Nocturnal wandering and violence. *Sleep* 18:740–48. [aAR]

Gutwein, B. M. & Fishbein, W. (1980a) Paradoxical sleep and memory (I): Selective alterations following enriched and impoverished environmental rearing. *Brain Research Bulletin* 5:9–12. [WF, aRPV]

(1980b) Paradoxical sleep and memory (II): Sleep circadian rhythmicity following enriched and impoverished environmental rearing. *Brain Research Bulletin* 5:105–109. [WF, aRPV]

Habib, M. & Sigiru, A. (1987) Pure topographical disorientation: A definition and anatomical basis. *Cortex* 23:73–85. [aMS]

Haiak, G., Klingelhofer, J., Scholz-Varszeg, M., Matzander, G., Sander, D., Conrad, B. & Ruther, E. (1994) Relationship between cerebral blood flow velocities and electrical activity in sleep. *Sleep* 17:11–19. [aJAH]

Hagan, J. J., Leslie, R. A., Patel, S., Evans, M. L., Wattam, T. A., Holmes, S.,

# References

Benham, C. D., Taylor, S. G., Routledge, C., Hemmati, P., Munton, R., Ashmeade, T. E., Shah, A. S., Hatcher, J. P., Hatcher, P. D., Jones, D. N. C., Smith, M. I., Piper, D. C., Hunter, A. J., Porter, R. A. & Upton, N. (1999) Orexin A activates locus coeruleus cell firing and increases arousal in the rat. *Proceedings of the National Academy of Sciences USA* 96:10911–16. [RS-P]

Hagan, J. J., Salamone, J. D., Simpson, J., Iversen, S. D. & Morris, R. G. (1988) Place navigation in rats is impaired by lesions of medial septum and diagonal band but not nucleus basalis magnocellularis. *Behavioural Brain Research* 27:9–20. [aRPV]

Hajak, G., Klingelhofer, J., Schulz-Varszegi, M., Matzander, G., Sander, D., Conrad, B. & Ruther, E. (1994) Relationship between blood flow velocities and cerebral electrical activity in sleep. *Sleep* 17:11–19. [CF]

Hall, C. S. (1966) Studies of dreams collected in the laboratory and at home. *Institute of Dream Research Monograph Series* (No. 1). Privately printed. [GWD]

(1967) Caveat lector. *Psychoanalytic Review* 54:655–61. [GWD]

(1981) Do we dream during sleep? Evidence for the Goblot hypothesis. *Perceptual and Motor Skills* 53:239–46. [IF]

Hall, C. S., Bukolz, E. & Fishburne, C. J. (1992) Imagery and the acquisition of motor skills. *Canadian Journal of Sports Science* 17:19–27. [aAR]

Hall, C. S. & Domhoff, G. W. (1963) Aggression in dreams. *International Journal of Social Psychiatry* 9:259–67. [aAR]

(1964) Friendliness in dreams. *Journal of Social Psychology* 62:309–14. [aAR]

Hall, C. S. & Van de Castle, R. I. (1966) *The content analysis of dreams.* Appleton-Century-Crofts/Meredith. [AC, RCo, GWD, aJAH, LM, JM, aTAN, arAR, aRPV]

Hamann, S. P., Ely, T. D., Grafton, S. T. & Kilts, C. D. (1999) Amygdala activity related to enhanced memory for pleasant and aversive stimuli. *Nature Neuroscience* 2:289–93. [aAR, AR]

Hansen, C. H. & Hansen, R. D. (1988) Finding the face in the crowd: An anger superiority effect. *Journal of Personality and Social Psychology* 54:917–24. [LM]

(1994) Automatic emotion: Attention and facial efference. In: *The heart's eye: Emotional influences in perception and attention,* ed. P. M. Neidenthal & S. Kitayama. Academic Press. [LM]

Hardman, J. G., Limbird, L. E., Molinoff, P. B., Ruddon, R. W. & Goodman, A. G., eds. (1996) *Goodman and Gilman's, The pharmacological basis of therapeutics, 9th edition.* McGraw-Hill. [aRPV]

Harlow, J. & Roll, S. (1992) Frequency of day residue in dreams of young adults. *Perceptual and Motor Skills* 74:832–34. [aJAH]

Harris, I. (1948) Observations concerning typical anxiety dreams. *Psychiatry* 11:301–309. [rAR]

Hars, B. & Hennevin, E. (1987) Impairment of learning by cueing during postlearning slow-wave sleep in rats. *Neuroscience Letters* 79:290–94. [RG, aTAN]

Hars, B., Hennevin, E. & Pasques, P. (1985) Improvement of learning by cueing during postlearning paradoxical sleep. *Behavioural Brain Research* 18:241–50. [aRPV]

Hartmann, E. (1966a) Reserpine: Its effect on the sleep dream cycle. *Psychopharmacologica* 9:242–17. [rJAH]

(1966b) The psychophysiology of free will. In: *Psychoanalysis: A general psychology,* ed. R. Lowenstein, L. Newman & A. Solnit. International Universities Press. [aJAH]

(1967) *The biology of dreaming.* Charles C. Thomas. [EH]

(1968) The effect of four drugs on sleep patterns in man. *Psychopharmacology* 12:346–53. [LJW]

(1970) A note on the nightmare. *International Psychiatry Clinics* 7:192–97. [LJW]

(1973) *The functions of sleep.* Yale University Press. [EH]

(1976) Long-term administration of psychotropic drugs: Effects on human sleep. In: *Pharmacology of sleep,* ed. R. L. Williams & I. Karacan. Wiley. [LJW]

(1978) The biochemistry of the nightmare: Possible involvement of dopamine. *Sleep Research* 7:186. [aJAH]

(1982) From the biology of dreaming to the biology of the mind. *The Psychoanalytic Study of the Child* 37:303–35. [aJAH]

(1984) *The nightmare: The psychology and biology of terrifying dreams.* Basic Books. [RG, RLe, aAR]

(1991) *Boundaries in the mind.* Basic Books. [EH]

(1995) Making connections in a safe place: Is dreaming psychotherapy? *Dreaming* 5:213–28. [aAR]

(1996a) Outline for a theory on the nature and functions of dreaming. *Dreaming* 6:147–70. [MBl, EH, aAR]

(1996b) We do not dream of the three R's: A student and implications. *Sleep Research* 25:136. [EH]

(1998) *Dreams and nightmares: The new theory on the origin and meaning of dreams.* Plenum Press. [MBl, RG, EH, RLe, WHM, arAR, MSc]

(2000a) Thought people and dream people: Individual differences on the waking

to dreaming continuum. In: *Individual differences and conscious experience,* ed. D. Wallace. John Benjamins. [EH]

(2000b) We do not dream of the three R's: Implications for the nature of dreaming mentation. *Dreaming* 10:103–10. [MBl, EH]

Hartmann, E. & Cravens, J. (1973) The effects of long term administration of psychotropic drugs on human sleep: III. The effects of amitriptyline. *Psychopharmacologia* 33:185–202. [aRPV]

Hartmann, E., Kunzendorf, R., Rosen, R. & Grace, N. (1998a) Contextualizing images in dreams and daydreams. *Sleep* 21S:279. [EH]

Hartmann, E., Rosen, R. & Grace, N. (1998b) Contextualizing images in dreams: More frequent and more intense after trauma. *Sleep* 21S:284. [EH]

Hartmann, E., Russ, D., Oldfield, M., Falke, R. & Skoff, B. (1980) Dream content: Effects of L-DOPA. *Sleep Research* 9:153. [aJAH, MSc, arMS]

Hartmann, E., Russ, D., Oldfield, M., Sivan, I. & Cooper, S. (1987) Who has nightmares? *Archives of General Psychiatry* 44:49–56. [RLe]

Hartmann, E. & Stern, W. C. (1972) Desynchronized sleep deprivation: Learning deficit and its reversal by increased catecholamines. *Physiology and Behavior* 8:585–87. [aRPV]

Hasegawa, I., Fukushima, T., Ihara, T. & Miyashita, Y. (1998) Callosal window between prefrontal cortices: Cognitive interaction to retrieve long-term memory. *Science* 281:814–18. [JSA]

Hasselmo, M. (1999) Neuromodulation: Acetylcholine and memory consolidation. *Trends in Cognitive Sciences* 3:351–59. [RWG, aJAH, RS]

Hasselmo, M. & Bower, J. M. (1993) Acetylcholine and memory. *Trends in Neuroscience* 16:218–22. [RS]

Hauri, P. (1970) Evening activity, sleep mentation, and subjective sleep quality. *Journal of Abnormal Psychology* 76:270–75. [aJAH, aTAN]

Hauri, P., Sawyer, J. & Rechtschaffen, A. (1967) Dimensions of dreaming: A factored scale for rating dream reports. *Journal of Abnormal Psychology* 22:16–22. [aJAH]

Hebb, D. O. (1949) *Organization of behavior.* Wiley. [WF, TJM]

Hécean, H. & Albert, M. (1978) *Human neuropsychology.* Wiley. [aMS]

Hecker, R. & Mapperson, B. (1997) Dissociation of visual and spatial processing in working memory. *Neuropsychologia* 35:599–603. [rTAN]

Heimer, L. & Alheid, G. (1999) Basal forebrain organization: Emerging concepts. In: *Encyclopedia of neuroscience,* ed. G. Adelman & B. H. Smith. Elsevier. [rJAH]

Heiss, W.-D., Pawlik, G., Herholz, K., Wagner, R. & Weinhard, K. (1985) Regional cerebral glucose metabolism in man during wakefulness, sleep, and dreaming. *Brain Research* 327:362–66. [AAB, aJAH, aMS]

Helekar, S. A. (1999) On the possibility of universal neural coding of subjective experience. *Consciousness and Cognition* 8:423–46. [rTAN]

Henane, R., Buguet, A., Roussel, B. & Bittel, J. (1977) Variations in evaporation and body temperatures during sleep in man. *Journal of Applied Psychology* 42:50–55. [aTAN]

Hendricks, J. C., Morrison, A. R. & Mann, G. L. (1982) Different behaviors during paradoxical sleep without atonia depend on pontine lesion site. *Brain Research* 239:81–105. [ARM]

Henik, A., Rafal, R. & Rhodes, D. (1994) Endogenously generated and visually guided saccades after lesions of the human frontal eye fields. *Journal of Cognitive Neuroscience* 6:400–11. [aJAH]

Henley, K. & Morrison, A. R. (1974) A re-evaluation of the effects of lesions of the pontine tegmentum and locus coeruleus on phenomena of paradoxical sleep in the cat. *Acta Neurobiologia Experimentalis* 34:215–32. [ARM]

Hennevin, E. & Hars, B. (1992) Second-order conditioning during sleep. *Psychobiology* 20:166–76. [aTAN]

Hennevin, E., Hars, B. & Bloch, V. (1989) Improvement of learning by mesencephalic reticular stimulation during postlearning paradoxical sleep. *Behavioral and Neural Biology* 51:291–306. [aTAN]

Hennevin, E., Hars, B. & Maho, C. (1995a) Memory processing in paradoxical sleep. *Sleep Research Society Bulletin* 1:44–50. [RS, aRPV]

Hennevin, E., Hars, B., Maho, C. & Bloch, V. (1995b) Processing of learned information in paradoxical sleep: Relevance for memory. *Behavioural Brain Research* 69:125–35. [aTAN, aRPV]

Henry, G. M., Weingartner, H. & Murphy, D. L. (1973) Influence of affective states and psychoactive drugs on verbal learning and memory. *American Journal of Psychiatry* 130:966–71. [aRPV]

Hepler, D. J., Olton, D. S., Wenk, G. L. & Coyle, J. T. (1985) Lesions in nucleus basalis magnocellularis and medial septal area of rats produce qualitatively similar memory impairments. *Journal of Neuroscience* 5:866–73. [aRPV]

Hepp, K., Henn, V., Vili, T. & Cohen, B. (1989) Brainstem regions related to saccade generation. In: *The neurobiology of saccadic eye movements,* ed. R. H. Wurtz & M. E. Goldberg. Elsevier. [aJAH]

Herculano, H. L., Munk, M. H., Neuenschwander, S. & Singer, W. (1999) Precisely synchronized oscillatory firing patterns require electroencephalographic activation. *Journal of Neuroscience* 19(10):3992–4010. [RWG]

Herman, J. H. (1992) The reproductive properties of dreams:

Evidence for cortical modulation of brain stem generators. In: *The neuropsychology of dreaming sleep*, ed. J. Antrobus & M. Bertini. Erlbaum. [JH, aJAH]

Herman, J. H., Barker, D. R. & Roffwarg, H. P. (1983) Similarity of eye movement characteristics in REM sleep and the awake state. *Psychophysiology* 20:537–43. [JH, aJAH]

Herman, J. H., Ellman, S. J. & Roffwarg, H. P. (1978) The problem of NREM recall re-examined. In: *The mind in sleep*, ed. A. M. Arkin, J. S. Antrobus & S. J. Ellman. Erlbaum. [aJAH, TLK, arTAN]

Herman, J. H., Herman, M., Boys, R., Perser, L., Taylor, M. E. & Roffwarg, H. P. (1984) Evidence for a directional correspondence between eye movements and dream imagery in REM sleep. *Sleep* 7:52–63. [aJAH, VSR]

Herman, J. H., Roffwarg, H. P., Taylor, M. E., Boys, R. M., Steigman, K. B. & Barker, D. R. (1981) Saccadic velocity in REM sleep dreaming, normal visual activity and total darkness. *Psychophysiology* 8:188. [aJAH]

Hernandez-Peon, A., Chavez-Zbarra, G., Morgans, P. J. & Timo-Iaria, C. (1963) Cholinergic pathways involved in sleep and emotional behavior. *Experimental Neurology* 8:93–111. [PJM]

Hertz, J., Krogh, A. & Palmer, R. G. (1991) *Introduction to the theory of neural computation. Lecture notes, vol. 1.* Addison-Wesley. [TJM]

Hill, C. E., Diemer, R., Hess, S., Hillyer, A. & Seeman, R. (1993) Are the effects of dream interpretation on session quality, insight, and emotions due to the dream itself, to projection, or to the interpretation process? *Dreaming* 3:269–80. [MBl, aAR]

Hillered, L. & Persson, L. (1995) Parabanic acid for monitoring of oxygen radical activity in the injured human brain. *NeuroReport* 6(13):1816–20. [CMP]

Hindmarch, I. & Bhatti, J. Z. (1988) Psychopharmacological effects of sertraline in normal, healthy volunteers. *European Journal of Clinical Pharmacology* 35:221–23. [aRPV]

Hindmarch, I., Shillingford, J. & Shillingford, C. (1990) The effects of sertraline on psychomotor performance in elderly volunteers. *Journal of Clinical Psychiatry* 51:34–36. [aRPV]

Hinton, G. E. & Sejnowski, T. J. (1986) Learning and relearning in Boltzmann machines. In: *Parallel distributed processing, vol. 1*, ed. David E. Rumelhart, James. L. NcClelland, and the PDP Research Group. A Bradford Book/MIT Press. [TJM]

Hinton, G. & Shallice, T. (1991) Lesioning an attractor network: Investigations of acquired dyslexia. *Psychological Review* 98:74–95. [JSA]

Hirai, N., Uchida, S., Maehara, T., Okubo, Y. & Shimizu, H. (1999) Enhanced gamma (30–150 Hz) frequency in the human medial temporal lobe. *Neuroscience* 90:1149–55. [aJAH]

Hobson, J. A. (1988a) Homeostasis and heteroplasticity: Functional significance of behavioral state sequences. In: *Clinical physiology of sleep*. American Physiological Association. [OF, aJAH]

(1988b) *The dreaming brain: How the brain creates both the sense and the nonsense of dreams*. Basic Books. [RBA, GWD, IF, JAH, arJAH, TLK, MM, aTAN, aAR, arMS, MS, aRPV, LJW]

(1989) *Sleep*. Scientific American Library. [aJAH]

(1990) Activation, input source, and modulation: A neurocognitive model of the state of the brain-mind. In: *Sleep and cognition*, ed. R. R. Bootzin, J. F. Kihlstrom & D. L. Schacter. American Psychological Association. [JH, arJAH]

(1992a) A new model of the brain-mind state: Activation level, input source, and mode of processing (AIM). In: *The neuropsychology of sleep and dreaming*, ed. J. S. Antrobus & M. Bertini. Erlbaum. [JH, arJAH, aTAN, aMS, RS]

(1992b) Sleep and dreaming: Induction and mediation of REM sleep by cholinergic mechanisms. *Current Opinions in Neurobiology* 2:759–63. [aJAH, LJW]

(1994) *The chemistry of conscious states*. Little, Brown. [aJAH, aAR, aMS]

(1997a) Consciousness as a state-dependent phenomenon. In: *Scientific approaches to the question of consciousness*, ed. J. Cohen & J. Schooler. Erlbaum. [aJAH]

(1997b) Dreaming as delirium: A mental status analysis of our nightly madness. *Seminars in Neurology* 17:121–28. [EH, aJAH, aRPV]

(1999a) *Consciousness*. Scientific American Library. [aAR]

(1999b) *Dreaming as delirium*. MIT Press. [rJAH]

(1999c) The new neuropsychology of sleep: Implications for psychoanalysis. *Neuropsychoanalysis* 1:157–83. [rMS]

Hobson, J. A., Alexander, J. & Fredrickson, C. J. (1969) The effect of lateral geniculate lesions on phasic activity of the cortex during desynchronized sleep in the rat. *Brain Research* 14:607–21. [RCo]

Hobson, J. A. & Brazier, M. A. B., eds. (1981) *The reticular formation revisited*. Raven Books. [aJAH]

Hobson, J. A., Datta, S., Calvo, J. M. & Quattrochi, J. (1993) Acetylcholine as a brain state modulator: Triggering and long-term regulation of REM sleep. *Progress in Brain Research* 98:389–404. [aJAH]

Hobson, J. A., Goldfrank, F. & Snyder, F. (1965) Respiration and mental activity in sleep. *Journal of Psychiatric Research* 3:79–90. [aJAH, arTAN]

Hobson, J. A., Hoffman, E., Helfand, R. & Kostner, D. (1987) Dream bizarreness

and the activation-synthesis hypothesis. *Human Neurobiology* 6:157–64. [aJAH]

Hobson, J. A., Lydic, R. & Baghdoyan, H. (1986) Evolving concepts of sleep cycle generation: From brain centers to neuronal populations. *Behavioral and Brain Sciences* 9:371–400. [aJAH, aMS, MS]

Hobson, J. A. & McCarley, R W. (1977) The brain as a dream state generator: An activation-synthesis hypothesis of the dream process. *American Journal of Psychiatry* 134:1335–48. [JAB, JAC, AC, RCo, CG, JH, aJAH, JAH, AK, aTAN, aAR, arMS, aRPV, GWV]

Hobson, J. A., McCarley, R. W. & Wyzinki, P. W. (1975) Sleep cycle oscillation: Reciprocal discharge by two brainstem neuronal groups. *Science* 189:55–58. [CG, arJAH, EKP]

Hobson, J. A. & Pace-Schott, E. F. (1999) Reply to Solms, Braun and Reiser. *Neuropsychoanalysis* 1:206–24. [arJAH, JAH]

Hobson, J. A., Pace-Schott, E. F. & Stickgold, R. (2000) Consciousness: Its vicissitudes in waking and sleep – an integration of recent neurophysiological and neuropsychological evidence. In: *The new cognitive neurosciences, second edition*, ed. M. Gazzaniga. MIT Press. [RCo, aJAH, EFP-S]

Hobson, J. A., Pace-Schott, E. F., Stickgold, R. & Kahn, D. (1998a) To dream or not to dream? Relevant data from new neuroimaging and electrophysiological studies. *Current Opinion in Neurobiology* 8:239–44. [aJAH, EFP-S]

Hobson, J. A. & Steriade, M. (1986) The neuronal basis of behavioral state control. In: *Handbook of physiology – the nervous system, vol. IV*, ed. F. E. Bloom. American Physiological Society. [aJAH]

Hobson, J. A. & Stickgold, R. ( 1994a) Dreaming: A neurocognitive approach. *Consciousness and Cognition* 3:1–15. [aJAH, aTAN]

(1994b) The conscious state paradigm: A neurocognitive approach to waking, sleeping and dreaming. In: *The cognitive neurosciences*, ed. M. Gazzaniga. MIT Press. [aJAH]

(1995) The conscious state paradigm: A neurocognitive approach to waking, sleeping, and dreaming. In: *The cognitive neurosciences*, ed. M. S. Gazzaniga. MIT Press. [aTAN]

Hobson, J. A., Stickgold, R. & Pace-Schott, E. (1998b) The neuropsychology of REM sleep dreaming. *NeuroReport* 9:R1–14. [MBo, JAC, GWD, CG, aJAH, JAH, AK, aMS, MS, arRPV]

Hobson, J. A., Stickgold, R., Pace-Schott, E. F. & Leslie, K. R. (1998c) Sleep and vestibular adaptation: Implications for function in microgravity. *Journal of Vestibular Research* 8:81–94. [JAC, aJAH]

Hodes, R. & Dement, W. C. (1964) Depression of electrically induced reflexes ("H-reflexes") in man during low voltage EEG sleep. *Electroencephalography and Clinical Neurophysiology* 17:617–29. [aJAH]

Hoelscher, T. J., Klinger, E. & Barta, S. G. (1981) Incorporation of concern- and nonconcern-related verbal stimuli into dream content. *Journal of Abnormal Psychology* 90:88–91. [aAR]

Hofle, N., Paus, T., Reutens, D., Fiset, P., Gotman, J., Evans, A. C. & Jones, B. E. (1997) Regional cerebral blood flow changes as a function of delta and spindle activity during slow wave sleep in humans. *Journal of Neuroscience* 17:4800–808. [AAB, RCo, CF, CG, arJAH, EFP-S]

Hökfelt, T., Ljungdahl, A., Fuxe, K. & Johansson, O. (1974) Dopamine nerve terminals in the limbic cortex: Aspects of the dopamine hypothesis of schizophrenia. *Science* 184:177–79. [CG]

Holmes, C. J. & Jones, B. E. (1994) Importance of cholinergic, GABAergic, serotonergic and other neurons in the medial medullary reticular formation for sleep-wake states studies by cytotoxic lesions in the cat. *Neuroscience* 62:1179–200. [aJAH]

Holmes, C. J., Mainville, L. S. & Jones, B. E. (1994) Distribution of cholinergic, GABAergic and serotonergic neurons in the medial medullary reticular formation and their projections studied by cytotoxic lesions in the cat. *Neuroscience* 62:1155–78. [aJAH]

Honda, T. & Semba, K. (1994) Serotonergic synaptic input to cholinergic neurons in the rat mesopontine tegmentum. *Brain Research* 47:299–306. [aJAH]

Hong, C. C. H., Gillin, J. C., Dow, B. M., Wu, J. & Buchsbaum, M. S. (1995) Localized and lateralized cerebral glucose metabolism associated with eye movements during REM sleep and wakefulness: A positron emission tomography (PET) study. *Sleep* 18:570–80. [CG, aJAH, EFP-S, arMS]

Hong, C. C. H., Potkin, S. G., Antrobus, J. S., Dow, B. M., Callaghan, G. M. & Gillin, J. C. (1997) REM sleep eye movement counts correlate with visual imagery in dreaming: A pilot study. *Psychophysiology* 34:377–81. [IF, aJAH]

Hopfield, J. J. (1982) Neural networks and physical systems with emergent collective computational abilities. *Proceedings of the National Academy of Sciences USA* 79:2554–58. [TJM]

Hoppe, K. (1977) Split brains and psychoanalysis. *Psychoanalytic Quarterly* 46:220–44. [arMS]

Hori, T., Hayashi, M. & Morikawa, T. (1994) Topography and coherence analysis of hypnagogic EEG. In: *Sleep onset: Normal and abnormal processes*, ed. R. D. Ogilivie & J. R. Harsh. American Psychological Association. [aTAN]

## References

Horne, J. A. (1976) Recovery sleep following different visual conditions during total sleep deprivation in man. *Biological Psychology* 4:107–18. [aRPV]
(1988) *Why we sleep: The function of sleep in humans and other mammals.* Oxford University Press. [aRPV]

Horne, J. A. & McGrath, M. J. (1984) The consolidation hypothesis for REM sleep function: Stress and other confounding factors – a review. *Biological Psychology* 18:165–84. [WF, aRPV]

Horne, J. A. & Walmsley, B. (1976) Daytime visual load and the effects upon human sleep. *Psychophysiology* 13:115–20. [aRPV]

Horner, R. L., Sanford, L. D., Annis, D., Pack, A. I. & Morrison, A. R. (1997) Serotonin at the laterodorsal tegmental nucleus suppresses rapid eye movement sleep in freely behaving rats. *Journal of Neuroscience* 17:7541–52. [aJAH]

Horvath, T. L., Peyron, C., Diano, S., Ivanov, A., Aston-Jones, G., Kilduff, T. S. & van den Pol, A. N. (1999) Hypocretin (orexin) activation and synaptic innervation of the locus coeruleus noradrenergic system. *Journal of Comparative Neurology* 415:145–59. [RS-P]

Hoshi, Y., Mizukami, S. & Tamora, M. (1994) Dynamic features of hemodynamic and metabolic changes in the human brain during all-night sleep as revealed by near-infrared spectroscopy. *Brain Research* 652:257–62. [aJAH]

Hrdy, S. B. (1977) Infanticide as a primate reproductive strategy. *American Scientist* 65:40–49. [aAR]

Hu, B., Steriade, M. & Deschenes, M. (1989) The cellular mechanisms of thalamo-ponto-geniculo-occipital (PGO) waves. *Neuroscience* 31:25–35. [aJAH]

Hu, D. G, Griesbach, G. & Amsel, A. (1997) Development of vicarious trial-and-error behavior in odor discrimination learning in the rat: Relation to hippocampal function? *Behavior and Brain Research* 86(1):67–70. [EEK]

Huang-Hellinger, F. R., Breiter, H. C., McCormack, G., Cohen, M. S., Kwong, K. K., Sutton, J. P., Savoy, R. L., Weisskoff, R. M., Davis, T. L., Baker, J. R., Belliveau, J. W. & Rosen, B. R. (1995) Simultaneous functional magnetic resonance imaging and electrophysiological recording. *Human Brain Mapping* 3:13–23. [aJAH]

Hublin, C., Kaprio, J., Partinen, M. & Koskenvuo, M. (1999a) Nightmares: Familial aggregation and association with psychiatric disorders in a nationwide twin cohort. *American Journal of Medical Genetics (Neuropsychiatric Genetics)* 88:329–36. [TAN, aAR, MSc]
(1999b) Limits of self-report in assessing sleep terrors in a population survey. *Sleep* 22:89–93. [aAR]

Huerta, P. T. & Lisman, J. E. (1993) Heightened synaptic plasticity of hippocampal CA1 neurons during a cholinergically induced rhythmic state. *Nature* 364:723–25. [aRPV]

Hulse, S. H., Egeth, H. & Deese, J. (1980) *The psychology of learning, 5th edition.* McGraw-Hill. [ARM]

Humphrey, M. & Zangwill, O. (1951) Cessation of dreaming after brain injury. *Journal of Neurology, Neurosurgery and Psychiatry* 14:322–25. [aMS]

Humphrey, N. (1980) Nature's psychologists. In: *Consciousness and the physical world,* ed. B. Josephson & V. Ramachandran. Pergamon Press. [NH]
(1983) Dreaming and being dreamed. In: *Consciousness regained: Chapters in the development of mind,* N. Humphrey. Oxford University Press. [NH]
(1986) A book at bedtime? In: *The inner eye.* Faber & Faber. [NH]

Hunt, H. T. (1982) Forms of dreaming. *Perceptual and Motor Skills* 54 (Monograph Supplement I-V54):559–633. [aJAH, HTH]
(1989) *The multiplicity of dreams: Memory, imagination, and consciousness.* Yale University Press. [aJAH, HTH, TLK, aTAN, aAR]
(1991) Dreams as literature/science: An essay. *Dreaming* 1:235–42. [aJAH]
(1995) *On the nature of consciousness.* Yale University Press. [HTH]

Hunt, H. T., Ruzycki-Hunt, K., Pariak, D. & Belicki, K. (1993) The relationship between dream bizarreness and imagination: Artifact or essence? *Dreaming* 3:179–99. [aJAH, HTH, arTAN, EFP-S]

Hunt, W. K., Sanford, L. D., Ross, R. J., Morrison, A. R. & Pack, A. I. (1998) Elicited ponto-geniculo-occipital waves and phasic suppression of diaphragm activity in sleep and wakefulness. *Journal of Applied Psychology* 84:2106–14. [aTAN]

Huttenlocher, P. R. (1961) Evoked and spontaneous activity in single units of medial brainstem during natural sleep and waking. *Journal of Neurophysiology* 24:451–68. [aJAH]

Hurwitz, T. D., Mahowald, M. W., Kuskowski, M. & Engdahl, B. E. (1998) Polysomnographic sleep is not clinically impaired in Vietnam combat veterans with chronic posttraumatic stress disorder. *Biological Psychiatry* 44:1066–73. [arAR]

Husserl, E. (1965) *Phenomenology and the crisis of philosophy,* trans. Q. Laurer. Harper & Row. [rTAN]

Ikeda, K. & Morotomi, T. (1997) Reversed discriminatory responses of heart rate during human REM sleep. *Sleep* 20:942–47. [aTAN]

Ilomoto, S. & Panksepp, J. (1999) The role of nucleus accumbens dopamine in motivated behavior: A unifying interpretation with special reference to reward-seeking. *Brain Research Reviews* 31:6–41. [JP]

Imeri, L., DeSimoni, M. G., Gigho, R., Clavenna, A. & Mancia, M. (1994) Changes in the serotonergic system during the sleep-wake cycle: Simultaneous polygraphic and voltametric recordings in hypothalamus using a telemetry system. *Neuroscience* 58:353–58. [aJAH]

Imon, H., Ito, K., Dauphin, L. & McCarley, R. W. (1996) Electrical stimulation of the cholinergic anterodorsal tegmental nucleus elicits scopolamine-sensitive excitatory postsynaptic potentials in medial pontine reticular formation neurons. *Neuroscience* 74:393–401. [aJAH]

Inanaga, K. (1998) Frontal midline theta rhythm and mental activity. *Psychiatry and Clinical Neurosciences* 52:555–66. [aJAH]

Inglis, W. L. & Semba, K. (1996) Colocalization of ionotropic glutamate receptor subunits with NADPH-diaphorase-containing neurons in the rat mesopontine tegmentum. *Journal of Comparative Neurology* 368:17–32. [aJAH]

Inglis, W. L. & Winn, P. (1995) The pedunculopontine tegmental nucleus: Where the striatum meets the reticular formation. *Progress in Neurobiology* 47:1–29. [aJAH]

Inman, D. J., Silver, S. M. & Doghramji, K. (1990) Sleep disturbance in post-traumatic stress disorder: A comparison with non-PTSD insomnia. *Journal of Traumatic Stress* 3:429–37. [aAR]

Inoue, S., Honda, K., Kimura, M. & Zhang, S.-Q. (1999a) Endogenous sleep substances and REM sleep. In: *Rapid eye movement sleep,* ed. B. N. Mallick & S. Inoue. Marcel Dekker. [aJAH]

Inoue, S., Saha, U. K. & Musha, T. (1999b) Spatio-temporal distribution of neuronal activities and REM sleep. In: *Rapid eye movement sleep,* ed. B. N. Mallick & S. Inoue. Marcel Dekker. [aJAH]

Itil, T. M. (1976) Discrimination between some hypnotic and anxiolytic drugs by computer-analyzed sleep. In: *Pharmacology of sleep,* ed. R. L. Williams & I. Karacan. Wiley. [LJW]

Ito, M. (1987) Oculomotor system, mechanisms. In: *Encyclopedia of neuroscience,* ed. G. Adelman. Birkhauser. [aJAH]

Ives, J. R., Thomas, R., Jakob, P. M., Lovblad, K. O., Matheson, J., Scammel, T., Edelman, R. R., Warah, S. & Schomer, D. L. (1997) Technique and methodology for recording/monitoring the subject's sleep stage during "quiet" functional magnetic resonance imaging (fMRI). *Sleep Research* 26:665. [aJAH]

Iwakiri, H., Matsuyama, K. & Mori, S. (1993) Extracellular levels of serotonin in the medial pontine reticular formation in relation to sleep-wake cycle in cats: A microdialysis study. *Neuroscience Research* 18:157–70. [aJAH]

Izard, C. E. (1991) *The psychology of emotions.* Plenum Press. [LM]

Jackson, F. (1986) What Mary didn't know. *The Journal of Philosophy* 93(5):291–95. [KG]

Jacobs, B. L. (1986) Single unit activity of locus coeruleus neurons in behaving animals. *Progress in Neurobiology* 27:183–94. [aJAH]

Jacobs, B. L. & Azmita, E. C. (1992) Structure and function of the brain serotonin system. *Physiological Reviews* 72:165–229. [arJAH]

Jacobs, L., Feldman, M. & Bender, M. (1972) Are the eye movements of dreaming sleep related to the visual images of dreams? *Psychophysiology* 9:393–401. [aJAH]

Jacobson, A., Kales, A., Lehmann, D. & Zweizig, J. R. (1965) Somnambulism: All-night EEG studies. *Science* 148:975–77. [DL]

Janz, D. (1974) Epilepsy and the sleep-waking cycle. In: *Handbook of clinical neurology,* ed. P. Vinken & G. Bruyn. Elsevier. [aMS]

Jasper, A. H. & Tessier, J. (1971) Acetylcholine liberation from cerebral cortex during paradoxical sleep (REM). *Science* 172:601–602. [CG, aJAH]

Jeannerod, M. (1994) The representing brain: Neural correlates of motor intention and imagery. *Behavioral and Brain Sciences* 17:187–245. [aAR]
(1995) Mental imagery in the motor context. *Neuropsychologia* 33:1419–32. [aAR]

Jeannerod, M. & Mouret, J. (1963) Recherches sur les mechanismes des movements des yeux observes au cours de la vielle et du sommeil. *Pathologie-Biologie* 11:1053–60. [aJAH]

Jeannerod, M., Mouret, J. & Jouvet, M. (1965) Etude de la motoricite oculaire au cours de la phase paradoxale du sommeil chez le chat. *Electro-encephalography and Clinical Neurophysiology* 18:554–66. [aJAH]

Jenkins, J. G. & Dallenbach, K. M. (1924) Obliviscence during sleep and waking. *American Journal of Psychology* 35:605–12. [aRPV]

Jessen, F., Erb, M., Klose, U., Lotze, M., Grodd, W. & Heun, R. (1999) Activation of human language processing brain regions after the presentation of random letter strings demonstrated with event-related functional magnetic resonance imaging. *Neuroscience Letters* 270:13–16. [EFP-S]

Jimenez-Capdeville, M. E. & Dykes, R. W. (1996) Changes in cortical acetylcholine release in the rat during day and night: Differences between motor and sensor areas. *Neuroscience* 71:567–79. [aJAH]

Johnson, L. & Lobin, A. (1966) Spontaneous electrodermal activity during waking and sleeping. *Psychophysiology* 3:8–17. [aJAH]

Johnson, V., Spinweber, C., Gomez, S. & Matteson, L. (1990) Daytime sleepiness, performance, mood, nocturnal sleep: The effect of benzodiazepine and caffeine on their relationship. *Sleep* 13:121–35. [MK]

Jouvet, D. (1979) Elimination of paradoxical sleep by lesions of the pontine gigantocellular tegmental field in the cat. *Neuroscience Letters* 13:285–93. [aMS]

(1991) Paradoxical sleep and its chemical/structural substrates in the brain. *Neuroscience* 40:637–56. [aJAH, BEJ, aRPV]

(1993) The organization of central cholinergic systems and their functional importance in sleep-waking states. *Progress in Brain Research* 98:61–71. [aJAH]

(1994) Reticular formation cytoarchitecture, transmitters and projections. In: *The nervous system of the rat*, ed. G. Paxinos. Academic Press. [aJAH]

(1998) The neural basis of consciousness across the sleep-waking cycle. In: *Consciousness: At the frontiers of neuroscience, advances in neurology*, vol. 77, ed. H. H. Jasper, L. Descarries, V. F. Castellucci & S. Rossignol. Lippincott-Raven. [BEJ, RS, arRPV] Also in: *Advances in Neurology* 77:75–94. [aJAH, BEJ, aRPV]

Jones, B. E. & Cuello, A. C. (1989) Afferents to the basal forebrain cholinergic cell area from pontomesencephalic-catecholamine, serotonin, and acetylcholine-neurons. *Neuroscience* 31:37–61. [arJAH, RS]

Jones, B. E. & Muhlethaler, M. (1999) Cholinergic and GABAergic neurons of the basal forebrain. In: *Handbook of behavioral state control: Molecular and cellular mechanisms*, ed. R. Lydic & H. A. Baghdoyan. CRC Press. [arJAH]

Jouvet, M. (1962) Recherches sur les structures nerveuses et les mécanismes responsables des differentes phases du sommeil physiologique. *Archives Italiennes de Biologie* 100:125–206. [CG, aJAH, ARM, aMS]

(1964) Cataplexie et sommeil paradoxical réflexes. *Comptes Rendus des Séances de La Société de Biologie et de Ses Filiales* 159:83–87. [ARM]

(1967) Neurophysiology of the states of sleep. *Physiological Reviews* 47:117–77. [aTAN]

(1969) Biogenic amines and the states of sleep. *Science* 163:32–41. [RS-P]

(1973) Essai sun le revu. *Archives Italiennes de Biologie* 111:564–76. [aJAH]

(1975) The function of dreaming: A neurophysiologist's point of view. In: Handbook of psychobiology, ed. M. S. Gazzaniga & C. Blakemore. Academic Press. [BEJ]

(1978) Does a genetic programming of the brain occur during paradoxical sleep? In: *Cerebral correlates of conscious experience*, ed. P. A. Buser & A. Rougeul-Buser. Elsevier. [JAB]

(1980) Paradoxical sleep and the nature-nurture controversy. *Progress in Brain Research* 53:331–46. [aAR]

(1999) *The paradox of sleep: The story of dreaming*. MIT Press. [arJAH, rAR]

Jouvet, M. & Delorme, F. (1965) Locus coeruleus et sommeil paradoxal. *Comptes Rendus des Séances de La Société de Biologie et de Ses Filiales* 159:895–99. [ARM, VSR]

Jouvet, M. & Michel, F. (1959) Correlation electromyographiques du sommeil chez le chat decortique et mesencephalique chronique. *Comptes Rendus des Seances de la Societe de Biologie et de Ses Filiates* 153:422–25. [aJAH]

Jouvet, M., Michel, F. & Mounier, D. (1960) Analyse electroencephalographique comparee du sommeil physiologique chez le chat et chez l'homme. *Revue Neurologique* 103:189–204. [aTAN]

Jouvet-Mounier, D., Astic, L. & Lacote, D. (1970) Ontogenesis of the states of sleep in rat, cat, and guinea pig during the first postnatal month. *Developmental Psychobiology* 2:216–39. [JMS]

Joy, R. M. & Prinz, P. N. (1969) The effect of sleep altering environments upon the acquisition and retention of a conditioned avoidance response in the rat. *Physiology and Behavior* 4:809–14. [aRPV]

Jung, C. G. (1933) *Modern man in search of a soul*. Harcourt. [JM, aAR]

(1959) *Archetypes of the collective unconscious*, trans. R. F. C. Hull. *The collected works of C. G. Jung, vol. 9(1)*. Bollingen Series XX. Princeton University Press. [JBP]

Jus, A., Jus, K., Villeneuve, A., Pires, A., Lachane, R., Fortier, J. & Villeneuve, R. (1973) Studies on dream recall in chronic schizophrenic patients after prefrontal lobotomy. *Biological Psychiatry* 6:275–93. [arMS]

Kagan, J. (1998) *Three seductive ideas*. Harvard University Press. [LJW]

Kahan, T. L. (1994) Measuring dream self-reflectiveness: A comparison of two approaches. *Dreaming* 4:177–93. [arJAH, TLK]

Kahan, T. L. & LaBerge, S. (1994) Lucid dreaming as metacognition: Implications for cognitive science. *Consciousness and Cognition* 3:246–64. [arJAH, TLK]

(1996) Cognition and metacognition in dreaming and waking: Comparisons of first and third-person ratings. *Dreaming* 6:235–49. [TLK, SL, rTAN]

Kahan, T. L., LaBerge, S., Levitan, L. & Zimbardo, P. (1997) Similarities and differences between dreaming and waking cognition: An exploratory study. *Consciousness and Cognition* 6:132–47. [MBl, TLK, SL, rTAN]

Kahana, M. J., Sekuler, R., Caplan, J. B., Kirschen, M. & Madsen, J. R. (1999) Human theta oscillations exhibit task dependence during virtual maze navigation. *Nature* 399:781–84. [rJAH]

Kahn, A., Dan, B., Groswasser, J., Franco, P. & Sottiaux, M. (1996) Normal sleep architecture in infants and children. *Journal of Clinical Neurophysiology* 13:184–97. [IF]

Kahn, D. & Hobson, J. A. (1993) Self-organization theory of dreaming. *Dreaming* 3:151–78. [aJAH]

(1994) Unpredictability and meaning in self-organized dreaming. Paper presented to the Association for the Study of Dreams, Leiden, NL, June 1994. [EH]

Kahn, D., Pace-Schott, E. F. & Hobson, J. A. (1997) Consciousness in waking and dreaming: The roles of neuronal oscillation and neuromodulation in determining similarities and differences. *Neuroscience* 78(1):13–38. [RCo, arJAH, aTAN, EFP-S, HSP, AR]

Kahn, D., Stickgold, R., Pace-Schott, E. F. & Hobson, J. A. (2001) Dreaming and waking consciousness: A character recognition study. *Journal of Sleep Research* 9:317–25. [rJAH]

Kahn, E., Fisher, C. & Edwards, A. (1991) Night terrors and anxiety dreams. In: *The mind in sleep: Psychology and psychophysiology*, ed. S. J. Ellman & J. S. Antrobus. Wiley. [aTAN]

Kahneman, D., Slovic, P. & Tversky, A. (1982) *Judgment under uncertainty: Heuristics and biases*. Cambridge University Press. [LM]

Kaijima, M., Da Costa-Rochette, L., Dodd, R. H. & Naquet, R. (1984) Hypnotic action of ethyl-B-carboline-3-carboxylate, a benzodiazepine receptor antagonist, in cats. *Electroencephalography and Clinical Neurophysiology* 58:277–81. [LJW]

Kajimura, N., Uchiyama, M., Takayama, Y., Uchida, S., Uema, S., Uema, T., Kato, M., Sekimoto, M., Watanabe, T., Nakajima, T., Horikoshi, M., Okawa, M. & Takahashi, K. (1999) Activity of midbrain reticular formation and neocortex during the progression of human non-rapid eye movement sleep. *The Journal of Neuroscience* 19:10065–73. [EFP-S]

Kales, A., Hoedemaker, F. S., Jacobson, A., Kales, J. D., Paulson, M. J. & Wilson, T. E. (1967) Mentation during sleep: REM and NREM recall reports. *Perceptual and Motor Skills* 24:555–60. [aJAH, aTAN]

Kales, A. & Jacobson, A. (1967) Mental activity during sleep: Recall studies, somnambulism, and effects of rapid eye movement deprivation and drugs. *Experimental Neurology Supplement* 4:81–91. [LJW]

Kales, A., Jacobson, A., Paulson, M. J., Kales, J. D. & Walter, R. D. (1966) Somnambulism: Psychophysiological correlates. I. All-night EEG studies. *Archives of General Psychiatry* 14:586–94. [aTAN]

Kales, A. & Vgontzas, A. N. (1995) Sleep disturbances as side effects of therapeutic drugs. In: *Pharmacology of sleep*, ed. A. Kales. Springer. [LJW]

Kalivas, P. W. & Barnes, C. D. (1993) *Limbic motor circuits and neuropsychiatry*. CRC Press. [rJAH, EFP-S]

Kallmeyer, R. J. & Chang, E. C. (1998) What makes dreams positive or negative: Relations to fundamental dimensions of positive and negative mood. *Perceptual and Motor Skills* 86:219–23. [MBl]

Kamiya, J. (1961) Behavioral, subjective and physiological aspects of drowsiness and sleep. In: *Functions of varied experience*, ed. D. W. Fiske & S. R. Maddi. Dorsey Press. [JSA, aJAH]

(1962) Behavioral and physiological concomitants of dreaming. Progress report to National Institute of Mental Health. [aTAN]

Kamiya, J. & Fong, S. (1962) Dream reporting from NREM sleep as related to respiration rate. Meeting of the Association for the Psychophysiological Study of Sleep, Chicago. [rTAN]

Kamminer, H. & Lavie, P. (1991) Sleep and dreaming in holocaust survivors: Dramatic decrease in dream recall in well-adjusted survivors. *Journal of Nervous and Mental Disease* 179:664–69. [MK]

Kamodi, A., Williams, J. A., Hutcheson, B. & Reiner, P. B. (1992) Membrane properties of mesopontine cholinergic neurons studied with the whole-cell patch-clamp technique: Implications for behavioral state control. *Journal of Neurophysiology* 68:1359–72. [aJAH]

Kanamori, N., Sakai, K. & Jouvet, M. (1980) Neuronal activity specific to paradoxical sleep in the ventromedial medullary reticular formation. *Brain Research* 189:251–55. [CG]

Kandel, E. R. (1991) Disorders of thought: Schizophrenia. In: *Principles of neural science*, ed. E. Kandel, J. Schwartz & T. Jessel. Appleton & Lange. [aMS]

(1998) A new intellectual framework for psychiatry. *American Journal of Psychiatry* 155(4):457–69. [MM]

Kang, Y. & Kitai, S. T. (1990) Electrophysiology properties of pedunculopontine neurons and their postsynaptic responses following stimulation of substantia nigra reticulata. *Brain Research* 535:79–95. [aJAH]

Kanwisher, N., McDermott, J. & Chun, M. M. (1997) The fusiform face area: A module in human extrastriate cortex specialized for face perception. *Journal of Neuroscience* 17:4302–11. [aJAH]

Kapur, S. & Remington, G. (1996) Serotonin-dopamine interaction and its relevance to schizophrenia. *American Journal of Psychiatry* 153:466–76. [aJAH]

Karacan, I. (1982) Evaluation of nocturnal penile tumescence and impotence. In: *Sleeping and waking disorders: Indications and techniques*, ed. C. Guilleminault. Addison–Wesley. [rTAN]

Karacan, I., Goodenough, D. R., Shapiro, A. & Starker, S. (1979) Erection cycle during sleep in relation to dream anxiety. In: *The content analysis of verbal behavior*, ed. L. A. Gottschalk. Spectrum. [aTAN] Also in: *Archives of General Psychiatry* 15:183–89. [aJAH]

## References

Karacan, I., Hursch, C. J., Williams, R. L. & Thornby, J. I. (1972) Some characteristics of nocturnal penile tumescence in young adults. *Archives of General Psychiatry* 26:351–56. [aTAN]

Karacan, I., Salis, P. J., Thornby, J. I. & Williams, R. L. (1976) The ontogeny of nocturnal penile tumescence. *Waking and Sleeping* 1:27–44. [rTAN]

Kardiner, A. (1932) The bio-analysis of the epileptic reaction. *Psychoanalytic Quarterly* 1:375–83. [aMS]

Karni, A. & Sagi, D. (1991) Where practice makes perfect in texture discrimination: Evidence for primary visual cortex plasticity. *Proceedings of the National Academy of Science USA* 88:4966–70. [rJAH]

(1993) The time course of learning a visual skill. *Nature* 365:250–52. [JB, arRPV]

Karni, A., Tanne, D., Rubenstein, B. S., Askenasy, J. J. M. & Sagi, D. (1994) Dependence on REM-sleep of overnight improvement of a perceptual skill. *Science* 265:679–82. [JB, rJAH, aTAN, MSc, CS, RS, arRPV]

Karni, A., Weisberg, J., Lalonde, F. &Ungerleider, L. G. (1995) Slow changes in primary and secondary visual cortex associated with perceptual skill learning: An fMRI study. LPP/NIMH, National Institutes of Health. [rJAH]

Kasamatsu, T. (1970) Maintained and evoked unit activity in the mesencephalic reticular formation of the freely behaving cat. *Experimental Neurology* 28:450–70. [aJAH]

Kaufman, L. S. & Morrison, A. R. (1981) Spontaneous and elicited PGO spikes in rats. *Brain Research* 214:61–72. [CG]

Kavanau, J. L. (1997) Memory, sleep and the evolution of mechanisms of synaptic efficacy maintenance. *Neuroscience* 79:7–44. [aAR]

Kay, D. C., Blackburn, A. B., Buckingham, J. A. & Karacan, I. (1976) Human pharmacology of sleep. In: *Pharmacology of sleep*, ed. R. L. Williams & I. Karacan. Wiley. [LJW]

Kayama, Y., Otha, M. & Jodo, E. (1992) Firing of "possibly" cholinergic neurons in the rat laterodorsal tegmental nucleus during sleep and wakefulness. *Brain Research* 569:210–20. [aTAN, CMP]

Kayed, K. (1995) Narcolepsy and hypnagogic hallucinations. *Impuls* 3:48–54. [aJAH]

Kellaway, P. & Frost, J. (1983) Biorhythmic modulation of epileptic events. In: *Recent advances in epilepsy I*, ed. T. Pedley & B. Meldrum. Chuchill Livingstone. [aMS]

Kellner, R., Neidhardt, J., Krakow, B. & Pathak, D. (1992) Changes in chronic nightmares after one session of desensitization or rehearsal instructions. *American Journal of Psychiatry* 149:659–63. [aAR]

Kerr, J. S., Fairweather, D. B. & Hindmarch, I. (1993) Effects of fluoxetine on psychomotor performance, cognitive function and sleep in depressed patients. *International Clinical Psychopharmacology* 8:341–43. [aRPV]

Kerr, J. S., Fairweather, D. B., Mahendran, R. & Hindmarch, I. (1992) The effects of paroxetine, alone and in combination with alcohol on psychomotor performance and cognitive function in the elderly. *International Clinical Psychopharmacology* 7:101–108. [aRPV]

Kerr, N. H. (1993) Mental imagery, dreams, and perception. In: *Dreaming as cognition*, ed. C. Cavallero & D. Foulkes. Harvester Wheatsheaf. [CCa]

Kerr, N. H., Foulkes, D. & Jurkovic, G. (1978) Reported absence of visual dream imagery in a normally sighted subject with Turner's syndrome. *Journal of Mental Imagery* 2:247–64. [aMS]

Kesner, R. P., Crutcher, K. A. & Measom, M. O. (1986) Medial septal and nucleus basalis magnocellularis lesions produce order memory deficits in rats which mimic symtomatology of Alzheimer's disease. *Neurobiology of Aging* 7:287–95. [aRPV]

Khateb, A., Fort, P., Pegna, A., Jones, B. E. & Winthaler, M. (1995) Cholinergic nucleus basalis neurons are excited by histamine in vitro. *Neuroscience* 69:495–506. [aJAH]

Kim, J. (1998) *Mind in a physical world*. MIT Press. [AR]

Kinchla, R. A. (1992) Attention. *Annual Review of Psychology* 43:711–42. [aJAH]

King, K. B. (1997) Psychologic and social aspects of cardiovascular disease. *Annals of Behavioral Medicine* 19:264–70. [TAN]

Kinney, I., Kramer, M. & Bonnet, M. (1981) Dream incorporation of meaningful names. *Sleep Research* 10:157. [MK]

Kinon, B. J. & Lieberman, J. A. (1996) Mechanism of action of atypical antipsychotic drugs: A critical analysis. *Psychopharmacology* 124:2–34. [CG]

Kinsbourne, M. (1971) Cognitive deficit: Experimental analysis. In: *Psychobiology*, ed. J. McGaugh. Academic Press. [MS]

Kirsch, I. (1985) Response expectancy as a determinant of experience and behavior. *American Psychologist* 1189–202. [MBl]

(1998) Volition as a believed-in imagining. In: *Believed-in imaginings*, ed. J. de Rivers & T. R. Sarbin. *American Psychological Association*. [MBl]

Kirschbaum, C., Wolf, O. T., May, M., Wippich, W. & Hellhammer, D. H. (1996) Stress–and treatment-induced elevations of cortisol levels associated with impaired declarative memory in healthy adults. *Life Science* 58:1475–83. [JB]

Kitahama, K., Valatx, J.-L. & Jouvet, M. (1976) Apprentissage d'un labyrinthe en y chez deux souches de souris. Effets de la privation instrumentale et pharmacologique du sommeil. *Brain Research* 108:75–86. [CS]

Kiyono, S., Seo, M. L. & Shibagaki, M. (1981) Effects of rearing environments upon sleep waking patterns in rats. *Physiology and Behavior* 26:391–94. [aRPV]

Kleiner, S. & Bringmann, A. (1996) Nucleus basalis magnocellularis and pedunculopontine tegmental nucleus: Control of the slow EEG waves in rats. *Archives Italiennes de Biologie* 134:153–67. [aJAH]

Klingelhofer, J., Haiak, G., Matzander, G., Schulz-Varszegi, M., Sandler, D., Ruther, E. & Conrad, B. (1995) Dynamics of cerebral blood flow velocities during normal human sleep. *Clinical Neurology and Neurosurgery* 97:142–48. [aJAH]

Klinger, E. (1990) *Daydreaming*. Jeremy P. Tarcher. [EH, aAR]

(1996) Emotional influences on cognitive processing, with implications for theories of both. In: *The psychology of action: Linking cognition and motivation to behavior*, ed. P. Gollwitzer & J. Bargh. Guilford. [EH]

Knegtering, H., Eijck, M. & Huijsman, A. (1994) Effects of antidepressants on cognitive functioning of elderly patients: A review. *Drugs and Aging* 5:192–99. [aRPV]

Knudson, R. & Minier, S. (1999) The on-going significance of significant dreams. *Dreaming* 9:235–45. [HTH]

Kocsis, B. & Vertes, R. P. (1994) Characterization of neurons of the supramammillary nucleus and mammillary body that discharge rhythmically with the hippocampal theta rhythm in the rat. *Journal of Neuroscience* 14:7040–52. [aRPV]

(1997) Phase relations of rhythmic neuronal firing in the supramammillary nucleus and mammillary body to the hippocampal theta activity in urethane anesthetized rats. *Hippocampus* 7:204–14. [aRPV]

Kodama, T. & Honda, Y. (1996) Acetylcholine releases of mesopontine PGO-on cells in the lateral geniculate nucleus in sleep-waking cycle and serotonergic regulation. *Progress in Neuro-Psychopharmacology and Biological Psychiatry* 20:1213–27. [aJAH]

Kodama, T., Takahashi, Y. & Honda, Y. (1990) Enhancement of acetylcholine release during paradoxical sleep in the dorsal tegmental field of the cat brain stem. *Neuroscience Letters* 114:277–82. [aJAH]

Koga, E. (1965) A new method of EEG analysis and its application to the study of sleep. *Folia Psychiatric Neurology/Japan* 19:269–78. [IF]

Kohonen, T. (1984) *Self-organization and associative memory*. Springer-Verlag. [TJM]

Kolb, B. & Whishaw, I. Q. (1996) *Fundamentals of human neuropsychology, fourth edition*. W. H. Freeman. [aJAH]

Kondo, T., Antrobus, J. & Fein, G. (1989) Later REM activation and sleep mentation. *Sleep Research* 18:147. [TLK, aMS]

Korovkin, P. P. (1975) *Inequalities*. MIR Publishers. [TJM]

Koshland, D. E. (1974) The chemotactic response as a potential model for neuronal systems. In: *The neurosciences, third study program*. MIT Press. [TJM]

Kosslyn, S. M. (1994) *Image and brain*. MIT Press. [aMS]

Kosslyn, S. M. & Koenig, O. (1992) *Wet mind: The new cognitive neuroscience*. The Free Press. [aJAH]

Kosslyn, S. M., Pascual-Leone, A., Felician, O., Camposano, S., Keenan, J. P., Thompson, W. L., Ganis, G., Sukel, K. E. & Alpert, N. M. (1999) The role of area 17 in visual imagery: Convergent evidence from PET and rTMS. *Science* 284:167–70. http://www.sciencemag.org/cgi/content/abstract/284/5411/167 [JAB, rJAH]

Kosslyn, S. M. & Thompson, W. L. (2000) Shared mechanisms in visual imagery and visual perception: Insights from cognitive neuroscience. In: *The new cognitive neurosciences, 2nd edition*, ed. M. S. Gazzaniga. MIT Press. [RCo]

Kosten, T. R., Frank, J. B., Dan, E., McDougle, C. J. & Giller, E. L., Jr. (1991) Pharmacology for posttraumatic stress disorder using phenelzine or imipramine. *Journal of Nervous and Mental Diseases* 179:366–70. [LJW]

Koukkou, M. & Lehmann, D. (1983) Dreaming: The functional state-shift hypothesis, a neuropsychophysiological model. *British Journal of Psychiatry* 142:221–31. [TLK, DL, MSc]

(1987) An information-processing perspective of psychophysiological measurements. *Journal of Psychophysiology* 1:109–12, 219–20. [DL]

(1993) A model of dreaming and its functional significance: The state-shift hypothesis. In: *The functions of dreaming*, ed. A. Moffitt, M. Kramer & R. Hoffmann. State University of New York Press. [RCo]

Koulack, D. (1991) *To catch a dream: Explorations of dreaming*. State University of New York Press. [RLe, JM]

Koulack, D. & Goodenough, D. (1976) Dream recall and dream recall failure: An arousal-retrieval model. *Psychological Bulletin* 83:975–84. [IF]

Koulack, D. & Schultz, K. J. (1974) Task performance after awakenings from different stages of sleep. *Perceptual and Motor Skills* 39:792–94. [aTAN]

Kovalzon, V. M. & Tsibulsky, V. L. (1984) REM-sleep deprivation, stress and emotional behavior in rats. *Behavioural Brain Research* 14:235–45. [AC, JP, aRPV]

Kowatch, R. A., Schnoll, S. S., Knisely, J. S., Green, D. & Elswick, R. K. (1992)

Electroencephalographic sleep and mood during cocaine withdrawal. *Journal of Addictive Diseases* 11:21–45. [rJAH]

Koyama, Y., Kayama, Y. & Sakai, K. (1999) Different physiological properties of two populations of PS-on neurons in the mesopontine tegmentum. In: *Rapid eye movement sleep,* ed. B. N. Mallick & S. Inoue. Marcel Dekker. [aJAH]

Krakow, B., Germain, A., Tandberg, D., Koss, M., Schrader, R., Hollifield, M., Cheng, D. & Edmond, T. (2000) Sleep breathing and sleep movement disorders masquerading as insomnia in sexual-assault survivors. *Comprehensive Psychiatry* 41:49–56. [TAN]

Krakow, B., Kellner, R., Pathak, D. & Lambert, L. (1995a) Imagery rehearsal treatment for chronic nightmares. *Behavior Research and Therapy* 33:837–43. [aAR]

(1996) Long term reduction of nightmares with imagery rehearsal treatment. *Behavioural and Cognitive Psychotherapy* 24:135–48. [aAR]

Krakow, B., Tandberg, D., Scriggins, L. & Barey, M. (1995b) A controlled comparison of self-rated sleep complaints in acute and chronic nightmare sufferers. *The Journal of Nervous and Mental Disease* 183:623–27. [aAR]

Kramer, M. (1991) The nightmare: A failure in dream function. *Dreaming* 1(4):277–85. [aAR]

(1993) The selective mood regulatory function of dreaming: An update and revision. In: *The functions of dreaming,* ed. A. Moffitt, M. Kramer & R. Hoffman. State University of New York Press. [aJAH, MK, RLe, WHM, aAR]

(2000) Dreams and psychopathology. In: *Principles and practices of sleep, 3rd edition,* ed. M. Kryger, T. Roth & W. Dement. W. B. Saunders. [MK]

Kramer, M., Hlasny, R., Jacobs, G. & Roth, T. (1976) Do dreams have meaning? An empirical inquiry. *American Journal of Psychiatry* 133:778–81. [MK]

Kramer, M. & Kinney, L. (1988) Sleep patterns in trauma victims with disturbed dreaming. *Psychiatric Journal of the University of Ottawa* 13:12–16. [TAN]

Kramer, M., Moshiri, A. & Scharf, M. (1982) The organization of mental content in and between the waking and dream state. *Sleep Research* 11:106. [MK]

Kramer, M., Roehrs, T. & Roth, T. (1972) The relationship between sleep and mood. *Sleep Research* 1:193. [MK]

Kramer, M. & Roth, T. (1979) The stability and variability of dreaming. *Sleep* 1:336–51. [MK]

Kramer, M., Roth, T., Arand, D. & Bonnet, M. (1981) Waking and dreaming mentation: A test of their interrelationship. *Neuroscience Letters* 22:83–86. [MK]

Kramer, M., Roth, T. & Trinder, J. (1975) Dreams and dementia: A laboratory exploration of dream recall and dream content in chronic brain syndrome patients. *International Journal of Aging and Human Development* 6:169–78. [aJAH, MK]

Kramer, M., Schoen, L. & Kinney, L. (1984) The dream experience in dream disturbed Vietnam veterans. In: *Post traumatic stress disorders: Psychological and biological sequellae,* ed. B. Van Der Kolk. American Psychological Association. [MK]

Kramer, M., Winget, C. & Whitman, R. (1971) A city dreams: A survey approach to normative dream content. *American Journal of Psychiatry* 127:1350–56. [MK]

Krauthamer, G. M., Gronwerg, B. S. & Krein, H. (1995) Pontine cholinergic neurons of the peduncolopontine tegmental nucleus projecting to the superior colliculus consist of sensory responsive and unresponsive populations which are functionally distinct from other mesopontine neurons. *Neuroscience* 69:507–17. [aJAH]

Krech, D., Rosenzweig, M. R. & Bennett, E. L. (1962) Relations between brain chemistry and problem solving among rats in enriched and impoverished environments. *Journal of Comparative and Physiological Psychology* 55:801–807. [aRPV]

Kremen, I. (1961) Dream reports and rapid eye movements. Unpublished doctoral dissertation, Harvard University, Cambridge, Mass. [aTAN]

Kreuger, J. M., Obal, F. & Fang, J. (1999) Humoral regulation of physiological sleep: Cytokines and GHRH. *Journal of Sleep Research* 8 (Supplement 1):53–59. [aJAH]

Krieckhaus, E. E. (1988) Preoperative training provides no protection against lesion-induced decrements in explicit processes. In: *Preoperative events: Their effects on behavior following brain damage,* ed. J. Schulkin. Erlbaum. [EEK]

(1999) Consideration of the drive properties of the mammillary bodies solves the "fornix problem." *Behavioral and Brain Sciences* 22:456–58. [EEK]

(submitted) Cognitive control of stress and alarm: A neuropharmacological model. *Behavioral and Brain Sciences.* [EEK]

Krieckhaus, E. E., Donahoe, J. W. & Morgan, M. A. (1992) Paranoid schizophrenia may be caused by dopamine hyperactivity of CA1 hippocampus. *Biological Psychiatry* 31:560–70. [EEK]

Krippner, S. (1981) Access to hidden reserves of the unconscious through dreams in creative problem solving. *Journal of Creative Behavior* 15:11–23. [RLe]

Kryger, M. H., Roth, T. & Dement, W. C. (1994) *Principles and practice of sleep medicine, second edition.* W. B. Saunders. [JFP]

Kubin, L., Reignier, C., Tojima, H., Taguchi, O., Pack, A. I. & Davies, R. O. (1994) Change in serotonin level in the hypoglossal nucleus region during carbachol-induced atonia. *Brain Research* 645:291–302. [aJAH]

Kubin, L., Tojima, H., Reignier, C., Pack, A. I. & Davies, R. O. (1996) Interaction of serotonergic excitatory drive to hypoglossal motorneurons with carbachol-induced, REM-sleep-like atonia. *Sleep* 19:187–95. [aJAH]

Kuboyama, T., Hori, A., Sato, T., Nikami, T., Yamaki, T. & Veda, S. (1997) Changes in cerebral blood flow velocity in healthy young men during overnight sleep and while awake. *Electroencephalography and Clinical Neurophysiology* 102:125–31. [aJAH]

Kuhlo, W. & Lehmann, D. (1964) Das Einschlaferleben und seine neurophysiologischen korrelate. *Archiv fur Psychiatrie und Zeitschrift f.d.ges. Neurologies* 205:687–716. [rTAN]

Kuiken, D. & Sikora, S. (1993) The impact of dreams on waking thoughts and feelings. In: *The functions of dreaming,* ed. A. Moffitt, M. Kramer & R. Hoffman. State University of New York Press. [aTAN, MSc]

Kumar, V. M., Datta, S. & Singh, B. (1989) The role of reticular activating system in altering medial preoptic neuronal activity in anaesthetized rats. *Brain Research Bulletin* 22:1031–37. [arJAH]

Kunzendorf, R., Hartmann, E., Cohen, R. & Cutler, J. (1997) Bizarreness of the dreams and daydreams reported by individuals with thin and thick boundaries. *Dreaming* 7:265–71. [EH]

Kunzendorf, R, & Maurer, J. (1989) Hypnotic attenuation of the "boundaries" between emotional, visual, and auditory sensations. *Imagination, Cognition and Personality* 8(3):225–34. [EH]

Kupfer, D. J. & Bowers, M. B., Jr. (1972) REM sleep and central monoamine oxidase inhibition. *Psychopharmacologia* 27:183–90. [aRPV]

Kupfer, D. J., Ehlers, C. L., Frank, E. L., Grochocinski, V. J., McEachran, A. B. & Buhari, A. (1994) Persistent effects of antidepressants: EEG sleep studies in depressed patients during maintenance treatment. *Biological Psychiatry* 35:781–93. [aRPV]

Kupfer, D. J., Ehlers, C. L., Pollock, B. G., Nathan, R. S. & Perel, J. M. (1989) Clomipramine and EEG sleep in depression. *Psychiatry Research* 30:165–80. [aRPV]

Kupfer, D. J., Perel, J. M., Pollock, B. G., Nathan, R. S., Grochocinski, V. J., Wilson, M. J. & McEachran, A. B. (1991) Fluvoxamine versus desipramine: Comparative polysomnographic effects. *Biological Psychiatry* 29:23–40. [aRPV]

Kupfer, D. J., Spiker, D. G., Coble, P. A. & McPartland, R. J. (1979) Amitriptyline and EEG sleep in depressed patients I: Drug effect. *Sleep* 1:149–59. [aRPV]

Kupfer, D. J., Spiker, D. G., Rossi, A., Coble, P. A., Shaw, D. & Ulrich, R. (1982) Nortriptyline and EEG sleep in depressed patients. *Biological Psychiatry* 17:535–46. [aRPV]

Kupfer, D. J. & Thase, M. E. (1983) The use of the sleep laboratory in the diagnosis of affective disorders. *Psychiatric Clinics of North America* 6(1):3–24. [GL]

Kutas, M. (1990) Event-related brain potential (ERP) studies of cognition during sleep. In: *Sleep and cognition,* ed. R. R. Bootzin, J. F. Kihlstrom & D. L. Schacter. American Psychological Association. [aTAN]

LaBerge, S. P. (1985) *Lucid dreaming.* Ballantine/J. P. Tarcher. [SL, aAR]

(1990) Lucid dreaming: Psychophysiological studies of consciousness during REM sleep. In: *Sleep and cognition,* ed. R. R. Bootzin, J. F. Kihlstrom & D. L. Schacter. American Psychological Association. [aJAH, TLK, SL]

(1992) Physiological studies of lucid dreaming. In: *The neuropsychology of dreaming sleep,* ed. J. Antrobus & M. Bertini. Erlbaum. [aJAH]

(1998) Dreaming and consciousness. In: *Toward a science of consciousness II,* ed. S. Hameroff, A. Kaszniak & S. Scott. MIT Press. [SL]

LaBerge, S. P., Kahan, T. & Levitan, L. (1995) Cognition in dreaming and waking. *Sleep Research* 24A:239. [SL]

LaBerge, S. P., Nagel, L. E., Dement, W. C. & Zarcone, V. P. (1981) Lucid dreaming verified by volitional communication during REM sleep. *Perceptual and Motor Skills* 52:727–32. [aJAH, SL, aTAN]

LaBerge, S. P. & Zimbardo, P. G. (2000) Smooth tracking eye movements discriminate both dreaming and perception from imagination. Paper presented at the Toward a Science of Consciousness Conference, Tucson, AZ, April 10–15, 2000. http://www.imprint-academic.demon.co.uk/Tucson4.03.10.html [SL]

Labruzza, A. L. (1978) The activation-synthesis hypothesis of dreams: A theoretical note. *American Journal of Psychiatry* 135:1536–38. [aTAN, aMS]

Lai, Y. Y. & Siegel, J. M. (1992) Pontomedullary glutamate receptors mediating locomotion and muscle tone suppression. *Journal of Neuroscience* 11:2931–37. [aJAH]

(1999) Muscle atonia and REM sleep. In: *Rapid eye movement sleep,* ed. B. N. Mallick & S. Inoue. Marcel Dekker. [aJAH]

Lairy, G. C., Barros de Ferreira, M. & Goldsteinas, L. (1967) Les phases intermédiaires du sommeil. In: *Abnormalities of sleep in man: Proceedings of the XVth European Meeting on Electroencephalography,* ed. H. Gastaut, E. Lugaresi, G. Berti Ceroni & G. Coccogna. Aulo Gaggi. [CG, HTH, aTAN]

Lairy, G. C. & Salzarulo, P. (1975) Concluding remarks. In: *The experimental study of human sleep: Methodological problems*, ed. G. C. Lairy & P. Salzarulo. Elsevier.   [PS]

Lamping, D. L., Spring, B. & Gelenberg, A. J. (1984) Effects of two antidepressants on memory performance in depressed outpatients: A double-blind study. *Psychopharmacology* 84:254–61.   [aRPV]

Landers, J. (1992) Reconstructing ancient populations. In: *The Cambridge encyclopedia of human evolution*, ed. S. Jones, R. Martin & D. Pilbeam. Cambridge University Press.   [aAR]

Landolt, H.-P., Dijk, D.-J., Achermann, P. & Borbely, A. A. (1996) Effect of age on the sleep EEG: Slow-wave activity and spindle frequency activity in young and middle-aged men. *Brain Research* 738:205–12.   [rTAN]

Lansky, M. (1995) *Posttraumatic nightmares*. Analytic Press.   [RLe]

Larkin, J. H. & Simon, H. A. (1987) Why a diagram is (sometimes) worth ten thousand words. *Cognitive Science* 11(1):65–100.   [WJC]

Larson, J. D. & Foulkes, D. (1969) Electromyogram suppression during sleep, dream recall, and orientation time. *Psychophysiology* 5:548–55.   [CG, MS, EFP-S, GWV]

Larson, J. D. & Lynch, G. (1986) Induction of synaptic potentiation in hippocampus by patterned stimulation involves two events. *Science* 232:985–88.   [aRPV]

(1988) Role of N-methyl-D-aspartate receptors in the induction of synaptic potentiation by burst stimulation patterned after the hippocampal theta-rhythm. *Brain Research* 441:111–18.   [aRPV]

Larson, J. D., Wong, D. & Lynch, G. (1986) Patterned stimulation at the theta frequency is optimal for the induction of hippocampal long-term potentiation. *Brain Research* 368:347–50.   [aRPV]

Lasaga, J. I. & Lasaga, A. M. (1973) Sleep learning and progressive blurring of perception during sleep. *Perceptual and Motor Skills* 37:51–62.   [aTAN]

Lavie, P. (1974a) BRAC and spiral after-effect. *Behavioral Biology* 11:373–78.   [aTAN]

(1974b) Differential effects of REM and non-REM awakenings on the spiral after effect. *Physiological Psychology* 2:107–108.   [aJAH]

(1990) Penile erections in a patient with nearly total absence of REM: A follow-up study. *Sleep* 13:276–78.

(1994) Dreaming and REM sleep – commentary. *World Federation of Sleep Research Societies Newsletter* 3:14–15.   [aTAN]

Lavie, P. & Giora, Z. (1973) Spiral after-effect durations following awakening from REM sleep and NREM sleep. *Perception and Psychophysics* 14(1):19–20.   [aJAH, aTAN]

Lavie, P., Katz, N., Pillar, G. & Zinger, Y. (1998) Elevated awaking thresholds during sleep: Characteristics of chronic war-related posttraumatic stress disorder patients. *Biological Psychiatry* 44:1060–65.   [MBl]

Lavie, P., Matanya, Y. & Yehuda, S. (1984) Cognitive asymmetries after wakings from REM and NONREM sleep in right-handed females. *International Journal of Neuroscience* 23:111–15.   [aTAN]

Lavie, P., Pratt, H., Scharf, B., Peled, R. & Brown, J. (1984) Localized pontine lesion: Nearly total absence of REM sleep. *Neurology* 34:118–20.   [aMS, aRPV]

Lavie, P., Schnall, R. P., Sheffy, J. & Shlitner, A. (2000) Peripheral vasoconstriction during REM sleep detected by a new plethysmographic method. *Nature Medicine* 6:606.   [rTAN]

Lavie, P. & Segal, J. (1989) Twenty-four-hour structure of sleepiness in morning and evening persons investigated by ultrashort sleep-wake cycle. *Sleep* 12:522–28.   [WHM]

Lavie, P. & Sutter, D. (1975) Differential responding to the beta movement after waking from REM and NONREM sleep. *American Journal of Psychology* 88:595–603.   [aJAH, aTAN]

Lavie, P. & Tzischinsky, O. (1984) Cognitive asymmetries after waking from REM and NONREM sleep: Effects of delayed testing and handedness. *International Journal of Neuroscience* 23:311–15.   [aTAN]

Lawson, M. L., Crewther, S. G. & Crewther, D. (1999) Temporal limitations of information processing in global and local attention: The effect of information content. *Australian and New Zealand Journal of Opthalmology* 27:261–64.   [RCo]

Leconte, P., Hennevin, E. & Bloch, V. (1974) Duration of paradoxical sleep necessary for the acquisition of conditioned avoidance in the rat. *Physiology and Behavior* 13:675–81.   [JP, aRPV]

LeDoux, J. E. (1996/1998) *The emotional brain*. Simon & Schuster, 1996.   [GWD, arJAH] Touchstone, 1998.   [aAR]

(1998) Fear and the brain: Where have we been, and where are we going? *Biological Psychiatry* 44(12):1229–38.   [aAR]

Lee, J. H., Bliwise, D. L., Lebret-Bories, E., Guilleminault, C. & Dement, W. C. (1993) Dream-disturbed sleep in insomnia and narcolepsy. *Journal of Nervous and Mental Diseases* 181:320–24.   [LJW]

Lee, L. H., Friedman, D. B. & Lydic, R. (1995) Respiratory nuclei share synaptic connectivity with pontine reticular regions regulating REM sleep. *American Journal of Physiology* 268:L251–62.   [RLy]

Lehmann, D., Grass, P. & Meier, B. (1995) Spontaneous conscious covert cognition states and brain electric spectral states in canonical correlations. *International Journal of Psychophysiology* 19:41–52.   [DL, aTAN]

Lehmann, D. & Koukkou, M. (1984) Physiological and mental processes during sleep: A model of dreaming. In: *Psychology of dreaming*, ed. M. Bosinelli & P. Cicogna. Cooperativa Libraria Universitaria Editrice.   [arTAN]

Lehmann, D., Meier, B., Meier, C. A., Mita, T. & Skrandies, W. (1983) Sleep onset mentation related to short epoch EEG spectra. *Sleep Research* 12:180.   [DL]

Lehmann, D., Strik, W. K., Henggeler, B., Koenig, T. & Koukkou, M. (1998) Brain electric microstates and momentary conscious mind states as building blocks of spontaneous thinking: I. Visual imagery and abstract thoughts. *International Journal of Psychophysiology* 29:1–11.   [DL]

Lehmann, H. & Hanrahan, G. (1954) Chlorpromazine, a new inhibiting agent for psychomotor excitement. *Archives of Neurology* 71:227–37.   [aMS]

Lejune, M., Decker, C. & Sanchez, X. (1994) Mental rehearsal in table tennis performance. *Perceptual and Motor Skills* 79:627–41.   [aAR]

Leonard, C. S. & Llinas, R. R. (1990) Electrophysiology of mammalian pedunculopontine and laterodorsal tegmental neurons in vitro: Implications for the control of REM sleep. In: *Brain cholinergic mechanisms*, ed. M. Steriade & D. Biesold. Oxford Science.   [aJAH]

(1994) Serotonergic and cholinergic inhibition of mesopontine cholinergic neurons controlling REM sleep: An in vitro electrophysiological study. *Neuroscience* 59:309–30.   [aJAH]

Leonard, T. O. & Lydic, R. (1995) Nitric oxide synthase inhibition decreases pontine acetylcholine release. *NeuroReport* 6:1525–29.   [RLy]

(1997) Pontine nitric oxide modulates acetylcholine release, rapid eye movement sleep generation, and respiratory rate. *Journal of Neuroscience* 17:774–85.   [aJAH, RLy]

(1999) Nitric oxide, a diffusable modulator of physiological traits and behavioral states. In: *Rapid eye movement sleep*, ed. B. N. Mallick & S. Inoue. Marcel Dekker.   [aJAH]

Lepkifker, E., Dannon, P. N., Iancu, I., Ziv, R. & Kotler, M. (1995) Nightmares related to fluoxetine treatment. *Clinical Neuropharmacology* 18:90–94.   [aJAH]

Leslie, K. & Ogilvie, R. (1996) Vestibular dreams: The effect of rocking on dream mentation. *Dreaming* 6:1–16.   [aJAH]

Leung, L. S., Shen, B. & Kaibara, T. (1992) Long-term potentiation induced by pattern stimulation of the commissural pathway to hippocampal CA1 region in freely moving rats. *Neuroscience* 48:63–74.   [aRPV]

Leutgeb, S. & Mizumori, S. J. Y. (1999) Excitotoxic septal lesions result in spatial memory deficits and altered flexibility of hippocampal single-unit representations. *Journal of Neuroscience* 19:6661–72.   [aRPV]

Levin, D. T. & Simons, D. J. (1997) Failure to detect changes to attended objects in motion pictures. *Psychonomic Bulletin and Review* 4(4):501–506.   [SL]

Levin, R. (1990a) Ego boundary impairment and thought disorder in frequent nightmare sufferers. *Psychoanalytic Psychology* 7:529–43.   [RLe]

(1990b) Psychoanalytic theories on the function of dreaming: A review of the empirical dream research. In: *Empirical studies of psychoanalytic theories, vol. 3*, ed. J. Masling. Analytic Press.   [RLe]

(1994) Prevalence of sleep and dreaming characteristics of frequent nightmare subjects in a university population. *Dreaming* 4:127–37.   [RLe]

(1998) Nightmares and schizotypy. *Psychiatry* 61:206–16.   [RLe]

Levin, R. & Livingstone, G. (1991) Concordance between two measures of dream bizarreness. *Perceptual and Motor Skills* 72:837–38.   [aTAN]

Levin, R. & Raulin, M. (1991) Preliminary evidence for the proposed relationship between frequent nightmares and schizotypal symptomatology. *Journal of Personality Disorders* 5:8–14.   [RLe]

Levine, J. B. (1991) The role of culture in the representation of conflict in dreams. A comparison of Bedouin, Irish and Israeli children. *Journal of Cross-Cultural Psychology* 22(4):472–90.   [arAR]

Levi-Strauss, C. (1962) *The savage mind*. Weidenfeld & Nicolson.   [NH]

Levitan, L. & LaBerge, S. (1993) Dream times and remembrances. *NightLight* 5(4):9–14.   [SL]

Lewicki, P., Czyzewska, M. & Hill, T. (1997) Cognitive mechanisms for acquiring "experience": The dissociation between conscious and nonconscious cognition. In: *Scientific approaches to consciousness*, ed. J. D. Cohen & J. W. Schooler. Erlbaum.   [aAR]

Lewicki, P., Hill, T. & Bizot, E. (1988) Acquisition of procedural knowledge about a pattern of stimuli that cannot be articulated. *Cognitive Psychology* 20:24–37.   [MBl]

Lewin, I. & Glaubman, H. (1975) The effect of REM deprivation: Is it detrimental, beneficial, or neutral? *Psychophysiology* 12:349–53.   [aRPV]

Leyhausen, P. (1979) *Cat behavior*. Garland Press.   [ARM]

Li, X. Y., Greene, R. W., Rainnie, D. G. & McCarley, R. W. (1997) Dual modulation of nicotine in DR neurons. *Sleep Research* 26:22.   [aJAH]

Liljequist, R., Linnoila, M. & Mattila, M. J. (1974) Effect of two weeks treatment with chlorimipramine and nortriptyline, alone or in combination with alcohol on learning and memory. *Psychopharmacologia* 39:181–86.   [aRPV]

Lin, J. S., Hou, Y., Sakai, K. & Jouvet, M. (1996) Histaminergic descending inputs to the mesopontine tegmentum and their role in the control of cortical activation and wakefulness in the cat. *The Journal of Neuroscience* 16:1523–37. [aJAH]

Lin, J. S., Kitahama, P., Fort, P., Panula, P., Denny, R. M. & Jouvet, M. (1993) Histaminergic system in the cat hypothalamus with reference to type B monoamine oxidase. *Journal of Comparative Neurology* 330:405–20. [aJAH]

Lin, J. S., Luppi, P. H., Salvert, D., Sakai, K. & Jouvet, M. (1986) Histamine-containing neurons in the cat hypothalamus. *Comptes Rendus de l'Academie des Sciences – Series iii, Sciences de la Vie* 303:371–76. [aJAH]

Lin, J. S., Sakai, K. & Jouvet, M. (1988) Evidence for histaminergic arousal mechanisms in the hypothalamus of cats. *Neuropharmacology* 27:111–22. [aJAH]

(1994) Hypothalamo-preoptic histaminergic projections in sleep-wake control in the cat. *European Journal of Neuroscience* 6:618–25. [aJAH]

Lin, J. S., Vanni-Mercier, G. & Jouvet, M. (1997) Histaminergic ascending and descending projections in the cat, a double immunocytochemical study focused on basal forebrain cholinergic cells and dorsal raphe nucleus serotoninergic neurons. *Sleep Research* 26:24. [aJAH]

Lin, L., Faraco, J., Li, R., Kadotani, H., Rogers, W., Lin, X., Qiu, X., deJong, P. J., Nishini, S. & Mignot, E. (1999) The sleep disorder canine narcolepsy is caused by a mutation in the hypocretin (orexin) receptor 2 gene. *Cell* 98:365–76. [aJAH, RS-P]

Linden, E. R., Bern, D. & Fishbein, W. (1975) Retrograde amnesia: Prolonging the fixation phase of memory consolidation by paradoxical sleep deprivation. *Physiology and Behavior* 14:409–12. [WF, aRPV]

Linden, R. D., Campbell, K. B., Hamel, G. & Picton, T. W. (1985) Human auditory steady state evoked potentials during sleep. *Ear and Hearing* 6:167–74. [aTAN]

Link, S. (1990) Modeling imageless thought: The relative judgment theory of numerical comparisons. *Journal of Mathematical Psychology* 34:2–41. [TJM]

Linnoila, M., Johnson, J., Dubyoski, T., Ross, R., Buchsbaum, M., Potter, W. Z. & Weingartner, H. (1983) Effects of amitriptyline, desipramine and zimeldine, alone and in combination with ethanol, on information processing and memory in healthy volunteers. *Acta Scandinavica* 68 (Suppl. 308):175–81. [aRPV]

Lisman, J. E. (1999) Relating hippocampal circuitry to function: Recall of memory sequences by reciprocal dentate-CA3 interactions. *Neuron* 22(2):233–42. [RWG]

Liu, C., Ding, J. M., Faiman, L. E. & Gillette, M. U. (1997) Coupling of muscarinic cholinergic receptors and CGMP in nocturnal regulation of the suprachiasmatic circadian clock. *Journal of Neuroscience* 17:659–66. [aJAH]

Livingston, G. & Levin, R. (1991) The effects of dream length on the relationship between primary process in dreams and creativity. *Dreaming* 1:301–309. [aJAH, RLe]

Livingstone, M. S. & Hubel, D. H. (1987) Psychophysical evidence for separate channels for the perception of form, color, movement, and depth. *The Journal of Neuroscience* 7:3416–68. [rTAN]

Llinas, R. & Pare, D. (1991) Of dreaming and wakefulness. *Neuroscience* 44:521–35. [aJAH, ARM]

Llinas, R. & Ribary, U. (1993) Coherent 40-Hz oscillation characterizes dream state in humans. *Proceedings of the National Academy of Sciences USA* 90:2078–81. [RCo, aJAH, EFP-S]

Llinas, R., Ribary, U., Joliot, M. & Wang, X. J. (1994) Content and context in thalamocortical binding. In: *Temporal coding in the brain,* ed. G. Buzsaki, R. R. Llinas, U. Ribary, M. Joliot & X. J. Wang. Springer-Verlag. [aJAH]

Lloyd, S. R. & Cartwright, R. D. (1995) The collection of home and laboratory dreams by means of an instrumental response technique. *Dreaming* 5:63–73. [aJAH, aTAN]

Lockwood, A. H., Murphy, B. W. & Khalak, H. (1997) Attentional systems and the allocation of cerebral resources in reading and grammatical tasks. *International Journal of Neuroscience* 91:241–52. [RCo]

Lovblad, K. O., Thomas, R., Jakob, P. M., Scammell, T., Bassetti, C., Griswold, M., Ives, J., Matheson, J., Edelman, R. R. & Warach, S. (1999) Silent functional magnetic resonance imaging demonstrates focal activation in rapid eye movement sleep. *Neurology* 53:2193–95. [aJAH, EFP-S]

Lu, J., Shiromani, P. & Saper, C. B. (1999) Retinal input to the sleep-active ventrolateral preoptic nucleus in the rat. *Neuroscience* 93:209–14. [aJAH]

(2000) Effects of ibotenic acid lesion of the VLPO on sleep. *Journal of Neuroscience* 20:3830. [RS-P]

Lubin, A., Nute, C., Naitoh, P. & Martin, W. B. (1973) EEG delta activity during sleep as a damped ultradian rhythm. *Psychophysiology* 10:27–35. [IF]

Lubow, R. E. (1989) *Latent inhibition and conditioned attention theory.* Cambridge University Press. [JBP]

Luebke, J. L., Greene, R. W., Semba, K., Kamodi, A., McCarley, R. W. & Reiner, P. B. (1992) Serotonin hyperpolarizes cholinergic low threshold burst neurons in the rat laterodorsal tegmental nucleus in vitro. *Proceedings of the National Academy of Sciences USA* 89:743–47. [aJAH]

Luebke, J. L., McCarley, R. W. & Greene, R. W. (1993) Inhibitory action of the acetylcholine agonist carbachol on neurons of the rat laterodorsal tegmental nucleus in the vitro brainstem slice. *Journal of Neuroscience* 70:2128–35. [aJAH]

Lugaresi, E., Medori, R., Montagna, P., Baruzzi, A., Cortelli, P., Lugaresi, A., Tinuper, P., Zucconi, M. & Gambetti, P. (1986) Fatal familial insomnia and dysautonomia with selective degeneration of thalamic nuclei. *New England Journal of Medicine* 315:997–1003. [aMS]

Lukas, S. E., Dorsey, C. M., Mello, N. K. & Mendelson, J. H. (1996) Reversal of sleep disturbance in cocaine– and heroin-dependent men during chronic buprenorphine treatment. *Experimental and Clinical Psychopharmacology* 4:413–20. [rJAH]

Luppi, P.-H., Gervasoni, D., Peyron, C., Rampon, C., Barbagli, B., Boissard, R. & Fort, P. (1999a) Norepinephrine and REM sleep. In: *Rapid eye movement sleep,* ed. B. N. Mallick & S. Inoue. Marcel Dekker. [aJAH]

Luppi, P.-H., Peyron, C., Rampon, C., Gervasoni, D., Barbagli, B., Boissard, R. & Fort, P. (1999b) Inhibitory mechanisms in the dorsal raphe nucleus and locus coeruleus during sleep. In: *Handbook of behavioral state control: Molecular and cellular mechanisms,* ed. R. Lydic & H. A. Baghdoyan. CRC Press. [aJAH]

Lutzenberger, W., Pulvermuller, F., Ebert, T. & Birnbaumer, N. (1995) Visual stimulation alters local 40 Hz responses in humans: An EEG study. *Neuroscience Letters* 183:39–42. [aJAH]

Lyamin, O. I., Manger, P. R., Mukhametov, L. A., Siegel, J. M. & Shuaib, A. (2000) Rest and activity states in a gray whale. *Journal of Sleep Research* 9:261–67. [JMS]

Lydic, R. & Baghdoyan, H. A. (1993) Pedunculopontine stimulation alters respiration and increases ACh release in the pontine reticular formation. *American Journal of Physiology* 264:R544–54. [RLy]

Lydic, R. & Baghdoyan, H. A., eds. (1999) *Handbook of behavioral state control: Molecular and cellular mechanisms.* CRC Press. [aJAH]

Lydic, R., Baghdoyan, H. A., Hibbard, L., Bonyak, E. V., DeJoseph, M. R. & Hawkins, R. A. (1991a) Regional brain glucose metabolism is altered during rapid eye movement sleep in the cat: A preliminary study. *Journal of Comparative Neurology* 304:517–29. [aJAH]

Lydic, R., Baghdoyan, H. A. & Lorinc, Z. (1991b) Microdialysis of cat pons reveals enhanced acetylcholine release during state dependent respiratory depression. *American Journal of Physiology* 261:R766–70. [aJAH, RLy]

Lydic, R., McCarley, R. & Hobson, J. A. (1983) The time-course of dorsal raphe discharge, PGO waves, and muscle tone averaged across multiple sleep cycles. *Brain Research* 274:365–70. [aJAH]

(1987) Serotonin neurons and sleep. II. Time course of dorsal raphe discharge, PGO waves and behavioral states. *Archives Italiennes de Biologie* 126:1–28. [aJAH]

Lyman, R., Kwan, S. & Chao, W. (1938) Left occipito-parietal tumour with observations on alexia and agraphia in Chinese and in English. *Chinese Medical Journal* 54:491–516. [aMS]

Mach, E. (1906/1959) *The analysis of sensations, and the relation of the physical to the psychical.* (5th edition, 1959). Dover. [aMS]

Mack, A. & Rock, I. (1998) *Inattentional blindness.* MIT Press. [SL]

Mackay, D. M. (1966) Cerebral organization and the conscious control of action. In: *Brain conscious experience,* ed. J. C. Eccles. Springer-Verlag. [TJM]

Mackiewicz, M., Veasey, S. C., Ro, M. & Pack, A. I. (1997) Spatial and temporal variations in the enzymatic activity of adenosine deaminase in the cat CNS in relation to the sleep-wake cycle. *Society for Neuroscience Abstracts* 23:20. [aJAH]

Macrae, D. & Trolle, E. (1956) The defect of function in visual agnosia. *Brain* 79:94–110. [aMS]

Maddock, R. J. (1999) The retrosplenial cortex and emotion: New insights from functional neuroimaging of the human brain. *Trends in Neurosciences* 22:310–16. [aJAH]

Madsen, P. C. (1993) Blood flow and oxygen uptake in the human brain during various states of sleep and wakefulness. *Acta Neurologica Scandinavia* 88 (Suppl. 148):9–19. [aMS]

Madsen, P. C., Holm, S., Vorstup, S., Friberg, L., Lassen, N. A. & Wildschiodtz, L. F. (1991a) Human regional cerebral blood flow during rapid eye movement sleep. *Journal of Cerebral Blood Flow and Metabolism* 11:502–507. [CG, aJAH, aMS]

Madsen, P. C., Schmidt, J. F., Wildschiodtz, L. F., Holm, S., Vorstup, S. & Lassen, N. A. (1991b) Cerebral O2 metabolism and cerebral blood flow in humans during deep and rapid-eye-movement sleep. *Journal of Applied Physiology* 70:2597–601. [aJAH, aMS]

Madsen, P. C. & Vorstrup, S. (1991) Cerebral blood flow and metabolism during sleep. *Cerebrovascular and Brain Metabolism Reviews* 3:281–96. [aJAH, aMS]

Maho, C. & Bloch, V. (1992) Responses of hippocampal cells can be conditioned during paradoxical sleep. *Brain Research* 581:115–22. [aTAN]

Mahowald, M. W. & Rosen, G. M. (1990) Parasomnias in children. *Pediatrician* 17:21–31. [aTAN]

Mahowald, M. W. & Schenck, C. H. (1991) Status dissociatus – a perspective on states of being. *Sleep* 14:69–79. [rTAN]

(1992/1999) Dissociated states of wakefulness and sleep. *Neurology* 42 (Suppl. 6):44–52. [aTAN, aAR] Also in: *Handbook of behavioral state control: Mollecular and cellular mechanisms,* ed. R. Lydic & H. A. Baghdoyan. CRC Press (1999). [aJAH]

(1994) REM sleep behavior disorder. In: *Principles and practice of sleep medicine, 2nd edition,* ed. M. H. Kryger, T. Roth & W. C. Dement. WB Saunders. [aTAN]

Mahowald, M. W., Woods, S. R. & Schenck, C. H. (1998) Sleeping dreams, waking hallucinations and the central nervous system. *Dreaming* 8:89–102. [aJAH, JFP, aAR]

Maier, S. J., Grahn, R. E., Lalman, B. A., Sutton, L. C., Wiertelak, E. P. & Watkins, L. R. (1993) The role of the amygdala and the dorsal raphe nucleus in mediating the behavioral consequences of inescapable shock. *Behavioral Neuroscience* 106:377–89. [EEK]

Maier Faber, B. (1988) *Psychophysiologishce Faktoren der REM-Traumerinnerung.* Doctoral dissertation, University of Zürich. [MSc]

Mainen, Z. F. & Sejnowski, T. J. (1996) Influence of dendritic structure on firing pattern in model neocortical neurons. *Nature* 382:363–66. [TJM]

Malcolm, N. (1956) Dreaming and skepticism. *Philosophical Review* 65:14–37. [rAR]

(1959) *Dreaming.* Routledge & Kegan Paul. [rAR]

Mallick, B. N., Fahringer, H. M., Wu, M. F. & Siegel, J. M. (1991) REM sleep deprivation reduces auditory evoked inhibition of dorsolateral pontine neurons. *Brain Research* 552:333–37. [aTAN]

Mallick, B. N., Kaur, S., Jha, S. K. & Siegel, J. M. (1999) Possible role of GABA in the regulation of REM sleep with special reference to REM-OFF neurons. In: *Rapid eye movement sleep,* ed. B. N. Mallick & S. Inoue. Marcel Dekker. [aJAH]

Mallick, B. N., Nitz, D., Fahringer, H. & Siegel, J. M. (1997) GABA release in the basal forebrain / medial septal region across the sleep cycle. *Sleep Research* 26:26. [aJAH]

Mallick, B. N. & Inoue, S., eds. (1999) *Rapid eye movement sleep.* Marcel Dekker. [aJAH]

Maloney, K. J. & Jones, B. E. (1997) C-FOS expression in cholinergic, GABAergic and monoaminergic cell groups during paradoxical sleep deprivation and recovery. *Society for Neuroscience Abstracts* 23:2131. [aJAH]

Mamelak, A. N. & Hobson, J. (1989a) Dream bizarreness as the cognitive correlate of altered neuronal behavior in REM sleep. *Journal of Cognitive Neuroscience* 1:201–22. [aJAH, aTAN]

(1989b) Nightcap: A home-based sleep monitoring system. *Sleep* 12:157–66. [aJAH]

Mamelak, M. (1991) A model for narcolepsy. *Canadian Journal of Psychology* 45:194–220. [aJAH]

Manaye, K. F., Zweig, R., Wu, D., Hersh, L. B., De Lacalle, S., Saper, C. B. & German, D. C. (1999) Quantification of cholinergic and select non-cholinergic mesopontine neuronal populations in the human brain. *Neuroscience* 89:759–70. [aJAH]

Mancia, M. (1995) One possible function of sleep: To produce dreams. *Behavioral Brain Research* 69:203–206. [aJAH]

Mancia, M. & Marini, G. (1997) Thalamic mechanisms in sleep control. In: *Sleep and sleep disorders: From molecule to behavior,* ed. O. Hayaishi & S. Inoue. Academic Press. [aJAH]

Mandai, O., Guerrin, A., Sockeel, P., Dujardin, K. & Leconte, P. (1989) REM sleep modifications following a morse code learning session in human. *Physiology and Behavior* 46:639–42. [MSc]

Manfridi, A. & Mancia, M. (1996) Desynchronized (REM) sleep inhibition induced by carbachol microinjections into the nucleus basalis of Meynert is mediated by the glutamatergic system. *Experimental Brain Research* 109:174–78. [aJAH]

Mann, C., Simmons, J., Wilson, C., Engel, J. & Bragin, A. (1997) EEG in human hippocampus, amygdala and entorhinal cortex during REM and NREM sleep. *Sleep Research* 26:27. [aJAH]

Manor, Y., Koch, C. & Segev, I. (1991) Effect of geometrical irregularities on propagation delay in axonal trees. *Biophysical Journal* 60:1424–37. [TJM]

Maquet, P. (1995) Sleep function(s) and cerebral metabolism. *Behavioural Brain Research* 69:75–83. [aJAH]

(2000) Functional neuroimaging of sleep by positron emission tomography. *Journal of Sleep Research* 9:207–231. [aJAH]

Maquet, P., Degueldre, C., Delfiore, G., Aerts, J., Peters, J.-M., Luxen, A. & Franck, G. (1997) Functional neuroanatomy of human slow wave sleep. *The Journal of Neuroscience* 17:2807–12. [CP, aJAH, EFP-S]

Maquet, P., Dive, D., Salmon, E., Sadzot, B., Franco, G., Poirier, R. & Franck, C. (1990) Cerebral glucose utilization during sleep-wake cycle in man determined by positron emission tomography and [18F]-2-fluoro-2 deoxy-D-glucose method. *Brain Research* 513:136–43. [aJAH, aMS]

(1992) Cerebral glucose utilization during stage 2 sleep in man. *Brain Research* 571:149–53. [aJAH]

Maquet, P. & Franck, G. (1997) REM sleep and the amygdala. *Molecular Psychiatry* 2:195–96. [aJAH, EFP-S]

Maquet, P., Peters, J.-M., Aerts, J., Delfiore, G., Degueldre, C., Luxen, A. & Franck, G. (1996) Functional neuroanatomy of human rapid-eye-movement sleep and dreaming. *Nature* 383(6596):163–66. [JSA, JAB, RCo, arJAH, EFP-S, aAR, arMS, aRPV]

Maquet, P. & Phillips, C. (1998) Functional brain imaging of human sleep. *Journal of Sleep Research* 7 (Supplement 1):42–47. [aJAH]

(1999) Rapid eye movement sleep: From cerebral metabolism to functional brain mapping. In: *Rapid eye movement sleep,* ed. B. N. Mallick & S. Inoue. Marcel Dekker. [aJAH]

Marczynski, G. T. (1983) Algorithm for calculating theoretical probabilities of patterns generated by sequential inequality testing. *International Journal of Biomedical Computing* 14:463–86. [TJM]

Marczynski, T. J. (1993) Slow potential changes in the human brain. In: *Proceedings of NATO Symposium,* ed. C. McCallum & S. H. Curry. Plenum Press. [TJM]

Marczynski, T. J., Burns, L. L., Livezey, G. T., Vimal, L. P. & Chen, E. (1984) Sleep and purposive behavior: Inverse deviations from randomness of neuronal firing patterns and the feline thalamus. A new form of homeostasis? *Brain Research* 298:75–90. [TJM]

Marczynski, T. J., Burns, L. L. & Monley, C. A. (1992) Empirically derived model of the role of sleep in associative learning and recuperative processes. *Neural Networks* 5:371–402. [TJM]

Marczynski, T. J. & Sherry, C. J. (1971) A new analysis of trains of increasing and decreasing neuronal spike intervals treated as self-adjusting sets of ratios. *Brain Research* 35:533–38. [TJM]

Marczynski, T. J., Wei, J. Y., Burns, L. L., Choi, S. Y., Chen, E. & Marczynski, G. T. (1982) Visual attention and neuronal firing patterns in the feline pulvinar nucleus of thalamus. *Brain Research Bulletin* 8:565–80. [TJM]

Margraf, J. (1996) *Lehrbuch der Verhaltenstherapie.* Springer. [MSc]

Marini, G., Gritti, I. & Mancia, M. (1992) Enhancement of tonic and phasic events of rapid eye movement sleep following bilateral ibotenic acid injections into centralis lateralis thalamic nucleus of cats. *Neuroscience* 48:877–88. [aJAH]

Marioti, M. & Formenti, A. (1990) Somatosensory transmission through the ventroposterolateral thalamic nucleus during sleep and wakefulness. In: *The diencephalon and sleep,* ed. M. Mancia & G. Marini. Raven Press. [CMP]

Markand, O. & Dyken, M. (1976) Sleep abnormalities in patients with brain stem lesions. *Neurology* 26:769–76. [aMS, aRPV]

Markowitsch, H. J. (1996) Neuropsychologie des menschlichen Gedächtnisses. *Spektrum der Wissenschaft* 9:52–61. [MSc]

Markowitz, J. (1991) Fluoxetine and dreaming. *Journal of Clinical Psychiatry* 52:432. [GL]

Marks, G. A. & Birabil, C. G. (1998) Enhancement of rapid eye movement sleep in the rat by cholinergic and adenosinergic agonists infused into the pontine reticular formation. *Neuroscience* 86:29–37. [aJAH]

Marks, G. A., Roffwarg, H. P. & Shaffery, J. P. (1999) Neuronal activity in the lateral geniculate nucleus associated with ponto-geniculo-occipital waves lacks lamina specificity. *Brain Research* 815:21–28. [RCo]

Marks, G. A. & Shaffery, J. P. (1996) A preliminary study of sleep in the ferret, *Mustela putorius furo:* A carnivore with an extremely high proportion of REM sleep. *Sleep* 19:83–93. [JMS]

Marks, I. M. & Nesse, R. M. (1994) Fear and fitness: An evolutionary analysis of anxiety disorders. *Ethology and Sociobiology* 15:247–61. [aAR]

Marrosu, F., Portas, C., Mascia, M. S., Casu, M. A., Fa, M., Giagheddu, M., Imperato, A. & Gessa, G. L. (1995) Microdialysis measurement of cortical and hippocampal acetylcholine release during sleep-wake cycle in freely moving cats. *Brain Research* 671:329–32. [CG, aJAH, EKP]

Marten, K. & Psarakos, S. (1995) Using self-view television to distinguish between self-examination and social behavior in the bottlenose dolphin (*Tursiops truncatus*). *Consciousness and Cognition* 4:205–24. [aJAH]

Mathews, A. & MacLeod, C. (1985) Selective processing of threat cues in anxiety states. *Behavioral Research and Therapy* 23:563–69. [LM]

Mathur, R. & Douglas, N. J. (1995) Frequency of EEG arousals from nocturnal sleep in normal subjects. *Sleep* 18:330–33. [RDO]

Matsukawa, M., Ogawa, M., Nakadate, K., Maeshima, T., Ichitani, Y., Kawai, N. & Okado, N. (1997) Serotonin and acetylcholine are crucial to maintain hippocampal synapses and memory acquisition in rats. *Neuroscience Letters* 230:13–16. [CG]

Maudhuit, C., Jolas, T., Chastanet, M., Hamon, M. & Adrien, J. (1996) Reduced inhibitory potency of serotonin reuptake blockers on central serotoninergic neurons in rats selectively deprived of rapid eye movement sleep. *Biological Psychiatry* 40(10):1000–1007. [GL]

Maurer, D., Lewis, T. L., Brent, H. P. & Levin, A. V. (1999) Rapid improvement in the acuity of infants after visual input. *Science* 286:108–10. [TJM]

Mazzoni, G., Gori, S., Formicola, G., Gneri, G., Massetani, R., Murri, L. &

Salzarulo, P. (1999) Word recall correlates with sleep cycles in elderly subjects. *Journal of Sleep Research* 8:185–88. [GM, PS]

McCarley, R. W. (1994) Dreams and the biology of sleep. In: *Principles and practice of sleep medicine, 2nd edition,* ed. M. H. Kryger, T. Roth & W. C. Dement. WB Saunders. [aTAN]

McCarley, R. W., Greene, R. W., Rannie, D. & Portas, C. M. (1995) Brainstem neuromodulation and REM sleep. *Seminars in Neuroscience* 7:341–54. [aJAH]

McCarley, R. W. & Hobson, J. A. (1970) Cortical unit activity in desynchronized sleep. *Science* 167:901–903. [CG]

(1975) Neuronal excitability modulation over the sleep cycle: A structural and mathematical model. *Science* 189:58–60. [JH, aJAH, aMS]

(1977) The neurobiological origins of psychoanalytic dream theory. *American Journal of Psychiatry* 134:1211–21. [JH, arJAH]

(1979) The form of dreams and the biology of sleep. In: *Handbook of dreams. Research, theory and applications,* ed. B. B. Wolman. WB Saunders/Van Nostrand Reinhold. [arTAN]

McCarley, R. W. & Hoffman, E. (1981) REM sleep dreams and the activation-synthesis hypothesis. *American Journal of Psychiatry* 138:904–12. [arJAH]

McCarley, R. W. & Massaquoi, S. G. (1986) A limit cycle mathematical model of the REM sleep generator. *American Journal of Physiology* 251 (*Regulatory Integrative Comparative Physiology* 20):R1011-R29. [aJAH]

McCarley, R. W. & Sinton, C. M. (2000) Neuroanatomical and neurophysiological aspects of sleep: Basic science and clinical relevance. *Seminars in Clinical Neuropsychiatry* 5(1):6–19. [AK]

McCarley, R. W., Strecker, R. E., Porkka-Hieskanen, T., Thakkar, M., Bjorkum, A. A., Portas, C. M., Rannie, D. G. & Greene, R. W. (1997) Modulation of cholinergic neurons by serotonin and adenosine in the control of REM and NREM sleep. In: *Sleep and sleep disorders: From molecule to behavior,* ed. O. Hayaishi & S. Inoue. Academic Press. [aJAH]

McCarley, R. W., Winkelman, J. W. & Duffy, H. (1983) Human cerebral potentials associated with REM sleep rapid eye movements: Links to PGO waves and waking potentials. *Brain Research* 274:359–64. [CG, aJAH]

McCarthy, G., Puce, A., Gore, J. C. & Truett, A. (1997) Face specific processing in the human fusiform gyrus. *Journal of Cognitive Neuroscience* 9:605–10. [aJAH]

McClelland, J. L., McNaughton, B. L. & O'Reilly, R. C. (1995) Why there are complementary learning systems in the hippocampus and neocortex: Insights from the successes and failures of connectionist models of learning and memory. *Psychological Review* 102:419–57. [rJAH]

McCormick, D. A. (1990) Cellular mechanisms of cholinergic control of neocortical and thalamic neuronal excitability. In: *Brain cholinergic systems,* ed. M. Steriade & D. Biesold. Oxford University Press. [aJAH]

McCormick, D. A. & Bal, T. (1997) Sleep and arousal: Thalamocortical mechanisms. *Annual Review of Neuroscience* 20:185–215. [CMP, HSP]

McCormick, D. A. & Williamson, A. (1991) Modulation of neuronal firing mode in cat and guinea pig LGN by histamine: Possible cellular mechanisms of histaminergic control of arousal. *Journal of Neuroscience* 11:3188–99. [aJAH]

McDonald, D. G., Schiet, W. W., Frajiet, R. E., Shallenberger, H. D. & Edwards, P. J. (1975) Studies of information processing in sleep. *Psychophysiology* 12:624–29. [aTAN]

McFarlane, A. C. (1987) Posttraumatic phenomena in a longitudinal study of children following a natural disaster. *Journal of the American Academy of Child and Adolescent Psychiatry* 26(5):764–69. [aAR]

McGinn, C. (1989) Can we solve the mind-body problem? *Mind* 98:349–66. [OF]

(1991) *The problem of consciousness.* Blackwell. [KG]

McGinty, D. & Harper, R. (1976) Dorsal raphe neurons: Depression of firing during sleep in cats. *Brain Research* 101:569–75. [CG, aJAH, EKP]

McGinty, D., Szymusiak, R. & Thompson, D. (1994) Preoptic/anterior hypothalamic warming increases EEG delta frequency activity within non-rapid eye movement sleep. *Brain Research* 667:273–77. [aJAH]

McGrath, M. J. & Cohen, D. B. (1978) REM sleep facilitation of adaptive waking behavior: A review of the literature. *Psychological Bulletin* 85(1):24–57. [aTAN, CS, aRPV]

McGuinn, C. (1999) *The mysterious flame.* Basic Books. [MK]

McNair, D. M., Kahn, R. J., Frankenthaler, L. M. & Faldetta, L. L. (1984) Amoxapine and amitriptyline II. Specificity of cognitive effects during brief treatment of depression. *Psychopharmacology* 83:134–39. [aRPV]

McNally, R. (1987) Preparedness and phobias: A review. *Psychological Bulletin* 101:283–303. [RLe, LM]

Mealey, L., Daood, C. & Krage, M. (1996) Enhanced memory for faces associated with potential threat. *Ethology and Sociobiology* 17:119–28. [LM]

Meehl, P. (1989) Schizotaxia revisited. *Archives of General Psychiatry* 46:935–44. [RLe]

Meindl, R. S. (1992) Human populations before agriculture. In: *The Cambridge encyclopedia of human evolution,* ed. S. Jones, R. Martin & D. Pilbeam. Cambridge University Press. [aAR]

Mellman, T. A. (1997) Psychobiology of sleep disturbances in posttraumatic stress disorder. *Annals of the New York Academy of Sciences* 821:142–49. [TAN]

Mellman, T. A., Kulick-Bell, R., Ashlock, L. E. & Nolan, B. (1995) Sleep events among veterans with combat-related posttraumatic stress disorder. *American Journal of Psychiatry* 152:110–15. [TAN]

Mellman, T. A., Nolan, B., Hebding, J., Kulick-Bell, R. & Dominguez, R. (1997) A polysomnographic comparison of veterans with combat-related PTSD, depressed men, and non-ill controls. *Sleep* 20:46–51. [TAN]

Mendlewicz, J., Kempenaers, C. & de Maertelaer, V. (1991) Sleep EEG and amitryptiline treatment in depressed inpatients. *Biological Psychiatry* 30:691–702. [aRPV]

Merikle, P. M., Joordens, S. & Stolz, J. A. (1995) Measuring the relative magnitude of unconscious influence. *Consciousness and Cognition* 4:422–39. [GM]

Merrill, W. (1987) The Raramuri stereotype of dreams. In: *Dreaming,* ed. B. Tedlock. [KG]

Merritt, J. M., Stickgold, R., Pace-Schott, E., Williams, J. & Hobson, J. A. (1994) Emotion profiles in the dreams of men and women. *Consciousness and Cognition* 3:46–60. [JAC, GWD, arJAH, LM]

Mesulam, M.-M. (1998) From sensation to cognition. *Brain* 121:1013–52. [RC, aJAH]

Mesulam, M.-M., Hersh, L. B., Mash, D. C. & Geula, C. (1992) Differential cholinergic innervation within functional subdivisions of the human cerebral cortex: A choline acetyltransferase study. *Journal of Comparative Neurology* 318:316–28. [rJAH]

Metcalfe, J. & Jacobs, W. J. (1998) Emotional memory. The effects of stress on "cool" and "hot" memory systems. *The Psychology of Learning and Motivation* 38:187–222. [EEK, aAR, AR]

Metcalfe, J. & Mischel, W. (1999) A hot/cool-system analysis of delay of gratification: Dynamics of willpower. *Psychological Review* 106(1):3–19. [EEK]

Metherate, R., Cox, C. L. & Ashe, J. H. (1992) Cellular bases of neocortical activation: Modulation of neural oscillations by the nucleus basalis and endogenous acetylcholine. *Journal of Neuroscience* 12:4701–11. [arJAH]

Metzger, L. J., Orr, S. P., Lasko, N. B., Berry, N. J. & Pitman, R. K. (1997) Evidence for diminished P3 amplitudes in PTSD. *Annals of the New York Academy of Sciences* 821:499–503. [TAN]

Meyer, J. S., Ishikawa, Y., Hata, T. & Karacan, I. (1987) Cerebral blood flow in normal and abnormal sleep and dreaming. *Brain and Cognition* 6:266–94. [aJAH]

M'Harzi, M. & Jarrard, L. E. (1992) Effects of medial and lateral septal lesions on acquisition of a place and cue radial maze task. *Behavioural Brain Research* 49:159–65. [aRPV]

Michel, F. & Sieroff, E. (1981) Une approche anatomo-clinique des deficits de l'imagerie oreirique, est-elle possible? In: *Sleep: Proceedings of an international colloquium.* Carlo Erba Formitala. [aMS]

Mignot, E. & Nishino, S. (1999) Narcolepsy. In: *Handbook of behavioral state control: Molecular and cellular mechanisms,* ed. R. Lydic & H. A. Baghdoyan. CRC Press. [aJAH]

Miller, A. M., Obermeyer, W. & Benca, R. (1997) The superior colliculus is involved in paradoxical sleep induction by lights-off stimulation in albino rats. *Sleep Research* 26:739. [aJAH]

Miller, G. A. (1956) The magical number seven, plus or minus two: Some limits on our capacity for processing information. *Psychological Review* 63:81–97. [JBP]

Miller, J. D., Farber, J., Gatz, P., Roffwarg, H. & Germain, D. C. (1983) Activity of mesencephalic dopamine and non-dopamine neurons across stages of sleep and waking in the rat. *Brain Research* 273:133–41. [CG, arJAH, EKP, rMS]

Miller, L., Drew, W. G. & Schwartz, I. (1971) Effect of REM sleep deprivation on retention of a one-trial passive avoidance response. *Perceptual and Motor Skills* 33:118. [aRPV]

Mirmiran, M. (1983) "Oneiric" behavior during active sleep induced by bilateral lesions of the pontine tegmentum in juvenile rats. In: *Sleep 1987,* ed. W. P. Koella. Karger. [CG]

Mirmiran, M., van den Dungen, H. & Uylings, H. B. M. (1982) Sleep patterns during rearing under different environmental conditions in juvenile rats. *Brain Research* 233:287–98. [aRPV]

Miro, E., Villanueva, Y., Del-Rio-Portilla, Y., Perez-Garci, E., Guevara, M. A. & Corsi-Cabrera, M. (1999) EEG changes underlying cognitive alterations during dreaming. *Sleep Research Online* 2 (Supplement 1):28. [aJAH]

Mitler, M. M., Van Den Hoed, J., Carskadon, M., Richardson, G., Park, R., Guilleminault, C. & Dement, W. C. (1979) REM sleep episodes during the multiple sleep latency test in narcoleptic patients. *Electroencephalography and Clinical Neurophysiology* 46:479–81. [aJAH]

Miyauchi, S., Takino, R. & Azakami, M. (1990) Evoked potentials during REM sleep reflect dreaming. *Electroencephalography and Clinical Neurophysiology* 76:19–28. [aJAH]

Miyauchi, S., Takino, R., Fukuda, H. & Torii, S. (1987) Electrophysiological evidence for dreaming: Human cerebral potentials associated with rapid eye movements during REM sleep. *Electroencephalography and Clinical Neurophysiology* 66:383–90. [CG, aJAH]

Mizumori, S. J., Perez, G. M., Alvarado, M. C., Barnes, C. A. & McNaughton, B. L. (1990) Reversible inactivation of the medial septum differentially affects two forms of learning in rats. *Brain Research* 528:12–20. [aRPV]

Moffitt, A. (1995) Dreaming: Functions and meanings. *Impuls* 3:18–31. [aJAH]

Moffitt, A., Hoffmann, R., Wells, R., Armitage, R., Pigeau, R. & Shearer, J. (1982) Individual differences among pre–and post-awakening EEG correlates of dream reports following arousals from different stages of sleep. *Psychiatric Journal of the University of Ottawa* 7:111–25. [TLK, aTAN]

Moffitt, A., Kramer, M. & Hoffmann, R., eds. (1993) *The functions of dreaming.* State University of New York Press. [EH, MK, RLe]

Mogilnicka, E., Przewlocka, B., van Luijtelaar, E. L. J. M., Klimek, V. & Coenen, A. M. L. (1986) Effects of REM sleep deprivation on central alpha 1- and beta-adrenoceptors in rat brain. *Pharmacology, Biochemistry and Behavior* 25:329–32. [AC]

Molinari, S. & Foulkes, D. (1969) Tonic and phasic events during sleep: Psychological correlates and implications. *Perceptual and Motor Skills* 29:343–68. [aJAH, MK, aTAN, EFP-S]

Monaco, A. P., Baghdoyan, H. A., Nelson, J. P. & Hobson, J. A. (1984) Cortical wave amplitude and eye movement direction are correlated in REM sleep but not in waking. *Archives Italiennes de Biologie* 122:213–23. [JSA, aJAH]

Monroe, L. J., Rechtschaffen, A., Foulkes, D. & Jensen, J. (1965) Discriminability of REM and NREM reports. *Journal of Personality and Social Psychology* 2:456–60. [aJAH, aTAN, aMS]

Montangero, J. (1991) How can we define the sequential organization of dreams? *Perceptual and Motor Skills* 73:1059–73. [aJAH]

(1993) Dream, problem-solving, and creativity. In: *Dreaming as cognition,* ed. C. Cavallero & D. Foulkes. Harvester Wheatsheaf. [aAR]

(1999) *Rêve et cognition.* Mardaga. [JM]

Monti, J. M. (1993) Involvement of histamine in the control of the waking state. *Life Sciences* 53:1331–38. [aJAH]

Monti, J. M., Jantos, H., Silveria, R., Reyes-Parada, M., Scorza, C. & Prunell, G. (1994) Depletion of brain seratonin by 5, 7-DHT: Effects on the 8–OH-DPAT –induced changes of sleep and waking in the rat. *Psychopharmacology* 115:273–77. [aJAH]

Monti, J. M. & Monti, D. (1999) Functional role of seratonin 5-HT1 and 5-HT2 receptor in the regulation of REM sleep. In: *Rapid eye movement sleep,* ed. B. N. Mallick & S. Inoue. Marcel Dekker. [aJAH]

Montplaisir, J., Petit, D., Lorrain, D., Gauthier, S. & Nielsen, T. A. (1995) Sleep in Alzheimer's disease: Further considerations on the role of brainstem and forebrain cholinergic populations in sleep-wake mechanisms. *Sleep* 18:145–48. [rTAN]

Moore, H., Fadel, J., Sarter, M. & Bruno, J. P. (1999) Role of accumbens and cortical dopamine receptors in the regulation of cortical acetylcholine release. *Neuroscience* 88:811–22. [arJAH]

Moradi, A. R., Doost, H. T. N., Taghavi, M. R., Yule, W. & Dalgleish, T. (1999) Everyday memory deficits in children and adolescents with PTSD: Performance on the Rivermead Behavioural Memory Test. *Journal of Child Psychology and Psychiatry and Allied Disciplines* 40:357–61. [TAN]

Morales, F. R. & Chase, M. H. (1978) Intracellular recording of lumbar motoneuron membrane potential during sleep and wakefulness. *Experimental Neurology* 62:821–27. [CG]

Morden, B., Conner, R., Mitchell, G., Dement, W. & Levine, S. (1967) Effects of rapid eye movement (REM) sleep deprivation on shock-induced fighting. *Physiology and Behavior* 3:425–32. [HS]

Morris, M., Bowers, D., Chatterjee, A. & Heilman, K. (1992) Amnesia following a discrete basal forebrain lesion. *Brain* 115:1827–47. [aMS]

Morrison, A. R. (1979) Brainstem regulation of behavior during sleep and wakefulness. In: *Progress in psychobiology and physiological psychology, vol. 8,* ed. J. M. Sprague & A. N. Epstein. Academic Press. [ARM]

(1982) Central activity states: Overview. In: *The neural basis of behavior,* ed. A. L. Beckman. Spectrum. [VSR]

(1983a) A window on the sleeping brain. *Scientific American* 248:86–94. [aAR]

(1983b) Paradoxical sleep and alert wakefulness: Variations on a theme. In: *Sleep disorders: Basic and clinical research,* ed. M. Chase & E. Weitzman. S. P. Medical and Scientific Books. [HTH]

Morrison, A. R. & Bowker, R. M. (1975) The biological significance of PGO spikes in the sleeping cat. *Acta Neurobiologiae Experimentalis* 35:821–40. [MBl, CG]

Morrison, A. R., Mann, G. L. & Hendricks, J. C. (1981) The relationship of excessive exploratory behavior in wakefulness to paradoxical sleep without atonia. *Sleep* 4:247–57. [ARM]

Morrison, A. R. & Pompeiano, O. (1966) Vestibular influences during sleep. IV. Functional relations between vestibular nuclei and lateral geniculate nucleus during desynchronized sleep. *Archives Italiennes de Biologie* 104:425–58. [JH]

(1970) Vestibular influences during sleep. VI. Vestibular control of autonomic functions during the rapid eye movements of desynchronized sleep. *Archives Italiennes de Biologie* 108:154–80. [JH]

Morrison, A. R. & Reiner, P. B. A. (1985) A dissection of paradoxical sleep. In: *Brain mechanisms of sleep,* ed. D. J. McGinty, R. Drucker-Colin, A. R. Morrison & P. L. Parmeggiani. Raven Press. [aJAH, ARM]

Morrison, A. R., Sanford, L. D., Ball, W. A., Mann, G. L. & Ross, R. J. (1995) Stimulus-elicited behavior in rapid eye movement sleep without atonia. *Behavioral Neurosciences* 109:972–79. [JH, ARM]

Morrison, A. R., Sanford, L. D. & Ross, R. J. (1999) Initiation of REM sleep: Beyond the brainstem. In: *Rapid eye movement sleep,* ed. B. N. Mallick & S. Inoue. Marcel Dekker/Narosa. [arJAH, ARM]

Morrison, J. H., Moliver, M. E., Grzanna, M. E. & Coyle, J. T. (1981) The intracortical trajectory of the coeruleo-cortical projection in the rat: A tangentially organized cortical afferent. *Neuroscience* 6:139–58. [FD]

Moruzzi, G. & Magoun, H. W. (1949) Brainstem reticular formation and activation of the EEG. *Electroencephalography and Clinical Neurophysiology* 1:455–73. [aJAH, RS-P]

Moscovitz, C., Moses, H. & Klawans, H. L. (1978) Levodopa-induced psychosis: A kindling phenomenon. *American Journal of Psychiatry* 135:669–75. [rJAH]

Moscowitz, E. & Berger, R. J. (1969) Rapid eye movements and dream imagery: Are they related? *Nature* 224:613–14. [IF, aJAH]

Moss, C. S. (1972) *Recovery with aphasia: The aftermath of my stroke.* Illinois University Press. [aMS]

Mouret, J. (1964) *Les movements oculaires au cours du sommeil paradoxal.* Imprimerie des Beaux Arts. [CG]

Mouret, J., Jeannerod, M. & Jouvet, M. D. (1963) L'activite electrique du systeme visuel au cours de la phase paradoxale du sommeil chez le chat. *Journal of Physiology (Paris)* 55:305–306. [aJAH]

Moyer, R. S. & Landauer, T. K. (1967) Time required for judgment of numerical inequality. *Nature* 21(5):1519–20. [TJM]

Mukhametov, L. A., Supin, A. Y. & Polyakova, I. G. (1977) Interhemispheric asymmetry of the electroencephalographic sleep patterns in dolphins. *Brain Research* 134:581–84. [JMS]

Mukhametov, L. M. (1984) Sleep in marine mammals. In: *Sleep mechanisms,* ed. A. Borbely & J. L. Valatx. *Experimental Brain Research* (Suppl 8):227–38. [GL]

(1987) Unihemispheric slow-wave sleep in the Amazonian dolphin, *Inia geoffrensis. Neuroscience Letters* 79:128–32. [aTAN, JMS]

Müller, F. (1892) Ein Beitrag zur Kenntniss der Seelenblindheit. *Archiv für Psychiatrie und Nervenkrankenheiten* 24:856–917. [aMS]

Müller, M. M., Bosch, J., Elbert, T., Krieter, A. K., Sosa, M. V., Sosa, P. V. & Rockstroh, B. (1996) Visually induced gamma-based responses in human electroencephalographic activity: A link to animal studies. *Experimental Brain Research* 112:96–102. [aJAH]

Murck, H., Guldner, J., Colla-Muller, M., Frieboes, R.-M., Schier, T., Weidemann, K., Holsboer, F. & Steiger, A. (1996) VIP decelerates non-REM-REM cycles and modulates hormone secretion during sleep in men. *American Journal of Physiology* 271:R905–11. [aJAH]

Murri, L., Arena, R., Siciliano, G., Mazzotta, R. & Muratorio, A. (1984) Dream recall in patients with focal cerebral lesions. *Archives of Neurology* 41:183–85. [aMS]

Murri, L., Massetani, R., Siciliano, G. & Arena, R. (1985) Dream recall after sleep interruption in brain-injured patients. *Sleep* 8:356–62. [aMS]

Muzio, J. N., Roffwarg, H. P., Anders, C. B. & Muzio, L. G. (1972) Retention of rote learned meaningful verbal material and alteration in the normal sleep EEG pattern. *Psychophysiology* 9:108. [aRPV]

Nadel, L. (1994) Multiple memory systems: What and why, an update. In: *Memory systems 1994,* ed. D. L. Schacter & E. Tulving. MIT Press. [aJAH]

Nadel, L. & Moscovitch, M. (1997) Memory consolidation, retrograde amnesia and the hippocampal complex. *Current Opinion in Neurobiology* 7:217–27. [aRPV]

Nader, K. (1996) Children's traumatic dreams. In: *Trauma and dreams,* ed. D. Barrett. Harvard University Press. [arAR]

Nader, K., Pynoos, R., Fairbanks, L. & Frederick, C. (1990) Children's PTSD reactions one year after a sniper attack at their school. *American Journal of Psychiatry* 147:1526–30. [aAR]

Nagel, T. (1986) *The view from nowhere.* Oxford University Press. [KG]

Nakamura, Y., Goldberg, L. J., Chandler, S. H. & Chase, M. H. (1978) Intracellular analysis of trigeminal motoneuron activity during sleep in the cat. *Science* 199:204–206. [CF]

Nakano, S., Tsuji, S., Matsunaga, K. & Murai, Y. (1995) Effect of sleep stage on somatosensory evoked potentials by median nerve stimulation. *Electroencephalography and Clinical Neurophysiology* 96:385–89. [aTAN]

Nambu, T., Sakurai, T., Mizukami, K., Hosoya, Y., Yamagisawa, M. & Goto, K. (1999) Distribution of orexin neurons in the adult rat brain. *Brain Research* 827:243–60. [RS-P]

Namgung, U., Matsuyama, S. & Routtenberg, A. (1997) Long-term potentiation activates the GAP-43 promoter: Selective participation of hippocampal mossy cells. *Proceedings of the National Academy of Sciences USA* 94:11678–80. [TJM]

Natale, V. & Battaglia, D. (1990) Temporal dating of autobiographical memories associated to REM and NREM dreams. *Imagination, Cognition and Personality* 10:279–84. [aTAN]

Nausieda, P., Weiner, W., Kaplan, L., Weber, S. & Klawans, H. (1982) Sleep disruption in the course of chronic levodopa therapy: An early feature of the levodopa psychosis. *Clinical Neuropharmacology* 5:183–94. [rJAH, aMS]

Nauta, W. J. H. (1946) Hypothalamic regulation of sleep in rats. An experimental study. *Journal of Neurophysiology* 9:285–316. [RS-P]

(1966) Projections of the lentiform nucleus in the monkey. *Brain Research* 1:3–42. [EEK]

Nauta, W. J. H. & Feirtag, M. (1986) *Fundamental neuroanatomy.* W. H. Freeman. [EEK]

Naville, F. & Brantmay, H. (1935) Contribution à l'étude des équivalents épileptiques chez les enfant. *Archives Suisses de Neurologie et de Psychiatrie* 35:96–122. [aMS]

Neal, P. (1988) *As I am.* Century. [aMS]

Neidhart, E. J., Krakow, B., Kellner, R. & Pathak, D. (1992) The beneficial effects of one treatment session and recording of nightmares on chronic nightmare sufferers. *Sleep* 15:470–73. [aAR]

Neisser, U. (1967) *Cognitive psychology.* Appleton-Century-Crofts. [PC]

Nelson, J. P., McCarley, R. W. & Hobson, J. A. (1983) REM sleep burst neurons, PGO waves, and eye movement information. *Journal of Neurophysiology* 50:784–97. [RCo, JH, aJAH]

Nesse, R. (1997) An evolutionary perspective on panic disorder and agoraphobia. In: *The maladapted mind: Classic readings in evolutionary psychopathology,* ed. S. Baron-Cohen. Psychology Press. [aAR]

Nesse, R. & Williams, G. (1997) Are mental disorders diseases? In: *The maladapted mind: Classic readings in evolutionary psychopathology,* ed. S. Baron-Cohen. Psychology Press. [aAR]

Nicholson, A. N., Belyavin, A. J. & Pascoe, P. A. (1989) Modulation of rapid eye movement sleep in humans by drugs that modify monoaminergic and purinergic transmission. *Neuropsychopharmacology* 2:131–43. [aJAH, aRPV]

Nicholson, A. N. & Pascoe, P. A. (1986) 5-hydroxytryptamine and noradrenaline uptake inhibition: Studies on sleep in man. *Neuropharmacology* 25:1079–83. [aRPV]

(1988) Studies on the modulation of the sleep-wakefulness continuum in man by fluoxetine, a 5-HT uptake inhibitor. *Neuropharmacology* 27:597–602. [aRPV, LJW]

(1991) Presynaptic alpha-adrenoreceptor function and sleep in man: Studies with clonidine and idazoxan. *Neuropharmacology* 30:367–72. [aJAH]

Nick, T. A. & Ribera, A. B. (2000) Synaptic activity modulates presynaptic excitability. *Nature Neuroscience* 3(2):142–49. [WF]

Nielsen, J. (1955) Occipital lobes, dreams and psychosis. *Journal of Nervous and Mental Diseases* 121:50–52. [aMS]

Nielsen, T. A. (1992) A self-observational study of spontaneous hypnagogic imagery using the upright napping procedure. *Imagination, Cognition and Personality* 11:353–66. [arTAN]

(1995) Describing and modeling hypnagogic imagery using a systematic self-observation procedure. *Dreaming* 5:75–94. [arTAN]

(1999) Mentation during sleep. The NREM/REM distinction. In: *Handbook of behavioral state control. Cellular and molecular mechanisms,* ed. R. Lydic & H. A. Baghdoyan. CRC Press. [aJAH, JAH, aTAN, EFP-S, aMS]

Nielsen, T. A. & Carrier, J. (2000) Unpublished findings. Centre d'étude du sommeil, Hôpital du Sacré Coeur, Montréal, Québec. [rTAN]

Nielsen, T. A. & Chénier, V. (1999) Variations in EEG coherence as an index of the affective content of dreams from REM sleep: Relationships with face imagery. *Brain and Cognition* 41:200–12. [rTAN]

Nielsen, T. A., Deslauriers, D. & Baylor, G. W. (1991) Emotions in dream and waking event reports. *Dreaming* 1:287–300. [aJAH]

Nielsen, T. A., Germain, A. & Ouellet, L. (1995) Atonia-signalled hypnagogic imagery: Comparative EEG mapping of sleep onset transitions, REM sleep and wakefulness. *Sleep Research* 24:133. [arTAN]

Nielsen, T. A., Kuiken, D., Hoffmann, R. & Moffitt, A. (2001) REM and NREM sleep mentation differences: A question of story structure? *Sleep and Hypnosis (in press).* [aTAN]

Nielsen, T. A., Kuiken, D. L., Moffitt, A., Hoffman, R. & Newell, R. (1983) Comparisons of the story structure of stage REM and stage 2 mentation reports. *Sleep Research* 12:181. [aTAN]

Nielsen, T. A. & Montplaisir, J. (1994) Dreaming and REM sleep – commentary. *World Federation of Sleep Research Societies Newsletters* 3:15–16. [aTAN]

Nielsen, T. A., Raymond, I., Bessette, P., Faucher, B. & Germain, A. (1999a) Qualitative differences in social interactions between REM and NREM mentation reports. Paper presented at the Sixteenth International Conference of the Association for the Study of Dreams, Santa Cruz, CA, July 6–10, 1999. [aTAN]

Nielsen, T. A. & Zadra, A. L. (2000) Dreaming disorders. In: *Principles and practice of sleep medicine, 3rd edition,* ed. M. Kryger, T. Roth & W. C. Dement. W. B. Saunders. [rTAN]

Nielsen, T. A., Zadra, A. L. & Fukuda, K. (1999b) Changes in the typical dreams of Japanese students over 40 years. Paper presented at the Sixteenth International Conference of the Association for the Study of Dreams, Santa Cruz, CA, July 6–10. [AG]

Nielsen, T. A., Zadra, A. L., Germain, A. & Montplaisir, J. (1998) The 55 typical dreams questionnaire: Assessment of 200 sleep patients. *Sleep* 21(Supplement):286. [AG]

(1999c) The typical dreams of sleep patients: Consistent profile with 284 new cases. *Sleep* 22 (Suppl. 1):S177–78. [AG]

Nietzsche, F. (1992) On the geneaology of morals. In: *Basic writings of Nietzsche,* trans. and ed. W. Kaufmann. Modern Library. [GL]

Niiyama, Y., Fujiwara, R., Satoh, N. & Hishikawa, Y. (1994) Endogenous components of event-related potential appearing during NREM stage 1 and REM sleep in man. *International Journal of Psychophysiology* 17:165–74. [aTAN]

Niiyama, Y., Sekine, A., Fushimi, M. & Hishikawa, Y. (1997) Marked suppression of cortical auditory evoked response shortly before the onset of REM sleep. *NeuroReport* 8:3303–308. [aJAH]

(1998) Cortical activity of REM sleep often occurs earlier than other physiological phenomena. *Psychiatry and Clinical Neuroscience* 52:152–54. [aTAN]

Niiyama, Y., Shimizo, T., Hu, M. & Hiohikawa, Y. (1988) Phasic EEG activities associated with rapid eye movements during REM sleep in man. *Electroencephalography and Clinical Neurophysiology* 70:396–403. [aJAH]

Nikles, C. D., Brecht, D. L., Klinger, E. & Bursell, A. L. (1998) The effects of current-concern–and nonconcern-related suggestions on nocturnal dream content. *Journal of Personality and Social Psychology* 75:242–55. [RLe, aAR]

Nillson, L. (1999) *Neurosurgery* 45(5):1176. [CMP]

Nillson, L., Hillered, L., Ponten, U. & Ungerstedt, U. (1990) Changes in cortical extracellular levels of energy-related metabolites and amino acids following concussive brain injury in rats. *Journal of Cerebral Blood Flow Metabolism* 10(5):631–37. [CMP]

Nishino, S. & Mignot, E. (1997) Pharmacological aspects of human and canine narcolepsy. *Progress in Neurobiology* 52:27–78. [aJAH]

Nitz, D. & Siegel, J. M. (1997) GABA release in the locus coeruleus as a function of sleep/wake state. *Neuroscience* 78:795–801. [aJAH]

Nofzinger, E. A., Mintun, M. A., Wiseman, M. B., Kupfer, D. J. & Moore, R. Y. (1997) Forebrain activation in REM sleep: An FDG PET study. *Brain Research* 770:192–201. [JSA, arJAH, EAN, EFP-S, HS, aMS, aRPV]

Nofzinger, E. A., Mintun, M. A., Price, J., Meltzer, C. C., Townsend, D., Buysse, D. J., Reynolds, C. F., Dachille, M., Martzzie, J., Kupfer, D. J. & Moore, R. Y. (1998) A method for the assessment of the functional neuroanatomy of human sleep using FDG PET. *Brain Research Protocols* 2(3):191–98. [EAN]

Nofzinger, E. A., Nichols, T. E., Meltzer, C. C., Price, J., Steppe, D. A., Miewald, J. M., Kupfer, D. J. & Moore, R. Y. (1999) Changes in forebrain function from waking to REM sleep in depression: Preliminary analyses of [18F]-FDG PET studies. *Psychiatry Research: Neuroimaging Section* 91(2):59–78. [EAN, EFP-S]

Nofzinger, E. A., Price, J. C., Meltzer, C. C., Buysse D. J., Villemagne, V. L., Miewald, J. M., Sembrat, R. C., Steppe, D. A. & Kupfer, D. J. (2000) Towards a neurobiology of dysfunctional arousal in depression: The relationship between beta EEG power and regional cerebral glucose metabolism during NREM sleep. *Psychiatry Research: Neuroimaging* 98(2):71–91. [EAN]

Noguchi, Y., Yamada, T., Yeh, M., Matsubar, M., Kokubun, Y., Kawada, J., Shiraish, G. & Kajimoto, S. (1995) Dissociated changes of frontal and parietal somatosensory-evoked potentials in sleep. *Neurology* 45:154–60. [aTAN]

Nordby, H., Hugdahl, K., Stickgold, R., Bronnick, K. S. & Hobson, J. A. (1996) Event-related potentials (ERPs) to deviant auditory stimuli during sleep and waking. *NeuroReport* 7:1082–86. [aTAN]

Noyes, G. R. (1950) *Poetical works of Dryden, New edition revised and enlarged, the Cambridge edition.* Houghton Mifflin. [aJAH]

Nunes, G. P., Jr., Tufik, S. & Nobrega, J. N. (1994) Autoradiographic analysis of D1 and D2 dopaminergic receptors in rat brain after paradoxical sleep deprivation. *Brain Research Bulletin* 34:453–56. [aJAH]

Nunez, A. (1996) Unit activity of rat basal forebrain neurons – relationship to cortical activity. *Neuroscience* 72:757–66. [aJAH]

Nykamp, K., Rosenthal, L., Folkerts, M., Roehrs, T., Guido, P. & Roth, T. (1998) The effects of REM sleep deprivation on the level of sleepiness/alertness. *Sleep* 21:609–14. [MBl]

Oakman, S. A., Faris, P. L., Cozzari, C. & Hartmann, B. K. (1999) Characterization of the extent of pontomesencephalic cholinergic neurons' projections to the thalamus: Comparison with projections to midbrain dopaminergic groups. *Neuroscience* 94:529–47. [arJAH]

Oatley, K. (1999) Why fiction may be twice as true as fact: Fiction as cognitive and emotional simulation. *Review of General Psychology* 3:101–17. [JBP]

# References

Obal, F. & Krueger, J. M. (1999) Hormones and REM sleep. In: *Rapid eye movement sleep*, ed. B. N. Mallick & S. Inoue. Marcel Dekker/Narosa. [aJAH, GL]

Obal, F., Opp, M., Cady, B., Johansen, L. & Krueger, J. M. (1989) Vasoactive intestinal peptide and peptide histidine methionine elicit selective increases in REM sleep in rabbits. *Brain Research* 490:292–300. [aJAH]

Occhionero, M., Cicogna, P., Natale, V., Esposito, M. J. & Bosinelli, M. (1998) A comparison of mental activity during slow wave and REM sleep. 14th ESRS Congress, Madrid. *Journal of Sleep Research* 7 (Suppl. 2):190. [MBo, MO]

Occhipinti, S. & Siegal, M. (1994) Reasoning about food and contamination. *Journal of Personality and Social Psychology* 66:243–53. [LM]

Ogawa, T., Satoh, T. & Takagi, K. (1967) Sweating during night sleep. *Japanese Journal of Physiology* 17:135–48. [aTAN]

Ogilvie, R. D., Hunt, H., Sawicki, C. & Samahalski, J. (1980) The REM/nonREM versus the tonic-phasic model of sleep. *Sleep Research* 9:21. [aJAH, EFP-S]

(1982) Psychological correlates of spontaneous MEMA during sleep. *Sleep* 11:11–27. [aJAH]

Ogilvie, R. D., Wilkinson, R. T. & Allison, S. (1989) The detection of sleep onset: Behavioral, physiological, and subjective convergence. *Sleep* 12:458–74. [RDO]

O'Hara, B. F., Edgar, D. M., Wiler, S. W., Cao, V. H., Clegg, D. A., Miller, J. D., Heller, H. C. & Kilduff, T. S. (1997) Nicotonic and muscarinic receptors in the developing and adult circadian system. *Society for Neuroscience Abstracts* 23:312. [aJAH]

Ohman, A. (1993) Fear and anxiety as emotional phenomena: Clinical phenomenology, evolutionary perspectives, and information processing mechanisms. In: *Handbook of emotions*, ed. M. Lewis & J. M. Haviland. Guilford Press. [LM]

Okabe, S. & Kubin, L. (1997) Role of 5HT1 receptors in the control of hypoglossal motorneurons in vivo. *Sleep* 18 (Suppl. 10):S150–53. [aJAH]

Okuma, T., Fukuma, E. & Kobayashi, K. (1975) "Dream detector" and comparison of laboratory and home dreams collected by REMP-awakening technique. *Advances in Sleep Research* 2:223–31. [aJAH]

Oleksenko, A. I., Mukhametov, L. M., Polykova, I. G., Supin, A. Y. & Kovalzon, V. M. (1992) Unihemispheric sleep deprivation in bottlenose dolphins. *Journal of Sleep Research* 1:40–44. [JMS]

Olive, M. F., Seidel, W. F. & Edgar, D. M. (1998) Compensatory sleep responses to wakefulness induced by the dopamine autoreceptor antagonist (-)DS121. *Journal of Pharmacology and Experimental Therapeutics* 285:1073–83. [aJAH]

Olson, D. R., Astington, J. W. & Harris, P. L. (1988) Introduction. In: *Developing theories of the mind*, ed. J. W. Astington, P. L. Harris & D. R. Olson. Cambridge University Press. [JFP]

Oniani, T. N. (1984) Does paradoxical sleep deprivation disturb memory trace consolidation? *Physiology and Behavior* 33:687–92. [AC]

Oniani, T. N. & Lortkipanidze, N. D. (1985) Effect of paradoxical sleep deprivation on learning and memory. In: *Neurophysiology of motivation, memory and sleep–wakefulness cycle, vol. 4*, ed. T. Oniani. Metzniereba. [VSR]

Onoe, H. & Sakai, K. (1995) Kainate receptors: A novel mechanism in paradoxical (REM) sleep generation. *NeuroReport* 6:353–56. [aJAH]

Orem, J. (1980) *Physiology in sleep*, ed. J. Orem. Academic Press. [aJAH]

(2000) Physiology in sleep. In: *Principles and practice of sleep medicine*, eds., M. H. Kryger, T. Roth & W. C. Dement. W. B. Saunders. [aJAH]

Orlinsky, D. (1962) Psychodynamic and cognitive correlates of dream recall – A study of individual differences. Unpublished doctoral dissertation, University of Chicago. [aTAN]

Orr, S. P., Lasko, N. B., Metzger, L. J. & Pitman, R. K. (1997) Physiologic responses to non-starting tones in Vietnam veterans with post-traumatic stress disorder. *Psychiatry Research* 73:103–107. [TAN]

Orr, W. J., Dozier, J. E., Green, L. & Cromwell, R. L. (1968) Self-induced waking: Changes in dreams and sleep patterns. *Comprehensive Psychiatry* 9:499–506. [aJAH]

Osaka, T. & Matsumura, H. (1993) Noradrenergic inputs to sleep-related neurons in the preoptic area from the locus coeruleus and ventrolateral medulla in the rat. *Neuroscience Research* 191:39–50. [aJAH]

Osorio, I. & Daroff, R. (1980) Absence of REM and altered NREM sleep in patients with spinocerebellar degeneration and slow saccades. *Annals of Neurology* 7:277–80. [aMS, aRPV]

Ostow, M. (1954) Psychodynamic disturbances in patients with temporal lobe disorder. *Journal of the Mount Sinai Hospital* 20:293–308. [aMS]

Ostrow, M. (2000) Freud under analysis: An exchange. *The New York Review of Books XLVII* (3):46–48. [OF]

Oswald, I. & Adam, K. (1986) Effects of paroxetine on human sleep. *British Journal of Clinical Pharmacology* 22:97–99. [aRPV]

Oswald, I. & Evans, J. (1985) On serious violence during sleep-walking. *British Journal of Psychiatry* 147:688–91. [aAR]

Oswald, I. & Priest, R. G. (1965) Five weeks to escape the sleeping-pill habit. *British Medical Journal* 2:1093–95. [LJW]

Otto, T., Eichenbaum, H., Wiener, S & Wible, C. (1991) Learning-related patterns of CA1 spike trains parallel stimulation parameters optimal for inducing hippocampal long-term potentiation. *Hippocampus* 1:181–92. [rJAH]

Pace-Schott, E. F., Gersh, T., Silvestri, R., Stickgold, R., Salzmann, C. & Hobson, J. A. (1998) The Nightcap can detect sleep quality changes caused by selective serotonin reuptake inhibitor (SSRI) treatment in normal subjects. *Sleep* 21 (Supplement):102. [aJAH]

(1999) Effects of selective serotonin reuptake inhibitors (SSRI) on dreaming in normal subjects. *Sleep* 22 (Supplement):S172. [arJAH]

(2000) Enhancement of subjective intensity of dream features in normal subjects by the SSRIs paroxetine and fluvoxamine. *Sleep* 23 (Supplement):A167. [rJAH]

(2001) SSRI treatment suppresses dream recall frequency but increases subjective dream intensity in normal subjects. Journal of Sleep Research 10. (in press)

Pace-Schott, E. F., Kaji, J., Stickgold, R. & Hobson, J. A.(1994) Nightcap measurement of sleep quality in self-described good and poor sleepers. *Sleep* 17:688–92. [aJAH]

Pace-Schott, E. F. & Hobson, J. A. (2000) Basic mechanisms of sleep: New evidence on the neuroanatomy and neuromodulation of the NREM-REM cycle. In: *ACNP fifth generation of progress*, eds., D. Charney, J. Coyle, K. Davis & C. Nemeroff (in press). [arJAH]

Pace-Schott, E. F., Stickgold, R. & Hobson, J. A. (1997a) Memory processes within dreaming: An affirmative probe for intra-state dreaming and waking memory events. *Sleep Research* 26:276. [aJAH]

(1997b) Memory processes within dreaming: Methodological issues. *Sleep Research* 26:277. [aJAH]

Pagel, J. F. (1999a) A "dream" can be gazpacho – searching for a definition of dream. *Dream Time* 16(1&2):6–28. [JFP]

(1999b) *The definition of dreaming*. Dream Section Working Group. *American Association of Sleep Medicine Bulletin* 6(4):34. [JFP]

Paiva, T. & Guimaraes, E. (1999) Dream content and EEG in normal subjects. *Sleep Research Online* 2 (Supplement 1):274. [aJAH]

Pal, P. K., Calne, S., Samii, A. & Fleming, J. A. E. (1999) A review of normal sleep and its disturbances in Parkinson's disease. *Parkinsonism Related Disorders* 5:1–17. [EKP]

Palfai, T., Wichlinski, L., Brown, H. A. & Brown, O. M. (1986) Effects of amnesic doses of reserpine or syrosingopine on mouse brain acetylcholine levels. *Pharmacology Biochemistry and Behavior* 24:1427–60. [LJW]

Panksepp, J. (1982) Toward a general psychobiological theory of emotions. *Behavioral and Brain Sciences* 5:407–22. [LM]

(1985) Mood changes. In: *Handbook of clinical neurology 45*, ed. P. Vinken, G. Bruyn & H. Klawans. Elsevier. [aMS]

(1993) Neurochemical control of moods and emotions: Amino acids to neuropeptides. In: *The handbook of emotions*, ed. M. Lewis & J. Haviland. Guilford. [JP]

(1998a/1999) *Affective neuroscience: The foundations of human and animal emotions*. Oxford University Press, 1998. [JP, arAR, arMS] Academic Press, 1999. [LM]

(1998b) The periconscious substrates of consciousness: Affective states and the evolutionary origins of the self. *Journal of Consciousness Studies* 5:566–82. [JP]

(2000) Affective consciousness and the instinctual motor system: The neural sources of sadness and joy. In: *The cauldron of consciousness: Motivation, affect and self–organization. Advances in consciousness research*, ed. R. Ellis & N. Newton. John Benjamins. [JP]

Panula, P., Pirvola, U., Auvinen, S. & Airaksinen, M. S. (1989) Histamine-immunoreactive fibers in the rat brain. *Neuroscience* 28:585–610. [aJAH]

Papa, M., Sellitti, S. & Sadile, A. G. (2000) Remodeling of neural networks in the anterior forebrain of an animal model of hyperactivity and attention deficits as monitored by molecular imaging probes. *Neuroscience and Biobehavioral Reviews* 24:149–56. [RCo]

Pare, D., Curro-Dossi, R. & Steriade, M. (1991) Three types of inhibitory postsynaptic potentials generated by interneurons in the anterior thalamic complex of cat. *Journal of Neurophysiology* 66:1190–204. [EEK]

Pare, D. & Llinas, R. (1995) Conscious and pre-conscious processes as seen from the standpoint of sleep-waking cycle neurophysiology. *Neuropsychologia* 33(9):1155–68. [CMP]

Parmeggiani, P. L. (1968) Telencephalo-diencephalic aspects of sleep mechanisms. *Brain Research* 7:350–59. [CF]

(1994) The autonomic nervous system in sleep. In: *Principles and practice of sleep medicine, 2nd edition*, ed. M. H. Kryger, T. Roth & W. C. Dement. W. B. Saunders. [rTAN]

Parmeggiani, P. L. & Morrison, A. R. (1990) Alterations in autonomic functions during sleep. In: *Central regulation of autonomic functions*, ed. A. D. Loewy & M. Spyer. Oxford University Press. [ARM]

Parnavelas, J. G. (1990) Neurotransmitters in the cerebral cortex. *Progress in Brain Research* 85:13–29. [EKP]

Parra, P., Gulyas, A. J. & Miles, R. (1998) How many subtypes of inhibitory cells in the hippocampus? *Neuron* 20(5):983–93. [RWG]

Partridge, M. (1950) *Pre-frontal leucotomy: A survey of 300 cases personally followed for 1–3 years.* Blackwell. [aMS]

Pascual-Leone, A., Hamilton, R., Tormos, J. M., Keenan, J. P. & Catala, M. D. (1999) Neuroplasticity in the adjustment to blindness. In: *Neuronal plasticity: Building a bridge from the laboratory to the clinic*, ed. J. Grafman & Y. Christen. Springer–Verlag. [aJAH]

Passingham, R. (1993) *The frontal lobes and voluntary action.* Oxford University Press. [aAR]

Passouant, P., Besset, A., Billard, M. & Negre, C. (1978) Effect of piribedil on nocturnal sleep. *Revue d'Electroencephalographie et de Neurophysiologie Clinique* 8:326–34. [rJAH]

Passouant, P., Cadhilac, J. & Billard, M. (1975) Withdrawal of the paradoxical sleep by the clomipramine, electrophysiological, histochemical and biochemical study. *International Journal of Neurology* 10:186–97. [aRPV]

Paus, T., Petrides, M., Evans, A. C. & Meyer, E. (1993) Role of the human anterior cingulate cortex in the control of oculomotor, manual, and speech responses: A positron emission tomography study. *Journal of Neurophysiology* 70:453–69. [aJAH]

Pavlides, C., Greenstein, Y. J., Grudman, M. & Winson, J. (1988) Long-term potentiation in the dentate gyrus is induced preferentially on the positive phase of theta-rhythm. *Brain Research* 439:383–87. [aRPV]

Pavlides, C. & Winson, J. (1989) Influences of hippocampal place cell firing in the awake state on the activity of these cells during subsequent sleep episodes. *Journal of Neuroscience* 9:2907–18. [aRPV]

Pavlov, I. P. (1923) "Innere Hemmung" der bedingten Reflexe an der Schlaf – ein und derselbe Prozess. *Skandinavische Archive für Physiologie* 44:42–58. [MSt]

Pazos, A., Probst, A. & Palacios, J. M. (1987) Serotonin receptors in the human brain. IV. Autoradiographic mapping of serotonin-2 receptors. *Neuroscience* 21:123–39. [EKP]

Pearlman, C. (1971) Latent learning impaired by REM sleep deprivation. *Psychonomic Science* 25:135–36. [aRPV]
(1973) Posttrial REM sleep: A critical period for consolidation of shuttlebox avoidance. *Animal Learning and Behavior* 1:49–51. [RG]
(1978) Interference with taste familiarization by several drugs in rats. *Behavioral Biology* 24:307–16. [aRPV]
(1979) REM sleep and information processing: Evidence from animal studies. *Neuroscience and Biobehavioural Reviews* 3:57–68. [CS, aRPV]

Pearlman, C. & Becker, M. (1973) Brief posttrial REM sleep deprivation impairs discrimination learning in rats. *Physiologial Psychology* 1:373–76. [aRPV]
(1974) REM sleep deprivation impairs bar-press acquisition in rats. *Physiology and Behavior* 13:813–17. [aRPV]

Peder, M., Elomaa, E. & Johansson, G. (1986) Increased aggression after rapid eye movement sleep deprivation in Wistar rats is not influenced by reduction of dimensions of enclosure. *Behavioral and Neural Biology* 45:287–91. [HS]

Pekala, R. J. (1991) *Quantifying consciousness: An empirical approach.* Plenum. [HTH]

Peña-Casanova, J., Roig-Rovira, T., Bermudez, A. & Tolosa-Sarro, E. (1985) Optic aphasia, optic apraxia, and loss of dreaming. *Brain and Language* 26:63–71. [aMS]

Penfield, W. (1938) The cerebral cortex in man. I: The cerebral cortex and consciousness. *Archives of Neurology and Psychiatry* 40:417–42. [arMS]
(1975) *The mystery of the mind.* Princeton University Press. [PC, aAR, WHM]

Penfield, W. & Erickson, T. (1941) *Epilepsy and cerebral localization.* Charles Thomas. [HSP, arMS]

Penfield, W. & Perot, P. (1963) The brain's record of auditory and visual experience. *Brain* 86:595–696. [PC]

Penfield, W. & Rasmussen, T. (1955) *The cerebral cortex of man.* Macmillan. [aMS]

Perlis, L. & Nielsen, T. A. (1993) Mood regulation, dreaming and nightmares: Evaluation of a desensitization function. *Dreaming* 3:243–57. [aJAH]

Perry, E. K. & Perry, R. H. (1995) Acetylcholine and hallucinations: Disease-related compared to drug-induced alterations in human consciousness. *Brain and Cognition* 28:240–58. [aJAH, aMS, LJW]

Perry, E. K., Walker, M., Grace, J. & Perry, R. H. (1999) Acetylcholine in mind: A neurotransmitter correlate of consciousness. *Trends in Neurosciences* 22:273–80. [arJAH, EKP]

Persson, L. & Hillered, L. (1992) Chemical monitoring of neurosurgical care patients using intracerebral microdialysis. *Journal of Neurosurgery* 76(1):72–80. [CMP]

Peselow, E. D., Corwin, J., Fieve, R. R., Rotrosen, J. & Cooper, T. B. (1991) Disappearance of memory deficits in outpatient depressives responding to imipramine. *Journal of Affective Disorders* 21:173–83. [aRPV]

Pessah, M. A. & Roffwarg, H. P. (1972) Spontaneous middle ear muscle activity in man: A rapid eye movement sleep phenomenon. *Science* 178:773–76. [CG]

Peterson, J. B. (1999) *Maps of meaning: The architecture of belief.* Routledge. [JBP]

Petit, L., Orssud, C., Tzourio, N., Crinello, F., Berthoz, A. & Mazoyer, B. (1996) Functional anatomy of a prelearned sequence of horizontal saccades in humans. *The Journal of Neuroscience* 16:3714–26. [aJAH]

Petrie, K. J., Booth, R. J. & Pennebaker, J. W. (1998) The immunological effects of thought suppression. *Journal of Personality and Social Psychology* 75:1264–72. [JBP]

Peyron, C., Tighe, D. K., Van den Pol, A. N., De Lecea, L., Heller, H. C., Sutcliffe, J. G. & Kilduff, T. S. (1998) Neurons containing hypocretin (orexin) project to multiple neuronal systems. *Journal of Neuroscience* 18:9996–10015. [RS-P]

*Physicians Desk Reference* (1999) Medical Economics Company. [rJAH]

Piccione, P., Thomas, S., Roth, T. & Kramer, M. (1976) Incorporations of the laboratory situation in dreams. *Sleep Research* 5:120. [MK]

Picton, T. W., Hillyard, S. A., Krausz, H. I. & Galombos, R. (1974) Human auditory evoked potentials. I. Evaluation of components. *Electroencephalography and Clinical Neurophysiology* 36:179–90. [aTAN]

Piehler, R. (1950) Über das Traumleben leukotomierter (Vorläufige Mitteilung). *Nervenärzt* 21:517–21. [aMS]

Pierrot-Deseilligny, C., Rivaud, S., Gaynard, B., Muri, R. & Vermersch, A. I. (1995) Cortical control of saccades. *Annals of Neurology* 37:557–67. [aJAH]

Piper, D. C., Smith, M. I., Upton, N. & Hunter, A. J. (1999) The effect of the novel neuropeptide, Orexin-A, on the sleep-wake cycle of the rat. *Sleep Research Online* (Supplement 1):73. [aJAH]

Pirolli, A. & Smith, C. (1989) REM sleep deprivation in humans impairs learning of a complex task. *Sleep Research* 18:375. [WHM]

Pishkin, V., Fishkin, S. M., Shurley, J. T., Lawrence, B. E. & Lovallo, W. R. (1978) Cognitive and psychophysiologic response to doxepin and chlordiazepoxide. *Comprehensive Psychiatry* 19:171–78. [aRPV]

Pitman, R. K., Shalev, A. Y. & Orr, S. P. (2000) Posttraumatic stress disorder: Emotion, conditioning, and memory. In: *The new cognitive neurosciences, 2nd edition*, ed. M. S. Gazzaniga. MIT Press. [rAR]

Pivik, R. T. (1971) *Mental activity and phasic events during sleep.* Doctoral dissertation, Stanford University. [arTAN]
(1978) Tonic states and phasic events in relation to sleep mentation. In: *The mind in sleep: Psychology and psychophysiology*, ed. A. M. Arkin, J. S. Antrobus & S. J. Ellman. Erlbaum. [aTAN]
(1986) Sleep physiology and psychophysiology. In: *Psychophysiology: Systems, processes and applications*, ed. M. G. H. Coles, E. Donchin & S. W. Porger. Guilford Press. [GWV]
(1991) Tonic states and phasic events in relation to sleep mentation. In: *The mind in sleep: Psychology and psychophysiology, 2nd edition*, ed. S. J. Ellman & J. S. Antrobus. Wiley. [RCo, aJAH, arTAN, EFP-S, aAR, MS, GWV]
(1994) The psychophysiology of sleep. In: *Principles and practices of sleep medicine, 2nd edition*, ed. M. H. Kryger, T. Roth & W. C. Dement. WB Saunders. [aTAN]

Pivik, R. T. & Foulkes, D. (1968) NREM mentation: Relation to personality, orientation time, and time of night. *Journal of Consulting and Clinical Psychology* 32:144–51. [aJAH, aTAN, GWV]

Pivik, R. T., McCarley, R. & Hobson, J. A. (1977) Eye-movement-associated discharge in brain stem neurons during desynchronized sleep. *Brain Research* 121:59–76. [aJAH, aMS]

Plihal, W. & Born, J. (1997) Effects of early and late nocturnal sleep on declarative and procedural memory. *Journal of Cognitive Neuroscience* 9(4):534–47. [JB, rJAH, aTAN, MSc, CS, RS, arRPV]
(1999a) Effects of early and late nocturnal sleep on priming and spatial memory. *Psychophysiology* 36:571–82. [JB, rRPV]
(1999b) Memory consolidation in human sleep depends on inhibition of glucocorticoid release. *NeuroReport* 10:2741–47. [JB]

Plihal, W., Pietrowsky, R. & Born, J. (1999) Dexamethasone blocks sleep induced improvement of declarative memory. *Psychoneuroendocrinology* 24:313–31. [JB, MSc]

Plum, F. & Posner, J. B. (1966) *The diagnosis of stupor and coma.* F. A. Davis. [aRPV]

Plutchik, R. (1980) *Emotion: A psychoevolutionary synthesis.* Harper & Row. [LM]

Poe, G. R., Skaggs, W. E., Barnes, C. A. & McNaughton, B. L. (1997) Theta phase precession remnants in REM sleep. *Society of Neuroscience Abstracts* 22:505. [RS]

Poggio, T. (1990) A theory of how the brain might work. *Cold Spring Harbor Symposium on Quantitative Biology* 55:899–910. [TJM]

Pokorny, A. D. (1978) Sleep disturbances, alcohol, and alcoholism. In: *Sleep disorders: Diagnosis and treatment*, ed. R. L. Williams & I. Karacan. Wiley. [aJAH]

Pompeiano, O. (1967a) The neurobiological mechanisms of the postural and motor events during desynchronized sleep. *Proceedings of the Association for Research of Nervous Mental Disorders* 45:351–423. [aJAH]
(1967b) Sensory inhibition during motor activity in sleep. In: *Neurophysiological basis of normal and abnormal motor activities*, ed. M. D. Yahr & D. P. Purpura. Raven Press. [aJAH]
(1980) Cholinergic activation of reticular and vestibular mechanisms controlling

posture and eye movements. In: *The reticular formation revisited*, ed. J. A. Hobson & M. A. B. Brazier. Raven Press. [aJAH]

Pompeiano, O. & Morrison, A. R. (1965) Vestibular influences during sleep. I. Abolition of the rapid eye movements of desynchronized sleep following vestibular lesions. *Archives Italiennes de Biologie* 103:569–95. [JH]

(1966) Vestibular origin of the rapid eye movements during desynchronized sleep. *Experientia* 22:60–61. [JH]

Pompeiano, O. & Valentinuzzi, M. (1976) A mathematical model for the mechanism of rapid eye movements induced by an anticholinesterase in the decerebrate cat. *Archives Italiennes de Biologie* 114:103–54. [JH]

Pope, R. A. (1973) Psychological correlates of theta burst activity in sleep onset. Unpublished master's thesis, University of Wyoming. [rTAN]

Popper, K. R. (1959) *The logic of scientific discovery*. Harper & Row. [WF]

Porkka-Heiskanen, T., Strecker, R. E., Stenberg, D., Bjorkum, A. A. & McCarley, R. W. (1997a) GABA and adenosine inhibit the dorsal raphe nucleus and increase REM sleep as studied by microdialysis. *Sleep Research* 26:35. [aJAH]

Porkka-Heiskanen, T., Strecker, R. E., Thakkar, M. & McCarley, R. W. (1997b) Brain extracellular adenosine levels during sleep-waking and prolonged wakefulness. *Society for Neuroscience Abstracts* 23:312. [aJAH]

Portas, C. M., Bjorvatn, B., Fagerland, S., Gronli, J., Mundal, V., Sorensen, E. & Ursin, R. (1998) On-line detection of extracellular levels of serotonin in dorsal raphe nucleus and frontal cortex over the sleep/wake cycle in the freely moving rat. *Neuroscience* 83:807–14. [aJAH]

Portas, C. M., Krakow, K., Allan, P., Josephs, O. & Frith, C. (1999) Processing auditory stimuli across the sleep-wake cycle: A functional MRI study in humans. *Sleep Research Online* 2 (Supplement 1):265. [aJAH, CMP]

Portas, C. M. & McCarley, R. W. (1994) Behavioral state-related changes of extracellular serotonin concentration in the dorsal raphe nucleus: A microdialysis study in the freely moving cat. *Brain Research* 648:306–12. [aJAH]

Portas, C. M., Rees, G., Howseman, A. M., Josephs, O., Turner, R. & Frith, C. D. (1999) A specific role for the thalamus in mediating the interaction of attention and arousal in humans. *Journal of Neuroscience* 18:8879–989. [aJAH]

Portas, C. M., Thakkar, M., Rainnie, D. G., Greene, R. W. & McCarley, R. W. (1997) Role of adenosine in behavioral state modulation: A microdialysis study in the freely moving cat. *Neuroscience* 79:225–35. [aJAH]

Portas, C. M., Thakkar, M., Rainnie, D. G. & McCarley, R. W. (1996) Microdialysis perfusion of 8-hydroxy-2-(di-N-Propylamine) tetralin (8-OH-DPAT) in the dorsal raphe nucleus decreases serotonin release and increases rapid eye movement sleep in the freely moving cat. *Journal of Neuroscience* 16:2820–28. [aJAH]

Porte, H. S. (1996) REMS reconsidered. *Sleep Research* 25:139. [aJAH]

(1997) Slower eye movement in sleep. *Sleep Research* 26:253. [aTAN]

(2000) Cognition in Stage 2 sleep parallels neural network state. *Sleep* 23:A175. [HSP]

Porte, H. S. & Hobson, J. A. (1986) Bizarreness in REM and NREM reports. *Sleep Research* 15:81. [aJAH, aTAN]

(1996) Physical motion in dreams: One measure of three theories. *Journal of Abnormal Psychology* 105:329–35. [arJAH, aTAN]

Posner, M. I. (1994a) Attention in cognitive neuroscience: An overview. In: *The cognitive neurosciences*, ed. M. Gazzaniga. MIT Press. [aJAH]

(1994b) Attention: The mechanisms of consciousness. *Proceedings of the National Academy of Sciences USA* 91:7398–403. [RCo]

Posner, J., Bye, A., Dean, K., Peck, A. W. & Whiteman, P. D. (1985) The disposition of bupropion and its metabolites in healthy male volunteers after single and multiple doses. *European Journal of Clinical Pharmacology* 29:97–103. [aJAH]

Posner, M. I. & Digirolamo, G. J. (2000) Attention in cognitive neuroscience: An overview. In: *The new cognitive neurosciences, 2nd edition*, ed. M. S. Gazzaniga. MIT Press. [RCo]

Posner, M. I. & Rothbart, M. K. (1998) Attention, self-regulation and consciousness. *Philosophy Transcripts of the Royal Society of London: B Biological Sciences* 353:1915–27. [RCo]

Post, R. M., Gerner, R. H. & Carmen, J. S. (1978) Effects of dopaminergic agonist, piribedil, in depressed patients. *Archives of General Psychiatry* 35:609–15. [aJAH]

Post, R. M., Gillin, J. C., Goodwin, F. K. & Wyatt, J. R. (1974) The effect of orally administered cocaine on sleep of depressed patients. *Psychopharmacology* 37:59–66. [arJAH]

Poucet, B., Herrmann, T. & Buhot, M. C. (1991) Effects of short-lasting inactivations of the ventral hippocampus and medial septum on long-term and short-term acquisition of spatial information in rats. *Behavioural Brain Research* 44:53–65. [aRPV]

Pradhan, N., Sadasivan, P. K., Chatterji, S. & Narayana Dutt, D. (1995) Patterns of attractor dimension of sleep EEG. *Computers in Biology and Medicine* 25:455–62. [AC]

Price, L. J. & Kremen, I. (1980) Variations in behavioral response threshold within the REM period of human sleep. *Psychophysiology* 17(2):133–40. [aJAH, aTAN]

Prospero-Garcia, O., Jiminez-Anguiano, A. & Drucker-Colin, R. (1993) The combination of VIP and atropine induces REM sleep in cats rendered insomniac by PCPA. *Neuropsychopharmacology* 8:387–90. [aJAH]

Prospero-Garcia, O., Navarro, L., Murillo-Rodriguez, E., Sanchez-Alvarez, M., Guzman-Marin, R., Mendez-Diaz, M., Gomez-Chavarin, M., Jiminez-Anguiano, A. & Drucker-Colin, R. (1999) Cellular and molecular changes occurring during REM sleep. In: *Rapid eye movement sleep*, ed. B. N. Mallick & S. Inoue. Marcel Dekker. [aJAH]

Prudom, A. E. & Klemm, W. R. (1973) Electrographic correlates of sleep behavior in a primitive mammal, the armadillo, *Dasypus novemcinctus*. *Physiology and Behavior* 10:275–82. [JMS]

Punamäki, R.-L. (1997) Determinants and mental health effects of dream recall among children living in traumatic conditions. *Dreaming* 7:235–63. [aAR]

(1998) The role of dreams in protecting psychological well-being in traumatic conditions. *International Journal of Behavioral Development* 22:559–88. [aAR]

Purcell, S., Moffitt, A. & Hoffmann, R. (1993) Waking, dreaming, and self-regulation. In: *The functions of dreaming*, ed. A. Moffitt, M. Kramer & R. Hoffmann. State University of New York Press. [TLK]

Purcell, S., Mullington, J., Moffitt, A., Hoffmann, R. & Pigeau, R. (1986) Dream self-reflectiveness as a learned cognitive skill. *Sleep* 9:423–37. [arJAH, aTAN]

(1993) Waking, dreaming and self-regulation. In: *The function of dreaming*, ed. A. Moffitt, M. Kramer & R. Hoffmann. State University of New York Press. [rJAH]

Putkonen, P. T. S. & Putkonen, A. R. (1971) Suppression of paradoxical sleep (PS) following hypothalamic defence reactions in cats during normal conditions and recovery from PS deprivation. *Brain Research* 26:333–47. [HS]

Pylyshyn, Z. W. (1989) Computing in cognitive science. In: *Foundations of cognitive science*, ed. M. I. Posner. MIT Press. [JSA, aJAH]

Python, A., de Saint Hilaire, Z. & Gaillard, J. M. (1996) Effects of a D2 receptor agonist RO 41–9067 alone and with clonidine on sleep parameters in the rat. *Pharmacology, Biochemistry and Behavior* 53:291–96. [aJAH]

Rabinowicz, A. L., Salas, E., Beserra, F., Leiguarda, R. C. & Vazquez, S. E. (1997) Changes in regional cerebral blood flow beyond the temporal lobe in unilateral temporal lobe epilepsy. *Epilepsia* 38:1011–14. [aJAH, EFP-S]

Rados, R. & Cartwright, R. D. (1982) Where do dreams come from? A comparison of presleep and REM sleep thematic content. *Journal of Abnormal Psychology* 91:433–36. [aJAH]

Rafal, R., Henik, A. & Smith, J. (1991) Extrageniculate contributions to reflex visual orienting in normal humans: A temporal hemifield advantage. *Journal of Cognitive Neuroscience* 3:322–28. [aJAH]

Rafal, R. & Robertson, L. (1994) The neurology of visual attention. In: *The cognitive neurosciences*, ed. M. Gazzaniga. MIT Press. [aJAH]

Rafal, R., Smith, J., Kranitz, A., Cohen, A. & Brennan, C. (1990) Extrageniculate vision in hemianopic humans: Saccade inhibition by signals in the blind field. *Science* 250:118–21. [aJAH]

Rajkowski, J., Silakov, V., Ivanova, S. & Aston-Jones, G. (1997) Locus coeruleus (LC) neurons in monkey are quiescent in paradoxical sleep (PS). *Society for Neuroscience Abstracts* 23:828.2. [aJAH]

*Random House Dictionary of the English Language* (1983) Unabridged second edition, ed. S. B. Flexner. Random House. [JFP]

Rannie, D. G., Grunze, H. C., McCarley, R. W. & Greene, R. W. (1994) Adenosine inhibition of mesopontine cholinergic neurons: Implications for EEG arousal. *Science* 263:689–92. [aJAH]

Rannie, D. G., McCarley, R. W. & Greene, R. W. (1997) Adenosine modulation of glutamatergic transmission in the laterodorsal tegmentum via an action at presynaptic receptors. *Sleep Research* 26:39. [aJAH]

Raskin, A., Friedman, A. S. & DiMascio, A. (1983) Effects of chlorpromazine, imipramine, diazepam and phenelzine on psychomotor and cognitive skills of depressed patients. *Psychopharmacology Bulletin* 19:649–52. [aRPV]

Rasmussen, D. D., Clow, K. & Szerb, J. C. (1994) Modification of neocortical acetylcholine release and electroencephalogram desynchronization due to brainstem stimulation by drugs applied to the basal forebrain. *Neuroscience* 60:665–77. [aJAH]

Rasmussen, K., Heym, J. & Jacobs, B. L. (1984) Activity of serotonin-containing neurons in nucleus centralis superior of freely moving cats. *Experimental Neurology* 83:302–17. [CG]

Rasmussen, K., Morilak, D. A. & Jacobs, B. L. (1986) Single unit activity of locus coeruleus neurons in the freely moving cat. I. During naturalistic behaviors and in response to simple and complex stimuli. *Brain Research* 371:324–34. [aJAH]

Rauch, S. L. & Renshaw, P. F. (1995) Clinical neuroimaging in psychiatry. *Harvard Review of Psychiatry* 2:297–312. [aJAH]

Raymond, I., Nielsen, T., Bessette, P., Faucher, B. & Germain, A. (1999)

Character incorporation higher in mentation reports from late night REM than in reports from late night stage 2 sleep. *Sleep* 22 (Supplement 1):S132. [aJAH]

Razmy, A. & Shapiro, C. M. (2000) Interactions of sleep and Parkinson's disease. *Seminars in Clinical Neuropsychiatry* 5(1):20–32. [AK]

Reader, T. A., Ferron, A., Descarries, L. & Jasper, H. H. (1979) Modulatory role for biogenic amines in the cerebral cortex. Microiontophoretic studies. *Brain Research* 160:217–29. [CG]

Rechtschaffen, A. (1973) The psychophysiology of mental activity during sleep. In: *The psychophysiology of thinking*, ed. F. J. McGuigan & R. A. Schoonover. Academic Press. [aJAH, aMS, MS, GWV]

(1978) The single-mindedness and isolation of dreams. *Sleep* 1:97–109. [arJAH, aTAN, aAR]

(1994) Dreaming and REM sleep – commentary. *World Federation of Sleep Research Societies Newsletters* 3:16–18. [aTAN]

(1998) Current perspectives on the function of sleep. *Perspectives in Biology and Medicine* 41:359–90. [aRPV]

Rechtschaffen, A. & Bergmann, B. M. (1995) Sleep deprivation in the rat by the disk-over-water method. *Behavioural Brain Research* 69:55–63. [aRPV]

(1999) Sleep stage priorities in rebounds from sleep deprivation: A response to Feinberg. *Sleep* 22:1025–30. [IF]

Rechtschaffen, A. & Buchignani, C. (1992) The visual appearance of dreams. In: *The neuropsychology of sleep and dreaming*, ed. J. S. Antrobus & M. Bertini. Erlbaum. [aJAH, aAR]

Rechtschaffen, A., Goodenough, D. & Shapiro, A. (1962) Patterns of sleep talking. *Archives of General Psychiatry* 7:418–26. [aTAN]

Rechtschaffen, A. & Kales, A., eds. (1968) *A manual of standardized terminology, techniques and scoring system for sleep stages of human subjects.* Brain Information Service/Brain Research Institute, University of California at Los Angeles. [aJAH, JFP, aMS, MS, RS] Public Health Service, U.S. Government Printing Office. [RCo, GWD, rMS] HEW Neurological Information Network. [rTAN]

Rechtschaffen, A. & Maron, L. (1964) The effect of amphetamine on the sleep cycle. *Electroencephalography and Clinical Neurophysiology* 16:438–45. [LJW]

Rechtschaffen, A., Molinari, S., Wabon, R. & Wincor, M. Z. (1970) Extraocular potentials: A possible indicator of PGO activity in the human. *Psychophysiology* 7:336. [CG]

Rechtschaffen, A., Verdone, P. & Wheaton, J. (1963a) Reports of mental activity during sleep. *Canadian Psychiatric Association Journal (Candadian Psychiatry)* 8:409–14. [AC, aJAH, aTAN]

Rechtschaffen, A., Vogel, G. & Shaikun, G. (1963b) Interrelatedness of mental activity during sleep. *Archives of General Psychiatry* 9:536–47. [aTAN]

Rechtschaffen, A., Watson, R., Wincor, M. Z., Molinari, S. & Barta, S. G. (1972) The relationship of phasic and tonic periorbital EMG activity to NREM mentation. *Sleep Research* 1:114. [aJAH, MS, GWV]

Reed, G. (1972) *The psychology of anomalous experience: A cognitive approach.* Hutchinson. [PC]

Reiner, P. B. (1986) Correlational analysis of central noradrenergic activity and sympathetic tone in behaving cats. *Brain Research* 378:86–96. [aJAH]

Reiner, P. B. & McGeer, E. G. (1987) Electrophysiological properties of cortically projecting histamine neurons of the rat hypothalamus. *Neuroscience Letters* 73:43–47. [aJAH]

Reinsel, R. A. & Antrobus, J. S. (1992) Lateralized task performance after awakening from sleep. In: *The neuropsychology of sleep and dreaming*, ed. J. S. Antrobus & M. Bertini. Erlbaum. [aTAN]

Reinsel, R., Antrobus, J. S. & Wollman, M. (1992) Bizarreness in dreams and waking fantasy. In: *The neuropsychology of sleep and dreaming*, ed. J. S. Antrobus & M. Bertini. Erlbaum. [JSA, GWD, aJAH]

Reinsel, R., Wollman, M. & Antrobus, J. (1986) Effects of environmental context and cortical activation on thought. *Journal of Mind and Behavior* 7:250–76. [MBl, GWD, aJAH]

Reiser, B. J., Black, J. B. & Abelson, R. P. (1985) Knowledge structures in the organization and retrieval of autobiographical memories. *Cognitive Psychology* 17:89–137. [PC]

Reiser, M. F. (1997) The art and science of dream interpretation: Isakower revisited. *Journal of the American Psychoanalytic Association* 45:891–907. [RG]

Rempel-Clower, N. L. & Barbas, H. (1998) Topographic organization of connections between the hypothalamus and prefrontal cortex in the rhesus monkey. *Journal of Comparative Neurology* 398:393–419. [rJAH]

Resnick, J., Stickgold, R., Rittenhouse, C. D. & Hobson, J. A. (1994) Self-representation and bizarreness in children's dream reports collected in the home setting. *Consciousness and Cognition* 3:30–45. [aJAH, aAR]

Revonsuo, A. (1994a) In search of the science of consciousness. In: *Consciousness in philosophy and cognitive neuroscience*, ed. A. Revonsuo & M. Kamppinen. Erlbaum. [rAR]

(1994b) The "multiple drafts" model and the ontology of consciousness. *Behavioral and Brain Sciences* 17:177–78. [rAR]

(1995) Consciousness, dreams, and virtual realities. *Philosophical Psychology* 8:35–58. [arAR, AR]

(1997) How to take consciousness seriously in cognitive neuroscience. *Communication and Cognition* 30:185–206. [aAR]

(1999a) Towards a cognitive neuroscience of consciousness. In: *Towards a science of consciousness III*, ed. S. Hameroff, A. Kaszniak & D. Chalmers. MIT Press. [aAR]

(1999b) Binding and the phenomenal unity of consciousness. *Consciousness and Cognition* 8(2):173–85. [AR]

(2000) Prospects for a scientific research program on consciousness. In: *Neural correlates of consciousness*, ed. T. Metzinger. MIT Press. [rAR]

Revonsuo, A. & Salmivalli, C. (1995) A content analysis of bizarre elements in dreams. *Dreaming* 5(3):169–87. [aJAH, arAR, AR]

Revonsuo, A. & Valli, K. (2000) Dreaming and consciousness: Testing the threat simulation theory of the function of dreaming. *Psyche* 6, http://psyche.cs.monash.edu.au/v6/psyche-6-08-revonsuo.html

Riemann, D., Low, H., Schredl, M., Wiegand, M., Dippel, B. & Berger, M. (1990) Investigations of morning and laboratory dream recall and content in depressive patients during baseline conditions and under antidepressive treatment with trimipramine. *Psychiatric Journal of the University of Ottawa* 15:93–99. [aTAN]

Rime, B. (1995) Mental rumination, social sharing, and the recovery from emotional exposure. In: *Emotion, disclosure and health*, ed. J. W. Pennebaker. American Psychological Association. [JBP]

Ritchie, D. (1959) *Stroke: A diary of recovery.* Faber & Faber. [aMS]

Rittenhouse, C. D., Broadley, D., Stickgold, R. & Hobson, J. A. (1993) Increased semantic priming upon awakenings from REM sleep. *Sleep Research* 22:97. [aJAH]

Rittenhouse, C. D., Stickgold, R. & Hobson, J. A. (1994) Constraints on the transformation of characters, objects, and settings in dream reports. *Consciousness and Cognition* 3:100–13. [GWD, arJAH, AR]

Robbins, P. & Houshi, F. (1983) Some observations on recurrent dreams. *Bulletin of the Menninger Clinic* 47:262–65. [aAR, AZ]

Robbins, T. W. & Everitt, B. J. (1995) Arousal systems and attention. In: *The cognitive neurosciences*, ed. M. S. Gazzaniga. MIT Press. [aJAH]

Robinson, D. L. & Peterson, S. E. (1992) The pulvinar and visual salience. *Trends in Neuroscience* 15:127–32. [RCo]

Rochlin, A., Hoffmann, R. & Armitage, R. (1998) EEG correlates of dream recall in depressed outpatients and healthy controls. *Dreaming* 8:109–23. [aJAH, aTAN]

Rodin, E., Mulder, D., Faucett, R. & Bickford, R. (1955) Psychologic factors in convulsive disorders of focal origin. *Archives of Neurology* 74:365–74. [aMS]

Roediger, H. L. & McDermott, K. B. (1995) Creating false memories: Remembering words not presented in lists. *Journal of Experimental Psychology: Learning, Memory and Cognition* 21:803–14. [PC]

Roehrs, T. A., Johansen, C. E., Schuh, K., Warbasse, L. & Roth, T. (1998) Effects of cocaine and its abstinence on sleep and daytime sleepiness. *Sleep* 21 (Supplement):265. [rJAH]

Roffwarg, H. P. & Belenky, G. (1996) Review of Hong et al. (1995) Localized and lateralized cerebral glucose metabolism associated with eye movements during REM sleep and wakefulness: A positron emission tomography (PET) study. *Sleep* 18:570–80. *World Federation of Sleep Research Societies Newsletter* 5:44. [aJAH]

Roffwarg, H. P., Dement, W. C. & Fisher, C. (1964) Preliminary observations of the sleep-dream pattern in neonates, infants, children and adults. In: *Problems of sleep and dreams in children*, ed. E. Harms. Pergamon Press. [rTAN]

Roffwarg, H. P., Dement, W. C., Muzio, J. N. & Fisher, C. (1962) Dream imagery: Relationship to rapid eye movements of sleep. *Archives of General Psychiatry* 7:235–58. [aJAH, SL, aTAN]

Roffwarg, H. P., Herman, J. H., Bowe-Anders, C. & Tauber, E. S. (1978) The effects of sustained alterations of waking visual input on dream content. In: *The mind in sleep: Psychology and psychophysiology*, ed. A. M. Arkin, J. S. Antrobus & S. J. Ellman. Erlbaum. [aJAH]

Roffwarg, H. P., Muzio, J. N. & Dement, W. C. (1966) Ontogenetic development of the human sleep-dream cycle. *Science* 152:604–19. [CG, rTAN, RDO, rRPV]

Rogan, M. T., Staubli, U. V. & LeDoux, J. E. (1997) Fear conditioning induces associative long-term potentiation in the amygdala. *Nature* 390:604–607. [ARM]

Roitt, I., Brostoff, J. & Male, D. (1998) *Immunology, 5th edition.* Mosby. [aAR]

Role, L. & Kelly, J. (1991) The brain stem: Cranial nerve nuclei and the monoaminergic systems. In: *Principles of neural science*, ed. E. Kandel, J. Schwartz & T. Jessel. Appleton & Lange. [aMS]

Roschke, J., Prentice-Cuntz, T., Wagner, P., Mann, K. & Frank, C. (1996) Amplitude frequency characteristics of evoked potentials during sleep: An analysis of the brain's transfer properties in depression. *Biological Psychiatry* 40:736–43. [aTAN]

Rose, G. M. & Dunwiddie, T. V. (1986) Induction of hippocampal long-term potentiation using physiologically patterned stimulation. *Neuroscience Letters* 69:244–48. [aRPV]

Rosen, J., Reynolds, C. F., Yeager, A. L., Houck, P. R. & Hurwitz, L. (1991) Sleep disturbances in survivors of the Nazi holocaust. *American Journal of Psychiatry* 148:62–66. [rAR]

Rosenblatt, S. I., Antrobus, J. S. & Zimler, J. P. (1992) The effect of postawakening differences in activation on the REM-NREM report effect and recall information from films. In: *The neuropsychology of sleep and dreaming*, ed. J. S. Antrobus & M. Bertini. Erlbaum. [JSA, aJAH]

Rosenfield, I. (1992) *The strange, familiar, and forgotten*. Vintage Books. [WJC]

Rosenlicht, N. & Feinberg, I. (1977) The relation of dreamlike elements to narrative length at sleep onset and REM awakenings. *Sleep Research* 26:254. [aJAH]

Rosenlicht, N., Maloney, T. & Feinberg, I. (1994) Dream report length is more dependent on arousal level than prior REM duration. *Brain Research Bulletin* 34:99–101. [CCa, RCo, IF, aJAH, MSc]

Ross, J. S. & Shua-Haim, J. R. (1998) Aricept-induced nightmares in Alzheimer's disease: 2 case reports. *Journal of American Geriatrics Society* 46:119–20. [rJAH, EKP]

Ross, R. J., Ball, W. A., Dinges, D. F., Kribbs, N. B., Morrison, A. R., Silver, S. M. & Mulvaney, F. D. (1994) Rapid eye movement sleep disturbance in posttraumatic stress disorder. *Biological Psychiatry* 35:195–202. [MBl, arAR, LJW]

Ross, R. J., Ball, W. A., Sullivan, K. A. & Caroff, S. N. (1989) Sleep disturbance as the hallmark of posttraumatic stress disorder. *American Journal of Psychiatry* 146:697–707. [TAN, rAR]

Rossi, A. M., Fuhrman, A. & Solomon, P. (1964) Sensory deprivation, arousal and rapid eye movement correlates of some effects. *Perceptual and Motor Skills* 19:447–51. [aJAH]

Rotenberg, V. S. (1982) *The adaptive function of sleep*. Nauka. [rTAN]
   (1984) Search activity in the context of psychosomatic disturbances, of brain monoamines and REM sleep function. *Pavlovian Journal of Biological Science* 19:1–15. [VSR]
   (1988) Functional deficiency of REM sleep and its role in the pathogenesis of neurotic and psychosomatic disturbances. *Pavlovian Journal of Biological Science* 23:1–3. [VSR]
   (1992) Sleep and memory 1: The influence of different sleep stages on memory. *Neuroscience and Biobehavioral Reviews* 16:497–502. [VSR]
   (1993a) REM sleep and dreams as mechanisms for the recovery of search activity. In: *The functions of dreaming*, ed. A. Moffitt, M. Kramer & R. Hoffman. State University of New York Press. [RG, VSR]
   (1993b) The estimation of sleep quality in different stages and cycles of sleep. *Journal of Sleep Research* 2:17–20. [aTAN]
   (1994a) The revised monoamine hypothesis: Mechanism of antidepressant treatment in the context of behavior. *Integrative Physiological and Behavioral Science* 29:182–88. [VSR]
   (1994b) An integrative psychophysiological approach to brain hemisphere functions in schizophrenia. *Neuroscience and Biobehavioral Reviews* 18:487–95. [VSR]

Rotenberg, V. S. & Arshavsky, V. V. (1979) Search activity and its impact on experimental and clinical pathology. *Activitas Nervosa Superior (Prague)* 21:105–15. [VSR]

Rotenberg, V. S., Kochubey, B. & Shachnarovich, V. (1975) The dynamic of GSR in different stages and cycles of night sleep. *Journal of High Nervous Activity* 4:858–60. [VSR]

Roth, B. (1978) Narcolepsy and hypersomnia. In: *Sleep disorders: Diagnosis and treatment*, ed. R. L. Williams & I. Karacan. Wiley. [aJAH]

Roth, M. T., Fleegal, M. A., Lydic, R. & Baghdoyan, H. A. (1996) Pontine acetylcholine release is regulated by muscarinic autoreceptors. *NeuroReport* 7:3069–72. [aJAH]

Roth, R. H. & Elsworth, J. D. (1999) Dopamine systems. In: *Encyclopedia of neuroscience*, ed. G. Adelman & B. H. Smith. Elsevier. [rJAH]

Roth, T. & Ancoli-Israel, S. (1999) Daytime consequences and correlates of insomnia in the United States: Results of the 1991 National Sleep Foundation survey, II. *Sleep* 22 (Suppl. 2):354–58. [MK]

Roth, T., Zorick, F., Wittig, R., McLenaghan, A. & Roehrs, T. (1982) The effects of doxepin HCl on sleep and depression. *Journal of Clinical Psychiatry* 43:366–68. [aRPV]

Rothman, T., Grayson, H. & Ferguson, J. (1962) A comparative investigation of isocarboxazid and imipramine in depressive syndromes. *Journal of Neuropsychiatry* 130:234–40. [aRPV]

Roussy, F., Camirand, C., Foulkes, D., De Koninck, J, Loftis, M. & Kerr, N. H. (1996) Does early-night REM dream content reliably reflect presleep state of mind? *Dreaming* 6:121–30. [aJAH]

Roussy, F., Gonthier, I., Raymond, I., Mercier, P. & DeKoninck, J. (1997) Further attempts at matching REM dream content with waking ideation. *Sleep Research* 26:255. [aJAH]

Rowley, J., Stickgold, R. A. & Hobson, J. A. (1998) Eye movement and mental activity at sleep onset. *Consciousness and Cognition* 7:67–84. [aJAH, EFP-S]

Rye, D. B. (1997) Contributions of the peduculopontine region to normal and altered REM sleep. *Sleep* 20:757–88. [aJAH]

Sacks, O. (1985/1987) *The man who mistook his wife for a hat*. Duckworth, 1985. [aMS] Harper & Row, 1987. [WJC]
   (1990) *Awakenings*. Harper Collins. [aMS]
   (1991) Neurological dreams. *MD* February:29–32. [aMS]
   (1992) *Migraine*. University of California Press. [aAR]
   (1995) *An anthropologist on mars*. Picador. [aMS]

Sacks, O. & Wasserman, R. (1987) The case of the colorblind painter. *New York Review of Books* 34(18):25–34. [aMS]

Sagales, T. & Domino, E. F. (1973) Effects of stress and REM sleep deprivation on the patterns of avoidance learning and brain acetylcholine in the mouse. *Psychopharmacologia* 29:307–15. [aRPV]

Sagot, J. C., Amoros, C., Candas, V. & Libert, J. P. (1987) Sweating responses and body temperatures during nocturnal sleep in humans. *American Journal of Physiology* 21:R462–70. [aTAN]

Sahlins, M. (1977) *Stone age economics*. Tavistock Press. [NH]

Sakai, F., Meyer, J. S., Karacan, I., Derman, S. & Yamamato, M. (1980) Normal human sleep: Regional cerebral hemodynamics. *Annals of Neurology* 7:471–78. [aJAH]

Sakai, K. (1988) Executive mechanisms of paradoxical sleep. *Archives Italinnes de Biologie* 126:239–57. [aJAH]

Sakai, K., El Mansari, M. & Jouvet, M. (1990) Inhibition by carbachol microinjections of presumptive cholinergic PGO-on neurons in freely moving cats. *Brain Research* 527:213–23. [aJAH]

Sakai, K. & Jouvet, M. (1980) Brain stem PGO-on cells projecting directly to the cat dorsal lateral geniculate nucleus. *Brain Research* 194:500–505. [aJAH]

Sakai, K. & Koyama, Y. (1996) Are there cholinergic and non-cholinergic paradoxical sleep generation in the cat. *European Journal of Neuroscience* 9:415–23. [aJAH]

Sakai, K. & Onoe, H. (1997) Critical role for M3 muscarinic receptors in paradoxical sleep generation in the cat. *European Journal of Neuroscience* 9:415–23. [aJAH]

Sakurai, T., Amemiya, A., Ishii, M., Matsuzaki, I., Chemelli, R. M., Tanaka, H., Williams, S. C., Richardson, J. A., Kozlowski, G. P., Wilson, S., Arch, J. R., Buckingham, R. E., Haynes, A. C., Carr, S. A., Annan, R. S., McNulty, D. E., Liu, W. S., Terrett, J. A., Elshourbagy, N. A., Bergsma, D. J. & Yanagisawa, M. (1998) Orexins and orexin receptors: A family of hypothalamic neuropeptides and G protein-couple receptors that regulate feeding behavior. *Cell* 92:573–85. [RS-P]

Saletu, B., Frey, R., Krupka, M., Anderer, P., Grunberger, J. & See, W. R. (1991) Sleep laboratory studies on the single-dose effects of serotonin reuptake inhibitors paroxetine and fluoxetine on human sleep and awakening qualities. *Sleep* 14:439–47. [aRPV]

Saletu, B. & Grunberger, J. (1988) Drug profiling by computed electroencephalography and brain maps, with special consideration of sertraline and its psychometric effects. *Journal of Clinical Psychiatry* 49:S59–71. [aRPV]

Saletu, B., Grunberger, J. & Rajna, P. (1983) Pharmaco-EEG profiles of antidepressants: Pharmacodynamic studies with fluvoxamine. *British Journal of Clinical Pharmacology* 15:369–84. [aRPV]

Saletu, B., Grunberger, J., Rajna, P. & Karobath, M. (1980) Clovoxamine and fluvoxamine-2 biogenic amine re-uptake inhibiting antidepressants: Quantitative EEG, psychometric and pharmacokinetic studies in man. *Journal of Neural Transmission* 49:63–86. [aRPV]

Salin-Pascual, R. J. & Jiminez-Anguiano, A. (1995) Vesamicol, an acetylcholine uptake blocker in presynaptic vesicles suppresses rapid eye movement (REM) sleep in the rat. *Psychopharmacology* 121:485–87. [aJAH]

Salisbury, D. F. (1994) Stimulus processing awake and asleep: Similarities and differences in electrical CNS responses. In: *Sleep onset. Normal and abnormal processes*, ed. R. D. Ogilvie & J. R. Harsh. American Psychological Association. [aTAN]

Sallanon, M., Denoyer, M., Kitahama, K., Auber, C., Gay, N. & Jouvet, M. (1989) Long-lasting insomnia induced by preoptic neuronal lesions and its transient reversal by muscimol injection into the posterior hypothalamus in the cat. *Neuroscience* 32:669–83. [aJAH]

Sallinen, M., Kaartinen, J. & Lyytinen, H. (1996) Processing of auditory stimuli during tonic and phasic periods of REM sleep as revealed by event-related potentials. *Journal of Sleep Research* 5:220–28. [aJAH]

Salloway, S., Malloy, P. & Cummings, J. L. (1997) *The neuropsychology of limbic and subcortical disorders*. American Psychiatric Press. [rJAH]

Salzarulo, P. (1995) Opening remarks: Perspectives on the relationship between cognitive processes and sleep disturbances. *Journal of Sleep Research* 4:1. [PS]

Salzarulo, P. & Cipolli, C. (1974) Spontaneously recalled verbal material and its linguistic organization in relation to different stages of sleep. *Biological Psychology* 2:47–57. [PS]

(1979) Linguistic organization and cognitive implications of REM and NREM sleep-related reports. *Perceptual and Motor Skills* 49:767–77. [aJAH, PS]

Salzarulo, P., Cipolli, C., Lairy, G. C. & Pecheux, M. G. (1973) L'etude psychophysiologique de l'activité mentale du sommeil: Analyse critique des méthodes et théories. *Evolutión Psychiatrique* 38:33–70. [PS]

Salzarulo, P., Formicola, G., Lombardo, P., Gori, S., Rossi, L., Murri, L. & Cipolli, C. (1997) Functional uncertainty, aging and memory processes during sleep. *Acta Neurologica Belgica* 97:118–22. [PS]

Salzarulo, P., Lairy, G. C., Bancaud, J. & Munari, C. (1975) Direct depth recording of the striate cortex during REM sleep in man: Are there PGO potentials? *Electroencephalography and Clinical Neurophysiology* 38:199–202. [aJAH]

Slazman, C., Miyawaki, E. K., le Bars, P. & Kerrihard, T. N. (1993) Neurobiologic basis of anxiety and its treatment. *Harvard Review of Psychiatry* 1:197–206. [aJAH]

Sanchez, R. & Leonard, C. S. (1996) NMDA-receptor-mediated synaptic currents in guinea pig laterodorsal tegmental neurons in vitro. *The Journal of Neuroscience* 76:1101–11. [aJAH]

Sandyk, R. (1997) Treatment with weak electromagnetic fields restores dream recall in a Parkinsonian patient. *International Journal of Neuroscience* 96:86–95. [EKP]

Sanford, L. D., Hunt, W. K., Ross, R. J., Morrison, A. R. & Pack, A. I. (1998a) Microinjections into the pedunculopontine tegmentum: Effects of the GABA A antagonist, bicuculline, on sleep, PGO waves and behavior. *Archives Italiennes de Biologie* 136:205–14. [aJAH]

Sanford, L. D., Hunt, W. K., Ross, R. J., Pack, A. I. & Morrison, A. R. (1998b) Central administration of a 5-HT2 receptor agonist and antagonist: Lack of effect on rapid eye movement sleep and PGO waves. *Sleep Research Online* 1:80–86. [aJAH]

Sanford, L. D., Kearney, K., McInerney, B., Horner, R. L., Ross, R. J. & Morrison, A. R. (1997a) Rapid eye movement sleep (REM) is not regulated by 5-HT2 receptor mechanisms in the laterodorsal tegmental nucleus. *Sleep Research* 26:127. [aJAH]

Sanford, L. D., Morrison, A. R., Ball, W. A., Ross, R. J. & Mann, G. L. (1992a) Spontaneous phasic activity in the brain: Differences between waves in lateral geniculate and central lateral nuclei across sleep states. *Journal of Sleep Research* 1:258–64. [aTAN]

(1992b) Varying expressions of alerting mechanisms in wakefulness and across sleep states. *Electroencephalography and Clinical Neurophysiology* 82:458–68. [JH, aTAN]

(1993) The amplitude of elicited PGO waves: A correlate of orienting. *Electroencephalography and Clinical Neurophysiology* 86:438–45. [RCo, JH, rJAH ARM]

Sanford, L. D., Morrison, A. R. & Ross, R. J. (1994a) Effects of auditory stimulation on phenomena of rapid eye movement sleep. In: *Environment and physiology*, ed. B. N. Mallick & R. Singh. Narosa. [RCo]

Sanford, L. D., Ross, R. J. & Morrison, A. R. (1995a) Serotonergic mechanisms in the amygdala terminate REM sleep. *Sleep Research* 24:54. [aJAH]

Sanford, L. D., Ross, R. J., Seggos, A. E., Morrison, A. R., Ball, W. A. & Mann, G. L. (1994b) Central administration of two 5-HT receptor agonists: Effect on REM sleep and PGO waves. *Pharmacology, Biochemistry and Behavior* 49:93–100. [aJAH]

Sanford, L. D., Ross, R. J., Tejani-Butt, S. M. & Morrison, A. R. (1995b) Amygdaloid control of alerting and behavioral arousal in rats: Involvement of serotonergic mechanisms. *Archives Italiennes de Biologie* 134:81–89. [aJAH]

Sanford, L. D., Silvestri, A. J., Ross, R. J. & Morrison, A. R. (in press) Influence of fear conditioning on elicited ponto-geniculo-occipital waves and rapid eye movement sleep. *Archives Italiennes de Biologie*. [ARM]

Sanford, L. D., Tejani-Butt, S. M., Ross, R. J. & Morrison, A. R. (1996) Elicited PGO waves in rats: Lack of 5-HT1A inhibition in putative pontine generator region. *Pharmacology, Biochemistry and Behavior* 53:323–27. [aJAH]

Sanford, L. D., Tidikis, D. E., Ross, R. J. & Morrison, A. R. (1997b) Carbachol microinjections into the amygdala of rats shortens REM sleep. *Sleep Research* 26:128. [aJAH]

Santamaria, J., Pujol, M., Orteu, N., Solanas, A., Cardenal, C., Santacruz, P., Chimeno, E. & Moon, P. (2000) Unilateral thalamic stroke does not decrease ipsilateral sleep spindles. *Sleep* 23:333–39. [rJAH]

Saper, C. B. & Loewy, A. D. (1980) Efferent connections of the parabrachial nucleus in the rat. *Brain Research* 197:291–317. [aJAH]

Saper, C. B., Sherin, J. E. & Elmquist, J. K. (1997) Role of the ventrolateral preoptic area in sleep induction. In: *Sleep and sleep disorders: From molecule to behavior*, ed. O. Hayaishi & S. Inoue. Academic Press. [aJAH]

Saredi, R., Baylor, G. W., Meier, B. & Strauch, I. (1997) Current concerns and REM-dreams: A laboratory study of dream incubation. *Dreaming* 7:195–208. [aJAH]

Sarbin, T. R. (1981) Hypnosis: A fifty year perspective. *Contemporary Hypnosis* 8:1–15. [NST]

Sasaki, Y., Fukuda, K., Takeuchi, T., Inugami, M. & Miyasita, A. (2000) Sleep onset REM period appearance rate is affected by REM propensity in circadian rhythm in normal nocturnal sleep. *Clinical Neurophysiology* 111:428–33. [rTAN, RDO]

Sastre, J. P., Buda, C. P., Kitahama, K. & Jouvet, M. (1996) Importance of the ventrolateral region of the periaqueductal gray and adjacent tegmentum as studied by muscimol microinjection in the cat. *Neuroscience* 74:415–26. [aJAH]

Sastre, J. P. & Jouvet, M. (1979) Le comportement onirique du chat. *Physiology and Behavior* 22:979–89. [JP]

Sastry, B. S. R. & Phillis, J. W. (1976) Depression of rat cerebral cortical neurons by H1 and H2 histamine receptor agonists. *European Journal of Pharmacology* 38:269–73. [CG]

Sato, S., McCutchen, C., Graham, B., Freeman, A., von Albertini-Carletti, T. & Alling, D. W. (1997) Relationship between muscle tone changes, sawtooth waves and rapid eye movements during sleep. *Electroencephalography and Clinical Neurophysiology* 103:627–32. [aJAH, EFP-S]

Sauvageau, A, Nielsen, T. A. & Montplaisir, J. (1998) Effects of somatosensory stimulation on dream content in gymnasts and control participants: Evidence of vestibulomotor adaptation in REM sleep. *Dreaming* 8:125–34. [aJAH]

Scarone, S., Spoto, G., Penati, G., Canger, R. & Moja, E. A. (1976) A study of the EEG sleep patterns and the sleep and dream experience of a group of schizophrenic patients treated with sulpiride. *Arzneimittel-Forschung* 26:1626–28. [rJAH]

Schacter, D. L. (1996) *Searching for memory*. Basic Books. [aAR]

Schacter, D. L., Chiu, C.-Y. P. & Ochsner, K. N. (1993) Implicit memory: A selective review. *Annual Review of Neuroscience* 16:159–82. [GM]

Schacter, D. L. & Tulving, E. (1994) *Memory systems 1994*. MIT Press. [RS]

Schanfald, D., Pearlman, C. & Greenberg, R. (1985) The capacity of stroke patients to report dreams. *Cortex* 21:237–47. [aMS]

Scharf, B., Moskowitz, C., Lupton, M. & Klawans, H. (1978) Dream phenomena induced by chronic Levodopa therapy. *Journal of Neural Transmission* 43:143–51. [aMS]

Schenck, C. H. (1993) REM sleep behavior disorder. In: *Encyclopedia of sleep and dreaming*, ed. M. A. Carskadon. Macmillan. [aAR]

Schenck, C. H., Bundlie, S. R., Ettiger, M. G. & Mahowald, M. W. (1986) Chronic behavioral disorders of human REM sleep: A new category of parasomnia. *Sleep* 9:293–308. [aAR]

Schenck, C. H., Hurwitz, T. D. & Mahowald, M. W. (1993) REM sleep behavior disorder: An update on a series of 96 patients and a review of the world literature. *Journal of Sleep Research* 2:224–31. [aJAH]

Schenck, C. H. & Mahowald, M. W. (1995) A polysomnographically documented case of adult somnambulism with long-distance automobile driving and frequent nocturnal violence: Parasomnia with continuing danger as a noninsane automatism? *Sleep* 18:765–72. [aAR]

(1996) REM sleep parasomniacs. *Neurological Clinics* 14:697–720. [aJAH]

Schenk, G. K., Giller, W. & Rauft, W. (1981) Double-blind comparisons of a selective seratonin uptake inhibitor, zimelidine, and placebo on quantified EEG parameters and psychological variables. *Acta Psychiatrica Scandinavica* 63:303–13. [aRPV]

Schibler, U. & Tafti, M. (1999) Molecular approaches towards the isolation of sleep-related genes. *Journal of Sleep Research* 8 (Supplement 1):1–10. [aJAH]

Schindler, R. (1953) Das Traumleben der Leukotomierten. *Wiener Zeitschrift fur Nervenheilkunde* 6:330. [aMS]

Schneider, A. & Domhoff, G. W. (1995) *The quantitative study of dreams* (updated 2000). [Web site]. Available at: http://www.dreamresearch.net/ [GWD]

Schoen, L., Kramer, M. & Kinney, L. (1984) Auditory thresholds in the dream disturbed. *Sleep Research* 13:102. [MK]

Scholz, U. J., Bianchi, A. M., Cerutti, S. & Kubicki, S. (1997) Vegetative background of sleep – spectral analysis of the heart rate variability. *Physiology and Behavior* 62:1037–43. [aJAH]

Schredl, M. (1999) Dream recall: Research, clinical implications and future and directions. *Sleep and Hypnosis* 1:99–108, A1-A4. [MSc]

Schredl, M. & Doll, E. (1997) Autogenic training and dream recall. *Perceptual and Motor Skills* 84:1305–306. [aTAN]

(1998) Emotions in diary dreams. *Consciousness and Cognition* 7:634–46. [MBl, aJAH, LM, MSc]

Schredl, M. & Engelhardt, H. (2000) Dreaming and psychopathology: Dream recall and dream content of psychiatric inpatients. *Journal of Abnormal Psychology*. (submitted). [MSc]

Schredl, M., Frauscher, S. & Shendi, A. (1995) Dream recall and visual memory. *Perceptual and Motor Skills* 81:256–58. [aJAH]

Schredl, M., Leins, M.-L., Weber, B. & Heuser, I. (2000) The effect of donepezil on sleep in elderly healthy subjects. *Experimental Gerontology*. (submitted). [MSc]

Schredl, M. & Montasser, A. (1996) Dream recall: State or trait variable? Part I: Model, theories, methodology and trait factors. *Imagination, Cognition and Personality* 16:181–210. [MSc]

(1997) Dream recall: State or trait variable? Part II: State factors, investigations and final conclusions. *Imagination, Cognition and Personality* 16:227–61. [aTAN, MSc]

Schredl, M. & Pallmer, R. (1998) Geschlechtsunterschiede in Angstträumen von SchülerInnen. *Praxis der Kinderpsychologie und Kinderpsychiatrie* 47:463–76. [MSc]

Schredl, M., Schäfer, G., Hofmann, F. & Jacob, S. (1999) Dream, content and personality: Thick vs. thin boundaries. *Dreaming* 9:257–63. [MSc]

Schredl, M., Schröder, A. & Löw, H. (1996) Traumerleben von älteren Menschen – Teil 2: Empirische Studie und Diskussion. *Zeitschrift für Gerontopsychologie und psychiatrie* 9:43–53. [MSc]

Schredl, M., Weber, B. & Heuser, I. (1998) REM-Schlaf und Gedachtnis. *Psychologische Beiträge* 40:340–49. [MSc]

Schultz, G. & Melzack, R. (1991) The Charles Bonnet syndrome: "Phantom visual images." *Perception* 20:809–25. [aAR]

Schultz, H. & Salzarulo, P. (1993) Methodological approaches to problems in experimental dream research: An introduction. *Journal of Sleep Research* 2:1–3. [PS]

Schwartz, B. A. (1968) Afternoon sleep in certain hypersomnolent states: "Intermediate sleep." *Electroencephalography and Clinical Neurophysiology* 24:569–81. [rTAN]

Schwartz, D. G., Weinstein, L. N. & Arkin, A. M. (1978) Qualitative aspects of sleep mentation. In: *The mind in sleep: Psychology and psychophysiology*, ed. A. M. Arkin, J. S. Antrobus & S. J. Ellman. Erlbaum. [aTAN]

Schwarz, C. & Thier, P. (1999) Binding of signals relevant to action: Towards a hypothesis of the functional role of the pontine nuclei. *Trends in Neurosciences* 22:443–51. [aJAH]

Scrima, L. (1984) Dream sleep and memory: New findings with diverse implications. *Integrative Psychiatry* 2:211–16. [WHM]

Segal, M. & Bloom, F. E. (1974) The action of norepinephrine in the rat hippocampus: I. Iontophoretic studies. *Brain Research* 72:79–97. [CG]

Sei, H., Sakai, K., Kanamori, N., Salvert, D., Vanni-Mercier, G. & Jouvet, M. (1994) Long-term variations of arterial blood pressure during sleep in freely moving cats. *Physiology and Behavior* 4:673–79. [CG]

Seidel, W. F., Dement, W. C., Mignot, E. & Edgar, D. M. (1997) Non-REM sleep recovery: Comparison of dopamine uptake inhibitors with other wake-promoting agents. *Sleep Research* 26:629. [aJAH]

Selden, N. R., Gitelman, D. R., Salamon-Murayama, N., Parrish, T. B. & Mesulam, M. M. (1998) Trajectories of cholinergic pathways within the cerebral hemispheres of the human brain. *Brain* 121:2249–57. [FD]

Seligman, M. E. P. & Yellen, A. (1987) What is a dream? *Behaviour Research and Therapy* 25:1–24. [JAB, aJAH, aTAN]

Selye, H. (1937) Studies on adaptation. *Endocrinology* 21(2):169–89. [IF]

Semba, K. (1999) The mesopontine cholinergic system: A dual role in REM sleep and wakefulness. In: *Handbook of behavioral state control: Molecular and cellular mechanisms*, ed. R. Lydic & H. A. Baghdoyan. CRC Press. [arJAH]

Semba, K. & Fibiger, H. C. (1992) Afferent connections of the laterodorsal and the pedunculopontine tegmental nuclei in the rat: A retro–and anterograde transport and immunohistochemical study. *Journal of Comparative Neurology* 323:387–410. [aJAH]

Semba, K., Reiner, P. B., McGeer, E. G. & Fibinger, H. C. (1988) Brainstem afferents to the magnocellular basal forebrain studied by axonal transport, immunohistochemistry and electrophysiology. *Journal of Comparative Neurology* 26:433–53. [aJAH]

Serafetinides, E. A. (1991) Comparison of eye movement patterns of dreaming and arousal in sleep. *International Journal of Neuroscience* 58:269–70. [aJAH]

Shaffery, J. P., Oksenberg, A. & Marks, G. A. (1996) Rapid eye movement sleep deprivation in kittens amplifies LGN cell-size disparity induced by monocular deprivation. *Developmental Brain Research* 97:51–61. [rJAH]

Shaffery, J. P., Roffwarg, H. P., Speciak, S. G. & Marks, G. A. (1999) Ponto-geniculo-occipital-wave suppression amplifiers lateral geniculate nucleus cell-size changes in monocularly deprived kittens. *Brain Research* 114(1):109–19. [rJAH]

Shakespeare, W. (1959/1986) A midsummer night's dream. In: *The complete works of William Shakespeare*, ed. S. Wells & G. Taylor. Clarendon Press. [JFP]

Shapiro, A., Goodenough, D. R., Biederman, I. & Sleser, I. (1964) Dream recall and the physiology of sleep. *Journal of Applied Psychology* 19:776–83. [rTAN]

Shapiro, A., Goodenough, D. R. & Gryler, R. B. (1963) Dream recall as a function of method of awakening. *Psychosomatic Medicine* 25:174–80. [RG, aTAN]

Shapiro, A., Goodenough, D. R., Lewis, H. B. & Sleser, I. (1965) Gradual arousal from sleep: A determinant of thinking reports. *Psychosomatic Medicine* 27:342–49. [RG, aTAN]

Shapiro, C. M. (1998) Fatigue: How many types and how common? *Journal of Psychosomatic Research* 45(1):1–3. [AK]

Shapiro, C. M., Catherall, J. R., Montgomery, I., Raab, G. M. & Douglas, N. J. (1986) Do asthmatics suffer bronchoconstriction during rapid eye movement sleep? *British Medical Journal* 292:1161–64. [AK]

Sharf, B., Moskovitz, C., Lupton, M. D. & Klawans, H. L. (1978) Dream phenomena induced by chronic levodopa therapy. *Journal of Neural Transmission* 43:143–51. [aJAH]

Sharpley, A. L. & Cowen, P. J. (1995) Effect of pharmacologic treatments on the sleep of depressed patients. *Biological Psychiatry* 37:85–98. [aRPV]

Sharpley, A. L., Williamson, D. J., Attenburrow, M. E. J., Pearson, G., Sargent, P. & Cowen, P. J. (1996) The effects of paroxetine and nefazodone on sleep: A placebo controlled study. *Psychopharmacology* 126:50–54. [aRPV]

Shatz, C. J. (1990) Impulse activity and the patterning of connections during CNS development. *Neuron* 5:745–56. [TJM]

(1992) The developing brain. *Scientific American* 26:61–67. [TJM]

Shea, T. B. (1994) Delivery of anti-GAP-43 antibodies into neuroblastoma cells reduces growth cone size. *Biochemical and Biophysical Communications* 203:459–64. [TJM]

Shen, J., Barnes, C. A., Wenk, G. L. & McNaughton, B. L. (1996) Differential effects of selective immunotoxic lesions of medial septal cholinergic cells on spatial working and reference memory. *Behavioral Neuroscience* 110:1181–86. [aRPV]

Shen, J., Kudrimoti, H. S., McNaughton, B. L. & Barnes, C. A. (1998) Reactivation of neuronal ensembles in hippocampal dentate gyrus during sleep after spatial experience. *Journal of Sleep Research* 7 (Supplement 1):6–16. [JP]

Sherin, J. E., Elmquist, J. K., Torrealba, F. & Saper, C. B. (1998) Innervation of histaminergic tuberomammillary neurons by GABAergic and galaninergic neurons in the ventrolateral preoptic nucleus of the rat. *Journal of Neuroscience* 18:4705–21. [aJAH, RS-P]

Sherin, J. E., Shiromani, P. J., McCarley, R. W. & Saper, C. B. (1996) Activation of ventrolateral preoptic neurons during sleep. *Science* 271:216–19. [aJAH, RS-P]

Shevrin, H. & Fisher, C. (1967) Changes in the effects of a waking subliminal stimulus as a function of dreaming and nondreaming sleep. *Journal of Abnormal Psychology* 72:362–68. [aJAH, HS]

Shimizu, A., Takehashi, H., Sumitsuji, N., Tanaka, M., Yoshida, M. & Kaneko, Z. (1977) Memory retention of stimulations during REM and NREM stages of sleep. *Electroencephalography and Clinical Neurophysiology* 43:658–65. [CCi, aTAN]

Shipley, J. E., Kupfer, D. J., Dealy, R. S., Griffin, S. J., Coble, P. A., McEachran, A. B. & Grochocnski, V. J. (1984) Differential effects of amitriptyline and of zimelidine on the sleep electroencephalogram of depressed patients. *Clinical Pharmacology Therapeutics* 36:251–59. [aRPV]

Shiromani, P. J. (1998) Sleep circuitry, regulation and function: Lesson from c-fos, leptin and timeless. In: *Progress in psychobiology and physiological psychology*, ed. A. Morrison & C. Fluharty. Academic Press. [RS-P]

Shiromani, P. J., Gutwein, B. M. & Fishbein, W. (1979) Development of learning and memory in mice after brief paradoxical sleep deprivation. *Physiology and Behavior* 22:971–78. [aRPV]

Shiromani, P. J., Malik, M., Winston, S. & McCarley, R. W. (1995) Time course of fos-like immunoreactivity associated with cholinergically induced REM sleep. *Journal of Neuroscience* 15:3500–508. [aJAH]

Shiromani, P. J., Scammell, T., Sherin, J. E. & Saper, C. B. (1999) Hypothalamic regulation of sleep. In: *Handbook of behavioral state control: Molecular and cellular mechanisms*, ed. R. Lydic & H. A. Baghdoyan. CRC Press. [arJAH]

Shiromani, P. J., Winston, S. & McCarley, R. W. (1996) Pontine cholinergic neurons show fos-like immunoreactivity associated with cholinergically induced REM sleep. *Molecular Brain Research* 38:77–84. [aJAH]

Shoemaker, P. J. (1996) Hardwired for news: Using biological and cultural evolution to explain the surveillance function. *Journal of Communication* 46:32–47. [LM]

Shouse, M. N. & Siegel, J. M. (1992) Pontine regulation of REM sleep components in cats: Integrity of the pedunculopontine tegmentum (PPT) is important for phasic events but unnecessary for atonia during REM sleep. *Brain Research* 571:50–63. [aJAH]

Shuman, S. L., Capece, M. L., Baghdoyan, H. A. & Lydic, R. (1995) Pertussis toxin-sensitive G proteins mediate cholinergic regulation of sleep and breathing. *American Journal of Physiology* 269:R308–17. [RLy]

Shurley, J., Serafetinides, E., Brooks, R., Elsner, R. & Kenney, D. (1969) Sleep in cetaceans. I. The pilot whale (*Globicephala scammoni*). *Psychophysiology* 6:230. [JMS]

Siegel, J. M. (1989) Brainstem mechanisms generating REM sleep. In: *Principles and practice of sleep medicine*, ed. M. H. Krieger, T. Roth & W. C. Dement. W. B. Saunders. [PS]

(1990) Mechanisms of sleep control. *Journal of Clinical Neurophysiology* 7(1):49–65. [RWG]

(1994) Brainstem mechanisms generating REM sleep. In: *Principles and practice of sleep medicine, second edition*, ed. M. Kryger, T. Roth & W. Dement. W. B. Saunders. [aJAH]

(1997) Monotremes and the evolution of REM sleep. *Sleep Research Society Bulletin* 4:31–32. [MBl]

(1998) The evolution of REM sleep. In: *Handbook of behavioral state control. Cellular and molecular mechanisms,* ed. R. Lydic & H. A. Baghdoyan. CRC Press LLC. [aTAN]

Siegel, J. M., Manger, P. R., Nienhuis, R., Fahringer, H. M. & Pettigrew, J. D. (1996) The Echidna *Tachyglossus aculeatus* combines REM and non-REM aspects in a single sleep state: Implications for the evolution of sleep. *Journal of Neuroscience* 16:3500–506. [aTAN]

(1997) The platypus has REM sleep. *Sleep Research* 26:177. [JMS]

Siegel, J. M., Manger, P. R., Nienhuis, R., Fahringer, H. M., Shalita, T. & Pettigrew, J. D. (1999) Sleep in the platypus. *Neuroscience* 91:391–400. [JMS]

Silbersweig, D. A., Stern, E., Frith, C., Cahill, C., Holmes, A., Grootoonk, S., Seaward, J., McKenna, P., Chua, S. E., Schnorr, L., Jones, T. & Frackowiak, R. S. J. (1995) A functional neuroanatomy of hallucinations in schizophrenia. *Nature* 378:176–79. [aJAH, EFP-S]

Silvestri, R., Pace-Schott, E. F., Gersh, T., Stickgold, R., Salzman, C. & Hobson, J. A. (1998) Changes in sleep and dreaming during SSRI treatment and withdrawal in a home setting. *Journal of Sleep Research* 7 (Supplement 2):250. [aJAH]

(in press). Effects of flouoxamine and paroxetine on sleep structure in normal subjects: A home based Nightcap evaluation during drug administration and withdrawal. Journal of Clinical Psychiatry.

Singer, J. L. (1966) *Daydreaming.* Random House. [aAR]

(1975) *The inner world of daydreaming.* Harper and Row. [EH]

(1988) Sampling ongoing consciousness and emotional experience: Implications for health. In: *Psychodynamics and cognition,* ed. M. Horowitz. University of Chicago Press. [aAR]

(1993) Experimental studies of ongoing conscious experience. In: *Experimental and theoretical studies of consciousness (Ciba Foundation Symposium 174:100–22).* Wiley. [PC]

Singh, S. & Mallick, B. N. (1996) Mild electrical stimulation of pontine tegmentum around locus coeruleus reduces rapid eye movement sleep in rats. *Neuroscience Research* 24:227–35. [aJAH]

Sippel, J. M., Giraud, G. D. & Holden, W. E. (1999) Nasal administration of the nitric oxide synthase inhibitor L-NAME induces daytime somnolence. *Sleep* 22:786–88. [aJAH]

Sira, J. ben (180 B.C./1993) Qumran "Dead Sea" Scrolls, Apocrypha, Sirach 34.3. In: The new Oxford annotated Bible with the Apocrypha, ed. H. G. May & B. C. Metzger. Oxford University Press.

Sitaram, N., Moore, A. M. & Gillin, J. C. (1978a) The effect of physostigmine on normal human sleep and dreaming. *Archives of General Psychiatry* 35:1239–43. [FD, arJAH]

Sitaram, N., Moore, A. & Gillin, J. C. (1978b) Experimental acceleration and slowing of REM sleep ultradian rhythm by cholinergic agonist and antagonist. *Nature* 274:490–92. [aJAH, aMS]

Sitaram, N., Wyatt, R., Dawson, S. & Gillin, J. C. (1976) REM sleep induction by physostigmine infusion during sleep. *Science* 191:1281–83. [aJAH, aMS]

Skaggs, W. E. & McNaughton, B. L. (1996) Replay of neuronal firing sequences in rat hippocampus during sleep following spatial experience. *Science* 271:1870–73. [rRPV]

Sloan, M. A. (1972) The effects of deprivation of rapid eye movement (REM) sleep on maze learning and aggression in the albino rat. *Journal of Psychiatric Research* 9:101–11. [aRPV]

Slover, G. P. T., Morris, R. W., Stroebel, C. F. & Patel, M. K. (1987) Case study of psychophysiological diary: Infradian rhythms. In: *Advances in chronobiology,* ed. J. E. Pauley & L. E. Scheving. Alan R. Liss. [aTAN]

Smiley, J. F. & Goldman-Rakic, P. S. (1993) Heterogenous targets of dopamine synapses in monkey prefrontal cortex demonstrated by serial section electron microscopy: A laminar analysis using the silver-enhanced diaminobenzidine sulfide (SEDS) immunolabeling technique. *Cerebral Cortex* 3:223–38. [CG]

Smiley, J. F., Subramaniam, M. & Mesulam, M.-M. (1999) Monoaminergic-cholinergic interactions in the primate basal forebrain. *Neuroscience* 93:817–29. [arJAH]

Smith, C. (1985) Sleep states and learning: A review of the animal literature. *Neuroscience and Biobehavioral Reviews* 9:157–68. [rJAH, ARM, aTAN, JP, CS, aRPV]

(1993) REM sleep and learning: Some recent findings. In: *The functions of dreaming,* ed. A. Moffitt, M. Kramer & R. Hoffmann. State University of New York Press. [rJAH, WHM, CS]

(1995) Sleep states and memory processes. *Behavioural Brain Research* 69:137–45. [JAB, rJAH, ARM, aTAN, aAR, CS, aRPV]

(1996) Sleep states, memory processes and synaptic plasticity. *Behavioural Brain Research* 78:49–56. [CS, aRPV]

(1999) Cognitive aspects of learning and their relationship to states of sleep. *Sleep Research Online* 2:789. [CS]

Smith, C. & Butler, S. (1982) Paradoxical sleep at selective times following training is necessary for learning. *Physiology and Behavior* 29:469–73. [CS, aRPV]

Smith, C., Conway, J. M. & Rose, G. M. (1998) Brief paradoxical sleep deprivation impairs reference, but not working, memory in the radial arm maze task. *Neurobiology of Learning and Memory* 69:211–17. [RS]

Smith, C. & Gisquet-Verrier, P. (1996) Paradoxical sleep deprivation and sleep recording following training in a brightness discrimination avoidance task in Sprague-Dawley rats: Paradoxical effects. *Neurobiology of Learning and Memory* 66:283–94. [RDO, CS]

Smith, C. & Kelly, G. (1988) Paradoxical sleep deprivation applied two days after the end of training retards learning. *Physiology and Behavior* 43:213–16. [arRPV]

Smith, C. & Lapp, L. (1991) Increases in number of REMs and REM density in humans following an intensive learning period. *Sleep* 14:325–30. [aJAH, CS, aRPV]

Smith, C. & MacNeill, C. (1993) A paradoxical sleep-dependent window for memory 53–56 h after the end of avoidance training. *Psychobiology* 21:109–12. [arRPV]

(1994) Impaired motor memory for a pursuit rotor task following Stage 2 sleep loss in college students. *Journal of Sleep Research* 3:206–13. [aTAN, CS, RS]

Smith, C. & Rose, G. M. (1996) Evidence for a paradoxical sleep window for place learning in the Morris water maze. *Physiology and Behavior* 59:93–97. [aRPV]

(1997) Posttraining paradoxical sleep in rats is increased after spatial learning in the Morris water maze. *Behavioral Neuroscience* 111:1197–204. [RS, aRPV]

Smith, C., Tenn, C. & Annett, R. (1991) Some biochemical and behavioral aspects of the paradoxical sleep window. *Canadian Journal of Psychology* 45:115–24. [CS]

Smith, C. & Weeden, K. (1990) Post-training REMs coincident auditory stimulation enhances memory in humans. *Psychiatric Journal of the University of Ottawa* 15(2):85–90. [CCi, aTAN]

Smith, D. S. & Skene, J. H. (1997) A transcription-dependent switch controls competence of adult neurons for distinct modes of axon growth. *The Journal of Neurosciences* 17:646–58. [TJM]

Smith, E. E. & Jonides, J. (1999) Storage and executive processes in the frontal lobes. *Science* 283:1657–61. [RDO]

Smith, G. S., Reynolds, C. F., III, Pollock, B., Derbyshire, S., Nofzinger, E. A., Dew, M. A., Houck, R. R., Milko, D., Meltzer, C. C. & Kupfer, D. J. (1999) Cerebral glucose metabolic response to combined total sleep deprivation and antidepressant treatment in geriatric depression. *American Journal of Psychiatry* 156:683–89. [EFP-S]

Smith, R. C. (1984) The meaning of dreams: The need for a standardized dream report. *Psychiatry Research* 13:267–74. [aTAN]

Snyder, F. (1965) The organismic state associated with dreaming. In: *Psychoanalysis and current biological thought,* ed. N. Greenfield & W. Lewis. University of Wisconsin Press. [aTAN, rRPV]

(1966) Toward an evolutionary theory of dreaming. *American Journal of Psychiatry* 123:121–36. [RCo, aJAH, RDO, aAR]

(1967) In quest of dreaming. In: *Experimental studies of dreaming,* ed. H. A. Witkin & H. B. Lewis. Random House. [aJAH]

(1968) Quelques hypothèses au sujet de la contribution du sommeil avec mouvements oculaires à la survivance des mammifères. In: *Rève et conscience,* ed. P. Wertheimer. Presses Universitaires de France. [CG]

(1970) The phenomenology of dreaming. In: *The psychodynamic implications of the physiological studies on dreams,* ed. L. Madow & L. H. Snow. Charles S. Thomas. [GWD, arJAH, aAR]

Snyder, F., Karacan, I., Thorp, U. & Scott, J. (1968) Phenomenology of REM dreaming. *Psychophysiology* 4:375. [GWV]

Snyder, H. (1958) Epileptic equivalents in children. *Pediatrics* 18:308–18. [aMS]

Soh, K., Morita, Y. & Sei, H. (1992) Relationship between eye movement and oneiric behavior in cats. *Physiology and Behavior* 52:553–58. [aJAH]

Solms, M. (1991) *Anoneira and the neuropsychology of dreams.* Doctoral dissertation, University of the Witwatersrand, Johannesburg. [rMS]

(1995) New findings on the neurological organization of dreaming: Implications for psychoanalysis. *Psychoanalytic Quarterly* 64:43–67. [SL, MM, aTAN]

(1997a) *The neuropsychology of dreams: A clinico-anatomical study.* Erlbaum. [RBA, MBo, RCo, GWD, FD, OF, EH, arJAH, JAH, TLK, AK, MK, SL, aTAN, CMP, aAR, arMS]

(1997b) What is consciousness? *Journal of the American Psychoanalytic Association* 45:681–778. [rMS]

(1999a) Paper presented at the Third International Congress of the World Federation of Sleep Research Societies (WFSRS), Dresden, Germany, October 5–9, 1999. [rTAN]

(1999b) Commentary on *The new neuropsychology of sleep: Implications for psychoanalysis* by J. A. Hobson. *Neuropsychoanalysis* 1:183–95. [arJAH, rMS]

(2000) Ongoing commentary on J. A. Hobson's, "The new neuropsychology of sleep: Implications for psychoanalysis." *Neuropsychoanalysis* 2. (in press). www.neuro–psa.com [aJAH, rMS]

Sparks, D. L. & Groh, J. M. (1994) The superior colliculus: A window for viewing issues in integrative neuroscience. In: *The cognitive neuroscience*, ed. M. Gazzaniga. MIT Press. [aJAH]

Sparks, D. L. & Hartwich-Young, R. (1989) The deep layers of the superior colliculus. In: *The neurobiology of saccadic eye movements*, ed. R. H. Wurtz & M. E. Goldberg. Elsevier Science. [aJAH]

Sparks, P. D. (1998) *The role of the septal complex in the modulation of emotion.* Doctoral dissertation, New York University. [EEK]

Spinks, A. & Mealey, L. (submitted) Linguistic biases for words representing threat? [LM]

Spitta, H. (1882) *Die Schlaf–und Traumzustände der menschlichen Seele mit besonderer Berücksichtigung ihres Verhältnisses zu den psychischen Alienationen.* Verlag und Druck von Franz Fues. (Original edition 1878). [rTAN]

Spitzer, M., Walder, S. & Clarenbach, P. (1993) Aktivierte assoziative Netzwerke im REM-Schlaf: Semantische Bahnungseffekte nach dem Aufwecken aus verschiedenen Schlafstadien. In: *Schlafmedizin*, ed. K. Meier-Ewert & E. Rühle. Fischer. [MSc]

Spring, B., Gelenberg, A. J., Garvin, R. & Thompson, S. (1992) Amitriptyline, clovoxamine and cognitive function: A placebo-controlled comparison in depressed outpatients. *Psychopharmacology* 108:3227–32. [aRPV]

Squire, L. R., Amaral, D. G. & Press, G. A. (1990) Magnetic resonance imaging of the hippocampal formation and mammillary nuclei distinguish medial temporal lobe and diencephalic amnesia. *Journal of Neuroscience* 10:3106–17. [EEK]

Staba, R. J., Fox, I. J., Behnke, E. J., Fields, T. A., MacDonald, K. A., Bragin, A., Fried, I., Simmons, J., Mann, C., Wilson, C. L. & Engel, J., Jr. (1998) Single unit recordings in human hippocampus and entorhinal cortex during sleep. *Society for Neuroscience Abstracts* 24:920. [aJAH]

Stackman, R. W. & Walsh, T. J. (1995) Distinct profile of working memory errors following acute or chronic disruption of the cholinergic septohippocampal pathway. *Neurobiology of Learning and Memory* 64:226–36. [aRPV]

Staner, L., Kerkhofs, M., Detroux, D., Leyman, S., Linkowski, P. & Mendlewicz, J. (1995) Acute, subchronic and withdrawal sleep EEG changes during treatment with paroxetine and amitriptyline: A double-blind randomized trial in major depression. *Sleep* 18:470–77. [aRPV]

Stanley, G. B., Li, F. F. & Dan, Y. (1999) Reconstruction of natural scenes from ensemble responses in the lateral geniculate nucleus. *Journal of Neuroscience* 19:8036–42. [RCo, aJAH]

Staubli, U. & Lynch, G. (1987) Stable hippocampal long-term potentiation elicited by 'theta' pattern stimulation. *Brain Research* 435:227–34. [aRPV]

Steiger, A. & Holsboer, F. (1997) Neuropeptides and human sleep. *Sleep* 20:1038–52. [aJAH]

Steinfels, G. F., Heym, J., Strecker, R. E. & Jacobs, B. L. (1983) Behavioral correlates of dopaminergic unit activity in freely moving cats. *Brain Research* 258:217–28. [aJAH, JP]

Steininger, J. L., Alam, M. N., Szymusiak, R. & McGinty, D. (1996) State dependent discharge of tuberomammilary neurons in the rat hypothalamus. *Sleep Research* 25:28. [aJAH]

Steininger, T. L., Wainer, B. H., Blakely, R. D. & Rye, D. B. (1997) Serotonergic dorsal raphe nucleus projections to the cholinergic and noncholinergic neurons of the pedunculopontine tegmental region: A light and electron microscopic anterograde tracing and immunohistochemical study. *Journal of Comparative Neurology* 382:302–22. [aJAH]

Stepanski, E. J. (1993) Toothgrinding. In: *Encyclopedia of sleep and dreaming*, ed. M. A. Carskadon. Macmillan. [rAR]

Steriade, M. (1991) Alertness, quiet sleep, dreaming. In: *Cerebral cortex. Vol. 9: Normal and altered states of function*, ed. A. Peters & E. G. Jones. Plenum Press. [MSt]

  (1996) Arousal: Revisiting the reticular activating system. *Science* 272:225–26. [RCo, aJAH]

  (1997) Synchronized activities of coupled oscillators in the cerebral cortex and thalamus at different levels of vigilance. *Cerebral Cortex* 7:583–604. [aJAH]

  (1999a) Coherent oscillations and short-term plasticity in corticothalamic networks. *Trends in Neuroscience* 22:337–45. [MSt]

  (1999b) Cellular substrates of oscillations in corticothalamic systems during states of vigilance. In: *Handbook of behavioral state control: Molecular and cellular mechanisms*, eds., R. Lydic, H. A. Baghdoyan. CRC Press. [aJAH]

  (2000) Corticothalamic resonance, states of vigilance and mentation. *Neuroscience* 101:243–76. [aJAH]

Steriade, M. & Amzica, F. (1996) Intracortical and corticothalamic coherency of fast (30–40Hz) spontaneous cortical rhythms during brain activation. *Proceedings of the National Academy of Science USA* 93:2533–38. [RCo]

  (1998) Coalescence of sleep rhythms and their chronology in corticothalamic networks. *Sleep Research Online* 1:1–10. [RCo]

Steriade, M., Amzica, F. & Contreras, D. (1996) Synchronization of fast (30–40 Hz) spontaneous cortical rhythms during brain activation. *Journal of Neuroscience* 16:392–417. [MSt]

Steriade, M. & Biesold, D., eds. (1990) *Brain cholinergic systems.* Oxford University Press. [aJAH]

Steriade, M. & Buzsaki, G. (1990) Parallel activation of thalamic and cortical neurons by brainstem and basal forebrain cholinergic systems. In: *Brain cholinergic systems*, ed. M. Steriade & D. Biesold. Oxford University Press. [arJAH]

Steriade, M. & Contreras, D. (1995) Relations between cortical and thalamic cellular events during transition from sleep patterns to paroxysmal activity. *Journal of Neuroscience* 15:623–42. [MSt]

Steriade, M., Contreras, D. & Amzica, F. (1994) Synchronized sleep oscillations and their paroxysmal developments. *Trends in Neuroscience* 17:199–208. [aJAH]

Steriade, M., Contreras, D., Dossi, C. & Nunez, A. (1993a) The slow (<1 Hz) oscillation in reticular thalamic and thalamocortical neurons: Scenario of sleep rhythm generation in interacting thalamic and neocortical networks. *Journal of Neuroscience* 13:3284–99. [aJAH]

Steriade, M., Deschênes, M., Wyzinski, P. & Hallé, J. Y. (1974) Input-output organization of the motor cortex during sleep and waking. In: *Basic sleep mechanisms*, ed. O. Petre-Quadens & J. Schlag. Academic Press. [MSt]

Steriade, M. & Hobson, J. A. (1976) Neuronal activity during the sleep-waking cycle. *Progress in Neurobiology* 6:155–376. [aJAH]

Steriade, M. & McCarley, R. W. (1990a) *Brainstem control of wakefulness and sleep.* Plenum Press. [CG, RWG, aJAH, aRPV]

  (1990b) Dreaming. In: *Brainstem control of wakefulness and sleep*, ed. M. Steriade & R. W. McCarley. Plenum Press. [aTAN]

Steriade, M., McCormick, D. A. & Sejnowski, T. (1993b) Thalamocortical oscillations in the sleeping and aroused brain. *Science* 262:679–84. [RWG, aJAH]

Steriade, M., Nuñez, A. & Amzica, F. (1993c) A novel slow (<1 Hz) oscillation of neocortical neurons in vivo: Depolarizing and hyperpolarizing components. *Journal of Neuroscience* 13:3252–65. [aJAH]

  (1993d) Intracellular analysis of relation between the slow (<1 Hz) neocortical oscillation and other sleep rhythms of the electroencephalogram. *Journal of Neuroscience* 13:3266–83. [aJAH, MSt]

Steriade, M., Paré, D., Bouhassira, D., Deschênes, M. & Oakson, G. (1989) Phasic activation of lateral geniculate and perigeniculate thalamic neurons during sleep with ponto-geniculo-occipital waves. *Journal of Neuroscience* 9:2215–29. [aTAN, MSt]

Steriade, M., Pare, D., Datta, S., Oakson, G. & Curro-Dossi, R. (1990) Different cellular types in mesopontine cholinergic nuclei related to ponto-geniculo-occipital waves. *Journal of Neuroscience* 10:2560–79. [aJAH]

Steriade, M., Pare, D., Parent, A. & Smith, Y. (1988) Projections of cholinergic and non-cholinergic neurons of the brainstem core to relay and associational thalamic nuclei in the cat and macaque monkey. *Neuroscience* 25:47–67. [RCo, aJAH]

Steriade, M., Ropert, N., Kitsikis, A. & Oakson, G. (1980) Ascending activating neuronal networks in midbrain reticular core and related rostral systems. In: *The reticular formation revisited*, ed. J. A. Hobson & M. A. B. Brazier. Raven Press. [arJAH]

Steriade, M., Timofeev, I. & Grenier, F. (1999) Intracellular activity of various neocortical cell-classes during the natural wake-sleep cycle. *Society of Neuroscience Abstract* 25:1661. [MSt]

Steriade, M., Timofeev, I., Grenier, F. & Dürmüller, N. (1998) Role of thalamic and cortical neurons in augmenting responses and self-sustained activity: Dual intracellular recordings in vivo. *Journal of Neuroscience* 16:6425–43. [MSt]

Stern, W. C. (1971) Acquisition impairments following rapid eye movement sleep deprivation in rats. *Physiology and Behavior* 7:345–52. [aRPV]

Stern, W. C., Forbes, W. B. & Morgane, P. J. (1974) Absence of ponto-geniculo-occipital (PGO) spikes in rats. *Physiology and Behavior* 12:293–96. [CG]

Stern, W. C., Morgane, P. & Bronzino, J. (1972) LSD: Effects on sleep patterns and spiking activity in the lateral geniculate nucleus. *Brain Research* 41:199–204. [aTAN]

Stevens, D. R., Gerber, V., McCarley, R. W. & Greene, R. W. (1996) Glycine-mediated inhibitory postsynaptic potentials in the medial pontine reticular formation of the rat in vitro. *Neuroscience* 73:791–96. [aJAH]

Stickgold, R. (1998) Sleep: Off-line memory reprocessing. *Trends in Cognitive Science* 2:484–92. [arJAH, EFP-S, CS, RS, rRPV]

Stickgold, R., James, L. & Hobson, J. A. (2000c) Visual discrimination learning requires sleep after training. *Nature Neuroscience* 3:1237–38. [aJAH]

Stickgold, R., Malia, A. & Hobson, J. A. (1999a) Sleep onset memory reprocessing and Tetris. *Journal of Cognitive Neuroscience* 11. (in press). [aJAH]

  (2000a) Hypnagogic dreams in normals and amnesiacs. *Journal of Cognitive Neuroscience* 12:S29. [rJAH]

Stickgold, R., Malia, A., Maguire, D., Roddenberry, D. & O'Connor, M. (2000d) Replaying the game: Hypnagogic images in normals and amnesics. *Science* 290:350–53. [aJAH]

Stickgold, R., Pace-Schott, E. F. & Hobson, J. A. (1994a) A new paradigm for dream research: Mentation reports following spontaneous arousal from REM and NREM sleep recorded in a home setting. *Consciousness and Cognition* 3:16–29. [RCo, IF, arJAH, aTAN, RDO]

(1997a) Subjective estimates of dream duration and dream recall process. *Sleep Research* 26:279. [PC, aJAH]

Stickgold, R., Rittenhouse, C. & Hobson, J. A. (1994b) Dream splicing: A new technique for assessing thematic coherence in subjective reports of mental activity. *Consciousness and Cognition* 3:114–28. [arJAH]

Stickgold, R., Sangodeyi, F. & Hobson, J. A. (1997b) Judges cannot identify thematic coherence in dream reports with discontinuities. *Sleep Research* 26:278. [aJAH]

Stickgold, R., Schirmer, B., Patel, V., Whidbee, D. & Hobson, J. A. (1998a) Visual discrimination learning: Both NREM and REM are required. *Sleep 21* (Supplement):256. [aJAH]

Stickgold, R., Scott, L., Fosse, R. & Hobson, J. A. (2001) Brain-mind states: I. Longitudinal field study of wake-sleep factors influencing mentation report length. Sleep 24:171–179. [rJAH]

Stickgold, R., Scott, L., Malia, A., Maher, E., Bennett, D. & Hobson, J. A. (1998b) Longitudinal collection of mentation reports over wake-sleep states. *Sleep 21* (Supplement):280. [arJAH]

Stickgold, R., Scott, L., Rittenhouse, C. & Hobson, J. A. (1999b) Sleep induced changes in associative memory. *Journal of Cognitive Neuroscience* 11:182–93. [arJAH, EFP-S, RS]

Stickgold, R., Whidbee, D., Schirmer, B., Patel, V. & Hobson, J. A. (2000b) Visual discrimination task improvement: A multi-step process occurring during sleep. *Journal of Cognitive Neuroscience* 12:246–54. [JB, arJAH, RS, arRPV]

Stirling, R. & Dunlop, S. A. (1995) The dance of the growth cones – where to, next. *Trends in the Neurosciences* 18:11–15. [TJM]

Stock, J. B. & Surette, M. G. (1996) Chemotaxis. In: *Escherechia coli and salmonellas, cellular and molecular biology, vol. 1,* ed. F. C. Neidhardt, R. Curtiss, J. L. Riley, M. O. Schaechter & H. E. Umbarger. American Society for Microbiology Press. [TJM]

Stoddard, F. J., Chedekel, D. S. & Shakun, L. (1996) Dreams and nightmares of burned children. In: *Trauma and dreams,* ed. D. Barrett. Harvard University Press. [aAR]

Stones, M. J. (1977) Memory performance after arousal from different sleep stages. *British Journal of Psychology* 68:177–81. [aJAH, aTAN]

Stoyva, J. M. (1961) The effects of suggested dreams on the length of rapid eye movement periods. Unpublished doctoral dissertation, University of Chicago. [aTAN]

(1965) Finger electromyographic activity during sleep: Its relation to dreaming in deaf and normal subjects. *Journal of Abnormal Psychology* 70:343–49. [aJAH, aTAN]

Strauch, I. (1996) Animal characters in dreams and fantasies of children. *ASD Newsletter* 13(1):11–13. [aAR]

Strauch, I. & Meier, B. (1996) *In search of dreams. Results of experimental dream research.* State University of New York Press. [MBl, GWD, MK, JM, aTAN, aAR, MSc]

Strecker, R. E., Porkka-Heiskanen, T., Bjorkum, A. A. & McCarley, R. W. (1997a) Adenosine actions on the dorsal raphe nucleus: Altered sleep-waking pattern. *Society for Neuroscience Abstracts* 23:1065. [aJAH]

Strecker, R. E., Porkka-Heiskanen, T., Bjorkum, A. A., Thakkar, M. & McCarley, R. W. (1997b) Is adenosine a physiological sleep factor? Evidence from microdialysis studies. *Sleep Research* 26:46. [aJAH]

Strub, R. & Geschwind, N. (1983) Localization in Gerstmann syndrome. In: *Localization in neuropsychology,* ed. A. Kertesz. Academic Press. [MS]

Strube, M. J. (1990) Psychometric principles: From physiological data to psychological constructs. In: *Principles of psychophysiology. Physical, social, and inferential elements,* ed. J. T. Cacioppo & L. G. Tassinary. Cambridge University Press. [aTAN]

Stuss, D. & Benson, D. F. (1983) Frontal lobe lesions and behavior. In: *Localization in neuropsychology,* ed. A. Kertesz. Academic Press. [aMS]

Suchecki, D. & Tufik, S. (2000) Social stability attenuates the stress in the modified multiple platform method for paradoxical sleep deprivation in the rat. *Physiology and Behavior* 68: 309–16. [AC]

Sudo, Y., Suhara, T., Honda, T., Nakajima, T., Okubo, Y., Suzuki, K., Nakashima, Y., Yoshikawa, K., Okauchi, T., Sasaki, Y. & Matsushita, M. (1998) Muscarinic cholinergic receptors in human narcolepsy: A PET study. *Neurology* 51:1297–302. [aJAH]

Susic, V. (1976) Potentiation of ketamine effects on the spiking activity in the lateral geniculate nucleus by rapid eye movement (REM) sleep deprivation. *Archives Internationales de Psychologie et de Biochemie* 84:229–34. [aTAN]

Sutton, J. P., Breiter, H. C., Caplan, J. B., Huang-Hellinger, F. R., Kwong, K. K., Hobson, J. A. & Rosen, B. R. (1996) Human brain activation during REM sleep detected by fMRI. *Society for Neuroscience Abstracts* 22:690. [aJAH]

Sutton, J. P., Caplan, J. B., Breiter, H. C., Huang-Hellinger, F. R., Kwong, K. K., McCormack, G., Hobson, J. A., Makris, N. & Rosen, B. R. (1997) Functional MRI study of human brain activity during NREM sleep. *Society for Neuroscience Abstracts* 23:21. [aJAH]

Sutton, J. P. & Hobson, J. A. (1994) State-dependent sequencing and learning. In: *Computation in neurons and neural systems,* ed. F. H. Eeckman. Kluwer. [aJAH]

Sutton, J. P., Holmes, J., Caplan, J., Rudnick, L., Kwong, K. K., Breiter, H. C., Huang-Hellinger, F. R., McCormack, G., Hobson, J. A., Makris, N., van der Gaag, C. & Rosen, B. R. (1998) Investigation of human sleep using simultaneous fMRI and polysomnography. *Sleep 21* (Supplement):102. [aJAH]

Sutton, J. P., Rittenhouse, C. D., Pace-Schott, E. F., Merritt, M., Stickgold, R. & Hobson, J. A. (1994a) Emotion and visual imagery in dream reports: A narrative graphing approach. *Consciousness and Cognition* 3:89–100. [arJAH]

Sutton, J. P., Rittenhouse, C. D., Pace-Schott, E. F., Stickgold, R. & Hobson, J. A. (1994b) Graphing and quantifying discontinuity in narrative dream reports. *Consciousness and Cognition* 3:61–88. [arJAH]

Sweeny, J. A., Minton, M. A., Kwee, S., Wiseman, M. B., Brown, D. L., Rosenberg, D. R. & Carl, J. R. (1996) Positive emission tomography study of voluntary saccadic eye movements and spatial working memory. *Journal of Neurophysiology* 75:454–68. [aJAH]

Swerdlow, N. R. & Geyer, M. A. (1998) Using an animal model of deficient sensorimotor gating to study the pathophysiology and new treatments of schizophrenia. *Schizophrenia Bulletin* 24:285–301. [RCo]

Szymusiak, R. (1995) Magnocellular nuclei of the basal forebrain: Substrates of sleep and arousal regulation. *Sleep* 18:478–500. [arJAH, RS-P]

Szymusiak, R., Alam, M. N., Steininger, T. L. & McGinty, D. (1998) Sleep-waking discharge patterns of ventrolateral preoptic/anterior hypothalamic neurons in rats. *Brain Research* 803:178–88. [RS-P]

Tagaya, H., Takahashi, K., Yamamoto, R., Ogata, S., Shiotsuka, S., Yamashita, Y., Maki, A., Yamamoto, T., Koizumi, H., Hirasawa, H., Ikawa, M., Koyama, K., Kitamura, N. & Atsumi, Y. (1999) Hemodynamic differences between Non-REM and REM sleep – Functional mapping of the occipital cortex by 24-channel NIRS imaging. *Sleep Research Online* 2 (Supplement 1):90. [aJAH]

Tagney, J. (1973) Sleep patterns related to rearing rats in enriched and impoverished environments. *Brain Research* 53:353–61. [aRPV]

Tait, R. & Silver, R. C. (1989) Coming to terms with major negative life events. In: *Unintended thoughts,* ed. J. S. Uleman & J. A. Bargh. Guilford Press. [JBP]

Takeuchi, T., Ferrelli, A. V., Murphy, T. I., Wrong, A., Veenhof, W., Lazic, S. & Ogilvie, R D. (1999a) Should we dismiss the hypothesis "REM sleep is necessary to produce dreams"? Prediction of dream recall by EEG activity in sleep onset REM periods elicited in normal sleepers. *Sleep Research* 22 (Supplement 1):S130–31. [aJAH, RDO]

Takeuchi, T., Ogilvie, R. D., Ferrelli, A. V., Murphy, T. I., Yamamoto, Y. & Inugami, M. (1999b) Dreams are not produced without REM sleep mechanisms. Poster session presented at the Third International Congress of World Federation of Sleep Research Society, Dresden, Germany. *Sleep Research Online* 2 (Supplement 1):279. [CG, aJAH, RDO]

Tallon-Baudry, C. & Bertrand, O. (1999) Oscillatory gamma activity in humans and its role in object representation. *Trends in Cognitive Sciences* 3:151–62. [aJAH]

Tallon-Baudry, C., Bertrand, O., Delpuech, C. & Pernier, J. (1996) Stimulus specificity of phase-locked and non-phase locked 40 Hz visual responses in humans. *Journal of Neuroscience* 16:4240–49. [aJAH]

(1997) Oscillatory gamma-band (30–70 Hz) activity induced by a visual search task in humans. *Journal of Neuroscience* 17:722–34. [aJAH]

Tallon-Baudry, C., Bertrand, O., Peronnet, F. & Pernier, J. (1998) Induced gamma-band activity during the delay of a visual short-term memory task in humans. *Journal of Neuroscience* 18:4244–54. [aJAH]

Tedlock, B. (1987a) Dreaming and dream research. In: *Dreaming,* ed. B. Tedlock. Cambridge University Press. [KG]

(1987b) *Dreaming: Anthropological and psychological perspectives,* ed. B. Tedlock. Cambridge University Press. [HTH]

Tees, R. C. (1999) Penfield, Wilder. In: *The MIT encyclopedia of the cognitive sciences,* ed. R. W. Wilson & F. C. Keil. MIT Press. [rJAH]

Tehovnik, E. J., Lee, K. & Schiller, P. H. (1994) Stimulation-evoked saccades from the dorsomedial frontal cortex of the rhesus monkey following lesions of the frontal eye fields and superior colliculus. *Experimental Brain Research* 98:179–90. [aJAH]

Tessier-Lavigne, M. & Goodman, C. S. (1996) The molecular biology of axon guidance. *Science* 274:1123–33. [TJM]

Teuber, H.-L. (1955) Physiological psychology. *Annual Review of Psychology* 6:267–96. [aMS]

Thakkar, M., Portas, C. & McCarley, R. W. (1996) Chronic low amplitude electrical stimulation of the laterodorsal tegmental nucleus of freely moving cats increases REM sleep. *Brain Research* 723:223–27. [aJAH, RLy]

Thakkar, M., Strecker, R. E. & McCarley, R. W. (1997) The 5HT1A agonist 8-OH-DPAT inhibits REM-on neurons but has no effect on waking and REM-on neurons: A combined microdialysis and unit recording study. *Sleep Research* 26:52. [aJAH]

(1998) Behavioral state control through differential serotonergic inhibition in the mesopontine cholinergic nuclei: A simultaneous unit recording and microdialysis study. *Journal of Neuroscience* 18:5490–97. [aJAH]

Thase, M. E. (1998) Depression, sleep, and antidepressants. *Journal of Clinical Psychiatry* 59 (Suppl. 4):55–65. [aRPV]

Tholey, P. (1991) Applications of lucid dreaming in sports. *Lucidity* 10:431–39. [AZ]

Thomayer, J. (1897) Sur la signification de quelques rêves. *Revue Neurologique* 5:98–101. [aMS]

Thompson, D. F. & Pierce, D. R. (1999) Drug-induced nightmares. *The Annals of Pharmacotherapy* 33:93–98. [CG, arJAH]

Thompson, N. S. (1981) Toward a falsifiable theory of evolution. In: *Perspectives in ethology, vol. 4*, ed. P. P. G. Bateson & P. H. Klopfer. Plenum. [NST]

Thompson, P. J. (1991) Antidepressants and memory: A review. *Human Psychopharmacology* 6:79–90. [aRPV]

Thompson, P. J. & Trimble, M. R. (1982) Non-MAOI antidepressant drugs and cognitive functions: A review. *Psychological Medicine* 12:539–48. [aRPV]

Thorpe, M. J. (1990) *The international classification of sleep disorders: Diagnostic and coding manual.* Allen Press. [JFP]

Tighilet, B. & Lacour, M. (1996) Distribution of histaminergic axonal fibres in the vestibular nuclei of the cat. *NeuroReport* 7:873–78. [aJAH]

Tilley, A. J. (1979) Sleep learning during stage 2 and REM sleep. *Biological Psychology* 9:155–61. [aTAN]

Tilley, A. J., Brown, S., Donald, M., Ferguson, S., Piccone, J., Plasto, K. & Statham, D. (1992) Human sleep and memory processes. In: *Sleep, arousal, and performance*, ed. R. J. Broughton & R. D. Ogilvie. Birkanser. [WHM]

Tilley, A. J. & Empson, J. A. C. (1978) REM sleep and memory consolidation. *Biological Psychology* 6:292–300. [aRPV]

Timofeev, I., Contreras, D. & Amzica, F. (1996) Synaptic responsiveness of cortical and thalamic neurons during various phases of the slow oscillation in cat. *Journal of Physiology London* 494:265–78. [MSt]

Timofeev, I. & Steriade, M. (1996) Low-frequency rhythms in the thalamus of intact-cortex and decorticated cats. *Journal of Neurophysiology* 76:4152–68. [MSt]

Tolman, E. C. (1938) The determiners of behavior at a choice point. *Psychological Review* 45:1–41. [EEK]

Tononi, G. & Edelman, G. M. (1998) Consciousness and complexity. *Science* 282:1846–51. [aJAH]

Tononi, G. & Pompeiano, O. (1995) Pharmacology of the cholinergic system. In: *Pharmacology of sleep*, ed. A. Kales. Springer. [LJW]

Tononi, G., Pompeiano, O. & Cirelli, C. (1991) Suppression of desynchronized sleep through microinjection of the alpha-2 adrenergic agonist clonidine in the dorsal pontine tegmentum of the cat. *Pflügers Archiv* 418:512–18. [aJAH]

Tooby, J. (1999) The most testable concept in biology, Part I. *Human Behavior and Evolution Society Newsletter* 8(2):1–4. [LM]

Tooby, J. & Cosmides, L. (1992) The psychological foundations of culture. In: *The adapted mind: Evolutionary psychology and the generation of culture*, ed. J. Barkow, L. Cosmides & J. Tooby. Oxford University Press. [aAR]

(1995) Mapping the evolved functional organization of mind and brain. In: *The cognitive neurosciences*, ed. M. S. Gazzaniga. MIT Press. [aAR]

Toth, M. A. (1971) A new method for detecting eye movement in sleep. *Psychophysiology* 7:516–22. [rTAN, VSR]

Toussaint, M., Luthringer, R., Schaltenbrand, N., Nicholas, A., Jacqmin, A., Carelli, G., Gresser, J., Muzet, A. & Macher, J. P. (1997) Changes in EEG power density during sleep laboratory adaptation. *Sleep* 20:1201–1207. [aTAN]

Tracy, R. L. & Tracy, L. N. (1973)) Reports of mental activity from sleep stages 2 and 4. *Sleep Research* 2:125. [aTAN]

(1974) Reports of mental activity from sleep stage 2 and 4. *Perceptual and Motor Skills* 38:647–48. [aJAH]

Trampus, M., Ferri, N., Adami, M. & Ongini, E. (1993) The dopamine receptor agonists, A68930 and SKF 38393, induce arousal and suppress REM sleep in the rat. *European Journal of Pharmacology* 235:83–87. [aJAH]

Tranel, D., Bechara, A. & Damasio, A. R. (2000) Decision making and the somatic marker hypothesis. In: *The new cognitive neurosciences, second edition*, ed. M. Gazzaniga. MIT Press. [rJAH]

Traub, R. D., Whittington, M. A., Buhl, E. H., Jefferys, J. G. R. & Faulkner, H. J. (1999) On the mechanism of the gamma to beta frequency shift in neuronal oscillations induced in rat hippocampal slices by tetanic stimulation. *The Journal of Neuroscience* 19(3):1088–105. [GL]

Trinder, J. & Kramer, M. (1971) Dream recall. *American Journal of Psychiatry* 128:76–81. [MK]

Trulson, M. E. & Jacobs, B. L. (1979) Raphe unit activity in freely moving cats: Correlation with level of behavioral arousal. *Brain Research* 163:135–50. [aJAH]

Trulson, M. E. & Preussler, D. W. (1984) Dopamine-containing ventral tegmental area neurons in freely moving cats: Activity during the sleep-waking cycle and effects of stress. *Experimental Neurology* 83:367–77. [CG, JP]

Trulson, M. E., Preussler, D. W. & Howell, A. G. (1981) Activity of the substantia nigra across the sleep-wake cycle in freely moving cats. *Neuroscience Letters* 26:183–88. [arJAH]

Tufik, S., Lindsey, C. J. & Carlini, E. A. (1978) Does REM sleep deprivation induce a supersensitivity of dopaminergic receptors in the rat brain? *Pharmacology* 16:98–105. [aJAH]

Tulving, E. (1994) Organization of memory: Quo vadis? In: *The cognitive neurosciences*, ed. M. Gazzaniga. MIT Press. [aJAH]

Tulving, E., Markowitsch, H. J., Craik, F. M., Habib, R. & Houle, S. (1996) Novelty and familiarity activation in PET studies of memory encoding and retrieval. *Cerebral Cortex* 6:71–79. [aJAH]

Tulving, E. & Thompson, D. H. (1973) Encoding specificity and retrieval process in episodic memory. *Psychological Review* 80:352–73. [aJAH]

Turner, V. W. (1982) *From ritual to theatre: The human seriousness of play. Performance studies series, Volume 1.* PAJ Publications. [aJAH]

Tzavaras, A. (1967) *Contribution a l'etude de l'agnosie des physiognomies.* Thesis, Faculté de Médecine de Paris. [aMS]

Uchida, S., Takizawa, Y., Hirai, N. & Ishiguro, M. (1997) High frequency EEG oscillations in human REM sleep. *Sleep Research* 26:57. [aJAH]

Ullman, M. (1959) The adaptive significance of the dream. *The Journal of Nervous and Mental Disease* 129:144–49. [aAR]

Ungerleider, L. G. & Mishkin, M. (1982) Two cortical visual systems. In: *Analysis of visual behavior*, ed. D. J. Ingle, M. A. Goodale & J. W. Mansfield. MIT Press. [EEK]

Valldeoriola, F., Nobbe, F. A. & Tolosa, E. (1997) Treatment of behavioral disturbances in Parkinson's disease. *Journal of Neural Transmission* 51 (Supplement):175–204. [rJAH]

Valldeoriola, F., Santamaria, J., Graus, F. & Tolosa, E. (1993) Absence of REM sleep, altered NREM sleep and supranuclear horizontal gaze paralysis caused by a lesion of the pontine tegmentum. *Sleep* 16:184–88. [aRPV]

Van de Castle, R. L. (1970) *The psychology of dreaming.* General Learning Press. [aAR]

(1983) Animal figures in fantasy and dreams. In: *New perspectives on our lives with companion animals*, ed. A. H. Katcher & A. M. Beck. University of Pennsylvania Press. [aAR]

(1994) *Our dreaming mind.* Ballantine. [aRPV]

Van de Castle, R. L. & Hauri, P. (1970) Psychophysiological correlates of NREM mentation. *Psychophysiology* 7:330. [rTAN]

Van den Hout, M. A., Zijlstra, B. C. & Merckelbach, H. (1989) Selective recall of surprising visual scenes: An experimental note on Seligman and Yellen's theory of dreams. *Behavior Research and Therapy* 27:199–201. [RCo]

van der Kolk, B., Blitz, R., Burr, W., Sherry, S. & Hartmann, E. (1984) Nightmares and trauma: A comparison of nightmares after combat with lifelong nightmares in veterans. *American Journal of Psychiatry* 141:187–90. [rAR]

Vanderwolf, C. H. (1969) Hippocampal electrical activity and voluntary movement in the rat. *Electroencephalography and Clinical Neurophysiology* 26:407–18. [aRPV]

Vanderwolf, C. H. & Robinson, T. E. (1981) Reticulo-cortical activity and behavior: A critique of arousal theory and a new synthesis. *Behavioral and Brain Sciences* 4:459–514. [HTH]

Van Hilten, J. J., Weggeman, M., Van der Velde, E. A., Kerkhof, G. A., Van Dijk, J. G. & Roos, R. A. C. (1993) Sleep, excessive daytime sleepiness and fatigue in Parkinson's disease. *Journal of Neural Transmitters* [P-D Section] 5:235–44. [EKP]

van Hulzen, Z. J. M. & Coenen, A. M. L. (1979) Selective deprivation of paradoxical sleep and consolidation of shuttle-box avoidance. *Physiology and Behavior* 23:821–26. [AC, arRPV]

(1980) The pendulum technique for paradoxical sleep deprivation in rats. *Physiology and Behavior* 25:807–11. [aRPV]

(1982) Effects of paradoxical sleep deprivation on two-way avoidance acquisition. *Physiology and Behavior* 29:581–87. [WF, JP, aRPV]

(1984) Photically evoked potentials in the visual cortex following paradoxical sleep deprivation. *Physiology and Behavior* 32:557–63. [AC]

van Luijtelaar, E. L. J. M. & Coenen, A. M. L. (1985) Paradoxical sleep deprivation and the immobility response in the rat: Effects of desipramine and phentolamine. *Sleep* 8:49–55. [AC]

(1986) Electrophysiological evaluation of three paradoxical sleep deprivation techniques in rats. *Physiology and Behavior* 36:603–609. [aRPV]

Vanni-Mercier, G. & Debilly, G. (1998) A key role for the caudoventral pontine tegmentum in the simultaneous generation of eye saccades in bursts and associated ponto-geniculo-occipital waves during paradoxical sleep in the cat. *Neuroscience* 86:571–85. [aJAH]

Vanni-Mercier, G., Debilly, G., Lin, J. S. & Pelisson, D. (1996) The caudoventral pontine tegmentum is involved in the generation of high velocity eye saccades in bursts during paradoxical sleep in the cat. *Neuroscience Letters* 213:127–31. [aJAH]

Vanni-Mercier, G., Pelisson, D., Goffart, L., Sakai, K. & Jouvet, M. (1994) Eye saccade dynamics during paradoxical sleep in the cat. *European Journal of Neuroscience* 6:1298–306. [aJAH]

Vanni-Mercier, G., Sakai, K. & Jouvet, M. (1984) Waking-state specific neurons in the caudal hypothalamus of the cat. *Comptes Rendus de l' Academie des Sciences* 298:195–200. [CG, aJAH, EKP]

Vanni-Mercier, G., Sakai, K. & Lin, J. S. (1989) Mapping of the cholinoceptive brainstem structures responsible for the generation of paradoxical sleep in the cat. *Archives Italiennes de Biologie* 127:133–64. [aJAH]

Van Stegeren, A. H., Everaerd, W., Cahill, L., McGaugh, J. L. & Gooren, L. J. (1998) Memory for emotional events: Differential effects of centrally versus peripherally acting beta-blocking agents. *Psychopharmacology* 138:305–10. [JB]

Van Sweden, B., Van Dijk, J. C. & Cackcbcke, J. F. V. (1994) Auditory information procession in sleep: Late cortical potentials in an oddball paradigm. *Neuropsychobiology* 29:152–56. [aTAN]

Van Twyver, H. & Allison, T. (1974) Sleep in the armadillo dasypus novemcinctus at moderate and low ambient temperatures. *Brain and Behavioral Evolution* 2:107–20. [JMS]

Vasar, V., Appelberg, B., Rimon, R. & Selvaratnam, J. (1994) The effect of fluoxetine on sleep: A longitudinal, double-blind polysomnographic study of healthy volunteers. *International Clinical Psychopharmacology* 9:203–206. [aRPV]

Velazquez-Moctezuma, J., Gillin, J. C. & Shiromani, P. J. (1989) Effect of specific M1, M2 muscarinic receptor agonists on REM sleep generation. *Brain Research* 503:128–31. [aJAH]

Velazquez-Moctezuma, J., Shaluta, M., Gillin, J. C. & Shiromani, P. J. (1991) Cholinergic antagonists and REM sleep generation. *Brain Research* 543:175–79. [aJAH]

Venkatakrishna-Bhatt, H., Bures, J. & Buresova, O. (1978) Differential effects of paradoxical sleep deprivation on acquisition and retrieval of conditioned taste aversion in rats. *Physiology and Behavior* 20:101–108. [aRPV]

Verdone, P. (1965) Temporal reference of manifest dream content. *Perceptual and Motor Skills* 20:1253–68. [aTAN]

Vertes, R. P. (1979) Brain stem gigantocellular neurons: Patterns of activity during behavior and sleep in the freely moving rat. *Journal of Neurophysiology* 42:214–28. [aRPV]

(1981) An analysis of ascending brain stem systems involved in hippocampal synchronization and desynchronization. *Journal of Neurophysiology* 46:1140–59. [CG, aRPV]

(1984) Brainstem control of the events of REM sleep. *Progress in Neurobiology* 22:241–88. [aJAH, aRPV]

(1986a) Brainstem modulation of the hippocampus. Anatomy, physiology and significance. In: *The hippocampus, vol. 4,* ed. R. L. Isaacson & K. H. Pribram. Plenum Press. [aRPV]

(1986b) A life-sustaining function for REM sleep: A theory. *Neuroscience and Biobehavioral Reviews* 10:371–76. [arRPV]

(1988) Brainstem afferents to the basal forebrain in the rat. *Neuroscience* 24:907–35. [aRPV]

(1990) Brainstem mechanisms of slow-wave sleep and REM sleep. In: *Brainstem mechanisms of behavior,* ed. W. R. Klemm & R. P. Vertes. Wiley. [aRPV]

(1992) PHA-L analysis of projections from the supramammillary nucleus in the rat. *The Journal of Comparative Neurology* 326:595–622. [aRPV]

(1995) Memory consolidation in REM sleep: Dream on. *Sleep Research Society Bulletin* 1:27–32. [aRPV]

(1996) Reply to William Fishbein's follow-up commentary on the issue of memory consolidation and REM sleep. *Sleep Research Society Bulletin* 2(4):56–57. [WF]

Vertes, R. P. & Kocsis, B. (1997) Brainstem-diencephalo-septohippocampal systems controlling the theta rhythm of the hippocampus. *Neuroscience* 81(4):893–926. [GL, aRPV]

Vertes, R. P. & Martin, G. F. (1988) Autoradiographic analysis of ascending projections from the pontine and mesencephalic reticular formation and the median raphe nucleus in the rat. *The Journal of Comparative Neurology* 275:511–41. [aRPV]

Vertes, R. P. & Perry, G. W. (1993) Sudden infant death syndrome: A theory. *Neuroscience and Biobehavioral Reviews* 17:305–12. [rRPV]

Violani, C., Bertini, M., Di Stefano, L. & Zoccolotti, P. (1983) Dream recall as a function of patterns of hemispheric activation. *Sleep Research* 12:190. [aTAN]

Vogel, G. W. (1975) A review of REM sleep deprivation. *Archives of General Psychiatry* 32:749–61. [aJAH, aRPV]

(1978a) An alternative view of the neurobiology of dreaming. *American Journal of Psychiatry* 135:1531–35. [GWD, CG, AK, aTAN, aMS, GWV]

(1978b) Sleep-onset mentation. In: *The mind in sleep,* ed. A. M. Arkin, J. S. Antrobus & S. J. Ellman. Erlbaum. [aTAN]

(1979) A motivational function of REM sleep. In: *The functions of sleep,* ed. R. Drucker-Collin, M. Shkurovich & M. Sterman. Academic Press. [HS]

(1983) Evidence for REM sleep deprivation as the mechanism of action of antidepressant drugs. *Progress in Neuropsychopharmacology and Biological Psychiatry* 7:343–49. [aRPV]

(1991) Sleep-onset mentation. In: *The mind in sleep: Psychology and psychophysiology, 2nd edition,* ed. S. J. Ellman & J. S. Antrobus. Wiley. [RCo, aJAH, aTAN, GWV]

(1993) Incorporation into dreams. In: *Encyclopedia of sleep and dreaming,* ed. M. A. Carskadon. Macmillan. [aAR]

Vogel, G. W., Barrowclough, B. & Giesler, D. (1972) Limited discriminability of REM and sleep onset reports and its psychiatric implications. *Archives of General Psychiatry* 26:449–55. [aMS]

Vogel, G. W., Buffenstein, A., Mintner, K. & Hennessey, A. (1990) Drug effects on REM sleep and on endogenous depression. *Neuroscience and Biobehavioral Reviews* 14:49–63. [aJAH, aRPV, LJW]

Vogel, G. W., Foulkes, D. & Trosman, H. (1966) Ego functions and dreaming during sleep onset. *Archives of General Psychiatry* 14:238–48. [aTAN]

Vogel, G. W., Thurmond, A., Gibbons, P., Sloan, K., Boyd, M. & Walker, M. (1975) REM sleep reduction effects on depressed syndromes. *Archives of General Psychiatry* 32:765–77. [RC]

von der Malsburg, C. (1999) The what and why of binding: The modeler's perspective. *Neuron* 24:95–104. [TJM]

Wagenaar, W. A. (1986) My memory: A study of autobiographical memory over six years. *Cognitive Psychology* 18:225–52. [PC]

(1994) Is memory self-serving? In: *The remembering self: Construction and accuracy in the self-narrative,* ed. U. Neisser & R. Fivush. Cambridge University Press. [PC]

Wagner, A. D., Schacter, D. C., Rotte, M., Koutstaal, W., Maril, A., Dale, A. M., Rosen, B. R. & Buckner, R. L. (1998) Building memories: Remembering and forgetting of verbal experiences as predicted by brain activity. *Science* 281:1188–91. [aJAH]

Wainer, B. H. & Mesulum, M. M. (1990) Ascending cholinergic pathways in the rat brain. In: *Brain cholinergic systems,* ed. M. Steriade & D. Biesold. Oxford University Press. [aJAH]

Walsh, T. J., Herzog, C. D., Gandhi, C., Stackman, R. W. & Wiley, R. G. (1996) Injection of IgG 192-saporin into the medial septum produces cholinergic hypofunction and dose-dependent working memory deficits. *Brain Research* 726:69–79. [aRPV]

Wang, X. J. (1999) Synaptic basis of cortical persistent activity: The importance of NMDA receptors to working memory. *Journal of Neuroscience* 19(21):9587–603. [RWG]

Wapner, W., Judd, T. & Gardner, H. (1978) Visual agnosia in an artist. *Cortex* 14:343–64. [aMS]

Warburton, D. M. (1979) Neurochemical basis of consciousness. In: *Chemical influences on behaviour,* ed. K. Brown & S. Cooper. Academic Press. [EKP]

Ware, J. C., Brown, F. W., Moorad, P. J., Jr., Pittard, J. T. & Cobert, B. (1989) Effects on sleep: A double blind study comparing trimipramine to imipramine in depressed insomnia patients. *Sleep* 12:537–49. [aRPV]

Warot, D., Berlin, I., Patat, A., Durrieu, G., Zieleniuk, I. & Puech, A. J. (1996) Effects of befloxatone, a reversible selective monoamine oxidase-A inhibitor on psychomotor function and memory in healthy subjects. *Journal of Clinical Pharmacology* 36:942–50. [aRPV]

Waterman, D. (1992) *Rapid eye movement sleep and dreaming. Studies of age and activation.* Doctoral dissertation, University of Amsterdam. [arTAN]

Waterman, D., Elton, M. & Kenemans, J. L. (1993) Methodological issues affecting the collection of dreams. *Journal of Sleep Research* 2:8–12. [aJAH, aTAN]

Watson, E., Bakos, L., Compton, P. & Gawin, F. (1992) Cocaine use and withdrawal: The effect on sleep and mood. *American Journal of Drug and Alcohol Abuse* 18:21–28. [rJAH]

Webster, H. H. & Jones, B. E. (1988) Neurotoxic lesions of the dorsolateral pontomesencephalic tegmentum-cholinergic cell area in the cat. II: Effects on sleep-waking states. *Brain Research* 458:285–302. [aJAH]

*Webster's third new international dictionary* (1993) Unabridged edition, ed. P. B. Gove. Merriam Webster. [JFP]

Wehr, T. A. (1992) Brain warming function for REM sleep. *Neuroscience and Biobehavioral Reviews* 16:372–97. [GL]

Weinberger, D. R. (1995) Neurodevelopmental perspectives on schizophrenia. In:

*Pscyhopharmacology: The fourth generation of progress*, ed. F. E. Bloom. Raven Press. [aJAH]

Weinstein, L. N., Schwartz, D. G. & Arkin, A. M. (1991) Qualitative aspects of sleep mentation. In: *The mind in sleep*, ed. S. J. Ellman & J. S. Antrobus. Wiley. [aJAH]

Weinstein, L. N., Schwartz, D. G. & Ellman, S. J. (1991) Sleep mentation as affected by REM deprivation: A new look. In: *The mind in sleep*, ed. S. J. Ellman & J. S. Antrobus. Wiley. [aTAN]

Weiss, T. & Adey, W. R. (1965) Excitability changes during paradoxical sleep in the rat. *Experientia* 21:292–93. [aTAN]

Weisz, R. & Foulkes, D. (1970) Home and laboratory dreams collected under uniform sampling conditions. *Psychophysiology* 6:588–96. [aJAH, rAR]

Werth, E., Achermann, P. & Borbély, A. A. (1996) Brain topography of the human sleep EEG: Antero-posterior shifts of spectral power. *NeuroReport* 8:123–27. [AAB]

(1997) Fronto-occipital EEG power gradients in human sleep. *Journal of Sleep Research* 6:102–12. http://www.swetsnet.nl/fulltexts/8520014.ipdirect. 749077.2017033.pdf [AAB]

Wesensten, N. J. & Badia, P. (1988) The p300 component in sleep. *Physiology and Behavior* 44:215–20. [aTAN]

Westlake, P. R. (1970) The possibilities of neural holographic process within the brain. *Kybernetik* 7:129–53. [TJM]

Whitman, R., Kramer, M. & Baldridge, B. (1963a) Which dream does the patient tell? *Archives of General Psychiatry* 8:277–82. [MK]

(1963b) Experimental study of supervision of psychotherapy. *Archives of General Psychiatry* 9:529–35. [MK]

Whitman, R., Pierce, C., Maas, J. & Baldridge, B. (1962) The dreams of the experimental subject. *Journal of Nervous and Mental Disease* 134:431–39. [aJAH]

Whitty, C. & Lewin, W. (1957) Vivid day-dreaming: An unusual form of confusion following anterior cingulectomy. *Brain* 80:72–76. [aMS]

Wichlinski, L. J. (1996) Possible involvement of an endogenous benzodiazepine receptor ligand of the inverse agonist type in the regulation of rapid-eye movement (REM) sleep: An hypothesis. *Progress in Neuro-Psychopharmacology and Biological Psychiatry* 20:1–44. [LJW]

Wiener, N. (1921) A new theory of measurement: A study in the logic of mathematics. *Proceedings of the London Mathematical Society* 19:181–205. [TJM]

Wilbrand, H. (1887) *Die Seelenblindheit als Herderscheinung und ihre Beziehung zur Alexie und Agraphie*. Bergmann. [aMS]

(1892/1996) Ein Fall von Seelenblindheit und Hemianopsie mit Sectionsbefund. *Deutsche Zeitschrift fur Nervenheilkunde* 2:361–87. (1996 English edition, trans. M. Solms, K. Kaplan-Solms & J. Brown. In: *Classic cases in neuropsychology*, ed. C. Code, C.-W. Wallesch, Y. Joannette & A. Lecours. Erlbaum). [aMS]

Williams, J. A., Comisarow, J., Day, J., Fibinger, H. C. & Reiner, P. B. (1994) State-dependent release of acetylcholine in rat thalamus measured by microdialysis. *The Journal of Neuroscience* 14:5236–42. [aJAH]

Williams, J. A., Merritt, J., Rittenhouse, C. & Hobson, J. A. (1992) Bizarreness in dreams and fantasies: Implications for the activation-synthesis hypothesis. *Consciousness and Cognition* 1:172–85. [arJAH]

Williams, J. A., Vincent, S. R. & Reiner, P. B. (1997) Nitric oxide production in rat thalamus changes with behavioral state, local depolarization and brainstem stimulation. *The Journal of Neuroscience* 17:420–27. [aJAH]

Williams, S. L., Kinney, P. J. & Falbo, J. (1989) Generalization of therapeutic changes in agoraphobia: The role of perceived self-efficacy. *Journal of Consulting and Clinical Psychology* 57:436–42. [JBP]

Williams, S. L., Kinney, P. J., Harap, S. T. & Liebmann, M. (1997) Thoughts of agoraphobic people during scary tasks. *Journal of Abnormal Psychology* 106:511–20. [JBP]

Williamson, D. E., Dahl, R. E., Birmaher, B., Goetz, R. R., Nelson, B. & Ryan, N. D. (1995) Stressful life events and EEG sleep in depressed and normal control adolescents. *Biological Psychiatry* 37:859–65. [MBl, aAR]

Wilmer, H. A. (1996) The healing nightmare: War dreams of Vietnam veterans. In: *Trauma and dreams*, ed. D. Barrett. Harvard University Press. [arAR]

Wilson, C. L., James, M. L., Behnke, E. J., Fried, I., Bragin, A., Simmons, G., Mahan, C., Engel, J. & Maidment, N. T. (1997) Direct measures of extracellular serotonin change in the human forebrain during waking. *Society for Neuroscience Abstracts* 23:2130. [arJAH]

Wilson, M. A. & McNaughton, B. L. (1994) Reactivation of hippocampal ensemble memories during sleep. *Science* 265:676–79. [MBl, rJAH, ARM, JP, MSt, arRPV]

Winget, C. & Kramer, M. (1979) *Dimensions of the dream*. University of Florida Press. [MK]

Winget, C., Kramer, M. & Whitman, R. (1972) Dreams and demography. *Canadian Psychiatric Association Journal* 17:203–208. [MK]

Winson, J. (1972) Interspecies differences in the occurrences of theta. *Behavioral Biology* 7:479–87. [aRPV]

(1978) Loss of hippocampal theta rhythm results in spatial memory deficit in the rat. *Science* 201:160–63. [aRPV]

(1985) *Brain and psyche*. Doubleday. [aRPV]

(1990) The meaning of dreams. *Scientific American* 262:86–96. [JAB, BEJ, aAR, aRPV]

(1993) The biology and function of rapid eye movement sleep. *Current Opinion in Neurobiology* 3:243–48. [RCo, TAN, aAR, aRPV]

Witkin, H. A. (1969) Influencing dream content. In: *Dream psychology and the new biology of dreaming*, ed. M. Kramer, R. M. Whitman, B. J. Baldridge & P. H. Ornstein. C. C. Thomas. [aJAH]

Witkin, H. A. & Lewis, H. B. (1967) Presleep experiences and dreams. In: *Experimental studies of dreaming*, ed. H. A. Witkin & H. B. Lewis. Random House. [aJAH]

Wolfe, J. & Schlesinger, L. K. (1997) Performance of PTSD patients on standard tests of memory. Implications for trauma. *Annals of the New York Academy of Sciences* 821:208–18. [TAN]

Wollman, M. C. & Antrobus, J. S. (1986) Sleeping and waking thought: Effects of external stimulation. *Sleep* 9:438–48. [aJAH]

(1987) Cortical arousal and mentation in sleeping and waking subjects. *Brain and Cognition* 6:334–46. [aTAN]

Wolpert, E. A. (1960) Studies in psychophysiology of dreams II. An electromyographic study of dreaming. *Archives of General Psychiatry* 2:121–31. [aTAN]

Wolpert, E. A. & Trosman, H. (1958) Studies in psychophysiology of dreams I. Experimental evocation of sequential dream episodes. *American Association Archives of Neurology and Psychiatry* 79:603–606. [aJAH, aTAN]

Wood, J. M., Sebba, D. & Domino, G. (1989) Do creative people have more bizarre dreams? A reconsideration. *Imagination, Cognition and Personality* 9:3–16. [aTAN]

Woolf, N. J. (1996) Global and serial neurons form a hierarchically arranged interface proposed to underlie memory and cognition. *Neuroscience* 74:625–51. [aJAH]

(1997) A possible role for cholinergic neurons of the basal forebrain and pontomesencephalon in consciousness. *Consciousness and Cognition* 6:574–96. [aJAH]

Wrangham, R. & Peterson, D. (1996) *Demonic males. Apes and the origins of human violence*. Houghton Mifflin. [aAR]

Wu, J. C., Buchsbaum, M. S., Gillin, J. C., Tang, C., Cadwell, S., Weigand, M., Najafi, A., Klein, E., Hazen, J. K. & Bunney, W. E., Jr. (1999) Prediction of antidepressant effects of sleep deprivation by metabolic rates in the ventral anterior cingulate and the medial prefrontal cortex. *American Journal of Psychiatry* 156:1149–58. [EFP-S]

Wu, J. C. & Bunney, W. E. (1990) The biological basis of an antidepressant response to sleep deprivation and relapse: Review and hypothesis. *American Journal of Psychiatry* 147(1):14–21. [GL]

Wurtz, R. H. & Munoz, D. P. (1994) Role of monkey superior colliculus in control of saccades and fixation. In: *The cognitive neurosciences*, ed. M. Gazzaniga. MIT Press. [aJAH]

Wyatt, R. J., Fram, D. H., Buchfinder, R. & Snyder, F. (1971a) Treatment of intractable narcolepsy with a monoamine oxidase inhibitor. *The New England Journal of Medicine* 285:987–91. [aRPV]

Wyatt, R. J., Fram, D. H., Kupfer, D. J. & Snyder, F. (1971b) Total prolonged drug-induced REM sleep suppression in anxious-depressed patients. *Archives of General Psychiatry* 24:145–55. [aRPV]

Wyatt, R. J., Kupfre, D. J., Scott, J., Robinson, D. S. & Snyder, F. (1969) Longitudinal studies of the effect of monoamine oxidase inhibitors on sleep in man. *Psychopharmacologia* 15:236–44. [aRPV]

Xi, M. C., Morales, F. R. & Chase, M. H. (1997) GABAergic synaptic transmission in the nucleus pontis oralis: A mechanism controlling the generation of active sleep. *Society for Neuroscience Abstracts* 23:1066. [aJAH]

Xiang, Z., Huguenard, J. R. & Prince, D. A. (1998) Cholinergic switching within neocortical inhibitory networks. *Science* 281(5379):985–88. [RWG]

Xie, Z., Stickgold, R. A., Pace-Schott, E. F. & Hobson, J. A. (1996) Visual discrimination learning task increases REM sleep. *Society for Neuroscience Abstracts* 22:359. [aJAH]

Yaguez, L., Nagel, D., Hoffman, H., Canavan, A. G., Wist, E. & Homberg, V. (1998) A mental route to motor learning: Improving trajectorial kinematics through imagery training. *Behavioural Brain Research* 90:95–106. [aAR]

Yamamoto, K., Mamelak, A. N. & Quattrochi, J. J. (1990a) A cholinoceptive desynchronized sleep induction zone in the anterdorsal pontine tegmentum: Locus of the sensitive region. *Neuroscience* 39:279–93. [aJAH]

Yamamoto, K., Mamelak, A. N., Quattrochi, J. J. & Hobson, J. A. (1990b) A cholinoceptive desynchronized sleep induction zone in the anterodorsal pontine tegmentum: Spontaneous and drug-induced neuronal activity. *Neuroscience* 39:295–304. [aJAH]

Yamuy, J., Fung, S. J., Xi, M., Morales, F. R. & Chase, M. H. (1999) Hypoglossal motoneurons are postsynaptically inhibited during carbachol-induced rapid eye movement sleep. *Neuroscience* 94:11–15. [aJAH]

Yamuy, J., Morales, F. R. & Chase, M. H. (1995) Induction of rapid eye movement sleep by the microinjection of nerve growth factor into the pontine reticular formation of the cat. *Neuroscience* 66(1):9–13. [aJAH, aTAN]

Yaroush, R., Sullivan, M. J. & Ekstrand, B. R. (1971) Effects of sleep on memory: II. Differential effect of the first and second half of the night. *Journal of Experimental Psychology* 88:361–66. [JB, rRPV]

Youngblood, B. D., Zhou, J., Smagin, G. N., Ryan, D. H. & Harris, R. B. S. (1997) Sleep deprivation by the "flower pot" technique and spatial reference memory. *Physiology and Behavior* 61:249–56. [aRPV]

Yue, G. & Cole, C. J. (1992) Strength increases from the motor program: Comparison of training with maximal voluntary and imagined muscle contractions. *Journal of Neurophysiology* 67:1114–23. [aAR]

Zadra, A. L. (1996) Recurrent dreams and their relation to life events and well-being. In: *Trauma and dreams,* ed. D. Barrett. Harvard University Press. [rAR, AZ]

Zadra, A. L. & Nielsen, T. A. (1997) Typical dreams: A comparison of 1958 versus 1996 student samples. *Sleep Research* 26:280. [AG]

(1999) The 55 typical dreams questionnaire: Consistencies across student samples. *Sleep* 22 (Suppl. 1):S175. [AG]

Zadra, A. L., Nielsen, T. A. & Donderi, D. C. (1998) Prevalence of auditory, olfactory, and gustatory experiences in home dreams. *Perceptual and Motor Skills* 87(3):819–26. [aJAH, aAR]

Zadra, A. L., O'Brien, S. A. & Donderi, D. C. (1997–1998) Dream content, dream recurrence and well-being: A replication with a younger sample. *Imagination, Cognition and Personality* 17(4):293–311. [aAR, AZ]

Zadra, A. L. & Pihl, R. O. (1997) Lucid dreaming as a treatment for recurrent nightmares. *Psychotherapy and Psychosomatics* 66(1):50–55. [aAR, AZ]

Zagrodzka, J., Hedberg, C. E., Mann, G. L. & Morrison, A. R. (1998) Contrasting expressions of aggressive behavior released by lesions of the central nucleus of the amygdala during wakefulness and rapid eye movement sleep without atonia. *Behavioral Neuroscience* 112:589–602. [ARM]

Zeki, S. (1993) *A vision of the brain.* Blackwell. [aMS]

Zepelin, H. (1993) Mammals. In: *Encyclopedia of sleep and dreaming,* ed. M. A. Carskadon. Macmillan. [rAR]

(1994) Mammalian sleep. In: *Principles and practice of sleep medicine,* ed. M. H. Kryger, T. Roth & W. C. Dement. W. B. Saunders. [JMS]

Zhang, J., Obal, F., Zheng, T., Fang, J., Taishi, P. & Krueger, J. M. (1999) Intrapreoptic microinjection of GHRH or its antagonist alters sleep in rats. *Journal of Neuroscience* 19:2187–94. [aJAH]

Zhou, W. & King, W. M. (1996) Binocular eye movements not coordinated during REM sleep. *Experimental Brain Research* 117:153–60. [aJAH]

Zimmerman, J. T., Stoyva, J. M. & Reite, M. L. (1978) Spatially re-arranged vision and REM sleep: A lack of effect. *Biological Psychiatry* 13:301–16. [aRPV]

Zimmerman, W. B. (1970) Sleep mentation and auditory awakening thresholds. *Psychophysiology* 6:540–49. [IF, aJAH, TLK, aTAN, MO, EFP-S, aMS]

Zubenko, M. D., Moossy, J., Martinez, J., Rao, G., Rosen, J. & Kopp, V. (1991) Neuropathological and neurochemical correlates of psychosis in primary dementia. *Archives of Neurology* 48:619–24. [EKP]

# Postscript: Recent findings on the neurobiology of sleep and dreaming

**Edward F. Pace-Schott**

*Harvard Medical School*

## I. Overview

It is important that persons seeking to theorize on the neural bases of dreaming be grounded in the most current findings on the neurobiology of sleep from the level of its molecular and cellular neurophysiology through the macroscopic regional activity patterns that more directly inform the study of its neuropsychology and phenomenology. In this volume, Hobson et al. have provided a primer on these neurobiological topics with a focus on the REM-NREM cycle and, in less detail, on the sleep-wake cycle. Hobson et al. (sections 3.1 & 3.2) review literature postdating the extensive reviews provided by Hobson and Steriade (1986) and Steriade and McCarley (1990) and approximately predating the year 2000. Far more extensive reviews focusing on specific neurochemical systems and anatomical networks can be found in the contributions to two recent books edited by Lydic and Baghdoyan (1999) and Mallick and Inoue (1999) as well as in the third edition of Kryger et al. (2000). An overview of more recent (approximately year 2000) literature can be found in Jones (2000), Pace-Schott and Hobson (2002), and Saper (2000). The current section is intended to briefly review the most recent (2000–2001) literature on the neurobiology of sleep most relevant to dream science in order to maximize the reference value of this volume for the contemporary (mid-2002) student of this discipline.

## II. New findings on the cellular neurophysiology of sleep

Although the prominence of cholinergic and aminergic neuronal populations in the control of the REM-NREM cycle is well accepted, the intricate modulation of the physiological components of these cardinal sleep stages by a wide variety of neurotransmitter substances continues to be investigated, with many new findings being recently reported. In the following section, we will briefly summarize some of these findings and refer the reader to major reviews of each neurochemical system in Lydic and Baghdoyan (1999), Mallick and Inoue (1999), and other recent publications.

### A. Reciprocal interaction between brainstem aminergic and cholinergic populations

An updated model of the reciprocal interaction model of the REM/NREM sleep cycle is provided in the Hobson et al. target article, section 3.2 (see Figures 3B and 4). Recent findings are supportive of this model and emphasize that, in the initiation and augmentation of REM, intermediate synaptic steps may intervene at both the level of REM-on mesopontine neurons and REM-off pontine aminergic nuclei. Evidence for a glutamatergic-cholinergic self-excitatory loop in the PPT/LDT is reviewed in Hobson et al. section 3.2. Luppi et al. (1999b) reviews evidence that GABAergic inputs to both the dorsal raphe nucleus and the noradrenergic locus coeruleus may be the final synaptic step responsible for the shutting down of serotonergic and noradrenergic REM-off cells during REM sleep. Moreover, there is now evidence for a GABAergic "switch" between waking and REM located in the nucleus pontis oralis (Xi et al. 2001).

The roles of aminergic and cholinergic neurotransmitters in the REM/NREM and sleep/wake cycles continue to be actively investigated (e.g., Sakai & Crochet 2001; Sakai et al. 2001; Kubin 2001). A review of recent findings is provided in Hobson et al. (target article), Jones (2000), and Pace-Schott and Hobson (2002), with greater detail available on the following: cholinergic systems (Semba 1999; Capece et al. 1999); noradrenergic systems (Luppi et al. 1999a); and serotonergic systems (Jacobs & Fornal 1999).

### B. Neurotransmitter and neuromodulatory systems

**1. Dopaminergic systems.** Given the key role assigned to dopamine (DA) in reward-based (Solms, target article) and psychotomimetic (Gottesmann, commentary) theories of dreaming, elucidation of this important CNS catecholamine's role in sleep will have major implications for theories of dream biology. There have not been comprehensive recent reviews of DA in sleep and dreaming; however the beginnings of such can be found in Gottesmann (1999), Hobson et al. (target: section 3.2.3, response: section HR4.3.2), Hobson and Pace-Schott (1999), and Rye (1997).

Recently, there has been increased interest in DA as evidenced by an entire symposium on DA in sleep at the 2001 Association of Professional Sleep Societies meetings at which D. M. Edgar, E. Mignot, D. B. Rye, and E. R. De-Long presented provocative recent findings on this topic. Renewed interest in DA's role in sleep arises, in part, from the dopaminergic action of agents used to treat narcolepsy, including the newest agent modafinil (Wisor et al. 2001) as well as the efficacy of dopaminergic agents in treating sleep-related movement disorders such as restless leg syndrome (e.g., Allen & Early 2001).

Although, as is often noted, the dopaminergic neurons of the midbrain do not alter their firing rate with changes in behavioral state (see Hobson et al. section 3.2.3), GABAergic neurons in the dopaminergic ventral tegmental area (VTA) have recently been shown, compared to waking, to decrease firing during slow wave sleep and to increase firing during REM with a possible concomitant role in state-dependent changes in cortical arousal (Lee et al. 2001). Moreover, an increasing complexity of cholinergic-dopaminergic interactions is emerging from studies such as those showing a muscarinic inhibition versus excitation of midbrain dopaminergic neurons (Fiorillo & Williams 2000) and regulation of striatal dopamine release by nicotinic receptors (Zhou et al. 2001). Perhaps most interesting from the point of view of sleep and dreaming is the description by Freeman et al. (2001) of a "mesothalamic" (in addition to mesostriatal, mesolimbic, and mesocortical) dopamine system whereby axon collaterals of dopaminergic neurons of the mesostriatal system synapse on motor and limbic-related thalamic nuclei, thus linking the mesostriatal dopaminergic system directly to ascending systems involved in behavioral state control.

The connection of dopaminergic neurotransmission and behavioral state is also suggested by recent receptor-ligand binding PET findings of reduced striatal dopaminergic terminals in patients with idiopathic REM behavior disorder (RBD), a sleep disorder now thought to herald Parkinsonian neurodegeneration, a hypo-dopaminergic condition (Albin et al. 2000). Because RBD is associated with vivid dreaming, this finding might be seen as evidence against a dopaminergic origin of normal dreaming as proposed by Solms (target article). However, the striatal DA system is not directly implicated in Solms's hypothesis. These findings suggest that endogenous dopamine may indeed affect the sleep/waking cycle in some ways and that the role of dopamine in sleep will be a topic of continuing interest to dream science.

**2. GABAergic systems.** The highly heterogenous sleep-related roles of GABA (Hobson et al. section 3.2.3) have recently been reviewed by Jones and Muhlethaler (1999), Jones (2000), Luppi et al. (1999a), and Pace-Schott and Hobson (2002). GABAergic inhibition plays both facilitory and inhibitory roles in both the sleep-wake and REM-NREM cycles.

For example, GABAergic inhibition (e.g., by the reticular nucleus of the thalamus) plays a central role in generating the intrinsic thalamocortical oscillations of slow wave sleep hypothesized to underlie NREM sleep mentation, sleep maintenance, and neuroplastic events associated with learning and memory (Steriade 2000, 2001, this volume-commentary). Similarly, GABA, along with the neuropeptide galanin, plays a key role in sleep onset and maintenance via the suppression of wake-active regions of the posterior hypothalamus and brainstem [for recent reviews, see Salin Pascual et al. (2001), Saper et al. (2001), Shiromani et al. (1999), and Szymusiak et al. (2001)]. Moreover, most hypnotic drugs enhance GABAergic neurotransmission via their action at GABA$_A$ receptors (Lancel 1999).

In the case of REM sleep, GABAergic inhibition of aminergic REM-off populations has been recently reviewed for both the locus coeruleus (Luppi et al. 1999b; Mallick et al.

1999) and the dorsal raphe (Luppi et al. 1999b). Recent findings in the locus coeruleus suggest that local GABAergic populations are excited by cholinergic afferents and, in turn, these inhibit the noradrenergic REM-off cells (Mallick et al. 2001).

Paradoxically, however, GABA also can suppress cholinergic mesopontine REM-on networks [reviewed in Datta (1999) and Rye (1997)], and in humans, GABAergic hypnotics suppress REM sleep as well as the delta frequency oscillations of deep NREM (Lancel 1999). Moreover, in another region of the brain, the nucleus pontis oralis of the pontine brainstem, a GABAergic mechanism may promote wakefulness at the expense of REM sleep (Xi et al. 1999, 2001).

These seeming paradoxes illustratrate an important point with regard to the ubiquitous and fast-acting amino acid neurotransmitters. These substances play many highly site-specific roles in the CNS whose net results on behavioral state and cortical arousal can contrast with their activating or inhibiting effects at the microscopic level (e.g., via disinhibitory processes). As a microscopically excitatory example, the thalamocortical oscillations of NREM sleep operate at the synaptic level in part via glutamatergic mechanisms (Steriade 2000).

**3. Orexinergic systems.** Perhaps the most dramatic recent finding in sleep biology has been the discovery of a new excitatory wake-promoting neuromodulatory system originating in orexin (or hypocretin)-producing cells of the lateral hypothalamus. Since the original reports that deficits in this system underlie animal models of narcolepsy (Chemelli et al. 1999; Lin et al. 1999), a multitude of reports on its role in human as well as animal sleep have appeared. [For reviews, see Kilduff and Peyron (2000) and Siegel et al. (2001).]

In the rat, intracerebroventricularly injected orexin (Piper et al. 2000) as well as microdialysis perfusion of orexin into the basal forebrain (Thakkar et al. 2001) has been shown to increase wakefulness in a dose-dependent manner. Moreover, in the rat, the densest projection of orexinergic cells is to the locus coeruleus (Horvath et al. 1999) whose noradrenergic output favors the cortical arousal of waking but opposes the arousal associated with REM (Hobson et al., target article). Projections from the suprachiasmatic nucleus of rat to orexinergic cells of the lateral hypothalamus have recently been described suggesting that orexin may also play a role in arousal associated with the circadian regulation of the sleep-wake cycle (Abrahamson et al. 2001).

A distribution pattern of orexinergic neurons (in the perifornical, lateral, and medial hypothalamus) and their projection pattern (densest to the locus coeruleus) similar to those in animals has now been described in humans (Moore et al. 2001). Moreover, compared to normals, human narcoleptics display a large reduction in number of lateral hypothalamic orexinergic neurons (Peyron et al. 2000; Thannickal et al. 2000) as well as a CSF orexin deficiency (Nishino et al. 2000).

**4. Other neurotransmitter and neuromodulatory systems.** Although by no means less important to the REM-NREM and sleep-wake cycles, recent findings on other neurotransmitter and neuromodulatory systems in the past year have been less dramatic than the discovery of orexin, less voluminous than, say, reports on GABA, or less relevant to the current topic, dreaming, than studies on dopamine. Recent reviews of these systems will, therefore, be cited here, interspersed with a few examples of pertinent new findings.

Many reports focus on **adenosine,** the putative endogenous somnogen whose accumulation in the brain during prolonged wakefulness may constitute a physiological basis of homeostatic sleep need (Porkka-Heiskanen et al. 1997) or "process S" (Borbely 2001). Recently, a remarkable degree of site-specificity in the sleep-promoting effect of sleep deprivation or experimentally induced adenosine accumulation has been demonstrated in the basal forebrain of the cat (Strecker et al. 2000; Porkka-Heiskanen et al. 2000).

A **histaminergic** arousal system originates in the tuberomammilary nucleus of the posterior hypothalamus and innervates the entire forebrain and brainstem regions known to be involved in behavioral state control [for recent reviews see Salin-Pascual et al. (2001), Saper (2000), Saper et al. (1997, 2001), and Shiromani et al. (1999)].

Since the discovery that **glycine,** another inhibitory neurotransmitter, regulates postsynaptic inhibition of somatic motoneurons during REM atonia (Chase et al. 1989), details of this mechanism continue to be reported [for recent reviews, see Gotttesmann (1997), Lai and Siegel (1999), and Rye (1997)]. As an example of recent findings, an important role for glycine (as well as for GABA) in the neuronal inhibition underlying the sensory blockade of REM sleep has been described by Soja et al. (2001). This afferent blocking role of glycine, thus, parallels its role in the gating of motor output during REM.

As noted above, **glutamate,** the most ubiquitous CNS excitatory neurotransmitter, may widely interact with cholinergic and cholinoceptive neurons to generate the exponential increase of mesopontine and pontine reticular activity associated with REM sleep activation [for a review, see Semba (1999)]. The glutamatergic roles in REM sleep atonia have been reviewed by Lai and Siegel (1999) and Rye (1997), and its role in the thalamocortical oscillations of NREM is described in Steriade (1999, 2000, 2001, this volume-commentary). One notable recent finding is that the Purkinje cell responsivity to glutamate is greatly reduced during NREM sleep (relative to waking) in the cerebellum (Andre & Arrighi 2001), a brain structure that is underresearched with respect to variations with behavioral state but is generating growing interest in regard to psychopathology and emotion (e.g., Schmahmann 2000).

**Nitric oxide,** a diffusable gaseous transmitter that can enhance synaptic release of neurotransmitters such as acetylcholine and enhance capillary vasodilation may play a role in maintaining the cholinergically mediated REM sleep state (Leonard & Lydic 1999; Leonard et al. 2002). For a recent review, see Leonard and Lydic (1999).

A wide variety of **neuropeptides** have been implicated in regulation of the sleep-wake and REM-NREM cycles. These include: galanin (Saper et al. 1997); orexin (see earlier discussion); vasoactive intestinal polypeptide [reviewed in Steiger and Holsboer (1997)]; and numerous hormones [reviewed in Krueger et al. (1999) and Obal and Krueger (1999)] and cytokines, some of which, like adenosine, have been proposed as endogenous somnogens [reviewed in Inoue et al. (1999) and Krueger and Fang (1999)]. As one example, corticotropin releasing factor (CRF) participates in the ponto-medullary control of REM atonia (Lai & Siegel 1999), and receptors to this substance have recently also

been identified on important arousal-related aminergic and cholinergic systems in the pontine brainstem and basal forebrain (Sauvage & Steckler 2001).

Much of the future excitement in sleep research, as in all of basic and clinical neuroscience, is likely to move well beyond the neurotransmitter and its receptors toward the roles of intracellular **second messengers** (e.g., Capece et al. 1999) as well as **gene transcription factors and other intranuclear systems** [for reviews, see Bentivoglio and Grassi-Zucconi (1999), Prospero-Garcia et al. (1999), Schibler and Tafti (1999), and Tononi and Cirelli (2001a)].

Recent examples of important molecular findings in sleep research include the following:

1. The discovery of the role of orexin in sleep regulation (see earlier discussion) arose from identification of narcolepsy-associated genes in animal models (Chemelli et al. 1999; Lin et al. 1999).

2. Molecular mechanisms by which sleep may promote the neuroplasticity underlying learning and memory have recently been proposed (Graves et al. 2001; Tononi & Cirelli 2001b).

3. Immediate early genes, such as the gene coding for *c-fos,* modulate transcription of other genes (Bentivoglio & Grassi-Zucconi 1999; Cirelli & Tononi 2000; Tononi & Cirelli 2001a) and are selectively expressed during waking in most brain neurons (Cirelli & Tononi 2000; Tononi & Cirelli, 2001a). One notable exception, the sleep-dependent expression of *c-fos* in the ventrolateral preoptic (VLPO) area of the hypothalamus, formed the basis of the discovery that this region was the hypothalamic sleep portion of a sleep-wake switch [for a review, see Saper et al. (2001)].

4. An extensive screening of the rat genome for genes selectively upregulated during sleep or waking has shown state-dependent changes in fewer than 1% of genes, a majority of these selective upregulate during wakefulness, and among these, as might be expected, are genes coding for elements of synaptic neurotransmission (Tononi & Cirelli 2001a).

5. As one recent example of important molecular findings in sleep research, binding of adenosine to A1 receptors in the basal forebrain has been shown to increase DNA binding of a specific transcription factor, NF-kB, thereby potentially extending knowledge on the physiological basis of sleep debt into the cell nucleus (Basheer et al. 2001).

## III. Selected new findings on the physiology of sleep

### A. Electrophysiology

Among the many new discoveries on the electrophysiology of sleep, the elucidation of the details of corticothalamic interactions in NREM by Steriade's group using animal models is of particular interest to the debate in this volume on the similarities versus differences between REM and NREM dreaming. This pattern has been recently described by Steriade in several extensive reviews (Steriade 1999, 2000, 2001; see also his commentary in this volume) as follows: In the cat, as presumably also in humans, disfacilitation due to decreased ascending arousal and GABAergic inhibition by thalamic reticular nucleus neurons suppresses firing of thalamocortical relay neurons following sleep onset and allows the emergence of underlying thalamocortical oscillatory rhythms. Thalamic neurons exhibit burst firing patterns first in spindle (12–14 Hz) and later in delta (1–4 Hz) frequencies as NREM deepens from Stage 2 to delta sleep. The cortex constrains these spindle- and delta-wave-generating thalamocortical bursts by a newly described slow (<1 Hz) oscillation recently also described in humans (Achermann & Borbely 1997). This slow oscillation occurs among neurons within the cortex during NREM, which takes the form of a slow (0.5–1 Hz) alternation of a quiescent hyperpolarization phase followed by intense spike trains during a depolarized phase. Steriade proposes that in NREM just prior to REM, renewed thalamic relay input to the cortex (which is normally absent during NREM) in the form of the PGO waves (which are known to precede the full complement of REM features) impinge on cortical neurons that are still in the slow oscillation mode and are thus primed to emit spike trains, which could be experienced as vivid NREM mental imagery (Steriade 2000, 2001). This could account for some aspects of the observed activation of the striate cortex in slow wave sleep (Hofle et al. 1997) and for some of the imagery ascribed to NREM mentation (e.g., Cavallero et al. 1992). This would, however, only apply to the immediate pre-REM period. Notably, this neuronal pattern presumably underlying some aspects of NREM mentation and imagery – an intracortical alternation of activity and quiescence (Steriade 2000) in the absence of subcortical activation or distinct cortical activity outside V1 (Braun et al. 1997) – differs markedly from the pattern of subcortical brainstem and limbic activation with highly selective cortical activity (and excluding V1) presumed to underlie REM sleep dreaming (reviewed in Hobson et al., target article).

The preceding focus on slow wave sleep relates to the REM-NREM mentation controversy in this volume and is not meant to minimize the importance of recent electrophysiological studies focusing on sleep in humans, for example:

1. discovery of a reliable electrophysiological correlate of homeostatic sleep drive, slow wave activity (Borbely 2001; Borbely & Achermann 2000), as well as a marker of increasing homeostatic sleep pressure during extended waking, theta/low-frequency alpha (Aeschbach et al. 2001);

2. the alpha band in REM (e.g., Cantero et al. 2000);

3. ERP studies on cognitive processing during sleep (e.g., Atienza et al. 2000);

4. EEG correlates of slow eye movements in the sleep-wake transition (De Gennaro et al. 2000);

5. localization of source generators of human sleep spindles (Anderer et al. 2001);

6. the association of gamma frequency oscillations with the brain-activated states of REM and waking and, within REM, with phasic versus tonic periods (Gross & Gotman 1999);

7. relating to the unique features of dream mentation attributable to hypofrontality in REM, a loss of synchrony in the gamma range between the frontal cortex and posterior perceptual areas in REM compared to NREM and waking has been hypothesized to contribute to the bizarreness of REM sleep dreaming (Perez-Garci et al. 2001).

Similarly, in animal models, there have been a wealth of new electrophysiological findings in addition to the work of Steriade's laboratory described earlier. For example, possible roles of the newly described rat equivalent of the feline PGO wave (Datta et al. 1998) have now been advanced (Datta 2000; see also the later discussion herein). Similarly, synchronization between two cardinal indicators of REM sleep, hippocampal theta and PGO waves, has recently been described in the cat (Karashima et al. 2001). Behavioral-state related changes in regions of the subcortex that were previously poorly studied in relation to sleep, such as the cerebellum (e.g., Andre & Arrighi 2001), are also continually emerging.

## B. Neuromodulatory effects on cognition

The neuromodulatory basis of altered dream cognition has been elucidated by some recent findings on specific neuromodulatory systems known to vary with sleep state. A recent human fMRI study by Furey et al. (2000) may offer a partial solution as to why the cholinergic enhancement of REM sleep promotes hallucinatory experience and illogical thought while at the same time, in waking, acetylcholine is essential to the ability of the brain to respond appropriately to demands placed on attention (Baxter & Chiba 1999), and ACh-deficient states promote hallucinosis (Perry et al. 1999). In a visual working memory task, these workers found that cholinergic enhancement using physostigmine increased recruitment of posterior (extrastriate) perceptual areas of putative working memory networks in the brain while, at the same time, decreasing recruitment of dorsal prefrontal "executive" working memory areas, presumably due to reduced demand, resulting from the cholinergic enhancement of the efficiency of upstream posterior areas (Furey et al. 2000). Therefore, during dreaming, a cholinergic bias might enhance extrastriate activity while diminishing frontal activity as is seen in the neuroimaging studies of REM (Braun et al. 1997, 1998; Maquet et al. 1996). In contrast, noradrenergic neurotransmission, at its minimum in REM, continues to be described as enhancing memory (e.g., Gibbs & Summers 2000) and attention (e.g., Usher et al. 1999) and as having a specific relationship to the executive areas of the prefrontal cortex (Aston-Jones et al. 2000). Recently, a tight interdependence of attention and working memory has been shown in a variety of experimental contexts (Awh & Jonides 2001; de Fockert et al. 2001). Because both of these cognitive modules are, at least in large part, subserved by prefrontal areas (Goldberg 2001), one might speculate that impairment of either of them might impact the other, further promoting the characteristic features of dreaming such as discontinuities (impaired attention) and amnesia (impaired working memory). For a recent argument for a dopaminergic contribution to normative dreaming, see Solms (2001) and Gottesmann (2000). In addition, recent studies have begun to look at specific dream effects of pharmacologically induced neuromodulatory changes via drugs such as the SSRIs (Pace-Schott et al. 2001).

## C. Circadian rhythms

Recent years have also shown some dramatic findings on the circadian control of the sleep-wake cycle. First, using a forced desynchrony protocol, it has been shown that the endogenous human circadian rhythm has a nearly 24-hr periodicity in contrast to the widely held notion that the endogenous rhythm slightly exceeds the length of the astronomical day (Czeisler et al. 1999).

Recent findings have begun to elucidate the genetic and neural bases for the circadian regulation of arousal. First, genetic control of the mammalian circadian pacemaker in cells of the suprachiasmatic nucleus (SCN) of the hypothalamus has now been described [see Reppert and Weaver (2001) for a review]. This system allows precisely timed translation and transcription of a group of identified genes utilizing feedback inhibition and promotion via their protein products [see Reppert and Weaver (2001) for full details]. Second, melanopsin has been identified as the circadian photopigment present in the retinal ganglionic cells, which are themselves now believed to be the primary circadian photoreceptor of the retina (Hannibal et al. 2002). Third, findings on the translation of this circadian signal to hypothalamic systems controlling various biological rhythms have now begun to appear (e.g., Lu et al. 2001; Aston Jones et al. 2001).

For example, projections from the suprachiasmatic nucleus to hypothalamic orexinergic cells were described earlier (Abrahamson et al. 2001). In addition, retrograde tracer studies using pseudorabies virus have shown a circuit extending from the suprachiasmatic nucleus through the dorsomedial nucleus of the hypothalamus to the locus coeruleus, whose noradrenergic output is a well-known promoter of arousal (Aston-Jones et al. 2001). Interestingly, this circuit may also recruit orexinergic neurons in the dorsomedial nucleus of the hypothalamus (Aston-Jones et al. 2001). Finally, the neuronal basis for circadian control of REM sleep propensity, described experimentally in humans by Czeisler et al. (1980), is now being elucidated in animal models (e.g., Wurts & Edgar 2000). Similarly, an ultradian rhythm in daytime arousal with a 3–4-hr. periodicity has recently been described (Chapotot et al. 2000).

### D. Sleep and neuroplasticity

The role of REM sleep in memory consolidation and other forms of neuroplasticity is controversial, and arguments can be raised opposing such a role for REM as detailed in the target article by Vertes and Eastman. Specifically, Vertes and Eastman suggest that REM deprivation (REMD) effects on learning in animal models may represent generalized stress-related short-term performance deficits resulting from the experimental REMD techniques traditionally employed by investigators. Similarly, they note that the well-documented REM suppressant effects of the three major classes of antidepressants are not associated with memory deficits in humans treated with these agents. These authors propose the alternative hypothesis that REM provides a periodic CNS stimulation during mammalian sleep, which promotes rapid recovery of the brain from the disfacilitated conditions of slow wave sleep. Nonetheless, publications in 2000–1 have continued to report empirical relationships between sleep and memory in both humans and animals [for a complete review see Peigneux et al. (2001a)], and Stickgold et al. (2001) have offered specific rebuttals to Vertes and Eastman's arguments against sleep-dependent learning. Reviews of work preceding the late 1990s can be found in Peigneux et al. (2001a), Smith (1985, 1993, 1995), and Stickgold (1998).

Since early reports on the replay during sleep of neuronal firing patterns recorded in the rat hippocampus during prior waking (Wilson & McNaughton 1994), there have been additional animal studies reporting on how sleep might promote consolidation of learning as well as the emergence of a theoretical framework for explaining how memories are consolidated via a neocortical-hippcampal dialog [Buzsáki 1996; for reviews, see Peigneux et al. (2001a, 2001b) and Stickgold (1998)]. For example, Siapas and Wilson (1998) have shown a temporal correlation of fast (200Hz) hippocampal "ripples" and cortical NREM sleep spindles suggesting transfer of hippocampally encoded short-term stores into longer-term cortical stores. Similarly, Poe et al. (2000) have shown that the experience-dependent firing of hippocampal place cells during REM sleep occurs in relationship to REM theta oscillations with a changing pattern over time suggestive of early long-term potentiation followed by later long-term depression as the putative transfer of hippocampal information to the neocortex is completed. Louie and Wilson (2001) have demonstrated reactivation of hippocampal neurons during REM occurring with the neurons firing in the same temporal sequence and over an equivalent time scale as they fired during prior waking. Steriade (2000) postulates that the alternating hyperpolarization and burst firing of neocortical neurons during NREM may play a role in neuroplasticity associated with learning and memory. Datta (2000) has shown an increase in pontine P-waves (considered a rodent equivalent of the feline PGO wave) following a period of avoidance task training as well as a correlation between P-wave density and degree of learning, suggesting that this aspect of REM sleep in the rat favors consolidation of learning in the forebrain structures targeted by such waves. Similarly, Sanford et al. (2001) have shown that fear conditioning increases the amplitude of elicited PGO waves in REM sleep, further suggesting a role for PGO waves in experience-dependent neuroplasticity. Moreover, rats exposed to rich sensorimotor experiences during waking display elevated levels of the plasticity-associated immediate-early gene, zif-268, during subsequent REM and slow wave sleep (Ribeiro et al. 1999).

Recent studies on a visual discrimination task (Karni & Sagi 1991) have shown a requirement for sleep in order for procedural learning to occur (Stickgold et al. 2000a). Subjects who were sleep deprived on the night following training, but were allowed two subsequent nights of unrestricted recovery sleep, showed no significant task improvement (Stickgold et al. 2000a). In addition, specific requirements have been shown for REM sleep (Karni et al. 1994; Stickgold et al. 2000b) and slow wave sleep (Gais et al. 2000; Stickgold et al. 2000b) on the night following training in order for next-day improvement on the task to be observed. These findings have been conceptualized in terms of an earlier theory based on animal studies (Giuditta et al. 1995) in which a two-step process involving both REM and SWS was hypothesized (Gais et al. 2000; Stickgold 1998; Stickgold et al. 2000b). Most dramatically, a recent $H_2{}^{15}O$ PET study by Maquet et al. (2000) has shown in humans, as has been shown in the rat by Louie and Wilson (2001), a reactivation during REM sleep of the same brain areas activated during previous waking performance of a cognitive task. These findings have since been extended to include demonstration of experience-dependent changes in parieto-frontal functional connectivity during REM sleep following task training (Laureys et al. 2001).

A number of recent reviews and theoretical pieces have addressed the question of how neuronal activity during sleep might favor the molecular changes leading to long-term consolidation of information acquired in waking (Graves et al. 2001; Sejnowski & Destexe 2000; Tononi & Cirelli 2001b). Moreover, recent depth electrode findings in humans have begun to examine unit activity in hippocampal and surrounding cortical areas during sleep (e.g., Staba et al. 2000), while experimental studies in humans have begun to dissect the contribution of neocortical versus hippocampal systems for information processing during sleep (Stickgold et al. 2000b).

## IV. New findings on the neuroimaging of sleep

### A. Normative sleep

Since the landmark PET neuroimaging studies of Braun et al. (1997, 1998), Maquet et al. (1996, 1997), and Nofzinger et al. (1997), additional functional neuroimaging studies have corroborated their findings. In the transition from wake to NREM, studies by Andersson et al. (1998), Bootzin et al. (1998), Buchsbaum et al. (2001), Kajimura et al. (1999), Nofzinger et al. (2000), and a recent comprehensive review of the literature by Maquet (2000) have all confirmed the previously described pattern of decreased thalamic, prefrontal, and multimodal parietal association cortex blood flow accompanying the onset and deepening of NREM sleep [see also reviews in Buchsbaum et al. (2001) and Schwartz and Maquet (2002)].

In the case of REM, the general pattern of lateral prefrontal deactivation and limbic midline activation has also been replicated (however, see discussion of [18]FDG PET findings). For example, frontal deactivation has now been described in the first fMRI study of REM sleep (Lovblad et al. 1999). Newer studies continue to report that portions of the ventromedial, limbic-related prefrontal cortices, and closely associated medial subcortex and cortex reactivate in REM following their deactivation relative to waking in NREM (Buchsbaum et al. 2001; Nofzinger et al. 1999, 2001; Smith et al. 1999; Wu et al. 1999, 2001). Nofzinger et al. (1999) have termed this area the "anterior paralimbic REM activation area." This area has recently been extensively studied as a biological indicator of responsivity to antidepressant therapy (Nofzinger et al. 1999; Wu et al. 1999, 2001). For example, patients responding to the antidepressant effect of one night's total sleep deprivation (TSD) show higher medial prefrontal activation during baseline waking and a greater decrease in medial prefrontal activation following TSD than did nonresponders (Smith et al. 1999; Wu et al. 1999, 2001). Similarly, unlike normals, depressed patients did not show an increase in medial prefrontal activation from waking to REM (Nofzinger et al. 1999) but, following successful treatment with bupropion, this relative medial prefrontal activation from wake to REM was restored via a decrease in waking medial prefrontal activity (Nofzinger et al. 2001). [For a recent review on the functions of a major component of this anterior paralimbic REM activation area, the anterior cingulate cortex, see Paus (2001).]

In addition, a recent study using near infrared spectroscopy (a new method of imaging surface cortical blood flow) has reported strong activation of the visual cortex during REM, which these authors hypothesize reflects ongoing dream imagery (Igawa et al. 2001). With regard to the search for PGO-like activity in humans, a recent $H_2^{15}O$ PET study has shown a REM-related blood flow pattern suggestive of PGO activity in the human (Peigneux et al. 2001b).

### B. Prefrontal cortical deactivation, sleep, and dreaming

A sleep-related deactivation of prefrontal cortices has emerged as a descriptive and explanatory factor in findings on sleep physiology and psychology extending from sleep onset, through the NREM-REM cycle of normal sleep, to the postsleep awakening processes. Frontal deactivation characterizes the transition from waking to NREM as measured by $H_2^{15}O$ PET (Braun et al. 1997; Maquet et al. 1997), 18FDG PET (Nofzinger et al. 2000), or quantitative EEG (Werth et al. 1997; Finelli et al. 2001; Borbely 2001). Moreover, the frontal cortex has been shown to be particularly sensitive to sleep debt independently of circadian factors (Cajochen et al. 2001). Deactivation proceeds with the deepening of NREM sleep (Hofle et al. 1997) and in $H_2^{15}O$ PET studies, is maintained in the transition from NREM to REM (Braun et al. 1997, 1998; Maquet et al. 1996). Moreover, there is now evidence of a further decoupling of the frontal cortex from the rest of the brain in the transition from NREM to REM (Perez-Garci et al. 2001). Upon awakening, the PFC lags behind other regions in achieving waking levels of activation (Balkin et al. 1999). However, the failure to find such widespread lateral frontal deactivation in REM relative to waking using [18]FDG PET (Nofzinger et al. 1997) continues to be reported (Buchsbaum et al. 2001), and future research needs to reconcile [18]FDG PET findings with the results using other neuroimaging and quantitative EEG methods.

In a striking parallel to these findings on normative sleep, neuroimaging studies have found profound effects of one night's total sleep deprivation on prefrontal areas (Drummond et al. 1999, 2000; Petiau et al. 1998; Smith et al. 1999; Thomas et al. 2001; Wu et al. 1999, 2001), which sometimes parallel measured declines in prefrontal task performance (Drummond 1999; Petiau et al. 1998; Thomas et al. 2001). Moreover, one night's acute total sleep deprivation has been shown to degrade performance on neuropsychological tasks believed to be subserved by the prefrontal cortex (Blagrove et al. 1995, 1996; Harrison & Horne 1997, 1998, 1999, 2000; Harrison et al. 2000; Herskovitch et al. 1980; Horne 1988, 1993; May & Kline 1987; Wimmer et al. 1992) although some populations may not show such deficits (Binks et al. 1999; Pace-Schott et al. 2002a, 2002b). Two very recent reviews have summarized evidence that sleep disruption results in impairments of prefrontal functioning (Beebe & Gozal 2002; Harrison & Horne 2000b). Beebe and Gozal (2002) summarize the growing body of evidence showing a link between a characteristic group of neuropsychological deficits in executive functions and physiological stressors associated with obstructive sleep apnea, which include sleep disruption as well as cardiovascular and blood metabolic changes. Harrison and Horne (2000b) provide a comprehensive review of the studies on sleep deprivation and executive deficits with a specific focus on executive skills relating to decision making.

These converging findings, along with the lack of dream effects following lesions of the frontal cortex (Doricchi & Violani 1992; Solms 1997) suggest that frontal deactivation must play a key role in any neurocognitive theory of dreaming. This holds equally true whether they are REM sleep-based theories (Hobson et al., target article; Nielsen, target article) or those based on a generalized picture of the brain in sleep (Solms 1997, and target article).

The concept of sleep as involving disengagement of prefrontal systems from posterior unimodal and multimodal associative sensory systems potentially holds great value for explaining dream phenomenology. This is especially compelling if one imagines that posterior systems contribute to powerful integrative synthetic cognitive capacities of the brain in sleep. These posterior systems, along with certain medial subcortical and prefrontal areas (see discussion later), may be capable of generating not only a virtual reality with a rich (albeit visually biased) sensory pallette but also with a capacity for generating an organizing narrative, albeit one lacking in full self-reflective awareness, executive control of much of one's behavior, selective attention, and, most importantly, lacking in full reality testing. One is *usually* not aware that one is dreaming. Although, as noted by Kahan and LaBerge (1994), there remains a large degree of here-and-now self-reflection and control in dreams, that is, one reflects on one's behavior and decides on one's actions *within the dream's reality*, what is missing, except in the case of full lucidity (a fragile and transient condition), is the awareness that one is dreaming and an awareness of a reality outside of the dream – most often, one's bedroom.

It is likely that this qualitative change in consciousness, from dreaming to waking, is subserved by the reengaged functions of certain prefrontal neural systems, possibly the more dorsolateral ones. The dorsolateral portion of the prefrontal cortex has been argued to be the terminus and executive motor output center of the "dorsal stream" of sensory processing (Goldman-Rakic et al. 2000), in which "where" or spatial information from various sensory modalities is processed [for reviews of dorsal and ventral cortical processing streams, see Mesulam (1998, 2000); for reviews of frontal executive systems, see Goldberg (2001), Fuster (1997), and Roberts et al. (1996)]. Interestingly, a posterior integrative node of this dorsal system is the temporo-parieto-occipital junction, including Brodmann's areas 39 and 40, the angular and supramarginal gyri of the inferior parietal lobe (Mesulam 1998, 2000) – the very area whose sole destruction has been shown to be sufficient to produce global cessation of dreaming (Doricchi & Violani 1992; Solms 1997).

The ventral stream, the "what" or object identification stream, is also argued to have dedicated termini in the prefrontal cortex (Goldman-Rakic et al. 2000) and includes visual associative areas that remain active in REM (Braun et al. 1998), are associated with waking hallucinosis (Ffytche et al. 1998), and may account for the visual hallucinosis of dreams as evidenced by the occurrence of nonvisual dreaming when they are destroyed (Solms 1997). Ventral portions of the visual associative cortex include face processing regions in the fusiform gyri (Haxby et al. 2000; Kanwisher 2000), while ventral portions of the prefrontal cortex include limbic-related regions subserving social judgment (Damasio 1996; Tranel et al. 2000). Moreover, the anterior paralimbic REM activation area (Nofzinger et al. 1999) is that portion of the subcortex and cortex subserving instinct and emotion (Damasio et al. 2000), functions often deployed in our interactions with others. Dreams are highly

social phenomena (e.g., Kahn et al. 2000, 2002), and it is interesting to speculate that just as dreams are visually biased when one considers the full range of sensory modalities, they may correspondingly be "people-biased" when one considers the full range of objects available in the world, one's memory, and one's imagination.

The prefrontal cortex has been suggested to be the area of the brain that makes us distinctively human (Goldberg 2001). It is the region proportionately most developed in the great apes, including the human, compared to other mammals and primates (Goldberg 2001; Semendeferi et al. 2002). Given that other mammals have well-developed posterior systems (albeit lacking temporal language areas), one might speculate that oneiric behavior in animals (Jouvet 1999) as well as their waking consciousness itself may resemble, in some ways, our non-self-reflective dream life!

## V. New findings from neuroimaging of conditions related to sleep and dreaming

In addition to sleep-wake and sleep stage comparisons, cognitive neuroscientists have begun looking at other global states of consciousness using neuroimaging techniques. In some of these studies, the concept of hypofrontality, as in sleep, emerges as an explanatory principle of nonnormative mental states in which executive function may be weakened. Such findings encourage the student of sleep and dream science to take a closer look at neuroimaging studies of altered waking states. These include states sometimes considered sleep-like, which can be induced in normals such as meditation, hypnosis, quiet waking, and anaesthesia as well as the symptomatology of neuropsychiatric disorders such as epilepsy, schizophrenia, and autism. A sampling of recent findings in this area will be addressed here, aiming not to provide an exhaustive review but rather to suggest to the reader the wealth of evidence now emerging from the cognitive neuroscience of waking, which might inform our understanding of dreaming.

### A. Meditation and religious experience

Meditation is particularly informative because it appears that its corresponding regional cerebral activation pattern resembles sleep to a greater or lesser degree depending on the specific method used to produce this state. Although most meditative practices result in the subjective experience of the "relaxation response" (Benson 1979) objectively measurable by autonomic (e.g., Lazar et al. 2000) or EEG (e.g., Lou et al. 1999) indices, some techniques promote a REM-like hypofrontality while others do not.

In a recent fMRI study by Lazar et al. (2000), subjects practiced a technique based on closely attending to one's breathing and silently generating one mantra to each inhalation and another to each exhalation. The resulting regional activation pattern (when subtractively compared to either a control task or an earlier stage of this meditation), unlike NREM sleep, activated anterior midline attentional structures (e.g., the anterior cingulate and the medial frontal gyrus) and, unlike REM, activated portions of the dorsolateral prefrontal cortex (e.g., BA9). Moreover, unlike sleep, activation of parietal attention-related (BA7), motor (BA4), premotor (BA6), and somatosensory (BA3) cortex as well as arousal-related midbrain structures was increased. These authors suggest that this activation pattern reflects a hyperattentive state associated with the concentration required by this technique of meditation.

Lou et al. (1999) studied yoga meditation using $H_2^{15}O$ PET. In contrast to the attentive, volitional method of Lazar et al. (2000), their subjects listened to an audiotape that guided them through experiences of attention to regions of the body, abstract joy, visualization of an outdoor scene, and visualization of an abstract symbol of the self. Lou et al. (1999) report activation patterns corresponding to brain regions known to process these sorts of experiences such as parietal and motor areas for bodily foci, left hemispheric activation for abstract joy [in accordance with the well-known, lateralization of affective valence, see Davidson & Irwin (1999), which, however, Lou et al. do not note], activation of extrastriate occipital and parietal areas by visualization, and parietal activation by the abstract image of the self. Notably, all these meditative periods were subtractively different from normal waking consciousness by a lesser activation of prefrontal, anterior cingulate, temporal, and inferior parietal cortices as well as of arousal-related subcortical areas (pons, thalamus). These authors remark on the similarity of the brain activation pattern they observed during meditation to that of REM sleep attributing, however, the absence of anterior cingulate activation to the emotionally bland character of their meditation compared to dreaming. In terms of the hypofrontality under discussion here, their technique lacked the volitional effort of Lazar et al.'s technique and, correspondingly, also lacked prefrontal activation despite robust activity of posterior systems.

Two more recent studies of meditation and religious experience have, similarly to the Lazar study, showed activation of prefrontal circuits. Using SPECT, Newberg et al. (2001) studied Tibetan Buddhist meditators and report increased regional blood flow to cingulate, orbitofrontal, and dorsolateral prefrontal cortices as well as thalamic increases during meditation compared with baseline. As in Lazar et al.'s study, Newberg et al. interpret this activation pattern as reflecting an effortful attending to the meditative practice. A somewhat different pattern of prefrontal activation was noted by Azari et al. (2001), who studied, with $H_2^{15}O$ PET, fundamental Christians attending to a biblical psalm that reliably evoked, for these individuals, a subjective religious experience. In these subjects, the religious experience was accompanied by activation of dorsolateral and dorsomedial cortices but not limbic-related ventral areas. These authors suggest that, for their subjects, religious experience was an attributional, thought-based state versus a state of affective or autonomic arousal that, in contrast, is a general characteristic of REM sleep dreaming (Hobson et al., target article).

From these studies, therefore, it appears that although religious experiences can occur in dreams, waking religious experience or meditation does not resemble the hypofrontality and accompanying executive deficits experienced during sleep and dreaming. A notable exception to this occurs when a great degree of passivity or visual imagery characterizes the meditative practice, as in the case of Lou et al. (1999).

### B. Hypnosis

Unlike the majority of the studies of meditation just discussed, but with some similarity to REM sleep, Maquet et al. (1999), using $H_2^{15}O$ PET, have reported that hypnotized subjects show increased activation of posterior cortices with no corresponding increase in frontal areas. The hypnotized state following its induction (indexed objectively by slow roving eye movements) was compared subtractively to a baseline alert condition of imagining, with audio prompts, pleasant autobiographical memories. After hypnotic induction, subjects showed mostly left-sided increases (left-sided except when otherwise noted) in activity of: extrastriate occipital (BA18, bilaterally; 19; 37), inferior parietal (BA40), motor and premotor (BA4; 6; 43), deep ventrolateral prefrontal (BA45) and anterior cingulate (right-sided, BA24; 32) cortical activation but decreased temporal (BA20; 21, bilaterally; 22R; 38; 39), medial prefrontal (8; 9; 10), premotor (right-sided BA6; 8), and posterior cingulate/precuneal (BA39; 7) cortices. Although this pattern bears some similarity to REM in that posterior visual association areas are more active and some frontal areas are less active, these authors note the lack of subcortical limbic activation seen in REM and emphasize the similarity of this cortical pattern to waking visualization (Maquet et al. 1999). They also note the complete absence of EEG signs of Stage 2 NREM.

A second recent $H_2^{15}O$ PET study of hypnosis, by Rainville et al. (1999), reports a generally similar pattern to the Maquet study with the interesting additional recruitment of frontal cortices by the addition of an analgesic suggestion to the hypnotic state. Interestingly, in this study, as in the Maquet et al. (1999) study, widespread increases in extrastriate occipital (bilateral BA18; 19) regional cerebral blood flow (rCBF) were also reported, even though no explicit instructions to visualize were given. These authors note that this occipital activity correlated positively with delta EEG activity, a correlation they note is also seen in slow wave sleep (Hofle et al. 1997). Noting that such an increase also correlates with prolonged performance of an auditory vigilance task (Paus et al. 1997), Rainville et al. (1999) speculate that the increase in occipital rCBF may be a general correlate of relaxation or decreased arousal, may be explained as a decrease in cross-modal suppression, and may be permissive of spontaneous visual imagery. Moving more anteriorly, during hypnosis compared to baseline, Rainville et al. (1999) note a decrease in posterior parietal/precuneal/posterior cingulate rCBF (BA7; 40; 31), which they attribute to decreased attention to extrapersonal and somatic stimuli and an increase in right anterior cingulate (BA24) rCBF. The addition of analgesic suggestion (which they subtractively dissociate from the painful experimental stimuli alone) caused widespread increase in left frontal rCBF, which they interpret as engagement of executive processes to maintain a verbal suggestion and reinterpret a noxious perception accordingly.

Compared with sleep (see Table 2, Hobson et al., target article), hypnosis produces a pattern of activation with some similarities to NREM [occipital rCBF correlation with delta activity; Rainville et al. (1999)], REM (extrastriate and anterior cingulate activation, both studies), or sleep generally (precuneal and posterior cingulate deactivation, both studies), but also some distinct differences, including a lack of the subcortical limbic activation of REM and, most notably, the engagability of the frontal cortex by volitional activity (Rainville et al. 1999).

Interestingly, a pathological waking condition possibly related to hypnosis, depersonalization dissociation, is characterized mainly by occipital and parietal changes when compared to normal waking (Simeon et al. 2000) and thus also fails to mimic conditons seen in sleep versus waking. In contrast, sleepwalking, an abnormal NREM sleep condition, is characterized by reactivation of thalamocingulate pathways in the persistence of generalized cortical inactivation (Bassetti et al. 2000).

### C. Conscious resting state

Another interesting comparison to both sleep and altered waking states is the elusive notion of a "baseline" conscious resting state (Binder et al. 1999). In an fMRI study in which such a state was defined by the absence of attending to a perceptual or a semantic processing task, a distinctive network of left hemispheric heteromodal areas appear to be more active during the resting and semantic processing conditions compared to the perceptual processing task (Binder et al. 1999). These included the inferior parietal lobe, posterior cingulate, and dorsal and ventromedial prefrontal cortices. Binder et al. (1999) suggest that these are components of a "conceptual processing" network that, in the absence of perceptual input, sorts, organizes, and prioritizes the contents of consciousness, including recent perceptual inputs, toward ends of adaptive value to the individual. Such a state clearly differs from REM sleep in that prefrontal and other hereteromodal are activated while visual association areas are not.

### D. Anesthesia

A recent $H_2^{15}O$ PET study by Finset et al. (1999) has investigated brain changes during the loss of consciousness induced by the anesthetic propofol. Correlating with increasing levels of anesthetic, these authors report decreasing rCBF in the thalamus, parietal (BA7; 39; 40), posterior cingulate (BA30; 31), ventral prefrontal (BA11) with a small number of regional increases in the anterior cingulate (BA24), temporal pole (BA20), and cerebellum. These authors note the particularly strong correlation of increasing anesthetic levels and thalamic deactivation and argue for a similarity between the progressive deactivation of the reticulothalamic system seen in descending NREM and that seen in the induction of anesthesia (Finset et al. 1999).

### E. Emotional experience

The activation pattern associated with waking emotional experience is relevant to the current discussion because:

1. Dreaming can be a highly emotional state (Hobson et al., target article).

2. REM sleep dreaming is accompanied by activation of limbic and paralimbic structures (Braun et al. 1997, 1998; Maquet et al. 1996; Nofzinger et al. 1997).

3. REM-associated limbic activation may have functional significance related to normal and abnormal emotional regulation (Maquet & Franck 1997; Nofzinger et al. 1997, 1999; Wu et al. 1999, 2001).

4. Dreaming itself may have functional significance for emotional regulation (Cartwright et al. 1998; Kramer 1993).

Interestingly, in neuroimaging paradigms using script-driven imagery to generate specific emotional states in normals, the expected subcortical/cortical limbic/paralimbic activations are sometimes accompanied by deactivation of certain brain areas, which are also deactivated relative to waking in REM sleep, such as portions of the lateral prefrontal cortex and the posterior cingulate (Damasio et al. 2000; Liotti et al. 2000; Mayberg et al. 1999). Mayberg et al. (1999) and Liotti et al. (2000) have suggested that such patterns represent a limbic-cortical reciprocal inhibition, which may underlie reciprocal relationships between strong emotions and cognitive functions such as attention in both normal feeling states and depression and anxiety disorders. For example, decreased waking activity of ventral anterior cingulate areas (e.g., subgenual area 25) accompanies successful treatment of depression by either sleep deprivation (Nofzinger et al. 1999; Wu et al. 1999, 2001) or pharmacological treatments (Mayberg et al. 1999; Nofzinger et al. 2001). Such normal regional reciprocity might be speculatively linked to the observed activity patterns of REM sleep either by hypothesizing a suppression of executive areas by limbic activation or a release/disinhibition of limbic activity by executive deactivation.

### F. Can regional cerebral activity patterns closely mimicking sleep occur in normals during waking?

From the studies described, the answer to this question appears to be negative. Although the progression of anesthesia appears to closely resemble NREM (as does coma), it cannot be considered waking. Among the studies of waking states, the characteristic activity pattern of REM sleep (high activation of extrastriate cortex and Nofzinger's anterior paralimbic REM activation area accompanied by dorsolateral prefrontal deactivation) is not seen in passive meditation, hypnosis, or depersonalization (lacking anterior paralimbic activation) or concentrative meditation, hypnosis with suggestion, or the conscious resting state (recruitment of prefrontal areas). Nevertheless, some very interesting parallels occur, such as the prominent extrastriate activation seen in the relaxed states of passive meditation and hypnosis and anterior cingulate activation seen in concentrative meditation, hypnosis, and even anesthesia. Combined with the distinct neuroimaging (see Hobson et al., target article) and electrophysiological differences (see earlier discussion) between REM and other states, it appears safe to speculate that REM sleep represents a truly unique brain state and encourages its study in relation to the unique cognitive features of dreaming with which it is correlated (albeit quantitatively and statistically versus exclusively) (Hobson et al., target article). Furthermore, the uniqueness of the REM (and, for that matter, the NREM) cerebral activation pattern argues strongly against the view sometimes expressed that the dream state constitutes an omnipresent baseline that emerges in the absence of cognitive-perceptual input (e.g., Reinsel et al. 1992).

### G. Neuropsychiatric disorders

**1. Schizophrenia, hallucinosis, and delusions.** The utility of comparing dreaming to any particular neuropsychiatric disorder, as opposed to any attempt to equate the two global states, is to gain a better understanding of the brain mechanism of particular cognitive phenomena by considering their putative neural substrates in two different states, one generated by normative behavioral state alternation and the other by pathophysiological processes (Hobson 1999, 2001). In the case of schizophrenia, such features include "positive symptoms," such as hallucinosis and delusionality, as well as "negative symptoms," such as prominent executive deficits (Mitchell et al. 2001).

A recent review of fMRI studies of schizophrenia (Mitchell et al. 2001) cites evidence that auditory hallucinations arise from a failure of executive dorsolateral prefrontal systems to monitor auditory processing in middle and superior temporal gyri in addition to abnormalities within the temporal cortex itself. Such failure of monitoring as well as of inhibitory influences of frontal areas over temporal regions during endogenous word generation is believed to account for misperception of internally generated stimuli as having an external origin (Mitchell et al. 2001). Interestingly, Fosse et al. (2001) have shown that as one progresses through the sleep-wake and REM-NREM cycles, hallucination and thought follow a reciprocal pattern of frequency in mentation reports. Therefore, the predominance of hallucination during REM, the time in the sleep cycle when the relative activity difference between a hypoactive prefrontal cortex and an active visual association area may be greatest, may have a formally similar cortical mechanism to visual hallucinosis in psychosis.

The possible connection between the limbic activation of REM sleep and dream delusionality (see Hobson et al., target article) and the delusionality of schizophrenia is suggested in a recent theoretical paper by Mujica-Parodi et al. (2000). These authors suggest that, rather than a failure of logical reasoning processes, the delusionality of schizophrenia arises when the "acquisition and evaluation of inputs" is distorted by excessively affect-driven cognition. In the case of schizophrenia, this includes biases in the acquisition of information or deficits in the ability to "filter" information (Mujica-Parodi et al. 2000). In either case, inappropriate inferences arise not from faulty logic but from having defective inputs that one's reasoning capacities then must then explain (Mujica-Parodi et al. 2000). Defective inputs may be perceptual, in which case a hallucination may occur, or a failure to attend to differences between objects or concepts, in which case illogical connections or ideas of reference arise (Mujica-Parodi et al. 2000). This sort of error bears a striking resemblance to the bizarre transformations of dreaming that have been shown to be constrained by visual similarity between dream images irrespective of their logical equivalence (Rittenhouse et al. 1994). Moreover, the similar condition of distorted or uncertain perceptual input in the presence of strong affective biases might, on the one hand, lead to delusions (such as misperceived threat) in the schizophrenic (Mujica-Parodi et al. 2000), while, in the normal dreamer, it might lead to irrational ad hoc explanations of dream objects and events (Williams et al. 1992).

The relationship of dopamine to psychotic symptoms and disorders has been well known since the advent of traditional neuroleptics in the treatment of schizophrenia [see Hobson (2001) for a historical account] and the use of L-DOPA and other dopamine agonists in the treatment of Parkinsonism [see Hobson & Pace-Schott (1999) and Solms (1997) for recent reviews]. Recent findings continue to stress the role of mesolimbic and mesocortical dopamine systems in psychotic disorders [for recent reviews, see Tzschentke (2001) and Moore et al. (1999)]. However, a purely dopaminergic hypothesis of psychosis is problematic. For example, some dopamine agonists fail to provoke psychotic symptoms in normals or in schizophrenics (e.g.,

Depatie & Lal 2001). Moreover, the roles of other neurotransmitters in psychotic disorders have become increasingly evident since the advent of the atypical neuroleptics, which possess both serotonergic and dopaminergic (as well as noradrenergic and cholinergic) properties (e.g., Ashby & Wang 1996). For example, both glutamatergic (Olney et al. 1999; Coyle 1996) and serotonergic (Aghajanian & Marek 2000; Bantick et al. 2001) neurotransmission may play key roles in psychotic disorders via their interraction with each other (Aghajanian & Marek 2000) as well as with dopaminergic systems (Kapur & Remington 1996; Vollenweider et al. 2000). Therefore, although dopamine clearly plays a role in the generation of waking psychosis, and, speculatively, the psychotomimetic properties of dreaming (e.g., Gottesmann 2000), it is clearly a complex interaction of multiple neurotransmitter systems that produces the resultant hallucinations and delusions in either state (Hobson 2001; Vollenweider 1998).

**2. Autism.** Autism presents a particularly instructive example of frontal deficits in psychopathology because its specific pattern of prefrontal deficits may, in some sense, be a mirror image of deficits in dreaming. While autistic patients may show deficits similar to those seen in orbitofrontal and ventromedial prefrontal disorders while preserving some of the putative capacities of dorsolateral prefrontal regions (Stone et al. 1998), the dreamer may show pronounced deficits in functions subserved by the dorsolateral prefrontal cortex (e.g., logical reasoning and working memory) and dorsolateral hypoperfusion (Hobson et al., target article) but a preserved capability for social reasoning and forming a "theory of mind" [the ability to attribute mental and affective states to others (Baron-Cohen 1995)].

There is evidence that the capacity to form a theory of mind (Baron-Cohen 1995; Fletcher et al. 1995; Goel et al. 1995), as well as its more refined sequelae in the social competencies of adult humans, is subserved at least in part by ventral (e.g., orbitofrontal) and medial (e.g. BA 32) portions of the prefrontal cortex (Damasio 1996; Tranel et al. 2000). These are anatomically or functionally proximal to areas that: (1) are selectively reactivated in REM sleep (Braun et al. 1997, 1998; Maquet et al. 1996; Nofzinger et al. 1997, 1999, 2000, 2001); (2) are shown to be sensitive to sleep deprivation (Smith et al. 1999; Wu et al. 1999); and (3) when lesioned, are associated with profound augmentation (anterior limbic) or attenuation (ventromedial prefrontal disconnective) of dreaming (Solms 1997).

Recent studies by Roger Godbout's laboratory (Godbout et al. 1998, 2000; Daoust et al. 2001a, 2001b) on sleep and dreaming in patients with Asperger's syndrome (a pervasive developmental disorder thought by some to be related to high functioning autism) suggest another potential avenue for investigating the neuropsychological basis of dreaming complementary to the lesion studies of Solms (1997) and Doricchi and Violani (1992). It is notable that in our dreams we preserve our capacity to attribute mental states to others (i.e., the characters in our dreams), that is, to form a "theory of mind" (Pace-Schott 2001). This is especially notable in that other executive capacities, such as logical reasoning, and more basic building blocks of executive capacities, such as working memory and attention, are impaired during dreaming (Hobson et al., target article).

Patients with autism and Asperger's syndrome show both deficient theory of mind capacities (Baron-Cohen 1995)

and highly attenuated dreaming (Craig & Baron-Cohen 1998; Daoust et al. 2001a, 2001b; Godbout et al. 1998). This attenuation has been hypothesized to reflect theory of mind deficits in individuals with autism spectrum disorders (Daoust et al. 2001a; Godbout et al. 1998). If impoverishment of dreaming in the autistic spectrum disorders proves to be generally found, it will provide a fascinating parallel to reports of global cessation of dreaming in ventromedial prefrontal disconnective lesion patients (Solms 1997).

**3. Hypofrontality, psychopathology, and dreaming.** An imbalance between prefrontal and posterior cortical or limbic activation has emerged as an explanatory principle for the symptomatology of a wide variety of neuropsychiatric disorders. For example, mesolimbic activation in the face of impaired prefrontal function is a prominent theory of both positive and negative symptoms in schizophrenia (Weinberger 1995). Similarly, suppression of prefrontal areas by limbic hyperactivity has been suggested in affective disorders (Mayberg et al. 1999). In addition, the ictal phase of epilepsy (during which hallucinations occur) is characterized by frontal hypoperfusion in the presence of hyperfusion of more posterior seizure loci (Rabinowicz 1997). Prefrontal areas have also been shown to be affected by delirium (Lerner & Rosenstein 2000; Trzepacz 2000), the psychosis most formally similar to dreaming (Hobson 1999). Frontal deficits are widely hypothesized to underlie specific deficits in autism (Stone et al. 1998; also see earlier discussion). Finally, Schwartz and Maquet (2002) note that certain bizarre dream features involving character misidentification bear marked similarity to neurological conditions such as Fregoli syndrome or Capgras delusion, both of which often involve a degree of frontal pathology. Therefore, the concept of hypofrontality, often in the presence of accentuated limbic or posterior perceptual cortical activity, may offer a convergent explanatory principle linking dreaming with a wide variety of psychopathological conditions.

## Conclusions and future directions

One organizing premise of this brief update is that the combined study of phenomenology, cognitive neuroscience, and neurochemistry across a wide variety of normal (e.g., dreaming, waking), alterable (e.g., meditation, hypnosis), and psychopathological states may provide a synergistic understanding of the phenomenology of any one of these states, in this case, dreaming. For example, one such hypothesis suggested here is that the deactivation of prefrontal cortical areas and, possibly, a reciprocal intensification of cortical or subcortical limbic and posterior perceptual cortical activity may underlie a wide variety of dream-like states as well as normal dreaming itself. An extremely fruitful future direction may be the neurobiological study of waking states as close as possible to dreaming (e.g., hypnotic suggestion of intensified imagery) as well as dream states as close as possible to waking (e.g., full lucidity in experienced practitioners of lucid dreaming).

The preceding update is intended to give the reader a flavor of the immensely rich and exciting convergent findings emerging in the fields of sleep and dream science, cognitive neuroscience, and clinical neuroscience. As detailed in Hobson et al.'s target article and response to commentary as well as Pace-Schott's commentary, emerging technologies in basic and clinical neuroscience will undoubtedly provide many breakthroughs in the coming years. Therefore, one can expect updates, such as this one, to be needed on a regular basis in order for the student of dream science to remain fully grounded in all the explanatory possibilities offered by current neurobiology.

This work was supported by NIDA RO1-DA11744-01A1. I wish to thank Mark T. Blagrove, Ellen Carlin, Stevan Harnad, J. Allan Hobson, Mark Solms, Robert Stickgold, and Robert P. Vertes for their critical readings of this piece and their helpful suggestions.

## References

Abrahamson, E. E., Leak, R. K. & Moore, R. Y. (2001) The suprachiasmatic nucleus project to posterior hypothalamic arousal systems. *NeuroReport* 12:435–40.

Achermann, P. & Borbely, A. A. (1997) Low frequency (<1 Hz ) oscillations in the human sleep electroencephalogram. *Neuroscience* 81:213–22.

Aeschbach, D., Postolache, T. T., Sher, L., Matthews, J. R., Jackson, M. A. & Wehr, T. A. (2001) Evidence from the waking electroencephalogram that short sleepers live under higher homeostatic sleep pressure than long sleepers. *Neuroscience* 102:493–502.

Aghajanian, G. K. & Marek, G. J. (2000) Serotonin model of schizophrenia: Emerging role of glutamate mechanisms. *Brain Research—Brain Research Reviews* 31:302–12.

Albin, R. L., Koeppe, R. A., Chervin, R. D., Consens, F. B., Wernette, K., Frey, K. A. & Aldrich, M. S. (2000) Decreased striatal dopaminergic innervation in REM sleep behavior disorder. *Neurology* 55:1410–12.

Allen R. P. & Early, C. J. (2001) Restless legs syndrome: A review of clinical and pathophysiologic features. *Journal of Clinical Neurophysiology* 18:128–47.

Anderer, P., Klosch, G., Gruber, G., Trenker, E., Pacual-Marqui, R. D., Zeitlhofer, J., Barbanoj, M. J., Rappelsberger, P. & Saletu, B. L. (2001) Low resolution brain electromagnetic tomography revealed simultaneously active frontal and parietal sleep spindle sources in the human cortex. *Neuroscience* 103:581–92.

Andersson, J., Onoe, H., Hetta, J., Lindstrom, K., Valind, S., Lilja, A., Sundin, A., Fasth, K. J., Westerberg, C. Broman, J. E., Watanabe, Y. & Langstrom, B. (1998) Brain networks affected by synchronized sleep visualized by positron emission tomography. *Journal of Cerebral Blood Flow and Metabolism* 18:701–15.

Andre, P. & Arrighi, P. (2001) Modulation of purkinje cell response to glutamate during the sleep-waking cycle. *Neuroscience* 105:731–46.

Ashby, C. R. & Wang, R. Y. (1996) Pharmacological actions of the atypical antipsychotic drug clozapine: A review. *Synapse* 24:349–94.

Aston-Jones, G., Chen, S., Zhu, Y. & Oshinsky, M. L. (2001) A neural circuit for circadian regulation of arousal. *Nature Neuroscience* 4: 732–38.

Aston-Jones, G., Rajkowski, J. & Cohen, J. (2000) Locus coeruleus and regulation of behavioral flexibility and attention. *Progress in Brain Research* 126:165–82.

Atienza, M., Cantero, J. L. & Gomez, C. M. (2000) Decay time of the auditory sensory memory trace during wakefulness and REM sleep. *Psychophysiology* 37:485–493.

Awh, E. & Jonides, J. (2001) Overlapping mechanisms of attention and spatial working memory. *Trends in Cognitive Sciences* 5:126.

Azari, N. P., Nickeo, J., Wunderlich, G., Niedeggen, M., Hefter, H., Tellmann, L., Herzog, H., Stoerig, P., Birnbacher, D., & Seitz, R. J. (2001) Neural correlates of religious experience. *European Journal of Neuroscience* 13:1649–52.

Balkin, T. J., Braun, A. R., Wesensten, N. J., Varga, M., Baldwin, P., Carson, R. E., Belenky, G. & Herskovitch, P. (1999) Bidirectional changes in regional cerebral blood flow across the first 20 minutes of wakefulness. *Sleep Research Online* 2(Supplement 1):6.

Bantick, R. A., Deakin, J. F., & Grasby, P. M. (2001) The 5-HT1A receptor in schizophrenia: A promising target for novel atypical neuroleptics? *Journal of Psychopharmacology* 15: 37–46.

Baron-Cohen, S. (1995) *Mindblindness: An essay on autism and theory of mind.* MIT Press.

Basheer, R., Rannie, D. G., Porkka-Heiskanen, T., Ramesh, V. & McCarley, R. W. (2001) Adenosine, prolonged wakefulness, and A1-activated NF-kB DNA binding in the basal forebrain of the rat. *Neuroscience* 104:731–9.

Bassetti, C., Vella, S., Donati, F., Wielepp, P. & Weder, B. (2000) SPECT during sleepwalking. *Lancet* 356:484–5.

Baxter, M. G. & Chiba, A. A. (1999) Cognitive functions of the basal forebrain. *Current Opinion in Neurobiology* 9:178–83.

Beebe, D. W. & Gozal, D. (2002) Obstructive sleep apnea and the prefrontal cortex: Towards a comprehensive model linking nocturnal upper airway obstruction to daytime cognitive and behavioral deficits. *Journal of Sleep Research* 22:1–16.

Benson, H. (1979) *The mind/body effect.* Simon and Schuster.

Bentivoglio, M. & Grassi-Zucconi, G. (1999) Immediate early gene expression in sleep and wakefulness. In: *Handbook of behavioral state control: Molecular and cellular mechanisms*, ed. R. Lydic & H. A. Baghdoyan. CRC Press.

Binder, J. R., Frost, J. A., Hammeke, T. A., Bellgowan, P. S. F., Rao, S. M. & Cox, R. W. (1999) Conceptual processing during the conscious resting state: A functional MRI study. *Journal of Cognitive Neuroscience* 11:80–93.

Binks, P. G., Waters, W. F. & Hurry, M. (1999) Short-term total sleep deprivation does not selectively impair higher cortical functioning. *Sleep* 22:328–34.

Blagrove, M. (1996) Effects of length of sleep deprivation on interrogative suggestibility. *Journal of Experimental Psychology, Applied* 2:48–9.

Blagrove, M., Alexander, C. A. & Horne, J. A (1995) The effects of chronic sleep reduction on the performance of cognitive tasks sensitive to sleep deprivation. *Applied Cognitive Psychology* 9:21–40.

Bootzin, R. R., Hubbard, T. L., Reiman, E. M., Bandy, D., Yun, L. S. & Munzlinger, T. (1998) Brain regions preferentially affected during different stages of sleep and wakefulness. *Sleep* 21(Supplement):272.

Borbely, A. A. (2001) From slow waves to sleep homeostasis: New perspectives. *Archives Italiennes de Biologie* 139:53–61.

Borbely, A. A. & Achermann, P. (2000) Sleep homeostasis and models of sleep regulation. In: *Principles and practice of sleep medicine*, ed. M. H. Kryger, T. Roth & W. C. Dement. W. B. Saunders.

Braun, A. R., Balkin, T. J., Wesensten, N. J., Carson, R. E., Varga, M., Baldwin, P., Selbie, S., Belenky, G. & Herscovitch, P. (1997) Regional cerebral blood flow throughout the sleep-wake cycle. *Brain* 120:1173–97.

Braun, A. R., Balkin, T. J., Wesensten, N. J., Gwadry, F., Carson, R. E., Varga, M., Baldwin, P., Belenky, G. & Herscovitch, P. (1998) Dissociated pattern of activity in visual cortices and their projections during human rapid eye-movement sleep. *Science* 279:91–5.

Buchsbaum, M. S., Hazlett, E. A., Wu, J. & Bunney, W. E. (2001) Positron emission tomography with deoxyglucose-F18 imaging of sleep. *Neuropsychopharmacology* 25:S50–6.

Buzsáki, G. (1996) The hippocampo-neocortical dialogue. *Cerebral Cortex* 6:81–92.

Cajochen, C., Knoblauch, V., Krauchi, K., Renz, R. & Wirz-Justice, A. (2001) Dynamics of frontal EEG activity, sleepiness and body temperature under high and low sleep pressure. *NeuroReport* 12:2277–81.

Cantero, J. L., Atienza, M. & Salas, R. (2000) Spectral features of EEG alpha activity in human REM sleep: Two variants with different functional roles? *Sleep* 23:746–50.

Capece, M. C., Baghdoyan, H. A. & Lydic, R. (1999) New directions for the study of cholinergic REM sleep generation: Specifying pre- and post-synaptic mechanisms. In: *Rapid eye movement sleep*, ed. B. N. Mallick & S. Inoue. Marcel Dekker.

Cartwright, R., Luten, A., Young, M., Mercer, P. & Bears, M. (1998) Role of REM sleep and dream affect in overnight mood regulation: A study of normal volunteers. *Psychiatry Research* 81:1–8.

Cavallero, C., Cicogna, P., Natale, V., Occhionero, M. & Zito, A. (1992) Slow wave sleep dreaming. *Sleep* 15:562–6.

Chapotot, F., Jouny, C., Muzet, A., Buguet, A. & Brandenberger, G. (2000) High frequency waking EEG: Reflection of a slow ultradian rhythm in daytime arousal. *NeuroReport.* 11:2223–7.

Chase, M. H., Soja, P. J. & Morales, F. R. (1989) Evidence that glycine mediates the post synaptic potentials that inhibit lumbar motorneurons during the atonia of active sleep. *The Journal of Neuroscience* 9:743–51.

Chemelli, R. M., Willie, J. T., Sinton, C. M., Elmquist, J. L., Scammell, T., Lee, C., Richardson, J. A., Williams, S. C., Xiong, Y., Kisanuki, Y., Fitch, T. E., Nakazato, M., Hammer, R. E., Saper, C. B. & Yanagisawa, M. (1999) Narcolepsy in orexin knockout mice: Molecular genetics of sleep regulation. *Cell* 98:437–51.

Cirelli, C. & Tononi, G. (2000) On the functional significance of c-fos induction during the sleep/waking cycle. *Sleep* 23:453–69.

Coyle, J. (1996) The glutamatergic dysfunction hypothesis for schizophrenia. *Harvard Review of Psychiatry* 3:241–53.

Craig, J. & Baron-Cohen, S. (1998) The hypothesis of the theory of the spirit: Do autistic children speak of their dreams? *Psychologie Francaise.* 43:169–76.

Czeisler, C. A., Duffy, J. F., Shanahan, T. L., Brown, E. N., Mitchell, J. F., Rimmer, D. W., Ronda, J. M., Silva, E. J., Allan, J. S., Emens, J. S., Dijk, D. & Kronauer, R. E. (1999) Stability, precision, and near 24 hour period of the human circadian pacemaker. *Science* 284:2177–81.

Czeisler, C. A., Zimmerman, J. C., Ronda, J. M., Moore-Ede, M. C. & Weitzman, E. D. (1980) Timing of REM sleep is coupled to the circadian rhythm of body temperature in man. *Sleep* 2:329–46.

Damasio, A. R. (1996) The somatic marker hypothesis and the possible functions of the prefrontal cortex. *Philosophical Transactions of the Royal Society of London* 351:1413–20.

Damasio, A. R., Grabowski, T. J., Bechara, A., Damasio, H., Ponto, L. L. B., Parvisi, J. & Hichwa, R. D. (2000) Subcortical and cortical brain activity during the feeling of self-generated emotions. *Nature Neuroscience* 3:1049–56.

Daoust, A., Limoges, E., Mottron L. A. & Godbout R. (2001a) Dream content analysis in Asperger's syndrome. *Sleep* 24(Abstract Supplement):A184.

Daoust, A., Limoges, E., Mottron L. A. & Godbout R. (2001b) Spectral analysis of REM sleep EEG in Asperger's syndrome. *Sleep* 24(Abstract Supplement):A121.

Datta, S. (1999) PGO wave generation: Mechanism and functional significance. In: *Rapid eye movement sleep*, ed. B. N. Mallick & S. Inoue. Marcel Dekker.

Datta, S. (2000) Avoidance task training potentiates phasic pontine-wave density in the rat: A mechanism for sleep-dependent plasticity. *Journal of Neuroscience*, 20, 8607–13.

Datta, S., Siwek, D. F., Patterson, E. H. & Cipolloni, P. B. (1998) Localization of pontine PGO wave generation sites and their anatomical projections in the rat. *Synapse* 30:409–23.

Davidson, R. J. & Irwin, W. (1999) The functional neuroanatomy of emotion and affective style. *Trends in Cognitive Sciences* 3:11–21.

de Fockert, J. W., Rees, G., Frith, C. D. & Lavie, N. (2001) The role of working memory in visual selective attention. *Science* 291:1803–6.

De Gennaro, L., Ferrara, M., Ferlazzo, F. & Bertini, M. (2000) Slow eye movements and EEG power spectra during wake-sleep transition. *Clinical Neurophysiology* 111:2107–15.

Depatie, L. & Lal, S. (2001) Apomorphine and the dopamine hypothesis of schizophrenia: A dilemma? *Journal of Psychiatry and Neuroscience.* 26:203–20.

Doricchi, F. & Violani, C. (1992) Dream recall in brain-damaged patients: A contribution to the neuropsychology of dreaming through a review of the literature. In: *The neuropsychology of sleep and dreaming*, ed. J. S. Antrobus & M. Bertini. Lawrence Erlbaum Associates.

Drummond, S. P. A., Brown, G. G., Stricker, J. L., Buxton, R. B., Wong, E. C. & Gillin, J. C. (1999) Sleep deprivation-induced reduction in cortical functional response to serial subtraction. *NeuroReport* 10:3745–8.

Drummond, S. P. A., Brown, G. G., Gillin, J. C., Stricker, J. L., Wong, E. C. & Buxton, R. B. (2000) Altered brain response to verbal learning following sleep deprivation. *Nature* 403:655–7.

Ffytche, D. H., Howard, R. J., Brammer, M. J., David, A., Woodruff, P. & Williams, S. (1998) The anatomy of conscious vision: An fMRI study of visual hallucinations. *Nature Neuroscience* 1:738–42.

Finelli, L. A., Borbely, A. A. & Achermann, P. (2001) Functional topography of the human nonREM sleep electroencephalogram. *European Journal of Neuroscience* 13:2282–90.

Finset, P., Paus, T., Daloze, T., Plourde, G., Meuret, P., Bonhomme, V., Hajj-Ali, N., Backman, S. B. & Evans, A. C. (1999) Brain mechanisms of propofol-induced loss of consciousness in humans: A positron emission tomographic study. *The Journal of Neuroscience* 19:5506–13.

Fiorillo, C. D. & Williams, J. T. (2000) Cholinergic inhibition of ventral midbrain dopamine neurons. *Journal of Neuroscience* 20:7855–60.

Fletcher, P. C., Happe, F., Frith, U., Baker, S. C., Dolan, R. J., Frackowiak, R. S. J. & Frith, C. (1995) Other minds in the brain: A functional imaging study of "theory of mind" in story comprehension. *Cognition* 57:109–28.

Fosse, R., Stickgold, R. & Hobson, J. A (2001) Brain-mind states: Reciprocal variation in thoughts and hallucinations. *Psychological Science* 12:30–6.

Freeman, A., Ciliax, B., Bakay, R., Daley, J., Miller, R. D., Keating, G., Levey, A. & Rye, D. (2001) Nigrostriatal collaterals to thalamus degenerate in Parkinsonian animal models. *Annals of Neurology* 50:321–9.

Furey, M. L., Pietrini, P. & Haxby, J. V. (2000) Cholinergic enhancement and increased selectivity of perceptual processing during working memory. *Science*, 290:2315–19.

Fuster, J. A. (1997) *The prefrontal cortex, third edition.* Lippincott-Raven.

Gais, S., Plihal, W., Wagner, U. & Born, J. (2000) Early sleep triggers memory for early visual discrimination skills. *Nature Neuroscience, 3*, 1335–9.

Gibbs, M. E. & Summers, R. J. (2000) Separate roles for B2 and B3 adrenoceptors in memory consolidation. *Neuroscience* 95:913–22.

Giuditta, A., Ambrosini, M. V., Montagnese, P., Mandile, P., Cotugno, M., Zucchoni, G. & Vescia, S. (1995) The sequential hypothesis on the function of sleep. *Behavioral Brain Research* 69:157–66.

Godbout, R., Bergeron, C., Stip, E. & Mottron, L. (1998) A laboratory study of sleep and dreaming in a case of Asperger's syndrome. *Dreaming* 8:75–88.

Godbout, R., Bergeron, C., Limoges, E., Mottron L. & Stip, E. (2000) A laboratory study of sleep in Asperger's syndrome. *NeuroReport* 11:127–30.

Goel, V., Graffman, J., Sadato, N. & Hallett, M. (1995) Modeling other minds. *NeuroReport* 6:1741–6.

Goldberg, E. (2001) *The executive brain: Frontal lobes and the civilized mind.* Oxford Univrsity Press.

Goldman-Rakic, P. S., O'Scalaidhe, S. P. & Chafee, M. V. (2000) Domain specificity in cognitive systems. In: *The new cognitive neurosciences, second edition,* ed. M. Gazzaniga. MIT Press.

Gottesmann C. (1997) Introduction to the neurophysiological study of sleep: Central regulation of skeletal and ocular activities. *Archives Italiennes de Biologie* 135:279–314.

Gottesmann, C. (1999) Neurophysiological support of consciousness during waking and sleep. *Progress in Neurobiology* 59:469–508.

Gottesmann, C. (2000) Hypothesis for the neurophysiology of dreaming. *Sleep Research Online* 3:1–4.

Graves, L., Pack, A. & Abel, T. (2001) Sleep and memory: A molecular perspective. *Trends in Neurosciences* 24:237–43.

Gross, D. W. & Gotman, J. (1999) Correlation of high-frequency oscillations with the sleep-wake cycle and cognitive activity in humans. *Neuroscience* 94:1005–18.

Hannibal, J., Hindersson, P., Knudsen, S. M., Georg, B. & Fahrenkrug, J. (2002) The photopigment melanopsin is exclusively present in pituitary adenylate cyclase-activating polypeptide-containing retinal ganglion cells of the retinohypothalamic tract. *Journal of Neuroscience* 22:RC191 (1–7).

Harrison, Y. & Horne, J. (1997) Sleep deprivation affects speech. *Sleep* 20:871–7.

Harrison, Y. & Horne, J. (1998) Sleep loss impairs short and novel language tasks having a prefrontal focus. *Journal of Sleep Research* 7:95–100.

Harrison, Y. & Horne, J. (1999) One night of sleep loss impairs innovative thinking and flexible decision making. *Organizational Behavior and Human Decision Processes* 78:128–45.

Harrison, Y. & Horne, J. (2000a) Sleep loss and temporal memory. *The Quarterly Journal of Experimental Psychology* 53A:271–9.

Harrison, Y. & Horne, J. (2000b) The impact of sleep deprivation on decision making: A review. *Journal of Experimental Psychology* 6:236–49.

Harrison, Y., Horne, J. & Rothwell, A. (2000) Prefrontal neuropsychological effects of sleeep deprivation in young adults – A model for healthy aging? *Sleep* 23:1067–73.

Haxby, J. V., Hoffman, E. A. & Gobbini, M. I. (2000) The distributed human neural system for face perception. *Trends in Cognitive Sciences* 4:223–33.

Herskovitch, J., Stuss, D. & Broughton, R. (1980) Changes in cognitive processing following short term cumulative partial sleep deprivation and recovery oversleeping. *Journal of Clinical Neuropsychology* 2:301–19.

Hobson, J. A. (1999) *Dreaming as delirium.* MIT Press.

Hobson, J. A. (2001) *The dream drug store.* MIT Press.

Hobson, J. A. & Pace-Schott, E. F. (1999) Reply to Solms, Braun and Reiser. *Neuropsychoanalysis* 1:206–24.

Hobson, J. A. & Steriade, M. (1986) The neuronal basis of behavioral state control. In: *Handbook of physiology – The nervous system, Vol. IV,* ed. F. E. Bloom. American Physiological Society.

Hofle, N., Paus, T., Reutens, D., Fiset, P., Gotman, J., Evans, A. C. & Jones, B. E. (1997) Regional cerebral blood flow changes as a function of delta and spindle activity during slow wave sleep in humans. *The Journal of Neuroscience* 17:4800–8.

Horne, J. A. (1988) Sleep loss and "divergent" thinking ability. *Sleep* 11:528–36.

Horne, J. A. (1993) Human sleep, sleep loss and behavior: Implications for the prefrontal cortex and psychiatric disorder. *British Journal of Psychiatry* 162:413–19.

Horvath, T. L., Peyron, C., Diano, S., Ivanov, A., Aston-Jones, G., Kilduff, T. S. & van den Pol, A. N. (1999) Hypocretin (orexin) activation and synaptic innervation of the locus coeruleus noradrenergic system. *Journal of Comparative Neurology* 415:145–59.

Igawa, M., Atsumi, Y., Takahashi, K., Shiotsuka, S., Hirasawa, H., Ryusei, Y., Maki, A., Yamashita, Y. & Koizuumi, H. (2001) Activation of visual cortex in REM sleep measured by 24-channel NIRS imaging. *Psychiatry and Clinical Neuroscience* 55:187–8.

Inoue, S., Honda, K., Kimura, M. & Zhang, S.-Q. (1999) Endogenous sleep substances and REM sleep. In: *Rapid eye movement sleep,* ed. B. N. Mallick & S. Inoue. Marcel Dekker.

Jacobs, B. L. & Fornal, C. A. (1999) An integrative role for serotonin in the central nervous system. In: *Handbook of behavioral state control: Molecular and cellular mechanisms,* ed. R. Lydic & H. A. Baghdoyan. CRC Press.

Jones, B. E. (2000) Basic mechanisms of sleep-wake states. In: *Principles and practice of sleep medicine,* ed. M. H. Kryger, T. Roth & W. C. Dement. W. B. Saunders.

Jones, B. E. & Muhlethaler, M. (1999) Cholinergic and GABAergic neurons of the basal forebrain. In: *Handbook of behavioral state control: Molecular and cellular mechanisms,* ed. R. Lydic & H. A. Baghdoyan. CRC Press.

Jouvet, M. (1999) *The paradox of sleep: The story of dreaming.* MIT Press.

Kahan, T. L. & LaBerge, S. (1994) Lucid dreaming as metacognition: Implications for cognitive science. *Consciousness and Cognition* 3:246–64.

Kahn, D., Stickgold, R., Pace-Schott, E. F. & Hobson, J. A. (2000) Dreaming and waking consciousness: A character recognition study. *Journal of Sleep Research* 9:317–25.

Kahn, D., Stickgold, R., Pace-Schott, E. F. & Hobson, J. A. (2002). Emotion and cognition: Feeling and character identification in dreaming. *Consciousness and Cognition* 11:34–50.

Kajimura, N., Uchiyama, M., Takayama, Y., Uchida, S., Uema, S., Uema, T., Kato, M., Sekimoto, M., Watanabe, T., Nakajima, T., Horikoshi, M., Okawa, M. & Takahashi, K. (1999) Activity of midbrain reticular formation and neocortex during the progression of human non-rapid eye movement sleep. *The Journal of Neuroscience* 19:10065–73.

Kanwisher N. (2000) Domain specificity in face perception. *Nature Neuroscience* 3:759–63.

Kapur, S. & Remington, G. (1996) Serotonin-dopamine interaction and its relevance to schizophrenia. *American Journal of Psychiatry* 153:466–76.

Karashima, A., Nakamura, K., Watanabe, M., Sato, N., Nakao, M., Katayama, N. & Yamamoto, M. (2001) Synchronization between hippocampal theta waves and PGO waves during REM sleep. *Psychiatry and Clinical Neurosciences* 55:189–90.

Karni, A. & Sagi, D. (1991) Where practice makes perfect in texture discrimination: Evidence for primary visual cortex plasticity. *Proceedings of the National Academy of Science of the United States of America* 88:4966–70.

Karni, A., Tanne, D., Rubenstein, B. S., Askenasy, J. J. M. & Sagi, D. (1994) Dependence on REM sleep of overnight improvement of a perceptual skill. *Science* 265:679–82.

Kilduff, T. S. & Peyron, C. (2000) The hypocretin/orexin ligand-receptor system: Implications for sleep and sleep disorders. *Trends in Neurosciences* 23:359–65.

Kramer, M. (1993) The selective mood regulatory function of dreaming: An update and revision. In; *The functions of dreaming,* ed. A. Moffitt, M. Kramer & R. Hoffman. State University of New York Press.

Krueger, J. M. & Fang, J. (1999) Cytokines and sleep regulation. In: *Handbook of behavioral state control: Molecular and cellular mechanisms,* ed. R. Lydic & H. A. Baghdoyan. CRC Press.

Krueger, J. M., Obal, F. & Fang, J. (1999) Humoral regulation of physiological sleep: Cytokines and GHRH. *Journal of Sleep Research* 8(Supplement 1): 53–9.

Kryger, M. H., Roth, T. & Dement, W. C., eds.(2000) *Principles and practice of sleep medicine.* W. B. Saunders.

Kubin, L. (2001) Carbachol models of REM sleep: Recent developments and new directions. *Archives Italiennes de Biologie* 139:37–51.

Lai, Y. Y. & Siegel, J. M. (1999) Muscle atonia and REM sleep. In: *Rapid eye movement sleep,* ed. B. N. Mallick & S. Inoue. Marcel Dekker.

Lancel, M. (1999) Role of GABA$_A$ receptors in the regulation of sleep: Initial sleep responses to peripherally administered modulators and agonists. *Sleep* 22:33–42.

Laureys, S., Peigneux, P., Phillips, C., Fuchs, S., Degueldre, C., Aerts, J., Del Fiore, G., Luxen, A., Franck, G., Van Der Linden, M., Cleeremans, A., Smith, C. & Maquet, P. (2001) Experience-dependent changes in cerebral functional connectivity during human rapid eye movement sleep. *Neuroscience* 105: 521–5.

Lazar, S. W., Bush, G., Gollub, R. L., Fricchione, G. L., Khalsa, G. & Benson, H. (2000) Functional brain mapping of the relaxation response and meditation. *NeuroReport* 11:1581–5.

Lee, R.-S., Steffensen, S. C. & Hendriksen, S. J. (2001) Discharge profiles of ventral tegmental area GABA neurons during movement, anaesthesia and the sleep-wake cycle. *The Journal of Neuroscience* 21:1757–66.

Leonard, T. O. & Lydic, R. (1999) Nitric oxide, a diffusible modulator of physiological traits and behavioral states. In: *Rapid eye movement sleep,* ed. B. N. Mallick & S. Inoue. Marcel Dekker.

Leonard, T. O., Michaelis, E. K. & Mitchell, K. M. (2001) Activity-dependent nitric oxide concentration dynamics in the laterodorsal tegmental nucleus in vivo. *Journal of Neurophysiology* 86:2159–72.

Lerner, D. M. & Rosenstein, D. L. (2000) Neuroimaging in delirium and related conditions. *Seminars in Clinical Neuropsychiatry* 5:98–112.

Lin, L., Faraco, J., Li, R., Qiu, X., deJong, P. J., Nishino, S. & Mignot, E. (1999) The sleep disorder canine narcolepsy is caused by a mutation in the hypocretin (orexin) receptor 2 gene. *Cell* 98:365–76.

Liotti, M., Mayberg, H. S., Brannan, S. K., McGinnis, S., Jerabek, P. & Fox, P. T. (2000) Differential limbic-cortical correlates of sadness and anxiety in healthy subjects: Implications for affective disorders. *Biological Psychiatry* 48:30–42.

Lou, H. C., Kjaer, T. W., Friberg, L., Wildschiodtz, G., Holm, S. & Nowak, M. (1999) A $^{15}$O-H$_2$O PET study of meditation and the resting state of normal consciousness. *Human Brain Mapping* 7:98–105.

Louie, K. & Wilson, M. A. (2001) Temporally structured replay of awake hippocampal ensemble activity during rapid eye movement sleep. *Neuron* 29:145–56.

Lovblad K. O., Thomas, R., Jakob, P. M., Scammell, T., Bassetti, C., Griswold, M., Ives, J., Matheson, J., Edelman, R. R. & Warach, S. (1999) Silent functional magnetic resonance imaging demonstrates focal activation in rapid eye movement sleep. *Neurology* 53:2193–5.

Lu, J., Zhang, Y.-H., Chou, T. C., Gaus, S. E., Elmquist, J. K., Shiromani, P. & Saper, C. (2001) Contrasting effects of ibotenate lesions of the paraventricular nucleus and subparaventricular zone on sleep-wake cycle and temperature regulation. *Journal of Neuroscience* 21:4864–74.

Luppi, P.-H., Gervasoni, D., Peyron, C., Rampon, C., Barbagli, B., Boissard, R. & Fort, P. (1999a) Norepinephrine and REM sleep. In: *Rapid eye movement sleep,* ed. B. N. Mallick & S. Inoue. Marcel Dekker.

Luppi, P.-H., Peyron, C., Rampon, C., Gervasoni, D., Barbagli, B., Boissard, R. & Fort P. (1999b) Inhibitory mechanisms in the dorsal raphe nucleus and locus coeruleus during sleep. In: *Handbook of behavioral state control: Molecular and cellular mechanisms,* ed. R. Lydic & H. A. Baghdoyan. CRC Press.

Lydic, R. & Baghdoyan, H. A., eds. (1999) *Handbook of behavioral state control: Molecular and cellular mechanisms.* CRC Press.

Mallick, B. N. & Inoue, S., ed. (1999) *Rapid eye movement sleep.* Marcel Dekker.

Mallick, B. N., Kaur, S., Jha, S. K. & Siegel, J. M. (1999) Possible role of GABA in the regulation of REM sleep with special reference to REM-OFF neurons. In: *Rapid eye movement sleep,* ed. B. N. Mallick & S. Inoue. Marcel Dekker.

Mallick, B. N., Kaur, S. & Saxena, R. N. (2001) Interactions between cholinergic and GABAergic neurotransmitters in and around the locus coeruleus for the induction and maintenance of rapid eye movement sleep. *Neuroscience* 104:467–85.

Maquet, P. (2000) Functional neuroimaging of normal human sleep by positron emission tomography. *Journal of Sleep Research* 9:207–31.

Maquet, P. & Franck, G. (1997) REM sleep and the amygdala. *Molecular Psychiatry* 2:195–6.

Maquet, P., Degueldre, C., Delfiore, G., Aerts, J., Peters, J. M., Luxen, A. & Franck, G. (1997) Functional neuroanatomy of human slow wave sleep. *The Journal of Neuroscience* 17:2807–12.

Maquet, P., Faymonville, M. E., Degueldre, C., Delfiore, G., Franck, G., Luxen, A. & Lamy, M. (1999) Functional neuroanatomy of hypnotic state. *Biological Psychiatry* 45:327–33.

Maquet, P., Laureys, S., Peigneux, P., Fuchs, S., Petiau, C., Phillips, C., Aerts, J., Del Fiore, G., Degueldre, C., Meulemans, T., Luxen, A., Franck, G., Van Der Linden, M., Smith, C. & Cleeremans, A. (2000) Experience-dependent changes in cerebral activation during human REM sleep. *Nature Neuroscience, 3,* 831–6.

Maquet, P., Peters, J. M., Aerts, J., Delfiore, G., Degueldre, C., Luxen, A. & Franck, G. (1996) Functional neuroanatomy of human rapid-eye-movement sleep and dreaming. *Nature* 383:163–6.

May, J. & Kline, P. (1987) Measuring the effects upon cognitive abilities of sleep loss during continuous operations. *British Journal of Psychology* 78:443–55.

Mayberg, H. S., Liotti, M., Brannan, S. K., McGinnis, S., Mahurin, R. K., Jerabek, P., Silva, J. A., Tekell, J. L., Martin, C. C., Lancaster, J. L. & Fox, P. T. (1999) Reciprocal limbic-cortical function and negative mood: Converging PET findings in depression and normal sadness. *American Journal of Psychiatry* 156:675–82.

Mesulam, M.-M. (1998) From sensation to cognition. *Brain* 121:1013–52.

Mesulam, M.-M. (2000) *Principles of behavioral and cognitive neurology, second edition.* Oxford University Press.

Mitchell, R. L. C., Elliot, R. & Woodruff, W. R. (2001) FMRI and cognitive dysfunction in schizophrenia. *Trends in Cognitive Sciences* 5:71–81.

Moore, H., West, A. R. & Grace A. A. (1999) The regulation of forebrain dopamine transmission: Relevance to the pathophysiology and psychopathology of schizophrenia. *Biological Psychiatry* 46:40–55.

Moore, R. Y., Abrahamson, E. A. & Van Den Pol, A. (2001) The hypocretin neuron system: An arousal system in the human brain. *Archives Italiennes de Biologie* 139:2195–205.

Mujica-Parodi, L. R., Malaspina, D. & Sackheim, H. A. (2000) Logical processing, affect and delusional thought in schizophrenia. *Harvard Review of Psychiatry* 8:73–83.

Newberg, A., Alavi, A., Baime, M., Pourdehnad, M., Santanna, J. & d'Aquili, E. (2001) The measurement of regional cerebral blood flow during the complex cognitive task of meditation: A preliminary SPECT study. *Psychiatry Research: Neuroimaging* 106:113–22.

Nishino, S., Ripley, B., Overeem, S., Lammers, G. J. & Mignot, E. (2000) Hypocretin (orexin) deficiency in human narcolepsy. *Lancet* 355:39–40.

Nofzinger, E. A., Berman, S., Fasiczka, A., Miewald, J. M., Meltzer, C. C., Price, J. C., Sembrat, R. C., Wood, A. & Thase, M. E. (2001) Effects of bupropion SR on anterior paralimbic function during waking and REM sleep in depression: Preliminary findings using [18F] FDG PET. *Psychiatry Research* 106:95–111.

Nofzinger, E. A., Mintun, M. A., Wiseman, M. B., Kupfer, D. J. & Moore, R. Y. (1997) Forebrain activation in REM sleep: An FDG PET study. *Brain Research* 770:192–201.

Nofzinger, E. A., Nichols, T. E., Meltzer, C. C., Price, J., Steppe, D. A., Miewald,

J. M., Kupfer, D. J. & Moore, R. Y. (1999) Changes in forebrain function from waking to REM sleep in depression: Preliminary analysis of [18F] FDG PET studies. *Psychiatry Ressearch: Neuroimaging* 91:59–78.

Nofzinger, E. A., Price, J., Meltzer, C. C., Buysse, D. J., Villemagne, V. L., Miewald, J. M., Sembrat, R. C., Steppe, D. A. & Kupfer, D. J. (2000) Towards a neurobiology of dysfunctional arousal in depression: The relationship between beta EEG power and regional cerebral glucose metabolism during NREM sleep. *Psychiatry Research: Neuroimaging* 98:71–91.

Obal, F. & Krueger, J. M. (1999) Hormones and REM sleep. In: *Rapid eye movement sleep,* ed. B. N. Mallick & S. Inoue. Marcel Dekker.

Olney, J. W., Newcomer, J. W. & Farber, N. B. (1999) NMDA receptor hypofunction model of schizophrenia. *Journal of Psychiatric Research* 33:523–33.

Pace-Schott, E. F. (2001) "Theory of mind," social cognition and dreaming. *Sleep Research Society Bulletin* 7:33–6.

Pace-Schott, E. F., Gersh, T., Silvestri-Hobson, R., Stickgold, R., Salzman, C. & Hobson, J. A. (2001) SSRI treatment suppresses dream recall frequency but increases subjective dream intensity in normal subjects. *Journal of Sleep Research* 10:129–42.

Pace-Schott, E. F. & Hobson, J. A. (2002) Basic mechanisms of sleep: New evidence on the neuroanatomy and neuromodulation of the NREM-REM cycle. In: *American College of Neuropsychopharmacology, Fifth Generation of Progress,* ed. D. Charney, J. Coyle, K. Davis & C. Nemeroff. Lippincott, Williams & Wilkins.

Pace-Schott, E. F., Hutcherson, C. A., Bemporad, B., Stickgold, R., Kumar, A. & Hobson, J. A. (2002a) Healthy young male adults are resistant to sleep-deprivation induced deficits in dorsolateral prefrontal function. *Sleep 25, Supplement* (in press).

Pace-Schott, E. F., Hutcherson, C. A., Bemporad, B., Stickgold, R., Kumar, A. & Hobson, J. A. (2002b) Healthy young male adults are resistant to sleep-deprivation induced deficits in ventromedial/orbital prefrontal function. *Sleep 25, Supplement* (in press).

Paus, T. (2001) Primate anterior cingulate cortex: Where motor control, drive and cognition interface. *Nature Reviews: Neuroscience* 2:417–24.

Paus, T., Zatorre, R. J., Hofle, N., Caramanos, Z., Gotman, J., Petrides, M. & Evans, A. (1997) Time-related changes in neural systems underlying attention and arousal during the perfoermance of an auditory vigilance task. *Journal of Cognitive Neuroscience* 9:392–408.

Peigneux, P., Laureys, S., Delbeuck, X. & Maquet, P. (2001a). Sleeping brain, learning brain. The role of sleep for memory systems. *NeuroReport* 12:A111–24.

Peigneux, P., Laureys, S., Fuchs, S., Delbeuck, X., Delgueldre, C., Aerts, J., Delfiore, G., Luxen, A. & Maquet, P. (2001b) Generation of rapid eye movements during pardoxical sleep in humans. *NeuroImage* 14:701–8.

Perez-Garci, E., del Rio-Portilla, Y., Guevara, M. A., Arce, C. & Corsi-Cabrera, M. (2001) Paradoxical sleep is characterized by uncoupled gamma activity between frontal and perceptual cortical regions. *Sleep* 24:118–26.

Perry, E., Walker, M., Grace, J. & Perry, R. (1999) Acetylcholine in mind: A neurotransmitter correlate of consciousness. *Trends in Neurosciences* 22:273–80.

Petiau, C., Harrison, Y., Delfiore, G., Degueldre, A., Luxen, J., Horne, J. & Maquet, P. (1998) Modification of fronto-temporal connectivity during a verb generation task after 30 hr total sleep deprivation: A PET study. *Journal of Sleep Research* 7 (Supplement2):208.

Peyron, C., Faraco, J., Rogers, W., Ripley, B., Overeem, S., Charnay, Y., Nevsimalova, S., Aldrich, M., Reynolds, D., Albin, R., Li, R., Hungs, M., Pedrazzoli, M., Padigaru, M., Kucherlapati, M., Fan, J., Maki, R., Lammers, G. J., Bouras, C., Kucherlapati, R., Nishino, S. & Mignot, E. (2000) A mutation in a case of early onset narcolepsy and a generalized absence of hypocretin peptides in human narcoleptic brains. *Nature Medicine* 6:991–7.

Piper, D. C., Upton, N., Smith, M. J. & Hunter, A. J. (2000) The novel brain neuropepotide, orexin-A, modulates the sleep-wake cycle of rats. *European Journal of Neuroscience* 12:726–30.

Poe, G. R., Nitz, D. A., McNaughton, B. L. & Barnes, C. A. (2000) Experience dependent phase reversal of hippocampal neuron firing during REM sleep. *Brain Research* 855:176–80.

Porkka-Heiskanen, T., Strecker, R. E. & McCarley, R. W. (2000) Brain site specificity of extracellular adenosine concentration changes during sleep deprivation and spontaneous sleep: An *in vivo* microdialysis study *Neuroscience* 99:507–17.

Porkka-Heiskanen, T., Strecker, R. E., Thakkar, M., Bjorkum, A. A., Greene, R. W. & McCarley, R. W. (1997) Adenosine: A mediator of the sleep-inducing effects of prolonged wakefulness. *Science* 276:1265–8.

Prospero-Garcia, O., Navarro, L., Murillo-Rodriguez, E., Sanchez-Alvarez, M., Guzman-Marin, R., Mendez-Diaz, M., Gomez-Chavarin, M., Jiminez-Anguiano, A. & Drucker-Colin, R. (1999) Cellular and molecular changes occurring during REM sleep. In: *Rapid eye movement sleep,* ed. B. N. Mallick & S. Inoue. Marcel Dekker.

Rabinowicz, A. L., Salas, E., Beserra, F., Leiguarda, R. C. & Vazquez, S. E. (1997) Changes in regional cerebral blood flow beyond the temporal lobe in unilateral temporal lobe epilepsy. *Epilepsia* 38:1011–14.

Rainville, P., Hofbauer, R. K., Paus, T., Duncan, G. H., Bushnell, M. C. & Price, D. D. (1999) Cerebral mechanisms of hypnotic induction and suggestion. *Journal of Cognitive Neuroscience* 11:110–25.

Reinsel, R., Antrobus, J. & Wollman, M. (1992) Bizarreness in dreams and waking fantasy. In: *The neuropsychology of sleep and dreaming*, ed. J. S. Antrobus & M. Bertini. Lawrence Erlbaum Associates.

Reppert, S. M. & Weaver, D. R. (2001) Molecular analysis of mammalian circadian rhythms. *Annual Review of Physiology* 63:647–76.

Ribeiro, S., Goyal, V., Mello, C. V. & Pavlides, C. (1999) Brain gene expression during REM sleep depends on prior waking experience. *Learning and Memory* 6:500–8.

Rittenhouse, C. D., Stickgold, R. & Hobson J. A. (1994) Constraints on the transformation of characters and objects in dream reports. *Consciousness and Cognition* 3:100–13.

Roberts, A. C., Robbins, T. W. & Weiskrantz, L., eds. (1996) Executive and cognitive functions of the prefrontal cortex. *Philosophical Transactions of the Royal Society of London, B* 351:1387–1527.

Rye, D. B. (1997) Contributions of the peduculopontine region to normal and altered REM sleep. *Sleep* 20:757–88.

Sakai, K. & Crochet, S. (2001) Differentiation of presumed serotonergic dorsal raphe neurons in relation to behavior and sleep wake states. *Neuroscience* 2001:1141–55.

Sakai, K., Crochet, S. & Onoe, H. (2001) Pontine structures and mechanisms involved in the generation of paradoxical (REM) sleep. *Archives Italiennes de Biologie* 139:93–107.

Salin-Pascual, R., Gerashchenko, D., Greco, M., Blanco-Centurion, C. & Shiromani, P. J. (2001) Hypothalamic regulation of sleep. *Neuropsychopharmacology* 25:S21–7.

Sanford, L. D., Silvestri, A. J., Ross, R. J. & Morrison, A. R. (2001) Influence of fear conditioning on elicited ponto-geniculo-occipital waves and rapid eye movement sleep. *Archives Italiennes de Biologie* 139:169–83.

Saper, C. B. (2000) Brain stem modulation of sensation, movement and consciousness. In: *Principles of neural science, fourth edition*, ed. E. R. Kandel, J. H. Schwartz & T. M. Jessell. McGraw-Hill.

Saper, C. B., Sherin, J. E. & Elmquist, J. K. (1997) Role of the ventrolateral preoptic area in sleep induction. In: *Sleep and sleep disorders: From molecule to behavior*, ed. O. Hayaishi, & S. Inoue. Academic Press.

Saper, C. B., Chou, T. C. & Scammell, T. E. (2001) The sleep switch: Hypothalamic control of sleep and wakefulness. *Trends in Neurosciences* 24:726–31.

Sauvage, M. & Steckler T. (2001) Detection of corticotropin-releasing hormone receptor 1 immunoreactivity in cholinergic, dopaminergic and noradrenergic neurons of the murine basal forebrain and brainstem nuclei-potential implications for arousal and attention. *Neuroscience* 104:643–52.

Schibler, U. & Tafti, M. (1999) Molecular approaches towards the isolation of sleep-related genes. *Journal of Sleep Research* 8(Supplement 1):1–10.

Schmahmann, J. D. (2000) The role of the cerebellum in affect and psychosis. *Journal of Neurolinguistics* 13:189–214.

Schwartz, S. & Maquet, P. (2002) Sleep imaging and the neuropsychological assessment of dreams. *Trends in Cognitive Sciences* 6:23–30.

Sejnowski, T. J. & Destexhe, A. (2000) Why do we sleep? *Brain Research* 886:208–23.

Semba, K. (1999) The mesopontine cholinergic system: A dual role in REM sleep and wakefulness. In: *Handbook of behavioral state control: Molecular and cellular mechanisms*, ed. R. Lydic & H. A. Baghdoyan. CRC Press.

Semendeferi, K., Lu., A., Schenker, N. & Damasio, H. (2002) Humans and great apes share a large frontal cortex. *Nature Neuroscience* 5:272–6.

Shiromani, P. J., Scammell, T., Sherin, J. E. & Saper, C. B. (1999) Hypothalamic regulation of sleep. In: *Handbook of behavioral state control: Molecular and cellular mechanisms*, ed. R. Lydic & H. A. Baghdoyan. CRC Press.

Siapas, A. G. & Wilson, M. A. (1998) Coordinated interactions between hippocampal ripples and cortical spindles during slow wave sleep. *Neuron* 21:1123–8.

Siegel, J., Moore, R., Thannickal, T. & Nienhuis, R. (2001) A brief history of hypocretin/orexin and narcolepsy. *Neuropsychopharmacology* 25:S14–20.

Simeon, D., Guralnik, O., Hazlett, E. A., Spiegel-Cohen, J., Hollander, E. & Buchsbaum, M. S. (2000) Feeling unreal: A PET study of depersonalization disorder. *American Journal of Psychiatry* 157:1782–88.

Smith, C. (1985) Sleep states and learning: A review of the animal literature. *Neuroscience Biobehavioral Review* 9:157–68.

Smith, C. (1993) REM sleep and learning: Some recent findings. In: *The Functions of Dreaming*, ed. A. Moffitt, M. Kramer & R. Hoffman. State University of New York Press.

Smith, C. (1995) Sleep states and memory processes. *Behavioural Brain Research* 69:137–45.

Smith, G. S., Reynolds, C. F., Pollock, B., Derbyshire, S., Nofzinger, E., Dew, M. A., Houck, R. R., Milko, D., Meltzer, C. C. & Kupfer, D. J. (1999) Cerebral glucose metabolic response to combined total sleep deprivation and antidepressant treatment in geriatric depression. *American Journal of Psychiatry* 156:683–9.

Soja, P. J., Pang, W., Taepavarapruk, N., Cairns, B. E. & McErlane, S. A. (2001) On the reduction of spontaneous and glutamate driven spinocerebellar and spinoreticular tract neuronal activity during active sleep. *Neuroscience* 104:199–206.

Solms, M. (1997) *The neuropsychology of dreams: A clinico-anatomical study*. Lawrence Erlbaum Associates.

Solms, M. (2002) Dreaming: Cholinergic and dopaminergic hypotheses. In: *Neurochemistry of consciousness: Transmitters in Mind*, ed. E. Perry, H. Ashton & A. Young. John Benjamins Publishing Company.

Staba, R. J., Wilson, C. L., Bragin, A., Fried, I. & Engel, J. (2000) Single unit firing properties recorded in human hippocampus and entorhinal cortex during sleep. *Society for Neuroscience Abstracts* 26:703.

Steiger, A. & Holsboer, F. (1997) Neuropeptides and human sleep. *Sleep* 20:1038–52.

Steriade, M. (1999) Coherent oscillations and short-term plasticity in corticothalamic networks. *Trends in Neurosciences* 22:337–45.

Steriade, M. (2000) Corticothalamic resonance, states of vigilance and mentation. *Neuroscience* 101:243–76.

Steriade, M. (2001) Active neocortical processes during quiescent sleep. *Archives Italiennes de Biologie* 139:37–51.

Steriade, M. & McCarley, R. W. (1990) *Brainstem control of wakefulness and sleep*. Plenum.

Stickgold, R. (1998) Sleep: Off-line memory reprocessing. *Trends in Cognitive Science* 2:484–92.

Stickgold, R., Hobson, J. A., Fosse, R. & Fosse, M. (2001) Sleep, learning and dreams: Offline memory reprocessing. *Science* 294:1052–7.

Stickgold, R., James, L. & Hobson, J. A. (2000a) Visual discrimination learning requires sleep after training. *Nature Neuroscience* 3:1237–8.

Stickgold, R., Malia, A., Maguire, D., Roddenberry, D. & O'Connor, M. (2000b) Replaying the game: Hypnagogic images in normals and amnesiacs. *Science* 290:350–3.

Stickgold, R., Whidbee, D., Schirmer, B., Patel, V. & Hobson, J. A. (2000c) Visual discrimination task improvement: A multi-step process occurring during sleep. *Journal of Cognitive Neuroscience* 12:246–54.

Stone, V. E., Baron-Cohen, S. & Knight, R. T. (1998) Frontal lobe contributions to theory of mind. *Journal of Cognitive Neuroscience* 10:640–56.

Strecker, R. E., Moriarty, S., Thakkar, M. M., Porkka-Heiskanen, T., Basheer, R., Dauphin, L. J., Rannie, D. G., Portas, C., Greene, R. W. & McCarley R. W. (2000) Adenosinergic modulation of basal forebrain and preoptic/anterior hypothalamic neuronal activity in the control of behavioral state. *Behavioral Brain Research* 115:183–204.

Szymusiak, R., Steininger, T., Alam, N. & McGinty, D. (2001) Preoptic area sleep regulating mechanisms. *Archives Italiennes de Biologie* 139:77–92.

Thakkar, M. M., Ramesh, V., Strecker, R. E. & McCarley, R. W. (2001) Microdialysis perfusion of orexin-A in the basal forebrain increases wakefulness in freely behaving rats. *Archives Italiennes de Biologie* 139:313–28.

Thannickal, T. C., Moore, R. Y., Nienhuis, R., Ramanathan, L., Gulyani, S., Aldrich, M., Cornford, M. & Siegel J. M. (2000) Reduced number of hypocretin neurons in human narcolepsy. *Neuron* 27:469–74.

Thomas, M., Sing, H., Belenky, G., Holcomb, H., Mayberg, H., Dannals, R., Wagner, H., Thorne, D., Popp, K., Rowland, L., Welsh, A., Balwinski, S. Redmond, D. (2001) Neural bases of alertness and cognitive performance impairments during sleepiness. I. Effects of 24h of sleep deprivation on waking human regional brain activity. *Journal of Sleep Research* 9:335–52.

Tononi, G. & Cirelli, C. (2001a) Modulation of brain gene expression during sleep and wakefulness: A review of recent findings. *Neuropsychopharmacology* 25:S28–35.

Tononi, G. & Cirelli, C. (2001b) Some considerations on sleep and neural plasticity. *Archives Italiennes de Biologie* 139:221–41.

Tranel, D., Bechara, A. & Damasio, A. R. (2000) Decision making and the somatic marker hypothesis. In: *The new cognitive neurosciences, second edition*, ed. M. Gazzaniga. MIT Press.

Trzepacz, P. T. (2000) Is there a final common pathway in delirium? Focus on acetylcholine and dopamine. *Seminars in Clinical Neuropsychiatry* 5:132–48.

Tzschentke, T. M. (2001) Pharmacology and behavioral pharmacology of the mesocortical dopamine system. *Progress in Neurobiology* 63:241–320.

Usher, M., Cohen, J. D., Servan-Schreiber, D., Rajlowski, J. & Aston-Jones, G. (1999) The role of locus coeruleus in the regulation of cognitive performance. *Science* 283:549–54.

Vollenweider, F. X. (1998) Advances and pathophysiological models of hallucinogenic drug actions in humans: A preamble to schizophrenia research. *Pharmacopsychiatry* 31:92–103.

Vollenweider, F. X., Vontobel, P., Oye, I., Hell, D. & Leenders, K. L. (2000) Effects of (S)-ketamine on striatal dopamine: An (11C) raclopride PET study of a model of psychosis in humans. *Journal of Psychiatric Research* 34:35–43.

Weinberger, D. R. (1995) Neurodevelopmental perspectives on schizophrenia. In: *Psychopharmacology: The fourth generation of progress,* ed. F. E. Bloom. Raven Press.

Werth, E., Achermann P. & Borbely A. A. (1997) Fronto-occipital EEG power gradients in human sleep. *Journal of Sleep Research* 6:102–12.

Williams, J., Merritt, J., Rittenhouse, C. & Hobson, J. A. (1992) Bizarreness in dreams and fantasies: Implications for the activation-synthesis hypothesis. *Consciousness and Cognition* 1:172–85.

Wilson, M. A. & McNaughton, B. L. (1994) Reactivation of hippocampal ensemble memories during sleep. *Science* 265:676–9.

Wimmer, F., Hofmann, R. F., Bonato, R. A. & Moffitt, A. R. (1992) The effect of sleep deprivation on divergent thinking and attention processes. *Journal of Sleep Research* 1: 223–30.

Wisor, J. P., Nishino, S., Sora, I., Uhl, G. H., Mignot, E. & Edgar, D. M. (2001) Dopaminergic role in stimulant induced wakefulness. *The Journal of Neuroscience* 21:1787–94.

Wu, J. C., Buchsbaum, M. & Bunney, W. E. (2001) Clinical neurochemical implications of sleep deprivation's effects on the anterior cingulate of depressed responders. *Neurpsychopharmacology* 25:S74–8.

Wu, J., Buchsbaum, M. S., Gillin, J. C., Tang, C., Cadwell, S., Weigand, M., Najafi, A., Klein, E., Hazen, J. K. & Bunney, W. E. Jr. (1999) Prediction of antidepressant effects of sleep deprivation by metabolic rates in the ventral anterior cingulate and the medial prefrontal cortex. *American Journal of Psychiatry* 156:1149–58.

Wurts, S. & Edgar, D. (2000) Circadian and homeostatic control of eye movement (REM) sleep: Promotion of REM tendency by the suprachiasmatic nucleus. *The Journal of Neuroscience.* 20:4300–10.

Xi, M. C., Morales, F. R. & Chase, M. H. (1999) Evidence that wakefulness and REM sleep are controlled by a GABAergic pontine mechanism. *Journal of Neurophysiology* 82:2015–19.

Xi, M. C., Morales, F. R. & Chase, M. H. (2001) Induction of wakefulness and inhibition of active sleep by GABAergic processes in the nucleus pontis oralis. *Archives Italiennes de Biologie* 139:125–45.

Zhou, F-M., Liang, Y. & Dani, J. A. (2001) Endogenous nicotinic cholinergic activity regulates dopamine release in the striatum. *Nature Neuroscience* 4:1224–9.

# Index

Page references to tables, figures, and notes are followed by "t," "f," or "n" respectively (e.g., 4t or 63f).